T0180912

Lecture Notes in Artificial Intelligence 12457

Subseries of Lecture Notes in Computer Science

More information about this subseries at http://www.springer.com/series/1244

Frank Hutter · Kristian Kersting ·
Jefrey Lijffijt · Isabel Valera (Eds.)

Machine Learning and Knowledge Discovery in Databases

European Conference, ECML PKDD 2020
Ghent, Belgium, September 14–18, 2020
Proceedings, Part I

Editors
Frank Hutter
Albert-Ludwigs-Universität
Freiburg, Germany

Kristian Kersting
TU Darmstadt
Darmstadt, Germany

Jefrey Lijffijt
Ghent University
Ghent, Belgium

Isabel Valera
Saarland University
Saarbrücken, Germany

ISSN 0302-9743 ISSN 1611-3349 (electronic)
Lecture Notes in Artificial Intelligence
ISBN 978-3-030-67657-5 ISBN 978-3-030-67658-2 (eBook)
https://doi.org/10.1007/978-3-030-67658-2

LNCS Sublibrary: SL7 – Artificial Intelligence

This Springer imprint is published by the registered company Springer Nature Switzerland AG
The registered company address is: Gewerbestrasse 11, 6330 Cham, Switzerland

Preface

This edition of the European Conference on Machine Learning and Principles and Practice of Knowledge Discovery in Databases (ECML PKDD 2020) is one that we will not easily forget. Due to the emergence of a global pandemic, our lives changed, including many aspects of the conference. Because of this, we are perhaps more proud and happy than ever to present these proceedings to you.

ECML PKDD is an annual conference that provides an international forum for the latest research in all areas related to machine learning and knowledge discovery in databases, including innovative applications. It is the leading European machine learning and data mining conference and builds upon a very successful series of ECML PKDD conferences.

Scheduled to take place in Ghent, Belgium, due to the SARS-CoV-2 pandemic, ECML PKDD 2020 was the first edition to be held fully virtually, from the 14th to the 18th of September 2020. The conference attracted over 1000 participants from all over the world. New this year was a joint event with local industry on Thursday afternoon, the AI4Growth industry track. More generally, the conference received substantial attention from industry through sponsorship, participation, and the revived industry track at the conference.

The main conference programme consisted of presentations of 220 accepted papers and five keynote talks (in order of appearance): Max Welling (University of Amsterdam), Been Kim (Google Brain), Gemma Galdon-Clavell (Eticas Research & Consulting), Stephan Günnemann (Technical University of Munich), and Doina Precup (McGill University & DeepMind Montreal).

In addition, there were 23 workshops, nine tutorials, two combined workshop-tutorials, the PhD Forum, and a discovery challenge.

Papers presented during the three main conference days were organized in four different tracks:

- Research Track: research or methodology papers from all areas in machine learning, knowledge discovery, and data mining;
- Applied Data Science Track: papers on novel applications of machine learning, data mining, and knowledge discovery to solve real-world use cases, thereby bridging the gap between practice and current theory;
- Journal Track: papers that were published in special issues of the journals *Machine Learning* and *Data Mining and Knowledge Discovery*;
- Demo Track: short papers that introduce a new system that goes beyond the state of the art, accompanied with a video of the demo.

We received a record number of 687 and 235 submissions for the Research and Applied Data Science Tracks respectively. We accepted 130 (19%) and 65 (28%) of these. In addition, there were 25 papers from the Journal Track, and 10 demo papers (out of 25 submissions). All in all, the high-quality submissions allowed us to put together an exceptionally rich and exciting program.

The Awards Committee selected research papers that were considered to be of exceptional quality and worthy of special recognition:

- Data Mining best paper award: "Revisiting Wedge Sampling for Budgeted Maximum Inner Product Search", by Stephan S. Lorenzen and Ninh Pham.
- Data Mining best student paper award: "SpecGreedy: Unified Dense Subgraph Detection", by Wenjie Feng, Shenghua Liu, Danai Koutra, Huawei Shen, and Xueqi Cheng.
- Machine Learning best (student) paper award: "Robust Domain Adaptation: Representations, Weights and Inductive Bias", by Victor Bouvier, Philippe Very, Clément Chastagnol, Myriam Tami, and Céline Hudelot.
- Machine Learning best (student) paper runner-up award: "A Principle of Least Action for the Training of Neural Networks", by Skander Karkar, Ibrahim Ayed, Emmanuel de Bézenac, and Patrick Gallinari.
- Best Applied Data Science Track paper: "Learning to Simulate on Sparse Trajectory Data", by Hua Wei, Chacha Chen, Chang Liu, Guanjie Zheng, and Zhenhui Li.
- Best Applied Data Science Track paper runner-up: "Learning a Contextual and Topological Representation of Areas-of-Interest for On-Demand Delivery Application", by Mingxuan Yue, Tianshu Sun, Fan Wu, Lixia Wu, Yinghui Xu, and Cyrus Shahabi.
- Test of Time Award for highest-impact paper from ECML PKDD 2010: "Three Naive Bayes Approaches for Discrimination-Free Classification", by Toon Calders and Sicco Verwer.

We would like to wholeheartedly thank all participants, authors, PC members, area chairs, session chairs, volunteers, co-organizers, and organizers of workshops and tutorials for their contributions that helped make ECML PKDD 2020 a great success. Special thanks go to Vicky, Inge, and Eneko, and the volunteer and virtual conference platform chairs from the UGent AIDA group, who did an amazing job to make the online event feasible. We would also like to thank the ECML PKDD Steering Committee and all sponsors.

October 2020

Tijl De Bie
Craig Saunders
Dunja Mladenić
Yuxiao Dong
Frank Hutter
Isabel Valera
Jefrey Lijffijt
Kristian Kersting
Georgiana Ifrim
Sofie Van Hoecke

Organization

General Chair

Tijl De Bie — Ghent University, Belgium

Research Track Program Chairs

Frank Hutter — University of Freiburg & Bosch Center for AI, Germany
Isabel Valera — Max Planck Institute for Intelligent Systems, Germany
Jefrey Lijffijt — Ghent University, Belgium
Kristian Kersting — TU Darmstadt, Germany

Applied Data Science Track Program Chairs

Craig Saunders — Amazon Alexa Knowledge, UK
Dunja Mladenić — Jožef Stefan Institute, Slovenia
Yuxiao Dong — Microsoft Research, USA

Journal Track Chairs

Aristides Gionis — KTH, Sweden
Carlotta Domeniconi — George Mason University, USA
Eyke Hüllermeier — Paderborn University, Germany
Ira Assent — Aarhus University, Denmark

Discovery Challenge Chair

Andreas Hotho — University of Würzburg, Germany

Workshop and Tutorial Chairs

Myra Spiliopoulou — Otto von Guericke University Magdeburg, Germany
Willem Waegeman — Ghent University, Belgium

Demonstration Chairs

Georgiana Ifrim — University College Dublin, Ireland
Sofie Van Hoecke — Ghent University, Belgium

Nectar Track Chairs

Jie Tang	Tsinghua University, China
Siegfried Nijssen	Université catholique de Louvain, Belgium
Yizhou Sun	University of California, Los Angeles, USA

Industry Track Chairs

Alexander Ypma	ASML, the Netherlands
Arindam Mallik	imec, Belgium
Luis Moreira-Matias	Kreditech, Germany

PhD Forum Chairs

Marinka Zitnik	Harvard University, USA
Robert West	EPFL, Switzerland

Publicity and Public Relations Chairs

Albrecht Zimmermann	Université de Caen Normandie, France
Samantha Monty	Universität Würzburg, Germany

Awards Chairs

Danai Koutra	University of Michigan, USA
José Hernández-Orallo	Universitat Politècnica de València, Spain

Inclusion and Diversity Chairs

Peter Steinbach	Helmholtz-Zentrum Dresden-Rossendorf, Germany
Heidi Seibold	Ludwig-Maximilians-Universität München, Germany
Oliver Guhr	Hochschule für Technik und Wirtschaft Dresden, Germany
Michele Berlingerio	Novartis, Ireland

Local Chairs

Eneko Illarramendi Lerchundi	Ghent University, Belgium
Inge Lason	Ghent University, Belgium
Vicky Wandels	Ghent University, Belgium

Proceedings Chair

Wouter Duivesteijn	Technische Universiteit Eindhoven, the Netherlands

Sponsorship Chairs

Luis Moreira-Matias	Kreditech, Germany
Vicky Wandels	Ghent University, Belgium

Volunteering Chairs

Junning Deng	Ghent University, Belgium
Len Vande Veire	Ghent University, Belgium
Maarten Buyl	Ghent University, Belgium
Raphaël Romero	Ghent University, Belgium
Robin Vandaele	Ghent University, Belgium
Xi Chen	Ghent University, Belgium

Virtual Conference Platform Chairs

Ahmad Mel	Ghent University, Belgium
Alexandru Cristian Mara	Ghent University, Belgium
Bo Kang	Ghent University, Belgium
Dieter De Witte	Ghent University, Belgium
Yoosof Mashayekhi	Ghent University, Belgium

Web Chair

Bo Kang	Ghent University, Belgium

ECML PKDD Steering Committee

Andrea Passerini	University of Trento, Italy
Francesco Bonchi	ISI Foundation, Italy
Albert Bifet	Télécom Paris, France
Sašo Džeroski	Jožef Stefan Institute, Slovenia
Katharina Morik	TU Dortmund, Germany
Arno Siebes	Utrecht University, the Netherlands
Siegfried Nijssen	Université catholique de Louvain, Belgium
Michelangelo Ceci	University of Bari Aldo Moro, Italy
Myra Spiliopoulou	Otto von Guericke University Magdeburg, Germany
Jaakko Hollmen	Aalto University, Finland
Georgiana Ifrim	University College Dublin, Ireland
Thomas Gärtner	University of Nottinghem, UK
Neil Hurley	University College Dublin, Ireland
Michele Berlingerio	IBM Research, Ireland
Elisa Fromont	Université de Rennes 1, France
Arno Knobbe	Universiteit Leiden, the Netherlands
Ulf Brefeld	Leuphana Universität Luneburg, Germany
Andreas Hotho	Julius-Maximilians-Universität Würzburg, Germany

Program Committees

Guest Editorial Board, Journal Track

Ana Paula Appel	IBM Research – Brazil
Annalisa Appice	University of Bari Aldo Moro
Martin Atzmüller	Tilburg University
Anthony Bagnall	University of East Anglia
James Bailey	The University of Melbourne
José Luis Balcázar	Universitat Politècnica de Catalunya
Mitra Baratchi	University of Twente
Srikanta Bedathur	IIT Delhi
Vaishak Belle	The University of Edinburgh
Viktor Bengs	Paderborn University
Batista Biggio	University of Cagliari
Hendrik Blockeel	KU Leuven
Francesco Bonchi	ISI Foundation
Ilaria Bordino	UniCredit R&D
Ulf Brefeld	Leuphana Universität Lüneburg
Klemens Böhm	Karlsruhe Institute of Technology
Remy Cazabet	Claude Bernard University Lyon 1
Michelangelo Ceci	University of Bari Aldo Moro
Loïc Cerf	Universidade Federal de Minas Gerais
Laetitia Chapel	IRISA
Marc Deisenroth	Aarhus University
Wouter Duivesteijn	Technische Universiteit Eindhoven
Tapio Elomaa	Tampere University
Stefano Ferilli	University of Bari Aldo Moro
Cesar Ferri	Universitat Politècnica de València
Maurizio Filippone	EURECOM
Germain Forestier	Université de Haute-Alsace
Marco Frasca	University of Milan
Ricardo José Gabrielli Barreto Campello	University of Newcastle
Esther Galbrun	University of Eastern Finland
Joao Gama	University of Porto
Josif Grabocka	University of Hildesheim
Derek Greene	University College Dublin
Francesco Gullo	UniCredit
Tias Guns	VUB Brussels
Stephan Günnemann	Technical University of Munich
Jose Hernandez-Orallo	Universitat Politècnica de València
Jaakko Hollmén	Aalto University
Georgiana Ifrim	University College Dublin
Mahdi Jalili	RMIT University
Szymon Jaroszewicz	Polish Academy of Sciences

Giovanni Stilo	Università degli Studi dell'Aquila
Mahito Sugiyama	National Institute of Informatics
Andrea Tagarelli	University of Calabria
Chang Wei Tan	Monash University
Nikolaj Tatti	University of Helsinki
Alexandre Termier	Univ. Rennes 1
Marc Tommasi	University of Lille
Ivor Tsang	University of Technology Sydney
Panayiotis Tsaparas	University of Ioannina
Steffen Udluft	Siemens
Celine Vens	KU Leuven
Antonio Vergari	University of California, Los Angeles
Michalis Vlachos	University of Lausanne
Christel Vrain	LIFO, Université d'Orléans
Jilles Vreeken	Helmholtz Center for Information Security
Willem Waegeman	Ghent University
Marcel Wever	Paderborn University
Stefan Wrobel	Univ. Bonn and Fraunhofer IAIS
Yinchong Yang	Siemens AG
Guoxian Yu	Southwest University
Bianca Zadrozny	IBM
Ye Zhu	Monash University
Arthur Zimek	University of Southern Denmark
Albrecht Zimmermann	Université de Caen Normandie
Marinka Zitnik	Harvard University

Area Chairs, Research Track

Cuneyt Gurcan Akcora	The University of Texas at Dallas
Carlos M. Alaíz	Universidad Autónoma de Madrid
Fabrizio Angiulli	University of Calabria
Georgios Arvanitidis	Max Planck Institute for Intelligent Systems
Roberto Bayardo	Google
Michele Berlingerio	IBM
Michael Berthold	University of Konstanz
Albert Bifet	Télécom Paris
Hendrik Blockeel	Katholieke Universiteit Leuven
Mario Boley	MPI Informatics
Francesco Bonchi	Fondazione ISI
Ulf Brefeld	Leuphana Universität Lüneburg
Michelangelo Ceci	Università degli Studi di Bari Aldo Moro
Duen Horng Chau	Georgia Institute of Technology
Nicolas Courty	Université de Bretagne Sud/IRISA
Bruno Cremilleux	Université de Caen Normandie
Andre de Carvalho	University of São Paulo
Patrick De Causmaecker	Katholieke Universiteit Leuven

Nicola Di Mauro	Università degli Studi di Bari Aldo Moro
Tapio Elomaa	Tampere University
Amir-Massoud Farahmand	Vector Institute & University of Toronto
Ángela Fernández	Universidad Autónoma de Madrid
Germain Forestier	Université de Haute-Alsace
Elisa Fromont	Université de Rennes 1
Johannes Fürnkranz	Johannes Kepler University Linz
Patrick Gallinari	Sorbonne University
Joao Gama	University of Porto
Thomas Gärtner	TU Wien
Pierre Geurts	University of Liège
Manuel Gomez Rodriguez	MPI for Software Systems
Przemyslaw Grabowicz	University of Massachusetts Amherst
Stephan Günnemann	Technical University of Munich
Allan Hanbury	Vienna University of Technology
Daniel Hernández-Lobato	Universidad Autónoma de Madrid
Jose Hernandez-Orallo	Universitat Politècnica de València
Jaakko Hollmén	Aalto University
Andreas Hotho	University of Würzburg
Neil Hurley	University College Dublin
Georgiana Ifrim	University College Dublin
Alipio M. Jorge	University of Porto
Arno Knobbe	Universiteit Leiden
Dragi Kocev	Jožef Stefan Institute
Lars Kotthoff	University of Wyoming
Nick Koudas	University of Toronto
Stefan Kramer	Johannes Gutenberg University Mainz
Meelis Kull	University of Tartu
Niels Landwehr	University of Potsdam
Sébastien Lefèvre	Université de Bretagne Sud
Daniel Lemire	Université du Québec
Matthijs van Leeuwen	Leiden University
Marius Lindauer	Leibniz University Hannover
Jörg Lücke	University of Oldenburg
Donato Malerba	Università degli Studi di Bari "Aldo Moro"
Giuseppe Manco	ICAR-CNR
Pauli Miettinen	University of Eastern Finland
Anna Monreale	University of Pisa
Katharina Morik	TU Dortmund
Emmanuel Müller	University of Bonn
Sriraam Natarajan	Indiana University Bloomington
Alfredo Nazábal	The Alan Turing Institute
Siegfried Nijssen	Université catholique de Louvain
Barry O'Sullivan	University College Cork
Pablo Olmos	University Carlos III of Madrid
Panagiotis Papapetrou	Stockholm University

Andrea Passerini	University of Turin
Mykola Pechenizkiy	Technische Universiteit Eindhoven
Ruggero G. Pensa	University of Torino
Francois Petitjean	Monash University
Claudia Plant	University of Vienna
Marc Plantevit	Université Claude Bernard Lyon 1
Philippe Preux	Université de Lille
Rita Ribeiro	University of Porto
Celine Robardet	INSA Lyon
Elmar Rueckert	University of Lübeck
Marian Scuturici	LIRIS-INSA de Lyon
Michèle Sebag	Univ. Paris-Sud
Thomas Seidl	Ludwig-Maximilians-Universität Muenchen
Arno Siebes	Utrecht University
Alessandro Sperduti	University of Padua
Myra Spiliopoulou	Otto von Guericke University Magdeburg
Jerzy Stefanowski	Poznań University of Technology
Yizhou Sun	University of California, Los Angeles
Einoshin Suzuki	Kyushu University
Acar Tamersoy	Symantec Research Labs
Jie Tang	Tsinghua University
Grigorios Tsoumakas	Aristotle University of Thessaloniki
Celine Vens	KU Leuven
Antonio Vergari	University of California, Los Angeles
Herna Viktor	University of Ottawa
Christel Vrain	University of Orléans
Jilles Vreeken	Helmholtz Center for Information Security
Willem Waegeman	Ghent University
Wendy Hui Wang	Stevens Institute of Technology
Stefan Wrobel	Fraunhofer IAIS & Univ. of Bonn
Han-Jia Ye	Nanjing University
Guoxian Yu	Southwest University
Min-Ling Zhang	Southeast University
Albrecht Zimmermann	Université de Caen Normandie

Area Chairs, Applied Data Science Track

Michelangelo Ceci	Università degli Studi di Bari Aldo Moro
Tom Diethe	Amazon
Faisal Farooq	IBM
Johannes Fürnkranz	Johannes Kepler University Linz
Rayid Ghani	Carnegie Mellon University
Ahmed Hassan Awadallah	Microsoft
Xiangnan He	University of Science and Technology of China
Georgiana Ifrim	University College Dublin
Anne Kao	Boeing

Javier Latorre	Apple
Hao Ma	Facebook AI
Gabor Melli	Sony PlayStation
Luis Moreira-Matias	Kreditech
Alessandro Moschitti	Amazon
Kitsuchart Pasupa	King Mongkut's Institute of Technology Ladkrabang
Mykola Pechenizkiy	Technische Universiteit Eindhoven
Julien Perez	NAVER LABS Europe
Xing Xie	Microsoft
Chenyan Xiong	Microsoft Research
Yang Yang	Zhejiang University

Program Committee Members, Research Track

Moloud Abdar	Deakin University
Linara Adilova	Fraunhofer IAIS
Florian Adriaens	Ghent University
Zahra Ahmadi	Johannes Gutenberg University Mainz
M. Eren Akbiyik	IBM Germany Research and Development GmbH
Youhei Akimoto	University of Tsukuba
Ömer Deniz Akyildiz	University of Warwick and The Alan Turing Institute
Francesco Alesiani	NEC Laboratories Europe
Alexandre Alves	Universidade Federal de Uberlândia
Maryam Amir Haeri	Technische Universität Kaiserslautern
Alessandro Antonucci	IDSIA
Muhammad Umer Anwaar	Mercateo AG
Xiang Ao	Institute of Computing Technology, Chinese Academy of Sciences
Sunil Aryal	Deakin University
Thushari Atapattu	The University of Adelaide
Arthur Aubret	LIRIS
Julien Audiffren	Fribourg University
Murat Seckin Ayhan	Eberhard Karls Universität Tübingen
Dario Azzimonti	Istituto Dalle Molle di Studi sull'Intelligenza Artificiale
Behrouz Babaki	Polytechnique Montréal
Rohit Babbar	Aalto University
Housam Babiker	University of Alberta
Davide Bacciu	University of Pisa
Thomas Baeck	Leiden University
Abdelkader Baggag	Qatar Computing Research Institute
Zilong Bai	University of California, Davis
Jiyang Bai	Florida State University
Sambaran Bandyopadhyay	IBM
Mitra Baratchi	University of Twente
Christian Beecks	University of Münster
Anna Beer	Ludwig Maximilian University of Munich

Liang Chen	Sun Yat-sen University
Dawei Cheng	Shanghai Jiao Tong University
Wei Cheng	NEC Labs America
Wen-Hao Chiang	Indiana University - Purdue University Indianapolis
Feng Chong	Beijing Institute of Technology
Pantelis Chronis	Athena Research Center
Victor W. Chu	The University of New South Wales
Xin Cong	Institute of Information Engineering, Chinese Academy of Sciences
Roberto Corizzo	UNIBA
Mustafa Coskun	Case Western Reserve University
Gustavo De Assis Costa	Instituto Federal de Educação, Ciência e Tecnologia de Goiás
Fabrizio Costa	University of Exeter
Miguel Couceiro	Inria
Shiyao Cui	Institute of Information Engineering, Chinese Academy of Sciences
Bertrand Cuissart	GREYC
Mohamad H. Danesh	Oregon State University
Thi-Bich-Hanh Dao	University of Orléans
Cedric De Boom	Ghent University
Marcos Luiz de Paula Bueno	Technische Universiteit Eindhoven
Matteo Dell'Amico	NortonLifeLock
Qi Deng	Shanghai University of Finance and Economics
Andreas Dengel	German Research Center for Artificial Intelligence
Sourya Dey	University of Southern California
Yao Di	Institute of Computing Technology, Chinese Academy of Sciences
Stefano Di Frischia	University of L'Aquila
Jilles Dibangoye	INSA Lyon
Felix Dietrich	Technical University of Munich
Jiahao Ding	University of Houston
Yao-Xiang Ding	Nanjing University
Tianyu Ding	Johns Hopkins University
Rui Ding	Microsoft
Thang Doan	McGill University
Carola Doerr	Sorbonne University, CNRS
Xiao Dong	The University of Queensland
Wei Du	University of Arkansas
Xin Du	Technische Universiteit Eindhoven
Yuntao Du	Nanjing University
Stefan Duffner	LIRIS
Sebastijan Dumancic	Katholieke Universiteit Leuven
Valentin Durand de Gevigney	IRISA

Saso Dzeroski	Jožef Stefan Institute
Mohamed Elati	Université d'Evry
Lukas Enderich	Robert Bosch GmbH
Dominik Endres	Philipps-Universität Marburg
Francisco Escolano	University of Alicante
Bjoern Eskofier	Friedrich-Alexander University Erlangen-Nürnberg
Roberto Esposito	Università di Torino
Georgios Exarchakis	Institut de la Vision
Melanie F. Pradier	Harvard University
Samuel G. Fadel	Universidade Estadual de Campinas
Evgeniy Faerman	Ludwig Maximilian University of Munich
Yujie Fan	Case Western Reserve University
Elaine Faria	Federal University of Uberlândia
Golnoosh Farnadi	Mila/University of Montreal
Fabio Fassetti	University of Calabria
Ad Feelders	Utrecht University
Yu Fei	Harbin Institute of Technology
Wenjie Feng	The Institute of Computing Technology, Chinese Academy of Sciences
Zunlei Feng	Zhejiang University
Cesar Ferri	Universitat Politècnica de València
Raul Fidalgo-Merino	European Commission Joint Research Centre
Murat Firat	Technische Universiteit Eindhoven
Francoise Fogelman-Soulié	Tianjin University
Vincent Fortuin	ETH Zurich
Iordanis Fostiropoulos	University of Southern California
Eibe Frank	University of Waikato
Benoît Frénay	Université de Namur
Nikolaos Freris	University of Science and Technology of China
Moshe Gabel	University of Toronto
Ricardo José Gabrielli Barreto Campello	University of Newcastle
Esther Galbrun	University of Eastern Finland
Claudio Gallicchio	University of Pisa
Yuanning Gao	Shanghai Jiao Tong University
Alberto Garcia-Duran	Ecole Polytechnique Fédérale de Lausanne
Eduardo Garrido	Universidad Autónoma de Madrid
Clément Gautrais	KU Leuven
Arne Gevaert	Ghent University
Giorgos Giannopoulos	IMSI, "Athena" Research Center
C. Lee Giles	The Pennsylvania State University
Ioana Giurgiu	IBM Research - Zurich
Thomas Goerttler	TU Berlin
Heitor Murilo Gomes	University of Waikato
Chen Gong	Shanghai Jiao Tong University
Zhiguo Gong	University of Macau

Hongyu Gong	University of Illinois at Urbana-Champaign
Pietro Gori	Télécom Paris
James Goulding	University of Nottingham
Kshitij Goyal	Katholieke Universiteit Leuven
Dmitry Grishchenko	Université Grenoble Alpes
Moritz Grosse-Wentrup	University of Vienna
Sebastian Gruber	Siemens AG
John Grundy	Monash University
Kang Gu	Dartmouth College
Jindong Gu	Siemens
Riccardo Guidotti	University of Pisa
Tias Guns	Vrije Universiteit Brussel
Ruocheng Guo	Arizona State University
Yiluan Guo	Singapore University of Technology and Design
Xiaobo Guo	University of Chinese Academy of Sciences
Thomas Guyet	IRISA
Jiawei Han	University of Illinois at Urbana-Champaign
Zhiwei Han	fortiss GmbH
Tom Hanika	University of Kassel
Shonosuke Harada	Kyoto University
Marwan Hassani	Technische Universiteit Eindhoven
Jianhao He	Sun Yat-sen University
Deniu He	Chongqing University of Posts and Telecommunications
Dongxiao He	Tianjin University
Stefan Heidekrueger	Technical University of Munich
Nandyala Hemachandra	Indian Institute of Technology Bombay
Till Hendrik Schulz	University of Bonn
Alexander Hepburn	University of Bristol
Sibylle Hess	Technische Universiteit Eindhoven
Javad Heydari	LG Electronics
Joyce Ho	Emory University
Shunsuke Horii	Waseda University
Tamas Horvath	University of Bonn and Fraunhofer IAIS
Mehran Hossein Zadeh Bazargani	University College Dublin
Robert Hu	University of Oxford
Weipeng Huang	Insight
Jun Huang	University of Tokyo
Haojie Huang	The University of New South Wales
Hong Huang	UGoe
Shenyang Huang	McGill University
Vân Anh Huynh-Thu	University of Liège
Dino Ienco	INRAE
Siohoi Ieng	Institut de la Vision
Angelo Impedovo	Università “Aldo Moro” degli studi di Bari

Anisio Lacerda	Centro Federal de Educação Tecnológica de Minas Gerais
Patricia Ladret	GIPSA-lab
Fabrizio Lamberti	Politecnico di Torino
James Large	University of East Anglia
Duc-Trong Le	University of Engineering and Technology, VNU Hanoi
Trung Le	Monash University
Luce le Gorrec	University of Strathclyde
Antoine Ledent	TU Kaiserslautern
Kangwook Lee	University of Wisconsin-Madison
Felix Leibfried	PROWLER.io
Florian Lemmerich	RWTH Aachen University
Carson Leung	University of Manitoba
Edouard Leurent	Inria
Naiqi Li	Tsinghua-UC Berkeley Shenzhen Institute
Suyi Li	The Hong Kong University of Science and Technology
Jundong Li	University of Virginia
Yidong Li	Beijing Jiaotong University
Xiaoting Li	The Pennsylvania State University
Yaoman Li	CUHK
Rui Li	Inspur Group
Wenye Li	The Chinese University of Hong Kong (Shenzhen)
Mingming Li	Institute of Information Engineering, Chinese Academy of Sciences
Yexin Li	Hong Kong University of Science and Technology
Qinghua Li	Renmin University of China
Yaohang Li	Old Dominion University
Yuxuan Liang	National University of Singapore
Zhimin Liang	Institute of Computing Technology, Chinese Academy of Sciences
Hongwei Liang	Microsoft
Nengli Lim	Singapore University of Technology and Design
Suwen Lin	University of Notre Dame
Yangxin Lin	Peking University
Aldo Lipani	University College London
Marco Lippi	University of Modena and Reggio Emilia
Alexei Lisitsa	University of Liverpool
Lin Liu	Taiyuan University of Technology
Weiwen Liu	The Chinese University of Hong Kong
Yang Liu	JD
Huan Liu	Arizona State University
Tianbo Liu	Thomas Jefferson National Accelerator Facility
Tongliang Liu	The University of Sydney
Weidong Liu	Inner Mongolia University
Kai Liu	Colorado School of Mines

Shiwei Liu	Technische Universiteit Eindhoven
Shenghua Liu	Institute of Computing Technology, Chinese Academy of Sciences
Corrado Loglisci	University of Bari Aldo Moro
Andrey Lokhov	Los Alamos National Laboratory
Yijun Lu	Alibaba Cloud
Xuequan Lu	Deakin University
Szymon Lukasik	AGH University of Science and Technology
Phuc Luong	Deakin University
Jianming Lv	South China University of Technology
Gengyu Lyu	Beijing Jiaotong University
Vijaikumar M.	Indian Institute of Science
Jing Ma	Emory University
Nan Ma	Shanghai Jiao Tong University
Sebastian Mair	Leuphana University Lüneburg
Marjan Mansourvar	University of Southern Denmark
Vincent Margot	Advestis
Fernando Martínez-Plumed	Joint Research Centre - European Commission
Florent Masseglia	Inria
Romain Mathonat	Université de Lyon
Deepak Maurya	Indian Institute of Technology Madras
Christian Medeiros Adriano	Hasso-Plattner-Institut
Purvanshi Mehta	University of Rochester
Tobias Meisen	Bergische Universität Wuppertal
Luciano Melodia	Friedrich-Alexander Universität Erlangen-Nürnberg
Ernestina Menasalvas	Universidad Politécnica de Madrid
Vlado Menkovski	Technische Universiteit Eindhoven
Engelbert Mephu Nguifo	Université Clermont Auvergne
Alberto Maria Metelli	Politecnico di Milano
Donald Metzler	Google
Anke Meyer-Baese	Florida State University
Richard Meyes	University of Wuppertal
Haithem Mezni	University of Jendouba
Paolo Mignone	Università degli Studi di Bari Aldo Moro
Matej Mihelčić	University of Zagreb
Decebal Constantin Mocanu	University of Twente
Christoph Molnar	Ludwig Maximilian University of Munich
Lia Morra	Politecnico di Torino
Christopher Morris	TU Dortmund University
Tadeusz Morzy	Poznań University of Technology
Henry Moss	Lancaster University
Tetsuya Motokawa	University of Tsukuba
Mathilde Mougeot	Université Paris-Saclay
Tingting Mu	The University of Manchester
Andreas Mueller	NYU
Tanmoy Mukherjee	Queen Mary University of London

Ksenia Mukhina	ITMO University
Peter Müllner	Know-Center
Guido Muscioni	University of Illinois at Chicago
Waleed Mustafa	TU Kaiserslautern
Mohamed Nadif	University of Paris
Ankur Nahar	Indian Institute of Technology Jodhpur
Kei Nakagawa	Nomura Asset Management Co., Ltd.
Haïfa Nakouri	University of Tunis
Mirco Nanni	KDD-Lab ISTI-CNR Pisa
Nicolo' Navarin	University of Padova
Richi Nayak	Queensland University of Technology
Mojtaba Nayyeri	University of Bonn
Daniel Neider	MPI SWS
Nan Neng	Institute of Information Engineering, Chinese Academy of Sciences
Stefan Neumann	University of Vienna
Dang Nguyen	Deakin University
Kien Duy Nguyen	University of Southern California
Jingchao Ni	NEC Laboratories America
Vlad Niculae	Instituto de Telecomunicações
Sofia Maria Nikolakaki	Boston University
Kun Niu	Beijing University of Posts and Telecommunications
Ryo Nomura	Waseda University
Eirini Ntoutsi	Leibniz University Hannover
Andreas Nuernberger	Otto von Guericke University of Magdeburg
Tsuyoshi Okita	Kyushu Institute of Technology
Maria Oliver Parera	GIPSA-lab
Bruno Ordozgoiti	Aalto University
Sindhu Padakandla	Indian Institute of Science
Tapio Pahikkala	University of Turku
Joao Palotti	Qatar Computing Research Institute
Guansong Pang	The University of Adelaide
Pance Panov	Jožef Stefan Institute
Konstantinos Papangelou	The University of Manchester
Yulong Pei	Technische Universiteit Eindhoven
Nikos Pelekis	University of Piraeus
Thomas Pellegrini	Université Toulouse III - Paul Sabatier
Charlotte Pelletier	Univ. Bretagne Sud
Jaakko Peltonen	Aalto University and Tampere University
Shaowen Peng	Kyushu University
Siqi Peng	Kyoto University
Bo Peng	The Ohio State University
Lukas Pensel	Johannes Gutenberg University Mainz
Aritz Pérez Martínez	Basque Center for Applied Mathematics
Lorenzo Perini	KU Leuven
Matej Petković	Jožef Stefan Institute

Bernhard Pfahringer	University of Waikato
Weiguo Pian	Chongqing University
Francesco Piccialli	University of Naples Federico II
Sebastian Pineda Arango	University of Hildesheim
Gianvito Pio	University of Bari "Aldo Moro"
Giuseppe Pirrò	Sapienza University of Rome
Anastasia Podosinnikova	Massachusetts Institute of Technology
Sebastian Pölsterl	Ludwig Maximilian University of Munich
Vamsi Potluru	JP Morgan AI Research
Rafael Poyiadzi	University of Bristol
Surya Prakash	University of Canberra
Paul Prasse	University of Potsdam
Rameshwar Pratap	Indian Institute of Technology Mandi
Jonas Prellberg	University of Oldenburg
Hugo Proenca	Leiden Institute of Advanced Computer Science
Ricardo Prudencio	Federal University of Pernambuco
Petr Pulc	Institute of Computer Science of the Czech Academy of Sciences
Lei Qi	Iowa State University
Zhenyue Qin	The Australian National University
Rahul Ragesh	PES University
Tahrima Rahman	The University of Texas at Dallas
Zana Rashidi	York University
S. S. Ravi	University of Virginia and University at Albany – SUNY
Ambrish Rawat	IBM
Henry Reeve	University of Birmingham
Reza Refaei Afshar	Technische Universiteit Eindhoven
Navid Rekabsaz	Johannes Kepler University Linz
Yongjian Ren	Shandong University
Zhiyun Ren	The Ohio State University
Guohua Ren	LG Electronics
Yuxiang Ren	Florida State University
Xavier Renard	AXA
Martí Renedo Mirambell	Universitat Politècnica de Catalunya
Gavin Rens	Katholiek Universiteit Leuven
Matthias Renz	Christian-Albrechts-Universität zu Kiel
Guillaume Richard	EDF R&D
Matteo Riondato	Amherst College
Niklas Risse	Bielefeld University
Lars Rosenbaum	Robert Bosch GmbH
Celine Rouveirol	Université Sorbonne Paris Nord
Shoumik Roychoudhury	Temple University
Polina Rozenshtein	Aalto University
Peter Rubbens	Flanders Marine Institute (VLIZ)
David Ruegamer	LMU Munich

Matteo Ruffini	ToolsGroup
Ellen Rushe	Insight Centre for Data Analytics
Amal Saadallah	TU Dortmund
Yogish Sabharwal	IBM Research - India
Mandana Saebi	University of Notre Dame
Aadirupa Saha	IISc
Seyed Erfan Sajjadi	Brunel University
Durgesh Samariya	Federation University
Md Samiullah	Monash University
Mark Sandler	Google
Raul Santos-Rodriguez	University of Bristol
Yucel Saygin	Sabancı University
Pierre Schaus	UCLouvain
Fabian Scheipl	Ludwig Maximilian University of Munich
Katerina Schindlerova	University of Vienna
Ute Schmid	University of Bamberg
Daniel Schmidt	Monash University
Sebastian Schmoll	Ludwig Maximilian University of Munich
Johannes Schneider	University of Liechtenstein
Marc Schoenauer	Inria Saclay Île-de-France
Jonas Schouterden	Katholieke Universiteit Leuven
Leo Schwinn	Friedrich-Alexander-Universität Erlangen-Nürnberg
Florian Seiffarth	University of Bonn
Nan Serra	NEC Laboratories Europe GmbH
Rowland Seymour	Univeristy of Nottingham
Ammar Shaker	NEC Laboratories Europe
Ali Shakiba	Vali-e-Asr University of Rafsanjan
Junming Shao	University of Science and Technology of China
Zhou Shao	Tsinghua University
Manali Sharma	Samsung Semiconductor Inc.
Jiaming Shen	University of Illinois at Urbana-Champaign
Ying Shen	Sun Yat-sen University
Hao Shen	fortiss GmbH
Tao Shen	University of Technology Sydney
Ge Shi	Beijing Institute of Technology
Ziqiang Shi	Fujitsu Research & Development Center
Masumi Shirakawa	hapicom Inc./Osaka University
Kai Shu	Arizona State University
Amila Silva	The University of Melbourne
Edwin Simpson	University of Bristol
Dinesh Singh	RIKEN Center for Advanced Intelligence Project
Jaspreet Singh	L3S Research Centre
Spiros Skiadopoulos	University of the Peloponnese
Gavin Smith	University of Nottingham
Miguel A. Solinas	CEA
Dongjin Song	NEC Labs America

Jiong Wang	Institute of Information Engineering, Chinese Academy of Sciences
Xiaobao Wang	Tianjin University
Shuheng Wang	Nanjing University of Science and Technology
Jihu Wang	Shandong University
Haobo Wang	Zhejiang University
Xianzhi Wang	University of Technology Sydney
Chao Wang	Shanghai Jiao Tong University
Jun Wang	Southwest University
Jing Wang	Beijing Jiaotong University
Di Wang	Nanyang Technological University
Yashen Wang	China Academy of Electronics and Information Technology of CETC
Qinglong Wang	McGill University
Sen Wang	University of Queensland
Di Wang	State University of New York at Buffalo
Qing Wang	Information Science Research Centre
Guoyin Wang	Chongqing University of Posts and Telecommunications
Thomas Weber	Ludwig-Maximilians-Universität München
Lingwei Wei	University of Chinese Academy of Sciences; Institute of Information Engineering, CAS
Tong Wei	Nanjing University
Pascal Welke	University of Bonn
Yang Wen	University of Science and Technology of China
Yanlong Wen	Nankai University
Paul Weng	UM-SJTU Joint Institute
Matthias Werner	ETAS GmbH, Bosch Group
Joerg Wicker	The University of Auckland
Uffe Wiil	University of Southern Denmark
Paul Wimmer	University of Lübeck; Robert Bosch GmbH
Martin Wistuba	University of Hildesheim
Feijie Wu	The Hong Kong Polytechnic University
Xian Wu	University of Notre Dame
Hang Wu	Georgia Institute of Technology
Yubao Wu	Georgia State University
Yichao Wu	SenseTime Group Limited
Xi-Zhu Wu	Nanjing University
Jia Wu	Macquarie University
Yang Xiaofei	Harbin Institute of Technology, Shenzhen
Yuan Xin	University of Science and Technology of China
Liu Xinshun	VIVO
Taufik Xu	Tsinghua University
Jinhui Xu	State University of New York at Buffalo
Depeng Xu	University of Arkansas
Peipei Xu	University of Liverpool

Yichen Xu	Beijing University of Posts and Telecommunications
Bo Xu	Donghua University
Hansheng Xue	Harbin Institute of Technology, Shenzhen
Naganand Yadati	Indian Institute of Science
Akihiro Yamaguchi	Toshiba Corporation
Haitian Yang	Institute of Information Engineering, Chinese Academy of Sciences
Hongxia Yang	Alibaba Group
Longqi Yang	HPCL
Xiaochen Yang	University College London
Yuhan Yang	Shanghai Jiao Tong University
Ya Zhou Yang	National University of Defense Technology
Feidiao Yang	Institute of Computing Technology, Chinese Academy of Sciences
Liu Yang	Tianjin University
Chaoqi Yang	University of Illinois at Urbana-Champaign
Carl Yang	University of Illinois at Urbana-Champaign
Guanyu Yang	Xi'an Jiaotong - Liverpool University
Yang Yang	Nanjing University
Weicheng Ye	Carnegie Mellon University
Wei Ye	Peking University
Yanfang Ye	Case Western Reserve University
Kejiang Ye	SIAT, Chinese Academy of Sciences
Florian Yger	Université Paris-Dauphine
Yunfei Yin	Chongqing University
Lu Yin	Technische Universiteit Eindhoven
Wang Yingkui	Tianjin University
Kristina Yordanova	University of Rostock
Tao You	Northwestern Polytechnical University
Hong Qing Yu	University of Bedfordshire
Bowen Yu	Institute of Information Engineering, Chinese Academy of Sciences
Donghan Yu	Carnegie Mellon University
Yipeng Yu	Tencent
Shujian Yu	NEC Laboratories Europe
Jiadi Yu	Shanghai Jiao Tong University
Wenchao Yu	University of California, Los Angeles
Feng Yuan	The University of New South Wales
Chunyuan Yuan	Institute of Information Engineering, Chinese Academy of Sciences
Sha Yuan	Tsinghua University
Farzad Zafarani	Purdue University
Marco Zaffalon	IDSIA
Nayyar Zaidi	Monash University
Tianzi Zang	Shanghai Jiao Tong University
Gerson Zaverucha	Federal University of Rio de Janeiro

Program Committee Members, Applied Data Science Track

Deepak Ajwani	Nokia Bell Labs
Nawaf Alharbi	Kansas State University
Rares Ambrus	Toyota Research Institute
Maryam Amir Haeri	Technische Universität Kaiserslautern
Jean-Marc Andreoli	Naverlabs Europe
Cecilio Angulo	Universitat Politècnica de Catalunya
Stefanos Antaris	KTH Royal Institute of Technology
Nino Antulov-Fantulin	ETH Zurich
Francisco Antunes	University of Coimbra
Muhammad Umer Anwaar	Technical University of Munich
Cristian Axenie	Audi Konfuzius-Institut Ingolstadt/Technical University of Ingolstadt
Mehmet Cem Aytekin	Sabancı University
Anthony Bagnall	University of East Anglia
Marco Baldan	Leibniz University Hannover
Maria Bampa	Stockholm University
Karin Becker	UFRGS
Swarup Ranjan Behera	Indian Institute of Technology Guwahati
Michael Berthold	University of Konstanz
Antonio Bevilacqua	Insight Centre for Data Analytics
Ananth Reddy Bhimireddy	Indiana University Purdue University - Indianapolis
Haixia Bi	University of Bristol
Wu Bin	Zhengzhou University
Thibault Blanc Beyne	INP Toulouse
Andrzej Bobyk	Maria Curie-Skłodowska University
Antonio Bonafonte	Amazon
Ludovico Boratto	Eurecat
Massimiliano Botticelli	Robert Bosch GmbH
Maria Brbic	Stanford University
Sebastian Buschjäger	TU Dortmund
Rui Camacho	University of Porto
Doina Caragea	Kansas State University
Nicolas Carrara	University of Toronto
Michele Catasta	Stanford University
Oded Cats	Delft University of Technology
Tania Cerquitelli	Politecnico di Torino
Fabricio Ceschin	Federal University of Paraná
Jeremy Charlier	University of Luxembourg
Anveshi Charuvaka	GE Global Research
Liang Chen	Sun Yat-sen University
Zhiyong Cheng	Shandong Artificial Intelligence Institute

João Mendes-Moreira	University of Porto
Ioannis Mitros	Insight Centre for Data Analytics
Elena Mocanu	University of Twente
Hebatallah Mohamed	Free University of Bozen-Bolzano
Roghayeh Mojarad	Université Paris-Est Créteil
Mirco Nanni	KDD-Lab ISTI-CNR Pisa
Juggapong Natwichai	Chiang Mai University
Sasho Nedelkoski	TU Berlin
Kei Nemoto	The Graduate Center, City University of New York
Ba-Hung Nguyen	Japan Advanced Institute of Science and Technology
Tobias Nickchen	Paderborn University
Aastha Nigam	LinkedIn Inc
Inna Novalija	Jožef Stefan Institute
Francisco Ocegueda-Hernandez	National Oilwell Varco
Tsuyoshi Okita	Kyushu Institute of Technology
Oghenejokpeme Orhobor	The University of Manchester
Aomar Osmani	Université Sorbonne Paris Nord
Latifa Oukhellou	IFSTTAR
Rodolfo Palma	Inria Chile
Pankaj Pandey	Indian Institute of Technology Gandhinagar
Luca Pappalardo	University of Pisa, ISTI-CNR
Paulo Paraíso	INESC TEC
Namyong Park	Carnegie Mellon University
Chanyoung Park	University of Illinois at Urbana-Champaign
Miquel Perelló-Nieto	University of Bristol
Nicola Pezzotti	Philips Research
Tiziano Piccardi	Ecole Polytechnique Fédérale de Lausanne
Thom Pijnenburg	Elsevier
Valentina Poggioni	Università degli Studi di Perugia
Chuan Qin	University of Science and Technology of China
Jiezhong Qiu	Tsinghua University
Maria Ramirez-Loaiza	Intel Corporation
Manjusha Ravindranath	ASU
Zhaochun Ren	Shandong University
Antoine Richard	Georgia Institute of Technology
Kit Rodolfa	Carnegie Mellon University
Mark Patrick Roeling	Technical University of Delft
Soumyadeep Roy	Indian Institute of Technology Kharagpur
Ellen Rushe	Insight Centre for Data Analytics
Amal Saadallah	TU Dortmund
Carlos Salort Sanchez	Huawei
Eduardo Hugo Sanchez	IRT Saint Exupéry
Markus Schmitz	University of Erlangen-Nuremberg/BMW Group
Ayan Sengupta	Optum Global Analytics (India) Pvt. Ltd.
Ammar Shaker	NEC Laboratories Europe

Manali Sharma	Samsung Semiconductor Inc.
Jiaming Shen	University of Illinois at Urbana-Champaign
Dash Shi	LinkedIn
Ashish Sinha	IIT Roorkee
Yorick Spenrath	Technische Universiteit Eindhoven
Simon Stieber	University of Augsburg
Hendra Suryanto	Rich Data Corporation
Raunak Swarnkar	IIT Gandhinagar
Imen Trabelsi	National Engineering School of Tunis
Alexander Treiss	Karlsruhe Institute of Technology
Rahul Tripathi	Amazon
Dries Van Daele	Katholieke Universiteit Leuven
Ranga Raju Vatsavai	North Carolina State University
Vishnu Venkataraman	Credit Karma
Sergio Viademonte	Vale Institute of Technology, Vale SA
Yue Wang	Microsoft Research
Changzhou Wang	The Boeing Company
Xiang Wang	National University of Singapore
Hongwei Wang	Shanghai Jiao Tong University
Wenjie Wang	Emory University
Zirui Wang	Carnegie Mellon University
Shen Wang	University of Illinois at Chicago
Dingxian Wang	East China Normal University
Yoshikazu Washizawa	The University of Electro-Communications
Chrys Watson Ross	University of New Mexico
Dilusha Weeraddana	CSIRO
Ying Wei	The Hong Kong University of Science and Technology
Laksri Wijerathna	Monash University
Le Wu	Hefei University of Technology
Yikun Xian	Rutgers University
Jian Xu	Citadel
Haiqin Yang	Ping An Life
Yang Yang	Northwestern University
Carl Yang	University of Illinois at Urbana-Champaign
Chin-Chia Michael Yeh	Visa Research
Shujian Yu	NEC Laboratories Europe
Chung-Hsien Yu	University of Massachusetts Boston
Jun Yuan	The Boeing Company
Stella Zevio	LIPN
Hanwen Zha	University of California, Santa Barbara
Chuxu Zhang	University of Notre Dame
Fanjin Zhang	Tsinghua University
Xiaohan Zhang	Sony Interactive Entertainment
Xinyang Zhang	University of Illinois at Urbana-Champaign
Mia Zhao	Airbnb
Qi Zhu	University of Illinois at Urbana-Champaign

Hengshu Zhu	Baidu Inc.
Tommaso Zoppi	University of Florence
Lan Zou	Carnegie Mellon University

Program Committee Members, Demo Track

Deepak Ajwani	Nokia Bell Labs
Rares Ambrus	Toyota Research Institute
Jean-Marc Andreoli	NAVER LABS Europe
Ludovico Boratto	Eurecat
Nicolas Carrara	University of Toronto
Michelangelo Ceci	Università degli Studi di Bari Aldo Moro
Tania Cerquitelli	Politecnico di Torino
Liang Chen	Sun Yat-sen University
Jiawei Chen	Zhejiang University
Zhiyong Cheng	Shandong Artificial Intelligence Institute
Silvia Chiusano	Politecnico di Torino
Henggang Cui	Uber ATG
Tiago Cunha	University of Porto
Chris Develder	Ghent University
Nat Dilokthanakul	Vidyasirimedhi Institute of Science and Technology
Daizong Ding	Fudan University
Kaize Ding	ASU
Xiaowen Dong	University of Oxford
Fuli Feng	National University of Singapore
Enrique Frias-Martinez	Telefónica Research and Development
Zuohui Fu	Rutgers University
Chen Gao	Tsinghua University
Thomas Gärtner	TU Wien
Derek Greene	University College Dublin
Severin Gsponer	University College Dublin
Xinyu Guan	Xi'an Jiaotong University
Junheng Hao	University of California, Los Angeles
Ziniu Hu	University of California, Los Angeles
Chao Huang	University of Notre Dame
Hong Huang	UGoe
Neil Hurley	University College Dublin
Guillaume Jacquet	Joint Research Centre - European Commission
Di Jiang	WeBank
Song Jiang	University of California, Los Angeles
Jihed Khiari	Johannes Kepler Universität Linz
Mark Last	Ben-Gurion University of the Negev
Thach Le Nguyen	The Insight Centre for Data Analytics
Vincent Lemaire	Orange Labs
Camelia Lemnaru	Universitatea Tehnică din Cluj-Napoca
Bowen Liu	Stanford University

Yin Lou	Ant Financial
Yao Ma	Michigan State University
Brian Mac Namee	University College Dublin
Susan Mckeever	Technological University Dublin
Edgar Meij	Bloomberg L.P.
Gabor Melli	Sony PlayStation
Decebal Constantin Mocanu	University of Twente
Elena Mocanu	University of Twente
Luis Moreira-Matias	Kreditech
Latifa Oukhellou	IFSTTAR
Chanyoung Park	University of Illinois at Urbana-Champaign
Julien Perez	NAVER LABS Europe
Chuan Qin	University of Science and Technology of China
Jiezhong Qiu	Tsinghua University
Zhaochun Ren	Shandong University
Kit Rodolfa	Carnegie Mellon University
Robert Ross	Technological University Dublin
Parinya Sanguansat	Panyapiwat Institute of Management
Niladri Sett	University College Dublin
Manali Sharma	Samsung Semiconductor Inc.
Jerzy Stefanowski	Poznań University of Technology
Luis Teixeira	Fraunhofer Portugal AICOS
Roberto Trasarti	ISTI-CNR, Pisa
Anton Tsitsulin	University of Bonn
Grigorios Tsoumakas	Aristotle University of Thessaloniki
Marco Turchi	Fondazione Bruno Kessler
Vishnu Venkataraman	Credit Karma
Yue Wang	Microsoft Research
Xiang Wang	National University of Singapore
Dingxian Wang	East China Normal University
Ying Wei	The Hong Kong University of Science and Technology
Le Wu	Hefei University of Technology
Jian Xu	Citadel
Carl Yang	University of Illinois at Urbana-Champaign
Yang Yang	Northwestern University
Haiqin Yang	Ping An Life
Chung-Hsien Yu	University of Massachusetts Boston
Chuxu Zhang	University of Notre Dame
Xiaohan Zhang	Sony Interactive Entertainment

Sponsors

amazon | science

Google

Qualcomm
AI research

visitgent

gentcongres

Invited Talks Abstracts

Interpretability for Everyone

Been Kim

Google Brain

Abstract. In this talk, I will share some of my reflections on the progress made in the field of interpretable machine learning. We will reflect on where we are going as a field, and what are the things that we need to be aware of to make progress. With that perspective, I will then discuss some of my work on 1) sanity checking popular methods and 2) developing more lay person-friendly interpretability methods.

Bio: Been Kim is a senior research scientist at Google Brain. Her research focuses on building interpretable machine learning—making ML understandable by humans for more responsible AI. The vision of her research is to make humans empowered by machine learning, not overwhelmed by it. She gave a talk at the G20 meeting on the digital economy summit in Argentina in 2019. Her work called TCAV received a UNESCO Netexplo award for "breakthrough digital innovations with the potential of profound and lasting impact on the digital society". This work was also a part of the CEO's keynote at Google I/O 2019. She gave talks on the topic at ICML in 2017, CVPR and MLSS at University of Toronto in 2018, and Lawrence Berkeley National Laboratory in 2019. She was a co-workshop Chair of ICLR 2019, and has been an area chair at conferences including NIPS, ICML, ICLR, and AISTATS, in 2018. She received her Ph.D. from MIT.

Building Knowledge for AI Agents with Reinforcement Learning

Doina Precup

McGill University and DeepMind Montreal

Abstract. Reinforcement learning allows autonomous agents to learn how to act in a stochastic, unknown environment, with which they can interact. Deep reinforcement learning, in particular, has achieved great success in well-defined application domains, such as Go or chess, in which an agent has to learn how to act and there is a clear success criterion. In this talk, I will focus on the potential role of reinforcement learning as a tool for building knowledge representations in AI agents whose goal is to perform continual learning. I will examine a key concept in reinforcement learning, the value function, and discuss its generalization to support various forms of predictive knowledge. I will also discuss the role of temporally extended actions, and their associated predictive models, in learning procedural knowledge. In order to tame the possible complexity of exploring in order to build knowledge, reinforcement learning agents can use the concepts of intents (i.e., intended consequences of actions) and affordances (which capture knowledge about the states in which an action is applicable). Finally, I will discuss the challenge of how to evaluate reinforcement learning agents whose goal is not just to control their environment, but also to build knowledge about their world.

Bio: Doina Precup splits her time between McGill University, where she co-directs the Reasoning and Learning Lab in the School of Computer Science, and DeepMind Montreal, where she has led the research team since its formation in October 2017. Her research interests are in the areas of reinforcement learning, deep learning, time series analysis, and diverse applications of machine learning in health care, automated control, and other fields. She became a senior member of the Association for the Advancement of Artificial Intelligence in 2015, Canada Research Chair in Machine Learning in 2016, Senior Fellow of the Canadian Institute for Advanced Research in 2017, and received a Canada CIFAR AI (CCAI) Chair in 2018. Dr. Precup is also involved in activities supporting the organization of Mila and the wider Montreal and Quebec AI ecosystem.

Algorithmic Auditing: How to Open the Black Box of ML

Gemma Galdon-Clavell

Eticas Research & Consulting

Abstract. As algorithms proliferate, so do concerns over how they work and their impacts. In the last few years, laws and regulations have been drafted calling for AI to be transparent, accountable, and explainable. However, it is still unclear how these principles can be translated into practices, and the standards that will guide the algorithms of tomorrow are still in the making. In this presentation, Gemma Galdon, founder of Eticas Consulting, will share the experience of auditing algorithms as a way to open the black box of AI processes. Through the discussion around this specific tool and the experiences of the Eticas team in auditing algorithms, the many sources of bias and inefficiencies in ML systems will be presented.

Bio: Dr. Gemma Galdon-Clavell is a policy analyst working on surveillance, social, legal and ethical impacts of technology, smart cities, privacy, security policy, resilience and policing. She is a founding partner at Eticas Research & Consulting and a researcher at the Universitat de Barcelona's Sociology Department. She completed her PhD on surveillance, security and urban policy in early 2012 at the Universitat Autònoma de Barcelona, where she also received an MSc in Policy Management, and was later appointed Director of the Security Policy Programme at the Universitat Oberta de Catalunya (UOC). Previously, she worked at the Transnational Institute, the United Nations' Institute for Training and Research (UNITAR) and the Catalan Institute for Public Security. She teaches topics related to her research at several foreign universities, mainly in Latin America, and is a member of the IDRC-funded Latin American Surveillance Studies Network. Additionally, she is a member of the international advisory board of Privacy International and a regular analyst on TV and radio and in print media. Her recent academic publications tackle issues related to the proliferation of surveillance in urban settings, urban security policy and community safety, security and mega events, the relationship between privacy and technology, and smart cities.

Amortized and Neural Augmented Inference

Max Welling

University of Amsterdam

Abstract. Amortized inference is the process of learning to perform inference from many related tasks. The Variational Autoencoder (VAE) employs an amortized inference network, otherwise known as the encoder. Amortization is a powerful concept that also applies to learning itself (learning to learn or meta learning) and to optimization (learning to optimize with reinforcement learning). In this talk we will develop hybrid amortized methods that combine classical learning, inference and optimization algorithms with learned neural networks. The learned neural network augments or corrects the classical solution, or, conversely the neural network is fed useful information computed by a classical method. We further extend these ideas to amortized causal inference, where we learn from data to discover the causal relations in data. Neural augmentation is applied to problems in MRI image reconstruction, LDPC decoding and MIMO detection.

Bio: Max Welling is a research chair in Machine Learning at the University of Amsterdam and a VP Technologies at Qualcomm. He has a secondary appointment as a senior fellow at the Canadian Institute for Advanced Research (CIFAR). He is co-founder of "Scyfer BV", a university spin-off in deep learning which was acquired by Qualcomm in summer 2017. In the past he held postdoctoral positions at Caltech ('98–'00), UCL ('00–'01) and U. Toronto ('01–'03). He received his PhD in '98 under the supervision of Nobel laureate Prof. G. 't Hooft. Max Welling has served as associate editor in chief of IEEE TPAMI from 2011–2015 (impact factor 4.8). He has served on the board of the NIPS foundation since 2015 (the largest conference in machine learning) and was program chair and general chair of NIPS in 2013 and 2014 respectively. He was also program chair of AISTATS in 2009 and ECCV in 2016 and general chair of MIDL 2018. He has served on the editorial boards of JMLR and JML and was an associate editor for Neurocomputing, JCGS and TPAMI. He received multiple grants from Google, Facebook, Yahoo, NSF, NIH, NWO and ONR-MURI including an NSF career grant in 2005. He was the recipient of the ECCV Koenderink Prize in 2010. Welling is on the board of the Data Science Research Center in Amsterdam, he directs the Amsterdam Machine Learning Lab (AMLAB), and co-directs the Qualcomm-UvA deep learning lab (QUVA) and the Bosch-UvA Deep Learning lab (DELTA). Max Welling has over 250 scientific publications in machine learning, computer vision, statistics and physics and an h-index of 62.

Can You Trust Your GNN? – Certifiable Robustness of Machine Learning Models for Graphs

Stephan Günnemann

Technical University of Munich

Abstract. Graph neural networks have achieved impressive results in various graph learning tasks and they have found their way into many applications such as molecular property prediction, cancer classification, fraud detection, and knowledge graph reasoning. Despite their proliferation, studies of their robustness properties are still very limited – yet, in domains where graph learning methods are often used the data is rarely perfect and adversaries (e.g., on the Web) are common. Specifically, in safety-critical environments and decision-making contexts involving humans, it is crucial to ensure the GNN's reliability. In my talk, I will shed light on the aspect of robustness for state-of-the art graph-based learning techniques. I will highlight the unique challenges and opportunities that come along with the graph setting and I will showcase the method's vulnerabilities. Based on these insights, I will discuss different principles allowing us to certify robustness, giving us provable guarantees about the GNN's behavior, and ways to improve their reliability.

Bio: Stephan Günnemann is a Professor at the Department of Informatics, Technical University of Munich. His main research focuses on reliable machine learning techniques, specifically targeting graphs and temporal data. His works on subspace clustering on graphs as well as adversarial robustness of graph neural networks have received the best research paper awards at ECML-PKDD and KDD. Stephan acquired his doctoral degree at RWTH Aachen University, Germany in the field of computer science. From 2012 to 2015 he was an associate of Carnegie Mellon University, USA. Stephan has been a visiting researcher at Simon Fraser University, Canada, and a research scientist at the Research & Technology Center of Siemens AG. In 2017 he became a Junior Fellow of the German Computer Science Society. He currently acts as scientific advisor for the Fraunhofer Society to build up a new institute for Cognitive Systems focusing on safe and reliable AI. Stephan has been a (senior) PC member/area chair at conferences including NeurIPS, ICML, KDD, ECML-PKDD, AAAI, and WWW.

Contents – Part I

Pattern Mining

Clustering

Privacy and Fairness

(Social) Network Analysis and Computational Social Science

Dimensionality Reduction and Autoencoders

Domain Adaptation

Pattern Mining

Maximum Margin Separations in Finite Closure Systems

Florian Seiffarth[1], Tamás Horváth[1,2,3](✉), and Stefan Wrobel[1,2,3]

[1] Department of Computer Science, University of Bonn, Bonn, Germany
{seiffarth,horvath,wrobel}@cs.uni-bonn.de
[2] Fraunhofer IAIS, Schloss Birlinghoven, Sankt Augustin, Germany
[3] Fraunhofer Center for Machine Learning, Sankt Augustin, Germany

Abstract. Monotone linkage functions provide a measure for proximities between elements and subsets of a ground set. Combining this notion with Vapnik's idea of support vector machines, we extend the concepts of maximal closed set and half-space separation in finite closure systems to those with maximum margin. In particular, we define the notion of margin for finite closure systems by means of monotone linkage functions and give a greedy algorithm computing a maximum margin closed set separation for two sets efficiently. The output closed sets are maximum margin half-spaces, i.e., form a partitioning of the ground set if the closure system is Kakutani. We have empirically evaluated our approach on different synthetic datasets. In addition to binary classification of finite subsets of the Euclidean space, we considered also the problem of vertex classification in graphs. Our experimental results provide clear evidence that maximal closed set separation with maximum margin results in a much better predictive performance than that with arbitrary maximal closed sets.

Keywords: Closure systems · Maximum margin separations · Monotone linkages · Binary classification

1 Introduction

Motivated by different applications of *finite closure systems*, including e.g. closed itemset mining [12], inductive logic programming [11], and formal concept analysis [5], in [14] we studied the algorithmic properties of *half-space* and *maximal closed set* separation in this kind of set systems. One of our results in [14] is a greedy algorithm, which takes as input two sets and returns two *disjoint* maximal closed sets containing them if their closures are disjoint. It is shown in [14] that this greedy algorithm provides an algorithmic characterization of the special class of *Kakutani* closure systems [2,9]. That is, for any two sets it returns two complementary *half-spaces* containing them if and only if the closures of the input sets are disjoint, where a half-space is a closed set such that its complement is also closed. For the case that the separating maximal closed sets or half-spaces are not unique, the greedy algorithm returns one of them selected arbitrarily.

F. Hutter et al. (Eds.): ECML PKDD 2020, LNAI 12457, pp. 3–18, 2021.
https://doi.org/10.1007/978-3-030-67658-2_1

This is similar to Rosenblatt's perceptron algorithm [13], which fulfills also the minimum requirement the output hyperplane to separate the input point sets. A major drawback of such unconstrained solutions is that they provide no control of *overfitting*. This problem has been addressed by Vapnik and his co-authors' work on *support vectors machines* (SVM) [1], which have become a well-established tool within machine learning for its well-founded theory and excellent predictive performance on a broad range of real-world problems. In particular, SVM resolve the problem of overfitting by separating the data points in an inner product feature space by the hyperplane maximizing the minimum of the distances to the sets of positive and negative examples.

Motivated by the same problem as SVM, in this work we adapt the idea of maximum margin hyperplanes to the binary separation problems studied in [14] for *finite closure systems*. We stress that our adaptation does *not* generalize SVM to finite closure systems. While in case of SVM the inner product induces a distance, in case of finite closure systems the ground set is typically *not* a metric space. To overcome this problem, we assume that the closure systems are provided by some *weak* measure of proximity defined by means of *monotone linkage functions* [10]. While this kind of functions strongly generalize distance functions (e.g., they are not required to fulfill symmetry or the triangle inequality), they preserve the *anti-monotonicity* of distances. That is, the linkage from a point to a set is anti-monotonic for set inclusion. Similarly to SVM, this feature is essential for the separation problems considered in this work. A second issue is how to define margins for closed set and half-space separations in finite closure systems. While there are different equivalent characterizations of maximum margins for SVM, it turns out that their equivalence does not hold when adapting them to abstract closure systems equipped with monotone linkage functions. In particular, in contrast to SVM, the linkage of the set of positive examples to a half-space can be different from that of the negative examples to the complementary half-spaces for *all* half-spaces. We therefore define the *margin* by the smallest linkage from the closures of the input sets to the complementary half-spaces. Furthermore, we generalize this concept to arbitrary closed set separations as well.

Using these notions, we formulate the computational problems of finding closed set and half-space separations *maximizing* the margin in finite closure systems equipped with monotone linkage functions. This problem preserves several key features of SVM for abstract closure systems. For the above problem we give another *greedy* algorithm and prove that it is correct and requires a linear number of evaluations of the underlying closure operator and linkage function. We also show that for Kakutani closure systems [9], the algorithm always returns a half-space separation of the input sets with maximum margin if and only if the closures of the two training sets are disjoint.

We experimentally evaluated the predictive performance of our algorithm on various synthetic datasets. Our empirical results concerning point classification in Euclidean spaces show that our algorithm clearly outperforms the greedy separation algorithm in [14]. In addition, we carried out several experiments with vertex classification in trees and also in random graphs using the shortest path closure operator. Similarly to the other experiments, our algorithm consistently outperformed the greedy algorithm in [14]. For space limitations we omit further applications dealing, among others, with finite lattices, in particular, with formal concept and subsumption lattices in inductive logic programming.

The rest of the paper is organized as follows. In Sect. 2 we collect the necessary concepts and fix the notation. In Sect. 3 we introduce our notion of margin defined by means of monotone linkage functions for finite closure systems. In Sect. 4 we present our greedy algorithm solving closed set and half-space separation with maximum margin and prove some of its basic formal properties. In Sect. 5 we report our experimental results. Finally, in Sect. 6, we conclude and formulate some problems for further study.

2 Preliminaries

In this section we collect some basic notions concerning *closure systems* (see, e.g., [3,9]) and *linkage functions* (see, e.g., [6]) and fix the notation.

Closure Systems. The power set of a set E is denoted by 2^E. A *set system* over a ground set E is a pair $(E, \mathcal{C})$, where $\mathcal{C} \subseteq 2^E$; $(E, \mathcal{C})$ is a *closure system* if it fulfills the axioms: (i) $E \in \mathcal{C}$ and (ii) $X \cap Y \in \mathcal{C}$ for all $X, Y \in \mathcal{C}$. Unless otherwise stated, by closure systems we always mean *finite* closure systems, i.e., $|E| < \infty$. It is a well-known fact (see, e.g., [3]) that closure systems give rise to closure operators and vice versa. More precisely, a *closure operator* over E is a function $\rho : 2^E \to 2^E$ satisfying

i) $X \subseteq \rho(X)$, (*extensivity*)
ii) $\rho(X) \subseteq \rho(Y)$ whenever $X \subseteq Y$, (*monotonicity*)
iii) $\rho(\rho(X)) = \rho(X)$ (*idempotency*)

for all $X, Y \subseteq E$. The following characterization is standard (see, e.g., [3]):

Proposition 1. *Let* $(E, \mathcal{C})$ *be a closure system and* $\rho : 2^E \to 2^E$ *be the map defined by* $\rho(X) = \bigcap\{C \in \mathcal{C} : X \subseteq C\}$ *for all* $X \subseteq E$. *Then* ρ *is a closure operator and* $\mathcal{C} = \{C \subseteq E : \rho(C) = C\}$. *Conversely, let* ρ *be a closure operator over* E. *Then* $(E, \mathcal{C}_\rho)$ *with* $\mathcal{C}_\rho = \{C \subseteq E : \rho(C) = C\}$ *is a closure system.*

The elements of $\mathcal{C}$ will be referred to as *closed* sets. We use the notation $\mathcal{C}_\rho$ to indicate that the closure system is defined by the closure operator ρ.

We will have a special interest in the following closure systems over Euclidean spaces, graphs, and lattices[1].

1. (*finite convex hulls in* $\mathbb{R}^d$) Let E be a finite subset of $\mathbb{R}^d$ for some $d > 0$. Then the function $\alpha : 2^E \to 2^E$ defined by

$$\alpha(X) = \text{conv}(X) \cap E \tag{1}$$

for all $X \subseteq E$ is a closure operator over E, where $\text{conv}(\cdot)$ denotes the *convex hull* operator on $\mathbb{R}^d$.

[1] For space limitation, the applications concerning closure systems over finite lattices will be discussed in the long version of this paper.

2. (*shortest path closure in graphs* [4]) Let $G = (V, E)$ be a graph with vertex set V and edge set E. Then $(V, \mathcal{C}_\gamma)$ is a closure system if
$$V' \in \mathcal{C}_\gamma \iff V(P) \subseteq V' \tag{2}$$
for all $V' \subseteq V$, $u, v \in V'$, and $P \in \mathcal{S}_{u,v}$, where $\mathcal{S}_{u,v}$ is the set of all shortest paths connecting u and v in G and $V(P)$ denotes the set of vertices in P.
3. (*closed sets in lattices* [15]) Let $(L; \leq)$ be a finite lattice. Then the function $\lambda : 2^L \to 2^L$ defined by
$$\lambda : L' \mapsto \{x \in L \mid \inf L' \leq x \leq \sup L'\} \tag{3}$$
for all $L' \subseteq L$ is a closure operator, where $\inf L'$ (resp. $\sup L'$) is the greatest lower bound or bottom (resp. least upper bound or top) element of L'.

The primary focus of this work is on maximum margin separation in closure systems. To formulate this problem in Sect. 3, we recall some definitions concerning separations in finite closure systems from [14]. More precisely, let $(E, \mathcal{C})$ be a closure system and $A, B \subseteq E$. Then A and B are

(i) *separable* in $(E, \mathcal{C})$ if there are disjoint closed sets C_A, C_B in $\mathcal{C}$ such that $A \subseteq C_A$ and $B \subseteq C_B$,
(ii) *maximal closed set separable* in $(E, \mathcal{C})$ if there are disjoint closed sets C_A, C_B in $\mathcal{C}$ such that $A \subseteq C_A$ and $B \subseteq C_B$, and there are no disjoint closed sets $C'_A \supseteq C_A$ and $C'_B \supseteq C_B$ such that at least one of the containments is proper,
(iii) *half-space separable* if there are $C, C^c \in \mathcal{C}$ such that $A \subseteq C$ and $B \subseteq C^c$, where $C^c = E \setminus C$.

Regarding (iii) above, a closed set $C \in \mathcal{C}$ is a *half-space* if its complement C^c is also closed. Finally, a closure system $(E, \mathcal{C})$ is *Kakutani* [9] if and only if all pairs of disjoint closed sets in $\mathcal{C}$ are half-space separable. It follows from the definitions that no separation is possible in $(E, \mathcal{C})$ if $\emptyset \notin \mathcal{C}$. Therefore, in the rest of the paper we always assume that the empty set is also closed, i.e., it is an element of the underlying closure system.

Monotone Linkage Functions. To adapt Vapnik's idea of maximum margin separation to (abstract) finite closure systems, we need some additional formal tool to quantify the closeness between subsets of the ground set. Such an abstract measure for the proximity between elements and subsets of a ground set is provided by *monotone linkage functions* introduced by Mullat [10]. This kind of functions preserve an important elementary property of distances from points to sets in metric spaces and can therefore be regarded as a very general "distance" concept. More precisely, a *monotone linkage function* over a set E is a map $l : 2^E \times E \to \mathbb{R}$ such that
$$X \subseteq Y \implies l(X, e) \geq l(Y, e)$$
holds for all $X, Y \subseteq E$ and $e \in E$. That is, l is *anti-monotone* w.r.t. set containment, which is an essential property satisfied by distances as well. Thus, all

distances give rise to monotone linkage functions; the converse is, however, not true. Note that by applying monotone linkage functions to singletons in the first argument, we obtain a pairwise proximity between the elements of the ground set. However, in contrast to metric spaces, the definition does not imply symmetry, i.e., $l(\{x\}, y)$ is not necessarily equal to $l(\{y\}, x)$. Furthermore, $l(X, e)$ is not required to be zero for $e \in X$.

There are several examples of monotone linkage functions on finite and infinite ground sets. Below we recall some of the most popular ones to illustrate the concept (c.f. [6] for further examples). The proof that the functions below are all monotone linkage is left to the reader.

(i) (*monotone linkage in* $\mathbb{R}^d$) For any distance D on $\mathbb{R}^d$, define $l : 2^{\mathbb{R}^d} \times \mathbb{R}^d \to \mathbb{R}$ by $l : (X, e) \mapsto \inf_{x \in X} \{D(x, e)\}$ for all $X \subseteq \mathbb{R}^d$ and $e \in \mathbb{R}^d$.
(ii) (*monotone linkage in (weighted) graphs*) For a (weighted) graph $G = (V, E)$ define $l : 2^V \times V \to R$ by $l : (X, e) \mapsto \min_{x \in X}\{d(x, e)\}$ for all $X \subseteq V$, where d denotes the (weighted) length of a (weighted) shortest path between vertices.
(iii) (*monotone linkage in graphs by maximum degree on induced subgraphs*) For a graph $G = (V, E)$, define $l : 2^V \times V \to \mathbb{R}$ by $l : (X, v) \mapsto \min_{x \in X}(\delta(v) - \delta_{G[X]}(x))$ for all $X \subseteq V$ and $v \in V$, where $G[X]$ is the subgraph of G induced by X, $\delta(v)$ the degree of v in G, and $\delta_{G[X]}(x)$ the degree of x in $G[X]$.

Monotone linkage functions have been studied intensively by Kempner [6–8] in the context of *clustering* over set systems and convex geometries. As mentioned above, we will apply them for defining margins in *arbitrary* finite closure systems. For this purpose, we will use the following notion many times in what follows. A *monotone linkage closure system* (MLCS) is a triple $(E, \mathcal{C}_\rho, l)$, where $(E, \mathcal{C}_\rho)$ is a closure system and l is a monotone linkage function on E. We will always assume that the closure operator and the linkage function are given implicitly by *oracles* under the usual complexity assumption. That is, for all $X \subseteq E$ and $e \in E$, $\rho(X)$ and $l(X, e)$ are returned in *unit time* by the oracles.

3 Maximum Margin Separations in MLCSs

Our main goal in this paper is to adapt Vapnik's idea [1] of *maximum margin* separating hyperplanes to finite closure systems. That is, given subsets A and B of some inner product (feature) space $\mathcal{F}$, in case of *support vector machines* (SVM) [1] we are interested in the hyperplane H^* having *maximum* distance to the two sets, i.e., which satisfies

$$d(A \cup B, H) \leq d(A \cup B, H^*) \tag{4}$$

for all hyperplanes H, where for all $X, Y \subseteq \mathcal{F}$, $d(X, Y) = \min_{y \in Y} d(X, y)$ with d being the distance induced by the underlying inner product. It is a well-known fact that if A and B are separable by a hyperplane, then H^* is *unique*; H^* is also referred to as the *maximum margin separating hyperplane*, where the *margin* of a separating hyperplane H is defined by

$$\mu(A, B) = d(A, H) + d(B, H) \ . \tag{5}$$

A key property of the margin is that it is *anti-monotone* w.r.t. set inclusion, i.e., $\mu(A', B') \leq \mu(A, B)$ for all $A' \supseteq A$ and $B' \supseteq B$. Note that (4) implies

$$d(A, H^*) = d(B, H^*) \ .$$

Clearly, the above definitions are *not* (directly) applicable to maximum margin separation in closure systems because we do not assume E to be an inner product or a metric space and have therefore no measure in general for the distance from a point $e \in E$ to a subset $X \subseteq E$. Furthermore, while the notion of half-spaces in $\mathbb{R}^d$ has been generalized to closure systems, for hyperplanes there is no analogous definition. Hence, to be in a position to define margins, we need some suitable functions for the abstraction of "closeness" from a point to a subset of the ground set. They should *generalize* metrics, but *preserve* the anti-monotonic property above at the same time.

The class of *monotone linkage functions* [10] defined in Sect. 2 fulfill both of these requirements. In addition to generality and anti-monotonicity, they have some further properties making this class an attractive candidate for our purpose. In particular, monotone linkage functions assume neither symmetry nor the triangle inequality.

To adapt the ordinary definition of margins to MLCSs, note that if a hyperplane $H \subseteq \mathbb{R}^d$ separates A and B, then (5) is equivalent to

$$\begin{aligned} \mu(A, B) &= d(A, H_2) + d(B, H_1) \\ &= d(\mathrm{conv}(A), H_2) + d(\mathrm{conv}(B), H_1), \end{aligned} \tag{6}$$

where $H_1 \supseteq A$ and $H_2 \supseteq B$ are the closed half-spaces defined by H (i.e., $H \subseteq H_1, H_2$). That is, in case of SVM, the margin given by a hyperplane H separating A and B is defined by the sum of the distances from the *convex hull* of A to the half-space H_2 containing B and from that of B to H_1 containing A.

Analogously to distances in metric spaces, we first extend linkage functions from sets to elements to those from sets to sets. Formally, for a linkage function l on E and subsets $X, Y \subseteq E$, we define the linkage l from X to Y by $l(X, Y) = \min_{y \in Y} l(X, y)$. Note that this extended definition preserves anti-monotonicity, i.e., $l(X', Y) \leq l(X, Y)$ holds whenever $X' \supseteq X$. Let H, H^c be half-spaces of an MLCS $(E, \mathcal{C}_\rho, l)$ and $A \subseteq H, B \subseteq H^c$ for some $A, B \subseteq E$. Then, by analogy with (6), our *first* definition of the *margin* of the half-space separation of A, B by H, H^c is

$$\mu^+_{H,H^c}(A, B) = l(\rho(A), H^c) + l(\rho(B), H). \tag{7}$$

While the above adaptation of the ordinary notion of margins to MLCSs is relatively natural, the generalization is less obvious for *maximum* margin half-space separations. This is because for SVM there are two *equivalent* properties characterizing maximum margin hyperplanes H^* defining the closed half-spaces $H_1 \supseteq A$ and $H_2 \supseteq B$:

(i) H^* maximizes $\mu(A, B)$ such that $d(\text{conv}(A), H_2) = d(\text{conv}(B), H_1)$.
(ii) H^* maximizes $\min\{d(\text{conv}(A), H_2), d(\text{conv}(B), H_1)\}$.

That is, the maximum margin hyperplane by (i) lies in the "middle" between the convex hulls of A and B; by (ii) it maximizes the minimum of the distances from the two convex hulls. While (i) and (ii) are equivalent in case of SVM, the situation is different for MLCSs as shown in the proposition below.

Proposition 2. *There exists an MLCS* $(E, \mathcal{C}_\rho, l)$ *and subsets* $A, B \subseteq E$ *such that* $\mu^+_{H_1,H_1^c}(A, B) \neq \mu^+_{H_2,H_2^c}(A, B)$, *where*

$$H_1 = \underset{H,H^c \in \mathcal{C}_\rho}{\arg\max}\, \mu^+_{H,H^c}(A, B) \text{ subject to } l(\rho(A), H^c) = l(\rho(B), H)$$
$$H_2 = \underset{H,H^c \in \mathcal{C}_\rho}{\arg\max} \min\{l(\rho(A), H^c), l(\rho(B), H)\} \ .$$

Proof. Consider MLCS $(E, \mathcal{C}, l)$ with $E = \{a, b, c, d\}$ and $\mathcal{C} = \{X \subseteq E : |X| \neq 3\}$. The monotone linkage function is defined by

$$l(\{a\}, b) = l(\{b\}, a) = l(\{b\}, d) = 3, \quad l(\emptyset, e) = 3 \text{ for all } e \in E, \quad l(\{a\}, c) = 2,$$
$$l(\{a\}, d) = l(\{b\}, c) = 1, \quad l(X, e) = 0 \text{ for all other } X \subseteq E \text{ and } e \in E$$

It can be easily checked that $(E, \mathcal{C})$ is a closure system and l fulfills the anti-monotonicity property. For $A = \{a\}, B = \{b\}$ there exist exactly two different separating half-spaces of size 2, i.e., $H_1 = \{a, c\}$ and $H_2 = \{a, d\}$. Using the definition of linkage on sets it follows $l(A, H_1^c) = l(B, H_1) = 1$. Moreover, $l(A, H_2^c) = 2$ and $l(B, H_2) = 3$. Thus, H_1 fulfills the first property and H_2 the second one, by noting that $2 = \min\{l(A, H_2^c), l(B, H_2)\} > \min\{l(A, H_1^c), l(B, H_1)\} = 1$. The claim then follows by $2 = \mu^+_{H_1,H_1^c}(A, B) \neq \mu^+_{H_2,H_2^c}(A, B) = 5$.

Thus, for an MLCS $(E, \mathcal{C}, l)$, maximizing the margin as defined in (7) subject to $l(\rho(A), H^c) = l(\rho(B), H)$ is *not* equivalent to maximizing

$$\mu_{H,H^c}(A, B) := \min\{l(\rho(A), H^c), l(\rho(B), H)\} \tag{8}$$

over *all* half-space separations of A and B in $(E, \mathcal{C}_\rho, l)$ (see, also, Fig. 1).

Since our primary interest is in classification, we prefer the definition in (ii) above and will accordingly focus on maximizing the margin defined by (8). Note that our definition of margin differs from that in SVM, as it involves only one part of the ordinary one.

Until now we have concentrated on half-space separations. In case of MLCSs, two sets with disjoint closures are, however, not always half-space separable. This motivates the relaxed concept of *maximal* closed set separation [14]. Fortunately, the above definition of margin can be extended naturally to arbitrary closed

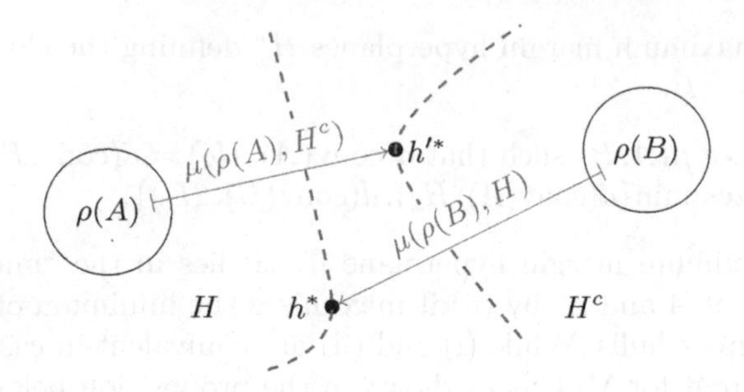

Fig. 1. Margins $\mu(\rho(A), H^c)$, $\mu(\rho(B), H)$ between closed sets $\rho(A), \rho(B)$ and half-spaces H, H^c together with the support elements h^*, h'^*.

sets. More precisely, for an MLCS $(E, \mathcal{C}_\rho, l)$, let $A, B \subseteq E$ and $C_A, C_B \in \mathcal{C}_\rho$ with $A \subseteq C_A$ and $B \subseteq C_B$. Then the *margin* for C_A and C_B is defined by

$$\mu_{C_A,C_B}(A, B) := \min\{l(\rho(A), C_A^c), l(\rho(B), C_B^c)\} \ . \tag{9}$$

Similarly to half-spaces, the definition takes only one part of the effective margin into account. Note that (8) is the special case of (9) for $C_A = H$ and $C_B = H^c$. We now show that the anti-monotonicity of monotone linkages extends to margins in MLCS. This property is essential for separations.

Lemma 3. *Let $(E, \mathcal{C}_\rho, l)$ be an MLCS, $A \subseteq A' \subseteq E$, $B \subseteq B' \subseteq E$, and $C_A \supseteq A', C_B \supseteq B'$ disjoint closed sets. Then $\mu_{C_A,C_B}(A, B) \geq \mu_{C_A,C_B}(A', B')$.*

Proof. This follows directly from the definition of margin in (9) and the anti-monotonicity of monotone linkage functions.

Moreover, maximizing the disjoint closed sets C_A and C_B in Lemma 3 maximizes the margin at the same time, as we show in the following lemma.

Lemma 4. *Let $C_A \subseteq C'_A$ and $C_B \subseteq C'_B$ be closed sets of an MLCS $(E, \mathcal{C}_\rho, l)$ with $C'_A \cap C'_B = \emptyset$ and $A \subseteq C_A, B \subseteq C_B$. Then $\mu_{C_A,C_B}(A, B) \leq \mu_{C'_A,C'_B}(A, B)$.*

Proof. From the definition of monotone linkages between sets it follows that $l(X, Y) \geq l(X, Y')$ whenever $Y \subseteq Y'$. Hence, by $C_A^c \supseteq C_A'^c$ and $C_B^c \supseteq C_B'^c$ we have

$$\begin{aligned}
\mu_{C_A,C_B}(A, B) &= \min\{l(\rho(A), C_A^c), l(\rho(B), C_B^c)\\
&\leq \min\{l(\rho(A), C_A'^c), l(\rho(B), C_B'^c)\\
&= \mu_{C'_A,C'_B}(A, B) \ .
\end{aligned}$$

Given a half-space separation of A, B with $A \subseteq H$ and $B \subseteq H^c$, similarly to SVM we can define the *support elements* by h^* and h'^* satisfying $l(\rho(A), H^c) = l(\rho(A), h'^*)$ and $l(\rho(B), H) = l(\rho(B), h^*)$, respectively. For example, in case of maximum margin separating half-spaces in trees, there are exactly two support elements corresponding to the two half-spaces.

4 The Maximum Margin Algorithm

Using (8) and (9) for the definition of margins for half-space and closed set separations, we are ready to formulate the separation problems in MLCS $(E, \mathcal{C}_\rho, l)$:

MAXIMUM MARGIN HALF-SPACE SEPARATION (MMHSS) PROBLEM: *Given* non-empty subsets A, B of E, *find* a half-space $H \in \mathcal{C}_\rho$ with $A \subseteq H, B \subseteq H^c$ that maximizes the margin, i.e., $H = \underset{H_1, H_1^c \in \mathcal{C}_\rho}{\arg\max}\ \mu_{H_1, H_1^c}(A, B)$, if A and B are half-space separable; o/w return "No".

MAXIMUM MARGIN CLOSED SET SEPARATION (MMCSS) PROBLEM: *Given* non-empty subsets A, B of E, *find* disjoint closed sets $C_A, C_B \in \mathcal{C}_\rho$ with $A \subseteq C_A, B \subseteq C_B$ that maximize the margin, i.e., for all other disjoint closed sets $C'_A \supseteq A, C'_B \supseteq B$ it holds that $\mu_{C_A, C_B}(A, B) \geq \mu_{C'_A, C'_B}(A, B)$, if $\rho(A) \cap \rho(B) = \emptyset$; o/w return "No".

Remark 5. The MMHSS problem is a special case of the MMCSS problem for $C_A = H, C_B = H^c$. Moreover, Lemma 4 implies that for any maximum margin closed set separation there exists a maximal closed set separation of the same margin. The converse is, however, not true in general.

We solve the above problems by Algorithm 1, which is based on an adaptation of the greedy algorithm in [14]. The input to the algorithm is an MLCS $(E, \mathcal{C}_\rho, l)$ together with two sets $A, B \subseteq E$ of training examples. We assume that $\mathcal{C}_\rho$ is given by the closure operator ρ, which returns the closure for any $X \subseteq E$ in unit time. Similarly, for any $X \subseteq E$ and $e \in E$, $l(X, e)$ is returned by another oracle in unit time. Accordingly, we measure the complexity of Algorithm 1 in terms of the number of closure operator calls and linkage function evaluations.

In Lines 1–4, the closures of A, B are calculated and checked for disjointness. In particular, if they are not disjoint, the algorithm terminates with "NO", as in this case A and B are not separable by closed sets. Thus, the algorithm is correct for this case. Consider the case that $\rho(A) \cap \rho(B) = \emptyset$. For this case, all elements not contained in the union of the closures of A and B are first collected in F and sorted then by their minimum linkage from these two closed sets (Lines 5–6). The elements f in F will be processed one by one in this order and then immediately removed, potentially together with other untreated elements (Line 13). In particular, if the linkage from the closure of A to f is not greater than that of B or the current closed set C_B containing B cannot be extended by f, we expand the current closed set $C_A \supseteq A$ with f if it does not violate the disjointness with C_B (see Lines 9–10). Otherwise, we extend C_B by f, if $\rho(C_B \cup \{f\})$ remains disjoint with C_A (Lines 11–12). We then remove f and all other elements from F (Line 13) that have been added to C_A or to C_B in Line 10 or 12.

An example of the algorithm to the case that $(E, \mathcal{C}_\rho, l)$ is defined over graphs with the shortest path closure operator is given in Fig. 2. We now show that Algorithm 1 is correct (Theorem 6) and efficient (Theorem 8). Furthermore, in case of Kakutani closure systems, the sets C_A, C_B returned in Line 15 form complementary half-spaces with maximum margin whenever $\rho(A) \cap \rho(B) \neq \emptyset$ (Corollary 7).

Algorithm 1: Maximum Margin Separation

Input: a finite MLCS $(E, \mathcal{C}_\rho, l)$ and sets $A, B \subseteq E$
Output: *maximum* margin closed sets $C_A, C_B \in \mathcal{C}_\rho$ with $A \subseteq C_A$ and $B \subseteq C_B$ if $\rho(A) \cap \rho(B) = \emptyset$; "No" otherwise

1 $\overline{A}, C_A \leftarrow \rho(A); \overline{B}, C_B \leftarrow \rho(B)$;
2 **if** $C_A \cap C_B \neq \emptyset$ **then**
3 | **return** No;
4 **end**
5 $F \leftarrow E \setminus \{C_A \cup C_B\}$;
6 compute $\min\{l(\overline{A}, f), l(\overline{B}, f)\}$ for all $f \in F$ and sort F by these values;
7 **while** $F \neq \emptyset$ **do**
8 | take the smallest element $f \in F$;
9 | **if** $(l(\overline{A}, f) \leq l(\overline{B}, f) \vee \rho(C_B \cup \{f\}) \cap C_A \neq \emptyset) \wedge \rho(C_A \cup \{f\}) \cap C_B = \emptyset$ **then**
10 | | $C_A \leftarrow \rho(C_A \cup \{f\})$;
11 | **else if** $\rho(C_B \cup \{f\}) \cap C_A = \emptyset$ **then**
12 | | $C_B \leftarrow \rho(B_B \cup \{f\})$;
13 | $F \leftarrow F \setminus (C_A \cup C_B \cup \{f\})$;
14 **end**
15 **return** C_A, C_B

Theorem 6. *Algorithm 1 solves the MMCSS problem correctly.*

Proof. Let $(E, \mathcal{C}_\rho, l)$ be an MLCS and $A, B \subseteq E$. By construction, the algorithm returns "No" only for the case that $\rho(A) \cap \rho(B) \neq \emptyset$, i.e., when A and B are not separable in $\mathcal{C}_\rho$, implying the correctness for this case. Otherwise, the closed sets $C_A \supseteq A, C_B \supseteq B$ returned are disjoint and hence, form a *separation* of A and B. They are *maximal*, as only such elements of E are discarded that violate the disjointness condition. All such elements can be removed ultimately from F, as they do not have to be reconsidered again for the monotonicity of ρ.

Regarding optimality, suppose for contradiction that there are other disjoint closed sets $C'_A \supseteq A, C'_B \supseteq B$ such that

$$\mu_{C'_A, C'_B}(A, B) > \mu_{C_A, C_B}(A, B). \tag{10}$$

For symmetry, we can assume w.l.o.g. that there is an $e^* \in C_A^c$ such that

$$\min\{l(\rho(A), C_A^c), l(\rho(B), C_B^c)\} = l(\rho(A), e^*),$$

i.e., $\mu_{C_A, C_B}(A, B) = l(\rho(A), e^*)$. Then, by (9) and (10) we have

$$l(\rho(A), e^*) < \min\{l(\rho(A), C_A'^c), l(\rho(B), C_B'^c)\} \tag{11}$$

implying $l(\rho(A), e^*) < l(\rho(A), C_A'^c)$. Thus, $e^* \notin C_A'^c$ and hence $e^* \in C'_A \subseteq C_B'^c$. But then, together with (11), we have $l(\rho(A), e^*) < l(\rho(B), e^*)$.

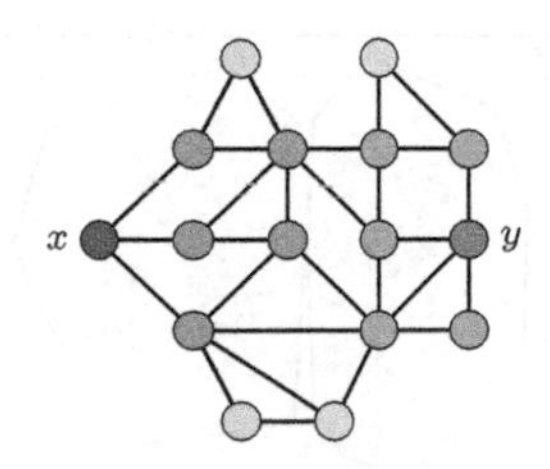

Fig. 2. Maximum margin half-space separation of x and y defined by the shortest path closure. Brighter nodes are added later to the respective class. The maximum margin between $\{x\}$ and $\{y\}$ is 2 for the linkage defined by weight 1 for all edges (see Sect. 2).

We prove that $e^* \in C'_A$ and $e^* \in C^c_A$ contradicts the assumptions. Conditions $C'_A \cap C'_B = \emptyset$ and $e^* \in C'_A$ imply that $\rho(\rho(A) \cup \{e^*\}) \cap \rho(B) = \emptyset$. Since $e^* \notin C_A$, e^* has not been added to C_A, though $l(\rho(A), e^*) < l(\rho(B), e^*)$. But this can happen only if there is a non-empty set $G \subseteq F$ such that for all $g \in G$, g is before e^* in F, i.e., $\min\{l(\rho(A), g), l(\rho(B), g)\} \leq l(\rho(A), e^*)$. Assume there is a $g \in G$ such that $g \in C_A$, but $g \notin C'_A$. Then $g \in C'^c_A$ and thus,

$$\begin{aligned}\mu_{C'_A, C'_B}(A, B) &= \min\{l(\rho(A), C'^c_A), l(\rho(B), C'^c_B)\} \\ &\leq \min\{l(\rho(A), g), l(\rho(B), g)\} \\ &\leq l(\rho(A), e^*) \\ &= \mu_{C_A, C_B}(A, B)\end{aligned}$$

contradicting (10). Hence, for all $g \in G$, $g \in C_A$ implies $g \in C'_A$. In a similar way we have that $g \in C_B$ implies $g \in C'_B$ for all $g \in G$.

Since $e \in C^c_A$, $e^* \notin C_A$. There are two possible cases: (i) $e^* \in \rho(\rho(B) \cup G_B) \subseteq C'_B$, where $G_B \subseteq G$ is the set of elements added to $\rho(B)$. But this contradicts $e^* \in C'^c_B$. (ii) At the step e^* is considered for adding to C_A, there are disjoint subsets $G_A, G_B \subseteq G$ already added to $\rho(A)$ and $\rho(B)$, respectively, such that

$$\rho(\rho(\rho(A) \cup G_A) \cup \{e^*\}) \cap \rho(\rho(B) \cup G_B) \neq \emptyset.$$

But then, for $G_A \subseteq C'_A$ and $G_B \subseteq C'_B$ and for the monotonicity of ρ, we have $C'_A \cap C'_B \neq \emptyset$, as $e^* \in C'_A$; a contradiction.

Corollary 7. *For all MLCSs $(E, \mathcal{C}_\rho, l)$, Algorithm 1 solves the MMHSS-problem correctly if $(E, \mathcal{C}_\rho)$ is Kakutani.*

Proof. It is a direct implication of Theorem 6, as maximal disjoint closed sets are always half-spaces in any Kakutani closure system.

Theorem 8. *Algorithm 1 requires at most $2 \cdot |E \setminus (\rho(A) \cup \rho(B))|$ evaluations of l and $2 \cdot |E \setminus (\rho(A) \cup \rho(B))| + 2$ calls of ρ.*

Proof. To sort F, we evaluate l twice for all $f \in F$ with $|F| = |E \setminus (\rho(A) \cup \rho(B))|$. The closure is calculated twice to determine the closures of the input sets (Line 1) and twice for all $f \in F$ in the worst case (Lines 9 and 11).

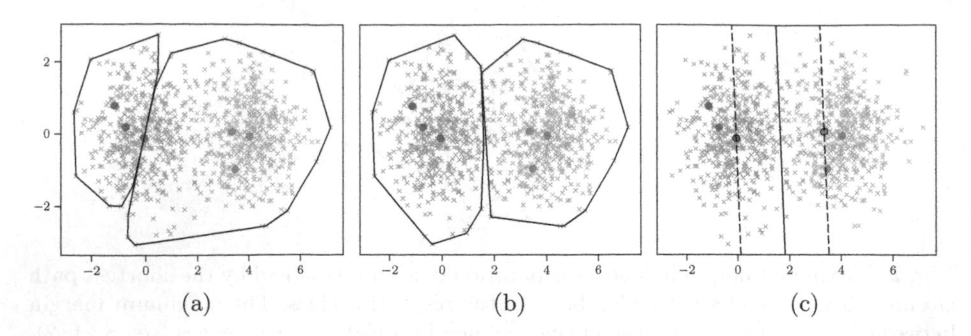

Fig. 3. Comparison of greedy separation (a), maximum margin separation (b) and ordinary support vector machines (c). (Color figure online)

5 Empirical Evaluations

Potential applications of maximum margin closed set and half-space separation in finite closure systems include, among others, graphs, lattices (e.g., in inductive logic programming [11], formal concept analysis [5], and itemset mining [12]), and finite point sets. For space limitations, we consider only two such applications in this short version[2]. The first one, discussed in Sect. 5.1, is concerned with the separation of finite point sets in $\mathbb{R}^d$. For this task, we compare Algorithm 1 to the greedy algorithm in [14] as well as to ordinary SVM on synthetic datasets. The other application described in Sect. 5.2 deals with vertex classification in random trees and graphs of different sizes and edge densities. For this task, we compare the predictive performance of our algorithm to that of the greedy algorithm in [14].

5.1 Binary Classification in Finite Point Sets

In this section we consider point separation in MLCSs over finite subsets of $\mathbb{R}^d$. The closure systems used in these experiments are given by the traces of convex hulls as defined by (1) in Sect. 2; the linkage function by means of the Euclidean distance. Our experimental results reported below show that the predictive performance of maximum margin separation in this kind of closure systems is comparable to that of SVM and that it outperforms the greedy separation algorithm in [14] on finite synthetic point sets in $\mathbb{R}^2, \mathbb{R}^3$ and $\mathbb{R}^4$ that are half-space separable. All datasets consist of two blobs, each with 500 points, such that the two classes are half-space separable[3]. In addition to the quantitative results below, for one of the random datasets from $\mathbb{R}^2$ we visualize the output obtained by the three algorithms (see Fig. 3). We selected three (in accordance to the VC-dimension of half-spaces in $\mathbb{R}^2$) random training examples for each class (denoted

[2] The source code and the data used in the experiments reported in this section are available at https://github.com/fseiffarth/MaxMarginSeparations.

[3] For a detailed description of these synthetic datasets, the reader is referred to [14].

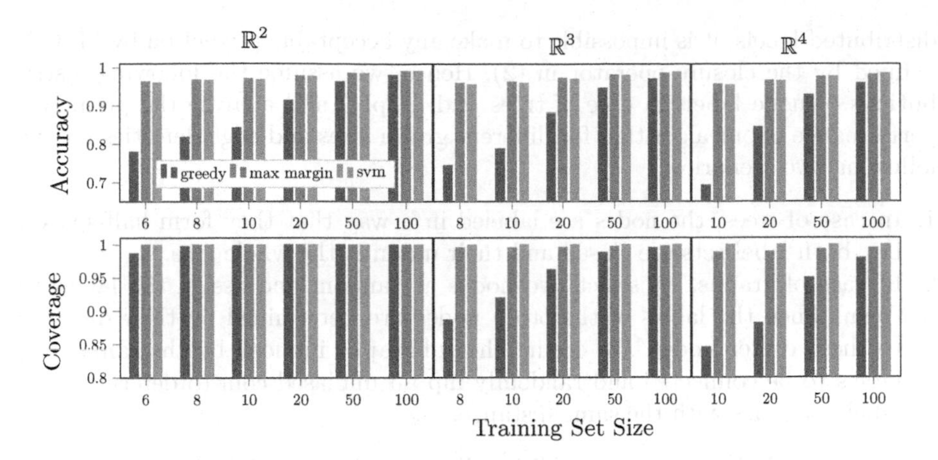

Fig. 4. Accuracy and coverage of greedy separation, maximum margin separation, and SVM for point set classification in $\mathbb{R}^2, \mathbb{R}^3, \mathbb{R}^4$.

by dark blue resp. dark red). The class labels are indicated by light blue and light red. The predictions are given by the convex hulls for the two greedy algorithms and by the separating hyperplane for SVM.

For each of the training set sizes (see the values of the x-axes of Fig. 4), we generated 1,000 binary labeled random sets as indicated above. Figure 4 shows the averaged accuracy (top row) and coverage (bottom row) for the three algorithms. The results obtained clearly show that maximum margin closed set separation outperforms the greedy separation algorithm in [14] in predictive performance, especially on small training set sizes. Furthermore, at least on the random datasets we used, it is also comparable to ordinary SVM by emphasizing that our definition is *not* a generalization of SVM; it is only an adaption of the idea of maximum margin separation to finite closure systems. The accuracy of the greedy algorithm strongly depends on the training set size and the dimension of the space, while the accuracy of the maximum margin algorithm is constantly above 0.9. Regarding the coverage, for which a similar behavior can be observed, note that finite point sets in $\mathbb{R}^d$ are not half-space separable by MLCSs in general. While the average coverage for the greedy algorithm drops below 0.85 in case of $\mathbb{R}^4$ and 10 training samples, the maximum margin algorithm has an average coverage above 0.95 for all training set sizes. By definition, SVM always achieve a coverage of 1.

5.2 Vertex Classification in Random Graphs

For tree and graph data we always consider the *shortest path* closure defined in (2), together with the monotone linkage function for weighted graphs as defined in Sect. 2. In case of graphs, we are interested in binary node predictions of random connected graphs. Of course, the distribution of the labels in the graphs plays an important role in the prediction. Clearly, in case of randomly distributed labels, it is impossible to make any acceptable prediction by MLCSs defined by the closure operator in (2). Hence, we assume the following distributions of node labels in case of trees and graphs, and analyze the predictive performance of our algorithm for different graph sizes and edge densities for the following two scenarios:

1. In case of trees, the nodes are labeled in a way that they form half-spaces, i.e., both label sets are closed and their union is the whole tree.
2. In case of graphs, we select two nodes at random and assign the labels to them. Then the labels of the other nodes are determined by their distance to these center nodes. We ensure the subgraphs induced by the same class labels to be connected and randomly flip an unbiased coin to determine the label for nodes with the same distance.

Moreover, in both cases we additionally use only graph labelings with nearly balanced class sizes, i.e., the minimum size of a class is at least 25% of the total size. In case of trees, we look at random trees of different sizes, ranging from $1,000$ to $20,000$ (see Fig. 5a). For each tree size and training sample size (see the x-axis of Fig. 5a), we generated $1,000$ binary labeled random trees in the above way. Then, for each run of the algorithm on a tree, $x/2$ training examples have been drawn at random from each of the two label sets for the input, where x is the x-axis value in Fig. 5a. For evaluation, we run the greedy algorithm from [14] and the maximum margin closed set separation algorithm on the training sets to predict the class labels of the unseen examples. The average accuracy, over all $1,000$ random trees is displayed in Fig. 5a. As a baseline, we take the percentage defined by the majority class. Note that trees induce Kakutani closure systems and hence the coverage is always 1. One can see that with increasing training set size, the accuracy increases up to more than 0.95 in case of maximum margin separation and 10 training samples. Moreover, the maximum margin separation leads to better accuracy compared to the greedy separation, especially for small training sample sizes. Somewhat surprisingly, the tree size has no significant impact on the predictive performance.

In case of graphs with different edge densities[4], we generated $1,000$ random graphs for each edge density (see the x-axis values in Fig. 5b) and assigned the nodes to one of the two classes as described above. The random graphs were generated from random trees by adding additional random edges until the required edge density has been reached. For each run of our algorithm, we selected 1 or 2 nodes from each label class at random for training such that their closures do not intersect. The accuracy results are shown in Fig. 5b. We present also the coverage values, as the underlying MLCSs are not Kakutani in general. For increasing edge density, the accuracy decreases to 0.8 in case of 4 training samples and to 0.75 in case of 2 training samples for the edge density of 1.2. For edge density 1.5, there are no obvious changes in the accuracy. This can be explained by the fact that the coverage decreases to approximately 0.38 in case of an edge density of 1.5.

[4] The edge density is the number of edges minus 1 divided by the number of nodes.

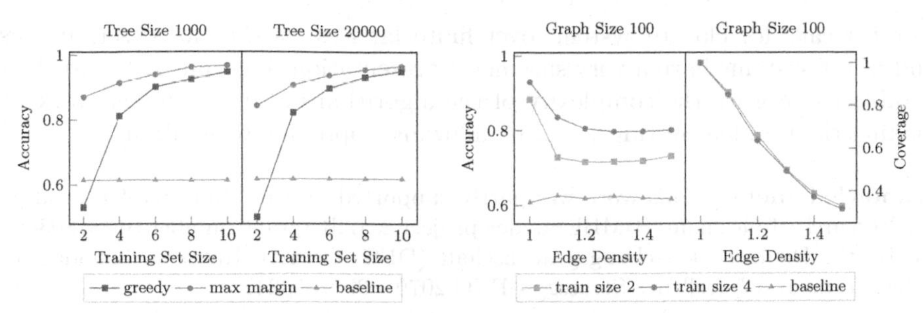

(a) Comparison of greedy algorithm and maximum margin algorithm for node prediction task in random trees.

(b) Accuracy and coverage of maximum margin separation in random graphs of different edge densities with 2 and 4 training samples.

Fig. 5. Accuracy of vertex prediction in random trees and random graphs of different sizes and edge densities.

6 Concluding Remarks

We adapted the idea of maximum margin separation in inner product feature spaces to abstract finite closure systems equipped with monotone linkage functions. Although not all properties of ordinary maximum margin separation could be preserved in this way, the anti-monotonicity property, a key feature of maximum margin separation, remains valid. Combining this concept with half-space and maximal closed set separation, we presented a simple greedy algorithm and proved that it computes a closed set separation with maximum margin correctly, using a linear number of closure operator calls and linkage function evaluations. In addition, for Kakutani closure systems the output closed sets are always complementary maximum margin half-spaces if the closures of the input sets are disjoint. Our experimental results on synthetic data clearly show that the maximum margin separation algorithm presented in this work outperforms the greedy algorithm in [14] both on point classification and vertex prediction in random trees.

We mention some interesting questions raised by this work. In contrast to ordinary SVM, the maximum margin separating half-spaces are not unique in general in Kakutani MLCSs. It would be important to *characterize* the class of Kakutani MLCSs from the point of view of *uniqueness*. In particular, how does the structure of the closure system interact with the linkage function in such a characterization, if it exists at all. Another important issue is the complexity of the maximum margin separation algorithm. Although the number of closure operator calls and linkage function evaluations is linear in the cardinality of the ground set, the algorithm is practically infeasible for MLCSs over very large ground sets (e.g., the set of vertices of the web graph). The question towards this direction is therefore to identify practically interesting classes of MLCSs for which the algorithm has sublinear complexity. For example, one can show that the greedy algorithm in [14] requires only a logarithmic number of closure operator calls for closure systems over finite lattices, as the closed sets in this kind of set systems have a very succinct representation. Last but not least, is it possible to improve the complexity of the algorithm presented in this work by relaxing the problem settings, e.g., by allowing approximate solutions.

Acknowledgments. This work was partly supported by the Ministry of Education and Research of Germany (BMBF) under project ML2R (grant number 01/S18038C) and by the Deutsche Forschungsgemeinschaft (DFG, German Research Foundation) under Germany's Excellence Strategy - EXC 2070 - 390732324.

References

1. Boser, B.E., Guyon, I.M., Vapnik, V.N.: Training algorithm for optimal margin classifiers. In: Proceedings of the Fifth Annual ACM Workshop on Computational Learning Theory, COLT 1992, pp. 144–152. ACM, New York (1992)
2. Chepoi, V.: Separation of two convex sets in convexity structures. J. Geom. **50**(1–2), 30–51 (1994)
3. Davey, B.A., Priestley, H.A.: Introduction to Lattices and Order, 2nd edn. Cambridge University Press, Cambridge (2002)
4. Farber, M., Jamison, R.E.: Convexity in graphs and hypergraphs. SIAM J. Algebr. Discrete Methods **7**(3), 433–444 (1986)
5. Ganter, B., Stumme, G., Wille, R. (eds.): Formal Concept Analysis. LNCS (LNAI), vol. 3626. Springer, Heidelberg (2005). https://doi.org/10.1007/978-3-540-31881-1
6. Kempner, Y., Mirkin, B., Muchnik, I.: Monotone linkage clustering and quasi-concave set functions. Appl. Math. Lett. **10**(4), 19–24 (1997)
7. Kempner, Y., Levit, V.E.: Duality between quasi-concave functions and monotone linkage functions. Discrete Math. **310**(22), 3211–3218 (2010)
8. Kempner, Y., Muchnik, I.: Clustering on antimatroids and convex geometries. WSEAS Trans. Math. **2**(1), 54–59 (2003)
9. Kubiś, W.: Separation properties of convexity spaces. J. Geometry **74**(1–2), 110–119 (2002)
10. Mullat, J.E.: Extremal subsystems of monotonic systems I. Avtomatica i Telemekhanika **5**, 130–139 (1976)
11. Nienhuys-Cheng, S.-H., de Wolf, R.: Foundations of Inductive Logic Programming. LNCS, vol. 1228. Springer, Heidelberg (1997). https://doi.org/10.1007/3-540-62927-0
12. Pasquier, N., Bastide, Y., Taouil, R., Lakhal, L.: Efficient mining of association rules using closed itemset lattices. Inf. Syst. **24**(1), 25–46 (1999)
13. Rosenblatt, F.: The perceptron: a probabilistic model for information storage and organization in the brain. Psychol. Rev. **65**(6), 386–408 (1958)
14. Seiffarth, F., Horváth, T., Wrobel, S.: Maximal closed set and half-space separations in finite closure systems. In: Brefeld, U., Fromont, E., Hotho, A., Knobbe, A., Maathuis, M., Robardet, C. (eds.) ECML PKDD 2019. LNCS (LNAI), vol. 11906, pp. 21–37. Springer, Cham (2020). https://doi.org/10.1007/978-3-030-46150-8_2
15. van de Vel, M.: Binary convexities and distributive lattices. Proc. Lond. Math. Soc. **s3-48**(1), 1–33 (1984)

Discovering Outstanding Subgroup Lists for Numeric Targets Using MDL

Hugo M. Proença[1(✉)], Peter Grünwald[1,2], Thomas Bäck[1], and Matthijs van Leeuwen[1]

[1] Leiden University, Leiden, Netherlands
{h.manuel.proenca,T.H.W.Baeck,m.van.leeuwen}@liacs.leidenuniv.nl
[2] CWI, Amsterdam, Netherlands
Peter.Grunwald@cwi.nl

Abstract. The task of subgroup discovery (SD) is to find interpretable descriptions of subsets of a dataset that stand out with respect to a target attribute. To address the problem of mining large numbers of redundant subgroups, subgroup set discovery (SSD) has been proposed. State-of-the-art SSD methods have their limitations though, as they typically heavily rely on heuristics and/or user-chosen hyperparameters.

We propose a dispersion-aware problem formulation for subgroup set discovery that is based on the minimum description length (MDL) principle and subgroup lists. We argue that the best subgroup list is the one that best summarizes the data given the overall distribution of the target. We restrict our focus to a single numeric target variable and show that our formalization coincides with an existing quality measure when finding a single subgroup, but that—in addition—it allows to trade off subgroup quality with the complexity of the subgroup. We next propose SSD++, a heuristic algorithm for which we empirically demonstrate that it returns outstanding subgroup lists: non-redundant sets of compact subgroups that stand out by having strongly deviating means and small spread.

Keywords: Pattern mining · Subgroup discovery · The MDL principle

1 Introduction

Subgroup discovery [2,9] (SD) is the task of discovering subsets of the data that stand out with respect to a given target. It has a wide range of applications in many different domains [17]. For example, insurance companies could use it for fraud detection, where a found subgroup '*provider* = HospitalX ∧ *care* = leg in cast → *average*(*claim*) = \$2829.50' might indicate that a certain health care provider claims much more for certain care than others.

Since its conception subgroup discovery has been developed for various types of data and targets, e.g., nominal, numeric, and multi-label [11] targets. In this paper we limit the scope to attribute-value data with a numeric target, i.e., each data point is a row with exactly one value for each attribute and a single, numeric target label, as is also considered in the regular regression setting.

F. Hutter et al. (Eds.): ECML PKDD 2020, LNAI 12457, pp. 19–35, 2021.
https://doi.org/10.1007/978-3-030-67658-2_2

Related Work. Subgroup discovery traditionally focused on mining the top-k subgroups, based on their individual qualities. This approach has two major drawbacks: 1) its focus on quality measures that only take into account the centrality measure of the subgroup, such as the mean or median, and 2) the *pattern explosion*, i.e., typically large amounts of redundant patterns are found.

In response to the centrality problem of numeric targets, dispersion-aware measures—that allow for efficient mining of the top-k patterns—were proposed [4], but these do not address the second drawback, i.e., the pattern explosion.

To address this drawback, methods for *subgroup set discovery* (SSD) have emerged. While SD aims on ranking the quality of subgroups regardless of how they cover the data together, SSD aims at finding good quality subgroups that together describe different regions of the data with minimum overlap between those. However, most of the SSD methods focus on binary target variables [3,5,10]. For the setting with a numerical target variable, three approaches have been proposed:

1) Sequential covering: CN2-SD [10], originally introduced for nominal targets, can be directly applied to numeric targets. The idea is to iteratively find the subgroup with the highest quality, removing the data covered by that subgroup, and repeating this process until no further subgroups are found. This is virtually the same as mining a list of subgroups and therefore closest to our approach.

2) Diverse Subgroup Set Discovery (DSSD) [12]: DSSD uses a diverse beam search to find a non-redundant set of high-quality subgroups. It is based on a two-step approach that first mines a large pool of subgroups based on their individual qualities and then selects subgroups from that pool that maximize quality while penalizing for overlap. DSSD relies on tunable hyperparameters for the search and overlap penalization, which strongly influence the results.

3) Subjectively interesting Subgroup Discovery (SISD) [13]: This approach finds the subjectively most interesting subgroup with regard to the prior knowledge of the user, based on an information-theoretic framework. By successively updating the prior knowledge based on the found subgroups, it iteratively mines a diverse set of subgroups that are also dispersion-aware.

Apart from the limitations already mentioned, all three approaches lack a *global* formalization of the optimal set of subgroups for a given dataset and instead employ a sequential approach for which the stopping criteria, such as the total number of patterns to be found, need to be manually defined.

Contributions.[1] We introduce a principled approach for dispersion-aware subgroup set discovery that builds on recent work [18,22] that uses the minimum description length (MDL) principle [8,19] for pattern-based modelling. The MDL principle states that the best model is the one that compresses the data and model best and is ideally suited for model selection tasks where the goal is to find succinct and descriptive models—such as is the case in subgroup discovery.

[1] The **extended version** of this work is available on arXiv [16].

Table 1. First 4 subgroups of a subgroup list obtained by SSD++ on the *Hotel booking* dataset with target *lead days*—number of days in advance the bookings were done (this case study is discussed in Sect. 6). *Description* contains information regarding client bookings, n the number of instances covered, $\hat{\mu}$ and $\hat{\sigma}$ are the mean and standard deviation in days, and *overlap* is the percentage of the subgroup description that is covered by subgroups that come before in the list, i.e., how independently can the subgroups be interpreted. The last line represents the dataset overall probability distribution. * The n of the dataset is the total number of instances in the dataset.

s	**description** of client bookings	n	$\hat{\mu}$	$\hat{\sigma}$	overlap
1	month = 9 & customer_type = Transient-Party & meal = Half Board & country = GBR & adults ≥ 2	22	533	34	–
2	month $\in [7, 9]$ & market_segment = Groups & weekend_nights = 1 & distribution_channel = Direct	29	336	~ 0	0%
3	month = 9 & week_nights =4 & distribution_channel = Corporate	16	343	3	0%
4	week_nights = 0 & deposit_type = Refundable & repeated_guest = no & adults≥ 2	20	9	~ 0	0%
dataset overall distribution		18 550*	92	99	–

Our three main contributions are: *1)* A formalization of subgroup set discovery for numeric targets using the MDL principle. To this end we devise a model class based on probabilistic rule lists. This probabilistic approach not only enables MDL-based model selection, naturally identifying compact subgroup lists, but also takes into account the dispersion (or spread) of the target value. By mining an ordered list of subgroups rather than an unordered set, we avoid the problem of a single instance being covered by multiple subgroups. This comes at the cost of slightly reduced interpretability, as the subgroups always need to be considered in order, but note that the still often-used sequential covering approach effectively identifies subgroup lists as well. *2)* Derivations that show how our formalization relates to both an existing subgroup quality measure and Bayesian testing, and—based on these insights—a novel evaluation measure for subgroup lists. *3)* SSD++, a heuristic algorithm that finds a set of non-redundant patterns according to our MDL-based problem formulation.

Example. To illustrate how our MDL-based problem formulation naturally defines a succinct and non-redundant set of subgroups for a given dataset, without the need to define the desired diversity or number of patterns in advance, we show an example subgroup list as obtained by our approach on the *Hotel booking* dataset (see Table 1 for the details and in depth explanation in Sect. 6). Our method identifies a detailed list of booking descriptions from which we show here the first four subgroups, each consisting of a short description that clearly represent different sub-populations of the data, i.e., different types of client bookings.

2 Subgroup Discovery with Numeric Targets

Consider a dataset $D = (X, Y) = \{(\mathbf{x}_1, y_1), (\mathbf{x}_2, y_2), ..., (\mathbf{x}_n, y_n)\}$. Each example $(\mathbf{x}_i, y_i)$ is composed of a numeric target value y_i and an instance of values of the explanatory variables $\mathbf{x}_i = (x_{i1}, x_{i2}, ..., x_{ik})$. Each instance value x_{ij} is associated to variable v_j and the total number of values in an instance is $k = |V|$ values, one for each variable v_j in V, which represents the set of all explanatory variables present in X. The domain of a variable v_j, denoted $\mathcal{X}_j$, can be one of three types: numeric, binary, or nominal (with > 2values). Y is a vector of values y_i of the numeric target variable with domain $\mathcal{Y} = \mathbb{R}$.

Subgroups. A subgroup, denoted by s, consists of a *description* (also intent) that defines a *cover* (also extent), i.e., a subset of dataset D.

Subgroup Description. A description a is a Boolean function over all explanatory variables V. Formally, it is a function $a : \mathcal{X}_1 \times \cdots \mathcal{X}_{|V|} \mapsto \{false, true\}$. In our case, a description a is a conjunction of conditions on V, each specifying a specific value or interval on a variable. The domain of possible conditions depends on the type of a variable: numeric variables support *greater and less than* $\{\geq, \leq\}$; binary and categorical support *equal to* $\{=\}$. The size of a pattern a, denoted $|a|$, is the number of variables it contains. In Table 1, subgroup 1 has description of size $|a| = 5$, where two of those conditions are $\{\text{meal} = \text{Half Board}\}$ and $\{\text{adult} \geq 2\}$; on a categorical and a numerical variable, respectively.

Subgroup Cover. The cover is the bag of instances from D where the subgroup description holds true. Formally, it is defined by $D_a = \{(\mathbf{x}, y) \in D \mid a \sqsubseteq \mathbf{x}\}$, where we use $a \sqsubseteq \mathbf{x}$ to denote $a(\mathbf{x}) = true$. Further, let $|D_a|$ denote the coverage of the subgroup, i.e., the number of instances it covers.

Interpretation as Probabilistic Rule. As D_a encompasses both the explanatory variables and the target variable, the effect of a on the target variable can be interpreted as a probabilistic rule $a \mapsto \hat{f}_a(Y)$ that associates the antecedent a to its corresponding target values in Y through the empirical distribution of their values $\hat{f}_a(y)$. Note that in general $\hat{f}_a(Y)$ can be described by a statistical model and corresponding statistics $\hat{\Theta}$, e.g., a normal distribution $\mathcal{N}$ with given mean $\hat{\mu}$ and standard deviation $\hat{\sigma}$.

Revisiting the subgroup list in Table 1, the description and corresponding statistics for the third subgroup are $a = \{\text{month} = 9 \ \& \ \text{week\_nights} = 4 \ \& \ \text{distribution\_channel} = \text{Corporate}\}$ and $\hat{\Theta}_a = \{\hat{\mu} = 343; \hat{\sigma} = 3\}$, respectively, and together represent the following rule:

$$\text{IF } a \sqsubseteq \mathbf{x} \text{ THEN lead time} \sim \mathcal{N}(\mu = 343; \sigma = 3)$$

where $\mathcal{N}(\mu; \sigma)$ is the probability density function of a normal distribution.

Quality Measures. To assess the quality (or interestingness) of a subgroup description a, a measure that scores subsets D_a needs to be chosen. The measures used vary depending on the target and task [2], but for a numeric target it usually has two components: 1) representativeness of the subgroup in the data, based on coverage $|D_a|$; and 2) a function of the difference between a statistic of the empirical target distribution of the pattern, $\hat{f}_a(Y)$, and the overall empirical target distribution of the dataset, $\hat{f}_d(Y)$. The latter corresponds to the statistics estimated over the whole data, e.g., in Table 1 it is $\hat{\Theta}_d = \{\hat{\mu} = 92; \hat{\sigma} = 99\}$ and it is estimated over all 18 550 instances of the dataset.

The general form of a quality measure to be maximized is

$$q(a) = |D_a|^{\alpha} g(\hat{f}_a(Y), \hat{f}_d(Y)),\ \alpha \in [0,1], \tag{1}$$

where α allows to control the trade-off between coverage and the difference of the distributions, and $g(\hat{f}_a(y), \hat{f}_d(y))$ is a function that measures how different the subgroup and dataset distributions are. The most adopted quality measure is the Weighted Relative Accuracy (WRAcc) [2], with $\alpha = 1$ and $g(\hat{f}_a(Y), \hat{f}_d(Y)) = \hat{\mu}_a - \hat{\mu}_d$ (the difference between averages of subgroup and dataset).

Subgroup Set Discovery. Subgroup set discovery [12] is the task of finding a set of high-quality, non-redundant subgroups that together describe all substantial deviations in the target distribution. That is, given a quality function Q for subgroup sets and the set of all possible subgroup sets $\mathcal{S}$, the task is to find that subgroup set $S^* = \{s_1, \ldots, s_k\}$ given by $S^* = \arg\max_{S \in \mathcal{S}} Q(S)$.

Ideally this measure should 1) *be global*, i.e., for a given dataset it should be possible to compare subgroup set qualities regardless of subgroup set size or coverage; 2) *maximize the individual qualities* of the subgroups; and 3) *minimize redundancy* of the subgroup set, i.e., the subgroups covers should overlap as little as possible while ensuring 2.

3 MDL-Based Subgroup Set Discovery

In this section we formalize the task of subgroup set discovery as a model selection problem using the Minimum Description Length (MDL) principle [8,19]. To this end we first need to define an appropriate model class $\mathcal{M}$; as we will explain next, we use *subgroup lists* as our models. The model selection problem should then be formalized using a two-part code [8], i.e.,

$$M^* = \underset{M \in \mathcal{M}}{\arg\min} L(D, M) = \underset{M \in \mathcal{M}}{\arg\min} \left[L(Y \mid X, M) + L(M) \right], \tag{2}$$

where $L(Y \mid X, M)$ is the encoded length, in bits[2], of target Y given explanatory data X and model M, and $L(M)$ is the encoded length, in bits, of the model. Intuitively, the best model M^* is that model that results in the best trade-off between how well the model compresses the target data and the complexity of that model—thus minimizing redundancy and automatically selecting the best subgroup list size. This formulation is similar to those previously used for two-view association discovery and multi-class classification [18,21]. We will first describe the details of the model class and then the required length functions.

[2] To obtain code lengths in bits, all logarithms in this paper are to the base 2.

3.1 Model Class: Subgroup Lists

Although Eq. (2) provides a *global* criterion that enables the comparison of subgroup sets of different sizes, subgroups are descriptions of *local* phenomena and we require each *individual subgroup to have high quality.*

We can accomplish this by using *subgroup lists* as models; see Eq. (3). Specifically, as we are only interested in finding subgroups for which the target deviates from the overall distribution, we assume y values to be distributed according to $\hat{f}_d$ by default (last line in Eq. (3)). For each region in the data for which the target distribution deviates from that distribution and a description exists, a subgroup specifying a different distribution $\hat{f}_a$ is added to the list.

We model the empirical distributions $\hat{f}$ by normal distributions, as those capture the two properties of interest, i.e., centre and spread, while being robust to cases where f violates the normality assumption [8]. We thus define $\hat{f}_{\hat{\mu},\hat{\sigma}}(y) = (2\pi\hat{\sigma})^{-1/2} \exp \frac{(y-\hat{\mu})^2}{2\hat{\sigma}^2}$, where $\hat{\mu}$ and $\hat{\sigma}$ are the estimated mean and standard deviation, respectively. These statistics can be easily estimated using the maximum likelihood estimator, so that a pattern a establishes a rule of the form $\text{IF } a \sqsubseteq \mathbf{x} \text{ THEN } \mathcal{N}(\hat{\mu}_i, \hat{\sigma}_i)$. Combining subgroup distributions $\hat{f}_{a,\hat{\mu}_a,\hat{\sigma}_a}$ with estimated dataset distribution $\hat{f}_{d,\hat{\mu}_d,\hat{\sigma}_d}$, this leads to a subgroup list M given by

$$\begin{aligned} \text{subgroup 1} &: \text{IF } a_1 \sqsubseteq \mathbf{x} \text{ THEN } \hat{f}_{a_1,\hat{\mu}_1,\hat{\sigma}_1}(y) \\ &\;\vdots \\ \text{subgroup k} &: \text{ELSE IF } a_k \sqsubseteq \mathbf{x} \text{ THEN } \hat{f}_{a_k,\hat{\mu}_k,\hat{\sigma}_k}(y) \\ \text{dataset} &: \text{ELSE } \hat{f}_{d,\hat{\mu}_d,\hat{\sigma}_d}(y) \end{aligned} \tag{3}$$

This corresponds to a probabilistic rule list with $k = |S|$ subgroups and a last (default) rule which is fixed to the overall empirical distribution $\hat{f}_{d,\hat{\mu},\hat{\sigma}}$ [18]. Fixing the distribution of this last 'rule' is crucial and differentiates a subgroup list from rule lists as used in classification and/or regression, as this enforces the discovery of a set of subgroups that individually all have target distributions that substantially deviate from the overall target distribution.

3.2 Model Encoding

The next step is to define the two length functions; we start with $L(M)$. Following the MDL principle [8], we need to ensure that 1) all models in the model class, i.e., all subgroup lists for a given dataset, can be distinguished; and 2) larger code lengths are assigned to more complex models. To accomplish the former we encode all elements of a model that can change, while for the latter we resort to two different codes: when a larger value represents a larger complexity we use the universal code for integers [8], denoted[3] $L_{\mathbb{N}}$, and when we have no prior knowledge but need to encode an element from a set we choose the uniform code.

[3] $L_{\mathbb{N}}(i) = \log k_0 + \log^* i$, where $\log^* i = \log i + \log \log i + \ldots$ and $k_0 \approx 2.865064$.

Specifically, the encoded length of a model M over variables V is given by

$$L(M) = L_{\mathbb{N}}(|S|) + \sum_{a_i \in S} \left[L_{\mathbb{N}}(|a_i|) + \log \binom{|V|}{|a_i|} + \sum_{v \in a_i} L(v) \right], \tag{4}$$

where we first encode the number of subgroups $|S|$ using the universal code for integers, and then encode each subgroup description individually. For each description, first the number $|a_i|$ of variables used is encoded, then the set of variables using a uniform code over the set of all possible combinations of $|a_i|$ from $|V|$ variables, and finally the specific condition for a given variable. As we allow variables of three types, the latter is further specified by

$$L(v_{bin}) = \log 2\,;\; L(v_{nom}) = \log |\mathcal{X}_v|\,;\; L(v_{num}) = \log N(n_{cut}), \tag{5}$$

where the code for each variable type assigns code lengths proportional to the number of possible partitions of the variable's domain. Note that this seems justified, as more partitions implies more potential spurious associations with the target that we would like to avoid. For `binary` variables only two conditions are possible, while for `nominal` variables this is given by the size of the domain. For `numeric` variables it equals the number of possible combinations $N(n_{cut})$, as there can be conditions with one (e.g. $x \leq 2$) or two operators (e.g. $1 \leq x \leq 2$), which is a function of the number of possible subsets generated by n_{cut} cut points. Note that we here assume that equal frequency binning is used, which means that knowing X and n_{cut} is sufficient to determine the cut points.

3.3 Data Encoding

The remaining length function is that of the target data given the explanatory data and model, $L(Y \mid X, M)$. For this we first observe that for any given subgroup list of the form of Eq. (3), *any individual instance* $(\mathbf{x}_i, y_i)$ *is 'covered' by only one subgroup.* That is, the cover of a subgroup a_i, denoted D_i, depends on the order of the list and is given by the instances where its description occurs minus those instances covered by previous subgroups:

$$D_i = \{X_i, Y_i\} = \{(\mathbf{x}, y) \in D \mid a_i \sqsubseteq \mathbf{x} \wedge \left(\bigwedge_{\forall j<i} a_j \not\sqsubseteq \mathbf{x} \right) \}. \tag{6}$$

Next, let $n_i = |D_i|$ be the number of instances covered by a subgroup (also known as *usage*). For a given subgroup a_i, we then estimate

$$\hat{\mu}_i = \frac{1}{n_i} \sum_{y \in Y_i} y \tag{7}$$

$$\hat{\sigma}_i^2 = \frac{1}{n_i} \sum_{y \in Y_i} (y - \hat{\mu}_i)^2, \tag{8}$$

where $\hat{\sigma}_i^2$ is the biased estimator such that the estimate times n_i equals the Residual Sum of Squares, i.e., $n_i\hat{\sigma}_i^2 = \sum_{y\in Y_i}(y - \hat{\mu}_i)^2 = RSS_a$.

Given the above, we can separately encode the covers of the individual subgroups, but we first show how to encode the target values not covered by any subgroup.

Encoding Target Values Not Covered by Any Subgroup. The target values not covered by any subgroup, given by $Y_d = \{(\mathbf{x}, y) \in D \mid \forall_{a_i \in M} a_i \not\sqsubseteq \mathbf{x}\}$, are covered by the default dataset 'rule' and distribution at the end of a subgroup list. As $\hat{f}_{d,\hat{\mu}_d,\hat{\sigma}_d}$ is known and constant for a given dataset, one can simply encode the instances using this (normal) distribution, resulting in encoded length

$$L(Y_d \mid \hat{\mu}_d, \hat{\sigma}_d) = \frac{n_d}{2}\log 2\pi + \frac{n_d}{2}\log \hat{\sigma}_d^2 + \left[\frac{1}{2\hat{\sigma}_d^2}\sum_{y\in Y_d}(y - \hat{\mu}_d)^2\right] \text{le}, \tag{9}$$

where le $= \log e$. The first two terms are normalizing terms of a normal distribution, while the last term represents the Residual Sum of Squares (RSS) normalized by the variance of the data. Note that when $Y_d = Y$, i.e., the whole dataset target, RSS is equal to $n_d\sigma_d$ and the last term reduces to le$n_d/2$.

Encoding Target Values Covered by a Subgroup. In contrast to the previous case, here *we do not know a priori the statistics defining the probability distribution corresponding to the subgroup*, i.e., $\hat{\mu}$ and $\hat{\sigma}$ are not given by the model and thus both need to be encoded. For this we resort to the Bayesian encoding of a normal distribution with mean μ and standard deviation σ unknown, which was shown to be asymptotically optimal [8]. An optimal code length is simply given by the negative logarithm of a probability, and the optimal Bayesian probability for Y_i is given by

$$P_{Bayes}(Y_i) = \int_{-\infty}^{+\infty}\int_{0}^{+\infty} (2\pi\sigma)^{-\frac{n_i}{2}} \exp - \frac{\sum_{y\in Y_i}(y-\mu)^2}{2\sigma^2} w(\mu,\sigma)\, \mathrm{d}\mu\, \mathrm{d}\sigma, \tag{10}$$

where $w(\mu, \sigma)$ is the prior on the parameters, which needs to be chosen.

The MDL principle requires the encoding to be as unbiased as possible for any values of the parameters, which leads to the use of uninformative priors. The most uninformative prior is Jeffrey's prior, which is $1/\sigma^2$ and therefore constant for any value of μ and σ, but unfortunately its integral is undefined, i.e., $\int\int \sigma^{-2}\,\mathrm{d}\sigma\,\mathrm{d}\mu = \infty$. Thus, we need to 1) constrain the parameter space and 2) make the integral finite, which we will do next in consecutive steps.

One of the best ways to constrain the parameter space without biasing it, is by multiplying Jeffrey's prior by a normal prior on the effect size, i.e., $\rho = \mu/\sigma \sim \mathcal{N}(0, \tau)$ [20]. We then still need to describe τ though; the most uninformative choice would be to use an inverse-chi-squared distribution, which would be equivalent to using a Cauchy prior on the effect size [20]. Unfortunately, this would lead to an open integral, which would render the approach infeasible for cases—like ours—where many probabilities need to be computed. The second best option is to fix $\tau = 1$, which gives a tractable formula that is equivalent to introducing a virtual point and converges[4] to the Bayes Information Criterion (BIC) for large n. This is the best we can do and we proceed with this option.

Now, given the prior defined by $\rho = \mu/\sigma \sim \mathcal{N}(0,1)$, the remaining question is how we can make the integral over the prior finite. The most common solution, which we also employ, is to use k data points from Y_i, denoted Y_i^k, to create a proper conditional prior $w(\mu, \sigma \mid Y_i^k)$. As there are only two unknown parameters, we only need two points hence $k = 2$ [7,8]. Consequently, we first encode Y_i^2 with a non-optimal code that is readily available—here the encoding with the dataset distribution of Eq. (9)—and then use the Bayesian rule to derive the total encoded length of Y_i as

$$L(Y_i) = -\log \frac{P_{Bayes}(Y_i)}{P_{Bayes}(Y_i^2)} P(Y_i^2 \mid \mu_d, \sigma_d) = L_{Bayes}(Y_i) + L_{cost}(Y_i^2), \tag{11}$$

where $L_{cost}(Y_i^2) = L(Y_i^2 \mid \mu_d, \sigma_d) - L_{Bayes}(Y_i^2)$ is the extra cost incurred by encoding two points non-optimally. After some re-writing[5] we obtain the encoded length of the y values covered by a subgroup Y_i as

$$\begin{aligned} L(Y_i) &= L_{Bayes}(Y_i) + L_{cost}(Y_i^2) \\ &= 1 + \frac{n_i}{2}\log\pi - \log\Gamma\left(\frac{n_i}{2}\right) + \frac{1}{2}\log(n_i+1) + \frac{n_i}{2}\log n\hat{\sigma}_a^2 + L_{cost}(Y_i^2), \end{aligned} \tag{12}$$

where Γ is the Gamma function that extends the factorial to the real numbers ($\Gamma(n) = (n-1)!$ for integer n) and $\hat{\mu}_i$ and $\hat{\sigma}_i$ are the statistics of Equations (7) and (8), respectively. Note that for Y_i^2 any two unequal values (otherwise $\hat{\sigma}_2 = 0$ and $L_{Bayes}(Y_i^2) = \infty$) can be chosen from Y_i, thus we choose them such that they minimize $L_{cost}(Y_i^2)$. Finally, the total encoded size of Y is given by

$$L(Y \mid X, M) = \sum_{i \in M} L(Y_i) + L(Y_d \mid \mu_d, \sigma_d). \tag{13}$$

3.4 Properties and Quality Measure for Subgroup Lists

We next show[6] that the proposed data encoding is an instance of the classical definition of a quality measure as given by Eq. (1), and is tightly related to both an existing quality measure and the Bayesian two-sample t-test.

First, we show that Eq. (12)—with mean and variance unknown—converges, for large n, to Eq. (9)—with mean and variance known—plus an additional term. Using the Stirling approximation of $\Gamma(n+1) \sim \sqrt{2\pi n}\left(\frac{n}{e}\right)^n$ leads to

$$L(Y_i) \sim \frac{n_i}{2}\log 2\pi + \frac{n_i}{2}\log\hat{\sigma}_i^2 + \frac{n_i}{2}\mathrm{le} + \log\frac{n_i}{e}, \tag{14}$$

[4] See proof in Appendix 2 of the extended version [16].

[5] The full derivation of the Bayesian encoding and an in-depth explanation are given in Appendix 1 of the extended version [16].

[6] Derivations are given in Appendix 4 of the extended version [16].

where $\log \frac{n}{e}$ is equal to the penalty term of BIC and similar to the usual MDL complexity of a distribution [8].

Now, we can show that minimizing our MDL criterion is equivalent to maximizing a subgroup discovery quality function of the form Eq. (1). Focusing on the case where $S = \{s_1\}$ contains only one subgroup with statistics $\hat{\Theta}_1 = \{\hat{\mu}_1, \hat{\sigma}_1\}$, we start with $L(Y \mid X, M)$ (Eq. (2)), multiply it by minus one to make it a maximization problem, and add a constant $L(Y \mid \hat{\mu}_d, \hat{\sigma}_d)$, i.e., the encoded size of the whole target Y using the overall distribution dataset, to obtain

$$
\begin{aligned}
L(Y \mid \hat{\Theta}_d) - L(Y \mid X, M) &\sim n_i \left[\log \frac{\hat{\sigma}_d}{\hat{\sigma}_i} + \frac{\hat{\sigma}_i^2 + (\mu_1 - \mu_2)^2}{2\sigma_d^2} \text{le} - \frac{\text{le}}{2} \right] - \log(n_i) - L(S) \\
&= n_i D_{KL}(\hat{\Theta}_a; \hat{\Theta}_d) - \log(n_i) - L(S),
\end{aligned}
\tag{15}
$$

where $\hat{\Theta}_a = \{\hat{\mu}_d, \hat{\sigma}_d\}$ and $n_i D_{KL}(\hat{\Theta}_a; \hat{\Theta}_d)$ is the usage-weighted Kullback-Leibler divergence between the normal distributions specified by the respective parameter vectors. This shows that *finding the MDL-optimal subgroup is equivalent to finding the subgroup that maximizes the weighted Kullback-Leibler (WKL) divergence*, an existing subgroup discovery quality measure [11] that was previously used for nominal targets, plus a term that defines the complexity of the subgroup. Moreover, note that Eq. (15) is equivalent to the Bayesian two-sample t-test [6] plus the complexity of the model, which plays the role of penalizing for multiple hypothesis testing. Finally, our measure is part of the family of *dispersion-corrected* subgroup quality measures, as it takes into account both the centrality and the spread of the target values [4].

Quality Measure for Subgroup Lists. Based on the previous, we naturally extend the KL-based measure for individual subgroups to subgroup lists and propose the Sum of Weighted Kullback-Leibler (SWKL) divergences:

$$
\text{SWKL}(S) = \sum_{a \in S} n_i D_{KL}(\hat{\Theta}_a; \hat{\Theta}_d) = \sum_{a_i \in S} n_i \left[\log \frac{\hat{\sigma}_d}{\hat{\sigma}_i} + \frac{\hat{\sigma}_i^2 + (\hat{\mu}_i - \hat{\mu}_d)^2}{2\hat{\sigma}_d^2} \text{le} - \frac{\text{le}}{2} \right]
\tag{16}
$$

An advantage of this measure is that it can not only be used for numeric targets, but for any type of probabilistic model. Note that computing SWKL is straightforward for subgroup lists as obtained by most methods, including ours, but not for subgroup sets as instances can be covered by multiple subgroups.

4 The SSD++ Algorithm

As the problem of finding an MDL-optimal list of subgroups is unfeasible, we propose a heuristic approach (as is common in MDL-based pattern mining [18,22]) based on Separate-and-Conquer (SaC) to construct the list, and beam-search to generate the subgroups to add at each iteration of SaC. The first reason for using greedy search to add one subgroup at the time, is its transparency, as it adds at

Algorithm 1: SSD++ algorithm

Data: Dataset D, number of cut points n_{cut}, beam width w_b, depth max. d_{max}
Result: Subgroup list S
1 $M \leftarrow [\Theta_d(Y)]$;
2 **repeat**
3 $\quad Cands \leftarrow BeamSearch(M, D, w_b, n_{cut}, d_{max})$;
4 $\quad s \leftarrow \arg\max_{\forall s' \in Cands} : \delta L(D, M \oplus s')$;
5 $\quad M \leftarrow M \oplus s$;
6 **until** $\delta L(D, M \oplus s') \leq 0, \forall s' \in Cands$;

each iteration the locally best subgroup found by the beam search. Beam-search, on the other hand, was empirically shown, in the context of subgroup discovery for numeric targets, to be very competitive in terms of quality when compared to a complete search with an associated speedup improvement [14]. Also, its straightforward implementation allows to easily extend this framework to other types of targets, not just numeric. To quantify the quality of annexing $\oplus$ a subgroup s at the end (after all the other subgroups) of model M, we use the *normalized gain* $\delta L(M \oplus s) = (L(D, M) - L(D, M \oplus s))/n_s$, which was first introduced in the classification setting and proved to outperform its non-normalized version in that setting [18]. For a detailed empirical comparison of normalized gain and its non-normalized version please refer to Appendix 6 [16].

Algorithm 1 presents SSD++, a greedy algorithm that iteratively adds subgroups to an empty subgroup list until no more compression can be gained, where compression is measured in terms of normalized gain of adding a subgroup s.

The *beam search algorithm* starts by discretizing all variables depending on their subsets, i.e. categorical and binary with the operator *equal to* (=) and numeric by generating all subsets with n_{cut} points. At each iteration the w_b subgroups that maximize the selected gain are chosen and will be expanded with all discretized variables until the maximum depth d_{max} of the description is achieved.

The *SSD++ algorithm* [15] takes as input the dataset D, and the beam search parameters, namely the number of cut points n_{cut}, the width of the beam w_b, and the maximum depth of search d_{max}. The algorithm starts by adding the dataset empirical distribution to the model (Ln 1). Then, while there is a subgroup that improves compression (Ln 6), it keeps iterating over three steps: 1) generating the candidates using beam search (Ln 3); 2) finding the subgroup that maximizes the normalized gain (Ln 4); and 3) adding that subgroup to the end of the model, i.e., after all the existing subgroups in the model (Ln 5). The beam search returns the best subgroup according to the data not covered by any subgroup in the model M and its parameters (w_b, n_{cut}, d_{max}). When there is no subgroup that improves compression (non-positive gain) the while loop stops and the subgroup list is returned. Note that beam search is used at each iteration, instead of only once at the beginning, as it can converge to local optima, and would thus bias our search to the top-k subgroups instead of the best at each iteration.

5 Experiments

We evaluate SSD++[7] by comparing it to 1) a classical top-k mining algorithm, as a baseline of a non-diverse method, and 2) the sequential covering algorithm, henceforth called top-k and seq-cover respectively, which are both available in the implementation of the DSSD algorithm[8].

DSSD and SISD will not be compared due to two interconnected issues: 1) the lack of a *global* definition of the optimal set for a dataset; 2) the absence of a definition for the interaction between subgroups that overlap. The first issue has as a natural consequence that none of the methods have a clear stopping criteria as the definition of when a set describes the data well is not available, apart from the user-specified hyperparameter 'number of subgroups'. Added to this, both issues give rise to the question of how to measure the interaction of subgroups in the region of their overlap from a model (global) perspective, i.e., they could behave as an additive or a multiplicative mixture of their probabilities for example. These issues hamper the comparison with both methods as they do not have a clear stopping criteria and a formulation of their overlap interaction, of which the latter is necessary for our proposed measure SWKL. On the other hand, a direct use of SWKL assuming a list formulation, i.e. ordering them and removing the overlap, will always rate them lower, which was corroborated with our initial experiments. Moreover, we do not compare with prediction algorithms that generate rules for regression, such as RIPPER or CART, as the rules generated aim at making the best prediction possible, and not the highest difference from the dataset distribution, as shown theoretically in Appendix 5 [16].

Data. We use a set of 16 benchmark datasets from the Keel[9] repository commonly used for subgroup discovery. The complete description of the datasets is given in Table 2; the datasets were chosen to be diverse, ranging from 297 to 22 784 instances and from 2 to 40 variables.

Hyperparameter Selection. *SSD++:* the algorithm admits as hyperparameters: the width of the beam w_b; number of cut points n_{cut}; and maximum depth of search d_{max}. By varying these parameters over the datasets the results can be seen in Appendix 7 [16] and it was concluded that: 1) no descriptions of size much greater than 5 are found; 2) after $n_{cut} = 5$ (the default value for seq-cover) the subgroups returned are virtually the same but with numerical values refined; 3) for most datasets the quality of the subgroup list stabilizes beyond $w_b = 100$. Thus, for the rest of our experiments the parameters are set accordingly.

Top-k: the software used here is the top-k subgroups implemented in DSSD, which is equivalent to most top-k subgroup miners. As it is common with top-k miners a depth-first search is used for small datasets $|D| \leq 2000$ and a beam

[7] For the implementation of SSD++ and to reproduce the experiments see Proença [15].

[8] http://www.patternsthatmatter.org/software.php#dssd/.

[9] http://www.keel.es/.

Table 2. Dataset properties: number of instances, and variables.

Dataset	$\|D\|$	categorical	numerical	Dataset	$\|D\|$	categorical	numerical
cholesterol	297	7	5	wizmir	1 461	0	9
baseball	337	4	12	abalone	4 177	0	8
autoMPG8	392	0	6	puma32h	8 192	0	32
dee	365	0	6	ailerons	13 750	0	40
ele-1	495	0	2	elevators	16 599	0	18
forestFires	517	0	12	bikesharing	17 379	2	10
concrete	1 030	0	8	california	20 640	0	8
treasury	1 049	0	15	house	22 784	0	16

search for the rest. For the quality measure it uses the Weighted Kullback-Leibler without dispersion, i.e., $WKL_{\mu}(s) = n_s/\hat{\sigma}_d(\hat{\mu_d} - \hat{\mu_s})^2$ as described in Appendix 3 [16], as the algorithm does not accept its dispersion-aware version used in Eq. (16). Also, as it does not have a termination criteria, the k number of subgroups returned is selected as the number of subgroups found by SSD++.

Seq-cover: to ensure fairness the same beam search hyperarameters as SSD++ are used, i.e., $d_{max} = 5$, $w_b = 100$, $n_{cut} = 5$. As quality measure it uses the Weighted Kullback-Leibler without dispersion for the same reasons as top-k.

5.1 Subgroup List Quality

The results can be seen in Table 3, and Figs. 1 and 2. The algorithms are compared in terms of Sum of Weighted Kullback-Leibler (SWKL) of Eq. (16) for the quality of the list, number of subgroups $|S|$, average number of variables per description $|a|$, standard deviation of the first subgroup $\tilde{\sigma}_{top1}$, runtime and average Jaccard index of the lists. Note that $\tilde{\sigma}_{top1}$ shows the most important characteristic first found by each miner. In the case of the averaged Jaccard index it is computed based on the average of the Jaccard index between the 1-vs-1 covers (when considered independently) of the subgroups in the list, i.e., for the case of a list of 4 subgroups, 6 values are averaged.

From Table 3 we see that SSD++ obtains the best score in terms of our proposed measure SWKL for 12 out of 16 datasets. As expected the top-k algorithm obtains a lower score for all datasets except for one. This supports that our proposed measure SWKL gives weight to subgroup sets that cover different parts of the dataset. Also, in terms of the dispersion of the first subgroup its value is lower for 80% of the cases. In terms of the number of rules and compared with seq-cover, SSD++ tends to find fewer subgroups for smaller datasets ($|D| \leq 10\,000$), and more for larger datasets. For the latter, the experiments showed that on average each subgroup covers more than 100 instances per subgroup. In terms of the number of variables per description, it tends to find more compact descriptions than top-k and seq-cover.

In terms of runtime, as per Fig. 1, SSD++ has a similar performance to seq-cover for small sample sizes ($|D| \leq 1000$) and 10 times slower for larger sizes.

Table 3. Performance results of {Summed Weighted Kullback-Leibler Divergence (SWKL) divided by number of examples; standard deviation of the first subgroup normalized by σ_d; number of subgroups; average number of conditions per subgroup description} per dataset for each algorithm.

	top-k				seq-cover				SSD++			
datasets	SWKL	$\tilde{\sigma}_{top1}$	$\|S\|$	$\|a\|$	SWKL	$\tilde{\sigma}_{top1}$	$\|S\|$	$\|a\|$	SWKL	$\tilde{\sigma}_{top1}$	$\|S\|$	$\|a\|$
cholesterol	0.14	1.49	1	5	**0.84**	**1.51**	33	4	0.11	1.99	1	3
baseball	0.25	0.85	8	5	1.69	0.82	26	4	**1.92**	**0.22**	8	2
autoMPG8	0.48	0.54	10	5	1.36	0.54	22	3	**1.65**	**0.18**	10	2
dee	0.49	0.47	8	5	**1.47**	0.50	20	4	1.33	**0.44**	8	2
ele-1	0.29	**1.06**	9	3	1.14	**1.06**	22	3	**1.25**	1.33	9	2
forestFires	0.58	6.84	23	5	2.85	6.84	57	4	**3.80**	**0.03**	23	3
concrete	0.25	0.78	19	5	**1.27**	0.65	35	4	**1.27**	**0.34**	19	3
treasury	0.42	0.70	31	5	2.41	0.68	25	3	**3.73**	**0.05**	31	2
wizmir	0.77	0.31	22	5	2.17	0.31	26	4	**2.73**	**0.16**	22	2
abalone	0.23	0.59	25	5	0.48	0.59	118	3	**0.71**	**0.45**	25	3
puma32h	0.55	0.59	42	5	**1.48**	0.59	76	5	1.42	**0.30**	42	3
ailerons	0.24	1.23	19	2	1.04	1.23	101	4	**1.58**	**1.10**	197	4
elevators	0.25	**1.44**	141	4	0.84	**1.44**	157	4	**1.30**	**1.44**	160	4
bikesharing	0.27	1.09	127	5	1.24	1.09	91	4	**1.68**	**0.07**	127	4
california	0.19	0.90	163	4	0.70	0.90	135	5	**1.15**	**0.84**	163	4
house	0.19	**1.59**	280	5	0.91	**1.59**	145	4	**2.08**	2.18	280	5

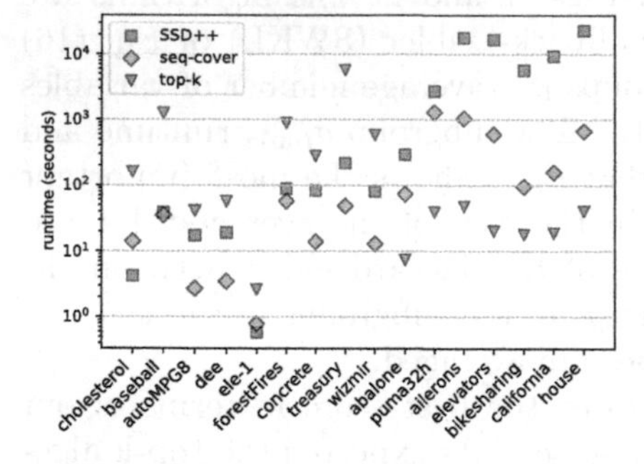

Fig. 1. Runtime in seconds per algorithm and dataset.

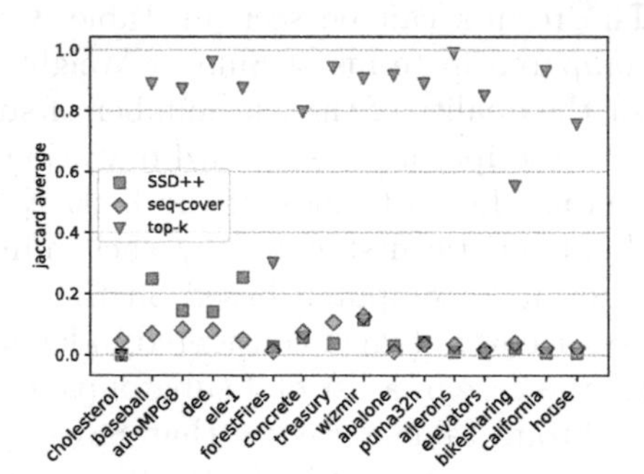

Fig. 2. Average overlap of subgroups in a list per dataset and algorithm.

This can, in part, be explained, by the larger number of subgroups found for these datasets—from 1.2 to 4 times more. Figure 2 shows that for small datasets the overlap is larger than for seq-cover, while for larger datasets our formulation tends to have similar level of overlap.

6 Case Study: Hotel Bookings

To test the usefulness of our method we applied it to the problem of understanding the type of clients that make a hotel booking based on how much time in advance (lead time in days) this was done. To this end we used the "Hotel booking demand dataset" [1], and analysed the data referent to a *resort hotel* in the year of 2016. The first four subgroups of a total of 260 obtained with SSD++ can be seen in Fig. 1 (in Sect. 1) and its subgroups versus the dataset in Fig. 3. Only the first 4 subgroups are shown here for clarity, and given that greedy search is used, they are also the 4 most interesting subgroups.

The results show us a detailed picture of the dataset and at first glance, one notices that most subgroups cover a small number of instances. Nevertheless, this is normal as they represent highly defined subgroups, with a very different mean and an almost zero standard deviation, compared with the dataset $\hat{\mu}_d = 92$ and $\hat{\sigma}_d = 99$. As an example, subgroup 1 has an average lead time circa 6 times higher than the dataset distribution, together with a standard deviation that is 3 times smaller. This subgroup seems to represent a group of people that travelled together from Great Britain and all chose the same type of booking, while with some slight days of difference in their bookings. Another interesting subgroup is the 4th which shows that there is a group of around 20 similar bookings for groups of 2 or more adults done with only 9 days before arrival when the deposit type is refundable. If one would follow the whole subgroup list one would have a complete summary of the bookings done.

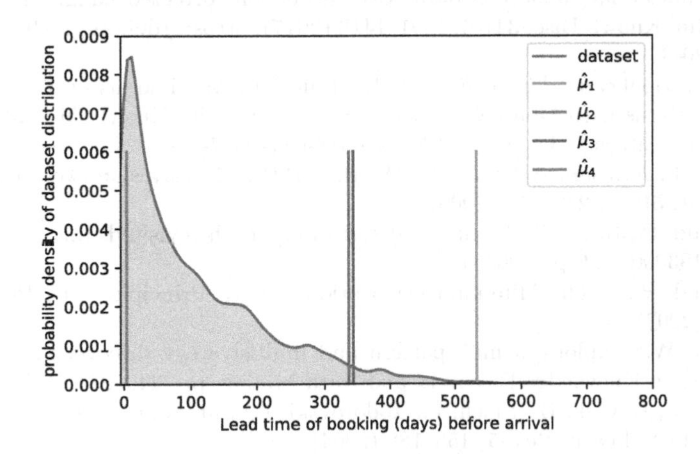

Fig. 3. Kernel density estimation of the dataset distribution and mean value of the first 4 found subgroups.

7 Conclusions

We introduced a dispersion-aware problem formulation for subgroup set discovery based on subgroup lists, the MDL principle, and Bayesian statistics. We proved our formulation to be equivalent to an existing subgroup quality measure for the case of finding the single best subgroup, and showed a relationship to Bayesian testing. Based on these insights we proposed a new evaluation measure for subgroup lists, the sum of Weighted Kullback-Leibler divergences (SWKL).

To find good subgroup lists we introduced SSD++, a greedy algorithm that we empirically evaluated on 16 datasets and compared against state-of-the-art algorithms. SSD++ was shown to outperform the other methods in terms of both our proposed measure and subgroup set complexity as quantified by subgroup and/or description sizes, and discovers subgroups with small standard deviation.

Acknowledgment. This work is part of the research programme Indo-Dutch Joint Research Programme for ICT 2014 with project number 629.002.201, SAPPAO, which is financed by the Netherlands Organisation for Scientific Research.

References

1. Antonio, N., de Almeida, A., Nunes, L.: Hotel booking demand datasets. Data Brief **22**, 41–49 (2019)
2. Atzmueller, M.: Subgroup discovery. Wiley Interdisc. Rev. Data Min. Knowl. Disc. **5**(1), 35–49 (2015)
3. Belfodil, A., et al.: FSSD-a fast and efficient algorithm for subgroup set discovery. In: Proceedings of DSAA 2019 (2019)
4. Boley, M., Goldsmith, B.R., Ghiringhelli, L.M., Vreeken, J.: Identifying consistent statements about numerical data with dispersion-corrected subgroup discovery. Data Min. Knowl. Disc. **31**(5), 1391–1418 (2017). https://doi.org/10.1007/s10618-017-0520-3
5. Bosc, G., Boulicaut, J.F., Raïssi, C., Kaytoue, M.: Anytime discovery of a diverse set of patterns with Monte Carlo tree search. Data Min. Knowl. Disc. **32**(3), 604–650 (2018). https://doi.org/10.1007/s10618-017-0547-5
6. Gönen, M., Johnson, W.O., Lu, Y., Westfall, P.H.: The Bayesian two-sample t test. Am. Stat. **59**(3), 252–257 (2005)
7. Grünwald, P., Roos, T.: Minimum description length revisited. Int. J. Math. Ind. **11**(1), 1930001 (29 p.) (2019)
8. Grünwald, P.D.: The Minimum Description Length Principle. MIT Press, Cambridge (2007)
9. Klösgen, W.: Explora: a multipattern and multistrategy discovery assistant. In: Advances in Knowledge Discovery and Data Mining, pp. 249–271 (1996)
10. Lavrač, N., Kavšek, B., Flach, P., Todorovski, L.: Subgroup discovery with CN2-SD. J. Mach. Learn. Res. **5**, 153–188 (2004)
11. van Leeuwen, M.: Maximal exceptions with minimal descriptions. Data Min. Knowl. Disc. **21**(2), 259–276 (2010). https://doi.org/10.1007/s10618-010-0187-5
12. van Leeuwen, M., Knobbe, A.: Diverse subgroup set discovery. Data Min. Knowl. Disc. **25**(2), 208–242 (2012). https://doi.org/10.1007/s10618-012-0273-y
13. Lijffijt, J., Kang, B., Duivesteijn, W., Puolamaki, K., Oikarinen, E., De Bie, T.: Subjectively interesting subgroup discovery on real-valued targets. In: 2018 IEEE ICDE, pp. 1352–1355. IEEE (2018)
14. Meeng, M., Knobbe, A.: For real: a thorough look at numeric attributes in subgroup discovery. Data Min. Knowl. Disc. **35**(1), 158–212 (2021)

15. Proença, H.M. : HMProenca/SSDpp-numeric: v2020.06.0 (2020). https://github.com/HMProenca/SSDpp-numeric. Archived at https://doi.org/10.5281/zenodo.3901236
16. Proença, H.M., Grünwald, P., Bäck, T., van Leeuwen, M.: Discovering outstanding subgroup lists for numeric targets using MDL. Preprint arXiv:2006.09186 (2020)
17. Proença, H.M., Klijn, R., Bäck, T., van Leeuwen, M.: Identifying flight delay patterns using diverse subgroup discovery. In: 2018 SSCI, pp. 60–67. IEEE (2018)
18. Proença, H.M., van Leeuwen, M.: Interpretable multiclass classification by MDL-based rule lists. Inf. Sci. **512**, 1372–1393 (2020)
19. Rissanen, J.: Modeling by shortest data description. Automatica **14**(5), 465–471 (1978)
20. Rouder, J.N., Speckman, P.L., Sun, D., Morey, R.D., Iverson, G.: Bayesian t tests for accepting and rejecting the null hypothesis. Psychon. Bull. Rev. **16**(2), 225–237 (2009)
21. Van Leeuwen, M., Galbrun, E.: Association discovery in two-view data. IEEE Trans. Knowl. Data Eng. **27**(12), 3190–3202 (2015)
22. Vreeken, J., Van Leeuwen, M., Siebes, A.: KRIMP: mining itemsets that compress. Data Min. Knowl. Disc. **23**(1), 169–214 (2011). https://doi.org/10.1007/s10618-010-0202-x

A Relaxation-Based Approach for Mining Diverse Closed Patterns

Arnold Hien[2], Samir Loudni[2,3](✉), Noureddine Aribi[1], Yahia Lebbah[1], Mohammed El Amine Laghzaoui[1], Abdelkader Ouali[2], and Albrecht Zimmermann[2]

[1] University of Oran1, Lab. LITIO, 31000 Oran, Algeria
[2] Normandie University, UNICAEN, CNRS – UMR GREYC, Normandie, France
samir.loudni@imt-atlantique.fr
[3] TASC (LS2N-CNRS), IMT Atlantique, 44307 Nantes, France

Abstract. In recent years, pattern mining has moved from a slow-moving repeated three-step process to a much more agile iterative/user-centric mining model. A vital ingredient of this framework is the ability to *quickly* present a set of *diverse* patterns to the user. In this paper, we use constraint programming (well-suited to user-centric mining due to its rich constraint language) to efficiently mine a diverse set of closed patterns. Diversity is controlled through a threshold on the Jaccard similarity of pattern occurrences. We show that the Jaccard measure has no monotonicity property, which prevents usual pruning techniques and makes classical pattern mining unworkable. This is why we propose anti-monotonic lower and upper bound relaxations, which allow effective pruning, with an efficient branching rule, boosting the whole search process. We show experimentally that our approach significantly reduces the number of patterns and is very efficient in terms of running times, particularly on dense data sets.

1 Introduction

The original data analysis model based on pattern mining consists of three steps in a kind of *multi-waterfall cycle*: 1) a user chooses the values of one or several mining parameters, 2) an underlying engine extracts patterns (often taking not inconsiderable time to do so), and 3) the user sifts through a (potentially very large) set of result patterns and interprets them, using their insights to return to the first step and repeat the cycle.

Recently, this approach has been challenged by an increasing focus on *user-centered*, *interactive*, and *anytime* pattern mining [14]. This new paradigm stresses that users should be presented quickly with patterns likely to be interesting to them, and typically affect later iterations of the mining process by giving feedback. A powerful framework for taking a variety of user feedback into account is pattern mining via constraint programming (CP). Much of the current focus in this domain is on user-centered/interactive mining, particularly the ability to elicit and exploit user feedback [9,14,18]. An important aspect of requesting such feedback is that the user be quickly presented with *diverse*

F. Hutter et al. (Eds.): ECML PKDD 2020, LNAI 12457, pp. 36–54, 2021.
https://doi.org/10.1007/978-3-030-67658-2_3

results. If patterns are too similar to each other, deciding which one to prefer can become challenging, and if they appear in several successive iterations, it eventually becomes a slog. Similarly, a method that produces diverse results but takes a long time to do so, risks that the user checks out of the process. Older work on diversity either post-process patterns derived from the process described above [5,12,21], use heuristics [20] or view it purely from the point of view of speeding up the extraction process [8]. Recent work, on the other hand, pushes diversity constraints into the mining process itself [3,4]. At the algorithmic level, additional user-specified constraints often require new implementations to filter out the patterns violating or satisfying the user's constraints, which can be computationally infeasible for large databases.

In the last decade, data mining has been combined with constraint programming to model various data mining problems [2,6,13,19]. The main advantage of CP for pattern mining is its declarativity and flexibility, which include the ability to incorporate new user-specified constraints without the need to modify the underlying system. Moreover, CP allows to define flexible search strategies[1]. In this paper, we propose to add to the literature on *explicitly* taking the diversity of patterns (in terms of the data instances they describe) into account and to use an exhaustive process to find candidates for inclusion into a result set. To achieve this, we use the widely accepted Jaccard index to compare patterns and formulate a diversity constraint, which has no monotonicity property, implying limited pruning during search. To cope with this problem, we propose two anti-monotonic relaxations: (*i*) A lower bound relaxation, which allows to prune non-diverse items during search. This is integrated in our constraint programming based approach through a new global constraint taking into account diversity with its filtering algorithms (aka, propagators); (*ii*) An upper bound relaxation to find items ensuring diversity. This is exploited through a new branching rule, boosting the search process towards diverse patterns. We demonstrate the performance of our proposed method experimentally, comparing to the state-of-the-art in CP-based closed pattern mining.

2 Preliminaries

2.1 Itemset Mining

Let $\mathcal{I} = \{1, ..., n\}$ be a set of n *items*, an *itemset* (or pattern) P is a non-empty subset of $\mathcal{I}$. The language of itemsets corresponds to $\mathcal{L}_\mathcal{I} = 2^\mathcal{I} \backslash \emptyset$. A transactional dataset $\mathcal{D}$ is a bag (or multiset) of transactions over $\mathcal{I}$, where each *transaction* t is a subset of $\mathcal{I}$, i.e., $t \subseteq \mathcal{I}$; $\mathcal{T} = \{1, ..., m\}$ a set of m *transaction* indices. An itemset P *occurs* in a transaction t, iff $P \subseteq t$. The *cover* of P in $\mathcal{D}$ is the set of transactions in which it occurs: $\mathcal{V}_\mathcal{D}(P) = \{t \in \mathcal{D} \mid p \subseteq t\}$. The *support* of P in $\mathcal{D}$ is the size of its cover: $sup_\mathcal{D}(P) = |\mathcal{V}_\mathcal{D}(P)|$. An itemset P is said to be *frequent* when its support exceeds a user-specified minimal threshold θ, $sup_\mathcal{D}(P) \geq \theta$.

[1] Opposed to more rigid search in classical pattern mining algorithms, which often rely on exploiting the properties of a particular constraint.

Given $S \subseteq \mathcal{D}$, $items(S)$ is the set of common items belonging to all transactions in S: $items(S) = \{i \in \mathcal{I} \mid \forall t \in S, i \in t\}$. The *closure* of an itemset P, denoted by $Clos(P)$, is the set of common items that belong to all transactions in $\mathcal{V}_\mathcal{D}(P)$: $Clos(P) = \{i \in \mathcal{I} \mid \forall t \in \mathcal{V}_\mathcal{D}(P), i \in t\ \}$. An itemset P is said to be *closed* iff $Clos(P) = P$. Constraint-based pattern mining aims at extracting all patterns P of $\mathcal{L}_\mathcal{I}$ satisfying a selection predicate c (called *constraint*) which is usually called *theory* [5]: $Th(c)$. A common example is the frequency measure leading to the minimal support constraint, which can be combined with the closure constraint to mine closed frequent itemsets.

Example 1. Figure 1 shows the itemset lattice derived from a toy dataset with five items and 100 transactions. As the figure shows, there exist 26 frequent closed itemsets with $\theta = 7$.

Most constraint-based mining algorithms take advantage of monotonicity which offers pruning conditions to safely discard non-promising patterns from the search space. Several frameworks exploit this principle to mine with a monotone or an anti-monotone constraint. Other classes of constraints have also been considered [15,16]. However, for constraints that are not anti-monotone, pushing them into the discovery algorithm might lead to less effective pruning phases. Thus, we propose in this paper to exploit the *witness* concept introduced in [11] to handle such constraints. A witness is a single itemset on which we can test whether a constraint holds and derive information about properties of other itemsets.

Definition 1 (Witness). *Let P, Q itemsets, and $C : \mathcal{I} \mapsto \{true, false\}$, then $W, P \subseteq W \subseteq P \cup Q$, is called a positive (negative) witness iff $\forall P', P \subseteq P' \subseteq P \cup Q : C(W) = true \Rightarrow C(P') = true$ ($C(W) = false \Rightarrow C(P') = false$).*

2.2 Diversity of Itemsets

The Jaccard index is a classical similarity measure on sets. We use it to quantify the overlap of the covers of itemsets.

Definition 2 (Jaccard index). *Given two itemsets P and Q, the Jaccard index is the relative size of the overlap of their covers: $Jac(P, Q) = \frac{|\mathcal{V}_\mathcal{D}(P) \cap \mathcal{V}_\mathcal{D}(Q)|}{|\mathcal{V}_\mathcal{D}(P) \cup \mathcal{V}_\mathcal{D}(Q)|}$.*

A lower Jaccard indicates low similarity between itemset covers, and can thus be used as a measure of diversity between pairs of itemsets.

Definition 3 (Diversity/Jaccard constraint). *Let P and Q be two itemsets. Given the Jac measure and a diversity threshold J_{max}, we say that P and Q are pairwise diverse iff $Jac(P, Q) \leq J_{max}$. We will denote this constraint by c_{Jac}.*

Our aim is to push the Jaccard constraint during pattern discovery to prune non-diverse itemsets. To achieve this, we maintain a *history* $\mathcal{H}$ of extracted pairwise diverse itemsets during search and constrain the next mined itemsets to respect a maximum Jaccard constraint with all itemsets already included in $\mathcal{H}$. This problem can be formalized as follows.

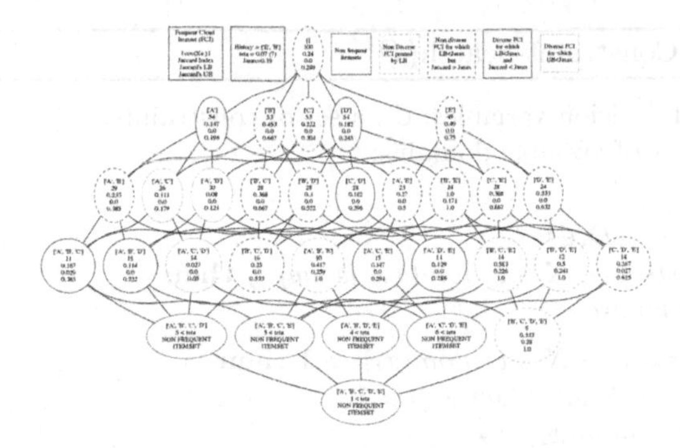

Fig. 1. The powerset lattice of frequent closed itemsets ($\theta = 7$) for the dataset $\mathcal{D}$ of Example 1.

Definition 4 (k diverse frequent itemsets). *Given a current history $\mathcal{H} = \{H_1, \ldots, H_k\}$ of k pairwise diverse frequent closed itemsets, the Jac measure and a diversity threshold J_{max}, the task is to mine new itemsets P such that $\forall H \in \mathcal{H}$, $Jac(P, H) \leq J_{max}$.*

Example 2. The lattice in Fig. 1 depicts the set of diverse FCIs (marked with blue and green solid line circles) with $J_{max} = 0.19$ and $\mathcal{H} = \{BE\}$. ACE is a diverse FCI (i.e., $Jac(ACE, BE) = 0.147 < 0.19$).

Proposition 1. *Let P, Q and P' be three itemsets s.t. $P \subset P'$. $Jac(P, Q)$ may be smaller, equal or greater than $Jac(P', Q)$.*

Based on the above proposition, the anti-monotonicity of the maximum Jaccard constraint does not hold, which disables pruning. Thus, instead of solving the problem of Definition 4 directly, we introduce bounds in Sect. 3 that allow us to prune the search space using a relaxation of the Jaccard constraint. The appeal of this approach is that we are able to infer monotone and anti-monotone properties from this relaxation.

2.3 Constraint Programming (CP)

Constraint programming [10] is a powerful paradigm which offers a generic and modular approach to model and solve combinatorial problems. A CP model consists of a set of variables $X = \{x_1, \ldots, x_n\}$, a set of domains D mapping each variable $x_i \in X$ to a finite set of possible values $dom(x_i)$, and a set of constraints $\mathcal{C}$ on X. A constraint $c \in \mathcal{C}$ is a relation that specifies the allowed combinations of values for its variables $X(c)$. An assignment on a set $Y \subseteq X$ of variables is a mapping from variables in Y to values in their domains. A solution is an assignment on X satisfying all constraints. Constraint solvers typically use backtracking search to explore the search space of partial assignments. Algorithm 1 provides a general overview of a CP solver. At each node of the search tree, procedure *Constraint-Search* selects an unassigned variable (line 8) according

Algorithm 1: Constraint-Search(D)

```
1  In: X : a set of decision variables; C : a set of constraints;
2  InOut: D : a set of variable domains;
3  begin
4      D ← Filtering(D, C)
5      if there exists x_i ∈ X s.t. dom(x_i) is empty then
6          return failure
7      if there exists x_i ∈ X s.t. |dom(x_i)| > 1 then
8          Select x_i ∈ X s.t. |dom(x_i)| > 1
9          forall v ∈ dom(x_i) do
10             Constraint-Search(Dom ∪ {x_i → {v}})
11     else
12         output solution D
```

to user-defined heuristics and assigns it a value (line 9). It backtracks when a constraint cannot be satisfied, i.e. when at least one domain is empty (line 5). A solution is obtained (line 12) when each domain $dom(x_i)$ is reduced to a singleton and all constraints are satisfied. The main concept used to speed up the search is constraint propagation by *Filtering algorithms*. At each assignment, constraint filtering algorithms prune the search space by enforcing local consistency properties like *domain consistency*. A constraint c on $X(c)$ is domain consistent, if and only if, for every $x_i \in X(c)$ and every $v \in dom(x_i)$, there is an assignment satisfying c such that $(x_i = v)$. Global constraints are families of constraints defined by a relation on any number of variables [10].

2.4 A CP Model for Frequent Closed Itemset Mining

The first constraint programming model for frequent closed itemset mining (FCIM) was introduced in [6]. It is based on reified constraints to connect item variables to transaction variables. The first global constraint CLOSEDPATTERNS for mining frequent closed itemsets was proposed in [13]. The global constraint COVERSIZE for computing the exact size of the cover of an itemset was introduced in [19]. It offers more flexibility in modeling problems. We present the global constraint CLOSEDPATTERNS.

Global Constraint ClosedPatterns. Most declarative methods use a vector x of Boolean variables $(x_1, \ldots, x_{|\mathcal{I}|})$ for representing itemsets, where x_i represents the presence of the item $i \in \mathcal{I}$ in the itemset. We will use the following notations: $x^+ = \{i \in \mathcal{I} \mid dom(x_i) = \{1\}\}$ the present items, $x^- = \{i \in \mathcal{I} \mid dom(x_i) = \{0\}\}$ the absent items and $x^* = \{i \in \mathcal{I} \mid i \notin x^+ \cup x^-\}$ the set of non assigned items.

Definition 5 (ClosedPatterns). *Let x be a vector of Boolean variables, θ a support threshold and $\mathcal{D}$ a dataset. The global constraint* CLOSEDPATTERNS$_{\mathcal{D},\theta}(x)$ *holds if and only if x^+ is a closed frequent itemset w.r.t. the threshold θ.*

Definition 6 (Closure extension [22]**).** *A non-empty itemset P is a closure extension of Q iff $\mathcal{V}_{\mathcal{D}}(P \cup Q) = \mathcal{V}_{\mathcal{D}}(Q)$.*

Filtering of ClosedPatterns. [13] also introduced a complete filtering algorithm for ClosedPatterns based on three rules. The first rule filters 0 from $dom(x_i)$ if $\{i\}$ is a closure extension of x^+ (see Definition 6). The second rule filters 1 from $dom(x_i)$ if the itemset $x^+ \cup \{i\}$ is infrequent w.r.t. θ. Finally, the third rule filters 1 from $dom(x_i)$ if $\mathcal{V}_{\mathcal{D}}(x^+ \cup \{i\})$ is a subset of $\mathcal{V}_{\mathcal{D}}(x^+ \cup \{j\})$ where j is an absent item, i.e. $j \in x^-$.

To show the strength and the flexibility of the CP approach in taking into account user's constraints, we formulate a CP model to extract more specific patterns using the following four global constraints: $\mathcal{C} = \{\text{ClosedP}_{\mathcal{D},\theta}(X), \text{AtLeast}(X, lb), \text{Knapsack}(X, z, w), \text{Regular}(X, DFA)\}$ The AtLeast constraint enforces that at least lb variables in X are assigned to 1; the Knapsack constraint restricts a weighted linear sum to be no more than a given capacity z, i.e. $\sum_i w_i X_i \leq z$; the Regular constraint imposes that X is accepted by deterministic finite automaton (DFA), which recognizes a regular expression.

Example 3. Let us consider the example of Fig. 1 using $lb = 2$, $w = \langle 8, 7, 5, 14, 16 \rangle$, $z = 20$, $\theta = 7$ and the regular expression $0^* \, 1^+ \, 0^*$ ensuring items' contiguity. Solving this CP model provides the solution: $Th(\mathcal{C}) = \{\{AB\}, \{BC\}, \{CD\}, \{ABC\}\}$.

3 A CP Model for Mining Diverse Frequent Closed Itemsets

We present our approach for computing diverse FCIs. The key idea is to compute an approximation of the set of diverse FCIs by defining two bounds on the Jaccard index that allow us to reduce the search space. All the proofs are given in the Supp. material [1].

3.1 Problem Reformulation

Proposition 1 states that the Jaccard constraint is neither monotonic nor anti-monotonic. So, we propose to approximate the theory of the original constraint c_{Jac} by a larger collection corresponding to the solution space of its *relaxation* c^r_{Jac}: $Th(c_{Jac}) \subseteq Th(c^r_{Jac})$. The key idea is to formulate a relaxed constraint having suitable monotonicity properties in order to exploit them for search space reduction. More precisely, we want to exploit *upper and lower bounding* operators to derive a monotone relaxation and an anti-monotone one of c_{Jac}.

Definition 7 (Problem reformulation). *Given a current history $\mathcal{H} = \{H_1, \ldots, H_k\}$ of extracted k pairwise diverse frequent closed itemsets, a diversity threshold J_{max}, a lower bound LB_J and an upper bound UB_J on the Jaccard index, the relaxed problem consists of mining candidate itemsets P such that $\forall H \in \mathcal{H}$, $LB_J(P, H) \leq J_{max}$. When $UB_J(P, H) \leq J_{max}$, for all $H \in \mathcal{H}$, the Jaccard constraint is fully satisfied.*

3.2 Jaccard Lower Bound

Let us now formalize how to compute the lower bound and how to exploit it. To arrive at a lower bound for the Jaccard value between two itemsets, we need to consider the situation where the *overlap* between them has become as small as possible, while the coverage that is proper to each itemset remains as large as possible.

Definition 8 (Proper cover). *Let P and Q be two itemsets. The proper cover of P w.r.t. Q is defined as* $\mathcal{V}_Q^{pr}(P) = \mathcal{V}_\mathcal{D}(P) \backslash \{\mathcal{V}_\mathcal{D}(P) \cap \mathcal{V}_\mathcal{D}(Q)\}$.

The lowest possible Jaccard would reduce the numerator to 0, which is however not possible under the minimum support threshold θ. The denominator, on the other hand, consists of $|\mathcal{V}_\mathcal{D}(H)|$ (which cannot change) and the part of P's coverage that does not overlap with H, i.e. $\mathcal{V}_H^{pr}(P)$.

Proposition 2 (Lower bound). *Consider a member pattern H of the history* $\mathcal{H}$. *Let P an itemset encountered during search such that* $sup_\mathcal{D}(P) \geq \theta$, *and* $\mathcal{V}_H^{pr}(P)$ *be the proper cover of P w.r.t. H.* $LB_J(P, H) = \frac{\theta - |\mathcal{V}_H^{pr}(P)|}{|\mathcal{V}_\mathcal{D}(P)| + |\mathcal{V}_\mathcal{D}(H)| + |\mathcal{V}_H^{pr}(P)| - \theta}$ *is a lower bound of* $Jac(P, H)$.

The lower bound on the Jaccard index enables us to discard some non-diverse itemsets, i.e., those with an LB_J value greater than J_{max} are *negative witnesses.*

Example 4. The set of all non diverse FCIs with a lower bound value greater than J_{max} are marked in Fig. 1 with orange line circles.

Proposition 3 (Monotonicity of LB_J). *Let* $H \in \mathcal{H}$ *be an itemset. For any two itemsets* $P \subseteq Q$, *the relationship* $LB_J(P, H) \leq LB_J(Q, H)$ *holds.*

Property 3 establishes an important result to define a pruning condition based on the monotonicity of the lower bound (cf. Sect. 3.4). If $LB_J(P, H) > J_{max}$, then no itemset $Q \supseteq P$ will satisfy the Jaccard constraint (because LB_J is a lower bound), rendering the constraint itself *anti-monotone.* So, we can safely prune Q.

3.3 Jaccard Upper Bound

As our relaxation approximates the theory of the Jaccard constraint, i.e. $Th(c_{Jac}) \subseteq Th(c_{Jac}^r)$, one could have itemsets P such that $LB_J(P, H) < J_{max}$ but $Jac(P, H) > J_{max}$ (see itemsets marked with blue dashed line circles in Fig. 1). To tackle this case, we define an upper bound on the Jaccard index to evaluate the satisfaction of the Jaccard constraint, i.e., those with $UB_J(P, H) \leq J_{max}$, $\forall H \in \mathcal{H}$, are *positive witnesses.*

To derive the upper bound, we need to follow the opposite argument as for the lower bound: the highest possible Jaccard will be achieved if $\mathcal{V}_\mathcal{D}(H) \cap \mathcal{V}_\mathcal{D}(P)$ stays unchanged but the set $\mathcal{V}_H^{pr}(P)$ is reduced as much possible (under the minimum support constraint). If the intersection is greater than or equal θ, in the worst case scenario (leading to the highest Jaccard), a future P' covers only transactions in the intersection. If not, the denominator needs to contain a few elements of $\mathcal{V}_H^{pr}(P)$, $\theta - |\mathcal{V}_\mathcal{D}(H) \cap \mathcal{V}_\mathcal{D}(P)|$, to be exact.

Proposition 4 (Upper bound). *Given a member pattern H of the history $\mathcal{H}$, and an itemset P such that $supp_\mathcal{D}(P) \geq \theta$. $UB_J(P,H) = \frac{|\mathcal{V}_\mathcal{D}(H) \cap \mathcal{V}_\mathcal{D}(P)|}{|\mathcal{V}_P^{pr}(H)| + \max\{\theta, |\mathcal{V}_\mathcal{D}(H)| \cap |\mathcal{V}_\mathcal{D}(P)|\}}$ is an upper bound of $Jac(P,H)$.*

Example 5. The set of all diverse FCIs with UB_J values less than J_{max} are marked with green line circles in Fig. 1.

Our upper bound can be exploited to evaluate the Jaccard constraint during mining. More precisely, in the enumeration procedure, if the upper bound of the current candidate itemset P is less than J_{max}, then c_{Jac} is fully satisfied. Moreover, if the upper bound is *monotonically decreasing* (or anti-monotonic), then all itemsets Q derived from P are also diverse (see Proposition 5).

Proposition 5 (Anti-monotonicity of UB_J). *Let H be a member pattern of the history $\mathcal{H}$. For any two itemsets $P \subseteq Q$, the relationship $UB_J(P,H) \geq UB_J(Q,H)$ holds.*

3.4 The Global Constraint ClosedDiversity

This section presents our new global constraint ClosedDiversity that exploits the LB relaxation to mine pairwise diverse frequent closed itemsets.

Definition 9 (ClosedDiversity). *Let x be a vector of Boolean item variables, $\mathcal{H}$ a history of pairwise diverse frequent closed itemsets (initially empty), θ a support threshold, J_{max} a diversity threshold and $\mathcal{D}$ a dataset. The* ClosedDiversity$_{\mathcal{D},\theta}(x, \mathcal{H}, J_{max})$ *global constraint holds if and only if: (1) x^+ is closed; (2) x^+ is frequent, $supp_\mathcal{D}(x^+) \geq \theta$; (3) x^+ is diverse, $\forall H \in \mathcal{H}, LB_J(x^+, H) \leq J_{max}$.*

Initially, the history $\mathcal{H}$ is empty. Our global constraint allows to incrementally update $\mathcal{H}$ with diverse FCIs encountered during search. Condition (3) expresses a necessary condition ensuring that x^+ is diverse. Indeed, one could have $LB_J(x^+, H) \leq J_{max}$ but $Jac(x^+, H) > J_{max}$. Thus, we propose in Sect. 4 to exploit our UB relaxation to guarantee the satisfaction of the Jaccard constraint.

The propagator for ClosedDiversity exploits the filtering rules of ClosedPatterns (see Sect. 2.4). It also uses our LB relaxation to remove items i that cannot belong to a solution containing x^+. We denote by x^-_{Freq} the set of items filtered by the rule of infrequent items and by x^-_{Div} the set of items filtered by our LB rule.

Algorithm 2: Filtering for CLOSEDDIVERSITY

```
1  In: θ, J_max : frequency and diversity thresholds; H : history of solutions
      encountered during search;
2  InOut: x = {x_1 ... x_n} : Boolean item variables;
3  begin
4      if ( |V_D(x⁺)| < θ ∨ !PGrowth_LB(x⁺, H, J_max)) then return false;
5      foreach i ∈ x* do
6          if (|V_D(x⁺ ∪ {i})| < θ) then
7              dom(x_i) ← dom(x_i) − {1}; x⁻_Freq ← x⁻_Freq ∪ {i}; x* ← x* \ {i};
               continue;
8          if (|V_D(x⁺ ∪ {i})| = |V_D(x⁺)|) then
9              dom(x_i) ← dom(x_i) − {0}; x⁺ ← x⁺ ∪ {i};    x* ← x* \ {i};
10         if (! PGrowth_LB(x⁺ ∪ {i}, H, J_max)) then
11             dom(x_i) ← dom(x_i) − {1}; x⁻_Div ← x⁻_Div ∪ {i}; x* ← x* \ {i};
               continue;
12         foreach k ∈ (x⁻_Freq ∪ x⁻_Div) do
13             if (V_D(x⁺ ∪ {i}) ⊆ V_D(x⁺ ∪ {k})) then
14                 dom(x_i) ← dom(x_i) − {1}
15                 if k ∈ x⁻_Freq then x⁻_Freq ← x⁻_Freq ∪ {i};
16                 else x⁻_Div ← x⁻_Div ∪ {i};
17                 x* ← x* \ {i}; break;
18     return true;
19 Function PGrowth_LB(x, H, J_max): Boolean
20     foreach H ∈ H do
21         if (LB_J(x, H) > J_max) then return false
22     return true
```

Proposition 6 (ClosedDiversity Filtering rule). *Given a history $\mathcal{H}$ of pairwise diverse frequent closed itemsets, a partial assignment on x, and a free item $i \in x^*$, $x^+ \cup \{i\}$ cannot lead to a diverse itemset if one of the two cases holds: 1) if $\exists H \in \mathcal{H}$ s.t. $LB_J(x^+ \cup \{i\}, H) > J_{max}$, then we remove* 1 *from $dom(x_i)$. 2) if $\exists k \in x^-_{Div}$ s.t. $\mathcal{V}_\mathcal{D}(x^+ \cup \{i\}) \subseteq \mathcal{V}_\mathcal{D}(x^+ \cup \{k\})$, then $LB_J(x^+ \cup \{i\}, H) > LB_J(x^+ \cup \{k\}, H) > J_{max}$, thus we remove* 1 *from $dom(x_i)$.*

Algorithm. The propagator for CLOSEDDIVERSITY is presented in Algorithm 2. It takes as input the variables x, the support threshold θ, the diversity threshold J_{max} and the current history $\mathcal{H}$ of pairwise diverse frequent closed itemsets. It starts by computing the cover of the itemset x^+ and checks if x^+ is either infrequent or not diverse (see function $\mathcal{P}Growth_{LB}$), if so the constraint is violated and a fail is returned (line 4). Algorithm 2 extends the filtering rules of CLOSEDPATTERNS (see Sect. 2.4) by examining the diversity condition of the itemset $x^+ \cup \{i\}$ (see Proposition 3). For each element $H \in \mathcal{H}$, the function $\mathcal{P}Growth_{LB}(x^+ \cup \{i\}, \mathcal{H}, J_{max})$ computes the value of $LB_J(x^+ \cup \{i\}, H)$ and tests if there exists an H s.t. $LB_J(x^+ \cup \{i\}, H) > J_{max}$ (lines 20–21). If so, we return *false* (line 21) because $x^+ \cup \{i\}$ cannot lead to a diverse itemset w.r.t. $\mathcal{H}$, remove 1 from $dom(x_i)$ (line 11), update x^-_{Div} and x^* and we continue with the next free item. Otherwise, we return true. Second, we remove 1 from each free item variable $i \in x^*$ such that its cover is a superset of the cover of an absent item $k \in (x^-_{Freq} \cup x^-_{Div})$ (lines 12–17). The LB filtering rule associated to the case $k \in x^-_{Div}$ is a new rule taking its originality from the reasoning made on absent items.

Proposition 7 (Consistency and time complexity). *Algorithm 2 enforces Generalized Arc Consistency (GAC) (a.k.a. domain consistency [10]) in* $\mathcal{O}(n^2 \times m)$.

4 Using Witnesses and the Estimated Frequency Within the Search

In this section, we show how to exploit the witness property and the estimated frequency so as to design a more informed search algorithm.

Positive Witness. During search, we compute incrementally the $UB(x^+ \cup \{i\}, H)$ of any extension of the partial assignment x^+ with a free item i. If, for each $H \in \mathcal{H}$, this upper bound is less or equal to J_{max}, then c_{Jac} is fully satisfied and $x^+ \cup \{i\}$ is a *positive witness*. Moreover, thanks to the anti-monotonicity of UB_J (see Proposition 5), all supersets of $x^+ \cup \{i\}$ will satisfy the Jaccard constraint.

Estimated Frequency. The frequency of an itemset can be computed as the cardinality of the intersection of its items' cover: $sup_{\mathcal{D}}(x^+) = |\cap_{i \in x^+} \mathcal{V}_{\mathcal{D}}(i)|$, the intersection between 2 covers being performed by a bitwise-AND. To limit the number of intersections, we use an estimation of the frequency of each item $i \in \mathcal{I}$ w.r.t the set of present items x^+, denoted $eSup_{\mathcal{D}}(i, x^+)$. This estimation constitutes a *lower bound* of $|\mathcal{V}_{\mathcal{D}}(x^+ \cup \{i\})|$. Interestingly, if $eSup_{\mathcal{D}}(i, x^+) \geq \theta$ then $|\mathcal{V}_{\mathcal{D}}(x^+ \cup \{i\})| \geq \theta$, meaning that the intersection between covers is performed only if $eSup_{\mathcal{D}}(i, x^+) < \theta$, thereby leading to performance enhancement. In addition, we argue that the estimated support is an interesting heuristic to reinforce the witness branching rule. Indeed, branching on the variable having the minimum estimated support (using the lower bound of the real support) will probably activate our filtering rules (see Algorithm 2), thus reducing the search space. It will be denoted as MINCOV variable ordering heuristic.

We propose Algorithm 3 as a branching procedure (returns the next variable to branch on). When the search begins, for each item $i \in x^*$, its estimated frequency is initialized to $eSup_{\mathcal{D}}(i, \emptyset) = |\mathcal{V}_{\mathcal{D}}(i)|$. Once an item j has been added to the partial solution, the estimated frequencies of unbound items must be updated (see lines 4–9). Thus, we first find the variable x^{es} having the minimal estimated support (line 4). Next, each item $i \in x^* \setminus \{x^{es}\}$ may lose some support, but no more than $|\mathcal{V}_{\mathcal{D}}(x^+)| - |\mathcal{V}_{\mathcal{D}}(x^+ \cup \{x^{es}\})|$, since some removed transactions may

Algorithm 3: Branching for ClosedDiversity

1 **In**: J_{max} : diversity thresholds; $\mathcal{H}$: history of solutions ;
2 **Out**: First witness index or x^{es} as the item with the smallest estimated support
3 **begin**
4 $x^{es} \leftarrow argmin_{i \in x^*}(eSup_{\mathcal{D}}(i, x^+))$;
5 diff $\leftarrow (|\mathcal{V}_{\mathcal{D}}(x^+)| - |\mathcal{V}_{\mathcal{D}}(x^+ \cup \{x^{es}\})|)$;
6 **foreach** $i \in x^* \setminus \{x^{es}\}$ **do**
7 $eSup_{\mathcal{D}}(i, x^+ \cup \{x^{es}\}) \leftarrow eSup_{\mathcal{D}}(i, x^+)$ – diff;
8 **if** $(eSup_{\mathcal{D}}(i, x^+ \cup \{x^{es}\}) < \theta)$ **then**
9 $eSup_{\mathcal{D}}(i, x^+ \cup \{x^{es}\}) \leftarrow |\mathcal{V}_{\mathcal{D}}(x^+ \cup \{x^{es}\}) \cap \mathcal{V}_{\mathcal{D}}(i)|$;
10 **foreach** $i \in x^*$ **do**
11 **if** $(\mathcal{P}Growth_{UB}(x^+ \cup \{i\}, \mathcal{H}, J_{max}))$ **then**
12 **return** $\langle i, \mathbf{true} \rangle$;
13 **return** $\langle x^{es}, \mathbf{false} \rangle$
14 **Function** $\mathcal{P}Growth_{UB}(x^+ \cup \{j\}, \mathcal{H}, J_{max})$: *Boolean*
15 **foreach** $H \in \mathcal{H}$ **do**
16 **if** $(UB_J(x^+ \cup \{j\}, H) > J_{max})$ **then**
17 **return false**
18 **return true**

not contain i (line 5). Using this upper bound (denoted by *diff*), the estimated frequency of i is updated and set to $eSup_{\mathcal{D}}(i, x^+)$ – *diff* (lines 6–9). As indicated above, if $eSup_{\mathcal{D}}(i, x^+) \geq \theta$ then $|\mathcal{V}_{\mathcal{D}}(x^+ \cup \{i\})| \geq \theta$. Otherwise, we have to compute the right support by performing the intersection between covers (line 9). It is important to stress that the branching variable x^{es} will be returned (line 13) only if no positive witness is found (lines 10–12). Finally, the function $\mathcal{P}Growth_{UB}(x^+ \cup \{i\}, \mathcal{H}, J_{max})$ allows to test whether the current instantiation x^+ can be extended to a witness itemset using the free item $\{i\}$. It returns true if the upper bound of the current itemset x^+ when adding one item $\{i\}$ is less than J_{max} for all $h \in \mathcal{H}$ (lines 15–17). Here, the Jaccard constraint is fully satisfied and thus, we return the item $\{i\}$ with the witness flag set to *true*. This information will be supplied to the search engine (line 12) to accelerate solutions certification. We will denote by FirstWitCov, our variable ordering heuristic that branches on the first free item satisfying the witness property.

Exploring the Witness Subtree. Let N be the node associated to the current itemset x^+ extended to a free item $\{i\}$. When the node N is detected as a positive witness during the branching, all supersets derived from N will also satisfy the Jaccard constraint. As these patterns are more likely to have similar covers, so a rather high Jaccard between them, we propose a simple strategy which avoids a complete exploration of the witness sub-tree rooted at N. Thus, we generate the first closed diverse itemset from N, add it to the current history and continue the exploration of the remaining search space using the new history. With a such strategy we have no guarantee that the closed itemset added to the history have the best Jaccard. But this strategy is fast.

5 Related Work

The question of mining sets of diverse patterns has been addressed in the recent literature, both to offer more interesting results and to speed up the mining process. Van Leeuwen *et al.* propose populating the beam for subgroup discovery not purely with the *best* partial patterns to be extended but to take coverage overlap into account [20]. Beam search is heuristic, as opposed to our exhaustive approach and since they mine all patterns at the same time, diverse partial patterns can still lead to a less diverse final result. Dzyuba *et al.* propose using *XOR* constraints to partition the transaction set into disjoint subsets that are small enough to be efficiently mined using either a CP approach or a dedicated itemset miner [8]. Their focus is on efficiency, which they demonstrate by approximating the result set of an exhaustive operation. While they discuss pattern sets, they limit themselves to a strict non-overlap constraint on coverages. In [4], the authors propose using Monte Carlo Tree Search and upper confidence bounds to direct the search towards interesting regions in the lattice given the already explored space. While MCTS is necessarily randomized, it allows for anytime mining. The authors of [3] consider sets of subgroup descriptions as *disjunctions* of such patterns. Using a greedy algorithm exploiting upper bounds, the authors propose to iteratively extract up to k subgroup descriptions (similarly to our work). Notably, this approach requires a target attribute *and* a target value to focus on while our approach allows for unsupervised mining.

Earlier work has treated reducing redundancy as a post-processing step, e.g. [12] where a number of redundancy measures such as entropy are exploited in exhaustive search and the number of patterns in the set limited, [7] where the constraint-based itemset mining constraint is adapted to the pattern set settings, [5], which exploit bounds on predicting the presence of patterns from the patterns already included in $\mathcal{H}$ in a heuristic algorithm, or [21], which exploits the MDL principle to minimize redundancy among itemsets (and, in later work, sequential patterns). All of those methods require a potentially rather costly first mining step, and none exploits the Jaccard measure. As discussed in Sect. 2.1, there exist a number of constraint properties that allow for pruning, and Kifer *et al.*'s witness concept unifies them and discusses how to deal with constraints that do not have monotonicity properties [11]. The way to proceed in such a case is establishing *positive* and *negative* witnesses for the constraint, something we have done for the maximum pairwise Jaccard constraint. A rarely discussed aspect is that witnesses are closely related to CP since every witness enforces/forbids the inclusion of certain domain values.

6 Experiments and Results

The experimental evaluation is designed to address the following questions: (1) How (in terms of CPU-times and # of patterns) does our global constraint (denoted ClosedDiv) compare to the ClosedPatterns global constraint (denoted ClosedP) and the approach of Dzyuba et al. [8] (denoted Flexics)? (2) How do the resulting diverse FCIs compare qualitatively with those resulting from ClosedP and Flexics? (3) How far is the distance between the Jaccard index and the upper/lower bounds.

Experimental Protocol. Experiments were carried out on classic UCI data sets, available at the FIMI repository (https://fimi.ua.ac.be/data). We selected several real-world data sets, their characteristics (name, number of items $|\mathcal{I}|$, number of transactions $|\mathcal{T}|$, density ρ) are shown in the first column of Table 1. We selected data sets of various size and density. Some data sets, such as Hepatitis and Chess, are very dense (resp. 50% and 49%). Others, such as T10 and Retail, are very sparse (resp. 1% and 0.06%). The implementation of the different global constraints and their constraint propagators were carried out in the `Choco` solver [17] version 4.0.3, a Java library for constraint programming. The source code is publicly available[2]. Experiments were conducted on AMD Opteron 6174, 2.2 GHz with a RAM of 256 GB and a time limit of 24 h. The default maximum heap size allowed by the JVM is 30 GB. We have selected for every data set frequency thresholds to have different numbers of frequent closed itemsets ($|Th(c)| \leq 15000$, $30000 \leq |Th(c)| \leq 10^6$, and $|Th(c)| > 10^6$). The only exception are the very large and sparse data sets Retail and Pumsb, where we do not find a large number of solutions. We used the ClosedP CP model as a baseline to determine suitable thresholds used with the ClosedDiv CP model. To evaluate the quality of a set of patterns in terms of diversity, we measured the average ratio of exclusive pattern coverages: $ECR(P_1, ..., P_k) = avg_{1 \leq i \leq k}(\frac{sup_{\mathcal{D}}(P_i) - |\mathcal{V}_{\mathcal{D}}(P_i) \cap \bigcup_{j \neq i} \mathcal{V}_{\mathcal{D}}(P_j)|}{sup_{\mathcal{D}}(P_i)})$.

(a) Comparing ClosedDiv with ClosedP and Flexics. Table 1 compares the performance of the two CP models for various values of θ on different data sets. Here, we report the CPU time (in seconds), the number of extracted patterns, and the number of nodes explored during search. This enables to evaluate the amount of inconsistent values pruned by each approach (filtering algorithm). We use MinCov as variable ordering heuristic. The maximum diversity threshold J_{max} is set to 0.05. First, the results highlight the great discrepancy between the two models with a distinctly lower number of patterns generated by ClosedDiv (in the thousands) in comparison to ClosedP (in the millions). On dense and moderately dense data sets (from Chess to Mushroom), the discrepancy is greatly amplified, especially for small values of θ. For instance, on Chess, the number of patterns for ClosedDiv is reduced by 99% (from $\sim 50 \cdot 10^6$ solutions to 393) for θ equal to 15%. The density of the data sets provides an appropriate explanation for the good performance of ClosedDiv. As the number of closed

[2] https://github.com/lobnury/ClosedDiversity.

Table 1. ClosedDiv (J_{max} = 0.05) vs ClosedP. For columns #Patterns and #Nodes, the values in bold indicate a reduction more than 20% of the total number of patterns and nodes." – " is shown when time limit is exceeded. OOM : Out Of Memory. (1): ClosedP (2): ClosedDiv

Dataset $\|\mathcal{I}\| \times \|\mathcal{T}\|$ $\rho(\%)$	$\theta(\%)$	#Patterns (1)	#Patterns (2)	Time (s) (1)	Time (s) (2)	#Nodes (2)	#Nodes (2)
Chess	20	22,808,625	**96**	2838.30	**5.87**	45,617,249	**436**
75 × 3196	15	50,723,131	**393**	5666.03	**75.40**	101,446,261	**1,855**
49.33%	10	OOM	**4,204**	OOM	**3825.29**	OOM	**18,270**
Hepatitis	30	83,048	**12**	9.64	**0.09**	166,095	**29**
68 × 137	20	410,318	**57**	42.00	**0.57**	820,635	**162**
50.00%	10	1,827,264	**2,270**	**169.59**	**76.91**	3,654,527	**5,256**
Kr-vs-kp	30	5,219,727	**17**	682.94	**0.74**	10,439,453	**82**
73 × 3196	20	21,676,719	**96**	2100.79	**5.64**	43,353,437	**448**
49.32%	10	OOM	**4,120**	OOM	**3035.49**	OOM	**17,861**
Connect	30	460,357	**18**	1666.14	**14.81**	920,713	**77**
129 × 67557	18	2,005,476	**197**	5975.44	**573.66**	4,010,951	**900**
33.33%	15	3,254,780	**509**	9534.07	**1989.35**	6,509,559	**2,188**
Heart-cleveland	10	12,774,456	**3,496**	1308.63	**257.39**	25,548,911	**7,977**
95 × 296	8	23,278,687	**12,842**	**2278.97**	2527.38	46,557,373	**28,221**
47.37%	6	43,588,346	58,240	**4126.84**	46163.06	87,176,691	124,705
Splice1	10	1,606	**422**	**6.55**	25.25	3,211	**843**
287 × 3190	5	31,441	**8,781**	**117.15**	5616.47	62,881	**17,594**
20.91%	2	589,588	-	**1179.55**	-	1,179,175	-
Dataset $\|\mathcal{I}\| \times \|\mathcal{T}\|$ $\rho(\%)$	$\theta(\%)$	#Patterns (1)	#Patterns (2)	Time (s) (1)	Time (s) (2)	#Nodes (2)	#Nodes (2)
Mushroom	5	8,977	**727**	**10.02**	60.70	17,953	**1,704**
112 × 8124	1	40,368	**12,139**	**34.76**	12532.95	80,735	**25,154**
18.75%	0.5	62,334	**27,768**	**50.05**	64829.06	124,667	**56,873**
T40I10D100K	8	138	127	**75.91**	447.20	275	253
942 × 100000	5	317	288	**331.47**	1561.34	633	575
4.20%	1	65,237	7,402	**5574.31**	58613.88	130,473	14,887
Pumsb	40	-	**4**	-	**57.33**	-	**16**
2113 × 49046	30	-	**15**	-	**267.72**	-	**64**
3.50%	20	-	**52**	-	**852.39**	-	**250**
T10I4D100K	5	11	11	**1.73**	6.31	21	21
870 × 100000	1	386	**361**	**434.25**	3125.06	771	722
1.16%	0.5	1,074	**617**	**881.31**	7078.90	2,147	**1,257**
BMS1	0.15	1,426	**609**	**11362.71**	68312.38	2,851	**1,220**
497 × 59602	0.14	1,683	**668**	**11464.93**	68049.00	3,365	**1,339**
0.51%	0.12	2,374	**823**	**13255.79**	79704.88	4,747	**1,651**
Retail	5	17	**13**	**10.74**	33.44	33	25
16470 × 88162	1	160	**111**	**297.21**	1625.73	319	227
0.06%	0.4	832	**528**	**6073.53**	31353.23	1,663	**1,093**

patterns increases with the density, redundancy among these patterns increases as well. On very sparse data sets, ClosedDiv still outputs fewer solutions than ClosedP but the difference is less pronounced. This is explainable by the fact that on these data sets, where we have few solutions, almost all patterns are diverse.

Second, regarding runtime, ClosedDiv exhibits different behaviours. On dense data sets ($\rho \geq 30\%$), ClosedDiv is more efficient than ClosedP and up to an order of magnitude faster. On Chess (resp. Connect), the speed-up is 1455 (resp. 112) for $\theta = 30\%$. For instances resulting in between 500 and 5000 diverse FCIs, the speed-up is up to 5. This good performance of ClosedDiv is mainly due to the strength of the LB filtering rule that provides the CP solving process with more propagation to remove more inconsistent values in the search space. In addition, the number of nodes explored by ClosedDiv is always small comparing to ClosedP. These results support our previous observations. The only exception is Heart-cleveland for which ClosedDiv is slower (especially for values of $\theta \leq 8\%$). This is mainly due to the relative large number of diverse patterns ($\geq$12000), which induces higher lower bound computational overhead. We observe the same behaviour on the two moderately dense data sets Splice1 and Mushroom. On sparse data sets, ClosedDiv can take significantly more time to extract all diverse FCIs. This can be explained by the fact that on these instances almost all FCIs are diverse w.r.t. lower bound (on average about 70% for Retail and 39% for BMS1, see Table 1). Thus, non-solutions are rarely filtered, while the lower bound overhead greatly penalizes the CP solving process. On the very large Pumsb data set, finally, our approach is very efficient while ClosedP fails to complete the extraction.

Finally, Fig. 2a compares ClosedDiv with Flexics (two variants) for various values of θ on different data sets: GFlexics, which uses CP4IM [6] as an oracle to enumerate the solutions, and EFlexics, a specialized variant, based on Eclat [23]. We run WeightGen with values of $\kappa \in \{0.1, 0.5, 0.9\}$ [9]. For each instance, we fixed the number of samples to the number of solutions returned by ClosedDiv. We report results corresponding to the best setting of parameter κ. First, ClosedDiv largely dominates GFlexics, being more than an order of magnitude faster. Second, while EFlexics is faster than GFlexics, our approach is almost always ranked first, illustrating its usefulness for mining diverse patterns in an anytime manner.

(b) Impact of Varying J_{max}. We varied J_{max} from 0.1 to 0.7. The minimum support θ is fixed for each data set (indicated after '-'). Figure 2 shows detailed results. As expected, the greater J_{max}, the longer the CPU time. In fact, the size of the history $\mathcal{H}$ grows rapidly with the increase of J_{max}. This induces significant additional costs in the lower and upper bound computations. Moreover, when J_{max} becomes sufficiently large, the LB filtering of ClosedDiv occurs rarely since the lower bound is almost always below the J_{max} value (see Fig. 3). Despite the hardness of some instances (i.e., $J_{max} \geq 0.35$), our CP approach is able to complete the extraction for almost all values of J_{max}. The only exception are the large and dense data sets, where ClosedDiv fails to complete the extraction

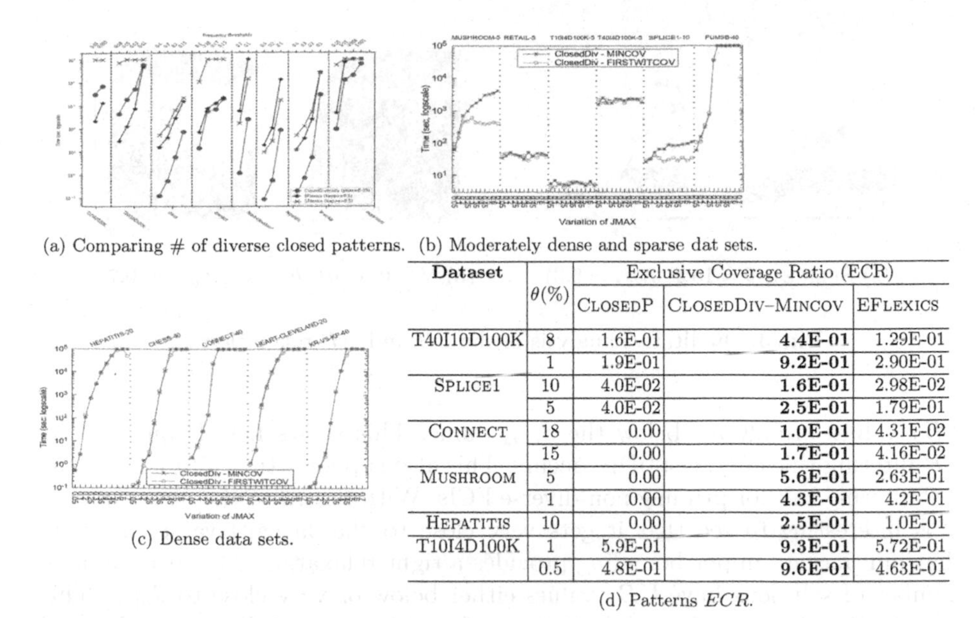

(a) Comparing # of diverse closed patterns. (b) Moderately dense and sparse dat sets.

(c) Dense data sets.

Dataset	θ(%)	Exclusive Coverage Ratio (ECR)		
		CLOSEDP	CLOSEDDIV–MINCOV	EFLEXICS
T40I10D100K	8	1.6E-01	**4.4E-01**	1.29E-01
	1	1.9E-01	**9.2E-01**	2.90E-01
SPLICE1	10	4.0E-02	**1.6E-01**	2.98E-02
	5	4.0E-02	**2.5E-01**	1.79E-01
CONNECT	18	0.00	**1.0E-01**	4.31E-02
	15	0.00	**1.7E-01**	4.16E-02
MUSHROOM	5	0.00	**5.6E-01**	2.63E-01
	1	0.00	**4.3E-01**	4.2E-01
HEPATITIS	10	0.00	**2.5E-01**	1.0E-01
T10I4D100K	1	5.9E-01	**9.3E-01**	5.72E-01
	0.5	4.8E-01	**9.6E-01**	4.63E-01

(d) Patterns *ECR*.

Fig. 2. CPU-time analysis (MINCOV vs FIRSTWITCOV and CLOSEDDIV vs FLEXICS) and patterns discrepancy analysis.

within the time limit for $J_{max} \geq 0.45$. However, in practice, the user will only be interested in small values of J_{max} because the diversity of patterns is maximal and the number of patterns returned becomes manageable.

Figures 2b and 2c also compare the resolution time of our CP model using the two variable ordering heuristics MINCOV and FIRSTWITCOV. First, on dense data sets, both heuristics perform similarly, with a slight advantage for FIRSTWITCOV. On these data sets, the number of witness patterns mined remains very low ($\leq$100), thus the benefits of FIRSTWITCOV is limited (see Supp. material). On moderately dense data sets (MUSHROOM and SPLICE1), FIRSTWITCOV is very effective; on MUSHROOM it is up 10 times faster than MINCOV for J_{max} equal to 0.7. On these data sets, the number of witness patterns extracted is relatively high compared to dense ones. In this case, FIRSTWITCOV enables to guide the search to find diverse patterns more quickly. On sparse data sets, no heuristic clearly dominates the other. When regarding the number of diverse patterns generated (see Supp. material), we observe that FIRSTWITCOV returns less patterns on moderately dense and sparse data sets, while on dense data sets the number of diverse patterns extracted remains comparable.

(c) Qualitative Analysis of the Proposed Relaxation. In this section, we shed light on the quality of the relaxation of the Jaccard constraint. Figure 3a shows, for a particular instance SPLICE1 with $J_{max} = 0.3$, the evolution of the LB_J and UB_J of the solutions found during search. Here, the solutions are sorted according to their UB_J. Concerning the lower bound, one can observe that the

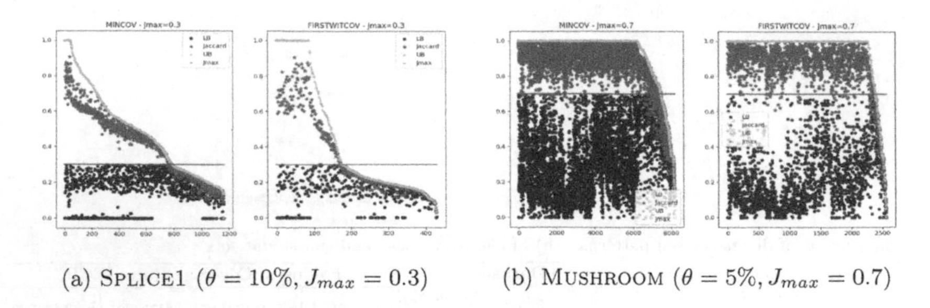

(a) SPLICE1 ($\theta = 10\%$, $J_{max} = 0.3$) (b) MUSHROOM ($\theta = 5\%$, $J_{max} = 0.7$)

Fig. 3. Qualitative analysis of the LB and UB relaxations.

LB_J values are always below the J_{max} value. This shows how frequently the LB filtering rule of CLOSEDDIV occurs. This also supports the suitability of the LB filtering rule for pruning non-diverse FCIs. With regard to the upper bound, it is interesting to see that it gets very close to the Jaccard value, meaning that our Jaccard upper bounding provides a tight relaxation. Moreover, a large number of solutions have UB_J values either below or very close to J_{max}. This is indicative of the quality of the patterns found in terms of diversity. We recall that when $UB_J < J_{max}$, all partial assignments can immediately be extended to diverse itemsets, thanks to the anti-monotonicity property of our UB (see Proposition 5). We observe the same behaviour on MUSHROOM with $J_{max} = 0.7$ (see Fig. 3b). Finally, we can see that FIRSTWITCOV allows to quickly discover solutions of better quality in terms of UB_J and Jaccard values compared to MINCOV. This demonstrates the interest and the strength of our UB_J branching rule to get diverse patterns.

(d) Qualitative Analysis of Patterns. Figure 2d compares CLOSEDDIV with CLOSEDP and EFLEXICS in terms of the ECR measure, which should be as high as possible. Due to the huge number of patterns generated by CLOSEDP, a random sample of $k = 10$ solutions of all patterns is considered. Reported values are the average over 100 trials. ECR penalises overlap, and thus having two similar patterns is undesirable. According to ECR, leveraging Jaccard in CLOSEDDIV clearly leads to pattern sets with more diversity among the patterns. This is indicative of patterns whose coverage are (approximately) mutually exclusive. This should be desirable for an end-user tasked with exploring and interpreting the set of returned patterns.

7 Conclusions

In this paper, we showed that mining diverse patterns using a maximum Jaccard constraint cannot be modeled using an anti-monotonic constraint. Thus, we have proposed (anti-) monotonic lower and upper bound relaxations, which allow to make pruning effective, with an efficient branching rule, boosting the whole search process. The proposed approach is introduced as a global constraint called ClosedDiv where diversity is controlled through a threshold on the Jaccard similarity of pattern occurrences. Experimental results on UCI datasets demonstrate that our approach significantly reduces the number of patterns, the set of patterns is diverse and the computation time is lower compared to ClosedP global constraint, particularly on dense data sets.

References

1. Supplementary Material, June 2020. https://github.com/lobnury/ClosedDiversity
2. Belaid, M., Bessiere, C., Lazaar, N.: Constraint programming for mining borders of frequent itemsets. In: Proceedings of IJCAI 2019, Macao, China, pp. 1064–1070 (2019)
3. Belfodil, A., et al.: Fssd-a fast and efficient algorithm for subgroup set discovery. In: Proceedings of DSAA, pp. 91–99 (2019)
4. Bosc, G., Boulicaut, J.F., Raïssi, C., Kaytoue, M.: Anytime discovery of a diverse set of patterns with Monte Carlo tree search. Data Min. Knowl. Disc. **32**(3), 604–650 (2018)
5. Bringmann, B., Zimmermann, A.: The chosen few: on identifying valuable patterns. Proc. ICDM **2007**, 63–72 (2007)
6. De Raedt, L., Guns, T., Nijssen, S.: Constraint programming for itemset mining. In: 14th ACM SIGKDD, pp. 204–212 (2008)
7. De Raedt, L., Zimmermann, A.: Constraint-based pattern set mining. In: 7th SIAM SDM, pp. 237–248. SIAM (2007)
8. Dzyuba, V., van Leeuwen, M., De Raedt, L.: Flexible constrained sampling with guarantees for pattern mining. Data Min. Knowl. Disc. **31**(5), 1266–1293 (2017). https://doi.org/10.1007/s10618-017-0501-6
9. Dzyuba, V., van Leeuwen, M.: Interactive discovery of interesting subgroup sets. In: Tucker, A., Höppner, F., Siebes, A., Swift, S. (eds.) IDA 2013. LNCS, vol. 8207, pp. 150–161. Springer, Heidelberg (2013). https://doi.org/10.1007/978-3-642-41398-8_14
10. Hoeve, W., Katriel, I.: Global constraints. In: Handbook of Constraint Programming, pp. 169–208. Elsevier Science Inc., (2006)
11. Kifer, D., Gehrke, J., Bucila, C., White, W.: How to quickly find a witness. In: Boulicaut, J.-F., De Raedt, L., Mannila, H. (eds.) Constraint-Based Mining and Inductive Databases. LNCS (LNAI), vol. 3848, pp. 216–242. Springer, Heidelberg (2006). https://doi.org/10.1007/11615576_11
12. Knobbe, A.J., Ho, E.K.Y.: Pattern teams. In: Fürnkranz, J., Scheffer, T., Spiliopoulou, M. (eds.) PKDD 2006. LNCS (LNAI), vol. 4213, pp. 577–584. Springer, Heidelberg (2006). https://doi.org/10.1007/11871637_58
13. Lazaar, N., et al.: A global constraint for closed frequent pattern mining. In: Proceedings of the 22nd CP, pp. 333–349 (2016)
14. Leeuwen, M.: Interactive data exploration using pattern mining. In: Holzinger, A., Jurisica, I. (eds.) Interactive Knowledge Discovery and Data Mining in Biomedical Informatics. LNCS, vol. 8401, pp. 169–182. Springer, Heidelberg (2014). https://doi.org/10.1007/978-3-662-43968-5_9
15. Ng, R.T., Lakshmanan, L.V.S., Han, J., Pang, A.: Exploratory mining and pruning optimizations of constrained association rules. In: Proceedings of ACM SIGMOD, pp. 13–24 (1998)

16. Pei, J., Han, J., Lakshmanan, L.V.S.: Mining frequent item sets with convertible constraints. In: Proceedings of ICDE, pp. 433–442 (2001)
17. Prud'homme, C., Fages, J.G., Lorca, X.: Choco Solver Documentation (2016)
18. Puolamäki, K., Kang, B., Lijffijt, J., De Bie, T.: Interactive visual data exploration with subjective feedback. In: Frasconi, P., Landwehr, N., Manco, G., Vreeken, J. (eds.) ECML PKDD 2016. LNCS (LNAI), vol. 9852, pp. 214–229. Springer, Cham (2016). https://doi.org/10.1007/978-3-319-46227-1_14
19. Schaus, P., Aoga, J.O.R., Guns, T.: CoverSize: a global constraint for frequency-based itemset mining. In: Beck, J.C. (ed.) CP 2017. LNCS, vol. 10416, pp. 529–546. Springer, Cham (2017). https://doi.org/10.1007/978-3-319-66158-2_34
20. Van Leeuwen, M., Knobbe, A.: Diverse subgroup set discovery. Data Min. Knowl. Disc. **25**(2), 208–242 (2012)
21. Vreeken, J., Van Leeuwen, M., Siebes, A.: Krimp: mining itemsets that compress. Data Min. Knowl. Disc. **23**(1), 169–214 (2011)
22. Wang, J., Han, J., Pei, J.: CLOSET+: searching for the best strategies for mining frequent closed itemsets. In: Proceedings of the Ninth KDD, pp. 236–245. ACM (2003)
23. Zaki, M., Parthasarathy, S., Ogihara, M., Li, W.: New algorithms for fast discovery of association rules. In: Proceedings of KDD 1997, Newport Beach, California, USA, August 14–17, pp. 283–286. AAAI Press (1997)

OMBA: User-Guided Product Representations for Online Market Basket Analysis

Amila Silva(✉), Ling Luo, Shanika Karunasekera, and Christopher Leckie

School of Computing and Information Systems, The University of Melbourne, Parkville, VIC, Australia
{amila.silva,ling.luo,karus,caleckie}@unimelb.edu.au

Abstract. Market Basket Analysis (MBA) is a popular technique to identify associations between products, which is crucial for business decision making. Previous studies typically adopt conventional frequent itemset mining algorithms to perform MBA. However, they generally fail to uncover rarely occurring associations among the products at their most granular level. Also, they have limited ability to capture temporal dynamics in associations between products. Hence, we propose OMBA, a novel representation learning technique for Online Market Basket Analysis. OMBA jointly learns representations for products and users such that they preserve the temporal dynamics of product-to-product and user-to-product associations. Subsequently, OMBA proposes a scalable yet effective online method to generate products' associations using their representations. Our extensive experiments on three real-world datasets show that OMBA outperforms state-of-the-art methods by as much as 21%, while emphasizing rarely occurring strong associations and effectively capturing temporal changes in associations.

Keywords: Market Basket Analysis · Online learning · Item representations · Transaction Data

1 Introduction

Motivation. Market Basket Analysis (MBA) is a technique to uncover relationships (i.e., association rules) between the products that people buy as a basket at each visit. MBA is widely used in today's businesses to gain insights for their business decisions such as product shelving and product merging. For instance, assume there is a strong association (i.e., a higher probability to buy together) between product p_i and product p_j. Then both p_i and p_j can be placed on the same shelf to encourage the buyer of one product to buy the other.

Typically, the interestingness of the association between product p_i and product p_j is measured using *Lift*: the *Support* (i.e., probability of occurrence) of p_i and p_j together divided by the product of the individual *Support* values of p_i and p_j as if they are independent. MBA attempts to uncover the sets of products with high *Lift* scores. However, it is infeasible to compute the *Lift* measure

F. Hutter et al. (Eds.): ECML PKDD 2020, LNAI 12457, pp. 55–71, 2021.
https://doi.org/10.1007/978-3-030-67658-2_4

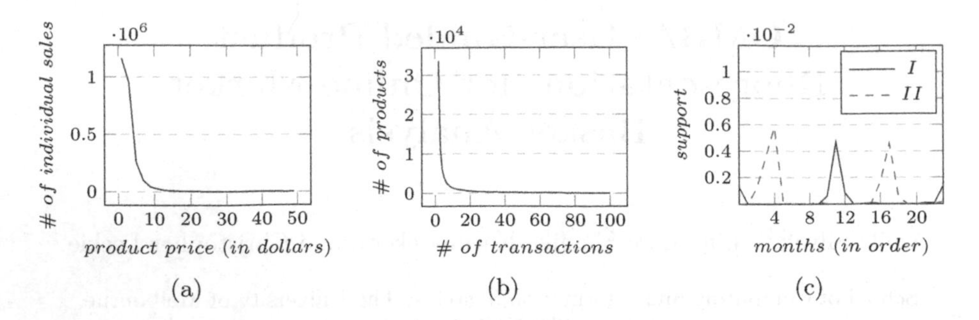

Fig. 1. (a) Number of products' sales with respect to the products' prices; (b) number of products with respect to their appearances in the transactions; and (c) temporal changes in the *Support* of : (I) Valentine Gifts and Decorations; and (II) Rainier Cherries. The plots are generated using Complete Journey dataset

between all product combinations for a large store in today's retail industry, as they offer a broad range of products. For example, Walmart offers more than 45 million products as of 2018[1]. Well-known MBA techniques [2,7,9] can fail to conduct accurate and effective analysis for such a store for the following reasons.

Research Gaps. First, MBA is typically performed using frequent itemset mining Algorithms [2,7,9,12] , which produce itemsets whose *Support* is larger than a predefined *minimum Support* value. However, such frequency itemset mining algorithms fail to detect associations among the products with low *Support* values (e.g., expensive products, which are rarely bought, as shown in Fig. 1a), but which are worth analysing. To further elaborate, almost all these algorithms compute *Support* values of itemsets, starting from the smallest sets of size one and gradually increasing the size of the itemsets. If an itemset P_i fails to meet the *minimum Support* value, all the supersets of P_i are pruned from the search space as they cannot have a *Support* larger than the *Support* of P_i. Nevertheless, the supersets of P_i can have higher *Lift* scores than P_i. Subsequently, the selected itemsets are further filtered to select the itemsets with high *Lift* scores as associated products. For instance, *'Mexican Seasoning Mixes'* and *'Tostado Shells'* have 0.012 and 0.005 support values in Complete Journey dataset (see Sect. 5) respectively. Because of the low support of the latter, conventional MBA techniques could fail to check the association between these two products despite them having a strong association with a $Lift$ score of 43.45. Capturing associations of products that have lower *Support* values is important due to the power law distribution of products' sales [13], where a huge fraction of the products have a low sales volume (as depicted in Fig. 1b).

Second, most existing works perform MBA at a coarser level, which groups multiple products, due to two reasons: (1) data sparsity at the finer levels, which requires a lower *minimum Support* value to capture association rules; and (2) large numbers of different products at the finer levels. Both aforementioned

[1] https://bit.ly/how_many_products_does_walmart_grocery_sell_july_2018.

reasons substantially increase the computation time of conventional association rule mining techniques. As a solution to this, several previous works [11,19] attempt to initially find strong associations at a coarser level, such as groups of products, and further analyse only the individual products in those groups to identify associations at the finer levels. Such approaches do not cover the whole search space at finer levels, and thus fail to detect the association rules that are only observable at finer levels. In addition, almost all the conventional approaches consider each product as an independent element. This is where representation learning based techniques, which learn a low-dimensional vector for each unit[2] such that they preserve the semantics of the units, have an advantage. Thus, representation learning techniques are useful to alleviate the cold-start problem (i.e., detecting associations that are unseen in the dataset) and data sparsity. While there are previous works on applying representation learning for shopping basket recommendation, *none of the previous representation learning techniques has been extended to mine association rules between products.* Moreover, user details of the shopping baskets could be useful to understand the patterns of the shopping baskets. Users generally exhibit repetitive buying patterns, which could be exploited by jointly learning representations for users and products in the same embedding space. While such user behaviors have previously been studied in the context of personalized product recommendation [15,21], they have not been exploited in the context of association rule mining of products.

Third, conventional MBA techniques consider the set of transactions as a single batch to generate association rules between products, despite that the data may come in a continuous stream. The empirical studies in [13,18] show that the association rules of products deviate over time due to various factors (e.g., seasonal variations and socio-economic factors). To further illustrate this point, Fig. 1c shows the changes in sales of two product categories over time at a retailer. As can be seen, there are significant variations of the products' sales. Consequently, the association rules of such products vary over time. Conventional MBA techniques fail to capture these temporal variations. As a solution to this problem in [13,18], the transactions are divided into different time-bins and conventional MBA techniques are used to generate association rules for different time bins from scratch. Such approaches are computationally and memory intensive; and ignore the dependencies between consecutive time bins.

Contribution. In this paper, we propose a novel representation learning based technique to perform Online Market Basket Analysis (OMBA), which: (1) *jointly learns representations for products and users such that the representations preserve their co-occurrences, while emphasizing the products with higher selling prices (typically low in Support) and exploiting the semantics of the products and users*; (2) *proposes an efficient approach to perform MBA using the learned product representations, which is capable of capturing rarely occurring and unseen associations among the products*; and (3) *accommodates online updating for product representations, which adapts the model to the temporal changes, without overfitting to the recent information or storing any historical records.* The code

[2] "Units" refers to the attribute values (could be products or users) of the baskets.

for all the algorithms presented in this paper and the data used in the evaluation are publicly available via https://bit.ly/2UwHfr0.

2 Related Work

Conventional MBA Techniques. Typically, MBA is performed using conventional frequent itemset mining algorithms [2,7,9,12], which initially produce the most frequent (i.e., high *Support*) itemsets in the dataset, out of which the itemsets with high *Lift* scores are subsequently selected as association rules. These techniques mainly differ from each other based on the type of search that they use to explore the space of itemsets (e.g., a depth-first search strategy is used in [12], and the work in [2] applies breadth-first search). As elaborated in Sect. 1, these techniques have the limitations: (1) inability to capture important associations among rarely bought products, which covers a huge fraction of products (see Fig. 1b); (2) inability to alleviate sparsity at the finer product levels; and (3) inability to capture temporal dynamics of association rules. In contrast, OMBA produces association rules from the temporally changing representations of the products to address these limitations.

Representation Learning Techniques. Due to the importance of capturing the semantics of products, there are previous works that adopt representation learning techniques for products [5,6,10,15,21,22]. Some of these works address the *next basket recommendation task*, which attempts to predict the next shopping basket of a customer given his/her previous baskets. Most recent works [5,22] in this line adopt recurrent neural networks to model long-term sequential patterns in users' shopping baskets. However, our task differs from these works by concentrating on the MBA task, which models the shopping baskets without considering the order of the items in each. The work proposed in [15], for *product recommendation tasks*, need to store large knowledge graphs (i.e, co-occurrences matrices), thus they are not ideal to perform MBA using transaction streams. In [6,10,21], word2vec language model [17] has been adopted to learn product representations in an online fashion. Out of them, Triple2Vec [21] is the most similar and recent work for our work, which jointly learns representation for products and users by adopting word2vec. However, Triple2Vec learns two embedding spaces for products such that each embedding space preserves semantic similarity (i.e., second-order proximity) as illustrated in Fig. 2. Thus, the products' associations cannot be easily discovered from such an embedding space. In contrast, the embedding

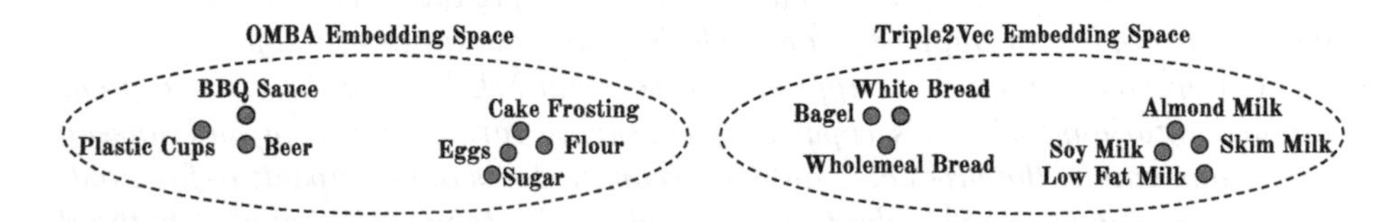

Fig. 2. An illustration of the embedding spaces of OMBA, which preserves complementarity, and Triple2Vec, which only preserves the semantic similarity

space of OMBA mainly preserves the complementarity (i.e., first-order proximity). Thus, the associated products are closely mapped in the embedding space as illustrated in Fig. 2. Moreover, none of the aforementioned works proposes a method to generate association rules from products' embeddings, which is the ultimate objective of MBA.

Online Learning Techniques.OMBA learns online representations for products, giving importance to recent information to capture the temporal dynamics of products' associations. However, it is challenging to incrementally update representations using a continuous stream without overfitting to the recent records [23]. When a new set of records arrives, sampling-based online learning approaches in [20,23] sample a few historical records to augment the recent records and the representations are updated using the augmented corpus to alleviate overfitting to recent records. However, sampling-based approaches need to retain historical records. To address this limitation, a constraint-based approach is proposed in [23], which imposes constraints on embeddings to preserve their previous embeddings. However, sampling-based approaches are superior to constraint-based approaches. In this work, we propose a novel adaptive optimization technique to accommodate online learning that mostly outperforms sampling-based approaches without storing any historical records.

3 Problem Statement

Let $B = \{b_1, b_2, ..., b_N, ...\}$ be a continuous stream of shopping transactions (i.e., baskets) at a retailer that arrive in chronological order. Each basket $b \in B$ is a tuple $< t^b, u^b, P^b >$, where: (1) t^b is the timestamp of b; (2) u^b is the user id of b; and (3) P^b is the set of products in b.

The problem is to learn the embeddings V_P for $P = \bigcup^{\forall b} P^b$ (i.e., the set of unique products in B) such that the embedding v_p of a product $p \in P$:

1. is a d-dimensional vector ($d << |P|$), where $|P|$ is the number of different products in B;
2. preserves the associations (i.e., co-occurrences) among products;
3. is continuously updated as new transactions (B_Δ) arrive to incorporate the latest information.

Market Basket Analysis (MBA) uncovers the associations among products in the form of *association rule*: $P_i \Rightarrow p_j$, "Consumers who buy the products in P_i are likely to buy the product p_j", where $P_i \subset P$ and $p_j \in P \backslash P_i$. The embedding learning module in OMBA is designed such that p_j and the products in P_i are mapped close together in the embedding space.

To quantitatively evaluate the embeddings that are learned for MBA, we focus on the ***intra-basket item retrieval task***, which is to retrieve the true product in a basket given the other attributes (i.e., user and other products) of the same basket. The embeddings should reflect the accurate associations between products to give better performance for this task.

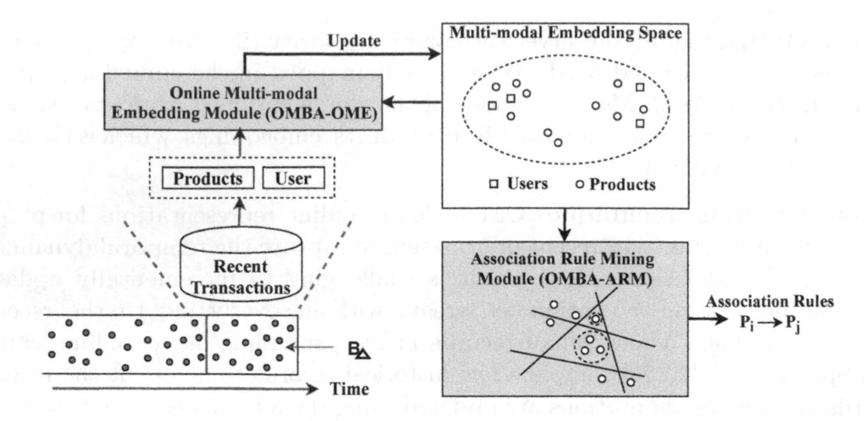

Fig. 3. OMBA consists of: (1) OMBA-OME to learn online product embeddings; and (2) OMBA-ARM to generate association rules using the embeddings

4 OMBA

4.1 Overview of OMBA

OMBA consists of two modules as depicted in Fig. 3: (1) Online Multi-Modal Embedding (OMBA-OME) module, which learns the representations for products; and (2) Association Rule Mining (OMBA-ARM) module, which generates the association rules using the representations from OMBA-OME.

To learn representations for products, OMBA-OME jointly embeds all the products and users into the same latent space. We conduct an empirical analysis using three datasets, which verifies that there is a significant similarity of the shopping baskets of a user (due to space limitations, more details of the empirical analysis are given in [4]). Thus, mapping users along with the products into a single embedding space allows us to exploit the user-specific patterns in shopping baskets to generalize the product representations.

The learning process of OMBA-OME proceeds in an online fashion, in which the embeddings are updated incrementally for the arrival of each of the new records B_{Δ}. Such an approach avoids the unnecessary cost of learning representations from scratch for each new arrival. However, this approach could lead to overfitting to recent information while abruptly forgetting (i.e., catastrophic forgetting) the previously learned information. To address this issue, we propose an online learning method in Sect. 4.2, which incorporates recent records effectively while alleviating the catastrophic forgetting without storing any historical records. Subsequently, our OMBA-ARM module adopts a clustering approach on the products' embedding space to extract association rules among the products.

4.2 Online Multi-modal Embedding (OMBA-OME)

OMBA-OME learns the embeddings for units such that the units of a given basket b can be recovered by looking at b's other units. Formally, we model the likelihood for the task of recovering unit $z \in b$ given the other units b_{-z} of b as:

$$p(z|b_{-z}) = \exp(s(z, b_{-z}))/\sum_{z' \in X} \exp(s(z', b_{-z})) \tag{1}$$

X is the set of units (could be user or product) of type z, and $s(z, b_{-z})$ is the similarity score between z and b_{-z}. We define $s(z, b_{-z})$ as $v_z^\top h_z$ where,

$$h_z = \begin{cases} (v_{u^b} + v_{\hat{P^b}})/2 & \text{if } z \text{ is a product} \\ v_{\hat{P^b}} & \text{if } z \text{ is a user} \end{cases} \tag{2}$$

Value-Based Weighting. The representation for the context products ($v_{\hat{P^b}}$) in a basket is computed based on a novel weighting scheme. This weighting scheme emphasizes learning better representations for higher value items from their rare occurrences by assigning higher weight for them considering their selling price. Formally, the term $v_{\hat{P^b}}$ in Equation 2 is computed as:

$$v_{\hat{P^b}} = \frac{\sum_{x \in P_{b_{-z}}} g(x) v_x}{\sum_{x \in P_{b_{-z}}} g(x)} \tag{3}$$

where $g(x)$ function returns a weight for product x based on its selling price, $SV(x)$. The function g is computed as follows:

- Assuming the number of appearances of x follows a power-law distribution with respect to its selling price $SV(x)$ (see Fig. 1a), a power-law formula (i.e., $y = cx^{-k}$) is fitted to the curve in Fig. 1a, which returns the probability of product x appearing in a basket b, $p(x \in b)$, as $1.3 * SV(x)^{-2.3}$ (the derivation of this formula is presented in [4]);
- Then, the function g is computed as:

$$g(x) = \frac{1}{p(x \in b)} = \frac{1}{1.3 * min(SV(x), 10)^{-2.3}} \tag{4}$$

The function g is clipped for the products with selling price > 10 (10 is selected as the point at which the curve in Fig. 1a reaches a plateau) to avoid the issue of exploding gradient.

Adaptive Optimization. Then, the final loss function is the negative log likelihood of recovering all the attributes of the recent transactions B_Δ:

$$O_{B_\Delta} = -\sum_{b \in B_\Delta} \sum_{z \in b} p(z|b_{-z}) \tag{5}$$

The objective function above is approximated using negative sampling (proposed in [17]) for efficient optimization. Then for a selected record b and unit $z \in b$, the loss function based on the reconstruction error is:

$$L = -\log(\sigma(s(z, b_{-z}))) - \sum_{n \in N_z} \log(\sigma(-s(n, b_{-z}))) \tag{6}$$

where N_z is the set of randomly selected negative units that have the type of z and $\sigma(.)$ is sigmoid function.

We adopt a novel optimization strategy to optimize the loss function, which is designed to alleviate overfitting to the recent records and the frequently appearing products in the transactions.

For each basket b, we compute the intra-agreement Ψ_b of b's attributes as:

$$\Psi_b = \frac{\sum_{z_i, z_j \in b, z_i \neq z_j} \sigma(v_{z_i}^\top v_{z_j})}{\sum_{z_i, z_j \in b, z_i \neq z_j} 1} \tag{7}$$

Then the adaptive learning rate of b is calculated as,

$$lr_b = \exp(-\tau \Psi_b) * \eta \tag{8}$$

where η denotes the standard learning rate and τ controls the importance given to Ψ_b. If the representations have already overfitted to b, then Ψ_b takes a higher value. Consequently, a low learning rate is assigned to b to avoid overfitting. In addition, the learning rate for each unit z in b is further weighted using the approach proposed in AdaGrad [8] to alleviate the overfitting to frequent items. Then, the final update of the unit z at the tth timestep is:

$$z_{t+1} = z_t - \frac{lr_b}{\sqrt{\sum_{i=0}^{t-1} \left(\frac{\partial L}{\partial z}\right)_i^2 + \epsilon}} \left(\frac{\partial L}{\partial z}\right)_t \tag{9}$$

4.3 Association Rule Mining (OMBA-ARM)

The proposed OMBA-OME module learns representations for products such that the frequently co-occurring products (i.e., products with strong associations) are closely mapped in the embedding space. Accordingly, the proposed OMBA-ARM module clusters in the products' embedding space to detect association rules between products. The product embeddings from the OMBA-OME module are in a higher-dimensional embedding space. Thus, Euclidean-distance based clustering algorithms (e.g., K-Means) suffers from the curse-of-dimensionality issue. Hence, we adopt a Locality-Sensitive Hashing (LSH) algorithm based on random projection [3], which assigns similar hash values to the products that are close in the embedding space. Subsequently, the products with the most similar hash values are returned as strong associations. Our algorithm is formally elaborated as the following five steps:

1. Create $|F|$ different hash functions such as $F_i(v_p) = sgn(f_i \cdot v_p^\top)$, where $i \in \{0, 1, \dots, |F| - 1\}$ and f_i is a d-dimensional vector from a Gaussian distribution $N \sim (0, 1)$, and $sgn(.)$ is the sign function. According to the Johnson-Lindenstrauss lemma [14], such hash functions approximately preserve the distances between products in the original embedding space.
2. Construct an $|F|$-dimensional hash value for v_p (i.e., a product embedding) as $F_0(v_p) \oplus F_1(v_p) \oplus \dots \oplus F_{|F|-1}(v_p)$, where $\oplus$ defines the concatenation operation. Compute the hash value for all the products.

3. Group the products with similar hash values to construct a hash table.
4. Repeat steps (1), (2), and (3) $|H|$ times to construct $|H|$ hash tables to mitigate the contribution of bad random vectors.
5. Return the sets of products with the highest collisions in hash tables as the products with strong associations.

$|H|$ and $|F|$ denote the number of hash tables and number of hash functions in each table respectively.

Finding Optimal Values for $|H|$ and $|F|$. In this section, the optimal values for $|F|$ and $|H|$ are found such that they guarantee that the rules generated from OMBA-ARM have higher *Lift*. Initially, we form two closed-form solutions to model the likelihood to have a strong association between product x and product y based on: (1) OMBA-ARM; and (2) LIFT.

(1) Based on OMBA-ARM, products x and y should collide at least in a single hash table to have a strong association. Thus, the likelihood to have a strong association between product x and product y is[3]:

$$\begin{aligned} p(x \Rightarrow y)_{omba} &= 1 - (1 - p(sgn(v_x \cdot f) = sgn(v_y \cdot f))^{|F|})^{|H|} \\ &= 1 - (1 - (1 - \frac{arccos(v_x \cdot v_y)}{\pi})^{|F|})^{|H|} \end{aligned} \tag{10}$$

where $sgn(x) = \{1 \ if \ x \geq 1; -1 \ otherwise\}$. $p(sgn(v_x \cdot f) = sgn(v_y \cdot f))$ is the likelihood to have a similar sign for products x and y with respect to a random vector. v_x and v_y are the normalized embeddings of product x and y respectively.

(2) Based on Lift, the likelihood to have a strong association between products x and y can be computed as (see footnote 3):

$$p(x \Rightarrow y)_{lift} = \frac{Lift(y,x)_{train}}{Lift(y,x)_{train} + |N_z| * Lift(y,x)_{noise}} = \sigma(A * v_x \cdot v_y) \tag{11}$$

where $Lift(y,x)_{train}$ and $Lift(y,x)_{noise}$ are the *Lift* scores of x and y calculated using the empirical distribution of the dataset and the noise distribution (to sample negative samples) respectively. $|N_z|$ is the number of negative samples for each genuine sample in Eq. 6, and A is a scalar constant.

Then, the integer solutions for parameters $|F| = 4$ and $|H| = 11$, and real solutions for $A = 4.3$ are found such that $p(x \Rightarrow y)_{omba} = p(x \Rightarrow y)_{lift}$. Such a selection of hyper-parameters theoretically guarantees that the rules produced from OMBA-ARM are higher in *Lift*, which is a statistically well-defined measure for strong associations.

The advantages of our OMBA-ARM module can be listed as follows: (1) it is simple and efficient, which simplifies the expensive comparison of all the products in the original embedding space ($O(dN^2)$ complexity) to $O(d|F||H|N)$,

[3] Detailed derivations of Eq. 10 and Eq. 11 are presented in [4].

Table 1. Descriptive statistics of the datasets

Datasets	# Users	# Items	# Transactions	# Baskets
Complete journey (CJ)	2,500	92,339	2,595,733	276,483
Ta-Feng (TF)	9,238	7,973	464,118	77,202
Instacart (IC)	206,209	49688	33,819,306	3,421,083

where $N \gg |F||H|$; (2) it approximately preserves the distances in the original embedding space, as the solutions for $|F|(=4)$ and $|H|(=11)$ satisfy the Johnson–Lindenstrauss lemma [14] ($|F||H| \gg log(N)$); (3) it outperforms underlying Euclidean norm-based clustering algorithms (e.g., K-means) for clustering data in higher dimensional space, primarily due to the curse of dimensionality issue [1]; and (4) theoretically guarantees that the rules have higher *Lift*.

5 Experimental Methodology

Datasets. We conduct our experiments using three publicly available datasets:

- Complete Journey Dataset (CJ) contains household-level transactions at a retailer by 2,500 frequent shoppers over two years.
- Ta-Feng Dataset (TF) includes shopping transactions of the Ta-Feng supermarket from November 2000 to February 2001.
- InstaCart Dataset (IC) contains the shopping transactions of Instacart, an online grocery shopping center, in 2017.

The descriptive statistics of the datasets are shown in Table 1. As can be seen, the datasets have significantly different statistics (e.g., TF has a shorter collection period, and IC has a larger user base), which helps to evaluate the performance in different environment settings. The time-space is divided into 1-day time windows, meaning that new records are received by the model once per day.

Baselines. We compare OMBA with the following methods:

- Pop recommends the most popular products in the training dataset.
- Sup recommends the product with the highest *Support* for a given context.
- Lift recommends the product that has the highest *Lift* for a given context.
- NMF [16] performs Non-Negative Matrix Factorization on the user-product co-occurances matrix to learn representations for users and products.
- Item2Vec [6] adopts word2vec for learning the product representations by considering a basket as a sentence and a product in the basket as a word.
- Prod2Vec [10] aggregates all the baskets related to a single user as a product sequence and apply word2vec to learn representations for products.
- Triple2Vec [21] joinlty learns representations for products and users such that they preserve the triplets in the form of $<user, item_1, item_2>$, which are generated using shopping baskets.

Also, we compare with a few variants of OMBA.

- OMBA-No-User does not consider users. The comparison with OMBA-No-User highlights the importance of users.
- OMBA-Cons adopts SGD optimization with the constraint-based online learning approach proposed in [23].
- OMBA-Decay and OMBA-Info adopt SGD optimization with the sampling-based online learning methods proposed in [23] and [20] respectively.

Parameter Settings. All the representation learning based techniques share three common parameters (default values are given in brackets): (1) the latent embedding dimension d (300), (2) the SGD learning rate η (0.05), (3) the negative samples $|N_z|$ (3) as appeared in Eq. 6, and (4) the number of epochs N (50). We set $\tau = 0.1$ after performing grid search using the CJ dataset (see [4] for a detailed study of parameter sensitivity). For the specific parameters of the baselines, we use the default parameters mentioned in their original papers.

Evaluation Metric. Following the previous work [23], we adopt the following procedure to evaluate the performance for the *intra-basket item retrieval task.* For each transaction in the test set, we select one product as the target prediction and the rest of the products and the user of the transaction as the context. We mix the ground truth target product with a set of M negative samples (i.e., products) to generate a candidate pool to rank. M is set to 10 for all the experiments. Then the size-$(M + 1)$ candidate pool is sorted to get the rank of the ground truth. The average similarity of each candidate product to the context of the corresponding test instance is used to produce the ranking of the candidate pool. Cosine similarity is used as the similarity measure of OMBA, Triple2Vec, Prod2Vec, Item2Vec and NMF. POP, Sup, and Lift use popularity, support, and lift scores as similarity measures respectively.

If the model is well trained, then higher ranked units are most likely to be the ground truth. Hence, we use three different evaluation metrics to analyze the ranking performance: (1) Mean Reciprocal Rank (MRR) $= \frac{\sum_{q=1}^{Q} 1/rank_i}{|Q|}$; (2) Recall@k (R@k) $= \frac{\sum_{q=1}^{Q} min(1, \lfloor k/rank_i \rfloor)}{|Q|}$; and (3) Discounted Cumulative Gain (DCG) $= \frac{\sum_{q=1}^{Q} 1/log_2(rank_i+1)}{|Q|}$, where Q is the set of test queries and $rank_i$ refers the rank of the ground truth label for the i-th query. $\lfloor . \rfloor$ is the floor operation. A good ranking performance should yield higher values for all three metrics. We randomly select 20 one-day query windows from the second half of the period for each dataset, and all the transactions in the randomly selected time windows are used as test instances. For each query window, we only use the transactions that arrive before the query window to train different models. Only OMBA and its variants are trained in an online fashion and all the other baselines are trained in a batch fashion for 20 repetitions.

Table 2. The comparison of different methods for *intra-basket item retrieval.* Each model is evaluated 5 times with different random seeds and the mean value for each model is presented. Recall values for different k values are presented in [4] due to space limitations

	Results for *intra-basket item retrieval*									Memory complexity
Dataset	CJ			IC			TF			
Metric	MRR	R@1	DCG	MRR	R@1	DCG	MRR	R@1	DCG	
Pop	0.2651	0.095	0.4295	0.2637	0.07841	0.4272	0.2603	0.0806	0.4247	$O(P + \|B_{max}\|)$
Sup	0.3308	0.1441	0.4839	0.3009	0.1061	0.4634	0.3475	0.1646	0.4972	$O(P^2 + \|B_{max}\|)$
Lift	0.5441	0.3776	0.6477	0.4817	0.2655	0.6170	0.4610	0.2981	0.5868	$O(P(P+1) + \|B_{max}\|)$
NMF	0.1670	0.0000	0.3565	0.5921	0.3962	0.6922	0.4261	0.2448	0.5591	$O(P(U+P) + \|B_{max}\|)$
Item2Vec	0.4087	0.2146	0.5457	0.5159	0.2929	0.6137	0.3697	0.1782	0.5149	$O(k(2*P) + \|B\|)$
Prod2Vec	0.4234	0.2275	0.5575	0.5223	0.3222	0.6363	0.3764	0.1854	0.5201	$O(k(2*P) + \|B\|)$
Triple2Vec	0.5133	0.3392	0.6269	0.6169	0.4348	0.7095	0.478	0.3040	0.5990	$O(k(U + 2*P) + \|B\|)$
OMBA-No-User	0.4889	0.3091	0.6004	0.5310	0.3384	0.6405	0.3873	0.2013	0.5298	$O(kP + \|B_{max}\|)$
OMBA-Cons	0.4610	0.2742	0.5843	0.5942	0.3393	0.6998	0.3996	0.2031	0.5324	$O(k(U+P) + \|B_{max}\|)$
OMBA-Decay	0.5984	0.4221	0.6948	0.7117	0.5442	0.7860	0.402	0.2186	0.5387	$O(k(U+P) + \|B_{max}\|/(1-e^{-\tau}))$
OMBA-Info	0.5991	0.4275	0.6937	**0.7482**	0.5852	**0.8027**	0.4046	0.2205	0.5421	$O(k(U+P) + \|B_{max}\|/(1-e^{-\tau}))$
OMBA	**0.6013**	**0.4325**	**0.6961**	0.7478	**0.5859**	0.8025	**0.5166**	**0.3466**	**0.6293**	$O(k(U+P) + \|B_{max}\|)$

Table 3. (a) 5 nearest neighbours in the embedding space of OMBA and Triple2Vec for a set of target products; (b) MRR for *intra-basket item retrieval* using the test queries, such that the ground truth's price > x (in dollars). All the results are calculated using CJ

(a)

Target Product	5 nearest products by OMBA	5 most nearest products by Triple2Vec
Layer Cakes	Cake Candles, Cake Sheets, Cake Decorations, Cake Novelties, Birthday/Celebration Layer	Cheesecake, Fruit/Nut Pies, Cake Candles, Flags, Salad Ingredients
Frozen Bread	Sauces, Eggs, Peanut Butter, Non Carbohydrate Juice, Pasta/Ramen	Frozen Breakfast, Breakfast Bars, Popcorn, Gluten Free Bread, Cookies/Sweet Goods
Turkey	Beef, Meat Sauce, Oil/Vinegar, Cheese, Ham	Ham, Beef, Cheese, Chicken, Lunch Meat
Authentic Indian Foods	Grain Mixes, Organic Pepper, Other Asian Foods, Tofu, South American Wines	Other Asian Foods, German Foods, Premium Mums, Herbs & Fresh Others, Bulb Sets

(b)

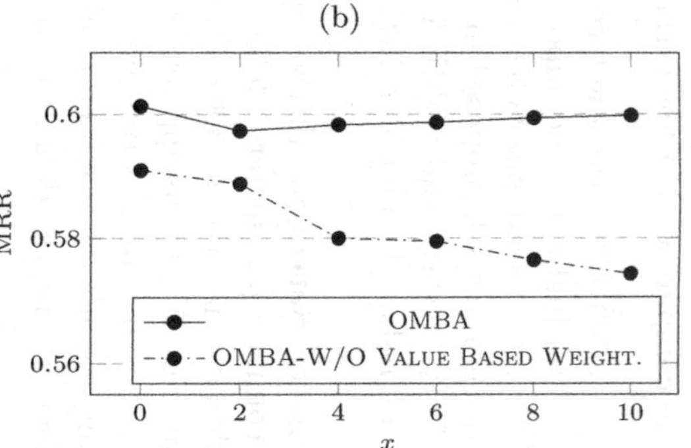

6 Results

Intra-basket Item Retrieval Task. Table 2 shows the results collected for the *intra-basket item retrieval* task. OMBA and its variants show significantly better results than the baselines, outperforming the best baselines on each dataset by as much as 10.51% for CJ, 21.28% for IC, and 8.07% for TF in MRR.

(1) Capture Semantics of Products. LIFT is shown to be a strong baseline. However, it performs poorly for IC, which is much sparser with a large userbase compared to the other datasets. This observation clearly shows the importance of using representation learning based techniques to overcome data sparsity by capturing the semantics of products. We further analyzed the wrongly predicted test instances of LIFT, and observed that most of these instances are products with significant seasonal variations such as "Rainier Cherries" and "Valentine Gifts and Decorations" (as shown in Fig. 1c). This shows the importance of capturing temporal changes of the products' associations.

Out of the representation learning-based techniques, TRIPLE2VEC is the strongest baseline, despite being substantially outperformed by OMBA. As elaborated in Sect. 2, TRIPLE2VEC mainly preserves semantic similarity between products. This could be the main reason for the performance difference between TRIPLE2VEC and OMBA. To validate that, Table 3 lists the nearest neighbours for a set of products in the embedding spaces from OMBA and TRIPLE2VEC. Most of the nearest neighbours in TRIPLE2VEC are semantically similar to the target products. For example, 'Turkey' has substitute products like 'Beef' and 'Chicken' as its nearest neighbours. In contrast, the embedding space of OMBA mainly preserve complementarity. Thus, multiple related products to fulfill a specific need are closely mapped. For example, 'Layer Cake' has neighbours related to a celebration (e.g., Cake Decorations and Cake Candles).

(2) Value-based Weighting. To validate the importance of the proposed value-based weighting scheme in OMBA, Table 3 shows the deviation of the performance for *intra-basket item retrieval* with respect to the selling price of the ground truth products. With the proposed value-based weighting scheme, OMBA accurately retrieves the ground truth products that have higher selling prices. Thus, we can conclude that the proposed value-based weighting scheme is important to learn accurate representations for rarely occurring products.

(3) Online Learning. Comparing the variants of OMBA, OMBA substantially outperforms OMBA-NO-USER, showing that users are important to model products' associations. OMBA's results are comparable (except 27.68% performance boost for TF) with sampling-based online learning variants of OMBA (i.e., OMBA-DECAY and OMBA-INFO), which store historical records to avoid overfitting to recent records. Hence, the proposed adaptive optimization-based online learning technique in OMBA achieves the performance of the state-of-the-art online learning methods (i.e., OMBA-DECAY and OMBA-INFO) in a memory-efficient manner without storing any historical records.

Association Rule Mining. To compare the association rules generated from OMBA, we generate the association rules for the CJ dataset using: (1) Apriori

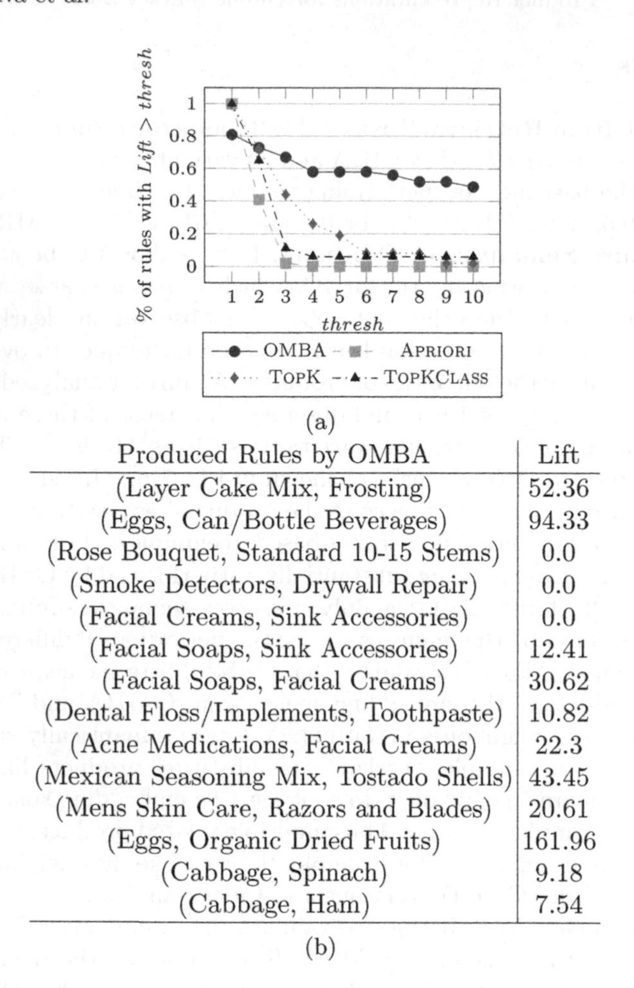

Produced Rules by OMBA	Lift
(Layer Cake Mix, Frosting)	52.36
(Eggs, Can/Bottle Beverages)	94.33
(Rose Bouquet, Standard 10-15 Stems)	0.0
(Smoke Detectors, Drywall Repair)	0.0
(Facial Creams, Sink Accessories)	0.0
(Facial Soaps, Sink Accessories)	12.41
(Facial Soaps, Facial Creams)	30.62
(Dental Floss/Implements, Toothpaste)	10.82
(Acne Medications, Facial Creams)	22.3
(Mexican Seasoning Mix, Tostado Shells)	43.45
(Mens Skin Care, Razors and Blades)	20.61
(Eggs, Organic Dried Fruits)	161.96
(Cabbage, Spinach)	9.18
(Cabbage, Ham)	7.54

(b)

Fig. 4. (a) The percentage of top 100 rules that have *Lift* scores > *thresh*; and (b) A few examples for the top 100 rules produced by OMBA

Algorithm [2] with *minimum Support = 0.0001*; (2) TopK Algorithm [9] with *k = 100*; and (3) TopKClass Algorithm[4] with *k = 100* and *consequent list*= {*products appear in the rules from* OMBA}. Figure 4a shows *Lift* scores of the top 100 rules generated from each approach.

(1) Emphasize Rarely Occurring Associations. As can be seen, at most 20% of the rules generated from the baselines have *Lift* scores greater than 5, while around 60% of the rules of OMBA have scores greater than 5. This shows that OMBA generates association rules that are strong and rarely occurring in the shopping baskets (typically have high *Lift* scores due to low *Support* values), which are not emphasized in conventional MBA techniques.

[4] https://bit.ly/spmf_TopKClassAssociationRules.

Table 4. Association Rules generated by OMBA-ARM on different days

	Mostly associated products by OMBA-ARM	
Target product	At month 11	At month 17
Valentine gifts and decorations	Rose consumer bunch, Apple juice, Ladies casual shoes	Peaches, Dried plums, Drain care
Mangoes	Blueberries, Dry snacks, Raspberries	Dry snacks, Red cherries, Tangelos
Frozen bread	Ham, Fluid milk, Eggs	Peanut butter, Eggs, Ham

(2) Discover Unseen Associations. However, Fig. 4a shows that some of the rules from OMBA have *Lift* scores less than 1 (around 20% of the rules). Figure 4b lists some of the rules generated from OMBA, including a few with $Lift < 1$. The qualitative analysis of these results shows that OMBA can even discover unseen associations in the dataset (i.e., associations with *Lift*= 0) as shown in Fig. 4b, by capturing the semantics of product names. The associations with *Lift* = 0 like "(Rose Bouquet, Standard 10–15 Stems)" and "(Smoke Detectors, Drywall Repair)" can be interpreted intuitively, but they would be omitted by conventional count-based approaches. The ability to capture the semantics of products enables OMBA to mitigate the *cold-start problem* and recommend products to customers that they have never purchased. The role of the OMBA-OME module on capturing products' semantics can be further illustrated using the rule "(Facial Creams, Sink Accessories)", which is unseen in the CJ dataset. However, both "Sink Accessories" and "Facial Creams" frequently co-occurred with "Facial Soaps", with 12.41 and 30.02 *Lift* scores. Hence, the representations for "Sink Accessories" and "Facial Creams" learnt by OMBA-OME are similar to the representation of "Facial Soaps" to preserve their co-occurrences. This leads to similar representations for "Sink Accessories" and "Facial Creams" and generates the rule. In contrast, conventional count-based approaches (e.g., Apriori, H-Mine, and TopK) are unable to capture these semantics of product names as they consider each product as an independent unit.

(3) Temporal Variations of Associations. Moreover, Table 4 shows that OMBA adopts associations rules to the temporal variations. For example, most of the associated products for 'Mangoes' at month 11 are summer fruits (e.g., 'Blueberries' and 'Raspberries'); 'Mangoes' are strongly associated with winter fruits (e.g., 'Cherries' and 'Tangelos') at Month 17. However, everyday products like 'Frozen Bread' show consistent associations with other products. These results further signify the importance of learning online product representation.

In summary, the advantages of OMBA are three-fold: (1) OMBA explores the whole products' space effectively to detect rarely-occurring non-trivial strong associations; (2) OMBA considers the semantics between products when learning associations, which alleviates the cold-start problem; and (3) OMBA learns products' representations in an online fashion, thus is able to capture the temporal changes of the products' associations accurately.

7 Conclusion

We proposed a scalable and effective method to perform Market Basket Analysis in an online fashion. Our model, OMBA, introduced a novel online representation learning technique to jointly learn embeddings for users and products such that they preserve their associations. Following that, OMBA adapted an effective clustering approach in high dimensional space to generate association rules from the embeddings. Our results show that OMBA manages to uncover non-trivial and temporally changing association rules at the finest product levels, as OMBA can capture the semantics of products in an online fashion. Nevertheless, extending OMBA to perform MBA for multi-level products and multi-store environments could be a promising future direction to alleviate data sparsity at the finest product levels. Moreover, it is worthwhile to explore sophisticated methods to incorporate users' purchasing behavior to improve product representations.

Acknowledgment. This research was financially supported by Melbourne Graduate Research Scholarship and Rowden White Scholarship.

References

1. Aggarwal, C.C., Hinneburg, A., Keim, D.A.: On the surprising behavior of distance metrics in high dimensional space. In: Van den Bussche, J., Vianu, V. (eds.) ICDT 2001. LNCS, vol. 1973, pp. 420–434. Springer, Heidelberg (2001). https://doi.org/10.1007/3-540-44503-X_27
2. Agrawal, R., Srikant, R., et al.: Fast algorithms for mining association rules. In: Proceedings of VLDB (1994)
3. Andoni, A., Indyk, P.: Near-optimal hashing algorithms for approximate nearest neighbor in high dimensions. In: Proceedings of FOCS (2006)
4. Silva, A., Luo, L., Karunasekera, S., Leckie, C.: Supplementary Materials for OMBA: On Learning User-Guided Item Representations for Online Market Basket Analysis (2020). https://bit.ly/30yGMbx
5. Bai, T., Nie, J.Y., Zhao, W.X., Zhu, Y., Du, P., Wen, J.R.: An attribute-aware neural attentive model for next basket recommendation. In: Proceedings of SIGIR (2018)
6. Barkan, O., Koenigstein, N.: Item2vec: neural item embedding for collaborative filtering. In: Proceedings of MLSP (2016)
7. Deng, Z.H., Lv, S.L.: Fast mining frequent itemsets using nodesets. Expert Syst. Appl. **41**, 4505 (2014)
8. Duchi, J., Hazan, E., Singer, Y.: Adaptive subgradient methods for online learning and stochastic optimization. In: Proceedings of COLT (2010)
9. Fournier-Viger, P., Wu, C.-W., Tseng, V.S.: Mining top-K association rules. In: Kosseim, L., Inkpen, D. (eds.) AI 2012. LNCS (LNAI), vol. 7310, pp. 61–73. Springer, Heidelberg (2012). https://doi.org/10.1007/978-3-642-30353-1_6

10. Grbovic, M., et al.: E-commerce in your inbox: product recommendations at scale. In: Proceedings of SIGKDD (2015)
11. Han, J., Fu, Y.: Discovery of Multiple-Level Association Large Databases. In: Proceedings of VLDB (1995)
12. Han, J., Pei, J., Yin, Y., Mao, R.: Mining frequent patterns without candidate generation: a frequent-pattern tree approach. Data Min. Knowl. Disc. **8**, 53 (2004). https://doi.org/10.1023/B:DAMI.0000005258.31418.83
13. Hariharan, S., Kannan, M., Raguraman, P.: A seasonal approach for analysis of temporal trends in retail marketing using association rule mining. Int. J. Comput. Appl. **71**, 10–15 (2013)
14. Johnson, W.B., Lindenstrauss, J.: Extensions of Lipschitz mappings into a Hilbert space. Contemp. Math. **26**, 189 (1984)
15. Le, D.T., Lauw, H.W., Fang, Y.: Basket-sensitive personalized item recommendation. In: Proceedings of IJCAI (2017)
16. Lee, D.D., Seung, H.S.: Algorithms for non-negative matrix factorization. In: Proceedings of NIPS (2001)
17. Mikolov, T., Sutskever, I., Chen, K., Corrado, G.S., Dean, J.: Distributed representations of words and phrases and their compositionality. In: Proceedings of NIPS (2013)
18. Papavasileiou, V., Tsadiras, A.: Time variations of association rules in market basket analysis. In: Iliadis, L., Maglogiannis, I., Papadopoulos, H. (eds.) AIAI/EANN -2011. IAICT, vol. 364, pp. 36–44. Springer, Heidelberg (2011). https://doi.org/10.1007/978-3-642-23960-1_5
19. Agarwal, R., Mittal, M.: Inventory classification using multilevel association rule mining. Int. J. Decis. Support Syst. Technol. **11**(2), 1–12 (2019)
20. Silva, A., Karunasekera, S., Leckie, C., Luo, L.: USTAR: Online multimodal embedding for modeling user-guided spatiotemporal activity. In: Proceedings of IEEE-Big Data (2019)
21. Wan, M., Wang, D., Liu, J., Bennett, P., McAuley, J.: Representing and recommending shopping baskets with complementarity, compatibility and loyalty. In: Proceedings of CIKM (2018)
22. Yu, F., Liu, Q., Wu, S., Wang, L., Tan, T.: A dynamic recurrent model for next basket recommendation. In: Proceedings of SIGIR (2016)
23. Zhang, C., et al.: React: online multimodal embedding for recency-aware spatiotemporal activity modeling. In: Proceedings of SIGIR (2017)

Clustering

Online Binary Incomplete Multi-view Clustering

Longqi Yang[1,2], Liangliang Zhang[3(✉)], and Yuhua Tang[1(✉)]

[1] State Key Laboratory of High Performance Computing, College of Computer, National University of Defense Technology, Changsha 410073, China
{yanglongqi19,yhtang}@nudt.edu.cn

[2] Artificial Intelligence Research Center, National Innovation Institute of Defense Technology, Beijing 100089, China

[3] Beijing Institute of System Engineering, Beijing, China
vermouthlove@hotmail.com

Abstract. Multi-view clustering has attracted considerable attention in the past decades, due to its good performance on the data with multiple modalities or from diverse sources. In real-world applications, multi-view data often suffer from incompleteness of instances. Clustering on such multi-view data is called incomplete multi-view clustering (IMC). Most of the existing IMC solutions are offline and have high computational and memory costs especially for large-scale datasets. To tackle these challenges, in this paper, we propose a Online Binary Incomplete Multi-view Clustering (OBIMC) framework. OBIMC robustly learns the common compact binary codes for incomplete multi-view features. Moreover, the cluster structures are optimized with the binary codes in an online fashion. Further, we develop an iterative algorithm to solve the resultant optimization problem with linear computational complexity and theoretically prove its convergence. Experiments on four real datasets demonstrate the efficiency and effectiveness of the proposed OBIMC method. As indicated, our algorithm significantly and consistently outperforms some state-of-the-art algorithms with much less running time.

1 Introduction

Clustering is a fundamental task in machine learning and data mining. Nowadays, in many real-world applications, a tremendous quantity of data is continuously generated from diverse sources or with multiple modalities, which are referred to as multi-view data. Multi-view clustering aims to provide improved performance on multi-view data in comparison with single-view data by considering the complementary information [13].

The recently proposed multi-view clustering (MVC) [1] methods focused on the *complete* multi-view data, which can be roughly divided into three categories: multi-view spectral clustering [6,7,14,18], multi-view subspace clustering [3,23,25] and multi-view matrix factorization [9,21]. The previously mentioned methods share a common assumption that all the views are complete, meaning that all the instances appear in individual views and correspond to each other.

F. Hutter et al. (Eds.): ECML PKDD 2020, LNAI 12457, pp. 75–90, 2021.
https://doi.org/10.1007/978-3-030-67658-2_5

However, this assumption is too strong as one may have only one or two language versions for a multilingual document; some samples may lose visual or audio information due to sensor failures. We call such data incomplete multi-view data. The research along this line is termed as *incomplete multi-view clustering (IMC)* (or *partial multi-view clustering*).

Based on the existing work, the challenges for IMC problems include: (1) How to deal with incomplete views? and (2) How to handle large-scale incomplete multi-view data (maybe online/streaming data)? The existing IMC methods learn a consensus representation for all views and conduct imputation on missing values. The widely used imputation algorithms include zero-filling, mean value filling, k-nearest-neighbor filling and expectation-maximization (EM) filling. Some advanced algorithms have recently been proposed to perform matrix imputation [16,22]. Whereas, most of existing IMC methods are validated on small datasets [8,10,16,19,20,26]. There are few existing studies on large-scale IMC problems. One direction is to cluster in an online fashion [5,15] which reduces the data size in each process. However, their clustering performances are sensitive to the data quality of individual views. Another direction is to learn a common representation that is often much smaller in dimensions than the original data. The recent advancement of binary code learning [4,24] provides a preferable solution to overcome the above difficulties. The key idea of binary code learning is to encode the original features by a set of compact binary codes in Hamming space. Unfortunately, [24] is not applicable to IMC problem.

In the following sections, we start with a brief review of a related work. Then, we propose our Online Binary Incomplete Multi-view Clustering (OBIMC)

Table 1. Summary of the notations

Notation	Description
N	Total number of instances
n_v	Total number of views
d_v	Dimensionality of features in the v-th view
$\boldsymbol{X}^v$	Data Matrix for v-th view
$\phi(\boldsymbol{X}^v)$	Anchor-based representation for the v-th view
$\boldsymbol{W}^v$	Diagonal instance weight matrix for the v-th view
$\boldsymbol{A}^v$	Anchors for the v-th view
$\boldsymbol{a}_i^v$	The i-th anchor for the v-th view
$\hat{\boldsymbol{a}}_i^v$	Completion i-th anchor for the v-th view
$\boldsymbol{B}$	Common binary coding
$\boldsymbol{C}$	Clustering centroids
$\boldsymbol{G}$	Clustering indicators
$\boldsymbol{U}^v$	Mapping matrix for the v-th view
m	The number of anchors
k	The number of clusters
l	Code length

approach and report the experimental results. Finally, we conclude the paper. Table 1 summarizes the notations used throughout the paper.

2 Preliminaries

In this paper, we build upon the work of Binary Multi-view Clustering (BMVC) [24], which is a state-of-the-art offline MVC method.

Given a set of input multi-view matrices $\mathcal{X} = [\boldsymbol{X}^1, \ldots, \boldsymbol{X}^{n_v}]$, where $\boldsymbol{X}^v \in \mathbb{R}^{d_v \times N}$ is the data matrix of the v-th view. d_v and N denote the dimensionality and the number of data points in $\boldsymbol{X}^v$, respectively. BMVC partitions the N data instances into k clusters as follows:

Step 1. Randomly select m instances as anchors (denoting as $\boldsymbol{A}^v = \{\boldsymbol{a}_i^v\}_{i=1}^m$) from multi-view data $\mathcal{X} = [\boldsymbol{X}^1, \ldots, \boldsymbol{X}^{n_v}]$.
Step 2. Construct the m-dimensional anchor-based representation $\phi(\boldsymbol{X}^v)$ from raw features $\boldsymbol{X}^v$, $\mathbb{R}^{d_v} \to \mathbb{R}^m$.
Step 3. Learn a l-dimensional consensus binary code $\boldsymbol{B}$ from each view $\phi(\boldsymbol{X}^v)$ with the mapping matrix $\boldsymbol{U}^v$, $(\boldsymbol{B} \approx \boldsymbol{U}^v\phi(\boldsymbol{X}^v))$, $\mathbb{R}^m \to \{-1, 1\}^l$.
Step 4. Perform binary clustering on the consensus binary code $\boldsymbol{B}$ by minimizing $\|\boldsymbol{B} - \boldsymbol{C}\boldsymbol{G}\|_F^2$.
Step 5. Integrate binary code learning (step 3) and binary clustering (step 4) into a framework.
Step 6. Optimize the unified framework with an iterative algorithm, then obtain the indicator $\boldsymbol{G}$.

Instead of fusion on the participation of each view, BMVC learned a common representation of input data, which is apart from the most existing methods. Moreover, BMVC integrated binary code learning and binary clustering into a unified framework. Thus, the clustering on the data-dependent binary code resulting the state-of-the-art performance. However, applying BMVC in a real-world incomplete setting might be limited by the following aspects:

- How to partition incomplete multi-view data?
- How to learn a consensus binary representation from the incomplete multi-view data?

The first question is the key question in our study. We will propose our solutions to these questions in the next section.

3 Proposed Method

This section presents the proposed OBIMC method together with its optimization algorithm.

3.1 Anchor-Based Representation

Given a set of input incomplete multi-view matrices $\mathcal{X} = [\boldsymbol{X}^1, \ldots, \boldsymbol{X}^{n_v}]$, where $\boldsymbol{X}^v \in \mathbb{R}^{d_v \times N}$ is the data matrix of the v-th view. d_v and N denote the dimensionality and the number of data points in $\boldsymbol{X}^v$, respectively. We assign 0 and NaN("not a number") to the unobserved instance and anchor, respectively.

Anchor Generation. The common strategy of anchor generation includes k-means and random selection. We adopt random selection for its lower computational complexity. Another reason is that it fails to conduct k-means on incomplete multi-view data. Thus, we randomly select m anchors from incomplete data, denoting as $\boldsymbol{A}^v = \{\boldsymbol{a}_i^v\}_{i=1}^m$.

Anchor Completion. Take the v-th view as an example, the anchor-based representation of i-th instance $\boldsymbol{x}_i^v$ is

$$\phi(\boldsymbol{x}_i^v) = [\exp(-\|\boldsymbol{x}_i^v - \boldsymbol{a}_1^v\|^2/\sigma), \cdots, \exp(-\|\boldsymbol{x}_i^v - \boldsymbol{a}_m^v\|^2/\sigma)]^T, \tag{1}$$

where $\phi(\boldsymbol{x}_i^v) \in \mathbb{R}^m$ indicates the m-dimensional nonlinear embedding, σ is the pre-defined kernel width. The main limitation of calculation of the above anchor-based representation $\phi(\boldsymbol{X}^v)$ is the incompleteness of anchors.

Given the m incomplete anchors, we now focus on completing those NaNs. To do so, we define a completion operator on $\boldsymbol{a}_i^v$:

$$\hat{\boldsymbol{a}}_i^v = \mathcal{P}(\boldsymbol{a}_i^v) = \begin{cases} \boldsymbol{a}_i^v, & \text{if } i\text{-th anchor is observed;} \\ \boldsymbol{a}_{\text{ave}}^v, & \text{otherwise.} \end{cases}$$

where $\boldsymbol{a}_{\text{ave}}^v = \sum_{i=1}^N \boldsymbol{x}_i^v / (\#\text{observed instances})$.

Anchor-Based Representation. After filling the unobserved anchor with the average value of the observed instances, we obtain the new representation of the observed instances.

$$\phi(\boldsymbol{x}_i^v) = [\exp(-\|\boldsymbol{x}_i^v - \hat{\boldsymbol{a}}_1^v\|^2/\sigma), \cdots, \exp(-\|\boldsymbol{x}_i^v - \hat{\boldsymbol{a}}_m^v\|^2/\sigma)]^T. \tag{2}$$

Note that the zeros-filling of unobserved instance x_i^v may introduce bias to 0, we will fix this in the binary code learning phase.

3.2 Binary Code Learning

The goal of binary code learning is to find a common compact code to represent the incomplete multi-view data. In detail, we project the features from different views into a Hamming space, i.e. $\mathbb{R}^m \rightarrow \{-1, 1\}^l$.

Binary Code Learning for Individual View. To learn a l-bit binary code, we utilize a mapping matrix $\boldsymbol{U}^v \in \mathbb{R}^{m \times l}$ to project $\phi(\boldsymbol{X}^v)$ onto the l-dimensional space, together with the binary hash function for individual view as $sgn(\boldsymbol{U}^v \phi(\boldsymbol{X}^v))$, where $sgn(\cdot)$ is an element-wise sign operator. The binary code for v-th view is $\boldsymbol{B}^v = sgn(\boldsymbol{U}^v \phi(\boldsymbol{X}^v))$.

Common Binary Code Learning. We introduce a set of diagonal matrices $\boldsymbol{W}^v \in \mathbb{R}^{N\times N}$ to indicate the incompleteness of input data:

$$\boldsymbol{W}^v_{j,j} = \begin{cases} 1, \text{if } j\text{-th instance is observed in } v\text{-th view;} \\ 0, \text{otherwise.} \end{cases}$$

As mentioned above, the zero-filling unobserved features might introduce bias to 0, which lead to incorrect encodings, finally resulting in poor clustering performance. We lower the weight of unobserved instances via $\boldsymbol{W}^v$. It turns to the minimization of $\sum_{v=1}^{n_v} \|(\boldsymbol{B} - \boldsymbol{B}^v)\boldsymbol{W}^v\|_F^2$. To handle the *sgn* function, we adopt continuous relaxation, i.e., directly dropping the sign function ($sgn(z) \approx z$), and obtain the following loss:

$$\min_{\boldsymbol{U}^v,\boldsymbol{B}} \sum_{v=1}^{n_v} \| \left(\boldsymbol{B} - \boldsymbol{U}^v\phi(\boldsymbol{X}^v)\right) \boldsymbol{W}^v\|_F^2. \tag{3}$$

Code Balance Regularization. In order to produce the code in which the variance of each bit is maximized and pairwise uncorrelated. We optimize the following regularization

$$\begin{aligned} \max \quad & var[\boldsymbol{B}] = var[\boldsymbol{U}^v\phi(\boldsymbol{X}^v)\boldsymbol{W}^v] = \frac{1}{N}tr[\boldsymbol{U}^v\phi(\boldsymbol{X}^v)\boldsymbol{W}^v(\boldsymbol{W}^v)^T\phi^T(\boldsymbol{X}^v)(\boldsymbol{U}^v)^T], \\ s.t. \quad & \boldsymbol{B}\boldsymbol{B}^T = I. \end{aligned}$$

We relax the bit uncorrelation condition $\boldsymbol{B}\boldsymbol{B}^T = I$ and control the scales of $\boldsymbol{U}^v$. And then, we define the regularizer as follows

$$R(\boldsymbol{U}^v) = -\gamma tr(\boldsymbol{U}^v\phi(\boldsymbol{X}^v)\boldsymbol{W}^v\phi^T(\boldsymbol{X}^v)(\boldsymbol{U}^v)^T) + \beta\|\boldsymbol{U}^v\|_F^2, \tag{4}$$

where β, γ are the parameters. Recall the definition of $\boldsymbol{W}^v$, we have $\boldsymbol{W}^v = \boldsymbol{W}^v(\boldsymbol{W}^v)^T$.

Objective Function. We write the objective function for common binary code learning as follows:

$$\min_{\boldsymbol{U}^v,\boldsymbol{B}} \sum_{v=1}^{n_v} \left(\|\boldsymbol{B} - \boldsymbol{U}^v\phi(\boldsymbol{X}^v)\boldsymbol{W}^v\|_F^2 + R(\boldsymbol{U}^v)\right) \quad s.t.\boldsymbol{B} \in \{-1,1\}^{l\times N} \tag{5}$$

3.3 Binary Clustering

Binary clustering aims to factorize the binary representation $\boldsymbol{B}$ into the binary centroids $\boldsymbol{C}$ and indicator matrix $\boldsymbol{G}$. Due to the equivalence of the conventional k-means clustering and matrix factorization [2], we construct the objective function as:

$$\min_{\boldsymbol{C},\boldsymbol{G}} = \|\boldsymbol{B} - \boldsymbol{C}\boldsymbol{G}\|_F^2, \quad s.t.\boldsymbol{C}^T\boldsymbol{1} = \boldsymbol{0}, \boldsymbol{C} \in \{-1,1\}^{l\times k}, \boldsymbol{G} \in \{0,1\}^{k\times N}, \sum_j \boldsymbol{g}_{ji} = 1. \tag{6}$$

The first constraint on clustering centers ($\boldsymbol{C}^T\mathbf{1} = \mathbf{0}$) aims to maximize the information of each bit via the bit balance condition, which means that each bit has about 50% chance of being 1 or −1.

Until now, we have introduced the *binary code learning* (Eq.(5)) and *binary clustering* (Eq.(6)), the overall objective function is:

$$\min_{\boldsymbol{U}^v,\boldsymbol{B},\boldsymbol{C},\boldsymbol{G}} \sum_{v=1}^{n_v} (\| (\boldsymbol{B} - \boldsymbol{U}^v\phi(\boldsymbol{X}^v))\boldsymbol{W}^v\|_F^2 + R(\boldsymbol{U}^v)) + \lambda\|\boldsymbol{B} - \boldsymbol{C}\boldsymbol{G}\|_F^2,$$
$$s.t.\quad \boldsymbol{B} \in \{-1,1\}^{l\times N}, \boldsymbol{C}^T\mathbf{1} = \mathbf{0}, \boldsymbol{C} \in \{-1,1\}^{l\times k}, \boldsymbol{G} \in \{0,1\}^{k\times N}, \sum_j \boldsymbol{g}_{ji} = 1.$$

where λ is the regularization parameter.

3.4 Online Binary Incomplete Multi-view Clustering

We propose to solve the above optimization problem in an online fashion with low computational complexity. We assume that the data of each view is reaching by chunks, whose size is s. The objective function $\mathcal{L}$ can be decomposed as:

$$\begin{aligned}
&\min\mathcal{L}_{\text{OBIMC}}(\boldsymbol{U}^v, \boldsymbol{B}_t, \boldsymbol{C}, \boldsymbol{G}_t)\\
&= \sum_{v=1}^{n_v}\sum_{t=1}^{\lceil N/s\rceil} (\| (\boldsymbol{B}_t - \boldsymbol{U}^v\phi(\boldsymbol{X}_t^v))\boldsymbol{W}_t^v\|_F^2 - \gamma g_t(\boldsymbol{U}^v)) + \sum_{v=1}^{n_v}\beta\|\boldsymbol{U}^v\|_F^2\\
&+ \lambda\sum_{t=1}^{\lceil N/s\rceil} \|\boldsymbol{B}_t - \boldsymbol{C}\boldsymbol{G}_t\|_F^2\\
&s.t. \boldsymbol{B}_t \in \{-1,1\}^{l\times s}, \boldsymbol{C}^T\mathbf{1} = \mathbf{0}, \boldsymbol{C} \in \{-1,1\}^{l\times k}, \boldsymbol{G}_t \in \{0,1\}^{k\times s}, \sum_j \boldsymbol{g}_{ji} = 1,
\end{aligned}\tag{7}$$

where $g_t(\boldsymbol{U}^v) = tr\left(\boldsymbol{U}^v\phi(\boldsymbol{X}_t^v)\boldsymbol{W}_t^v\phi^T(\boldsymbol{X}_t^v)\boldsymbol{U}^v\right)$.

3.5 Optimization

The solution to problem (7) is non-trivial as it involves a mixed binary integer program with discrete constraints. Next up, we introduce an alternating optimization algorithm to iteratively update each variable while fixing others. The subproblem of $\boldsymbol{U}^v$ and $\boldsymbol{B}_t$ could be solved with the close-form solution. The optimization converges fast with the help of a convergence guaranteed algorithm for subproblem $\boldsymbol{C}$ and $\boldsymbol{G}_t$.

Subproblem of $\{\boldsymbol{U}^v\}_{v=1}^{n_v}$. With $\boldsymbol{B}_t$, $\boldsymbol{C}$ and $\boldsymbol{G}_t$ fixed, to optimize $\boldsymbol{U}^v$ for specific view v at time t, we only need to minimize the following objective:

$$\mathcal{F}^{(t)}(\boldsymbol{U}^v) = \sum_{i=1}^{t}(\| (\boldsymbol{B}_i - \boldsymbol{U}^v\phi(\boldsymbol{X}_i^v))\boldsymbol{W}_i^v\|_F^2 - \gamma g_i(\boldsymbol{U}^v)) + \beta\|\boldsymbol{U}^v\|_F^2$$

which has a closed-form solution obtained by setting the derivation $\frac{\partial \mathcal{F}^{(t)}(\boldsymbol{U}^v)}{\partial \boldsymbol{U}^v} = 0$. The updating rule is as follows:

$$\boldsymbol{U}^v = \sum_{i=1}^{t} \boldsymbol{B}_i \boldsymbol{W}_i^v \phi^T(\boldsymbol{X}_i^v)((1-\frac{\gamma}{N})\sum_{i=1}^{t} \phi(\boldsymbol{X}_i^v)\boldsymbol{W}_i^v \phi^T(\boldsymbol{X}_i^v) + \beta \boldsymbol{I})^{-1}. \quad (8)$$

For the sake of convenience, we introduce two terms $\boldsymbol{Q}_t^v$ and $\boldsymbol{R}_t^v$ as follows:

$$\boldsymbol{Q}_t^v = \sum_{i=1}^{t} \boldsymbol{B}_i \boldsymbol{W}_i^v \phi^T(\boldsymbol{X}_i^v), \quad \boldsymbol{R}_t^v = \sum_{i=1}^{t} \phi(\boldsymbol{X}_i^v)\boldsymbol{W}_i^v \phi^T(\boldsymbol{X}_i^v).$$

Consequently, the update rule of $\boldsymbol{U}^v$ can be rewritten as:

$$\boldsymbol{U}^v = \boldsymbol{Q}_t^v \left((1-\frac{\gamma}{N})\boldsymbol{R}_t^v + \beta \boldsymbol{I}\right)^{-1}. \quad (9)$$

Then, when new chunk comes, the matrices $\boldsymbol{Q}_t^v$ and $\boldsymbol{R}_t^v$ can be updated easily as follows:

$$\begin{aligned} \boldsymbol{Q}_t^v &= \boldsymbol{Q}_{t-1}^v + \boldsymbol{B}_t \boldsymbol{W}_t^v \phi^T(\boldsymbol{X}_t^v), \\ \boldsymbol{R}_t^v &= \boldsymbol{R}_{t-1}^v + \phi(\boldsymbol{X}_t^v)\boldsymbol{W}_t^v \phi^T(\boldsymbol{X}_t^v). \end{aligned} \quad (10)$$

Subproblem of $\boldsymbol{B}_t$. Similarly, problem (7) writes w.r.t $\boldsymbol{B}_t$:

$$\begin{aligned} &\min_{\boldsymbol{B}_t} \sum_{v=1}^{n_v} \| (\boldsymbol{B}_t - \boldsymbol{U}^v \phi(\boldsymbol{X}_t^v))\boldsymbol{W}_t^v \|_F^2 + \lambda \|\boldsymbol{B}_t - \boldsymbol{C}\boldsymbol{G}_t\|_F^2 \\ &= tr\left[\boldsymbol{B}_t \boldsymbol{B}_t^T \left(\sum_{v=1}^{n_v} \boldsymbol{W}_t^v + \lambda \boldsymbol{I}\right)\right] - 2tr\left[\boldsymbol{B}_t^T \left(\sum_{v=1}^{n_v} \boldsymbol{U}^v \phi(\boldsymbol{X}_t^v)\boldsymbol{W}_t^v + \lambda \boldsymbol{C}\boldsymbol{G}_t\right)\right] + const, \\ &s.t. \boldsymbol{B}_t \in \{-1,1\}^{l \times s}. \end{aligned} \quad (11)$$

where *const* means the constant value w.r.t. $\boldsymbol{B}$. Since $tr(\boldsymbol{B}_t \boldsymbol{B}_t^T) = tr(\boldsymbol{B}_t^T \boldsymbol{B}_t) = sl$ is a constant, problem (11) can be rewritten as

$$\min_{\boldsymbol{B}_t} -2tr\left[\boldsymbol{B}_t^T \left(\sum_{v=1}^{n_v} \boldsymbol{U}^v \phi(\boldsymbol{X}_t^v)\boldsymbol{W}_t^v + \lambda \boldsymbol{C}\boldsymbol{G}_t\right)\right], s.t. \boldsymbol{B}_t \in \{-1,1\}^{l \times s}. \quad (12)$$

Eq.(12) has a closed-form solution:

$$\boldsymbol{B}_t = sgn\left(\sum_{v=1}^{n_v} \boldsymbol{U}^v \phi(\boldsymbol{X}_t^v)\boldsymbol{W}_t^v + \lambda \boldsymbol{C}\boldsymbol{G}_t\right). \quad (13)$$

Subproblem of $\boldsymbol{C}$ *and* $\boldsymbol{G}_t$. By removing the irrelevant terms, we obtain the following problem:

$$\begin{aligned} \min \quad & \mathcal{F}^{(t)}(\boldsymbol{C}) = \sum_{i=1}^{t} \|\boldsymbol{B}_i - \boldsymbol{C}\boldsymbol{G}_i\|_F^2, \\ s.t. \quad & \boldsymbol{C}^T \boldsymbol{1} = \boldsymbol{0}, \boldsymbol{C} \in \{-1,1\}^{l \times k}, \boldsymbol{G}_i \in \{0,1\}^{k \times s}, \sum_j \boldsymbol{g}_{ji} = 1. \end{aligned}$$

Algorithm 1: Online Binary Incomplete Multi-view Clustering

Input: Incomplete Multi-view features $\{\boldsymbol{X}^v\}$, code length l, number of clusters k, weight matrices $\{\boldsymbol{W}^v\}$, m, β, γ, λ.

Output: Binary representation $\boldsymbol{B}$, cluster centroid $\boldsymbol{C}$ and binary cluster indicator $\boldsymbol{G}$;

```
1  Initial: Randomly selected m anchor points from each view;
2  R_0^v = 0, Q_0^v = 0 for each view v;
3  for t = 1 : ⌈N/s⌉ do
4      Calculate φ(X_t^v) by Eq. (2) and normalize them to have zero-centered mean;
5      repeat
6          for v = 1 : n_v do
7              Update U^v by Eq.(9);
8          end
9          Update B_t by Eq.(13);
10         repeat
11             Update C by Eq.(15);
12             Update G_t by Eq.(16);
13         until Convergence of Eq.(14);
14     until Convergence of Eq.(7);
15     Update R_t^v and Q_t^v by Eq.(10);
16 end
```

Equation(14) is equivalent to the above problem with sufficiently large $\rho > 0$.

$$\begin{aligned} \min \quad & \mathcal{F}^{(t)}(\boldsymbol{C}) = \sum_{i=1}^{t} \|\boldsymbol{B}_i - \boldsymbol{C}\boldsymbol{G}_i\|_F^2 + \rho\|\boldsymbol{C}^T\mathbf{1}\|_F^2, \\ s.t. \quad & \boldsymbol{C} \in \{-1,1\}^{l\times k}, \boldsymbol{G}_i \in \{0,1\}^{k\times s}, \sum_j \boldsymbol{g}_{ji} = 1. \end{aligned} \tag{14}$$

We solve Eq.(14) iteratively by keeping the discrete constraints during optimizations following the adaptive discrete proximal linearized optimization (ADPLM) [17,24].

C-step: With $\boldsymbol{G}_t$ fixed, we have

$$\min \mathcal{F}^{(t)}(\boldsymbol{C}) = \sum_{i=1}^{t} -2tr(\boldsymbol{B}_i{}^T\boldsymbol{C}\boldsymbol{G}_i) + \rho\|\boldsymbol{C}^T\mathbf{1}\|_F^2 + const, \quad s.t. \boldsymbol{C} \in \{-1,1\}^{l\times k}.$$

According to the rule of ADPLM, we update $\boldsymbol{C}$ in the $(p+1)$-th iteration by

$$\boldsymbol{C}^{p+1} = \operatorname{sgn}\left(\boldsymbol{C}^p - \frac{1}{\mu}\nabla\mathcal{F}^{(t)}\left(\boldsymbol{C}^p\right)\right). \tag{15}$$

where $\nabla\mathcal{F}^{(t)}(\boldsymbol{C})$ is the gradient of $\mathcal{F}^{(t)}(\boldsymbol{C})$. We set $\mu^p \in (L, 2L)$, where L is the Lipschitz constant.

$\boldsymbol{G}_t$-Step: Similarly, when fixing $\boldsymbol{C}$ and $\boldsymbol{B}_t$, the problem w.r.t. $\boldsymbol{G}_t$ turns into

$$\min_{\boldsymbol{G}_t} \|\boldsymbol{B}_t - \boldsymbol{C}\boldsymbol{G}_t\|_F^2, \quad s.t. \boldsymbol{G}_t \in \{0,1\}^{k\times s}, \sum_j \boldsymbol{g}_{ji} = 1.$$

Inspired by *K-means*, let $\boldsymbol{D} \in \mathbb{R}^{s\times k}$ record the hamming distance between binary code (the column of $\boldsymbol{B}_t$) and cluster centroid (the column of $\boldsymbol{C}$), i.e. $\boldsymbol{d}_{ij} = H(\boldsymbol{b}_i, \boldsymbol{c}_j^{p+1})$, where $H(\boldsymbol{b}_i, \boldsymbol{c}_j)$ is the hamming distance between i-th binary code $\boldsymbol{b}_i$ and the j-th cluster centroid $\boldsymbol{c}_j$. It is worth noting that the calculations are much faster in hamming space than in Euclidean space. The optimal solution of the indicator matrix $\boldsymbol{G}_t$ at indices (i, j) can be easily obtained by

$$\boldsymbol{g}_{ij}^{p+1} = \begin{cases} 1, \, j = \arg\min_s \boldsymbol{d}_{is} \\ 0, \;\; \text{otherwise} \, . \end{cases} \tag{16}$$

Convergence Condition. The convergence of problem in our study is defined as $(\text{obj}^{(t-1)} - \text{obj}^{(t)})/\text{obj}^{(t)} < 1\text{e}-4$.

We have so far presented the whole optimization procedures for the problem (7). The entire optimization procedure for OBIMC is summarized in Algorithm 1.

3.6 Complexity

We discuss the computational complexity of OBIMC in the four aspects: optimizing $\boldsymbol{U}^v$, $\boldsymbol{B}_t$, $\boldsymbol{C}$, and $\boldsymbol{G}_t$, respectively. In general, $l \ll N$ and $m \ll N$, p is empirically set from 5 to 10. The time complexity of OBIMC is $\mathcal{O}(Nlm^2n_v)$, which is linear to $\mathcal{O}(N)$. Owing to the economic online optimization, the space complexity of OBIMC is $\mathcal{O}(s(l+k) + n_v lm + lk)$, which is linear to $\mathcal{O}(s)$, where s is the size of the data chunks.

3.7 Convergence

Theorem 1. *The proposed optimization algorithm monotonically decreases the value of $\mathcal{L}(\boldsymbol{U}, \boldsymbol{B}, \boldsymbol{C}, \boldsymbol{G})$ in each optimization step.*

Proof. As shown in Algorithm (1), the optimization of OBIMC can be divided into three subproblems. It is obvious that $\{\boldsymbol{U}^v, \boldsymbol{B}_t\}$ generated via the process (9) and (13) are the exact minimum points of the subproblems (8) and (12), respectively. According to the theoretical analysis in [17], (15) has an analytical solution by using ADPLM [24]. Based on the learning scheme of k-means, $\{\boldsymbol{C}, \boldsymbol{G}_t\}$ computed by using ADPLM and (16) are the optimal solution of subproblem (14). As a result, the value of the objective function $\mathcal{L}(\boldsymbol{U}, \boldsymbol{B}, \boldsymbol{C}, \boldsymbol{G})$ in (7) is decreasing in each iteration of the proposed optimization algorithm. Thus, by finding the optimal solution to each subproblem alternatively, our algorithm can at least find a locally optimal solution.

Table 2. Summary of the datasets.$\{n_v, k, N, d_v\}$: number of {views, clusters, instances, features} in each view, respectively.

Dataset	n_v	k	N	$d_v(v = 1, \ldots, n_v)$
Digit	6	10	2000	6, 47, 64, 76, 216, 240
Caltech101	5	102	9144	40, 254, 512, 928, 1984
NUS-WIDE-Obj	5	31	30000	65, 74, 129, 145, 226
Reuters	5	6	111740	11547, 15506, 21531, 24983, 34279

4 Experiment

4.1 Datasets and Baselines

Datasets. We evaluate the proposed algorithm on four widely used complete multi-view benchmark data sets, including Digit[1], Caltech101[2], NUS-WIDE-Obj[3], and Reuters[4]. The details of these datasets are shown in Table 2.

Baselines. We consider the following algorithms as baselines: **CoSC** [7], **Multi-NMF** [9], **OMVC** [15], **OPIMC** [5], **BMVC** [24]. **BIMC** is the offline version of proposed **OBIMC**. Note that CoSC, MultiNMF and BMVC only work for complete multi-view data, we first fill the unobserved instance in each incomplete view with zero.

4.2 Experiment Setting

To simulate the incomplete view setting, we follow the approach in [11] to generate the incomplete multi-view data. In detail, we first randomly select round(ϵN) samples, where round($\cdot$) is a rounding function. For each selected sample, we generate a random vector $\boldsymbol{e} = (e_1, \ldots, e_{n_v}) \in [0,1]^{n_v}$ and a scalar $e_0 (e_0 \in [0,1])$. The p-th view will be present for this sample, if $e_p \geq e_0$ is satisfied. If none of $\{e_i\}_{i=1}^{n_v}$ satisfies the condition, we will repeat the procedure and make sure that at least one view would be presented. Moreover, OBIMC does not require a complete view across all the samples. We finally generate the incomplete multi-view dataset according to the vectors above mentioned. In summary, ϵ controls the percentage of samples that have incomplete views. We compare these algorithms with respect to different incomplete rate: $\epsilon \in \{0.1, 0.2, \ldots, 0.9\}$.

We report the experimental results on clustering accuracy (ACC) and normalized mutual information (NMI) [12]. The running time of all the methods are reported in this paper. For the fair comparison, we use the codes provided by the corresponding authors with the default or optimal parameter settings in

[1] http://archive.ics.uci.edu/ml/datasets/Multiple+Features.
[2] http://www.vision.caltech.edu/Image_Datasets/Caltech101/.
[3] http://lms.comp.nus.edu.sg/research/NUS-WIDE.htm.
[4] http://archive.ics.uci.edu/ml/machine-learning-databases/00259/.

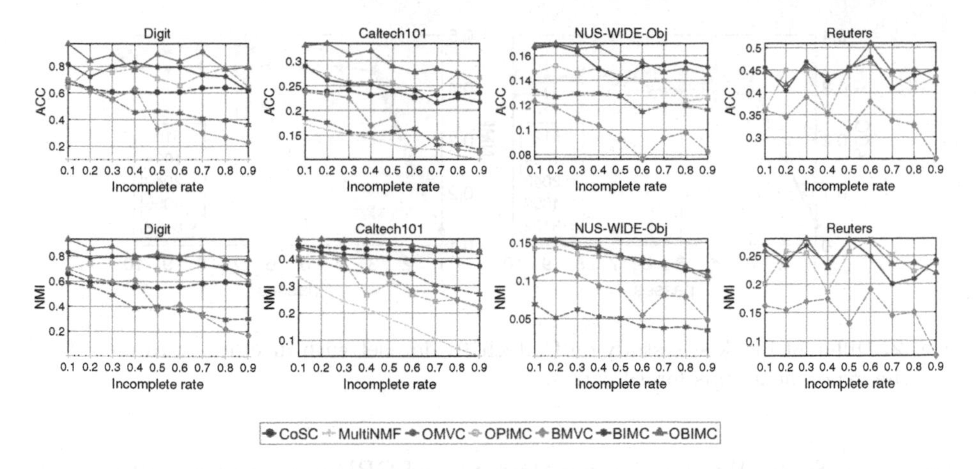

Fig. 1. ACC and NMI comparison with various incomplete rates on different datasets. The chunk size s is set as 512 for Digit dataset and s is set as 4906 for Caltech101, NUS-WIDE-obj and Reuters.

their original papers. The performances of all approaches are evaluated over 10 independent runs. All the experiments are implemented using Matlab 2016b on a standard PC with an Intel i3-2310M 2.1 GHz CPU and 16 GB RAM.

4.3 Experimental Results

Figure 1 demonstrates the performance results in terms of ACC and NMI. From Fig. 1 we make the following observations:

- Our OBIMC outperforms the baselines in all the incomplete rate settings. As the incomplete rate increases, the clustering performance of all the methods drops.
- The baseline of spectral clustering (CoSC) is inferior to our model in the Digit and Caltech101 dataset. However, CoSC is not capable for large datasets, due to its high computational complexity.
- For two online IMC baselines (OPIMC and OMVC), OPIMC outperforms OMVC in all datasets. One reason is that the average value filling of OMVC results in a large deviation.
- Our OBIMC outperforms the offline binary coding baseline BMVC markedly. In this work, we complete the missing anchors with the average imputation and optimize the objective function in an online fashion. This shows that our method is successful in dealing with missing instances.
- The recently online baseline OPIMC performs better than other baselines in most datasets, but worse than our method. This shows that our fusion strategy is more stable than the one in OPIMC.

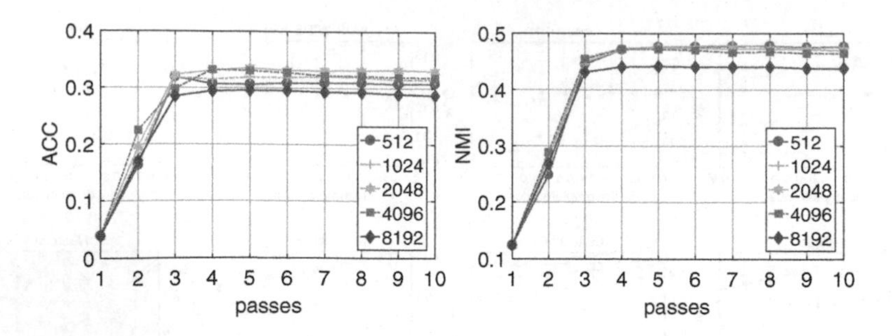

Fig. 2. Different block size study on Caltech101 dataset with incomplete rate $\epsilon = 0.1$, and the experiment runs for 10 passes.

Block Size Study. We report the performance of OBIMC on Caltech101 dataset (incomplete rate = 0.1) with different block sizes in the range of $s \in \{512, 1024, 2048, 4096, 8192\}$. From Fig. 2, we can see that the proposed OBIMC is insensitive to the block size. It is worth noting that the parameters are fixed in this comparison. By setting $s = 8192$, the performance drops a little bit. We set $s = 512$ for small-size dataset (i.e, $N \leq 5000$) and $s = 4096$ for large-scale dataset.

Parameter Study. In the proposed framework, there are four parameters to be tuned, i.e. β, γ, λ, and m. In the experiments, we first employ the grid search strategy to find the best choices for all parameters on the Caltech101 dataset. When applying to a large dataset, we only fine-tune these parameters on a subset of $10,000$ instances from the corresponding dataset. From Fig. 3(a) and 3(b), we can see that the best clustering results are established when the values of β and γ are smaller than 1 but larger than 10^{-6}. From Fig. 3(c) and 3(d), we can observe that our method is insensitive to the value of parameters, which can be selected over a wide range. In this study, we set $m = 256$, $\lambda = 10^{-4}$.

Running Time Analysis. Although BMVC, OPIMC and OBIMC are all with linear time complexity, the time comparison demonstrates clearer advantages for binary clustering methods, which are much faster than the real-valued ones. The main reason is that they benefit from the highly efficient distance calculation using the Hamming metric rather than the Euclidean distance measurement. Particularly, our OBIMC costs much less time than all the other relevant methods. For example, on Caltech101 dataset, binary clustering BMVC is 31.80 times faster than OMVC, while the speed-up of our OBIMC is more obvious by a margin of 58.34 times. Comparing with the-state-of-the-art online method OPIMC on five datasets, the speed-up of our OBIMC is $1.29, 13.40, 4.50, 2.11, 1.83$ times, respectively. All these observations prove the efficiency and effectiveness of our model.

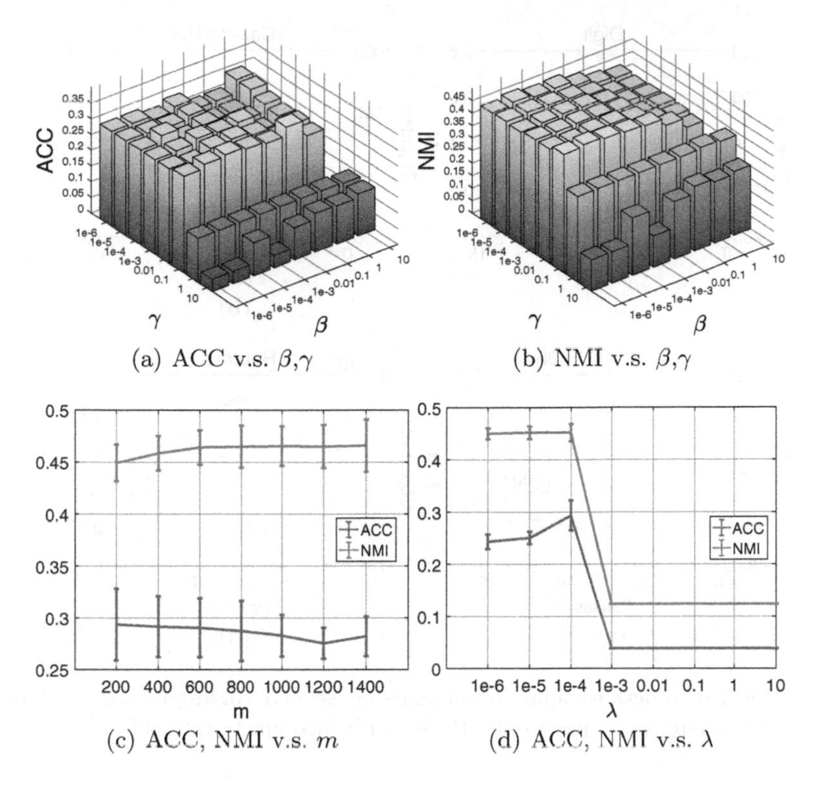

(a) ACC v.s. β,γ

(b) NMI v.s. β,γ

(c) ACC, NMI v.s. m

(d) ACC, NMI v.s. λ

Fig. 3. Parameters (β,γ,λ,m) study on Caltech101 dataset.

Table 3. Run time for different methods (in seconds), 'N/A-T' means that no result returns after 1 h, 'N/A-M' means out of memory error.

Method	*Digit*	*Caltech*101	NUSWIDEObj	Reuters
CoSC	28.87	1908.08	N/A-M	N/A-M
MultiNMF	2.45	241.52	67.00	N/A-T
OMVC	31.83	169.20	168.20	N/A-T
OPIMC	0.36	4.68	20.84	43.09
BMVC	1.31	5.32	12.07	47.11
OBIMC	**0.28**	**2.90**	**4.63**	**20.45**

Convergence Study. Figure 4 demonstrates the convergence curve and corresponding NMI performance for five datasets with an incomplete rate of 50%. The objective function value monotonically decreases as the training goes on. When the loss converges, the NMI performance gets stable values.

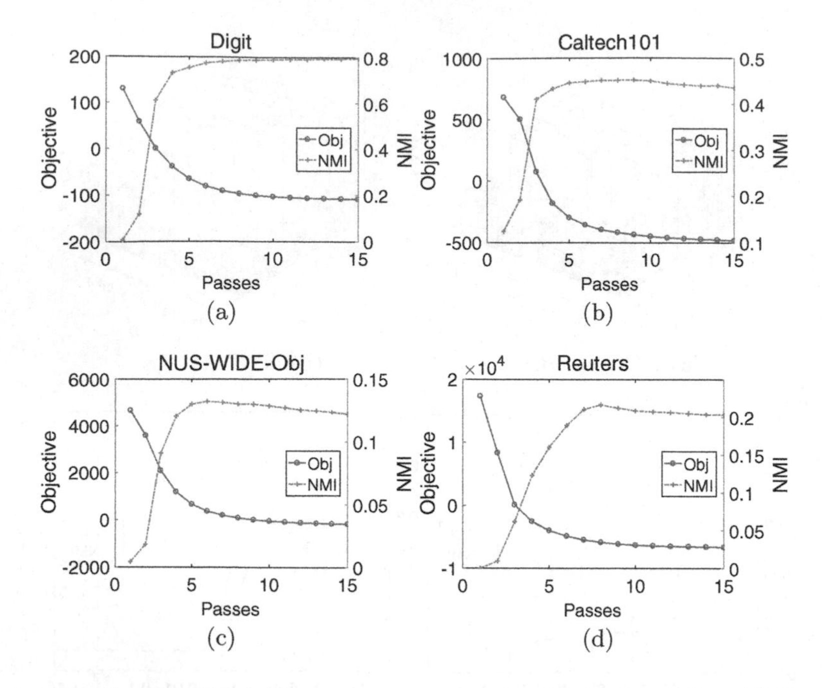

Fig. 4. Objective function value convergence curve and corresponding NMI performance curve vs number of passes of OBIMC with incomplete rate 50%.

5 Conclusion

In this paper, we propose a novel method to solve the challenging problem of incomplete multi-view clustering on large-scale data. In this work, we develop an anchor-based representation for incomplete multi-view data, the robust binary representations and binary cluster structures are jointly learned, which can effectively integrate the complementary information from multiple views. Our method deals with the large-scale data in an online fashion with the help of two global statistics. Moreover, an effective alternating optimization algorithm with guaranteed convergence is proposed to ensure the high-quality binary solutions. Experiments performed on the four real-world multi-view datasets show the clear superiority of OBIMC in comparison with the state-of-the-art clustering methods. In the future, we will study how to extend this binary coding based online approach to semi-supervised setting.

Acknowledgment. This work is supported by the Major Research plan of the National Natural Science Foundation of China (Grant No.91648204), National Key Research and Development Project (Grant No.2017YFB1300203), and National Natural Science Foundation of China (Grant No.61973313).

References

1. Bickel, S., Scheffer, T.: Multi-view clustering. In: IEEE International Conference on Data Mining (2004)
2. Ding, C.H.Q., He, X., Simon, H.D., Jin, R.: On the equivalence of nonnegative matrix factorization and k-means — spectral clustering. Lawrence Berkeley National Laboratory (2005)
3. Gao, H., Nie, F., Li, X., Huang, H.: Multi-view subspace clustering. In: IEEE International Conference on Computer Vision (2015)
4. Gong, Y., Lazebnik, S., Gordo, A., Perronnin, F.: Iterative quantization: a procrustean approach to learning binary codes for large-scale image retrieval. IEEE Trans. Pattern Anal. Mach. Intell. **35**(12), 2916–2929 (2013)
5. Hu, M., Chen, S.: One-pass incomplete multi-view clustering. In: Proceedings of the Thirty-Third AAAI Conference on Artificial Intelligence, pp. 3838–3845 (2019)
6. Kumar, A., Iii, H.D.: A co-training approach for multi-view spectral clustering abhishek kumar. In: International Conference on International Conference on Machine Learning (2011)
7. Kumar, A., Rai, P., Daumé, H.: Co-regularized multi-view spectral clustering. In: International Conference on Neural Information Processing Systems (2011)
8. Li, S., Jiang, Y., Zhou, Z.: Partial multi-view clustering. In: AAAI Conference on Artificial Intelligence, vol. 3, pp. 1968–1974, January 2014
9. Liu, J., Wang, C., Gao, J., Han, J.: Multi-view clustering via joint nonnegative matrix factorization. In: Proceedings of the 2013 SIAM International Conference on Data Mining, pp. 252–260 (2013). https://doi.org/10.1137/1.9781611972832.28. https://epubs.siam.org/doi/abs/10.1137/1.9781611972832.28
10. Liu, X., Li, M., Wang, L., Dou, Y., Yin, J., Zhu, E.: Multiple kernel k-means with incomplete kernels. In: Proceedings of the Thirty-First AAAI Conference on Artificial Intelligence, pp. 2259–2265. AAAI'2017, AAAI Press (2017). http://dl.acm.org/citation.cfm?id=3298483.3298565
11. Liu, X., Zhu, X., Li, M., Wang, L., Gao, W.: Late fusion incomplete multi-view clustering. IEEE Trans. Pattern Anal. Mach. Intell. **41**(10), 2410–2423 (2018)
12. Manning, C.D., Raghavan, P., Schütze, H.: Introduction to Information Retrieval. Cambridge University Press, Cambridge (2008)
13. Nie, F., Cai, G., Li, J., Li, X.: Auto-weighted multi-view learning for image clustering and semi-supervised classification. IEEE Trans. Image Process. **27**(3), 1501–1511 (2017)
14. Ren, P., et al.: Robust auto-weighted multi-view clustering. In: Proceedings of the Twenty-Seventh International Joint Conference on Artificial Intelligence, IJCAI-18, pp. 2644–2650, July 2018. https://doi.org/10.24963/ijcai.2018/367
15. Shao, W., He, L., Lu, C.T., Yu, P.S.: Online multi-view clustering with incomplete views. In: International Conference on Big Data, pp. 1012–1017 (2016)
16. Shao, W., He, L., Yu, P.S.: Multiple incomplete views clustering via weighted nonnegative matrix factorization with $\ell_{2,1}$ regularization. In: ECML PKDD, pp. 318–334 (2015)
17. Shen, F., Zhou, X., Yang, Y., Song, J., Shen, H.T., Tao, D.: A fast optimization method for general binary code learning. IEEE Trans. Image Process. **25**(12), 5610–5621 (2016)

18. Tao, Z., Liu, H., Li, S., Ding, Z., Fu, Y.: From ensemble clustering to multi-view clustering. In: Proceedings of the Twenty-Sixth International Joint Conference on Artificial Intelligence, IJCAI-17, pp. 2843–2849 (2017). https://doi.org/10.24963/ijcai.2017/396
19. Wang, H., Zong, L., Liu, B., Yang, Y., Zhou, W.: Spectral perturbation meets incomplete multi-view data. In: Proceedings of the Twenty-Eighth International Joint Conference on Artificial Intelligence, IJCAI-19, pp. 3677–3683. International Joint Conferences on Artificial Intelligence Organization, July 2019. https://doi.org/10.24963/ijcai.2019/510
20. Wen, J., Zhang, Z., Xu, Y., Zhong, Z.: Incomplete multi-view clustering via graph regularized matrix factorization. In: Proceedings of the European Conference on Computer Vision (ECCV) (2018)
21. Xiao, C., Nie, F., Huang, H.: Multi-view k-means clustering on big data. In: International Joint Conference on Artificial Intelligence (2013)
22. Xu, C., Tao, D., Xu, C.: Multi-view learning with incomplete views. IEEE Trans. Image Process. **24**(12), 5812–5825 (2015)
23. Yang, W., Xuemin, L., Lin, W., Wenjie, Z., Qing, Z., Xiaodi, H.: Robust subspace clustering for multi-view data by exploiting correlation consensus. IEEE Trans. Image Process. **24**(11), 3939–49 (2015)
24. Zhang, Z., Liu, L., Shen, F., Shen, H.T., Shao, L.: Binary multi-view clustering. IEEE Trans. Pattern Anal. Mach. Intell. **41**(7), 1774–1782 (2018)
25. Zhao, H., Ding, Z., Fu, Y.: Multi-view clustering via deep matrix factorization. In: AAAI Conference on Artificial Intelligence (2017). https://aaai.org/ocs/index.php/AAAI/AAAI17/paper/view/14647
26. Zhao, H., Liu, H., Fu, Y.: Incomplete multi-modal visual data grouping. In: Proceedings of the Twenty-Fifth International Joint Conference on Artificial Intelligence, IJCAI-16, pp. 2392–2398 (2016)

Utilizing Structure-Rich Features to Improve Clustering

Benjamin Schelling[1,2,3](✉), Lena Greta Marie Bauer[4], Sahar Behzadi[3], and Claudia Plant[3,4]

[1] MCML, Munich, Germany
[2] Ludwig-Maximilians-Universität München, Munich, Germany
[3] Faculty of Computer Science, University of Vienna, Vienna, Austria
benjamin.schelling@univie.ac.at, sahar.behzadi@univie.ac.at
[4] ds:UniVie, Vienna, Austria
{lena.bauer,claudia.plant}@univie.ac.at

Abstract. For successful clustering, an algorithm needs to find the boundaries between clusters. While this is comparatively easy if the clusters are compact and non-overlapping and thus the boundaries clearly defined, features where the clusters blend into each other hinder clustering methods to correctly estimate these boundaries. Therefore, we aim to extract features showing clear cluster boundaries and thus enhance the cluster structure in the data. Our novel technique creates a condensed version of the data set containing the structure important for clustering, but without the noise-information. We demonstrate that this transformation of the data set is much easier to cluster for k-means, but also various other algorithms. Furthermore, we introduce a deterministic initialisation strategy for k-means based on these structure-rich features.

1 Introduction

Clustering is the task of grouping data points and dividing a data set into the correct partitioning. For this, it is most important that the clusters can be distinguished from each other and the boundaries between them are easily found. Useless information can make this difficult for many if not all clustering approaches, causing a faulty analysis of the data, as the clusters can not be separated. To battle this difficulty we propose an approach that extracts the features with the highest amount of information relevant for clustering and constructs a condensed data set, which is easier to cluster.

Features where the clusters do not blend into each other, are those important for clustering. If the clusters are well separated, it is possible to determine boundaries between them, enabling a good clustering result. Concentrations of data points show that there are clusters present and a meaningful partitioning can be found. They also reveal where to draw the boundaries.

Let us demonstrate this using the running example shown in Fig. 1, consisting of three non-overlapping Gaussian clusters and a third feature made up of uniform noise. The first axis of the running example is shown in Fig. 2. It contains hardly any information useful for clustering as it is unimodal, meaning that

F. Hutter et al. (Eds.): ECML PKDD 2020, LNAI 12457, pp. 91–107, 2021.
https://doi.org/10.1007/978-3-030-67658-2_6

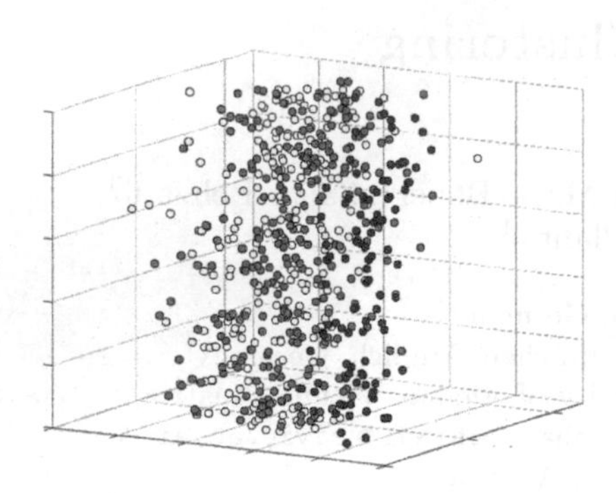

Fig. 1. The running example in a 3D scatterplot.

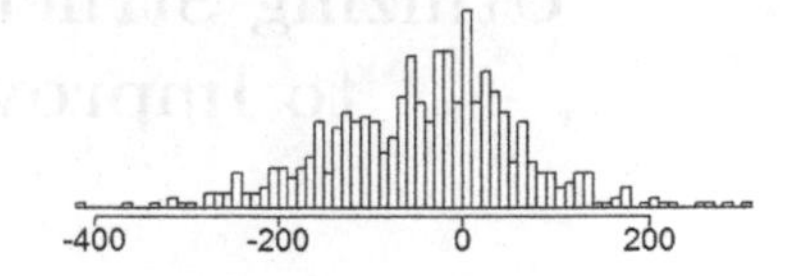

Fig. 2. Histogram of the first axis of the running example.

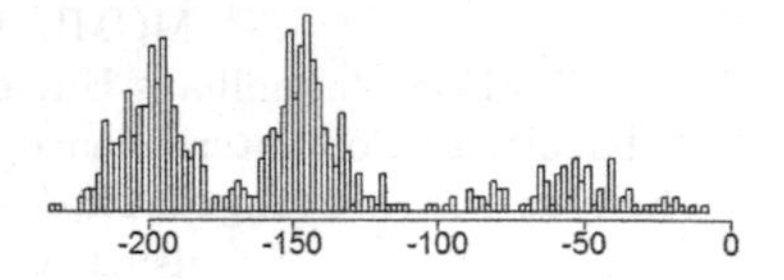

Fig. 3. Histogram of the feature with the highest dip value.

the clusters overlap and have no clear boundaries. In contrast, the projection of the data set on a feature that has clear boundaries between clusters is shown in Fig. 3. It is far more relevant for clustering, as it shows where the data set should be partitioned. This feature contains all the information necessary for clustering, as the clusters are almost perfectly separated in this projection. K-means, for instance, finds the correct partitioning based solely on this feature. However, this projection is not an arbitrary one, it is the projection with the maximal possible dip value in this specific data set. The dip test [8] is a parameter-free statistical test developed in the1980 s which measures how much a one-dimensional sample deviates from uni-modality. Multi-modality indicates clear cluster boundaries, making the dip test a useful tool to estimate how much information relevant for clustering is contained in a feature.

Our approach DipExt (dip test based extraction) searches for and extracts structure-rich features. DipExt creates a lower-dimensional representation containing the structure of the data set in a condensed form. The 2 features with the highest dip values, as found by our approach, can be seen in Fig. 4. The second feature is not needed regarding the clustering goal, but we include it for

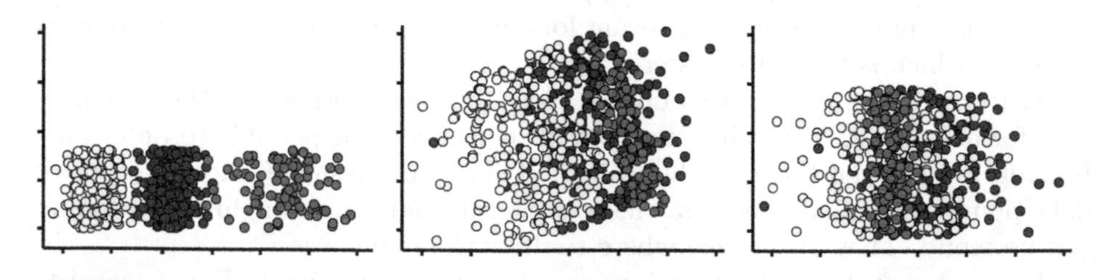

Fig. 4. The running example in 2D with our method (left), PCA (middle) and ICA (right).

visualisation-purposes. Also shown are the 2D representations of the data set obtained by PCA and ICA. The difference is clear: While our approach produces a lower-dimensional representation which is trivial to cluster, PCA and ICA do not. In contrast to them, our method explicitly searches for *structure*, i.e. information relevant for clustering, while PCA searches for variance and ICA for statistical independence. Neither of the latter two guarantees that the found features are important for clustering. Features with high dip values, on the other hand, contain separated concentrations of data points, i.e. densely grouped data points. Therefore, they most likely carry the information relevant for clustering.

Our goal is to find a lower-dimensional transformation of the data which contains the cluster-information. DipExt extracts the features with structure and combines them, creating a condensed form of the data, which contains only the information relevant for clustering. After these features are found we re-scale them according to their relevance for clustering, i.e., with their dip value. This scaled subset of features comprises the information from the data processed for clustering, so that, for example, k-means can find the correct clustering more easily. Additionally, we also present an initialisation strategy for k-means called DipInit (dip test based Initialisation), to ensure that k-means converges to a suitable optimum. By clustering features with high dip values first, DipInit ensures that these features have a higher impact on the clustering result. Thus DipInit makes full use of structure-rich features. It is deterministic and based on our assumption that the dip value can determine the importance of a feature for clustering. It is highly compatible with DipExt and makes full use of its characteristics, but it is also very competent on its own.

1.1 Related Work

DipExt finds a subset of the features, extracting those relevant for clustering. Thus, it could be counted as a Subspace Clustering-method (see [12] for an introduction to subspace clustering). The difference between DipExt and many Subspace Clustering methods is that they generally look for a separate subspace for every cluster, while DipExt looks for a common subspace for all clusters. The goal to find one optimal subspace, valid for all clusters, is a more recent trend in subspace clustering (and, therefore, not mentioned in [12]). This trend is sometimes referred to as "cluster-aware" [26] or **"cluster-friendly"** [27] **subspace clustering**/dimensionality reduction. Examples are SubKMeans [17] or FossClu [6]. SubKMeans separates the features into a "cluster-friendly" subspace and one which does not contain features important for clustering [17]. FossClu proceeds similarly and removes unimodal features based on its objective to create an optimal subspace. Autoencoder-based approaches like DCN [27] are at the forefront of this subspace clustering trend. DCN explicitly looks for a "k-means friendly space" via its objective function. DCN, along with DEC [25] and IDEC [7], was one of the first deep learning-based methods combining dimensionality reduction with clustering. They transform the data non-linearly, making the data difficult to interpret and sometimes lead to extremely distorted representations.

The best known "general" dimensionality reduction methods are PCA and ICA. The main difference between this "general" dimensionality reduction and "cluster-friendly" dimensionality reduction is the consideration of cluster structure in the found subspace. While "cluster-friendly" subspace clustering methods try to find a subspace suitable explicitly for clustering, PCA and ICA are general methods used in many areas of Data Mining. Further examples are, t-SNE [13], which tries to preserve the neighbourhoods of data points while projecting to smaller dimensionalities, or UMAP [15], both of which are non-linear.

The dip test [8] was first used in data mining in DipMeans [10], to estimate the number of clusters. In SkinnyDip [16] it is used to cluster noisy data. Recently, it has been generalised to higher dimensions in [22] and [3]. However, in [22] it is not used as a clustering algorithm, but as a criterion of whether clustering makes sense. Both these generalisations effectively apply the one-dimensional dip test with a criterion to select the next data point in a multi-dimensional data set. DipTransformation [19] reshapes the data set to an easier clusterable form making use - as will we - of DipScaling [20]. It transforms the data by changing the position of the data points to make clusters more compact. It does not, however, change the dimensionality of the data. DipScaling is closly related to normalisation methods like Z-transformation or min-max-normalisation. It can be also subsumed under feature weighting methods, like EWKM [9], which assign a "weight" for the importance for clustering to a feature.

As stated, part of our approach is an initialisation strategy for k-means - DipInit. K-means is very sensitive to its initialisation as this determines to which local optimum k-means converges. The most common strategies are k-means++ [1] and random initialisation. They are, like most other strategies (see [2] for an overview), based on random effects. Contrary to that, DipInit is deterministic. It employs the dip test to start clustering on features, where the clusters are well separated, allowing k-means to find better optima.

1.2 Contributions

- We present a deterministic method, DipExt, which extracts the features with the highest level of structure in a data set and rescales them. It creates a condensed version of the data set, containing the information needed for clustering. This version is far easier to cluster for k-means, on which we focus, but also for other methods, as we will show.
- This condensed version has often a considerably smaller dimensionality. It is possible to get an arbitrary dimensionality as specified by the user, or to find it **automatically**. We demonstrate that our approach is stable in regard to choosing the dimensionality.
- Initialisation-strategies for k-means mostly include random components. **DipInit** is an initialisation strategy which is **deterministic** and makes use of the specific properties of the subspace found by DipExt. It puts the main focus on features with high dip-values, helping to converge to a better local optimum.

2 The Algorithm(s)

2.1 The Dip Test - How Much Structure is in a Feature?

As stated, the goal is to find the features with the highest dip value. First, we cover a few basics about the dip test.

The dip test is a statistical tool developed in the1980 s by Hartigan & Hartigan to measure multi-modality. It is a well established and **parameter-free** test to measure the probability of a univariate sample to be multi-modal. Essentially, it tells the user whether a sample has multiple peaks/modes in it, like the one shown in Fig. 3 or if there is only one peak like in Fig. 2. It does this by estimating how much the sample deviates from a uni-modal/uniform distribution. This deviation value - we refer to it as dip value - lies in the range $(0, 0.25]$. A dip value close two 0 indicates that the sample is uni-modal, while a value close to 0.25 suggests the presence of multiple modes. The dip test itself has a runtime of $\mathcal{O}(n)$, but needs sorted input and thus the total runtime is $\mathcal{O}(n \cdot \log(n))$. Due to the space restrictions, it is not possible to describe the dip test in detail. An introduction can be found in [16].

In our case, we search for the feature with the highest dip value in a multi-dimensional data set. Thus, we measure the dip value for a projection vector v. More specific, we project the data D onto the vector v via the scalar product $\cdot$, i.e, $f = D \cdot v$, to get the feature f in the direction of v. This projection f is now univariate and we can compute its dip value. The dip test is scaling invariant, thus, it is enough to compute the dip value for the vectors in the unit sphere with $||v|| = 1$. The question is: How do we find the projection vector in the unit sphere with the maximal dip value?

2.2 DipExt - Extracting Features with Structure

The dip landscape on the unit sphere is highly complicated for most data sets. For both the running example and the Skinsegmentation data set, we can plot this dip landscape as both data sets are 3-dimensional, making their unit sphere 2-dimensional. The dip landscapes are shown in Fig. 5 as a heatmap of the dip

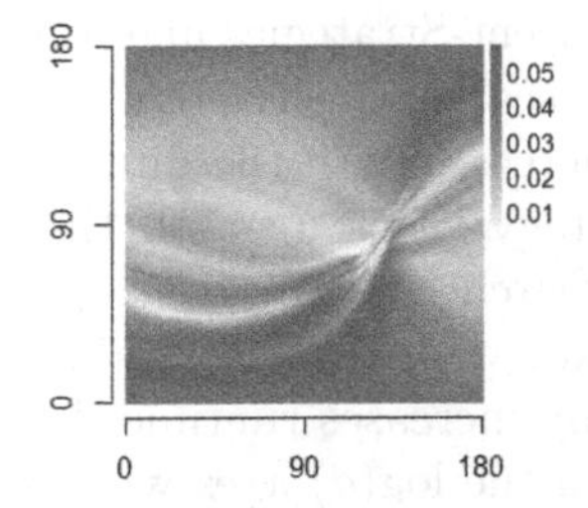

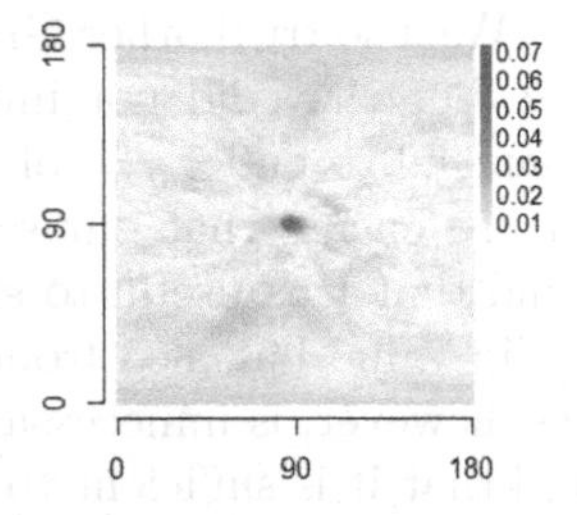

Fig. 5. The dip landscape for the Skinsegmentation data set (left) and the running example (right). The dip values were computed for every vector on the 2D unit sphere.

values of the polar coordinates of the unit vectors. These landscapes are clearly not trivial, thus, finding the maximal dip value dip_{max} of the projection vector v_{max} is not straightforward.

Krause et al. showed in [11] that the dip is continuous almost everywhere and, more importantly, that a gradient for a projection vector can be computed. In a d-dimensional data set D, the dip value of a changing projection vector $v \in \mathbb{R}^d$ changes smoothly and, thus, the partial derivative of the dip value for the projection vector v is given by:[1]

$$\frac{\partial\, dip(v)}{\partial v_i} = \begin{cases} -\frac{i_3-i_1}{n} \cdot \frac{v\cdot(\beta_i\gamma-\gamma_i\beta)}{(v\cdot\gamma)^2}, & \eta > 0 \\ \frac{i_3-i_1}{n} \cdot \frac{v\cdot(\beta_i\gamma-\gamma_i\beta)}{(v\cdot\gamma)^2}, & \eta \leq 0 \end{cases}, \tag{1}$$

with $\cdot$ the scalar product, $\beta = (\beta_1, \ldots, \beta_d) = x_{i_2} - x_{i_1}$, $\gamma = (\gamma_1, \ldots, \gamma_d) = x_{i_3} - x_{i_1}$. The indices i_1, i_2, i_3 give the *modal triangle* and $x_{i_1}, x_{i_2}, x_{i_3} \in D$ are corresponding to the indices. The height h of the modal triangle fulfils $h = 2 \cdot dip$. The value $\eta = i_2 - i_1 - (i_3 - i_1)(v \cdot \beta)/(v \cdot \gamma)$ merely ensures that the gradient points in the correct direction. Please refer to [11] for a thorough explanation of the technical details.

Since the gradient can be computed, we can make use of **Gradient Descent**-approaches to search for v_{max}. Naive Gradient Descent is unsuited for our needs, as it quickly converges to a local optimum, which the dip landscape is full of, and finding v_{max} becomes unlikely. We need to ensure that our search strategy does not stop too soon and keeps on looking, even if the dip value slightly decreases after a step. This is obtained with a momentum term, which keeps the search from changing direction too fast:

$$\begin{aligned} w_t &= m \cdot w_{t-1} + s \cdot \nabla dip(v) \\ v_{t+1} &= v_t + w_t \end{aligned} \tag{2}$$

The momentum m is set to 0.95 and step-size s to 0.1 for all experiments in this paper. Both of these are common values for these parameters (they are, e.g., used in [24]) and worked well for various data sets. $\nabla dip(v)$ is simply the gradient of the dip, i.e., the direction in which the dip value for the projection vector increases, as computed in Eq. (1). As a starting point, we chose the axis with the highest dip value. This momentum-based search strategy is very capable of finding v_{max}. We also tried other Gradient Descent-Strategies like ADAM, NAG or AMSGrad, but they did not improve the search effectively. The area that should be searched to find v_{max}, of course, increases with the dimensionality of the data set. To ensure that our search strategy keeps up with this increased area, we found that it is useful to start the search not only from the axis with the highest dip value but also from other axes with high dip values. Starting from all axes, however, is unnecessary and only increases runtime. Experiments demonstrated that it is sufficient to start from the $\log(d)$ axes with the highest dip values to ensure that a wider area is searched without being to quickly satisfied with a local optimum.

[1] We follow the argument given in [16] in regard to the explicit form of the derivative.

Table 1. The maximal dip values as found by us compared to overall maximal values.

Data set	SKIN	BANK	IRIS	USER	BRST	FRST	MICE	AIBO	PROX	MOTE	DIAT
Brute	0.048	0.089	0.124	0.070	0.065	0.116	0.146	0.096	0.195	0.067	0.165
DipExt	0.046	0.085	0.124	0.065	0.043	0.107	0.076	0.090	0.192	0.064	0.162

Now, when v_{max} is found, we can extract the projection $f_{max} = D \cdot v_{max}$ of the data onto v_{max} from the data D and continue the search on the orthogonal complement of D with respect to v_{max}. Thus, f_{max} is stored as the projection with the highest dip value, i.e., the feature most important for clustering. The Gradient Descent strategy is repeated on the orthogonal complement of v_{max}, which has now a dimensionality of $d - 1$. After enough features are found (we cover in Sect. 2.4 what "enough" entails), they are combined to a new condensed data set.

To show that our strategy is very capable in finding these high dip values, we also searched for them via brute force. We computed the highest dip value for various real world data sets and compared them to the highest dip value as found by us. The results can be seen in Table 1. We got extremely close to the optimal values with our strategy on almost all data sets. So, even if other parameter values or Gradient Descent-strategies are chosen, one could not realistically hope to find better dip values. The two data sets, where our strategy was sub-optimal will be covered later on.

2.3 DipScaling - Scaling Features According to Their Relevance

DipExt has extracted the features with the highest dip values. As an example, the condensed form of the Banknote-Authentication data set is shown in Fig. 6.

The clusters (the data points are coloured according to the ground truth) are perfectly separated. This is impressive, as it is not an easy-to-cluster-data set (as can be seen in the experiments). A scatter plot, as well as the 2D-representation

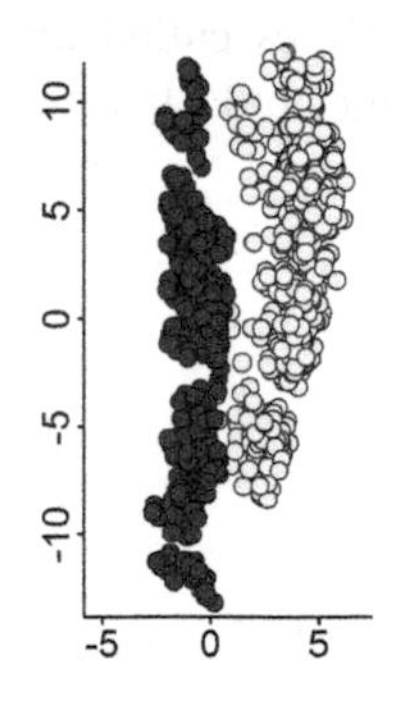

Fig. 6. The 2D extraction of the Banknote-data set.

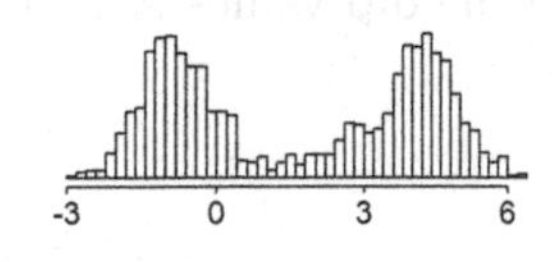

Fig. 7. The first feature.

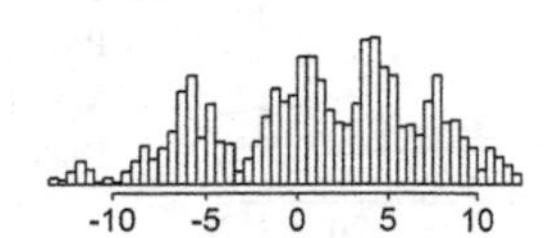

Fig. 8. The second feature.

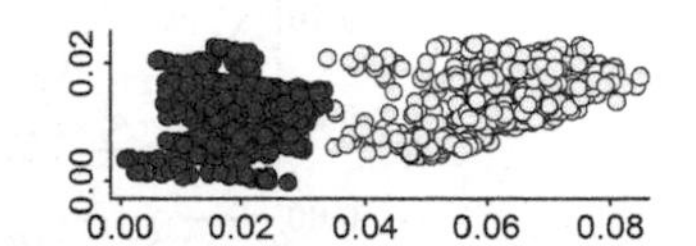

Fig. 9. The extraction after rescaling it with DipScaling.

of the data set with various methods, can be found in the Download, showing the difficulty involved with understanding this data set. Separating the clusters is challenging, but DipExt found a representation, where a correct partitioning is possible. However, even with a perfect initialisation, k-means will not find a good clustering. The axes, i.e., the two features with the highest dip values as found by DipExt (shown as histograms in Fig. 7 and 8) are scaled as they were in the original data set, causing a terrible result for k-means. We can bypass this predicament, by rescaling the axes according to their relevance for clustering. For this, DipScaling [20] is used. The result can be seen in Fig. 9. The data set is now very easy to cluster for k-means.

DipScaling computes the dip values of the axes and rescales the axes with them. Essentially, it executes min-max-normalisation and the new maximum of the axis is its dip value. The effect of this transformation is that the importance of the axes for k-means changes. Axes with very low dip values have a small range, i.e., the values are somewhat similar, thus, they do not influence the computation of the centres in the k-means update step exceedingly. An axis with a high dip value, on the other hand, is scaled rather large causing it to have more influence in determining the centres. This is in line with our assumption of a high dip value signalling a high importance for clustering.

2.4 How Many Features?

A question we left open before was when to stop extracting features. After the feature with the maximal dip value dip_{max} has been extracted, DipExt continues on the orthogonal complement of this feature to find the feature with the next highest dip value. It is possible to continue until DipExt has extracted all available features, but this is clearly unsatisfactory. In Fig. 10, the dip values of the extracted features for various data sets with a dimensionality larger than 25 are plotted. The behaviour is roughly the same for all of them. They start with a somewhat high dip value (the absolute value depends on the data set, thus, dip_{max} can differ greatly), a few features with higher dip values might still be found, before the values start to drop to a somewhat constant base level. One possibility would be to apply the Elbow-method (sometimes called knee-method) to find the point where the dip values start to change, but this would

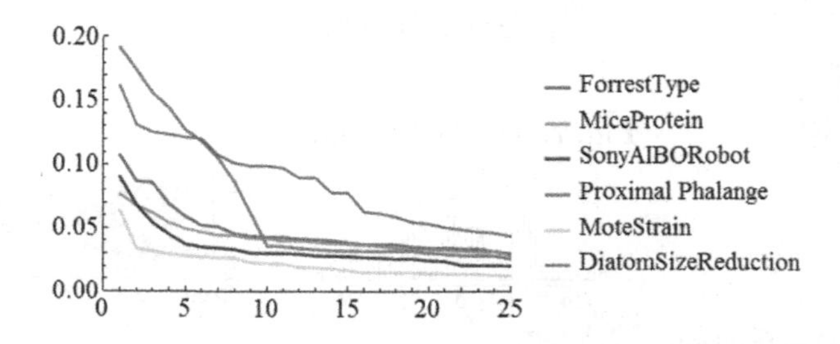

Fig. 10. The DipValues of data sets with dimensionality >25.

mean extracting features only to get their dip value, without using the features later on. This is an overhead, which we are not willing to accept. Instead, a heuristic which we found to work well is to simply search for features until a feature has a dip value smaller than $dip_{max}/2$ and stop after that. Consider the following explanation for this strategy: The features found by DipExt are scaled by DipScaling, thus, a feature with only half the range compared to others will have a small impact on clustering. It will change the position of the cluster centres minimally, if at all. Since its effect is negligible, it might as well be left out. This heuristic manages to keep the dimensionality of the condensed data set small and, furthermore, tailors DipExt to the specific dip landscape of a data set. We discuss it in more detail in Sect. 3.1.

2.5 DipInit - Ensuring the Correct Optimum

In the "Banknote-Authentication"-data set it is enough to apply DipScaling to ensure that k-means converges to a good optimum. In general, however, it is not clear to which optimum k-means converges as it is highly sensitive to the initialisation. The main assumption for this paper is, that a high dip value determines the relevance for clustering. Based on this assumption, we create a new initialisation strategy that is tailored to DipExt.

We have seen before how much cluster information can be contained in a single feature and how much a feature with a high dip value can reveal about the correct boundaries of a cluster (compare Fig. 2 to 3). The obvious conclusion is to use this for the initialisation of k-means by clustering the axes with high dip values first. The basic shape of the clustering is thus determined by the feature with the highest dip value, which "knows" the most about the boundaries of the clusters. Features with a smaller dip value are brought in afterwards when the basic shape is fixed.

We show the effect of this initialisation on the DipExt-version of the Iris-data set (shown in Fig. 11a). Since DipExt reduced the dimensionality to 2, we can plot it directly. Figure 11b shows a histogram of the feature with the highest dip value. This one-dimensional feature is now clustered first. It is sorted, split into equally large parts all containing $\frac{n}{k}$ many data points (equal-frequency binning) and then k-means is executed. To this 1D-data set, the axis with the second-highest dip value is now added. We now have a 2D-data set with the labels from the clustering of the 1D-data set (Fig. 11c). K-means resumes and we get the result as shown in Fig. 11d. If there were more axes, they would be iteratively added and, each time, k-means would resume.

The basic shape of the clustering is fixed, when the first axis is clustered (Fig. 11c). Adding the second axis leads to a few changes at the border of the red and yellow cluster, as the Voronoi cells of k-means slightly change. The borders of the Voronoi cells are included in Fig. 11c and Fig. 11d. The final result gets us very close to the ground truth shown in Fig. 11a. As we wanted, the features with the highest dip value determine the principal form of the clustering, while less structured features, which are added to the data consecutively, have a smaller influence on the clustering and mainly improve on details. An advantage of

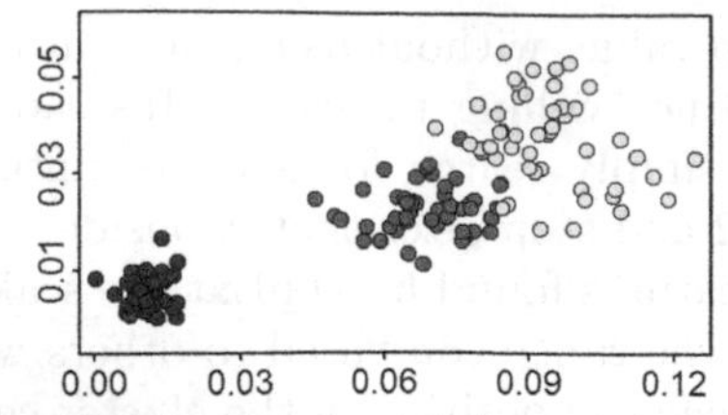

(a) The 2D reduction of the Iris data set. Colours correspond to the ground truth.

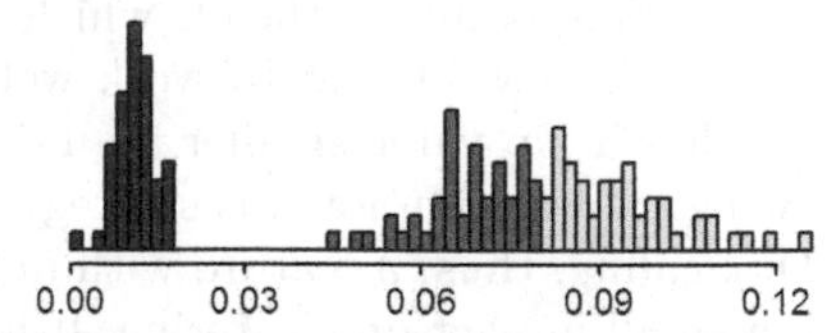

(b) The feature with the highest dip value is split in k parts with equally many data points, and then k-means is executed.

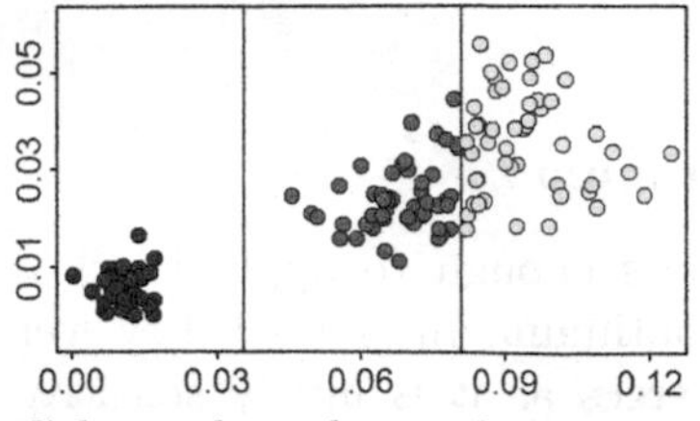

(c) Colours show the results if only the first axis as shown in (b) is clustered.

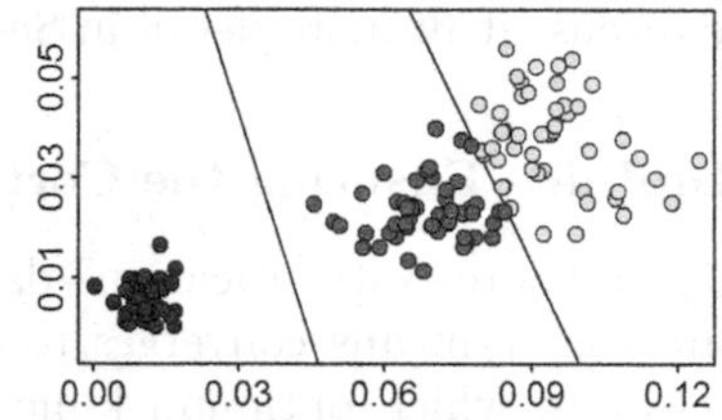

(d) Running k-means on (c) with the preliminary clusters leads to the final clustering.

Fig. 11. The DipInit initialisation strategy on the reduced Iris-data set.

DipInit is that it is **deterministic**. Contrary to most initialisation strategies for k-means, which contain some random elements (see [2] for an overview), every step of DipInit is performed without any decision left to chance (Pseudo code for DipInit can be found in the Download). A further advantage of this initialisation is that it makes full use of the characteristics of DipExt. DipInit needs the features with the highest dip values, which DipExt extracted. It can be used as a stand-alone initialisation for k-means, where it is more than competitive (see Download for details), but it truly shines in combination with DipExt.

3 Experiments

Sourcecode: The source code, datasets, labels, parameters for our and all compared methods can be found under the following links:

https://doi.org/10.6084/m9.figshare.12063252.v1
https://dm.cs.univie.ac.at/research/downloads

With DipExt we extract the subset of the features with the highest dip values. DipScaling rescales these features into a more fitting range. DipInit ensures that k-means converges to an optimum, which makes sense from the point of our main assumption. The question is how well this approach fares in comparison to other methods. To answer this, we compared our approach on 11 data sets to a wide range of algorithms. We chose the compared methods from cluster-aware subspace clustering (SubKmeans, FossClu, DEC, IDEC), transformation approaches

Table 2. Experimental results given in Adjusted Mutual Information (AMI) [23]. DipExt in combination with DipInit. For random methods the average of 100 runs is shown. Correct k is always given. Cases which were aborted after 1h or failed due to non-trivial implementation bugs are marked with —. Best result bold.

Data set	SKIN	BANK	IRIS	USER	BRST	FRST	MICE	AIBO	PROX	MOTE	DIAT
Orig. dim.	3	4	4	5	10	27	70	77	80	84	345
Red. dim.	2	2	2	4	9	5	7	5	8	2	14
DipExt	**0.48**	**0.81**	**0.88**	**0.61**	**0.50**	**0.83**	**0.54**	**0.60**	**0.49**	**0.45**	**0.80**
k-means	0.02	0.03	0.68	0.27	0.27	0.68	0.29	0.35	0.45	0.30	0.76
SubKMeans	0.01	0.01	0.65	0.21	0.41	0.49	0.34	0.00	**0.49**	0.40	0.76
FossClu	0.27	0.44	0.74	0.49	0.29	0.55	0.26	0.35	—	—	—
DEC	0.11	0.08	0.61	0.20	0.34	0.54	0.22	0.25	0.42	0.12	0.44
IDEC	0.01	0.11	0.58	0.22	0.33	0.50	0.22	0.20	0.34	0.12	0.58
DipTrans.	0.39	0.68	0.83	0.50	0.46	0.68	0.46	0.16	0.48	0.26	0.67
EWKM	0.03	0.08	0.78	0.24	0.28	0.08	0.30	0.01	0.45	0.12	0.70
Z-trans.	0.01	0.01	0.63	0.20	0.44	0.51	0.33	0.23	0.47	0.35	0.74
min-max n.	0.02	0.02	0.69	0.27	0.45	0.70	0.35	0.35	0.48	0.35	0.71
SkinnyDip	—	0.34	0.55	0.29	0.24	0.48	0.19	—	0.45	0.00	—
DipMeans	—	0.25	0.55	0.00	0.00	0.00	0.00	0.00	0.45	0.00	0.75
PCA	0.01	0.01	0.63	0.21	0.44	0.50	0.32	0.21	0.46	0.35	—
ICA	0.24	0.07	0.55	0.16	0.37	0.63	0.48	0.32	0.43	0.37	0.79
t-SNE	—	0.73	0.34	0.05	0.01	0.00	0.04	0.00	0.00	0.24	0.05
UMAP	0.00	0.40	0.75	0.37	0.29	0.75	0.49	0.13	0.46	0.37	0.78
DBSCAN	—	0.20	0.30	0.01	0.27	0.06	0.17	0.00	0.00	0.08	0.25
EM	0.19	0.00	0.84	0.45	0.27	0.71	0.29	0.34	0.45	0.06	0.75
Spec. Clust.	—	0.03	0.59	0.28	0.41	0.74	0.43	0.04	0.47	0.00	0.71
SingleLink	—	0.12	0.70	0.26	0.14	0.41	0.18	0.04	0.03	0.03	0.02
k-means++	0.02	0.03	0.69	0.27	0.16	0.66	0.30	0.39	0.45	0.30	0.74

(DipTransformation, EWKM, Z-transformation, min-max-normalisation), dip-based methods (SkinnyDip, DipMeans) and general dimensionality reduction methods (PCA, ICA, t-SNE, UMAP) and standard methods (DBSCAN, EM, Spectral Clustering, SingleLink, k-means++). We compare to a wide range of methods, to show that the results we obtain are highly competitive. This is also the reason why standard methods like DBSCAN are included. Most of the approaches (DEC, SubKmeans, t-SNE, ...) are combined with k-means, but for some data sets, k-means might simply be the wrong framework and, e.g., DBSCAN a far better choice. To show that our results are competitive, these very different methods are also included.

The data sets - SkinSegmentation (SKIN), Banknote-Authentication (BANK), Iris (IRIS), Userknowledge (USER), BreastTissue (BRST), Forrest-type (FRST), Mice Protein (MICE), SonyAIBORobot (AIBO), ProximalPhalanxOutlineAgeGroup (PROX), MoteStrain (MOTE) and DiatomSizeReduction (DIAT) - are all publicly available. They range greatly in size (BRST n = 106, SKIN n = 245057) and dimensionality (SKIN d = 3, DIAT d = 345), which shows that our approach is not per se restricted in these regards. Links to the data sets can be found with the Source Code. The results of these experiments can be seen in Table 2. As an evaluation metric to measure the quality of clustering we used Adjusted Mutual Information (AMI)[23] with the max-normalisation. AMI ranges from 1.0 (perfect clustering result) to 0.0 (purely random cluster assignments). Please keep in mind, that different normalisations can lead to (slightly) different results. Apart from AMI, Normalized Mutual Information (NMI) is also frequently used to evaluate clustering results. However, using NMI instead of AMI has little influence. The values are very similar, with DipExt again being the best method on all data sets.

The results in Table 2 paint a very clear picture. Our approach is the method with the best results on all data sets. It is particularly noteworthy for us that we succeeded in improving the quality of clustering for k-means by more than 0.26 in AMI on average. BANK, as an example, was improved from 0.03 to 0.81 in AMI. We managed to extract a subspace where clusters can be far easier found. The condensed data set makes the information relevant for clustering more easily accessible than SubKMeans, PCA, DipTransformation, ... were able to do so. Furthermore, the condensed version of the data set has now a considerably smaller dimensionality: MOTE has now 2 instead of 84 dimensions, DIAT 14 instead of 345.

Parameters: Due to the large number of compared methods, listing all parameters would be rather expansive. They can instead be found in the Download.

As we have seen, our approach reaches very impressive results. Now, there are a few topics which we wish to discuss in regard to the stability of our dimensionality-criterion, runtime and the applicability of DipExt to methods besides k-means.

3.1 How Much Structure Is Enough Structure?

As stated, we extract features until one is found with a dip value smaller than $dip_{max}/2$. We changed this threshold to $dip_{max}/3$ and $dip_{max}/1.5$ to see what the effect of this threshold is and found that the results were barely influenced by it (see Table 3). Only three data sets had a change ≥ 0.02 in AMI, and they would essentially still be among the best clustering results. We can see from Fig. 12 that the main difference in these thresholds was the number of features DipExt extracts. Here comes another advantage of DipScaling into effect, which shows how well it fits DipExt and DipInit. DipScaling rescales the axes depending on their dip values, thus, features with low dip values have a smaller impact on

Table 3. Comparing stopping-criterion $dip_{max}/2$ to /3 and /1.5.

	SKIN	BANK	IRIS	USER	BRST	FRST	MICE	AIBO	PROX	MOTE	DIAT
/1.5	0.48	0.81	0.88	0.61	0.54	0.83	0.46	0.54	0.48	0.45	0.81
/2	0.48	0.81	0.88	0.61	0.50	0.83	0.54	0.60	0.49	0.45	0.80
/3	0.48	0.81	0.86	0.61	0.50	0.81	0.54	0.60	0.49	0.45	0.79

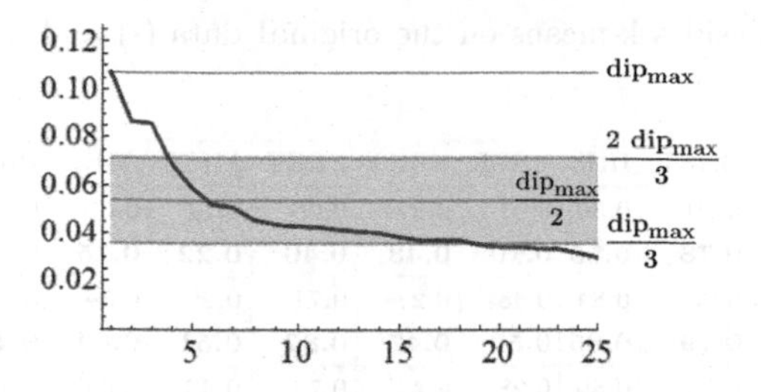

Fig. 12. The dip values for the 25 features with maximal dip values for FRST.

clustering, making DipExt more stable in regards to the number of features extracted. Furthermore, DipInit ensures that the features with high dip values are clustered first, fixing the principal shape of clustering, such that features with low dip values have a smaller impact on clustering. The combination of these effects makes our approach extremely stable in regard to the number of features extracted. It makes barely a difference if there are 5 or 15 features extracted by DipExt. The result will be almost the same.

3.2 Runtime

A more detailed calculation of the runtime as well as the pseudocode can be found in the Download. The results show an $\mathcal{O}(n \cdot \log(n))$-dependency on the number of data points n, as well as an $\mathcal{O}(d \cdot \log(d))$-dependency on the dimensionality of the data set d for our approach. This behaviour can also be seen in Fig. 13, showing that these estimates are correct. We also find that DipExt and DipInit are competitive to many state-of-the-art methods, even though the code is currently not optimised in regard to runtime. Please see the Download for details.

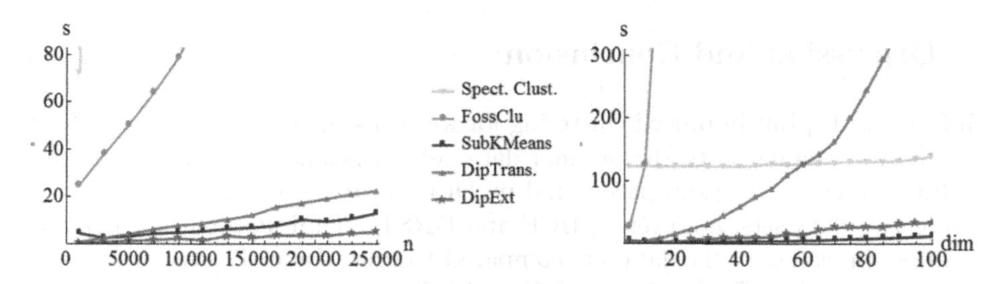

Fig. 13. Runtime relative to the dimensionality of the data set d (right). Dimensionality ranging from 5 to 100 with a data set size of ≈1.500 data points. Runtime relative to the data set size n (left). Data size ranging from 1.000 to 25.000 data points with a constant dimensionality of 5.

Table 4. Methods besides k-means on the original data (-) and in combination with DipExt (+).

Data set	SKIN	BANK	IRIS	USER	BRST	FRST	MICE	AIBO	PROX	MOTE	DIAT
DBSCAN	–	0.20	0.30	0.01	0.27	0.06	0.17	0.00	0.00	0.08	0.25
DBSCAN+	–	**0.78**	**0.55**	**0.10**	**0.43**	**0.40**	**0.22**	**0.15**	**0.45**	**0.28**	**0.76**
EM	0.19	0.00	0.84	0.45	0.27	0.71	0.29	0.34	**0.45**	0.06	0.75
EM+	**0.50**	**0.79**	**0.85**	**0.52**	**0.47**	**0.82**	**0.51**	**0.64**	**0.45**	**0.34**	**0.76**
Spec. Clust.	–	0.03	0.59	0.28	0.41	0.74	0.43	0.04	0.47	0.00	**0.71**
Spec. Clust.+	–	**0.89**	**0.68**	**0.53**	**0.52**	**0.81**	**0.52**	**0.56**	**0.48**	**0.12**	0.70
SingleLink	–	0.12	0.70	0.26	0.14	0.41	0.18	0.04	0.03	0.03	**0.02**
SingleLink+	–	**0.90**	**0.78**	**0.49**	**0.37**	**0.48**	**0.35**	**0.09**	**0.48**	**0.39**	0.01

3.3 Methods Besides K-Means

DipExt has been created with k-means in mind. However, there is no reason why it should not also be used in combination with other methods. To test whether it is compatible, we applied 4 of the most often used clustering algorithms to the subset of features extracted by DipExt. The results (shown in Table 4) are clear. All methods profit massively from DipExt: 38 of 41 results improved showing an average increase in clustering quality of 0.23 in AMI. We want to emphasise that some of these results (e.g. Spectral Clustering on BANK with 0.89 in AMI) are now better than all other listed in Table 2. This shows that DipExt is not bound to k-means. Furthermore, it emphasises our claim that DipExt creates a condensed version of the data set containing all the information relevant for clustering. It removes distracting features containing mostly noise information which are detrimental to clustering, i.e., features with uni-modal distributions. The extracted features contain clearer borders between clusters making the separation of them easier. Thus, for a wide range of methods, the subset of features found by DipExt is easier to cluster than the original data set.

4 Discussion and Conclusion

DipExt and DipInit improved clustering for k-means by more than 0.26 in AMI on average. Compared to the original data sets, k-means in combination with DipInit can cluster the subspace found by DipExt far better.

For two data sets, however – MICE and BRST – DipExt could not find the maximal dip values of the data sets compared to brute force. The result of brute force improved MICE from 0.55 to 0.62 in AMI and did not affect BRST. Using brute force on the other data sets improved some of their clustering results as well, if only by a small value. These observations are further indications of the strong connection between dip value and success of clustering. The higher the dip values are, the better the clustering results will be. The clusters can now be better separated, corroborating our assumption. However, the improvement of brute force compared to DipExt is only very small, as DipExt is already highly capable of finding the features with the maximal dip values as we have seen in Sect. 2.2/Table 1. At best, only a slightly better dip value can be found. This shows that DipExt is very close to the optimal subspace, which can be extracted in our framework on these data sets.

DipExt can be considered as a "cluster-friendly subspace clustering"-method. Many of these methods, though, focus only on one method, mostly k-means, and try to find a subspace suited for it. While DipExt is also highly compatible with k-means, especially with the DipInit initialisation, it is not limited to it, contrary to, e.g., SubKMeans. DipExt extracts features and transforms them into a low-dimensional representation of the data, which is far easier to cluster for a wide range of methods. We demonstrated this for various standard methods in Sect. 3.3/Table 4, where we showed that their results could be greatly improved. These results are now so good that they often outperform other "cluster-friendly subspace clustering"-methods. Considering the difference in the approach of the standard methods, we conclude that this improvement also applies to a wide range of other clustering approaches. This shows that DipExt creates a "clustering-friendly subspace" that is not limited to one method.

There are, of course, limitations to our method. Multimodal or interlocking clusters pose a problem. It is also possible to create pathological cases where the clusters are positioned such that no univariate projection reveals anything relevant about the cluster structure. We have further found that very small datasets (<100 data points) are not suitable for DipExt. It seems that this is not enough data points to make the dip test statistically significant. For many data sets, though, DipExt and DipInit are interesting tools for clustering. DipInit is one of the few deterministic initialisation strategies for k-means. K-means is probably the most often used clustering algorithm, widely applied in all areas of data-driven research. However, since the results of k-means depend on the initialisation, one either runs it once and might have a mediocre result, or runs it multiple times and has to choose one of them. With DipInit, k-means finds one optimum which is in accord with our assumptions and which most likely, following our experiments, is well above average. DipExt is very useful for exploratory data analysis. It extracts only those features where the cluster boundaries are well-defined, creating a condensed data set which is far easier to understand (see IRIS in Fig. 11a or BANK in Fig. 9). This transformation can now be far better clustered, not only by k-means but by various other methods as well. DipExt may not be limited to clustering. As DipExt allows for a better separation of clusters, combining it with methods for estimating the number of clusters is a possible application. One could also consider the combination of DipExt with classification-methods. We intend to explore possible applications for DipExt and potential improvements further in the future.

References

1. Arthur, D., Vassilvitskii, S.: k-means++: the advantages of careful seeding, SODA (2007)
2. Celebi, M., Kingravi, H., Vela, P.: A comparative study of efficient initialisation methods for the K-Means clustering algorithm. Expert Syst. Appl. **40**(1), 200–210 (2013)
3. Chronis, P., Athanasiou, S., Skiadopoulos, S.: Automatic clustering by detecting significant density dips in multiple dimensions. In: ICDM (2019)
4. Dempster, A.P., Laird, N.M., Rubin, D.B.: Maximum-Likelihood from incomplete data via the EM algorithm. J. Royal Stat. Soc. **39**(1), 1–22 (1977)
5. Ester, M., Kriegel, H.-P., Sander, J., Xu, X.: A density-based algorithm for discovering clusters in large spatial databases with noise. In: KDD (1996)
6. Goebl, S., He, X., Plant, C., Böhm, C.: Finding the optimal subspace for clustering. In: ICDM (2014)
7. Guo, X., Gao, L., Liu, X., Yin, J.: Improved deep embedded clustering with local structure preservation. In: IJCAI (2017)
8. Hartigan, J.A., Hartigan, P.M.: The dip test of unimodality. Ann. Stat. **131**, 70–84 (1985)
9. Jing, L., Ng, M.K., Huang, J.Z.: An entropy weighting k-means algorithm for subspace clustering of high-dimensional sparse data. In: TKDE (2007)
10. Kalogeratos, A., Likas, A.: Dip-means: an incremental clustering method for estimating the number of clusters. In: NIPS (2012)
11. Krause, A., Liebscher, V.: Multimodal projection pursuit using the dip statistic, Preprint-Reihe Mathematik (2005)
12. Kriegel, H.P., Kröger, P., Zimek, A.: Clustering high-dimensional data: a survey on subspace clustering, pattern-based clustering, and correlation clustering. In: TKDD (2009)
13. Maaten, L., Hinton, G.: Visualizing data using t-SNE. J. Mach. Learn. Res. (2008)
14. MacQueen, J.B.: Some methods for classification and analysis of multivariate observations. In: Berkeley Symposium on Math. Stat. and Prob. (1967)
15. McInnes, L., Healy, J., Melville, J.: Uniform manifold approximation and projection for dimension reduction. arXiv:1802.03426 (2018)
16. Maurus, S., Plant, C.: Skinny-dip: clustering in a sea of noise. In: KDD (2016)
17. Mautz, D., Ye, W., Plant, C., Böhm, C.: Towards an optimal subspace for k-means. In: KDD (2017)
18. Ng, A., Jordan, M., Weiss, Y.: On spectral clustering: analysis and an algorithm. In: NIPS (2002)
19. Schelling, B., Plant, C.: DipTransformation: enhancing the structure of a dataset and thereby improving clustering. In: ICDM (2018)
20. Schelling, B., Plant, C.: Dataset-transformation: improving clustering by enhancing the structure with DipScaling and DipTransformation. In: KAIS (2019)
21. Sibson, R.: SLINK: an optimally efficient algorithm for the single-link cluster method. Comput. J. **16**(1), 30–34 (1973)
22. Siffer, A., Fouque, P.A., Termier, A., Largouet, C.: Are your data gathered? In: KDD (2018)
23. Vinh, N.X., Bailey, J.: Information theoretic measures for clusterings comparison: variants, properties, normalization and correction for chance. JMLR **11**, 2837–2854 (2011)

24. Wu, H., Gu, X.: Max-Pooling dropout for regularization of convolutional neural networks. In: ICONIP (2015)
25. Xie, J., Girshick, R., Farhadi, A.: Unsupervised deep embedding for clustering analysis. In: ICML (2016)
26. Yang, B., Fu, X., Sidiropoulos, N.: Learning from hidden traits: joint factor analysis and latent clustering. IEEE Trans. Signal Process. (2017)
27. Yang, B., Fu, X., Sidiropoulos, N., Hong, M.: Towards K-means-friendly spaces: simultaneous deep learning and clustering. In: ICML (2017)

Simple, Scalable, and Stable Variational Deep Clustering

Lele Cao(✉), Sahar Asadi(✉), Wenfei Zhu, Christian Schmidli, and Michael Sjöberg

King Digital Entertainment, Activision Blizzard Group, Stockholm, Sweden
{lele.cao,sahar.asadi,wenfei.zhu,christian.schmidli, michael.sjoberg}@king.com

Abstract. Deep clustering (DC) has become the state-of-the-art for unsupervised clustering. In principle, DC represents a variety of unsupervised methods that jointly learn the underlying clusters and the latent representation directly from unstructured datasets. However, DC methods are generally poorly applied due to high operational costs, low scalability, and unstable results. In this paper, we first evaluate several popular DC variants in the context of industrial applicability using eight empirical criteria. We then choose to focus on variational deep clustering (VDC) methods, since they mostly meet those criteria except for simplicity, scalability, and stability. To address these three unmet criteria, we introduce four generic algorithmic improvements: initial γ-training, periodic β-annealing, mini-batch GMM (Gaussian mixture model) initialization, and inverse min-max transform. We also propose a novel clustering algorithm S3VDC (simple, scalable, and stable VDC) that incorporates all those improvements. Our experiments show that S3VDC outperforms the state-of-the-art on both benchmark tasks and a large unstructured industrial dataset without any ground truth label. In addition, we analytically evaluate the usability and interpretability of S3VDC.

Keywords: Deep clustering · Deep embedding · Variational deep clustering · Gaussian mixture model · User profiling

1 Introduction

Clustering algorithms aim to group a set of data points into clusters such that: 1) points within each cluster are similar, and 2) points from different clusters are dissimilar. Traditional clustering methods like k-means and Gaussian mixture models (GMM) are, however, highly dependent on the input data; hence, they are ineffective when the input dimensionality is very high. Formerly, feature extraction methods such as dimension reduction and representation learning were extensively applied as a step prior to clustering. Recent works show that optimizing clustering jointly with feature extraction using deep neural networks (DNNs) yields superior results [12,15,21]. This paradigm is usually termed deep

F. Hutter et al. (Eds.): ECML PKDD 2020, LNAI 12457, pp. 108–124, 2021.
https://doi.org/10.1007/978-3-030-67658-2_7

clustering (DC). Despite the high evaluation scores achieved by DC approaches reported on a few benchmark datasets (Table 1), DC has not shown consistent success on large dynamic industrial datasets. We empirically discovered that the following properties are crucial to the *industrial applicability* of DCs. This motivates the active adoption of unsupervised DC algorithms for exploring large-scale industrial datasets.

P1. Truly unsupervised: unavailability of ground-truth label should not significantly compromise the credibility of model performance.
P2. Access to latent embedding: this is typically important to understand the feature space. In addition, the learned embedding can be utilized as input to other machine learning (ML) applications.
P3. Indicator of the optimal number of clusters: there should be a natural way to determine the best number of clusters without any label guidance.
P4. Generative: the model allows generating samples from the underlying data distribution to enable tasks such as data augmentation and cluster interpretation.
P5. Learnable cluster weights: the optimal cluster weights (the relative likelihood of each cluster) should be learned without enforcing any constant prior.
P6. *Simplicity*: once implemented (in a preferably end-to-end manner, and independent of any pretrained model or sequential steps) and verified, the manual operational cost stays constantly and continuously at a low level.
P7. *Scalability*: the computational complexity and memory efficiency remain largely invariant to both the size of datasets and the number of clusters.
P8. *Stability*: favorable results with low variance should be guaranteed from different trials with random initialization so that the frequent model iteration process (subsequent training on newly accumulated data) becomes easier and more efficient.

We examine the state-of-the-art DC methods on P1–P5 in Sect. 2 qualitatively and show that variational deep clustering (VDC) meets these properties the best. Section 3 presents a unified introduction to the mainstream VDC algorithms. In Sect. 4, we discuss four common problems of VDCs that violate P6–P8, and we propose generic solutions to address each of those problems. Finally, Sect. 5 presents a comprehensive experimental and exploratory analysis. The main contributions of this work are:

- We identify essential properties for the industrial applicability of DC. In addition, we present a qualitative evaluation of state-of-the-art DC algorithms and recommend VDC model family as the most suitable approach.
- We propose practical and generic solutions to address the *simplicity*, *scalability*, and *stability* problems in VDC baselines, resulting in an industrial-friendly VDC algorithm: S3VDC (simple, scalable, and stable VDC)[1].
- Our experimental results show that S3VDC outperforms baseline VDCs on scalability and stability. Furthermore, our exploratory analysis demonstrates how other industrial "nice-to-have" properties can be fulfilled.

[1] Source code: https://github.com/king/s3vdc.

Table 1. Qualitative evaluation of DC algorithms with respect to properties P1–P8 (Sect. 1). "✓" means that the method has the property by design. "✗" denotes that the model prohibits the property. "**H**" (high) and "**L**" (low) denote the required effort to satisfy the property. "-" indicates cases where we could not conclude with reasonable effort. NMI (normalized mutual information) is a supervised clustering metric.

DC methods vs. preferred properties		*P1	*P2	*P3	*P4	*P5	*P6	*P7	*P8	MNIST Dataset	
										Accuracy	NMI
* DNN based DC	DEC [24]	✓	✓	H	✗	H	L	H	H[a]	0.84	0.80
	DCEC [12]	✓	✓	H	✗	H	L	H	L[a]	0.88	0.88
	JULE [25]	✓	✓	H	✗	H	H	H	–	–	0.91
	DEPICT [10]	✓	✓	H	✗	H	H	H	H	0.96	0.91
	IMSAT [14]	L	L	H	✗	H	L	L	H	0.98	–
* GAN based DC	InfoGAN [4]	L	L	H	✓	H	L	✓	H	0.89 [21]	0.86 [21]
	Sub-GAN [18]	✓	✓	H	✓	H	L	H	H	0.85	–
	ClusterGAN [21]	✓	L	H	✓	H	L	✓	H	0.95	0.89
	ClusterGAN [9]	✓	L	H	✓	H	H	H	–	0.96	0.92
* VAE based DC (VDC)	M1+M2 [16][b]	L	L	L	✓	H	L	✓	H	–	–
	M1+2 [8][c]	✓	✓	L	✓	H	L	L	H	0.83	0.80
	DLGMM [22][b]	L	✓	L	✓	H	H	L	H	–	–
	GMVAE [7]	✓	✓	L	✓	H	H	H	H	0.82	–
	VaDE [15]	✓	✓	L	✓	✓	H[a]	H[a]	H[a]	0.94	–

[a] Evaluated using the open-source implementation of the respective method.
[b] Strictly, these methods can not be regarded as unsupervised DC approaches.
[c] We use "M1 + 2" to refer to the simplified and unsupervised VDC [8] adapted from the original M1 + M2 [16].

2 Review and Evaluation of Deep Clustering (DC)

DC models fall into three primary categories according to their network architecture: DNN, generative adversarial network (GAN), and variational auto-encoder (VAE) based DC, among which GAN and VAE are deep generative approaches; thus, they can capture high dimensional probability distributions, impute missing data, and deal with multi-modal outputs. Table 1 presents a qualitative evaluation of the state-of-the-art DC methods with respect to properties P1–P8.

The DNN-based category includes DC approaches that apply multi-layer perceptron (MLP), deep belief network (DBN), or convolutional neural network (CNN). Deep embedded clustering (DEC) [24] is the first well-known DC method. It learns the representations sequentially using stacked auto-encoder (AE), initializes clusters with k-means, and finetunes the encoder with a clustering loss. [12] proposed DCEC, which improves DEC by preserving its decoder and employing CNN as a feature extractor. Among other influential methods in this category are JULE [25], DEPICT [10], and IMSAT [14]. Lack of P4 (generating new samples) and P5 (modeling cluster weights) is the primary issue with DNN-based DCs due to their underlying architecture (see Table 1).

GAN-based methods contain a system of two DNNs (generator and discriminator) that play a zero-sum game [11]. InfoGAN [4], a highly cited GAN-based

DC baseline, optimizes the mutual information of latent variables constructed by a mixture of Gaussian (MoG). In [18], a novel clustering model, Sub-GAN, was proposed. Recently, two GAN-based DC approaches [9,21], both named after ClusterGAN, were proposed: [21] considers discrete-continuous mixtures to sample noise variable, while [9] utilizes conditional entropy minimization as clustering loss combined with a few tricks to enable self-paced and balanced training. GAN-based methods may suffer from catastrophic model collapse as well as unstable training (P8). Moreover, they require additional discriminator networks (P6). They also struggle to determine the optimal number of clusters and their weights (P3, P5).

VAE is a generative variant of AE that forces latent variables to follow a predefined distribution. In Table 1, M1 + M2 [16] and DLGMM [22] are two mixture VAEs, which may be applied to unsupervised clustering tasks with modest adaptations. M1 + M2 tackles semi-supervised classification problems using hierarchical stochastic layers that are fundamentally cumbersome to train. DLGMM uses MoG to approximate VAE posterior but does not model the cluster variable. Recent work directly combines VAE with GMMs, producing a few well-known VAE-based DC (a.k.a. VDC) approaches: GMVAE [7], M1 + 2 [8], and VaDE [15]. VDC methods satisfy more industrial properties compared to DNN and GAN based ones except for simplicity, scalability, and stability (P6–P8). The next sections focus on making VDC approaches comply with P6–P8.

3 Variational Deep Clustering (VDC) Algorithms

VDCs try to generatively model an unlabeled dataset $\mathbf{X} = \{\mathbf{x}^{(n)}\}_{n=1}^{N}$ under the constraint of C clusters; N denotes the total number of samples in $\mathbf{X}$. For the sake of conciseness, we use the general term $\mathbf{x} \in \mathbb{R}^D$ to denote any D-dimensional data sample when we walk-through the VDC methods mentioned in Sect. 2: M1 + 2 [8], GMVAE [7], and VaDE [15].

3.1 Generative Processes

The generative step of VDCs starts with sampling a cluster c from a categorical distribution $p(c)$ parameterized by $\boldsymbol{\pi} \in \mathbb{R}_+^C$:

$$c \sim Cat(\boldsymbol{\pi}), \qquad \text{s.t.:} \sum_{c=1}^{C} \pi_c = 1 \;, \tag{1}$$

where π_c is the prior probability (i.e. cluster weights) for the c-th cluster. Note that VaDE treats π_c as trainable GMM parameters while others set them uniformly to C^{-1}. Next, a latent vector $\mathbf{z}$ is chosen from $p(\mathbf{z})$ via one of

$$\text{M1+2:}\ \ \mathbf{z} \sim \mathcal{N}(0, I), \tag{2}$$

$$\text{GMVAE:}\ \ \mathbf{z}' \sim \mathcal{N}(0, \mathbf{I}), \mathbf{z} \sim \mathcal{N}(\boldsymbol{\mu}_c(\mathbf{z}'), \boldsymbol{\sigma}_c^2(\mathbf{z}')), \tag{3}$$

$$\text{VaDE:}\ \ \mathbf{z} \sim \mathcal{N}(\boldsymbol{\mu}_c, \boldsymbol{\sigma}_c^2 \mathbf{I}), \tag{4}$$

where M1 + 2 uses a single multivariate Gaussian $\mathcal{N}(\cdot)$, which may lower the upper bound of its performance; $\mathbf{I}$ is an identity matrix; $\boldsymbol{\mu}_c$ and $\boldsymbol{\sigma}_c^2$ are the mean and variance of the Gaussian for the c-th cluster, respectively. VaDE initializes those as global parameters with a GMM. In GMVAE, if we marginalize out $\mathbf{z}'$, $\mathbf{z}$ is an arbitrary distribution parameterized by functions $\boldsymbol{\mu}_c(\cdot)$ and $\boldsymbol{\sigma}_c^2(\cdot)$ approximated by neural networks (NNs). Thus, it does not scale well to large C values due to the larger network size required to represent each Gaussian.

VDCs then compute $\boldsymbol{\mu}_x$ (and also $\boldsymbol{\sigma}_x^2$ if $\mathbf{x}$ is real-valued) via functions approximated by density NNs parameterized by $\boldsymbol{\theta}$, i.e. $f_{\boldsymbol{\theta}}(\mathbf{z}; c)$ for M1+2 or $f_{\boldsymbol{\theta}}(\mathbf{z})$ otherwise. Finally, a sample $\mathbf{x}$ is selected from:

$$\mathbf{x} \sim \mathcal{N}(\boldsymbol{\mu}_x, \boldsymbol{\sigma}_x^2 \mathbf{I}) \quad \text{or} \quad \mathcal{B}(\boldsymbol{\mu}_x), \tag{5}$$

where $\mathcal{B}(\cdot)$ is a Bernoulli distribution parameterized by $\boldsymbol{\mu}_x$ when $\mathbf{x}$ is binary. For simplicity, we proceed with the assumption that sample $\mathbf{x}$ is binary.

3.2 Variational Lower Bound Objectives

VDCs are usually optimized through DNNs, so the stochastic gradient variational Bayes (SGVB) estimator and the reparameterization trick [6] can be used to maximize the log-evidence lower bound (ELBO). As shown previously, M1+2 considers the generative model $p(\mathbf{x}, \mathbf{z}, c) = p(\mathbf{x}|\mathbf{z}, c)p(\mathbf{z})p(c)$, and the posterior $p(\mathbf{z}, c|\mathbf{x})$ is approximated with a tractable mean-field distribution $q(\mathbf{z}, c|\mathbf{x}) = q(\mathbf{z}|\mathbf{x})q(c|\mathbf{x})$. So, its ELBO is

$$\begin{aligned}\mathcal{L}_{\text{M1+2}} &= \mathbb{E}_{q(\mathbf{z},c|\mathbf{x})}[\ln(p(\mathbf{x}, \mathbf{z}, c)/q(\mathbf{z}, c|\mathbf{x}))] \\ &= \mathbb{E}_{q(\mathbf{z},c|\mathbf{x})}[\ln p(\mathbf{x}|\mathbf{z}, c)] - \mathcal{D}(q(c|\mathbf{x})||p(c)) - \mathcal{D}(q(\mathbf{z}|\mathbf{x})||p(\mathbf{z})),\end{aligned} \tag{6}$$

where the first term is the negative reconstruction loss, and the terms followed are Kullback-Leibler (KL) divergence functions, noted as $\mathcal{D}(\cdot||\cdot)$, which regularize the categorical and Gaussian distributions. Similarly, VaDE uses the generative model $p(\mathbf{x}, \mathbf{z}, c) = p(\mathbf{x}|\mathbf{z})p(\mathbf{z}|c)p(c)$, hence the ELBO is

$$\begin{aligned}\mathcal{L}_{\text{VaDE}} &= \mathbb{E}_{q(\mathbf{z},c|\mathbf{x})}[\ln p(\mathbf{x}|\mathbf{z})] - \mathcal{D}(q(\mathbf{z}, c|\mathbf{x})||p(\mathbf{z}, c)) \\ &= \mathbb{E}_{q(\mathbf{z},c|\mathbf{x})}[\ln p(\mathbf{x}|\mathbf{z})] - \mathcal{D}(q(c|\mathbf{x})||p(c)) - \mathcal{D}(q(\mathbf{z}|\mathbf{x})||p(\mathbf{z}|c)),\end{aligned} \tag{7}$$

where the first part is the reconstruction term and the rest regularize the latent posterior $q(\mathbf{z}, c|\mathbf{x})$ to lie on a manifold of MoG prior $p(\mathbf{z}, c)$. GMVAE considers the generative model $p(\mathbf{x}, \mathbf{z}, \mathbf{z}', c) = p(\mathbf{x}|\mathbf{z})p(\mathbf{z}|\mathbf{z}', c)p(\mathbf{z}')p(c)$ and a mean-field posterior proxy $q(\mathbf{z}, \mathbf{z}', c|\mathbf{x}) = q(\mathbf{z}|\mathbf{x})q(\mathbf{z}'|\mathbf{x})p(c|\mathbf{z}, \mathbf{z}')$, thus the ELBO can be written as

$$\begin{aligned}\mathcal{L}_{\text{GMVAE}} &= \mathbb{E}_{q(\mathbf{z},\mathbf{z}',c|\mathbf{x})}[\ln(p(\mathbf{x}, \mathbf{z}, \mathbf{z}', c)/q(\mathbf{z}, \mathbf{z}', c|\mathbf{x}))] \\ &= \mathbb{E}_{q(\mathbf{z},\mathbf{z}',c|\mathbf{x})}[\ln p(\mathbf{x}|\mathbf{z})] \\ &\quad - \mathcal{D}(p(c|\mathbf{z}, \mathbf{z}')||p(c)) - \mathcal{D}(q(\mathbf{z}'|\mathbf{x})||p(\mathbf{z}')) - \mathcal{D}(q(\mathbf{z}|\mathbf{x})||p(\mathbf{z}|\mathbf{z}', c)),\end{aligned} \tag{8}$$

where the three terms denote the reconstruction error and regularizers of c-prior, $\mathbf{z}'$-prior, and conditional prior, respectively. Universally in ELBO objectives of

VDC methods, $p(\mathbf{x}|\mathbf{z})$, $q(\mathbf{z}|\mathbf{x})$, and $q(\mathbf{z}'|\mathbf{x})$ are respectively modelled with DNNs approximating functions of $f_{\boldsymbol{\theta}}(\mathbf{z})$, $g_{\boldsymbol{\phi}}(\mathbf{x})$, and $g_{\boldsymbol{\phi}'}(\mathbf{x})$, where $\boldsymbol{\theta}$, $\boldsymbol{\phi}$, and $\boldsymbol{\phi}'$ are their trainable parameters.

4 Simple, Scalable, and Stable VDC

In this section, we discuss four frequently encountered problems that violate the *simplicity*, *scalability*, and *stability* properties (P6–P8 in Sect. 1), and propose generic solutions with pertinence, which lead to a holistic algorithm S3VDC.

4.1 Initial γ-Training: Reproduce with Milder Volatility

The necessity of VDC reproducibility is beyond doubt. However, we found that the clustering results of VDCs are heavily affected by randomness (e.g., in parameter initialization and data pipeline implementation), which agrees with the pattern reported by [20]. The evaluation of VDCs towards the stability property (P8) in Table 1 shows that the objective functions of VDC have little impact on stability. A simple remedy is to pretrain networks $f_{\boldsymbol{\theta}}(\mathbf{z})$, $g_{\boldsymbol{\phi}}(\mathbf{x})$, and $g_{\boldsymbol{\phi}'}(\mathbf{x})$ using a stacked AE that has the same network architecture and reconstruction loss [15]. Once pretrained, the weights are copied to the original VDC model before finetuning the fully fleshed ELBO objectives. Unfortunately, this practice improves the stability at the cost of simplicity due to requiring 1) multiple sequential steps, 2) higher space complexity, and 3) maintaining multiple pretrained models for different datasets.

The naïve ELBO optimization targets for VDCs consist of two parts (see Eqs. (6–8)): 1) a reconstruction term that guarantees data reconstruction capability of the latent representation $\mathbf{z}$, and 2) several KL regularization terms to leverage the prior knowledge. Authors of [5] concluded that higher impact of regularizers pushes VAE to posterior collapse, while less regularization makes VAE more interchangeable with a plain AE. Consequently, we propose an initial "γ-training" phase, indicated by the straight red line in Fig. 1, that trains VDCs with a much lower emphasis (weighted by $0 < \gamma \ll 1$) on the regularizer terms for the first T_γ mini-batch steps:

$$\mathcal{L}_\gamma = \mathbb{E}_{q(\cdot|\hat{\mathbf{x}})}[\ln p(\mathbf{x}|\cdot)] - \gamma[\mathcal{D}(q(\cdot|\hat{\mathbf{x}})||\cdot) + \ldots], \; T_\gamma \geq t > 0, \tag{9}$$

where t is the index of the current training step. We also apply the denoising mechanism, in which $\mathbf{x}$ is corrupted into $\hat{\mathbf{x}} \in \hat{\mathbf{X}}$ by adding noise from a zero-mean random normal distribution. In this way, VDCs reconstruct a "repaired" input from a corrupted version. With this, pretraining is fused into the VDC optimization processes, which brings not only *stable* results but also *simple* maintenance.

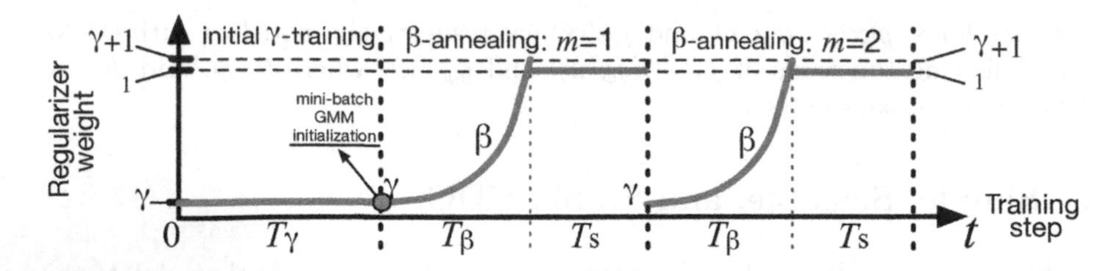

Fig. 1. An overview of the training process in S3VDC algorithm: initial γ-training, mini-batch GMM initialization, and periodic β-annealing.

4.2 Periodic β-Annealing: Improve Latent Disentanglement

We empirically observed that VDCs are vulnerable to model degeneration where the optimization focuses too much on KL regularizers from a certain point and never escapes, leading to uninformative $\mathbf{z}$ that jeopardizes the learning of disentangled latent representations. The intuitive solution is to carefully schedule the regularization weight (a.k.a. regularizer annealing). Similar modification is made to VAEs [2,13,19]. The key to designing the annealing policy lies between the intricate balance of two seemingly contradictory factors: 1) the provision of meaningful $\mathbf{z}$ before training $f_{\boldsymbol{\theta}}(\mathbf{z})$, and 2) continuously leveraging better $\mathbf{z}$ distribution. To that end, we propose a periodic polynomial annealing strategy, "periodic β-annealing", after optimizing Eq. (9) for T_γ steps. As illustrated in Fig. 1, any period m has two phases: 1) β-annealing for T_β steps optimizing the following objective, and 2) training with naïve VDC objectives for T_{s} steps.

$$\begin{aligned}
&\mathcal{L}_\beta = \mathbb{E}_{q(\cdot|\hat{\mathbf{x}})}[\ln p(\mathbf{x}|\cdot)] - \beta[\mathcal{D}(q(\cdot|\hat{\mathbf{x}})||\cdot) + \ldots] \;, \\
&\text{s.t.: } \beta = \gamma + \left[\frac{t - T_\gamma - (m-1)(T_\beta + T_{\mathrm{s}})}{T_\beta}\right]^u \;, \\
&\qquad T_\gamma + (m-1)(T_\beta + T_{\mathrm{s}}) + T_\beta \geq t > T_\gamma + (m-1)(T_\beta + T_{\mathrm{s}}) \;.
\end{aligned} \tag{10}$$

β represents the scheduled weight of KL regularizers during the first phase in each period; $m \in \{1 \ldots M\}$ denotes the m-th annealing period. To *stabilize* the training as much as possible, u should be selected to ensure both a smooth transition from γ-training and a mild increase of β. We empirically find that periodic β-annealing improves latent disentanglement in practice, therefore, it further improves the clustering *stability*.

4.3 Mini-batch GMM Initialization: Scale to Large Datasets

For VDCs to work on large-scale datasets, their performance should be largely invariant to both the total number of samples N and the number of clusters C. M1 + 2 meets both conditions at the expense of restricting to using a single Gaussian. GMVAE uses one NN to approximate a tuple ($\boldsymbol{\mu}_c$, $\boldsymbol{\sigma}_c^2$), so its complexity increases linearly with C. As a result, M1 + 2 and GMVAE have much inferior performance compared to VaDE (see accuracy scores in Table 1). On the other hand, the computational bottleneck of VaDE lies in the step where π_c, $\boldsymbol{\mu}_c$ and $\boldsymbol{\sigma}_c^2$ are initialized using all N samples. To improve the *scalability* of VaDE, we merely take $k \times L$ ($\ll N$) Monte Carlo samples from the entire dataset to initialize GMM (represented with an anchor "●" in Fig. 1); L denote the mini-batch size, hence the name "mini-batch GMM initialization". Section 5.3 shows how robust this simplification is in relation to the value of $k \in \{1, 2, \ldots\}$. As an auxiliary study, we also tried to update π_c, $\boldsymbol{\mu}_c$ and $\boldsymbol{\sigma}_c^2$ (using $k \times L$ samples for every GMM initialization) before each β-annealing period with a momentum factor (similar to [3]), yet it did not offer observable uplift of clustering accuracy.

4.4 Inverse Min-Max Transform: Avoid NaN Losses

NaN loss values are not uncommon when training VAE and VDC models using ML frameworks like Tensorflow and PyTorch. Spinning up a large number of parallel training sessions to guarantee sufficient successful ones is a waste of computing resources. Worst of all, there is no easy way to escape from NaN loss if pretraining is considered as a standalone step, which is the situation we end up with when experimenting with certain VDC models such as [15].

Long story short, the curse of NaN losses originates from the terms of $q(c|\mathbf{x})$ and $q(c|\mathbf{z}, \mathbf{z}')$ in Eqs. (6), (7), and (8). The authors of [15] illustrate that $q(c|\mathbf{x})$ can be approximated with $q(c|\mathbf{z})$ in VaDE:

$$q(c|\mathbf{x}) \approx p(c|\mathbf{z}) \equiv \frac{p(c)e^{\ln p(\mathbf{z}|c)}}{\sum_{c'=1}^{C} p(c')e^{\ln p(\mathbf{z}|c')}}, \tag{11}$$

where $p(\mathbf{z}|c) = \mathcal{N}(\mathbf{z}|\boldsymbol{\mu}_c, \boldsymbol{\sigma}_c^2\mathbf{I})$. For the sake of numerical stability, we formulate $p(\mathbf{z}|c)$ as $e^{\ln p(\mathbf{z}|c)}$ instead. The log probability $\ln p(\mathbf{z}|c)$ is directly calculated with

$$\ln p(\mathbf{z}|c) - \ln \frac{1}{\boldsymbol{\sigma}_c} - \frac{1}{2}\left[\ln(2\pi) + \frac{\boldsymbol{\mu}_c^2}{\boldsymbol{\sigma}_c^2} - \frac{2\boldsymbol{\mu}_c}{\boldsymbol{\sigma}_c^2} + \frac{\mathbf{z}^2}{\boldsymbol{\sigma}_c^2}\right], \tag{12}$$

whose operators are all element-wise. If the model becomes extremely confident that a sample $\mathbf{x}$ (represented by latent embedding $\mathbf{z}$) does not belong to cluster c, then the value of $\ln p(\mathbf{z}|c)$ can be rather small, which potentially leads to numerical overflow when computing $e^{\ln p(\mathbf{z}|c)}$. Pretraining generally pushes $\ln p(\mathbf{z}|c)$ to either 0 or $-\infty$, resulting in a much higher probability of having NaN loss values. GMVAE also suffers from this problem due to the formulation of $p(c|\mathbf{z}, \mathbf{z}')$:

$$p(c|\mathbf{z}, \mathbf{z}') = \frac{p(c)e^{\ln p(\mathbf{z}|\mathbf{z}', c)}}{\sum_{c'=1}^{C} p(c')e^{\ln p(\mathbf{z}|\mathbf{z}', c')}}, \tag{13}$$

where $p(\mathbf{z}|\mathbf{z}', c) = \mathcal{N}(\mathbf{z}|\boldsymbol{\mu}_c(\mathbf{z}'), \boldsymbol{\sigma}_c^2(\mathbf{z}')\mathbf{I})$. M1 + 2 largely refrain from this problem by replacing the categorical distribution with a continuous one approximated by NN. The naïve fix of limiting the range of input values and initial model weights is suboptimal since it does not address the root cause of NaN loss.

Algorithm 1: S3VDC: Simple, Scalable, and Stable VDC.

Input: dataset $\mathbf{X}$ and noisy dataset $\hat{\mathbf{X}}$ with N samples
Output: Trained $\boldsymbol{\theta}$, $\boldsymbol{\phi}$, $\boldsymbol{\pi}$, $\boldsymbol{\mu}_c$, $\boldsymbol{\sigma}_c^2$
Parameter: L, γ, T_γ, T_β, T_s, M, λ, C, k, $f_{\boldsymbol{\theta}}(\cdot)$, $g_{\boldsymbol{\phi}}(\cdot)$

1 **for** $(t=1; t \le T_\gamma; t++)$ **do**
2 Perform γ-training using objective in Equation (9) w.r.t. Equation (7): $\mathcal{L}_\gamma = \mathbb{E}_{q(\mathbf{z},c|\hat{\mathbf{x}})}[\ln p(\mathbf{x}|\mathbf{z})] - \gamma \mathcal{D}(q(\mathbf{z},c|\hat{\mathbf{x}})||p(\mathbf{z},c))$;
3 Initialize $\boldsymbol{\pi}$, $\boldsymbol{\mu}_c$, $\boldsymbol{\sigma}_c^2$ with GMM using $k \times L$ samples;
4 **for** $(m=1; m \le M; m++)$ **do**
5 $T_m = T_\gamma + (m-1)(T_\beta + T_s)$;
6 **for** $(t = T_m + 1; t \le T_m + T_\beta + T_s; t++)$ **do**
7 Calculate $\ln p(\mathbf{z}|c)$ with Equation (12);
8 Use Equation (14) to calculate: $\widetilde{\ln}\, p(\mathbf{z}|c) = \frac{\lambda[\ln p(\mathbf{z}|c) - \min(\ln p(\mathbf{z}|c))]}{\max(\ln p(\mathbf{z}|c)) - \min(\ln p(\mathbf{z}|c))}$;
9 Calculate $q(c|\hat{\mathbf{x}})$ with Equation (11): $q(c|\hat{\mathbf{x}}) = \frac{p(c)e^{\widetilde{\ln}\, p(\mathbf{z}|c)}}{\sum_{c'=1}^{C} p(c')e^{\widetilde{\ln}\, p(\mathbf{z}|c')}}$;
10 **if** $t \le T_m + T_\beta$ **then**
11 Calculate $\beta = \gamma + [(t - T_m)/T_\beta]^3$;
12 Perform β-annealing using Equation (10) w.r.t. Equation (7): $\mathcal{L}_\beta = \mathbb{E}_{q(\mathbf{z},c|\hat{\mathbf{x}})}[\ln p(\mathbf{x}|\mathbf{z})] - \beta[\mathcal{D}(q(c|\hat{\mathbf{x}})||p(c)) + \mathcal{D}(q(\mathbf{z}|\hat{\mathbf{x}})||p(\mathbf{z}|c))]$;
13 **else**
14 Optimize denoising Equation (7): $\mathcal{L}_{\text{VaDE}} = \mathbb{E}_{q(\mathbf{z},c|\hat{\mathbf{x}})}[\ln p(\mathbf{x}|\mathbf{z})] - \mathcal{D}(q(c|\hat{\mathbf{x}})||p(c)) + \mathcal{D}(q(\mathbf{z}|\hat{\mathbf{x}})||p(\mathbf{z}|c))$;
15 **return** $\boldsymbol{\theta}$, $\boldsymbol{\phi}$, $\boldsymbol{\pi}$, $\boldsymbol{\mu}_c$, $\boldsymbol{\sigma}_c^2$

Since each mini-batch training step uses L samples, for unification, we use $\mathbf{V} \in \mathbb{R}^-_{c \times L}$ to denote the actual matrix of $\ln p(\mathbf{z}|c)$ and $\ln p(\mathbf{z}|\mathbf{z}', c)$ in VaDE and GMVAE, respectively. If we manage to eliminate overly small elements in $\mathbf{V}$, the `NaN` loss can also be prevented. One intuitive solution is clipping the values in $\mathbf{V}$, but it can easily halt the training due to the vanishing gradient. We instead propose an "inverse min-max transform" operator that re-scales each element in $\mathbf{V}$ to the range of $[-\lambda, 0]$:

$$\widetilde{\mathbf{V}} = \lambda[\mathbf{V} - \min(\mathbf{V})]/[\max(\mathbf{V}) - \min(\mathbf{V})] . \tag{14}$$

The functions $\min(\cdot)$ and $\max(\cdot)$ return the minimum and maximum element from a matrix, respectively. λ is a scaling factor, to which we suggest to assign a value between 20 and 50. With this, the properties of *stability* and *simplicity* of VDC training are attained by guaranteeing the convergence while entirely preventing the occurrence of `NaN` losses.

4.5 The Proposed Holistic Algorithm: S3VDC

The solutions presented above (Sect. 4.1–4.4) are generic for VDCs. However, for concreteness, we choose to focus on VaDE due to its ability to finetune the cluster weights $\boldsymbol{\pi}$. We propose S3VDC (simple, scalable, and stable VDC) that incorporates all of the introduced solutions to a vanilla VaDE. Algorithm 1 illustrates the details of S3VDC, where the lines 2, 12, and 14 respectively represent γ-training, β-annealing, and vanilla VaDE optimization. We empirically discover that applying the inverse min-max transform to line 13 and 14 is mandatory, while it is optional for γ-training (line 2).

Like all VDC models, S3VDC is optimized with stochastic mini-batch steps, thus there is no known formal guarantee on the number of steps required to achieve convergence. However, the time complexity of the t-th mini-batch optimization step (i.e. line 1 and 6 in Algorithm 1), $\mathcal{O}_t$, is dominated by the amount of matrix multiplications in the NN architecture. For a standard o_l-layered MLP with o_n neurons in each layer, $\mathcal{O}_t$ is largely $\mathcal{O}(L \times o_l \times o_n^2)$. Besides, line 3 in Algorithm 1 is a special training step that finds optimal initial values for $\boldsymbol{\pi}$, $\boldsymbol{\mu}_c$, $\boldsymbol{\sigma}_c^2$ using GMM EM (expectation-maximization) algorithm[2]; and the computational complexity of each EM step is approximately $\mathcal{O}_{\text{GMM}}(k \times L \times C \times d_{\mathbf{z}}^3)$, where $d_{\mathbf{z}}$ is the dimension of $\mathbf{z}$. Since GMM initialization is only executed once, the overall time complexity of each training step is equivalent to $\mathcal{O}_t$; but $\mathcal{O}_{\text{GMM}}$ may become a critical training bottleneck when a large number of samples (e.g. $k \times L > 1 \times 10^6$) are used for GMM initialization. We will investigate the optimal $k \times L$ in Sect. 5.3. Despite the implementation differences in various software libraries, the space complexity of S3VDC primarily consists of three parts: mini-batch samples ($\mathcal{O}(L \times D)$), DNN (such as weights, gradients, and intermediate cache that are often consistent among VDCs as long as they share the same network architecture), and GMM initialization ($\mathcal{O}(k \times L \times C \times d_{\mathbf{z}}^2)$ [26]).

5 Experiments and Explorations

In Sect. 4.1 and 4.4, we discussed *simplicity* (P6). This section intends to evaluate *scalability* (P7), *stability* (P8), and other industrial properties (P1–P5) of S3VDC using four unstructured datasets (Table 2). We zero-center the original inertial signals in InertialHAR [1] dataset, but use MNIST [17] and Fashion [23] as they are. King10M dataset contains the daily behaviour of almost ten million game players that play Candy Crush Soda Saga (CCSS)[3]. Each sample contains counters (aggregated per day over a period of 30 days commencing from a randomly picked date) for eight types of in-game actions (e.g. win a level, buy an item, send a message, etc.) from an individual player. The individual players in King10M dataset are randomly sampled from the entire population of CCSS game players.

[2] Specifically the implementation in https://github.com/scikit-learn.

[3] https://king.com/game/candycrushsoda.

Table 2. Dataset specifications (top) and S3VDC hyper-parameters (bottom).

Specification	InertialHAR [1]	MNIST [17]	Fashion [23]	King10M
# sample	10,299	70,000	70,000	9,666,892
# feature[a]	9×128	28×28	28×28	8×30
# cluster: C	6	10	10	N/A (6)[b]
dimension of $\mathbf{z}$: $d_{\mathbf{z}}$	4	8	6	10
mini-batch size: L	1,024	128	64	1,024
GMM initialization steps: k	5	200	400	750
initial learning rate[c]	1.5×10^{-3}	2×10^{-3}	2×10^{-3}	1.5×10^{-3}
γ-training weight: γ	5×10^{-6}	5×10^{-4}	5×10^{-3}	5×10^{-4}
γ-training steps: T_γ	6×10^{3}	1×10^{5}	1×10^{5}	2.5×10^{4}
β-annealing steps: T_β	2.5×10^{3}	9×10^{3}	9×10^{3}	4.5×10^{3}
static steps: T_s	5×10^{2}	1×10^{3}	1×10^{3}	5×10^{2}
# period: m	2	10	10	6

[a] Each sample is transformed into a shape of $L \times 28 \times 28 \times$(#channel) via zero padding and resizing (interpolation). InertialHAR and King10M treat each timeseries as a channel.
[b] King10M has no label, thus we set $C=6$ based on the results in Sect. 5.4 (Table 5).
[c] Exponential learning rate decay with a terminating value of 1×10^{-6}.

Table 3. Comparison of clustering accuracy and stability over benchmark datasets.

Dataset	DCEC[a]		VaDE[a]		S3VDC	
	Accuracy	STD	Accuracy	STD	Accuracy	STD
InertialHAR	0.5594	0.0682	0.5900	0.0590	**0.6705**	**0.0130**
MNIST	0.8690	0.0890	0.8400	0.1060	**0.9360**	**0.0181**
Fashion	0.5180	0.0380	0.5500	0.0420	**0.6053**	**0.0091**

[a] For DCEC, we initialize k-means using the same number of samples as S3VDC. For VaDE, we do not use any pre-trained model and discard the runs with `NaN` losses.

All experiments are carried out on the Google cloud platform with the hardware scale tier `BASIC-GPU`[4]. The noise added to input follows a Gaussian distribution with a standard deviation (STD) of 5×10^{-9}. We determine the hyper-parameters (lower part of Table 2) using a discrete random search with early stopping, which is carried out using an in-house developed ML platform. During that process, λ in Equation (14) is uniformly set to 50, and GMM initialization takes 1×10^4 EM iteration steps; we also discovered that setting $u = 3$ in Eq. (10) generally works well on all of the datasets presented in Table 2. The CNN architecture is the same as [12]. We used the same setting as S3VDC for VaDE and DCEC when applicable for a fair comparison. All reported scores are averaged over five trials with randomly initialized model weights.

[4] https://cloud.google.com/ml-engine/docs/machine-types.

InertialHAR

	Acc.	CH	Sil.
Accuracy (Acc.)	1.00		
Calinski-Harabasz (CH)	0.54	1.00	
Silhouette Score (Sil.)	0.36	0.21	1.00

MNIST

	Acc.	CH	Sil.
Accuracy (Acc.)	1.00		
Calinski-Harabasz (CH)	-0.37	1.00	
Silhouette Score (Sil.)	-0.77	0.60	1.00

Fashion

	Acc.	CH	Sil.
Accuracy (Acc.)	1.00		
Calinski-Harabasz (CH)	-0.15	1.00	
Silhouette Score (Sil.)	-0.36	0.30	1.00

Fig. 2. Pair-wise correlation of clustering metrics for S3VDC.

5.1 Clustering Stability and Accuracy

The stability is measured by STD of accuracy, which is calculated by finding the best mapping between cluster assignments and labels. Table 3 illustrates that S3VDC not only reaches the state-of-the-art clustering accuracy but also obtains a substantial decrease in STD. This guarantees extremely stable clustering results while maintaining high accuracy with respect to the target labels. The accuracy of VaDE and DCEC reported in this work is different than that in [15] and [12] due to (1) they use the same hyper-parameters as S3VDC for comparability, and (2) no pre-trained model is used in our experiments for simplicity.

5.2 Unsupervised Model Selection

Unlike the labeled benchmark datasets, it is much harder to evaluate clustering results without any ground truth label. To that end, we investigated the feasibility of model selection using unsupervised metrics and latent disentanglement. We considered all successful trials from hyper-parameter search.

Correlation Among Clustering Metrics: Figure 2 illustrates how much supervised metric (accuracy) and unsupervised metrics (Calinski-Harabasz index and Silhouette score) agree by calculating the pair-wise Spearman rank correlation for the selected metrics. Calinski-Harabasz (CH) index is defined as the ratio between within-cluster dispersion and between-cluster dispersion, which is higher when clusters are dense and well-separated. Silhouette score measures how similar a sample is to its own cluster compared to other clusters. This score ranges from -1 to 1, where higher values indicate better clustering quality. We observe that supervised and unsupervised metrics do not correlate in the same way on different datasets, yet unsupervised metrics almost always agree. So, in addition to unsupervised clustering metrics, examining the latent disentanglement becomes inevitable for unsupervised model selection.

Disentanglement of Latent Embedding: In Fig. 3, we demonstrate that the disentanglement degree (visualized together with the predicted clusters) of the learned latent embedding $\mathbf{z}$ is an indicator of clustering quality. The left column contains our selected S3VDC models; and the S3VDC models in the middle column are obtained during the hyper-parameter search and considered sub-optimal. The right-most column of Fig. 3 represents the best VaDE models obtained on each dataset. It can be seen that disentanglement degree sometimes leads to selecting a different model than what unsupervised metrics

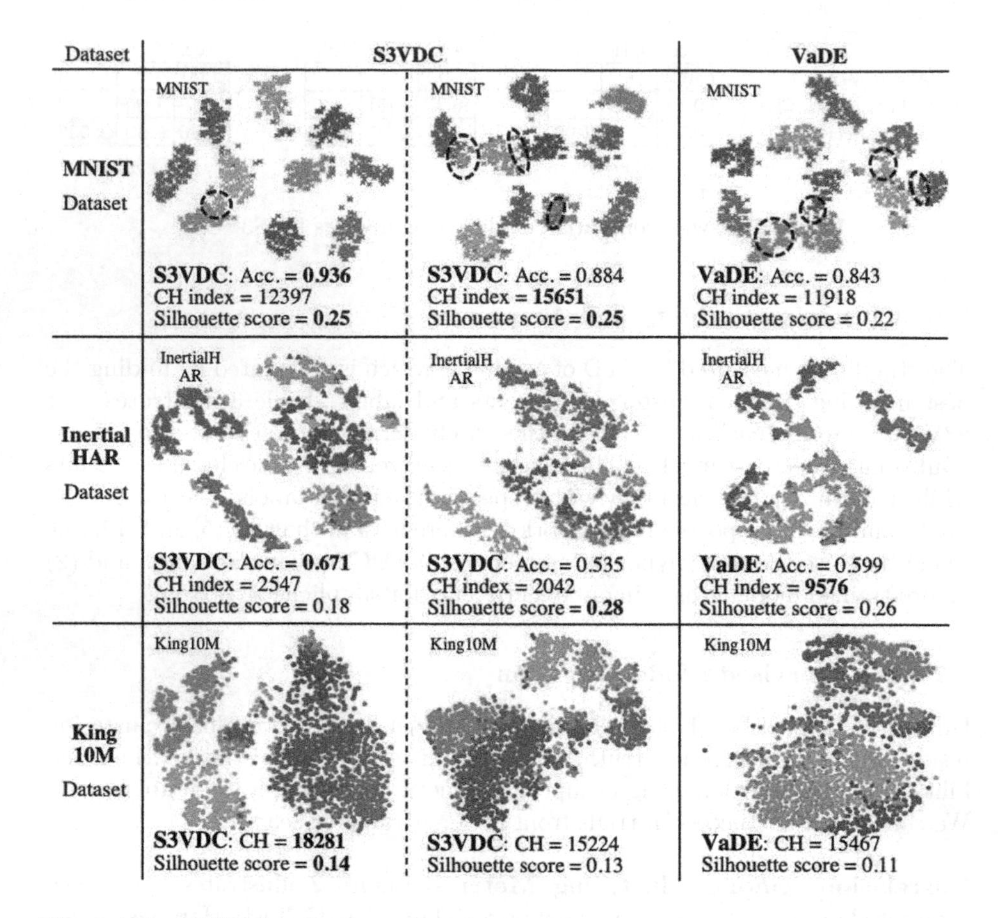

Fig. 3. The t-SNE visualization of **z** colored with predicted clusters. To enable VaDE training on King10M, we apply the same mini-batch GMM initialization as S3VDC. (Color figure online)

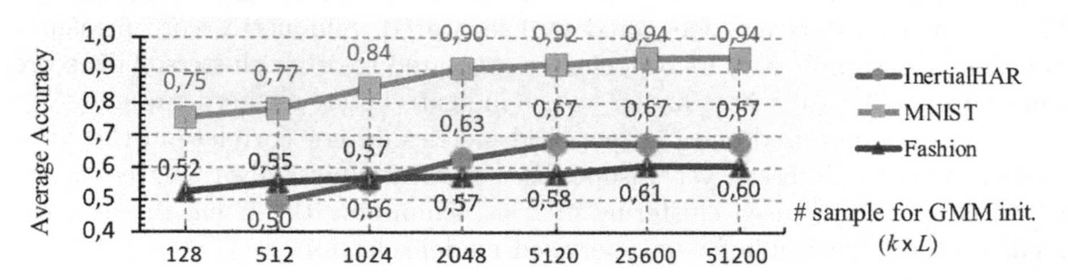

Fig. 4. S3VDC accuracy in relation to the number of samples for GMM initialization.

might suggest; for an instance, we would never select the left-most S3VDC model for InertialHAR dataset by only look at either CH index or Silhouette score. Thus, we used a combination of unsupervised clustering metrics and latent disentanglement to choose the best hyper-parameters for S3VDC on King10M dataset.

Table 4. The average time consumption (seconds) of GMM initialization.

VDC Models	InertialHAR	MNIST	Fashion	King10M
VaDE	0.66 s	2.44 s	2.42 s	Failed to initialize
S3VDC[a]	0.34 s (50%)	0.98 s (37%)	1.04 s (37%)	9.15 s (8%)

[a] The minimum percentage of dataset required to achieve the accuracy plateau is indicated in parentheses. VaDE uses the entire dataset for GMM initialization.

Table 5. The S3VDC marginal likelihood $-\ln p(\mathbf{x})$ vs. the target cluster number C.

Dataset	C = 4	C = 5	C = 6	C = 7	C = 8	C = 9	C = 10	C = 11	C = 12
InertialHAR	23.89	23.81	**22.97**	23.54	24.05	–	–	–	–
MNIST	–	–	–	–	89.61	89.09	**89.08**	89.09	89.32
Fashion	–	–	–	58.46	58.35	**57.97**	57.98	58.01	58.17
King10M	39.77	39.94	**39.68**	39.95	40.11	–	–	–	–

5.3 Training Scalability

In Sect. 4.3, we proposed to use only a Monte Carlo subset (of size $k \times L$) of the entire dataset to initialize the global GMM parameters. Figure 4 shows that the clustering accuracy of S3VDC monotonically increases with the value of $k \times L$, and saturates at different places. Observed from Table 4, it is generally sufficient to use less than half of the dataset for GMM initialization, which leads to improved time complexity and training scalability. Specifically, King10M merely needs about 8% of all samples to reach the saturated Silhouette and CH scores, while VaDE keeps failing at the full-scale GMM initialization step.

5.4 Optimal Number of Clusters

To determine the optimal number of clusters, C, without using any label information, we report the Monte Carlo estimated marginal likelihood $-\ln p(\mathbf{x})$ for S3VDC over the test sets (~10% of each dataset) for different values of C. From Table 5, we find that $-\ln p(\mathbf{x})$ often reaches its minimum around the "preferred" C on benchmark datasets. The true number of clusters can become less obvious if hard-to-disentangle clusters exist (e.g., Fashion dataset visualized in Fig. 3).

5.5 Model Interpretation Without Label

When evaluating the performance of S3VDC over large unlabeled datasets like King10M, we suggest to simultaneously look at 1) unsupervised metrics, 2) the joint visualization of latent disentanglement and cluster prediction, and 3) the marginal likelihood. Accordingly, we selected the best S3VDC model (see bottom-left in Fig. 3) trained on King10M. To assign meaningful labels to all six clusters for King10M, we first generate 1,000 CCSS game players by sampling the latent embedding from each cluster using the learned S3VDC model, then we plot them in a three-dimensional KPI (key performance indicator) space for each cluster. As shown in Fig. 5, the three KPIs describe 1) engagement (horizontal axis): the level of commitment to the gaming activities; 2) monetization (vertical axis): the amount of in-game spending; and 3) aptitude (anchor color): the probability of winning game levels. The KPI calculation (from the generated time-series) and the actual values are considered to be sensitive proprietary data and therefore removed from the plot. We observe that clusters 1, 2, and 6 contain slightly more active players; clusters 2 and 6 represent mostly players with high aptitude, while clusters 1 and 3 have mainly low-aptitude players; the ranking of in-game spending is 6 > 1,2,4 > 3,5. Moreover, cluster 3 and 6 respectively have the highest and lowest cluster weights.

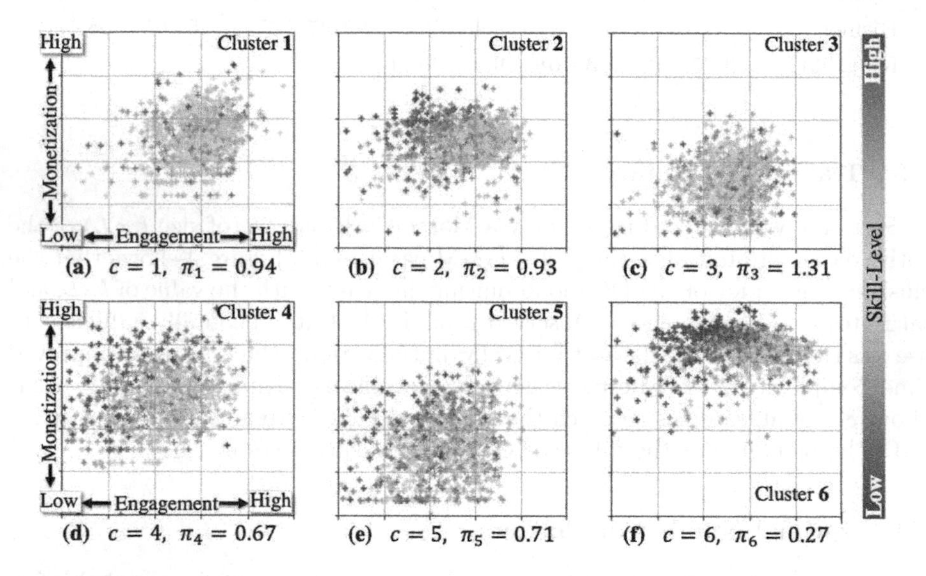

Fig. 5. Interpret the generated player data from a S3VDC model (trained on King10M dataset) towards three player KPIs: engagement, revenue, and aptitude.

6 Conclusion

In this paper, we challenge the state-of-the-art in DC solutions in their applicability in industry. To that end, we argue for eight properties that we believe DC methods need to hold to be able to fill in the gap of applicability. To address this gap, we propose, S3VDC, a simple stable and scalable VDC algorithm. Our experiments show that S3VDC outperforms the state-of-the-art VDC in terms of stability on standard datasets. In addition, we show that the proposed method is scalable since it requires half of the samples needed for initialization compared to the state-of-the-art VDC. Finally, we experimentally demonstrate how

S3VDC facilitates unsupervised model selection and interpretation. Future work includes a sensitivity analysis of hyper-parameters in S3VDC, ablation study, and utilizing generative processes for quantitative evaluation.

References

1. Anguita, D., Ghio, A., Oneto, L., Parra, X., Reyes-Ortiz, J.L.: A public domain dataset for human activity recognition using smartphones. In: ESANN (2013)
2. Bowman, S.R., Vilnis, L., Vinyals, O., Dai, A.M., Jozefowicz, R., Bengio, S.: Generating sentences from a continuous space. In: Proceedings of CoNLL (2016)
3. Cao, L., Sun, F., Kotagiri, R., Huang, W., Cheng, W., Liu, X.: Real-time recurrent tactile recognition: momentum batch-sequential echo state networks. IEEE Trans. Syst. Man Cybern. Syst. **50**(4), 1350–1361 (2020)
4. Chen, X., Duan, Y., Houthooft, R., Schulman, J., Sutskever, I., Abbeel, P.: InfoGAN: Interpretable representation learning by information maximizing generative adversarial nets. In: Proceedings of NIPS, pp. 2172–2180 (2016)
5. Chou, J.: Generated loss and augmented training of MNIST variational auto encoder. arXiv preprint arXiv:1904.10937 (2019)
6. Diederik, P.K., Welling, M., et al.: Auto-encoding variational Bayes. In: ICLR (2014)
7. Dilokthanakul, N., et al.: Deep unsupervised clustering with Gaussian mixture variational autoencoders. arXiv preprint arXiv:1611.02648 (2016)
8. Figueroa, J.A., Rivera, A.R.: Is simple better?: revisiting simple generative models for unsupervised clustering. In: NIPS Workshop on Bayesian Deep Learning (2017)
9. Ghasedi, K., Wang, X., Deng, C., Huang, H.: Balanced self-paced learning for generative adversarial clustering network. In: CVPR, pp. 4391–4400 (2019)
10. Ghasedi Dizaji, K., Herandi, A., Deng, C., Cai, W., Huang, H.: Deep clustering via joint convolutional autoencoder embedding and relative entropy minimization. In: Proceedings of CVPR, pp. 5736–5745 (2017)
11. Goodfellow, I., et al.: Generative adversarial nets. In: Proceedings of NIPS, pp. 2672–2680 (2014)
12. Guo, X., Liu, X., Zhu, E., Yin, J.: Deep clustering with convolutional autoencoders. In: Liu, D., Xie, S., Li, Y., Zhao, D., El-Alfy, E.S. (eds.) Neural Information Processing. ICONIP 2017. Lecture Notes in Computer Science, vol. 10635, pp. 373–382. Springer, Cham (2017). https://doi.org/10.1007/978-3-319-70096-0_39
13. Higgins, I., Matthey, L., Pal, A., Burgess, C., Glorot, X., Botvinick, M., Mohamed, S., Lerchner, A.: beta-VAE: learning basic visual concepts with a constrained variational framework. In: Proceedings of ICLR, p. 6 (2017)
14. Hu, W., Miyato, T., Tokui, S., Matsumoto, E., Sugiyama, M.: Learning discrete representations via information maximizing self-augmented training. In: Proceedings of ICML, pp. 1558–1567 (2017)
15. Jiang, Z., Zheng, Y., Tan, H., Tang, B., Zhou, H.: Variational deep embedding: an unsupervised and generative approach to clustering. In: IJCAI, pp. 1965–1972 (2017)
16. Kingma, D.P., Mohamed, S., Rezende, D.J., Welling, M.: Semi-supervised learning with deep generative models. In: Proceedings of NIPS, pp. 3581–3589 (2014)
17. LeCun, Y., Bottou, L., Bengio, Y., Haffner, P., et al.: Gradient-based learning applied to document recognition. Proc. IEEE **86**(11), 2278–2324 (1998)

18. Liang, J., Yang, J., Lee, H.Y., Wang, K., Yang, M.H.: Sub-GAN: an unsupervised generative model via subspaces. In: Proceedings of ECCV, pp. 698–714 (2018)
19. Liu, X., Gao, J., Celikyilmaz, A., Carin, L., et al.: Cyclical annealing schedule: a simple approach to mitigating KL vanishing. In: Proceedings of NAACL (2019)
20. Locatello, F., et al.: Challenging common assumptions in the unsupervised learning of disentangled representations. In: Proceedings of ICML, pp. 4114–4124 (2019)
21. Mukherjee, S., Asnani, H., Lin, E., Kannan, S.: ClusterGAN: latent space clustering in generative adversarial networks. In: AAAI, vol. 33, pp. 4610–4617 (2019)
22. Nalisnick, E., Hertel, L., Smyth, P.: Approximate inference for deep latent Gaussian mixtures. In: NIPS Workshop on Bayesian Deep Learning, vol. 2 (2016)
23. Xiao, H., Rasul, K., Vollgraf, R.: Fashion-MNIST: a novel image dataset for benchmarking machine learning algorithms. arXiv preprint arXiv:1708.07747 (2017)
24. Xie, J., Girshick, R., Farhadi, A.: Unsupervised deep embedding for clustering analysis. In: Proceedings of ICML, pp. 478–487 (2016)
25. Yang, J., Parikh, D., Batra, D.: Joint unsupervised learning of deep representations and image clusters. In: Proceedings of CVPR, pp. 5147–5156 (2016)
26. Zhou, Y., Rangarajan, A., Gader, P.D.: A Gaussian mixture model representation of endmember variability in hyperspectral unmixing. IEEE Trans. Image Process. **27**(5), 2242–2256 (2018)

Gauss Shift: Density Attractor Clustering Faster Than Mean Shift

Richard Leibrandt(✉) and Stephan Günnemann

Technical University of Munich, Boltzmannstr. 3, 85748 Garching, Germany
r.leibrandt@tum.de, guennemann@in.tum.de
http://www.daml.in.tum.de

Abstract. Mean shift is a popular and powerful clustering method. While techniques exist that improve its absolute runtime, no method has been able to effectively improve its quadratic time complexity with regard to dataset size. To enable development of an alternative, faster method that leads to the same results, we first contribute the formal cluster definition, which mean shift implicitly follows. Based on this definition we derive and contribute Gauss shift – a method that has linear time complexity. We quantify the characteristics of Gauss shift using synthetic datasets with known topologies. We further qualify Gauss shift using real-life data from active neuroscience research, which is the most comprehensive description of any subcellular organelle to date.

Supplementary material: www.daml.in.tum.de/gauss-shift.

Keywords: Clustering · Density attractor clustering · Gauss shift · mean shift · Efficiency · Optimization · Local search · Neuroscience

1 Introduction

Clustering methods assign data objects to groups, called clusters. Objects within a cluster are similar to each other, while objects from different clusters are dissimilar. Clustering is used in diverse fields, such as social analytics, molecular dynamics, crystallography, and airplane flight path analysis, so it comes to no surprise that there are many different notions of what constitutes a cluster. Additionally, for a single notion there are usually multiple formal cluster definitions and sometimes algorithms are implemented without providing an explicit, algorithm-free cluster definition. None of those notions/definitions is better or worse per se; they are just differently useful in different situations. A proficient user will know about different notions/definitions and apply them accordingly.

One popular notion is that objects of the same cluster must be close to a common representative (e.g. a Euclidean vector in the case of k-means or a normal distribution in the case of Gaussian mixture models). Another popular notion is that objects of the same cluster are connected by higher density, while objects in different clusters are separated by lower density. A common interpretation is that the density calculated from the data objects approximates the probability density function (PDF) of the underlaying, unknown distribution from which

F. Hutter et al. (Eds.): ECML PKDD 2020, LNAI 12457, pp. 125–142, 2021.
https://doi.org/10.1007/978-3-030-67658-2_8

the data objects are drawn. Examples of density-based clustering methods are DBSCAN [7] and mean shift [4]. While for a modified version of DBSCAN, called DBSCAN* [2], a formal cluster definition can be derived from density-contour clusters [9, p. 205] (which we present in Definition 1), so far, no formal cluster definition has been provided for mean shift and algorithms that implicitly use the same definition.

However, we are convinced that a formal, mathematical definition of what constitutes a cluster is required for rigorous data *science*. Such a definition has to forgo a description of a method that creates respective clusters from data objects. Otherwise, we intrinsically commit the fallacy of first executing the method and retrospectively claiming the results are what was aimed for. Not only is such a definition essential to be able to reason about the results theoretically, but also to enable researchers to develop alternative methods that create clusters based on the same definition. Researchers should provide algorithms that improve *effectiveness*, *efficiency* or *versatility* for each definition separably or new definitions.

Contributions. With Definition 3, we provide a rigorous, formal cluster definition for Euclidean spaces that is implicitly used by mean shift and similar algorithms. For comparison purposes, we also provide Definition 1 (implicitly used by DBSCAN* [2]), which is more concise and complete than previous definitions.

Based on Definition 3, we provide a new method that improves on *efficiency*: Gauss shift has linear time complexity with regard to dataset size, in contrast to the quadratic time complexity of the state-of-the-art, and superlinear local convergence rate, in contrast to the linear convergence rate of the standard method mean shift. We focus on clustering data objects in multivariate real coordinate Euclidean vector spaces, with $\mathbb{X}$ being its underlying set. $F \in \mathbb{N}$ is the dimension of $\mathbb{X} = \mathbb{R}^F$ and the dimension of each object $\boldsymbol{o}_i \in \mathbb{X}$. The objects make up the dataset $\mathcal{O}$ – a multiset of size $|\mathcal{O}|$. Gauss shift is quantified using synthetic datasets with known topologies and further qualified using real-life data from neuroscientific research conducted from 2015 to 2020, the most comprehensive description of any subcellular organelle (a subunit within a biological cell, that has a specific function) to date.

Notation. Blackboard bold font denotes infinite sets (e.g. natural numbers $\mathbb{N}$, real numbers $\mathbb{R}$, or $\mathbb{X}$) and respective superscripts denote the Cartesian product (e.g. $\mathbb{R}^2 = \mathbb{R} \times \mathbb{R}$). The natural numbers include 0. Calligraphy font denotes finite (multi)sets (e.g. $\mathcal{O}$). Cardinality is denoted, e.g., with $|\mathcal{O}|$. Non-bold font indicates univariate objects, bold font indicates multivariate objects. The n-th derivative ($n \in \mathbb{N}$) of function ϱ with respect to $\boldsymbol{x} \in \mathbb{R}^F$ is denoted with $\nabla_{\boldsymbol{x}}^n \varrho(\boldsymbol{x}) = \nabla_{\boldsymbol{x}}^n(\varrho(\boldsymbol{x}))$ (no directional derivative), with respect to the entire input before function evaluation is denoted with $\nabla^n \varrho(\boldsymbol{x}) = (\nabla^n \varrho)(\boldsymbol{x})$. The same applies to functions $\varrho(y)$ of univariate vectors $y \in \mathbb{R}^1$. Definiteness of matrices is denoted with $\succ 0$ (positive), $\prec 0$ (negative), and positive semi-definiteness with $\succeq 0$.

Mean shift refers to the complete non-blurring, non-hierarchical clustering algorithm, not the step between local search iterations, which we denote as *mean shift step*. Analogous, the same applies to the terms *Gauss shift* and *Gauss shift step*. Slanted font introduces terminological terms, italic font indicates emphasis.

2 Model Structure: Defining Density Clustering

For the general notion of density-based clusters, we provide two specific definitions. E.g. DBSCAN* [2] follows the first definition, mean shift the second (see Fig. 1).

Definition 1. (Density superlevel clustering (DSC)). *Given a Euclidean space* $\mathbb{X}$*, over which a density function* $\varrho\colon \mathbb{X} \to \mathbb{R}$ *is defined based on the set of data objects* $\mathcal{O} \subset \mathbb{X}$*, then a cluster is a connected component (maximal connected subset) of a superlevel set of the density function. The superlevel set of a density function* $\varrho\colon \mathbb{X} \to \mathbb{R}$ *is defined as* $\{\boldsymbol{x} \in \mathbb{X} \mid \varrho(\boldsymbol{x}) \geq c\}$ *for a cut value* $c \in \mathbb{R}$*. Data objects that are elements of the same connected component belong to the same cluster. Each object* $\boldsymbol{o}_i \in \mathcal{O}$ *outside every cluster is considered noise.*

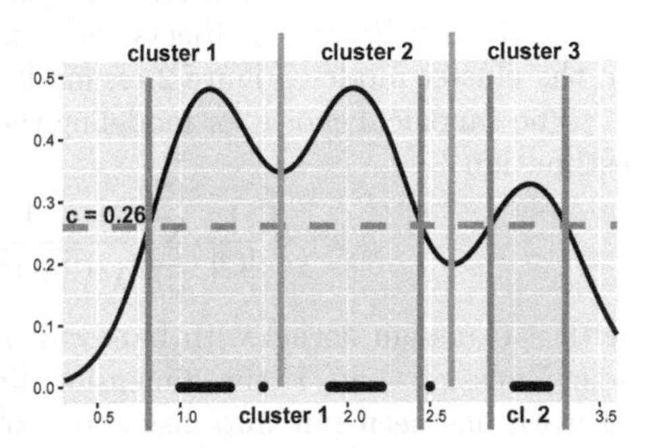

Fig. 1. Different definitions lead to different results: DSC produces two clusters for $c = 0.26$, excluding some objects. DAC devides all objects into three clusters.

Local search optimization is a class of greedy algorithms that move iteratively from solution point to solution point in such a way that subsequent points improve a target function compared to the previous point by applying local changes, until a convergence criterion is met. All solution points are elements of the set $\mathbb{X}$ (the *search space*), and subsequent points are required to be in the neighborhood of the point. Using the definition of local search, we state a second cluster definition.

Definition 2. (Local search optimization). *Given a target function* $\varrho(\boldsymbol{x})$*,* $\boldsymbol{x} \in \mathbb{X}$ *over a topological space* $(\mathbb{X}, \tau)$ *and an objective regarding the target function (e.g. maximization for mode searching), then local search optimization is defined as the family of functions* $L = (l_t)_{t\in\mathcal{T}}$ *with* $t, T \in \mathbb{N} \cup \{\infty\}, \mathcal{T} = \{t \mid 1 \leq t \leq T\}$*, such that* $\boldsymbol{x}_{t+1} = l_t(\boldsymbol{x}_t) \in \mathcal{N}_t(\boldsymbol{x}_t)$ *induces a sequence* $\{\boldsymbol{x}_t\}_{t=1}^{T}$ *(with* $\boldsymbol{x}_t \in \mathbb{X}$*) for neighborhoods* $\mathcal{N}_t(\boldsymbol{x})$ *over* $\mathbb{X}$ *such that* $\varrho(\boldsymbol{x}_{t+1})$ *is closer to the objective than* $\varrho(\boldsymbol{x}_t)$ *(e.g.*$\varrho(\boldsymbol{x}_{t+1}) > \varrho(\boldsymbol{x}_t)$ *in case of maximization). From this follows, with the initial candidate solution* $\boldsymbol{x}_1 \in \mathbb{X}$*, that* $\boldsymbol{x}_{t+1} = l_t(\boldsymbol{x}_t) = (l_t \circ \cdots \circ l_1)(\boldsymbol{x}_1)$*. The local search is terminating with a final candidate solution* $\boldsymbol{x}_T$ *(*$T < \infty$*) if no* $\varrho(\boldsymbol{x}_{T+1})$ *exists that is closer to the objective than* $\boldsymbol{x}_T$*, otherwise it is non-terminating (*$T = \infty$*). An attractor* $\boldsymbol{a} \in \mathbb{X}$ *is a limit for a local search induced sequence. The basin of attraction of* $\boldsymbol{a}$ *is the non-empty set* $\{\boldsymbol{x} \mid \lim_{t\to T} (l_t \circ \cdots \circ l_1)(\boldsymbol{x}) = \boldsymbol{a}\}$.

Definition 3. (Density attractor clustering (DAC)). *Given a Euclidean space $\mathbb{X}$, over which a density function $\varrho\colon \mathbb{X} \to \mathbb{R}$ is defined based on the set of data objects $\mathcal{O} \subset \mathbb{X}$, then, using ϱ as the target function of local searches for mode searching with infinitesimal neighborhoods, a cluster is the basin of attraction of a single attractor of these local searches initialized with elements of $\mathbb{X}$. Data objects that are elements of the same basin of attraction belong to the same cluster.*

2.1 Density Attractor Clustering Using Kernel Density Estimation

Clustering based on Definition 3 requires two major design decisions: How to model the density of the objects and what algorithm to use to assign the objects to the cluster modes. Figure 2 exemplifies the method.

The canonical choice for modeling the density function is using kernel density estimation

$$\varrho(\boldsymbol{x}) = \frac{1}{|\mathcal{O}|} \cdot \sum_{i=1}^{|\mathcal{O}|} \frac{1}{\sqrt{(2 \cdot \pi)^F} \cdot \omega^F} \cdot \exp\left(\frac{\|\boldsymbol{x} - \boldsymbol{o}_i\|_2^2}{-2 \cdot \omega^2}\right) \tag{1}$$

with a Gaussian kernel with bandwidth $\omega \in \mathbb{R}$, with $\boldsymbol{o}_i$ being the mode of the i-th Gaussian, and F the dimensionality of the data. A considerable amount of work has been put into discussing different kernels, bandwidth choices, and approaches in modeling the density function. We refer to the literature on why kernel density estimation with a Gaussian kernel is considered to be the first choice in most cases, as the modeling of the density function will not be a topic of this publication.

The canonical choice for assigning each object $\boldsymbol{o}_i$ to a cluster mode is to apply a local search for the maximum of the local density initialized with the object's location $\boldsymbol{o}_i$ and assign objects whose local search converges to the same mode to the same cluster set. To our knowledge, this is the only approach for multivariate data. This paper omits the index for the local search of a particular object. The index i, referring to an object $\boldsymbol{o}_i$, only enumerates symbols used during that local search.

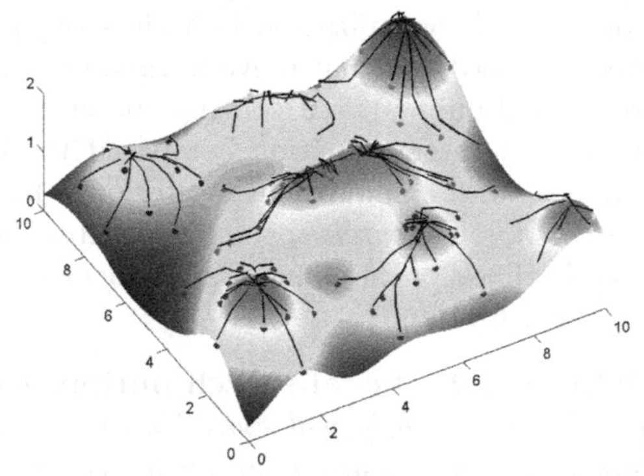

Fig. 2. Density attractor clustering: objects are colored by clustering, lines show local search trajectories. (Color figure online)

3 Theory: Methods for Solving Density Attractor Clustering

Different local search classes exist, of which we discuss four that approximate the target function locally. The first two classes are well researched and usually use Taylor approximation. While mean shift search belongs to a third class, we show that its update rule Eq. (7) is by no means arbitrary, but implicitly uses Taylor approximation as well. This also lays the foundation to develop our more advanced methods.

For Euclidean search spaces, a single iteration step of a local search method can be expressed as

$$\boldsymbol{x}_{t+1} = \boldsymbol{x}_t + \eta_t \cdot \boldsymbol{p}_t\ , \quad \eta_t \in \mathbb{R}\ , \quad \boldsymbol{p}_t \in \mathbb{R}^F\ , \quad t, T \in \mathbb{N}\ , \quad 1 \leq t \leq T\ , \tag{2}$$

where $\frac{\boldsymbol{p}_t}{\|\boldsymbol{p}_t\|_2}$ is the search direction, η_t is the step rate, and the step size is $\eta_t \cdot \|\boldsymbol{p}_t\|_2 = \|\boldsymbol{x}_{t+1} - \boldsymbol{x}_t\|_2$. Addressing the framework of local search, $\mathcal{N}_t(\boldsymbol{x}_t)$ is determined by the step size. Since Definition 3 expects $\mathcal{N}$ to be infinitesimal, any line search with a positive step size is an approximation that meets Definition 3 only if the step size is not too large.

Different classes of local search exist to determine $\boldsymbol{p}_t$ and η_t: (i) Trust region optimization (TROpt) methods first determine the *maximal* step size. Then they choose an element of the resulting subset of the search space (the *trust region*) as the next solution, thus determining the step direction (and the effective step size) implicitly. (ii) Line search optimization (LSOpt) methods first determine the search direction in which to improve. Then they choose an element of the search space along this direction from the current solution (the *line search*) as the next solution, thus determining the step size implicitly. (iii) Closed-form optimization (CFOpt) methods determine $\boldsymbol{p}_t$ and η_t explicitly with a closed-form expression without an additional subroutine. (iv) Simultaneous search optimization (SSOpt) methods determine $\boldsymbol{p}_t$ and η_t simultaneously using a numerical subprocess. TROpt and LSOpt are classical approaches, while some advances in CFOpt and SSOpt are more recent.

In order to choose $\boldsymbol{x}_{t+1}$ from the trust region in TROpt or $\boldsymbol{p}_t$ in LSOpt, many methods construct polynomial local models. From Taylor's theorem we get the Taylor expansion

$$\varrho(\boldsymbol{x}_t + \eta_t \cdot \boldsymbol{p}_t) = \varrho(\boldsymbol{x}_t) + \eta_t \cdot \boldsymbol{p}_t^\intercal \cdot \nabla\varrho(\boldsymbol{x}_t) + 0.5 \cdot \eta_t^2 \cdot \boldsymbol{p}_t^\intercal \cdot \nabla^2\varrho(\boldsymbol{x}_t) \cdot \boldsymbol{p}_t + \sum_{i=3}^{\infty} \frac{(\eta_t \cdot \boldsymbol{p}_t \cdot \nabla)^i \varrho(\boldsymbol{x}_t)}{i!} \tag{3}$$

with which we approximate ϱ polynomially around $\boldsymbol{x}_t$ by aborting Eq. (3) after one of the summands. Particularly useful are models built by aborting either after the second or the third summand, thus using up to the first (1^{st}OD) or the second order derivative (2^{nd}OD).

We denote optimization methods using *up to* the 1^{st}OD or 2^{nd}OD as 1^{st}OD or 2^{nd}OD methods (not necessarily building *polynomial* models from these). If a 1^{st}OD method builds *linear* local models, then it uses the gradient as $\boldsymbol{p}_t$ and is called *gradient ascent*.

In the context of density attractor clustering, the standard algorithm for the local search is mean shift search, which is a 1^{st}OD CFOpt gradient ascent method, as we show in Sect. 3.1. This motivates our CFOpt method in Sect. 3.2 which constructs a *non-polynomial* model. To our knowledge, it is the first 2^{nd}OD CFOpt method. We refine it into a 2^{nd}OD SSOpt method in Sect. 3.3.

3.1 Deriving Mean Shift Search as 1^{st}OD CFOpt Gradient Ascent Method with Adaptive Step Rate from Taylor Series

For gradient ascent, we locally approximate the density function by aborting after the second summand of Eq. (3), thus building a linear local model at $\boldsymbol{x}_t$. To find the step direction, we solve

$$\max_{\boldsymbol{p}_t}(\ \varrho(\boldsymbol{x}_t + \eta_t \cdot \boldsymbol{p}_t)\ )\,, \text{(here) subject to } \|\boldsymbol{p}_t\|_2 = 1\,. \tag{4}$$

Since $\boldsymbol{p}_t^\mathsf{T} \cdot \nabla\varrho(\boldsymbol{x}_t) = \|\boldsymbol{p}_t\|_2 \cdot \|\nabla\varrho(\boldsymbol{x}_t)\|_2 \cdot \cos(\phi) = \|\nabla\varrho(\boldsymbol{x}_t)\|_2 \cdot \cos(\phi)$, where ϕ is the angle between $\boldsymbol{p}_t$ and $\nabla\varrho(\boldsymbol{x}_t)$, the maximum is attained for $\cos(\phi) = 1$ and $\boldsymbol{p}_t = \frac{\nabla\varrho(\boldsymbol{x}_t)}{\|\nabla\varrho(\boldsymbol{x}_t)\|_2}$ must apply. Finally, now omitting the normalization, we obtain $\boldsymbol{x}_{t+1} = \boldsymbol{x}_t + \eta_t \cdot \nabla\varrho(\boldsymbol{x}_t)$ as the gradient ascent.

For the following theorem and the proof, we first define the density function based on kernel density estimation for an arbitrary kernel function k as

$$\varrho(\boldsymbol{x}) = \frac{1}{|\mathcal{O}|} \cdot \sum_{i=1}^{|\mathcal{O}|} \frac{1}{\omega^F} \cdot \hat{k}_i(\boldsymbol{x})\ , \quad (5) \quad \hat{k}_i(\boldsymbol{x}) = k\left(\left\|\frac{\boldsymbol{x} - \boldsymbol{o}_i}{\omega}\right\|_2\right)\,. \tag{6}$$

Theorem 4. *The mode searching method "mean shift search" that uses the standard update rule with a kernel h (left in Eq. (7), $\hat{h}_i(\boldsymbol{x}) = h(\|(\boldsymbol{x} - \boldsymbol{o}_i)/\omega\|_2)$) is equivalent to a 1^{st}OD CFOpt gradient ascent method (right in Eq. (7)), where the target function is a density function calculated by kernel density estimation with an (unnormalized) kernel $k(y) = -\int y \cdot h(y)\ \mathrm{d}y$, namely*

$$\boldsymbol{x}_{t+1} = \frac{\sum_{i=1}^{|\mathcal{O}|} \hat{h}_i(\boldsymbol{x}_t) \cdot \boldsymbol{o}_i}{\sum_{i=1}^{|\mathcal{O}|} \hat{h}_i(\boldsymbol{x}_t)} \overset{(5)}{=} \boldsymbol{x}_t + \underbrace{\frac{\omega^{2+F} \cdot |\mathcal{O}|}{\sum_{i=1}^{|\mathcal{O}|} \hat{h}_i(\boldsymbol{x}_t)}}_{\eta_t} \cdot \underbrace{\nabla\varrho(\boldsymbol{x}_t)}_{\boldsymbol{p}_t}\,. \tag{7}$$

Here, $\boldsymbol{x}_{t+1} - \boldsymbol{x}_t$ is the mean shift step. (Proof is available in the suppl. material.)

3.2 Deriving Focused Gauss Shift Search as 2ndOD CFOpt Method with Adaptive Step Rate from Taylor Series

Unfortunately, gradient ascent is sensitive to (local) poor scaling [18, p. 27]. Poor scaling means that changes to $\boldsymbol{x}$ in a certain direction produce much larger variations in $\varrho(\boldsymbol{x})$ than changes in another direction. We can eliminate this issue by taking the 2ndOD into account: We locally approximate $\varrho(\boldsymbol{x})$ by aborting Eq. (3) after the third summand, thus building a quadratic local model at $\boldsymbol{x}_t$. To get the search direction, we assume for now that

$$\nabla^2\varrho(\boldsymbol{x}_t) \prec 0 \,, \tag{8}$$

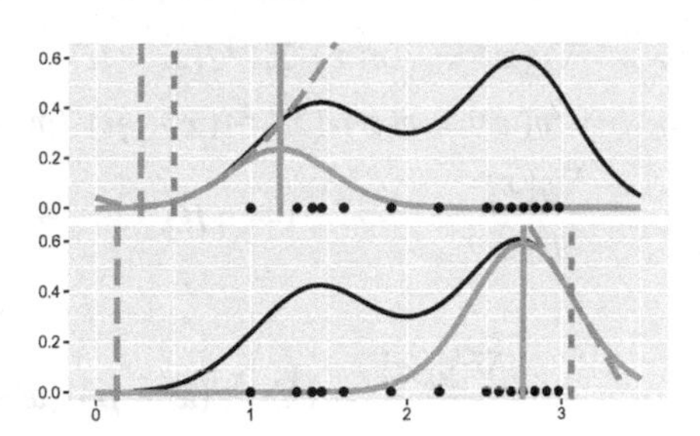

Fig. 3. Gaussian models (curves) surpass quadratic ones (curves). **Top:** If $\mathbf{H}_t \succ 0$ at $\boldsymbol{x}_t$ (⁝), Newton's step (¦) minimizes unintentionally, while Gauss shift step (|) still maximizes. **Bottom:** $\mathbf{H}_t \prec 0$ at $\boldsymbol{x}_t$ (⁝), yet Newton's step (¦) overleaps all modes, since its step size is too large, while Gauss shift step (|) converges very quickly to the mode.

solve Eq. (4) (here without the constraint) by setting the derivative of ϱ equal zero

$$0 \stackrel{!}{=} \nabla_{\boldsymbol{p}_t}\, \varrho(\boldsymbol{x}_t + \eta_t \cdot \boldsymbol{p}_t)$$
$$\stackrel{(3)}{=} \eta_t \cdot \nabla\varrho(\boldsymbol{x}_t) + \eta_t^2 \cdot \nabla^2\varrho(\boldsymbol{x}_t) \cdot \boldsymbol{p}_t \iff \boldsymbol{p}_t \quad = -\left(\eta_t \cdot \nabla^2\varrho(\boldsymbol{x}_t)\right)^{-1} \cdot \nabla\varrho(\boldsymbol{x}_t)$$

and receive $\boldsymbol{x}_{t+1} = \boldsymbol{x}_t - \left(\nabla^2\varrho(\boldsymbol{x}_t)\right)^{-1} \cdot \nabla\varrho(\boldsymbol{x}_t)$, which is the update rule of the CFOpt method *Newton's method*. Here, $\boldsymbol{x}_{t+1}$ is the mode of the *quadratic* local model. Re-introducing a step rate and using the notation $\mathbf{H}_t$ for the inverse Hessian, we receive the update rule of *Dampened Newton's method*

$$\boldsymbol{x}_{t+1} = \boldsymbol{x}_t - \eta_t \cdot \mathbf{H}_t \cdot \nabla\varrho(\boldsymbol{x}_t) \;, \quad \mathbf{H}_t \coloneqq \left(\nabla^2\varrho(\boldsymbol{x}_t)\right)^{-1} \;, \quad -\mathbf{H}_t \cdot \nabla\varrho(\boldsymbol{x}_t) = \boldsymbol{p}_t \;. \tag{9}$$

This method usually produces a much better step (*Newton's step*) than gradient ascent and – consequently – mean shift search. However, Dampened Newton's method requires, just as Newton's method, $\nabla^2\varrho(\boldsymbol{x}_t) \prec 0$.

Motivated by this and by considering the fact that $\varrho(\boldsymbol{x})$ is constructed from a sum of Gaussians, we build a *Gaussian* local model from ϱ and its first two derivatives instead of a polynomial model. The advantage is that the Gaussian model g improves on the approximation of ϱ and thus relaxes the requirement (8). With the model parameters $\alpha \in \mathbb{R}, \boldsymbol{\mu} \in \mathbb{R}^F, \boldsymbol{\Sigma} \in \mathbb{R}^{F\times F}$, where the symmetric matrix $\boldsymbol{\Sigma}$ has to be invertible, we have the model equations (omitting t for now)

$$g(\boldsymbol{x}) = \alpha \cdot \exp(-0.5 \cdot (\boldsymbol{x} - \boldsymbol{\mu})^\intercal \cdot \boldsymbol{\Sigma}^{-1} \cdot (\boldsymbol{x} - \boldsymbol{\mu})) \tag{10}$$

$$\Longrightarrow \quad \nabla g(\boldsymbol{x}) = -g(\boldsymbol{x}) \cdot \boldsymbol{\Sigma}^{-1} \cdot (\boldsymbol{x} - \boldsymbol{\mu})$$

$$\Longrightarrow \quad \nabla^2 g(\boldsymbol{x}) = g(\boldsymbol{x}) \cdot (\boldsymbol{\Sigma}^{-1} \cdot (\boldsymbol{x} - \boldsymbol{\mu}) \cdot (\boldsymbol{x} - \boldsymbol{\mu})^\intercal - \mathbf{I}) \cdot \boldsymbol{\Sigma}^{-1} \qquad \text{which lead to}$$

$$\frac{\nabla g(\boldsymbol{x})}{g(\boldsymbol{x})} = -\boldsymbol{\Sigma}^{-1} \cdot (\boldsymbol{x} - \boldsymbol{\mu}) \quad \Longleftrightarrow \quad \boldsymbol{\mu} = \boldsymbol{x} + \boldsymbol{\Sigma} \cdot \frac{\nabla g(\boldsymbol{x})}{g(\boldsymbol{x})} \tag{11}$$

$$\begin{aligned} \frac{\nabla^2 g(\boldsymbol{x})}{g(\boldsymbol{x})} &= (\boldsymbol{\Sigma}^{-1} \cdot (\boldsymbol{x} - \boldsymbol{\mu}) \cdot (\boldsymbol{x} - \boldsymbol{\mu})^\intercal - \mathbf{I}) \cdot \boldsymbol{\Sigma}^{-1} \\ &\overset{(11)}{=} \left(\boldsymbol{\Sigma}^{-1} \cdot \left(-\boldsymbol{\Sigma} \cdot \frac{\nabla g(\boldsymbol{x})}{g(\boldsymbol{x})}\right) \cdot \left(-\boldsymbol{\Sigma} \cdot \frac{\nabla g(\boldsymbol{x})}{g(\boldsymbol{x})}\right)^\intercal - \mathbf{I}\right) \cdot \boldsymbol{\Sigma}^{-1} \\ &\overset{\boldsymbol{\Sigma} = \boldsymbol{\Sigma}^\intercal}{=\!=} \frac{\nabla g(\boldsymbol{x}) \cdot (\nabla g(\boldsymbol{x}))^\intercal}{g(\boldsymbol{x})^2} - \boldsymbol{\Sigma}^{-1} \end{aligned}$$

$$\Longleftrightarrow \qquad \boldsymbol{\Sigma}^{-1} = \frac{\nabla g(\boldsymbol{x}) \cdot (\nabla g(\boldsymbol{x}))^\intercal - \nabla^2 g(\boldsymbol{x}) \cdot g(\boldsymbol{x})}{g(\boldsymbol{x})^2} \tag{12}$$

Since g shall approximate ϱ in the neighborhood $\mathcal{N}_t(\boldsymbol{x}_t)$, we impose the conditions

$$g(\boldsymbol{x}_t) \overset{!}{=} \varrho(\boldsymbol{x}_t)\,, \nabla g(\boldsymbol{x}_t) \overset{!}{=} \nabla \varrho(\boldsymbol{x}_t)\,, \nabla^2 g(\boldsymbol{x}_t) \overset{!}{=} \nabla^2 \varrho(\boldsymbol{x}_t) \tag{13}$$

at $\boldsymbol{x}_t$. By substituting g with ϱ and $\boldsymbol{x}$ with $\boldsymbol{x}_t$ in Eq. (12), plugging Eq. (12) into Eq. (11) and setting the next candidate solution $\boldsymbol{x}_{t+1}$ to be the mean of the Gaussian local model, we receive

Theorem 5. *The mode searching method "focused Gauss shift search", defined by its update rule (formulations are equivalent)*

$$\begin{aligned} \boldsymbol{x}_{t+1} &= \boldsymbol{x}_t + \left(\nabla \varrho(\boldsymbol{x}_t) \cdot (\nabla \varrho(\boldsymbol{x}_t))^\intercal / \varrho(\boldsymbol{x}_t) - \nabla^2 \varrho(\boldsymbol{x}_t)\right)^{-1} \cdot \nabla \varrho(\boldsymbol{x}_t) \\ &= \boldsymbol{x}_t - \left(1 + \frac{(\nabla \varrho(\boldsymbol{x}_t))^\intercal \cdot \mathbf{H}_t \cdot \nabla \varrho(\boldsymbol{x}_t)}{\varrho(\boldsymbol{x}_t) - (\nabla \varrho(\boldsymbol{x}_t))^\intercal \cdot \mathbf{H}_t \cdot \nabla \varrho(\boldsymbol{x}_t)}\right) \cdot \mathbf{H}_t \cdot \nabla \varrho(\boldsymbol{x}_t)\,, \end{aligned} \tag{14}$$

is a 2nd*OD CFOpt method, which (a) search direction is parallel to the Newton's method, and which (b) adaptive step rate relaxes the requirement* (8). *Here,* $\boldsymbol{x}_{t+1} - \boldsymbol{x}_t$ *is the focused Gauss shift step.* (The proof is available in the supplementary material.)

3.3 Deriving Smoothing Gauss Shift Search as an NthOD SSOpt Method

Though we relaxed the requirement (8) with the Gaussian local model, we did not eliminate it yet. Additionally, Eq. (14) shares the issue with Newton's step that the step size can become too large to meet Definition 3 here if the eigenvalues of $\boldsymbol{\Sigma}_t^{-1}$ are very small or $(\nabla \varrho(\boldsymbol{x}_t))^\intercal \cdot \mathbf{H}_t \cdot \nabla \varrho(\boldsymbol{x}_t) \approx \varrho(\boldsymbol{x}_t)$. We remedy this and additionally increase the reliability of each Gauss shift step by generalizing *focused* Gauss shift search to *smoothing* Gauss shift search: First we recognize that the conditions (13) are met if Eq. (14) is reformulated to

$$\alpha_t, \boldsymbol{x}_{t+1}, \boldsymbol{\Sigma}_t^{-1} = \underset{\alpha_t^\star, \boldsymbol{\mu}_t^\star, \boldsymbol{\Sigma}_t^{-1\star}}{\operatorname{argmin}} \left(\delta(\boldsymbol{x}_t) * \sum_{d=0}^{2} \left\| \nabla^d g_t^\star(\boldsymbol{x}_t) - \nabla^d \varrho(\boldsymbol{x}_t) \right\|_2^2 \right)$$

where δ is the Dirac delta function – the neutral element of the binary operation of convolution $*$, $g_t^\star(\boldsymbol{x}_t) = \alpha_t^\star \cdot \exp\Big(-0.5\cdot(\boldsymbol{x}_t - \boldsymbol{\mu}_t^\star)^\intercal \cdot \boldsymbol{\Sigma}_t^{-1\star} \cdot (\boldsymbol{x} - \boldsymbol{\mu}_t^\star)\Big)$ is from Eq. (10), and $d \in \mathbb{N}$. Since δ can be formulated with the limit of a normalized Gaussian $w_\lambda(y) = (\sqrt{2\cdot\pi}\cdot\lambda)^{-F} \cdot \exp(-0.5\cdot y^2/\lambda^2)$ and the convolution as an integration, we get

$$\begin{pmatrix} \alpha_t \\ \boldsymbol{x}_{t+1} \\ \boldsymbol{\Sigma}_t^{-1} \end{pmatrix} = \underset{\alpha_t^\star, \boldsymbol{\mu}_t^\star, \boldsymbol{\Sigma}_t^{-1\star}}{\operatorname{argmin}} \left(\int_{\boldsymbol{x}\in\mathbb{R}^F} \lim_{\lambda\to 0} \Big(w_\lambda(\|\boldsymbol{x} - \boldsymbol{x}_t\|_2) \Big) \cdot \sum_{d=0}^{2} \left\| \nabla^d g_t^\star(\boldsymbol{x}) - \nabla^d \varrho(\boldsymbol{x}) \right\|_2^2 \, \mathrm{d}\boldsymbol{x} \right).$$

The generalization is performed by allowing an arbitrary locality $\lambda \in \mathbb{R}$ and an arbitrary set of derivatives $\{\nabla^d \mid 0 \le d \le D \in \mathbb{N}\}$ to fit the local model. D can be chosen arbitrarily and can be small for reasons discussed in the next section. We obtain the model parameters by solving

$$\begin{pmatrix} \alpha_t \\ \boldsymbol{x}_{t+1} \\ \boldsymbol{\Sigma}_t^{-1} \end{pmatrix} = \underset{\alpha_t^\star, \boldsymbol{\mu}_t^\star, \boldsymbol{\Sigma}_t^{-1\star}}{\operatorname{argmin}} \left(\int_{\boldsymbol{x}\in\mathbb{R}^F} w_\lambda(\|\boldsymbol{x} - \boldsymbol{x}_t\|_2) \cdot \sum_{d=0}^{D} \left\| \nabla^d g_t^\star(\boldsymbol{x}) - \nabla^d \varrho(\boldsymbol{x}) \right\|_2^2 \mathrm{d}\boldsymbol{x} \right) \quad (15)$$

subject to $\alpha_t \overset{!}{>} 0$ and $\boldsymbol{\Sigma}_t^{-1} \overset{!}{\succeq} 0$. No inversion of $\boldsymbol{\Sigma}_t \overset{!}{\succ} 0$ is required, since we optimize $\boldsymbol{\Sigma}_t^{-1}$ directly, so *semi*-definiteness of $\boldsymbol{\Sigma}_t^{-1}$ is sufficient. Using Eq. (15) during the mode search effectively fits the local model to ϱ, evaluated at the entire search space instead of just $\boldsymbol{x}_t$. We still obtain a *local* model, since the weight function w: $\mathbb{R} \to \mathbb{R}$ emphasizes residuals closer to $\boldsymbol{x}_t$ over those farther away. In the context of local search, w determines $\mathcal{N}_t(\boldsymbol{x}_t)$ and makes sure ϱ is well approximated in $\mathcal{N}_t(\boldsymbol{x}_t)$. The resulting overall clustering method is called *smoothing Gauss shift*.

4 Algorithm: Smoothing Gauss Shift with Linear Time Complexity

Obviously, solving Eq. (15) is intractable. Thus, for implementation, we sample support vectors $\boldsymbol{s}_j \in \mathbb{R}^F$ $(j, J \in \mathbb{N}, 1 \le j \le J)$, instead of integrating over the entire $\mathbb{R}^F$, and receive

$$\alpha_t, \boldsymbol{x}_{t+1}, \boldsymbol{\Sigma}_t^{-1} = \underset{\alpha_t^\star, \boldsymbol{\mu}_t^\star, \boldsymbol{\Sigma}_t^{-1\star}}{\operatorname{argmax}} \left(\sum_{j=1}^{J} \left(w_\lambda(\|\boldsymbol{s}_j - \boldsymbol{x}_t\|_2) \cdot \sum_{d=0}^{D} \left\| \nabla^d g_t^\star(\boldsymbol{s}_j) - \nabla^d \varrho(\boldsymbol{s}_j) \right\|_2^2 \right) \right)$$

$$\text{subject to } \alpha_t \overset{!}{>} 0 \text{ and } \boldsymbol{\Sigma}_t^{-1} \overset{!}{\succeq} 0 \quad (16)$$

which is a non-linear least-squares optimization problem. It can be solved, e.g., with sequential least squares programming [13] or semidefinite programming (SDP) for full $\mathbf{\Sigma}_t^{-1}$ or trust region algorithms [1,25] for diagonal $\mathbf{\Sigma}_t^{-1}$. SDP is enabled by the fact that *semi*-definiteness is sufficient for $\mathbf{\Sigma}_t^{-1}$. Open source implementations can be found, among others, at SciPy [24]. We refer to the rich literature on discussions on solvers, since it is not a topic of this publication.

Mean shift has a time complexity of $O\big(\Phi\cdot\bar{T}\cdot\bar{\Omega}\cdot F\big)$ and focused Gauss shift of $O\big(\Phi\cdot\bar{T}\cdot\bar{\Omega}\cdot F^2\big)$, where Φ denotes the number of mode searches, $\bar{T}$ the average number of mode search steps per mode search, $\bar{\Omega}$ the average number of objects used per mode search step, and F the dataset dimension; without approximation $\Phi=|\mathcal{O}|$, $\bar{\Omega}=|\mathcal{O}|$. The number of $\nabla^d\varrho$-calls equals $\Phi\cdot\bar{T}$. *Smoothing* Gauss shift improves on this: Reusing the support vectors for each mode search step allows us to calculate the J support vectors beforehand ($J \ll |\mathcal{O}|$). This makes the number of $\nabla^d\varrho$-calls – now J – independent of Φ and thus independent of $|\mathcal{O}|$, which is why Gauss shift has linear time complexity in the number of data objects. The time complexity is $O\Big(J\cdot\bar{\Omega}\cdot F^{\max(\{1,\,D\})}+\Phi\cdot\bar{T}\cdot F^{D+\Psi}\Big)$, where the number of residuals is of order $O\big(F^D\big)$, and $\Psi\in[1,2]$ depends on the degree of freedom of $\mathbf{\Sigma}_t^{-1\star}$ (e.g. if $\mathbf{\Sigma}_t^{-1\star}$ is non-diagonal, $\Psi>1$). Our paper is the first to introduce this decoupling. Hereafter, we assume that this technique is used and $\Psi=1$.

The support vectors can be received quickly, e.g. by rounding each $\boldsymbol{o}_i$ to the nearest vertex on a rectilinear grid and using the result as support vectors. Note that using support vectors is *not* to be confused with the approach to pre-cluster objects into prototypes in a first stage and then cluster the prototypes with DAC in a second stage. The difference is that pre-clustering approximates $\nabla^d\varrho$ and changes the initials of the mode searches, while Eq. (16) uses *every* object $\boldsymbol{o}_i$ to calculate $\nabla^d\varrho$ exactly and uses the original objects as initials. The support vectors simply specify the residuals which are calculated during the local search, thus completely meeting Definition 3.

One might consider using support vectors to construct linear or quadratic local models. However, a linear model would not provide a step size and we would have to determine η_t in an additional (costly) subroutine, while a quadratic local model would choose too large step sizes in the tails of ϱ and thus would not meet Definition 3 (see Sect. 3.2, especially Fig. 3).

As mentioned, D can be chosen to be small. This is justified by the principles of numerical differentiation, which state that higher order derivatives can be sufficiently well approximated using multiple lower order derivatives or function values. By implication, information regarding the former is contained within a set of the later. This means that even if we use low order derivatives $\nabla\varrho$ or only ϱ at each $\boldsymbol{s}_j$ ($D=1$ or $D=0$), we still – in effect – preserve a 2^{nd}OD model and thus keep its fast rate of local convergence, which is typically quadratic for methods using $\nabla^2\varrho(\boldsymbol{x})$ [18, p. 23].

5 Review of the State-of-the-Art

In order to increase the speed of DAC, the number of kernel evaluations needs to be reduced. This can be achieved by (i) decreasing $\bar{\Omega}$ by approximating $\nabla^d \varrho$ or distances, or (ii) decreasing the number of calls to $\nabla^d \varrho$. The number of $\nabla^d \varrho$-calls ($\Phi \cdot \bar{T}$ for mean shift) can be decreased by reducing Φ or $\bar{T}$, or by ensuring that, in the first place, Φ does not influence the number of $\nabla^d \varrho$-calls via decoupling. In the case of decoupling, $\nabla^d \varrho$ are still called at some point and (other) computations are performed per mode search step, so reducing $\bar{\Omega}$, Φ, and $\bar{T}$ still improves speed.

Approximating $\nabla^d \varrho$ and Distances. When approximating the $\nabla^d \varrho$, one option is to **ignore** every $\boldsymbol{o}_i$ that is (i) outside a certain range of the current $\boldsymbol{x}_t$ [3, method *ms2*], which equals using a truncated kernel, or is (ii) not one of the k-nearest neighbors of $\boldsymbol{x}_t$, or (iii) to randomly ignore a percentage of $\mathcal{O}$ (which equals mini-batch optimization). Note that *ms2*, leveraging that images pixels are arranged on a grid, only works for image segmentation. It can incur large errors [3]. These techniques can be used in Gauss shift when calculating $\nabla^d \varrho$ or, in smoothing Gauss shift, to approximate the distances between $\boldsymbol{x}_t$ and the $\boldsymbol{s}_j$.

Another option, originally called *sparse EM* [17], is to use all objects when calculating the initial $\nabla^d \varrho(\boldsymbol{x}_1)$ [10] and possibility at few other steps $\boldsymbol{x}_{t'}$ [3, *ms3*], but to otherwise **freeze** some of the distances $\|\boldsymbol{x}_t - \boldsymbol{o}_i\|_2$ for subsequent $\boldsymbol{x}_t$. A distance of an object $\boldsymbol{o}_i$ to the candidate solution is frozen if $\boldsymbol{o}_i$ is (i) outside a certain range of $\boldsymbol{x}_{t'}$ [3, *ms3*], is (ii) not one of the k-nearest neighbors of $\boldsymbol{x}_{t'}$ [10], or (ii) randomly chosen distances are frozen. The speedup of *ms3* is similar or slightly worse than for *ms2*, but the results are more reliable [3]. These techniques can be used in Gauss shift to approximate the distances between $\boldsymbol{x}_t$ and the $\boldsymbol{s}_j$.

A pre-clustering stage can be used either to approximate $\nabla^d \varrho$ [11] and decrease $\bar{\Omega}$, or each initial $\boldsymbol{x}_1$ [6] and decrease Φ. Pre-clustering can also be done per mode search step, aggregating groups of distant objects before calculating $\nabla^d \varrho$, possibly using the *fast multipole method* on top [14,28]. A considerable disadvantage is, however, that this methodology cannot be used for non-standard kernel densities [15]. These techniques can be used in Gauss shift when calculating $\nabla^d \varrho$.

The above approaches can be made fast by using spatial indexing using tree data structures, but this is only efficient for small ω. For an ω in the magnitude that is required to produce few clusters, a considerable percentage of the objects is needed to calculate the $\nabla^d \varrho$ correctly. In this case, most of the tree branches must be explored and the time complexity of tree structures is no longer $O(|\mathcal{O}| \cdot log(|\mathcal{O}|))$, but $O(|\mathcal{O}|^2)$ [5].

Gauss shift still requires calculating the distance for each pair of $\boldsymbol{s}_j$ and $\boldsymbol{o}_i$ for determining $\nabla^d \varrho$, as well as between each $\boldsymbol{x}_t$ and the $\boldsymbol{s}_j$ during the mode search. Therefore, the mentioned techniques are orthogonal to ours and can be integrated into Gauss shift, just as they have been integrated into mean shift. However, these methods only improve the absolute runtime, not the time complexity.

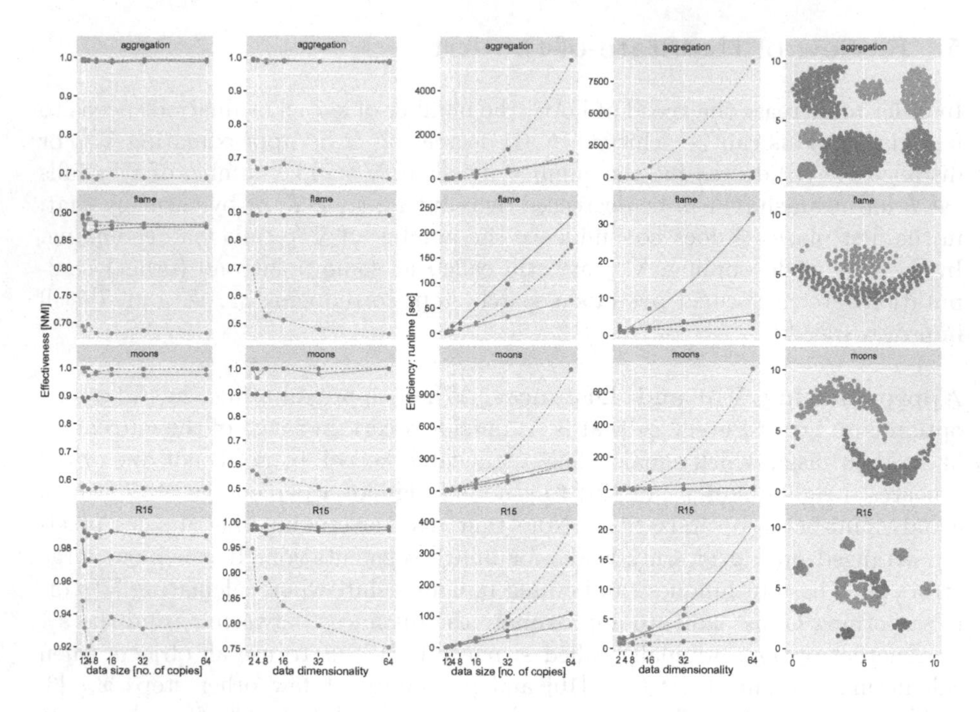

Fig. 4. Quantifying experiments for mean shift, *ms4*, and Gauss shift for $D = 1$, $D = 0$: While other methods have *quadratic* time complexity, Gauss shift has *linear* complexity with regard to dataset size and retains excellent effectiveness. Gauss shift $D = 0$ has linear complexity also regarding dataset dimensionality.

Techniques to Reduce $\bar{T}$. In order to accelerate the linear convergence of mean shift search, it is possible to switch between a mean shift step and a Newton's step [3, *ms4*]. While the authors of *ms4* claim speedup factors of 1.5 to 6 (larger than for *ms2* or *ms3*), we could not reproduce this. Tracking trajectories of previous mode searches can be used to predict the limit of the current mode search [3, *ms1*]. The idea is orthogonal to our method and can be used additionally. Like *ms2*, *ms1* can only be used for image segmentation. Again, these methods only improve the absolute runtime, not the time complexity class.

Related Methods for Non-Euclidean Spaces. The DAC methods mentioned so far work in Euclidean space. DAC methods for metric spaces assign the next $\boldsymbol{x}_{t+1}$ to be either the locality-weighted medoid [22], the closest $\boldsymbol{o}_i$ with a higher density [23], or the Tukey median [21]. (Clustering by density peaks [20] is a repetition of quick shift [23].) A method for manifolds utilizes the geometric median [26]. These methods are not primarily designed for Euclidean space and have runtime complexities ranging from $O(|\mathcal{O}|^2)$ to $O(|\mathcal{O}|^3)$ with larger runtime constants than mean shift.

Related Non-DAC Methods. *Blurring mean shift* performs a mean shift step for every $\boldsymbol{o}_i$ simultaneously and moves every $\boldsymbol{o}_i$, thus changing the target function, with each step. In contrast to mean shift, this method and its hierarchical version [5] do not adhere to DAC, but to a notion of iteratively smoothing the density until a stability of entropy is reached [4].

While DBSCAN* has quadratic runtime complexity, improvements of DBSCAN exist [8,12]. But, as mentioned, these methods do not adhere to Definition 3.

6 Experiments

Methods. The standard method to solve the DAC problem is mean shift, which is a reasonably fast algorithm. Most of the techniques and methods in Sect. 5 can (i) be used together with our method, (ii) are known to be very slow, or (ii) do not follow Definition 3. The first group is not competing against our method, the second group is already known to be slower than mean shift, and the third group solves a differently defined problem. Comparing against a different definition can make sense, but only when having an application in mind or at least a (personal and subjective) expectation. The only method that does not fall into any of those groups is *ms4*, so we are going to compare against it in addition to mean shift. We put smoothing Gauss shift with $D=1$ to the test (as the main contender) and a variant with $D=0$, for reasons we will explain in the following.

All experiments were performed on a regular Lenovo Thinkpad T530 office notebook.

6.1 Quantitative Experiments

Setup. The well-known synthetic benchmark datasets *aggregation* ($|\mathcal{O}|=788$), *flame* ($|\mathcal{O}| = 240$), *moons* ($|\mathcal{O}| = 422$), and *R15* ($|\mathcal{O}| = 600$) serve as basis. They are displayed in the last column in Fig. 4, where the coloring indicates the reference labels we use when measuring effectiveness.

Two series of experiments were conducted. In the first series ($|\mathcal{O}|$-exp.), the basis datasets were copied for a certain number of times, concatenated, and added with Gaussian noise (see columns 1,3 in Fig. 4). E.g. for *aggregation*, this amounts to up to 50,432 data objects for a multiple of 64. In the second series (F-exp.), the basis datasets were linearly mapped into a Euclidean space of higher dimension and added with Gaussian noise (see columns 2,4 in Fig. 4).

Effectiveness. To measure effectiveness (columns 1,2 in Fig. 4), we calculated the geometrically normalized mutual information (NMI) between the clustering results obtained by the methods and the labels. NMI is a standard measurement metric used in the literature for comparing clustering results – 1 is a perfect match, 0 the opposite. Since it is known that mean shift solves the DAC problem exactly, an alternative method can use its results for comparison. Therefore, the main goal is not to have an NMI as high as possible, but to have the same NMI as mean shift. *ms4* is able to reproduce the results of mean shift sufficiently only for low-dimensional *R15*. The reason is that the Newton's step often takes step sizes that are too large. Since we implemented *ms4* faithfully, we suspect that either important details were not published or it is working particularly well for image segmentation – the sole published use case [3]. In contrast, Gauss shift always reproduces the results of mean shift – virtually exactly for $D=1$ and close for $D=0$.

Efficiency for Increasing $|\mathcal{O}|$. To validate the claims of Sect. 4, the runtime is displayed in column 3 in Fig. 4. While the time complexity of mean shift and *ms4* is quadratic, it is linear for Gauss shift. For $D=1$, the runtime is always better than for mean shift and nearly always better than for *ms4*.

Efficiency for Increasing F. For increasing F, the column 4 in Fig. 4 shows that Gauss shift with $D=1$ has quadratic time complexity, as foreshadowed in Sect. 4, since the time complexity depends on F with order $O(F^{D+1})$ (with $\Psi=1$ as mentioned). Indeed, for $D=0$ Gauss shift displays linear time complexity. As expected, the time complexity of mean shift is linear and for *ms4* quadratic, due to the need to invert a $F\times F$ matrix during the Newton's step. Gauss shift does not surpass mean shift in each mode search step, due to mean shift's simplicity and (relative) independence of F. However, this is not such an issue in real-life: (i) For high-dimensional datasets, objects are usually located in low-dimensional manifolds. After preprocessing, the problem becomes low-dimensional in practical applications. (ii) If this is not the case, clustering is most likely meaningless because the curse of dimensionality renders distances meaningless. (iii) Even if this is not the case for a particular dataset, apart from *ms4*, the runtime is low anyway, so, here, runtime is not an urgent issue in the first place. (iv) Finally, while datasets for clustering are usually not as high-dimensional as in the F-exp., they may very well become as large as in the $|\mathcal{O}|$-exp., yet the runtimes of the $|\mathcal{O}|$-exp. already dwarf the runtimes of the F-exp. (ignoring *ms4*). So mean shift would only have a small advantage in the (unrealistic) case of a high-dimensional dataset with few objects. From all that follows that the runtime superiority of Gauss shift for large datasets (regardless of dimension) is a huge success, while mean shift's advantage for small, high-dimensional datasets is mostly irrelevant.

6.2 Real-Life Case Study: Clustering Dendritic Spines for Neuroscience

Background. The Institute of Neuro- & Sensory Physiology at the University Medical Center Göttingen uses advanced super-resolution fluorescence microscopy to research the morphology of spines, which are membranous protrusions from neuron dendrites, which receive input from axons at the synapse [19,27]. In 2019 we started collaborating with the institute to analyze data collected from 2015 to 2018. Among other things, we applied DAC to a dendrite dataset, which is the most comprehensive description of any subcellular organelle to date, featuring nanoscale localizations of various proteins. Figure 5 displays one of the recorded microscopy images. From these images, vectors were extracted using domain-specific feature engineering. Regarding the neuroscientific insights, medical details, and further analyses, we refer to the related upcoming publications. Here, we only discuss results concerning this publication: The dataset discussed above describes 19,571 mushroom spines which can be expressed as vectors in nine-dimensional Euclidean space. We compare the clustering results of mean shift and Gauss shift, omitting *ms4* due to inadequate results.

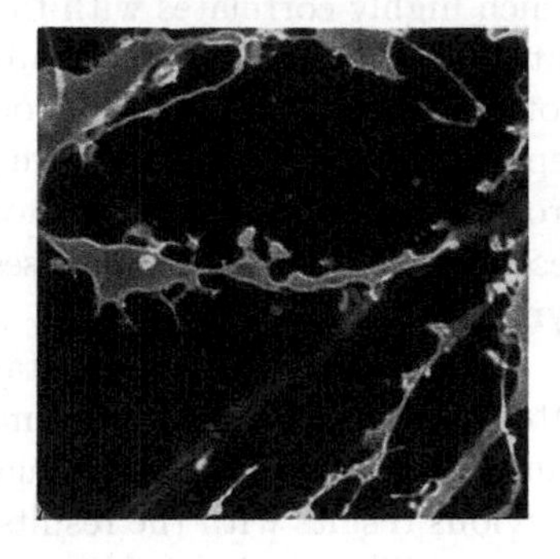

Fig. 5. Dendrites with spines

Results. For appropriate ω, both methods cluster 95.42% of the spines into the same clusters, as displayed in Fig. 6 (top). While the results are virtually the same, mean shift takes up to 66% more time than Gauss shift. This is about what we would expect, considering that for *R15* with $|\mathcal{O}| = 19,200$, mean shift took about 80%–170% more time, but here we are in $\mathbb{R}^9$ instead of $\mathbb{R}^2$, substantiating that being fast for high $|\mathcal{O}|$ is more important than for high F.

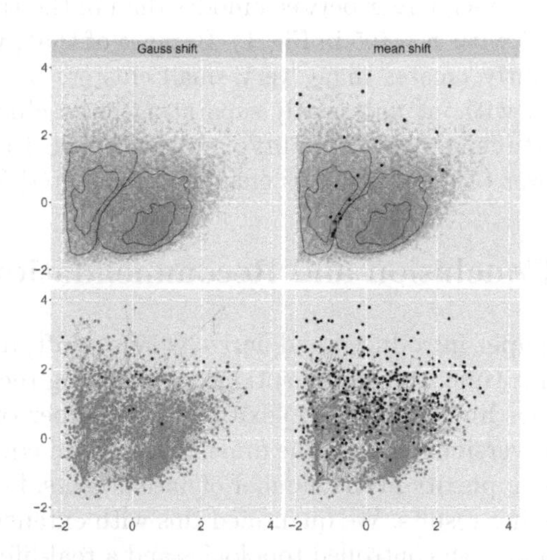

Fig. 6. First two principal components of the spines data. **Top:** Appropriate ω for each method (contours only illustrate the structure). **Bottom:** Decrease in ω leads to a rapid increase of spurious clusters (black) for mean shift only (lines represent paths taken by mode searches, only 8.4% of objects are displayed for clarity). (Color figure online)

The results deliver neuroscientific insights: The first principal component (PC) represents the size and intensity of the Homer postsynaptic density (PSD), which highly correlates with the size of the postsynaptic membrane, which indicates the strength of the synapse. Stronger synapses are the result of long-term potentiation, a well-known process that occurs during learning. The second PC represents the size of the entire postsynapse. Figure 6 suggests the existence of a group with stronger (red) and one with weaker (green) synapses, and, for further research, that the "red" synapses were more involved in learning than the "green" synapses.

The next experiment substantiates that higher-dimensional real-life data can often be clustered in lower dimensions, further emphasizing the greater importance of Gauss shift's advantage over mean shift's advantage: We compared the previous results with the results of clustering only in the first two principal components. The results overlap with each other in 93% of the cases. However, now in the lower dimensional Euclidean space, our method is 14 times faster than mean shift.

Behavior across ω Ranges. While tuning ω is not a core subject of this publication, this spine data reveals an additional interesting and beneficial property of Gauss shift: Over a range of ω, Gauss shift and mean shift produce a similar collection of *dominant* clusters, however mean shift produces many additional spurious clusters. This can become an issue: When using scale-space filtering to tune ω, a search is performed for the largest range of ω in which the number of clusters stays stable [16]. Real-life datasets often have distributions ϱ whose value is much larger between modes than in the (multivariate) tails (e.g. compare $\boldsymbol{x} = 1.5$ with $\boldsymbol{x} = 0.5$ in Fig. 1). Because of that, when ω is decreased, mean shift constantly creates many very small clusters in the tails, usually with just one object, without necessarily separating large clusters. This wrongly suggests to the ω-tuner that no stability can be reached. In contrast, Gauss shift produces far fewer of these spurious clusters, as depicted in Fig. 6.

7 Conclusion and Recommendation

This paper introduced and derived Gauss shift, a method to solve DAC, and, for the first time, the aforementioned decoupling technique. We showed that Gauss shift has linear time complexity in the number of data objects and – depending on the version – also in the dimensionality. In contrast, mean shift has quadratic time complexity in the number of data objects. Both methods return very similar clustering results. We quantified this with extensive analyses with four synthetic datasets with controlled topologies and a real-life case study. We also found that Gauss shift produces fewer spurious clusters than mean shift. Mean shift can only be faster for very small datasets with high dimensions (a rare case in practice). In all other cases, we recommend using Gauss shift.

References

1. Branch, M.A., Coleman, T.F., Li, Y.: A subspace, interior, and conjugate gradient method for large-scale bound-constrained minimization problems. SIAM J. Sci. Comput. **21**(1), 1–23 (1999)
2. Campello, R.J.G.B., Moulavi, D., Sander, J.: Density-based clustering based on hierarchical density estimates. In: Pei, J., Tseng, V.S., Cao, L., Motoda, H., Xu, G. (eds.) PAKDD 2013. LNCS (LNAI), vol. 7819, pp. 160–172. Springer, Heidelberg (2013). https://doi.org/10.1007/978-3-642-37456-2_14
3. Carreira-Perpiñán, M.Á.: Acceleration strategies for Gaussian mean-shift image segmentation. In: IEEE CVPR 2006, vol. 1, pp. 1160–1167 (2006)
4. Carreira-Perpiñán, M.Á.: A review of mean-shift algorithms for clustering (2015)
5. DeMenthon, D., Megret, R.: Spatio-temporal segmentation of video by hierarchical mean shift analysis. Center for Automation Research, University of Maryland, College Park (2002)
6. van der Ende, D., Thiery, J.M., Eisemann, E.: Accelerated mean shift for static and streaming environments. In: The 4th International Conference on Data Analytics (2015)
7. Ester, M., Kriegel, H.P., Sander, J., Xu, X.: A density-based algorithm for discovering clusters in large spatial databases with noise. In: 2nd International Conference on Knowledge Discovery and Data Mining (KDD) (1996)
8. Gionis, A., et al.: Comparison of enhanced DBSCAN algorithms: a review. In: ICIATE, vol. 5 (2017)
9. Hartigan, J.A.: Clustering Algorithms. JWS, Hoboken (1975)
10. Hinneburg, A., Gabriel, H.-H.: DENCLUE 2.0: fast clustering based on kernel density estimation. In: R. Berthold, M., Shawe-Taylor, J., Lavrač, N. (eds.) IDA 2007. LNCS, vol. 4723, pp. 70–80. Springer, Heidelberg (2007). https://doi.org/10.1007/978-3-540-74825-0_7
11. Hinneburg, A., Keim, D.A.: An efficient approach to clustering in large multimedia databased with noise. In: Proceedings of the 4th KDD, pp. 58–65 (1998)
12. Khan, K., Rehman, S.U., Aziz, K., Fong, S., Sarasvady, S.: DBSCAN: past, present and future. In: 5th ICADIWT, pp. 232–238 (2014)
13. Kraft, D.: A software package for sequential quadratic programming. Technical report, DFVLR-FB 88-28, DLR German Aerospace Center - Institute for Flight Mechanics, Cologne, Germany (1988)
14. Lee, D., Moore, A.W., Gray, A.G.: Dual-tree fast Gauss transforms. In: NIPS vol. 18, pp. 747–754 (2006)
15. Leibrandt, R., Günnemann, S.: Making kernel density estimation robust towards missing values in highly incomplete multivariate data without imputation. In: Proceedings of the SIAM SDM 2018 (2018)
16. Leung, Y., et al.: Clustering by scale-space filtering. IEEE TPAMI **22**(12), 1396–1410 (2000)
17. Neal, R.M., Hinton, G.E.: A view of the EM algorithm that justifies incremental, sparse, and other variants. In: Jordan, M.I. (eds.) Learning in Graphical Models. NATO ASI Series (Series D: Behavioural and Social Sciences), vol. 89, pp. 355–368. Springer, Dordrecht (998). https://doi.org/10.1007/978-94-011-5014-9_12
18. Nocedal, J., Wright, S.J.: Numerical Optimization. Springer, New York (2000)
19. Richter, K., et al.: Glyoxal as an alternative fixative to formaldehyde in immunostaining and super-resolution microscopy. EMBO J. **37**(1), 139–159 (2018)

20. Rodriguez, A., Laio, A.: Clustering by fast search and find of density peaks. Science **344**(6191), 1492–1496 (2014)
21. Shapira, L., Avidan, S., Shamir, A.: Mode-detection via median-shift. In: IEEE 12th International Conference on Computer Vision, pp. 1909–1916 (2009)
22. Sheikh, Y.A., Khan, E.A., Kanade, T.: Mode-seeking by Medoidshifts. In: IEEE 11th International Conference on Computer Vision, pp. 1–8 (2007)
23. Vedaldi, A., Soatto, S.: Quick shift and kernel methods for mode seeking. In: Forsyth, D., Torr, P., Zisserman, A. (eds.) ECCV 2008. LNCS, vol. 5305, pp. 705–718. Springer, Heidelberg (2008). https://doi.org/10.1007/978-3-540-88693-8_52
24. Virtanen, P., et al.: SciPy 1.0: fundamental algorithms for scientific computing in Python. Nat. Methods **17**, 261–272 (2020)
25. Voglis, C., Lagaris, I.E.: A rectangular trust region dogleg approach for unconstrained and bound constrained nonlinear optimization. In: 6th WSEAS AMATH (2004)
26. Wang, Y., Huang, X., Wu, L.: Clustering via geometric median shift over Riemannian manifolds. Inf. Sci. **220**, 292–305 (2013)
27. Wilhelm, B., et al.: Composition of isolated synaptic boutons reveals the amounts of vesicle trafficking proteins. Science **344**(6187), 1023–1028 (2014)
28. Yang, C., Duraiswami, R., Gumerov, N.A., Davis, L.: Improved fast Gauss transform and efficient kernel density estimation. In: IEEE 9th International Conference on Computer Vision, vol. 1, pp. 664–671 (2003)

Privacy and Fairness

Privacy-Preserving Decision Trees Training and Prediction

Adi Akavia[1(✉)], Max Leibovich[1], Yehezkel S. Resheff[2], Roey Ron[2], Moni Shahar[3], and Margarita Vald[2,4(✉)]

[1] University of Haifa, Haifa, Israel
adi.akavia@gmail.com, max.fhe.phd@gmail.com
[2] Intuit Inc., Petah Tikva, Israel
hezi.resheff@gmail.com, roey1rg@gmail.com
[3] Facebook Inc., Tel Aviv-Yafo, Israel
monishahar@gmail.com
[4] Tel Aviv University, Tel Aviv-Yafo, Israel
margarita.vald@cs.tau.ac.il

Abstract. In the era of cloud computing and machine learning, data has become a highly valuable resource. Recent history has shown that the benefits brought forth by this data driven culture come at a cost of potential data leakage. Such breaches have a devastating impact on individuals and industry, and lead the community to seek privacy preserving solutions. A promising approach is to utilize Fully Homomorphic Encryption (FHE) to enable machine learning over encrypted data, thus providing resiliency against information leakage. However, computing over encrypted data incurs a high computational overhead, thus requiring the redesign of algorithms, in an "FHE-friendly" manner, to maintain their practicality.

In this work we focus on the ever-popular tree based methods (e.g., boosting, random forests), and propose a new privacy-preserving solution to training and prediction for trees. Our solution employs a low-degree approximation for the step-function together with a lightweight interactive protocol, to replace components of the vanilla algorithm that are costly over encrypted data. Our protocols for decision trees achieve practical usability demonstrated on standard UCI datasets, encrypted with fully homomorphic encryption. In addition, the communication complexity of our protocols is independent of the tree size and dataset size in prediction and training, respectively, which significantly improves on prior works.

Keywords: Fully homomorphic encryption · Privacy preserving machine learning · Decision trees · Training · Prediction

---

The first author thanks the Israel Science Foundation (grant 3380/19) and Israel National Cyber Directorate via the Haifa, BIU and Tel-Aviv cyber centers for their support. The authors wish to thank Yaron Sheffer for helpful discussions.

F. Hutter et al. (Eds.): ECML PKDD 2020, LNAI 12457, pp. 145–161, 2021.
https://doi.org/10.1007/978-3-030-67658-2_9

1 Introduction

The ubiquity of data collected by products and services is often regarded as the key to the so called *AI revolution*. User and usage information is aggregated across individuals, to drive smart products, personalized experience, and automation. In order to achieve these goals, stored data is accessed by multiple microservices each performing different calculations. However, these benefits come at a cost of a threat on privacy.

The public is constantly being informed of data breaches, events which impact privacy and safety of individuals and in turn have a large negative effect on the breached service providers. Whether the leakage is passwords, private pictures and messages, or financial information, it is becoming increasingly clear that drastic measures must be taken to safeguard data that is entrusted to corporations.

There are several approaches to safeguarding data and minimizing the impact of potential breaches. Most fundamentally, encryption of data at rest ensures that even if the entire database is stolen, the data is still safe. While this may have sufficed in the past, the rise of microservice-based architectures in the cloud resulted in a large number of applications having access to the cleartext (i.e., unencrypted) information, making the attack surface uncontrollably large. Ideally, we would like to allow all these applications to operate without ever being exposed to the actual information. Recent advances in the field of Homomorphic Encryption provide some hope of achieving this level of privacy.

Fully Homomorphic Encryption (FHE) [5,6,10,15,16,34] is a type of encryption that allows computation to be performed over encrypted data ("homomorphic computation"), producing an encrypted version of the result. Concretely, FHE supports addition and multiplication over encrypted data, and hence allows evaluating any polynomial. The downside of FHE is the heavy cost of the multiplication operation, which imposes computational limitations on the degree of the evaluated polynomial and the number of total multiplications.

Unfortunately, common computations are not "FHE-friendly" as their polynomial representation is of high degree, which is a major obstacle to the widespread deployment of computing over encrypted data in practice. In particular, machine learning models require complex calculations to train and predict, and adaptations must be made in order to make them practical with FHE. Previous work on machine learning with FHE focused mostly on training and evaluation of logistic regression models, e.g., [8,25], and on more complex models such as shallow neural networks e.g., [17,31]. While these are two widely used classes of models, they are far from encompassing the entire scope of broadly used machine learning methods. In practice, tree based models remain some of the most popular methods, ranging from single decision trees, to random forests and boosting.

A decision tree is a model used for prediction, *i.e.*, mapping a feature vector to a score or a label. The prediction is done by traversing a path from root to leaf, where the path is determined by a sequence of comparison operations "$x_i > \theta$" between a feature value x_i and a threshold θ (continuing to right-child if satisfied, left otherwise). Training is the process of producing a decision tree from a dataset of labeled examples, with the goal of yielding accurate predictions on new unlabeled data instances.

Any solution for decision trees over encrypted data would need to address how to perform the comparison operations over such data. For prediction over encrypted data, Bost *et al.* [4] instantiated the comparison component via an interactive protocol, yielding communication complexity proportional to the tree size; subsequent work [2,7,11,22,27,38,39,42], likewise, followed the interactive approach with communication complexity proportional to the tree size or depth, imposing a significant burden on the bandwidth. In the context of training, existing protocols [12,14,21,29,30,36,40,41,43] consider a multi-party setting where each party holds a cleartext subset of the dataset to be trained on, in contrast to our setting, where the data is encrypted and no entity in the system holds in it the clear. This leaves the question of training decision trees over encrypted data together with non-interactive prediction as an open problem.

Elaborating on the above, this work is motivated by the enterprise setting, with a primary goal of providing a privacy-preserving solution compatible with the existing enterprise architecture. In this architecture, data is stored encrypted in a centralized storage, called *data lake*, and used by multiple microservices (referred to as server) that perform computations on cleartext data decrypted with a key provided by the enterprise key-management service (KMS). The KMS is an entity holding enterprise secrets and keys and providing crypto-services to authorized entities, and thus must be safeguarded. As part of its safeguarding, the KMS is restricted to a lightweight and predefined functionality, in particular, it is prohibited from executing heavy or general purpose code. Our goal is to completely eliminate the microservices access to cleartext data, and replace it with computation over encrypted data producing an encrypted outcome (that may either be decrypted by the KMS or used in encrypted form for subsequent computations). The KMS may be employed for computation on cleartext data, provided it adheres to the aforementioned restrictions on the KMS, in particular, in must be lightweight.

Our Contribution. In this work we present the first protocols for privacy-preserving decision tree based training and prediction that attain all the following desirable properties (see Fig. 3 and 4 and Table 1 in Sect. 4):

1. *Prediction:* a non-interactive protocol on encrypted data.
2. *Training:* a d-round protocol between a server computing on encrypted data and the KMS, with communication complexity independent of the dataset size, where d is the constructed tree depth.
3. *Security:* provable privacy guarantees against an adversary who follows the protocol specification but may try to learn more information (semi-honest).
4. *Practical usability:* high accuracy comparable to the classical vanilla decision tree, fast prediction (seconds) and practical training (minutes to hours) demonstrated on standard UCI datasets encrypted with FHE.

Our Technique for Comparison Over Encrypted Data. We devise a low degree polynomial approximation for step functions by using the least squares method, and utilize our approximation for fast and accurate prediction and training over encrypted data. To achieve better accuracy in our algorithms and protocols, the approximation uses a weighting function that is zero in a window around the step and constant elsewhere. See Sect. 3.1.

Further Applications. Our training and prediction protocols can be employed in additional settings:

(a) Cross-entity: Our prediction protocol can trivially be used in settings where one company holds an unlabeled example, with the goal of learning the prediction result, and the other company holds a decision tree.
(b) Secure outsourcing: Both our protocols can be employed in settings where the client is the owner of example and tree in prediction (respectively, the dataset in training), and the server performs all computation (besides the lightweight KMS tasks performed by the client), resulting in protocols with lightweight client.
(c) Two-server model: Our training protocol can be employed, with slight modification, in settings where no KMS is available but rather an untrusted but non-colluding server; in this setting we guarantee privacy against both servers.

Terminology. Henceforth we use the more neutral "*client*" terminology rather than "KMS", in order to capture the aforementioned wide scope of applications.

Prior Work on Privacy-Preserving Decision Trees. For *prediction*, prior works considered the cross-entity setting, presenting interactive protocols with communication complexity proportional to the tree size in [2,4,7,11,22,27,38,42] or depth [39]. In contrast, our protocol is non-interactive.

For *training*, the prior works [12,14,21,29,30,36,40,41,43] considered multiparty computation settings, where multiple parties communicate to train a model on the union of their private individual datasets with the goal of preventing leakage on their private dataset. In particular, every example in the training dataset is visible in cleartext to at least one participant. Moreover, their communication complexity is proportional to the dataset size. In contrast, in our setting all data is encrypted and there is no data owner who sees cleartext data examples; furthermore, the communication complexity of our protocol is independent of the dataset size.

The technique of employing *low degree approximation* to speedup computing on encryption data was previously used for other functions, such as *ReLU*, *Sigmoid* and *Tanh*, see [3,9,20,24,26,28].

2 Preliminaries

Throughout the rest of the paper, we use the following notation and definitions. For $n \in \mathbb{N}$, let $[n]$ denote the set $\{1, \ldots, n\}$. A function $g : \mathbb{N} \to \mathbb{R}^+$ is *negligible* if it tends to zero faster than any inverse polynomial, *i.e.*, for all $c \in \mathbb{N}$ there exists $k_c \in \mathbb{N}$ such that for every $k > k_c$ it holds that $g(k) < k^{-c}$. We use $\mathsf{neg}(\cdot)$ to denote a negligible function if we do not need to specify its name. A L-dimensional binary vector $y = (y_1, \ldots, y_L)$ is a *1-hot encoding* of $\ell \in [L]$, if the ℓ'th entry is the only non-zero entry in y.

Finally, we use a standard notion of convergence of functions, as stated next.

Definition 1 (Uniform Convergence). *Let E be a set and let $(f_n)_{n \in \mathbb{N}}$ be a sequence of real-valued functions on E. We say that $(f_n)_{n \in \mathbb{N}}$ is* uniformly convergent *on E to a function f if for every $\epsilon > 0$ there exists a $n_0 \in \mathbb{N}$ such that for all $n \geq n_0$ and $x \in E$ it holds that $|f_n(x) - f(x)| < \epsilon$.*

Next we give cryptographic definitions for CPA-security, fully homomorphic encryption and privacy-preserving protocols, favoring readability over formality; see formal definitions in [18,23] and Definition 2.6.2 at [19] respectively.

In the setting of public key encryption, Alice shares with the world a *public encryption key pk* and keeps the corresponding secret key *sk*. Then any person, Bob, in hold of *pk* can encrypt messages and send the corresponding ciphertexts to Alice. Alice can decrypt these ciphertexts using her secret key. However, any eavesdropper Eve who doesn't know the secret key (but does know the public key) cannot learn any new information on the underlying message content from seeing such ciphertexts. Here, Alice, Bob and Eve are probabilistic polynomial time algorithms, where "polynomial" (similarly, "negligible") is measured with respect to a system parameter λ called the *security parameter*.

Definition 2 ((Informal) CPA-Secure Public-Key Encryption (PKE)). *A public-key encryption scheme with respect to message space $\mathcal{M}$ consists of polynomial-time algorithms* (Gen, Enc, Dec) *with the following properties.*

- **Syntax:**
  - *(Key generation:)* Gen *is a randomized algorithm that given the security parameter outputs a secret key sk and a public encryption key pk.*
  - *(Encryption:)* $\mathsf{Enc}_{pk}(m)$ *is a randomized algorithm that takes as input the public key pk and message $m \in \mathcal{M}$, and outputs a ciphertext ct.*
  - *(Decryption:)* $\mathsf{Dec}_{sk}(ct)$ *is a deterministic algorithm that takes as input the secret key sk and ciphertext ct, and outputs a decrypted message.*
- **Correctness:** *for any message $m \in \mathcal{M}$, $\mathsf{Dec}_{sk}(\mathsf{Enc}_{pk}(m)) = m$.*
- **CPA-security:** *For any probabilistic polynomial-time adversary* A *and any two equal length messages $m_0, m_1 \in \mathcal{M}$ of its choice there is a negligible function* neg *such that* A *cannot distinguish between $\mathsf{Enc}_{pk}(m_0)$ and $\mathsf{Enc}_{pk}(m_1)$ with probability greater than $\frac{1}{2} + \mathsf{neg}(\lambda)$.*

A fully homomorphic public-key encryption scheme (FHE) is a public-key encryption scheme that has an additional randomized efficient algorithm called Eval, which supports "homomorphic evaluations" on ciphertexts. Namely, given two ciphertexts $c_1 = \mathsf{Enc}(m_1)$ and $c_2 = \mathsf{Enc}(m_2)$ encrypting messages m_1 and m_2 respectively with FHE, it is possible to produce new ciphertexts c_3 and

c_4 that decrypt to $\mathsf{Dec}(c_3) = m_1 + m_2$ and $\mathsf{Dec}(c_4) = m_1 \times m_2$ respectively. The addition and multiplication of the messages m_1, m_2 (encoded as numbers) may be in Boolean arithmetic, a finite field arithmetic or over the reals. The *correctness* requirement is extended to hold with respect to any sequence of homomorphic evaluations performed on ciphertexts encrypted under *pk* using $\mathsf{Eval}_{pk}(\cdot)$. A fully homomorphic encryption scheme must satisfy an additional property called *compactness*, which only requires that the size of the ciphertext does not grow with the complexity of the sequence homomorphic operations.

Finally, the definition of privacy-preserving protocols is adapted from [19].

Definition 3 ((Informal)[19] Def. 2.6.2.: Privacy-Preserving Protocol). *A polynomial-time two-party protocol is* privacy-preserving *if each party cannot distinguish with probability greater than* $\frac{1}{2} + \mathsf{neg}(\lambda)$*, for some negligible function* neg*, whether the other party is executing the protocol (semi-honest) on input* x_0 *or* x_1*, for any two equal length inputs* $x_0, x_1 \in \{0,1\}^*$.

3 Decision Trees with Low Degree Approximation

In this section we present our algorithms for training and prediction of decision trees. The algorithms are tailored to being evaluated over encrypted data, in the sense of avoiding complexity bottlenecks of homomorphic evaluation.

The key component in our algorithms is a low degree polynomial approximation for the step function "$x < \theta$" (aka, soft-step function); See Sect. 3.1. The obtained low degree approximation is used to replace the step function at each tree node in our new prediction and training algorithms, presented in Sects. 3.2–3.3 respectively.

3.1 Low Degree Approximation of a Step Function

We construct a low-degree polynomial approximation of a step function. Specifically, we consider the step function $I_0 \colon \mathbb{R} \to \{0,1\}$ with threshold zero, defined by: $I_0(x) = 1$ if $x \geq 0$ and $I_0(x) = 0$ otherwise.

There are several convenient methods for replacing piece-wise continuous functions with limited-degree polynomial approximation. One approach is to consider the appropriate space of functions as a metric space, and then to find a polynomial of the desired degree that minimizes the deviation from the target function in this metric. Natural choices of metrics are the uniform error, integral square error, and integral absolute error. We aim to replace a step function with a *soft-step function*, *i.e.*, a polynomial approximation. In choosing these polynomials we opt for the mean square integral solution, due to its extendability. That is, the soft-step function would be the solution to the following optimization problem:

$$\phi = \min_{p \in P_n} \int_{-2}^{2} \left(I_0(x) - p(x)\right)^2 \, dx \tag{1}$$

where P_n is the set of polynomial functions of degree at most n over the reals. Setting the interval of the approximation to be $[-2, 2]$ is sufficient once we have pre-processed all data to be in the range $[-1, 1]$. A soft-step at $\theta \in [-1, 1]$ is of the form $\phi(x - \theta)$, and thus $x - \theta \in [-2, 2]$.

Theorem 4. *Let ϕ_n be the soft-step function obtained from Eq. 1 with P_n being the set of polynomial functions of degree at most n over the reals. For every $\epsilon > 0$ the sequence of functions $(\phi_n)_{n \in \mathbb{N}}$ uniformly converges to the step function I_0 on $[-2, 2] \setminus (-\epsilon, \epsilon)$.*

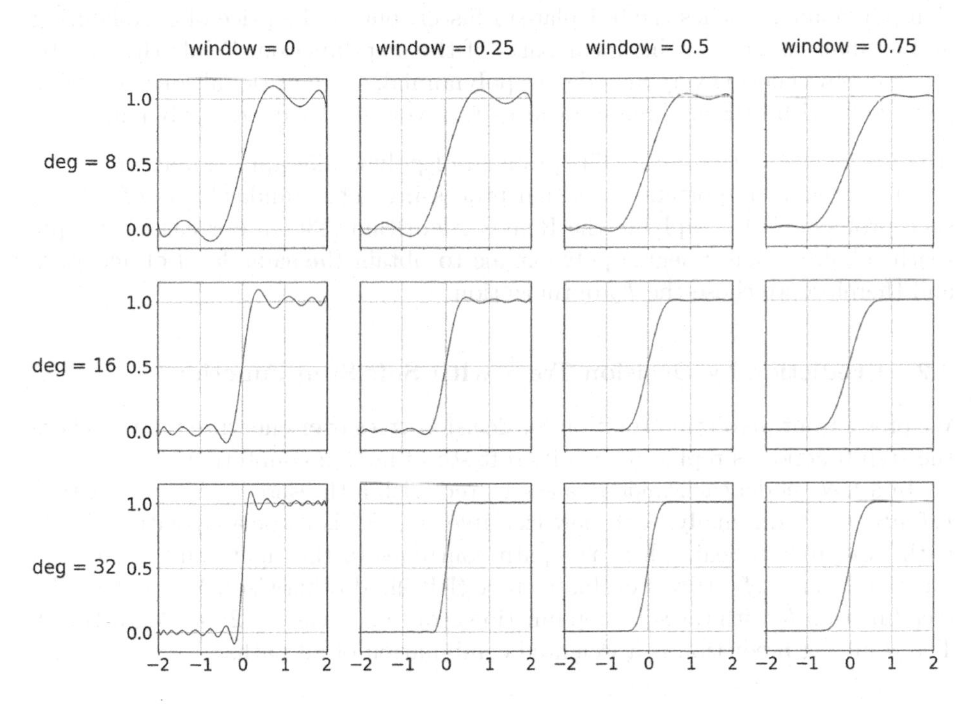

Fig. 1. Polynomial approximations of I_0 in $[-2, 2]$, with varying polynomial degrees (rows) and width of neglected window around the transition (columns)

However, in many cases the sensitivity to error in the approximation is not uniform over the domain. Errors at an interval around the threshold may harm the overall result of the algorithm less, compared to errors away from the threshold value. Adding an importance-weighting of the approximation interval leads to the following optimization problem:

$$\phi = \min_{p \in P_n} \int_{-2}^{2} (I_0(x) - p(x))^2 \; w(x) \, dx \tag{2}$$

with a weighting function $w(x) \geq 0 \; \forall x \in [-2, 2]$ and $\int_{-2}^{2} w(x) \, dx = 1$. We note that the unique solution to this problem is obtained by the projection of I_0 onto P_n, in the w-weighted norm (see Chapter II in [35]), or alternatively by applying polynomial regression (minimizing MSE) on a discrete set of points sampled proportionally to w.

Experiments with polynomials which are solutions to Eq. 2 with various weight functions and degrees show that by neglecting an interval around the threshold we are able to obtain tighter approximations to the linear phases of the step function. More specifically, by using weighting functions that are zero in a window around the step and constant otherwise, a trade-off is obtained between the slope of the transition and tightness of approximation at the edges of the interval (Fig. 1). For larger slopes (smaller neglected windows) the approximation polynomial reaches the 0–1 plateau faster, but at the price of overshooting and oscillations around the linear parts of the step function. While this can be remedied by choosing very high degree polynomial, computational considerations (especially with FHE arithmetic) lead us to favor smaller degree polynomials.

Approximating in L_∞-norm. The problem of polynomial approximation in L_∞-norm is of great importance in signal processing. The standard way of solving such problems is by applying the Remez Algorithm [33]. In our case, this approach required higher degree polynomials to obtain the same level of accuracy, and therefore we chose the L_2 optimization.

3.2 Prediction by Decision Trees with Soft-Step Function

We present our prediction solution for decision trees over encrypted data, where the step function is replaced with its soft-step function counterpart.

In a tree model each node v is associated with a threshold $v.\theta$ and a feature $v.feature$. Traditionally, each new example $x \in \mathbb{R}^k$ is associated with a single path from root to leaf, where the path continues to the right-child if the step function $I_{v.\theta}(x[v.feature])$ evaluates to 1 (left-child otherwise). We replace the step function I_0 with the soft-step function ϕ obtained via Eq. 2; see Algorithm 1. The resulting prediction is a weighted combination of all paths.

Algorithm 1. Tree Prediction Function

Let $\mathsf{T} = (V, E)$ be a decision tree, where each node v is a data structure containing the following fields: $feature$ and θ denote the feature and the threshold associated with the node; $leaf\_value$ is a 1-hot encoding of the value in case v is a leaf (a vector of dimension L, where L is the number of labels); $v.right$ and $v.left$ denote the right and left sub-trees of v, respectively. The variable v is initialized to $\mathsf{T}.root$.
We denote by ϕ the soft-step function from Equation 2 in Section 3.1.
Input: $x \in [-1, 1]^k$, where k is the number of features.

function Tree_Predict(v, x)
  if v is a leaf **then**
    return $v.leaf\_value$
  else
    return $\phi(x[v.feature] - v.\theta) \cdot$ Tree_Predict($v.right, x$) +
        $\phi(v.\theta - x[v.feature]) \cdot$ Tree_Predict($v.left, x$)
  end if
end function

Algorithm 1 outputs the same prediction as the traditional algorithm, provided the soft step function ϕ is of sufficiently high degree; See Theorem 5.

Theorem 5 (Path Convergence). *Let* T *be a decision tree. For a sample* $x \in [-1,1]^k$ *we denote by* T_x *the associated path of* x *in* T*, and set* $\Delta_x = \min_{v \in \mathsf{T}_x} \{|x[v.feature] - v.\theta|\}$. *For every* $\delta > 0$ *there exists* $n \in \mathbb{N}$ *such that for every sample* x *satisfying* $\Delta_x > \delta$*, the path* T_x *is equal to*

$$\underset{path=(root,\ldots,leaf)\in\mathsf{T}}{arg\ max} \prod_{v \in path \setminus leaf} \phi_n\big(g \cdot (x[v.feature] - v.\theta)\big)$$

where $g = -1$ *if* $v.left \in path$ *and* 1 *otherwise.*

We note that Algorithm 1 naturally extends to evaluation of base-classifiers for random forests and tree boosting, hence our method supports these common tree-based machine learning methods as well. Moreover Algorithm 1 can also be viewed as a "recipe" for converting an existing tree model to a polynomial form of low degree, which may be of independent interest.

Our prediction protocol on encrypted data is derived from Algorithm 1 by replacing each operation on cleartext data with the corresponding homomorphic operation on encrypted data. Since the algorithm consists solely of addition and multiplication operations, this is straightforward; specifically, we homomorphically evaluate the following formula on the *encrypted* input x:

$$\sum_{path=(root,\ldots,leaf)\in\mathsf{T}} leaf.leaf\_value \cdot \prod_{v \in path \setminus leaf} \phi\big(g \cdot (x[v.feature] - v.\theta)\big)$$

where notations here are as in Algorithm 1 and Theorem 5.

The resulting prediction protocol is privacy-preserving, as long as the FHE encryption is CPA-secure; see Theorem 6.

Theorem 6. *Let* m *be the tree size and* L *the number of labels. Suppose the input data is encrypted with a CPA-secure* FHE *scheme. Our prediction protocol is a non-interactive privacy-preserving protocol that executes* $O(m \cdot L)$ *basic homomorphic operations.*

Extension to Encrypted Trees. Algorithm 1 can be used in a setting where privacy of the tree is needed, i.e., where both the tree and the input data are encrypted. This requires k additional multiplications and additions per node, for k the number of features.

3.3 Training Decision Trees with Soft-Step Function

The standard training procedure considers splits that partition the training dataset $\mathcal{X}$ and builds a tree according to local objectives based on the number of examples of each label that flow to each side at a chosen, per node, split. In the training procedure, at each node, impurity scores are calculated for each potential split, then the feature and threshold are chosen in order to minimize some impurity measure, *e.g.*, the weighted Gini impurity (see Fig. 2).

Traditionally, the training procedure associates with each node a set of indicators $W = \{w_x\}_{x \in \mathcal{X}}$, so that w_x is a bit indicating if example x is participating in the training of a sub-tree rooted at this node, and W is updated for the children nodes as follows: for the chosen feature i^*, and threshold θ^*, the right sub-tree (respectively, left sub-tree)

$$\forall x \in \mathcal{X} : w_x^{\mathsf{right}} = w_x \cdot I_{\theta^*}(x[i^*]) \qquad (\text{resp. } w_x^{\mathsf{left}} = w_x \cdot (1 - I_{\theta^*}(x[i^*])) \,) \quad (3)$$

In our approach, to avoid the comparison operation that is expensive over encrypted data, we replace the step function I_0 by the low-degree polynomial approximation ϕ obtained via Eq. 2.

Algorithm 2. Tree Training Function

$(\mathcal{X}, \mathcal{Y})$ is a input dataset and W is a set of weights, s.t. $\forall x \in \mathcal{X}$: $w_x \in \mathbb{R}$ is the weight (initially 1) associated with the example $x \in [-1, 1]^k$ (at the current node). We denote by k and L the number of features and labels in the associated problem, respectively. We denote by S the set of considered thresholds. The parameter $\mathsf{maximal\_depth}$ is the depth of the trained tree, the variable *depth* is initialized to 0. We denote by ϕ the soft-step function from Equation 2. The function $\mathsf{Gini}(\cdot)$ in Figure 2 computes the weighted Gini impurity and returns the best threshold and feature.
Input: a set of n examples $\mathcal{X}$ and the corresponding labels $\mathcal{Y}$ where each $x \in \mathcal{X}$ is in $[-1, 1]^k$ and the corresponding $y_x \in \mathcal{Y}$ is a 1-hot encoding of the label.

1: **function** $\mathsf{Tree\_Train}((\mathcal{X}, \mathcal{Y}), w, depth)$
2: **if** reached $\mathsf{maximal\_depth}$ **then** ▷ leaf node
3: $label \leftarrow \arg\max_{\ell \in [L]} \sum_{x \in \mathcal{X}} w_x \cdot y_x$
4: **else** ▷ search best split for this node
5: **for each** feature i and **each** threshold θ **do**
6: $\mathsf{right}[i, \theta] \leftarrow \sum_{x \in \mathcal{X}} w_x \cdot \phi(x[i] - \theta) \cdot y_x$
7: $\mathsf{left}[i, \theta] \leftarrow \sum_{x \in \mathcal{X}} w_x \cdot \phi(\theta - x[i]) \cdot y_x$
8: **end for**
9: $i^*, \theta^* \leftarrow \mathsf{Gini}(\{\mathsf{right}[i, \theta], \mathsf{left}[i, \theta]\}_{i \in [k], \theta \in S})$
10: ▷ See Figure 2
11: $\forall x \in \mathcal{X} : w_x^{\mathsf{right}} \leftarrow w_x \cdot \phi(x[i^*] - \theta^*)$
12: $\mathsf{Tree\_Train}((\mathcal{X}, \mathcal{Y}), \{w_x^{\mathsf{right}}\}_{x \in \mathcal{X}}, depth + 1)$
13: ▷ build right-side sub-tree
14: $\forall x \in \mathcal{X} : w_x^{\mathsf{left}} \leftarrow w_x \cdot \phi(\theta^* - x[i^*])$
15: $\mathsf{Tree\_Train}((\mathcal{X}, \mathcal{Y}), \{w_x^{\mathsf{left}}\}_{x \in \mathcal{X}}, depth + 1)$
16: ▷ build left-side sub-tree
17: **end if**
18: **end function**

Notice that our approximated version of Eq. 3 has real valued weights instead of Boolean indicators. This means that *every example* reaches every node, and is evaluated at all nodes. This results in a soft partition of the data, where the two children nodes get a part of each data point, weighted differently rather than hard splitting the data (partitioning to right and left children at each split). In order to efficiently keep track of the weight of each data example at each node, we keep a weights set W while constructing the tree during training. All weights are initialized to 1, and recursively multiplied by the polynomial approximation at the current node before passing on to the children nodes. The details of the training algorithm are presented in Algorithm 2.

Looking ahead, we carefully divide the operations in Algorithm 2 to those where homomorphic evaluation on encrypted data is "efficient" vs. "costly". Concretely, addition, multiplication and evaluating the low degree polynomial ϕ are efficient, whereas computing the Gini Impurity (Fig. 2) involves the costly division and argmin operations.

Our training protocol on encrypted data is derived from Algorithm 2 by replacing operations on cleartext data with their corresponding homomorphic operations on encrypted data, except for computing the Gini Impurity (Line 9) and label (Line 3). These operations are computed with the aid of a "client" possessing the secret decryption key. Recall, for example, that in our enterprise use-case, the client and server are the KMS and microservice, respectively.

Procedure: Gini impurity computation.
We denote by k and L the number of features and labels in the associated problem, respectively. We denote by S the set of considered thresholds. The function computes the weighted Gini impurity and returns the best threshold and feature.
Given a set of L-dimensional vectors $\{\mathsf{right}[i,\theta], \mathsf{left}[i,\theta]\}_{i\in[k],\theta\in S}$ proceed as follows:

1. for each threshold $\theta \in S$, each feature $i \in [k]$, and $side \in \{\mathsf{right}, \mathsf{left}\}$ compute

$$\mathsf{total\_side}[i,\theta] \leftarrow \sum_{\ell\in L} \mathsf{side}[i,\theta][\ell]$$

$$\tilde{I}_G[i,\theta] = \sum_{side\in\{\mathsf{right},\mathsf{left}\}} \left(1 - \sum_{\ell\in L}\left[\frac{\mathsf{side}[i,\theta][\ell]}{\mathsf{total\_side}[i,\theta]}\right]^2\right) \cdot \mathsf{total\_side}[i,\theta]$$

2. return the selected feature and threshold, i.e., $i^*, \theta^* \leftarrow \arg\min_{i,\theta} \tilde{I}_G[i,\theta]$

Fig. 2. The weighted Gini impurity computation

Elaborating on the above, to compute the Gini Impurity (Line 9), the server sends the client *encrypted* aggregated values $\{\mathsf{right}[i,\theta], \mathsf{left}[i,\theta]\}_{i\in[k],\theta\in S}$, the client decrypts the ciphertexts, executes Fig. 2 (Gini) and sends back to the server the chosen θ^* and i^* in *encrypted* form. To compute the label (Line 3), the server sends to the client a *ciphertext* containing $\sum_{x\in\mathcal{X}} w_x \cdot y_x$, the client decrypts and computes arg max, then sends back the resulting *label* in *encrypted* form.

By design, the computational burden is almost fully on the server. In particular, the client and communication complexity is independent of the dataset size (n). Concretely, number of communication rounds is proportional to tree depth, and the client and communication complexity is proportional to the number of features, labels, and considered thresholds by the algorithm; See Theorem 7.

Theorem 7. *Let k, S, L, n be as in Algorithm 2. Let m and d be the number of nodes and depth of the trained tree, respectively. Suppose the dataset $(\mathcal{X}, \mathcal{Y})$ is encrypted with a CPA-secure* FHE *scheme. Our training protocol is a d-round privacy-preserving protocol whose complexity is dominated by the following: (1) server: $O(m \cdot |S| \cdot k \cdot L \cdot n)$ basic homomorphic operations, (2) communication: $O(m \cdot k \cdot |S| \cdot L)$ transmitted ciphertexts, (3) client: $O(m \cdot k \cdot |S| \cdot L)$ decryption/encryption operations.*

Remarks. The server is exposed only to FHE encrypted ciphertext, as it does not have the decryption key. Extensions to other impurity measures, such as *entropy*, can straightforwardly be instantiated and executed over encrypted data.

4 Implementation Details and Experimental Results

We empirically evaluated our decision trees algorithms and protocols for both accuracy and run-time performance on encrypted data in Sects. 4.1–4.2, respectively. Our evaluations are done with respect to a single decision tree, and can naturally be extended to random forests where trees are trained/evaluated in parallel, each on a separate CPU core. The employed soft-step function is a polynomial of degree 15, constructed via Eq. 2 with a weighting function $w\colon [-2, 2] \to [0, 1]$ defined to be zero in the interval $[-0.2, 0.2]$ and a constant positive value elsewhere. For training, we use thresholds on a 0.05 grid in the $[-1, 1]$ interval. We use standard UCI repository datasets [13] in our evaluation, ranging in size from very small (iris with 4 features and 150 examples) to the moderate size common in real-world applications (forest cover with 54 features and over half a million examples).

4.1 Accuracy of Our Decision-Tree Algorithms

The accuracy of our Algorithms Algorithms 1–2 is evaluated in comparison to standard trees, on the benchmark datasets. We use a 3-fold cross-validation procedure, where each dataset was randomly partitioned into three equal-size parts, and each of the three parts serves as a test-set for a classifier trained on the remaining two. The *overall accuracy* is calculated as the percentage of correct classification on test examples (each example in the data is taken exactly once, so the accuracy reported is simply the percentage of examples that were correctly classified).

We compared all four possible combinations of training and prediction according to our algorithms vs. the standard algorithms; see Fig. 3. We train trees up to depth 5, as customary when using random forests.

The results show an overall comparable accuracy, indicating that our Algorithms 1–2 are a valid replacement for standard decision trees in terms of accuracy.

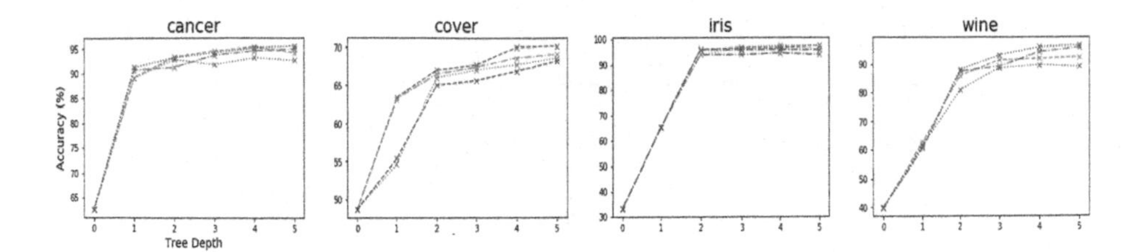

Fig. 3. Accuracy of ours vs. scikit-learn [32] algorithms on four UCI datasets and tree depth 0–5 (depth 0 is the majority-class baseline), in four execution modes: our training and prediction (red), our training and scikit-learn prediction (green), scikit-learn training and our prediction (orange), scikit-learn training and prediction (blue). (Color figure online)

4.2 Running-Time on Encrypted Data

We implemented our training and prediction protocols over data encrypted with CKKS homomorphic encryption scheme [10] in Microsoft SEAL v3.2.2 [37]. We set SEAL parameters to security-level 80 bits and irreducible polynomial degree 8192. We ran experiments with depth 4 trees.

Training Over Encrypted Datasets. Our experiments were executed on AWS x1.16xlarge as the server. The input examples are encrypted feature-by-feature, while packing 4096 values in each ciphertext; the associated labels (in 1-hot encoding) are likewise encrypted in a packed form.

The server run-time on encrypted UCI datasets (subs-sampled) ranged from minutes on small datasets to hours on medium ones; see Table 1. Executions on encrypted synthetic datasets containing up to 131,000 encrypted samples and (features,labels,splits) $= (8, 2, 21)$ exhibited a server run-time of under a day.

Table 1. Server run-time for training depth 4 decision trees on encrypted UCI datasets

Dataset name	# training examples	# features	# labels	Server training time (minutes)
Iris	100	4	3	19
Wine	119	13	3	59
Cancer	381	30	2	107
Digits	1,203	64	10	665
Cover	10,000	54	7	1220

The client (*e.g.*, KMS) run-time in all experiments ranged from seconds to under three minutes. The communication complexity is independent of the dataset size. See Fig. 4 for the number of transmitted ciphertexts. The ciphertext size is roughly 0.25 MB after compression.

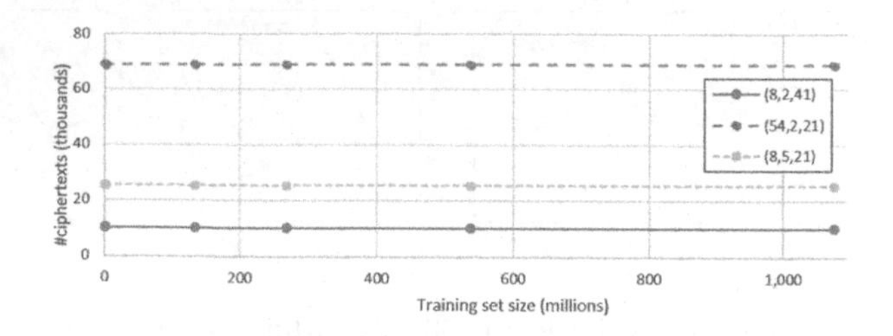

Fig. 4. Total number of transmitted ciphertexts vs. the training set size, when training a full tree of depth 4 on various (features, labels, splits) settings

Prediction over encrypted data was performed on encrypted examples and cleartext trees, and executed on a personal Intel Core i7-4790 CPU 16 GB memory computer (using a single core). The average run-time was 1.15 s.

5 Conclusions

In this work we present FHE-friendly algorithms and protocols for decision tree based prediction and training over encrypted data that are the first to attain all the following desired properties: non-interactive prediction, lightweight client and communication in training, and rigorous privacy guarantee. We ran extensive experiments on standard UCI and synthetic datasets, all encrypted with fully homomorphic encryption, demonstrating high accuracy comparable to standard algorithms on cleartext data, fast prediction (1.15 s), and feasible training (minutes to hours). Our protocols support real-life enterprise use-cases, and are well suited for offline execution, *e.g.* in nightly batched prediction. For formal details omitted from this extended abstract we refer the reader to [1].[1]

As a central technique we devised a soft-step function yielding a low-degree weighted approximation for the step function. This may have further applications, beyond tree-based models, including *neural networks* based prediction utilizing our soft-step function to replace *sigmoid* or *sign* activation to gain speedup.

[1] We remark that in [1] we prove privacy against malicious adversaries (stronger) and require circuit-private FHE. Here the adversary is semi-honest and our proof does not require circuit-privacy; our proof is the same as the proof of Theorem 2 in [1] except for encrypting random elements in $\mathcal{M}$ rather than executing Eval.

References

1. Akavia, A., Leibovich, M., Resheff, Y.S., Ron, R., Shahar, M., Vald, M.: Privacy-preserving decision tree training and prediction against malicious server. Cryptology ePrint Archive, Report 2019/1282 (2019)
2. Barni, M., Failla, P., Kolesnikov, V., Lazzeretti, R., Sadeghi, A.-R., Schneider, T.: Secure evaluation of private linear branching programs with medical applications. In: Backes, M., Ning, P. (eds.) ESORICS 2009. LNCS, vol. 5789, pp. 424–439. Springer, Heidelberg (2009). https://doi.org/10.1007/978-3-642-04444-1_26
3. Blatt, M., Gusev, A., Polyakov, Y., Rohloff, K., Vaikuntanathan, V.: Optimized homomorphic encryption solution for secure genome-wide association studies. Cryptology ePrint Archive, Report 2019/223 (2019). https://eprint.iacr.org/2019/223
4. Bost, R., Popa, R.A., Tu, S., Goldwasser, S.: Machine learning classification over encrypted data. In: NDSS, vol. 4324, p. 4325 (2015)
5. Brakerski, Z.: Fully homomorphic encryption without modulus switching from classical GapSVP. In: Safavi-Naini, R., Canetti, R. (eds.) CRYPTO 2012. LNCS, vol. 7417, pp. 868–886. Springer, Heidelberg (2012). https://doi.org/10.1007/978-3-642-32009-5_50
6. Brakerski, Z., Gentry, C., Vaikuntanathan, V.: (leveled) fully homomorphic encryption without bootstrapping. In: Innovations in Theoretical Computer Science 2012, Cambridge, MA, USA, 8–10 January 2012, pp. 309–325 (2012)
7. Brickell, J., Porter, D.E., Shmatikov, V., Witchel, E.: Privacy-preserving remote diagnostics. In: Proceedings of the 14th ACM Conference on Computer and Communications Security, pp. 498–507. ACM (2007)
8. Chen, H., et al.: Logistic regression over encrypted data from fully homomorphic encryption. BMC Med. Genomics **11**(4), 81 (2018)
9. Chen, H., et al.: Logistic regression over encrypted data from fully homomorphic encryption. BMC Med. Genomics **11**, 81 (2018). https://doi.org/10.1186/s12920-018-0397-z
10. Cheon, J.H., Kim, A., Kim, M., Song, Y.: Homomorphic encryption for arithmetic of approximate numbers. In: Takagi, T., Peyrin, T. (eds.) ASIACRYPT 2017. LNCS, vol. 10624, pp. 409–437. Springer, Cham (2017). https://doi.org/10.1007/978-3-319-70694-8_15
11. De Cock, M., et al.: Efficient and private scoring of decision trees, support vector machines and logistic regression models based on pre-computation. IEEE Trans. Dependable Secure Comput. **16**(2), 217–230 (2017)
12. Du, W., Zhan, Z.: Building decision tree classifier on private data. In: Proceedings of the IEEE International Conference on Privacy, Security and Data Mining, vol. 14, pp. 1–8. Australian Computer Society, Inc. (2002)
13. Dua, D., Graff, C.: UCI machine learning repository (2017)
14. Emekci, F., Sahin, O.D., Agrawal, D., El Abbadi, A.: Privacy preserving decision tree learning over multiple parties. Data Knowl. Eng. **63**(2), 348–361 (2007)
15. Fan, J., Vercauteren, F.: Somewhat practical fully homomorphic encryption. IACR Cryptology ePrint Archive 2012, 144 (2012)
16. Gentry, C.: A fully homomorphic encryption scheme. Ph.D. thesis. Stanford University (2009). https://crypto.stanford.edu/craig/
17. Gilad-Bachrach, R., Dowlin, N., Laine, K., Lauter, K., Naehrig, M., Wernsing, J.: CryptoNets: applying neural networks to encrypted data with high throughput and accuracy. In: International Conference on Machine Learning, pp. 201–210 (2016)

18. Halevi, S.: Homomorphic encryption. Tutorials on the Foundations of Cryptography. ISC, pp. 219–276. Springer, Cham (2017). https://doi.org/10.1007/978-3-319-57048-8_5
19. Hazay, C., Lindell, Y.: Efficient Secure Two-Party Protocols. ISC. Springer, Heidelberg (2010). https://doi.org/10.1007/978-3-642-14303-8
20. Hesamifard, E., Takabi, H., Ghasemi, M., Wright, R.: Privacy-preserving machine learning as a service. In: Proceedings on Privacy Enhancing Technologies 2018, pp. 123–142 (06 2018)
21. de Hoogh, S., Schoenmakers, B., Chen, P., op den Akker, H.: Practical secure decision tree learning in a teletreatment application. In: Christin, N., Safavi-Naini, R. (eds.) FC 2014. LNCS, vol. 8437, pp. 179–194. Springer, Heidelberg (2014). https://doi.org/10.1007/978-3-662-45472-5_12
22. Joye, M., Salehi, F.: Private yet efficient decision tree evaluation. In: Kerschbaum, F., Paraboschi, S. (eds.) DBSec 2018. LNCS, vol. 10980, pp. 243–259. Springer, Cham (2018). https://doi.org/10.1007/978-3-319-95729-6_16
23. Katz, J., Lindell, Y.: Introduction to Modern Cryptography. Chapman & Hall/CRC Cryptography and Network Security Series. Chapman & Hall/CRC, Boca Raton (2007)
24. Kim, A., Song, Y., Kim, M., Lee, K., Cheon, J.: Logistic regression model training based on the approximate homomorphic encryption. BMC Med. Genomics **11**, 23–31 (2018)
25. Kim, A., Song, Y., Kim, M., Lee, K., Cheon, J.H.: Logistic regression model training based on the approximate homomorphic encryption. BMC Med. Genomics **11**(4), 83 (2018)
26. Kim, M., Song, Y., Wang, S., Xia, Y., Jiang, X.: Secure logistic regression based on homomorphic encryption: design and evaluation. JMIR Med. Inf. **6**, e19 (2017)
27. Kiss, Á., Naderpour, M., Liu, J., Asokan, N., Schneider, T.: SoK: modular and efficient private decision tree evaluation. PoPETs **2019**(2), 187–208 (2019)
28. Kyoohyung, H., Hong, S., Cheon, J., Park, D.: Logistic regression on homomorphic encrypted data at scale. In: Proceedings of the AAAI Conference on Artificial Intelligence, vol. 33, pp. 9466–9471, July 2019
29. Lindell, Y., Pinkas, B.: Privacy preserving data mining. In: Bellare, M. (ed.) CRYPTO 2000. LNCS, vol. 1880, pp. 36–54. Springer, Heidelberg (2000). https://doi.org/10.1007/3-540-44598-6_3
30. Lory, P.: Enhancing the efficiency in privacy preserving learning of decision trees in partitioned databases. In: Domingo-Ferrer, J., Tinnirello, I. (eds.) PSD 2012. LNCS, vol. 7556, pp. 322–335. Springer, Heidelberg (2012). https://doi.org/10.1007/978-3-642-33627-0_25
31. Nandakumar, K., Ratha, N.K., Pankanti, S., Halevi, S.: Towards deep neural network training on encrypted data. In: IEEE Conference on Computer Vision and Pattern Recognition Workshops, CVPR Workshops 2019, Long Beach, CA, USA, 16–20 June 2019. p. 0. Computer Vision Foundation/IEEE (2019)
32. Pedregosa, F., et al.: Scikit-learn: machine learning in Python. J. Mach. Learn. Res. **12**, 2825–2830 (2011)
33. Remez, E.Y.: Sur la détermination des polynômes d'approximation de degré donnée. Comm. Soc. Math. Kharkov **10**(4163), 196 (1934)
34. Rivest, R.L., Adleman, L., Dertouzos, M.L.: On data banks and privacy homomorphisms. Found. Sec. Comput. **4**, 169–179 (1978)
35. Rivlin, T.J.: An Introduction to the Approximation of Functions. Courier Corporation, North Chelmsford (2003)

36. Samet, S., Miri, A.: Privacy preserving ID3 using Gini index over horizontally partitioned data. In: Proceedings of the 2008 IEEE/ACS International Conference on Computer Systems and Applications, AICCSA 2008, pp. 645–651. IEEE Computer Society, Washington, DC (2008)
37. Microsoft SEAL (release 3.3). Microsoft Research, Redmond (2019). https://github.com/Microsoft/SEAL
38. Tai, R.K.H., Ma, J.P.K., Zhao, Y., Chow, S.S.M.: Privacy-preserving decision trees evaluation via linear functions. In: Foley, S.N., Gollmann, D., Snekkenes, E. (eds.) ESORICS 2017. LNCS, vol. 10493, pp. 494–512. Springer, Cham (2017). https://doi.org/10.1007/978-3-319-66399-9_27
39. Tueno, A., Kerschbaum, F., Katzenbeisser, S.: Private evaluation of decision trees using sublinear cost. PoPETs **2019**(1), 266–286 (2019)
40. Vaidya, J., Clifton, C., Kantarcioglu, M., Patterson, A.S.: Privacy-preserving decision trees over vertically partitioned data. ACM Trans. Knowl. Disc. Data (TKDD) **2**(3), 14 (2008)
41. Wang, K., Xu, Y., She, R., Yu, P.S.: Classification spanning private databases. In: Proceedings, The Twenty-First National Conference on Artificial Intelligence and the Eighteenth Innovative Applications of Artificial Intelligence Conference, 16–20 July 2006, Boston, Massachusetts, USA, pp. 293–298. AAAI Press (2006)
42. Wu, D.J., Feng, T., Naehrig, M., Lauter, K.: Privately evaluating decision trees and random forests. Proc. Priv. Enhancing Technol. **2016**(4), 335–355 (2016)
43. Xiao, M.J., Huang, L.S., Luo, Y.L., Shen, H.: Privacy preserving ID3 algorithm over horizontally partitioned data. In: Proceedings of the Sixth International Conference on Parallel and Distributed Computing Applications and Technologies, PDCAT 2005, pp. 239–243. IEEE Computer Society, Washington, DC(2005)

Poisoning Attacks on Algorithmic Fairness

David Solans[1(✉)], Battista Biggio[2(✉)], and Carlos Castillo[1(✉)]

[1] Universitat Pomepu Fabra, Barcelona, Spain
{david.solans,carlos.castillo}@upf.edu, chato@acm.org
[2] Università degli Studi di Cagliari, Cagliari, Italy
battista.biggio@unica.it

Abstract. Research in adversarial machine learning has shown how the performance of machine learning models can be seriously compromised by injecting even a small fraction of poisoning points into the training data. While the effects on model accuracy of such poisoning attacks have been widely studied, their potential effects on other model performance metrics remain to be evaluated. In this work, we introduce an optimization framework for poisoning attacks against algorithmic fairness, and develop a gradient-based poisoning attack aimed at introducing classification disparities among different groups in the data. We empirically show that our attack is effective not only in the white-box setting, in which the attacker has full access to the target model, but also in a more challenging black-box scenario in which the attacks are optimized against a substitute model and then transferred to the target model. We believe that our findings pave the way towards the definition of an entirely novel set of adversarial attacks targeting algorithmic fairness in different scenarios, and that investigating such vulnerabilities will help design more robust algorithms and countermeasures in the future.

Keywords: Algorithmic discrimination · Algorithmic fairness · Poisoning attacks · Adversarial machine learning · Machine learning security

1 Introduction

Algoritmic Fairness is an emerging concern in computing science that started within the data mining community but has extended into other fields including machine learning, information retrieval, and theory of algorithms [12]. It deals with the design of algorithms and decision support systems that are non-discriminatory, i.e., that do not introduce an unjustified disadvantage for members of a group, and particularly that do not further place at a disadvantage members of an already disadvantaged social group. In machine learning, the problem that has been most studied to date is supervised classification, in which algorithmic fairness methods have been mostly proposed to fulfill criteria related

F. Hutter et al. (Eds.): ECML PKDD 2020, LNAI 12457, pp. 162–177, 2021.
https://doi.org/10.1007/978-3-030-67658-2_10

to parity (equality) [25]. Most of the methods proposed to date assume benevolence from the part of the data scientist or developer creating the classification model: she is envisioned as an actor trying to eliminate or reduce potential discrimination in her model.

The problem arises when dealing with malicious actors that can tamper with the model development, for instance by tampering with training data. Traditionally, *poisoning attacks* have been studied in *Adversarial Machine Learning.* These attacks are usually crafted with the purpose of increasing the misclassification rate in a machine learning model, either for certain samples or in an indiscriminate basis, and have been widely demonstrated in adversarial settings (see, e.g., [4]).

In this work, we show that an attacker may be able to introduce algorithmic discrimination by developing a novel poisoning attack. The purpose of this attacker is to create or increase a disadvantage against a specific group of individuals or samples. For that, we explore how analogous techniques can be used to compromise a machine learning model, not to drive its accuracy down, but with the purpose of adding algorithmic discrimination, or exaggerating it if it already exists. In other words, the purpose of the attacker will be to create or increase a disadvantage against a specific group of individuals or samples.

Motivation. The main goal of this paper is to show the potential harm that an attacker can cause in a machine learning system if the attacker can manipulate its training data. For instance, the developer of a criminal recidivism prediction tool [1] could sample training data in a discriminatory manner to bias the tool against a certain group of people. Similar harms can occur when training data is collected from public sources, such as online surveys that cannot be fully trusted. A minority of ill-intentioned users could *poison* this data to introduce defects in the machine learning system created from it. In addition to these examples, there is the unintentional setting, where inequities are introduced in the machine learning model as an undesired effect of the data collection or data labeling. For instance, human annotators could systematically make mistakes when assigning labels to images of people of a certain skin color [6].

The methods we describe on this paper could be used to model the potential harm to a machine learning system in the worst-case scenario, demonstrating the undesired effects that a very limited amount of wrongly labeled samples can cause, even if created in an unwanted manner.

Contributions. This work first introduces a novel optimization framework to craft poisoning samples that against algorithmic fairness. After this, we perform experiments in two scenarios: a "black-box" attack in which the attacker only has access to a set of data sampled from the same distribution as the original training data, but not the model nor the original training set, and a "white-box" scenario in which the attacker has full access to both. The effects of these attacks are measured using impact quantification metrics. The experiments show that by carefully perturbing a limited amount of training examples, an skilled attacker has the possibility of introducing different types of inequities for certain groups of individuals. This, can be done without large effects on the overall accuracy of the system, which makes these attacks harder to detect. To facilitate the reproducibility of the obtained results, the code generated for the experiments has been published in an open-source repository[1].

Paper Structure. The rest of this paper is organized as follows. Section 2, describes the proposed methodology to craft poisoning attacks for algorithmic fairness. Section 3 demonstrates empirically the feasibility of the new types of attacks on both synthetic and real-world data, under different scenarios depending on the attacker knowledge about the system. Section 4 provides further references to related work. Section 5 presents our conclusions.

2 Poisoning Fairness

In this section we present a novel gradient-based poisoning attack, crafted with the purpose of compromising algorithmic fairness, ideally without significantly degrading accuracy.

Notation. Feature and label spaces are denoted in the following with $\mathcal{X} \subseteq \mathbb{R}^{\mathsf{d}}$ and $\mathcal{Y} \in \{-1, 1\}$, respectively, with d being the dimensionality of the feature space. We assume that the attacker is able to collect some training and validation data sets that will be used to craft the attack. We denote them as $\mathcal{D}_{tr}$ and $\mathcal{D}_{val}$. Note that these sets include samples along with their labels. $L(\mathcal{D}_{\text{val}}, \theta)$ is used to denote the validation loss incurred by the classifier $f_\theta : \mathcal{X} \to \mathcal{Y}$, parametrized by θ, on the validation set $\mathcal{D}_{\text{val}}$. $\mathcal{L}(\mathcal{D}_{\text{tr}}, \theta)$ is used to represent the regularized loss optimized by the classifier during training.

2.1 Attack Formulation

Using the aforementioned notation, we can formulate the optimal poisoning strategy in terms of the following bilevel optimization:

$$\max_{\mathbf{x}_c} \ \mathcal{A}(\mathbf{x}_c, y_c) = L(\mathcal{D}_{\text{val}}, \theta^\star), \tag{1}$$

$$\text{s.t.} \ \theta^\star \in \arg\min_{\theta} \mathcal{L}(\mathcal{D}_{\text{tr}} \cup (\mathbf{x}_c, y_c), \theta), \tag{2}$$

$$\mathbf{x}_{\text{lb}} \preceq \mathbf{x}_c \preceq \mathbf{x}_{\text{ub}}. \tag{3}$$

The goal of this attack is to maximize a loss function on a set of untainted (validation) samples, by optimizing the poisoning sample $\mathbf{x}_c$, as stated in the outer optimization problem (Eq. 1). To this end, the poisoning sample is labeled as y_c and added to the training set $\mathcal{D}_{\text{tr}}$ used to learn the classifier in the inner optimization problem (Eq. 2). As one may note, the classifier $\theta^\star$ is learned on the poisoned training data, and then used to compute the outer validation loss. This highlights that there is an implicit dependency of the outer loss on the poisoning point $\mathbf{x}_c$ via the optimal parameters $\theta^\star$ of the trained classifier. In

[1] https://github.com/dsolanno/Poisoning-Attacks-on-Algorithmic-Fairness.

other words, we can express the optimal parameters $\theta^\star$ as a function of $\mathbf{x}_c$, i.e., $\theta^\star(\mathbf{x}_c)$. This relationship tells us how the classifier parameters change when the poisoning point $\mathbf{x}_c$ is perturbed. Characterizing and being able to manipulate this behavior is the key idea behind poisoning attacks.

Within this formulation, additional constraints on the feature representation of the poisoning sample can also be enforced, to make the attack samples stealthier or more difficult to detect. In this work we only consider a box constraint that requires the feature values of $\mathbf{x}_c$ to lie within some lower and upper bounds (in Eq. 3, the operator $\preceq$ enforces the constraint for each value of the feature vectors involved). This constraint allows us to craft poisoning samples that lie within the feature values observed in the training set. Additional constraints can be additionally considered, e.g., constraints imposing a maximum distance from an initial location or from samples of the same class, we leave their investigation to future work. Our goal here is to evaluate the extent to which a poisoning attack which is only barely constrained can compromise algorithmic fairness.

The bilevel optimization considered here optimizes one poisoning point at a time. To optimize multiple points, one may inject a set of properly-initialized attack points into the training set, and then iteratively optimize them one at a time. Proceeding on a greedy fashion, one can add and optimize one point at a time, sequentially. This strategy is typically faster but suboptimal (as each point is only optimized once, and may become suboptimal after injection and optimization of the subsequent points).

Attacking Algorithmic Fairness. We now define an objective function $\mathcal{A}(\mathbf{x}_c, y_c)$ in terms of a validation loss $L(\mathcal{D}_{\rm val}, \theta)$ that will allow us to compromise algorithmic fairness without significantly affecting classification accuracy. To this end, we consider the *disparate impact* criterion [3]. This criterion assumes data items, typically representing individuals, can be divided into unprivileged (e.g., people with a disability) and privileged (e.g., people without a disability), and that there is a positive outcome (e.g., being selected for a scholarship).

Although one might argue that there are several algorithmic fairness definitions [19] that could be used for this analysis, we selected this criterion for its particularity of being incorporated in legal texts in certain countries [10,26]. Apart of that, recent studies [11] show how fairness metrics are correlated in three clusters what means that targeting this criterion will also affect a set of other metrics with similar strength. In addition to this, authors of [2] used this metric to illustrate the first of the three historical fairness goals that have been used to define fairness metrics. Disparate impact is observed when the fraction of unprivileged people obtaining the positive outcome is much lower the fraction of privileged people obtaining the positive outcome. Formally, to avoid disparate impact:

$$D = \frac{P(\hat{Y} = 1 | G = u)}{P(\hat{Y} = 1 | G = p)} \geq 1 - \epsilon, \tag{4}$$

where $\hat{Y}$ is the predicted label, and $G = \{u, p\}$ a *protected attribute* denoting the group of unprivileged (u) and privileged (p) samples within a set $\mathcal{D}$. Disparate impact thus measures the ratio between the fractions of unprivileged and privileged samples that are assigned to the positive class. Typically, one sets $\epsilon \approx 0.2$ which suggests $D \geq 0.8$ for a fair classifier, as stated by the four-fifths rule of maximum acceptable disparate impact proposed by the US Equal Employment Opportunity Commission (EEOC) [10,26]. Thus, in general, we should have D values closer to one to improve fairness.

For our poisoning attack to work, we aim to minimize such a ratio, i.e., decreasing the fraction of unprivileged samples for which $\hat{y} = 1$, while increasing the fraction of privileged users which are assigned $\hat{y} = 1$. For numerical convenience, we choose to maximize the difference (instead of the ratio) between the mean loss computed on the unprivileged and the privileged samples:

$$L(\mathcal{D}_{\text{val}}, \theta) = \underbrace{\sum_{k=1}^{p} \ell(\mathbf{x}_k, y_k, \theta)}_{\text{unprivileged}} + \lambda \underbrace{\sum_{j=1}^{m} \ell(\mathbf{x}_j, y_j, \theta)}_{\text{privileged}} . \tag{5}$$

Note that the parameter λ here is set to p/m to balance the class priors (rather than dividing the first term by p and the second by m).

To minimize D, we would like to have unprivileged samples classified as negative (lower numerator) and privileged classified as positive (higher denominator). As we aim to maximize $L(\mathcal{D}_{\text{val}}, \theta)$, we can label the unprivileged samples as positive ($y_k = 1$), and the privileged samples as negative ($y_j = -1$). Maximizing this loss will enforce the attack to increase the number of unprivileged samples classified as negative and of privileged samples classified as positive.

In Fig. 1, we report a comparison of the attacker's loss $\mathcal{A}(\mathbf{x}_c, y_c) = L(\mathcal{D}_{\text{val}}, \theta^\star)$ as given by Eq. (5) and the disparate impact D, as a function of the attack point $\mathbf{x}c$ (with $y_c = 1$) in a bi-dimensional toy example. Each point in the plot represents the value of the function (either $\mathcal{A}$ or D computed on an untainted validation set) when the point $\mathbf{x}_c$ corresponding to that location is added to the training set. These plots show that our loss function provides a nice smoother approximation of the disparate impact, and that maximizing it correctly amounts to minimizing disparate impact, thus compromising algorithmic fairness.

2.2 Gradient-Based Attack Algorithm

Having defining our (outer) objective, we are now in the position to discuss how to solve the given bilevel optimization problem. Since our objective is differentiable, we can make use of existing gradient-based strategies to tackle this problem. In particular, we will use a simple gradient ascent strategy with projection (to enforce the box constraint of Eq. 3). The complete algorithm is given as Algorithm 1. In Fig. 1 we also report an example of how this algorithm is able to find a poisoning point that maximizes the attacker's loss.

Attack Initialization. An important remark to be made here is that *initialization* of the poisoning samples plays a key role. In particular, if we initialize the attack point as a point which is correctly classified by the algorithm, the attack will

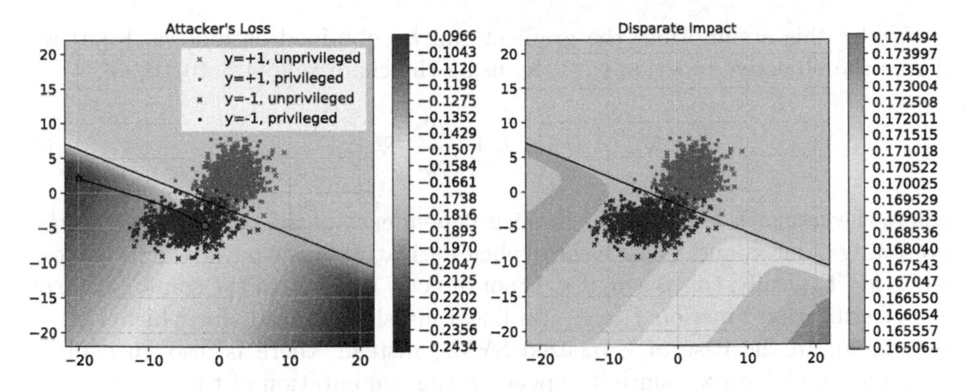

Fig. 1. Attacker's loss $\mathcal{A}(\mathbf{x}_c, y_c)$ (*left*) and disparate impact (*right*) as a function of the attack point $\mathbf{x}_c$ with $y_c = 1$, on a bi-dimensional classification task. Note how the attacker's loss provides a smoother approximation of the disparate impact, and how our gradient-based attack successfully optimizes the former, which amounts to minimizing disparate impact, compromising algorithmic fairness. (Color figure online)

Algorithm 1. Gradient-based poisoning attack

Require: $\mathbf{x}_c, y_c$: the initial location of the poisoning sample and its label; η: the gradient step size; $t > 0$: a small number.
Ensure: $\mathbf{x}'_c$: the optimized poisoning sample.
1: Initialize the attack sample: $\mathbf{x}'_c \leftarrow \mathbf{x}_c$
2: **repeat**
3: Store attack from previous iteration: $\mathbf{x}_c \leftarrow \mathbf{x}'_c$
4: Update step: $\mathbf{x}'_c \leftarrow \Pi\left(\mathbf{x}_c + \eta \nabla_{\mathbf{x}_c} \mathcal{A}\right)$, where Π ensures projection onto the feasible domain (i.e., the box constraint in Eq. 3).
5: **until** $|\mathcal{A}(\mathbf{x}'_c, y_c) - \mathcal{A}(\mathbf{x}_c, y_c)| \leq t$
6: **return** $\mathbf{x}'_c$

---

not even probably start at all. This is clear if one looks at Fig. 1, where we consider an attack point labeled as positive (red). If we had initialized the point in the top-right area of the figure, where positive (red) points are correctly classified, the point would have not even moved from its initial location, as the gradient in that region is essentially zero (the value of the objective is constant). Hence, for a poisoning attack to be optimized properly, a recommended strategy is to initialize points by sampling from the available set at random, but then flipping their label. This reduces the risk of starting from a flat region with null gradients [5,23].

Gradient Computation. Despite the simplicity of the given projected gradient-ascent algorithm, the computation of the poisoning gradient $\nabla_{\mathbf{x}_c} \mathcal{A}$ is more complicated. In particular, we do not only need the outer objective to be sufficiently smooth w.r.t. the classification function, but also the solution $\theta^\star$ of the inner optimization to vary smoothly with respect to $\mathbf{x}_c$ [4,5,8,18]. In general, we need $\mathcal{A}$ to be sufficiently smooth w.r.t. $\mathbf{x}_c$.

Under this assumption, the gradient can be obtained as follows. First, we derive the objective function w.r.t. $\mathbf{x}_c$ using the chain rule [4,5,16,18,23]:

$$\nabla_{\mathbf{x}_c}\mathcal{A} = \nabla_{\mathbf{x}_c}L + \frac{\partial\theta^\star}{\partial\mathbf{x}_c}^\top \nabla_\theta L, \tag{6}$$

where the term $\frac{\partial\theta^\star}{\partial\mathbf{x}_c}$ captures the implicit dependency of the parameters θ on the poisoning point $\mathbf{x}$, and $\nabla_{\mathbf{x}_c}L$ is the explicit derivative of the outer validation loss w.r.t. $\mathbf{x}_c$. Typically, this is zero if $\mathbf{x}_c$ is not directly involved in the computation of the classification function f, e.g., if a linear classifier is used (for which $f(\mathbf{x}) = \mathbf{w}^\top\mathbf{x} + b$). In the case of kernelized SVMs, instead, there is also an explicit dependency of L on $\mathbf{x}_c$, since it appears in the computation of the classification function f when it joins the set of its support vectors (see, e.g., [5,8]).

Under regularity of $\theta^\star(\mathbf{x}_c)$, the derivative $\frac{\partial\theta^\star}{\partial\mathbf{x}_c}$ can be computed by replacing the inner optimization problem in Eq. (2) with its equilibrium (Karush-Kuhn-Tucker, KKT) conditions, i.e., with the implicit equation $\nabla_\theta\mathcal{L}(\mathcal{D}_{\text{tr}} \cup (\mathbf{x}_c, y_c), \theta) \in \mathbf{0}$ [16,18]. By deriving this expression w.r.t. $\mathbf{x}_c$, we get a linear system of equations, expressed in matrix form as $\nabla_{\mathbf{x}_c}\nabla_\theta\mathcal{L} + \frac{\partial\theta^\star}{\partial\mathbf{x}}^\top \nabla^2_{\mathbf{w}}\mathcal{L} \in \mathbf{0}$. We can now compute $\frac{\partial\theta^\star}{\partial\mathbf{x}_c}$ from these equations, and substitute the result in Eq. (6), obtaining the required gradient:

$$\nabla_{\mathbf{x}_c}\mathcal{A} = \nabla_{\mathbf{x}_c}L - (\nabla_{\mathbf{x}_c}\nabla_\theta\mathcal{L})(\nabla^2_\theta\mathcal{L})^{-1}\nabla_\theta L. \tag{7}$$

These gradients can be computed for various classifiers (see, e.g., [8]). In our case, we simply need to compute the term $\nabla_\theta L$, to account for the specific validation loss that we use to compromise algorithmic fairness (Eq. 5).

Finally, in Fig. 2, we show how our poisoning attack modifies the decision function of a linear classifier to worsen algorithmic fairness on a simple bi-dimensional example. As one may appreciate, the boundary is slightly tilted, causing more unprivileged samples to be classified as negative, and more privileged samples to be classified as positive.

2.3 White-Box and Black-Box Poisoning Attacks

The attack derivation and implementation discussed throughout this section implicitly assumes that the attacker has full knowledge of the attacked system, including the training data, the feature representation, and the learning and classification algorithms. This sort of *white-box* access to the targeted system is indeed required to compute the poisoning gradients correctly and run the poisoning attack [4]. It is however possible to also craft *black-box* attacks against different classifiers by using essentially the same algorithm. To this end, one needs to craft the attacks against a *surrogate model*, and then check if these attack samples *transfer* successfully to the actual target model. Interestingly, in many cases these black-box transfer attacks have been shown to work effectively, provided that the surrogate model is sufficiently similar to the target ones [8,20]. The underlying assumption here is that it is possible to train the surrogate model

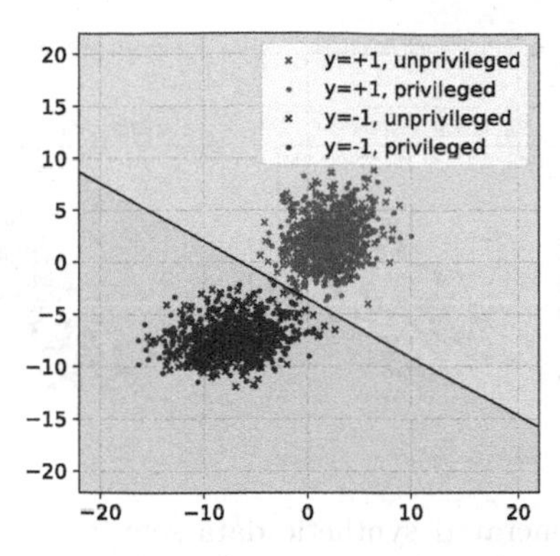

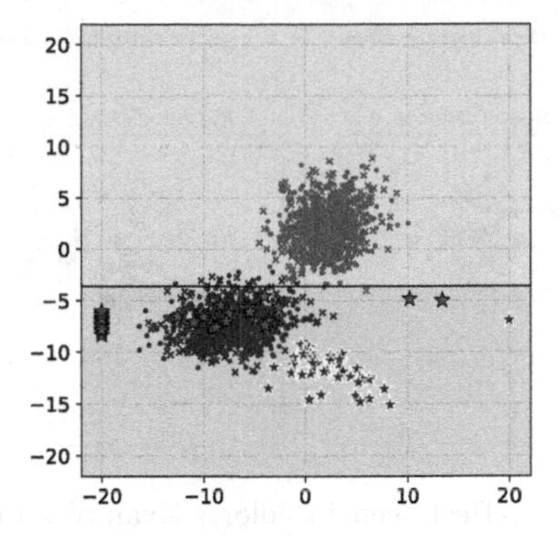

Fig. 2. Gradient-based poisoning attack against a logistic classifier, on a bi-dimensional classification task. The classification function and the corresponding decision regions are reported before (*left*) and after (*right*) injection of the poisoning samples (red and blue stars in the right plot). (Color figure online)

on samples drawn from the same distribution as those used by the target model, or that sufficient queries can be sent to the target model to reconstruct its behavior.

In our experiments we consider both white-box attacks and black-box transfer attacks to also evaluate the threat of poisoning fairness against weaker attackers that only possess limited knowledge of the target model. For black-box attacks, in particular, we assume that the attacker trains the substitute models on a training set sampled from the same distribution as that of the target models, but no queries are sent to the target classifiers while optimizing the attack.

3 Experiments

This section describes the obtained results for two different datasets, one synthetic set composed of 2000 samples, each of them having three features, one of them considered the sensitive attribute, not used for the optimization. The second dataset corresponds to one of the most widely used by the *Algorithmic Fairness* community, a criminal recidivism prediction dataset composed by more than 6000 samples, with 18 features describing each individuals. For each dataset, we consider both the white-box and the black-box attack scenarios described in Sect. 2.3.

3.1 Experiments with Synthetic Data

The first round of experiments uses synthetic data set to empirically test the impact of the attacks with respect to varying levels of disparity already found in the (unaltered) training data. Data is generated using the same approach of Zafar et al. [26]. Specifically, we generate 2,000 samples and assign them to binary class labels ($y = +1$ or $y = -1$) uniformly at random. Each sample

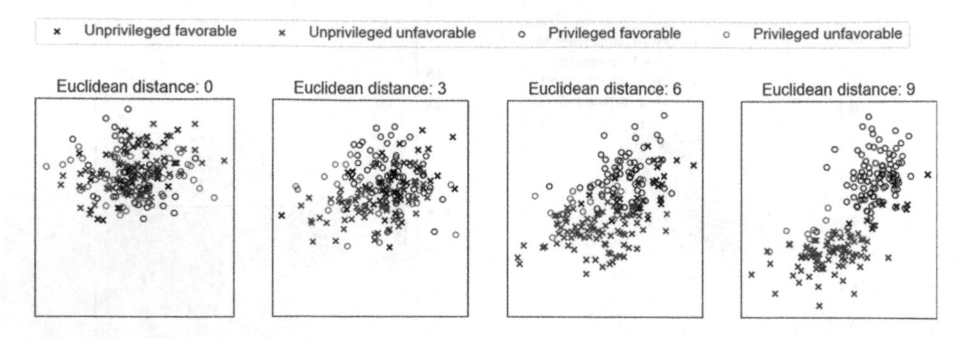

Fig. 3. (Best seen in color.) Examples of generated synthetic data sets for different values of the separation S between groups. Privileged elements ($z = +1$) are denoted by circles and unprivileged elements ($z = -1$) by crosses. Favorable labels ($y = +1$) are in green, while unfavorable labels ($y = -1$) are in red. (Color figure online)

is represented by a 2-dimensional feature vector created by drawing samples from two different Gaussian distributions: $p(x|y = +1) \sim N([2; 2], [5, 1; 1, 5]$ and $p(x|y = -1) \sim N([\mu_1; \mu_2], [10, 1; 1, 3])$ where μ_1, μ_2 are used to modify the euclidean distance S between the centroids of the distributions for the privileged and unprivileged groups so that different base rates [7] can be tested in the experiments. Then, a sample's sensitive attribute z is assigned by drawing from a Bernoulli distribution using $p(z = +1) = \frac{p(x'|y=+1)}{p(x'|y=+1)+p(x'|y=-1)}$ where $x' = [cos(\phi) - sin(\phi); sin(\phi), cos(\phi)]x$ corresponds to a rotated version of the feature vector x.

Using the generator we have described, datasets such as the ones as depicted in Fig. 3 can be obtained. In this figure, the feature vector x is represented in the horizontal and vertical axes, while the color represents the assigned label y (green means favorable, red means unfavorable) and the symbol the sensitive attribute z (circle means privileged, cross means unprivileged).

We generate multiple datasets by setting $S \in \{0, 1, 2, \ldots, 9\}$. We then split each dataset into training D_{tr} (50% of the samples), validation D_{val} (30%) and testing D_{test} (20%) subsets. In each run, a base or initial model $\mathcal{M}$ is trained. This model $\mathcal{M}$ corresponds to a Logistic Regression model in the first setting and to a Support Vector Machine with linear kernel in the second scenario. The regularization parameter C is automatically selected between $[0.5, 1, 5, 10]$ through cross validation. In the *White-Box* setting, the attack is optimized for $\mathcal{M}$ so that Eq. 1 is minimized in the training set D_{tr} and Eq. 3 is maximized in the validation set D_{val}. In the *Black-Box* setting, the attack is optimized against a surrogate model $\hat{\mathcal{M}}$, a Logistic Regression classifier, trained with another subset of data generated for the same value of the parameter S Each of these attacks generates a number of poisoning samples. The poisoned model is the result of retraining the original model with a training set that is the union of D_{tr} and the poisoned samples.

The attack performance is measured by comparing the model trained on the original training data with a model trained on the poisoned data. The evaluation is done according to the following metrics, which for each dataset are averaged over ten runs of each attack:

- **Accuracy.** The accuracy on test obtained by the poisoned model is similar and correlated with the accuracy obtained by a model trained on the original data. It is important to note that the separability of the generated data is also highly correlated with the separation between the groups in the data, creating this effect.
- **Demographic parity.** Measures the allocation of positive and negative classes across the population groups. Framed within the Disparate impact criteria that aims to equalize assigned outcomes across groups, this metric is formulated as:

$$P(\hat{Y} = 1|G = unprivileged) - P(\hat{Y} = 1|G = privileged)$$

  It tends to zero in a fair scenario and is bounded between $[1, -1]$ being -1 the most unfair setting. This metric is correlated with the *Disparate impact* metric introduced in Sect. 2 and has been selected for convenience in the visual representation of the results.
- **Average odds difference.** The average odds difference is a metric of disparate mistreatment, that attempts for Equalized odds [13], it accounts for differences in the performance of the model across groups. This metric is formulated as:

$$\frac{1}{2}[(FPR_p - FPR_u) + (TPR_p - TPR_u)]$$

  It gets value zero in a fair scenario and is bounded between $[1, -1]$ being -1 the most unfair setting.
- **FNR privileged.** False Negative Rate for the privileged group of samples.
- **FNR unprivileged.** False Negative Rate for the unprivileged group of samples.
- **FPR privileged.** False Positive Rate for the unprivileged group of samples.
- **FPR unprivileged.** False Positive Rate for the unprivileged group of samples.

Results shown on Fig. 4 show the obtained performance of the attacks for the generated data. In this figure, the horizontal axis is the separation S between classes in each of the ten datasets. Analyzing the results, we observe that the poisoned models increase disparities in comparison with a model created on the unaltered input data, across all settings. Additionally, they yield an increased FPR for the privileged group (privileged samples that actually have an unfavorable outcome are predicted as having a favorable one), increasing significantly the observed unfairness as measured by the fairness measurements. We note that the attacks also decrease the FNR of the unprivileged group (unprivileged samples that actually have a favorable outcome are predicted as having an unfavorable

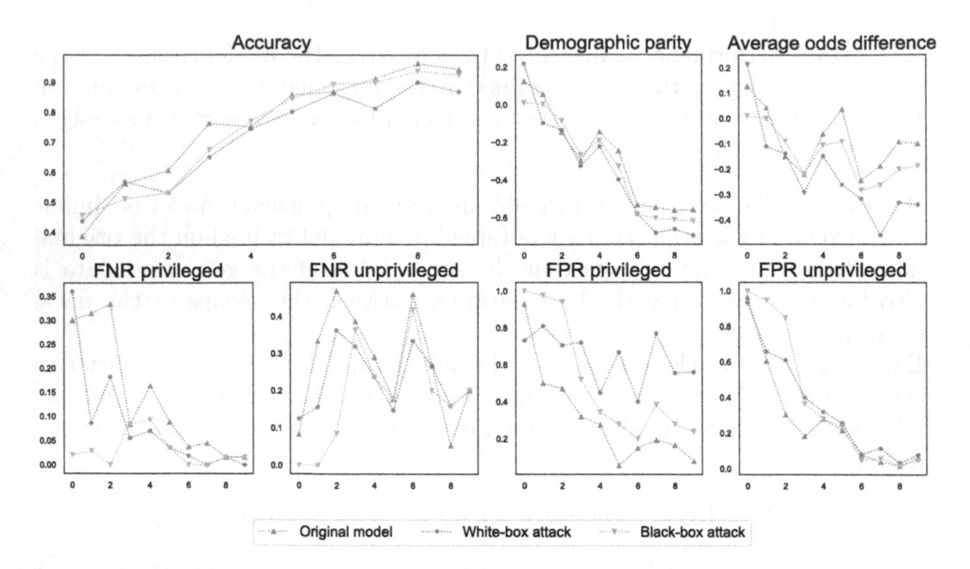

Fig. 4. Comparison of the original model against the model generated by the white-box attack and black-box attacks, for ten synthetic datasets generated by different separation parameters (S). Each data point is the average of ten runs of an attack. We observe that attacks have a moderate effect on the accuracy of the classifier, and can affect the classifier fairness (demographic parity and odds difference) to an extent that becomes more pronounced if the original dataset already has a large separation between classes (larger values of S).

one). This is most likely a consequence of the attack's objective of maintaining accuracy and show that this attack is not trivial. If the attack were only to increase disparities, it would also increase the FNR of the unprivileged group with a larger decrease in accuracy than what we observe. The decrease of FNR for the unprivileged group, however, is smaller than the increase of FPR for the privileged group, as the average odds difference plot shows, and hence the attack succeeds.

3.2 Experiments with Real Data

To demonstrate the attacks on real data, we use the COMPAS dataset released by ProPublica researchers [1], which is commonly used by researchers on Algorithmic Fairness. This dataset contains a prediction of criminal recidivism based on a series of attributes for a sample of 6,167 offenders in prison in Broward County, Florida, in the US. The attributes for each inmate include criminal history features such as the number of juvenile felonies and the charge degree of the current arrest, along with sensitive attributes: race and gender. For each individual, the outcome label ("recidivism") is a binary variable indicating whether he or she was rearrested for a new crime within two years of being released from jail.

We use this dataset for two different types of experiments. First, we show how the attacks demonstrated on synthetic data can also be applied to this data, and demonstrate the effect of varying the amount of poisoned samples, Second, we evaluate the transferability of the attack to other classification models.

White-Box and Black-Box Poisoning Attacks with Varying Amounts of Poisoned Samples. This experiment compares the original model against the model obtained under the two attack models.

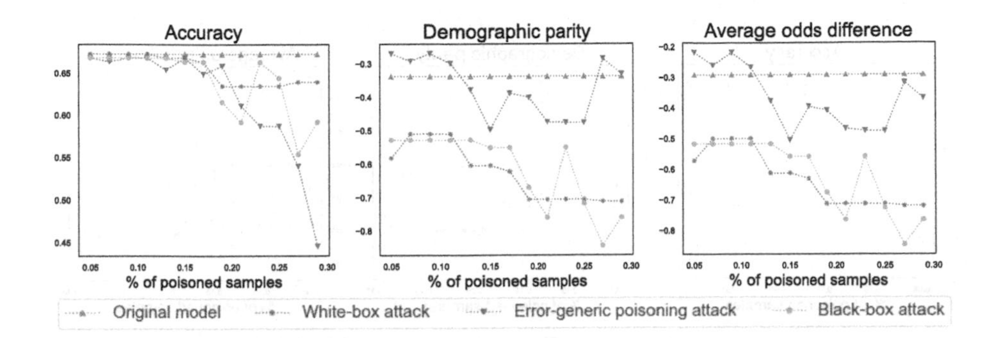

Fig. 5. Comparison of the original model against the model generated by a white-box attack and a black-box attack, for varying percentages of poisoned samples. The main difference between both types of attack is that the black-box attack starts having more noisy behaviour also drastically reducing the accuracy of the classifier (thus being more easily detectable) when the percentage of poisoned samples exceeds a certain threshold (about 20%).

Figure 5 shows the results, which are in line with the findings of the experiments on synthetic data. According to the obtained results, both types of poisoning attacks are is able to increase unfairness of the model with a more modest effect on the accuracy. Also, an interesting finding is the stability of the *White-Box* attack as opposite to the *Black-Box* attack. Whereas the first keeps the same trend with the growing number of samples, the later starts having a unstable and noisy behaviour after adding the 20% of samples, causing for some cases a more unfair model but also affecting the accuracy of the system in a manner that could be easily detected.

In Fig. 5 we also include an Error-Generic Poisoning Attack [8] for the Logistic Regression model, which is designed to decrease the accuracy of the resulting model. We observe that this type of generic adversarial machine learning attack does not affect the fairness of the classifier nearly as much as the attacks we have described on this paper.

As expected, computing the obtained performance for all the stated metrics, (Figure omitted for brevity) can be observed that the effect of any attack increases with the number of poisoned samples. In general, these attacks increase the False Negatives Rate (FNR) for the unprivileged samples, and increase the False Positives Rate (FPR) for the privileged samples.

Transferability of the Attack. We study how an attack would affect the performance of other type of models, simulating different scenarios of *Zero Knowledge* attacks.

Specifically, the attacks we perform is optimized for a Logistic Regression model, and its performance is tested for other models: (a) Gaussian Naive Bayes. (b) Decision Tree; (c) Random Forest; (d) Support Vector Machine with linear kernel; and (e) Support Vector Machine with Radial Basis Function (RBF) kernel.

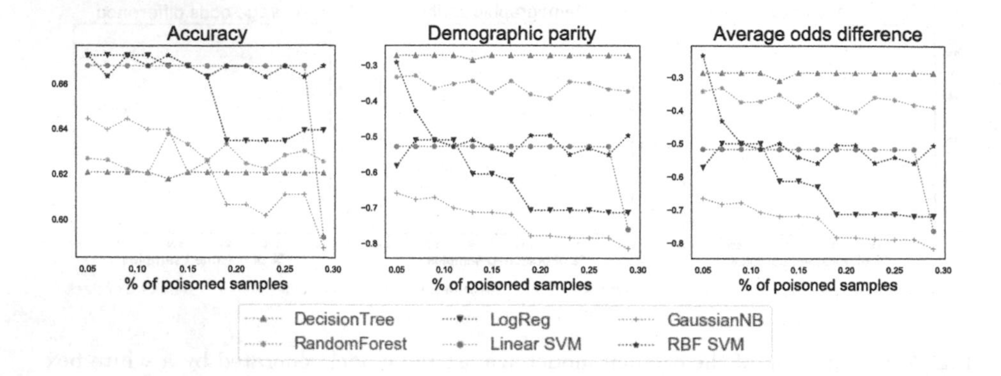

Fig. 6. Transferability of the attacks from Logistic Regression to other models.

Results are shown on Fig. 6, in which each data point corresponds to the average of five experimental runs. We observe that the attack optimized on a Logistic Regression classifier has a stronger effect on the Logistic Regression, Support Vector Machine (for both types of kernel tested) and Naive Bayes models. In contrast, while it can introduce unfairness through demographic disparity and average odds difference on a Decision Tree or Random Forest classifier, its effects are more limited.

4 Related Work

Adversarial Machine Learning Attacks. This work is based on Gradient-Based Optimization, an optimization framework widely used in the literature on Adversarial Machine Learning for crafting poisoning attacks [5,8,14,16,18]. Such framework is used to solve the bilevel optimization given by Eqs. (1)–(3), and requires computing the gradient of the classification function learned by the classifier. As a result, poisoning samples can be obtained by iteratively optimizing one attack point at a time [23].

Measuring Algorithmic Fairness. Many different ways of measuring algorithmic fairness have been proposed [19]. Among those that can be applied in an automatic classification context we find two main types: individual fairness metrics and group fairness metrics [12]. The former seek *consistency* in the sense that similar elements should be assigned similar labels [9]. The latter seek some form of *parity*, and in many cases can be computed from a contingency table indicating the number of privileged and unprivileged samples receiving a positive or negative outcome [21]. Popular group fairness metrics include disparate impact, equalized odds [13], and disparate mistreatment [24].

Optimization-Based Approaches to Increase Fairness. Algorithmic fairness can and often is compromised unintentionally, as discrimination in machine learning is often the result of training data reflecting discriminatory practices that may not be apparent initially [2]. When this is the case, training data can be modified by a type of poisoning attack, in which so-called "antidote" samples are added to a training set to reduce some measure of unfairness. One such approach proposes a method to be applied on recommender systems based on matrix factorization [22]; another is based in the Gradient-Based Optimization framework used in this work [15].

In addition to methods to mitigate unfairness by modifying training data (something known as a pre-processing method for algorithmic fairness [12]), other methods modify the learning algorithm itself to create, for instance, a fair classifier [24,26] In these works, the trade-off between accuracy and fairness is approached through an alternative definition of fairness based in covariance between the users' sensitive attributes and the signed distance between the feature vectors of misclassified users and the classifier decision boundary.

5 Conclusions and Future Work

The results show the feasibility of a new kind of adversarial attack crafted with the objective of increasing disparate impact and disparate mistreatment at the level of the system predictions. We have demonstrated an attacker effectively alter the algorithmic fairness properties of a model even if pre-existing disparities are present in the training data. This means that these attacks can be used to both introduce algorithmic unfairness, as well as for increasing it where it already exists. This can be done even without access to the specific model being used, as a surrogate model can be used to mount a black-box transfer attack.

Studying adversarial attacks on algorithmic fairness can help to make machine learning systems more robust. Additional type of models such neural nets and/or other data sets can be considered in the future to extend the work proposed here. Although experiments in this paper are done using a specific technique based on a poisoning attack, other techniques can be certainly considered. Other approaches such as causality-based techniques could be explored as future work.

Acknowledgments. This research was supported by the European Commission through the ALOHA-H2020 project. Also, we wish to acknowledge the usefulness of the Sec-ML library [17] for the execution of the experiments of this paper. C. Castillo thanks La Caixa project LCF/PR/PR16/11110009 for partial support. B. Biggio acknowledges that this work has been partly funded by BMK, BMDW, and the Province of Upper Austria in the frame of the COMET Programme managed by FFG in the COMET Module S3AI.

References

1. Angwin, J., Larson, J., Mattu, S., Kirchner, L.: Machine bias. There's software used across the country to predict future criminals. And it's biased against blacks (2016)
2. Barocas, S., Hardt, M.: Fairness in machine learning nips 2017 tutorial (2017). https://mrtz.org/nips17/#/
3. Barocas, S., Selbst, A.D.: Big data's disparate impact. Calif. L. Rev. **104**, 671 (2016)
4. Biggio, B., Roli, F.: Wild patterns: ten years after the rise of adversarial machine learning. Pattern Recogn. **84**, 317–331 (2018)
5. Biggio, B., Nelson, B., Laskov, P.: Poisoning attacks against support vector machines. In: Langford, J., Pineau, J. (eds.) 29th International Conference on Machine Learning, pp. 1807–1814. Omnipress (2012)
6. Buolamwini, J., Gebru, T.: Gender shades: intersectional accuracy disparities in commercial gender classification. In: Proceedings of the 1st Conference on Fairness, Accountability and Transparency, pp. 77–91 (2018)
7. Chouldechova, A.: Fair prediction with disparate impact: a study of bias in recidivism prediction instruments (2016)
8. Demontis, A., et al.: Why do adversarial attacks transfer? Explaining transferability of evasion and poisoning attacks. In: 28th USENIX Security Symposium (USENIX Security 2019). USENIX Association (2019)
9. Dwork, C., Hardt, M., Pitassi, T., Reingold, O., Zemel, R.: Fairness through awareness. In: Proceedings of the 3rd Innovations in Theoretical Computer Science Conference, pp. 214–226 (2012)
10. Feldman, M., Friedler, S., Moeller, J., Scheidegger, C., Venkatasubramanian, S.: Certifying and removing disparate impact (2015)
11. Friedler, S.A., Scheidegger, C., Venkatasubramanian, S., Choudhary, S., Hamilton, E.P., Roth, D.: A comparative study of fairness-enhancing interventions in machine learning (2018)
12. Hajian, S., Bonchi, F., Castillo, C.: Algorithmic bias: from discrimination discovery to fairness-aware data mining. In: Proceedings of the 22nd ACM SIGKDD International Conference on Knowledge Discovery and Data Mining, pp. 2125–2126 (2016)
13. Hardt, M., Price, E., Srebro, N.: Equality of opportunity in supervised learning (2016)
14. Jagielski, M., Oprea, A., Biggio, B., Liu, C., Nita-Rotaru, C., Li, B.: Manipulating machine learning: poisoning attacks and countermeasures for regression learning (2018)
15. Kulynych, B., Overdorf, R., Troncoso, C., Gürses, S.: POTs: protective optimization technologies. In: Proceedings of the 2020 Conference on Fairness, Accountability, and Transparency, January 2020). https://doi.org/10.1145/3351095.3372853
16. Mei, S., Zhu, X.: Using machine teaching to identify optimal training-set attacks on machine learners. In: 29th AAAI Conference on Artificial Intelligence (AAAI 2015) (2015)

17. Melis, M., Demontis, A., Pintor, M., Sotgiu, A., Biggio, B.: SecML: a python library for secure and explainable machine learning. arXiv preprint arXiv:1912.10013 (2019)
18. Muñoz-González, L., et al.: Towards poisoning of deep learning algorithms with back-gradient optimization. In: Thuraisingham, B.M., Biggio, B., Freeman, D.M., Miller, B., Sinha, A. (eds.) 10th ACM Workshop on Artificial Intelligence and Security, AISec 2017, pp. 27–38. ACM, New York (2017)
19. Narayanan, A.: Translation tutorial: 21 fairness definitions and their politics. In: Proceedings of the Conference on Fairness Accountability Transparency, New York, USA (2018)
20. Papernot, N., McDaniel, P.D., Goodfellow, I.J.: Transferability in machine learning: from phenomena to black-box attacks using adversarial samples. ArXiv e-prints abs/1605.07277 (2016)
21. Pedreschi, D., Ruggieri, S., Turini, F.: A study of top-k measures for discrimination discovery. In: Proceedings of the 27th Annual ACM Symposium on Applied Computing, SAC 2012, pp. 126–131. ACM, New York (2012). https://doi.org/10.1145/2245276.2245303
22. Rastegarpanah, B., Gummadi, K.P., Crovella, M.: Fighting fire with fire. In: Proceedings of the Twelfth ACM International Conference on Web Search and Data Mining - WSDM 2019 (2019). https://doi.org/10.1145/3289600.3291002
23. Xiao, H., Biggio, B., Brown, G., Fumera, G., Eckert, C., Roli, F.: Is feature selection secure against training data poisoning? In: Bach, F., Blei, D. (eds.) JMLR W&CP - Proceedings of the 32nd International Conference on Machine Learning (ICML), vol. 37, pp. 1689–1698 (2015)
24. Zafar, M.B., Valera, I., Gomez Rodriguez, M., Gummadi, K.P.: Fairness beyond disparate treatment & disparate impact. In: Proceedings of the 26th International Conference on World Wide Web - WWW 2017 (2017)
25. Zafar, M.B., Valera, I., Rodriguez, M., Gummadi, K., Weller, A.: From parity to preference-based notions of fairness in classification. In: Advances in Neural Information Processing Systems, pp. 229–239 (2017)
26. Zafar, M.B., Valera, I., Rodriguez, M.G., Gummadi, K.P.: Fairness constraints: mechanisms for fair classification (2015)

(Social) Network Analysis and Computational Social Science

SpecGreedy: Unified Dense Subgraph Detection

Wenjie Feng[1(✉)], Shenghua Liu[1(✉)], Danai Koutra[2], Huawei Shen[1], and Xueqi Cheng[1]

[1] Institute of Computing Technology, Chinese Academy of Sciences, Beijing 100190, China
wenchiehfeng.us@gmail.com, {liushenghua,shenhuawei,cxq}@ict.ac.cn
[2] University of Michigan, Ann Arbor, MI, USA
dkoutra@umich.edu

Abstract. How can we effectively detect fake reviews or fraudulent connections on a website? How can we spot communities that suddenly appear based on users' interaction? And how can we efficiently find the minimum cut in a big graph? All of these are related to the problem of finding dense subgraphs, an important primitive problem in graph data analysis with extensive applications across various domains.

We focus on formulating the problem of detecting the densest subgraph in real-world large graphs, and we theoretically compare and contrast several closely related problems. Moreover, we propose a unified framework for the densest subgraph detection (GenDS) and devise a simple and computationally efficient algorithm, SpecGreedy, to solve it by leveraging the graph spectral properties with a greedy approach. We conduct thorough experiments on 40 real-world networks with up to 1.47 billion edges from various domains, and demonstrate that our algorithm yields up to 58.6× speedup and achieves better or approximately equal-quality solutions for the densest subgraph detection compared to the baselines. Moreover, SpecGreedy scales linearly with the graph size and is proved effective in applications, such as finding collaborations that appear suddenly in a big, time-evolving co-authorship network.

Keywords: Dense subgraph detection · Graph pattern mining · Algorithm

1 Introduction

How can we capture the most contrast groups or communities in temporal or dynamic graphs—e.g. hot-topics or collaborations in the research community

W. Feng, S. Liu, H. Shen and X. Cheng—They are also with CAS Key Laboratory of Network Data Science & Technology, CAS, and University of Chinese Academy of Sciences, Beijing 100049, China.

F. Hutter et al. (Eds.): ECML PKDD 2020, LNAI 12457, pp. 181–197, 2021.
https://doi.org/10.1007/978-3-030-67658-2_11

that appear suddenly? How can we efficiently determine the minimum cut for a large graph? How can we find the most suspicious users based on their behaviors or spot the largest group with consensus opinion on controversial issues? All these real-world problems are related to the densest subgraph detection task.

Dense pattern mining in graphs is a key primitive task for extracting useful information and capturing underlying principles in relational data. It has benefited various application domains [16], such as capturing the functional groups in biology [30], traffic patterns in human behaviors and interactions [20], communities in social networks [25], anomaly detection in financial and other networks [1], and more. The densest subgraph problem has garnered significant interest in practice because it can be solved exactly in polynomial-time and has an adequate approximation in almost linear time. Goldberg's maximum flow algorithm [13] and Charikar's LP-based algorithm [7] provide the exact solution, and Charikar [7] proved that the simple greedy algorithm is guaranteed to find a result of quality better than the factor 2-approx. with linear time to the graph size. However, these algorithms still incur a prohibitive computational cost for the massive graphs that arise in modern data science applications, without considering the properties of real-world data.

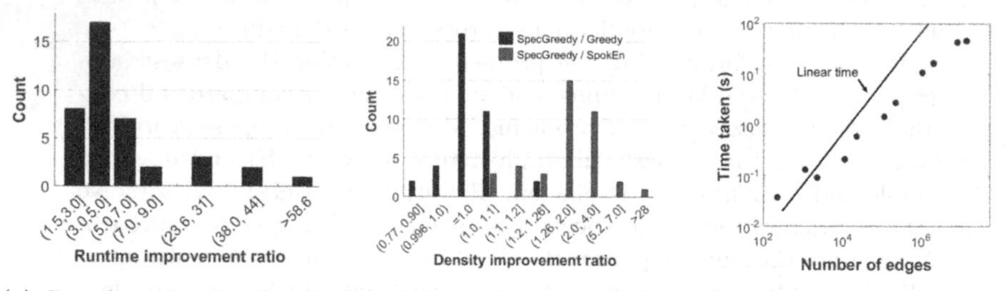

(a) Speedup statistic for the densest subgraph detection (b) Optimal densest subgraph density quality comparison (c) The linear scalability of SPECGREEDY

Fig. 1. Proposed algorithm SPECGREEDY is fast, effective, and scalable. (a) Our method detects the densest subgraphs (qualities in Fig. 1(b)) up to 58.6× faster than the widely-used GREEDY algorithm for various real-world datasets. (b) SPECGREEDY has better or comparable density quality compared with GREEDY and SPOKEN algorithm in the densest subgraph detection. It consistently outperforms SPOKEN for all graphs and finds up to 28 × denser subgraph; it obtains the same or denser (more than 1.26×) optimal density for most graphs compared with GREEDY, and 4 graphs with very close densities ($\geq 0.996\times$) and only 2 graphs with less than 0.9 density improvement. (c) The time taken of SPECGREEDY grows linearly with the size of graph.

To the best of our knowledge, there is no related work to study the connection of the above problems. Here we summarize the differences and relations for some well-known related problems, including detecting community with sparse cut and suspicious dense subgraphs. We also propose a unified formulation, *generalized densest subgraph (*GENDS*) problem*, which subsumes various application problems. This unification explicitly highlights those relations in a formal way and leads to a consistent method for solving these problems. We thus devise an efficient detection algorithm, SpecGreedy, that leverages the graph spectral properties and greedy peeling strategy to solve the generalized problem. With thorough experiments using 40 diverse real-world networks, we demonstrate that our algorithm is fast, highly effective, and scalable (linear to the number of edges, as shown in Fig. 1); it yields 58.6× speedup, and achieves almost equal or better quality, even for a very large graph with $1.47B$ edges. We also find interesting patterns, such as contrast collaboration dense patterns in DBLP co-authorship data.

Our main contributions include:

- **Theory & Correspondences:** We propose the generalized densest subgraph detection formulation, GenDS, to unify several related problems, and analyze the optimization in the principle of the spectral theory;
- **Algorithm:** We devise SpecGreedy, a fast and scalable algorithm to solve the unified GenDS problem;
- **Experiment:** We conduct thorough empirical analyses of various real-world graphs to verify the efficiency and linear-scalability of SpecGreedy. We also find some large contrast dense subgraphs in co-authorship relations.

Reproducibility: Our open-sourced code, the data used, and the supplement document are available at https://github.com/wenchieh/specgreedy.

2 Related Work

In this section, we summarize the related work on the densest subgraph problem and various methods for detecting dense subgraphs in different applications.

Finding the densest subgraph in the large input graph is a widely studied problem [16]. Generally speaking, the goal of such a problem is to find a set of nodes of a given input graph to maximize some notion of density. The so called *densest subgraph problem* (DSP) aims to find a subgraph that maximize the degree density, which is the average of the weights of all its edges. When the edge weights are non-negative, the densest subgraph can be identified optimally in polynomial time by using maximum flow algorithms [13]. However, obtaining the exact solution with maximum flow requires expensive computations despite the theoretical progress achieved in recent years, thus making it prohibitive for large graphs. Charikar [7] introduces a linear-programming formulation of the problem and shows that the greedy algorithm proposed by Asashiro et al. [4] produces a $1/2$-approximation of the optimum density in linear time. [21] proposes an optimization model for local community detection by extending the densest subgraph problem. A recent study [5] proposes a Greedy++ algorithm to improve the output quality of the subgraph over Charikar's greedy peeling algorithm [7] by drawing insights from the iterative approaches from convex optimization. However, when the edge weight can be negative, the above problem becomes NP-hard [27]. When restrictions on the size lower bound are specified, the *densest k-subgraph problem* (DkS) becomes NP-complete [2] and there does not exist any PTAS under a reasonable complexity assumption.

Another line of related research includes contrast graph pattern mining, which aims to discover subgraphs that manifest drastic differences between graphs. Yang et al. [31] proposed to detect the density contrast subgraphs which is equivalent to mining the densest subgraph from a "difference" graph, and employed a local search algorithm to find the solution. Tsourakakis et al. [27] focused on the risk-aversion dense subgraph pattern for a graph with small negative weights and extended the greedy algorithm for this case. [9] detects the k-oppositive cohesive groups by solving a quadratic optimization problem for signed networks. Also, dense subgraphs are used to detect communities [8,30] and anomaly [15,24]. Fraudar [15] proposed to use the greedy method that incorporates the suspiciousness of nodes and edges during optimization. SPOKEN [24] utilizes the "eigenspokes" pattern of community in the EE-plots produced by pairs of eigenvectors of a graph, which is applied to fraud detection.

Table 1. Symbols and definitions

Symbol	Definition
$\mathcal{G} = (V, E)$	Undirected graph with node set V and edge set $E \subseteq V \times V$
$\hat{\mathcal{G}} = (L \cup R, E)$	Bipartite graph with node set, L and R, and edge set $E \subseteq L \times R$
$\mathcal{G}_r = (V, E_r)$	Positive residual graph with node set V and residual edge set E_r
$\boldsymbol{x}, \boldsymbol{y}$	Indicator vector for the selected subset of nodes
$\boldsymbol{u}, \boldsymbol{v}$	Eigenvector or singular vector
$\mathbf{A}, \mathbf{L}$	Adjacency and Laplacian matrix of a graph
$\boldsymbol{d}, \mathbf{D}$	Node degree vector and its diagonal matrix, $d_i = \sum_j a_{ij}$
$\mathbf{I}$	Identity matrix of size $n \times n$
$\mathbf{D}_x$	Diagonal matrix for the vector $\boldsymbol{x}$

Besides, there are many works that utilize the spectral properties of the graph to detect communities [25] and dense subgraphs [3,22], and to partition the input graph [10].

3 Problem and Correspondences

Preliminaries and Definitions. Throughout the paper, vectors are denoted by boldface lowercase letters (e.g. $\boldsymbol{x}$), matrices are denoted by boldface uppercase letters (e.g. $\mathbf{A}$), and sets are denoted by uppercase letters(e.g. S, V). The operator $|\cdot|$ denotes the cardinality of a set or the number of non-zero (**nnz**) elements in a vector and $\|\cdot\|$ is the l_2 norm of a vector, $\lceil x \rceil \equiv \{1, \ldots, x\}$ for brevity. Table 1 gives the complete list of symbols we use in the paper.

Consider an undirected graph $\mathcal{G} = (V, E)$ with $|V| = n$. Let $S \subseteq V$ and $E(S)$ be the edges of subgraph $\mathcal{G}(S)$ induced by the subset S, i.e. $E(S) = \{e_{ij} :$

$v_i, v_j \in S \wedge e_{ij} \in E\}$. Let $\mathbf{A} = (a_{ij}) \in \mathbb{R}^{n \times n}$ be the adjacency matrix of $\mathcal{G}$ and $a_{ij} \geq 0$.

Given an indicator vector $\boldsymbol{x}$ of size n for the subset S, the average degree density of the subgraph $\mathcal{G}(S)$, being the mostly used density measure for the densest subgraph problem, is defined by Charikar [7] as

$$g(S) = \frac{|E(S)|}{|S|} = \frac{1}{2} \cdot \frac{\boldsymbol{x}^T \mathbf{A} \boldsymbol{x}}{\boldsymbol{x}^T \boldsymbol{x}}, \; \boldsymbol{x} \in \{0,1\}^n, \tag{1}$$

and avoids the trivial solution by limiting $|\boldsymbol{x}| \geq 1$. Generally, Hooi et al. [15] proposed to consider the node weight (some constant for each node) for the total mass, the density of $\mathcal{G}(S)$ is

$$g(S) = \frac{|E(S)| + \sum_{i \in V} c_i}{|S|} = \frac{\boldsymbol{x}^T \mathbf{A} \boldsymbol{x}}{2 \cdot \boldsymbol{x}^T \boldsymbol{x}} + \frac{\boldsymbol{x}^T \mathbf{D}_c \boldsymbol{x}}{\boldsymbol{x}^T \boldsymbol{x}} = \frac{1}{2} \frac{\boldsymbol{x}^T (\mathbf{A} + 2\mathbf{D}_c) \boldsymbol{x}}{\boldsymbol{x}^T \boldsymbol{x}}, \; \boldsymbol{x} \in \{0,1\}^n, \tag{2}$$

where $c_i \in \mathbb{R}^+$ is the weight of node i and $\mathbf{D}_c$ is the diagonal matrix of the weight vector $\boldsymbol{c} = [c_1, \ldots, c_n]$.

In addition to dense subgraphs within a single graph, we also consider the "contrast" patterns of cross-graphs, i.e., a subset of nodes that have significantly different edges or edge weights in two given graphs of the same nodeset, like the different snapshots of a dynamic graph.

Table 2. Summary for correspondence to problem GenDS

	Method	matrix P	matrix Q	Constraint
1	MinQuotientCut [10]	$\mathbf{A} \; - \; \mathbf{D} \; = -\mathbf{L}$	$\mathbf{I}$	$\lvert\boldsymbol{x}\rvert < n$
2	Charikar [7]	$\mathbf{A}$	$\mathbf{I}$	$\lvert\boldsymbol{x}\rvert \geq 1$
3	Fraudar [15]	$\mathbf{A} \; + \; 2 \; \mathbf{D}_w$	$\mathbf{I}$	
4	SparseCutDS[1] [21]	$\mathbf{A} \; \frac{2 \cdot \alpha}{2\alpha+1} \; \mathbf{D}$	$\mathbf{I}$	
5	TempDS [30]	$\mathbf{A}_t$	$\mathbf{A}_{t-1} \; + \; 2\,\mathbf{I} \; = \tilde{\mathbf{A}}_{t-1}$	
6	Risk-averse DS [27]	$\mathbf{A}^+ \; + \; \lambda_1 \; \mathbf{I} \; = \tilde{\mathbf{A}}^+$	$\mathbf{A}^- \; + \; \lambda_2 \mathbf{I} \; = \tilde{\mathbf{A}}^-$	
	GenDS[2]	$\mathbf{A} \; + \; 2 \; \mathbf{D}_c$	$\mathbf{A}' \; + \; \gamma\,\mathbf{I} \; = \tilde{\mathbf{A}}'$	

[1] The contrast subgraph pattern [19] equals to set $\alpha = 1$, and $\alpha = \frac{1}{2}$ is considered in [18] for community detection.

[2] Bipartite graphs can be transformed into an undirected graph as Lemma 9.

Generalized Densest Subgraph Problem. Therefore, we propose a generalized densest subgraph problem which subsumes various well-known formulations:

Problem 1 (GenDS: Generalized Densest Subgraph detection). **Given** a graph $\mathcal{G} = (V, E)$ and its contrast $\mathcal{G}' = (V, E')$ with $|V| = n$ nodes; **find** the optimal subset $S^* \subseteq V$ and $|S^*| \geq 1$ **such that**

$$S^* = \arg\max_{S \subseteq V, |S| \geq 1} g(S; \mathbf{P}, \mathbf{Q}) = \arg\max_{\boldsymbol{x} \in \{0,1\}^n, |\boldsymbol{x}| \geq 1} \frac{\boldsymbol{x}^T \mathbf{P} \boldsymbol{x}}{\boldsymbol{x}^T \mathbf{Q} \boldsymbol{x}}, \tag{3}$$

where matrices $\mathbf{P}$ and $\mathbf{Q}$ are related to $\mathcal{G}$ and $\mathcal{G}'$, that is, $\mathbf{P} = \mathbf{A} + 2\mathbf{D}_c$ and $\mathbf{Q} = \mathbf{A}' + \gamma \mathbf{I}$.

Here we define $\tilde{\mathbf{A}}' = \mathbf{A}' + \gamma \mathbf{I}$ as the *augmented adjacency matrix* of graph $\mathcal{G}'$. The denominator in Eq. (3) simultaneously considers the size of the node subset and the connections in the subgraph $\mathcal{G}(S)$. Specifically, if the contrast $\mathcal{G}'$ is an empty graph, $\mathbf{Q}$ degenerates to be a γ-scale identity matrix with only considering the size of the subgraph in GenDS. Note that $\mathbf{P}$ also becomes an augmented adjacency matrix of $\mathcal{G}$ as well if the node weights are equal, i.e., $c_i = c_1 > 0$.

As we show in Theorem 2, our proposed GenDS problem is more general and many dense subgraph-based formulations are special cases of it.

Theorem 2. GenDS *is a general framework for the MinQuotientCut, the densest subgraph detection (Charikar), Fraudar (suspicious dense subgraph),* SparseCutDS *(dense community with sparse cut),* TempDS *(temporal dense subgraph), and Risk-averse DS (consensus dense subgraph), and more.*

The following remarks provide detailed instantiations of GenDS for several problems. Table 2 summarizes the setting and provides the corresponding equation carefully aligned to highlight the correspondences to GenDS.

Remark 3. [MinQuotientCut] The optimal quotient cut ratio problem aims at partitioning the graph into two parts with minimum cut. Let the set of cut edges for S be $cut(S) = \{(u, v) \in E | u \in S, v \in V \setminus S\}$, its size can be formulated as

$$|cut(S)| = \sum_{e_{ij} \in E} a_{ij} (\boldsymbol{x}_i - \boldsymbol{x}_j)^2 = \boldsymbol{x}^T (\mathbf{D} - \mathbf{A}) \boldsymbol{x} = \boldsymbol{x}^T \mathbf{L} \boldsymbol{x},$$

where $\boldsymbol{x}_i = 1$ if $i \in S$, and $\boldsymbol{x}_i = 0$ otherwise. The cut ratio of S is $\frac{|cut(S)|}{\min\{|S|, |V \setminus S|\}}$. Without loss of generality, assuming S is the smaller set compared with its complement, we have the minimum cut ratio by maximizing $-\frac{\boldsymbol{x}^T \mathbf{L} \boldsymbol{x}}{\boldsymbol{x}^T \boldsymbol{x}}$, which corresponds to $\mathbf{P} = -\mathbf{L}$ with $\boldsymbol{c} = -\frac{\boldsymbol{d}}{2}$ and $\mathbf{Q} = \mathbf{I}$ with $\mathbf{A}' = \mathbf{0}$ and $\gamma = 1$.[1]

Remark 4. [Charikar] The densest subgraph detection problem as formulated in Eq. (1) corresponds to $\mathbf{P} = \mathbf{A}$ and $\mathbf{Q} = \mathbf{I}$ with ignoring the constant factor.[2]

[1] In the other setting with $\mathbf{Q} = \mathbf{D}$, this problem is also equivalent to set $\mathbf{P} = -\mathbf{D}^{-1/2} \mathbf{L} \mathbf{D}^{-1/2}$, i.e., the normalized Laplacian matrix of $\mathcal{G}$, and $\mathbf{Q} = \mathbf{I}$.

[2] [23,29] used $\tilde{\mathbf{A}}$ with different γ to explore the trade-off between density and size of final dense subgraphs with the domain-set based optimization method.

Remark 5. [Fraudar] The suspicious densest group detection problem treats the weights of nodes and edges as their suspiciousness score of nodes and edges, i.e. c_u and a_{ij} measure how individually suspicious the particular node u and edge e_{ij} is (can be determined by other information, like profile and text of content) resp. As Eq. (2) shows, it corresponds to $\mathbf{P} = \mathbf{A} + 2\mathbf{D}_c$ and $\mathbf{Q} = \mathbf{I}$, where the numerator $\boldsymbol{x}^T\mathbf{P}\boldsymbol{x}$ is the total suspiciousness of the subgraph ignoring the constant factor.

Remark 6. [SPARSECUTDS] SPARSECUTDS finds a community that is densely connected internally but sparsely connected to the rest of the graph, and it is optimized by maximizing the density while minimizing the average cut size [21]. With the formulation of the cut size in remark 3, the objective to be maximized by SPARSECUTDS is denoted as

$$g_\alpha(S) = \frac{|E(S)| - \alpha \cdot |cut(S)|}{|S|} = \frac{\boldsymbol{x}^T\left((\frac{1}{2}+\alpha)\mathbf{A} - \alpha\mathbf{D}\right)\boldsymbol{x}}{\boldsymbol{x}^T\boldsymbol{x}} = c \cdot \frac{\boldsymbol{x}^T\left(\mathbf{A} - \frac{2\alpha}{2\alpha+1}\mathbf{D}\right)\boldsymbol{x}}{\boldsymbol{x}^T\boldsymbol{x}},$$

where α controls the weight of the $|cut(S)|$ term and $c = \frac{1}{2} + \alpha$ is a constant. Thus, it corresponds to $\mathbf{P} = \mathbf{A} + 2\mathbf{D}_c$ with $\mathbf{D}_c = -\frac{\alpha}{2\alpha+1}\mathbf{D}$ and $\mathbf{Q} = \mathbf{I}$.

Remark 7. [TEMPDS] TEMPDS detects dense subgraphs with nodes S appearing at time t suddenly while having very few edges at time $t-1$ [30]. Let $\mathbf{A}_t$ and $\mathbf{A}_{t-1}$ be adjacency matrices of the snapshots of a temporal graph. Thus, $\boldsymbol{x}^T\mathbf{A}_t\boldsymbol{x}$ and $\boldsymbol{x}^T\mathbf{A}_{t-1}\boldsymbol{x}$ are twice the numbers of edges in corresponding subgraphs. By considering the size of subset S, the objective of TEMPDS can be formulated as:

$$g(S) = \frac{\boldsymbol{x}^T\mathbf{A}_t\boldsymbol{x}}{\boldsymbol{x}^T(\mathbf{A}_{t-1}+2\mathbf{I})\boldsymbol{x}} = \frac{\boldsymbol{x}^T\mathbf{A}_t\boldsymbol{x}}{\boldsymbol{x}^T\tilde{\mathbf{A}}_{t-1}\boldsymbol{x}}.$$

Remark 8. [Risk-averse DS] Given a graph $\mathcal{G}$, the positive entry a_{ij} of its adjacency matrix $\mathbf{A}$ represents the expected *reward* of the edge (u_i, u_j) and the negative entry is opposite to the *risk* of the edge, the absolute value $|a_{ij}|$ measures the strength. Then $\mathbf{A}$ can be written into $\mathbf{A} = \mathbf{A}^+ - \mathbf{A}^-$, where $\mathbf{A}^+$ is the *reward network* and composed of all positive edges in $\mathbf{A}$, that is, its entry $\mathbf{A}^+_{i,j} = \max(a_{ij}, 0)$; and $\mathbf{A}^-$ is the *opposition risk network* and its entry $\mathbf{A}^-_{i,j} = |\min(a_{ij}, 0)|$.

The Risk-averse dense subgraph detection problem finds a subgraph that has a large positive average degree and small negative average degree [27], it is formulated in GENDS format by setting $\mathbf{P} = \mathbf{A}^+ + 2\mathbf{D}_c$ with $\boldsymbol{c} = \frac{\gamma_1}{2}\mathbf{1}$ and $\mathbf{Q} = \mathbf{A}^- + \gamma_2\mathbf{I}$, where $\gamma_1, \gamma_2 \geq 0$ control the size of the subgraph by considering the contribution of the size of the subset S.

As for the densest subgraph detection in a bipartite graph $\hat{\mathcal{G}}$, it can be reduced to the GENDS framework by converting $\hat{\mathcal{G}}$ to be a monopartite graph as following.

Lemma 9. *Given a bipartite graph $\hat{\mathcal{G}} = (L \cup R, E)$ with $|L| + |R| = n$, the densest bipartite subgraph detection problem over $\hat{\mathcal{G}}$ corresponds to the setting that $\boldsymbol{x} = [\boldsymbol{y}, \boldsymbol{z}]$, where $\boldsymbol{y} \in \{0,1\}^{|L|}, \boldsymbol{z} \in \{0,1\}^{|R|}$, and $\mathbf{P}$, $\mathbf{Q} \in \mathbb{R}^{n \times n}$,*

$$\mathbf{P} = \begin{bmatrix} \mathbf{D}_{\boldsymbol{c}_L} & \frac{\mathbf{A}}{2} \\ \frac{\mathbf{A}^T}{2} & \mathbf{D}_{\boldsymbol{c}_R} \end{bmatrix} = \frac{1}{2}\begin{bmatrix} \mathbf{0} & \mathbf{A} \\ \mathbf{A}^T & \mathbf{0} \end{bmatrix} + \begin{bmatrix} \mathbf{D}_{\boldsymbol{c}_L} & \mathbf{0} \\ \mathbf{0} & \mathbf{D}_{\boldsymbol{c}_R} \end{bmatrix}, \ \mathbf{Q} = \begin{bmatrix} \mathbf{I}_{|L|} & \mathbf{0} \\ \mathbf{0} & \mathbf{I}_{|R|} \end{bmatrix} \tag{4}$$

where $\boldsymbol{c}_L$ and $\boldsymbol{c}_R$ are the node weight vectors for the nodesets L and R respectively, $\mathbf{I}_{|L|}$ is the identity matrix of size $|L| \times |L|$, and $\mathbf{I}_{|R|}$ is similar.

To avoid the trivial solution for the weighted graph (single edge with heavy weight), we can introduce column weights as $\mathbf{A} \cdot diag(\frac{1}{h(\mathbf{1}^T\mathbf{A})})$ for some function h, e.g. $h(x) = x^\alpha$ with $\alpha \in \mathbb{R}^+$ or $h(x) = \log(x + c)$ (c is a small constant to prevent the denominator becoming zero). Besides, we can use the motif-based high-order graphs [32] to recognize more complex and interesting dense patterns.

4 Theoretical Analysis

In this section, we connect the optimization of GenDS to the graph spectral theory, showing that we can efficiently approximate the solution by the skewness properties of the spectrum in real-world graphs, thus guide our algorithm design.

Given the graph $\mathcal{G}$ and its contrast $\mathcal{G}'$, we construct a *"positive residual"* graph $\mathcal{G}_r = (V, E_r)$ with $E_r = \{(u,v)|(u,v) \in E \wedge (u,v) \notin E'\}$, and its adjacency matrix is denoted as $\mathbf{A}_r = (\mathbf{P} - \mathbf{Q})^+$. Then the densest subgraph detection in $\mathcal{G}_r$ means that it maximizes the density in $\mathcal{G}$ while minimizes the connection in $\mathcal{G}'$. Thus, the objective function in Eq. (3) is reformulated as

$$S^* = \underset{S \subseteq V, |S| \geq 1}{\arg\max}\ g(S; \mathbf{P}, \mathbf{Q}) = \underset{\boldsymbol{x} \in \{0,1\}^n, |\boldsymbol{x}| \geq 1}{\arg\max} \frac{\boldsymbol{x}^T(\mathbf{P} - \mathbf{Q})^+\boldsymbol{x}}{\boldsymbol{x}^T\boldsymbol{x}} = \underset{\boldsymbol{x} \in \{0,1\}^n, |\boldsymbol{x}| \geq 1}{\arg\max} \frac{\boldsymbol{x}^T\mathbf{A}_r\boldsymbol{x}}{\boldsymbol{x}^T\boldsymbol{x}}. \tag{5}$$

We will use this transformation in the following theoretical optimality analysis.

Consider the optimization problem with a similar form as Eq. (5) defined in the real domain, which is formulated in the *Rayleigh quotient* manner, that is

$$R(\mathbf{A}_r, \boldsymbol{x}) = \frac{\boldsymbol{x}^T\mathbf{A}_r\boldsymbol{x}}{\boldsymbol{x}^T\boldsymbol{x}}, \boldsymbol{x} \in \mathbb{R}^n, \ \boldsymbol{x} \neq \mathbf{0}, \tag{6}$$

where $\mathbf{A}_r \in \mathbb{R}^{n \times n}$ is a symmetric matrix; $R(\mathbf{A}_r, c\boldsymbol{x}) = R(\mathbf{A}_r, \boldsymbol{x})$ for any non-zero scalar c. The objective of GenDS in Eq. (5) is a binary-variable special case.

The Rayleigh–Ritz Theorem [10] in the spectral theory gives the optimality of Eq. (6) with eigenvalues of $\mathbf{A}_r \in \mathbb{R}^{n \times n}$, that is,

Theorem 10 (Rayleigh–Ritz Theorem). *Let $\mathbf{A}_r$ be a symmetric matrix with eigenvalues $\lambda_1 \geq \ldots \geq \lambda_n$ and corresponding eigenvectors $\boldsymbol{u}_1, \ldots, \boldsymbol{u}_n$.*

Then[3]

$$\lambda_1 = \max_{\boldsymbol{x}\neq \boldsymbol{0}} R(\mathbf{A}_r, \boldsymbol{x}) = \max_{\boldsymbol{x}\in\mathbb{R}^n, \|\boldsymbol{x}\|=1} \boldsymbol{x}^T\mathbf{A}_r\boldsymbol{x} \Longrightarrow \boldsymbol{x} = \boldsymbol{u}_1$$
$$\lambda_n = \min_{\boldsymbol{x}\neq \boldsymbol{0}} R(\mathbf{A}_r, \boldsymbol{x}) = \min_{\boldsymbol{x}\in\mathbb{R}^n, \|\boldsymbol{x}\|=1} \boldsymbol{x}^T\mathbf{A}_r\boldsymbol{x} \Longrightarrow \boldsymbol{x} = \boldsymbol{u}_n. \quad (7)$$

In general, for $1 \leq k \leq n$, *let* $\mathcal{S}_k$ *denote the span of* $\boldsymbol{u}_1, \ldots, \boldsymbol{u}_k$ *(with* $\mathcal{S}_0 = \mathbf{0}$*), and let* $\mathcal{S}_k^\perp$ *denote the orthogonal complement of* $\mathcal{S}_k$. *Then*

$$\lambda_k = \max_{\boldsymbol{x}\neq\mathbf{0},\, \boldsymbol{x}\in\mathcal{S}_{k-1}^\perp} R(\mathbf{A}_r, \boldsymbol{x}) = \max_{\|\boldsymbol{x}\|=1,\, \boldsymbol{x}\in\mathcal{S}_{k-1}^\perp} \boldsymbol{x}^T\mathbf{A}_r\boldsymbol{x} \Longrightarrow \boldsymbol{x} = \boldsymbol{u}_k, \quad (8)$$

which means λ_k is the largest value of $R(\mathbf{A}_r, \boldsymbol{x})$ over the complement space $\mathcal{S}_{k-1}^\perp$.

With the analogy of eigenvalues and singular values of matrices, the latter achieve the optimality property that resembles those of Rayleigh quotient matrices [11]. To avoid the large magnitude negative eigenvalues for the real graphs [26], here we utilize the singular values and singular vectors instead in the following.

Let $\mathbf{A}_r = \mathbf{U\Sigma V}^T = \sum_{i=1}^r \sigma_i \boldsymbol{u}_i \boldsymbol{v}_i^T$ be the singular value decomposition of the matrix $\mathbf{A}_r$, the columns of $\mathbf{U}$ and $\mathbf{V}$ are called the left- and right- singular vectors respectively, i.e., $\mathbf{U} = [\boldsymbol{u}_1, \ldots, \boldsymbol{u}_r]$ and $\mathbf{V} = [\boldsymbol{v}_1, \ldots, \boldsymbol{v}_r]$. $\mathbf{\Sigma} = diag(\sigma_1, \ldots, \sigma_r)$ for singular values $\sigma_1 \geq \cdots \geq \sigma_r > 0$. Then, we also have the following representation regard to the GENDS problem,

Lemma 11. *The optimal solution for the* GENDS *in Eq.* (3) *can be written as*

$$S^* = \operatorname*{arg\,max}_{\boldsymbol{x}\in\{0,1\}^n, |\boldsymbol{x}|\geq 1} \frac{\boldsymbol{x}^T\mathbf{A}_r\boldsymbol{x}}{\boldsymbol{x}^T\boldsymbol{x}} = \operatorname*{arg\,max}_{|S|\geq 1} \frac{1}{|S|} \sum_{i=1}^n \sigma_i \left(\sum_{j\in S} \boldsymbol{u}_{ij}\right)\left(\sum_{j\in S} \boldsymbol{v}_{ij}\right) \quad (9)$$

where $\boldsymbol{u}_{ij}$ *and* $\boldsymbol{v}_{ij}$ *denote the* j*-th element of the singular vector* $\boldsymbol{u}_i$ *and* $\boldsymbol{v}_i$ *corresponding to the singular value* σ_i *resp. The optimal density value* $g_{opt} \leq \sigma_1$.

As for the bipartite graph case, given an asymmetric matrix $\mathbf{A}_r \in \mathbb{R}^{m\times n}$, we define the related quadratic optimization problem as

$$R(\mathbf{A}_r; \boldsymbol{x}, \boldsymbol{y}) = \frac{\boldsymbol{x}^T\mathbf{A}_r\boldsymbol{y}}{\boldsymbol{x}^T\boldsymbol{x} + \boldsymbol{y}^T\boldsymbol{y}}, \boldsymbol{x} \in \mathbb{R}^m, \boldsymbol{y} \in \mathbb{R}^n, \boldsymbol{x} \neq \mathbf{0}, \boldsymbol{y} \neq \mathbf{0}. \quad (10)$$

And we also obtain the following theorem that leads to a similar statement as Theorem 10. Thus, it helps to avoid constructing the big matrix ($\mathbb{R}^{(m+n)\times(m+n)}$) for the bipartite graph. The detailed proof is given in the supplement.

Theorem 12 (Bigraph Spectral). *Suppose* $\mathbf{A}_r$ *is an* $m \times n$ *matrix,* $\mathbf{A}_r = \mathbf{U\Sigma V}^T$ *is its singular value decomposition. For any vector* $\boldsymbol{x} \in \mathbb{R}^m, \boldsymbol{y} \in \mathbb{R}^n$,

$$\sigma_1 = \max_{\|\boldsymbol{x}\|=\|\boldsymbol{y}\|=1} \boldsymbol{x}^T\mathbf{A}_r\boldsymbol{y} \geq \max_{\boldsymbol{x}\neq 0, \boldsymbol{y}\neq 0} 2\cdot R(\mathbf{A}_r, \boldsymbol{x}, \boldsymbol{y}) \Longrightarrow \begin{matrix}\boldsymbol{x} = \boldsymbol{u}_1 \\ \boldsymbol{y} = \boldsymbol{v}_1\end{matrix}. \quad (11)$$

[3] The proof details of the theorem refer to [10].

In general, for $1 \leq k \leq r$*, let* $\mathcal{S}_k^U$, $\mathcal{S}_k^V$ *denote the span of* $\boldsymbol{u}_1, \ldots, \boldsymbol{u}_k$ *and* $\boldsymbol{v}_1, \ldots, \boldsymbol{v}_k$ *(with* $\mathcal{S}_0^U = \mathbf{0}, \mathcal{S}_0^V = \mathbf{0}$*), then*

$$\sigma_k = \max_{\substack{\|\boldsymbol{x}\|=\|\boldsymbol{y}\|=1 \\ \boldsymbol{x}\perp\mathcal{S}_{k-1}^U,\, \boldsymbol{y}\perp\mathcal{S}_{k-1}^V}} \boldsymbol{x}^T \mathbf{A}_r \boldsymbol{y} \geq \max_{\substack{\boldsymbol{x}\neq\mathbf{0},\boldsymbol{y}\neq\mathbf{0} \\ \boldsymbol{x}\perp\mathcal{S}_{k-1}^U,\, \boldsymbol{y}\perp\mathcal{S}_{k-1}^V}} 2 \cdot R(\mathbf{A}_r, \boldsymbol{x}, \boldsymbol{y}) \Longrightarrow \begin{matrix} \boldsymbol{x} = \boldsymbol{u}_k \\ \boldsymbol{y} = \boldsymbol{v}_k \end{matrix}.$$

Therefore, given a bipartite graph $\hat{\mathcal{G}} = (L \cup R, E)$ with the adjacency matrix $\mathbf{A} \in \mathbb{R}^{|L|\times|R|}$, we will have the similar properties as Lemma 11 as

Lemma 13. *For the densest bipartite subgraph detection in Fraudar with* $\mathbf{P} = diag([\mathbf{A}/2, \mathbf{A}^T/2])$ *and* $\boldsymbol{x}^T\mathbf{P}\boldsymbol{x} = |E(S)|$*, the optimal solution can be written as*

$$\begin{aligned} S^* = & \arg\max_{\boldsymbol{x}\in\{0,1\}^n, |\boldsymbol{x}|\geq 1} \frac{\boldsymbol{x}^T\mathbf{P}\boldsymbol{x}}{\boldsymbol{x}^T\boldsymbol{x}} = \arg\max_{\boldsymbol{y}\in\{0,1\}^{|L|},\, \boldsymbol{z}\in\{0,1\}^{|R|}, |\boldsymbol{y}|>0, |\boldsymbol{z}|>0} R(\mathbf{A}_r, \boldsymbol{y}, \boldsymbol{z}) \\ \leq & \arg\max_{S=\delta(\boldsymbol{y})\cup\delta(\boldsymbol{z}), |S|\geq 1} \frac{1}{|S|} \sum_{i=1} \sigma_i \left(\sum_{j\in\delta(\boldsymbol{y})} \boldsymbol{u}_{ij} \right) \left(\sum_{j\in\delta(\boldsymbol{z})} \boldsymbol{v}_{ij} \right), \end{aligned} \tag{12}$$

where $\boldsymbol{u}_{ij}, \boldsymbol{v}_{ij}$ *denote the* j*-th element of the singular vector* $\boldsymbol{u}_i$ *and* $\boldsymbol{v}_i$*, and the optimal density value* $g_{opt} \leq \sigma_1$*.*

Moreover, if the matrix $\mathbf{Q}$ is positive definite (i.e., $\boldsymbol{x}\mathbf{Q}\boldsymbol{x}^T > 0$ for any $\boldsymbol{x} \neq \mathbf{0}$) in GenDS, the Eq. (3) under the relaxation $\boldsymbol{x} \in \mathbb{R}^n$ is equivalent to the *generalized Rayleigh quotient*, its optimization reduces to the *generalized eigenvalue decomposition* problem; the *min-max principle* provides result about the optimality similar to Theorem 10. Due to the singularity of $\mathbf{Q}$ in the real scenario, we take the residual graph form $\mathcal{G}_r$ for approximation as discussed above.

Real-World Graph Properties. The sparsity and various power-laws are key components of the real-world networks gathered from the world-wide-web, social networks, E-commerce, on-line reviews, recommend systems, and more. Those primary properties contribute to the time and space-efficient computing or storage, and synthetically modeling the realistic networks. Various studies [12,17] have shown that most real-world graphs have a statistically significant power-law distribution with degree distribution, the distribution of "bipartite cores" (≈communities), a cutoff in the eigenvalue or singular values of the adjacency matrix and the Laplacian matrix, etc. Also, the distribution of eigenvector elements (indicators of "network value") associated with the top-ranked eigenvalues of the graph adjacency matrix is skewed [6].

Thus, based on the spectral formulation of GenDS, the skewness of singular values and components in singular vectors of real-world graphs guarantees that we can simply consider the top singular vectors and use a few of top-rank elements in them to efficiently construct the candidates for dense subgraphs and detect the optimal result, We will introduce this in more details in the following algorithm.

5 Algorithms and Complexity Analysis

In this section, we present our proposed method SpecGreedy for the generalized densest subgraph detection problem GenDS and provide analysis for its property.

We first review the related Charikar's peeling algorithm. It takes the entire original graph as the starting point, then greedily removes the node with the smallest degree from the graph, and returns the densest one among the shrinking sequence of subgraphs created by the procedure. It is guaranteed to return a solution of at least half of the optimum density, i.e., $g^* \geq \frac{1}{2} g_{opt}$. In addition, using the priority tree to manage the nodes in the peeling process, the complexity of the greedy algorithm is $O(|E| \log |V|)$.

However, the densest subgraphs usually have small sizes and are embedded in a large graph (background), which leads to many searches and update steps to obtain an approximation solution or even the candidates for Charikar's algorithm.

Implications of Theoretical Analysis: Lemma 11 and 13 show the upper bound of the optimal density, i.e., $g_{opt} \leq \sigma_1$, and the σ_k is the optimal value for the real space orthogonal to $\mathcal{S}_{k-1}$ $(k > 1)$ as Theorem 10 and 12; the formulation of S^* highlights that the real-value singular vectors provide some insight to find the optimal densest subgraph. Thus, these nodes in S^* will have higher importance in the singular vectors associating with the top-ranked singular values.

Considering the skewed distribution of the elements in a singular vector, we can construct some small nodeset candidates, which derive some subgraphs, with the top-ranked nodes based on the singular vectors to avoid detecting the densest subgraph from the whole graph, that is, $S_C = \{S_1, \ldots, S_k\}$ for some $1 \leq k < n$, where the candidate $S_i = \{j; \boldsymbol{u}_{ij} > \Delta_L, j \in \lceil |L| \rceil\} \cup \{j; \boldsymbol{v}_{ij} > \Delta_R, j \in \lceil |R| \rceil\}$ for the singular vectors $\boldsymbol{u}_i$ and $\boldsymbol{v}_i$, Δ_L and Δ_R are some pre-defined truncation thresholds; the optimal density for $\mathcal{G}(S_i)$ is $g_i \leq \sigma_i$. Here we determine the selection thresholds as $\Delta_L = 1/\sqrt{|L|}$ and $\Delta_R = 1/\sqrt{|R|}$[4] based on the re-formulation of the optimal solution in the Eq. (9) and Eq. (12).

Proposed Algorithm. Therefore, we propose SpecGreedy, which utilizes graph spectral properties and the greedy peeling strategy to solve the GenDS problem. Algorithm 1 summarizes our approach.

Given the adjacency matrix $\mathbf{A}_r$ of the positive residual graph $\mathcal{G}_r$, density metric g, and the top approximation rank k which controls the maximum size of the candidate set. SpecGreedy finds the top-k spectral decomposition of the matrix at first (Line 2) , then detects the possible densest subgraphs based on the top singular vectors. In each round, it constructs the candidate subset S_r based on the truncated singular vectors $\boldsymbol{u}_r$ and $\boldsymbol{v}_r$, then uses the *greedy algorithm* to search the densest subgraph for $\mathcal{G}(S_r)$ to maximize the density metric g. It checks

[4] If $\mathbf{A}_r$ is the symmetric matrix as in Eq. (9), $|L| = |R| = n$ and $\Delta_L = \Delta_R = 1/\sqrt{n}$.

Algorithm 1 SpecGreedy: General dense subgraph detection

Input: Matrix $\mathbf{A}_r$ of the positive residual $\mathcal{G}_r$; density metric g; top approx. rank k.
Output: The densest subgraph.
1: $S = \emptyset$
2: $[\mathbf{U}, \Sigma, \mathbf{V}] = \mathbf{SVD}(\mathbf{A}_r, k)$ ▷ Top-k spectral decomposition of $\mathbf{A}_r$
3: **for** $r \leftarrow 1, \ldots, k$ **do**
4: Construct the candidate node subset S_r based on $\boldsymbol{u}_r$ and $\boldsymbol{v}_r$, i.e.
$S_r = \{i : \boldsymbol{u}_{ri} > \frac{1}{\sqrt{|L|}}, i \in L\} \cup \{j : \boldsymbol{v}_{rj} > \frac{1}{\sqrt{|R|}}, j \in R\}$
5: $S_r^* \leftarrow$ Greedy($\mathcal{G}(S_r)$, g) ▷ Greedily remove nodes to maximize the metric g.
6: **if** $g(S_r^*) > g(S)$ **then** ▷ $g(S) = g_{cur}^*$
7: $S \leftarrow S_r^*$
8: **if** $g(S) > \sigma_{r+1}$ **then** ▷ Spectral early-stopping condition
9: **break**
10: **return** $\mathcal{G}(S)$.

the stop condition based on the next singular value for the current optimal result in Line 8 for early stopping.

How many subgraph candidates do we need to check? Let g_{cur}^* be the current detected optimal density with some off-the-shelf detection approaches, if there is some $1 < j \leq k$ satisfied that $g_{cur}^* \geq \sigma_j$, the optimal density then can be achieved based on the singular vectors is g_{cur}^* due to the decreasing-order of singular values ($\sigma_j > \sigma_{j+1}$) and the aforementioned upper-bound ($g_i \leq \sigma_i$). Finally, the subgraph with the optimal density is returned. It is worth mentioning that the power-law distribution nature of the eigenvalues and singular values of real-world graphs and the theoretical bounds of solutions (the exact or 1/2-approx. result) for detection approaches guarantee that the size of candidates will be very small.

Besides the pre-computing top-k spectral decomposition strategy in Line 2, we can use a lazy or online way to compute the $(r+1)$-th largest spectral decomposition result with the power method or the efficient Krylov subspace methods such as the Lanczos method [14]. In the experiment, we adopt an incremental decomposition way which gets the top-l singular values and singular vectors first, and if the stop condition in Line 8 is not satisfied, then get the further top-$(l+s)$ singular values and vectors with step-size s. This stepwise increasing-decomposition will continue until $l+s \geq k$ or the early-stopping condition holds. Moreover, we can use other densest subgraph detection approaches in Line 5 considering the enhancement of solution, e.g. Greedy++ [5] or the LP method.

Theorem 14 (Time Complexity). *The complexity of* SpecGreedy *algorithm is* $O(K \cdot |E| + K \cdot |E(\tilde{S})| \log |\tilde{S}|)$ *where* $\tilde{S} = \max_{|S_i|} S_i$ *and* K *is the top approximation rank.*

Ideally, $K = \min\{k, r_{opt} + 1\}$ where k is the input parameter and r_{opt} is the rank with optimal resultant density g^*. The complexity of computing a top eigenvector/singular vector in sparse graphs is linear, i.e., $O(|E(V)|)$, and the total complexity of the greedy algorithm in Line 5 is $O(|E(S)| \log |S|)$ for $\mathcal{G}(S)$.

Given the skewness of the top singular vectors in real-world graphs, we usually have $|\tilde{S}| \ll |V|$, making SpecGreedy a linear algorithm in the number of edges.

6 Experiments

We design experiments to answer the following questions:

1. **Q1. Efficiency:** How does our SpecGreedy compare to the state-of-the-art greedy algorithm for detecting the densest subgraph?
2. **Q2. Effectiveness:** How well does SpecGreedy work on real data, and perform on detecting the contrast dense subgraph and injected subgraphs?
3. **Q3. Scalability:** How does our method scale with the input graph size?

Data: We used a variety of datasets (40 in total) obtained from 5 popular network repositories, including 32 monopartite graphs and 8 bipartite ones, and 5 of them also have edge weights; the largest unweighted graph is the `soc-twitter` graph with roughly $1.47B$ edges, while the smallest unweighted graph has roughly $14.5K$ edges. Multiple edges, self-loops are removed, and directionality is ignored for directed graphs. The detailed information about those real-world networks is provided in the supplement.

Implementations: We implemented efficient dense subgraph detection algorithms for comparison. We implemented our algorithm, Greedy [7], SpokEn [24], and Fraudar [15] in Python; SpokEn actually detects the densest subgraph only based on the truncation of the singular vectors like our method. In all the experiments, we set the parameter of top approximation rank $k = 10$ and $l = s = 3$ for SpecGreedy. We ran all experiments on a machine with 2.4GHz Intel(R) Xeon(R) CPU, 64GB of main memory.

6.1 Q1. Efficiency

To answer Q1, we apply our method SpecGreedy and the baseline Greedy on 40 unweighted networks and compare their runtime.

Figure 1(a) shows the statistical information about the runtime improvement ratio of SpecGreedy compared with the Greedy algorithm for detecting the densest subgraphs; Fig. 2(a) illustrates more detailed information about the time taken of the two methods: for each network dataset, it provides the runtime of the two methods and the network size.

Observation: Our method runs faster than Greedy and achieves the same or comparable optimal densities as shown in Fig. 1(b). Among these varied-size datasets, SpecGreedy achieves 3.0–5.0× speedup for 17 of them, 1.5–3.0× for 8, and 5.0–7.0× for 7 graphs, and more than 58.6× for the `ca-DBLP2012` graph. As we can see, SpecGreedy is efficient for large graphs, e.g. 30× for `ca-DBLP-NET`, 25× for `cit-Patents`, and 3× speedup for `soc-twitter`.

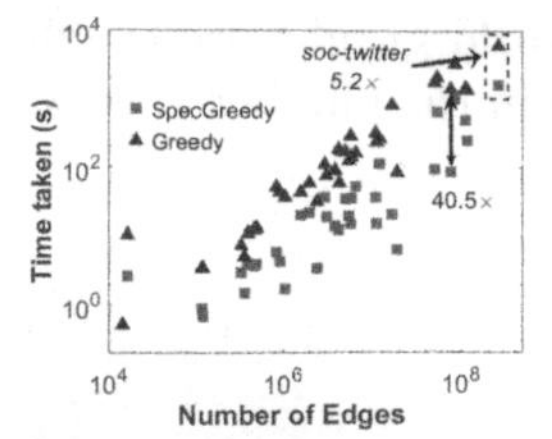

(a) The densest subgraph detection for all graphs.

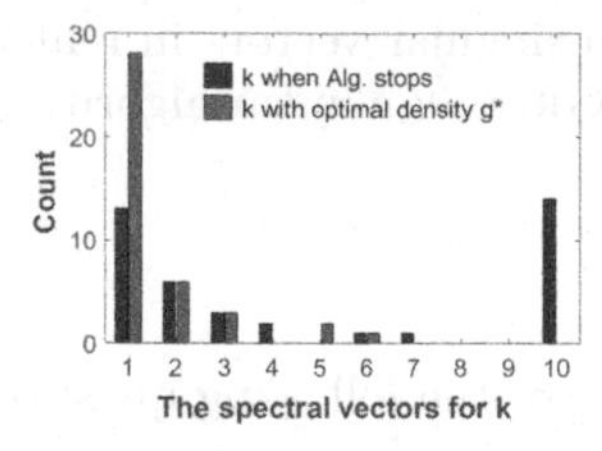

(b) The statistic of spectral vectors for k.

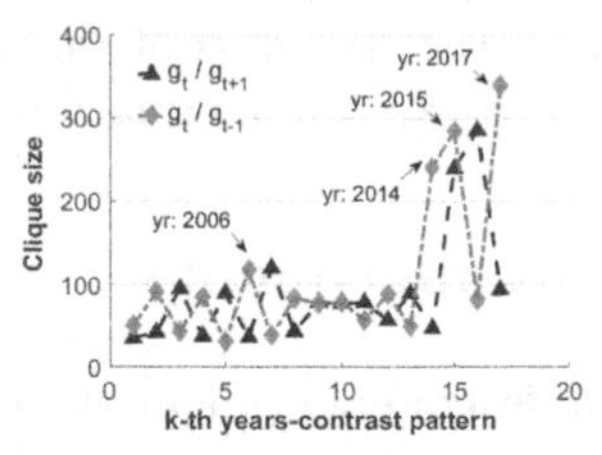

(c) Contrast patterns for the DBLP co-authorship graphs.

Fig. 2. The performance of SpecGreedy for the real-world graphs. (a) SpecGreedy runs faster than Greedy in all graphs for detecting the densest subgraph with the same or comparable density, achieves 58.6× speedup for *ca-DBLP2012* and about 3× for the largest graph *soc-twitter*. (b) The statistic information about k for spectral vectors. The densest subgraphs with optimal density $g*$ are achieved in the first singular vector for most of the datasets. The blue bars show the statistics of k when algorithm stops given the parameter $k = 10$. (c) The contrast patterns for DBLP co-authorship data in 2000 – 2017 with the positive residual $\mathcal{G}_r$ (very large cliques in 2017, 2015, and 2014). (Color figure online)

For the 5 weighted graphs, we observe similar results as above. SpecGreedy achieves 24–39× speedup for 3 of them and 11–17× for the rest. Greedy will have poor performance for the graph dominated by few edges with heavy weights due to it needs to peel each edge of the whole graph.

Figure 2(b) summarizes the statistics about spectral vectors k for obtaining the optimal density g^* and actual k when the algorithm stops. Larger k means taking more time for SVD and detection candidate subgraphs. We can see that the densest subgraphs with optimal density g^* are achieved in the first spectral vector for most of the datasets, the second one for 6 of the graphs, and only 3 graphs need to check more than 5 singular vectors. There are 26 graphs where SpecGreedy stops for the early-stopping condition, while the rest need to check all 10 singular vectors due to the small optimal density or flat power-law factor of singular values. Besides, we find that some subgraphs detected based on the top $k - 1$ vectors also cliques with a smaller size than the optimal one. So, the above heuristic observation and the power-law distribution of singular values contribute to the efficiency of SpecGreedy, and the small k is enough for good results.

6.2 Q2. Effectiveness

In this section, we verify that SpecGreedy detects high-quality densest subgraphs in real-world graphs and accurately detects injected subgraphs with different injection density. Moreover, focusing on a large-scale collaboration network, we show that SpecGreedy also finds significant contrast dense subgraphs.

Density Improvement. Following the setup we described in Q1, Fig. 1(b) shows the improvement ratio of optimal densities found by SpecGreedy compared to the Greedy and SpokEn algorithm. As we can see, SpecGreedy consistently outperforms SpokEn by detecting denser densest subgraphs for all real-world datasets. It even achieves more than 28.3× higher density for the `soc-twitter` graph, Also, SpecGreedy obtains the same or denser (more than 1.26×) optimal density for most graphs compared with Greedy; there are 4 graphs that the optimal densities detected by SpecGreedy have less than but very close ($\geq 0.996\times$) densities as detected by Greedy, and 2 graphs with less than 0.9 density improvement. So, utilizing the spectral distribution of the densest subgraph, SpecGreedy can improve the quality of solution of Greedy in most cases due to avoid arbitrary ties-break in graphs for removing in Greedy to some extent.

Injection Detection. We further evaluate the performance of SpecGreedy by performing a synthetic experiment where we inject dense subgraphs as ground truth. For a more realistic setting, we also added extra edges as 'camouflage' between the nodes in the selected injection subgraph and the remaining unselected nodes. We compared SpecGreedy, Greedy and SpokEn in terms of F measure in detecting the injected patterns, and reports the averaged F-score over 5 trials. Specifically, we injected a 600×600 subgraph with different injection densities to an amazon-Art review subgraph of size $4K \times 4K$, and we select the two different cases with background densities 2.7E-5 and 3.4E-5 for comparison. From the result, we observe that SpecGreedy achieves equally high accuracy as Greedy and is better than SpokEn, the detailed figures are provided in the supplement.

Case Study. As a case study, we also apply SpecGreedy on the DBLP co-authorship data [28] from 2000 to 2017 to identify interesting contrast dense patterns. Figure 2(c) shows the contrast dense subgraphs pattern detected by SpecGreedy with constructing the positive residual graphs $\mathcal{G}_r$. Those densest contrast subgraphs are all cliques of different sizes, which means the connections that form a clique only appear in $\mathcal{G}_t$ rather than $\mathcal{G}_{t-1}$ (or $\mathcal{G}_{t+1}$). As we can see, there are 3 extremely large cliques for 2017, 2015, and 2014, related to the publications in 'Brain network and Disease', 'Neurology and Medicine', and 'Physics' from some large collaborative groups of different disciplines.

6.3 Q3. Scalability

Figure 1(c) shows the linear scaling of SpecGreedy's running time in the number of edges of the graph. Here we used the `ca-Patents-AM` graph and randomly subsampled different proportions of the edges from it for detecting the densest subgraph. The slope parallel to the main diagonal indicates linear growth.

7 Conclusions

In this paper, we propose the generalized densest subgraph detection, GenDS, which unifies several well-known instances of related problems. We devise the SpecGreedy algorithm to solve the generalized problem based on graph spectral properties and a greedy peeling approach. Our main contributions are as follows.

- **Theory & Correspondences:** We propose the unified formulation for the densest subgraph detection from different applications, and analyze our proposed optimization problem by leveraging spectral theory.
- **Algorithm:** We devise a fast algorithm, SpecGreedy, to solve the GenDS.
- **Experiments:** The efficiency of SpecGreedy is verified on 40 real-world graphs. SpecGreedy runs linearly with the graph size and is effective in applications, like finding sudden bursts in research co-authorship relationships.

The quality guaranteed detection algorithm design and streaming graphs adaptation are also possible extension directions for this work.

Acknowledgments. This work was upported by the Strategic Priority Research Program of Chinese Academy of Sciences, Grant No. XDA19020400, NSF of China No. 61772498, U1911401, 61872206, 91746301, National Science Foundation under Grant No. IIS 1845491, and Army Young Investigator Award No. W911NF1810397.

References

1. Akoglu, L., Tong, H., Koutra, D.: Graph based anomaly detection and description: a survey. Data Min. Knowl. Discov. **29**(3), 626–688 (2015). https://doi.org/10.1007/s10618-014-0365-y
2. Andersen, R., Chellapilla, K.: Finding dense subgraphs with size bounds. In: Avrachenkov, K., Donato, D., Litvak, N. (eds.) WAW 2009. LNCS, vol. 5427, pp. 25–37. Springer, Heidelberg (2009). https://doi.org/10.1007/978-3-540-95995-3_3
3. Andersen, R., Cioaba, S.M.: Spectral densest subgraph and independence number of a graph. J. UCS **13**(11), 1501–1513 (2007)
4. Asahiro, Y., Iwama, K., Tamaki, H., Tokuyama, T.: Greedily finding a dense subgraph. J. Algorithms **34**(2), 203–221 (2000)
5. Boob, D., et al.: Flowless: Extracting densest subgraphs without flow computations. In: WWW 2020 (2020)
6. Chakrabarti, D., Zhan, Y., Faloutsos, C.: R-MAT: a recursive model for graph mining. In: SDM, pp. 442–446. SIAM (2004)
7. Charikar, M.: Greedy approximation algorithms for finding dense components in a graph. In: Jansen, K., Khuller, S. (eds.) APPROX 2000. LNCS, vol. 1913, pp. 84–95. Springer, Heidelberg (2000). https://doi.org/10.1007/3-540-44436-X_10
8. Chen, J., Saad, Y.: Dense subgraph extraction with application to community detection. In: IEEE TKDE (2010)
9. Chu, L., Wang, Z., Pei, J., Wang, J., Zhao, Z., Chen, E.: Finding gangs in war from signed networks. In: KDD, pp. 1505–1514. ACM (2016)

10. Fan, R.K.C.: Spectral graph theory. American Mathematical Society (1996)
11. Dax, A.: From eigenvalues to singular values: a review. APM **3**, 17 (2013)
12. Eikmeier, N., Gleich, D.F.: Revisiting power-law distributions in spectra of real world networks. In: KDD, pp. 817–826 (2017)
13. Goldberg, A.V.: Finding a maximum density subgraph. UCB (1984)
14. Golub, G.H., Van Loan, C.F.: Matrix Computations, vol. 3. JHU Press, Baltimore (2012)
15. Hooi, B., Song, H.A., Beutel, A., Shah, N., Shin, K., Faloutsos, C.: FRAUDAR: bounding graph fraud in the face of camouflage. In: SIGKDD, pp. 895–904 (2016)
16. Lee, V.E., Ruan, N., Jin, R., Aggarwal, C.: A survey of algorithms for dense subgraph discovery. In: Aggarwal, C., Wang, H. (eds.) Managing and Mining Graph Data. Advances in Database Systems, vol. 40, pp. 303–336. Springer, Boston (2010). https://doi.org/10.1007/978-1-4419-6045-0_10
17. Leskovec, J., Chakrabarti, D., Kleinberg, J., Faloutsos, C., Ghahramani, Z.: Kronecker graphs: an approach to modeling networks. JMLR **11**, 985–1042 (2010)
18. Li, Z., Zhang, S., Wang, R.-S., Zhang, X.-S., Chen, L.: Erratum: quantitative function for community detection. Phys. Rev. E **91**(1), 019901 (2015)
19. Liu, S., Hooi, B., Faloutsos, C.: A contrast metric for fraud detection in rich graphs. TKDE **31**(12), 2235–2248 (2018)
20. Liu, Y., Zhu, L., Szekely, P.A., Galstyan, A., Koutra, D.: Coupled clustering of time-series and networks. In: SDM, pp. 531–539. SIAM (2019)
21. Miyauchi, A., Kakimura, N.: Finding a dense subgraph with sparse cut. In: CIKM (2018)
22. Papailiopoulos, D., Mitliagkas, I., Dimakis, A., Caramanis, C.: Finding dense subgraphs via low-rank bilinear optimization. In: ICML, pp. 1890–1898 (2014)
23. Pavan, M., Pelillo, M.: Dominant sets and pairwise clustering. IEEE Trans. Pattern Anal. Mach. Intell. **29**(1), 167–172 (2006)
24. Prakash, B.A., Sridharan, A., Seshadri, M., Machiraju, S., Faloutsos, C.: EigenSpokes: surprising patterns and scalable community chipping in large graphs. In: Zaki, M.J., Yu, J.X., Ravindran, B., Pudi, V. (eds.) PAKDD 2010. LNCS (LNAI), vol. 6119, pp. 435–448. Springer, Heidelberg (2010). https://doi.org/10.1007/978-3-642-13672-6_42
25. Shen, H.-W., Cheng, X.-Q.: Spectral methods for the detection of network community structure: a comparative analysis. JSTAT **2010**(10), P10020 (2010)
26. Tsourakakis, C.E.: Fast counting of triangles in large real networks without counting: algorithms and laws. In: ICDM, pp. 608–617. IEEE (2008)
27. Tsourakakis, C.E., Chen, T., Kakimura, N., Pachocki, J.: Novel dense subgraph discovery primitives: risk aversion and exclusion queries. In: Brefeld, U., Fromont, E., Hotho, A., Knobbe, A., Maathuis, M., Robardet, C. (eds.) ECML PKDD 2019. LNCS (LNAI), vol. 11906, pp. 378–394. Springer, Cham (2020). https://doi.org/10.1007/978-3-030-46150-8_23
28. Wan, H., Zhang, Y., Zhang, J., Tang, J.: AMiner: search and mining of academic social networks. Data Intell. **1**(1), 58–76 (2019)
29. Wang, Z., Chu, L., Pei, J., Al-Barakati, A., Chen, E.: Tradeoffs between density and size in extracting dense subgraphs: a unified framework. In: ASONAM (2016)
30. Wong, S.W., Pastrello, C., Kotlyar, M., Faloutsos, C., Jurisica, I.: SDREGION: fast spotting of changing communities in biological networks. In: SIGKDD (2018)
31. Yang, Y., Chu, L., Zhang, Y., Wang, Z., Pei, J., Chen, E.: Mining density contrast subgraphs. In: ICDE, pp. 221–232. IEEE (2018)
32. Yin, H., Benson, A.R., Leskovec, J., Gleich, D.F.: Local higher-order graph clustering. In: KDD, pp. 555–564 (2017)

Networked Point Process Models Under the Lens of Scrutiny

Guilherme Borges(✉), Flavio Figueiredo, Renato M. Assunção, and Pedro O. S. Vaz-de-Melo

Universidade Federal de Minas Gerais, Belo Horizonte, Brazil
{guilherme.borges,flaviovdf,assuncao,olmo}@dcc.ufmg.br

Abstract. Recently, there has been much work on the use of Networked Point Processes (NPPs) to extract the latent network structure of timestamp data. Several models currently exist to capture implicit interactions in hospital visits, blog posts, e-mail messages, among others. The problem is that evaluating these solutions is not a trivial task. First, the methods have only been evaluated in a few datasets by a limited number of metrics. Second, and even worse, the evaluation metrics are often unsuitable for the typically sparse networks, which consequently lead to inconclusive results. To provide the community with a rigorous benchmark, in this paper we propose an empirical evaluation framework of NPP models in the task of network extraction. We reevaluate several models of the literature using our framework and compare the results to two null models designed for this task. In our discussion, we point out when some methods should be used depending on the expected efficacy, execution time, or dataset properties. Overall, we find that only three models show consistent significant results in real-world data.
Source Code: http://github.com/guilhermeresende/NPPs/.

Keywords: Temporal point processes · Benchmarking · Network inference

1 Introduction

Modeling the occurrence of events through temporal point processes is a necessity in many different fields. A temporal point process $\mathcal{P}$ is a stochastic process whose realizations consist of random event times $t_1, t_2, \cdots$, where each t_i is the time of the occurrence of the $i-$th event. Different point processes can be interrelated in such a way that events occurring in one process influence the occurrence of events in other processes [12]. In this case, the relationship among them can be encoded as an influence graph (or matrix) of Networked Point Processes (NPPs), where the nodes are the different processes, and the edge weights encode the influences from one node to the others [1,5]. Inferring this influence matrix via timestamp data only [41] is a challenging task that has been studied in a variety of settings, such as hospital visits of multiple patients [10], viewing records of TV programs [41], and the publication times of articles across different websites [1,45].

F. Hutter et al. (Eds.): ECML PKDD 2020, LNAI 12457, pp. 198–215, 2021.
https://doi.org/10.1007/978-3-030-67658-2_12

Following the growing popularity trend of Machine Learning (ML), several models for inferring the influence matrix of NPPs were recently proposed [1,2,15, 25,41,45,46]. One problem is that, similarly to what is being reported for other fields within ML [20,22,27,30,35–38], the rate of empirical advancement of NPP models is not being followed by a consistent increase in the level of empirical rigor. Models are being evaluated using only a handful of efficacy metrics at best [1,41,45,45], and most of these metrics are inadequate to assess sparse weighted networks, which is usually the case in NPPs. In addition, most studies evaluated their approaches using only a few real-world datasets, sometimes even only one [1]. Hyper-parameter sensitivity analyses are not conducted and almost no effort is shown to tune hyperparameters for baselines. Finally, very little is done to account for model limitations. Previous studies rarely report confidence intervals or comparisons with null models. As usual, in the field of ML, negative results are omitted in favor of wins. Sanity checks, such as verifying if the model is robust to false positives (i.e., i.i.d. data), are also ignored.

Motivated by these issues, and inspired by recent general guidelines for empirical rigor in the field of ML [35], we propose an empirical evaluation framework for NPP models. Our framework assumes that NPP models receive k series of timestamps (processes) as input and outputs the influence matrix among these k processes. This task is especially challenging because the input data is usually bursty [3], with sparse connectivity among processes [45], and may contain all sorts of noise (e.g., spams or bot messages) [44]. Also, the latent network structure may serve several purposes and, because of that, should be assessed and evaluated accordingly. While one may need the network to identify only the most influential nodes, another might require the whole network structure. Moreover, while some applications demand a low false-positive error rate, others are more concerned with false-negative errors. Thus, properly evaluating NPP models is not a trivial task.

Our proposed framework has five complementary fronts of evaluation, which can be applied to assess any data-driven model. First, we perform an analysis of how the likelihood of the model to the data correlates with evaluation metrics (e.g., NDCG). Second, we present a methodology to estimate an empirical upper-bound estimate based on the number of hyper-parameter configurations tested. Third, we evaluate the asymptotic complexity of training such methods and, fourth, we evaluate the models on seven real-world datasets using three complementary metrics. Finally, we assess the robustness of the models to false-positives via i.i.d. simulated data. In short, the main contributions of this work are:

- A comprehensive empirical evaluation framework for NPP models;
- A thorough comparison of state of the art NPP models;
- Guidelines for selecting NPP models for the task of network inference.

The rest of this work is organized as follows. In Sect. 2, we describe the related work. Next, in Sect. 3, we present the problem formulation and our empirical evaluation framework. Section 4 describes the results and, finally, in Sect. 5 we discuss the conclusions.

2 Related Work

In a recent position paper, Sculley *et al.* [35] made a sound statement about how empirical rigor is not keeping pace with advances in Machine Learning (ML). The good news is that, like us, many others are taking a step back and reviewing the current state of the art of different areas of ML through rigorous and thorough empirical evaluations [20,22,27,30,36–38]. A striking and common conclusion of these evaluations is that hyper-parameter tuning and testing multiple datasets can make traditional and most recent methods to perform equally. This was verified for generative adversarial networks [27], decision making [37], language models [30], information retrieval [22,42], and clinical prediction [20].

Inspired by these works, we conduct an extensive evaluation of NPP models for network inference. Most previous work on NPPs falls into three classes: those that explore Hawkes Processes [21,28,31,33], those that explore information cascades [7,16–18,32,39], and those that explore Wold processes [15]. We argue that NPP models have also reached a point where there is a need to take a step back and evaluate models using a rigorous empirical methodology. Often, previous efforts to extract latent networks from data rely on fixed hyper-parameters [1] and test their methods on only a handful of real datasets [9,14,29,43,45,46]. Also, previous works have employed a variety of distinct metrics and tools for validation, such as different simulated datasets, different problem definitions, and different error and ranking measures. Moreover, metrics do not usually transfer from one work to the other. For instance, [15] discussed how the Kendall correlations employed by [41,43,46] is unsuitable for sparse matrices.

3 The Evaluation Framework

We argue that a proper framework to evaluate the networked point process models must take into account: (1) a comparison with the state of the art models using (2) different metrics over (3) several datasets with (4) the proper use of confidence intervals and null models to test significance hypotheses. Throughout this section, we present the problem definition and our methodology to perform such evaluations.

3.1 Problem Definition

Consider multiple point processes $\mathcal{P}_a, \mathcal{P}_b, \cdots, \mathcal{P}_K$ observed simultaneously, where each event is associated to a single process: $\mathcal{P}_a = \{0 \leq t_{a_1} < t_{a_2} < \ldots\}$. Let $\mathcal{P} = \{1, \cdots, K\}$ be the set of all processes, where $|\mathcal{P}| = K$, i.e., we have a total of K processes or nodes. For each $a \in \mathcal{P}$, the counting process $N_a(t) = \sum_{i=1}^{|\mathcal{P}_a|} \mathbb{1}_{t_{a_i} \leq t}$ is the total number of events of $\mathcal{P}_a$ until time t. Finally, $N = |\bigcup_{a=1}^{K} \mathcal{P}_a|$ is is the total number of events considering the superimposed union of all events from all point processes.

Let $\mathcal{H}_a(t)$ be the history of the a-th point process up to time t, called the filtration of the process. A point process can be completely characterized by its conditional intensity function [12] defined as: $\lambda_a(t|\mathcal{H}_a(t)) = \lim_{h \to 0} \frac{\mathbb{P}(N_a(t+h) - N_a(t) > 0 | \mathcal{H}_a(t))}{h}$. Assuming no simultaneous events, this function

Table 1. Complexity and hyper-parameter ranges used for optimizing the models.

Method	Complexity (per iteration)	Hyper-parameter range
ExpKern	$O(NK^2)$	$\beta \in [0, 10]$; $C \in [0, 10^3]$; $pnlty \in [l_1, l_2, nuc, none]$; $solver \in [gd, agd]$; $\alpha \in [10^{-4}, 1]$;$tol \in [10^{-8}, 10^{-4}]$
HkEM	$\Omega(NK^2)$	$support \in [1, 200]$; $size \in [1, 400]$; $tol \in [10^{-8}, 10^{-4}]$
ADM4	$O(N^3K^2)$	$\beta \in [0, 10]$; $C \in [0, 10^3]$; $lasso \in [0, 1]$; $tol \in [10^{-8}, 10^{-4}]$
MLE-SGP	$O(MN^3K^2)$	$max - mean - f \in [1, 200]$; $\#f \in [1, 20]$; $\alpha \in [10^{-9}, 10^{-3}]$; $C \in [0, 10^3]$; $lasso \in [0, 1]$; $tol \in [10^{-8}, 10^{-4}]$
HC	$O(K^3)$	$H \in [1, 200]$; $C \in [0, 10^3]$; $solver \in [adam, ada, rmsp, adad]$; $pnlty \in [l_1, l_2, none]$; $\alpha \in [10^{-4}, 1]$; $tol \in [10^{-8}, 10^{-4}]$
GB	$O(N(\log(N) + \log(K)))$	$\beta \in [1, 10]$
NetInf	—	$\alpha \in [0, 1]$; $m \in [exp, powerlaw, rayleigh]$

defines the instantaneous probability of one event occurring at each time t given the entire previous history up to t. In a multivariate setting, we can extend this definition by conditioning the intensity $\lambda_a(t|\mathcal{H}_{\mathcal{P}}(t))$ on the history of all events $\mathcal{H}_{\mathcal{P}}(t)$. For instance, if we have three timestamps from two processes $t_{a1} < t_{b1} < t_{a2} < t$: $\mathcal{H}_{\mathcal{P}}(t) = \{t_{a1}, t_{b1}, t_{a2}\}$; $\mathcal{H}_a(t) = \{t_{a1}, t_{a2}\}$; and, $\mathcal{H}_b(t) = \{t_{b1}\}$. In other words, $\lambda_a(t|\mathcal{H}_a(t))$ captures the evolution of process $\mathcal{P}_a$ by focusing solely on $\mathcal{P}_a$'s history, $\lambda_a(t|\mathcal{H}_{\mathcal{P}}(t))$ states that a's evolution also depends on $\mathcal{P}_b$. This is our starting point for defining the *network structure* as follows.

Network Structure: Let $\mathcal{P}_a$ and $\mathcal{P}_b$ be two arbitrary processes in $\mathcal{P}$. Also, let $\mathcal{Q} = \mathcal{P} - \{\mathcal{P}_b\}$. When $\lambda_a(t|\mathcal{H}_{\mathcal{Q}}(t)) = \lambda_a(t|\mathcal{H}_{\mathcal{P}}(t))$ we can state that process $\mathcal{P}_b$ does not influence process $\mathcal{P}_a$. That is, the intensity function of a with or without $\mathcal{P}_b$ is, at least statistically, equivalent. When $\lambda_a(t|\mathcal{H}_{\mathcal{Q}}(t)) \neq \lambda_a(t|\mathcal{H}_{\mathcal{P}}(t))$, $\mathcal{P}_b$ influences $\mathcal{P}_a$, as $\mathcal{P}_b$'s past has some influence on the intensity function of $\mathcal{P}_a$. In other words, $\mathcal{P}_b$'s history impacts the instantaneous probability of one event occurring at t for process $\mathcal{P}_a$.

The literature commonly exploits two class of processes for modeling NPPs: Hawkes [19] and Wold [40] processes. In both classes, one may re-write the intensity as a sum of two factors, an exogenous (Poissonian) constant, μ_a, and an endogenous factor accounting for $\mathcal{H}_{\mathcal{P}}(t)$: $\lambda_a(t|\mathcal{H}_a(t)) = \mu_a + \sum_{b=1}^{K} \alpha_{ba}\phi_a(\mathcal{H}_b(t))$, where ϕ is some influence kernel. Following this definition, let $\hat{\boldsymbol{A}} = [\alpha_{ba}]$ be a K by K matrix capturing the network structure of the processes, with $\alpha_{ba} \geq 0$. Here, $\alpha_{ba} = 0$ when $\mathcal{P}_b$ does **not** influence $\mathcal{P}_a$ (i.e., $\alpha_{ba}\phi_a(\mathcal{H}_b(t)) = 0$). Conversely, $\mathcal{P}_b$ **does** influence $\mathcal{P}_a$ when $\alpha_{ba} \geq 0$. In other words, an edge exists in the latent graph when $\alpha_{ba} \neq 0$. Inferring $\hat{\boldsymbol{A}}$ is done using timestamps (or in some cases cascades) only.

3.2 Models

We evaluated six different state-of-the-art models, chosen to represent both recent and older methods (2011 to 2018), with different characteristics (e.g., employ Hawkes, cascades and Wold Processes), and open-source implementations available online[1]. We also evaluate a classic Hawkes process with an Exponential kernel [19]. The runtime complexity per learning iteration for each model is measured according to K and N.

The most commonly employed Hawkes model is one with an Exponential influence function [19], simply called **ExpKern**. We learn this model using the open source `tick` library. It has an asymptotic cost of $O(NK^2)$ (K^2 parameters updated for each N timestamp per learning iteration). Out of the recent methods, from 2011 onward, the earliest we evaluate is HawkesEM (**HkEM**) [25]. It is a non-parametric estimator for the Hawkes processes with a complexity lower bound of $\Omega(NK^2)$ per iteration[2], where $\Omega(.)$ is a lower bound. In 2013, Zhou et al. proposed **ADM4** [45], a parametric Hawkes model that employs an exponential kernel with a fixed decay and complexity of $O(N^3K^2)$. We also tested **MLE-SGP** [41], which uses a parametrization of the kernel as a sum of M basis functions (Gaussian functions in the author's experiments) with a complexity of $O(MN^3K^2)$, where M is the number of basis functions. More recently, Achab et al. proposed HawkesCumulants (**HC**) [1], a non-parametric model that uses a moment matching method to compute the causal matrix. Because of this, the method avoids estimating the kernels themselves and the per-iteration complexity does not depend on N, but only on K. The per iteration complexity is $O(K^3)$.

For Wold processes only one option was available, namely Granger-Busca (**GB**) [15]. The method is an EM algorithm with complexity of $O(N(\log(N) + \log(K)))$. For the cascades framework, we tested **NetInf**[3], one of the state-of-the-art models. Cascade models are usually employed in datasets different than the ones we explore in this work (i.e., require cascade information), so we chose to evaluate **NetInf** on the only dataset where this is possible (Memetracker). The authors of this model state that there is no closed form for run-time complexity as it depends on the underlying network structure.

Hyper-parameters were optimized via random search [4]. As we discuss in the next section, when extracting networks via point processes, cross-validation is not usually employed [1,15,45]. A ground truth matrix of source destination pairs is constructed and models are learned using only events from either sources or destinations. Thus, the most appropriate model for the dataset is selected using the log-likelihood function. When available in the source code, we present results for the learned models with the best log-likelihood. For **HC**, the

---

1 https://github.com/X-DataInitiative/tick. http://github.com/flaviovdf/granger-busca. http://snap.stanford.edu/netinf/.

2 The asymptotic upper bound was not reported by the authors.

3 Other models considered were MMEL [46] and Hawkes Conditional Law [2]. However, the first crashed consistently and the latter did not finish its execution on time.

Table 2. Number of processes and events for each dataset.

Dataset	# of processes	# of events
CollegeMsg	100	17750
sx-askubuntu	28	3335
sx-mathoverflow	29	2767
sx-superuser	33	4157
email-Eu-core	100	11220
wiki-talk	100	30442
Memetracker	25	177163

objective function [1] was used. Thus, we were able to properly optimize hyper-parameters for **HkEM**, **ADM4**, **ExpKern**, **GB** and for **HC**. The remaining method (**MLE-SGP**) does not have a log-likelihood function implemented. Thus, for this approach we keep the default values. Table 1 summarizes our models and hyper-parameters choices.

3.3 Ground Truth Data

In order to evaluate the models, we need a ground truth causal matrix $\boldsymbol{A}$. A common approach to construct this matrix is through datasets composed by (*source*, *destination*, *timestamp*) triples, which indicates that some interaction (e.g. messages) occurred from *source* to *destination* at time *timestamp*. The ground truth matrix is, thus, captured by the normalized number of interactions between (*source*, *destination*) pairs: $\boldsymbol{A}_{ij} = \#\text{events}(i \rightarrow j)/\#\text{events}$ where i is a source. To train the models on such triples, we remove the source column. Each process consists of destination nodes and their timestamps only. The models in these settings capture the causal notion that a received message can trigger other messages.

The most common dataset explored by these types of models is Memetracker [23], composed of publication times of articles across multiple web-domains. We selected the top 25 domains with the highest number of hyper-links during the month of January, 2009. Sources are webpages, and the ground truth edges arise when the source creates a hyperlink to a destination. Thus, the number of times the source domain cites the destination domain defines the whole matrix. Six other datasets were gathered from the Snap Network Repository [24] representing human communications in social networks, where sources and destinations are defined when one user contacts another. For each network, we selected the month with the highest number of events and only considered processes with more than 60 events each. If there were more than 100 processes that fulfilled this condition, we chose the top 100. Table 2 shows the number of processes and events for each dataset.

3.4 Metrics

We now describe the metrics used to evaluate the quality of the influence matrix inferred by the model. Let $\boldsymbol{A}$ be the $K \times K$ *ground truth* matrix and $\boldsymbol{A}_i$ the i_{th} row of this matrix. Similarly, we denote by $\hat{\boldsymbol{A}}$ the matrix generated by a model. We defined our metrics considering the rows of $\boldsymbol{A}_i$ and $\hat{\boldsymbol{A}}_i$, i.e., they capture the efficacy of the method from the viewpoint of a source node, or process $\mathcal{P}_i$.

Average Precision at n (AP@n). Initially, we calculate the Precision@n ($P@n$) as: $P@n(\boldsymbol{A}_i, \hat{\boldsymbol{A}}_i) = |\mathbb{T}_n(\boldsymbol{A}_i) \cap \mathbb{T}_n(\hat{\boldsymbol{A}}_i)|/n$, used in [15]. Here, $\mathbb{T}_n(\boldsymbol{A}_i)$ are the top n elements in $\boldsymbol{A}_i$ ordered by their value. The $P@n$ metric avoids the problem of sparsity, as it only considers the edges with the highest weight. However, it ignores the distribution of weights for the edges that are not in the top n. Average Precision at n ($AP@n$) aggregates the Precision at n for every possible n. By doing so, this metric solves the issue of choosing a specific n. However, equal weight is given to all of these choices. The NDCG, our next metric, mitigates this issue.

Normalized Discounted Cumulative Gain (NDCG). NDCG [11] is a measure of ranking quality that penalizes the errors according to the ranking. It is defined as $\frac{DCG_i}{IDCG_i}$, where for each row, $DCG_i(\boldsymbol{A}_i, \hat{\boldsymbol{A}}_i) = \sum_{\boldsymbol{A}_{ij} \in \boldsymbol{A}_i} \frac{2^{\boldsymbol{A}_{ij}}-1}{\log_2(pos^m(j)+1)}$ and $IDCG_i(\boldsymbol{A}_i, \hat{\boldsymbol{A}}_i) = \sum_{\boldsymbol{A}_{ij} \in \boldsymbol{A}_i} \frac{2^{\boldsymbol{A}_{ij}}-1}{\log_2(pos(j)+1)}$. $pos(j)$ captures the position of $\boldsymbol{A}_{ij}$ in the ranking of cell values $\boldsymbol{A}_{ij} \in \boldsymbol{A}_i$ and, similarly, $pos^m(j)$ is the position of $\hat{\boldsymbol{A}}_{ij}$ in the ranking of $\hat{\boldsymbol{A}}_{ij} \in \hat{\boldsymbol{A}}_i$. However, the NDCG does not penalize an overestimation of edge values if they are not too far up in the ranking. The NDCG varies from 0 to 1, the higher the better.

Normalized Root-Mean-Square Error (NRMSE). NRMSE considers the difference in values between two matrices, being defined as the root-mean-square error between the cells of the estimated and ground truth matrices, normalized by the range of values: $NRMSE(\boldsymbol{A}_i, \hat{\boldsymbol{A}}_i) = \frac{\sqrt{\sum_i (\boldsymbol{A}_{ij} - \hat{\boldsymbol{A}}_{ij})^2/K}}{\boldsymbol{A}_{imax} - \boldsymbol{A}_{imin}}$, where $\boldsymbol{A}_{imax}$ and $\boldsymbol{A}_{imin}$ are the maximum and minimum values for the vector $\boldsymbol{A}_i$, respectively.

Other metrics were used to evaluate NPPs, such as the average row rank correlation coefficient [1,15,45] and the relative error [1,15,41,45]. However, these metrics suffer from small sample sizes (non-zero columns in each rows) due to graph sparsity. It is expected that small samples lead to less statistical significance, thus we shall limit our discussion to the subset of complementary metrics that are able to reject a higher number null-hypothesis tests (e.g., we have a higher confidence that their values did not arise due to chance).

3.5 Null Models and Confidence Intervals

We also use a Null model composed of a random generation (or permutation) of rankings for the ranking metrics (e.g., AP@n). For the remaining metrics, we generate a random permutation of the values of the ground truth matrix. This matrix allows the null model to maintain the sparsity and usual range of values of the original ground truth matrix. These permutations are generated 1000 times and compared to the ground truth matrix using the metrics to create a confidence interval of 95% for each (metric, dataset) pair.

We also generated confidence intervals for the estimates calculated with the real datasets. As we do not know the true stochastic data generating mechanism, we used a bootstrap procedure adapted to point processes data [26,34]. The time axis $(0, T]$ was partitioned into 10 parts of equal length: $\mathcal{T} = (0, T/10], \ldots, (9T/10, T]$ and new pseudo-datasets were created by gluing together 10 temporal pieces sampling with replacements from $\mathcal{T}$. The estimates were calculated in the pseudo datasets, and this procedure was repeated 100 times to generate the sampling distributions from which confidence intervals were derived. In this paper, we present the first comparison of models under statistical significance tests.

3.6 Upper Bounds and Optimization

As with any ML task, an important step when comparing models is testing and choosing hyper-parameters. However, in the setting of Latent Network Extraction, we do not have validation sets to optimize hyper-parameters. That is, starting from an intensity function $\lambda_a(t)$ (see Sect. 3.1), we derive a likelihood for the entire dataset [12]. This is one of the most important results from Point Processes, and the intensity is a sufficient function for learning model parameters. Using the theory of Maximum Likelihood Estimation (MLE), we can argue that hyper-parameters should be chosen to maximize the likelihood for the datasets. However, which guarantees do we have that this model in-fact performs the best network extraction? Notice that without any edge, we cannot test the hyper-parameters on the validation data. It turns out that (from our empirical results in the next section) most methods do not present a correlation of likelihood with the aforementioned efficacy metrics. We now detail how we used the expected validation [13] alongside with this correlation to evaluate the NPP models.

Given a choice of n hyper-parameter configurations, we can compute the expected validation score (for a choice of evaluation metric) of the models by assuming that the graph is known. That is, we may simply pick the choice of hyper-parameters that maximizes some accuracy metric and not the likelihood. This score is captured via the equation:

$$\mathbb{E}[V_n^*|n] = \sum_{v} v\ P(V_n^* = v|n) = \sum_{v} v\ (\hat{P}(V_i \leq v)^n - \hat{P}(V_i < v)^n).$$

Here, V_n^* is the maximum score after testing n hyper-parameters configurations, v are the observed scores and V_i is the score for the i-th configuration, drawn i.i.d. (with random search, for example). Given that hyper-parameter configurations are i.i.d., $\hat{P}(V_i \leq v)^n$ captures the probability of observing n configurations below or equal to the threshold v. Similarly, $\hat{P}(V_i < v)^n$ captures the probability of these n configurations being below v. By iterating over all observed values of the given score v, this function is thus an expectation of the score given that:

$\hat{P}(V_i = v) = \hat{P}(V_i \leq v)^n - \hat{P}(V_i < v)^n$. Thus this value is simply an expected value of the score when testing n hyper-parameter configurations.

Conditioning on n is interesting as it measures the expected validation score as a function of the computational power set forth to optimize the model. More hyper-parameter tests imply a need of more computational resources. Thus, from this metric we can infer not only an **upper-bound** on the score, as the graph is known, but also have some notion of the expected computing cycles needed to achieve such a score. After computing the value of $\mathbb{E}[V_n^*|n]$, we analyze if there is a correlation between a higher likelihood and a higher value for the efficacy metrics. If a correlation is present, then we can argue that such an upper-bound is achievable by the method. If not, we can state that optimizing the likelihood will not lead to better results for that model. To have some notion of the variability of this estimate, we can also compute the variance $\mathbb{V}[V_n^*|n] = \mathbb{E}[V_n^{2*}|n] - \mathbb{E}[V_n^*|n]^2$ by computing the the expected value using v^2 in order to estimate $\mathbb{E}[V_n^{2*}|n]$.

3.7 Sensitivity to False Positives

When processes are known to be independent among themselves, LNE models must yield a zero matrix $\hat{\boldsymbol{A}}$ and we can test them for *false positives*. To do so, we simulate $K = 10$ independent Poisson (homogeneous and inhomogeneous) processes with $n = 1000$ events each. A Poisson process is independent of its history, and its intensity function is given by $\lambda(t|\mathcal{H}_a(t)) = \lambda(t)$. In this case, besides comparing $\hat{\boldsymbol{A}}$ with a zero matrix, the models should also infer the *baseline intensities* μ, which should be similar to the average $\lambda(t)$, or simply n/T.

4 Results

We start this section with a discussion about the empirical maximum values each model can achieve. Next, we evaluate the models' resilience to false positives using synthetic data. Experiments were run on an i7-6700 CPU with eight cores and 32 GB of RAM.

4.1 Upper Bounds and Optimization

Three factors are considered for assessing the accuracy of the models when learning the network structure:

1. **F1:** the best score achieved by the model when it is tuned specifically to improve the comparison metric over the ground truth network;
2. **F2:** the number of hyper-parameter configurations needed to be tested to achieve this maximum score;
3. **F3:** whether optimizing the model without using the ground truth network (i.e. using the likelihood or loss function over the timestamps) can improve the model's performance measured by the comparison metric.

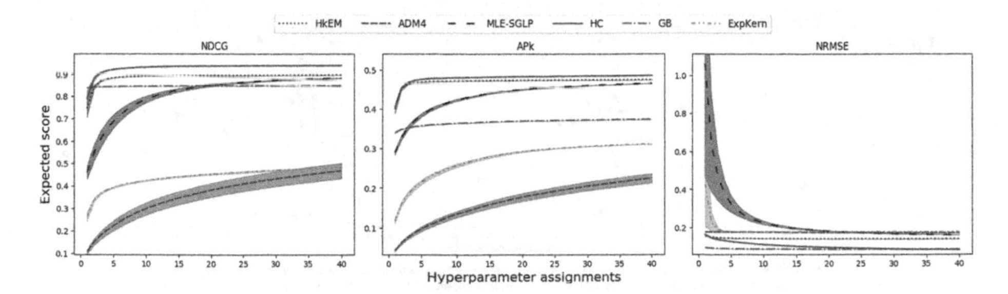

Fig. 1. Expected value of the best score according to the number of hyper-parameter configurations tested for the *sx-superuser* dataset.

These analyses serve to verify whether the models are well-defined, i.e., if optimizing the likelihood leads to better estimates of $\boldsymbol{A}$. Hyper-parameter optimization is done via random-search. Each step of the search defines a hyper-parameter configuration. For every dataset and hyper-parameter configuration, we computed the expected validation score $\mathbb{E}[V_n^*|n]$ and the variance $\mathbb{V}[V_n^*|n]$, which may vary according with the number of hyper-parameter configurations tested. In Fig. 1, we show the expected score for the models on the *sx-superuser* dataset as a function of the number of hyper-parameter configurations tested. The variance is shown as the shaded region. For each curve we can derive two values: the number of hyper-parameter tests necessary until the metric no longer improves (defined as an increase of less than 0.001) and the metric value at this step, which we refer as the **upper-bound** of the triple (method, dataset, metric).

In Fig. 2 we plot these two quantities for every triple (method, dataset, metric). We do not present results for **MLE-SGP** on the Memetracker dataset because it did not converge. Each method is determined by a color and each dataset by a symbol. The shaded region on the plots are determined by median values on the x and y-axis. When inside this region, we can state that the method and dataset pair is above the overall median for the corresponding axis considering the given metric. In the caption of the figure we describe how many times each method fell inside this region. Higher values of counts indicates that the method is performing well for (dataset, metric) pairs. In total, we have 7 datasets and 3 metrics (21 results), so the ideal method would be counted 21 times. In general, **HkEM**, **HC** and **GB** obtained the best upper bounds with few hyper-parameter tests, and the remaining methods need a larger amount of tests and obtain varying results. **HkEM** and **GB**, both with 13 wins, require few hyper-parameter tests to reach their upper-bound.

Recall that a model can only approximate the upper bound in practice if optimizing its likelihood (or loss function) improves the quality of the recovered network (**F1**). Thus, in Fig. 3 we show the Kendall-tau correlation (τ) between the value of the metric and the likelihood of each model when this value is achieved. Every point is a (metric value, likelihood) pair and each sub-figure refers to a (method, metric) pair. Due to space limitation, we only show these results for the *sx-superuser* dataset. Notice that only **HkEM** and **ADM4** have

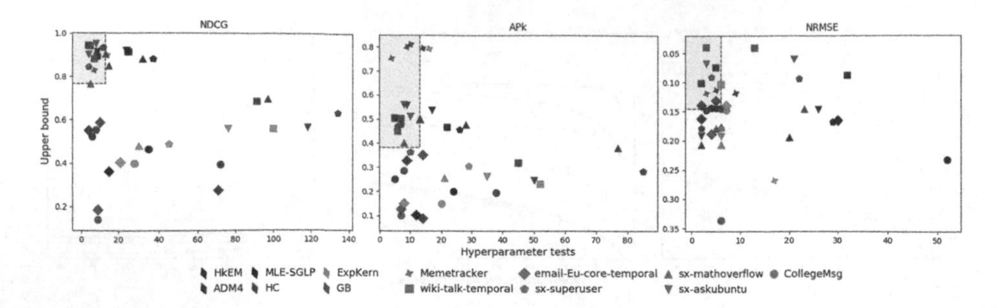

Fig. 2. Maximum expected value achieved vs number of hyper-parameter tests. The shaded area is where methods are above the median for both axis. Counts for how many times each model was in the shaded area: **HkEM**: 13, **ADM4**: 4, **MLE-SGP**: 0, **HC**: 12, **GB**: 13, **ExpKern**: 1.

high correlations regardless of the evaluation metric. This result is an initial indicator that these methods are better defined than the others. Being ill-defined means that optimizing the likelihood will not lead to better network recovery.

In Fig. 4, we summarize these results for all datasets by plotting the Kendall-tau correlation and the upper-bound for the models for all metrics and datasets. Note that **MLE-SGP** is not included in this analysis since no implemented likelihood function was available for testing. Similar to Fig. 2, we also present a shaded region indicating where models out-perform others (medians of the x and y-axis). Observe that **HkEM** tends to have high correlation values and usually outperforms others (11 wins), being followed by **ADM4** (7 wins). **GB** (2 wins), which was a good candidate before, regularly shows no correlation. The remaining models also performed poorly.

This second analysis argues in favor of **HkEM** and **ADM4**. Our results so far are based on a hypothetical setting in which we are able to measure validation scores. Some methods (e.g. **HC**) appear to be able to reach high metric values (e.g. NDCG) but the lack of correlation limits their applicability. To provide a more accurate evaluation, we next discuss the real-world setting where such optimization is not possible, that is, there is no ground truth and only the likelihood is available for hyper-parameter tuning.

4.2 Experimental Results for the LNE Task

In Fig. 5 we show the overall results of the models for the latent network extraction (LNE) task. Each plot contains the results for a given metric. The markers are the values models achieved when executed on the original timestamp data and the lines are the 95% confidence intervals (computed using bootstrap). Areas where the models are equivalent or worse than the Null model (95% confidence interval) are shaded in gray. In a few cases, the result on the original data lies marginally outside of the confidence interval. This is expected because Web datasets are bursty, and results may depend on specific periods of time left out

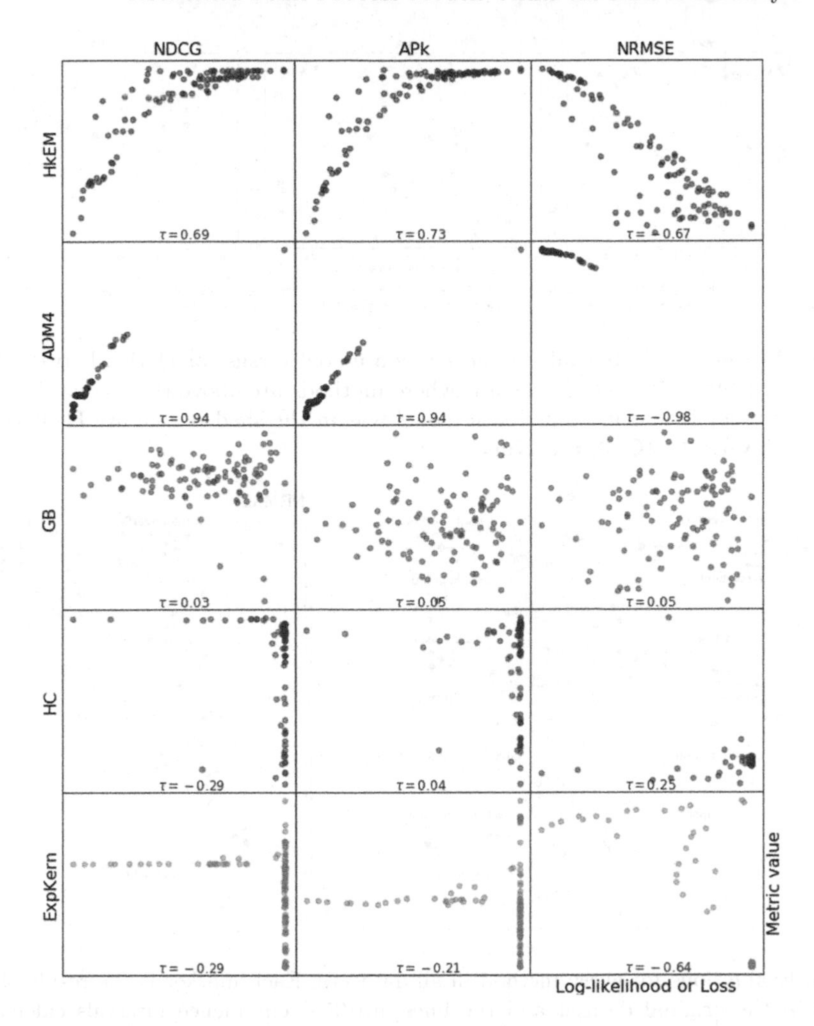

Fig. 3. Log-likelihood of the model (or loss function for **HC**) vs metric values when comparing the generated network with the ground truth for the *sx-superuser* dataset.

by the bootstrap. Hyper-parameters were optimized using the log-likelihood or loss function. For **MLE-SGP**, the default parameters were used.

Again, **HkEM** achieved the best results, being the best model in 9 out of the 21 combinations of dataset and metric, never overlapping with the null model's range. **ADM4**, **MLE-SGP**, **HC** and **GB** followed next, with each one getting the best value for 3 (dataset, metric) pairs. **ExpKern** was never the best model. These results can be well explained by the previous experiments. A model with good correlation and upper bound managed to get the best scores (**HkEM**), while models without a good correlation are difficult to optimize and tend to have erratic behaviour (**HC** and **ExpKern**).

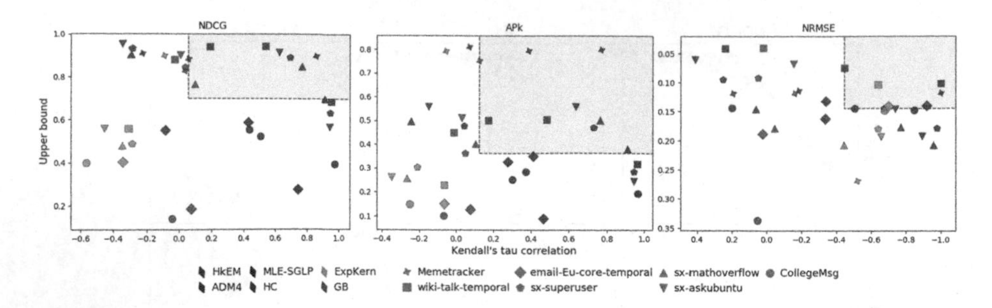

Fig. 4. Maximum expected value achieved by a model versus correlation between likelihood and metrics. The shaded area is where methods are above the median for both axis. Counts for how many times each model was in the shaded area are **HkEM**: 11, **ADM4**: 7, **GB**: 2, **HC**: 3, **ExpKern**: 2.

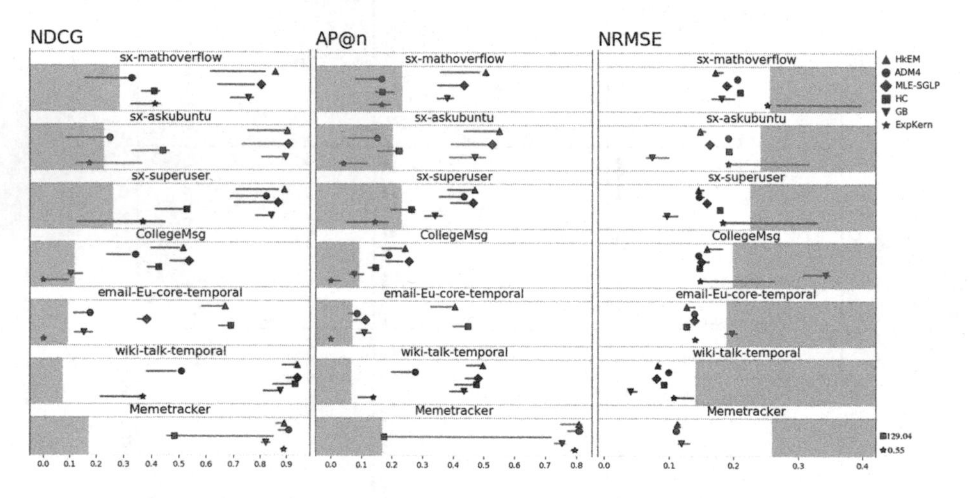

Fig. 5. Evaluations for every method in all datasets. Each marker is the result of the model in the original dataset and the lines are 95% confidence intervals calculated through bootstrap. The shaded regions correspond to the confidence interval of null models.

Considering the bootstrap intervals, **HkEM** was tied for the best model 14 times. **MLE-SGP** and **GB** were the best models 12 times, **HC** 8 times, **ADM4** 5 times, and **ExpKern** 2 times. Concerning the datasets, Memetracker seems to be the easiest to capture the network structure. Relationships defined from links created from one process to another may be more suitable for NPP models than when defined from message exchanges. On the other hand, datasets without self-excitation (*CollegeMsg* and *email-Eu-core*) are the hardest to infer the latent network.

Table 3. MSE for the baselines and Granger matrix on simulated Poisson processes.

Method/Metric	Homogeneous		Inhomogeneous	
	Baseline error	LN error	Baseline error	LN error
HkEM	1.00 (1.00, 1.00)	1.08 (1.06, 1.10)	1.00 (1.00, 1.00)	**3.47 (3.27, 3.67)**
ADM4	**0.02 (0.01, 0.03)**	**0.05 (0.03, 0.07)**	0.71 (0.68, 0.76)	6.03 (5.86, 6.16)
MLE-SGP	0.39 (0.30, 0.48)	1.61 (1.14, 2.12)	0.92 (0.89, 0.97)	20.11 (19.45, 20.79)
HC	2.44 (0.75, 37.32)	22.46 (14.48, 729.34)	2.46 (1.28, 3.60)	8.63 (5.47, 34.30)
GB	0.32 (0.31, 0.34)	–	**0.40 (0.39, 0.41)**	–
ExpKern	0.03 (0.02, 0.05)	0.07 (0.05, 0.09)	0.68 (0.65, 0.70)	4.50 (4.32, 4.67)

Our results so far argue in favor of **HkEM** and possibly **ADM4** as the best methods. Nevertheless, we point out that **GB** should not be ignored as it is the only method that executes on Web scale data [15] (see asymptotic cost in Sect. 3). Also, **HC** tends to have very high upper-bound values, but the lack of correlation leads the model to under-perform when used for the LNE task. This is not the case for **GB** as its efficacy appears to be less sensitive to the choice of hyper-parameters.

4.3 Comparison with Cascades Method

We now compare the methods when more information (e.g., cascade sets) are available for usage. In particular, we compare Hawkes and Wold models with **NetInf** [32], the only cascades model. Because **NetInf** requires cascade data, our analysis is limited for the Memetracker dataset. Also, since the problem formulation is slightly different, we processed the Memetrackler data in two different forms. In Memetracker, cascades are captured when short sentences are repeated over websites. Thus, to determine cascades, sentences with more than 3 words were extracted together with the timestamps when each different website first mentioned the sentence. The sequence of timestamps of a sentence form a cascade. While all the other models process only the sequence of timestamps on each website, **NetInf** also processes the cascade to which each timestamp belong. Only the top $K = 50$ websites in terms of hyperlink mentions were considered in our analysis. The ground truth is the hyperlink counts as described in Sect. 3.3, a total of 182,379 events.

It is also important to note that **NetInf** infers an unweighted network, thus the metrics that employ rankings (AP@n and NDCG) become meaningless. For this reason, we limit our comparison using NRSME. We found that **NetInf** achieves a worse score than most of the other Hawkes and Wold models, with a NRMSE of 0.398. Out of the other six methods, only **ExpKern** had a higher value (0.616). This result is quite remarkable as the Hawkes and Wold methods do not explore cascades. That is, they make use of less information than **NetInf**. Overall, these results point that exploring cascade data may not be a necessity in real world settings. Full results are presented in the Supplementary Information material [6].

4.4 False Positive Analysis (Simulations)

To test for false positives, we simulated $K = 10$ **homogeneous** Poisson processes with $\lambda(t) = 1$ for $N = 10,000$ steps. In this case, all models should infer baseline (or background) intensities $\mu_a = 1$ for each process a and $\hat{\boldsymbol{A}}_{ij} = 0\ \forall i, j$. Next, we simulate 10 **inhomogeneous** Poisson processes, where each process indexed by $a = \{1, 2, \cdots, 10\}$ has intensity $\lambda_a(t) = 1 + \sin(\omega_a t)$, where $\omega_a = \frac{1}{2a} \in \mathbb{R}$. For both experiments we run the models over the simulated timestamps and repeated this process 200 times to calculate the confidence intervals.

Table 3 shows the mean squared error (MSE) of the estimated baseline intensities (μ_a) with respect to the Poisson intensities ($\lambda(t)$), denoted by **baseline error**, and the MSE of the cells of the matrix with respect to the all zero matrix, denoted by **LN error**. The values between parenthesis are the bounds for the 95% confidence interval. Note that we do not calculate the squared error of the matrix estimated by **GB**, because it is a row stochastic matrix and, thus, would never generate a zeroed row.

Surprisingly, only **ADM4** and **ExpKern** were able to capture the *homogeneous* Poisson processes successfully, with both errors close to 0. The probable reason is that both models impose an exponential kernel on its intensity function, while all the other Hawkes methods try to approximate a kernel function, which can lead to overfitting. **GB** performed fairly well in estimating the baseline intensities, having the third smallest error, followed by **HkEM** and **MLE-SGP**. However, part of the events of these three models were still estimated to be caused by other processes. **HC** was the one with the worst performance, having the highest errors for the baseline and the matrix. Regarding the *inhomogeneous* Poisson processes, all models identified a dependence structure among the processes, which does not exist. **GB**, **ADM4** and **ExpKern** models had the smallest baseline errors. **HkEM** had the closest estimate for the network, followed by **ExpKern** and **ADM4**. Other Hawkes models had higher errors.

5 Conclusions

Laborious, rigorous evaluations and scrutiny is a necessity in any scientific field. As we have argued throughout this paper, the machine learning (ML) field is currently at a position where there is a need to revisit its state of the art, given the plethora of applications and scenarios involved. In particular, empirical rigor is critical to avoid spurious findings and to delineate the scope of ML models. In his original review of the "two cultures" [8], Breiman made an in-depth comparison between theory-based models and empirical prediction-based algorithms, concluding that researchers must move away from exclusive dependence on data models and adopt a more diverse set of tools. Similarly, machine learning researchers [20,22,27,30,35–38] are taking a step back and asking themselves if too much effort is put on sophisticated and complex ideas that, in practice, perform worse than traditional approaches.

This work serves as a methodological guide to new studies on network inference from networked point processes and corroborates with the warnings mentioned above, which asks for more empirical rigor in ML research. Based on our results, we argue that the classical **HkEM** model is the best general-purpose NPP model as it had superior results in most settings and positive correlations for likelihood and evaluation scores. **ADM4**, while generally less useful than **HkEM**, also presents strong positive correlations and is the best model for detecting false positives. Although **GB** does not present a correlation between likelihood and efficacy, it is the only method known to scale for full Web datasets [15], being recommended for these cases. The other models have not excelled in any of the five evaluation fronts.

Acknowledgements. Funding was provided by the authors' grants from CNPq, CAPES, and Fapemig. This project was also partially funded via CNPq's Universal Grant 421884/2018-5.

References

1. Achab, M., Bacry, E., Gaiffas, S., Mastromatteo, I., Muzy, J.F.: Uncovering causality from multivariate Hawkes integrated cumulants. In: ICML (2017)
2. Bacry, E., Muzy, J.-F.: Second order statistics characterization of Hawkes processes and non-parametric estimation. arXiv preprint arXiv:1401.0903 (2014)
3. Barabási, A.-L., Albert, R., Jeong, H.: Scale-free characteristics of random networks: the topology of the world-wide web. Physica A **281**(1–4), 69–77 (2000)
4. Bergstra, J., Bengio, Y.: Random search for hyper-parameter optimization. J. Mach. Learn. Res. **13**, 281–305 (2012)
5. Blundell, C., Beck, J., Heller, K.A.: Modelling reciprocating relationships with Hawkes processes. In: NeuRIPS, pp. 2600–2608 (2012)
6. Borges, G.R., Figueiredo, F., Assunção, R., Vaz-de Melo, P.O.: Networked point process models under the lens of scrutiny: Supplementary information material (2020). https://github.com/guilhermeresende/NPPs/. Accessed 24 June 2020
7. Bourigault, S., Lamprier, S., Gallinari, P.: Representation learning for information diffusion through social networks: an embedded cascade model. In: WWW (2016)
8. Breiman, L.: Statistical modeling: the two cultures. Stat. Sci. **16**(3), 199–231 (2001)
9. Chen, S., Shojaie, A., Shea-Brown, E., Witten, D.: The multivariate Hawkes process in high dimensions: beyond mutual excitation. arXiv preprint arXiv:1707.04928 (2017)
10. Choi, E., Du, N., Chen, R., Song, L., Sun, J.: Constructing disease network and temporal progression model via context-sensitive Hawkes process. In: ICDM (2015)
11. Croft, W.B., Metzler, D., Strohman, T.: Search Engines: Information Retrieval in Practice, vol. 520. Addison-Wesley, Reading (2010)
12. Daley, D.J., Vere-Jones, D.: An Introduction to the Theory of Point Processes, vol. 1. Springer, New York (2003). https://doi.org/10.1007/b9727710.1007/b97277
13. Dodge, J., Gururangan, S., Card, D., Schwartz, R., Smith, N.A.: Show your work: improved reporting of experimental results. In: EMNLP (2019)
14. Eichler, M., Dahlhaus, R., Dueck, J.: Graphical modeling for multivariate Hawkes processes with nonparametric link functions. J. Time Ser. Anal. **38**(2), 225–242 (2017)

15. Figueiredo, F., Borges, G., de Melo, P.O., Assunção, R.M.: Fast estimation of causal interactions using Wold processes. In: NeuRIPS, pp. 2975–2986 (2018)
16. Ghalebi, E., Mirzasoleiman, B., Grosu, R., Leskovec, J.: Dynamic network model from partial observations. In: NeuRIPS, pp. 9862–9872 (2018)
17. Gomez-Rodriguez, M., Leskovec, J., Krause, A.: Inferring networks of diffusion and influence. ACM Trans. Knowl. Discov. Data **5**(4), 1–37 (2012)
18. Gomez Rodriguez, M., Leskovec, J., Schölkopf, B.: Structure and dynamics of information pathways in online media. In: WSDM (2013)
19. Hawkes, A.G.: Point spectra of some mutually exciting point processes. J. Roy. Stat. Soc. Ser. B (Methodol.) **33**, 438–443 (1971)
20. Jie, M., et al.: A systematic review shows no performance benefit of machine learning over logistic regression for clinical prediction models. J. Clin. Epidemiol. **110**, 12–22 (2019)
21. Junuthula, R., Haghdan, M., Xu, K.S., Devabhaktuni, V.: The block point process model for continuous-time event-based dynamic networks. In: WWW (2019)
22. Kryscinski, W., Keskar, N.S., McCann, B., Xiong, C., Socher, R.: Neural text summarization: a critical evaluation. In: EMNLP-IJCNLP (2019)
23. Leskovec, J., Backstrom, L., Kleinberg, J.: Meme-tracking and the dynamics of the news cycle. In: KDD (2009)
24. Leskovec, J., Krevl, A.: SNAP Datasets. http://snap.stanford.edu/data
25. Lewis, E., Mohler, G.: A nonparametric EM algorithm for multiscale Hawkes processes. J. Nonparametric Stat. **1**(1), 1–20 (2011)
26. Loh, J., Stein, M.: Bootstrapping a spatial point process. Statistica Sinica **14**(1), 69–101 (2004)
27. Lucic, M., Kurach, K., Michalski, M., Bousquet, O., Gelly, S.: Are Gans created equal? A large-scale study. In: NeuRIPS (2018)
28. Mavroforakis, C., Valera, I., Gomez-Rodriguez, M.: Modeling the dynamics of learning activity on the web. In: WWW (2017)
29. Mei, H., Eisner, J.M.: The neural Hawkes process: a neurally self-modulating multivariate point process. In: NeuRIPS (2017)
30. Melis, G., Dyer, C., Blunsom, P.: On the state of the art of evaluation in neural language models. In: ICLR (2018)
31. Rizoiu, M.A., Xie, L., Sanner, S., Cebrian, M., Yu, H., Van Hentenryck, P.: Expecting to be hip: Hawkes intensity processes for social media popularity. In: WWW (2017)
32. Rodriguez, M.G., Balduzzi, D., Schölkopf, B.: Uncovering the temporal dynamics of diffusion networks. In: ICML (2011)
33. Santos, T., Walk, S., Kern, R., Strohmaier, M., Helic, D.: Self-and cross-excitation in stack exchange question & answer communities. In: WWW (2019)
34. Sarma, S.V., et al.: Computing confidence intervals for point process models. Neural Comput. **23**(11), 2731–2745 (2011)
35. Sculley, D., Snoek, J., Wiltschko, A., Rahimi, A.: Winner's curse? On pace, progress, and empirical rigor. In: ICLR (2018)
36. Strang, B., van der Putten, P., van Rijn, J.N., Hutter, F.: Don't rule out simple models prematurely: a large scale benchmark comparing linear and non-linear classifiers in OpenML. In: IDA (2018)
37. Riquelme, C., Tucker, G., Snoek, J.: Deep Bayesian bandits showdown: an empirical comparison of Bayesian deep networks for Thompson Sampling. In: ICLR (2018)
38. Vaswani, A., et al.: Attention is all you need. In: NeuRIPS (2017)
39. Wang, J., Zheng, V.W., Liu, Z., Chang, K.C.-C.: Topological recurrent neural network for diffusion prediction. In: ICDM (2017)

40. Wold, H.: On stationary point processes and Markov chains. Scand. Actuarial J. **1948**(1–2), 229–240 (1948)
41. Xu, H., Farajtabar, M., Zha, H.: Learning granger causality for Hawkes processes. In: ICML (2016)
42. Yang, W., Lu, K., Yang, P., Lin, J.: Critically examining the "neural hype": weak baselines and the additivity of effectiveness gains from neural ranking model. In: SIGIR (2019)
43. Yang, Y., Etesami, J., He, N., Kiyavash, N.: Online learning for multivariate Hawkes processes. In: NeuRIPS (2017)
44. Yi, L., Liu, B., Li., X.: Eliminating noisy information in web pages for data mining. In: KDD (2003)
45. Zhou, K., Zha, H., Song, L.: Learning social infectivity in sparse low-rank networks using multi-dimensional Hawkes processes. In: AISTATS (2013)
46. Zhou, K., Zha, H., Song, L.: Learning triggering kernels for multi-dimensional Hawkes processes. In: ICML (2013)

FB2vec: A Novel Representation Learning Model for Forwarding Behaviors on Online Social Networks

Li Ma[1], Mingding Liao[1], Xiaofeng Gao[1](✉), Guoze Zhang[2], Qiang Yan[2], and Guihai Chen[1]

[1] Shanghai Key Laboratory of Data Science, Department of Computer Science and Engineering, Shanghai Jiao Tong University, Shanghai, China
{mali-cs,uracil_forever}@sjtu.edu.cn, {gao-xf,gchen}@cs.sjtu.edu.cn
[2] Tencent, Shenzhen, China
{givenzhang,rolanyan}@tencent.com

Abstract. Representation learning in online social networks has been an important research task for better service, which targets at learning the low-dimensional vector representation for nodes in a network. There exists a kind of social network, which not only includes the topological structure and node attributes, but other information, such as the user behaviors. It is necessary to use these behaviors to learn the node representations. In this paper, we propose FB2vec to analyze forwarding behaviors and achieve better node representations. Moreover, an information intensity function based on the utility function is proposed to measure the possibility of forwarding behaviors. However, the intensity function can not reflect exact possibility value and only provide a relative intensity order. Therefore, we sample the intensity order pairs from datasets and train the intensity function to adapt original orders by an attribute-reserved siamese network. Extensive experiments demonstrate the effectiveness of FB2vec and the visualization of information intensity function indicates the rationality of FB2vec.

Keywords: User representation learning · Online social networks · Forwarding behaviors · Siamese networks

1 Introduction

Online social networks play an essential role in daily life with the rapid development of the internet and information technology. Network representation learning, which is to learn the low-dimension vector representations for network structure and node information [20], has shown great potential on the application of

This work was supported by the National Key R&D Program of China [2018YFB1004700]; the National Natural Science Foundation of China [61872238, 61972254]; the Tencent Joint Research Program, and the Open Project Program of Shanghai Key Laboratory of Data Science (No. 2020090600001).

F. Hutter et al. (Eds.): ECML PKDD 2020, LNAI 12457, pp. 216–231, 2021.
https://doi.org/10.1007/978-3-030-67658-2_13

online social networks. Most of the existing works of network representation learning for online social networks concentrate on analyzing the relationship of the network. However, the sparsity of the friendship network limits these applications. For example, Rochester used in [20], which is a Facebook network dataset, contains $4,563$ nodes but only $161,404$ links, which ignores the possibility that two nodes may be similar, but there is no link between them. However, there is such a specific network with a typical user behavior over online social networks. For example, the users from the online social networks WeChat could forward some articles into a group, although some of members of the group are not his or her friends. Compared with friendship networks, the networks with forwarding behaviors could be more dense because some of online social networks allow users to look up strangers' recommendation contents. So representation learning for forwarding behaviors is a potential research direction. However, few works pay attention to that because a forwarding behavior is a triple relationship and contains the information about forwarding amount, which makes it hard to be processed in a traditional network embedding model. Thus, the research on representation learning for forwarding behaviors is significant for data mining on online social networks.

Information dissemination is a common phenomenon in most of online social networks and the forwarding behavior is a form of it. In the process of information dissemination over social networks, a part of users generate contents, and the contents would be received and forwarded by other users, and then received by more users. Thus, forwarding behaviors could be described by three participants: a content generator, a content receiver and a content forwarder. There could be no direct link between a generator and a receiver, but a generator and a receiver could be connected by a forwarder. So the introduction of forwarding behaviors could break the limitation of friendship networks. Representation learning based on information dissemination could weaken the impact of the sparsity of friendship networks.

To get a better representation, many related works not only utilize the topological structure of the network, but the attributes of nodes, which could be beneficial for representation learning since they may crucially impact the connections and interactions with nodes [20]. However, these attribute-aware network embedding method could not be applied in the networks with information dissemination because there is more information. Information dissemination is affected by the content of information and the reactions of users to information, which could be measured by content influence and social influence correspondingly. The lack of researches on the information dissemination indicates excellent research prospects.

In this paper, we provide a representation learning model based on **F**orwarding **B**ehaviors called FB2vec. We focus on forwarding behaviors in the information dissemination process on online social networks and explore node embedding vectors to describe the information of users. In this model, users are classified into generators, forwarders and receivers, and they are encoded through multi-modal auto-encoder. In order to describe the forwarding behaviors of users, we innovatively design an information intensity function to measure the intensity of forwarding behaviors, which takes the embedding vectors of generators, forwards and receivers as input and takes the intensity of forwarding behaviors among users as output, and we use the siamese network to train the function. With the help of information intensity function, we can further promote the learning of embedding vectors of users. The experiments indicate that FB2vec outperforms baselines on the behavior prediction problem and similar user detection problem.

The main contributions of this paper can be summarized in these points:

1) We provide a novel model for representation learning for forwarding behaviors on online social networks (FB2vec) to utilize the additional dissemination information to promote the quality of the embedding vectors;
2) We propose a new method information intensity function, which combines the idea of network embedding and metric learning, to build the connection between embedding vectors and forwarding behaviors intensity;
3) Experiments prove that our model works better than state-of-the-art models both for behavior prediction task and similar user detection task, and parameter study shows that our model could measure the receivers' interest on forwarders.

2 Related Work

Network embedding methods aim at learning the low-dimensional latent vector representation of nodes in a network, and these representations can be used as features for a wide range of tasks on graphs. Generally, network embedding models can be categorized into random walk model, edge-based model, and matrix factorization model.

Random walk models are exploited to capture structural relationships between nodes by the random walk. DeepWalk [11] is the pioneering work of random walk based network embedding, with the follow-up researches such as node2vec [4] and MMDW [17].

Matrix factorization methods represent the connections between network vertices in the form of a matrix and use matrix factorization to obtain the embedding vectors. As a successful example, APNE [5] incorporates structure, content and label information of the network simultaneously. [14] provides a unified framework for matrix factorization based model and analyzes DeepWalk, LINE, PTE and Node2vec as examples. [13] proposes the algorithm of large-scale network embedding as sparse matrix factorization.

Edge-based methods directly learn vertex representations from vertex-vertex connections to overcome the high time-consuming of random walk based models. LINE [16] first focus on first-order proximity and second-order proximity to learn the node representation, which achieves a much higher speed than node2vec. Graph Convolutional Networks indicates that edge-based models become a mature scheme [8] which reflects the influence between embedding vectors through links in networks. SDNE [19] provides an auto-encoder based model to train the embedding vectors of two vertices of a link. TransNet [18] considers the embedding vectors of a link as the difference of embedding vectors of its vertices, which could be considered as an expansion of SDNE. DANE [2] designs a multi-modal autoencoder for joint training of structure network information and vertices attribute information, which is employed in our model. NECS [9] preserves the high-order proximity and incorporates the community structure in vertex representation learning. Especially, [1] showed the application potential of metric learning methods on network embedding, which inspires our model to employ metric learning methods to train the representation of forwarding behaviors.

3 Definition and Problem Statement

Information dissemination in the social networks is that a part of users generate contents, and the contents would be received and forwarded by other users, and then received by more users. So there are three views of users: generators, forwarders and receivers. More explanations of generators, forwarders, and receivers could be found in Sect. 5.1. Definition 1 elaborates the above phenomenon.

Definition 1 ***Information Dissemination.*** *Let $G = (V, E)$ denotes the structure of a social network, where V is the set of users and $E \subset V \times V$ is the relationships between users. $D = \{(u^g, v^f, w^r, c)\}$ is the set of information dissemination behaviors, where (u^g, v^f, w^r, c) means that content c, generated by generator u^g, is transmitted by forwarder v^f and then influences receiver w^r.*

The behaviors of receivers reflect the influence among users. In the view of the reason for forwarding, content influence and social influence are the main factors to promote users' forwarding behaviors. In other words, the receiver is affected by the generator, either because the receiver has interest on the content generated by the generator (content influence), or because the forwarders, the friends of the receiver (social influence) like the content. However, existing works of network embedding ignore the behaviors of users and fail to distinct content influence and social influence. Thus we define the representation learning for forwarding behaviors as Definition 2 to formulate the latent factors in these three views, inspired by traditional network embedding.

Definition 2 ***Representation Learning for Forwarding Behaviors.*** *Let $P^g, P^f, P^r \in R^{||V|| \times d}$ denote the generator embedding vector, forwarder embedding vector and receiver embedding vector correspondingly, where d is the length of embedding vector. Information intensity function $f(P^g_u, P^f_v, P^r_w)$ is defined to measure the comprehensive influence of the generator u and the forwarder v on the receiver w in the process of information dissemination and is used to the training of embedding vectors.*

4 Behavior Embedding via Siamese Network

In this section, we propose FB2vec, a novel behavior embedding model based on metric learning and multi-modal autoencoder, which is illustrated in Fig. 1. Overall, there are two branches for a user where the first branch captures the topological features and the second one captures the profile features. These two kinds of features of generators, forwarders and receivers are mapped to the low-dimensional space by multi-modal autoencoder. Specially, we design a requirement aggregating function to integrate generator behavior embedding and forwarder behavior embedding, which is inspired by a utility function in economics. Then the information intensity function based on the requirement aggregating function could measure the information dissemination intensity with providing a considerable order instead of the exact value of the possibility of dissemination. Therefore, we sample the dissemination behavior into pairwise intensity order and employ an attribute-reserved siamese network to train the information intensity function and representations of users.

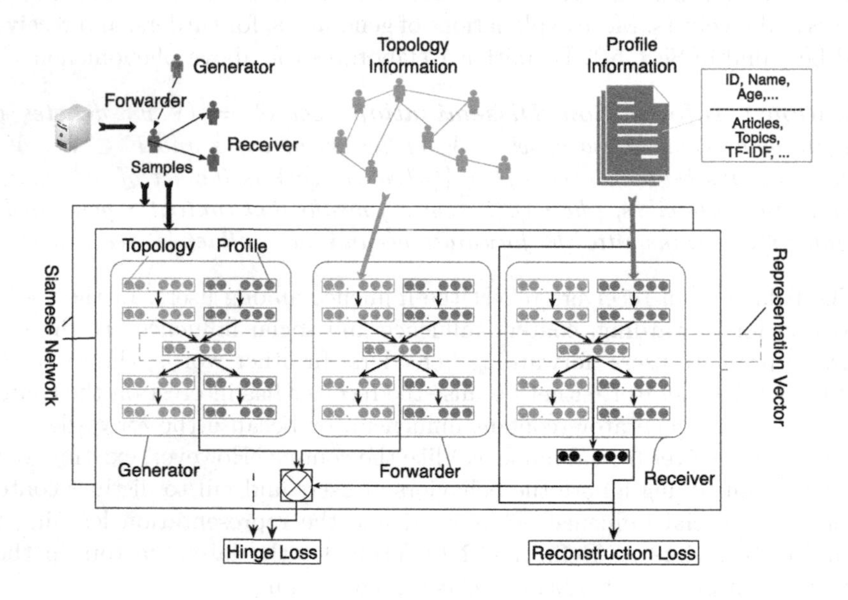

Fig. 1. Illustration of FB2vec

4.1 Attribute Preservation

The basic work for behavior embedding is to capture and preserve the topological structure of and attributes of users, which is solved by a multi-modal autoencoder. The two branches in Fig. 1 for each user (generator, forwarder or receiver) form a multi-modal autoencoder. All generators share the same autoencoder $\mathcal{AE}^g$ parameters. Similarly, one autoencoder $\mathcal{AE}^f$ is employed for all forwarders and

$\mathcal{AE}^r$ for all receivers. Let x_u be the topological structure features of u and y_u be the profile features of u. Take $\mathcal{AE}^g$ as an example. There are K layers in $\mathcal{AE}^g$ in Eq. (1)–(3). Equation (1) represents the multi-layer neural network for an encoder.

$$\begin{aligned}
h^g_{x,u,(1)} &= \sigma(W^g_{x,(1)} x_u + b^g_{x,(1)}),\\
&\ldots,\\
h^g_{x,u,(K)} &= \sigma(W^g_{x,(K)} h^g_{x,u,(K-1)} + b^{(g)}_{x,(K)}),\\
h^g_{y,u,(1)} &= \sigma(W^g_{y,(1)} y_u + b^g_{y,(1)}),\\
&\ldots,\\
h^g_{y,u,(K)} &= \sigma(W^g_{y,(K)} h^g_{y,u,(K-1)} + b^{(g)}_{y,(K)}).
\end{aligned} \tag{1}$$

Equation (2) maps topological structure features and attribute features of generator u into the embedding vectors.

$$P^g_u = \sigma(W^g[h^g_{x,u,(K)}, h^g_{y,u,(K)}] + b^g) \tag{2}$$

Equation (3) reconstructs features from embedding vector.

$$\begin{aligned}
h^g_{x,u,(-K)} &= \sigma(W^g_{x,(-K)} P^g_u + b^g_{x,(-K)}),\\
&\ldots,\\
h^g_{x,u,(-1)} &= \sigma(W^g_{x,(-1)} h^g_{x,u,(-2)} + b^g_{x,(-1)}),\\
h^g_{y,u,(-K)} &= \sigma(W^g_{y,(-K)} P^g_u + b^g_{y,(-K)}),\\
&\ldots,\\
h^g_{y,u,(-1)} &= \sigma(W^g_{y,(-1)} h^g_{y,u,(-2)} + b^g_{y,(-1)})
\end{aligned} \tag{3}$$

Here, $\theta_g = \{W^g_{x,(k)}, W^g_{y,(k)}, b^g_{x,(k)}, b^g_{y,(k)}\}$ are the model parameters. $\sigma(\cdot)$ denotes the non-linear activation function. The model parameters can be learned by minimizing the reconstruct error as Eq. (4), where $\otimes$ is the Hadamard product. Considering the possible sparsity of features, the number of non-zero elements in x or y may be far less than the number of zero elements. Thus, we impose a penalty function ϕ to make the autoencoder concentrate on the non-zero elements. When the features are sparse, $\phi(x) = |x|$, otherwise, $\phi(x) \equiv 1$.

$$\mathcal{L}^g = \sum_{u \in V} \|(\hat{x}^g_u - x^g_u) \otimes \phi(x^g_u)\| + \|(\hat{y}^g_u - y^g_u) \otimes \phi(y^g_u)\| \tag{4}$$

Similarly, model parameters θ_f for the forwarder autoencoder and model parameters θ_r for the receiver autoencoder can be learned by minimizing the reconstruct error $\mathcal{L}^f$ and $\mathcal{L}^r$, where x_v, y_v is the topological structure features and the profile features of forwarder v, x_w, y_w is the topological structure features and the profile features of receiver w:

$$\mathcal{L}^f = \sum_{v \in V} \|(\hat{x}^f_v - x^f_v) \otimes \phi(x^f_v)\| + \|(\hat{y}^f_v - y^f_v) \otimes \phi(y^f_v)\| \tag{5}$$

$$\mathcal{L}^r = \sum_{w \in V} \|(\hat{x}_w^r - x_w^r) \otimes \phi(x_w^r)\| + \|(\hat{y}_w^r - y_w^r) \otimes \phi(y_w^r)\| \tag{6}$$

Generally, FB2vec could make use of other kinds of neural networks such as convolutional neural networks for image features and recurrent neural networks for text features easily in the multi-modal autoencoder, which makes FB2vec become a general framework for any social networks.

4.2 Information Dissemination Intensity

Information dissemination intensity is expected to evaluate the intensity of influence on the receiver w under the joint action of the generator u and the forwarder v. Obviously, if the influence is higher, the higher the frequency of forwarding behavior (u, v, w, c) is. So we design a function, called information intensity function, that takes the embedding vectors of generators, forwarders and receivers as input, and outputs the information dissemination intensity. Then FB2vec trains the output of information dissemination intensity to match the real data, which in turn can promote the optimization of embedding vectors of generators, forwarders and receivers. In other words, the design of information intensity function is a fake task, which aims to help the training of users' representation.

Since the reactions of receivers are dependent on the source of information P_u^g (content requirement) and the reactions of their friends P_v^f (social requirement) in information dissemination, the information intensity function $f(P_u^g; P_v^f; P_w^r)$ should integrate the hidden influence factors of generator u and forwarder v into the context factors and calculate them with the hidden factors of receiver w. Assuming that the receiver w pursues the maximization of content requirement and social requirement as a rational individual, so we design the requirement aggregating function as Eq. (7) inspired by a kind of utility function [3], where P_1 refers to P_u^g, P_2 refers to P_v^f, β is the utility parameter and $\otimes$ is the Hadamard product. With the help of requirement aggregating function, the information dissemination intensity in the triple relationship could be measured. The information intensity function f is defined as Eq. (8), where $A \in R^{d \times d}$ is the embedding combination matrix. We have known that the reactions of users are dependent on the source of information P_u^g and the reactions of their friends P_v^f. We intuitively know that the distance of P_u^g and P_w^r reflects the intimacy between generators and receivers and the distance of P_f^g and P_w^r reflects the intimacy between forwarders and receivers. The choice of the utility function is reasonable because the reactions of a receiver depends on the two options (content and social requirement) with best utility.

$$g(P_1, P_2, \beta) = P_1^{\beta} \otimes P_2^{1-\beta} \tag{7}$$

$$f(P_u^g, P_v^f, P_w^r) = QAQ^T \text{where } Q = g(P_u^g, P_v^f, \beta(P_w^r)) - P_w^r \tag{8}$$

$$f(P_u^g, P_v^f, P_w^r) = \|g(P_u^g, P_v^f, \beta(P_w^r))B - P_w^r B\|^2 \tag{9}$$

In further discussion, information intensity function is equivalent to the Euclidean distance of linear transform of $g(P_u^g, P_v^f)$ and P_w^r, because Eq. (8)

can be transformed into Eq. (9) when A is semi-positive and $A = BB^T$. The platform-aware parameter A (as well as B) could be trained on different social networks, which indicates the application potential of FB2vec on different social networks with information dissemination and guarantees the generality of FB2vec. For a better description of users' patterns, β is replaced as the user-aware parameter $\beta(P_w^r)$ in $[0, 1]$, which is defined as the output of a multi-layer neural network with the input of the corresponding receiver behavior embedding vector P_w^r.

However, the utility function used in the Eq. (7) could only represent the order of preference factors instead of the exact value of latent factors [3], so the information intensity function in FB2vec is not suitable to measure the exact value, which makes it challenging to employ the ordinary network embedding model for parameter training. To solve the training problem, we explore the pairwise intensity order through a siamese network.

4.3 Pairwise Intensity Order

In the process of information dissemination, the users' behaviors are described as (u^g, v^f, w^v, c) in Definition 2. Without loss of generality, the behaviors of the same u^g, v^f and w^v are aggregated into $l(e_{u,v,w})$ where $e_{u,v,w}$ is short for the behavior (u^g, v^f, w^v, c) and $l(\cdot)$ is the number of $e_{u,v,w}$.

The information intensity function may not accurately describe the specific quantity of forwarding behaviors, but only the relative size relationship of quantity as it depends on the utility function. Therefore, we train the relationship between $l(e_{u,v,w})$ and the number of actual forwarding behaviors. That means the multi-modal autoencoder of generators, forwarders and receivers is used twice in each training, which forms a siamese network. FB2vec inputs two behaviors at the same time, one is the data extracted from the dataset, and the other is the negative sampling data for the same data, which is similar to the negative sampling in network embedding.

We randomly sample the generators, forwarders and receivers for each item $e_{u,v,w}$ to generate the confident negative samples. The sampling procedure is summarized as Algorithm 1. In detail, there are four categories of negative samples: negative samples nw^r for the receiver w^r, negative samples nv^f for the forwarder v^f, negative samples nu^g for the generator u^g and the forwarder v^f when $u^g = v^f$, and negative samples nu^g for the generator u^g when $u^g \neq v^f$. For example, the negative receiver sample nw^r means that the intensity from the generator u^g and the forwarder v^f to the receiver w^r is much larger than intensity from the u^g and v^f to nw^r. The negative samples for the generator u^g should be categorized into two classes by whether the forwarder is also the generator. Negative samples have to maintain consistency if the forwarder is also the generator.

Algorithm 1: Sampling intensity order relationship

Input: Graph G, Dissemination Behaviors D
Output: Sample Result $S[\cdot]$ for each $e_{u,v,w}$
1 Aggregate (u^g, v^f, w^r, c) into $l(e_{u,v,w})$;
2 **foreach** $e_{u,v,w}$ **do**
3 Randomly sample several nw^rs where $l(u^g, v^f, nw^r) \ll l(e_{u,v,w})$;
4 $S[e_{u,v,w}]$.add((u^g, v^f, nw^r));
5 Randomly sample several nv^rs where $l(u^g, nv^f, w^r) \ll l(e_{u,v,w})$;
6 $S[e_{u,v,w}]$.add((u^g, nv^f, w^r));
7 **if** $u^g = v^f$ **then**
8 Random sample several nu^rs where $l(nu^g, nu^g, w^r) \ll l(e_{u,v,w})$;
9 $S[e_{u,v,w}]$.add((nu^g, nu^g, w^r));
10 **else**
11 Random sample several nu^rs where $l(nu^g, v^f, w^r) \ll l(e_{u,v,w})$;
12 $S[e_{u,v,w}]$.add((nu^g, v^f, w^r));
13 return S;

4.4 Model Optimization

After the sampling procedure, there are several negative samples for each $e_{u,v,w} = (u^g, v^f, w^r, c)$. We employ the large margin strategy to distinguish each $e_{u,v,w}$ with their negative samples [21]. Large margin strategy only focuses on the relative distance between positive and negative items, which perfectly meets the requirement of FB2vec with the utility function. Therefore, the aim is to minimize Eq. (10), where m is the margin. The object function enforces each $f(e)$ to become smaller than $f(ne) - m$.

$$\mathcal{L}^{infl} = \sum_{e} \sum_{ne \in S[e]} \min\left(f(e) - f(ne) + m, 0\right) \tag{10}$$

Based on the mathematically information intensity function and pairwise intensity order, we propose a novel representation learning model FB2vec for forwarding behaviors with preservation of the topological and profile features on online social networks. The whole object function of FB2vec can be written as Eq. (11) where L^{reg} is the L_2 regularizer for all parameters, α and γ are the weighting factors. The object function is minimized by the root mean square propagation gradient method [7] to optimize the representation learning model.

$$\mathcal{L} = \mathcal{L}^{infl} + \alpha(\mathcal{L}^g + \mathcal{L}^f + \mathcal{L}^r) + \gamma L^{reg} \tag{11}$$

5 Experiments

To comprehensively evaluate the proposed FB2vec, we conduct experiments to show the efficiency of behavior embedding vectors in aspect of the behavior prediction problem and the similar user detection problem.

5.1 Experiment Setup

Dataset Description. We apply our model to a commercial dataset *WeChat Article* and a public social network retweets datasets *Sina Weibo* [22]. The details of datasets are the following:

- *WeChat Article*: This dataset is collected from Tencent WeChat, which is composed of the articles published by Official Accounts, the forwarding and reading behaviors of general users accounts. Official Accounts are special accounts for publishing contents and building dynamic services. In WeChat, only Official Accounts could be generators. A forwarder could forward the articles into WeChat groups or "Moments" to recommend them to its friends. These friends would see the brief introduction of articles and view the whole article if interested. If friends read the whole article, they would be treated as receivers.
- *Sina Weibo*: This dataset crawled from *Sina Weibo*, which is the largest online social network in China, contains about 177 thousand users and 23 million retweet behavior [22]. In *Sina Weibo*, generators, forwarders and receivers are all general users. When a user posts a tweet, it becomes a generator, and then other users would forward it as forwarders. If a user forwards the retweets from forwarders or the tweets from generators, it would be treated as a receiver.

We clean the data and extract dense forwarding subgraphs which are sampled through random walk methods for convenient model comparison. Through preprocessing, *WeChat Article* contains 90 thousand users, 40 thousand Official Accounts and 1.54 million forwarding behaviors. Weibo contains 250 thousand users and 422 thousand forwarding behaviors. We split 70% of forwarding behaviors as the training set.

Implementation Details. We use Tensorflow to implement the program. The profile features used in Weibo dataset is the verified status, gender, and tweet counts of users. The profile features used in *WeChat Article* are the one-hot vector of related article classes. For example, the profile features of a forwarder are the classes of articles it forwards. The topological features are 128 embedding vectors generated by DeepWalk [11].

The parameters of the model are set by preliminary experiments. The weighting factor α and γ are set to guarantee the same order of magnitude of the different components object function. For *Sina Weibo* dataset, we set $\alpha = 3 \times 10^{-2}$, $\gamma = 10^{-5}$, $d = 64$. For *WeChat Article* dataset, we set $\alpha = 3 \times 10^{-5}, \gamma = 10^{-5}, d = 64$. The negative sample number for each behavior is 24 (6 samples for each sample type) on two datasets. The neural network used in multi-modal autoencoder and used for β are all 3 layers, the node number of the hidden layer is 50 for network features, 25 for profile features and 10 for β. The training batch number is 2560. The maximum number of iteration step is set as 40.

Evaluation Protocol. We deploy FB2vec and baselines to *Behavior Prediction Problem* and *Similarity User Detection Problem*, and choose several comparative methods, including some significant or state-of-the-art models for network embedding and forwarding prediction, to evaluate the performance of FB2vec. Four standard metrics Precision, Recall, F1 score and AUC (Area Under Curve) [12] are employed to measure the performance.

Behavior Prediction Problem is to predict whether a user w is a receiver in the context of the generator u and the forwarder v. In *Sina Weibo* dataset, a user is a receiver if it forwards a tweet generated by u and forwarded by v. In *WeChat* Article dataset, a user is a receiver if it reads the whole article generated by u and forwarded by v. The problem is similar to the famous forwarding prediction problem [10]. We select 50 thousand triples (u, v, w) from the testing dataset as positive forwarding item and sample randomly 50 thousand triples (u, v, w) which does not appear in datasets as negative forwarding item. FB2vec solves the *Behavior Prediction Problem* by calculating the information intensity $f(\cdot)$ for each triple (u, v, w). If the information intensity is smaller than a threshold, w is a receiver in the context of the generator u and the forwarder v. The threshold is set as the median number of all information intensity in test data.

Similarity User Detection Problem is required to judge whether the given two users are similar or not, which is a common problem solved by network embedding methods. In our experiments, the similarity is defined by the ratio of common receivers as Eq. (12), where $D(u)$ means the set of all receivers of the generator u. We select 50 thousand pairs of generators by uniform sampling and select 50 thousand pairs of generators with at least one common receivers. Then generator pairs with the top 50% similarity are labeled as similar user pairs and the other 50% pairs are labeled as unsimilar user pairs. FB2vec solves *Similarity User Detection Problem* by computing the distance of generator embedding vector as Eq. (13). If the distance is smaller than a threshold, the pair of generators is predicted as a similar user pair. The threshold is set as the median number of all distances in test data. The threshold used in network embedding baseline models is set in the same way.

$$Sim(u_1, u_2) = \frac{2|D(u_1) \cap D(u_2)|}{|D(u_1)| + |D(u_2)|} \tag{12}$$

$$Dist(u_1, u_2) = -\left\|P^g_{u_1}B - P^g_{u_2}B\right\|^2 \tag{13}$$

The baseline models of the two tasks are DeepWalk, LINE, DANE and SDNE. The parameters for these four compared models are set as either the default settings suggested by the authors or tuned to find the best settings. Specially, we set all embedding vectors length as 128. Since network embedding model can not handle the triple relationship (u, v, w), we mask the generator u in training and testing datasets and apply network embedding models on the network connected between the receivers and their corresponding forwarders. Then network embedding models predict the links between forwarders and receivers.

We add another two baseline models, PMF [15] and OCCF [6], for behavior prediction. PMF is the basic one and OCCF is the start-of-the-art one considering node attributes. Both two models are matrix factorization model since other kinds of models are hard to implement because of the lack of available codes and the difficulty of feature requirements for most of the feature-based models. These models are to predict whether a message would be forwarded. We mask the forwarders in datasets and apply them to predict whether the contents of generators would be accepted by the receiver w.

5.2 Experiment Result

Behavior Prediction. We organize experimental results of FB2vec and other comparative methods for behavior prediction in Table 1. In the comparison of baselines, FB2vec outperforms on two datasets. In the *WeChat Article* dataset, FB2vec increases F1-score by from 3.9% to 33.5% and increases AUC by from 6.5% to 16.9%. In the *Sina Weibo* dataset, FB2vec increases F1-score by from 0.2% to 28.0% and increases AUC from 4.4% to 23.6%. The performances of all models on the *Sina Weibo* dataset is much worser than that on the *WeChat Article* because the network of the *WeChat Articl* is much denser.

Table 1. Comparison results: behavior prediction

Model	WeChat				Weibo			
	Precision	Recall	F1-score	AUC	Precision	Recall	F1-score	AUC
DeepWalk	0.791	0.772	0.781	0.861	0.551	0.577	0.562	0.627
LINE	0.701	0.714	0.707	0.847	0.552	0.505	0.528	0.593
SDNE	0.882	0.857	0.869	0.927	0.602	0.622	0.611	0.668
DANE	0.911	0.905	0.908	0.907	0.644	0.621	0.638	0.700
PMF	0.729	0.761	0.745	0.845	0.562	0.597	0.578	0.655
OCCF	0.897	0.916	0.907	0.921	0.658	**0.691**	0.674	0.702
FB2vec	**0.940**	**0.946**	**0.944**	**0.988**	**0.676**	0.676	**0.676**	**0.733**

Similar User Detection. The experimental results of FB2vec and other comparative methods for similar user detection are shown in Table 2. It shows that FB2vec outperforms on two datasets. In the *WeChat* dataset, FB2vec increases F1-score by from 4.1% to 41.4% and increases AUC by from 1.9% to 31.5%. In the *Sina Weibo* dataset, FB2vec increases F1-score by from 0.4% to 31.1% and increases AUC by from 1.1% to 8.5%.

Parameter Study. β is the utility parameter to represent the weight of generator factors (content influence) and forwarder factors (social influence). Users are more inclined to content influence when β approaches 0, while users are more inclined to social influence when β approaches 1. Figure 2 shows the number of

Table 2. Comparison results: similar user detection

Model	WeChat				Weibo			
	Precision	Recall	F1-score	AUC	Precision	Recall	F1-score	AUC
DeepWalk	0.653	0.667	0.659	0.738	0.600	0.603	0.601	0.703
LINE	0.582	0.599	0.591	0.700	0.523	0.531	0.527	0.701
DANE	0.791	0.815	0.803	0.903	0.684	**0.692**	0.688	0.752
SDNE	0.768	0.752	0.760	0.882	0.661	0.653	0.657	0.701
FB2vec	**0.831**	**0.840**	**0.836**	**0.921**	**0.701**	0.680	**0.691**	**0.761**

users with the distribution of β. In *Sina Weibo*, most of β ranges from 0.4 to 0.8 with a normal distribution. However, most of β in *WeChat Article* dataset ranges from 0.72 to 0.79, which is approximate to uniform distribution. These results indicate that content is more important than social influence in both two datasets. Moreover, the pattern of user behavior in WeChat is more similar than that in *Sina Weibo*.

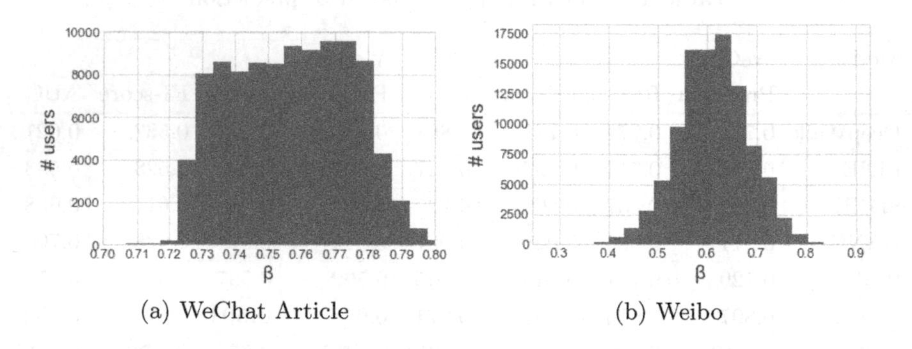

(a) WeChat Article (b) Weibo

Fig. 2. Distribution of β

Here, we visualize the information dissemination intensity index by three Official Accounts on *WeChat Article* dataset in Fig. 3, 4 and 5 as examples. The reason to visualize it on WeChat datasets instead of Weibo datasets is that Official Account product information specially and professionally and they tend to attract a specific group. We call these three Official Accounts as NewsAccount, MilitaryAccount, and EntertainAccount[1], because they are the famous Official Accounts with more than one million followers in the news section, military section and entertainment section correspondingly. The NewsAccount is the official accounts operated by one of the most famous Chinese newspaper, focusing on news of Chinese society and culture. MilitaryAccount is operated by the biggest official Chinese military newspaper. EntertainAccount is one of

[1] Their names are hidden because of the requirement of Tencent Company.

the biggest Official Accounts for anecdotes of entertainer, which are popular among women. Figure 3, 4 and 5 show their information intensity distribution. The x-axis shows age and the y-axis shows information intensity. Note that the lower information intensity represents higher attraction. The dots in the lower left quarter in Fig. 4(a) means that EntertainAccount attracts women and the younger more. Similarly, MilitaryAccount attracts men and the old more indicated by Fig. 5(b), and the preference of people with different ages and genders to NewsAccount is similar. The above phenomenon is matched with prior knowledge of these Official Accounts. Therefore, FB2vec satisfies the expectation of information dissemination intensity measure.

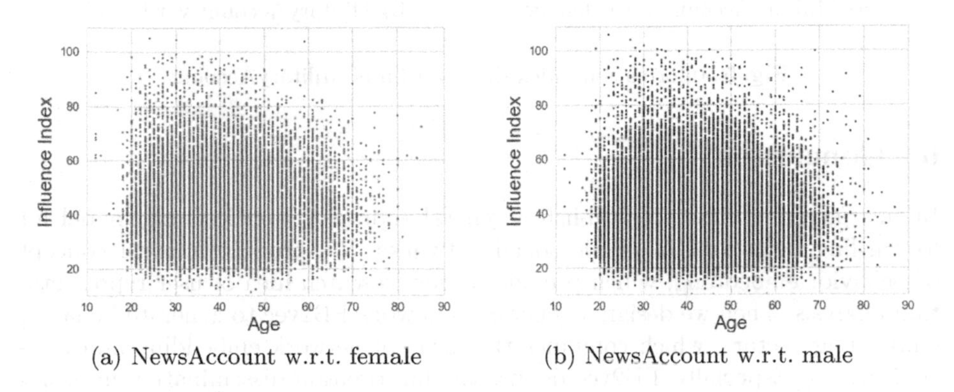

(a) NewsAccount w.r.t. female (b) NewsAccount w.r.t. male

Fig. 3. Information index distribution of NewsAccount

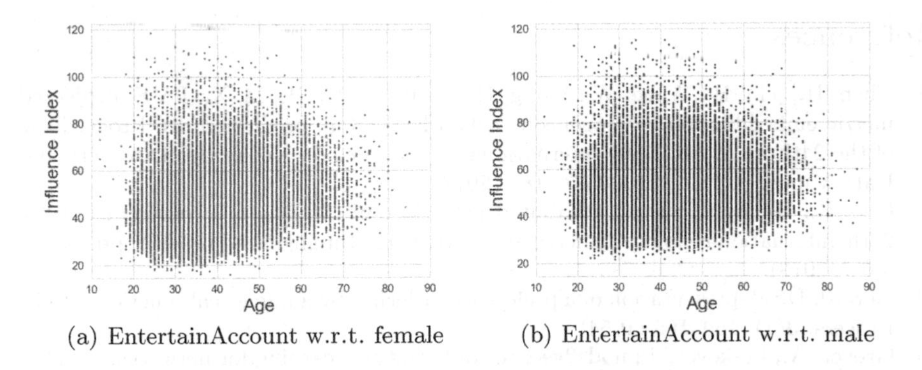

(a) EntertainAccount w.r.t. female (b) EntertainAccount w.r.t. male

Fig. 4. Information index distribution of EntertainAccount

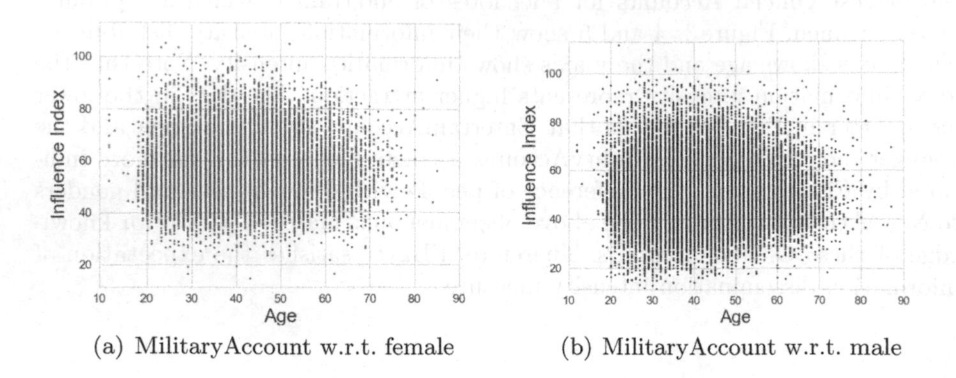

(a) MilitaryAccount w.r.t. female (b) MilitaryAccount w.r.t. male

Fig. 5. Information index distribution of MilitaryAccount

6 Conclusion

In this paper, we aim to establish a general representation learning model for forwarding behaviors on online social networks. We propose a novel concept of behavior embedding, which provides a new research idea of user representation analysis. Then we design a general framework FB2vec to generate behavior embedding vectors, which combines the ideas of network embedding and metric learning. Especially, FB2vec defines the information dissemination intensity function with the help of economic utility function and we train FB2vec by adapting the intensity order with real data. The experimental results on two datasets *Sina Weibo* and *WeChat Article* show that FB2vec outperforms baselines for different tasks. The parameter study proves that the FB2vec coincides with the prior knowledge of representation of behaviors.

References

1. Chen, H., Yin, H., Wang, W., Wang, H., Nguyen, Q.V.H., Li, X.: PME: projected metric embedding on heterogeneous networks for link prediction. In: Proceedings of the 24th ACM SIGKDD International Conference on Knowledge Discovery and Data Mining (KDD), pp. 1177–1186 (2018)
2. Gao, H., Huang, H.: Deep attributed network embedding. In: Proceedings of the 27th International Joint Conference on Artificial Intelligence (IJCAI), pp. 3364–3370 (2018)
3. Gerard, D.: Representation of a preference ordering by a numerical function. Decis. Process. **3**(1), 159–165 (1954)
4. Grover, A., Leskovec, J.: node2vec: scalable feature learning for networks. In: Proceedings of the 22nd ACM SIGKDD International Conference on Knowledge Discovery and Data Mining (KDD), pp. 855–864 (2016)
5. Guo, J., Xu, L., Huang, X., Chen, E.: Enhancing network embedding with auxiliary information: an explicit matrix factorization perspective. In: Pei, J., Manolopoulos, Y., Sadiq, S., Li, J. (eds.) DASFAA 2018. LNCS, vol. 10827, pp. 3–19. Springer, Cham (2018). https://doi.org/10.1007/978-3-319-91452-7_1

6. Jiang, B., et al.: Retweeting behavior prediction based on one-class collaborative filtering in social networks. In: Proceedings of the 39th International ACM SIGIR conference on Research and Development in Information Retrieval (SIGIR), pp. 977–980 (2016)
7. Kingma, D.P., Ba, J.: Adam: a method for stochastic optimization. In: 3rd International Conference on Learning Representations (ICLR), pp. 1–15 (2015)
8. Kipf, T.N., Welling, M.: Semi-supervised classification with graph convolutional networks. In: 5th International Conference on Learning Representations (ICLR), pp. 1–14 (2017)
9. Li, Y., Wang, Y., Zhang, T., Zhang, J., Chang, Y.: Learning network embedding with community structural information. In: Proceedings of the 28th International Joint Conference on Artificial Intelligence (IJCAI), pp. 2937–2943 (2019)
10. Liu, Y., Zhao, J., Xiao, Y.: C-RBFNN: a user retweet behavior prediction method for hotspot topics based on improved RBF neural network. Neurocomputing **275**, 733–746 (2018)
11. Perozzi, B., Al-Rfou, R., Skiena, S.: DeepWalk: online learning of social representations. In: Proceedings of the 20th ACM SIGKDD International Conference on Knowledge Discovery and Data Mining (KDD), pp. 701–710 (2014)
12. Powers, D.M.: Evaluation: from precision, recall and f-measure to ROC, informedness, markedness and correlation. J. Mach. Learn. Technol. **2**(1), 37–63 (2011)
13. Qiu, J., et al.: NetSMF: large-scale network embedding as sparse matrix factorization. In: The World Wide Web Conference (WWW), pp. 1509–1520 (2019)
14. Qiu, J., Dong, Y., Ma, H., Li, J., Wang, K., Tang, J.: Network embedding as matrix factorization: unifying DeepWalk, LINE, PTE, and Node2vec. In: Proceedings of the Eleventh ACM International Conference on Web Search and Data Mining (WSDM), pp. 459–467 (2018)
15. Salakhutdinov, R., Mnih, A.: Probabilistic matrix factorization. In: Advances in Neural Information Processing Systems 20, Proceedings of the Twenty-First Annual Conference on Neural Information Processing Systems (NeurIPS), pp. 1257–1264 (2007)
16. Tang, J., Qu, M., Wang, M., Zhang, M., Yan, J., Mei, Q.: LINE: large-scale information network embedding. In: Proceedings of the 24th International Conference on World Wide Web (WWW), pp. 1067–1077 (2015)
17. Tu, C., Zhang, W., Liu, Z., Sun, M.: Max-margin DeepWalk: discriminative learning of network representation. In: Proceedings of the Twenty-Fifth International Joint Conference on Artificial Intelligence (IJCAI), pp. 3889–3895 (2016)
18. Tu, C., Zhang, Z., Liu, Z., Sun, M.: TransNet: translation-based network representation learning for social relation extraction. In: Proceedings of the Twenty-Sixth International Joint Conference on Artificial Intelligence (IJCAI), pp. 2864–2870 (2017)
19. Wang, D., Cui, P., Zhu, W.: Structural deep network embedding. In: Proceedings of the 22nd ACM SIGKDD International Conference on Knowledge Discovery and Data Mining (KDD), pp. 1225–1234 (2016)
20. Wang, H., et al.: A united approach to learning sparse attributed network embedding. In: IEEE International Conference on Data Mining (ICDM), pp. 557–566 (2018)
21. Weinberger, K.Q., Saul, L.K.: Distance metric learning for large margin nearest neighbor classification. J. Mach. Learn. Res. **10**, 207–244 (2009)
22. Zhang, J., Tang, J., Li, J., Liu, Y., Xing, C.: Who influenced you? Predicting retweet via social influence locality. ACM Trans. Knowl. Discov. Data (TKDD) **9**(3), 25:1–25:26 (2015)

A Framework for Deep Quantification Learning

Lei Qi[(✉)], Mohammed Khaleel, Wallapak Tavanapong, Adisak Sukul, and David Peterson

Iowa State University, Ames, IA 50011, USA
{leiqi,mkhaleel,tavanapo,adisak,daveamp}@iastate.edu

Abstract. A quantification learning task estimates class ratios or class distribution given a test set. Quantification learning is useful for a variety of application domains such as commerce, public health, and politics. For instance, it is desirable to automatically estimate the proportion of customer satisfaction in different aspects from product reviews to improve customer relationships. We formulate the quantification learning problem as a maximum likelihood problem and propose the first end-to-end Deep Quantification Network (DQN) framework. DQN jointly learns quantification feature representations and directly predicts the class distribution. Compared to classification-based quantification methods, DQN avoids three separate steps: classification of individual instances, calculation of the predicted class ratios, and class ratio adjustment to account for classification errors. We evaluated DQN on four public datasets, ranging from movie and product reviews to multi-class news. We compared DQN against six existing quantification methods and conducted a sensitivity analysis of DQN performance. Compared to the best existing method in our study, (1) DQN reduces Mean Absolute Error (MAE) by about 35%. (2) DQN uses around 40% less training samples to achieve a comparable MAE.

Keywords: Quantification learning · Class distribution estimate · Deep learning

1 Introduction

In various problem domains, it is important to estimate class ratios (class prevalence) of a subject of interest. For instance, in commerce, knowing the prevalence of customer complaints in different aspects (e.g., packaging, durability, delivery and after sale service) of a product is key to improve the product and customer experience. In politics, knowing voters' proportional interest in different policy areas (e.g., healthcare, education) is useful to improve political candidates' campaign strategies to attract these voters. In healthcare, knowing the proportion of residents in different age groups affected by a specific disease is vital for

This work is partially supported in part by the NSF SBE Grant No. 1729775.

F. Hutter et al. (Eds.): ECML PKDD 2020, LNAI 12457, pp. 232–248, 2021.
https://doi.org/10.1007/978-3-030-67658-2_14

determining appropriate prevention and treatment responses. Research in quantification learning has been conducted in various disciplines, leading to different terminologies, such as prior probability shift [1], prevalence estimation [2], or class ratio estimation [3]. The earliest work dated back to the application in screening tests in epidemiology [4]. In this paper, we use class ratios and class distribution interchangeably.

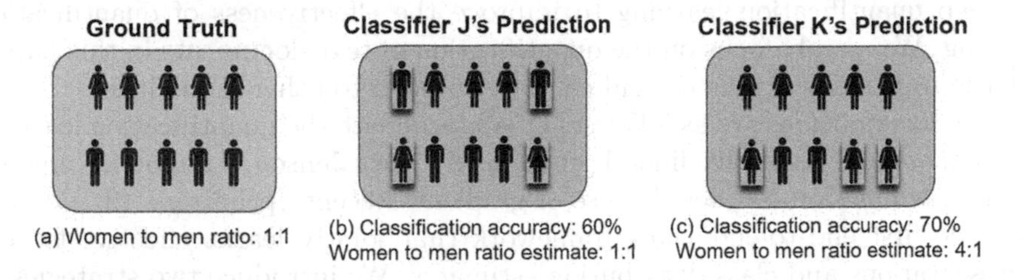

Fig. 1. Mismatch in performance when using classification for quantification

Forman [5] gave the first formal definition of quantification learning: "Given a labeled training set, induce a quantifier that takes an unlabeled test set as input and returns its best estimate of the class distribution." Mathematically, a quantification problem can be defined as follows. Let x_i be the *i-th* instance, and C is the set of class labels with the cardinality denoted as $|C|$. Given a training dataset $D = \{(x_1, y_1), (x_2, y_2), \ldots, (x_n, y_n)\}$, where $y_i \in C$ is the corresponding class label of x_i, and a test dataset $T = \left\{x_{n+1}, x_{n+2}, \cdots, x_{n+|T|}\right\}$, a quantifier outputs a vector $\widehat{P} = \left[\hat{p}_1, \hat{p}_2, \ldots, \hat{p}_{|C|}\right]$ and $\hat{p}_j$ is the predicted ratio of the number of instances in class j to the total number of instances in the test dataset. An instance can be a text document, an image, or an audio file, depending on the applications.

With a large volume of data, manual estimation of class distribution is prohibitively time-consuming and impractical. Although quantification and classification learning are both supervised learning tasks, quantification learning has been relatively under-explored mainly because it is seen as a trivial and straightforward post-processing step of classification [6]. Quantification learning focuses on correct predictions at the aggregate level while classification learning aims at predicting the class of individual instances. Furthermore, when using imperfect classifiers to conduct quantification, the more accurate classifier does not necessarily predict a more accurate class distribution. Figure 1 illustrates this important issue. Figure 1(a) shows the women to men ratio of 1:1 in the ground truth. Figure 1(b) demonstrates that classifier **J** (with classification accuracy of 60%) mis-predicting two females as males and vice versa is able to give accurate class ratios of 1:1 as in the ground truth. However, classifier **K** with a 10% higher classification accuracy than classifier **J** gives a much worse estimate of four times the number of women to the number of men.

Another major issue is that classification learning generally assumes that training data and test data are independent and identically distributed (i.i.d)

[7,8] whereas quantification learning does not make the i.i.d. assumption. When there is a difference between the class distribution of the training data and that of the test data, an over-estimate or under-estimate of the class distribution in the training data is likely to occur [5]. Therefore, quantification learning should be investigated in its own right. Existing works in quantification learning show some benefits of inducing a quantifier directly without classification [9,10], but none has used deep learning for quantification. In this paper, we design a framework for deep quantification learning to improve the effectiveness of quantification learning. We mainly focus on the quantification of text documents in this paper, but the framework is generic and can be extended to other modalities.

Our contributions are as follows: (1) We formulate the quantification learning problem as a maximum likelihood problem. We use a Jensen Shannon Divergence as the loss function to allow for use of gradient descent optimizers. (2) We propose the first end-to-end DQN framework that jointly learns effective feature representations and class distribution estimates. We introduce two strategies to select a set of documents (termed a *tuplet*) for training to investigate how well the induced quantifier generalizes to test sets with class distributions different from that of the training set. We examine five methods to extract the feature representations of tuplets. (3) We evaluated DQN variants on quantification tasks of text documents on four public datasets with 2, 4, and 20 classes using three metrics commonly used for measuring quantification errors: Mean Absolute Error, Relative Mean Absolute Error, and Kullback-Leibler Divergence. We performed a sensitivity analysis of DQN performance by varying tuplet sizes and training dataset sizes. (4) The highlights of our findings are as follows. Compared to the best existing method in our study, (i) DQN reduces Mean Absolute Error (MAE) by about 35%. (ii) DQN uses around 40% less training samples to achieve a comparable MAE. Therefore, DQN is more desirable since it reduces the manual labeling effort to create the training dataset significantly.

Reproducibility: Our source code is available at https://github.com/cml-cs-iastate/CyText.

2 Related Work

We categorize existing methods for quantification learning into two categories: classification-based quantification and direct quantification. We further divide the classification-based quantification methods into two sub-categories: (1) classify and count; (2) hybrid.

2.1 Classification-Based Quantification

2.1.1 Classify and Count Sub-category

Methods in this sub-category classify individual documents, count the number of documents predicted in each class, and calculate the class ratios. The methods differ in either the counting step or the post-processing steps.

- **Classify and Count (CC)**: Forman [5] and most authors published in this field used this basic CC method as the baseline. CC overestimates when the prevalence of a class in the test dataset is lower than the prevalence of the same class in the training data and underestimates when the prevalence of the class in the test dataset is higher [5]. To address this drawback, several methods of adjusting the predicted class ratios after CC have been proposed by [11,12].
- **Adjusted Classify and Count (ACC)**: Forman [11] proposed Eq. (1) to adjust the class ratios estimated by the CC method with *tpr* (true positive rate) and *fpr* (false positive rate) obtained from cross-validation for a binary class classification as follows:

$$p = \frac{\hat{p} - fpr}{tpr - fpr} \tag{1}$$

  where $\hat{p}$ is the predicted proportion of the positive class from a CC method and p is the adjusted proportion. According to Eq. (1), ACC is vulnerable to *tpr* and *fpr* values. For example, when *tpr* is close to *fpr*, the denominator is close to zero, which leads to a large value of p in Eq. (1). Hopkins and King generalized Eq. (1) for a multi-class classification task [13].
- **Probabilistic Classify and Count (PCC)**: Bella et al. [14] proposed probabilistic versions of Classify and Count (**PCC**) and adjusted Classify and Count (**PACC**) using the posterior probabilities output by a classifier for counting.

2.1.2 Hybrid Sub-category

This sub-category adapts traditional classification algorithms for quantification tasks by either introducing a new combined loss function or using an ensemble of quantification results. Milli et al. [15] proposed Quantification Trees, which use either Decision Tree or Random Forest for classification. They investigated two loss functions. Let FP_i and FN_i be the numbers of false positives and false negatives for a class i, respectively. One method calculates an error for the class i, $E_i = |FP_i - FN_i|$. The other method computes $E_i = |FP_i - FN_i| \times |FP_i + FN_i|$ where the first term represents the quantification error and the second term represents the classification error. Other methods in this category include [16,17]. Esuli and Sebastiani's quantifier [16] used SVMperf [18] as a classifier for quantification. Barranquero et al. [17] proposed Q-measure as a loss function, an analogy to F-measure for classification. Pérez-Gállego et al. proposed the first ensemble quantifier for binary quantification [19], which is inspired by the idea of ensemble learning for classification. Although the methods in the Hybrid category reduce quantification errors, they still rely on the classification of individual documents.

2.2 Direct Quantification

King and Lu [9] presented a non-parametric approach (denoted as ReadMe) to estimate the distribution of the cause-of-death without training a classifier.

Hopkins and King applied this method to estimate document category distributions [13]. The key assumption is that the proportions of the word patterns occurring in documents in each class are the same for both the test and labeled datasets. In simple terms, it means the same writing style for documents in each class in both test and labeled datasets. ReadMe works in iterations until the number of iterations reaches the user-specified value. In each iteration, it randomly selects k words to form word patterns and calculates the proportion of the word patterns used in documents of different classes. ReadMe uses these proportions to estimate the class distribution of the labeled dataset. The error of the estimate and the truth gets smaller with more iterations. González et al. proposed methods based on Hellinger Distance (HD) [10] between two distributions and it was used for quantification of image data. The method repeatedly generates a validation dataset V from the training dataset with a given prior probability. Then, it uses the distribution of a validation dataset $\hat{V}$ with the least HD value to the distribution of the test dataset as the estimated distribution.

In summary, the classification-based methods have these major drawbacks. (1) They require separate steps (feature extraction, classifier training, counting, and adjustment of the class ratios) that are not jointly optimized to reduce the quantification error. (2) Except for the rare case of 100% accurate classification, more accurate classifiers do not always lead to more accurate class distribution estimates. See Fig. 1 example. (3) For the direct quantification category, the HD-based method requires good feature representations of instances to begin with. For ReadMe, the randomly chosen word patterns may not be good features for estimating class ratios. Motivated by the nature of the quantification learning problem, the major drawbacks of the existing methods, and the success of deep learning in several tasks [20], we proposed DQN.

3 Problem Formulation of Quantification Learning

Recall in the introduction that given a labeled dataset D and a test dataset T, a quantifier estimates the class distribution of the entire dataset T. We partition the set of training instances $X = \{x_1, ..., x_n\}$ in D during training or T—the set of test instances during testing, into a number of tuplets. A tuplet is a set of m instances where $m \geq 1$. This enables the prediction of the class distribution for each tuplet and gives a reliable class distribution estimate by averaging the estimated class distributions of a large number of tuplets.

For ease of presentation, we consider a binary quantification problem and formulate the problem as a maximum likelihood problem as follows. Let the class ratio for a positive class be r; the ratio of the negative class is then $1 - r$. For training, the training instances in X are grouped into tuplets, and the corresponding class distribution of each tuplet is calculated from computing the ratio of the number of instances in the tuplet in each class to the total number of instances in the tuplet. Different methods for assigning the training instances into tuplets can be used. We introduce our methods in Sect. 4.2. Let $DT = \{(t_1, r_1), \ldots, (t_N, r_N)\}$ be a set of N tuplets generated from D with their corresponding class distribution, where $r_i \epsilon [0,1]$ is the corresponding class ratio of the positive class of the tuplet t_i. Let $TT = \{t_{N+1}, t_{N+2}, \ldots, t_{N+K}\}$ be a set of K tuplets generated from T. Given DT and TT, a quantifier outputs a vector $\widehat{P} = [\hat{p}, 1 - \hat{p}]$ where $\hat{p}$ is the predicted ratio for the positive class of TT. Therefore, DQN learns parameters of a function F that performs a complex mapping from t_k to r_k. We write it as $r_k \sim F(t_k, \Theta)$, where F is parameterized by the parameter set Θ. The conditional probability of r_k given t_k and Θ is written as $P(r_k|t_k, \Theta)$. We assume that the conditional probability of the class ratios in DT is the same as that in TT. The likelihood function under F is in Eq. (2) and the log-likelihood function $L(\Theta)$ is given in Eq. (3).

$$\prod_{k=1}^{N} P\left(r_k|t_k,\ \Theta\right) \tag{2}$$

$$LL(\Theta) = \sum_{k=1}^{N} \log\left[P\left(r_k|t_k, \Theta\right)\right] \tag{3}$$

Maximizing the log-likelihood is the same as minimizing the negative log-likelihood. We rewrite Eq. (3) as the negative log-likelihood as in Eq. (4).

$$NLL(\Theta) = -\sum_{k=1}^{N} \log\left[P\left(r_k|t_k, \Theta\right)\right] \tag{4}$$

We use Jensen-Shannon Divergence (JSD) [21] as our loss function to measure the similarity between the predicted and the true distributions. JSD has several good properties; its value range is [0, 1]; it is differentiable and it has the symmetric property. JSD has been applied in Generative Adversarial Networks (GAN) [22] as well as in several other research fields. Equation (4) with the JSD as the loss function can be solved to obtain an estimated parameter set $\hat{\Theta}$ using an optimizer such as gradient descent optimizers. For testing, we use the learned parameters to estimate the class distribution from TT.

4 Deep Quantification Network (DQN) Framework

4.1 DQN Framework

Our problem formulation defines a tuplet as a set of m instances; we call m tuplet size. Therefore, the first component of the framework is the tuplet generator that assigns input instances to tuplets. Ideally, a tuplet should be large enough to include samples of instances from all the classes in order to get a reasonable class distribution estimate from a single tuplet. The tuplet size is a hyper-parameter of DQN. On one extreme, when the tuplet size m is one (i.e., a tuplet with only one instance), DQN degrades to a classification-based method for quantification. On the other extreme, if the tuplet size is as large as the size of the training dataset, we are restricted to only one tuplet with a fixed class

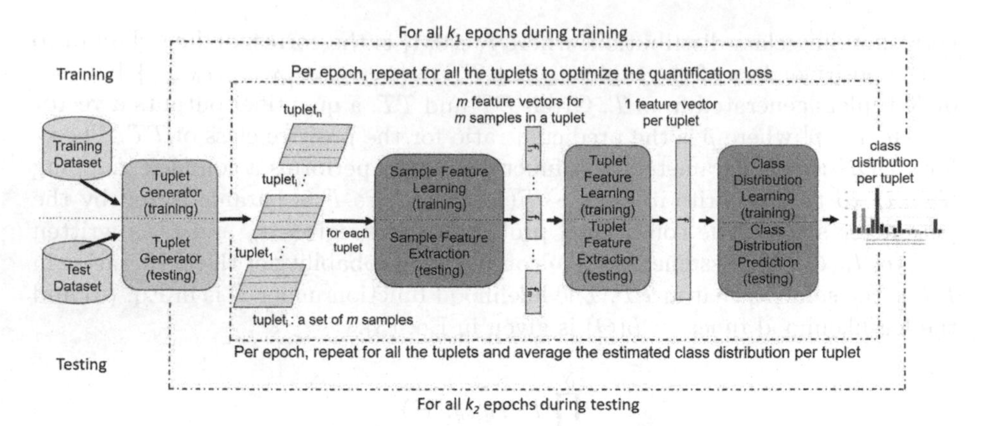

Fig. 2. DQN framework; the top half of the diagram shows the training process. The bottom half illustrates the test process; m is a DQN hyper-parameter; k_1 and k_2 are the numbers of epochs for training and testing, respectively.

distribution, which makes the trained model unable to perform well for other datasets. We describe our tuplet generation strategies in detail in Sect. 4.2. During training (illustrated in the top half of Fig. 2), the tuplet generator generates all the tuplets from all labeled instances in the training dataset as well as calculates the corresponding class distribution for each tuplet based on the class labels of the instances of the tuplet. In each epoch, the tuplet generator passes each tuplet to the sample feature learning component. This component is a layer of neurons to learn parameters of a function that extracts feature representation for each sample in a tuplet. Any deep learning architecture that is effective for the modality of the instances (e.g., LSTM for text documents, CNN for images) can be used. Since each tuplet has m instances, this component outputs m feature vectors f_1, ..., f_m. These vectors become the input to the tuplet feature learning component to extract a tuplet feature vector f. The feature vector is passed to one or more layers of neurons trained to estimate the class distribution of a tuplet such that the quantification loss between the predicted and the ground truth is optimized. During back-propagation, the weights and biases are updated based on a gradient descend algorithm as done in classification learning. The entire process is repeated until the desired number of epochs is reached.

During testing, for each epoch, the tuplet generator generates tuplets from all unlabeled instances in the test dataset. The second component extracts a feature vector for each sample in a tuplet using the pre-trained weights and biases. The third component extracts the tuplet feature vector for the last component that outputs the class distribution estimate of a tuplet. This process repeats for all the generated tuplets in this epoch and the arithmetic mean of the estimated class ratios for each class from all the tuplets is used as the class ratio estimate for the class. Subsequent epochs follow the same process and average all the estimated class distributions of all the epochs as the final class distribution estimate.

4.2 Tuplet Generation

We introduce two strategies for assigning training instances to tuplets: the random selection strategy and the Zipf distribution selection strategy. We use the random selection strategy to establish the baseline performance. We introduce the Zipf distribution [23] selection strategy to generate a variety of class distributions that might be much different from the underlying class distribution in the training dataset to enhance the generalization ability of the model for test datasets with different class distributions.

Random Selection Strategy: This strategy randomly selects m instances without replacement from the training dataset to form a tuplet. That is, an instance is assigned exclusively to one tuplet in each epoch. If the number of instances is not divisible by m, we have one incomplete tuplet for each run. We discard the incomplete tuplet. Note that the training instances that are not used in this run of tuplet generation may be used in a subsequent run of tuplet generation due to random selection. Although random sampling with replacement can also be considered, we are against it for two reasons. First, some instances may be selected many more times than other instances, creating a bias in training. Second, even when the number of training instances is divisible by m, we cannot guarantee that all training instances are selected per epoch unless we implement more constraints. We recommend using more epochs with our random selection without replacement strategy than using random selection with replacement.

Zipf Distribution Selection Strategy: Our goal is to prevent DQN from overfitting the class distribution in the training dataset and to generalize DQN for different class distributions of future test datasets. We propose to generate tuplets with different class distributions synthetically. Because many types of data in physical and social sciences can be approximated well with Zipf distribution [23], we chose the Zipf distribution. That is, we use Eq. (5) to calculate num_i, the number of instances in the i-th class in a tuplet.

$$num_i = \frac{m}{i^z * \sum_{j=1}^{|C|} j^z} \tag{5}$$

where $|C|$ is the number of classes and $z \in [0, 1]$ is the skew factor. When the skew factor value is zero, all classes have the same number of instances, i.e., uniform distribution. When the skew factor is one, a few classes have many instances while several classes have very few instances. After calculating the number of instances for each i-th class for a tuplet, we randomly select num_i instances without replacement for the class from the training dataset and assign them to the tuplet. The already assigned instances are not eligible for other tuplets in this epoch. The difference between the two strategies is the class distribution of each tuplet. With the Zipf distribution selection strategy, the class distributions of tuplets can vary significantly from the class distribution of the training dataset.

4.3 Sample Feature Learning/Extraction

To obtain feature representation of a tuplet, we first extract feature representation of each sample in the tuplet. To apply the DQN framework for a specific application, it is necessary to choose an appropriate neural network (NN) architecture suitable for the modality of the data and the application (e.g., CNN for images, 3D-CNN for videos). As we focus on the application of DQN on text documents in this paper, we choose Long Short-Term Memory (LSTM) [24] to learn effective feature representations for each sample (document) in a tuplet during training. LSTM can deal with variable length documents and is good at feature extraction of sequence data like text. This step outputs m fixed-length vectors: $f_1, \dots, f_m$, where $f_i \in \mathbb{R}^d$ and d is the number of elements in the feature vector f_i. During testing, the learned LSTM parameters are used to extract sample feature vectors.

4.4 Tuplet Feature Learning/Extraction

We study five alternatives to obtain a tuplet feature vector f from the feature vectors f_1, ..., f_m of the samples in the tuplet. Let $|f|$ denote the dimension of the tuplet feature vector f.

1. Concatenation (CAT): We concatenate f_1, ..., f_m one after another; therefore, $|f| = d * m$.
2. Average (AVG), Median (MED), and Maximum (MAX): we compute the column-wise arithmetic mean (or median, or maximum, respectively) for each dimension of the m feature vectors in a tuplet to obtain a unified feature vector for the tuplet. Therefore, $|f| = d$.
3. Additional neural network (NN) layers: We feed f_1 ,..., f_m to the additional NN layers such as a dense layer, or a convolutional layer. For these NN layers, we do not use any architecture that is impacted by the order of the samples in a tuplet (e.g., LSTM) since the order should not impact quantification results. The dimension of the tuplet feature vector f depends on the chosen NN architecture. During training, the parameters of the NN layers are learned. During testing, the learned parameters are used to extract one tuplet feature vector per tuplet.

4.5 Class Distribution Learning

The last component in Fig. 2 is a fully connected layer followed by a softmax layer. During training, this component takes the tuplet feature vector to learn a probability-valued vector, which is the estimated class distribution of the tuplet. As we mentioned in Sect. 2, we use JSD as the loss function between the estimated class distribution and the true class distribution during training. During testing, this component uses the learned parameters to estimate the class distribution.

4.6 Train and Test Algorithms

Training Algorithm (Algorithm 1 in Fig. 3): Given a set of labeled instances, denoted as D, the tuplet generator generates tuplets from D using one of the methods proposed in Sect. 4.2. Either method generates the number of tuplets per epoch of $\lfloor \frac{|D|}{m} \rfloor$.

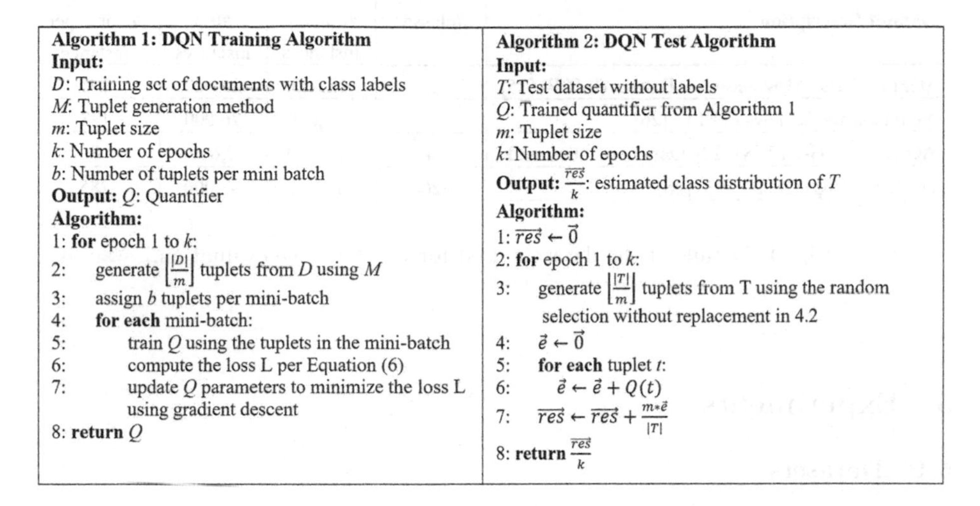

Algorithm 1: DQN Training Algorithm
Input:
D: Training set of documents with class labels
M: Tuplet generation method
m: Tuplet size
k: Number of epochs
b: Number of tuplets per mini batch
Output: Q: Quantifier
Algorithm:
1: **for** epoch 1 to k:
2: generate $\lfloor \frac{|D|}{m} \rfloor$ tuplets from D using M
3: assign b tuplets per mini-batch
4: **for each** mini-batch:
5: train Q using the tuplets in the mini-batch
6: compute the loss L per Equation (6)
7: update Q parameters to minimize the loss L using gradient descent
8: **return** Q

Algorithm 2: DQN Test Algorithm
Input:
T: Test dataset without labels
Q: Trained quantifier from Algorithm 1
m: Tuplet size
k: Number of epochs
Output: $\frac{\overrightarrow{res}}{k}$: estimated class distribution of T
Algorithm:
1: $\overrightarrow{res} \leftarrow \vec{0}$
2: **for** epoch 1 to k:
3: generate $\lfloor \frac{|T|}{m} \rfloor$ tuplets from T using the random selection without replacement in 4.2
4: $\vec{e} \leftarrow \vec{0}$
5: **for each** tuplet t:
6: $\vec{e} \leftarrow \vec{e} + Q(t)$
7: $\overrightarrow{res} \leftarrow \overrightarrow{res} + \frac{m * \vec{e}}{|T|}$
8: **return** $\frac{\overrightarrow{res}}{k}$

Fig. 3. Training and test algorithms for DQN. ***res*** and ***e*** are vectors; ***0*** is a zero vector.

The tuplets and their corresponding class distribution are given as input to the sample feature-learning component in mini-batches. We use Eq. (6) to calculate the overall loss L for all the tuplets in a mini-batch.

$$L = \sum_{i=0}^{b} JSD(r_i \| \hat{r_i}) + \lambda \sum_{\theta \in \theta} \theta^2 \tag{6}$$

where r_i is the real class distribution of a tuplet i from the training data, and $\hat{r_i}$ is the estimated class distribution of the same tuplet in a mini-batch. We update the hyper-parameters of Q using a gradient descent method. Finally, we obtain the trained quantifier Q. The last term $\lambda \sum_{\theta \in \Theta} \theta^2$ in Eq. (6) is the regularization term [20] to prevent overfitting and obtain a smooth model. λ is the weight decay and Θ is the set of all parameters in the model.

Testing Algorithm (Algorithm 2 in Fig. 3): Given the trained quantifier Q and a test dataset T, we use the random selection without replacement method introduced in Sect. 4.2 to generate tuplets from T and input them into the pre-trained quantifier Q. The test algorithm runs in multiple epochs. For each epoch, Lines 4–7 find the average estimate of class distributions of all the tuplets for the epoch and store it in the vector variable ***res***. The average is averaged again over the number of epochs in Line 8. Notice that in different epochs, the same instance has a chance to be assigned with other tuplets. Hence, using a sufficiently large number of epochs will produce different combinations of instances in tuplets. This is to have a reliable estimate of the underlying class distribution in the test data.

Dataset Description	#classes	Total #training instances	Total #test instances	Average #words per instance
Stanford Large Movie Review Dataset (IMDB) [25]	2	25,000	25,000	231
Yelp Polarity Reviews (YELP) [26]	2	560,000	76,000	133
AG News (AG-NEWS) [27]. Only news titles used	4	120,000	7,600	8
20 Newsgroups (20-NEWS) [28]	20	~16,000	~2,000	285

Fig. 4. Details of the datasets used for performance evaluation.

5 Experiments

5.1 Datasets

The details of four public balanced datasets are in Fig. 4. To evaluate the performance of a quantifier on test datasets with different class ratios as done in previous works [11], we created test datasets artificially from the original test datasets. For binary quantification tasks, we extracted instances with a prevalence of the positive class varying from 0.1 to 0.9 with the interval of 0.1. For multi-class quantification tasks, we synthetically created the test datasets with different class ratios using the Zipf distribution in Eq. (5) with different skew factors. We varied the skew factors from 0 (uniform distribution) to 0.9 (highly skewed distribution) with the interval of 0.1. For each prevalence or skew factor, we created ten different test datasets and compute the average of the quantification errors from the experiments on these datasets.

5.2 Compared Methods and Performance Metrics

We compared DQN with six existing methods across all the categories of existing works: **CC**, **ACC**, **PCC**, **PACC**, Hybrid Approach (**HA**), and **ReadMe** [13]. HA denotes a Hybrid method that uses the sum of equally weighted classification loss and quantification loss defined as $\frac{1}{|C|}\sum_{c\in C}|FPR_c - FNR_c|$ with a similar idea as that of Quantification Tree [15]. FPR_c and FNR_c are false positive and false negative rates for each class, respectively. The compared techniques were introduced in Sect. 2. ReadMe is the only existing direct quantification method. Both ReadMe and DQN do not require features to be known in advance. Therefore, we chose ReadMe to represent the existing work in the direct quantification category. For all the compared methods except ReadMe, we used the same LSTM architecture. This is to ensure that any performance difference does not come from different types of classifiers used. For ReadMe, we ran the original code [13] with their default parameters.

We used three commonly used metrics: Mean Absolute Error (MAE), Relative Mean Absolute Error (RMAE), and Kullback-Leibler Divergence (KLD) to quantify the errors [6]. Techniques that offer the lowest errors are most desirable.

5.3 Hyper-parameters Setting

Recall that DQN has four components (Fig. 2). For the tuplet generator using the Zipf distribution selection (Sect. 4.2), during training, we varied the Zipf skew parameter z value from 0.1 to 1.0 with an interval of 0.1. For the sample feature learning component (Sect. 4.3), we used one shared LSTM layer with 128 neurons and ReLU as the activation function. For the tuplet feature learning component that uses additional layers, we used one dense layer with 256 neurons to extract a high-level representation for a tuplet. Finally, the class distribution learning component (Sect. 4.4) had 256 nodes for the fully connected layer with the sigmoid function as the activation function and $|C|$ nodes for softmax layers where $|C|$ is the number of classes. We chose the hyper-parameter values for our training empirically. For DQN, we set the mini-batch size to 8 and the tuplet size to 100. We used stochastic gradient descent as the optimizer. We used the dropout rate and the recurrent dropout rate of 0.2. The learning rate was 1.0E−5. For the classification-based quantification methods, CC, ACC, PCC, PACC and HA, to make the comparison fair, we used LSTM as the classifier; we kept the same number of neurons, dropout rate and leaning rate as that of DQN, but used cross-entropy as the loss function for the classifier. The mini-batch size of the classifier was 64. The word embedding size was 150, the same for all the methods.

5.4 Experimental Results and Discussion

Figure 5 presents quantification errors on the four datasets when the quantifiers were trained on their respective entire training dataset. In all the experimental results, we used the following legends. DQN-R and DQN-Z denote DQN using the random selection and the selection with Zipf distribution to generate tuplets as previously discussed, respectively. The suffixes, CON, AVG, MED, MAX, NN denote the tuplet feature learning/extraction method. We have five findings as follows. (1) Our DQN variants perform better than all the other compared methods regardless of binary and multi-class quantification tasks. On average across the four datasets, DQN-Z-NN gives 35% lower MAE achieved by the best existing methods. See finding (5) below. DQN learns good feature representations for quantification and the tuplet generation strategy is able to utilize combinations of individual training instances to generate many tuplets for optimizing parameters to avoid overfitting with the training class distribution. (2) DQN-Z consistently gives lower quantification errors than DQN-R does in all

	Binary quantification						Multi-class quantification					
	IMDB			YELP			AG-NEWS			20-NEWS		
	MAE	RMAE	KLD	MAE	RMAE	KLD	MAE	RMAE	KLD	MAE	RMAE	KLD
CC	0.135	0.316	0.047	0.128	0.284	0.043	0.101	0.221	0.038	0.054	0.142	0.024
PCC	0.154	0.361	0.064	0.146	0.325	0.059	0.116	0.253	0.051	0.062	0.162	0.032
ACC	0.077	0.242	0.022	0.076	0.237	0.028	**0.051**	**0.150**	**0.014**	0.031	0.109	0.011
PACC	**0.068**	**0.214**	**0.017**	0.073	0.218	0.020	0.058	0.169	0.018	0.027	0.096	0.009
HA	0.074	0.233	0.020	**0.070**	**0.210**	**0.018**	0.056	0.163	0.016	0.030	0.105	0.010
ReadMe	0.072	0.226	0.019	0.075	0.236	0.028	0.052	0.153	0.013	**0.026**	**0.094**	**0.008**
DQN-R-CON	0.062	0.195	0.013	0.059	0.176	0.012	0.047	0.137	0.010	0.025	0.088	0.007
DQN-R-AVG	0.065	0.204	0.014	0.062	0.184	0.013	0.049	0.143	0.011	0.026	0.092	0.007
DQN-R-MED	0.064	0.201	0.014	0.061	0.181	0.013	0.048	0.141	0.011	0.026	0.090	0.007
DQN-R-MAX	0.055	0.173	0.010	0.052	0.156	0.009	0.041	0.121	0.008	0.022	0.078	0.005
DQN-R-NN	0.050	0.157	0.008	0.047	0.141	0.007	0.038	0.110	0.006	0.020	0.071	0.004
DQN-Z-CON	0.053	0.167	0.009	0.050	0.150	0.008	0.040	0.117	0.007	0.021	0.075	0.005
DQN-Z-AVG	0.056	0.176	0.011	0.052	0.156	0.009	0.042	0.123	0.009	0.022	0.079	0.006
DQN-Z-MED	0.055	0.173	0.010	0.053	0.158	0.010	0.041	0.121	0.008	0.022	0.078	0.005
DQN-Z-MAX	0.047	0.147	0.007	0.045	0.132	0.006	0.035	0.103	0.006	0.019	0.066	0.004
DQN-Z-NN	***0.044***	***0.138***	***0.006***	***0.042***	***0.124***	***0.005***	***0.033***	***0.097***	***0.005***	***0.018***	***0.062***	***0.003***
Error reduced	35%	36%	53%	40%	41%	72%	35%	35%	64%	31%	34%	63%

Fig. 5. Quantification errors on different quantification tasks.

four datasets. We trained DQN-Z with different class distributions of training tuplets so that it learns feature representations and parameter values for diverse class distributions that may occur in the test dataset. We recommend using DQN-Z to predict class distributions, especially in the applications that expect the class ratios to change significantly or periodically. (3) Fig. 5 also shows that DQN-Z-NN (with the dense layer of 256 neurons for tuplet feature learning) achieves the lowest quantification errors among all the variants. The additional dense layer can extract better tuplet feature representations. (4) Among the remaining tuplet feature extraction methods, MAX consistently gives the lowest error below those of CAT, AVG, and MED. (5) The best existing methods are as follows: PACC for IMDB, HA for YELP, ACC for AG-NEWS, and ReadMe for 20-NEWS. The 20-NEWS dataset has long formal documents similar to the datasets was originally investigated with ReadMe [13]. Nevertheless, both DQN-R and DQN-Z gave the lowest quantification errors in terms of three metrics among all the compared methods across all four datasets. The p-values of the two-tailed paired t-test (between the best of the existing methods and DQN-Z-NN) results are 9.64E−7, 3.34E−7, 3.38E−6, 3.31E−6 for IMDB, YELPS, AG-NEWS and 20-NEWS, respectively, indicating that the improvement made by DQN-Z-NN is statistically significant at the 95% confident interval.

5.5 Sensitivity Analysis

We demonstrate the impact of training dataset sizes and tuplet sizes on DQN using IMDB (binary quantification) and AG-NEWS (multi-class quantification). We used the same strategy as mentioned in Sect. 5.3 for creating test datasets artificially from the original test datasets to evaluate the performance of a quantifier on different class ratios and report the average.

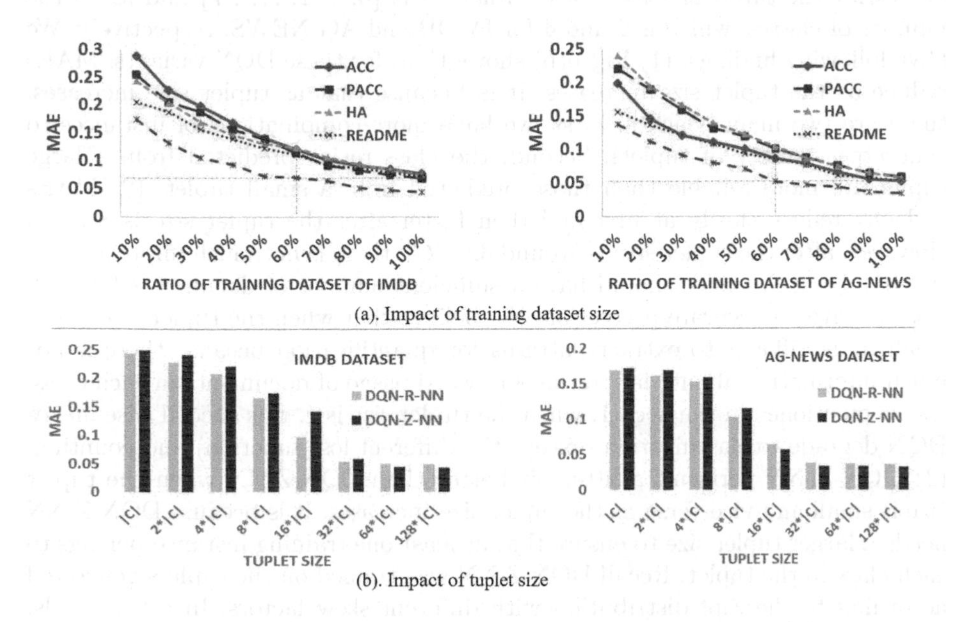

(a). Impact of training dataset size

(b). Impact of tuplet size

Fig. 6. Impact of training dataset size and tuplet size on DQN.

Impact of the Training Dataset Size on DQN: Our goal is to determine robustness of different methods to training data, which has practical impact on minimizing time-consuming manual labeling effort. Figure 6(a) shows that ACC and PACC perform consistently much better than CC and PCC, and DQN-Z-NN gives the lowest quantification errors. Therefore, we selected these methods, ACC, PACC, HA, ReadMe, and DQN-Z-NN to evaluate their performance under different training set sizes. We varied the percentage of the training dataset used to train the five quantifiers. We randomly selected $p\%$ documents from the entire training dataset to create the limited training dataset according to the specified percentage $p \in \{10, 20, \ldots, 100\}$. We trained each of the quantifiers using the limited training dataset and calculated MAE. Figure 6(a) shows that DQN consistently offers the lowest MAEs. For IMDB, ACC and PACC suffer the most from the limited training data. For AG-NEWS with short text documents, HA and PACC suffer the most. These methods rely on correct prediction of individual documents to estimate the class distribution. Both direct quantification methods (ReadMe and DQN) are less impacted by the decrease in the training dataset size. With around 60% of the entire training data, DQN reduces MAE significantly down to around 0.07 and 0.051 for IMDB and AG-NEWS, respectively. To offer the similar MAE, the other methods need 90% or more of the training data.

Impact of the Tuplet Size on DQN: We investigated DQN-R and DQN-Z with the NN tuplet feature learning when varying tuplet sizes on two datasets. We varied the tuplet size as $2^n * |C|$ where n is $\{0, 1, 2, \ldots, 7\}$ and $|C|$ is the number of classes, which is 2 and 4 for IMDB and AG-NEWS, respectively. We have following findings. (1) Fig. 6(b) shows that, for these DQN variants, MAEs reduce as the tuplet size increases. It is because that as tuplet size increases, there are two main benefits. First, we have more combinations of instances to generate a variety of tuplets. Second, the class ratios predicted from a large tuplet are more reliable than those predicted from a small tuplet. (2) MAEs of DQN reduce slowly at first and then faster after the tuplet size is $8 * |C|$. However, after the tuplet size is around $32 * |C|$, there is not much improvement. Intuitively, each tuplet should have a sufficient number of documents for each class to give an accurate prediction. We believe that when the tuplet size is too small, it is difficult to extract patterns for quantification because there is not much information about class ratios and word usage of documents in each class. As we mentioned in the Sect. 1, when the tuplet size is 1, this special case makes DQN degrade to classification (except the different loss function) and counting. (3) DQN-R-NN performs a litter bit better than DQN-Z-NN when the tuplet size is small and vice versa as the tuplet size increases. It is because DQN-Z-NN needs a larger tuplet size to ensure that at least one training instance belongs to each class in the tuplet. Recall DQN-Z-NN was trained on the tuplets generated according to the Zipf distribution with different skew factors. In other words, most of tuplets have a severe class imbalance. Therefore, we recommend using a large tuplet size when expecting a severe imbalance.

6 Conclusion and Future Work

We present the first attempt to use deep learning for quantification tasks to estimate the class distribution of a given dataset. We introduce DQN—a framework for deep quantification learning applicable for various quantification tasks. We present extensive evaluation results of DQN on four public datasets against six existing methods in all categories in the literature. DQN outperforms these methods on different types of text data and tasks, especially when the training dataset is small. We performed sensitivity analyses on important parameters of DQN. However, our work has some limitations. We did not evaluate DQN performance on other data modalities such as audio, image, and graph data. Nevertheless, we expect DQN to perform well when using an appropriate deep model for learning effective feature representations of the data. Second, interpretation of DQN models in reaching the predicted class distribution is challenging and critical to increase trust and transparency of the black-box quantification model. We will explore these issues in our future work.

References

1. Quionero-Candela, J., Sugiyama, M., Schwaighofer, A., Lawrence, N.D.: Dataset Shift in Machine Learning. The MIT Press, Cambridge (2009)

2. Barranquero, J., et al.: On the study of nearest neighbor algorithms for prevalence estimation in binary problems. Pattern Recognit. **46**(2), 472–82 (2013)
3. Asoh, H., et al.: A fast and simple method for profiling a population of twitter users. In: The Third International Workshop on Mining Ubiquitous and Social Environments. Citeseer (2012)
4. Buck, A.A., Gart, J.J.: Comparison of a screening test and a reference test in epidemiologic studies. II. A probabilistic model for the comparison of diagnostic tests. Am. J. Epidemiol. **83**(3), 593–602 (1966)
5. Forman, G.: Quantifying trends accurately despite classifier error and class imbalance. In: Proceedings of the 12th ACM SIGKDD International Conference on Knowledge Discovery and Data Mining (2006)
6. González, P., Castaño, A., Chawla, N.V., Coz, J.J.D.: A review on quantification learning. ACM Comput. Surv. (CSUR) **50**(5), 74 (2017)
7. Bishop, C.M.: Pattern Recognition and Machine Learning. Springer, Heidelberg (2006)
8. Hofer, V., Krempl, G.: Drift mining in data: a framework for addressing drift in classification. Comput. Stat. Data Anal. **57**(1), 377–391 (2013)
9. King, G., Lu, Y.: Verbal autopsy methods with multiple causes of death. Stat. Sci. **23**(1), 78–91 (2008)
10. González-Castro, V., Alaiz-Rodríguez, R., Fernández-Robles, L., Guzmán-Martínez, R., Alegre, E.: Estimating class proportions in boar semen analysis using the Hellinger distance. In: García-Pedrajas, N., Herrera, F., Fyfe, C., Benítez, J.M., Ali, M. (eds.) IEA/AIE 2010. LNCS (LNAI), vol. 6096, pp. 284–293. Springer, Heidelberg (2010). https://doi.org/10.1007/978-3-642-13022-9_29
11. Forman, G.: Counting positives accurately despite inaccurate classification. In: Gama, J., Camacho, R., Brazdil, P.B., Jorge, A.M., Torgo, L. (eds.) ECML 2005. LNCS (LNAI), vol. 3720, pp. 564–575. Springer, Heidelberg (2005). https://doi.org/10.1007/11564096_55
12. Forman, G.: Quantifying counts and costs via classification. Data Min. Knowl. Discov. **17**(2), 164–206 (2008)
13. Hopkins, D.J., King, G.: A method of automated nonparametric content analysis for social science. Am. J. Polit. Sci. **54**(1), 229–247 (2010)
14. Bella, A., Ferri, C., Hernández-Orallo, J., et al.: Quantification via probability estimators. In: 2010 IEEE International Conference on Data Mining, pp. 737–742. IEEE (2010)
15. Milli, L., Monreale, A., Rossetti, G., et al.: Quantification trees. In: ICDM, pp. 528–536. IEEE (2013)
16. Esuli, A., Sebastiani, F.: Optimizing text quantifiers for multivariate loss functions. ACM Trans. Knowl. Discov. Data **9**(4), 27:1–27:27 (2015)
17. Barranquero, J., et al.: Quantification-oriented learning based on reliable classifiers. Pattern Recognit. **48**(2), 591–604 (2015)
18. Joachims, T.: A support vector method for multivariate performance measures. In: Proceedings of the 22nd International Conference on Machine Learning (2005)
19. Pérez-Gállego, P., Quevedo, J.R., del Coz, J.J.: Using ensembles for problems with characterizable changes in data distribution: a case study on quantification. Inf. Fusion **34**, 87–100 (2017)
20. Goodfellow, I., Bengio, Y., Courville, A.: Deep Learning. MIT Press, Cambridge (2016)
21. Endres, D.M., Schindelin, J.E.: A new metric for probability distributions. IEEE Trans. Inf. Theory **49**(7), 1858–1860 (2003)

22. Goodfellow, I., et al.: Generative adversarial nets. In: NIPS, pp. 2672–2680 (2014)
23. Zipf, G.K.: Human Behavior and the Principle of Least Effort: An Introduction to Human Ecology. Ravenio Books, Cambridge (2016)
24. Hochreiter, S., Schmidhuber, J.: Long short-term memory. Neural Comput. **9**(8), 1735–1780 (1997)
25. Maas, A.L., et al.: Learning word vectors for sentiment analysis. In: ACL, pp. 142–150 (2011)
26. Zhang, X., et al.: Character-level convolutional networks for text classification. In: NIPS, pp. 649–657 (2015)
27. Lang, K.: NewsWeeder: learning to filter netnews. In: Machine Learning Proceedings, pp. 331–339 (1995)

PROMO for Interpretable Personalized Social Emotion Mining

Jason (Jiasheng) Zhang(✉) and Dongwon Lee

The Pennsylvania State University, State College, USA
{jpz5181,dongwon}@psu.edu

Abstract. Unearthing a set of users' collective emotional reactions to news or posts in social media has many useful applications and business implications. For instance, when one reads a piece of news on Facebook with dominating "angry" reactions, or another with dominating "love" reactions, she may have a general sense on how social users react to the particular piece. However, such a collective view of emotion is unable to answer the subtle differences that may exist among users. To answer the question "which emotion who feels about what" better, therefore, we formulate the *Personalized Social Emotion Mining (PSEM)* problem. Solving the PSEM problem is non-trivial in that: (1) the emotional reaction data is in the form of ternary relationship among user-emotion-post, and (2) the results need to be *interpretable*. Addressing the two challenges, in this paper, we develop an expressive probabilistic generative model, PROMO, and demonstrate its validity through empirical studies.

Keywords: Personalized Social Emotion Mining · Ternary relationship data · Probabilistic generative models

1 Introduction

It has become increasingly important for businesses to better understand their users and leverage the learned knowledge to their advantage. One popular method for such a goal is to mine users' digital footprints to unearth latent "emotions" toward particular products or services. In this paper, we focus on such a problem, known as *social emotion mining* (**SEM**) [24], to uncover latent emotions of social media posts, documents, or users. Note that SEM is slightly different from the conventional *sentiment analysis* (**SA**).

First and foremost, SEM is to unearth the latent emotion of a *reader* to a social post while SA is to detect underlying emotion of a *writer* of a social post or article. Therefore, for instance, knowing viewers' positive emotion toward a YouTube video for its frequent thumbs-up votes is the result of SEM, while knowing an underlying negative emotion of a news editorial is the result of SA. Second, in objectives, the goal of SEM is to find *which emotion people feel about what*, while SA aims to find *which emotion an author conveys through what* (referring to text in this work); Third, in methods, SEM learns the correlation between

F. Hutter et al. (Eds.): ECML PKDD 2020, LNAI 12457, pp. 249–265, 2021.
https://doi.org/10.1007/978-3-030-67658-2_15

topics of text and emotions while SA needs to search for or learn emotional structures, such as emotional words in text. For example, SEM is to predict the aggregated emotional reactions to a news article based on its content while an SA task may attempt to predict the emotions of customers based on the reviews that they wrote.

While useful, note that current SEM methods capture the emotional reactions of crowds as aggregated votes for different emotions. This brings two difficulties in understanding people's emotions in a finer granularity. On one hand, different people have their own ways to express similar emotions. For example, reading a sad story, some will react with "sad," but others may choose "love," both for expressing sympathy. When such emotions are aggregated, the cumulative emotions could be noisy. On the other hand, different people do not have to feel the same toward similar content. For instance, for the same Tweet, republican and democratic supporters may react opposite. In this case, no single emotion can represent the emotion of crowds well [3].

To close this gap, by observing personal difference as their common source, we formulate the novel *Personalized Social Emotion Mining* (**PSEM**) problem. Compared with SEM, here our goal is to find the hidden structures among the three dimensions, *users-emotions-posts*, in the emotional reactions data. The hidden structures are the clusters of different dimensions (i.e., groups of users, topics of posts) and the connections (i.e., the emotional reactions) among them. In short, we want to know *which emotion who feels about what.* This additional "who" helps reveal the differences in expression among users and also conserves the diversity in opinions. Such new knowledge in personalized emotion may help, for instance, business to provide personalized emotion based recommendation.

Note that it is nontrivial to model the ternary relationship of users-emotions-posts. More common relationship is binary (e.g., network, recommendation), where the closeness between the two entities (e.g., a user assigns 5 stars to an item) is modeled. The post dimension can be further expanded to the textual content of each post, which helps answer not only which post but also which topic triggers certain emotions. Moreover, because the task is to manifest the hidden structures of the data, the model applied should be well interpretable, which excludes the direct adoption of existing ternary relation learning methods [10,12,18,19,27,29,31,33]. To the best of our knowledge, no existing models can satisfy the two requirements simultaneously.

To solve these challenges, in this work, we propose the **PROMO** (PRobabilistic generative cO-factorization MOdel), that learns the hidden structures in emotional reaction data by co-factorizing both ternary relationships and post-word binary relationships with a probabilistic generative model. PROMO is both *interpretable* and *expressive.* We especially showcase the interpretability of PROMO by the hidden structures of the emotional reactions revealed from two real-world dataset, which is not available by existing methods. In addition, the empirical experiments using two supervised learning tasks validate PROMO by its improved performance against state-of-the-art competing methods.

2 Related Work

As PSEM is a newly formulated problem, in this section, we review related literature in three aspects: (1) the social emotion mining problem (SEM) is related to PSEM as a real-world problem; (2) the ternary relationship data modeling problem is one of the challenges of the PSEM problem; and (3) probabilistic generative models are commonly applied to textual data when interpretability is concerned.

Social Emotion Mining (SEM). Previous works on SEM [4,5,17,22,23,25, 30,35,38] attempt to find a mapping from online post content (i.e., textual) to the cumulative votes for different emotions from the users. Different forms of the representation of the emotional reactions are studied. For example, [5] uses the normalized cumulative votes for different emotions as the emotion distribution of a post, [4,17,25] focus on the dominating emotion tags for posts, which leads to a classification problem; [23,35] treat the emotional reactions as a ranking given the emotional tags and solve the label ranking problem with the imbalance challenge [35]. However, none of the existing works have added personalized views to SEM.

A few works have studied personalized emotion perception [2,28,34,36,37], which is to predict readers' perception of the emotional expressions in content. For example, given an image of flowers as content, the task is to predict whether a reader will tag keywords, expressing love or happiness. In such a case, the perception of emotion of objective content is less subjective compared with news content, such as political news in PSEM. As a result, methods used in personalized emotion perception [2,28,34,36,37], which does not explicitly consider the users-emotions-posts ternary relationship, are not a good fit to solve PSEM.

Ternary Relationship Data Modeling. One challenge of the PSEM problem is to model the ternary relationship data. Most previous methods are based on tensor factorization, such as Tucker decomposition [31] and CP (CANDECOMP/PARAFAC) decomposition [8]. Some are based on intuition from knowledge base, such as TransE [7], which still can be transformed into a factorization model [33]. A more advanced model based on neural network, the neural tensor network, is proposed in [29]. Such multi-relational learning methods have been applied to the tag recommendation [10], context-aware collaborative filtering [19], drug-drug interaction [18] and knowledge graph completion [7,26,29].

The addition of side information of post textual content brings more methods into our scope. For example, factorization machines [27] is a powerful method to model sparse interactions. When each user is treated as a task, multi-task learning methods [1,12] are also investigated for personalized sentiment classification. However, the models currently used in the multi-relational learning problem are *not* interpretable enough to manifest the hidden structures of the data to answer the PSEM problem.

Probabilistic Generative Models. Since the classical topic models PLSA [16] and LDA [6], probabilistic generative model becomes a popular tool to analyze text data. There exist other closely-related works using topic model to enhance recommendation system [11,14,32]. For example, [14] uses topic model to analyze the legislator-vote-bill network to find the ideal point of legislators. The difference with this work is that the vote relation is still one-dimensional. Generally, no existing probabilistic generative models are designed to model ternary relationships.

3 Problem Formulation

In a PSEM problem, data is in the form of tuples <user, emotion, post>, where posts are typically news represented by short textual messages, such as headlines. Thereafter, the post is also referred to as the document. In this work, the bag-of-word representation is taken for documents.

Formally, there are four sets of nodes, U users $\mathcal{U} = \{\mu \in [U]\}$, where $[U] = \{1, 2, ..., U\}$, D documents $\mathcal{D} = \{d \in [D]\}$, E emotion labels $\mathcal{E} = \{e \in [E]\}$, and V distinct words $\mathcal{V} = \{v \in [V]\}$. There are two kinds of relationships. $R_e \subseteq \mathcal{U} \times \mathcal{E} \times \mathcal{D}$ is the collection of user emotional reactions to documents, where each document has M_d emotional reactions $\{\epsilon_{dm} | m \in [M_d]\}$ from different users $\{u_{dm} | m \in [M_d]\}$; R_w is the document-word relationship. $R_w \subseteq \mathcal{D} \times \mathcal{V}$, where each document has N_d words $\{w_{dn} | n \in [N_d]\}$. The problem framework is visualized in Fig. 1. The annotation used is summarized in Table 1

Problem 1 (PSEM (Personalized Social Emotion Mining)). *Given users $\mathcal{U}$, documents $\mathcal{D}$, emotion labels $\mathcal{E}$ and vocabulary $\mathcal{V}$, find the hidden structures among them from relationships R_e and R_w.*

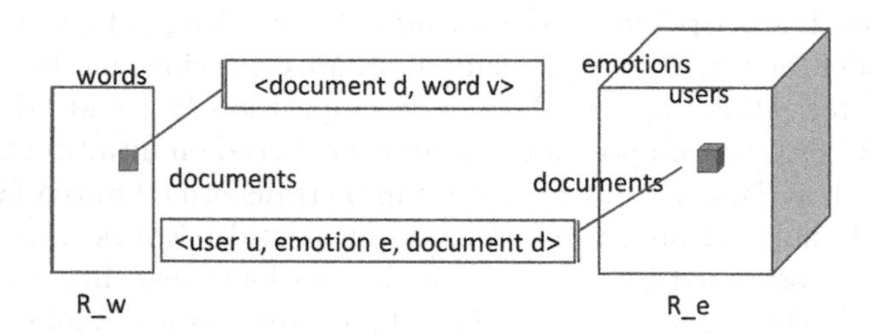

Fig. 1. PSEM data structure

4 Methodology

We propose a probabilistic generative co-factorization model (PROMO) to model the emotional reaction data. The probabilistic generative model itself provides a

Table 1. Annotation summary

Symbol	Description
R_e	The user-emotion-document relationship data
R_w	The document-word relationship data
$\mathcal{U}$	User set, of size $U = \|\mathcal{U}\|$, indexed by μ
$\mathcal{D}$	Document (post) set, of size $D = \|\mathcal{D}\|$, indexed by d
$\mathcal{E}$	Emotion label set, of size $E = \|\mathcal{E}\|$, indexed by e
$\mathcal{V}$	Vocabulary, of size $V = \|\mathcal{V}\|$, indexed by v
N_d	Number of words in the document d
M_d	Number of emotional reactions to the document d
u_{dm}	A user, as a U-dimension one-hot vector
ϵ_{dm}	An emotional reaction, as an E-dimension one-hot vector
w_{dn}	A word, as a V-dimension one-hot vector
K	Hyperparameter as the number of topics
G	Hyperparameter as the number of groups
θ	Corpus-wise topic distribution
ϕ	Topic-word distribution
ψ	User-group distribution
η	Group-topic-emotion distribution
z_d	Topic indicator, as a K-dimension one-hot vector
x_{dm}	Group indicator, as a G-dimension one-hot vector
α	Hyperparameter for the Dirichlet prior of θ
β	Hyperparameter for the Dirichlet prior of ϕ
ζ	Hyperparameter for the Dirichlet prior of ψ
γ	Hyperparameter for the Dirichlet prior of η

straightforward way to manifest the hidden structure of data, which meets the interpretability requirement. The two modules of PROMO to model the two relationships, R_w and R_e, are described separately, followed by the complete model description. In addition, the real-world interpretation of PROMO is discussed. After model construction, the inference algorithm is derived using stochastic variational inference. To valid the model, we show how to apply PROMO to two supervised learning tasks. Finally, the relationship between PROMO and existing models is discussed.

4.1 A Module for Short Documents

The most typical posts in the PSEM problem are news message posted by news channels in social media, such as CNN[1] in Facebook. These messages are usually

[1] www.facebook.com/cnn.

short. Adopting the idea that there is only a single topic for each such short document in social media [9], the document-word relationship R_w is modeled as below.

For each document d, a K-dimensional one-hot variable z_d is associated to d, representing the unique topic of d. z_d is generated by a corpus-wise topic distribution, which is a K-dimensional multinomial distribution parameterized by θ, that $\forall k, \theta^k \geq 0, \sum_k \theta^k = 1$. Without ambiguity, the corpus-wise topic distribution is referred to as its parameter θ, similarly for other distributions introduced in this work. θ is generated according to a symmetric Dirichlet distribution parameterized by α.

Consistently with conventional topic model, the topic z_d generates the n_d words of the document d, i.i.d., according to the topic-word distributions parameterized by ϕ. For each topic $k \in [K]$, the topic-word distribution is a multinomial distribution parameterized by ϕ_k, that $\forall v \in [V]\, \phi_k^v \geq 0, \sum_v \phi_k^v = 1$ and ϕ_k is generated according to a symmetric Dirichlet distribution parameterized by β.

4.2 A Module for Ternary Relationships

Extended from the module for documents, each document d is summarized by its topic z_d.

Inspired by the social norm and latent user group theories [12], we assume users form G groups. For each emotional reaction ϵ_{dm}, indexed as $m \in [M_d]$ in reactions to the document d, the user $u_{dm} \in [U]$ [2] belongs to one group, represented by G-dimensional one-hot variable x_{dm}. x_{dm} is generated by u_{dm}, according to the user-group distributions parameterized by ψ. For each user $\mu \in [U]$, the user-group distribution is a multinomial distribution parameterized by ψ_u, that $\forall g \in [G]\, \psi_\mu^g \geq 0, \sum_g \psi_\mu^g = 1$. In other words, the group x_{dm} to which a user u_{dm} belongs is generated i.i.d according to $\psi_{u_{dm}}$, when all reactions from a user are considered. For each user μ, ψ_μ is generated according to a symmetric Dirichlet distribution parameterized by ζ.

To model the emotional reactions from different users toward different documents, we assume that the users from the same group react the same to the documents of the same topic. Formally, each emotional reaction ϵ_{dm} is generated by the combination of the topic of the document d, z_d and the group of the user u_{dm}, according to the group-topic-emotion distributions parameterized by η. For each topic k and group g, the group-topic-emotion distribution is a multinomial distribution parameterized by η_{gk}, that $\forall e \in [E]\, \eta_{gk}^e \geq 0, \sum_e \eta_{gk}^e = 1$ and η_{gk} is generated according to a symmetric Dirichlet distribution parameterized by γ.

4.3 PROMO: PRobabilistic Generative cO-factorization MOdel

Our final PROMO model is made by combining two aforementioned modules. The annotations are summarized in Table 1. The graphic model representation and the generative process of PROMO is summarized below.

[2] The U-dimensional one-hot variable u_{dm} is also used as its index of the non-zero entry interchangeably, which is applied to all one-hot variables in this work.

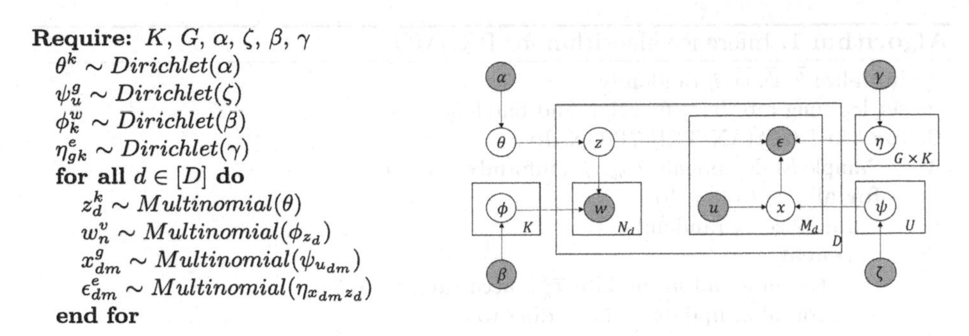

Fig. 2. Generative process and graphical model representation of PROMO

4.4 Interpretation

The discrete latent variable structure of PROMO gives clear translation of PSEM problem. *Which emotion who feels about what* is translated to *which emotion the user from which group feels about document about which topic*, which can be interpreted from PROMO model variables. "Users from which group" is the user-group distribution ψ; "document about which topic" is the document topic z; and η can be interpreted as which emotion a group of users will feel about which topic of documents, which carries the core hidden structure of the emotional reaction data, that is, the answer toward the PSEM problem (Fig. 2).

4.5 Inference

For the inference of PROMO with data, the stochastic variational inference [15] method is used. In the PSEM problem, the number of emotional reactions per document M can be very large (e.g., several thousands), which makes the collapsed Gibbs sampling [13], a more easily derivable inference method, too slow, due to the sequential sampling of each hidden variables that are not collapsed. With stochastic variational inference, within one document, the same type of latent variables can be inferred in parallel; and also the inference of each document within a batch can be trivially parallerized.

To approximate the intractable posterior distribution, we use a fully decomposable variational distribution q,

$$p(\theta, \psi, \phi, \eta, z, x | w, \epsilon; \Theta) \approx q(\theta, \psi, \phi, \eta, z, x | \overline{\theta}, \overline{\psi}, \overline{\eta}, \overline{z}, \overline{x}), \tag{1}$$

where Θ represents the set of hyperparameters $\{\alpha, \beta, \zeta, \gamma\}$, $\bar{\cdot}$ (e.g., $\overline{\theta}$, $\overline{\psi}$) are the parameters of q, the approximation is in terms of KL-divergence $D(q||p)$ and q can be decomposed as $q = q(\theta|\overline{\theta})q(\psi|\overline{\psi})q(\phi|\overline{\phi})q(\eta|\overline{\eta})q(z|\overline{z})q(x|\overline{x})$. For the variables as the parameters of multinomial distributions in PROMO, that is, θ, ψ, ϕ and η, the variational distributions are Dirichlet distributions; for the one-hot variables, z and x, the variational distributions are multinomial distributions.

Algorithm 1. Inference algorithm for PROMO

1: Initialize $\overline{\theta}$, $\overline{\psi}$, $\overline{\phi}$, $\overline{\eta}$ randomly.
2: set learning rate $lr(t)$ function and batch size bs.
3: **for** t in 1 **to** MAXITERATION **do**
4: Sample bs documents $\mathcal{D}_{batch}$ uniformly from data.
5: **for all** $d \in \mathcal{D}_{batch}$ **do**
6: Initialize $\overline{z}_d$ randomly.
7: **repeat**
8: For all g and m, update $\overline{x}^g_{dm}$ according to eq.3
9: For all k, update $\overline{z}^k_d$ according to eq.4
10: **until** converge
11: **end for**
12: **for all** par in $\{\overline{\theta}, \overline{\psi}, \overline{\phi}, \overline{\eta}\}$ **do**
13: Update par^* according to eq.5
14: Update $par \leftarrow (1 - lr(t))par + lr(t)par^*$
15: **end for**
16: **end for**

The inference task, calculating the posterior distribution p is then reduced to finding the best variational distributions,

$$(\overline{\theta}^*, \overline{\psi}^*, \overline{\phi}^*, \overline{\eta}^*, \overline{z}^*, \overline{x}^*) = argmin_{\overline{\theta}, \overline{\psi}, \overline{\phi}, \overline{\eta}, \overline{z}, \overline{x}}(D(q||p)). \tag{2}$$

The optimization in Eq. 2 is done by iteratively optimizing each parameter. Readers who are interested in derivation detail can refer to the previous stochastic variational inference works [6,15]. The update rules for those parameters are followed. Within each document d, for each emotional reaction indexed by m, the group distribution of the user u_{dm} is updated as

$$\overline{x}^g_{dm} \propto exp(\sum_{\mu} u^{\mu}_{dm} F(\overline{\psi}_{\mu})^g + \sum_k \sum_e \overline{z}^k_d \epsilon^e_{dm} F(\overline{\eta}_{gk})^e); \tag{3}$$

and the topic distribution of the document d is updated as

$$\overline{z}^k_d \propto exp(F(\overline{\theta})^k + \sum_n \sum_v w^v_{dn} F(\overline{\phi}_k)^v + \sum_m \sum_g \sum_e \overline{x}^g_{dm} \epsilon^e_{dm} F(\overline{\eta}_{gk})^e), \tag{4}$$

where $F(y)^l = \Psi(y^l) - \Psi(\sum_l y^l)$, with $\Psi()$ the digamma function. For corpus-level parameters, the updating rules are

$$\begin{aligned} \overline{\theta}^{k*} &= \alpha + \frac{D}{bs} \sum_{d \in \mathcal{D}_{batch}} \overline{z}^k_d, & \overline{\psi}^{g*}_{\mu} &= \zeta + \frac{D}{bs} \sum_{d \in \mathcal{D}_{batch}} \sum_m^{M_d} u^{\mu}_{dm} \overline{x}^g_{dm}, \\ \overline{\phi}^{v*}_k &= \beta + \frac{D}{bs} \sum_{d \in \mathcal{D}_{batch}} \sum_n^{N_d} w^v_{dn} \overline{z}^k_d, & \overline{\eta}^{e*}_{gk} &= \gamma + \frac{D}{bs} \sum_{d \in \mathcal{D}_{batch}} \sum_m^{M_d} \epsilon^e_{dm} \overline{x}^g_{dm} \overline{z}^k_d, \end{aligned} \tag{5}$$

where $\mathcal{D}_{batch}$ is the set of documents in a mini-batch and $bs = |\mathcal{D}_{batch}|$.

The complete stochastic variational inference algorithm for PROMO is shown in Algorithm 1.

4.6 Supervised Learning Tasks

In order to validate the ability of the PROMO model to reveal the hidden structures of the PSEM data, we propose two supervised learning tasks which the PROMO can be applied to. The two tasks test whether the hidden structures PROMO reveals can be generalizable to unseen data.

1. Warm-start emotion prediction. This task is to predict the emotional reaction given a user and a document. It is called warm-start, because both the user and the document exist in R_e for training. For each user $\mu \in [U]$, document $d \in [D]$ and without loss of generality, indexing the emotional reaction as m, the posterior probability of $p(\epsilon_{dm}|\mu, d, R_w, R_e; \Theta)$ can be calculated in PROMO as

$$\int dz_d dx_{dm} d\eta d\psi\, p(\epsilon_{dm}|z_d, x_{dm}, \eta) p(z_d, \eta|\mu, R_w, R_e; \Theta) p(x_{dm}, \psi|\mu, R_w, R_e; \Theta) \tag{6}$$

where the two posterior distributions of z_d and x_{dm} can be replaced with the fitted variational distributions as $p(z_d, \eta|\mu, R_w, R_e; \Theta) \approx q(z_d|\overline{z}_d)q(\eta|\overline{\eta})$ and $p(x_{dm}, \psi|\mu, R_w, R_e; \Theta) \approx p(x_{dm}|\mu, \psi)q(\psi|\overline{\psi})$. Finally, because of the one-hot property of z_d, ϵ_{dm} and x_{dm}, the posterior distribution $p(\epsilon_{dm}|\mu, d, R_w, R_e; \Theta)$ can be derived as

$$\prod_e (\sum_k \sum_g \overline{z}_d^k \langle\eta\rangle_{gk}^e \langle\psi\rangle_\mu^g)^{\epsilon_{dm}^e}, \tag{7}$$

where for any $f \in \{\theta, \psi, \phi, \eta\}$, $\langle f \rangle$ is the variational mean of f, which is $\langle f \rangle^l = \overline{f}^l / \sum_{l'} \overline{f}^{l'}$ for $f \sim Dirichlet(\overline{f})$.

2. Cold-start emotion prediction. With the module for documents, PROMO can be applied to predicting the emotional reaction given a user and a new document. For a new document d with words $w_d = \{w_{dn}|n \in [N_d]\}$, the posterior probability of $<\mu, \epsilon_{dm}, d>$ can be calculated followed the similar derivations for Eq. 6 and Eq. 7, as

$$p(\epsilon_{dm}|\mu, w_d, R_w, R_e; \Theta) \approx \prod_e (\sum_k \sum_g \widehat{z}_d^k \langle\eta\rangle_{gk}^e \langle\psi\rangle_\mu^g)^{\epsilon_{dm}^e}, \tag{8}$$

where $\widehat{z}_d$ is the estimated topic distribution of d, which can be calculated as $\widehat{z}_d^k \propto \prod_v \langle\phi\rangle_k^{v(\sum_n^{N_d} w_{dn}^v)} \langle\theta\rangle^k$. Compared with the warm-start prediction, Eq. 7, where $\overline{z}_d$ is inferred from both the words and also the emotional reactions of the document d, the cold-start prediction, Eq. 8, uses $\widehat{z}_d$, which is estimated only from the words w_d of the document d.

4.7 The Relation to Existing Models

In PROMO, we introduce a group-topic-emotion distribution η to address the challenge of the ternary relationship user-emotion-document. The reconstruction of R_e in the warm-start emotion prediction (Sect. 4.6), Eq. 7, can be translated as the factorization of the $R_e \in \{0,1\}^{U \times E \times D}$ into the document latent vectors $\overline{z} \in [0,1]^{D \times K}$, the user latent vectors $\langle \psi \rangle \in [0,1]^{U \times G}$ and the emotion interaction core $\langle \eta \rangle \in [0,1]^{G \times K \times E}$. It is equivalent in terms of expressive power to the RESCAL [26] model, a variant of Tucker decomposition, proposed for multi-relational learning. It can be proved by observing that any R_e constructed with the RESCAL model can also be constructed with PROMO (i.e., Eq. 7) by applying rescale factors to the corresponding vectors in RESCAL.

One of the differences between PROMO and RESCAL is that the document latent vector in PROMO, $\overline{z}$ are inferred from both R_w and R_e, while the counterpart in RESCAL only from R_e. Therefore, in warm-start prediction task, R_w serves as the regularization. On the other hand, the generative architecture grants PROMO better interpretability.

5 Experimental Validation

In this section, we apply PROMO to real-world data to answer two questions: (1) *Interpretability*: what can be revealed from emotional reactions data by PROMO? (2) *Validity*: Is PROMO a valid model for emotional reactions data? All code[3] and dataset[4] are publicly available.

5.1 Data Description

We crawled the emotional reactions data from the public posts published in news pages of Facebook. More specifically, we crawled posts and corresponding user emotional reactions from Fox News[5] page from May 13th to October 17th, 2016, and CNN[6] page from March 1, 2016 to July 14, 2017. As for documents, we use the post message, which is short and headline-like text, appearing in most posts of news pages; as for emotional reactions, we use the emoticon labels that users click for the posts. Besides the two data sets, we combine them and keep posts within the same publication period into a new data set.

We excludes posts without any words. For each document, we remove URL's and stop words; and all numbers are replaced with a unique indicator "NUMBER", as number in news title is often informative. For emotional reactions, clicks of "like" are excluded due to its ambiguous meaning [21]; after that, to eliminate noise, only documents and users with more than 10 emotional reactions are used, which is 10-core decomposition. The resulting data statistics is shown in Table 2.

[3] http://github.com/JasonLC506/PSEM.
[4] http://tiny.cc/ecml20.
[5] http://www.facebook.com/FoxNews/.
[6] http://www.facebook.com/cnn/.

5.2 Interpretability: What Can PROMO Reveal?

We apply PROMO to COMBINE data with $K = 7$ and $G = 5$. After the inference, we visualize and analyze the topic-word distribution ϕ and the group-topic-emotion distribution η to describe the hidden structures of the emotional reaction data, which is the answer to the PSEM problem.

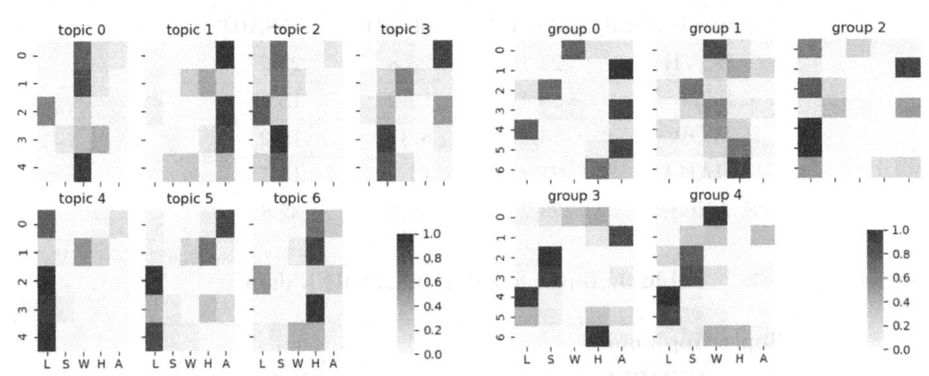

(a) Group-emotion distribution for each topic, where rows are groups, and columns are emotion labels

(b) Topic-emotion distribution for each group, where rows are topics, and columns are emotion labels

Fig. 3. The group-topic-emotion distributions are visualized as slices on the topic dimension and the group dimension, respectively. The marks of emotion labels are L for LOVE, S for SAD, W for WOW (surprise), H for HAHA, and A for ANGRY.

Topic Words. The posterior topic word distribution, $\langle\phi\rangle$, is visualized as top words for each topic in Table 3. We found four clusters of topics in terms of their top words. Topic 0 is about surprising news, topic 2 is about disaster news, topic 3 is about violence news and topic 1, 4, 5, 6 are all about political news. This phenomenon is due to the fact that the topic in PROMO is inferred from both R_w and R_e. Therefore, topics with similar topic-word distributions appear when their emotional reactions are highly different.

Emotional Reactions. First, the slices of the posterior group-topic-emotion distribution $\langle\eta\rangle$ on the topic dimension are visualized in Fig. 3a to show the user difference in emotional reactions within each single topic. For example, within topic 2, known as some disaster news from Table 3, cause most groups of users to react with SAD, while the users from the group 2 tend to react with LOVE. Similar example can be observed in topic 3, where two kinds of groups of users tend to use SAD and ANGRY, respectively. In addition, as an example of controversial topics, topic 5, related to political news, attracts LOVE from users in group 2, 3 and 4, but ANGRY from those in group 0.

Table 2. Data statistics

	CNN	Fox News	COMBINE
# users	869,714	355,244	524,888
# documents	23,236	3,934	10,981
# reactions	38,931,052	12,257,816	17,327,148
# emotion labels	5	5	5
# words (total)	300,135	66,241	158,104
% LOVE	23.4	26.8	25.9
% SAD	18.4	12.2	15.4
% WOW	13.1	8.3	10.2
% HAHA	21.4	14.2	15.7
% ANGRY	23.7	38.5	32.8

Table 3. Topics found on COMBINE data

Topic	Top words
0	NUMBER, world, one, new, years, life, found
1	clinton, hillary, trump, donald, said, j, NUMBER
2	NUMBER, one, police, people, died, said, us
3	NUMBER, police, people, said, one, new, killed
4	trump, j, donald, NUMBER, said, people, clinton
5	trump, clinton, hillary, j, donald, president, obama
6	trump, donald, j, clinton, hillary, NUMBER, said

Next, the slices of $\langle\eta\rangle$ on the group dimension are visualized in Fig. 3b to show the emotional reaction difference toward different topics from each single group of users. For example, users from group 0 tend to react with LOVE to the topic 4 but ANGRY to the topic 1, 3 and 5, though topics 1, 4 and 5 are similar in their topic-word distributions. Moreover, comparing group 1 and group 2 in general, users from group 1 generally prefer SAD, WOW and HAHA rather than emotions with more tension, such as LOVE and ANGRY, and vice versa for those from group 2. It shows that such results can also show the general difference between two groups of users.

5.3 Validity: Is PROMO Good for Emotional Reaction Data?

We answer this question by comparing the performance of PROMO and that of competing methods in two supervised learning tasks in real-world emotional reaction datasets.

Experimental Setting. For each data set, among all posts, 10% are randomly selected as the cold-start prediction test set, 5% as the validation set; Within the remaining posts, among all emotional reactions, 10% are randomly selected as the warm-start prediction test set, 5% as the validation set, and remaining as the training set for both tasks. The validation set is used for hyperparameter tuning and the convergence check for each task.

Evaluation Measures. We formulate both tasks as multi-class classification problems, given that there can be only one emotional reaction given a single <document, user> pair. As a result, macro AUC-precision-recall, $AUCPR$ is used as evaluation measure, which is known more robust in imbalanced case.

Warm-Start Emotion Prediction. We use tensor factorization based multi-relational learning methods as baselines to show that PROMO learns from text data. Besides, we use an adapted personalized emotion perception model, the factorization machine and a multi-task sentiment classification model to assert the necessity to explicitly consider the users-emotions-posts ternary relationship in PSEM.

- PROMO: the model proposed in this work.
- PROMO_NT: PROMO trained without text data of posts.
- RESCAL: model based on Tucker decomposition with relation dimension summed up with core tensor, which is similar to PROMO [26].
- CP: CANDECOMP/PARAFAC (CP) decomposition [8].
- MultiMF: separate matrix factorization for each relation type [21].
- NTN (Neural Tensor Network) [29]: a method combining tensor factorization and neural network to catch complex interactions among different dimensions.

The former five baselines (PROMO_NT, RESCAL, CP, MultiMF and NTN) are tensor factorization based multi-relational learning methods; while the followings make use of the textual content of the post.

- BPEP (Bayesian Personalized Emotion Perception): We adopt the Bayesian text domain module of the personalized Emotion perception work [28]. It builds naive Bayesian estimate of emotion distribution of each word for each user.
- FM (Factorization Machine): We use the rank-2 factorization machine [27] to model the interaction between textual content, posts and users. The feature is the concatenation of bag-of-word textual feature, post identifier and user identifier, both as one-hot vectors.
- MT-LinAdapt (Multi-Task Linear Model Adaptation) [12]: a state-of-the-art multi-task sentiment classification method that is shown to outperform its two-stage counterpart LinAdapt [1]. In our case, we adopt the setting that each user is a task. The bag-of-word textual feature is used as input for minimum comparison.
- CUE-CNN [2]: a state-of-the-art model for personalized sentiment classification, which employs convolutional neural network to capture textual content.
- CADEN [36]: a state-of-the-art model for personalized sentiment classification, which employs recurrent neural network to capture textual content.

Table 4. Results on warm-start emotion prediction task

	CNN	Fox News	COMBINE
PROMO	**0.809**	**0.779**	**0.791**
PROMO_NT	0.791	0.766	0.773
RESCAL	0.793	0.768	0.759
CP	0.776	0.753	0.760
MultiMF	0.748	0.697	0.698
NTN	0.805	0.685	0.733
BPEP	0.445	0.431	0.445
FM	0.785	0.774	0.773
MT-LinAdapt	0.741	0.750	0.733
CUE-CNN	0.787	0.706	0.736
CADEN	0.740	0.665	0.695

Table 5. Results on cold-start emotion prediction task

	CNN	Fox News	COMBINE
PROMO	0.601	**0.462**	**0.525**
PROMO_NG	0.520	0.404	0.424
BPEP	0.431	0.429	0.450
eToT	0.396	0.364	0.391
CUE-CNN	**0.625**	0.421	0.501
CADEN	0.565	0.388	0.453

All methods are adapted to use negative log likelihood as loss function for training. We implement RESCAL, CP, MultiMF, NTN, FM, MT-LinAdapt and CADEN using stochastic gradient descent with Adam [20]. We used validation set to search for the best hyperparameters for each model. For PROMO, $K = 50$ and $G = 64$, $\alpha = 1.0/K$, $\beta = 1.0$, $\gamma = 1.0$, $\zeta = 0.1$ and batch size is set to full batch; same setting for PROMO_NT, except $K = 30$ and $G = 16$.

As results shown in Table 4, PROMO consistently outperforms baseline methods in all data sets. From the comparison between PROMO and the tensor factorization based methods, we assert that PROMO learns from text data. On the other hand, the comparison between PROMO and BPEP, FM, MT-LinAdapt, CUE-CNN and CADEN supports that only text information, without explicitly considering the users-emotions-posts ternary relationship, is not enough in PSEM problem, which also distinguish PSEM problem from personalized sentiment analysis [1,12,36].

A detail check over the performance in different datasets provides some clue for the superiority of PROMO. From Table 4, in CNN dataset, the best of former five tensor factorization based methods, NTN, outperforms the latter three methods that focus more on textual information; while in Fox News and COMBINE datasets, the best of the latter three methods, FM, is comparable or better than the former five. This observation reveals the different contribution of users-emotions-posts interactions and textual information in different datasets (i.e., more contribution of the interactions for CNN while less in Fox News and COMBINE). However, the Bayesian architecture of PROMO provides an automatic balance between the two aspects, so that consistently outperforms others in all datasets.

Cold-Start Emotion Prediction. There are less existing works that can be applied to cold-start emotion prediction task. We test following competing models, besides PROMO, BPEP, CUE-CNN and CADEN described in the previous task.

- PROMO_NG: the PROMO with number of group set to 1, that is $G = 1$. This variant shows the situation when user difference is not considered.
- eToT: a probabilistic generative model for social emotion mining with temporal data, [38], excluding the temporal components, which is implemented using collapsed Gibbs sampling [13].

We used validation set to search for the best hyperparameters for each model. For PROMO, we set $K = 50$ and $G = 16$, $\alpha = 1.0/K$, $\beta = 0.01$, $\gamma = 100.0$, $\zeta = 0.1$ and batch size is set to full batch. For eToT, $K = 50$ is used.

As results shown in Table 5, PROMO outperforms other models in all but CNN dataset. Compared with that in warm-start task, models with more advanced textual feature extraction methods, i.e., CUE-CNN and CADEN perform much better. For example, CUE-CNN obtains a result even better than PROMO. However, those deep learning based model lost the interpretability of PROMO. In more detail, the improvement from PROMO_NG to PROMO supports the basic assumption of PSEM that users are different in emotional reaction. Besides, the comparable results between PROMO_NG and BPEP support the conclusion in the previous experiment, that only text information, as used in BPEP, may not be enough to describe user difference. Finally, the comparison between results of PROMO and eToT show that PROMO takes a superior probabilistic generative architecture for PSEM problem.

6 Conclusion

In this work, we formulate the novel Personalized Social Emotion Mining (PSEM) problem, to find the hidden structures of emotional reaction data. As a solution, then, we develop the PROMO (PRobabilistic generative cO-factorization MOdel), which is both well interpretable and expressive to address the PSEM problem. We showcase its interpretability by the meaningful hidden structures found by PROMO on a real-world data set. We also demonstrate that PROMO is a valid and effective model for emotional reactions data by showing its superiority against competing methods in two supervised learning tasks.

Acknowledgement. The authors would like to thank the anonymous referees and Noseong Park at GMU for their valuable comments. This work was supported in part by NSF awards #1422215, #1525601, #1742702, and Samsung GRO 2015 awards.

References

1. Al Boni, M., Zhou, K., Wang, H., Gerber, M.S.: Model adaptation for personalized opinion analysis. In: ACL, vol. 2, pp. 769–774 (2015)
2. Amir, S., Wallace, B.C., Lyu, H., Carvalho, P., Silva, M.J.: Modelling context with user embeddings for sarcasm detection in social media. In: SIGNLL (2016)

3. Arrow, K.J.: A difficulty in the concept of social welfare. J. Polit. Econ. **58**(4), 328–346 (1950)
4. Bai, S., Ning, Y., Yuan, S., Zhu, T.: Predicting reader's emotion on Chinese web news articles. In: Zu, Q., Hu, B., Elçi, A. (eds.) ICPCA/SWS 2012. LNCS, vol. 7719, pp. 16–27. Springer, Heidelberg (2013). https://doi.org/10.1007/978-3-642-37015-1_2
5. Bao, S., et al.: Joint emotion-topic modeling for social affective text mining. In: 2009 Ninth IEEE International Conference on Data Mining, pp. 699–704. IEEE (2009)
6. Blei, D.M., Ng, A.Y., Jordan, M.I.: Latent Dirichlet allocation. JMLR **3**, 993–1022 (2003)
7. Bordes, A., Usunier, N., Garcia-Duran, A., Weston, J., Yakhnenko, O.: Translating embeddings for modeling multi-relational data. In: NIPS, pp. 2787–2795 (2013)
8. Carroll, J.D., Chang, J.J.: Analysis of individual differences in multidimensional scaling via an n-way generalization of "Eckart-Young" decomposition. Psychometrika **35**(3), 283–319 (1970)
9. Ding, Z., Qiu, X., Zhang, Q., Huang, X.: Learning topical translation model for microblog hashtag suggestion. In: IJCAI, pp. 2078–2084 (2013)
10. Feng, W., Wang, J.: Incorporating heterogeneous information for personalized tag recommendation in social tagging systems. In: SIGKDD, pp. 1276–1284 (2012)
11. Gerrish, S., Blei, D.M.: How they vote: issue-adjusted models of legislative behavior. In: NIPS, pp. 2753–2761 (2012)
12. Gong, L., Al Boni, M., Wang, H.: Modeling social norms evolution for personalized sentiment classification. In: ACL, vol. 1, pp. 855–865 (2016)
13. Griffiths, T.L., Steyvers, M.: Finding scientific topics. PNAS **101**(suppl 1), 5228–5235 (2004)
14. Gu, Y., Sun, Y., Jiang, N., Wang, B., Chen, T.: Topic-factorized ideal point estimation model for legislative voting network. In: SIGKDD, pp. 183–192 (2014)
15. Hoffman, M.D., Blei, D.M., Wang, C., Paisley, J.: Stochastic variational inference. J. Mach. Learn. Res. **14**(1), 1303–1347 (2013)
16. Hofmann, T.: Probabilistic latent semantic analysis. In: UAI, pp. 289–296 (1999)
17. Jia, Y., Chen, Z., Yu, S.: Reader emotion classification of news headlines. In: NLP-KE 2009, pp. 1–6. IEEE (2009)
18. Jin, B., Yang, H., Xiao, C., Zhang, P., Wei, X., Wang, F.: Multitask dyadic prediction and its application in prediction of adverse drug-drug interaction. In: AAAI, pp. 1367–1373 (2017)
19. Karatzoglou, A., Amatriain, X., Baltrunas, L., Oliver, N.: Multiverse recommendation: n-dimensional tensor factorization for context-aware collaborative filtering. In: RecSys, pp. 79–86. ACM (2010)
20. Kingma, D.P., Ba, J.: Adam: a method for stochastic optimization. arXiv preprint arXiv:1412.6980 (2014)
21. Lee, S.Y., Hansen, S.S., Lee, J.K.: What makes us click "like" on Facebook? Examining psychological, technological, and motivational factors on virtual endorsement. Comput. Commun. **73**, 332–341 (2016)
22. Lei, J., Rao, Y., Li, Q., Quan, X., Wenyin, L.: Towards building a social emotion detection system for online news. Future Gener. Comput. Syst. **37**, 438–448 (2014)
23. Lin, K.H.Y., Chen, H.H.: Ranking reader emotions using pairwise loss minimization and emotional distribution regression. In: EMNLP, pp. 136–144 (2008)
24. Lin, K.H.Y., Yang, C., Chen, H.H.: What emotions do news articles trigger in their readers? In: SIGIR, pp. 733–734. ACM (2007)

25. Lin, K.H.Y., Yang, C., Chen, H.H.: Emotion classification of online news articles from the reader's perspective. In: WI-IAT 2008, vol. 1, pp. 220–226. IEEE (2008)
26. Nickel, M., Tresp, V., Kriegel, H.P.: A three-way model for collective learning on multi-relational data. In: ICML, vol. 11, pp. 809–816 (2011)
27. Rendle, S.: Factorization machines. In: ICDM, pp. 995–1000. IEEE (2010)
28. Rui, T., Cui, P., Zhu, W.: Joint user-interest and social-influence emotion prediction for individuals. Neurocomputing **230**, 66–76 (2017)
29. Socher, R., Chen, D., Manning, C.D., Ng, A.: Reasoning with neural tensor networks for knowledge base completion. In: NIPS, pp. 926–934 (2013)
30. Tang, Y.J., Chen, H.H.: Emotion modeling from writer/reader perspectives using a microblog dataset. In: Proceedings of IJCNLP Workshop on Sentiment Analysis Where AI Meets Psychology, pp. 11–19 (2011)
31. Tucker, L.R.: Some mathematical notes on three-mode factor analysis. Psychometrika **31**(3), 279–311 (1966)
32. Wang, E., Liu, D., Silva, J., Carin, L., Dunson, D.B.: Joint analysis of time-evolving binary matrices and associated documents. In: NIPS, pp. 2370–2378 (2010)
33. Yang, B., Yih, W.T., He, X., Gao, J., Deng, L.: Embedding entities and relations for learning and inference in knowledge bases. arXiv preprint arXiv:1412.6575 (2014)
34. Yang, Y., Cui, P., Zhu, W., Yang, S.: User interest and social influence based emotion prediction for individuals. In: Multimedia - MM 2013, pp. 785–788 (2013)
35. Zhang, J.J., Lee, D.: ROAR: robust label ranking for social emotion mining. In: AAAI (2018)
36. Zhang, L., Xiao, K., Zhu, H., Liu, C., Yang, J., Jin, B.: CADEN : a context-aware deep embedding network for financial opinions mining. In: ICDM, pp. 757–766 (2018)
37. Zhao, S., Yao, H., Gao, Y., Ding, G., Chua, T.S.: Predicting personalized image emotion perceptions in social networks. TAC **9**, 526–540 (2016)
38. Zhu, C., Zhu, H., Ge, Y., Chen, E., Liu, Q.: Tracking the evolution of social emotions: a time-aware topic modeling perspective. In: ICDM, pp. 697–706. IEEE (2014)

Progressive Supervision for Node Classification

Yiwei Wang[1(✉)], Wei Wang[1], Yuxuan Liang[1], Yujun Cai[2], and Bryan Hooi[1]

[1] School of Computing, National University of Singapore, Singapore, Singapore
{y-wang,wangwei,yuxliang,bhooi}@comp.nus.edu.sg
[2] Nanyang Technological University, Singapore, Singapore
yujun001@e.ntu.edu.sg

Abstract. Graph Convolution Networks (GCNs) are a powerful approach for the task of node classification, in which GCNs are trained by minimizing the loss over the final-layer predictions. However, a limitation of this training scheme is that it enforces every node to be classified from the fixed and unified size of receptive fields, which may not be optimal. We propose ProSup (Progressive Supervision), that improves the effectiveness of GCNs by training them in a different way. ProSup supervises all layers progressively to guide their representations towards the characteristics we desire. In addition, we propose a novel technique to reweight the node-wise losses, so as to guide GCNs to pay more attention to the nodes that are hard to classify. The hardness is evaluated progressively following the direction of information flows. Finally, ProSup fuses the rich hierarchical activations from multiple scales to form the final prediction in an adaptive and learnable way. We show that ProSup is effective to enhance the popular GCNs and help them to achieve superior performance on miscellaneous graphs.

Keywords: Graph Convolutional Networks · Progressive supervision · Node classification

1 Introduction

Node classification is a fundamental task on graph data, which aims to classify the nodes in an (attributed) graph [14]. For this task, Graph Convolutional Networks (GCNs) have achieved state-of-the-art performance [23]. Typically, GCNs follow a multi-layer structure (see Fig. 1(a)). Across layers, GCNs update node representations via the 'message-passing' mechanism, i.e., they aggregate the representations of each node and its neighbors to produce new ones at the next layer. Denote the subgraph contributing to a node's representation as its receptive field. From bottom to top, the receptive field expands gradually, which is generally a node's l-hop neighborhood at the lth layer [19].

For training GCNs, it is common to minimize the classification loss on the final-layer predictions. This training scheme is convenient, but not necessarily

F. Hutter et al. (Eds.): ECML PKDD 2020, LNAI 12457, pp. 266–281, 2021.
https://doi.org/10.1007/978-3-030-67658-2_16

ideal for effectiveness. One limitation is that it enforces GCNs to classify all nodes from the unified size of receptive fields, but nodes can have diverse 'appropriate' receptive fields for classification [24]. In a social network, for example, famous people include much noise with a small number of layers (hops), while freshmen may need more layers to include the relevant features. Another limitation is about the discriminativeness of learned features. [13] demonstrates that a classifier will perform better when trained on more discriminative features. However, minimizing the loss only on the final-layer predictions ignores the discriminativeness of hidden layers and may degrade the performance.

The central idea of this paper is to make predictions separately on each layer and progressively supervise hidden layers from bottom to top (see Fig. 1(b)). We encapsulate this idea in a new scheme, called ProSup (short for Progressive Supervision), that produces the side-outputs on hidden layers, and then fuse the multi-level, multi-scale activations to form the unified prediction. With our design, a node is classified at a side-output from an 'appropriate' receptive field instead of absorbing extra noise when propagating its representation to the final layer. Besides, our supervision over hidden layers acts as a kind of regularization. It regularizes hidden layers to produce discriminative and meaningful representations, rather than supervises only the final-layer representations. Note that ProSup can be incorporated into popular GCN architectures, e.g., GCN [11], LGCN [8], GraphSAGE [9], etc., for better effectiveness.

Each layer has some nodes easy to classify and some hard ones (see Fig. 2). To leverage the information of classification hardness, we propose a technique to progressively reweight the node-wise losses. Following the direction of information flows, we encourage GCNs to pay more attention to hard nodes. In principle, we give the nodes, that are classified incorrectly at a layer, larger weights at the next layer. This technique facilitates communication across GCN layers to mine the hard nodes progressively. In terms of gradient propagation, ProSup simultaneously minimizes the classification losses on the side-outputs and the fused prediction. The former propagates the local errors to the corresponding layer, while the latter directly supervises all layers globally. We use both of them to train the model holistically so as to produce discriminative representations and the finer fused prediction.

We evaluate ProSup on the node classification task using the Citeseer, Cora, Pubmed [14], Flickr [15], Yelp [25], and Reddit [9] datasets. Qualitatively, ProSup makes the class-specific representations more concentrated (see Fig. 4). We also observe quantitative improvements evaluated by test accuracy and F1-micro scores. Overall, ProSup improves the popular GCN [11], LGCN [8], GraphSAGE [9] and GraphSAINT [25] models by a significant margin, and enhances them to outperform the benchmark methods. As we analyze, our ProSup improves the effectiveness of GCNs without changing the time complexity.

2 Related Work

Due to the long history of Graph Neural Networks, we refer readers to [23] and [27] for a comprehensive review. The first work that proposes the convo-

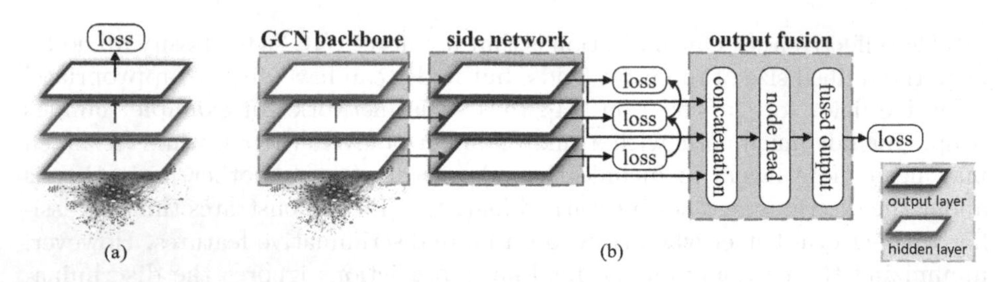

Fig. 1. ProSup applied to node classification. (a) Standard GCN takes an input graph and yields predictions on the final layer, where the loss is calculated. (b) ProSup supervises hidden layers progressively and fuses the outputs from multiple layers to produce the unified prediction. We minimize the classification losses on both the side-outputs and the unified output simultaneously.

lution operation on graph data is [2]. More recently, [11] and [7] speed up the graph convolution operations by introducing localized filters based on Chebyshev expansion. Specifically, [11] has made breakthrough advancements in the task of node classification. As a result, the model proposed in [11] is generally denoted as the vanilla GCN, or GCN. After [11], numerous GCN methods are proposed for better performance on node classification. There are two main lines of research in this field. The first one is to propose new GCN architectures to improve the model capacity. The second is to propose mini-batch techniques for GCNs to achieve better scalability without loss of effectiveness.

To improve the model capacity, [21,26], and [10] use the attention mechanism to better capture neighbor features by dynamically adjusting edge weights. Mixture Model Network (MoNet) [16] adopts a different approach to assign edge weights. It introduces node pseudo-coordinates to determine the relative position between a node and its neighbors, then defines a weight function to map the relative positions to edge weights. [28] utilizes the positive pointwise mutual information (PPMI) matrix to capture nodes co-occurrence information through random walks sampled from a graph. [12] combines PageRank with GCNs to enable efficient information propagation. [22] alternatively drives local network embeddings to capture global structural information by maximizing local mutual information. [4] proposes a non-uniform graph convolutional strategy, which learns different convolutional kernel weights for different neighboring nodes according to their semantic meanings. LGCN [8] ranks a node's neighbors based on node features. It assembles a feature matrix that consists of its neighborhood and sorts this feature matrix along each column. [18] proposes GMNN to model the dependency of object labels through statistical relational learning.

Another line of research focuses on scaling GCNs to large graphs efficiently and effectively. GraphSAGE [9] performs uniform node sampling on the previous layer neighbors. It enforces a pre-defined budget on the sample size, so as to bound the mini-batch computation complexity. [5] further restricts neighborhood size by requiring only two support nodes in the previous layer. Instead of

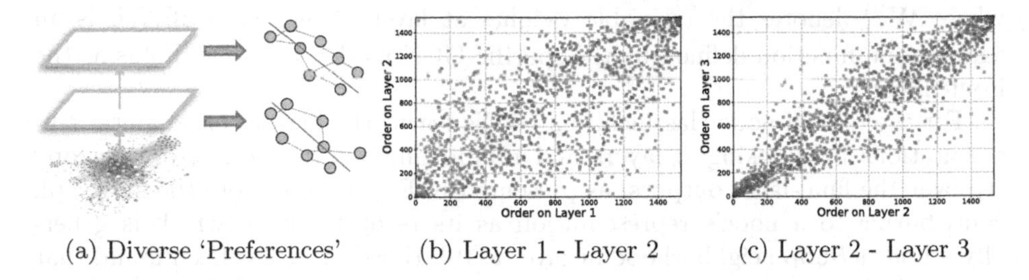

(a) Diverse 'Preferences' (b) Layer 1 - Layer 2 (c) Layer 2 - Layer 3

Fig. 2. GCN layers have diverse 'preferences' on node classification. (a) GCN layers produces various node representation, which lead to various optimal decision boundaries (red lines). (b, c) Descending order of the node-wise losses in `Citeseer`. (b) Each point represents a node. x-value is the node order on layer 1, and y is for layer 2. (c) is similar for layers 2 (x) and 3 (y). If the 'preferences' are the same, all the points in (b) and (c) concentrate on the line of $x = y$. (Color figure online)

sampling layers, ClusterGCN [6] and GraphSAINT [25] build mini-batches from subgraphs, so as to avoid the 'neighbor explosion' problem.

Our work is orthogonal to the two lines above in the sense that it neither alters the GCN architecture for better capacity nor introduces a new mini-batch technique. We propose a new scheme that takes a GCN architecture as the backbone to enhance its effectiveness without changing the time complexity. Meanwhile, existing mini-batch techniques can be applied to our scheme for better scalability. Our scheme is inspired by deeply-supervised nets [13], that perform deep supervision to 'guide' early classification results. We find that the favorable characteristics of feature representations provided by ProSup lead to more accurate predictions.

3 Methodology

In this section, we describe our proposed ProSup scheme for node classification. We first introduce the background and mathematical notations. Next, we introduce the modules of ProSup, including the GCN backbone, the side network, and output fusion, as illustrated in Fig. 1(b). Finally, we analyze why ProSup can improve the performance of GCNs.

3.1 Background and Motivation

We define a graph as $G = (\mathcal{V}, \mathcal{E})$, where $\mathcal{V}$ denotes the set of nodes, and $\mathcal{E}$ is the set of edges. The input feature vector of node i is $\mathbf{x}_i$, and the neighborhood of node i is $\mathcal{V}_i = \{j \in \mathcal{V} | (i, j) \in \mathcal{E}\}$. Suppose the representation of node i at layer l is $\mathbf{h}_i^{(l)}$. Typically, a GCN layer obtains $\mathbf{h}_i^{(l)}$ as:

$$\mathbf{h}_i^{(l)} = \mathsf{AGGREGATE}\left(\mathbf{h}_i^{(l-1)}, \left\{\mathbf{h}_j^{(l-1)}, j \in \mathcal{V}_i\right\}, \mathbf{W}^{(l)}\right), \tag{1}$$

where $\mathbf{W}^{(l)}$ denotes the trainable weights at layer l, and $\mathsf{AGGREGATE}$ is an aggregation function defined by the specific GCN model. $\mathbf{h}_i^{(0)} = \mathbf{x}_i$ holds at the input layer.

For the task of node classification, GCNs learns the high-level semantic representations by stacking L layers and minimizing the loss, e.g. cross entropy [1], over the final-layer outputs, as presented in Fig. 1(a). Denote the subgraph contributing to a node's representation as its receptive field, which is generally a node's l-hop neighborhood at layer l [19]. However, it has been found that increasing L does not necessarily improve the effectiveness of GCNs, even though it expands the receptive fields for classification. Popular GCN models observe empirically optimal performance with a small L, e.g. 2 or 3 [11]. A reason is that nodes have diverse 'appropriate' receptive field sizes for effective classification [24]. Larger receptive fields can introduce extra noise, while smaller ones miss valuable information. In addition, each GCN layer has some nodes easy to classify and others hard. Hence, a GCN layer may be effective at classifying some nodes, but not others. This 'preference' varies with l, i.e., various receptive fields, as shown in Fig. 2. Thus, minimizing the classification loss on the final-layer outputs, i.e., enforcing GCNs to classify all the nodes based on the fixed and unified receptive field of L-hop neighborhoods, inevitably introduces noise. The larger L is, the more noise is introduced, which may lead to decreased accuracy.

In our work, we propose our progressive supervision (ProSup) scheme to enhance GCN models for node classification. As depicted in Fig.1(b), the core idea is to supervise the feature maps on hidden layers progressively, and fuse the side-outputs from multiple levels and multiple scales to produce the unified output in an adaptive and learnable way. We combine the rich hierarchical responses, that allow each node to be classified at its 'appropriate' receptive fields, from different layers. For training, we minimize the classification losses over both the side-outputs and the fused output simultaneously. Next, we introduce the details of ProSup.

3.2 GCN Backbone and Side Network

We take an existing GCN model as the backbone, e.g., GCN, GAT, LGCN, etc., as shown in Fig. 1(b). This backbone typically has a multi-layer structure, of which the aggregation mechanism is introduced in Sect. 3.1. We construct a side network parallel to the GCN backbone for conducting progressive supervision. Following the notations in Eq. (1), we denote node representations in layer l as $\mathbf{H}^{(l)} \in \mathbb{R}^{N \times d_l}$, where $N = |\mathcal{V}|$ is the number of nodes, and d_l is the dimension of representations in layer l. As defined in Eq. (1), the representation of node i, $\mathbf{h}_i^{(l)}$, is the ith row of $\mathbf{H}^{(l)}$. Accordingly, the input feature vectors $\mathbf{x}_i$ form the rows of $\mathbf{X} = \mathbf{H}^{(0)}$. We produce the side-outputs at layer l as:

$$\mathbf{A}^{(l)} = \mathbf{H}^{(l)} \mathbf{W}_{side}^{(l)}, \quad \hat{\mathbf{Y}}^{(l)} = \sigma\left(\mathbf{A}^{(l)}\right), \tag{2}$$

where $\mathbf{A}^{(l)}$, $\mathbf{W}_{side}^{(l)} \in \mathbb{R}^{d_l \times C}$, and $\hat{\mathbf{Y}}^{(l)} \in \mathbb{R}^{N \times C}$ are the activation, trainable weights, and the prediction respectively. C is the number of classes. $\sigma(\cdot)$ is the nonlinear activation function used to produce probabilities, which is softmax for single-label classification and sigmoid for multi-label classification [1].

Unlike existing GCN architectures that produce outputs only at their final layer, we build an output layer after each convolutional layer. The receptive fields of these side-outputs become larger as l increases. Thus, we enable hidden layers to produce predictions by themselves, instead of only contributing to the final-layer prediction indirectly through feed-forward propagation. As a result, the nodes that are easier to classify with a smaller receptive field are classified earlier, rather than absorbing more noise through further aggregations. In addition, the hidden layers are expected to produce more discriminative representations through our direct supervision.

3.3 Output Fusion and Loss Function

To produce the unified prediction, we construct node-wise features by combining (e.g., concatenating) the activations from multiple layers, $\{\mathbf{A}^{(l)}\}_l$. Given the node-wise feature representation, we make node-wide class predictions using a light-weight node head, implemented as a multi-layer perception (MLP). The node head shares weights across all nodes, analogous to PointNet [17]:

$$\mathbf{A} = \delta\left(\mathsf{CONCAT}\left(\left\{\mathbf{A}^{(l)}\right\}_l\right)\mathbf{W}_{fuse}^{(1)}\right)\mathbf{W}_{fuse}^{(2)}, \quad \hat{\mathbf{Y}} = \sigma\left(\mathbf{A}\right), \tag{3}$$

where $\mathbf{W}_{fuse}^{(1)} \in \mathbb{R}^{LC \times d_{fuse}}, \mathbf{W}_{fuse}^{(2)} \in \mathbb{R}^{d_{fuse} \times C}$ are the learnable weights for fusing the multi-level activations, while $\mathbf{A}$ is the fused activation for the unified output. $\delta(\cdot)$ is the activation function, which we set as ReLU. During inference, we take $\hat{\mathbf{Y}}$ as the unified prediction. During training, we supervise both the side-outputs and the fused output. The loss on side-outputs is:

$$\mathcal{L}_{side} = -\sum_{l}\sum_{i \in \mathcal{V}_L}\sum_{c=1}^{C} Y_{ic} \log \hat{Y}_{ic}^{(l)}, \tag{4}$$

where $\mathcal{V}_L$ is the set of node indices that have labels in the training set, and Y_{ic} is the binary ground-truth label value of node i. $Y_{ic} = 1$ indicates node i belongs to class c, and $Y_{ic} = 0$, otherwise.

As for the fused prediction $\hat{\mathbf{Y}}$, we calculate the loss:

$$\mathcal{L}_{fuse} = -\sum_{i \in \mathcal{V}_L}\sum_{c=1}^{C} Y_{ic} \log \hat{Y}_{ic}. \tag{5}$$

Putting these together, we minimize the following objective function:

$$\mathcal{L} = \mathcal{L}_{fuse} + \alpha \mathcal{L}_{side}, \tag{6}$$

where α is the hyper-parameter for balancing $\mathcal{L}_{fuse}$ and $\mathcal{L}_{side}$.

From the perspective of gradient propagation, minimizing $\mathcal{L}_{side}$ over a side-output propagates the local errors to the corresponding layer, while minimizing $\mathcal{L}_{fuse}$ propagates the global errors to all layers simultaneously. The former supervises each layer to generate semantically meaningful predictions and regularizes GCNs to learn consistently discriminative representations across hidden layers. This endows the final prediction with the higher quality given the better hidden representations. Depending on the receptive field size, the side-output of any single layer may be coarse. For example, lower layers have only localized features, while higher layers may give over-smooth outputs [20]. Thus, we fuse the (coarse) side-outputs to form the unified (fine) output. This process of combining multi-level and multi-scale activations is learnable and adaptive, by which the final prediction of each node can be obtained from an 'appropriate' receptive field.

3.4 Progressively Re-weighting Hard Nodes

At each layer, some nodes are easy to classify, while others are hard, as shown in Fig. 2. Giving all the nodes the same weight in the loss function $\mathcal{L}_{side}$ can lose the information on the classification hardness. To address this limitation, we design a new scheme to progressively reweight the hard nodes. As shown in Fig. 1, on the losses computed on side-outputs, from bottom to top, we conduct a node re-weighting operation layer by layer. This bottom-up hierarchical mechanism is inspired by the sequential information flow across hidden layers of GCNs. For example, the third layer of the side network needs the feature map from the second layer to compute the classification results and loss. If the lower layers can inform the upper layers which nodes are hard to classify, the higher layers will pay more attention to these hard nodes. Through this communication, hard examples can be mined and classified more effectively from bottom to top, resulting in a better fused prediction.

Specifically, given $\hat{\mathbf{Y}}^{(l)}$ as the output from the l th layer of the side network, we denote the differences of node i between the prediction on layer l and the ground truth label as:

$$\xi_i^{(l)} = \sum_{c=1}^{C} \left| \hat{Y}_{ic}^{(l)} - Y_{ic} \right| \tag{7}$$

Then, we rewrite the loss function in Eq. (4) to:

$$\begin{aligned} \mathcal{L}_{side-reweight} = &- \sum_{i \in \mathcal{V}_L} \sum_{c=1}^{C} Y_{ic} \log \hat{Y}_{ic}^{(1)} \\ &- \sum_{l=2}^{L} \sum_{i \in \mathcal{V}_L} \frac{|\mathcal{V}_L|}{\sum_{i \in \mathcal{V}_L} |\xi_i^{(l-1)}|} |\xi_i^{(l-1)}| \sum_{c=1}^{C} Y_{ic} \log \hat{Y}_{ic}^{(l)}, \end{aligned} \tag{8}$$

Algorithm 1. ProSup: progressive supervision for enhancing Graph Convolutional Networks (GCNs) on node classification.

Input: Graph $G = (\mathcal{V}, \mathcal{E})$, Feature Matrix $\mathbf{X}$, a GCN backbone with the aggregation function $\mathsf{AGGREGATE}(\cdot)$, hyper-parameter α for balancing losses, d_{fuse} for the node head, ground-truth labels $\mathbf{Y}$, labeled nodes in the training set $\mathbf{V}_L$.
Output: Fused prediction $\hat{\mathbf{Y}}$, trained parameters of the GCN backbone $\left\{\mathbf{W}^{(l)}\right\}_l$, and trained parameters of ProSup, $\left\{\mathbf{W}^{(l)}_{side}\right\}_l$, $\mathbf{W}^{(1)}_{fuse}$, and $\mathbf{W}^{(2)}_{fuse}$.

1: Initialize all parameters.
2: **while** $\mathcal{L}$ does not converge **do**
3: $\quad \mathbf{H}^{(0)} \leftarrow \mathbf{X}$
4: $\quad$ **for** $l \leftarrow 1$ to L **do**
5: $\quad\quad$ **for** $i \leftarrow 1$ to $|\mathcal{V}|$ **do**
6: $\quad\quad\quad \mathbf{h}^{(l)}_i \leftarrow \mathsf{AGGREGATE}\left(\mathbf{h}^{(l-1)}_i, \left\{\mathbf{h}^{(l-1)}_j, j \in \mathcal{V}_i\right\}, \mathbf{W}^{(l)}\right)$
7: $\quad\quad$ **end for**
8: $\quad\quad \mathbf{A}^{(l)} \leftarrow \mathbf{H}^{(l)}\mathbf{W}^{(l)}_{side}$
9: $\quad\quad \hat{\mathbf{Y}}^{(l)} \leftarrow \sigma\left(\mathbf{A}^{(l)}\right)$
10: $\quad\quad$ **for** $i \leftarrow 1$ to $|\mathcal{V}_L|$ **do**
11: $\quad\quad\quad \xi^{(l)}_i \leftarrow \sum_{c=1}^{C}\left|\hat{Y}^{(l)}_{ic} - Y_{ic}\right|$
12: $\quad\quad$ **end for**
13: $\quad$ **end for**
14: $\quad \mathbf{A} \leftarrow \delta\left(\mathsf{CONCAT}\left(\left\{\mathbf{A}^{(l)}\right\}_l\right)\mathbf{W}^{(1)}_{fuse}\right)\mathbf{W}^{(2)}_{fuse}$
15: $\quad \hat{\mathbf{Y}} \leftarrow \sigma(\mathbf{A})$
16: $\quad$ Calculate $\mathcal{L}_{side}$ from Eq. (8)
17: $\quad \mathcal{L}_{fuse} \leftarrow -\sum_{i\in\mathcal{V}_L}\sum_{c=1}^{C} Y_{ic}\log\hat{Y}_{ic}$
18: $\quad \mathcal{L} \leftarrow \mathcal{L}_{fuse} + \alpha\mathcal{L}_{side}$
19: $\quad$ Back-propagation for minimizing $\mathcal{L}$
20: **end while**

where $\frac{|\mathcal{V}_L|}{\sum_i |\xi^{(l-1)}_i|}|\xi^{(l-1)}_i|$ is the normalized node weight. $\frac{|\mathcal{V}_L|}{\sum_i |\xi^{(l-1)}_i|}$ helps to ensure that, $\forall\, l$, the average loss weight of all the labeled nodes in the training set is

$$\frac{1}{|\mathcal{V}_L|}\sum_{i\in\mathcal{V}_L}\frac{|\mathcal{V}_L|}{\sum_{i\in\mathcal{V}_L}|\xi^{(l-1)}_i|}|\xi^{(l-1)}_i| = \frac{|\mathcal{V}_L|}{|\mathcal{V}_L|}\sum_{i\in\mathcal{V}_L}\frac{|\xi^{(l-1)}_i|}{\sum_{i\in\mathcal{V}_L}|\xi^{(l-1)}_i|} = 1.$$

Thus, with the progressive re-weighting scheme, we leverage the predictions of a shallower side network to weight loss terms in a deeper side network. This facilitates communication among hidden layers and provide better activations $\{\mathbf{A}^l\}_l$ for producing the final prediction effectively. We provide pseudo-code for ProSup in Algorithm 1.

3.5 Discussion

ProSup has several advantages over the classical GCN training scheme that minimizes the loss on the final-layer outputs. With ProSup, different nodes have

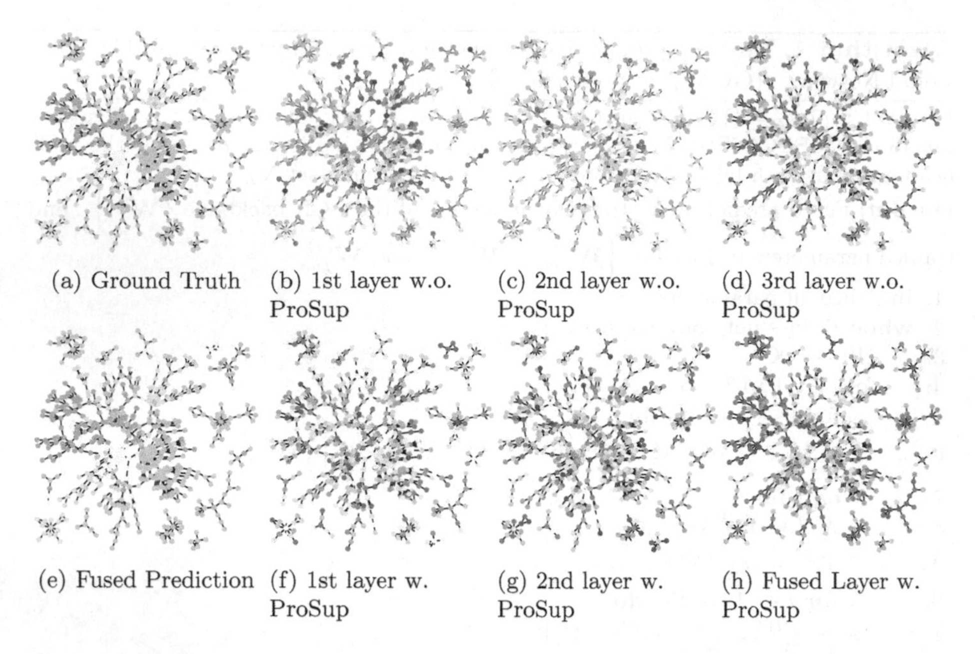

Fig. 3. ProSup enhances GCN to provide better predictions. In the first column, colors denote classes of nodes in the `Cora` dataset. (a) shows the ground truth labels. (e) is the fused prediction given by GCN with ProSup. In the other columns, colors denote the predicted score on the correct class. Colors closer to red mean that the score is closer to 1 (correct). Otherwise, the blue color corresponds to the score of 0 (incorrect). ProSup enhances GCN to provide hidden layers with better predictions, improving the final prediction. (Color figure online)

adaptively 'appropriate' receptive fields for classification, since the corresponding hidden layers can produce the outputs by themselves rather than pass the representations to the next layers, which may aggregate extra noise. Second, a discriminative classifier trained on highly discriminative features will perform better than that on less discriminative features [13]. By training the classifier constructed at each hidden layer of a GCN, we are able to progressively influence the weight update process of hidden layers to favor highly discriminative feature maps. This is a source of supervision that acts within the GCN at each layer: we expect the final classification predictions to be more reliable than the case of relying on backpropagation from the final layer alone.

Moreover, ProSup can also be seen as a form of regularization, as GCN learns consistently discriminative features among hidden layers instead of overfitting the training data by minimizing the classification loss solely on the final layer. Finally, in terms of explainability, ProSup provides a classifier for each hidden layer to endow their representations with clear semantic information, rather than treating GCNs as a black box with complicated and non-interpretable input-output relations. An example of node classification by GCN with ProSup is

Table 1. Statistics of the utilized datasets. 'm' stands for multi-label classification, while 's' for single-label.

Dataset	Citeseer	Cora	Pubmed	Flickr	Yelp	Reddit
#Nodes	3,327	2,708	19,717	89,250	716,847	232,965
#Edges	4,732	5,429	44,338	899,756	6,977,410	11,606,919
#Features	3,703	1,433	500	500	300	602
#Classes	7 (s)	6 (s)	3 (s)	7 (s)	100 (m)	41 (s)

presented in Fig. 3. The predictions on a hidden layer of GCN without ProSup are obtained by training a fully-connected layer using the learned representations.

4 Complexity Analysis

We train the model in the end-to-end style. The time complexity of GCN is $\mathcal{O}\left(|\mathcal{E}|\sum_{l=1}^{L} d_l + |\mathcal{V}|\sum_{l=1}^{L} d_{l-1}d_l\right)$. With the side network, we have the time complexity as $\mathcal{O}\left(|\mathcal{V}|\sum_{l=1}^{L} d_l C\right)$. In the output fusion module, we have the complexity $\mathcal{O}\left(|\mathcal{V}|LCd_{fuse}\right)$. Taking all the computation into consideration, we have the complexity of $\mathcal{O}\left(|\mathcal{E}|\sum_{l=1}^{L} d_l + |\mathcal{V}|\left(\sum_{l=1}^{L} d_l d_{l-1} + C\left(d_l + d_{fuse}\right)\right)\right)$. The complexity is linear to $|\mathcal{E}|$ and $|\mathcal{V}|$, same as in GCNs. Note that $C < d_l, \forall\ l$ holds generally. Thus, using ProSup to improve the effectiveness of GCNs does not increase their time complexity.

5 Experiment

In this section, we present the empirical improvements over various GCN architectures achieved by ProSup. We report the experimental results under both the transductive and inductive settings. In addition, we visualize the learned representations of GCN with ProSup compared with the GCN without ProSup. Finally, we conduct ablation studies to show the influence of different components of ProSup, as well as the sensitivity with respect to the hyper-parameters of ProSup.

We use standard benchmark datasets: Cora, Citeseer, Pubmed [14], Flickr [15], Yelp [25], and Reddit [9] for evaluation. The former three are citation networks, where each node is a document and each edge is a citation link. In Flickr, each node represents one image. An edge is built between two images if they share some common properties (e.g. same geographic location, same gallery, etc.). The Yelp dataset contains a social network, where an edge means that the connected users are friends. Reddit is collected from an online discussion forum where users comment in different topical communities. Two posts (nodes) are connected if some user comments on both posts. Each dataset contains an

Table 2. Test Accuracy (%) of transductive node classification. #Layers indicates the best-performing number of layers among 1 to 8. We conduct 100 trials with random weight initialization. The mean and standard derivations are reported.

Method	#Layers	Citeseer	#Layers	Cora	#Layers	Pubmed
GCN [11]	2	77.1 ± 1.4	2	88.3 ± 0.8	3	86.4 ± 1.1
GAT [21]	2	76.3 ± 0.8	3	87.6 ± 0.5	3	85.7 ± 0.7
JKNet-MaxPool [24]	1	77.4 ± 0.6	6	89.5 ± 0.8	3	86.5 ± 0.9
JKNet-Concat [24]	1	78.1 ± 0.9	6	89.1 ± 1.2	4	86.9 ± 1.3
JKNet-LSTM [24]	1	74.7 ± 0.8	1	86.1 ± 0.9	1	85.8 ± 1.2
LGCN [8]	2	77.5 ± 1.1	2	89.0 ± 1.2	2	86.5 ± 0.6
GMNN [18]	2	77.4 ± 1.5	2	88.7 ± 0.8	2	86.7 ± 1.0
ProSup + GCN	3	$\mathbf{79.3 \pm 1.2}$	4	$\mathbf{90.8 \pm 0.7}$	4	$\mathbf{88.2 \pm 0.8}$
ProSup + LGCN	3	$\mathbf{79.6 \pm 0.7}$	3	$\mathbf{91.2 \pm 0.8}$	3	$\mathbf{88.5 \pm 0.5}$

unweighted adjacency matrix and bag-of-words features. The statistics of these datasets are summarized in Table 1.

In the transductive setting, we have access to the features of all nodes but only the labels of nodes in the training set for training. In the inductive setting, both the features and labels of the nodes in the validation/testing set are unavailable during training.

For the hyper-parameters of the benchmarks, e.g. the number of hidden units, the optimizer, the learning rate, we set them as suggested by their authors. For the hyper-parameters of our ProSup, we set $d_{fuse} = d_{L-1}$ for the dimensionality of the output fusion module, and $\alpha = 1$ for the weight of the loss from side-outputs by default.

5.1 Transductive Node Classification

In the transductive settings, we take the popular GCN architectures of GCN [11], GAT [21], LGCN [8], JKNet [24], and GMNN [18] as the baselines for comparison. We split nodes in each graph into 60%, 20%, 20% for training, validation, and testing. We make 10 random splits and conduct the experiments for 100 trials with random weight initialization for each split.

We vary the number of layers from 1 to 8 for each model and choose the best performing number with respect to the validation set. The results are reported in Table 2. We observe that ProSup improves the test accuracy of GCN by 2.9% on Citeseer, 2.8% on Cora, 2.1% on Pubmed, and LGCN by 2.7% on Citeseer, 2.5% on Cora, and 2.3% on Pubmed respectively. As a result, ProSup enhances GCN and LGCN to outperform all the benchmark methods.

Taking a closer look, we observe that ProSup increases the number of layers that GCN and LGCN need to achieve their best performance. The reason is that more layers inevitably introduce extra noise for learning the final-layer representations. However, ProSup enables hidden layers to make outputs by themselves, so the node-wise predictions are made with the 'appropriate' receptive fields.

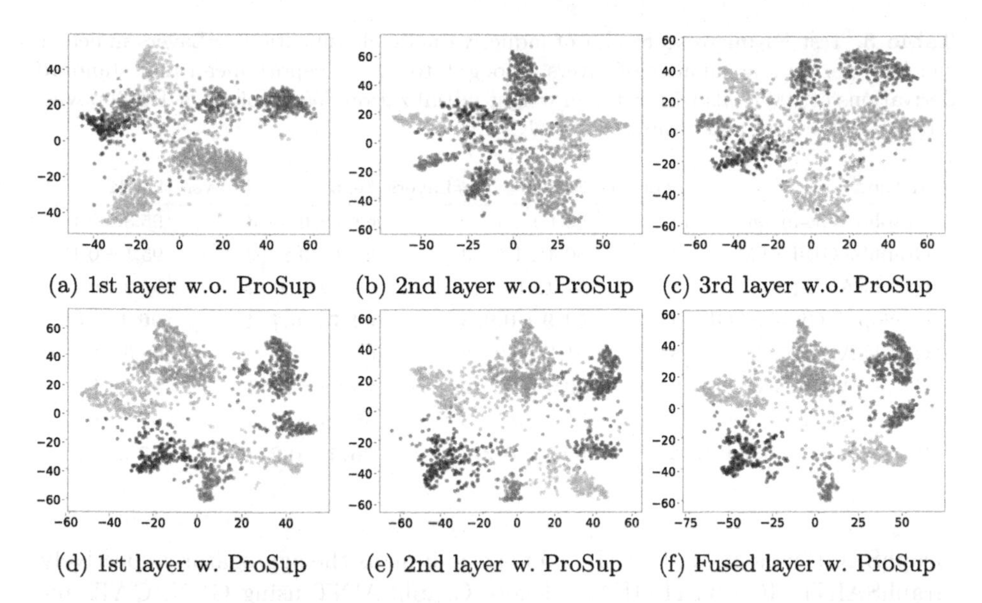

(a) 1st layer w.o. ProSup (b) 2nd layer w.o. ProSup (c) 3rd layer w.o. ProSup

(d) 1st layer w. ProSup (e) 2nd layer w. ProSup (f) Fused layer w. ProSup

Fig. 4. The learnt representations of the nodes in the `Cora` dataset (visualized by t-SNE [3]). Colors denote the ground-truth class labels. ProSup (lower row) makes class-specific representations more concentrated (than the upper row).

Besides, in terms of the fitting capacity, more layers provide a larger fitting capacity. In this case, minimizing the loss solely on the final-layer prediction leads to higher risks of overfitting, since it ignores the characteristics of representations of hidden layers. ProSup decreases this risk by regularizing GCNs to learn consistently discriminative features among its hidden layers.

Figure 4 presents the learned representations obtained by a 3-layer GCN and a 3-layer GCN with ProSup. We can observe that the hidden layers supported by ProSup learn more discriminative features consistently, thanks to the direct supervision applied to them. These highly discriminative features potentially help to produce better class predictions than less discriminative features.

5.2 Inductive Node Classification

In the inductive settings, we use the datasets Flickr, Yelp, Reddit with the fixed partition [25] for evaluation. These datasets are too large to be handled well by the full-batch implementations of GCN architectures. Hence, we use GraphSAGE [9] and GraphSAINT [25] as the benchmarks for comparison, which are more scalable. We implement ProSup with GraphSAGE-mean and GraphSAINT-GCN to observe whether ProSup can improve the performance of GCNs under the inductive setting.

We vary the number of layers of each method from 1 to 8 for each model and choose the best performing model with respect to the validation set. The results are reported in Table 3. GraphSAGE-mean/LSTM/pool denotes that

Table 3. Test F1-micro score (%) of inductive node classification. #Layers indicates the best-performing number of layers among 1 to 8. We report mean and standard derivations of 100 trials with random weight initialization. We implement ProSup with GraphSAGE-mean and GraphSAINT-GCN.

Method	#Layers	Flickr	#Layers	Yelp	#Layers	Reddit
GraphSAGE-mean	3	50.1 ± 1.1	2	63.4 ± 0.6	3	95.3 ± 0.1
GraphSAGE-LSTM	3	50.3 ± 1.3	3	63.2 ± 0.8	2	95.1 ± 0.1
GraphSAGE-pool	3	50.0 ± 0.8	3	63.1 ± 0.5	2	95.2 ± 0.1
ProSup + GraphSAGE	4	**51.9 ± 0.9**	3	**64.7 ± 0.7**	4	**96.1 ± 0.1**
GraphSAINT-GCN	3	51.1 ± 0.2	2	65.3 ± 0.3	3	96.6 ± 0.1
GraphSAINT-GAT	2	50.5 ± 0.1	2	65.1 ± 0.2	3	95.8 ± 0.0
GraphSAINT-JKNet	4	51.3 ± 0.5	3	65.3 ± 0.4	4	97.0 ± 0.1
ProSup + GraphSAINT	3	**52.8 ± 0.2**	3	**66.2 ± 0.2**	4	**97.3 ± 0.1**

GraphSAGE uses mean, LSTM, and max-pooling as the aggregator respectively. GraphSAINT-GCN/GAT/JKNet denote GraphSAINT using GCN, GAT, and JKNet as the base architecture respectively. We observe that ProSup improves the test F1-micro scores of GraphSAGE-mean by 3.6% on Flickr, 2.1% on Yelp, 0.8% on Reddit, and GraphSAINT-GCN by 3.3% on Flickr, 1.4% on Yelp, and 0.3% on Reddit respectively. As a result, ProSup enhances them to outperform the benchmark methods. Overall, the above results validate that ProSup is effective in improving the performance of the popular GCN models under both transductive and inductive settings.

5.3 Ablation Study

We conduct a number of ablations to analyze ProSup. First, we investigate the effects of the side network and the output fusion module. We compare the results on the Flickr dataset of GraphSAINT and its variants combined with our proposed techniques in Table 4. To obtain the classification results on hidden layers of GraphSAINT-GCN, we train a fully-connected layer over the learned hidden representations at every GCN layer. We observe that, with our side network, the prediction results are better on every layer, which demonstrates that our progressive supervision can act as a form of regularization, which not only makes the hidden representations more discriminative but also improves the quality of the final-layer predictions. Moreover, with our module of output fusion, the prediction performance of hidden layers is improved further, thanks to the additional global supervision. Finally, the fusion module fuses the coarse predictions from hidden layers to form the finer final prediction.

Table 5 presents the performance of GCNs with and without our technique of progressively re-weighting the hard nodes among hidden layers. Empirically, this technique consistently improves the classification performance of GCN and GraphSAGE over all datasets. This demonstrates that it is valuable to leverage

Table 4. F1-micro score (%) of inductive node classification on Flickr.

	GCN	GCN w. side network	GCN w. ProSup
Layer 1	47.3	51.3	51.7
Layer 2	48.6	51.6	51.9
Layer 3	**51.1**	**52.0**	52.2
Fused output	–	–	**52.8**

Table 5. Test Accuracy (%) of GCN on Citeseer, Cora, Pubmed and F1-micro score (%) of GraphSAGE-mean on Flickr, Yelp, Reddit.

	Citeseer	Cora	Pubmed	Flickr	Yelp	Reddit
GCN + ProSup w.o Reweighting	78.9	90.5	88.0	51.1	64.3	95.9
GCN + ProSup w. Reweighting	**79.3**	**90.8**	**88.2**	**51.9**	**64.7**	**96.1**

the hardness information implied by the predicted scores through different layers. The re-weighting technique encourages communication among hidden layers so as to improve the final predictions.

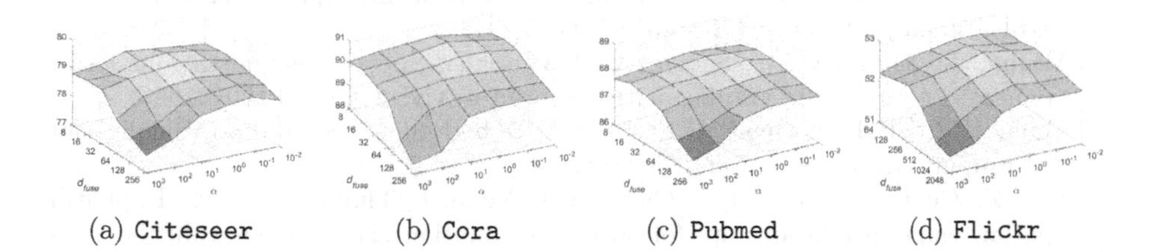

(a) Citeseer (b) Cora (c) Pubmed (d) Flickr

Fig. 5. Test accuracy (% on z-axis) of GCN with ProSup on Citeseer, Cora, Pubmed and F1-micro scores of GraphSAINT-GCN with ProSup on Flickr versus values of hyper-parameters d_{fuse} (x-axis) and α (y-axis).

Finally, we evaluate how sensitive our ProSup is to the selection of hyper-parameter values: d_{fuse} to control the dimensionality of the node head in the output fusion module, and α to adjust the weight of side-output loss. We visualize the results in Fig. 5. As we can see, the performance of GCN with ProSup is relatively smooth when parameters are within certain ranges. However, extremely large values of d_{fuse} and α result in low performance on all datasets, which should be avoided in practice. Moreover, increasing α from 0.01 to 1 improves the test accuracy/F1-micro scores on all datasets, demonstrating that the supervision progressively applied on hidden layers plays an important role in improving the performance of GCNs.

6 Conclusion

In this paper, we have developed a new graph convolutional network based node classifier and demonstrated its superior performance on datasets comprising documents, images, friendships, and online discussion posts. Our scheme builds on the idea of progressively supervising the hidden layers of GCNs. During training, every hidden layer participates in the loss calculation directly to learn discriminative features. In addition, We propose a reweighting technique to deal with hard nodes by giving them larger loss weights in a progressive way. This encourages GCNs to pay more attention to hard nodes and thus make predictions more effectively. Our method shows the effectiveness of predicting node-wise labels by combining multi-scale and multi-level responses in an adaptive and learnable way. Moreover, ProSup does not incur any change in the time complexity of GCNs, as we analyze. An interesting future direction is to extend our progressive supervision algorithm to other tasks on graph data, such as graph classification, link prediction, personalized recommendation, etc.

Acknowledgments. This paper is supported by NUS ODPRT Grant R252-000-A81-133 and Singapore Ministry of Education Academic Research Fund Tier 3 under MOEs official grant number MOE2017-T3-1-007.

References

1. Bishop, C.M.: Pattern Recognition and Machine Learning. Springer, Boston (2006). https://doi.org/10.1007/978-1-4615-7566-5
2. Bruna, J., Zaremba, W., Szlam, A., LeCun, Y.: Spectral networks and locally connected networks on graphs. arXiv preprint arXiv:1312.6203 (2013)
3. Buja, A., Cook, D., Swayne, D.F.: Interactive high-dimensional data visualization. J. Comput. Graph. Stat. **5**(1), 78–99 (1996)
4. Cai, Y., Ge, L., Liu, J., Cai, J., Cham, T.J., Yuan, J., Thalmann, N.M.: Exploiting spatial-temporal relationships for 3d pose estimation via graph convolutional networks. In: Proceedings of the IEEE International Conference on Computer Vision, pp. 2272–2281 (2019)
5. Chen, J., Zhu, J., Song, L.: Stochastic training of graph convolutional networks with variance reduction. arXiv preprint arXiv:1710.10568 (2017)
6. Chiang, W.L., Liu, X., Si, S., Li, Y., Bengio, S., Hsieh, C.J.: Cluster-GCN: an efficient algorithm for training deep and large graph convolutional networks. In: Proceedings of the 25th ACM SIGKDD International Conference on Knowledge Discovery & Data Mining, pp. 257–266 (2019)
7. Defferrard, M., Bresson, X., Vandergheynst, P.: Convolutional neural networks on graphs with fast localized spectral filtering. In: Advances in Neural Information Processing Systems, pp. 3844–3852 (2016)
8. Gao, H., Wang, Z., Ji, S.: Large-scale learnable graph convolutional networks. In: Proceedings of the 24th ACM SIGKDD International Conference on Knowledge Discovery & Data Mining, pp. 1416–1424 (2018)
9. Hamilton, W., Ying, Z., Leskovec, J.: Inductive representation learning on large graphs. In: Advances in Neural Information Processing Systems, pp. 1024–1034 (2017)

10. Haonan, L., Huang, S.H., Ye, T., Xiuyan, G.: Graph star net for generalized multi-task learning. arXiv preprint arXiv:1906.12330 (2019)
11. Kipf, T.N., Welling, M.: Semi-supervised classification with graph convolutional networks. arXiv preprint arXiv:1609.02907 (2016)
12. Klicpera, J., Bojchevski, A., Günnemann, S.: Predict then propagate: Graph neural networks meet personalized pagerank. arXiv preprint arXiv:1810.05997 (2018)
13. Lee, C.Y., Xie, S., Gallagher, P., Zhang, Z., Tu, Z.: Deeply-supervised nets. In: Artificial Intelligence and Statistics, pp. 562–570 (2015)
14. London, B., Getoor, L.: Collective classification of network data. Data Classif. Algorithms Appl. **399** (2014)
15. McAuley, J., Leskovec, J.: Image labeling on a network: using social-network metadata for image classification. In: Fitzgibbon, A., Lazebnik, S., Perona, P., Sato, Y., Schmid, C. (eds.) ECCV 2012. LNCS, vol. 7575, pp. 828–841. Springer, Heidelberg (2012). https://doi.org/10.1007/978-3-642-33765-9_59
16. Monti, F., Boscaini, D., Masci, J., Rodola, E., Svoboda, J., Bronstein, M.M.: Geometric deep learning on graphs and manifolds using mixture model CNNs. In: Proceedings of the IEEE Conference on Computer Vision and Pattern Recognition, pp. 5115–5124 (2017)
17. Qi, C.R., Su, H., Mo, K., Guibas, L.J.: PointNet: deep learning on point sets for 3D classification and segmentation. In: Proceedings of the IEEE Conference on Computer Vision and Pattern Recognition, pp. 652–660 (2017)
18. Qu, M., Bengio, Y., Tang, J.: GMNN: graph Markov neural networks. arXiv preprint arXiv:1905.06214 (2019)
19. Quan, P., Shi, Y., Lei, M., Leng, J., Zhang, T., Niu, L.: A brief review of receptive fields in graph convolutional networks. In: IEEE/WIC/ACM International Conference on Web Intelligence, vol. 24800, pp. 106–110. ACM (2019)
20. Rong, Y., Huang, W., Xu, T., Huang, J.: DropEdge: towards deep graph convolutional networks on node classification. In: International Conference on Learning Representations (2019)
21. Veličković, P., Cucurull, G., Casanova, A., Romero, A., Lio, P., Bengio, Y.: Graph attention networks. arXiv preprint arXiv:1710.10903 (2017)
22. Veličković, P., Fedus, W., Hamilton, W.L., Liò, P., Bengio, Y., Hjelm, R.D.: Deep graph infomax. arXiv preprint arXiv:1809.10341 (2018)
23. Wu, Z., Pan, S., Chen, F., Long, G., Zhang, C., Yu, P.S.: A comprehensive survey on graph neural networks. arXiv preprint arXiv:1901.00596 (2019)
24. Xu, K., Li, C., Tian, Y., Sonobe, T., Kawarabayashi, K.i., Jegelka, S.: Representation learning on graphs with jumping knowledge networks. arXiv preprint arXiv:1806.03536 (2018)
25. Zeng, H., Zhou, H., Srivastava, A., Kannan, R., Prasanna, V.: GraphSaint: graph sampling based inductive learning method. arXiv preprint arXiv:1907.04931 (2019)
26. Zhang, J., Shi, X., Xie, J., Ma, H., King, I., Yeung, D.Y.: GAAN: gated attention networks for learning on large and spatiotemporal graphs. arXiv preprint arXiv:1803.07294 (2018)
27. Zhou, J., et al.: Graph neural networks: a review of methods and applications. arXiv preprint arXiv:1812.08434 (2018)
28. Zhuang, C., Ma, Q.: Dual graph convolutional networks for graph-based semi-supervised classification. In: Proceedings of the 2018 World Wide Web Conference, pp. 499–508 (2018)

Modeling Dynamic Heterogeneous Network for Link Prediction Using Hierarchical Attention with Temporal RNN

Hansheng Xue[1], Luwei Yang[2], Wen Jiang[2], Yi Wei[2], Yi Hu[2], and Yu Lin[1(✉)]

[1] Research School of Computer Science, The Australian National University, Canberra, Australia
{hansheng.xue,yu.lin}@anu.edu.au
[2] Alibaba Group, Hangzhou, China
{luwei.ylw,wen.jiangw,yi.weiy}@alibaba-inc.com

Abstract. Network embedding aims to learn low-dimensional representations of nodes while capturing structure information of networks. It has achieved great success on many tasks of network analysis such as link prediction and node classification. Most of existing network embedding algorithms focus on how to learn static homogeneous networks effectively. However, networks in the real world are more complex, e.g.., networks may consist of several types of nodes and edges (called heterogeneous information) and may vary over time in terms of dynamic nodes and edges (called evolutionary patterns). Limited work has been done for network embedding of dynamic heterogeneous networks as it is challenging to learn both evolutionary and heterogeneous information simultaneously. In this paper, we propose a novel dynamic heterogeneous network embedding method, termed as DyHATR, which uses hierarchical attention to learn heterogeneous information and incorporates recurrent neural networks with temporal attention to capture evolutionary patterns. We benchmark our method on four real-world datasets for the task of link prediction. Experimental results show that DyHATR significantly outperforms several state-of-the-art baselines.

Keywords: Dynamic heterogeneous network · Hierarchical attention · Recurrent neural network · Temporal self-attention

1 Introduction

Network embedding is to encode network structures into non-linear space and represent nodes of networks as low-dimensional features [2,6]. It has been a popular and critical machine learning task with wide applications in many areas, such as recommender systems [28,35], natural language processing [9,32] and computational biology [10,22,24].

L. Yang—Equal Contribution.

F. Hutter et al. (Eds.): ECML PKDD 2020, LNAI 12457, pp. 282–298, 2021.
https://doi.org/10.1007/978-3-030-67658-2_17

Existing network embedding methods have achieved significant performance on many downstream tasks such as link prediction and node classification, most of these approaches focus on static homogeneous networks [13,25,30], static heterogeneous networks [3,7,34,37,40,41] or dynamic homogeneous networks [11,12,27,42,43]. However, many networks in the real-world are both dynamic and heterogeneous, which usually contain multiple types of nodes or edges [3,7] and the structure of networks may evolve over time [8,31]. For example, customer-product networks are usually heterogeneous with multiple node types to distinguish customers and products as well as dynamic with evolving nodes and edges to capture dynamic user activities.

Network embedding is challenging for dynamic heterogeneous networks to capture both heterogeneous information and evolutionary patterns. Dynamic networks are usually described as an ordered list of static network snapshots [4,19,20,23,29]. As nodes and edges may vary in different snapshots, network embedding over dynamic heterogeneous networks need not only to capture the structural information of static heterogeneous snapshots but also to learn the evolutionary patterns between consecutive snapshots.

Currently, limited work has been done for network embedding of dynamic heterogeneous networks. MetaDynaMix [21] integrates metapath-based topology features and latent representations to learn both heterogeneity and temporal evolution. Change2vec [1] focuses on measuring changes in snapshots instead of learning the whole structural information of each snapshot, and it also uses a metapath-based model to capture heterogeneous information. The above two methods both focus on short-term evolutionary information between adjacent snapshots of dynamic networks and thus become insufficient to capture long-term evolutionary patterns. More recently, Sajadmanesh et al. [26] uses a recurrent neural network model to learn long-term evolutionary patterns of dynamic networks on top of metapath-based models and proposes a non-parametric generalized linear model, NP-GLM, to predict continuous-time relationships. Yin et al. [39] propose a DHNE method, which learns both the historical and current heterogeneous information and models evolutionary patterns by constructing comprehensive historical-current networks based on consecutive snapshots. Then, DHNE performs metapath-based random walk and dynamic heterogeneous skip-gram model to capture representations of nodes. Kong et al. [17] introduce a dynamic heterogeneous information network embedding method called HA-LSTM. It uses graph convolutional networks to learn heterogeneous information networks and employs attention models and long-short time memory to capture evolving information over timesteps. Refer to Table 1 for a brief summary of existing network embedding methods.

To better capture the heterogeneity of static snapshots and model evolutionary patterns among consecutive snapshots. In this paper, we propose a novel dynamic heterogeneous network embedding method, named DyHATR. Specifically, DyHATR uses a hierarchical attention model to learn static heterogeneous snapshots and temporal attentive RNN model to capture evolutionary patterns. The contributions of this paper are summarized as:

- We propose a novel dynamic heterogeneous network embedding method, named DyHATR, which captures both heterogeneous information and evolutionary patterns.
- We use the hierarchical attention model (including node-level and edge-level attention) to capture the heterogeneity of static snapshots.
- We emphasize the evolving information of dynamic networks and use the temporal attentive GRU/LSTM to model the evolutionary patterns among continuous snapshots.
- We evaluate our method DyHATR on the task of link prediction. The results show that DyHATR significantly outperforms several state-of-the-art baselines on four real-world datasets.

Table 1. Summary of typical network embedding methods.

Embedding type	Method	Node type	Edge type	Dynamic
Static Homogeneous Network Embedding (SHONE)	DeepWalk [25]	Single	Single	No
	node2vec [13]	Single	Single	No
	LINE [30]	Single	Single	No
	GCN [16]	Single	Single	No
	GraphSAGE [14]	Single	Single	No
	GAT [33]	Single	Single	No
Static Heterogeneous Network Embedding (SHENE)	metapath2vec [7]	Multiple	Single	No
	HAN [34]	Multiple	Single	No
	HetGNN [40]	Multiple	Single	No
	MNE [41]	Single	Multiple	No
	GATNE-T [3]	Multiple	Multiple	No
Dynamic Homogeneous Network Embedding (DHONE)	DynGEM [12]	Single	Single	Yes
	DynamicTriad [43]	Single	Single	Yes
	dyngraph2vec [11]	Single	Single	Yes
	EvolveGCN [23]	Single	Single	Yes
	E-LSTM-D [4]	Single	Single	Yes
	DySAT [27]	Single	Single	Yes
Dynamic Heterogeneous Network Embedding (DHENE)	MetaDynaMix [21]	Multiple	Single	Yes
	HA-LSTM [17]	Multiple	Single	Yes
	change2vec [1]	Multiple	Multiple	Yes
	NP-GLM [26]	Multiple	Multiple	Yes
	DHNE [39]	Multiple	Multiple	Yes

2 Method

The main task of dynamic heterogeneous network embedding is how to capture heterogeneous information and temporal evolutionary patterns over dynamic networks simultaneously. We propose a novel dynamic heterogeneous network embedding method that uses the hierarchical attention model to capture the heterogeneity of snapshots and temporal attentive RNN model to learn the evolutionary patterns over evolving time. The whole framework of our proposed model, DyHATR, is shown in Fig. 1.

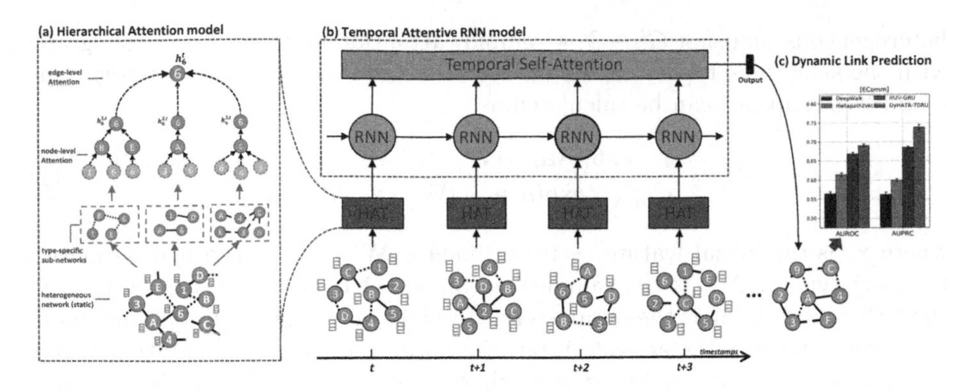

Fig. 1. The overall framework of DyHATR. The proposed model contains three parts, (a) Hierarchical Attention model, (b) Temporal Attentive RNN model and (c) Dynamic Link Prediction. The right histogram shows the performance improvement of DyHATR over DeepWalk, metapath2vec, and metapath2vec-GRU on the EComm dataset.

Dynamic heterogeneous networks are often described as an ordered list of observed heterogeneous snapshots $\mathbb{G} = \{G^1, G^2, ..., G^T\}$, where T is the number of continuous snapshots. $G^t = (V^t, E^t)$ represents the t-th snapshot, where V^t is the node set and E^t is the edge set. For each static heterogeneous snapshot, the node type set O and the edge type set R satisfies $|O| + |R| > 2$. The aim of dynamic network embedding is to learn a non-linear mapping function that encodes node $v \in V^t$ into a latent feature space $f : v \to \mathbf{z}_v^t$, where $\mathbf{z}_v^t \in \mathbb{R}^d, d \ll |V^t|$. For each snapshot G^t, the learned embedding matrix can be represented as $\mathbf{Z}^t \in \mathbb{R}^{|V^t| \times d}$, where d is the dimension of final embeddings. Note that each snapshot G^t is considered as a static heterogeneous network and we will first explain how DyHATR uses the hierarchical attention model to capture the heterogeneity of snapshots in the next section. Then the temporal attentive RNN for modeling evolutionary patterns across snapshots follows.

2.1 Hierarchical Attention for Heterogeneous Information

We introduce hierarchical attention model to capture heterogeneity information of static snapshots by splitting heterogeneous snapshots into several type-specific sub-networks according different edge types. The hierarchical attention model contains two components, node-level attention and edge-level attention. The illustration of hierarchical attention is shown in Fig. 1 (a).

Node-Level Attention. The node-level attention model aims to learn the importance weight of each node's neighborhoods and generate novel latent representations by aggregating features of these significant neighbors. For each static heterogeneous snapshot $G^t \in \mathbb{G}$, we employ attention models for every subgraph with the same edge type. The weight coefficient of node pair (i, j) for edge type r and t-th snapshot can be calculated as:

$$\alpha_{i,j}^{rt} = \frac{\exp(\sigma(\mathbf{a}_r^\top \cdot [\mathbf{W}^r \cdot \mathbf{x}_i || \mathbf{W}^r \cdot \mathbf{x}_j]))}{\sum_{k \in N_i^{rt}} \exp(\sigma(\mathbf{a}_r^\top \cdot [\mathbf{W}^r \cdot \mathbf{x}_i || \mathbf{W}^r \cdot \mathbf{x}_k]))}, \tag{1}$$

where $\mathbf{x}_i$ is the initial feature vector of node i, $\mathbf{W}^r$ is a transformation matrix for edge type r. N_i^{rt} is the sampled neighbors of node i with edge type r in the t-th snapshot; $\mathbf{a}_r$ is the parameterized weight vector of attention function in r-th edge type; σ is the activation function; and $||$ denotes the operation of concatenation. Then, aggregating the neighbors' latent embedding with calculated weight coefficients, we can obtain the final representation of node i at edge type r and t-th snapshot as

$$\hat{\mathbf{h}}_i^{rt} = \sigma(\sum_{j \in N_i^{rt}} \alpha_{ij}^{rt} \cdot \mathbf{W}^r \mathbf{x}_j), \tag{2}$$

where $\hat{\mathbf{h}}_i^{rt}$ is the aggregated embedding of node i for edge type r and t-th snapshot. To obtain stable and effective features, we employ the node-level attention model with multi-head mechanisms. Specifically, we parallelly run κ independent node-level attention models and concatenate these learned features as the output embedding. Thus, the multi-head attention representation of node i at edge type r and t-th snapshot can be described as

$$\mathbf{h}_i^{rt} = \text{Concat}(\hat{\mathbf{h}}^1, \hat{\mathbf{h}}^2, ..., \hat{\mathbf{h}}^\kappa), \tag{3}$$

where $\hat{\mathbf{h}}^\kappa$ is the simplified symbol of $\hat{\mathbf{h}}_i^{rt}$, which is the output from the κ-th node-level attention model, and κ is the number of attention heads. After the multi-head node-level attention model, we can obtain node representation sets with different edge types R and all snapshots T, $\{\mathbf{h}_1^{rt}, \mathbf{h}_2^{rt}, ..., \mathbf{h}_{|V^t|}^{rt}, \mathbf{h}_i^{rt} \in \mathbb{R}^{F'}, r \in R, t \in T\}$, where $F' \ll |V^t|$ is the dimension of the embedded features at the node-level attention model.

Edge-Level Attention. Node-level attention model can capture single edge-type-specific information. However, heterogeneous networks usually contain multiple types of edges. To integrate multiple edge-specific information for each node, we employ an edge-level attention model to learn the importance weight of different types of edges and aggregate these various type-specific information to generate novel embeddings.

Firstly, we input edge-specific embeddings into non-linear transformation functions to map into the same feature space $\sigma(\mathbf{W} \cdot \mathbf{h}_i^{rt} + \mathbf{b})$, where σ is the activation function (i.e., tanh function); $\mathbf{W}$ and $\mathbf{b}$ are learnable weight matrix and bias vector respectively. Both these parameters are shared across all different edge types and snapshots. Then, we measure the importance coefficients of inputting edge-specific embedding by calculating similarities between mapped edge-specific embedding and edge-level attention parameterized vector $\mathbf{q}$. The normalized weight coefficients of node i for edge type r and t-th snapshot can be represented as β_i^{rt}. Then, we can aggregate these edge-specific embeddings to generate the final representation features of node i at t-th snapshot:

$$\beta_i^{rt} = \frac{\exp(\mathbf{q}^\top \cdot \sigma(\mathbf{W} \cdot \mathbf{h}_i^{rt} + \mathbf{b}))}{\sum_{r \in R} \exp(\mathbf{q}^\top \cdot \sigma(\mathbf{W} \cdot \mathbf{h}_i^{rt} + \mathbf{b}))}, \qquad \mathbf{h}_i^t = \sum_{r=1}^{R} \beta_i^{rt} \cdot \mathbf{h}_i^{rt}, \tag{4}$$

After aggregating all edge-specific embeddings, we can obtain the final node embeddings for different snapshots, $\{\mathbf{h}^1, \mathbf{h}^2, ..., \mathbf{h}^T, \mathbf{h}^t \in \mathbb{R}^{|V^t| \times F}\}$, where $F \ll |V^t|$ is the dimension of embeddings outputted by edge-level attention model.

2.2 Temporal Attentive RNN for Evolutionary Patterns

The temporal evolutionary patterns show the appearance and disappearance of nodes and edges as time goes by. The hierarchical attention model of DyHATR can capture the heterogeneity of static snapshots effectively, but it cannot model patterns evolving over time. Recently, the introduction of the recurrent neural network (RNN) for the dynamic network embedding methods achieve promising performance [11,29,36]. In our paper, we extend the existing RNN model and propose a temporal attentive RNN model to capture deeper evolutionary patterns among continuous timestamps. The proposed temporal attentive RNN model mainly contains two parts, the recurrent neural network and the temporal self-attention model. The illustration of temporal attentive RNN model is shown in Fig. 1 (b).

Recurrent Neural Network Model. The structure of recurrent neural network enables the capability of modeling sequential information and learning evolutionary patterns among continuous snapshots. Two prominent and widely-used variants of RNN are used in our proposed model, Long short term memory (LSTM) and Gated-recurrent-unit (GRU).

Long short-term memory (LSTM) is an classical recurrent neural network model which can capture long-distance information of a sequence data and achieves promising performance on dynamic network learning tasks [15]. Through the above hierarchical attention model, we can obtain node embedding set for all snapshots $\{\mathbf{h}^1, \mathbf{h}^2, ..., \mathbf{h}^T, \mathbf{h}^t \in \mathbb{R}^{|V^t| \times F}\}$, where t is the t-th snapshot, $|V^t|$ represents the number of nodes in the t-th snapshot, and F indicates the dimension of each node embedding. Each LSTM unit computes the input vector $\mathbf{i}^t$, the forget gate vector $\mathbf{f}^t$, the output gate vector $\mathbf{o}^t$, the memory cell vector $\mathbf{c}^t$ and the state vector $\mathbf{s}^t$. The formulations of single LSTM unit are as follows:

$$\begin{aligned}
\mathbf{i}^t &= \sigma(\mathbf{W}_i \cdot [\mathbf{h}^t || \mathbf{s}^{t-1}] + \mathbf{b}_i), \\
\mathbf{f}^t &= \sigma(\mathbf{W}_f \cdot [\mathbf{h}^t || \mathbf{s}^{t-1}] + \mathbf{b}_f), \\
\mathbf{o}^t &= \sigma(\mathbf{W}_o \cdot [\mathbf{h}^t || \mathbf{s}^{t-1}] + \mathbf{b}_o), \\
\widetilde{\mathbf{c}^t} &= \tanh(\mathbf{W}_c \cdot [\mathbf{h}^t || \mathbf{s}^{t-1}] + \mathbf{b}_c), \\
\mathbf{c}^t &= \mathbf{f}^t \odot \mathbf{c}^{t-1} + \mathbf{i}^t \odot \widetilde{\mathbf{c}^t}, \\
\mathbf{s}^t &= \mathbf{o}^t \odot \tanh(\mathbf{c}^t),
\end{aligned} \tag{5}$$

where $t \in \{1, 2, ..., T\}$; $\mathbf{i}^t, \mathbf{f}^t, \mathbf{o}^t, \mathbf{c}^t \in \mathbb{R}^D$ are input, forget, output gates and memory cell respectively; $\mathbf{W}_i, \mathbf{W}_f, \mathbf{W}_o, \mathbf{W}_c \in \mathbb{R}^{D \times 2F}$ and $\mathbf{b}_i, \mathbf{b}_f, \mathbf{b}_o, \mathbf{b}_c \in \mathbb{R}^D$ are trainable parameters; σ is the activation function; $||$ denotes the operation of concatenation; and $\odot$ is an element-wise multiplication operator.

Gated-recurrent-unit (GRU) is a simpler variant of LSTM [5] and introduced to capture the long-term association in many dynamic network embedding models [23,36]. GRU units have fewer parameters to learn compared to LSTM units, and each GRU unit only contains two gates, update gate and reset gate. The state vector $\mathbf{s}^t \in \mathbb{R}^D$ for each input snapshot can be calculated as the following equations iteratively:

$$\begin{aligned}
\mathbf{u}^t &= \sigma(\mathbf{W}_u \cdot [\mathbf{h}^t || \mathbf{s}^{t-1}] + \mathbf{b}_u), \\
\mathbf{r}^t &= \sigma(\mathbf{W}_r \cdot [\mathbf{h}^t || \mathbf{s}^{t-1}] + \mathbf{b}_r), \\
\widetilde{\mathbf{s}^t} &= \tanh(\mathbf{W}_s \cdot [\mathbf{h}^t || (\mathbf{r}^t \odot \mathbf{s}^{t-1})] + \mathbf{b}_s), \\
\mathbf{s}^t &= (1 - \mathbf{u}^t) \odot \mathbf{s}^{t-1} + \mathbf{u}^t \odot \widetilde{\mathbf{s}^t},
\end{aligned} \tag{6}$$

where $t \in \{1, 2, ..., T\}$; $\mathbf{u}^t, \mathbf{r}^t \in \mathbb{R}^D$ are update and reset gates respectively; $\mathbf{W}_u, \mathbf{W}_r, \mathbf{W}_s \in \mathbb{R}^{D \times 2F}$ and $\mathbf{b}_u, \mathbf{b}_r, \mathbf{b}_s \in \mathbb{R}^D$ are trainable parameters; σ is the activation function; $||$ denotes the operation of concatenation; and $\odot$ is an element-wise multiplication operator. In the iterative equation of GRU, we use $\mathbf{r}^t \odot \mathbf{s}^{t-1}$ to model the latent features communicating from the previous state.

The outputs of the RNN model are denoted as $\{\mathbf{s}^1, \mathbf{s}^2, ..., \mathbf{s}^T, \mathbf{s}^t \in \mathbb{R}^{|V^t| \times D}\}$, where D is the dimension of each node after training the RNN model. Then, a traditional step may concatenate these state vectors $[\mathbf{s}_i^1 || \mathbf{s}_i^2 ||, ..., || \mathbf{s}_i^T]$ or take the last state $\mathbf{s}_i^T$ as the final embedding for node i. Note that the original state vectors (output from RNN) contain temporal information that allows us to capture time-related patterns but the traditional approach may not make full use of such temporal information. Therefore, we employ a temporal-level attention model on the output of the RNN model to capture the significant feature vectors.

Temporal Self-attention Model. Instead of concatenating all feature vectors together as the final embedding, we employ a temporal-level self-attention model to further capture the evolutionary patterns over dynamic networks. From the output of RNN model, we can obtain the embeddings of node i at different snapshots, $\{\mathbf{s}_i^1, \mathbf{s}_i^2, ..., \mathbf{s}_i^T, \mathbf{s}_i^t \in \mathbb{R}^D\}$, where D is the dimension of input embeddings. Assuming the output of temporal-level attention model for node i is $\{\mathbf{z}_i^1, \mathbf{z}_i^2, ..., \mathbf{z}_i^T\}, \mathbf{z}_i^t \in \mathbb{R}^d$, where d is the dimension of output embeddings.

The Scaled Dot-Product Attention is used in our model to learn the embeddings of each node at different snapshot. We pack representations from previous step for node i across time, $\mathbf{S}_i \in \mathbb{R}^{T \times D}$, as input. The corresponding output is denoted as $\mathbf{Z}_i \in \mathbb{R}^{T \times d}$. Firstly, the input is mapped into different feature space, $\mathbf{Q}_i = \mathbf{S}_i\mathbf{W}_q$ for queries, $\mathbf{K}_i = \mathbf{S}_i\mathbf{W}_k$ for keys, and $\mathbf{V}_i = \mathbf{S}_i\mathbf{W}_v$ for values, where $\mathbf{W}_q, \mathbf{W}_k, \mathbf{W}_v \in \mathbb{R}^{D \times d}$ are trainable parameters. The temporal-level attention model is defined as:

$$\mathbf{Z}_i = \mathbf{\Gamma}_i \cdot \mathbf{V}_i = \text{softmax}(\frac{(\mathbf{S}_i\mathbf{W}_q)(\mathbf{S}_i\mathbf{W}_k)^T}{\sqrt{d}} + \mathbf{M}) \cdot (\mathbf{S}_i\mathbf{W}_v), \tag{7}$$

where $\mathbf{\Gamma}_i \in \mathbb{R}^{T \times T}$ is the importance matrix; $\mathbf{M} \in \mathbb{R}^{T \times T}$ denotes the mask matrix. If $M_{uv} = -\infty$, it means the attention from snapshot u to v is off and the corresponding element in the importance matrix is zero, i.e., $\Gamma_i^{uv} = 0$. When snapshot satisfies $u \leq v$, we set $M_{u,v} = 0$; otherwise $M_{u,v} = -\infty$.

It is also feasible to use multi-head attention to improve the effectiveness and robustness of the model. The final embeddings of node i with κ' heads can be formulated as $\mathbf{Z}_i = \text{Concat}(\hat{\mathbf{Z}}_i^1, \hat{\mathbf{Z}}_i^2, ..., \hat{\mathbf{Z}}_i^{\kappa'})$. At this step, we obtain all of the embeddings of node i across all snapshots. But we will use embeddings of the last snapshot, denoted as $\mathbf{z}_i^T$, for optimization and downstream tasks.

2.3 Objective Function

The aim of DyHATR is to capture both heterogeneous information and evolutionary patterns among dynamic heterogeneous networks, we define the objective function that utilizes a binary cross-entropy function to ensure node u at the last snapshot have similar embedding features with its nearby nodes. The loss function of DyHATR is defined as:

$$L(\mathbf{z}_u^T) = \sum_{v \in N^T(u)} -\log(\sigma(<\mathbf{z}_u^T, \mathbf{z}_v^T>)) - Q \cdot \mathbb{E}_{v_n \sim P_n(v)} \log(\sigma(<-\mathbf{z}_u^T, \mathbf{z}_{v_n}^T>)) + \lambda \cdot L_p. \tag{8}$$

where σ is the activation function (i.e., sigmoid function); $< . >$ is the inner product operation; $N^T(u)$ is the neighbor with fixed-length random walk of node u at the last t-th snapshot; $P_n(v)$ denotes the negative sampling distribution and Q is the number of negative samples; L_p is the penalty term of the loss function to avoid over-fitting, i.e., L_2 regularization; λ is the hyper-parameters to control the penalty function. The pseudocode for DyHATR is shown in Algorithm 1.

Algorithm 1: The DyHATR algorithm.

Input: A sequence of snapshots $\mathbb{G} = \{G^1, G^2, ..., G^T\}$ with $G^t = (V^t, E^t)$, the dimension of output embeddings d, the number of multi-heads κ, initialization parameters;

Output: Node Embedding $\mathbf{Z}$

1 **for** each timestamp $t \in [1, 2, ...T]$ **do**
2   Construct edge-type-specific sub-networks set R;
3   **for** each edge-type $r \in R$ **do**
4   Obtain the weight coefficient of node pair (i, j), $\alpha_{i,j}^{rt}$ by Equation (1);
5   Obtain the representation of node i at edge type r and t-th snapshot, $\hat{\mathbf{h}_i^{rt}}$ by Equation (2);
6   The multi-head attention of node i, $\mathbf{h}_i^{rt} \leftarrow \text{Concat}(\hat{\mathbf{h}}^1, \hat{\mathbf{h}}^2, ..., \hat{\mathbf{h}}^\kappa)$;
7   **end**
8   Calculate the weight coefficient of node i for edge type r and t-th snapshot, β_i^{rt};
9   Obtain the representation of node i at t-th snapshot, $\mathbf{h}_i^t$ by Equation (4);
10 **end**
11 Obtain $\mathbf{s}_i^t$ by Equation (5) or (6) through GRU/LSTM ;
12 Get $\mathbf{Z}_i \in \mathbb{R}^{T \times d}$ by Equation (7) through temporal self-attention model;
13 Optimize the loss function $L(\mathbf{z}_u^T)$ by Equation (8);
14 **return** $\mathbf{Z}$;

3 Experiments

3.1 Datasets

We benchmark DyHATR on dynamic link prediction to evaluate the performance of network embedding. Four real-world dynamic network datasets are used in our experiments. The statistics of these datasets are listed in Table 2.

Twitter.[1] This social network is sampled from the Higgs Twitter dataset of the SNAP platform [18]. It reflects three kinds of user behaviors (re-tweeting, replying and mentioning) between 1st and 7th July 2012.

Math-Overflow.[2] This temporal network is collected from the stack exchange website Math Overflow and opened on the SNAP platform [18]. It represents three different types of interactions between users (answers to questions, comments to questions, comments to answer) within 2,350 days. In our experiments, we split this time span into 11 snapshots.

EComm.[3] This dataset, real-world heterogeneous bipartite graphs of the e-commerce, is extracted from the AnalytiCup challenge of CIKM-2019. EComm mainly records shopping behaviors of users within 11 daily snapshots from 10th

[1] http://snap.stanford.edu/data/higgs-twitter.html.
[2] http://snap.stanford.edu/data/sx-mathoverflow.html.
[3] https://tianchi.aliyun.com/competition/entrance/231719.

June 2019 to 20th June 2019, and it consists of two types of nodes (users and items) and four types of edges (click, buy, add-to-cart and add-to-favorite).

Alibaba.com. This dataset consists of user behavior logs collected from the e-commerce platform of Alibaba.com [38]. It mainly contains customers' activities records from 11th July 2019 to 21st July 2019, and consists of two types of nodes (users and items) and three types of activities (click, enquiry and contact).

Table 2. Statistics of four real-world datasets.

Dataset	#Nodes	#Edges	#Node types	#Edge types	#snapshots
Twitter	100,000	63,410	1	3	7
Math-overflow	24,818	506,550	1	3	11
EComm	37,724	91,033	2	4	11
Alibaba.com	16,620	93,956	2	3	11

3.2 Experimental Setup

The task of dynamic link prediction is to learn node representations over previous t snapshots and predict the links at $(t+1)$-th snapshot. Specifically, the previous t snapshots $\{G^1, ..., G^t\}$ are used to learn representations of nodes and the $(t+1)$-th snapshot G^{t+1} is the whole evaluation set. We randomly sample 20% of edges in the evaluation set (snapshot G^{t+1}) as the hold-out validation set to tune hyper-parameters. Then the remaining 80% of edges in the snapshot G^{t+1} are used for link prediction task. Among the remaining evaluation set, we further randomly select 25% edges and the rest 75% edges as training and test sets respectively for link prediction. Meanwhile, we randomly sample equal number of pairs of nodes without link as negative examples for training and test sets respectively. We use the inner product of the embedding features of two nodes as the feature of the link.

For this task, we train a Logistic Regression model as the classifier, which is similar with previous works [27,43]. We use the area under the ROC curve (AUROC) and area under the precision-recall curve (AUPRC) as evaluation metrics. We repeatedly run our model and baselines for five times and report the mean and standard deviation values.

The proposed DyHATR model is conducted on four real-world datasets using the Linux server with 6 Intel(R) Core(TM) i7-7800X CPU @3.50 GHz, 96 GB RAM and 2 NVIDIA TITAN Xp 12 GB. The codes of DyHATR are implemented in Tensorflow 1.14 and Python 3.6. In the DyHATR model, we sample 25 neighbors in each layer, the number of negative samples is 5, and use multi-head attention model with 4 or 8 heads to capture node feature embeddings. The dimension of the final embedding is 32. The stochastic gradient descent (SGD) and Adam optimizer are used in our proposed model to update and optimize parameters. In the dynamic link prediction part, we use a logistic regression classifier and evaluation metric functions from the scikit-learn library. DyHATR is freely available at https://github.com/skx300/DyHATR.

3.3 Baselines

We compare DyHATR with several state-of-the-art network embedding methods, including three for static homogeneous networks (DeepWalk [25], GraphSAGE [14] and GAT [33]), two for static heterogeneous networks (metapath2vec [7], MNE [41]), and three for dynamic homogeneous networks (DynamicTriad [43], dyngraph2vec [11] and DySAT [27]). For static network embedding methods, we firstly integrate all snapshots into a static network. Then, we apply them on the integrated static network to learn latent representations of nodes.

There exists five dynamic heterogeneous network embedding methods, MetaDynaMix [21], change2vec [1], HA-LSTM [17], DHNE [39] and NP-GLM [26]. Here, we mainly focus on DHNE and NP-GLM method. The representative algorithm NP-GLM mainly uses the idea of combining metapath2vec and RNN model to embed dynamic heterogeneous networks, we thus compare DyHATR with two methods, metapath2vec-GRU and metapath2vec-LSTM. To make a fair comparison, we set the same final embedding size as DyHATR, i.e. 32, for all baselines. The hyper-parameters for different baselines are all optimized specifically to be optimal.

Table 3. The experimental results on the task of dynamic link prediction. The bold value is the best performance achieved by DyHATR. The underline indicates the highest AUROC/AUPRC score achieved by baselines.

Methods	Twitter		Math-overflow		EComm		Alibaba.com	
	AUROC	AUPRC	AUROC	AUPRC	AUROC	AUPRC	AUROC	AUPRC
DeepWalk	0.520(0.040)	0.686(0.021)	0.714(0.002)	0.761(0.001)	0.564(0.006)	0.562(0.006)	0.546(0.019)	0.570(0.008)
GraphSAGE-mean	0.600(0.024)	0.765(0.016)	0.683(0.005)	0.712(0.003))	0.635(0.007)	0.642(0.008)	0.571(0.019)	<u>0.610(0.014)</u>
GraphSAGE-meanpool	0.562(0.009)	0.711(0.005)	0.671(0.007)	0.697(0.008)	0.600(0.018)	0.590(0.017)	0.552(0.015)	0.570(0.021)
GraphSAGE-maxpool	0.571(0.031)	0.731(0.021)	0.637(0.008)	0.660(0.009)	0.594(0.010)	0.587(0.009)	0.542(0.009)	0.573(0.008)
GraphSAGE-LSTM	0.546(0.016)	0.710(0.010)	0.672(0.005)	0.697(0.008)	0.599(0.013)	0.588(0.015)	0.567(0.006)	0.585(0.007)
GAT	0.587(0.019)	0.744(0.013)	<u>0.737(0.004)</u>	0.768(0.003)	0.648(0.003)	0.642(0.008)	<u>0.572(0.010)</u>	<u>0.610(0.008)</u>
metapath2vec	0.520(0.040)	0.686(0.021)	0.714(0.002)	0.761(0.001)	0.614(0.005)	0.599(0.004)	0.561(0.004)	0.601(0.007)
MNE	0.605(0.013)	0.729(0.014)	0.723(0.012)	0.766(0.010)	0.525(0.003)	0.535(0.006)	0.505(0.005)	0.523(0.010)
DynamicTriad	0.626(0.027)	0.785(0.013)	0.704(0.005)	0.759(0.003)	0.614(0.008)	<u>0.688(0.007)</u>	0.543(0.013)	0.598(0.011)
dyngraph2vec-AE	0.549(0.025)	0.724(0.021)	0.517(0.006)	0.565(0.005)	0.503(0.002)	0.509(0.006)	0.506(0.019)	0.534(0.014)
dyngraph2vec-AERNN	0.511(0.034)	0.706(0.031)	0.582(0.003)	0.598(0.003)	0.503(0.004)	0.501(0.004)	0.505(0.008)	0.554(0.045)
DySAT	<u>0.634(0.003)</u>	<u>0.796(0.005)</u>	0.506(0.008)	0.543(0.006)	0.512(0.002)	0.513(0.002)	0.521(0.013)	0.550(0.010)
metapath2vec-GRU	0.539(0.025)	0.744(0.018)	0.730(0.004)	0.770(0.004)	<u>0.668(0.004)</u>	0.685(0.003)	0.544(0.003)	0.576(0.004)
metapath2vec-LSTM	0.554(0.027)	0.754(0.017)	0.727(0.004)	<u>0.771(0.003)</u>	0.666(0.005)	0.683(0.008)	0.547(0.003)	0.578(0.008)
DHNE	0.552(0.005)	0.649(0.010)	0.678(0.001)	0.720(0.001)	0.547(0.003)	0.626(0.004)	0.515(0.001)	0.561(0.003)
DyHATR-TGRU	0.649(0.005)	0.805(0.007)	0.741(0.002)	0.781(0.001)	0.690(0.004)	**0.738(0.007)***	0.590(0.008)	0.602(0.008)
DyHATR-TLSTM	**0.660(0.010)***	**0.810(0.009)***	**0.751(0.001)***	**0.790(0.002)***	**0.696(0.005)***	0.734(0.005)	**0.605(0.003)***	**0.617(0.005)**

* Asterisks indicate the improvement over baselines achieved by DyHATR is significant (rank-sum p-value < 0.01).

3.4 Experimental Results

Task of Link Prediction. The experimental results are summarized in Table 3. Overall, DyHATR achieves the best performance on both AUROC and AUPRC metrics over four datasets among all baselines, including three typical dynamic heterogeneous network embedding methods (metapath2vec-GRU, metapath2vec-LSTM, and DHNE). The highest AUROC and AUPRC achieved by DyHATR on the EComm dataset are 0.696 and 0.738 respectively, which are significantly higher than the scores of baselines (0.668 for AUROC by metapath2vec-GRU and 0.688 for AUPRC by DynamicTrid). For Twitter, the AUROC and AUPRC score of DyHATR-TLSTM are 0.660 and 0.810 respectively, which are sightly higher than the second highest score achieved by DySAT (0.634 for AUROC and 0.796 for AUPRC). Besides, we also perform the rank-sum test to validate the significance of experimental results achieved by DyHATR. Note that dynamic homogeneous network embedding methods (such as DySAT and DynamicTriad) perform relatively better on the Twitter and Math-Overflow datasets than EComm and Alibaba.com datasets, probably because there are less number of node and edge types in Twitter and Math-Overflow than the other two datasets.

Effectiveness of Hierarchical Attention Model. The dynamic heterogeneous network can be represented as an ordered series of static heterogeneous snapshots. In our proposed DyHATR model, we use a hierarchical attention model to capture the heterogeneity of static snapshots. To evaluate the effectiveness of DyHATR on capturing heterogeneity of static snapshots, we compare the hierarchical attention model (HAT) with metapath2vec (m2v), a classic and widely-used heterogeneous network embedding method.

To make a full comparison, we adopt two different ways to integrate multiple snapshots. One is to merge snapshots into a single heterogeneous snapshot and run HAT/m2v representation learning methods. The other is to run HAT/m2v on each snapshot respectively and concatenate these embedding vectors for the downstream tasks. From the experimental results (Fig. 2), the hierarchical attention model outperforms metapath2vec on both Twitter and EComm datasets with two different integration methods, demonstrating the effectiveness of HAT on learning heterogeneous information of each snapshot.

Effectiveness of Temporal Attentive RNN Model. In our proposed method, DyHATR uses the temporal attentive GRU (TGRU) and LSTM (TSTM) model to learn the evolutionary patterns among continuous timestamps. To validate the effectiveness of this component. We have done an experiment by replacing/adding sub-component step by step. All methods use HAT to model each snapshot. The initial method is HAT-C, which concatenates the embedding vectors from each snapshot. Then the second method HAT-GRU/LSTM uses GRU/LSTM module to replace the concatenation to capture the sequential information. While the third one HAT-T uses temporal attention model instead. The final one HAT-TGRU/TLSTM, which is our proposed DyHATR, combines temporal attention model and GRU/LSTM. Figure 3 shows the experimental results on Twitter and EComm datasets. We can see that either GRU/LSTM or

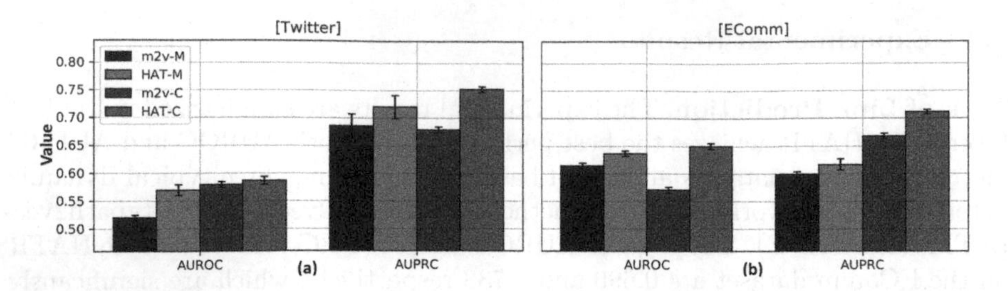

Fig. 2. The comparison results between hierarchical attention model (HAT) and metapath2vec (m2v). The x-axis shows two evaluation metrics (AUROC and AUPRC). The y-axis represents the evaluation value. HAT-M/m2v-M denotes merge snapshots into a single network and run HAT/m2v model; HAT-C/m2v-C represents run HAT/m2v first on each snapshot and concatenate these embeddings to predict dynamic links.

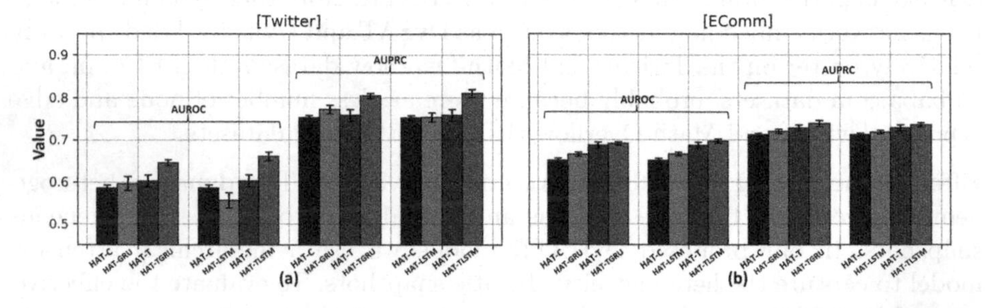

Fig. 3. The comparison results of different components in DyHATR. The x-axis shows models with different components. The y-axis represents the AUROC/AUPRC value. HAT-C denotes hierarchical attention model; HAT-T is the HAT model with temporal attention model; TGRU/TLSTM denotes temporal attentive GRU/LSTM model.

temporal attention model outperforms the naive concatenation. The combination of temporal attention model and GRU/LSTM further improve the performance better. This shows the superiority of temporal attentive RNN model on modeling evolving information.

Parameters Sensitivity. The number of multi-heads in both hierarchical attention (HAT) and temporal attention (TAT), and the dimension of final embedding output are critical parameters for our proposed DyHATR algorithm. Thus, we further analyze the effect of these parameters on the dynamic link prediction task. When we vary one parameter to check the sensitivity, the other parameters are kept fixed. Figure 4 (a) and (b) show the result on Twitter, the optimal numbers of multi-head for HAT and TAT are 4 and 8 respectively. Similarly, Fig. 4 (d) and (e) report 8 and 8 for EComm. Figure 4 (c) and (f) shows the performance respect to the dimension of final embedding output. When the dimension equals to 32, DyHATR achieves the highest AUROC value in our experiments.

Besides, we also compare the training time of DyHATR with metapath2vec-GRU/LSTM methods. For EComm, the training time of metapath2vec-GRU

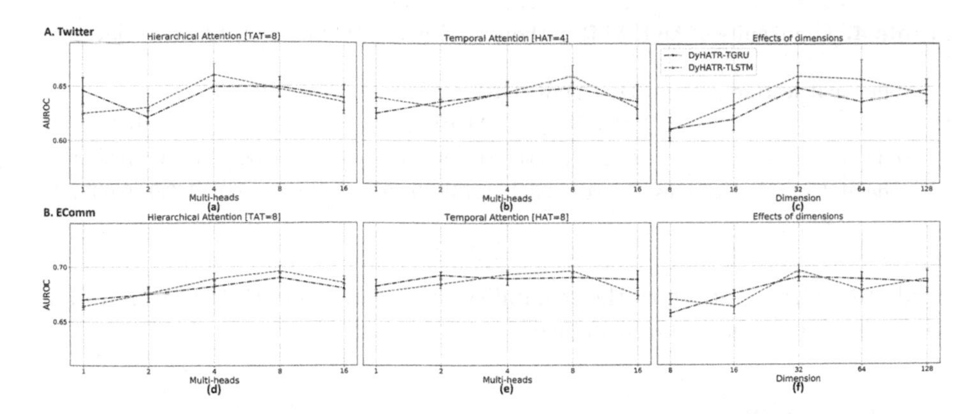

Fig. 4. The results of DyHATR varying different parameters on Twitter and EComm dataset. Upper row shows the results on Twitter dataset. Lower row reports the results on EComm dataset. (a) and (d) varies the number of multi-heads on hierarchical attention model; (b) and (e) varies the number of multi-heads on temporal attention model; (c) and (f) validates the effects of embedding dimensions. The x-axis shows the varying parameters and the y-axis represents the AUROC value.

and metapath2vec-LSTM are 475 s and 563 s respectively. DyHATR is slightly slower than meatpath2vec–based methods (579 s for DyHATR-TGRU and 595 s for DyHATR-TLSTM), because of the large scale of parameters in attention and RNN model.

Granularity of Snapshots. In this paper, we describe dynamic network as an ordered list of snapshots. The duration of each snapshot will affect the total number of snapshots. For example, the duration of snapshot in EComm is one day, which results in 10 snapshots. We find that the finer granularity of snapshots, which means the shorter of the duration, would improve the performance. When the duration is set to two days in EComm, there are 5 snapshots in total. The AUROC values are 0.640 and 0.625 for DyHATR-TGRU and DyHATR-TLSTM respectively, which are worse than that of 10 snapshots. Same situation is observed for Twitter, when we set duration as two days, there are 3 snapshots. The AUROC values are 0.580 and 0.570 for DyHATR-TGRU and DyHATR-TLSTM respectively, which are worse than that of 6 snapshots. One of the reasons is that the finer granularity is more beneficial to capture evolving patterns.

Effects of Attributes. Considering not all datasets contains node attributes, in previous experiments we only use node id embedding as the node input feature for convenience. However, it is easy to incorporate node inherent attributes as input feature. For example, in EComm dataset, users and items contain many attributes, such as gender, age, education, career, income, category, brand, prize and so on. One may simply process and concatenate these information as input features of nodes. Table 4 shows an experiment on EComm dataset using more

Table 4. The results of DyHATR without and with attributes on EComm dataset.

Node input feature	DyHATR-TGRU		DyHATR-TLSTM	
	AUROC	AUPRC	AUROC	AUPRC
Node id	0.690 (0.004)	0.738 (0.007)	0.696 (0.005)	0.734 (0.005)
Node id + other attributes	0.704 (0.002)	0.753 (0.006)	0.701 (0.004)	0.749 (0.005)

attributes. The results show the scalability of our proposed DyHATR and more attributes may improve performance.

4 Conclusions

Network embedding has recently been widely used in many domains and achieved significant progress. However, existing network embedding methods mainly focus on static homogeneous networks, which can not directly be applied to dynamic heterogeneous networks. In this paper, we propose a dynamic heterogeneous network embedding algorithm, termed as DyHATR, which can simultaneously learn both heterogeneous information and evolutionary patterns. DyHATR mainly contains two components, hierarchical attention for capturing heterogeneity and temporal attentive RNN model for learning evolving information overtime. Compared with other state-of-the-art baselines, DyHATR demonstrates promising performance on the task of dynamic links prediction. Besides, our proposed DyHATR is scalable to attributed dynamic heterogeneous networks and model inherent information of nodes.

References

1. Bian, R., Koh, Y.S., Dobbie, G., Divoli, A.: Network embedding and change modeling in dynamic heterogeneous networks. In: SIGIR, pp. 861–864 (2019)
2. Cai, H., Zheng, V.W., Chang, K.: A comprehensive survey of graph embedding: problems, techniques, and applications. IEEE TKDE **30**(09), 1616–1637 (2018)
3. Cen, Y., Zou, X., Zhang, J., Yang, H., Zhou, J., Tang, J.: representation learning for attributed multiplex heterogeneous network. In: SIGKDD, pp. 1358–68 (2019)
4. Chen, J., et al.: E-LSTM-D: a deep learning framework for dynamic network link prediction. IEEE Trans. Syst. Man Cybern. Syst. 1–14 (2019)
5. Cho, K., et al.: Learning phrase representations using RNN encoder-decoder for statistical machine translation. arXiv:abs/1406.1078 (2014)
6. Cui, P., Wang, X., Pei, J., Zhu, W.: A survey on network embedding. IEEE TKDE **31**(05), 833–852 (2019)
7. Dong, Y., Chawla, N.V., Swami, A.: Metapath2Vec: scalable Representation Learning for Heterogeneous Networks. In: SIGKDD, pp. 135–144 (2017)
8. Du, L., Wang, Y., Song, G., Lu, Z., Wang, J.: Dynamic network embedding : an extended approach for skip-gram based network embedding. In: IJCAI, pp. 2086–2092 (2018)

9. Fang, H., Wu, F., Zhao, Z., Duan, X., Zhuang, Y., Ester, M.: Community-based question answering via heterogeneous social network learning. In: AAAI, pp. 122–128 (2016)
10. Gligorijević, V., Barot, M., Bonneau, R.: deepNF: deep network fusion for protein function prediction. Bioinformatics **34**(22), 3873–3881 (2018)
11. Goyal, P., Chhetri, S.R., Canedo, A.: dyngraph2vec: capturing network dynamics using dynamic graph representation learning. Knowl. Based Syst. **187**, 104816 (2019)
12. Goyal, P., Kamra, N., He, X., Liu, Y.: Dyngem: Deep embedding method for dynamic graphs. arXiv preprint arXiv:1805.11273 (2018)
13. Grover, A., Leskovec, J.: Node2Vec: scalable feature learning for networks. In: SIGKDD, pp. 855–864 (2016)
14. Hamilton, W.L., Ying, R., Leskovec, J.: Inductive representation learning on large graphs. In: NIPS, pp. 1024–1034 (2017)
15. Hochreiter, S., Schmidhuber, J.: Long short-term memory. Neural Comput. **9**, 1735–1780 (1997)
16. Kipf, T.N., Welling, M.: Semi-supervised classification with graph convolutional networks. In: ICLR (2017)
17. Kong, C., Li, H., Zhang, L., Zhu, H., Liu, T.: Link prediction on dynamic heterogeneous information networks. In: CSoNet (2019)
18. Leskovec, J., Krevl, A.: SNAP Datasets: Stanford large network dataset collection, June 2014. http://snap.stanford.edu/data
19. Li, T., Zhang, J., Yu, P.S., Zhang, Y., Yan, Y.: Deep dynamic network embedding for link prediction. IEEE Access **6**, 29219–29230 (2018)
20. Lu, Y., Wang, X., Shi, C., Yu, P.S., Ye, Y.: Temporal network embedding with micro- and macro-dynamics. In: CIKM, p. 469–478 (2019)
21. Milani Fard, A., Bagheri, E., Wang, K.: Relationship prediction in dynamic heterogeneous information networks. In: Azzopardi, L., Stein, B., Fuhr, N., Mayr, P., Hauff, C., Hiemstra, D. (eds.) ECIR 2019. LNCS, vol. 11437, pp. 19–34. Springer, Cham (2019). https://doi.org/10.1007/978-3-030-15712-8_2
22. Nelson, W., Zitnik, M., Wang, B., Leskovec, J., Goldenberg, A., Sharan, R.: To embed or not: network embedding as a paradigm in computational biology. Front. Genet. **10**, 381 (2019)
23. Pareja, A., et al.: EvolveGCN: evolving graph convolutional networks for dynamic graphs. In: AAAI (2020)
24. Peng, J., Xue, H., Wei, Z., Tuncali, I., Hao, J., Shang, X.: Integrating multi-network topology for gene function prediction using deep neural networks. Brief. Bioinform. (2020)
25. Perozzi, B., Al-Rfou, R., Skiena, S.: DeepWalk: online learning of social representations. In: SIGKDD, pp. 701–710 (2014)
26. Sajadmanesh, S., Bazargani, S., Zhang, J., Rabiee, H.R.: Continuous-time relationship prediction in dynamic heterogeneous information networks. ACM TKDD **13**(4), 44:1–44:31 (2019)
27. Sankar, A., Wu, Y., Gou, L., Zhang, W., Yang, H.: Dynamic graph representation learning via self-attention networks. In: Workshop on Representation Learning on Graphs and Manifolds in ICLR (2019)
28. Shi, C., Hu, B., Zhao, W.X., Yu, P.S.: Heterogeneous information network embedding for recommendation. IEEE TKDE **31**(2), 357–370 (2019)
29. Singer, U., Guy, I., Radinsky, K.: Node embedding over temporal graphs. In: IJCAI, pp. 4605–4612 (2019)

30. Tang, J., Qu, M., Wang, M., Zhang, M., Yan, J., Mei, Q.: LINE: large-scale information network embedding. In: WWW, pp. 1067–1077 (2015)
31. Trivedi, R., Farajtabar, M., Biswal, P., Zha, H.: DyRep: learning representations over dynamic graphs. In: ICLR (2019)
32. Tu, C., Liu, H., Liu, Z., Sun, M.: CANE: context-aware network embedding for relation modeling. In: ACL, pp. 1722–1731, July 2017
33. Veličković, P., Cucurull, G., Casanova, A., Romero, A., Liò, P., Bengio, Y.: Graph attention networks. ICLR (2018)
34. Wang, X., et al.: Heterogeneous graph attention network. In: WWW, pp. 2022–2032 (2019)
35. Wen, Y., Guo, L., Chen, Z., Ma, J.: Network embedding based recommendation method in social networks. In: WWW, pp. 11–12 (2018)
36. Xu, D., Cheng, W., Luo, D., Liu, X., Zhang, X.: Spatio-temporal attentive RNN for node classification in temporal attributed graphs. In: IJCAI (2019)
37. Xue, H., Peng, J., Li, J., Shang, X.: Integrating multi-network topology via deep semi-supervised node embedding. In: CIKM, pp. 2117–2120 (2019)
38. Yang, L., Xiao, Z., Jiang, W., Wei, Y., Hu, Y., Wang, H.: Dynamic heterogeneous graph embedding using hierarchical attentions. In: Jose, J.M., et al. (eds.) ECIR 2020. LNCS, vol. 12036, pp. 425–432. Springer, Cham (2020). https://doi.org/10.1007/978-3-030-45442-5_53
39. Yin, Y., Ji, L., Zhang, J., Pei, Y.: DHNE: network representation learning method for dynamic heterogeneous networks. IEEE Access **7**, 134782–134792 (2019)
40. Zhang, C., Song, D., Huang, C., Swami, A., Chawla, N.V.: Heterogeneous graph neural network. In: SIGKDD, pp. 793–803 (2019)
41. Zhang, H., Qiu, L., Yi, L., Song, Y.: Scalable multiplex network embedding. In: IJCAI, pp. 3082–3088 (2018)
42. Zhang, Z., Cui, P., Pei, J., Wang, X., Zhu, W.: Timers: error-bounded SVD restart on dynamic networks. In: AAAI (2018)
43. Zhou, L.k., Yang, Y., Ren, X., Wu, F., Zhuang, Y.: Dynamic network embedding by modeling triadic closure process. In: AAAI (2018)

GIKT: A Graph-Based Interaction Model for Knowledge Tracing

Yang Yang[1], Jian Shen[1], Yanru Qu[2], Yunfei Liu[1], Kerong Wang[1], Yaoming Zhu[1], Weinan Zhang[1(✉)], and Yong Yu[1(✉)]

[1] Shanghai Jiao Tong University, Shanghai, China
{yyang,rockyshen,ymzhu,yyu}@apex.sjtu.edu.cn
{liuyunfei,wangkerong,wnzhang}@sjtu.edu.cn
[2] University of Illinois, Urbana-Champaign, USA
yanruqu2@illinois.edu

Abstract. With the rapid development in online education, *knowledge tracing* (KT) has become a fundamental problem which traces students' knowledge status and predicts their performance on new questions. Questions are often numerous in online education systems, and are always associated with much fewer skills. However, the previous literature fails to involve question information together with high-order question-skill correlations, which is mostly limited by data sparsity and multi-skill problems. From the model perspective, previous models can hardly capture the long-term dependency of student exercise history, and cannot model the interactions between student-questions, and student-skills in a consistent way. In this paper, we propose a Graph-based Interaction model for Knowledge Tracing (GIKT) to tackle the above problems. More specifically, GIKT utilizes graph convolutional network (GCN) to substantially incorporate question-skill correlations via embedding propagation. Besides, considering that relevant questions are usually scattered throughout the exercise history, and that question and skill are just different instantiations of knowledge, GIKT generalizes the degree of students' master of the question to the interactions between the student's current state, the student's history states, the target question, and related skills. Experiments on three datasets demonstrate that GIKT achieves the new state-of-the-art performance, with 2%–6% absolute AUC improvement.

Keywords: Knowledge tracing · Graph Neural Network · Information interaction

1 Introduction

In online learning platforms such as MOOCs or intelligent tutoring systems, *knowledge tracing* (KT) [6] is an essential task, which aims at tracing the knowledge state of students. At a colloquial level, KT solves the problem of predicting whether the students can answer the new question correctly according to their

F. Hutter et al. (Eds.): ECML PKDD 2020, LNAI 12457, pp. 299–315, 2021.
https://doi.org/10.1007/978-3-030-67658-2_18

previous learning history. The KT task has been widely studied and various methods have been proposed to handle it.

Existing KT methods [2,21,34] commonly build predictive models based on the skills that the target questions correspond to rather than the questions themselves. In the KT task, there exists several skills and lots of questions where one skill is related to many questions and one question may correspond to more than one skill, which can be represented by a relation graph such as the example shown in Fig. 1. Due to the assumption that skill mastery can reflect whether the students are able to answer the related questions correctly to some extent, it is a feasible alternative to make predictions based on the skills just like previous KT works.

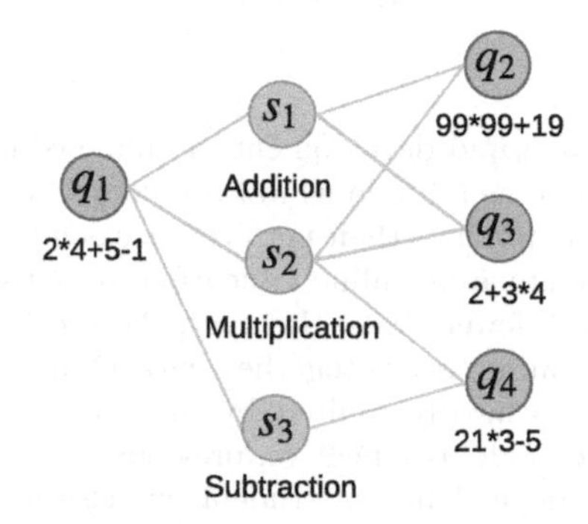

Fig. 1. Simple example of question-skill relation graph.

Although these pure skill-based KT methods have achieved empirical success, the characteristics of questions are neglected, which may lead to degraded performance. For instance, in Fig. 1, even though the two questions q_2 and q_3 share the same skills, their different difficulties may result in different probabilities of being answered correctly. To this end, several previous works [14] utilize the question characteristics as a supplement to the skill inputs. However, as the number of questions is usually large while many students only attempt on a small subset of questions, most questions are only answered by a few students, leading to the data sparsity problem [28]. Besides, for those questions sharing part of common skills (*e.g.* q_1 and q_4), simply augmenting the question characteristics loses latent inter-question and inter-skill information. Based on these considerations, it is important to exploit high-order information between the questions and skills.

In this paper, we first investigate how to effectively extract the high-order relation information contained in the question-skill relation graph. Motivated by the great power of Graph Neural Networks (GNNs) [10,13,26] to extract graph representations by aggregating information from neighbors, we leverage a graph convolutional network (GCN) to learn embeddings for questions and skills from high-order relations. Once the question and skill embeddings are aggregated, we can directly feed question embeddings together with corresponding answer embeddings as the input of KT models.

In addition to the input features, another key issue in KT is the model framework. Recent advances in deep learning simulate a fruitful line of deep KT works, which leverage deep neural networks to sequentially capture the changes of students' knowledge state. Two representative deep KT models are Deep Knowledge Tracing (DKT) [21] and Dynamic Key-Value Memory Networks (DKVMN) [34] which leverage Recurrent Neural Networks (RNN) [30] and Memory-Augmented Neural Networks (MANN) respectively to solve KT. However, they are notoriously unable to capture long-term dependencies in a question sequence [1]. To handle this problem, Sequential Key-Value Memory Networks (SKVMN) [1] proposes a hop-LSTM architecture that aggregates hidden states of similar exercises into a new state and Exercise-Enhanced Recurrent Neural Network with Attention mechanism (EERNNA) [25] uses the attention mechanism to perform weighted sum aggregation for all history states.

Instead of aggregating related hidden states into a new state for prediction directly, we take a step further towards improving long-term dependency capture and better modeling student's mastery degree. Inspired by SKVMN and EERNNA, we introduce a recap module on top of the recurrent layer to select several the most related hidden states according to the attention weight with the intention of noise reduction. Considering the mastery of the new question and its related skills, we generalize the interaction module and interact the relevant hidden states and the current hidden states with the aggregated question embeddings and skill embeddings. The generalized interaction module can better model student's mastery degree of question and skills. Besides, an attention mechanism is applied on each interaction to make final predictions, which automatically weights the prediction utility of all the interactions.

To sum up, in this paper, we propose an end-to-end deep framework, namely Graph-based Interaction for Knowledge Tracing (GIKT), for knowledge tracing. Our main contributions are summarized as follows: 1) By leveraging a graph convolutional network to aggregate question embeddings and skill embeddings, GIKT is capable to exploit high-order question-skill relations, which mitigates the data sparsity problem and the multi-skill issue. 2) By introducing a recap module followed by an interaction module, our model can better model the student's mastery degree of the new question and its related skills in a consistent way. 3) Empirically we conduct extensive experiments on three benchmark datasets and the results demonstrate that our GIKT outperforms the state-of-the-art baselines substantially.

2 Related Work

2.1 Knowledge Tracing

Existing knowledge tracing methods can be roughly categorized into two groups: traditional machine learning methods and deep learning methods. In this paper, we mainly focus on the deep KT methods.

Traditional machine learning KT methods mainly involve two types: Bayesian Knowledge Tracing (BKT) [6] and factor analysis models. BKT is a hidden Markov model which regards each skill as a binary variable and uses bayes rule to update state. Several works extends the vanilla BKT model to incorporate more information into it such as slip and guess probability [2], skill difficulty [19] and student individualization [18,33]. On the other hand, factor analysis models focus on learning general parameters from historical data to make predictions. Among the factor analysis models, Item Response Theory (IRT) [8] models parameters for student ability and question difficulty, Performance Factors Analysis (PFA) [20] takes into account the number of positive and negative responses for skills and Knowledge Tracing Machines [27] leverages Factorization Machines [24] to encode side information of questions and users into the parameter model.

Recently, due to the great capacity and effective representation learning, deep neural networks have been leveraged in the KT literature. Deep Knowledge Tracing (DKT) [21] is the first deep KT method, which uses recurrent neural network (RNN) to trace the knowledge state of the student. Dynamic Key-Value Memory Networks (DKVMN) [34] can discover the underlying concepts of each skill and trace states for each concept. Based on these two models, several methods have been proposed by considering more information, such as the forgetting behavior of students [16], multi-skill information and prerequisite skill relation graph labeled by experts [4] or student individualization [15]. GKT [17] builds a skill relation graph and learns their relation explicitly. However, these methods only use skills as the input, which causes information loss.

Some deep KT methods take question characteristices into account for predictions. Dynamic Student Classification on Memory Networks (DSCMN) [14] utilizes question difficulty to help distinguish the questions related to the same skills. Exercise-Enhanced Recurrent Neural Network with Attention mechanism (EERNNA) [25] encodes question embeddings using the content of questions so that the question embeddings can contain the characteristic information of questions, however in reality it is difficult to collect the content of questions. Due to the data sparsity problem, DHKT [28] augments DKT by using the relations between questions and skills to get question representations, which, however, fails to capture the inter-question and inter-skill relations. In this paper we use GCN to extract the high-order information contained in the question-skill graph. To handle the long term dependency issue, Sequential Key-Value Memory Networks (SKVMN) [1] uses a modified LSTM with hops to enhance the capacity of capturing long-term dependencies in an exercise sequence. And EERNNA [25] assumes that current student knowledge state is a weighted sum aggregation of all historical student states based on correlations between current question and historical questions. Our method differs from these two works in the way that they aggregate related hidden states into a new state for prediction, while we first select the most useful hidden states to reduce the effects of the noisy states, and then we perform pairwise interaction for prediction.

2.2 Graph Neural Networks

In recent years, graph data is widely used in deep learning models. However, the traditional neural network suffers from the complex non-Euclidean structure of graph. Inspired by CNNs, some works use the convolutional method for the graph-structure data [7,13]. Graph convolutional networks (GCNs) [13] is proposed for semi-supervised graph classification, which updates node representations based on itself and its neighbors. In this way, updated node representations contain attributes of neighbor nodes and information of high-order neighbors if multiple graph-convolutional layers are used. Due to the great success of GCNs, some variants are further proposed for graph data [10,26].

With the development of Graph Neural Networks (GNNs), many applications based on GNNs appear in various domains, such as natural language processing (NLP) [3,32], computer vision (CV) [9,22] and recommendation systems [23,29]. As GNNs help to capture high-order information, we use GCN in our GIKT model to extract relations between skills and questions into their representations. To the best of our knowledge, our method GIKT is the first work to model question-skill relations via graph neural network.

3 Preliminaries

Knowledge Tracing. In the knowledge tracing task, students sequentially answer a series of questions that the online learning platforms provide. After the students answer each question, a feedback of whether the answer is correct will be issued. Here we denote an exercise as $\boldsymbol{x}_i = (q_i, a_i)$, where q_i is the question ID and $a_i \in \{0, 1\}$ represents whether the student answered q_i correctly. Given an exercise sequence $\boldsymbol{X} = \{\boldsymbol{x}_1, \boldsymbol{x}_2, ..., \boldsymbol{x}_{t-1}\}$ and the new question q_t, the goal of KT is to predict the probability of the student correctly answering it $p(a_t = 1|\boldsymbol{X}, q_t)$.

Question-Skill Relation Graph. Each question q_i corresponds to one or more skills $\{s_1, ..., s_{n_i}\}$, and one skill s_j is usually related to many questions $\{q_1, ..., q_{n_j}\}$, where n_i and n_j are the number of skills related to question q_i and the number of questions related to skill s_j respectively. Here we denote the relations as a question-skill relation bipartite graph $\mathcal{G}$, which is defined as $\{(q, r_{qs}, s)|q \in \mathcal{Q}, s \in \mathcal{S}\}$, where $\mathcal{Q}$ and $\mathcal{S}$ correspond to the question and skill sets respectively. And $r_{qs} = 1$ if the question q is related to the skill s.

4 The Proposed Method GIKT

In this section, we will introduce our method in detail, and the overall framework is shown in Fig. 2. We first leverage GCN to learn question and skill representations aggregated on the question-skill relation graph, and a recurrent layer is used to model the sequential change of knowledge state. To capture long term dependency and exploit useful information comprehensively, we then design a recap module followed by an interaction module for the final prediction.

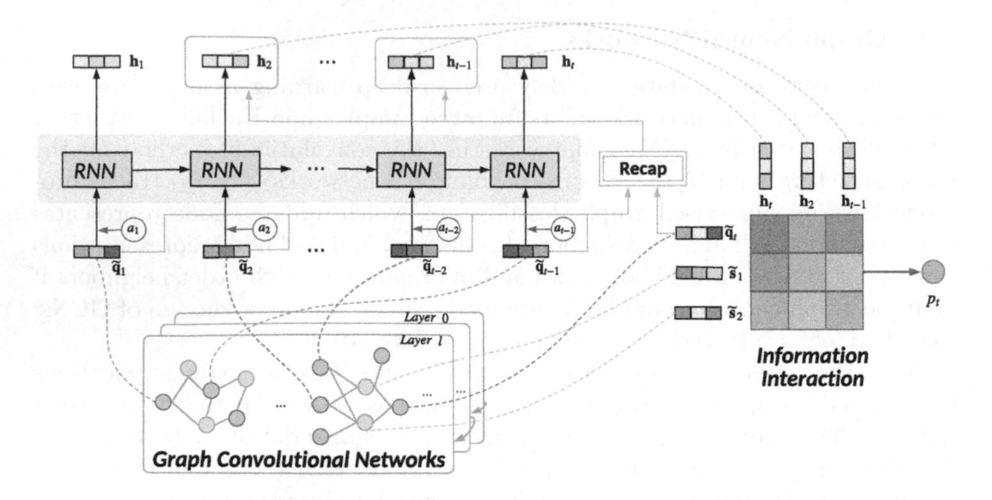

Fig. 2. An illustration of GIKT at time step t, where q_t is the new question. First we use GCN to aggregate question and skill embeddings. Then a recurrent neural network is used to model the sequential knowledge state $\mathbf{h}_t$. In recap module we select the most related hidden states of q_t, which corresponds to soft selection and hard selection implementation. The information interaction module performs pairwise interaction between the student's current state, the selected student's history states, the target question and related skills for the prediction p_t.

4.1 Embedding Layer

Our GIKT method uses embeddings to represent questions, skills and answers. Three embedding matrices $\mathbf{E}_s \in \mathbb{R}^{|\mathcal{S}|\times d}$, $\mathbf{E}_q \in \mathbb{R}^{|\mathcal{Q}|\times d}$, $\mathbf{E}_a \in \mathbb{R}^{2\times d}$ are denoted for look-up operation where d stands for the embedding size. Each row in $\mathbf{E}_s$ or $\mathbf{E}_q$ corresponds to a skill or a question. The two rows in $\mathbf{E}_a$ represent incorrect and correct answers respectively. For i-th row vector in matrices, we use $\mathbf{s}_i$, $\mathbf{q}_i$ and $\mathbf{a}_i$ to represent them respectively.

In our framework, we do not pretrain these embeddings and they are trained by optimizing the final objective in an end-to-end manner.

4.2 Embedding Propagation

From training perspective, sparsity in question data raises a big challenge to learn informative question representations, especially for those with quite limited training examples. From the inference perspective, whether a student can answer a new question correctly depends on the mastery of its related skills and the question characteristic. When he/she has solved similar questions before, he/she is more likely answer the new question correctly. In this model, we incorporate question-skill relation graph $\mathcal{G}$ to solve sparsity, as well as to utilize prior correlations to obtain better question representations.

Considering the question-skill relation graph is bipartite, the 1st hop neighbors of a question should be its corresponding skills, and the 2nd hop neighbors should be other questions sharing same skills. To extract the high-order information, we leverage graph convolutional network (GCN) [13] to encode relevant skills and questions into question embeddings and skill embeddings.

Graph convolutional network stacks several graph convolution layers to encode high-order neighbor information, and in each layer the node representations can be updated by embeddings of itself and neighbor nodes. Denote the representation of node i in the graph as $\mathbf{x}_i$ ($\mathbf{x}_i$ can represent skill embedding $\mathbf{s}_i$ or question embedding $\mathbf{q}_i$) and the set of its neighbor nodes as $\mathcal{N}_i$, then the formula of l-th GCN layer can be expressed as:

$$\mathbf{x}_i^l = \sigma\left(\frac{1}{|\mathcal{N}_i|} \sum_{j \in \mathcal{N}_i \cup \{i\}} \mathbf{w}^l \mathbf{x}_j^{l-1} + \mathbf{b}^l\right), \tag{1}$$

where $\mathbf{w}^l$ and $\mathbf{b}^l$ are the aggregate weight and bias to be learned in l-th GCN layer, σ is the non-linear transformation such as ReLU.

After embedding propagation by GCN, we get the aggregated embedding of questions and skills. We use $\widetilde{\mathbf{q}}$ and $\widetilde{\mathbf{s}}$ to represent the question and skill representation after embedding propagation. For easy implementation and better parallelization, we sample a fixed number of question neighbors (i.e., n_q) and skill neighbors (i.e., n_s) for each batch. And during inference, we run each example multiple times (sampling different neighbors) and average the model outputs to obtain stable prediction results.

4.3 Student State Evolution

For each history time t, we concatenate the question and answer embeddings and project to d-dimension through a non-linear transformation as exercise representations:

$$\mathbf{e}_t = \text{ReLU}(\mathbf{W}_1([\widetilde{\mathbf{q}_t}, \mathbf{a}_t]) + \mathbf{b}_1), \tag{2}$$

where we use $[,]$ to denote vector concatenation.

There may exist dependency between different exercises, thus we need to model the whole exercise process to capture the student state changes and to learn the potential relation between exercises. To model the sequential behavior of a student doing exercise, we use LSTM [11] to learn student states from input exercise representations:

$$\mathbf{i}_t = \sigma(\mathbf{W}_i[\mathbf{e}_t, \mathbf{h}_{t-1}, \mathbf{c}_{t-1}] + \mathbf{b}_i), \tag{3}$$

$$\mathbf{f}_t = \sigma(\mathbf{W}_f[\mathbf{e}_t, \mathbf{h}_{t-1}, \mathbf{c}_{t-1}] + \mathbf{b}_f), \tag{4}$$

$$\mathbf{o}_t = \sigma(\mathbf{W}_o[\mathbf{e}_t, \mathbf{h}_{t-1}, \mathbf{c}_{t-1}] + \mathbf{b}_o), \tag{5}$$

$$\mathbf{c}_t = \mathbf{f}_t \mathbf{c}_{t-1} + \mathbf{i}_t \tanh\left(\mathbf{W}_c[\mathbf{e}_t, \mathbf{h}_{t-1}] + \mathbf{b}_c\right), \tag{6}$$

$$\mathbf{h}_t = \mathbf{o}_t \tanh\left(\mathbf{c}_t\right), \tag{7}$$

where $\mathbf{h}_t$, $\mathbf{c}_t$, $\mathbf{i}_t$, $\mathbf{f}_t$, $\mathbf{o}_t$ represents hidden state, cell state, input gate, forget gate, output gate respectively. It is worth mentioning that this layer is important for capturing coarse-grained dependency like potential relations between skills, so we just learn a hidden state $\mathbf{h}_t \in \mathbb{R}^d$ as the current student state, which contains coarse-grained mastery state of skills.

4.4 History States Recap Module

In a student's exercise history, questions of relevant skills are very likely scattered in the long history. From another point, consecutive exercises may not follow a coherent topic. These phenomena raise challenges for LSTM sequence modeling in traditional KT methods: (i) As is well recognized, LSTM can hardly capture long-term dependencies in very long sequences, which means the current student state $\mathbf{h}_t$ may "forget" history exercises related to the new target question q_t. (ii) The current student state $\mathbf{h}_t$ considers more about recent exercises, which may contain noisy information for the new target question q_t. When a student is answering a new question, he/she may quickly recall similar questions he/she has done before to help him/her to understand the new question. Inspired from this behavior, we propose to select relevant history states[1] $\{\mathbf{h}_i | i \in [1, \dots, t-1]\}$ to better represent a student's ability on a specific question q_t, called history states recap module.

We develop two methods to find relevant history states. The first one is hard selection, i.e., we only consider the exercises sharing same skills with the new question:

$$\mathbf{I}_h = \{\mathbf{h}_i | \mathcal{N}_{q_i} = \mathcal{N}_{q_t}, i \in [1, .., t-1]\}, \tag{8}$$

Another method is soft selection, i.e., we learn the relevance between target question and history states through an attention network, and choose top-k states with highest attention scores:

$$\mathbf{I}_h = \{\mathbf{h}_i | R_{i,t} \leq k, i \in [1, .., t-1]\}, \tag{9}$$

where $R_{i,t}$ is the ranking of attention function $f(\widetilde{\mathbf{q}_i}, \widetilde{\mathbf{q}_t})$.

4.5 Generalized Interaction Module

Previous KT methods predict a student's performance mainly according to the interaction between student state $\mathbf{h}_t$ and question representation $\mathbf{q}_t$, i.e., $\langle \mathbf{h}_t, \mathbf{q}_t \rangle$. We generalize the interaction in the following aspects: (i) we use $\langle \mathbf{h}_t, \widetilde{\mathbf{q}_t} \rangle$ to represent the student's mastery degree of question q_t, $\langle \mathbf{h}_t, \widetilde{\mathbf{s}_j} \rangle$ to represent the student's mastery degree of the corresponding skill $s_j \in \mathcal{N}_{q_t}$, (ii) we generalize the interaction on current student state to history states, i.e., $\langle \mathbf{h}_i, \widetilde{\mathbf{q}_t} \rangle$ and $\langle \mathbf{h}_i, \widetilde{\mathbf{s}_j} \rangle$,

[1] We try other implementations like using history exercises (question-answer pair) instead of history states, and the results show using history states results in a better performance.

$\mathbf{h}_i \in \mathcal{I}_h$, which is equivalent to let the student to answer the target question in history timesteps.

Then we consider all above interactions for prediction, and define the generalized interaction module. In order to encourage relevant interactions and reduce noise, we use an attention network to learn bi-attention weights for all interaction terms, and compute the weighted sum as the prediction:

$$\alpha_{i,j} = \text{Softmax}_{i,j}(\boldsymbol{W}^T[\boldsymbol{f}_i, \boldsymbol{f}_j] + b) \tag{10}$$

$$p_t = \sum_{\boldsymbol{f}_i \in \mathbf{I}_h \cup \{\mathbf{h}_t\}} \sum_{\boldsymbol{f}_j \in \widetilde{\mathbf{N}}_{q_t} \cup \{\widetilde{\mathbf{q}_t}\}} \alpha_{i,j} g(\boldsymbol{f}_i, \boldsymbol{f}_j) \tag{11}$$

where p_t is the predicted probability of answering the new question correctly, $\widetilde{\mathbf{N}}_{q_t}$ represents the aggregated neighbor skill embeddings of q_t and we use inner product to implement function g. Similar to the selection of neighbors in relation graph, we set a fixed number of $\mathbf{I}_h$ and $\widetilde{\mathbf{N}}_{q_t}$ by sampling from these two sets.

4.6 Optimization

To optimize our model, we update the parameters in our model using gradient descent by minimizing the cross entropy loss between the predicted probability of answering correctly and the true label of the student's answer:

$$\mathcal{L} = -\sum_t (a_t \log p_t + (1 - a_t) \log (1 - p_t)). \tag{12}$$

5 Experiments

In this section, we conduct several experiments to investigate the performance of our model. We first evaluate the prediction error by comparing our model with other baselines on three public datasets. Then we make ablation studies on the GCN and the interaction module of GIKT to show their effectiveness in Sect. 5.5. Finally, we evaluate the design decisions of the recap module to investigate which design performs better in Sect. 5.6.

5.1 Datasets

To evaluate our model, the experiments are conducted on three widely-used datasets in KT and the detailed statistics are shown in Table 1.

- **ASSIST09**[2] was collected during the school year 2009–2010 from ASSISTments online education platform[3]. We conduct our experiments on "skill-builder" dataset. Following the previous work [31], we remove the duplicated records and scaffolding problems from the original dataset. This dataset has 3852 students with 123 skills, 17,737 questions and 282,619 exercises.

[2] https://sites.google.com/site/assistmentsdata/home/assistment-2009-2010-data/skill-builder-data-2009-2010.

[3] https://new.assistments.org/.

Table 1. Dataset statistics

	ASSIST09	ASSIST12	EdNet
#students	3,852	27,485	5000
#questions	17,737	53,065	12,161
#skills	123	265	189
#exercises	282,619	2,709,436	676,974
Questions per skill	173	200	147
Skills per question	1.197	1.000	2.280
Attempts per question	16	51	56
Attempts per skill	2,743	10,224	8,420

- **ASSIST12**[4] was collected from the same platform as ASSIST09 during the school year 2012–2013. In this dataset, each question is only related to one skill, but one skill still corresponds to several questions. After the same data processing as ASSIST09, it has 2,709,436 exercises with 27,485 students, 265 skills and 53,065 questions.
- **EdNet**[5] was collected by [5]. As the whole dataset is too large, we randomly select 5000 students with 189 skills, 12,161 questions and 676,974 exercises.

Note that for each dataset we only use the sequences of which the length is longer than 3 in the experiments as the too short sequences are meaningless. For each dataset, we split 80% of all the sequences as the training set, 20% as the test set. To evaluate the results on these datasets, we use the area under the curve (AUC) as the evaluation metric.

5.2 Baselines

In order to evaluate the effectiveness of our proposed model, we use the following models as our baselines:

- **BKT** [6] uses Bayesian inference for prediction, which models the knowledge state of the skill as a binary variable.
- **KTM** [27] is the latest factor analysis model that uses Factorization Machine to interact each feature for prediction. Although KTM can use many types of feature, for fairness we only use question ID, skill ID and answer as its side information in comparison.
- **DKT** [21] is the first method that uses deep learning to model knowledge tracing task. It uses recurrent neural network to model knowledge state of students.

[4] https://sites.google.com/site/assistmentsdata/home/2012-13-school-data-with-affect.

[5] https://github.com/riiid/ednet.

- **DKVMN** [34] uses memory network to store knowledge state of different concepts respectively instead of using a single hidden state.
- **DKT-Q** is a variant of DKT that we change the input of DKT from skills to questions so that the DKT model directly uses question information for prediction.
- **DKT-QS** is a variant of DKT that we change the input of DKT to the concatenation of questions and skills so that the DKT model uses question and skill information simultaneously for prediction.
- **GAKT** is a variant of the model Exercise-Enhanced Recurrent Neural Network with Attention mechanism (EERNNA) [25] as EERNNA utilizes question text descriptions but we can't acquire this information from public datasets. Thus we utilize our input question embeddings aggregated by GCN as input of EERNNA and follow its framework design for comparison.

5.3 Implementation Details

We implement all the compared methods with TensorFlow. The code for our method is available online[6]. The embedding size of skills, questions and answers are fixed to 100, all embedding matrices are randomly initialized and updated in the training process. In the implementation of LSTM, a stacked LSTM with two hidden layers is used, where the sizes of the memory cells are set to 200 and 100 respectively. In embedding propagation module, we set the maximal aggregate layer number $l = 3$. We also use dropout with the probability of 0.6 to avoid overfitting. All trainable parameters are optimized by Adam algorithm [12] with learning rate of 0.001 and the mini-batch size is set to 32. Other hyper-parameters are chosen by grid search, including the number of question neighbors in GCN, skill neighbors in GCN, related hidden states and skills related to the new question.

5.4 Overall Performance

Table 2 reports the AUC results of all the compared methods. From the results we observe that our GIKT model achieves the highest performance over three datasets, which verifies the effectiveness of our model. To be specific, our proposed model GIKT achieves at least 2% higher results than other baselines and especially in ASSIST12 our model achieves 5% higher results. Among the baseline models, traditional machine learning models like BKT and KTM perform worse than deep learning models, which shows the effectiveness of deep learning methods. DKVMN performs slightly worse than DKT on average as building states for each concept may lose the relation information between concepts. Besides, GAKT performs worse than our model, which indicates that exploiting high-order skill-question relations through selecting the most related information and performing interaction makes a difference.

[6] https://github.com/Rimoku/GIKT.

Table 2. The AUC results over three datasets. Among these models, BKT, DKT and DKVMN predict for skills, other models predict for questions. Note that "*" indicates that the statistically significant improvements over the best baseline, with p-value smaller than 10^{-5} in two-sided t-test.

Model	ASSIST09	ASSIST12	EdNet
BKT	0.6571	0.6204	0.6027
KTM	0.7169	0.6788	0.6888
DKVMN	0.7550	0.7283	0.6967
DKT	0.7561	0.7286	0.6822
DKT-Q	0.7328	0.7621	0.7285
DKT-QS	0.7715	0.7582	0.7428
GAKT	0.7684	0.7652	0.7281
GIKT	**0.7996***	**0.8278***	**0.7758***

On the other hand, we find that directly using questions as input may achieve superior performance than using skills. For the question-level model DKT-Q, it has comparable or better performance than DKT over ASSIST12 and EdNet datasets. However, DKT-Q performs worse than DKT in ASSIST09 dataset. The reason may be that the average number of attempts per question in ASSIST09 dataset is significantly less than other two datasets as observed in Table 1, which illustrates DKT-Q suffers from data sparsity problem. Besides, the AUC results of the model DKT-QS are higher than DKT-Q and DKT, except on ASSIST12 as it is a single-skill dataset, which indicates that considering question and skill information together improves overall performance.

5.5 Ablation Studies

To get deep insights on the effect of each module in GIKT, we design several ablation studies to further investigate on our model. We first study the influence of the number of aggregate layers, and then we design some variants of the interaction module to investigate their effectiveness.

Effect of Embedding Propagation Layer. We change the number of the aggregate layers in GCN ranging from 0 to 3 to show the effect of the high-order question-skill relations and the results are shown in Table 3. Specially, when the number of the layer is 0, it means the question embeddings and skill embeddings used in our model are indexed from embedding matrices directly.

From Table 3 we find that, compared to no embedding propagation layer cases, GIKT achieves better performance when the number of aggregate layers is non-zero, which validates the effectiveness of GCN. The results also imply that exploiting high-order relations contained in the question-skill graph is necessary for adequate results as the performance of adopting more layers is better than using less layers.

Table 3. Effect of the number of aggregate layers.

Layers	ASSIST09	ASSIST12	EdNet
0	0.7864	0.7805	0.7598
1	0.7815	0.7933	0.7583
2	0.7905	0.8218	0.7655
3	**0.7996**	**0.8278**	**0.7758**

Effect of Interaction Module. To verify the impact of interaction module in GIKT, we conduct ablation studies on four variants of our model. The details of the four settings are listed as below and the performance of them is shown in Table 4.

- **GIKT-RHS** (Remove History related hidden states and Skills related to the new question) For GIKT-RHS, we just use the current state of the student and the new question to perform interaction for prediction.
- **GIKT-RH** (Remove History related hidden states) For GIKT-RH, we only use the current state of the student to model mastery of the new question and related skills.
- **GIKT-RS** (Remove Skills related to the new question) For GIKT-RS, we do not model the mastery of skills related to the new question.
- **GIKT-RA** (Remove Attention in interaction module) GIKT-RA removes the attention mechanism after interaction, which treats each interaction pair as equally important and average the prediction scores directly in interaction part for prediction.

Table 4. Effect of information module

Model	ASSIST09	ASSIST12	EdNet
GIKT-RHS	0.7807	0.7664	0.7354
GIKT-RH	0.7798	0.768	0.7415
GIKT-RS	0.7966	0.8164	0.7745
GIKT-RA	0.7735	0.8018	0.7533
GIKT	**0.7996**	**0.8278**	**0.7758**

From Table 4 we have the following findings: Our GIKT model considering all interaction aspects achieve best performance, which shows the effectiveness of the interaction module. Meanwhile, from the results of GIKT-RH we can find that relevant history states can help better model the student's ability on the new question. Besides, the performance of GIKT-RS is slightly worse than GIKT, which implies that model mastery degree of question and skills simultaneously can further help prediction. Comparing the results of GIKT-RA with GIKT, the worse performance confirms the effectiveness of attention in interaction module, which distinguishes different interaction terms for better prediction results. By calculating different aspects of interaction and weighted sum for the prediction, information from different level can be fully interacted.

5.6 Recap Module Design Evaluation

To evaluate the detailed design of the recap module in GIKT, we conduct experiments of several variants. The details of the settings are listed as below and the performance of them is shown in Table 5.

- **GIKT-HE** (Hard select history Exercises) For GIKT-HE, we select the related exercises sharing the same skills.
- **GIKT-SE** (Soft select history Exercises) For GIKT-SE, we select history exercises according to the attention weight.
- **GIKT-HS** (Hard select hidden States) For GIKT-HS, we select the related hidden states of the exercises sharing the same skills.
- **GIKT-SS** (Soft select hidden States) For GIKT-SS, we select hidden states according to the attention weight. The reported results of GIKT in previous sections are taken by the performance of GIKT-SS.

Table 5. Results of different recap module design

Model	ASSIST09	ASSIST12	EdNet
GIKT-HE	0.786	0.7722	0.7426
GIKT-SE	0.7932	0.7788	0.7526
GIKT-HS	0.7797	0.7581	0.7426
GIKT-SS	**0.7996**	**0.8218**	**0.7643**

From Table 5 we find that the performance comparison between these four implementations is GIKT-SS > GIKT-SE > GIKT-HE > GIKT-HS. This result implies that using attention mechanism can achieve better selection coverage for better performance whether hidden states or exercises are selected. If we use the hard selection variant, selecting exercises performs better than selecting the hidden states. The reason may be that the exercises related to the new question contain less noise.

6 Conclusion

In this paper, we propose a framework to employ high-order question-skill relation graphs into question and skill representations for knowledge tracing. Besides, to model the student's mastery for the question and related skills, we design a recap module to select relevant history states to represent student's ability. Then we extend a generalized interaction module to represent the student's mastery degree of the new question and related skills in a consistent way. To distinguish relevant interactions, we use an attention mechanism for the prediction. The experimental results show that our model achieve better performance.

References

1. Abdelrahman, G., Wang, Q.: Knowledge tracing with sequential key-value memory networks. In: the 42nd International ACM SIGIR Conference (2019)
2. Baker, R.S.J., Corbett, A.T., Aleven, V.: More accurate student modeling through contextual estimation of slip and guess probabilities in Bayesian knowledge tracing. In: Woolf, B.P., Aïmeur, E., Nkambou, R., Lajoie, S. (eds.) ITS 2008. LNCS, vol. 5091, pp. 406–415. Springer, Heidelberg (2008). https://doi.org/10.1007/978-3-540-69132-7_44
3. Beck, D., Haffari, G., Cohn, T.: Graph-to-sequence learning using gated graph neural networks. arXiv preprint arXiv:1806.09835 (2018)
4. Chen, P., Lu, Y., Zheng, V.W., Pian, Y.: Prerequisite-driven deep knowledge tracing. In: 2018 IEEE International Conference on Data Mining (ICDM), pp. 39–48. IEEE (2018)
5. Choi, Y., et al.: EdNet: a large-scale hierarchical dataset in education. arXiv preprint arXiv:1912.03072 (2019)
6. Corbett, A.T., Anderson, J.R.: Knowledge tracing: modeling the acquisition of procedural knowledge. User Model. User-Adap. Inter. **4**(4), 253–278 (1994)
7. Defferrard, M., Bresson, X., Vandergheynst, P.: Convolutional neural networks on graphs with fast localized spectral filtering. In: Advances in Neural Information Processing Systems, pp. 3844–3852 (2016)
8. Ebbinghaus, H.: Memory: a contribution to experimental psychology. Annals Neurosci. **20**(4), 155 (2013)
9. Garcia, V., Bruna, J.: Few-shot learning with graph neural networks. arXiv preprint arXiv:1711.04043 (2017)
10. Hamilton, W., Ying, Z., Leskovec, J.: Inductive representation learning on large graphs. In: Advances in Neural Information Processing Systems, pp. 1024–1034 (2017)
11. Hochreiter, S., Schmidhuber, J.: Long short-term memory. Neural Comput. **9**(8), 1735–1780 (1997)
12. Kingma, D.P., Ba, J.: Adam: a method for stochastic optimization. arXiv preprint arXiv:1412.6980 (2014)
13. Kipf, T.N., Welling, M.: Semi-supervised classification with graph convolutional networks. arXiv preprint arXiv:1609.02907 (2016)
14. Minn, S., Desmarais, M.C., Zhu, F., Xiao, J., Wang, J.: Dynamic student classification on memory networks for knowledge tracing. In: Yang, Q., Zhou, Z.-H., Gong, Z., Zhang, M.-L., Huang, S.-J. (eds.) PAKDD 2019. LNCS (LNAI), vol. 11440, pp. 163–174. Springer, Cham (2019). https://doi.org/10.1007/978-3-030-16145-3_13
15. Minn, S., Yu, Y., Desmarais, M.C., Zhu, F., Vie, J.J.: Deep knowledge tracing and dynamic student classification for knowledge tracing. In: 2018 IEEE International Conference on Data Mining (ICDM), pp. 1182–1187. IEEE (2018)
16. Nagatani, K., Zhang, Q., Sato, M., Chen, Y.Y., Chen, F., Ohkuma, T.: Augmenting knowledge tracing by considering forgetting behavior. In: The World Wide Web Conference, pp. 3101–3107 (2019)

17. Nakagawa, H., Iwasawa, Y., Matsuo, Y.: Graph-based knowledge tracing: modeling student proficiency using graph neural network. In: IEEE/WIC/ACM International Conference on Web Intelligence, pp. 156–163. ACM (2019)
18. Pardos, Z.A., Heffernan, N.T.: Modeling individualization in a Bayesian networks implementation of knowledge tracing. In: De Bra, P., Kobsa, A., Chin, D. (eds.) UMAP 2010. LNCS, vol. 6075, pp. 255–266. Springer, Heidelberg (2010). https://doi.org/10.1007/978-3-642-13470-8_24
19. Pardos, Z.A., Heffernan, N.T.: KT-IDEM: introducing item difficulty to the knowledge tracing model. In: Konstan, J.A., Conejo, R., Marzo, J.L., Oliver, N. (eds.) UMAP 2011. LNCS, vol. 6787, pp. 243–254. Springer, Heidelberg (2011). https://doi.org/10.1007/978-3-642-22362-4_21
20. Pavlik Jr, P.I., Cen, H., Koedinger, K.R.: Performance factors analysis-a new alternative to knowledge tracing. Online Submission (2009)
21. Piech, C., et al.: Deep knowledge tracing. In: Advances in Neural Information Processing Systems, pp. 505–513 (2015)
22. Qi, X., Liao, R., Jia, J., Fidler, S., Urtasun, R.: 3d graph neural networks for RGBD semantic segmentation. In: Proceedings of the IEEE International Conference on Computer Vision, pp. 5199–5208 (2017)
23. Qu, Y., Bai, T., Zhang, W., Nie, J., Tang, J.: An end-to-end neighborhood-based interaction model for knowledge-enhanced recommendation. In: Proceedings of the 1st International Workshop on Deep Learning Practice for High-Dimensional Sparse Data, pp. 1–9 (2019)
24. Rendle, S.: Factorization machines. In: 2010 IEEE International Conference on Data Mining, pp. 995–1000. IEEE (2010)
25. Su, Y., et al.: Exercise-enhanced sequential modeling for student performance prediction. In: Thirty-Second AAAI Conference on Artificial Intelligence (2018)
26. Veličković, P., Cucurull, G., Casanova, A., Romero, A., Lio, P., Bengio, Y.: Graph attention networks. arXiv preprint arXiv:1710.10903 (2017)
27. Vie, J.J., Kashima, H.: Knowledge tracing machines: factorization machines for knowledge tracing. In: Proceedings of the AAAI Conference on Artificial Intelligence, vol. 33, pp. 750–757 (2019)
28. Wang, T., Ma, F., Gao, J.: Deep hierarchical knowledge tracing. In: Proceedings of the 12th International Conference on Educational Data Mining, EDM 2019, Montréal, Canada, 2–5 July 2019 (2019). https://drive.google.com/file/d/1wlW6zAi-l4ZAw8rBA_mXZ5tHgg6xKL00
29. Wang, X., He, X., Cao, Y., Liu, M., Chua, T.S.: KGAT: knowledge graph attention network for recommendation. In: Proceedings of the 25th ACM SIGKDD International Conference on Knowledge Discovery & Data Mining, pp. 950–958 (2019)
30. Williams, R.J., Zipser, D.: A learning algorithm for continually running fully recurrent neural networks. Neural Comput. **1**(2), 270–280 (1989)
31. Xiong, X., Zhao, S., Van Inwegen, E.G., Beck, J.E.: Going deeper with deep knowledge tracing. Int. Educ. Data Min. Soc. (2016)
32. Yao, L., Mao, C., Luo, Y.: Graph convolutional networks for text classification. In: Proceedings of the AAAI Conference on Artificial Intelligence, vol. 33, pp. 7370–7377 (2019)

33. Yudelson, M.V., Koedinger, K.R., Gordon, G.J.: Individualized Bayesian knowledge tracing models. In: Lane, H.C., Yacef, K., Mostow, J., Pavlik, P. (eds.) AIED 2013. LNCS (LNAI), vol. 7926, pp. 171–180. Springer, Heidelberg (2013). https://doi.org/10.1007/978-3-642-39112-5_18
34. Zhang, J., Shi, X., King, I., Yeung, D.Y.: Dynamic key-value memory networks for knowledge tracing. In: Proceedings of the 26th international conference on World Wide Web, pp. 765–774 (2017)

Dimensionality Reduction and Autoencoders

Simple and Effective Graph Autoencoders with One-Hop Linear Models

Guillaume Salha[1,2(✉)], Romain Hennequin[1], and Michalis Vazirgiannis[2,3]

[1] Deezer Research, Paris, France
research@deezer.com
[2] LIX, École Polytechnique, Palaiseau, France
[3] Athens University of Economics and Business, Athens, Greece

Abstract. Over the last few years, graph autoencoders (AE) and variational autoencoders (VAE) emerged as powerful node embedding methods, with promising performances on challenging tasks such as link prediction and node clustering. Graph AE, VAE and most of their extensions rely on multi-layer graph convolutional networks (GCN) encoders to learn vector space representations of nodes. In this paper, we show that GCN encoders are actually unnecessarily complex for many applications. We propose to replace them by significantly simpler and more interpretable linear models w.r.t. the direct neighborhood (one-hop) adjacency matrix of the graph, involving fewer operations, fewer parameters and no activation function. For the two aforementioned tasks, we show that this simpler approach consistently reaches competitive performances w.r.t. GCN-based graph AE and VAE for numerous real-world graphs, including all benchmark datasets commonly used to evaluate graph AE and VAE. Based on these results, we also question the relevance of repeatedly using these datasets to compare complex graph AE and VAE.

Keywords: Graphs · Autoencoders · Variational autoencoders · Graph convolutional networks · Linear encoders · Graph representation learning · Node embedding · Link prediction · Node clustering

1 Introduction

Graphs have become ubiquitous, due to the proliferation of data representing relationships or interactions among entities [10,39]. Extracting relevant information from these entities, called the *nodes* of the graph, is crucial to effectively tackle numerous machine learning tasks, such as link prediction or node clustering. While traditional approaches mainly focused on hand-engineered features [2,20], significant improvements were recently achieved by methods aiming at directly *learning* node representations that summarize the graph structure (see [10] for a review). In a nutshell, these *representation learning* methods aim at embedding nodes as vectors in a low-dimensional vector space in which nodes with structural proximity in the graph should be close, e.g., by leveraging random walk strategies [8,27], matrix factorization [4,24] or graph neural networks [11,17].

F. Hutter et al. (Eds.): ECML PKDD 2020, LNAI 12457, pp. 319–334, 2021.
https://doi.org/10.1007/978-3-030-67658-2_19

In particular, *graph autoencoders* (AE) [18,34,37] and *graph variational autoencoders* (VAE) [18] recently emerged as powerful node embedding methods. Based on encoding-decoding schemes, i.e. on the design of low dimensional vector space representations of nodes (*encoding*) from which reconstructing the graph (*decoding*) should be possible, graph AE and VAE models have been successfully applied to address several challenging learning tasks, with competitive results w.r.t. popular baselines such as [8,27]. These tasks include link prediction [9,18,25,30,35], node clustering [25,28,36], matrix completion for inference and recommendation [1,7] and molecular graph generation [14,22,23,33]. Existing models usually rely on graph neural networks (GNN) to encode nodes into embeddings. More precisely, most of them implement *graph convolutional networks* (GCN) encoders [7,9,12,13,18,25,28,30,32] with multiple layers, a model originally introduced by [17].

However, despite the prevalent use of GCN encoders in recent literature, the relevance of this design choice has never been thoroughly studied nor challenged. The actual benefit of incorporating GCNs in graph AE and VAE w.r.t. significantly simpler encoding strategies remains unclear. In this paper[1], we propose to tackle this important aspect, showing that GCN-based graph AE and VAE are often unnecessarily complex for numerous applications. Our work falls into a family of recent efforts questioning the systematic use of complex deep learning methods without clear comparison to less fancy but simpler baselines [5,21,31]. More precisely, our contribution is threefold:

- We introduce and study simpler versions of graph AE and VAE, replacing multi-layer GCN encoders by linear models w.r.t. the direct neighborhood (one-hop) adjacency matrix of the graph, involving a unique weight matrix to tune, fewer operations and no activation function.
- Through an extensive empirical analysis on 17 real-world graphs with various sizes and characteristics, we show that these simplified models consistently reach competitive performances w.r.t. GCN-based graph AE and VAE on link prediction and node clustering tasks. We identify the settings where simple linear encoders appear as an effective alternative to GCNs, and as first relevant baseline to implement before diving into more complex models. We also question the relevance of current benchmark datasets (Cora, Citeseer, Pubmed) commonly used in the literature to evaluate graph AE and VAE.
- We publicly release the code[2] of these experiments, for reproducibility and easier future usages.

This paper is organized as follows. After reviewing key concepts on graph AE, VAE and on multi-layer GCNs in Sect. 2, we introduce the proposed simplified graph AE and VAE models in Sect. 3. We present and interpret our experiments in Sect. 4, and we conclude in Sect. 5.

[1] A preliminary version of this work has been presented at the NeurIPS 2019 workshop on Graph Representation Learning [29].

[2] https://github.com/deezer/linear_graph_autoencoders.

2 Preliminaries

We consider a graph $\mathcal{G} = (\mathcal{V}, \mathcal{E})$ with $|\mathcal{V}| = n$ nodes and $|\mathcal{E}| = m$ edges. In most of this paper, we assume that $\mathcal{G}$ is undirected (we extend our approach to directed graphs in Sect. 4.2). A is the direct neighborhood (i.e. one-hop) binary adjacency matrix of $\mathcal{G}$.

2.1 Graph Autoencoders

Graph autoencoders (AE) [18,34,37] are a family of models aiming at mapping (*encoding*) each node $i \in \mathcal{V}$ of the graph $\mathcal{G}$ to a low-dimensional vector $z_i \in \mathbb{R}^d$, with $d \ll n$, from which reconstructing (*decoding*) the graph should be possible. The intuition of this encoding-decoding scheme is the following: if, starting from the node embedding, the model is able to reconstruct an adjacency matrix $\hat{A}$ close to the true one, then the low-dimensional vectors z_i should capture some important characteristics of the original graph structure.

Formally, the $n \times d$ matrix Z, whose rows are the z_i vectors, is usually the output of a graph neural network (GNN) [3,6,17] processing A. Then, to reconstruct the graph, most models stack an *inner product decoder* [18] to this GNN, i.e. we have $\hat{A}_{ij} = \sigma(z_i^T z_j)$ for all node pairs (i, j), with $\sigma(\cdot)$ denoting the sigmoid function: $\sigma(x) = 1/(1 + e^{-x})$. Therefore, the larger the inner product $z_i^T z_j$ in the embedding, the more likely nodes i and j are connected in $\mathcal{G}$ according to the AE. In a nutshell, we have:

$$\hat{A} = \sigma(ZZ^T) \text{ with } Z = \text{GNN(A)}. \tag{1}$$

Several recent works also proposed more complex decoders [9,30,32], that we consider as well in our experiments in Sect. 4. During the training phase, the GNN weights are tuned by gradient descent to iteratively minimize a *reconstruction loss* capturing the similarity between A and $\hat{A}$, formulated as a cross entropy loss [18]:

$$\mathcal{L}^{\text{AE}} = -\frac{1}{n^2} \sum_{(i,j) \in \mathcal{V} \times \mathcal{V}} \Big[A_{ij} \log \hat{A}_{ij} + (1 - A_{ij}) \log(1 - \hat{A}_{ij}) \Big], \tag{2}$$

In the above Eq. (2), some existing methods [18,28,30] also adopt a link reweighting strategy, to reinforce the relative importance of positive links ($A_{ij} = 1$) in the loss when dealing with sparse graphs.

2.2 Graph Variational Autoencoders

[18] also extended the *variational autoencoder* (VAE) framework from [16] to graph structures. Authors designed a probabilistic model involving a latent variable z_i of dimension $d \ll n$ for each node $i \in \mathcal{V}$, interpreted as node representations in a d-dimensional embedding space. In their approach, the inference

model, i.e. the *encoding* part of the graph VAE, is defined as:

$$q(Z|A) = \prod_{i=1}^{n} q(z_i|A) \text{ with } q(z_i|A) = \mathcal{N}(z_i|\mu_i, \text{diag}(\sigma_i^2)). \quad (3)$$

Gaussian means and variances parameters are learned using a GNN for each one, i.e. $\mu = \text{GNN}_\mu(A)$, with μ the $n \times d$ matrix stacking up d-dimensional mean vectors μ_i ; likewise, $\log \sigma = \text{GNN}_\sigma(A)$. Latent vectors z_i are samples drawn from these distributions. Then, a generative model aims at reconstructing (*decoding*) A, leveraging inner products with sigmoid activations:

$$p(A|Z) = \prod_{i=1}^{n} \prod_{j=1}^{n} p(A_{ij}|z_i, z_j) \text{ with } p(A_{ij} = 1|z_i, z_j) = \hat{A}_{ij} = \sigma(z_i^T z_j). \quad (4)$$

During training, GNN weights are tuned by iteratively maximizing a tractable variational lower bound (ELBO) of the model's likelihood [18]:

$$\mathcal{L}^{\text{VAE}} = \mathbb{E}_{q(Z|A)}\Big[\log p(A|Z)\Big] - \mathcal{D}_{KL}(q(Z|A)||p(Z)), \quad (5)$$

by gradient descent, with a Gaussian prior on the distribution of latent vectors, and using the *reparameterization trick* from [16]. $\mathcal{D}_{KL}(\cdot||\cdot)$ denotes the Kullback-Leibler divergence [19].

2.3 Graph Convolutional Networks

While the term *GNN encoder* is generic, a majority of successful applications of graph AE and VAE actually relied on multi-layer *graph convolutional networks (GCN)* [17] to encode nodes. This includes the seminal graph AE and VAE models from [18] as well as numerous extensions [7,9,12,13,25,28,30,32]. In a multi-layer GCN with L layers ($L \geq 2$), with input layer $H^{(0)} = I_n$ and output layer $H^{(L)}$ (with $H^{(L)} = Z$ for AE, and $H^{(L)} = \mu$ or $\log \sigma$ for VAE), embedding vectors are iteratively updated, as follows:

$$H^{(l)} = \text{ReLU}(\tilde{A}H^{(l-1)}W^{(l-1)}), \text{ for } l \in \{1, ...L-1\} \quad (6)$$
$$H^{(L)} = \tilde{A}H^{(L-1)}W^{(L-1)}$$

where $\tilde{A} = D^{-1/2}(A+I_n)D^{-1/2}$. D is the diagonal degree matrix of $A+I_n$, and $\tilde{A}$ is therefore its symmetric normalization. At each layer, each node averages representations from its neighbors (that, from layer 2, have aggregated representations from their own neighbors), with a ReLU activation: $\text{ReLU}(x) = \max(x, 0)$. $W^{(0)}, ..., W^{(L-1)}$ are weight matrices to tune; their dimensions can differ across layers.

GCNs became popular encoders for graph AE and VAE, thanks to their reduced computational complexity w.r.t. other GNNs [3,6], and notably the linear time complexity w.r.t. m of evaluating each layer [17]. Moreover, GCN models can also leverage node-level features, summarized in an $n \times f$ matrix X, in addition to the graph structure. In such setting, the input layer becomes $H^{(0)} = X$ instead of the identity matrix I_n.

3 Simplifying Graph AE and VAE with One-Hop Linear Encoders

Graph AE and VAE emerged as powerful node embedding methods with promising applications and performances [1,7,9,12–14,22,23,25,28,30,32,33,36]. However, while almost all recent efforts from the literature implement multi-layer GCN (or an other GNN) encoders, the question of the actual benefit of such complex encoding schemes w.r.t. much simpler strategies remains widely open. In the following two sections, we tackle this important problem, arguing that these encoders often bring unnecessary complexity and redundancy. We propose and study alternative versions of graph AE and VAE, learning node embeddings from linear models, i.e. from simpler and more interpretable encoders, involving fewer parameters, fewer computations and no activation function.

3.1 Linear Graph AE

In this paper, we propose to replace the multi-layer GCN encoder by a simple linear model w.r.t. the normalized one-hop adjacency matrix of the graph. In the AE framework, we set:

$$Z = \tilde{A}W, \text{ then } \hat{A} = \sigma(ZZ^T). \tag{7}$$

We refer to this model as *linear graph AE*. Embedding vectors are obtained by multiplying the $n \times n$ normalized adjacency matrix $\tilde{A}$, as defined in (6), by a unique $n \times d$ weight matrix W. We tune this matrix in a similar fashion w.r.t. graph AE [18], i.e. by iteratively minimizing a weighted cross-entropy loss capturing the quality of the reconstruction $\hat{A}$ w.r.t. the original matrix A, by gradient descent.

This encoder is a straightforward linear mapping. Each element of z_i is a weighted average from node i's direct one-hop connections. Contrary to multi-layer GCN encoders (as $L \geq 2$), it ignores higher-order information (k-hop with $k > 1$). Also, the encoder does not include any non-linear activation function. In Sect. 4, we will highlight the very limited impact of these two simplifications on empirical performances.

This encoder runs in a linear time w.r.t. the number of edges m using a sparse representation for $\tilde{A}$, and involves fewer matrix operations than a GCN. It includes nd parameters i.e. slightly fewer than the $nd + (L-1)d^2$ parameters required by a L-layer GCN with d-dim layers. However, as for standard graph AE, the inner-product decoder has a quadratic $O(dn^2)$ complexity, as it involves the multiplication of the two dense matrices Z and Z^T. We discuss scalability strategies in Sect. 4, where we also implement two alternative decoders.

Linear graph AE models can also leverage graph datasets that include node-level features vectors of dimension f, stacked up in an $n \times f$ matrix X. In such setting, the encoding step becomes:

$$Z = \tilde{A}XW, \tag{8}$$

where the weight matrix W is then of dimension $f \times d$.

3.2 Linear Graph VAE

We adopt a similar approach to replace the two multi-layer GCNs of standard graph VAE models by:

$$\mu = \tilde{A}W_{\mu} \text{ and } \log \sigma = \tilde{A}W_{\sigma}, \tag{9}$$

with $n \times d$ weight matrices W_{μ} and W_{σ}. Then:

$$\forall i \in \mathcal{V}, z_i \sim \mathcal{N}(\mu_i, \text{diag}(\sigma_i^2)), \tag{10}$$

with similar decoder w.r.t. standard graph VAE (Eq. (4)). We refer to this simpler model as *linear graph VAE*. During the learning phase, as standard graph VAE, we iteratively optimize the ELBO bound of equation (4), w.r.t. W_{μ} and W_{σ}, by gradient descent. When dealing with graphs that include node features X, we instead compute:

$$\mu = \tilde{A}XW_{\mu} \text{ and } \log \sigma = \tilde{A}XW_{\sigma}, \tag{11}$$

and weight matrices W_{μ} and W_{σ} are then of dimension $f \times d$. Figure 1 displays a schematic representation of the proposed linear graph AE and VAE models.

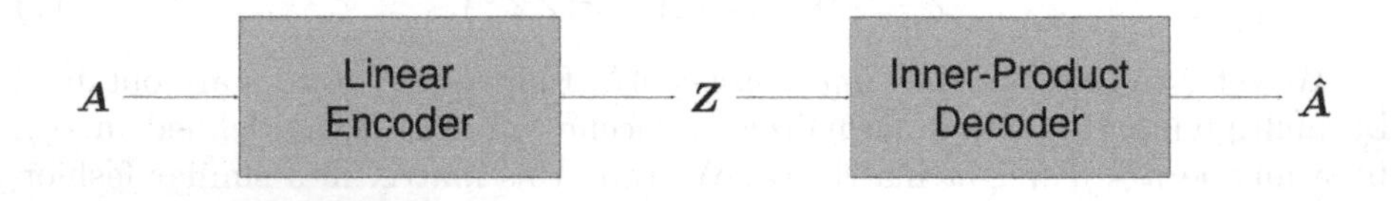

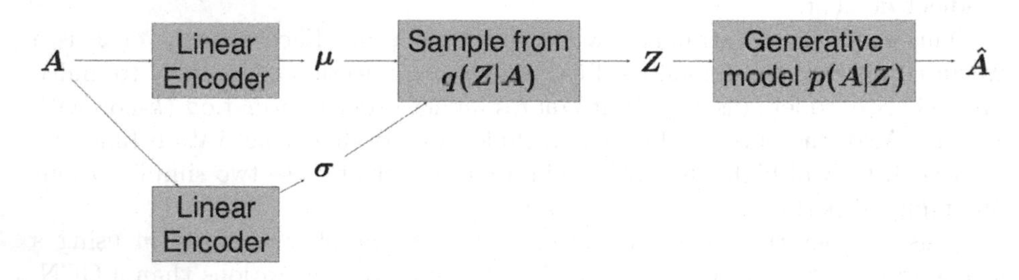

Fig. 1. Top: Linear Graph AE model. Bottom: Linear Graph VAE model.

3.3 Related Work

Our work falls into a family of research efforts aiming at challenging and questioning the prevalent use of complex deep learning methods without clear comparison to simpler baselines [5,21,31]. In particular, in the graph learning community, [38] recently proposed to simplify GCNs, notably by removing nonlinearities between layers and collapsing some weight matrices during training.

Their simplified model empirically rivals standard GCNs on several large-scale classification tasks. While our work also focuses on GCNs, we argue that the two papers actually tackle very different and complementary problems:

- [38] focus on *supervised* and semi-supervised settings. They consider the GCN as the model itself, optimized to classify node-level labels. On the contrary, we consider two *unsupervised* settings, in which GCNs are only a building part (the encoder) of a larger framework (the AE or the VAE), and where we optimize reconstruction losses from GCN-based embedding vectors (for AE) or from vectors drawn from distributions learned through two GCNs (for VAE).
- Our encoders only capture one-hop interactions: nodes aggregate information from their direct neighbors. On the contrary, [38] still rely on a stacked layers design that, although simplified, allows learning from higher-order interactions. Contrary to us, considering such relationships is crucial in their model for good performances (we explain in Sect. 4 that, in our setting, it would increase running times while bringing few to no improvement).

4 Empirical Analysis and Discussion

In this section, we propose an in-depth experimental evaluation of the proposed simplified graph AE and VAE models.

4.1 Experimental Setting

Tasks. We consider two learning tasks. Firstly, we focus on *link prediction*, as in [18] and most subsequent works. We train models on incomplete versions of graphs where 15% of edges were randomly removed. Then, we create validation and test sets from removed edges (resp. from 5% and 10% of edges) and from the same number of randomly sampled pairs of unconnected nodes. We evaluate the model's ability to classify edges from non-edges, using the mean *Area Under the Receiver Operating Characteristic (ROC) Curve* (AUC) and *Average Precision* (AP) scores on test sets, averaged over 100 runs where models were trained from 100 different random train/validation/test splits.

As a second task, we perform *node clustering* from the z_i vectors. When datasets include node-level ground-truth communities, we train models on complete graphs, then run k-means algorithms in embedding spaces. Then, we compare the resulting clusters to ground-truth communities via the mean *Adjusted Mutual Information (AMI)* scores computed over 100 runs.

Datasets. We provide experiments on 17 publicly available real-world graphs. For each graph, Tables 1 and 2 report the number of nodes n, the number of edges m, and the dimension f of node features (when available):

- We first consider the Cora, Citeseer and Pubmed citation graphs[3], with and without node features corresponding to f-dimensional bag-of-words vectors. These three graphs were used in the original experiments of [18] and then in the wide majority of recent works [9,12,13,25,26,28,30,32,35,36], becoming the *de facto* benchmark datasets for evaluating graph AE and VAE. Therefore, comparing linear and GCN-based models on these graphs was essential.
- We also report results on 14 alternative graphs. We consider four other citations networks: DBLP (see Footnote 5), Arxiv-HepTh (See Footnote 4), Patent[4] and a larger version of Cora (see Footnote 5), that we denote Cora-larger. We add the WebKD(see Footnote 3), Blogs (see Footnote 5) and Stanford[5] web graphs, where hyperlinks connect web pages, as well as two Google web graphs (a medium-size one (see Footnote 5), denoted Google, and a larger one (see Footnote 4), denoted Google-large). We complete the list with two social networks (Hamsterster (see Footnote 5) and LiveMocha (see Footnote 5)), the Flickr (see Footnote 5) image graph (nodes represent images, connected when sharing metadata), the Proteins (see Footnote 5) network of proteins interactions and the Amazon (see Footnote 5) products co-purchase network. We span a wide variety of real-world graphs of various natures, characteristics and sizes (from 877 to 2.7 million nodes, from 1 608 to 13.9 million edges).

Models. In all experiments, we compare the proposed simplified model to 2-layer and 3-layer GCN-based graph AE/VAE. We do not report performances of deeper models, due to significant scores deterioration. For a comparison to non-AE/VAE methods, which is out of the scope of this study, we refer to [18,28].

All models were trained for 200 epochs (resp. 300 epochs) for graphs with $n < 100\ 000$ (resp. $n \geq 100\ 000$). We thoroughly checked the convergence of all models, in terms of mean AUC performances on validation sets, for these epochs numbers. As [18], we ignored edges directions when initial graphs were directed. For Cora, Citeseer and Pubmed, we set identical hyperparameters w.r.t. [18] to reproduce their results, i.e. we had $d = 16$, 32-dim hidden layer(s) for GCNs, and we used Adam optimizer [15] with a learning rate of 0.01.

For other datasets, we tuned hyperparameters by performing grid search on the validation set. We adopted a learning rate of 0.1 for Arxiv-HepTh, Patent and Stanford; of 0.05 for Amazon, Flickr, LiveMocha and Google-large; of 0.01 for Blogs, Cora-larger, DBLP, Google and Hamsterster and Proteins (AE models); of 0.005 for WebKD (except linear AE and VAE where we used 0.001 and 0.01) and Proteins (VAE models). We set $d = 16$ (but we reached similar conclusions with $d = 32$ and 64), with 32-dim hidden layer(s) and without dropout.

Last, due to the prohibitive quadratic cost of reconstructing (*decoding*) the exact matrix $\hat{A}$ for large graphs (with $n \geq 100\ 000$), we adopted a simple stochastic sampling strategy for these graphs. At each training iteration, we estimated losses by only reconstructing a subgraph of 10 000 nodes from the original graph. These 10 000 nodes were randomly picked during training at each iteration.

[3] https://linqs.soe.ucsc.edu/data.
[4] http://snap.stanford.edu/data/index.html.
[5] http://konect.uni-koblenz.de/networks/.

Table 1. Link prediction on Cora, Citeseer and Pubmed benchmark datasets. Cells are grayed when linear graph AE/VAE are reaching competitive results w.r.t. standard GCN-based models (i.e. at least as good ± 1 standard deviation).

Model	Cora (n = 2 708, m = 5 429)		Citeseer (n = 3 327, m = 4 732)		Pubmed (n = 19 717, m = 44 338)	
	AUC (in %)	**AP (in %)**	**AUC (in %)**	**AP (in %)**	**AUC (in %)**	**AP (in %)**
Linear AE (ours)	83.19 ± 1.13	87.57 ± 0.95	77.06 ± 1.81	83.05 ± 1.25	81.85 ± 0.32	87.54 ± 0.28
2-layer GCN AE	84.79 ± 1.10	88.45 ± 0.82	78.25 ± 1.69	83.79 ± 1.24	82.51 ± 0.64	87.42 ± 0.38
3-layer GCN AE	84.61 ± 1.22	87.65 ± 1.11	78.62 ± 1.74	82.81 ± 1.43	83.37 ± 0.98	87.62 ± 0.68
Linear VAE (ours)	84.70 ± 1.24	88.24 ± 1.02	78.87 ± 1.34	83.34 ± 0.99	84.03 ± 0.28	87.98 ± 0.25
2-layer GCN VAE	84.19 ± 1.07	87.68 ± 0.93	78.08 ± 1.40	83.31 ± 1.31	82.63 ± 0.45	87.45 ± 0.34
3-layer GCN VAE	84.48 ± 1.42	87.61 ± 1.08	79.27 ± 1.78	83.73 ± 1.13	84.07 ± 0.47	88.18 ± 0.31
Model	**Cora, with features** (n = 2 708, m = 5 429, f = 1 433)		**Citeseer, with features** (n = 3 327, m = 4 732, f = 3 703)		**Pubmed, with features** (n = 19 717, m = 44 338, f = 500)	
	AUC (in %)	**AP (in %)**	**AUC (in %)**	**AP (in %)**	**AUC (in %)**	**AP (in %)**
Linear AE (ours)	92.05 ± 0.93	93.32 ± 0.86	91.50 ± 1.17	92.99 ± 0.97	95.88 ± 0.20	95.89 ± 0.17
2-layer GCN AE	91.27 ± 0.78	92.47 ± 0.71	89.76 ± 1.39	90.32 ± 1.62	96.28 ± 0.36	96.29 ± 0.25
3-layer GCN AE	89.16 ± 1.18	90.98 ± 1.01	87.31 ± 1.74	89.60 ± 1.52	94.82 ± 0.41	95.42 ± 0.26
Linear VAE (ours)	92.55 ± 0.97	93.68 ± 0.68	91.60 ± 0.90	93.08 ± 0.77	95.91 ± 0.13	95.80 ± 0.17
2-layer GCN VAE	91.64 ± 0.92	92.66 ± 0.91	90.72 ± 1.01	92.05 ± 0.97	94.66 ± 0.51	94.84 ± 0.42
3-layer GCN VAE	90.53 ± 0.94	91.71 ± 0.88	88.63 ± 0.95	90.20 ± 0.81	92.78 ± 1.02	93.33 ± 0.91

4.2 Results

Cora, Citeseer and Pubmed Benchmarks. Table 1 reports link prediction results for Cora, Citeseer and Pubmed. For standard graph AE and VAE, we managed to reproduce similar performances w.r.t. [18]. We show that linear graph AE and VAE models consistently reach competitive performances w.r.t. 2 and 3-layer GCN-based models, i.e. they are at least as good (±1 standard deviation). In some settings, linear graph AE/VAE are even slightly better (e.g. +1.25 points in AUC for linear graph VAE on Pubmed with features, w.r.t. 2-layer GCN-based graph VAE). These results emphasize the effectiveness of the proposed simple encoding scheme on these datasets, where the empirical benefit of multi-layer GCNs is very limited. In Table 3, we consolidate our results by reaching similar conclusions on the node clustering task. Nodes are documents clustered in respectively 6, 7 and 3 topic classes, acting as ground-truth communities. In almost all settings, linear graph AE and VAE rival their GCN-based counterparts (e.g. +4.31 MI points for linear graph VAE on Pubmed with features, w.r.t. 2-layer GCN-based graph VAE).

Table 2. Link prediction on alternative real-world datasets. Cells are grayed when linear graph AE/VAE are reaching competitive results w.r.t. standard GCN-based models (i.e. at least as good ± 1 standard deviation).

Model	WebKD (n = 877, m = 1 608)		WebKD, with features (n = 877, m = 1 608, f = 1 703)		Hamsterster (n = 1 858, m = 12 534)	
	AUC (in %)	AP (in %)	AUC (in %)	AP (in %)	AUC (in %)	AP (in %)
Linear AE (ours)	77.20 ± 2.35	83.55 ± 1.81	84.15 ± 1.64	87.01 ± 1.48	93.07 ± 0.67	94.20 ± 0.58
2-layer GCN AE	77.88 ± 2.57	84.12 ± 2.18	86.03 ± 3.97	87.97 ± 2.76	92.07 ± 0.63	93.01 ± 0.69
3-layer GCN AE	78.20 ± 3.69	83.13 ± 2.58	81.39 ± 3.93	85.34 ± 2.92	91.40 ± 0.79	92.22 ± 0.85
Linear VAE (ours)	83.50 ± 1.98	86.70 ± 1.53	85.57 ± 2.18	88.08 ± 1.76	91.08 ± 0.70	91.85 ± 0.64
2-layer GCN VAE	82.31 ± 2.55	86.15 ± 2.03	87.87 ± 2.48	88.97 ± 2.17	91.62 ± 0.60	92.43 ± 0.64
3-layer GCN VAE	82.17 ± 2.70	85.35 ± 2.25	89.69 ± 1.80	89.90 ± 1.58	91.06 ± 0.71	91.85 ± 0.77

Model	DBLP (n = 12 591, m = 49 743)		Cora-larger (n = 23 166, m = 91 500)		Arxiv-HepTh (n = 27 770, m = 352 807)	
	AUC (in %)	AP (in %)	AUC (in %)	AP (in %)	AUC (in %)	AP (in %)
Linear AE (ours)	90.11 ± 0.40	93.15 ± 0.28	94.64 ± 0.08	95.96 ± 0.10	98.34 ± 0.03	98.46 ± 0.03
2-layer GCN AE	90.29 ± 0.39	93.01 ± 0.33	94.80 ± 0.08	95.72 ± 0.05	97.97 ± 0.09	98.12 ± 0.09
3-layer GCN AE	89.91 ± 0.61	92.24 ± 0.67	94.51 ± 0.31	95.11 ± 0.28	94.35 ± 1.30	94.46 ± 1.31
Linear VAE (ours)	90.62 ± 0.30	93.25 ± 0.22	95.20 ± 0.16	95.99 ± 0.12	98.35 ± 0.05	98.46 ± 0.05
2-layer GCN VAE	90.40 ± 0.43	93.09 ± 0.35	94.60 ± 0.20	95.74 ± 0.13	97.75 ± 0.08	97.91 ± 0.06
3-layer GCN VAE	89.92 ± 0.59	92.52 ± 0.48	94.48 ± 0.28	95.30 ± 0.22	94.57 ± 1.14	94.73 ± 1.12

Model	LiveMocha (n = 104 103, m = 2 193 083)		Flickr (n = 105 938, m = 2 316 948)		Patent (n = 2 745 762, m = 13 965 410)	
	AUC (in %)	AP (in %)	AUC (in %)	AP (in %)	AUC (in %)	AP (in %)
Linear AE (ours)	93.35 ± 0.10	94.83 ± 0.08	96.38 ± 0.05	97.27 ± 0.04	85.49 ± 0.09	87.17 ± 0.07
2-layer GCN AE	92.79 ± 0.17	94.33 ± 0.13	96.34 ± 0.05	97.22 ± 0.04	82.86 ± 0.20	84.52 ± 0.24
3-layer GCN AE	92.22 ± 0.73	93.67 ± 0.57	96.06 ± 0.08	97.01 ± 0.05	83.77 ± 0.41	84.73 ± 0.42
Linear VAE (ours)	93.23 ± 0.06	94.61 ± 0.05	96.05 ± 0.08	97.12 ± 0.06	84.57 ± 0.27	85.46 ± 0.30
2-layer GCN VAE	92.68 ± 0.21	94.23 ± 0.15	96.35 ± 0.07	97.20 ± 0.06	83.77 ± 0.28	83.37 ± 0.26
3-layer GCN VAE	92.71 ± 0.37	94.01 ± 0.26	96.39 ± 0.13	97.16 ± 0.08	85.30 ± 0.51	86.14 ± 0.49

Model	Blogs (n = 1 224, m = 19 025)		Amazon (n = 334 863, m = 925 872)		Google-large (n = 875 713, m = 5 105 039)	
	AUC (in %)	AP (in %)	AUC (in %)	AP (in %)	AUC (in %)	AP (in %)
Linear AE (ours)	91.71 ± 0.39	92.53 ± 0.44	90.70 ± 0.09	93.46 ± 0.08	95.37 ± 0.05	96.93 ± 0.05
2-layer GCN AE	91.57 ± 0.34	92.51 ± 0.29	90.15 ± 0.15	92.33 ± 0.14	95.06 ± 0.08	96.40 ± 0.07
3-layer GCN AE	91.74 ± 0.37	92.62 ± 0.31	88.54 ± 0.37	90.47 ± 0.38	93.68 ± 0.15	94.99 ± 0.14
Linear VAE (ours)	91.34 ± 0.24	92.10 ± 0.24	84.53 ± 0.08	87.79 ± 0.06	91.13 ± 0.14	93.79 ± 0.10
2-layer GCN VAE	91.85 ± 0.22	92.60 ± 0.25	90.14 ± 0.22	92.33 ± 0.23	95.04 ± 0.09	96.38 ± 0.07
3-layer GCN VAE	91.83 ± 0.48	92.65 ± 0.35	89.44 ± 0.25	91.23 ± 0.23	93.79 ± 0.22	95.12 ± 0.21

Model	Stanford (n = 281 903, m = 2 312 497)		Proteins (n = 6 327, m = 147 547)		Google (n = 15 763, m = 171 206)	
	AUC (in %)	AP (in %)	AUC (in %)	AP (in %)	AUC (in %)	AP (in %)
Linear AE (ours)	97.73 ± 0.10	98.37 ± 0.10	94.09 ± 0.23	96.01 ± 0.16	96.02 ± 0.14	97.09 ± 0.08
2-layer GCN AE	97.05 ± 0.63	97.56 ± 0.55	94.55 ± 0.20	96.39 ± 0.16	96.66 ± 0.24	97.45 ± 0.25
3-layer GCN AE	92.19 ± 1.49	92.58 ± 1.50	94.30 ± 0.19	96.08 ± 0.15	95.10 ± 0.27	95.94 ± 0.20
Linear VAE (ours)	94.96 ± 0.25	96.64 ± 0.15	93.99 ± 0.10	95.94 ± 0.16	91.11 ± 0.31	92.91 ± 0.18
2-layer GCN VAE	97.60 ± 0.11	98.02 ± 0.10	94.57 ± 0.18	96.18 ± 0.33	96.11 ± 0.59	96.84 ± 0.51
3-layer GCN VAE	97.53 ± 0.13	98.01 ± 0.10	94.27 ± 0.25	95.71 ± 0.28	95.10 ± 0.54	96.00 ± 0.44

Table 3. Node clustering on graphs with communities. Cells are grayed when linear graph AE/VAE are reaching competitive results w.r.t. standard GCN-based models (i.e. at least as good ± 1 standard deviation).

Model	Cora (n = 2 708, m = 5 429)	Cora with features (n = 2 708, m = 5 429, f = 1 433)	Citeseer (n = 3 327, m = 4 732)	Citeseer with features (n = 3 327, m = 4 732, f = 3 703)
	AMI (in %)	AMI (in %)	AMI (in %)	AMI (in %)
Linear AE (ours)	26.31 ± 2.85	47.02 ± 2.09	8.56 ± 1.28	20.23 ± 1.36
2-layer GCN AE	30.88 ± 2.56	43.04 ± 3.28	9.46 ± 1.06	19.38 ± 3.15
3-layer GCN AE	33.06 ± 3.10	44.12 ± 2.48	10.69 ± 1.98	19.71 ± 2.55
Linear VAE (ours)	34.35 ± 1.42	48.12 ± 1.96	12.67 ± 1.27	20.71 ± 1.95
2-layer GCN VAE	26.66 ± 3.94	44.84 ± 2.63	9.85 ± 1.24	20.17 ± 3.07
3-layer GCN VAE	28.43 ± 2.83	44.29 ± 2.54	10.64 ± 1.47	19.94 ± 2.50

Model	Pubmed (n = 19 717, m = 44 338)	Pubmed with features (n = 19 717, m = 44 338, f = 500)	Cora-larger (n = 23 166, m = 91 500)	Blogs (n = 1 224, m = 19 025)
	AMI (in %)	AMI (in %)	AMI (in %)	AMI (in %)
Linear AE (ours)	10.76 ± 3.70	26.12 ± 1.94	40.34 ± 0.51	46.84 ± 1.79
2-layer GCN AE	16.41 ± 3.15	23.08 ± 3.35	39.75 ± 0.79	72.58 ± 4.54
3-layer GCN AE	23.11 ± 2.58	25.94 ± 3.09	35.67 ± 1.76	72.72 ± 1.80
Linear VAE (ours)	25.14 ± 2.83	29.74 ± 0.64	43.32 ± 0.52	49.70 ± 1.08
2-layer GCN VAE	20.52 ± 2.97	25.43 ± 1.47	38.34 ± 0.64	73.12 ± 0.83
3-layer GCN VAE	21.32 ± 3.70	24.91 ± 3.09	37.30 ± 1.07	70.56 ± 5.43

Alternative Graph Datasets. Table 2 reports link prediction results for all other graphs. Linear graph AE models are competitive in 13 cases out of 15, and sometimes even achieve better performances (e.g., +1.72 AUC points for linear graph AE on the largest dataset, Patent, w.r.t. 3-layer GCN-based graph AE). Moreover, linear graph VAE models rival or outperform GCN-based models in 10 cases out of 15. Overall, linear graph AE/VAE also achieve very close results w.r.t. GCN-based models in all remaining datasets (e.g. on Google, with a mean AUC score of 96.02% ± 0.14 for linear graph AE, only slightly below the mean AUC score of 96.66% ± 0.24 of 2-layer GCN-based graph AE). This confirms the empirical effectiveness of simple node encoding schemes, that appear as a suitable alternative to complex multi-layer encoders for many real-world applications. Regarding node clustering (Table 3), linear AE and VAE models are competitive on the Cora-larger graph, in which nodes are documents clustered in 70 topic classes. However, 2-layer and 3-layer GCN-based models are significantly outperforming on the Blogs graph, where political blogs are classified as either left-leaning or right-leaning (e.g. −23.42 MI points for linear graph VAE w.r.t. 2-layer GCN-based graph VAE).

Experiments on More Complex Decoders. So far, we compared different encoders but the (standard) inner-product decoder was fixed. As a robustness check, in Table 4, we report complementary link prediction experiments, on variants of graph AE/VAE with two more complex decoders from recent works:

- The Graphite model from [9], that still considers undirected graphs, but rely on an iterative graph refinement strategy inspired by low-rank approximations for decoding.
- The Gravity-Inspired model from [30], that provides an asymmetric decoding scheme (i.e. $\hat{A}_{ij} \neq \hat{A}_{ji}$). This model handles directed graphs. Therefore, contrary to previous experiments, we do not ignore edges directions when initial graphs were directed.

We draw similar conclusions w.r.t. Tables 1 and 2, consolidating our conclusions. For brevity, we only report results for the Cora, Citeseer and Pubmed graphs, where linear models are competitive, and for the Google graph, where GCN-based graph AE and VAE slightly outperform. We stress out that scores from Graphite [9] and Gravity [30] models are *not* directly comparable, as the former ignores edges directionalities while the latter processes directed graphs, i.e. the learning task becomes a *directed* link prediction problem.

When (Not) to Use Multi-layer GCN Encoders? Linear graph AE and VAE reach strong empirical results on all graphs, and rival or outperform GCN-based graph AE and VAE in a majority of experiments. These models are also significantly simpler and more interpretable, each element of z_i being interpreted as a weighted average from node i's direct neighborhood. Therefore, we recommend the systematic use of linear graph AE and VAE as a first baseline, before diving into more complex encoding schemes whose actual benefit might be unclear.

Moreover, from our experiments, we also conjecture that multi-layer GCN encoders *can* bring an empirical advantage when dealing with graphs with *intrinsic non-trivial high-order interactions.* Notable examples of such graphs include the Amazon co-purchase graph (+5.61 AUC points for 2-layer GCN VAE) and web graphs such as Blogs, Google and Stanford, in which two-hop hyperlinks connections of pages usually include relevant information on the global network structure. On such graphs, capturing this additional information tends to improve results, especially 1) for the probabilistic VAE framework, and 2) when evaluating embeddings via the node clustering task (20+ AMI points on Blogs for 2-layer GCN AE/VAE) which is, by design, a more *global* learning task than the quite *local* link prediction problem. On the contrary, in citation graphs, the relevance of two-hop links is limited. Indeed, if a reference A in an article B cited by some authors is relevant to their work, authors will likely also cite this reference A, thus creating a one-hop link. Last, while the impact of the graph *size* is unclear in our experiments (linear models achieve strong results even on large graphs, such as Patent), we note that graphs where multi-layer GCN encoders tend to outperform linear models are all relatively *dense*.

Table 4. Link prediction with Graphite and Gravity alternative decoding schemes. Cells are grayed when linear graph AE/VAE are reaching competitive results w.r.t. GCN-based models (i.e. at least as good ± 1 standard deviation).

Model	Cora (n = 2 708, m = 5 429)		Citeseer (n = 3 327, m = 4 732)	
	AUC (in %)	**AP (in %)**	**AUC (in %)**	**AP (in %)**
Linear Graphite AE (ours)	83.42 ± 1.76	87.32 ± 1.53	77.56 ± 1.41	82.88 ± 1.15
2-layer Graphite AE	81.20 ± 2.21	85.11 ± 1.91	73.80 ± 2.24	79.32 ± 1.83
3-layer Graphite AE	79.06 ± 1.70	81.79 ± 1.62	72.24 ± 2.29	76.60 ± 1.95
Linear Graphite VAE (ours)	83.68 ± 1.42	87.57 ± 1.16	78.90 ± 1.08	83.51 ± 0.89
2-layer Graphite VAE	84.89 ± 1.48	88.10 ± 1.22	77.92 ± 1.57	82.56 ± 1.31
3-layer Graphite VAE	85.33 ± 1.19	87.98 ± 1.09	77.46 ± 2.34	81.95 ± 1.71
Linear Gravity AE (ours)	90.71 ± 0.95	92.95 ± 0.88	80.52 ± 1.37	86.29 ± 1.03
2-layer Gravity AE	87.79 ± 1.07	90.78 ± 0.82	78.36 ± 1.55	84.75 ± 1.10
3-layer Gravity AE	87.76 ± 1.32	90.15 ± 1.45	78.32 ± 1.92	84.88 ± 1.36
Linear Gravity VAE (ours)	91.29 ± 0.70	93.01 ± 0.57	86.65 ± 0.95	89.49 ± 0.69
2-layer Gravity VAE	91.92 ± 0.75	92.46 ± 0.64	87.67 ± 1.07	89.79 ± 1.01
3-layer Gravity VAE	90.80 ± 1.28	92.01 ± 1.19	85.28 ± 1.33	87.54 ± 1.21
Model	**Pubmed** (n = 19 717, m = 44 338)		**Google** (n = 15 763, m = 171 206)	
	AUC (in %)	**AP (in %)**	**AUC (in %)**	**AP (in %)**
Linear Graphite AE (ours)	80.28 ± 0.86	85.81 ± 0.67	94.30 ± 0.22	95.09 ± 0.16
2-layer Graphite AE	79.98 ± 0.66	85.33 ± 0.41	95.54 ± 0.42	95.99 ± 0.39
3-layer Graphite AE	79.96 ± 1.40	84.88 ± 0.89	93.99 ± 0.54	94.74 ± 0.49
Linear Graphite VAE (ours)	79.59 ± 0.33	86.17 ± 0.31	92.71 ± 0.38	94.41 ± 0.25
2-layer Graphite VAE	82.74 ± 0.30	87.19 ± 0.36	96.49 ± 0.22	96.91 ± 0.17
3-layer Graphite VAE	84.56 ± 0.42	88.01 ± 0.39	96.32 ± 0.24	96.62 ± 0.20
Linear Gravity AE (ours)	76.78 ± 0.38	84.50 ± 0.32	97.46 ± 0.07	98.30 ± 0.04
2-layer Gravity AE	75.84 ± 0.42	83.03 ± 0.22	97.77 ± 0.10	98.43 ± 0.10
3-layer Gravity AE	74.61 ± 0.30	81.68 ± 0.26	97.58 ± 0.12	98.28 ± 0.11
Linear Gravity VAE (ours)	79.68 ± 0.36	85.00 ± 0.21	97.32 ± 0.06	98.26 ± 0.05
2-layer Gravity VAE	77.30 ± 0.81	82.64 ± 0.27	97.84 ± 0.25	98.18 ± 0.14
3-layer Gravity VAE	76.52 ± 0.61	80.73 ± 0.63	97.32 ± 0.23	97.81 ± 0.20

To conclude, we conjecture that denser graphs with intrinsic high-order interactions (e.g. web graphs) should be better suited than the sparse Cora, Citeseer and Pubmed citation networks, to evaluate and to compare complex graph AE and VAE models, especially on global tasks such as node clustering.

On k-hop linear encoders. While, in this work, we only learn from direct neighbors interactions, variants of our models could capture higher-order links by considering polynomials of the matrix A. For instance, we could learn embeddings from one-hop and two-hop links by replacing $\tilde{A}$ by the normalized version of $A + \alpha A^2$ (with $\alpha > 0$), or simply A^2, in the linear encoders of Sect. 3.

Our online implementation proposes such alternative. We observed few to no improvement on most of our graphs, consistently with our claim on the effectiveness of simple one-hop strategies. Such variants also tend to increase running times (see below), as A^2 is usually denser than A.

On Running Times. While this work put the emphasis on performance and not on training speed, we also note that linear AE and VAE models are 10% to 15% faster than their GCN-based counterparts. For instance, on an NVIDIA GTX 1080 GPU, we report a 6.03 s (vs 6.73 s) mean running time for training our linear graph VAE (vs 2-layer GCN graph VAE) on the featureless Citeseer dataset, and 791 s (vs 880 s) on the Patent dataset, using our sampling strategy from Sect. 4.1. This gain comes from the slightly fewer parameters and matrix operations required by one-hop linear encoders and from the sparsity of the one-hop matrix $\tilde{A}$ for most real-world graphs.

Nonetheless, as an opening, we point out that the problem of scalable graph autoencoders remains quite open. Despite advances on the encoder, the standard inner-product decoder still suffer from a $O(dn^2)$ time complexity. Our very simple sampling strategy to overcome this quadratic cost on large graphs (randomly sampling subgraphs to reconstruct) might not be optimal. Future works will therefore tackle these issues, aiming at providing more efficient strategies to scale graph AE and VAE to large graphs with millions of nodes and edges.

5 Conclusion

Graph autoencoders (AE), graph variational autoencoders (VAE) and most of their extensions rely on multi-layer graph convolutional networks (GCN) encoders to learn node embedding representations. In this paper, we highlighted that, despite their prevalent use, these encoders are often unnecessarily complex. In this direction, we introduced and studied significantly simpler versions of these models, leveraging one-hop linear encoding strategies. Using these alternative models, we reached competitive empirical performances w.r.t. GCN-based graph AE and VAE on numerous real-world graphs. We identified the settings where simple one-hop linear encoders appear as an effective alternative to multi-layer GCNs, and as first relevant baseline to implement before diving into more complex models. We also questioned the relevance of repeatedly using the same sparse medium-size datasets (Cora, Citeseer, Pubmed) to evaluate and to compare complex graph AE and VAE models.

References

1. Berg, R.V.D., Kipf, T.N., Welling, M.: Graph convolutional matrix completion. KDD Deep Learning Day (2018)
2. Bhagat, S., Cormode, G., Muthukrishnan, S.: Node classification in social networks. In: Aggarwal, C. (ed.) Social Network Data Analytics, pp. 115–148. Springer, Boston (2011). https://doi.org/10.1007/978-1-4419-8462-3_5
3. Bruna, J., Zaremba, W., Szlam, A., LeCun, Y.: Spectral networks and locally connected networks on graphs. In: International Conference on Learning Representations (2014)
4. Cao, S., Lu, W., Xu, Q.: GraRep: learning graph representations with global structural information. In: ACM International Conference on Information and Knowledge Management (2015)
5. Dacrema, M.F., Cremonesi, P., Jannach, D.: Are we really making much progress? A worrying analysis of recent neural recommendation approaches. In: ACM Conference on Recommender Systems (2019)
6. Defferrard, M., Bresson, X., Vandergheynst, P.: Convolutional neural networks on graphs with fast localized spectral filtering. In: Advances in Neural Information Processing Systems (2016)
7. Do, T.H., et al.: Matrix completion with variational graph autoencoders: application in hyperlocal air quality inference. In: IEEE International Conference on Acoustics, Speech and Signal Processing (2019)
8. Grover, A., Leskovec, J.: node2vec: scalable feature learning for networks. In: ACM SIGKDD International Conference on Knowledge Discovery and Data Mining (2016)
9. Grover, A., Zweig, A., Ermon, S.: Graphite: iterative generative modeling of graphs. In: International Conference on Machine Learning (2019)
10. Hamilton, W.L., Ying, R., Leskovec, J.: Representation learning on graphs: Methods and applications. IEEE Data Engineering Bulletin (2017)
11. Hamilton, W.L., Ying, Z., Leskovec, J.: Inductive representation learning on large graphs. In: Advances in Neural Information Processing Systems (2017)
12. Hasanzadeh, A., Hajiramezanali, E., Narayanan, K., Duffield, N., Zhou, M., Qian, X.: Semi-implicit graph variational auto-encoders. In: Advances in Neural Information Processing Systems (2019)
13. Huang, P.Y., Frederking, R., et al.: RWR-GAE: random walk regularization for graph auto encoders. arXiv preprint arXiv:1908.04003 (2019)
14. Jin, W., Barzilay, R., Jaakkola, T.: Junction tree variational autoencoder for molecular graph generation. In: International Conference on Machine Learning (2018)
15. Kingma, D.P., Ba, J.: Adam: a method for stochastic optimization. In: International Conference on Learning Representations (2015)
16. Kingma, D.P., Welling, M.: Auto-encoding variational Bayes. In: International Conference on Learning Representations (2014)
17. Kipf, T.N., Welling, M.: Semi-supervised classification with graph convolutional networks. In: International Conference on Learning Representations (ICLR 2017) (2016)
18. Kipf, T.N., Welling, M.: Variational graph auto-encoders. In: NeurIPS Workshop on Bayesian Deep Learning (2016)
19. Kullback, S., Leibler, R.A.: On information and sufficiency. Ann. Math. Stat. **22**(1), 79–86 (1951)
20. Liben-Nowell, D., Kleinberg, J.: The link-prediction problem for social networks. J. Am. Soc. Inform. Sci. Technol. **58**(7), 1019–1031 (2007)

21. Lin, J.: The neural hype and comparisons against weak baselines. ACM SIGIR Forum **52**(2), 40–51 (2019)
22. Liu, Q., Allamanis, M., Brockschmidt, M., Gaunt, A.: Constrained graph variational autoencoders for molecule design. In: Advances in Neural Information Processing Systems (2018)
23. Ma, T., Chen, J., Xiao, C.: Constrained generation of semantically valid graphs via regularizing variational autoencoders. In: Advances in Neural Information Processing Systems (2018)
24. Ou, M., Cui, P., Pei, J., Zhang, Z., Zhu, W.: Asymmetric transitivity preserving graph embedding. In: ACM SIGKDD International Conference on Knowledge Discovery and Data Mining (2016)
25. Pan, S., Hu, R., Long, G., Jiang, J., Yao, L., Zhang, C.: Adversarially regularized graph autoencoder for graph embedding. In: International Joint Conference on Artificial Intelligence (2018)
26. Park, J., Lee, M., Chang, H.J., Lee, K., Choi, J.: Symmetric graph convolutional autoencoder for unsupervised graph representation learning. arXiv preprint arXiv:1908.02441 (2019)
27. Perozzi, B., Al-Rfou, R., Skiena, S.: DeepWalk: online learning of social representations. In: ACM SIGKDD International Conference on Knowledge Discovery and Data Mining (2014)
28. Salha, G., Hennequin, R., Tran, V.A., Vazirgiannis, M.: A degeneracy framework for scalable graph autoencoders. In: International Joint Conference on Artificial Intelligence (2019)
29. Salha, G., Hennequin, R., Vazirgiannis, M.: Keep it simple: graph autoencoders without graph convolutional networks. In: NeurIPS Workshop on Graph Representation Learning (2019)
30. Salha, G., Limnios, S., Hennequin, R., Tran, V.A., Vazirgiannis, M.: Gravity-inspired graph autoencoders for directed link prediction. In: ACM International Conference on Information and Knowledge Management (2019)
31. Shchur, O., Mumme, M., Bojchevski, A., Günnemann, S.: Pitfalls of graph neural network evaluation. In: NeurIPS Workshop on Relational Representation Learning (2018)
32. Shi, H., Fan, H., Kwok, J.T.: Effective decoding in graph auto-encoder using triadic closure. In: AAAI Conference on Artificial Intelligence (2020)
33. Simonovsky, M., Komodakis, N.: GraphVAE: towards generation of small graphs using variational autoencoders. In: International Conference on Artificial Neural Networks (2018)
34. Tian, F., Gao, B., Cui, Q., Chen, E., Liu, T.Y.: Learning deep representations for graph clustering. In: AAAI Conference on Artificial Intelligence (2014)
35. Tran, P.V.: Multi-task graph autoencoders. arXiv preprint arXiv:1811.02798 (2018)
36. Wang, C., Pan, S., Long, G., Zhu, X., Jiang, J.: MGAE: marginalized graph autoencoder for graph clustering. In: ACM Conference on Information and Knowledge Management (2017)
37. Wang, D., Cui, P., Zhu, W.: Structural deep network embedding. In: ACM SIGKDD International Conference on Knowledge Discovery and Data Mining (2016)
38. Wu, F., Souza, A., Zhang, T., Fifty, C., Yu, T., Weinberger, K.: Simplifying graph convolutional networks. In: International Conference on Machine Learning (2019)
39. Wu, Z., Pan, S., Chen, F., Long, G., Zhang, C., Yu, P.S.: A comprehensive survey on graph neural networks. arXiv preprint arXiv:1901.00596 (2019)

Sparse Separable Nonnegative Matrix Factorization

Nicolas Nadisic[1(✉)], Arnaud Vandaele[1], Jeremy E. Cohen[2], and Nicolas Gillis[1]

[1] University of Mons, Mons, Belgium
{nicolas.nadisic,arnaud.vandaele,nicolas.gillis}@umons.ac.be
[2] Univ Rennes, Inria, CNRS, IRISA, Rennes, France
jeremy.cohen@irisa.fr

Abstract. We propose a new variant of nonnegative matrix factorization (NMF), combining separability and sparsity assumptions. Separability requires that the columns of the first NMF factor are equal to columns of the input matrix, while sparsity requires that the columns of the second NMF factor are sparse. We call this variant sparse separable NMF (SSNMF), which we prove to be NP-complete, as opposed to separable NMF which can be solved in polynomial time. The main motivation to consider this new model is to handle underdetermined blind source separation problems, such as multispectral image unmixing. We introduce an algorithm to solve SSNMF, based on the successive nonnegative projection algorithm (SNPA, an effective algorithm for separable NMF), and an exact sparse nonnegative least squares solver. We prove that, in noiseless settings and under mild assumptions, our algorithm recovers the true underlying sources. This is illustrated by experiments on synthetic data sets and the unmixing of a multispectral image.

Keywords: Nonnegative matrix factorization · Sparsity · Separability

1 Introduction

Nonnegative Matrix Factorization (NMF) is a low-rank model widely used for feature extraction in applications such as multispectral imaging, text mining, or blind source separation; see [6,8] and the references therein. Given a nonnegative data matrix $M \in \mathbb{R}_+^{m \times n}$ and a factorization rank r, NMF consists in finding two nonnegative matrices $W \in \mathbb{R}_+^{m \times r}$ and $H \in \mathbb{R}_+^{r \times n}$ such that $M \approx WH$. NMF can be formalized as the following optimization problem:

$$\min_{W \geq 0, H \geq 0} \|M - WH\|_F^2. \tag{1}$$

Electronic supplementary material The online version of this chapter (https://doi.org/10.1007/978-3-030-67658-2_20) contains supplementary material, which is available to authorized users.

F. Hutter et al. (Eds.): ECML PKDD 2020, LNAI 12457, pp. 335–350, 2021.
https://doi.org/10.1007/978-3-030-67658-2_20

In this paper, we use the Frobenius norm to measure the quality of the approximation. Although other measures are possible, the Frobenius norm is by far the most commonly used, because it assumes Gaussian noise (which is reasonable in many real-life applications) and allows for efficient computations [8].

One of the advantages of NMF over similar methods such as principal component analysis (PCA) is that the nonnegativity constraint favors a part-based representation [13], which is to say that the factors are more easily interpretable, in particular when they have a physical meaning. If each column of M represents a data point, then each corresponding column of H contains the coefficients to reconstruct it from the r atoms represented by the columns of W, since $M(:,j) \approx WH(:,j)$ for all j. Every data point is therefore expressed as a linear combination of atoms. For example, when using NMF for multispectral unmixing, a data point is a pixel, an atom is a specific material, and each column of H contains the abundance of these materials in the corresponding pixel; see Sect. 5.2 for more details. Geometrically, the atoms (columns of W) can be seen as r vertices whose convex hull contains the data points (columns of M), under appropriate scaling.

1.1 Separability

In general, computing NMF is NP-hard [19]. However, Arora et al. [2] proved that NMF is solvable in polynomial time under the *separability* assumption on the input matrix.

Definition 1. *A matrix M is r-separable if there exists a subset of r columns of M, indexed by $\mathcal{J}$, and a nonnegative matrix $H \geq 0$, such that $M = M(:,\mathcal{J})H$.*

Equivalently, M is r-separable if M has the form $M = W[I_r, H']\Pi$, where I_r is the identity matrix of size r, H' is a nonnegative matrix, and Π is a permutation. Separable NMF consists in selecting the right r columns of M such that M can be reconstructed perfectly. In other words, it consists in finding the atoms (columns of W) *among* the data points (columns of M).

Problem 1 (Separable NMF). Given a r-separable matrix M, find $W = M(:,\mathcal{J})$ with $|\mathcal{J}| = r$ and $H \geq 0$ such that $M = WH$.

Note that, if W is known, the computation of H is straightforward: it is a convex problem that can be solved using any nonnegative least squares (NNLS) solver (for example, it can be solved with the Matlab function `lsqnonneg`). However, the solution is not necessarily unique, unless W is full rank.

In the presence of noise, which is typically the case in real-life applications, this problem is called near-separable NMF and is also solvable in polynomial time given that the noise level is sufficiently small [2]. In this case, we are given a near-separable matrix $M \approx M(:,\mathcal{J})H$ where $|\mathcal{J}| = r$ and $H \geq 0$.

1.2 Successive Nonnegative Projection Algorithm

Various algorithms have been developed to tackle the (near-)separable NMF problem. Some examples are the successive projections algorithm (SPA) [1], the

Algorithm 1: SNPA

Input: A near-separable matrix $M \in \mathbb{R}^{m \times n}$, the number r of columns to be extracted, and a strongly convex function f with $f(0) = 0$ (by default, $f(x) = \|x\|_2^2$).

Output: A set of r indices $\mathcal{J}$, and a matrix $H \in \mathbb{R}_+^{r \times n}$ such that $M \approx M(:, \mathcal{J})H$.

1 Init $R \leftarrow M$
2 Init $\mathcal{J} = \{\}$
3 Init $t = 1$
4 **while** $R \neq 0$ & $t \leq r$ **do**
5 $\quad p = \text{argmax}_j\, f(R(:, j))$
6 $\quad \mathcal{J} = \mathcal{J} \cup \{p\}$
7 $\quad$**foreach** j **do**
8 $\quad\quad H^*(:, j) = \underset{h \in \Delta}{\text{argmin}}\, f(M(:, j) - M(:, \mathcal{J})h)$
9 $\quad\quad R(:, j) = M(:, j) - M(:, \mathcal{J})H^*(:, j)$
10 $\quad$ t = t + 1

fast canonical hull algorithm [12], or the successive nonnegative projections algorithm (SNPA) [7]. Such algorithms start with an empty matrix W and a residual matrix $R = M$, and then alternate between two steps: a greedy selection of one column of R to be added to W, and an update of R using M and the columns extracted so far. As SNPA was shown, both theoretically and empirically, to perform better and to be more robust than its competitors [7], it is the one we study here in detail. Moreover, SNPA is able to handle the underdetermined case when $\text{rank}(W) < m$ which will be key for our problem setting (see below for more details).

SNPA is presented in Algorithm 1. SNPA selects, at each step, the column of M maximizing a function f (which can be any strongly convex function such that $f(0) = 0$, and $f = ||.||_2^2$ is the most common choice). Then, the columns of M are projected onto the convex hull of the origin and the columns extracted so far, see step 8 where we use the notation

$$\Delta = \left\{h \mid h \geq 0, \sum_i h_i \leq 1\right\},$$

whose dimension is clear from the context. After r steps, given that the noise is sufficiently small and that the columns of W are vertices of $\text{conv}(W)$, SNPA is guaranteed to identify W. An important point is that SNPA requires the columns of H to satisfy $\|H(:, j)\|_1 \leq 1$ for all j, where $\|x\|_1 = \sum_i |x_i|$ is the ℓ_1 norm. This assumption can be made without loss of generality by properly scaling the columns of the input matrix to have unit ℓ_1 norm; see the discussion in [7].

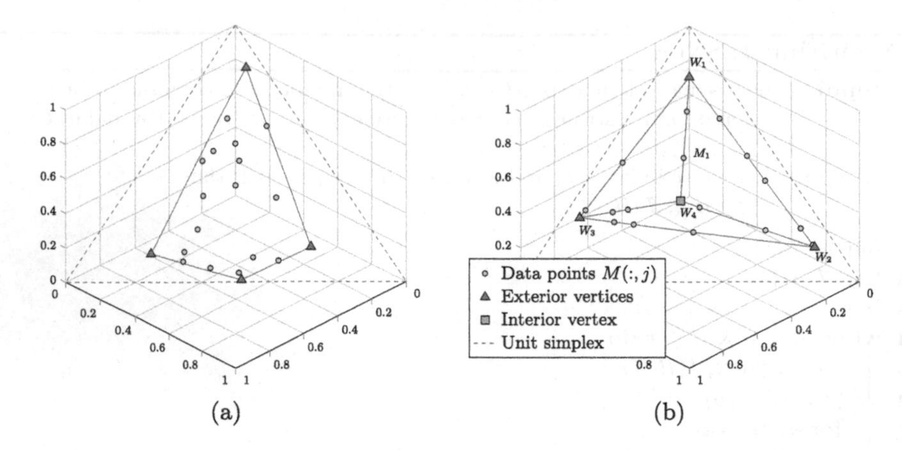

Fig. 1. On the left (a), all vertices are exterior, and SNPA is assured to identify them all. On the right (b), the data points are 2-sparse combinations of 4 points, one of which (vertex 4) is "interior" hence it cannot be identified with separable NMF.

1.3 Model Limitations

Unfortunately, some data sets cannot be handled successfully by separable NMF, even when all data points are linear combinations of a subset of the input matrix. In fact, in some applications, the columns of the basis matrix W, that is, the atoms, might not be vertices of $\text{conv}(W)$. This may happen when one seeks a matrix W which is not full column rank. For example, in multispectral unmixing, m is the number of spectral bands which can be smaller than r, which is the number of materials present in the image; see Sect. 5.2 for more details. Therefore, it is possible for some columns of W to be contained in the convex hull of the other columns, that is, to be additive linear combinations of others columns of W; see Fig. 1 for illustrations in three dimensions (that is, $m = 3$).

These difficult cases cannot be handled with separable NMF, because it assumes the data points to be linear combinations of vertices, so an "interior vertex" cannot be distinguished from another data point. However, if we assume the *sparsity* of the mixture matrix H, we may be able to identify these interior vertices. To do so, we introduce a new model, extending the approach of SNPA using additional sparsity constraints. We introduce in Sect. 2 a proper definition of this new problem, which we coin as sparse separable NMF (SSNMF). Before doing so, let use recall the literature on sparse NMF.

1.4 Sparse NMF

A vector or matrix is said to be *sparse* when it has few non-zero entries. Sparse NMF is one of the most popular variants of NMF, as it helps producing more interpretable factors. In this model, we usually consider column-wise sparsity of the factor H, meaning that a data point is expressed as the combination of only a few atoms. For example, in multispectral unmixing, the column-wise sparsity of H means that a pixel is composed of fewer materials than the total number of materials present in the image. When sparsity is an a priori knowledge on the structure of the data, encouraging sparsity while computing NMF is likely to reduce noise and produce better results.

Sparse NMF is usually solved by extending standard NMF algorithms with a regularization such as the ℓ_1 penalty [9,11], or constraints on some sparsity measure, like the one introduced in [10]. Recently, exact k-sparse methods based on the ℓ_0-"norm" have been used for NMF, using a brute-force approach [4], or a dedicated branch-and-bound algorithm [16]. They allow the explicit definition of a maximum number (usually noted k) of non-zero entries per column of H. These approaches leverage the fact that, in most NMF problems, the factorization rank r is small, hence it is reasonable to solve the k-sparse NNLS subproblems exactly.

1.5 Contributions and Outline

In this work, we study the SSNMF model from a theoretical and a practical point of view. Our contributions can be summarized as follows:

- In Sect. 2, we introduce the SSNMF model. We prove that, unlike separable NMF, SSNMF is NP-complete.
- In Sect. 3, we propose an algorithm to tackle SSNMF, based on SNPA and an exact sparse NNLS solver.
- In Sect. 4, we prove that our algorithm is correct under reasonable assumptions, in the noiseless case.
- In Sect. 5, experiments on both synthetic and real-world data sets illustrate the relevance and efficiency of our algorithm.

2 Sparse Separable NMF

We explained in the previous section why separable NMF does not allow for the identification of "interior vertices", as they are nonnegative linear combinations of other vertices. However, if we assume a certain column-wise sparsity on the coefficient matrix H, they may become identifiable. For instance, the vertex W_4 of Fig. 1b can be expressed as a combination of the three exterior vertices (W_1, W_2, and W_3), but not as a combination of any two of these vertices. Moreover, some data points cannot be explained using only pairs of exterior vertices, while they can be if we also select the interior vertex W_4.

2.1 Problem Statement and Complexity

Let us denote $\|x\|_0$ the number of non-zero entries of the vector x.

Definition 2. *A matrix M is k-sparse r-separable if there exists a subset of r columns of M, indexed by $\mathcal{J}$, and a nonnegative matrix $H \geq 0$ with $\|H(:,j)\|_0 \leq k$ for all j such that $M = M(:,\mathcal{J})H$.*

Definition 2 corresponds to Definition 1 with the additional constraint that H has k-sparse columns, that is, columns with at most k non-zero entries. A natural assumption to ensure that we can identify W (that is, find the set $\mathcal{J}$), is that the columns of W are not k-sparse combinations of any other columns of W; see Sect. 4 for the details.

Problem 2 (SSNMF). Given a k-sparse r-separable matrix M, find $W = M(:, \mathcal{J})$ with $|\mathcal{J}| = r$ and a column-wise k-sparse matrix $H \geq 0$ such that $M = WH$.

As opposed to separable NMF, given $\mathcal{J}$, computing H is not straightforward. It requires to solve the following ℓ_0-constrained optimization problem

$$H^* = \operatorname*{argmin}_{H \geq 0} f(M - M(:, \mathcal{J})H) \text{ such that } \|H(:, j)\|_0 \leq k \text{ for all } j. \quad (2)$$

Because of the combinatorial nature of the ℓ_0-"norm", this k-sparse projection is a difficult subproblem with $\binom{r}{k}$ possible solutions, which is known to be NP-hard [17]. In particular, a brute-force approach could tackle this problem by solving $\mathcal{O}(r^k)$ NNLS problems. However, this combinatorial subproblem can be solved exactly and at a reasonable cost by dedicated branch-and-bound algorithms, such as ARBORESCENT [16], given that r is sufficiently small, which is typically the case in practice. Even when k is fixed, the following result shows that no provably correct algorithm exists for solving SSNMF in polynomial time (unless P = NP):

Theorem 1. *SSNMF is NP-complete for any fixed $k \geq 2$.*

Proof. The proof is given in the supplementary material. Note that the case $k = 1$ is trivial since each data point is a multiple of a column of W.

However, in Sect. 4, we show that under a reasonable assumption, SSNMF can be solved in polynomial time when k is fixed.

2.2 Related Work

To the best of our knowledge, the only work presenting an approach to tackle SSNMF is the one by Sun and Xin (2011) [18]—and it does so only partially. It studies the blind source separation of nonnegative data in the underdetermined case. The problem tackled is equivalent to NMF in the case $m < r$. The assumptions used in this work are similar to ours, that is, separability and sparsity. However, the setup considered is less general than SSNMF because the sparsity assumption (on each column of H) is limited to $k = m - 1$, while the only case considered theoretically is the case $r = m + 1$ with only one interior vertex.

The proposed algorithm first extracts the exterior vertices using the method LP-BSS from [15], and then identifies the interior vertex using a brute-force geometric method. More precisely, they select an interior point, and check whether at least two of the $m - 1$ hyperplanes generated by this vertex with $m - 2$ of the extracted exterior vertices contain other data points. If it is the case, then

Algorithm 2: BRASSENS

Input: A k-sparse-near-separable matrix $M \in \mathbb{R}^{m \times n}$, and the desired sparsity level k.

Output: A set of r indices $\mathcal{J}$, and a matrix $H \in \mathbb{R}_+^{r \times n}$, such that $\|H(:,j)\|_0 \leq k$ for all j and $M = \tilde{M}(:,\mathcal{J})H$.

1 $\mathcal{J} = \text{SNPA}(M, \infty)$
2 $\mathcal{J}' = \text{kSSNPA}(M, \infty, \mathcal{J})$
3 **foreach** $j \in \mathcal{J}'$ **do**
4   **if** $\min_{\|h\|_0 \leq k, h \geq 0} f(M(:,j) - M(:,\mathcal{J}' \setminus j)h) > 0$ **then**
5     $\mathcal{J} = \mathcal{J} \cup \{j\}$
6 $H = \text{arborescent}(M, M(:,\mathcal{J}), k)$

they conclude that the selected point is an interior vertex, otherwise they select another interior point. For example, when $m = 3$, this method consists in constructing the segments between the selected interior point and all the exterior vertices. If two of these segments contain at least one data point, then the method stops and the selected interior point is chosen as the interior vertex. Looking at Fig. 1b, the only interior point for which two segments joining this point and an exterior vertex contain data points is W_4. Note that, to be guaranteed to work, this method requires at least two hyperplanes containing the interior vertex and $m-2$ exterior vertices to contain data points. This will not be a requirement in our method.

3 Proposed Algorithm: BRASSENS

In the following, we assume that the input matrix M is k-sparse r-separable. Our algorithm, called BRASSENS[1], is presented formally in Algorithm 2. On Line 1 we apply the original SNPA to select the exterior vertices; it is computationally cheap and ensures that these vertices are properly identified. The symbol ∞ means that SNPA stops only when the residual error is zero. For the noisy case, we replace the condition $R \neq 0$ by $R > \delta$, where δ is a user-provided noise-tolerance threshold.

Then, we adapt SNPA to impose a k-sparsity constraint on H: the projection step (Line 8 of Algorithm 1) is replaced by a k-sparse projection step that imposes the columns of H to be k-sparse by solving (2). We call *kSSNPA* this modified version of SNPA. Note that, if $k = r$, kSSNPA reduces to SNPA.

On Line 2 we apply kSSNPA to select candidate interior vertices. We provide it with the set $\mathcal{J}$ of exterior vertices so that they do not need to be identified again. kSSNPA extracts columns of M as long as the norm of the residual $\|M - M(:,\mathcal{J}')H\|_F$ is larger than zero. At this point, all vertices have been identified:

[1] It stands for BRASSENS Relies on Assumptions of Separability and Sparsity for Elegant NMF Solving.

the exterior vertices have been identified by SNPA, while the interior vertices have been identified by kSSNPA because we will assume that they are not k-sparse combinations of any other data points; see Sect. 4 for the details. Hence, the error will be equal to zero if and only if all vertices have been identified. However, some selected interior points may not be interior vertices, because the selection step of kSSNPA chooses the point that is furthest away from the k-*sparse hull* of the selected points, that is, the union of the convex hulls of the subsets of k already selected points. For example, in Fig. 1b, if W_1, W_2, and W_3 are selected, the k-sparse hull is composed of the 3 segments $[W_1, W_2]$, $[W_2, W_3]$, and $[W_3, W_1]$. In this case, although only point W_4 is a interior vertex, point M_1 is selected before point W_4, because it is located further away from the k-sparse hull.

On Lines 3 to 5, we apply a postprocessing to the selected points by checking whether they are k-sparse combinations of other selected points; this is a k-sparse NNLS problem solved with ARBORESCENT [16]. If they are, then they cannot be vertices and they are discarded, such as point M_1 in Fig. 1b which belongs to the segment $[W_1, W_4]$.

Note that this "postprocessing" could be applied directly to the whole data set by selecting data points as the columns of W if they are not k-sparse combinations of other data points. However, this is not reasonable in practice, as it is equivalent to solving n times a k-sparse NNLS subproblem in $n-1$ variables. The kSSNPA step can thus be interpreted as a safe screening technique, similarly as done in [5] for example, in order to reduce the number of candidate atoms from all the columns of M to a subset $\mathcal{J}'$ of columns. In practice, we have observed that kSSNPA is very effective at identifying good candidates points; see Sect. 5.1.

4 Analysis of BRASSENS

In this section, we first discuss the assumptions that guarantee BRASSENS to recover W given the k-sparse r-separable matrix M, and then discuss the computational complexity of BRASSENS.

4.1 Correctness

In this section, we show that, given a k-sparse r-separable matrix M, the BRASSENS algorithm provably solves SSNMF, that is, it is able to recover the correct set of indices $\mathcal{J}$ such that $W = M(:, \mathcal{J})$, under a reasonable assumption.

Clearly, a necessary assumption for BRASSENS to be able to solve SSNMF is that no column of W is a k-sparse nonnegative linear combinations of other columns of W, otherwise kSSNPA might set that column of W to zero, hence might not be able to extract it.

Assumption 1. *No column of W is a nonnegative linear combination of k other columns of W.*

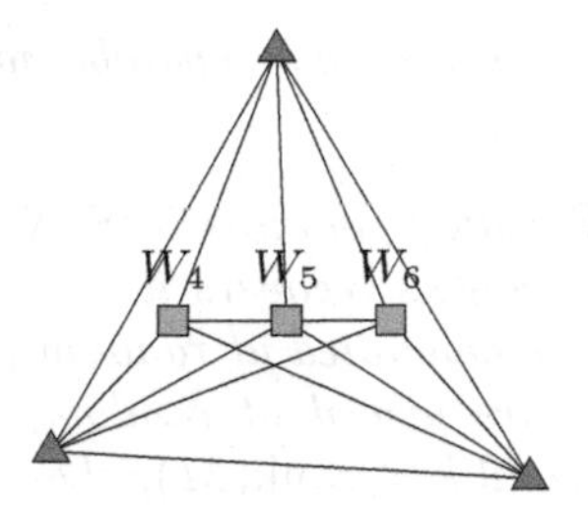

Fig. 2. There are three interior vertices. One of them (W_5) is a combination of the others two.

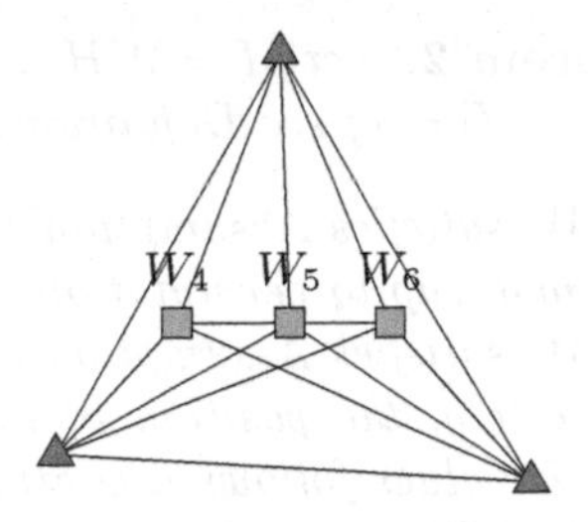

Fig. 3. There are two interior vertices. One of them (W_4) is a 2-sparse combination of data points (M_1 and M_2).

Interestingly, unlike the standard separable case (that is, $k = r$), and although it is necessary in our approach with BRASSENS, Assumption 1 is not necessary in general to be able to uniquely recover W. Take for example the situation of Fig. 2, with three aligned points in the interior of a triangle, so that $r = 6$, $m = 3$ and $k = 2$. The middle point of these three aligned points is a 2-sparse combination of the other two, by construction. If there are data points on each segment joining these three interior points and the exterior vertices, the only solution to SSNMF with $r = 6$ is the one selecting these three aligned points. However, Assumption 1 is a reasonable assumption for SSNMF.

Unfortunately, Assumption 1 is not sufficient for BRASSENS to provably recover W. In fact, we need the following stronger assumption.

Assumption 2. *No column of W is a nonnegative linear combination of k other columns of M.*

This assumption guarantees that a situation such as the one shown on Fig. 3 where one of the columns of W is a 2-sparse combination of two data points is not possible. In fact, in that case, if BRASSENS picks these two data points before the interior vertex W_4 in between them, it will not be able to identify W_4 as it is set to zero within the projection step of kSSNPA.

Interestingly, in the standard separable case, that is, $k = r$, the two assumptions above coincide; this is the condition under which SNPA is guaranteed to work. Although Assumption 2 may appear much stronger than Assumption 1, they are actually generically equivalent given that the entries of the columns of H are generated randomly (that is, non-zero entries are picked at random and follow some continuous distribution). For instance, for $m = 3$ and $k = 2$, it means that no vertex is on a segment joining two data points. If the data points are generated randomly on the segments generated by any two columns of W, the probability for the segment defined by two such data points to contain a column of W is zero. In fact, segments define a set of measure zero in the unit simplex.

We can now provide a recovery result for BRASSENS.

Theorem 2. *Let $M = WH$ with $W = M(:, \mathcal{J})$ be a k-sparse r-separable matrix so that $|\mathcal{J}| = r$; see Definition 2. We have that*

- *If W satisfies Assumption 2, then the factor W with r columns in SSNMF is unique (up to permutation and scaling) and* BRASSENS *recovers it.*
- *If W satisfies Assumption 1, the entries of H are generated at random (more precisely, the position of the non-zero entries are picked at random, while their values follows a continuous distribution) and $k < \text{rank}(M)$, then, with probability one, the factor W with r columns in SSNMF is unique (up to permutation and scaling) and* BRASSENS *recovers it.*

Proof. Uniqueness of W in SSNMF under Assumption 2 is straightforward: since the columns of W are not k-sparse combinations of other columns of M, they have to be selected in the index set $\mathcal{J}$. Otherwise, since columns of W are among the columns of M, it would not possible to reconstruct M exactly using k-sparse combinations of $M(:, \mathcal{J})$. Then, since all other columns are k-sparse combinations of the r columns of W (by assumption), no other columns needs to be added to $\mathcal{J}$ which satisfies $|\mathcal{J}| = r$.

Let us show that, under Assumption 2, BRASSENS recovers the correct set of indices $\mathcal{J}$. kSSNPA can only stop when all columns of W have been identified. In fact, kSSNPA stops when the reconstruction error is zero, while, under Assumption 2, this is possible only when all columns of W are selected (for the same reason as above). Then, the postprocessing will be able to identify, among all selected columns, the columns of W, because they will be the only ones that are not k-sparse combinations of other selected columns.

The second part of the proof follows from standard probabilistic results: since $k < \text{rank}(M)$, the combination of k data points generates a subspace of dimension smaller than that of $\text{col}(M)$. Hence, generating data points at random is equivalent to generating such subspaces at random. Since these subspaces form a space of measure zero in $\text{col}(M)$, the probability for these subspaces to contain a column of W is zero, which implies that Assumption 2 is satisfied with probability one.

4.2 Computational Cost

Let us derive an upper bound on the computational cost of BRASSENS. First, recall that solving an NNLS problem up to any precision can be done in polynomial time. For simplicity and because we focus on the non-polynomial part of BRASSENS, we denote $\bar{\mathcal{O}}(1)$ the complexity of solving an NNLS problem. In the worst case, kSSNPA will extract all columns of M. In each of the n iterations of kSSNPA, the problem (2) needs to be solved. When $|\mathcal{J}| = \mathcal{O}(n)$, this requires to solve n times (one for each column of M) a k-sparse least squares problem in $|\mathcal{J}| = \mathcal{O}(n)$ variables. The latter requires in the worst case $\mathcal{O}(n^k)$ operations by trying all possible index sets; see the discussion after (2). In total, kSSNPA will therefore run in the worst case in time $\bar{\mathcal{O}}(n^{k+2})$.

Therefore, when k is fixed (meaning that k is considered as a fixed constant) and under Assumption 2, BRASSENS can solve SSNMF in polynomial time. Note that this is not in contradiction with our NP-completeness results when k is fixed (Theorem 1) because our NP-completeness proof does not rely on Assumption 2.

In summary, to make SSNMF hard, we need either k to be part of the input, or the columns of W to be themselves k-sparse combinations of other columns of W.

5 Experiments

The code and data are available online[2]. All experiments have been performed on a personal computer with an i5 processor, with a clock frequency of 2.30GHz. All algorithms are single-threaded. All are implemented in Matlab, except the sparse NNLS solver ARBORESCENT, which is implemented in C++ with a Matlab MEX interface.

As far as we know, no algorithm other than BRASSENS can tackle SSNMF with more than one interior point (see Sect. 2.2) hence comparisons with existing works are unfortunately limited. For example, separable NMF algorithms can only identify the exterior vertices; see Sect. 1. However, we will compare BRASSENS to SNPA on a real multispectral image in Sect. 5.2, to show the advantages of the SSNMF model over separable NMF. In Sect. 5.1, we illustrate the correctness and efficiency of BRASSENS on synthetic data sets.

5.1 Synthetic Data Sets

In this section, we illustrate the behaviour of BRASSENS in different experimental setups. The generation of a synthetic data set is done as follows: for a given number of dimensions m, number of vertices r, number of data points n, and data sparsity k, we generate matrices $W \in \mathbb{R}_+^{m \times r}$ such that the last $r - m$ columns of W are linear combinations of the first m columns, and $H \in \mathbb{R}_+^{r \times n}$ such that $H = [I_r, H']$ and $\|H(:,j)\|_0 \leq k$ for all j. We use the uniform distribution in the interval [0,1] to generate random numbers (columns of W and columns of H), and then normalize the columns of W and H to have unit ℓ_1 norm. We then compute $M = WH$. This way, the matrix M is k-sparse r-separable, with r vertices, of which $r - m$ are interior vertices (in fact, the first m columns of W are linearly independent with probability one as they are generated randomly). We then run BRASSENS on M, with the parameter k, and no noise-tolerance. For a given setup, we perform 30 rounds of generation and solving, and we measure the median of the running time and the median of the number of candidates extracted by kSSNPA. This number of candidates corresponds to $|\mathcal{J}'|$ in Algorithm 2, that is, the number of interior points selected by kSSNPA as potential interior vertices. Note that a larger number of candidates only results in an increased computation time, and does not change the output of the algorithm which is guaranteed to extract all vertices (Theorem 2).

[2] https://gitlab.com/nnadisic/ssnmf.

Figure 4 shows the behaviour of BRASSENS when n varies, with fixed $m = 3$, $k = 2$, and $r = 5$. To the best of our knowledge, this case is not handled by any other algorithm in the literature. Both the number of candidates and the run time grow slower than linear. The irregularities in the plot are due to the high variance between runs. Indeed, if vertices are generated in a way that some segments between vertices are very close to each other, BRASSENS typically selects more candidates before identifying all columns of W.

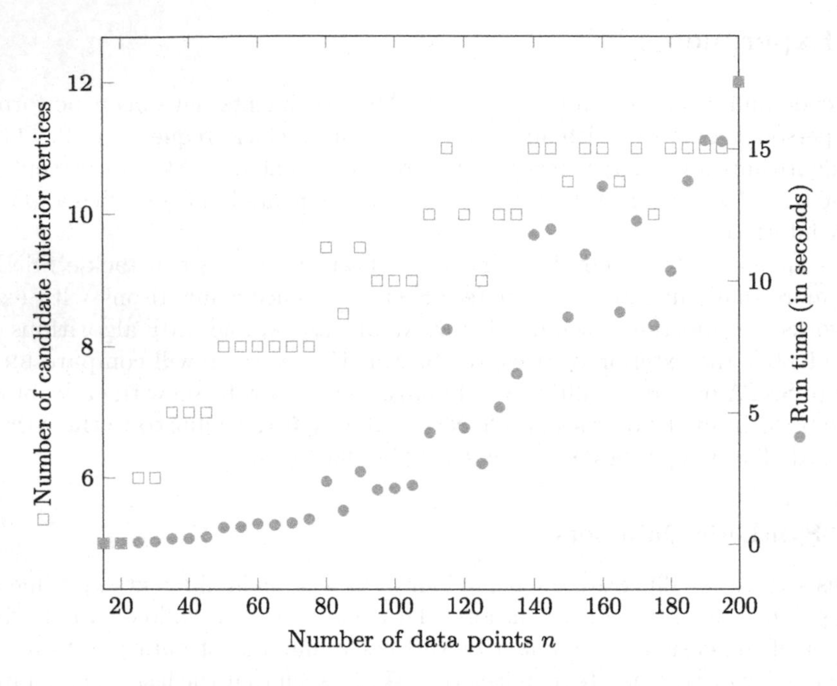

Fig. 4. Results for BRASSENS on synthetic data sets for different values of n, with fixed $m = 3$, $k = 2$, and $r = 5$ with 3 exterior and 2 interior vertices. The values showed are the medians over 30 experiments.

In Table 1 we compare the performance of BRASSENS for several sets of parameters. The number of candidates grows relatively slowly as the dimensions (m, n) of the problem increase, showing the efficiency of the screening performed by kSSNPA. However, the run time grows rather fast when the dimensions (m, n) grow. This is because, not only the number of NNLS subproblems to solve increase, but also their size.

In all cases, as guaranteed by Theorem 2, BRASSENS was able to correctly identify the columns of W. Again, as far as we know, no existing algorithms in the literature can perform this task.

Table 1. Results for BRASSENS on synthetic data sets (median over 30 experiments).

m	n	r	k	Number of candidates	Run time in seconds
3	25	5	2	5.5	0.26
4	30	6	3	8.5	3.30
5	35	7	4	9.5	38.71
6	40	8	5	13	395.88

To summarize, our synthetic experiments show the efficiency of the screening done by kSSNPA, and the capacity of BRASSENS to handle medium-scale data sets.

5.2 Blind Multispectral Unmixing

A multispectral image is an image composed of various wavelength ranges, called *spectral bands*, where every pixel is described by its *spectral signature*. This signature is a vector representing the amount of energy measured for this pixel in every considered spectral band. Multispectral images usually have a small number of bands (between 3 and 15). These bands can be included or not in the spectrum of visible light. In the NMF model, if the columns of M are the n pixels of the image, then its rows represent the m spectral bands.

The unmixing of a multispectral image consists in identifying the different materials present in that image. When the spectral signatures of the materials present in the image are unknown, it is referred to as *blind unmixing*. The use of NMF for blind unmixing of multispectral images relies on the linear mixing model, that is, the assumption that the spectral signature of a pixel is the linear combination of the spectral signatures of the material present in this pixel. This corresponds exactly to the NMF model, which is therefore able to identify both the materials present in the image (W) and the proportions/abundances of materials present in every pixel (H); see [3,14] for more details.

Let us apply BRASSENS to the unmixing of the well-known Urban satellite image [21], composed of 309×309 pixels. The original cleaned image has 162 bands, but we only keep 3 bands, namely the bands 2, 80, and 133 – these were obtained by selecting different bands with SPA applied on M^T – to obtain a data set of size $3 \times 94\,249$. The question is: can we still recover materials by using only 3 bands? (The reason for this choice is that this data set is well known and the ground truth is available, which is not the case of most multispectral images with only 3 bands.) We first normalize all columns of M so that they sum to one. Then, we run BRASSENS with a sparsity constraint $k = 2$ (this means that we assume that a pixel can be composed of at most 2 materials, which is reasonable for this relatively high resolution image) and a noise-tolerance threshold of 4%; this means that we stop SNPA and kSSNPA when $\|M - M(:, \mathcal{J})H\|_F \leq 0.04\|M\|_F$. BRASSENS extracts 5 columns of the input matrix. For comparison, we run SNPA with $r = 5$. Note that this setup corresponds to underdetermined blind unmixing, because $m = 3 < r = 5$. It would not be possible to tackle this problem using standard NMF algorithms (that would return a trivial solution such as $M = I_3 M$). It can be solved with SNPA, but SNPA cannot identify interior vertices.

SNPA extracts the 5 vertices in 3.8 s. BRASSENS extracts 5 vertices, including one interior vertex, in 33 s. The resulting abundance maps are showed in Fig. 5. They correspond to the reshaped rows of H, hence they show which pixel contains which extracted material (they are more easily interpretable than the spectral signatures contained in the columns of W). The materials they contain are given in Table 2, using the ground truth from [20]. We see that BRASSENS produces a better solution, as the materials present in the image are better separated: the first three abundance maps of BRASSENS are sparser and correspond to well-defined materials. The last two abundances maps of SNPA and of BRASSENS are similar but extracted in a different order. The running time of BRASSENS is reasonable, although ten times higher than SNPA.

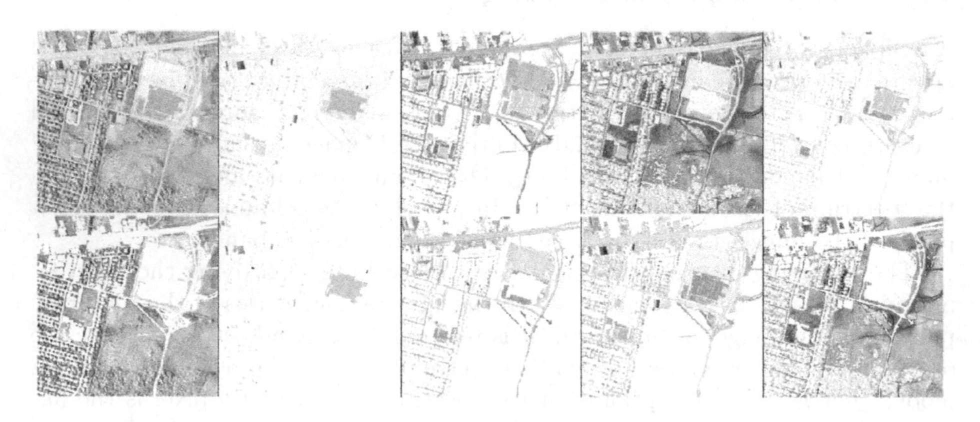

Fig. 5. Abundances maps of materials (that is, reshaped rows of H) extracted by SNPA (top) and by BRASSENS (bottom) in the Urban image with only 3 spectral bands.

Table 2. Interpretation of the unimixing results from Fig. 5.

Image	Materials extracted by SNPA	Materials extracted by BRASSENS
1	Grass + trees + roof tops	Grass + trees
2	Roof tops 1	Roof tops 1
3	Dirt + road + roof tops	Road
4	Dirt + grass	Roof tops 1 and 2 + road
5	Roof tops 1 + dirt + road	Dirt + grass

6 Conclusion

In this paper, we introduced SSNMF, a new variant of the NMF model combining the assumptions of separability and sparsity. We presented BRASSENS, an algorithm able to solve exactly SSNMF, based on SNPA and an exact sparse NNLS solver. We showed its efficiency for various setups and in the successful unmixing of a multispectral image. The present work provides a new way to perform underdetermined blind source separation, under mild hypothesis, and a new way to regularize NMF. It makes NMF identifiable even when atoms of W are nonnegative linear combinations of other atoms (as long as these combinations have sufficiently many non-zero coefficients). Further work includes the theoretical analysis of the proposed model and algorithm in the presence of noise.

Acknowledgments. The authors are grateful to the reviewers, whose insightful comments helped improve the paper. NN and NG acknowledge the support by the European Research Council (ERC starting grant No 679515), and by the Fonds de la Recherche Scientifique - FNRS and the Fonds Wetenschappelijk Onderzoek - Vlanderen (FWO) under EOS project O005318F-RG47.

References

1. Araújo, M.C.U., Saldanha, T.C.B., Galvão, R.K.H., Yoneyama, T., Chame, H.C., Visani, V.: The successive projections algorithm for variable selection in spectroscopic multicomponent analysis. Chemometr. Intell. Lab. Syst. **57**(2), 65–73 (2001)
2. Arora, S., Ge, R., Kannan, R., Moitra, A.: Computing a nonnegative matrix factorization – provably. In: Proceedings of the Forty-Fourth Annual ACM Symposium on Theory of Computing, pp. 145–162 (2012)
3. Bioucas-Dias, J.M., Plaza, A., Dobigeon, N., Parente, M., Du, Q., Gader, P., Chanussot, J.: Hyperspectral unmixing overview: geometrical, statistical, and sparse regression-based approaches. IEEE J. Sel. Topics Appl. Earth Observ. Remote Sens. **5**(2), 354–379 (2012)
4. Cohen, J.E., Gillis, N.: Nonnegative low-rank sparse component analysis. In: IEEE International Conference on Acoustics, Speech and Signal Processing (ICASSP), pp. 8226–8230 (2019)
5. El Ghaoui, L., Viallon, V., Rabbani, T.: Safe feature elimination in sparse supervised learning technical report no. Technical report, UC/EECS-2010-126, EECS Department, University of California at Berkeley (2010)
6. Fu, X., Huang, K., Sidiropoulos, N.D., Ma, W.K.: Nonnegative matrix factorization for signal and data analytics: identifiability, algorithms, and applications. IEEE Signal Process. Mag. **36**(2), 59–80 (2019)
7. Gillis, N.: Successive nonnegative projection algorithm for robust nonnegative blind source separation. SIAM J. Imag. Sci. **7**, 1420–1450 (2014)
8. Gillis, N.: The why and how of nonnegative matrix factorization. Regularization, optimization, kernels, and support vector machines **12**(257), 257–291 (2014)
9. Hoyer, P.O.: Non-negative sparse coding. In: Proceedings of the 12th IEEE Workshop on Neural Networks for Signal Processing, pp. 557–565 (2002)

10. Hoyer, P.O.: Non-negative matrix factorization with sparseness constraints. J. Mach. Learn. Res. **5**, 1457–1469 (2004)
11. Kim, H., Park, H.: Sparse non-negative matrix factorizations via alternating non-negativity-constrained least squares for microarray data analysis. Bioinformatics **23**(12), 1495–1502 (2007)
12. Kumar, A., Sindhwani, V., Kambadur, P.: Fast conical hull algorithms for near-separable non-negative matrix factorization. In: Proceedings of the 30th International Conference on Machine Learning (2013)
13. Lee, D.D., Seung, H.S.: Learning the parts of objects by non-negative matrix factorization. Nature **401**(6755), 788–791 (1999)
14. Ma, W.K., et al.: A signal processing perspective on hyperspectral unmixing: insights from remote sensing. IEEE Signal Process. Mag. **31**(1), 67–81 (2014)
15. Naanaa, W., Nuzillard, J.M.: Blind source separation of positive and partially correlated data. Sig. Process. **85**(9), 1711–1722 (2005)
16. Nadisic, N., Vandaele, A., Gillis, N., Cohen, J.E.: Exact sparse nonnegative least squares. In: IEEE International Conference on Acoustics, Speech and Signal Processing (ICASSP), pp. 5395–5399 (2020)
17. Natarajan, B.K.: Sparse approximate solutions to linear systems. SIAM J. Comput. **24**(2), 227–234 (1995)
18. Sun, Y., Xin, J.: Underdetermined sparse blind source separation of nonnegative and partially overlapped data. SIAM J. Sci. Comput. **33**(4), 2063–2094 (2011)
19. Vavasis, S.A.: On the complexity of nonnegative matrix factorization. SIAM J. Optim. **20**(3), 1364–1377 (2010)
20. Zhu, F.: Hyperspectral unmixing: ground truth labeling, datasets, benchmark performances and survey. arXiv preprint arXiv:1708.05125 (2017)
21. Zhu, F., Wang, Y., Xiang, S., Fan, B., Pan, C.: Structured sparse method for hyperspectral unmixing. ISPRS J. Photogram. Remote Sens. **88**, 101–118 (2014)

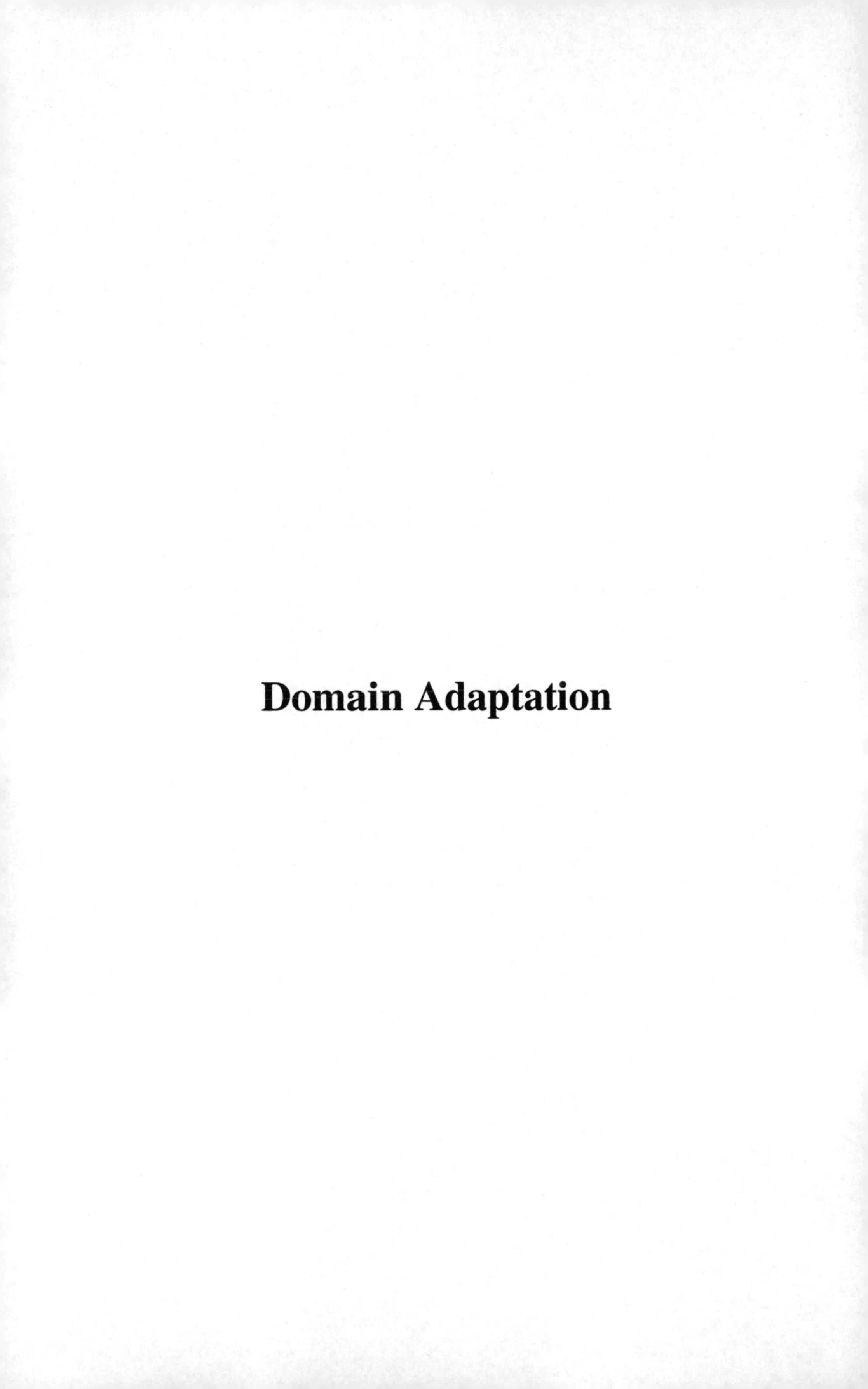

Domain Adaptation

Robust Domain Adaptation: Representations, Weights and Inductive Bias

Victor Bouvier[1,2(✉)], Philippe Very[3], Clément Chastagnol[4], Myriam Tami[1], and Céline Hudelot[1]

[1] CentraleSupélec, Mathématiques et Informatique pour la Complexité et les Systèmes, Université Paris-Saclay, 91190 Gif-sur-Yvette, France
{victor.bouvier,myriam.tami,celine.hudelot}@centralesupelec.fr

[2] Sidetrade, 114 Rue Gallieni, 92100 Boulogne-Billancourt, France
vbouvier@sidetrade.com

[3] Lend-Rx, 24 Rue Saint Dominique, 75007 Paris, France
philippe.very@lend-rxtech.com

[4] Alan, 117 Quai de Valmy, 75010 Paris, France
clement.chastagnol@alan.eu

Abstract. Unsupervised Domain Adaptation (UDA) has attracted a lot of attention in the last ten years. The emergence of Domain Invariant Representations (IR) has improved drastically the transferability of representations from a labelled source domain to a new and unlabelled target domain. However, a potential pitfall of this approach, namely the presence of *label shift*, has been brought to light. Some works address this issue with a relaxed version of domain invariance obtained by weighting samples, a strategy often referred to as Importance Sampling. From our point of view, the theoretical aspects of how Importance Sampling and Invariant Representations interact in UDA have not been studied in depth. In the present work, we present a bound of the target risk which incorporates both weights and invariant representations. Our theoretical analysis highlights the role of inductive bias in aligning distributions across domains. We illustrate it on standard benchmarks by proposing a new learning procedure for UDA. We observed empirically that weak inductive bias makes adaptation more robust. The elaboration of stronger inductive bias is a promising direction for new UDA algorithms.

Keywords: Unsupervised domain adaptation · Importance sampling · Invariant representations · Inductive bias

1 Introduction

Deploying machine learning models in the real world often requires the ability to generalize to *unseen samples i.e.* samples significantly different from those seen

P. Very—Work done when author was at Sidetrade.

F. Hutter et al. (Eds.): ECML PKDD 2020, LNAI 12457, pp. 353–377, 2021.
https://doi.org/10.1007/978-3-030-67658-2_21

during learning. Despite impressive performances on a variety of tasks, deep learning models do not always meet these requirements [3,14]. For this reason, *out-of-distribution generalization* is recognized as a major challenge for the reliability of machine learning systems [1,2]. Domain Adaptation (DA) [28,30] is a well-studied approach to bridge the gap between train and test distributions. In DA, we refer to train and test distributions as *source* and *target* respectively noted $p_S(x, y)$ and $p_T(x, y)$ where x are inputs and y are labels. The objective of DA can be defined as learning a good classifier on a poorly sampled target domain by leveraging samples from a source domain. Unsupervised Domain Adaptation (UDA) assumes that only unlabelled data from the target domain is available during training. In this context, a natural assumption, named *Covariate shift* [19,33], consists in assuming that the mapping from the inputs to the labels is conserved across domains, *i.e.* $p_T(y|x) = p_S(y|x)$. In this context, *Importance Sampling* (IS) performs adaptation by weighting the contribution of sample x in the loss by $w(x) = p_T(x)/p_S(x)$ [30]. Although IS seems natural when unlabelled data from the target domain is available, the covariate shift assumption is not sufficient to guarantee successful adaptation [5]. Moreover, for high dimensional data [12] such as texts or images, the shift between $p_S(x)$ and $p_T(x)$ results from non-overlapping supports leading to unbounded weights [20].

In this particular context, representations can help to reconcile non-overlapping supports [5]. This seminal idea, and the corresponding theoretical bound of the target risk from [5], has led to a wide variety of deep learning approaches [13,23,24] which aim to learn a so-called *domain invariant representation*:

$$p_S(z) \approx p_T(z) \tag{1}$$

where $z := \varphi(x)$ for a given non-linear representation φ. These assume that the *transferability* of representations, defined as the combined error of an ideal classifier, remains low during learning. Unfortunately, this quantity involves target labels and is thus intractable. More importantly, looking for strict invariant representations, $p_S(z) = p_T(z)$, hurts the transferability of representations [20,22,36,40]. In particular, there is a fundamental trade-off between learning invariant representations and preserving transferability in presence of label shift ($p_T(y) \neq p_S(y)$) [40]. To mitigate this trade-off, some recent works suggest to relax domain invariance by weighting samples [8,9,36,37]. This strategy differs with (1) by aligning a *weighted source* distribution with the target distribution:

$$w(z)p_S(z) \approx p_T(z) \tag{2}$$

for some weights $w(z)$. We now have two tools, w and φ, which need to be calibrated to obtain distribution alignment. Which one should be promoted? How weights preserve good transferability of representations?

While most prior works focus on the invariance error for achieving adaptation [13,23,24], this paper focuses on the transferability of representations. We show that weights allow to design an interpretable generalization bound where transferability and invariance errors are uncoupled. In addition, we discuss the role of inductive design for both the classifier and the weights in addressing the lack of labelled data in the target domain. Our contributions are the following:

1. We introduce a new bound of the target risk which incorporates both weights and domain invariant representations. Two new terms are introduced. The first is an *invariance error* which promotes alignment between a weighted source distribution of representations and the target distribution of representations. The second, named *transferability error*, involves labelling functions from both source and target domains.
2. We highlight the role of **inductive bias** for approximating the transferability error. First, we establish connections between our bound and popular approaches for UDA which use target predicted labels during adaptation, in particular Conditional Domain Adaptation [24] and Minimal Entropy [15]. Second, we show that the inductive design of weights has an impact on representation invariance.
3. We derive a new learning procedure for UDA. The particularity of this procedure is to only minimize the transferability error while controlling representation invariance with weights. Since the transferability error involves target labels, we use the predicted labels during learning.
4. We provide an empirical illustration of our framework on two DA benchmarks (**Digits** and **Office31** datasets). We stress-test our learning scheme by modifying strongly the label distribution in the source domain. While methods based on invariant representations deteriorate considerably in this context, our procedure remains robust.

2 Preliminaries

We introduce the *source* distribution *i.e.* data where the model is trained with supervision and the *target* distribution *i.e.* data where the model is tested or applied. Formally, for two random variables (X, Y) on a given space $\mathcal{X} \times \mathcal{Y}$, we introduce two distributions: the source distribution $p_S(x, y)$ and the target distribution $p_T(x, y)$. Here, labels are one-hot encoded *i.e.* $y \in [0,1]^C$ such that $\sum_c y_c = 1$ where C is the number of classes. The distributional shift situation is then characterized by $p_S(x, y) \neq p_T(x, y)$ [30]. In the rest of the paper, we use the index notation S and T to differentiate source and target terms. We define the hypothesis class $\mathcal{H}$ as a subset of functions from $\mathcal{X}$ to $\mathcal{Y}$ which is the composition of a representation class Φ and a classifier class $\mathcal{G}$, *i.e.* $\mathcal{H} = \mathcal{G} \circ \Phi$. For the ease of reading, given a classifier $g \in \mathcal{G}$ and a representation $\varphi \in \Phi$, we note $g\varphi := g \circ \varphi$. Furthermore, in the definition $z := \varphi(x)$, we refer indifferently to z, φ, $Z := \varphi(X)$ as the *representation*. For two given h and $h' \in \mathcal{H}$ and ℓ the L^2 loss $\ell(y, y') = ||y - y'||^2$, the risk in domain $D \in \{S, T\}$ is noted:

$$\varepsilon_D(h) := \mathbb{E}_D[\ell(h(X), Y)] \tag{3}$$

and $\varepsilon_D(h, h') := \mathbb{E}_D[\ell(h(X), h'(X))]$. In the seminal works [5,27], a theoretical limit of the target risk when using a representation φ has been derived:

Bound 1 (Ben David et al.). *Let* $d_{\mathcal{G}}(\varphi) = \sup_{g,g' \in \mathcal{G}} |\varepsilon_S(g\varphi, g'\varphi) - \varepsilon_T(g\varphi, g'\varphi)|$ *and* $\lambda_{\mathcal{G}}(\varphi) = \inf_{g\in\mathcal{G}}\{\varepsilon_S(g\varphi) + \varepsilon_T(g\varphi)\}$, $\forall g \in \mathcal{G}, \forall \varphi \in \Phi$:

$$\varepsilon_T(g\varphi) \leq \varepsilon_S(g\varphi) + d_{\mathcal{G}}(\varphi) + \lambda_{\mathcal{G}}(\varphi) \tag{4}$$

This generalization bound ensures that the target risk $\varepsilon_T(g\varphi)$ is bounded by the sum of the source risk $\varepsilon_S(g\varphi)$, the disagreement risk between two classifiers from representations $d_{\mathcal{G}}(\varphi)$, and a third term, $\lambda_{\mathcal{G}}(\varphi)$, which quantifies the ability to perform well in both domains from representations. The latter is referred to as the *adaptability* error of representations. It is intractable in practice since it involves labels from the target distribution. Promoting distribution invariance of representations, *i.e.* $p_S(z)$ close to $p_T(z)$, results on a low $d_{\mathcal{G}}(\varphi)$. More precisely:

$$d_{\mathcal{G}}(\varphi) \leq 2 \sup_{d\in\mathcal{D}} |p_S(d(z) = 1) - p_T(d(z) = 0)| \tag{5}$$

where $\mathcal{D}$ is the so-called set of *discriminators* or *critics* which verifies $\mathcal{D} \supset \{g \oplus g' : (g, g') \in \mathcal{G}^2\}$ where $\oplus$ is the XOR function [13]. Since the domain invariance term $d_{\mathcal{G}}(\varphi)$ is expressed as a supremal value on classifiers, it is suitable for domain adversarial learning with critic functions. Conversely, the adaptability error $\lambda_{\mathcal{G}}(\varphi)$ is expressed as an infremal value. This 'sup / inf' duality induces an unexpected trade-off when learning domain invariant representations:

Proposition 1 (Invariance hurts adaptability [20,40]**).** *Let* ψ *be a representation which is a richer feature extractor than* φ: $\mathcal{G} \circ \varphi \subset \mathcal{G} \circ \psi$. *Then,*

$$d_{\mathcal{G}}(\varphi) \leq d_{\mathcal{G}}(\psi) \text{ while } \lambda_{\mathcal{G}}(\psi) \leq \lambda_{\mathcal{G}}(\varphi) \tag{6}$$

As a result of Proposition 1, the benefit of representation invariance must be higher than the loss of adaptability, which is impossible to guarantee in practice.

3 Theory

To overcome the limitation raised in Proposition 1, we expose a new bound of the target risk which embeds a new trade-off between invariance and transferability (3.1). We show this new bound remains inconsistent with the presence of label shift (3.2) and we expose the role of weights to address this problem (3.3).

3.1 A New Trade-Off Between Invariance and Transferability

Core Assumptions. Our strategy is to express both the transferability and invariance as a supremum using Integral Probability Measure (IPM) computed on a critic class. We thus introduce a class of critics suitable for our analysis. Let $\mathcal{F}$ from $\mathcal{Z} \to [-1, 1]$ and $\mathcal{F}_C$ from $\mathcal{Z} \to [-1, 1]^C$ with the following properties:

- (A1) $\mathcal{F}$ and $\mathcal{F}_C$ are symmetric (*i.e.* $\forall f \in \mathcal{F}, -f \in \mathcal{F}$) and convex.
- (A2) $\mathcal{G} \subset \mathcal{F}_C$ and $\{\mathbf{f} \cdot \mathbf{f}' \ ; \ \mathbf{f}, \mathbf{f}' \in \mathcal{F}_C\} \subset \mathcal{F}$.

- (A3) $\forall \varphi \in \Phi$, $\mathbf{f}_D(z) \mapsto \mathbb{E}_D[Y|\varphi(X) = z] \in \mathcal{F}_C$.[1]
- (A4) For two distributions p and q on $\mathcal{Z}$, $p = q$ if and only if:

$$\mathrm{IPM}(p, q; \mathcal{F}) := \sup_{f \in \mathcal{F}} \{\mathbb{E}_p[f(Z)] - \mathbb{E}_q[f(Z)]\} = 0 \tag{7}$$

The assumption (A1) ensures that rather comparing two given $\mathbf{f}$ and $\mathbf{f}'$, it is enough to study the error of some $\mathbf{f}'' = \frac{1}{2}(\mathbf{f} - \mathbf{f}')$ from $\mathcal{F}_C$. This brings back a supremum on $\mathcal{F}_C^2$ to a supremum on $\mathcal{F}_C$. The assumption (A2), combined with (A1), ensures that an error $\ell(\mathbf{f}, \mathbf{f}')$ can be expressed as a critic function $f \in \mathcal{F}$ such that $f = \ell(\mathbf{f}, \mathbf{f}')$. The assumption (A3) ensures that $\mathcal{F}_C$ is rich enough to contain label function from representations. Here, $\mathbf{f}_D(z) = \mathbb{E}_D[Y|Z = z]$ is a vector of probabilities on classes: $f_D(z)_c = p_D(Y = c|Z = z)$. The last assumption (A4) ensures that the introduced IPM is a distance. Classical tools verify these assumptions *e.g.* continuous functions; here $\mathrm{IPM}(p, q; \mathcal{F})$ is the *Maximum Mean Discrepancy* [16] and one can reasonably believe that $\mathbf{f}_S$ and $\mathbf{f}_T$ are continuous.

Invariance and Transferability as IPMs. We introduce here two important tools that will guide our analysis:

- $\mathrm{INV}(\varphi)$, named *invariance error*, that aims at capturing the difference between source and target distribution of representations, corresponding to:

$$\mathrm{INV}(\varphi) := \sup_{f \in \mathcal{F}} \{\mathbb{E}_T[f(Z)] - \mathbb{E}_S[f(Z)]\} \tag{8}$$

- $\mathrm{TSF}(\varphi)$, named *transferability error*, that catches if the coupling between Z and Y shifts across domains. For that, we use our class of functions $\mathcal{F}_C$ and we compute the IPM of $Y \cdot \mathbf{f}(Z)$, where $\mathbf{f} \in \mathcal{F}_C$ and $Y \cdot \mathbf{f}(Z)$ is the scalar product[2], between the source and the target domains:

$$\mathrm{TSF}(\varphi) := \sup_{\mathbf{f} \in \mathcal{F}_C} \{\mathbb{E}_T[Y \cdot \mathbf{f}(Z)] - \mathbb{E}_S[Y \cdot \mathbf{f}(Z)]\} \tag{9}$$

A New Bound of the Target Risk. Using $\mathrm{INV}(\varphi)$ and $\mathrm{TSF}(\varphi)$, we can provide a new bound of the target risk:

Bound 2. *$\forall g \in \mathcal{G}$ and $\forall \varphi \in \Phi$:*

$$\varepsilon_T(g\varphi) \leq \varepsilon_S(g\varphi) + 6 \cdot \mathrm{INV}(\varphi) + 2 \cdot \mathrm{TSF}(\varphi) + \varepsilon_T(\mathbf{f}_T\varphi) \tag{10}$$

The proof is in Appendix A.1. In contrast with Bound 1 (Eq. 6), here two IPMs are involved to compare representations ($\mathrm{INV}(\varphi)$ and $\mathrm{TSF}(\varphi)$). A new term, $\varepsilon_T(\mathbf{f}_T\varphi)$, reflects the level of noise when fitting labels from representations. All the trade-off between invariance and transferability is embodied in this term:

[1] See Appendix A.1 for more details on this assumption.

[2] The scalar product between Y and $\mathbf{f}(Z)$ emerges from the choice of the L^2 loss.

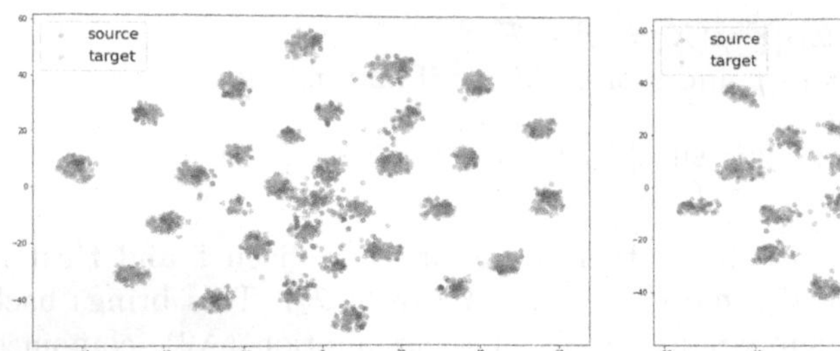

(a) $\lambda_{\mathcal{G}}(\varphi)$ adaptability in bound 1 from [5]. Inside class clusters, source and target representations are separated.

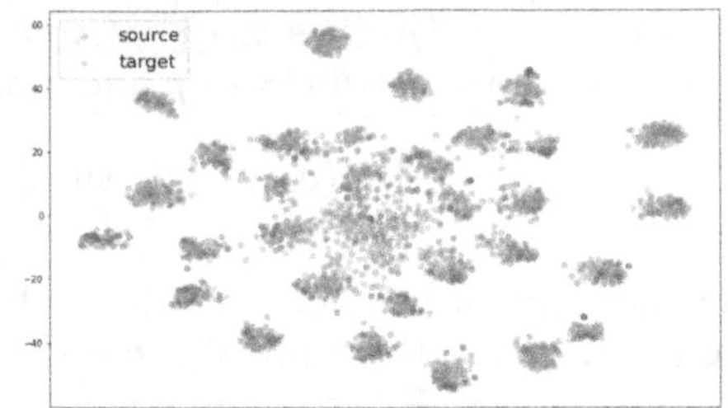

(b) TSF(φ) transferability from bound 2 (contribution). Inside class clusters, source and target representations are not distinguishable

Fig. 1. t-SNE [26] visualisation of representations when trained to minimize (a) adaptability error $\lambda_{\mathcal{G}}(\varphi)$ from [5], (b) transferability error TSF(φ) introduced in the present work. The task used is A→W of the **Office31** dataset. *Labels in the target domain are used during learning in this specific experiment.* For both visualisations of representations, we observe well-separated clusters associated to the label classification task. Inside those clusters, we observe a separation between source and target representations for $\lambda_{\mathcal{G}}(\varphi)$. That means that representations embed domain information and thus are not invariant. On the contrary, source and target representations are much more overlapping inside of each cluster with TSF(φ), illustrating that this new term is not conflictual with invariance.

Proposition 2. *Let ψ a representation which is a richer feature extractor than φ: $\mathcal{F} \circ \varphi \subset \mathcal{F} \circ \psi$ and $\mathcal{F}_C \circ \varphi \subset \mathcal{F}_C \circ \psi$. φ is more domain invariant than ψ:*

$$\mathrm{INV}(\varphi) \leq \mathrm{INV}(\psi) \text{ while } \varepsilon_T(\mathbf{f}_T^{\psi}\psi) \leq \varepsilon_T(\mathbf{f}_T^{\varphi}\varphi) \tag{11}$$

where $\mathbf{f}_T^{\varphi}(z) = \mathbb{E}_T[Y|\varphi(X) = z]$ and $\mathbf{f}_T^{\psi}(z) = \mathbb{E}_T[Y|\psi(X) = z]$. Proof in Appendix A.2.

Bounding the target risk using IPMs has two advantages. First, it allows to better control the invariance/transferability trade-off since $\varepsilon_T(\mathbf{f}_T\varphi) \leq \lambda_{\mathcal{G}}(\varphi)$. This is paid at the cost of $4 \cdot \mathrm{INV}(\varphi) \geq d_{\mathcal{G}}(\varphi)$ (see Proposition 7 in Appendix A.1). Second, $\varepsilon_T(\mathbf{f}_T\varphi)$ is source free and indicates whether there is enough information in representations for learning the task in the target domain at first. This means that TSF(φ) is only dedicated to control if aligned representations have the same labels across domains. To illustrate the interest of our new transferability error, we provide visualisation of representations (Fig. 1) when trained to minimize the adaptability error $\lambda_{\mathcal{G}}(\varphi)$ from Bound 1 and the transferability error TSF(φ) from Bound 2.

3.2 A Detailed View on the Property of Tightness

An interesting property of the bound, named tightness, is the case when $\mathrm{INV}(\varphi) = 0$ and $\mathrm{TSF}(\varphi) = 0$ simultaneously. The condition of tightness of the bound provides rich information on the properties of representations.

Proposition 3. $\text{INV}(\varphi) = \text{TSF}(\varphi) = 0$ *if and only if* $p_S(y, z) = p_T(y, z)$.

The proof is given in Appendix A.3. Two important points should be noted:

1. $\text{INV}(\varphi) = 0$ ensures that $p_S(z) = p_T(z)$, using (A4). Similarly, $\text{TSF}(\varphi) = 0$ leads to $p_S(y, z) = p_T(y, z)$. Since $p_S(y, z) = p_T(y, z)$ implies $p_S(z) = p_T(z)$, $\text{INV}(\varphi)$ does not bring more substantial information about representations distribution than $\text{TSF}(\varphi)$. More precisely, one can show that $\text{TSF}(\varphi) \geq \text{INV}(\varphi)$ noting that $Y \cdot \mathbf{f}(Z) = f(z)$ when $\mathbf{f}(z) = (f(z), ..., f(z))$ for $f \in \mathcal{F}$.
2. Second, the equality $p_S(y, z) = p_T(y, z)$ also implies that $p_S(y) = p_T(y)$. Therefore, in the context of label shift (when $p_S(y) \neq p_T(y)$), the transferability error cannot be null. This is a big hurdle since it is clearly established that most real world UDA tasks exhibit some label shift. This bound highlights the fact that representation invariance alone can not address UDA in complex settings such as the label shift one.

3.3 Reconciling Weights and Invariant Representations

Based on the interesting observations from [20,40] and following the line of study that proposed to relax invariance using weights [9,36–38], we propose to adapt the bound by incorporating weights. More precisely, we study the effect of modifying the source distribution $p_S(z)$ to a *weighted source* distribution $w(z)p_S(z)$ where w is a positive function which verifies $\mathbb{E}_S[w(Z)] = 1$. By replacing $p_S(z)$ by $w(z)p_S(z)$ (distribution referred as $w \cdot S$) in Bound 2, we obtain a new bound of the target risk incorporating both weights and representations:

Bound 3. $\forall g \in \mathcal{G}, \forall w : \mathcal{Z} \to \mathbb{R}^+$ *such that* $\mathbb{E}_S[w(z)] = 1$*:*

$$\varepsilon_T(g\varphi) \leq \varepsilon_{w \cdot S}(g\varphi) + 6 \cdot \text{INV}(w, \varphi) + 2 \cdot \text{TSF}(w, \varphi) + \varepsilon_T(\mathbf{f}_T\varphi)$$

where $\text{INV}(w, \varphi) := \sup_{f \in \mathcal{F}} \{\mathbb{E}_T[f(Z)] - \mathbb{E}_S[w(Z)f(Z)]\}$ *and* $\text{TSF}(w, \varphi) := \sup_{\mathbf{f} \in \mathcal{F}_C} \{ \mathbb{E}_T[Y \cdot \mathbf{f}(Z)] - \mathbb{E}_S[w(Z)Y \cdot \mathbf{f}(Z)]\}$.

As for the previous Bound 2, the property of tightness, *i.e.* when invariance and transferability are null simultaneously, leads to interesting observations:

Proposition 4. $\text{INV}(w, \varphi) = \text{TSF}(w, \varphi) = 0$ *if and only if* $w(z) = \frac{p_T(z)}{p_S(z)}$ *and* $\mathbb{E}_T[Y|Z = z] = \mathbb{E}_S[Y|Z = z]$*. The proof is given in Appendix A.4.*

This proposition means that the nullity of invariance error, *i.e.* $\text{INV}(w, \varphi) = 0$, implies distribution alignment, *i.e.* $w(z)p_S(z) = p_T(z)$. This is of strong interest since both representations and weights are involved for achieving domain invariance. The nullity of the transferability error, *i.e.* $\text{TSF}(w, \varphi) = 0$, implies that labelling functions, $\mathbf{f} : z \mapsto \mathbb{E}[Y|Z = z]$, are conserved across domains. Furthermore, the equality $\mathbb{E}_T[Y|Z] = \mathbb{E}_S[Y|Z]$ interestingly resonates with a recent line of work called *Invariant Risk Minimization* (IRM) [2]. Incorporating weights in the bound thus brings two benefits:

1. First, it raises the inconsistency issue of invariant representations in presence of label shift, as mentioned in Sect. 3. Indeed, tightness is not conflicting with label shift.
2. $\text{TSF}(w, \varphi)$ and $\text{INV}(w, \varphi)$ have two distinct roles: the former promotes domain invariance of representations while the latter controls whether aligned representations share the same labels across domains.

4 The Role of Inductive Bias

Inductive Bias refers to the set of assumptions which improves generalization of a model trained on an empirical distribution. For instance, a specific neural network architecture or a well-suited regularization are prototypes of inductive biases. First, we provide a theoretical analysis of the role of inductive bias for addressing the lack of labelling data in the target domain (4.1), which is the most challenging part of *Unsupervised* Domain Adaptation. Second, we describe the effect of weights to induce invariance property on representations (4.2).

4.1 Inductive Design of a Classifier

General Formulation. Our strategy consists in approximating target labels error through a classifier $\tilde{g} \in \mathcal{G}$. We refer to the latter as the inductive design of the classifier. Our proposition follows the intuitive idea which states that the best source classifier, $g_S := \arg\min_{g \in \mathcal{G}} \varepsilon_S(g\varphi)$, is not necessarily the best target classifier *i.e.* $g_S \neq \arg\min_{g \in \mathcal{G}} \varepsilon_T(g\varphi)$. For instance, a well-suited regularization in the target domain, noted $\Omega_T(g)$ may improve performance, *i.e.* setting $\tilde{g} := \arg\min_{g \in \mathcal{G}} \varepsilon_S(g\varphi) + \lambda \cdot \Omega_T(g)$ may lead to $\varepsilon_T(\tilde{g}\varphi) \leq \varepsilon_T(g_S\varphi)$. We formalize this idea through the following definition:

Definition 5 (Inductive design of a classifier). *We say that there is an inductive design of a classifier at level* $0 < \beta \leq 1$ *if for any representations* φ*, noting* $g_S = \arg\min_{g \in \mathcal{G}} \varepsilon_S(g\varphi)$*, we can determine* $\tilde{g}$ *such that:*

$$\varepsilon_T(\tilde{g}\varphi) \leq \beta \varepsilon_T(g_S\varphi) \tag{12}$$

We say the inductive design is $\beta-$*strong when* $\beta < 1$ *and weak when* $\beta = 1$.

In this definition, β does not depend of φ, which is a strong assumption, and embodies the strength of the inductive design. The closer to 1 is β, the less improvement we can expect using the inductive classifier $\tilde{g}$. We now study the impact of the inductive design of a classifier in our previous Bound 3. Thus, we introduce the approximated transferability error:

$$\widehat{\text{TSF}}(w, \varphi, \tilde{g}) = \sup_{\mathbf{f} \in \mathcal{F}_C} \{\mathbb{E}_T[\tilde{g}(Z) \cdot \mathbf{f}(Z)] - \mathbb{E}_S[w(Z)Y \cdot \mathbf{f}(Z)]\} \tag{13}$$

leading to a bound of the target risk where transferability is target labels free:

Bound 4 (Inductive Bias and Guarantee). *Let $\varphi \in \Phi$ and $w : \mathcal{Z} \to \mathbb{R}^+$ such that $\mathbb{E}_S[w(z)] = 1$ and a $\beta-$strong inductive classifier $\tilde{g}$ and $\rho := \frac{\beta}{1-\beta}$ then:*

$$\varepsilon_T(\tilde{g}\varphi) \leq \rho \left(\varepsilon_{w \cdot S}(g_{w \cdot S}\varphi) + 6 \cdot \mathrm{INV}(w, \varphi) + 2 \cdot \widehat{\mathrm{TSF}}(w, \varphi, \tilde{g}) + \varepsilon_T(\mathbf{f}_T\varphi) \right) \quad (14)$$

The proof is given in Appendix A.5. Here, the target labels are only involved in $\varepsilon_T(\mathbf{f}_T\varphi)$ which reflects the level of noise when fitting labels from representations. Therefore, transferability is now free of target labels. This is an important result since the difficulty of UDA lies in the lack of labelled data in the target domain. It is also interesting to note that the weaker the inductive bias ($\beta \to 1$), the higher the bound and vice versa.

The Role of Predicted Labels. Predicted labels play an important role in UDA. In light of the inductive classifier, this means that $\tilde{g}$ is simply set as $g_{w \cdot S}$. This is a weak inductive design ($\beta = 1$), thus, theoretical guarantee from Bound 4 is not applicable. However, there is empirical evidence that showed that predicted labels help in UDA [15,24]. It suggests that this inductive design may find some strength in the finite sample regime. A better understanding of this phenomenon is left for future work (See Appendix B). In the rest of the paper, we study this weak inductive bias by establishing connections between $\widehat{\mathrm{TSF}}(w, \varphi, g_S)$ and popular approaches of the literature.

Connections with Conditional Domain Adaptation Network. CDAN [24] aims to align the joint distribution $(\hat{Y}, Z)$ across domains, where $\hat{Y} = g_S\varphi(X)$ are estimated labels. It is performed by exposing the tensor product between $\hat{Y}$ and Z to a discriminator. It leads to substantial empirical improvements compared to *Domain Adversarial Neural Networks* (DANN) [13]. We can observe that it is a similar objective to $\widehat{\mathrm{TSF}}(w, \varphi, g_S)$ in the particular case where $w(z) = 1$.

Connections with Minimal Entropy. MinEnt [15] states that an adapted classifier is confident in prediction on target samples. It suggests the regularization: $\Omega_T(g) := H(\hat{Y}|Z) = \mathbb{E}_{Z \sim p_T}[-g(Z) \cdot \log g(Z)]$ where H is the entropy. If labels are smooth enough (*i.e.* it exists α such that $\frac{\alpha}{C-1} \leq \mathbb{E}_S[\hat{Y}|Z] \leq 1-\alpha$), MinEnt is a lower bound of transferability: $\widehat{\mathrm{TSF}}(w, \varphi, g_s) \geq \eta \left(H_T(g_S\varphi) - \mathrm{CE}_{w \cdot S}(Y, g_S\varphi) \right)$ for some $\eta > 0$ and $\mathrm{CE}_{w \cdot S}(g_S\varphi, Y)$ is the cross-entropy between $g_S\varphi$ and Y on $w(z)p_S(z)$ (see Appendix A.6).

4.2 Inductive Design of Weights

While the bounds introduced in the present work involve weights in the representation space, there is an abundant literature that builds weights in order to relax the domain invariance of representations [8,9,36,37]. We study the effect of inductive design of w on representations. To conduct the analysis, we consider there is a non-linear transformation ψ from $\mathcal{Z}$ to $\mathcal{Z}'$ and we assume that weights are computed in $\mathcal{Z}'$, *i.e.* w is a function of $z' := \psi(z) \in \mathcal{Z}'$. We refer to this as *inductive design of weights*. For instance, in the particular case where $\psi = g_S$, weights are designed as $w(\hat{y}) = p_T(\hat{Y} = \hat{y})/p_S(\hat{Y} = \hat{y})$ [9] where $\hat{Y} = g\varphi(X)$. In [24], *entropy conditioning* is introduced by designing weights $w(z') \propto 1 + e^{-z'}$ where $z' = -\frac{1}{C}\sum_{1 \leq c \leq C} g_{S,c} \log(g_{S,c})$ is the predictions entropy. The inductive design of weights imposes invariance property on representations:

Proposition 6 (Inductive design of w and invariance). *Let $\psi : \mathcal{Z} \to \mathcal{Z}'$ such that $\mathcal{F} \circ \psi \subset \mathcal{F}$ and $\mathcal{F}_C \circ \psi \subset \mathcal{F}_C$. Let $w : \mathcal{Z}' \to \mathbb{R}^+$ such that $\mathbb{E}_S[w(Z')] = 1$ and we note $Z' := \psi(Z)$. Then,* $\text{INV}(w, \varphi) = \text{TSF}(w, \varphi) = 0$ *if and only if:*

$$w(z') = \frac{p_T(z')}{p_S(z')} \quad \textit{and} \quad p_S(z|z') = p_T(z|z') \tag{15}$$

while both $\mathbf{f}_S^\varphi = \mathbf{f}_T^\varphi$ and $\mathbf{f}_S^\psi = \mathbf{f}_T^\psi$. The proof is given in Appendix A.7.

This proposition shows that the design of w has a significant impact on the property of domain invariance of representations. Furthermore, both labelling functions are conserved. In the rest of the paper we focus on weighting in the representation space which consists in:

$$w(z) = \frac{p_T(z)}{p_S(z)} \tag{16}$$

Since it does not leverage any transformations of representations ψ, we refer to this approach as a weak inductive design of weights. It is worth noting this inductive design controls naturally the invariance error *i.e.* $\text{INV}(w, \varphi) = 0$.

5 Towards Robust Domain Adaptation

In this section, we expose a new learning procedure which relies on weak inductive design of both weights and the classifier. This procedure focuses on the transferability error since the inductive design of weights naturally controls the invariance error. Our learning procedure is then a bi-level optimization problem, named RUDA (Robust UDA):

$$\begin{cases} \varphi^\star = \arg\min_{\varphi \in \Phi} \; \varepsilon_{w(\varphi)\cdot S}(g_{w\cdot S}\varphi) + \lambda \cdot \widehat{\text{TSF}}(w, \varphi, g_{w\cdot S}) \\ \quad \text{such that } w(\varphi) = \arg\min_{w} \text{INV}(w, \varphi) \end{cases} \tag{RUDA}$$

where $\lambda > 0$ is a trade-off parameter. Two discriminators are involved here. The former is a domain discriminator d trained to map 1 for source representations and 0 for target representations by minimizing a domain adversarial loss:

$$\mathcal{L}_{\text{INV}}(\theta_d|\theta_\varphi) = \frac{1}{n_S}\sum_{i=1}^{n_S} -\log(d(z_{S,i})) + \frac{1}{n_T}\sum_{i=1}^{n_T} -\log(1 - d(z_{T,i})) \tag{17}$$

where θ_d and θ_φ are respectively the parameters of d and φ, and n_S and n_T are respectively the number of samples in the source and target domains. Setting weights $w_d(z) := (1 - d(z))/d(z)$ ensures that $\text{INV}(w, \varphi)$ is minimal (See

Appendix C.2). The latter, noted $\mathbf{d}$, maps representations to the label space $[0,1]^C$ in order to obtain a proxy of the transferability error expressed as a domain adversarial objective (See Appendix C.1):

$$\mathcal{L}_{\text{TSF}}(\theta_\varphi, \theta_{\mathbf{d}}|\theta_d, \theta_g) = \inf_{\mathbf{d}} \left\{ \frac{1}{n_S} \sum_{i=1}^{n_S} -w_d(z_{S,i}) g(z_{S,i}) \cdot \log(\mathbf{d}(z_{S,i})) + \frac{1}{n_T} \sum_{i=1}^{n_T} -g(z_{T,i}) \cdot \log(1 - \mathbf{d}(z_{T,i})) \right\} \tag{18}$$

where $\theta_{\mathbf{d}}$ and θ_g are respectively parameters of $\mathbf{d}$ and g. Furthermore, we use the cross-entropy loss in the source weighted domain for learning θ_g:

$$\mathcal{L}_c(\theta_g, \theta_\varphi|\theta_d) = \frac{1}{n_S} \sum_{i=1}^{n_S} -w_d(z_{S,i}) y_{S,i} \cdot \log(g(z_{S,i})) \tag{19}$$

Finally, the optimization is then expressed as follows:

$$\begin{cases} \theta_\varphi^\star = \arg\min_{\theta_\varphi} \mathcal{L}_c(\theta_g, \theta_\varphi|\theta_d) + \lambda \cdot \mathcal{L}_{\text{TSF}}(\theta_\varphi, \theta_{\mathbf{d}}|\theta_d, \theta_g) \\ \theta_g = \arg\min_{\theta_g} \mathcal{L}_c(\theta_g, \theta_\varphi|\theta_d) \\ \theta_d = \arg\min_{\theta_d} \mathcal{L}_{\text{INV}}(\theta_d|\theta_\varphi) \end{cases} \tag{20}$$

Losses are minimized by stochastic gradient descent (SGD) where in practice $\inf_d$ and $\inf_{\mathbf{d}}$ are gradient reversal layers [13]. The trade-off parameter λ is pushed from 0 to 1 during training. We provide an implementation in `Pytorch` [29] based on [24]. The algorithm procedure is described in Appendix C.5.

6 Experiments

6.1 Setup

Datasets. We investigate two digits datasets: **MNIST** and **USPS** transfer tasks MNIST to USPS (M→U) and USPS to MNIST (U→M). We used standard train/test split for training and evaluation. **Office-31** is a dataset of images containing objects spread among 31 classes captured from different domains: **Amazon**, **DSLR** camera and a **Webcam** camera. **DSLR** and **Webcam** are very similar domains but images differ by their exposition and their quality.

Label Shifted Datasets. We stress-test our approach by investigating more challenging settings where the label distribution shifts strongly across domains. For the **Digits** dataset, we explore a wide variety of shifts by keeping only 5%, 10%, 15% and 20% of digits between 0 and 5 of the original dataset (refered as $\% \times [0 \sim 5]$). We have investigated the tasks U→M and M→U. For the **Office-31** dataset, we explore the shift where the object spread in classes 16 to 31 are duplicated 5 times (refered as $5 \times [16 \sim 31]$). Shifting distribution in the source domain rather than the target domain allows to better appreciate the drop in performances in the target domain compared to the case where the source domain is not shifted.

Comparison with the State-of-the-Art. For all tasks, we report results from DANN [13] and CDAN [24]. To study the effect of weights, we name our method RUDA when weights are set to 1, and RUDA$_w$ when weights are used. For the non-shifted datasets, we report a weighted version of CDAN (entropy conditioning CDAN+E [24]). For the label shifted datasets, we report IWAN [38], a weighted DANN where weights are learned from a second discriminator, and CDAN$_w$ a weighted CDAN where weights are added in the same setting than RUDA$_w$.

Training Details. Models are trained during 20.000 iterations of SGD. We report end of training accuracy in the target domain averaged on five random seeds. The model for the **Office-31** dataset uses a pretrained ResNet-50 [18]. We used the same hyper-parameters than [24] which were selected by importance weighted cross-validation [35]. The trade-off parameters λ is smoothly pushed from 0 to 1 as detailed in [24]. To prevent from noisy weighting in early learning, we used weight relaxation: based on the sigmoid output of discriminator $d(z) = \sigma(\tilde{d}(z))$, we used $d_\tau(z) = \sigma(\tilde{d}(z/\tau))$ and weights $w(z) = (1 - d_\tau(z))/d_\tau(z)$. τ is decreased to 1 during training: $\tau = \tau_{\min} + 2(\tau_{\max} - \tau_{\min})/(1 + \exp(-\alpha p))$ where $\tau_{\max} = 5, \tau_{\min} = 1$, $p \in [0, 1]$ is the training progress. In all experiments, α is set to 5 (except for 5% $\times$ [0 $\sim$ 5] where $\alpha = 15$, see Appendix C.3 for more details).

Table 1. Accuracy (%) on the **Office-31** dataset.

	Method	A→W	W→A	A→D	D→A	D→W	W→D	Avg
Standard	ResNet-50	68.4 ± 0.2	60.7 ± 0.3	68.9 ± 0.2	62.5 ± 0.3	96.7 ± 0.1	99.3 ± 0.1	76.1
	DANN	82.0 ± 0.4	67.4 ± 0.5	79.7 ± 0.4	68.2 ± 0.4	96.9 ± 0.2	99.1 ± 0.1	82.2
	CDAN	93.1 ± 0.2	68.0 ± 0.4	89.8 ± 0.3	70.1 ± 0.4	98.2 ± 0.2	100. ± 0.0	86.6
	CDAN+E	94.1 ± 0.1	69.3 ± 0.4	**92.9 ± 0.2**	**71.0 ± 0.3**	**98.6 ± 0.1**	**100. ± 0.0**	**87.7**
	RUDA	**94.3 ± 0.3**	**70.7 ± 0.3**	92.1 ± 0.3	70.7 ± 0.1	98.5 ± 0.1	100. ± 0.0	87.6
	RUDA$_w$	92.0 ± 0.3	67.9 ± 0.3	91.1 ± 0.3	70.2 ± 0.2	98.6 ± 0.1	100. ± 0.0	86.6
5 × [16 ∼ 31]	ResNet-50	72.4 ± 0.7	59.5 ± 0.1	79.0 ± 0.1	61.6 ± 0.3	97.8 ± 0.1	99.3 ± 0.1	78.3
	DANN	67.5 ± 0.1	52.1 ± 0.8	69.7 ± 0.0	51.5 ± 0.1	89.9 ± 0.1	75.9 ± 0.2	67.8
	CDAN	82.5 ± 0.4	62.9 ± 0.6	81.4 ± 0.5	65.5 ± 0.5	98.5 ± 0.3	99.8 ± 0.0	81.6
	RUDA	85.4 ± 0.8	66.7 ± 0.5	81.3 ± 0.3	64.0 ± 0.5	98.4 ± 0.2	99.5 ± 0.1	82.1
	IWAN	72.4 ± 0.4	54.8 ± 0.8	75.0 ± 0.3	54.8 ± 1.3	97.0 ±0.0	95.8 ±0.6	75.0
	CDAN$_w$	81.5 ± 0.5	64.5 ± 0.4	80.7 ± 1.0	65 ± 0.8	**98.7 ± 0.2**	99.9 ± 0.1	81.8
	RUDA$_w$	**87.4 ± 0.2**	**68.3 ± 0.3**	**82.9** ± 0.4	**68.8 ± 0.2**	**98.7** ± 0.1	**100.** ± 0.0	**83.8**

6.2 Results

Unshifted Datasets. On both **Office-31** (Table 1) and **Digits** (Table 2), RUDA performs similarly than CDAN. Simply performing the scalar product allows to achieve results obtained by multi-linear conditioning [24]. This presents a second advantage: when domains exhibit a large number of classes, *e.g.* in **Office-Home** (See Appendix), our approach does not need to leverage a random layer. It is interesting to observe that we achieve performances close to CDAN+E on **Office-31** while we do not use entropy conditioning. However, we observe

Table 2. Accuracy (%) on the **Digits** dataset.

Method	U→ M						M→U						Avg
Shift of [0 ∼ 5]	5%	10%	15%	20%	100%	Avg	5%	10%	15%	20%	100%	Avg	
DANN	41.7	51.0	59.6	69.0	94.5	63.2	34.5	51.0	59.6	63.6	90.7	59.9	63.2
CDAN	50.7	62.2	82.9	82.8	**96.9**	75.1	32.0	69.7	78.9	81.3	**93.9**	71.2	73.2
RUDA	44.4	58.4	80.0	84.0	95.5	72.5	34.9	59.0	76.1	78.8	93.3	68.4	70.5
IWAN	73.7	74.4	78.4	77.5	95.7	79.9	72.2	82.0	84.3	86.0	92.0	83.3	81.6
$CDAN_w$	68.3	78.8	84.9	**88.4**	96.6	83.4	69.4	80.0	83.5	87.8	93.7	82.9	83.2
$RUDA_w$	**78.7**	**82.8**	**86.0**	86.9	93.9	**85.7**	**78.7**	**87.9**	**88.2**	**89.3**	92.5	**87.3**	**86.5**

a substantial drop in performance when adding weights, but still get results comparable with CDAN in **Office-31**. This is a deceptive result since those datasets naturally exhibit label shift; one can expect to improve the baselines using weights. We did not observe this phenomenon on standard benchmarks.

Label Shifted Datasets. We stress-tested our approach by applying strong label shifts to the datasets. First, we observe a drop in performance for all methods based on invariant representations compared with the situation without label shift. This is consistent with works that warn the pitfall of domain invariant representations in presence of label shift [20,40]. RUDA and CDAN perform similarly even in this setting. It is interesting to note that the weights improve significantly RUDA results (+1.7% on **Office-31** and +16.0% on **Digits** both in average) while CDAN seems less impacted by them (+0.2% on **Office-31** and +10.0% on **Digits** both in average).

Should we Use Weights? To observe a significant benefit of weights, we had to explore situations with strong label shift *e.g.* 5% and 10% × [0 ∼ 5] for the **Digits** dataset. Apart from this cases, weights bring small gain (*e.g.* + 1.7% on **Office-31** for RUDA) or even degrade marginally adaptation. Understanding why RUDA and CDAN are able to address small label shift, without weights, is of great interest for the development of more robust UDA.

7 Related Work

This paper makes several contributions, both in terms of theory and algorithm. Concerning theory, our bound provides a risk suitable for domain adversarial learning with weighting strategies. Existing theories for non-overlapping supports [4,27] and importance sampling [11,30] do not explore the role of representations neither the aspect of adversarial learning. In [5], analysis of representation is conducted and connections with our work is discussed in the paper. The work [20] is close to ours and introduces a distance which measures support overlap between source and target distributions under covariate shift. Our analysis does not rely on such assumption, its range of application is broader.

Concerning algorithms, the covariate shift adaptation has been well-studied in the literature [17,19,35]. Importance sampling to address label shift has also been investigated [34], notably with kernel mean matching [39] and Optimal Transport [31]. Recently, a scheme for estimating labels distribution ratio with consistency guarantee has been proposed [21]. Learning domain invariant representations has also been investigated in the fold of [13,23] and mainly differs by the metric chosen for comparing distribution of representations. For instance, metrics are domain adversarial (Jensen divergence) [13,24], IPM based such as MMD [23,25] or Wasserstein [6,32]. Our work provides a new theoretical support for these methods since our analysis is valid for any IPM.

Using both weights and representations is also an active topic, namely for Partial Domain Adaptation (PADA) [9], when target classes are strict subset of the source classes, or Universal Domain Adaptation [37], when new classes may appear in the target domain. [9] uses an heuristic based on predicted labels for re-weighting representations. However, it assumes they have a good classifier at first in order to obtain cycle consistent weights. [38] uses a second discriminator for learning weights, which is similar to [8]. Applying our framework to Partial DA and Universal DA is an interesting future direction. Our work shares strong connections with [10] (authors were not aware of this work during the elaboration of this paper) which uses consistent estimation of true labels distribution from [21]. We suggest a very similar empirical evaluation and we also investigate the effect of weights on CDAN loss [24] with a different weighting scheme since our approach computes weights in the representation space. All these works rely on an assumption at some level, *e.g. Generalized Label Shift* in [10], when designing weighting strategies. Our discussion on the role of inductive design of weights may provide a new theoretical support for these approaches.

8 Conclusion

The present work introduces a new bound of the target risk which unifies weights and representations in UDA. We conduct a theoretical analysis of the role of inductive bias when designing both weights and the classifier. In light of this analysis, we propose a new learning procedure which leverages two weak inductive biases, respectively on weights and the classifier. To the best of our knowledge, this procedure is original while being close to straightforward hybridization of existing methods. We illustrate its effectiveness on two benchmarks. The empirical analysis shows that weak inductive bias can make adaptation more robust even when stressed by strong label shift between source and target domains. This work leaves room for in-depth study of stronger inductive bias by providing both theoretical and empirical foundations.

Acknowledgements. Victor Bouvier is funded by Sidetrade and ANRT (France) through a CIFRE collaboration with CentraleSupélec. Authors thank the anonymous reviewers for their insightful comments for improving the quality of the paper. This work was performed using HPC resources from the "Mésocentre" computing center of CentraleSupélec and École Normale Supérieure Paris-Saclay supported by CNRS and Région Île-de-France (http://mesocentre.centralesupelec.fr/).

A Proofs

We provide full proof of bounds and propositions presented in the paper.

A.1 Proof of Bound 2

We give a proof of Bound 2 which states:

$$\varepsilon_T(g\varphi) \leq \varepsilon_S(g\varphi) + 6 \cdot \mathrm{INV}(\varphi) + 2 \cdot \mathrm{TSF}(\varphi) + \varepsilon_T(\mathbf{f}_T\varphi) \tag{21}$$

First, we prove the following lemma:

Bound 5 (Revisit of Theorem 1). *$\forall g \in \mathcal{G}$:*

$$\varepsilon_T(g\varphi) \leq \varepsilon_S(g\varphi) + d_{\mathcal{F}_C}(\varphi) + \varepsilon_T(\mathbf{f}_S\varphi, \mathbf{f}_T\varphi) + \varepsilon_T(\mathbf{f}_T\varphi) \tag{22}$$

Proof. This is simply obtained using triangular inequalites:

$$\begin{aligned}\varepsilon_T(g\varphi) &\leq \varepsilon_T(\mathbf{f}_T\varphi) + \varepsilon_T(g\varphi, \mathbf{f}_T\varphi) \\ &\leq \varepsilon_T(\mathbf{f}_T\varphi) + \varepsilon_T(g\varphi, \mathbf{f}_S\varphi) + \varepsilon_T(\mathbf{f}_S\varphi, \mathbf{f}_T\varphi)\end{aligned}$$

Now using (A3) ($\mathbf{f}_S \in \mathcal{F}_C$):

$$|\varepsilon_T(g\varphi, \mathbf{f}_S\varphi) - \varepsilon_S(g\varphi, \mathbf{f}_S\varphi)| \leq \sup_{\mathbf{f}\in\mathcal{F}_C} |\varepsilon_T(g\varphi, \mathbf{f}\varphi) - \varepsilon_S(g\varphi, \mathbf{f}\varphi)| = d_{\mathcal{F}_C}(\varphi) \tag{23}$$

which shows that: $\varepsilon_T(g\varphi) \leq \varepsilon_S(g\varphi, \mathbf{f}_S) + d_{\mathcal{F}_C}(\varphi) + \varepsilon_T(\mathbf{f}_S\varphi, \mathbf{f}_T\varphi) + \varepsilon_T(\mathbf{f}_T\varphi)$ and we use the property of conditional expectation $\varepsilon_S(g\varphi, \mathbf{f}_S\varphi) \leq \varepsilon_S(g\varphi)$. □

Second, we bound $d_{\mathcal{F}_C}(\varphi)$.

Proposition 7. $d_{\mathcal{F}_C}(\varphi) \leq 4 \cdot \mathrm{INV}(\varphi)$.

Proof. We remind that $d_{\mathcal{F}_C}(\varphi) = \sup_{\mathbf{f},\mathbf{f}'\in\mathcal{F}_C} |\mathbb{E}_S[||\mathbf{f}\varphi(X) - \mathbf{f}'\varphi(X)||^2] - \mathbb{E}_T[||\mathbf{f}\varphi(X) - \mathbf{f}'\varphi(X)||^2]|$. Since (A1) ensures $\mathbf{f}' \in \mathcal{F}_C$, $-\mathbf{f}' \in \mathcal{F}_C$, then $\frac{1}{2}(\mathbf{f} - \mathbf{f}') = \mathbf{f}'' \in \mathcal{F}_C$ and finally $d_{\mathcal{F}_C}(\varphi) \leq 4 \sup_{\mathbf{f}''\in\mathcal{F}_C} |\mathbb{E}_S[||\mathbf{f}''\varphi||^2] - \mathbb{E}_T[||\mathbf{f}''\varphi||^2]|$. Furthermore, (A2) ensures that $\{||\mathbf{f}''\varphi||^2\} \subset \{f\varphi, f \in \mathcal{F}\}$ which leads finally to the announced result. □

Third, we bound $\varepsilon_T(\mathbf{f}_S\varphi, \mathbf{f}_T\varphi)$.

Proposition 8. $\varepsilon_T(\mathbf{f}_S\varphi, \mathbf{f}_T\varphi) \leq 2 \cdot \mathrm{INV}(\varphi) + 2 \cdot \mathrm{TSF}(\varphi)$.

Proof. We note $\Delta = \mathbf{f}_T - \mathbf{f}_S$ and we omit φ for the ease of reading

$$\begin{aligned}\varepsilon_T(\mathbf{f}_S, \mathbf{f}_T) &= \mathbb{E}_T[||\Delta||^2] \\ &= \mathbb{E}_T[\mathbf{f}_T \cdot \Delta] - \mathbb{E}_T[\mathbf{f}_S \cdot \Delta] \\ &= (\mathbb{E}_T[\mathbf{f}_T \cdot \Delta] - \mathbb{E}_S[\mathbf{f}_S \cdot \Delta]) + (\mathbb{E}_S[\mathbf{f}_S \cdot \Delta] - \mathbb{E}_T[\mathbf{f}_S \cdot \Delta])\end{aligned}$$

Since $\mathbf{f}_T$ does not intervene in $\mathbb{E}_S[\mathbf{f}_S \cdot \Delta] - \mathbb{E}_T[\mathbf{f}_S \cdot \Delta]$, we show this term behaves similarly than $\text{INV}(\varphi)$. First,

$$\begin{aligned}
\mathbb{E}_S[\mathbf{f}_S \cdot \Delta] - \mathbb{E}_T[\mathbf{f}_S \cdot \Delta] &\leq 2 \sup_{\mathbf{f} \in \mathcal{F}_C} \mathbb{E}_S[\mathbf{f}_S \cdot \mathbf{f}] - \mathbb{E}_T[\mathbf{f}_S \cdot \mathbf{f}] && \text{(Using(A1))} \\
&\leq 2 \sup_{\mathbf{f}, \mathbf{f}' \in \mathcal{F}_C} \mathbb{E}_S[\mathbf{f}' \cdot \mathbf{f}] - \mathbb{E}_T[\mathbf{f}' \cdot \mathbf{f}] && \text{(Using(A3))} \\
&\leq 2 \sup_{f \in \mathcal{F}} \mathbb{E}_S[f] - \mathbb{E}_T[f] && \text{(Using(A2))} \\
&= 2 \cdot \text{INV}(\varphi) && (24)
\end{aligned}$$

Second,

$$\mathbb{E}_T[\mathbf{f}_T \cdot \Delta] - \mathbb{E}_S[\mathbf{f}_S \cdot \Delta] \leq 2 \sup \mathbb{E}_T[\mathbf{f}_T \cdot \mathbf{f}] - \mathbb{E}_S[\mathbf{f}_S \cdot \mathbf{f}] = 2 \cdot \text{TSF}(\varphi) \quad \text{(Using(A1))}$$

which finishes the proof. □

Note that the fact $\mathbf{f}_S, \mathbf{f}_T \in \mathcal{F}_C$ is not of the utmost importance since we can bound:

$$\varepsilon_T(g\varphi) \leq \varepsilon_S(g\varphi, \hat{\mathbf{f}}_S) + d_{\mathcal{F}_C}(\varphi) + \varepsilon_T(\hat{\mathbf{f}}_S, \hat{\mathbf{f}}_T) + \varepsilon_T(\hat{\mathbf{f}}_T) \quad (25)$$

where $\hat{\mathbf{f}}_D = \arg\min_{\mathbf{f} \in \mathcal{F}_C} \varepsilon_D(v)$. The only change emerges in the transferability error which becomes:

$$\text{TSF}(w, \varphi) = \sup_{\mathbf{f} \in \mathcal{F}_C} \mathbb{E}_T[\hat{\mathbf{f}}_T\varphi \cdot \mathbf{f}\varphi] - \mathbb{E}_S[\hat{\mathbf{f}}_S\varphi \cdot \mathbf{f}\varphi] \quad (26)$$

A.2 Proof of the New Invariance Transferability Trade-Off

Proposition 9. *Let ψ a representation which is a richer feature extractor than φ: $\mathcal{F} \circ \varphi \subset \mathcal{F} \circ \psi$ and $\mathcal{F}_C \circ \varphi \subset \mathcal{F}_C \circ \psi$. Then, φ is more domain invariant than ψ:*

$$\text{INV}(\varphi) \leq \text{INV}(\psi) \text{ while } \varepsilon_T(f_T^{\psi}\psi) \leq \varepsilon_T(f_T^{\varphi}\varphi) \quad (27)$$

where $f_T^{\varphi}(z) = \mathbb{E}_T[Y|\varphi(X) = z]$ and $f_T^{\psi}(z) = \mathbb{E}_T[Y|\psi(X) = z]$.

Proof. First, $\text{INV}(\varphi) \leq \text{INV}(\psi)$ a simple property of the supremum. The definition of the conditional expectation leads to $\varepsilon_T(\mathbf{f}_T^{\psi}\psi) = \inf_{f \in \mathcal{F}_m} \varepsilon_T(f\psi)$ where $\mathcal{F}_m$ is the set of measurable functions. Since (A3) ensures that $\mathbf{f}_T^{\psi} \in \mathcal{F}_C$ then $\varepsilon_T(\mathbf{f}_T^{\psi}\psi) = \inf_{\mathbf{f} \in \mathcal{F}_C} \varepsilon_T(\mathbf{f}\psi)$. The rest is simply the use of the property of infremum. □

A.3 Proof of the Tightness of Bound 2

Proposition 10. $\text{INV}(\varphi) + \text{TSF}(\varphi) = 0$ *if and only if* $p_S(y, z) = p_T(y, z)$.

Proof. First, $\text{INV}(\varphi) = 0$ implies $p_T(z) = p_S(z)$ which is a direct application of (A4). Now $\text{TSF}(\varphi) = \sup_{\mathbf{f} \in \mathcal{F}_C} \mathbb{E}_S[\mathbf{f}_S(Z) \cdot \mathbf{f}(Z)] - \mathbb{E}_T[\mathbf{f}_T(Z) \cdot \mathbf{f}(Z)] = \sup_{\mathbf{f} \in \mathcal{F}_C} \mathbb{E}_S[\mathbf{f}_S(Z) \cdot \mathbf{f}(Z)] - \mathbb{E}_S[\mathbf{f}_T(Z) \cdot \mathbf{f}(Z)] = \sup_{\mathbf{f} \in \mathcal{F}_C} \mathbb{E}_S[(\mathbf{f}_S - \mathbf{f}_T)(Z) \cdot \mathbf{f}(Z)]$. For the particular choice of $\mathbf{f} = \frac{1}{2}(\mathbf{f}_S - \mathbf{f}_T)$ leads to $\mathbb{E}_S[||\mathbf{f}_S - \mathbf{f}_T||^2]$ then $\mathbf{f}_s = \mathbf{f}_T$, p_S almost surely. All combined leads to $p_S(y, z) = p_T(y, z)$. The converse is trivial. Note that $\text{TSF}(\varphi) = 0$ is enough to show $p_S(z) = p_T(z)$ by choosing $\mathbf{f}(z) = (f(z), ..., f(z))$ (C times $f(z)$) and $Y \cdot \mathbf{f}(Z) = f(Z)$ then $\text{TSF}(\varphi) \geq \sup_{f \in \mathcal{F}} \mathbb{E}_S[f(Z)] - \mathbb{E}_T[f(Z)]$. □

A.4 Proof of the Tightness of Bound 3

Proposition 11. $\text{INV}(w, \varphi) + \text{TSF}(w, \varphi) = 0$ *if and only if* $w(z) = \frac{p_T(z)}{p_S(z)}$ *and* $\mathbb{E}_T[Y|Z = z] = \mathbb{E}_S[Y|Z = z]$.

Proof. First, $\text{INV}(w, \varphi) = 0$ implies $p_T(z) = w(z)p_S(z)$ then which is a direct application of (A4). Now $\text{TSF}(w, \varphi) = \sup_{\mathbf{f} \in \mathcal{F}_C} \mathbb{E}_S[w(z)\mathbf{f}_S(Z) \cdot \mathbf{f}(Z)] - \mathbb{E}_T[\mathbf{f}_T(Z) \cdot \mathbf{f}(Z)] = \sup_{\mathbf{f} \in \mathcal{F}_C} \mathbb{E}_S[w(z)\mathbf{f}_S(Z) \cdot \mathbf{f}(Z)] - \mathbb{E}_S[w(z)\mathbf{f}_T(Z) \cdot \mathbf{f}(Z)] = \sup_{\mathbf{f} \in \mathcal{F}_C} \mathbb{E}_S[(\mathbf{f}_S - \mathbf{f}_T)(Z) \cdot \mathbf{f}(Z)]$. For the particular choice of $\mathbf{f} = \frac{1}{2}(\mathbf{f}_S - \mathbf{f}_T)$ leads to $\mathbb{E}_S[||\mathbf{f}_S - \mathbf{f}_T||^2]$ then $\mathbf{f}_s = \mathbf{f}_T$, p_T almost surely. The converse is trivial. □

A.5 Proof of Bound 4

Bound 6 (Inductive Bias and Guarantee). *Let* $\varphi \in \Phi$ *and* $w : \mathcal{Z} \to \mathbb{R}^+$ *such that* $\mathbb{E}_S[w(z)] = 1$ *and a* β*−strong inductive classifier* $\tilde{g}$*, then:*

$$\varepsilon_T(\tilde{g}\varphi) \leq \frac{\beta}{1 - \beta}\left(\varepsilon_{w \cdot S}(g_{w \cdot S}\varphi) + 6 \cdot \text{INV}(w, \varphi) + 2 \cdot \widehat{\text{TSF}}(w, \varphi, \tilde{g}) + \varepsilon_T(\mathbf{f}_T\varphi)\right)$$

Proof. We prove the bound in the case where $w = 1$, the general case is then straightforward. First, we reuse Bound 5 with a new triangular inequality involving the inductive classifier $\tilde{g}$:

$$\varepsilon_T(g\varphi) \leq \varepsilon_S(g\varphi) + d_{\mathcal{F}_C}(\varphi) + \varepsilon_T(\mathbf{f}_S\varphi, \tilde{g}\varphi) + \varepsilon_T(\tilde{g}\varphi, \mathbf{f}_T\varphi) + \varepsilon_T(\mathbf{f}_T\varphi) \quad (28)$$

where $\varepsilon_T(\tilde{g}\varphi, \mathbf{f}_T\varphi) \leq \varepsilon_T(\tilde{g}\varphi)$. Now, following previous proofs, we can show that:

$$\varepsilon_T(\mathbf{f}_S\varphi, \tilde{g}\varphi) \leq 2 \cdot \widehat{\text{TSF}}(\varphi, \tilde{g}) + 2 \cdot \text{INV}(\varphi) \quad (29)$$

Then,

$$\varepsilon_T(g\varphi) \leq \varepsilon_S(g\varphi) + 6 \cdot \text{INV}(\varphi) + 2 \cdot \widehat{\text{TSF}}(\varphi, \tilde{g}) + \varepsilon_T(\tilde{g}\varphi) + \varepsilon_T(\mathbf{f}_T\varphi) \quad (30)$$

This bound is true for any g and in particular for the best source classifier we have:

$$\varepsilon_T(g_S\varphi) \leq \varepsilon_S(g_S\varphi) + 6 \cdot \text{INV}(w, \varphi) + 2 \cdot \widehat{\text{TSF}}(w, \varphi, \tilde{g}) + \varepsilon_T(\tilde{g}\varphi) + \varepsilon_T(\mathbf{f}_T\varphi) \quad (31)$$

then the assumption of $\beta-$strong inductive bias is $\varepsilon_T(\tilde{g}\varphi) \leq \beta\varepsilon_T(g_S\varphi)$ which leads to

$$\varepsilon_T(g_S\varphi) \leq \varepsilon_S(g_S\varphi) + 6 \cdot \text{INV}(w,\varphi) + 2 \cdot \widehat{\text{TSF}}(w,\varphi,\tilde{g}) + \beta\varepsilon_T(g_S\varphi) + \varepsilon_T(\mathbf{f}_T\varphi) \quad (32)$$

Now we have respectively $\varepsilon_T(g_S\varphi)$ and $\beta\varepsilon_T(g_S\varphi)$ at left and right of the inequality. Since $1-\beta > 0$, we have:

$$\varepsilon_T(g_S\varphi) \leq \frac{1}{1-\beta}\left(\varepsilon_S(g_S\varphi) + 6 \cdot \text{INV}(w,\varphi) + 2 \cdot \widehat{\text{TSF}}(w,\varphi,\tilde{g}) + \varepsilon_T(\mathbf{f}_T\varphi)\right) \quad (33)$$

And finally:

$$\varepsilon_T(\tilde{g}\varphi) \leq \beta\varepsilon_T(g_S\varphi) \leq \frac{\beta}{1-\beta}\left(\varepsilon_S(g_S\varphi) + 6 \cdot \text{INV}(w,\varphi) + 2 \cdot \widehat{\text{TSF}}(w,\varphi,\tilde{g}) + \varepsilon_T(\mathbf{f}_T\varphi)\right) \quad (34)$$

finishing the proof. □

A.6 MinEnt [15] is a Lower Bound of Transferability

Proof. We consider a label smooth classifier $g \in \mathcal{G}$ *i.e.* there is $0 < \alpha < 1$ such that:

$$\frac{\alpha}{C-1} \leq g(z) \leq 1-\alpha \quad (35)$$

and we note $Y = g\varphi(X)$. One can show that:

$$\log\left(\frac{\alpha}{C-1}\right) \leq \log(g(z)) \leq \log(1-\alpha) \quad (36)$$

and finally:

$$1 \geq \frac{1}{\log(\frac{\alpha}{C-1})}\log(g(z)) \geq \frac{1}{\log(\frac{\alpha}{C-1})}\log(1-\alpha) \geq 0 \quad (37)$$

We choose as particular $\mathbf{f}$, $\mathbf{f}(z) = -\eta\log(g(z))$ with $\eta = -\log(\frac{\alpha}{|\mathcal{Y}|-1})^{-1} > 0$. The coefficient η ensures that $\mathbf{f}(z) \in [0,1]$ to make sure $\mathbf{f} \in \mathcal{F}_C$. We have the following inequalities:

$$\begin{aligned}\widehat{\text{TSF}}(w,\varphi,g) &\geq \eta \cdot (\mathbb{E}_T[-g(Z)\cdot\log(g(Z))] - \mathbb{E}_{w\cdot S}[-Y\log(g(Z))]) \\ &\geq \eta \cdot \left(H_T(\hat{Y}|Z) - \text{CE}_{w\cdot S}(Y, g(Z))\right)\end{aligned}$$

Interestingly, the cross-entropy is involved. Then, when using $\text{CE}_{w\cdot S}(Y,g(Z))$ as a proxy of $\varepsilon_{w\cdot S}(g\varphi)$, we can observe the following lower bound:

$$\text{CE}_{w\cdot S}(Y,g(Z)) + \widehat{\text{TSF}}(w,\varphi,g) \geq (1-\eta)\cdot\text{CE}_{w\cdot S}(Y,g(Z)) + \eta\cdot H_T(\hat{Y}|Z) \quad (38)$$

which is a trade-off between minimizing the cross-entropy in the source domain while maintaining a low entropy in prediction in the target domain (Fig. 2).

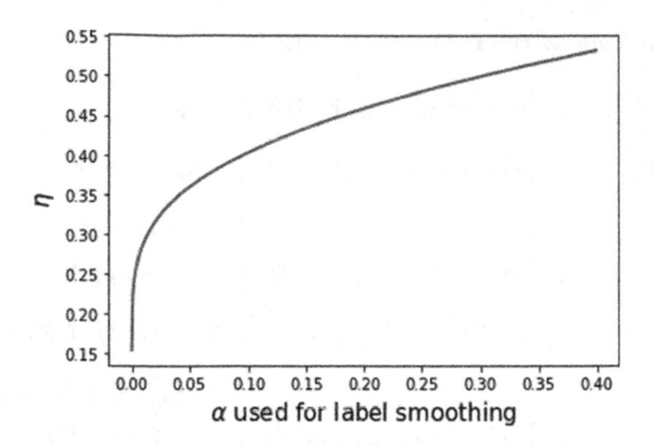

Fig. 2. We set $C = 31$ which is the number of classes in **Office31**. Label smoothing α leads naturally to a coefficient η which acts as a trade-off between cross-entropy minimization in the source domain and confidence in predictions in the target domain. This result follows a particular choice of the critic function in the transferability error introduced in this paper.

A.7 Proof of the Inductive Design of Weights

Proposition 12 (Inductive design of w and invariance). *Let $\psi : \mathcal{Z} \to \mathcal{Z}'$ such that $\mathcal{F} \circ \psi \subset \mathcal{F}$ and $\mathcal{F}_C \circ \psi \subset \mathcal{F}_C$. Let $w : \mathcal{Z}' \to \mathbb{R}^+$ such that $\mathbb{E}_S[w(Z')] = 1$ and we note $Z' := \psi(Z)$. Then,* $\mathrm{INV}(w, \varphi) = \mathrm{TSF}(w, \varphi) = 0$ *if and only if:*

$$w(z') = \frac{p_T(z')}{p_S(z')} \quad \text{and} \quad p_S(z|z') = p_T(z|z') \tag{39}$$

while both $\mathbf{f}_S^{\varphi} = \mathbf{f}_T^{\varphi}$ *and* $\mathbf{f}_S^{\psi} = \mathbf{f}_T^{\psi}$.

Proof. First,

$$\begin{aligned}
\mathrm{INV}(w, \varphi) &= \sup_{f \in \mathcal{F}} \mathbb{E}_S[w(Z')f(Z)] - \mathbb{E}_T[f(z)] && (40)\\
&\geq \sup_{f \in \mathcal{F}} \mathbb{E}_S[w(Z')f \circ \psi(Z)] - \mathbb{E}_T[f \circ \psi(z)] && (\mathcal{F} \circ \psi \subset \mathcal{F})\\
&= \sup_{f \in \mathcal{F}} \mathbb{E}_S[w(Z')f(Z')] - \mathbb{E}_T[f(z')] = 0 && (Z' = \psi(Z))
\end{aligned}$$

which leads to $w(z')p_S(z') = p_S(z')$ which is $w(z') = p_T(z')/p_S(z')$. Second, $\mathrm{INV}(w, \varphi) = 0$ also implies that $w(z')p_S(z) = p_T(z)$:

$$w(z') = \frac{p_T(z)}{p_S(z)} = \frac{p_T(z|z')}{p_S(z|z')}\frac{p_T(z')}{p_S(z')} = \frac{p_T(z|z')}{p_S(z|z')}w(z') \tag{41}$$

then $p_T(z|z') = p_S(z|z')$. Finally,

$$\begin{aligned}
\mathrm{TSF}(w,\varphi) &= \sup_{\mathbf{f}\in\mathcal{F}_C} \mathbb{E}_S[w(Z')Y\cdot\mathbf{f}(Z)] - \mathbb{E}_T[Y\cdot\mathbf{f}(Z)] && (42)\\
&= \sup_{\mathbf{f}\in\mathcal{F}_C} \mathbb{E}_{Z'\sim p_S}\left[w(Z')\mathbb{E}_{Z|Z'\sim p_S}[Y\cdot\mathbf{f}(Z)]\right] - \mathbb{E}_{Z'\sim p_T}\left[\mathbb{E}_{Z|Z'\sim p_T}[Y\cdot\mathbf{f}(Z)]\right] && (43)\\
&= \sup_{\mathbf{f}\in\mathcal{F}_C} \mathbb{E}_{Z'\sim p_S}\left[w(Z')\mathbb{E}_{Z|Z'\sim p_S}[Y\cdot\mathbf{f}(Z)]\right] - \mathbb{E}_{Z'\sim p_T}w(Z')\left[\mathbb{E}_{Z|Z'\sim p_S}[Y\cdot\mathbf{f}(Z)]\right]\\
& \qquad\qquad (w(z')p_S(z') = p_T(z'))\\
&= \sup_{\mathbf{f}\in\mathcal{F}_C} \mathbb{E}_{Z'\sim p_S}\left[w(Z')\left(\mathbb{E}_{Z|Z'\sim p_S}[Y\cdot\mathbf{f}(Z)] - \mathbb{E}_{Z|Z'\sim p_T}[Y\cdot\mathbf{f}(Z)]\right)\right] && (44)\\
&= \sup_{\mathbf{f}\in\mathcal{F}_C} \mathbb{E}_{Z'\sim p_S}\left[w(Z')\left(\mathbb{E}_{Z|Z'\sim p_S}[\mathbf{f}_S(Z)\cdot\mathbf{f}(Z) - \mathbf{f}_T(Z)\cdot\mathbf{f}(Z)]\right)\right]\\
& \qquad\qquad (p_S(z|z') = p_T(z|z'))\\
&= \sup_{\mathbf{f}\in\mathcal{F}_C} \mathbb{E}_{Z'\sim p_S}\left[w(Z')\left(\mathbb{E}_{Z|Z'\sim p_S}[\mathbf{f}_S(Z)\cdot\mathbf{f}(Z) - \mathbf{f}_T(Z)\cdot\mathbf{f}(Z)]\right)\right]\\
& \qquad\qquad (p_S(z|z') = p_T(z|z'))\\
&\geq 2\mathbb{E}_{Z'\sim p_S}\left[w(Z')\left(\mathbb{E}_{Z|Z'\sim p_S}[||\mathbf{f}_S(Z) - \mathbf{f}_T(Z)||^2]\right)\right] && (45)\\
&\geq 2\mathbb{E}_{Z'\sim p_T}\left[\left(\mathbb{E}_{Z|Z'\sim p_T}[||\mathbf{f}_S(Z) - \mathbf{f}_T(Z)||^2]\right)\right] && (46)\\
&\geq 2\mathbb{E}_{Z'\sim p_T}\left[\left(\mathbb{E}_{Z|Z'\sim p_T}[||\mathbf{f}_S(Z) - \mathbf{f}_T(Z)||^2]\right)\right] && (47)\\
&\geq 2\mathbb{E}_{Z\sim p_T}\left[||\mathbf{f}_S(Z) - \mathbf{f}_T(Z)||^2\right] && (48)
\end{aligned}$$

Which leads to $\mathbf{f}_S(z) = \mathbf{f}_T(z)$, $p_T(z)$ almost surely, then $\mathbb{E}_T[Y|Z] = \mathbb{E}_S[Y|Z]$ for $Z \sim p_T$. Now we finish by observing that:

$$\begin{aligned}
\mathrm{TSF}(w,\varphi) &= \sup_{\mathbf{f}\in\mathcal{F}_C} \mathbb{E}_S[w(Z')Y\cdot\mathbf{f}(Z)] - \mathbb{E}_T[Y\cdot\mathbf{f}(Z)] && (49)\\
&\geq \sup_{\mathbf{f}\in\mathcal{F}_C} \mathbb{E}_S[w(Z')Y\cdot\mathbf{f}\circ\psi(Z)] - \mathbb{E}_T[Y\cdot\mathbf{f}\circ\psi(Z)] && (50)\\
&\geq \sup_{\mathbf{f}\in\mathcal{F}_C} \mathbb{E}_S[w(Z')Y\cdot\mathbf{f}(Z')] - \mathbb{E}_T[Y\cdot\mathbf{f}(Z')] && (51)
\end{aligned}$$

which leads to $\mathbb{E}_S[Y|Z'] = \mathbb{E}_T[Y|Z']$ for $Z' \sim p_T$. The converse is trivial. □

B CDAN, DANN and TSF: An Open Discussion

In CDAN [24], authors claims to align conditional $Z|\hat{Y}$, by exposing the multilinear mapping of $\hat{Y}$ by Z, hence its name of Conditional Domain Adversarial Network. Here, we show this claim can be theoretically misleading:

Proposition 13. *If $\mathbb{E}[\hat{Y}|Z]$ is conserved across domains, i.e. g is conserved, and $\mathcal{D}$ and $\mathcal{D}_\otimes$ are infinite capacity set of discriminators, this holds:*

$$\mathrm{DANN}(\varphi) = \mathrm{CDAN}(\varphi) \tag{52}$$

Proof. First, let $d_\otimes \in \mathcal{D}_\otimes$. Then, for any $(\hat{y}, z) \sim p_S$ (similarly $\sim p_T$), $d(\hat{y}\otimes z) = d(g(z)\otimes z)$ since $\hat{y} = g(z) = \mathbb{E}[\hat{Y}|Z = z]$ is conserved across domains. Then $\tilde{d} : z \mapsto d_\otimes(g(z)\otimes z)$ is a mapping from $\mathcal{Z}$ to $[0, 1]$. Since $\mathcal{D}$ is the set of infinite capacity discriminators, $\tilde{d} \in \mathcal{D}$. This shows $\mathrm{CDAN}(\varphi) \leq \mathrm{DANN}(\varphi)$. Now we introduce $T : \mathcal{Y}\otimes\mathcal{Z} \to \mathcal{Z}$ such that $T(y\otimes z) = \sum_{1\leq c\leq|\mathcal{Y}|} y_c(y\otimes z)_{cr:(c+1)r} = z$

where $r = \dim(Z)$. The ability to reconstruct z from $\hat{y}\otimes z$ results from $\sum_c y_c = 1$. This shows that $\mathcal{D}_{\otimes} \circ T = \mathcal{D}$ and finally $\text{CDAN}(\varphi) \geq \text{DANN}(\varphi)$ finishing the proof.

This proposition follows two key assumptions. The first is to assume that we are in context of infinite capacity discriminators of both $\mathcal{Z}$ and $\mathcal{Y} \otimes \mathcal{Z}$. This assumption seems reasonable in practice since discriminators are multi-layer perceptrons. The second is to assume that $\mathbb{E}[\hat{Y}|Z]$ is conserved across domains. Pragmatically, the same classifier is used in both source and target domains which is verified in practice. Despite the empirical success of CDAN, there is no theoretical evidence of the superiority of CDAN with respect to DANN for UDA. However, our discussion on the role of inductive design of classifiers is an attempt to explain the empirical superiority of such strategies.

C More Training Details

C.1 From IPM to Domain Adversarial Objective

While our analysis holds for IPM, we recall the connections with $\mathsf{f}-$divergence, where domain adversarial loss is a particular instance, for comparing distributions. This connection is motivated by the furnished literature on adversarial learning, based on domain discriminator, for UDA. This section is then an informal attempt to transport our theoretical analysis, which holds for IPM, to $\mathsf{f}-$divergence. Given f a function defined on $\mathbb{R}^+$, continuous and convex, the $\mathsf{f}-$divergence between two distributions p and q: $\mathbb{E}_p[\mathsf{f}(p/q)]$, is null if and only if $p = q$. Interestingly, $\mathsf{f}-$divergence admits a 'IPM style' expression $\mathbb{E}_p[\mathsf{f}(p/q)] = \sup_f \mathbb{E}_p[f] - \mathbb{E}_q[\mathsf{f}^\star(f)]$ where $\mathsf{f}^\star$ is the convex conjugate of f. It is worth noting it is not a IPM expression since the critic is composed by $\mathsf{f}^\star$ in the right expectation. The domain adversarial loss [13] is a particular instance of $\mathsf{f}-$divergence (see [7] for a complete description in the context of generative modelling). Then, we informally transports our analysis on IPM distance to domain adversarial loss. More precisely, we define:

$$\text{INV}_{\text{adv}}(w, \varphi) := \log(2) - \sup_{d \in \mathcal{D}} \mathbb{E}_S[w(Z)\log(d(Z))] + \mathbb{E}_T[\log(1 - d(Z))] \tag{53}$$

$$\text{TSF}_{\text{adv}}(w, \varphi) := \log(2) - \sup_{\mathbf{d} \in \mathcal{D}_{\mathcal{Y}}} \mathbb{E}_S[w(Z)Y \cdot \log(\mathbf{d}(Z))] + \mathbb{E}_T[Y \cdot \log(1 - \mathbf{d}(Z))] \tag{54}$$

where $\mathcal{D}$ is the well-established domain discriminator from $\mathcal{Z}$ to $[0, 1]$, and $\mathcal{D}_{\mathcal{Y}}$ is the set of *label domain discriminator* from $\mathcal{Z}$ to $[0, 1]^C$.

C.2 Controlling Invariance Error with Relaxed Weights

In this section, we show that even if representations are not learned in order to achieve domain invariance, the design of weights allows to control the invariance error during learning. More precisely $w^\star(\varphi) = \arg\min_w \text{INV}(w, \varphi)$ has a closed form when given a domain discriminator d *i.e.* the following function from the representation space $\mathcal{Z}$ to $[0, 1]$:

$$d(z) := \frac{p_S(z)}{p_S(z) + p_T(z)} \tag{55}$$

Here, setting $w^\star(z) := (1 - d(z))/d(z) = p_T(z)/p_S(z)$ leads to $w(z)p_S(z) = p_T(z)$ and finally $\text{INV}(w^\star(\varphi), \varphi) = 0$. At early stage of learning, the domain discriminator d has a weak predictive power to discriminate domains. Using exactly the closed form $w^\star(z)$ may degrade the estimation of the transferability error. Then, we suggest to build relaxed weights $\tilde{w}_d$ which are pushed to $w^\star$ during training. This is done using temperature relaxation in the sigmoid output of the domain discriminator:

$$w_d^\tau(z) := \frac{1 - \sigma\left(\tilde{d}(z)/\tau\right)}{\sigma\left(\tilde{d}(z)/\tau\right)} \tag{56}$$

where $d(z) = \sigma(\tilde{d}(z))$; when $\tau \to 1$, $w_d(z, \tau) \to w^\star(z)$.

C.3 Ablation Study of the Weight Relaxation Parameter α

α is the rate of convergence of relaxed weights to optimal weights. We investigate its role on the task U→M. Increasing α degrades adaptation, excepts in the harder case ($5\% \times [0 \sim 5]$). Weighting early during training degrades representations alignment. Conversely, in the case $5\% \times [0 \sim 5]$, weights need to be introduced early to not learn a wrong alignment. In practice $\alpha = 5$ works well (except for $5\% \times [0 \sim 5]$ in **Digits**) (Fig. 3).

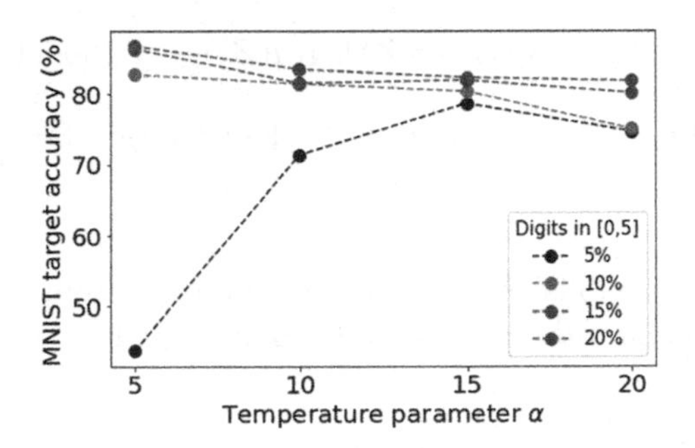

Fig. 3. Effect of α.

C.4 Additional Results on Office-Home Dataset

See Table 3.

Table 3. Accuracy (%) on **Office-Home** based on ReseNet-50.

Method	Ar→Cl	Ar→Pr	Ar→Rw	Cl→Ar	Cl→Pr	Cl→Rw	Pr→Ar	Pr→Cl	Pr→Rw	Rw→Ar	Rw→Cl	Rw→Pr	Avg
ResNet50	34.9	50.0	58.0	37.4	41.9	46.2	38.5	31.2	60.4	53.9	41.2	59.9	46.1
DANN	45.6	59.3	70.1	47.0	58.5	60.9	46.1	43.7	68.5	63.2	51.8	76.8	57.6
CDAN	49.0	69.3	74.5	54.4	66.0	68.4	55.6	48.3	75.9	68.4	55.4	80.5	63.8
CDAN+E	50.7	**70.6**	**76.0**	**57.6**	**70.0**	**70.0**	**57.4**	**50.9**	**77.3**	**70.9**	56.7	**81.6**	**65.8**
RUDA	**52.0**	67.1	74.4	56.8	69.5	69.8	57.3	**50.9**	77.2	70.5	**57.1**	81.2	64.9

C.5 Detailed Procedure

The code is available at https://github.com/vbouvier/ruda.

Algorithm 1. Procedure for Robust Unsupervised Domain Adaptation

Input: Source samples $(x_{S,i}, y_{S,i})_i$, Target samples $(x_{T,i}, y_{T,i})_i$, $(\tau_t)_t$ such that $\tau_t \rightarrow 1$, learning rates $(\eta_t)_t$, trade-off $(\alpha_t)_t$ such that $\alpha_t \rightarrow 1$, batch-size b

1: $\theta_g, \theta_\varphi, \theta_d, \theta_{\mathbf{d}}$ random initialization.
2: $t \leftarrow 0$
3: **while** stopping criterion **do**
4: $\quad \mathcal{B}_S \sim (x_i^s)$, $\mathcal{B}_T \sim (x_j^t)$ of size b.
5: $\quad \theta_d \leftarrow \theta_d - \eta_t \nabla_{\theta_d} \mathcal{L}_{\text{INV}}(\theta_d | \theta_\varphi; \mathcal{B}_S, \mathcal{B}_T)$
6: $\quad \theta_{\mathbf{d}} \leftarrow \theta_{\mathbf{d}} - \eta_t \nabla_{\theta_{\mathbf{d}}} \mathcal{L}_{\text{TSF}}(\theta_g, \theta_\varphi, \theta_{\mathbf{d}} | \theta_d, \tau_t)$
7: $\quad \theta_\varphi \leftarrow \theta_\varphi - \eta_t \nabla_{\theta_\varphi} (\mathcal{L}_c(\theta_g, \theta_\varphi | \theta_d, \tau_t) - \alpha_t \mathcal{L}_{\text{TSF}}(\theta_\varphi, \theta_{\mathbf{d}} | \theta_g, \theta_d, \tau_t))$
8: $\quad \theta_g \leftarrow \theta_g - \eta_t \nabla_{\theta_g} \mathcal{L}_c(\theta_g, \theta_\varphi | \theta_d, \tau_t)$
9: $\quad t \leftarrow t + 1$
10: **end while**

References

1. Amodei, D., Olah, C., Steinhardt, J., Christiano, P., Schulman, J., Mané, D.: Concrete problems in AI safety. arXiv preprint arXiv:1606.06565 (2016)
2. Arjovsky, M., Bottou, L., Gulrajani, I., Lopez-Paz, D.: Invariant risk minimization. arXiv preprint arXiv:1907.02893 (2019)
3. Beery, S., Van Horn, G., Perona, P.: Recognition in terra incognita. In: Proceedings of the European Conference on Computer Vision (ECCV), pp. 456–473 (2018)
4. Ben-David, S., Blitzer, J., Crammer, K., Kulesza, A., Pereira, F., Vaughan, J.W.: A theory of learning from different domains. Mach. Learn. **79**(1–2), 151–175 (2010)
5. Ben-David, S., Blitzer, J., Crammer, K., Pereira, F.: Analysis of representations for domain adaptation. In: Advances in Neural Information Processing Systems, pp. 137–144 (2007)

6. Bhushan Damodaran, B., Kellenberger, B., Flamary, R., Tuia, D., Courty, N.: DeepJDOT: deep joint distribution optimal transport for unsupervised domain adaptation. In: Proceedings of the European Conference on Computer Vision (ECCV), pp. 447–463 (2018)
7. Bottou, L., Arjovsky, M., Lopez-Paz, D., Oquab, M.: Geometrical insights for implicit generative modeling. In: Rozonoer, L., Mirkin, B., Muchnik, I. (eds.) Braverman Readings in Machine Learning. Key Ideas from Inception to Current State. LNCS (LNAI), vol. 11100, pp. 229–268. Springer, Cham (2018). https://doi.org/10.1007/978-3-319-99492-5_11
8. Cao, Y., Long, M., Wang, J.: Unsupervised domain adaptation with distribution matching machines. In: Thirty-Second AAAI Conference on Artificial Intelligence (2018)
9. Cao, Z., Ma, L., Long, M., Wang, J.: Partial adversarial domain adaptation. In: Proceedings of the European Conference on Computer Vision (ECCV), pp. 135–150 (2018)
10. Combes, R.T.D., Zhao, H., Wang, Y.X., Gordon, G.: Domain adaptation with conditional distribution matching and generalized label shift. arXiv preprint arXiv:2003.04475 (2020)
11. Cortes, C., Mansour, Y., Mohri, M.: Learning bounds for importance weighting. In: Advances in Neural Information Processing Systems, pp. 442–450 (2010)
12. D'Amour, A., Ding, P., Feller, A., Lei, L., Sekhon, J.: Overlap in observational studies with high-dimensional covariates. arXiv preprint arXiv:1711.02582 (2017)
13. Ganin, Y., Lempitsky, V.: Unsupervised domain adaptation by backpropagation. In: International Conference on Machine Learning, pp. 1180–1189 (2015)
14. Geva, M., Goldberg, Y., Berant, J.: Are we modeling the task or the annotator? An investigation of annotator bias in natural language understanding datasets. In: Proceedings of the 2019 Conference on Empirical Methods in Natural Language Processing and the 9th International Joint Conference on Natural Language Processing (EMNLP-IJCNLP), pp. 1161–1166 (2019)
15. Grandvalet, Y., Bengio, Y.: Semi-supervised learning by entropy minimization. In: Advances in Neural Information Processing Systems, pp. 529–536 (2005)
16. Gretton, A., Borgwardt, K.M., Rasch, M.J., Schölkopf, B., Smola, A.: A kernel two-sample test. J. Mach. Learn. Res. **13**(Mar), 723–773 (2012)
17. Gretton, A., Smola, A., Huang, J., Schmittfull, M., Borgwardt, K., Schölkopf, B.: Covariate shift by kernel mean matching. Dataset Shift Mach. Learn. **3**(4), 5 (2009)
18. He, K., Zhang, X., Ren, S., Sun, J.: Deep residual learning for image recognition. In: Proceedings of the IEEE Conference on Computer Vision and Pattern Recognition, pp. 770–778 (2016)
19. Huang, J., Gretton, A., Borgwardt, K., Schölkopf, B., Smola, A.J.: Correcting sample selection bias by unlabeled data. In: Advances in Neural Information Processing Systems, pp. 601–608 (2007)
20. Johansson, F., Sontag, D., Ranganath, R.: Support and invertibility in domain-invariant representations. In: The 22nd International Conference on Artificial Intelligence and Statistics, pp. 527–536 (2019)
21. Lipton, Z., Wang, Y.X., Smola, A.: Detecting and correcting for label shift with black box predictors. In: International Conference on Machine Learning, pp. 3122–3130 (2018)
22. Liu, H., Long, M., Wang, J., Jordan, M.: Transferable adversarial training: a general approach to adapting deep classifiers. In: International Conference on Machine Learning, pp. 4013–4022 (2019)

23. Long, M., Cao, Y., Wang, J., Jordan, M.I.: Learning transferable features with deep adaptation networks. In: Proceedings of the 32nd International Conference on International Conference on Machine Learning-vol. 37, pp. 97–105. JMLR.org (2015)
24. Long, M., Cao, Z., Wang, J., Jordan, M.I.: Conditional adversarial domain adaptation. In: Advances in Neural Information Processing Systems, pp. 1640–1650 (2018)
25. Long, M., Zhu, H., Wang, J., Jordan, M.I.: Deep transfer learning with joint adaptation networks. In: Proceedings of the 34th International Conference on Machine Learning-vol. 70, pp. 2208–2217. JMLR.org (2017)
26. Maaten, L.V.D., Hinton, G.: Visualizing data using T-SNE. J. Mach. Learn. Res. **9**(Nov), 2579–2605 (2008)
27. Mansour, Y., Mohri, M., Rostamizadeh, A.: Domain adaptation: learning bounds and algorithms. In: 22nd Conference on Learning Theory, COLT 2009 (2009)
28. Pan, S.J., Yang, Q.: A survey on transfer learning. IEEE Trans. Knowl. Data Eng. **22**(10), 1345–1359 (2009)
29. Paszke, A., et al.: Pytorch: an imperative style, high-performance deep learning library. In: Advances in Neural Information Processing Systems, pp. 8024–8035 (2019)
30. Quionero-Candela, J., Sugiyama, M., Schwaighofer, A., Lawrence, N.D.: Dataset Shift in Machine Learning. MIT Press, Cambridge (2009)
31. Redko, I., Courty, N., Flamary, R., Tuia, D.: Optimal transport for multi-source domain adaptation under target shift. arXiv preprint arXiv:1803.04899 (2018)
32. Shen, J., Qu, Y., Zhang, W., Yu, Y.: Wasserstein distance guided representation learning for domain adaptation. In: Thirty-Second AAAI Conference on Artificial Intelligence (2018)
33. Shimodaira, H.: Improving predictive inference under covariate shift by weighting the log-likelihood function. J. Stat. Plann. Inference **90**(2), 227–244 (2000)
34. Storkey, A.: When training and test sets are different: characterizing learning transfer. Dataset Shift Mach. Learn. **30**, 3–28 (2009)
35. Sugiyama, M., Krauledat, M., MÃžller, K.R.: Covariate shift adaptation by importance weighted cross validation. J. Mach. Learn. Res. **8**(May), 985–1005 (2007)
36. Wu, Y., Winston, E., Kaushik, D., Lipton, Z.: Domain adaptation with asymmetrically-relaxed distribution alignment. In: International Conference on Machine Learning, pp. 6872–6881 (2019)
37. You, K., Long, M., Cao, Z., Wang, J., Jordan, M.I.: Universal domain adaptation. In: Proceedings of the IEEE Conference on Computer Vision and Pattern Recognition, pp. 2720–2729 (2019)
38. Zhang, J., Ding, Z., Li, W., Ogunbona, P.: Importance weighted adversarial nets for partial domain adaptation. In: Proceedings of the IEEE Conference on Computer Vision and Pattern Recognition, pp. 8156–8164 (2018)
39. Zhang, K., Schölkopf, B., Muandet, K., Wang, Z.: Domain adaptation under target and conditional shift. In: International Conference on Machine Learning, pp. 819–827 (2013)
40. Zhao, H., Des Combes, R.T., Zhang, K., Gordon, G.: On learning invariant representations for domain adaptation. In: International Conference on Machine Learning, pp. 7523–7532 (2019)

Target to Source Coordinate-Wise Adaptation of Pre-trained Models

Luxin Zhang[1,2(✉)], Pascal Germain[2,3], Yacine Kessaci[1], and Christophe Biernacki[2]

[1] Worldline, Bezons, France
{luxin.zhang,yacine.kessaci}@worldline.com
[2] MODAL Team, Inria, Lille, France
{luxin.zhang,germain.pascal,christophe.biernacki}@inria.fr
[3] Université Laval, Quebec City, QC, Canada
pascal.germain@ift.ulaval.ca

Abstract. Domain adaptation aims to alleviate the gap between source and target data drawn from different distributions. Most of the related works seek either for a latent space where source and target data share the same distribution, or for a transformation of the source distribution to match the target one. In this paper, we introduce an original scenario where the former trained source model is directly reused on target data, requiring only finding a transformation from the target domain to the source domain. As a first approach to tackle this problem, we propose a greedy coordinate-wise transformation leveraging on optimal transport. Beyond being fully independent of the model initially learned on the source data, the achieved transformation has the following three assets: scalability, interpretability and feature-type free (continuous and/or categorical). Our procedure is numerically evaluated on various real datasets, including domain adaptation benchmarks and also a challenging fraud detection dataset with very imbalanced classes. Interestingly, we observe that transforming a small subset of the target features leads to accuracies competitive with "classical" domain adaptation methods.

Keywords: Domain adaptation · Optimal transport · Feature selection

1 Introduction

Traditional supervised machine learning algorithms assume the estimated model to be used on the data having the same underlying distribution as the training one. However, this assumption is not always valid. For example, a predictive model may have been trained on a dataset of users living in a specific country, and be used afterward on a dataset of users living in another geographical region.

Electronic supplementary material The online version of this chapter (https://doi.org/10.1007/978-3-030-67658-2_22) contains supplementary material, which is available to authorized users.

F. Hutter et al. (Eds.): ECML PKDD 2020, LNAI 12457, pp. 378–394, 2021.
https://doi.org/10.1007/978-3-030-67658-2_22

A common domain adaptation strategy used to mitigate these differences is to align the training (source) data and the test (target) data. However, most of these methods rely on training a predictor on the transformed source dataset.

Related Works. To mitigate the gap between source and target domains, former "classical" domain adaptation methods seek to minimize a discrepancy term between the source and target distributions. The minimization of Kullback-Leibler Divergence [14] and Maximum Mean Discrepancy [11] are among the most widely used techniques. The former re-weights source samples in the target domain, while the latter aligns source and target distributions in a latent space. Besides, a recently proposed correlation alignment method [15] aims to align the source domain and target domain second-order statistics. Nevertheless, one still needs to re-estimate a model on the transformed source data.

Deep learning approaches have also been proposed to tackle the domain adaptation problem. Methods like Deep Adaptation Networks [9] and Domain Adversarial Neural Networks [6] have achieved state-of-the-art results in computer vision tasks. However, it is known that these methods are not interpretable and requires careful tuning of hyperparameters.

Finally, prior works that leverage on optimal transport [3,4] focus mainly on the source to target transformation and are not scalable to a huge dataset. Hence, to the best of our knowledge, we are the first to tackle the target to source adaptation problem that is scalable and requires no retraining.

Contributions. In this paper, we propose a new domain adaptation perspective by transforming the target data into the source one, and applying directly the pre-trained model on the adapted data. We argue that this strategy is more suited to some real-life scenarios. Indeed, in an industrial context, a prediction system might have been developed and trained by employees or contractors that are no more available. This system being used as a "black box" predictor, it cannot be retrained on a new dataset. However, it is often mandatory to adapt such a system to different contexts, given that the industry wants to expand their business, or just because the data distribution naturally "drift" as time goes.

Therefore, we introduce an original scenario where the former trained model is directly reused on target data, requiring only finding a transformation from the target domain to the source domain. As a first approach to tackle this problem, we propose a greedy coordinate-wise transformation leveraging on optimal transport. Beyond being fully independent of the model initially learned on the source data, we design our approach with the following three characteristics in mind, as we want our method to be appealing for users in an industrial context.

1. Scalability: We want our transformation to be computable even on very large datasets.
2. Interpretability: We want to provide the user some information that can help him to interpret the nature of the target to source transformation.

3. Feature-type free: We want our transformation to apply to a variety of data types. In this paper, we address specifically the case where the data is a mix of numerical and categorical attributes.

Our domain adaptation method is applicable without any label, but we show that one can use few target labels to select the features to adapt. We show empirically that this feature selection scheme leads to competitive results with state-of-the-art domain adaptation methods that necessitate a training phase. Our experiments are performed on three real-life datasets.

2 Domain Adaptation and Optimal Transport

2.1 The Studied Domain Adaptation Framework

Let denote $(x, y) \in \mathcal{X} \times \mathcal{Y}$ a data point, where $\mathcal{X}$ and $\mathcal{Y}$ refer respectively to the input space and the output space. In this paper, $\mathcal{X}$ encompasses vectors of categorical and numerical types, possibly mixed together. Moreover, we focus on classification problems, with binary labels $\mathcal{Y} = \{0, 1\}$, but our method naturally extends to multilabel classification. In the supervised learning framework, one observes a training dataset $\{(x_i, y_i)\}_{i=1}^{m}$, and each observed training sample (x_i, y_i) is viewed as a realization of a random variable pair (X, Y) obeying a joint probability $P(X, Y)$. Thus, to learn a classifier, a supervised learning algorithm aims to model $P(Y|X)$ from the training dataset.

The domain adaptation framework differs from the standard supervised learning one by the fact that two distinct joint probabilities over the input-output space $\mathcal{X} \times \mathcal{Y}$ are considered, referred to as the *source domain* probability $P(X^s, Y^s)$ and the *target domain* probability $P(X^t, Y^t)$. The domain adaptation classification problem is to infer the target predictive model $P(Y^t|X^t)$ in the situation where most observed learning data are instances of (X^s, Y^s). More precisely, we stand hereafter in the semi-supervised domain adaptation setting where no target label is provided to the learner. We denote the source training points as $\mathbb{X}^s = \{x_1^s, \ldots, x_i^s, \ldots, x_m^s\} \in \mathcal{X}^m$, $\mathbb{Y}^s = \{y_1^s, \ldots, y_i^s, \ldots, y_m^s\} \in \mathcal{Y}^m$, and the target training points as $\mathbb{X}^t = \{x_1^t, \ldots, x_j^t, \ldots, x_n^t\} \in \mathcal{X}^n$. Of course, the challenging task of learning $P(Y^t|X^t)$ without target labels[1] is only achievable under the assumption that, despite $P(X^s, Y^s) \neq P(X^t, Y^t)$, both joint probabilities are "similar" to each other. Any rigorous domain adaptation study must characterize the underlying source-target similarity assumptions. In the domain literature, one common assumption is the so-called *covariate shift* setting [13], which describes the case where the source domain and target domain output conditional distributions coincide, that is $P(Y^s|X^s) = P(Y^t|X^t)$ whereas the input marginal distributions $P(X^s)$ and $P(X^t)$ differ. Alternatively, Courty et al. [4] assume that there exists a *mapping function* $\mathcal{T} : \mathcal{X} \to \mathcal{X}$ that models the domain drift from the source to the target, such that $P(Y^s|\mathcal{T}(X^s)) = P(Y^t|X^t)$. Our

[1] In the experiment section, we consider few target labels in order to perform model selection.

work builds on a sibling assumption, that is the existence of a mapping function $\mathcal{G} : \mathcal{X} \to \mathcal{X}$ that models the *domain drift* from the target to the source:

$$P(Y^s|X^s) = P(Y^t|\mathcal{G}(X^t)).$$

One could also look for a bijective mapping function such that $\mathcal{G}^{-1} = \mathcal{T}$, but this is not required by our analysis. Inspired by Courty et al. [4], we choose to leverage on *optimal transport* methods to empirically estimate $\mathcal{G}$ from the training dataset. Nonetheless, the source to target method of Courty et al. [4] requires computing the transportation map from $\mathbb{X}^s$ to $\mathbb{X}^t$, and training a learning model on the training dataset $\{(\mathcal{T}(x_i^s), y_i^s)\}_{i=1}^m$ afterward, while our method is based on the idea of using a pre-trained source model.

2.2 Optimal Transport for Domain Adaptation

The optimal transport problem was first introduced by Monge in the 18th century [10] and further developed by Kantorovich in the mid-20th [8]. Intuitively, the original Monge-Kantorovich problem looks for minimal effort to move masses of dirt to fill a given collection of pits. Optimal transport has been revisited over the past years to solve a variety of computational problems [12], including many machine learning ones [5]. It is naturally suited for domain adaptation problems [4], and it offers a principled method to transform a source distribution into a target one.

Let us now present the important optimal transport notions we rely upon. Even if we tailor the nomenclature to the context of our domain adaptation problem (*e.g.* we invoke "target" and "source" densities), the remaining part of this section is shared by the general optimal transport literature.

The training sets $\mathbb{X}^s$ and $\mathbb{X}^t$ provide discrete estimations of the input domain densities $P(X^s)$ and $P(X^t)$. Classically, we consider the empirical distributions as additions of Dirac functions. Denoting by δ_x the Dirac measure on $x \in \mathcal{X}$, we define the empirical estimation of the source domain distribution and the target domain distribution as

$$\hat{P}(X^s) = \sum_{i=1}^m w_i^s \delta_{x_i^s}\,, \text{ and } \hat{P}(X^t) = \sum_{j=1}^n w_j^t \delta_{x_j^t}\,,$$

where $\mathbb{W}^s = \{w_1^s, ..., w_i^s, ..., w_m^s\}$ and $\mathbb{W}^t = \{w_1^t, ..., w_j^t, ..., w_n^t\}$ are weights over the training points. Typically, we consider that the mass is uniformly distributed among each point, *i.e.* $w_i^s = \frac{1}{m}$ and $w_j^t = \frac{1}{n}$, but the framework allows reweighing the samples, such that

$$\sum_{i=1}^m w_i^s = \sum_{j=1}^n w_j^t = 1\,; \quad w_i^s, w_j^t \geq 0\,.$$

For simplicity, we assume that every x_i^s appears only once in $\mathbb{X}^s$, respectively x_j^t in $\mathbb{X}^t$. We then write $\hat{P}(X^s = x_i^s) = w_i^s$ and $\hat{P}(X^t = x_j^t) = w_j^t$.

Central to optimal transport methods is the notion of a *cost function* between a source point and a target point, denoted by

$$c : \mathcal{X} \times \mathcal{X} \to \mathbb{R} \,. \tag{1}$$

Moreover, $C \in R^{m \times n}$ denotes the *cost matrix* between source and target training points such that $C_{i,j} = c(x_i^s, x_j^t)$ corresponds to the cost of moving weight from $x_j^t \in \mathbb{X}^t$ to $x_i^s \in \mathbb{X}^s$. Based on these concepts, we present below the Kantorovich [8] formulation of the optimal transport problem in the discrete case.

Definition 1 (Kantorovich's discrete optimal transport problem). *The relationship between source and target examples is encoded as a joint probability* coupling matrix $\gamma \in \mathbb{R}_+^{m \times n}$, *where* $\gamma_{i,j}$ *corresponds to the* weight *to be* moved *from* $x_j^t \in \mathbb{X}^t$ *to* $x_i^s \in \mathbb{X}^s$. *The set of admissible coupling matrices is given by*

$$\varGamma = \left\{ \gamma \in \mathbb{R}_+^{m \times n} \;\middle|\; w_{i'}^s = \sum_{j=1}^{n} \gamma_{i',j} \;\; and \;\; w_{j'}^t = \sum_{i=1}^{m} \gamma_{i,j'} \right\}.$$

Then, the optimal coupling matrix γ^* *is obtained by solving*

$$\gamma^* = \operatorname*{argmin}_{\gamma \in \varGamma} \langle C, \gamma \rangle = \operatorname*{argmin}_{\gamma \in \varGamma} \sum_{i=1}^{m} \sum_{j=1}^{n} C_{i,j} \gamma_{i,j} \,. \tag{2}$$

In turns the transformation function $\mathcal{G}$ *is given by*

$$\mathcal{G}(x_j^t) = \operatorname*{argmin}_{x \in \mathcal{X}} \sum_{i=1}^{m} \gamma_{i,j}^* c(x, x_j^t) \,. \tag{3}$$

The solution $x \in \mathcal{X}$ *of Eq.* (3) *minimization problem is commonly referred to as the* barycenter *in the optimal transport literature.*

Equation (2) is a linear optimization problem, and algorithms such as the network simplex or dual ascent methods can be used to compute the solution [12]. The obtained joint probability γ^* reveals the allocation of mass from one domain to the other.

However, the computational complexity of this linear optimization problem is expensive; In the case where the number of target or source training points are equal ($m = n$), the computation time is $O(n^3)$. By adding a regularization term $H(\gamma) = \sum_{i,j} \gamma_{i,j} \log(\gamma_{i,j})$ to Eq. (2), Cuturi [5] achieve to reduce the computation complexity. The regularized optimization problem is expressed as

$$\gamma_\eta^* = \operatorname*{argmin}_{\gamma \in \varGamma} \langle C, \gamma \rangle + \eta H(\gamma) \,, \tag{4}$$

where $\eta > 0$ is a hyperparameter to be fixed. Altschuler et al. [1] show that, using the Sinkhorn iteration method [5], solving the regularized optimal transport of Eq. (4) requires about $O(n^2 \log(n))$ operations. This is still too expensive to apply to large learning problems.[2] Also, as mentioned before, such domain adaptation methods imply training a new model from the transformed source dataset.

To overcome these limitations of existing domain adaptation methods, we propose in the following section a coordinate-wise target to source domain adaptation method.

3 Target to Source Coordinate-Wise Domain Adaptation

In this section, we formalize our target to source domain adaptation problem and propose our adaptation method.

3.1 Formalization of Target to Source Transformation

The originality of our domain adaptation approach is to rely solely on a target to source transformation. That is, we consider that we have access to a source training dataset $\mathbb{X}^s$, a target dataset $\mathbb{X}^t$, and a pre-trained source predictor

$$h^s : \mathcal{X} \rightarrow \mathcal{Y}. \tag{5}$$

This predictor h^s might have been trained on the available source training dataset $\mathbb{X}^s$, but it could also originate from another dataset that is no more available. Also, no assumption is made on the nature of the predictor. For instance, it could be a neural network, a support vector machine, a decision tree, *etc.* Hence, we consider h^s as a "black box" predictor, and the goal is to predict the label of the samples in $\mathbb{X}^t$.

We consider that h^s has been trained to minimize a loss function $l: \mathcal{Y}\times\mathcal{Y}\rightarrow\mathbb{R}_+$. The loss on the source distribution is $R_s^l(h^s) = \mathbf{E}l\big(h^s(X^s), Y^s\big)$. In the absence of labeled data, we cannot assess the quality of this estimation. Nonetheless, we want to be able to provide a similar performance on the target distribution, according to the same metric. That is, we would like to find a target to source mapping $\mathcal{G} : \mathcal{X} \rightarrow \mathcal{X}$ such that h^s minimizes the loss on the transformed samples from the target distribution: $R_t^l(h^s \circ \mathcal{G}) = \mathbf{E}l\big(h^s(\mathcal{G}(X^t)), Y^t\big)$. The following proposition states sufficient conditions (Eqs. 6 and 7) for which our goal (Eq. 8) is achieved.

Proposition 1. *Under the assumption that*

$$P(Y^s) = P(Y^t), \tag{6}$$

if we can find a transformation $\mathcal{G} : \mathcal{X} \rightarrow \mathcal{X}$ *such that*

$$\forall x \in \mathcal{X}\,; \forall y \in \mathcal{Y} : \quad P(X^s = x|Y^s = y) = P(X^t = \mathcal{G}(x)|Y^t = y)\,, \tag{7}$$

[2] The number of transactions in fraud detection datasets as the ones used in the experiments of Sect. 4 is around ten million.

then for any $h^s : \mathcal{X} \to \mathcal{Y}$*, we have*

$$R_t^l(h^s \circ \mathcal{G}) = R_s^l(h^s)\,. \tag{8}$$

Proof. The proof is straightforward by noticing that Eqs. (6) and (7) imply

$$\forall x \in \mathcal{X}\,; \forall y \in \mathcal{Y}: \quad P(X^s = x, Y^s = y) = P(X^t = \mathcal{G}(x), Y^t = y)\,.$$

Since the $\mathcal{G}$-mapped target joint probability equals the source joint probability, $P(\mathcal{G}(X^t), Y^t) = P(X^s, Y^s)$, Eq. (8) is obtained by a change of variable. □

Albeit simple, this training-free target to source perspective on domain adaptation is—up to our knowledge—an unexplored problem. The landscape of possible methods to address this problem is certainly vast. In the remaining part of the paper, we propose and evaluate a variant of optimal transport methods.

3.2 Marginal Coordinate-Wise Optimal Transport

The target to source transformation explored in the current paper assumes that the output marginal distributions are equivalent, as stated by Eq. (6) of Proposition 1, and is motivated by the goal to find a mapping $\mathcal{G}$ complying Eq. (7). However, due to the lack of labeled data, it is not possible to directly estimate $\mathcal{G}$; we cannot properly estimate the conditional of the inputs (X^s or X^t) given the outputs (Y^s or Y^t). In these conditions, we relax the requirement of Eq. (7), and we seek for a function $\mathcal{G}$ belonging to the family of transformations that aligns the input marginal distributions:

$$P(X^s = x) = P(X^t = \mathcal{G}(x))\,. \tag{9}$$

Inspired by the previous works on optimal transport for domain adaptation introduced in Sect. 2.2, we choose $\mathcal{G}$ to minimize a transportation cost. Although it would be possible to define a cost function (see Eq. 1) between every $x_i^s \in \mathbb{X}^s$ and $x_j^t \in \mathbb{X}^t$ in order to find the Kantorovich discrete optimal transport problem (Definition 1), it would cause some issues that we aim to overcome:

- Solving Kantorovich's optimization problem (Eq. 2), or even its regularized version (Eq. 4), is computationally expensive on large datasets;
- In the case where the input space $\mathcal{X}$ contains mixed attributes, such a mix of numerical and categorical values, defining a cost function might be difficult;
- Even in the case where the input space contains exclusively numerical attributes (*e.g.* $\mathcal{X} \subseteq \mathbb{R}^d$), multidimensional distance metrics like Euclidean distance is not able to deal properly with the different scaling of each coordinate.
- As performing multidimensional optimal transport addresses the dependence across attributes, the solution is a large variance estimator of the optimal transformation in the case where the available data is not sufficiently abundant.

The proposed domain adaptation method is then performed by solving a sequence of one-dimensional optimal transport method. Doing so, we decompose the transformation $\mathcal{G}$ by feature-wise transformations $\mathcal{G}_k$:

$$\mathcal{G} = [\mathcal{G}_1, ..., \mathcal{G}_k, ..., \mathcal{G}_d] , \tag{10}$$

where d is less or equal to the number of features of the input space $\mathcal{X}$.

Each elementary transformation $\mathcal{G}_k$ solves the Kantorovich optimization problem (Eq. 2) on one feature only, thus the total computation is generally less expensive compared than the relaxed optimal transport problem (Eq. 4).[3] The distance measure can also be easily defined for each specific feature, especially when each of them has a different significance. Note that this feature by feature transformation is also robust to variation of scaling.

We denote by d' the number of transformations of numerical features. Without loss of generality, we refer the one-dimensional transformations of numerical features as $\mathcal{G}_1...\mathcal{G}_{d'}$, and to the transformations of categorical features as $\mathcal{G}_{d'+1}...\mathcal{G}_d$. The next two subsections detail how we process the numerical and the categorical features.

3.3 One-Dimensional Mapping of Numerical Features

The 1-D optimal transport on the real line has a closed-form solution [12] provided that the cost of moving one point to another is defined with respect to an ℓ^p norm:

$$\forall x^s, x^t \in \mathbb{R} , \; c^p_{num}(x^s, x^t) = |x^s - x^t|^p .$$

Different from the resolution of multidimensional optimal transport, in our 1-D scenario, we first need to sort the numerical features values in ascending order. We denote the obtained real-valued vector as

$$\forall k \in \{1, \ldots, d'\} , \; \mathbb{X}^s_k = \left(x^s_{k,1}, \ldots, x^s_{k,i}, \ldots, x^s_{k,m}\right) \in \mathbb{R}^m ,$$
$$\forall k \in \{1, \ldots, d'\} , \; \mathbb{X}^t_k = \left(x^t_{k,1}, \ldots, x^t_{k,j}, \ldots, x^t_{k,n}\right) \in \mathbb{R}^n ,$$

where $x^s_{k,i}$ is the k-th feature of a training point in $\mathbb{X}^s$, ranked at position i according to the sorted order: $x^s_{k,i} \leq x^s_{k,i+1}$ for all $i \in \{1, \ldots, m-1\}$. Then, the transformation function $\mathcal{G}_k$ is obtained by the following formula:

$$\mathcal{G}_k(x^t_{k,j}) = (F_s^{-1} \circ F_t)(x^t_{k,j}) , \tag{11}$$

where $F_t(x^t_{k,j})$ is the cumulative distribution function of $\hat{P}(X^t)$, obtained by counting the elements of vector $\mathbb{X}^t_k$ with values lesser or equal to $x^t_{k,j}$. This solution is also known as increasing arrangement. Note that the transformation $\mathcal{G}_k$ is a mapping that "moves" the target smallest value to the source smallest value ($\mathcal{G}_k(x^t_{k,1}) = x^s_{k,1}$), and the target largest value to the source largest value ($\mathcal{G}_k(x^t_{k,n}) = x^s_{k,m}$). For the intermediate target points ($1 < k < m$), the mapping

[3] See Sects. 3.3 and 3.4 for details.

is given by a barycenter in the source domain, according to Eq. (3) applied with the choice of the cost function c^p.

Recall that sorting a vector of n elements requires $O(n \log n)$ steps. For a specific attribute k, once vectors $\mathbb{X}_k^s$ and $\mathbb{X}_k^t$ are sorted, computing $\mathcal{G}(x_{k,j}^t)$ for every $x_{k,j}^t \in \mathbb{X}_k^t$ requires a single pass over both vectors that is $O(n)$ steps (provided $n \geq m$). Given that the number of numerical features to process is typically small compared to the number of instances ($d' \ll n$), the required $O(d'n \log n)$ computational time of our method is favorably compared to the typical $O(n^2 \log n)$ time of regularized optimal transport.

3.4 One-Dimensional Mapping of Categorical Features

Let $D_k = \{e_1^k, \ldots, e_{n_k}^k\}$ be the (non-ordered) set of values taken by a categorical feature, where $k \in \{d'+1, \ldots, d\}$ is the feature index, and n_k is the number of unique values in D_k. We denote the set of source and target values for categorical features by

$$\forall k \in \{d'+1, \ldots, d\}\,,\ \mathbb{X}_k^s = (x_{k,1}^s, \ldots, x_{k,i}^s, \ldots, x_{k,m}^s) \in D_k^m\,,$$
$$\forall k \in \{d'+1, \ldots, d\}\,,\ \mathbb{X}_k^t = (x_{k,1}^t, \ldots, x_{k,j}^t, \ldots, x_{k,n}^t) \in D_k^n\,.$$

As for numerical features, we propose to rely on Kantorovich's optimal transport to transform the target features into the source feature, but we cannot rely on a ℓ^p norm cost function as in Sect. 3.3. Instead, we need to define a cost function between the categorical values.

One could tailor this cost metric to the specificity of the learning problem at hand. In our experiments, we use a generic strategy that can be applied to any categorical features, by defining the cost in terms of the occurrence frequency [7]:

$$\forall e_l^k, e_r^k \in D_k\,,\ c_{cate}(e_l^k, e_r^k) = C_{l,r}^k = \begin{cases} 1 & \text{if } e_l^k = e_r^k\,, \\ \dfrac{1}{1 + \log(\frac{1}{v_l^k}) \log(\frac{1}{v_r^k})} & \text{otherwise,} \end{cases} \tag{12}$$

where $v_l^k \in (0, 1]$ is the frequency of occurrences of the value e_l^k for the k-th feature in $\mathbb{X}_k^s \cup \mathbb{X}_k^t$ (respectively $v_r^k \in (0, 1]$ is the frequency of the value e_r^k). In Eq. (12), we write $C_{l,r}^k$ for the entry of the cost matrix $C^k \in \mathbb{R}^{n_k \times n_k}$. Then, we state our optimal transport problem on a categorical feature in terms of the following coupling matrix $\gamma^k \in \mathbb{R}_+^{n_k \times n_k}$ in place of Eq. (2):

$$\gamma^k = \operatorname*{argmin}_{\gamma \in \Gamma^k} \langle C^k, \gamma \rangle = \operatorname*{argmin}_{\gamma \in \Gamma^k} \sum_{l=1}^{n_k} \sum_{r=1}^{n_k} C_{l,r}^k \gamma_{l,r}\,, \tag{13}$$

with

$$\Gamma^k = \left\{ \gamma \in \mathbb{R}_+^{n_k \times n_k} \;\middle|\; \frac{|\{i \mid x_{k,i}^s = e_l^k\}|}{m} = \sum_{j=1}^{n_k} \gamma_{l,j} \text{ and } \frac{|\{j \mid x_{k,j}^t = e_r^k\}|}{n} = \sum_{i=1}^{n_k} \gamma_{i,r} \right\}.$$

That is, we perform the optimal transport on the n_k categorical values instead of on the n source (and m target examples). Typically, $n_k \ll n$, and the computation is thus less expensive than the original problem. However, unlike numerical features where we can compute a barycenter thanks to Eq. (3), the barycenter for categorical features is difficult to define. We distinguish two strategies.

Numerical Embedding. In some case, a categorical feature has a numerical representation $\phi_k : D_k \to \mathbb{R}^{d_k}$ (for example, a real vector embedding like the common "Word2Vec" representation). In such cases, we use the barycenter of numerical representations as the adapted value:

$$\mathcal{G}_k(e_r^k) = \operatorname*{argmin}_{x \in \mathbb{R}^{d_k}} \sum_{l=1}^{n_k} \gamma_{l,r}^k \, c_{num}^p(x\,, \phi_k(e_r^k))\,. \tag{14}$$

Stochastic Mapping. More generally, we can define a stochastic transformation of categorical features. The probability of transforming one point $x_{k,j}^t$ to $x_{k,i}^s$ is

$$P(\mathcal{G}_k(e_r^k) = e_l^k) = \frac{\gamma_{l,r}^k}{\sum_{i=1}^{n_k} \gamma_{i,r}^k}\,. \tag{15}$$

Based on this method, to predict on a target example $x_j^t \in \mathbb{X}_k^t$ with the source predictor h^s (Eq. 5), we perform a Monte Carlo estimation by sampling the value of every categorical feature thanks to Eq. (15), and performing an average of the outputs predicted by h^s on these stochastic transformations.

3.5 Weakly Supervised Feature Selection

We have noticed in various experiments on different domain adaptation tasks that some features contribute more to domain adaptation than others. Instead of adapting all the features, the adaptation of well-selected features has better performance. In the scenario where few labeled target data are available, we propose to use them in order to select the features for which a transformation is beneficial.

The proposed feature selection scheme is a greedy algorithm. At initialization, no feature is adapted. Then, at each step of the process, we transform one feature of the target set. The selected feature is the one allowing the greatest accuracy increase on the small set of labeled target samples. The process is stopped when the accuracy improvement is no more significant according to the following criteria: we perform bootstrap sampling on the target label set, and we stop if more than one half of improvements are negative. Therefore, the remaining unselected features are unchanged when the target to source mapping is applied.

4 Experiments

In this section, we evaluate our methods on three datasets. The first one is the well-known sentiment classification task on Amazon review datasets. The second one is a fraud detection dataset that we collect from a Kaggle competition. The third one is the real-life industrial fraud detection dataset collected from a company which is one of the leaders of payment systems. Note that the methods that we have compared to are CORAL [15] which aligns the correlation matrices between domains, Deep Adaptation Networks (DAN) [9] which projects the source domain into a latent space and Domain Adversarial Neural Networks (DANN) [6] which generates the features that are both discriminating and invariant to the change of domains. Extended details concerning the experimental framework, additional results and related works are provided as supplementary material.[4]

4.1 Datasets

Amazon Reviews. This dataset contains reviews across different product categories from Amazon. Every review is a short text with associated score. According to the scores, reviews are labeled as positive or negative. Using supervised learning, one can build a sentiment classification model to estimate the different points of view of buyers. However, the model trained on one category does not generalize perfectly to another category. Some words appear frequently in reviews of one category but not the others.

The dataset we use here is the same as [2] and [6]. It is a class-balanced small dataset with 4 domains: Books (B), DVDs (D), Electronics (E) and Kitchen appliances (K). Each domain has 2000 training examples and around 4000 test examples, and features are bags-of-words. We generate a new feature representation using the mSDA unsupervised auto-encoder [2]. We generate two mSDA representations, one with all 5000 words dimensions, and another with only the most frequent 400 words dimensions. Similar to Chen et al. [2] we use a mSDA transformation with 5 layers. However, instead of stacking all hidden representations, we take only the hidden representation of the last layer.

Kaggle Fraud Detection. This is a public dataset from Kaggle IEEE-CIS Fraud Detection competition.[5] The objective is to predict the probability that an online transaction is fraudulent. The dataset contains transactions issued from different devices. We consider the mobile device as the source domain and the desktop device as the target domain. The dimension of the raw dataset is over 400 and contains missing values. Since the paper does not focus on the transformation of features with missing values, we remove all dimensions with more than 1% missing values and all transactions with missing values. The dimension of the dataset after preprocessing is 120 and the proportion of fraud is around 8%. We train the source models in a 4-folds cross-validation way and name the models from Model1 to Model4.

[4] The supplementary material, the code and data for the first two tasks are available on Github: https://github.com/marrvolo/CDA. Due to confidential reasons, the real-life fraud dataset is not shared.

[5] https://www.kaggle.com/c/ieee-fraud-detection.

Table 1. Amazon reviews results (prediction accuracy) using SVM on 5000 dimensions.

Domains	Source	No retrain		
		CORAL	OT	1D OT
$B \leftarrow D$	.8407	.8229	.7420	**.8438**
$B \leftarrow E$	.7692	.7283	.6856	**.8010**
$B \leftarrow K$	**.8445**	.7527	.7098	.8403
$D \leftarrow B$	**.8358**	.8132	.7433	.8351
$D \leftarrow E$	.8002	.7511	.7150	**.8355**
$D \leftarrow K$	**.8704**	.7806	.7248	.8639
$E \leftarrow B$	.7695	.7162	.6864	**.7854**
$E \leftarrow D$	.8084	.7406	.6979	**.8109**
$E \leftarrow K$	.8889	.8724	.7900	**.8970**
$K \leftarrow B$	.7870	.7357	.7025	**.8002**
$K \leftarrow D$	.8042	.7501	.7180	**.8145**
$K \leftarrow E$	**.8794**	.8560	.7863	.8785

Real Fraud Detection. This real-life fraud detection dataset consists of two domains: Belgian dataset and German dataset. The two datasets are real anonymous clients' transactions in production environments. They have the same mixture types of features, the number of categorical features is 8 and the number of numerical features is 23, numerical features are nearly all generated manually and are normalized between 0 and 1. We use 3 months of data, from July 2018 to October 2018. There are 180 million transactions for Belgian dataset and 90 million for German dataset.

4.2 Results

Amazon Reviews.[6] We first test our method on SVM model and compare our proposed method with two other methods that can be easily applied to a target to source adaptation problem. CORAL [15] aligns the correlations between domains and OT [4] is the optimal transport method in multidimensional space without any regularization. Here we use Euclidean distance as the cost metric. As shown in Table 1 (the left arrow shows the target to source direction of adaptation), our proposed 1D optimal transport outperforms the two others for this sentiment analysis task. Furthermore, we notice that the multidimensional optimal transport and CORAL consistently perform negative transfer. The potential reason is that the dataset size is not sufficiently large to accurately capture the multidimensional relations between attributes.

Table 2 reports experimental results using a neural network predictor. Our method is compared with CORAL, OT and with two deep adaptation methods

[6] See supplementary material for the results in the 400 dimensional dataset.

Table 2. Amazon reviews results (prediction accuracy) using neural networks on 5000 dimensions.

Domains	Source	Retrain		No retrain		
		DANN	DAN	CORAL	OT	1D OT
$B \leftarrow D$	.8276	.8382 ± .0041	**.8421 ± .0019**	.8179	.7200	.8374
$B \leftarrow E$	.7410	.7845 ± .0194	**.8256 ± .0020**	.7319	.6912	.8033
$B \leftarrow K$	.8267	.8361 ± .0049	**.8546 ± .0027**	.7589	.7222	.8412
$D \leftarrow B$	.8221	**.8353 ± .0021**	.8320 ± .0041	.7867	.7381	.8181
$D \leftarrow E$	.7919	.8073 ± .0119	**.8468 ± .0012**	.7268	.7081	.8185
$D \leftarrow K$	.8501	.8666 ± .0057	**.8691 ± .0016**	.7513	.7169	.8433
$E \leftarrow B$	.7711	.7743 ± .0078	**.8091 ± .0050**	.7146	.6860	.7905
$E \leftarrow D$	.8095	.7987 ± .0050	**.8148 ± .0052**	.7322	.6943	.8064
$E \leftarrow K$	.8904	.8853 ± .0057	.8911 ± .0012	.8723	.7991	**.8925**
$K \leftarrow B$	.7899	**.8022 ± .0019**	.7995 ± .0033	.7357	.7034	.7957
$K \leftarrow D$	.8095	.8213 ± .0018	**.8259 ± .0016**	.7481	.7052	.8139
$K \leftarrow E$	.8748	.8745 ± .0013	**.8749 ± .0010**	.8563	.7882	.8744

DAN [9] and DANN [6]. For a given dataset, the adaptation of CORAL and OT are deterministic, so no standard deviation is reported. The DANN and DAN models are trained with 30 different random states, and we report the average accuracy and associated standard deviation. On this 5000 dimensional dataset, the 1D optimal transport gets overall results comparable to the best adaptation method, while CORAL and OT still perform negative transfer.

Kaggle Fraud Detection.[7] As shown in Table 3, we evaluate our coordinate-wise domain adaptation method by transforming all features (1D OT), only numerical features (1D OT NUM) and only categorical features (1D OT CATE) in an unsupervised setting. We also compare the weakly supervised adaptation method by selecting significant features using few labeled target data to the retraining supervised (marked by **ws**) model. The column named "%n" shows the percentage of labeled target data that are available. The column "*d*" shows the number of adapted features on average. The reported metric is the area under the precision-recall curve (PR-AUC). Similar to experiment settings of Amazon reviews task, we repeat 30 times for no deterministic experiments (DANN and DAN) and report their standard deviations.

Table 3 first presents results using Gradient Boosting Decision Tree (GBDT) models. The coordinate-wise adaptation method on categorical features achieve the best performance among all unsupervised adaptation methods without retraining. The CORAL has a significant negative transfer on this dataset. Both weakly supervised methods (marked by **ws**) have improved the performance compared to the adaptation of all features. However, in the situation where 10% of

[7] See supplementary material for the results on Model3 and Model4.

Table 3. PR-AUC scores for domain adaptation models and non adaptative models (annotated by a † mark) on Kaggle Fraud Detection dataset.

GBDT model						
Retrain	Method	%n	Model1	d	Model2	d
YES	Train on target†	1	.5604 ± .0316	-	.5646 ± .0361	-
	Train on target†	10	**.7332 ± .0191**	-	**.7299 ± .0185**	-
NO	Source model†	0	.6712	-	.6625	-
	CORAL NUM	0	.6041	112	.5660	112
	1D OT NUM	0	.6497	112	.6615	112
	1D OT CATE	0	.6986	8	.7011	8
	1D OT	0	.6748	120	.6898	120
	1D OT (ws)	1	.6848 ± .0064	9 ± 3	.7075 ± .0089	13 ± 4
	1D OT (ws)	10	.7061 ± .0024	36 ± 4	.7159 ± .0017	42 ± 6

Neural network model						
Retrain	Method	%n	Model1	d	Model2	d
YES	Train on target†	1	.3660 ± .0745	-	.3960 ± .0584	-
	Train on target†	10	.5893 ± .0595	-	.5941 ± .0501	-
	DAN	0	.6489	120	.6372	120
	DANN	0	.6398	120	.6079	120
NO	Source model†	0	.5912	-	.5990	-
	CORAL	0	.5490	120	.5840	120
	1D OT NUM	0	.5872	112	.6008	112
	1D OT CATE	0	.6218	8	.6288	8
	1D OT	0	.6314	120	.6329	120
	1D OT (ws)	1	.6232 ± .0178	16 ± 6	.6202 ± .0218	15 ± 5
	1D OT (ws)	10	**.6561 ± .0100**	40 ± 8	**.6548 ± .0101**	36 ± 10

target labels are available, retraining a GBDT model gets the best performance. Compared to weakly supervised methods with 10% of labels, the method using only 1% target labels has comparable performance and less adapted features.

Regarding the neural networks source models, the coordinate-wise adaptation method on all features has the best performance among all unsupervised adaptation methods. In the case where few labeled target examples are available, the weakly supervised adaptation methods have selected the most significant features for domain adaptation and improved the prediction performance. Interestingly, the value of the standard deviation of selected features is smaller in the case of 1% of target labels in most of the experiments. This may due to the fact that selected features is far less than the ones in the case of 10% of target labels.

Figure 1 reveals the progression of the accuracy metric of our coordinate-wise domain adaptation method using a greedy search feature selection approach. These graphs are obtained using all target labels, in order to show that in an

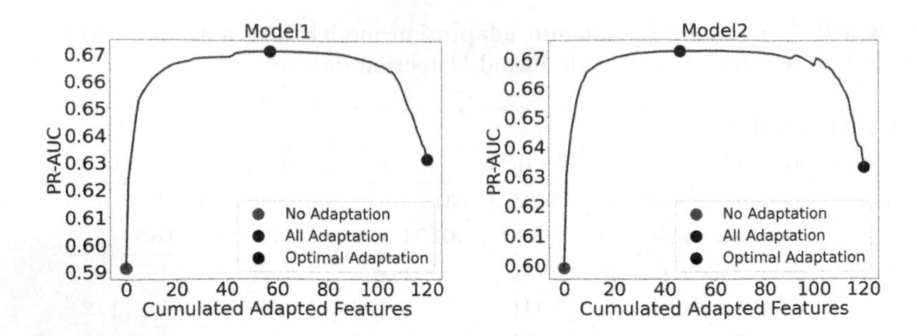

Fig. 1. Feature selection greedy algorithm with neural networks source model on Kaggle fraud detection task (idealized scenario involving all target labels).

Table 4. PR-AUC scores for domain adaptation models and non adaptative models (annotated by a † mark) on the Real Fraud Detection dataset.

GBDT model								
Retrain	Method	%n	July	d	August	d	September	d
YES	Train on target†	1	.0441 ± .0348	-	.0170 ± .0126	-	.0698 ± .0395	-
	Train on target†	10	.2554 ± .0683	-	.0870 ± .0592	-	.2923 ± .0854	-
NO	Source model†	0	.2595	-	.1546	-	.3840	-
	CORAL NUM	0	.2296	23	.1709	23	.3245	23
	1D OT NUM	0	.2323	23	.2357	23	.3684	23
	1D OT CATE	0	**.2705**	8	.1738	8	**.3886**	8
	1D OT	0	.2383	31	.2199	31	.3695	31
	1D OT (ws)	1	.2630 ± .0071	8 ± 5	.2097 ± .0235	9 ± 4	.3830 ± .0107	10 ± 5
	1D OT (ws)	10	.2697 ± .0043	14 ± 2	**.2515 ± .0165**	14 ± 3	.3837 ± .0112	15 ± 1

Neural network model								
Retrain	Method	%n	July	d	August	d	September	d
YES	Train on target†	1	.0412 ± .0380	-	.0171 ± .0163	-	.0486 ± .0417	-
	Train on target†	10	.2325 ± .0679	-	.0918 ± .0456	-	.2637 ± .0626	-
	DAN	0	.3073	31	.2010	31	.2881	31
	DANN	0	**.2849**	31	.1966	31	**.2945**	31
NO	Source model†	0	.2351	-	.1852	-	.2607	-
	CORAL	0	.1548	31	.1796	31	.1039	31
	1D OT NUM	0	.2341	23	.1985	23	.2552	23
	1D OT CATE	0	.2459	8	.1718	8	.2247	8
	1D OT	0	.2392	31	.1965	31	.2371	31
	1D OT (ws)	1	.2511 ± .0163	11 ± 4	.2009 ± .0154	10 ± 4	.2554 ± .0127	9 ± 4
	1D OT (ws)	10	.2599 ± .0160	14 ± 3	**.2100 ± .0181**	14 ± 3	.2591 ± .0153	14 ± 3

ideal case few feature transformations are required to achieve a good adaptation, and that adapting all features might hurt.

Real Fraud Detection. Similar performance can be observed on the real-life fraud detection adaptation task from Table 4. The models are evaluated in 3 different periods. Notice that in all adaptation tasks except the month of September, at least one of our proposed adaptation methods outperforms the source model.

The weakly supervised adaptation methods in GBDT and neural networks show that with only 1% of label information, the feature selection can achieve comparable performance as the one with 10% of label information and transform fewer features. Both weakly supervised adaptation methods improve the performance than the adaptation of all features. As expected, the standard deviation of adaptation performance with 1% of target labels slightly larger than that with 10% of target labels. Although, in the neural networks, the adaptation method like DAN and DANN have the best performance, they require the retraining of the model and the adaptation is performed in a latent space which is difficult to interpret. Moreover, these two methods are only for neural networks and our proposed methods are totally model independent.

5 Conclusion and Future Works

This article introduced a new target to source perspective for domain adaptation tasks. An unsupervised and a weakly supervised coordinate-wise transformation are proposed and achieve comparable results to the state-of-the-art methods. In addition, the proposed method is parameter-free and can be easily applied to various families of pre-trained models such as neural networks and decision trees. Although we have shown experimentally that transforming a small subset of target features leads to better predictions. As for future research, we aim to reveal the further relevance between these selected features and domain adaptation tasks.

Acknowledgements. This work was partially supported by the Canada CIFAR AI Chair Program.

References

1. Altschuler, J., Niles-Weed, J., Rigollet, P.: Near-linear time approximation algorithms for optimal transport via sinkhorn iteration. In: NeurIPS (2017)
2. Chen, M., Xu, Z.E., Weinberger, K.Q., Sha, F.: Marginalized denoising autoencoders for domain adaptation. In: ICML (2012)
3. Courty, N., Flamary, R., Habrard, A., Rakotomamonjy, A.: Joint distribution optimal transportation for domain adaptation. In: NeurIPS (2017)
4. Courty, N., Flamary, R., Tuia, D., Rakotomamonjy, A.: Optimal transport for domain adaptation. IEEE Trans. Pattern Anal. Mach. Intell. **39**(9), 1853–1865 (2016)
5. Cuturi, M.: Sinkhorn distances: lightspeed computation of optimal transport. In: NIPS (2013)
6. Ganin, Y., et al.: Domain-adversarial training of neural networks. JMLR **17**(1), 1–35 (2016)
7. Jones, K.S.: A statistical interpretation of term specificity and its application in retrieval. J. Doc. (1972)
8. Kantorovitch, L.: On the translocation of masses. Manag. Sci. **5**(1), 1–4 (1958)
9. Long, M., Cao, Y., Wang, J., Jordan, M.I.: Learning transferable features with deep adaptation networks. In: ICML (2015)

10. Monge, G.: Mémoire sur la théorie des déblais et des remblais. Histoire de l'Académie Royale des Sciences de Paris (1781)
11. Pan, S.J., Tsang, I.W., Kwok, J.T., Yang, Q.: Domain adaptation via transfer component analysis. IEEE Trans. Neural Networks **22**(2), 199–210 (2010)
12. Peyré, G., Cuturi, M.: Computational optimal transport. Found. Trends Mach. Learn. **11**(5–6), 355–607 (2019)
13. Shimodaira, H.: Improving predictive inference under covariate shift by weighting the log-likelihood function. J. Stat. Plann. Inference **90**(2), 227–244 (2000)
14. Sugiyama, M., Nakajima, S., Kashima, H., Buenau, P.V., Kawanabe, M.: Direct importance estimation with model selection and its application to covariate shift adaptation. In: NIPS (2008)
15. Sun, B., Feng, J., Saenko, K.: Correlation alignment for unsupervised domain adaptation. In: Csurka, G. (ed.) Domain Adaptation in Computer Vision Applications. ACVPR, pp. 153–171. Springer, Cham (2017). https://doi.org/10.1007/978-3-319-58347-1_8

Unsupervised Multi-source Domain Adaptation for Regression

Guillaume Richard[1,2](✉), Antoine de Mathelin[1], Georges Hébrail[2], Mathilde Mougeot[1,3], and Nicolas Vayatis[1]

[1] Centre Borelli, ENS Paris-Saclay, Cachan, France
guillaume.richard@ens-paris-saclay.fr
[2] EDF R&D, Palaiseau, France
[3] ENSIIE, Evry, France

Abstract. We consider the problem of unsupervised domain adaptation from multiple sources in a regression setting. We propose in this work an original method to take benefit of different sources using a weighted combination of the sources. For this purpose, we define a new measure of similarity between probabilities for domain adaptation which we call hypothesis-discrepancy. We then prove a new bound for unsupervised domain adaptation combining multiple sources. We derive from this bound a novel adversarial domain adaptation algorithm adjusting weights given to each source, ensuring that sources related to the target receive higher weights. We finally evaluate our method on different public datasets and compare it to other domain adaptation baselines to demonstrate the improvement for regression tasks.

Keywords: Domain adaptation · Adversarial · Multiple sources · Discrepancy

1 Introduction

In classical machine learning, one assumes that the source data used to train an algorithm comes from the same distribution as the target data it is applied to. This assumption is not true for many applications: for instance, a human activity recognition model trained on young people may not perform well when applied to older ones. Moreover, for many applications, different sources have different relations to the target domain. Including sources that are not related to the target may lead to negative transfer i.e. reduce the performance of adaptation on the target domain. Hence we consider in this paper the problem of unsupervised domain adaptation from multiple sources tackling the issue of adapting from a labelled source domain to a target domain with no labeled data.

Electronic supplementary material The online version of this chapter (https://doi.org/10.1007/978-3-030-67658-2_23) contains supplementary material, which is available to authorized users.

F. Hutter et al. (Eds.): ECML PKDD 2020, LNAI 12457, pp. 395–411, 2021.
https://doi.org/10.1007/978-3-030-67658-2_23

An abundant literature exists for unsupervised domain adaptation with a single source for classification: [3] introduced a single source adaptation bound for classification. It was later used in several works, notably adversarial methods of [10] and [19]. While those methods can be applied to regression, it is not theoretically founded and often fails in practice. [14] proposed a novel theoretical bound for regression using the notion of discrepancy between predictors. It is not easy to estimate in the general case but has led to several works using linear regression [1] to train GANs or kernels [8] for domain adaptation.

The main risk of adapting from multiple sources comes from one or several sources being detrimental to adaptation. It is particularly true with adversarial methods trying to match source and target domains. Then one wants to find a way to give high weights to sources the most related to the target. In [15] a weighting scheme is proposed assuming that the target distribution is a convex combination of the sources. A boosting method is used in [21] to derive weights. Recently, [22] extended previous bounds with a maximum over multiple sources leading to an algorithm giving high weights to sources far from the target. In [13] inter-relationships between sources are used to compute the weights.

There are two main contributions in this work: firstly, we prove new a bound for multi-source domain adaptation that is tighter than existing bounds in a regression task. It is based on a new measure of similarity between distributions which we call hypothesis-discrepancy. For a given predictor, it measures how another predictor can give different results on one of the domains while staying close on the other and can be computed with adversarial learning. The second main contribution is a new algorithm optimizing both representations and weights of each source for multi-source domain adaptation. To the best of the authors' knowledge, this is the first adversarial unsupervised domain adaptation tailored for regression. We conduct experiments on both synthetic and real-world datasets and improve on state of the art results for multi-source adversarial domain adaptation for regression.

2 Unsupervised Multiple Source Domain Adaptation with Hypothesis-Discrepancy

Setting. We first define the problem of Multi-Source Domain Adaptation (MSDA). We define K independent source domains $\mathcal{D}_k$ such that $\mathcal{D}_k = \{X_k, f_k\}$ where X_k is the input data with associated marginal distribution $X_k \sim p_k$ and f_k the true labelling function of the domain. Similarly, we define a target domain $\mathcal{D}_t = \{X_t, f_t\}$ with $X_t \sim p_t$. We assume that every input is in the same space $\mathcal{X}$ i.e. $X_k \in \mathcal{X}$ and $X_t \in \mathcal{X}$ which is the case of homogeneous transfer. The prediction task is the same for both domains i.e. $f_k : \mathcal{X} \to \mathcal{Y}$ and $f_t : \mathcal{X} \to \mathcal{Y}$ (f_k and f_t are supposed to be close to each other). For instance, $\mathcal{Y} \subset \mathbb{R}$ for regression or $\mathcal{Y} = \{0, 1\}$ for binary classification. We also consider a loss $L : \mathcal{Y} \times \mathcal{Y} \to \mathbb{R}^+$ and a hypothesis class $\mathcal{H}$ of hypotheses $h : \mathcal{X} \to \mathcal{Y}$. We also assume that the loss L is bounded over $\mathcal{Y}$ by $M = \sup_k sup_{x \in \mathcal{S}_k, h \in \mathcal{H}} L(h(x), f_k(x))$.

For two hypotheses h and h', we define $\epsilon_k(h, h') = \mathbb{E}_{x \sim p_k}[L(h(x), h'(x))]$ the average loss of two hypotheses over a the source domain $\mathcal{D}_k$ and $\epsilon_t(h, h') = \mathbb{E}_{x \sim p_t}[L(h(x), h'(x))]$ over the target domain. We also consider a labelled source sample $\mathcal{S}_k$ of size m with an associated empirical probability $\hat{p}_k$. Similarly, we consider a unlabelled target sample $\mathcal{S}_t$ of size n with an associated empirical probability of $\hat{p}_t$.

Objective. The goal of Domain Adaptation is to minimize the target risk $\epsilon_t(h, f_t) = \mathbb{E}_{x \sim p_t}[L(h(x), f_t(x))]$. In *unsupervised domain adaptation*, no label is available in the target task and we cannot directly estimate f_t. Consequently we want to leverage the information about the labels in the source domains f_k to adapt to the target domain. After defining the hypothesis-discrepancy we propose a new bound relating the target risk with a weighted combination of the source risks $\epsilon_k(h, f_k)$.

2.1 Hypothesis-Discrepancy

We introduce the concept of hypothesis-discrepancy:

Definition 1. *For two distributions P, Q over a set $\mathcal{X}$ and for a hypothesis class $\mathcal{H}$ over $\mathcal{X}$, for any $h \in \mathcal{H}$, the hypothesis-discrepancy (or HDisc) associated with h is defined as:*

$$HDisc_{\mathcal{H},L}(P, Q; h) = \max_{h' \in \mathcal{H}} |\mathbb{E}_{x \sim P}[L(h(x), h'(x))] - \mathbb{E}_{x \sim Q}[L(h(x), h'(x))]| \quad (1)$$

For any given $h \in \mathcal{H}$ hypothesis-discrepancy measures a similarity between two distributions. It is directly dependent on the hypothesis class $\mathcal{H}$ and the loss L and can be estimated with finite samples. In the definition, h' can be seen as a predictor that would be very close to h on the source domain but far on the target domain (or vice-versa). Using *HDisc*, we are able to show the following proposition for unsupervised single source domain adaptation:

Proposition 1. *If L is symmetric and follows the triangle inequality, then the following bound holds for any $k \in \{1, ..., K\}$,*

$$\epsilon_t(h, f_t) \leq \epsilon_k(h, f_k) + \eta_{\mathcal{H}}(f_k, f_t) + HDisc_{\mathcal{H},L}(p_t, p_k; h) \quad (2)$$

where

$$\eta_{\mathcal{H}}(f_k, f_t) = \min_{h_0 \in \mathcal{H}} [\epsilon_t(h_0, f_t) + \epsilon_k(h_0, f_k)]$$

Proof. See Appendix A

□

This bound gives a good intuition about the conditions under which domain adaptation can work. Indeed, the first term $\epsilon_k(h, f_k)$ corresponds to the error made by h on the source data. The third term is our hypothesis-discrepancy and characterizes the similarity between marginal probabilities over the input data. The second term $\eta_{\mathcal{H}}$ is the sum of the error made by the ideal hypothesis on both domains: it is small when the two labelling functions are close which is the general assumption of unsupervised domain adaptation [4]. As it involves f_t it cannot be controlled in unsupervised domain adaptation without access to labels in the target domain. It follows that, under the assumption that the two labelling functions are close, if the two other terms of the bound can be minimized, the target risk will also be minimized. Another strength of *HDisc* is that it is directly dependent on $\mathcal{H}$ and L: it can be used for any task including regression.

2.2 Multi-source Domain Adaptation Bound

When multiple sources are available, a straightforward idea would be to merge all the source domains into one and transform the problem to single-source domain adaptation where Proposition 1 applies. This solution is clearly not optimal as different source domains may have different relationships to the target one.

We propose to attribute weights to each source: we introduce the α-weighted source domain $\mathcal{D}_\alpha = \{p_\alpha, f_\alpha\}$ such that for $\alpha \in \Delta = \{\alpha \in \mathbb{R}^K; \alpha_k \geq 0, \sum_{k=1}^K \alpha_k = 1\}$, $f_\alpha : x \rightarrow (\sum_{k=1}^K \alpha_k p_k(x) f_k(x))/(\sum_{j=1}^K \alpha_j p_j(x))$ and $p_\alpha = \sum_{k=1}^K \alpha_k p_k$.

The α-weighted sample is $\mathcal{S}_\alpha = \bigcup_{k=1}^K \mathcal{S}_k$ with probabilities $\hat{p_\alpha}(x_i^{(k)}) = \alpha_k/m$.

Similarly, we consider an unlabeled target sample $\mathcal{S}_t = \{(x_1^{(t)}, ..., x_n^{(t)})\}$ where $x_i^{(t)} \overset{i.i.d}{\sim} p_t$. We define the sets $\mathcal{H}_k = \{g : x \rightarrow L(h(x), f_k(x)); h \in \mathcal{H}\}$. Moreover, the Rademacher complexity of a set $\mathcal{H}_k$ is defined as

$$\mathcal{R}_m(\mathcal{H}_k) = \mathbb{E}_{\mathcal{S}_k}[\mathbb{E}_\sigma[sup_{g \in \mathcal{H}_k} \sum_{i=1}^m \sigma_i g(x_i^{(k)})]]$$

where the expectations is taken over any sample $\mathcal{S}_k = \{x_k^{(1)}, ..., x_k^{(m)}\} \sim \hat{p_k}^{(m)}$ and σ_i are *iid* variables uniformly distributed over $\{-1, 1\}$ independent from $X_1, ..., X_K$.

Theorem 1. *Assuming that the loss L is symmetric and follows the triangle inequality, then for any hypothesis $h \in \mathcal{H}$, with probability $1 - \delta$ the following bound holds:*

$$\begin{aligned} \epsilon_t(h, f_t) \leq & \sum_{k=1}^K \alpha_k \hat{\epsilon_k}(h, f_k) + HDisc_{\mathcal{H},L}(p_t, p_\alpha) + \eta_{\mathcal{H},\alpha} \\ & + 2\sum_{k=1}^K \alpha_k \mathcal{R}_m(\mathcal{H}_k) + \|\alpha\|_2 M \sqrt{\frac{\log(1/\delta)}{2m}} \end{aligned} \tag{3}$$

where

- $\eta_{\mathcal{H},\alpha} = \min_{h_0 \in \mathcal{H}} [\epsilon_\alpha(h_0, f_\alpha) + \epsilon_t(h_0, f_t)]$
- $\mathcal{R}_m(\mathcal{H}_k)$ *is the Rademacher complexity of* $\mathcal{H}_k = \{h : x \to L(h(x), f_k(x)); h \in \mathcal{H}\}$.

Proof. We give a sketch of the proof. The full details can be found in Appendix B Using Proposition 1 with p_α, we get:

$$\epsilon_t(h, f_t) \leq \epsilon_\alpha(h, f_\alpha) + \eta_{\mathcal{H}}(f_t, f_\alpha) + HDisc_{\mathcal{H},L}(p_\alpha, p_k; h) \tag{4}$$

We then define $\phi = \epsilon_\alpha(h, f_\alpha) - \hat{\epsilon_\alpha}(h, f_\alpha)$. Using McDiarmid's inequality [17] for ϕ, we obtain that with probability $1 - \delta$,

$$\epsilon_\alpha(h, f_\alpha) \leq \hat{\epsilon_\alpha}(h, f_\alpha) + \mathbb{E}_{\hat{p_\alpha}}[\phi] + \|\alpha\|_2 M \sqrt{\frac{\log 1/\delta}{2m}}$$

Then one can show using the usual ghost sample argument of Rademacher complexity that:

$$\mathbb{E}_{\hat{p_\alpha}}[\phi] \leq 2 \sum_{k=1}^{K} \alpha_k \mathcal{R}_m(\mathcal{H}_k)$$

Noting that $\hat{\epsilon_\alpha}(h, f_\alpha) = \sum_{k=1}^{K} \alpha_k \hat{\epsilon_k}(h, f_k)$ concludes the proof.

□

Theorem 1 gives a theoretical analysis in the multi-source domain adaptation framework. The first term corresponds to the α-weighted source risks and can be controlled by learning h close to f_k. The second term connects the target distribution with the α-weighted source distribution. The third term is related to how different the labelling functions on the target and the source are and is expected to be small in unsupervised domain adaptation. The last two terms show the convergence rate of this bound and it was proven in [6] that $\mathcal{R}_m(\mathcal{H}_k) = \mathcal{O}(1/\sqrt{m})$ for some functions such as neural networks.

Then in order to adapt from sources $\mathcal{S}_1, ..., \mathcal{S}_k$ to the target, we need to minimize the hypothesis-discrepancy between the α-weighted domain and the target domain. We propose in Sect. 3 an algorithm to find ideal representations of the sources for adaptation.

3 Adversarial Algorithm for Multi-source Domain Adaptation

3.1 Optimization Objective: A Min-Max Problem

We now present the practical solution derived from Theorem 1. We introduce a feature extractor parametrized by θ $\phi_\theta : \mathcal{X} \to \mathcal{Z}$ and a class of predictor $\mathcal{H}_\mathcal{Z} : \mathcal{Z} \to \mathcal{Y}$. Given unlabeled target sample $X_t = \{x_1^{(t)}, ..., x_n^{(t)}\} \in \mathbb{R}^{n \times d}$ and K

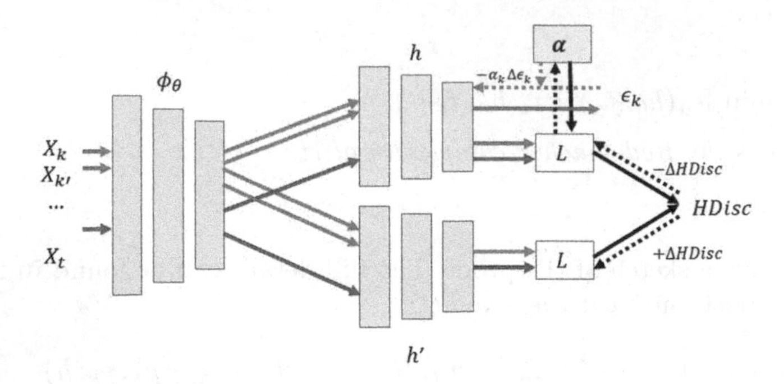

Fig. 1. MSDA: The adversarial scheme is similar to single-source with weights α. At each iteration, the weights α are updated.

labeled sources $X_k = \{x_1^{(k)}, ..., x_m^{(k)}\} \in \mathbb{R}^{m \times d}$ with labels $Y_k = \{y_1^{(k)}, ..., y_m^{(k)}\} \in \mathbb{R}^m$, we want to minimize the combination of the source risk and the hypothesis-discrepancy between the marginal weighted source distribution and target distribution as in Theorem 1.

Using the definition of *HDisc*, we formulate the following objective for our Adversarial Hypothesis-Discrepancy Multi-Source Domain Adaptation (**AHD-MSDA**):

$$\min_{\substack{\phi_\theta, h \in \mathcal{H} \\ \|\alpha\|_1 = 1}} \max_{h' \in \mathcal{H}} \left[\sum_{k=1}^{K} \alpha_k \epsilon_k (h \circ \phi_\theta, y_k) + \lambda \|\alpha\|_2 \right. \\ \left. + \left| \epsilon_t (h \circ \phi_\theta, h' \circ \phi_\theta) - \sum_{k=1}^{K} \alpha_k \epsilon_k (h \circ \phi_\theta, h' \circ \phi_\theta) \right| \right] \tag{5}$$

where λ is a hyperparameter. It was shown in [1] that if $\mathcal{H}_\mathcal{Z}$ is a subsect of μ_h-Lipschitz functions and ϕ_θ is continuous in θ then the discrepancy is continuous in θ and it also stands for hypothesis-discrepancy.

The first term of the objective forces h to be a good predictor on the source task. For any given h and h' the discrepancy term constrains both representations ϕ_θ and weights α to align domains. The term $\eta_\mathcal{H}$ is ignored in our objective and assumed to be small.

This objective involves a min-max formulation that is similar to the ones used in adversarial domain adaptation [10]. While computing the true solution of this min-max problem is still impossible in practice, we derive an alternate optimization algorithm in the next section.

3.2 Adversarial Domain Adaptation

We display the general structure of our algorithm in Fig. 1. Similarly to most other adversarial methods, we sequentially optimize differents parameters of our networks according to different objectives. At a given iteration, four losses are minimized sequentially:

Algorithm 1. Pseudo-algorithm for **AHD-MSDA**

Initialize $\alpha_k = \frac{1}{K}$, h, h' and θ randomly, choose learning rates η_h, η_θ and η_α
for $e = 1...epochs$ **do**
  Forward propagation
  $\epsilon_k = \frac{1}{m}\sum_{i=1}^{m} L(h^{(e)}(\phi_{\theta^{(e)}}(x_i^{(k)}), y_i^{(k)}))$
  $HDisc = \left|\epsilon_t(h^{(e)} \circ \phi_{\theta^{(e)}}, h'^{(e)} \circ \phi_{\theta^{(e)}}) - \sum_{k=1}^{K} \alpha_k \epsilon_k(h^{(e)} \circ \phi_{\theta^{(e)}}, h'^{(e)} \circ \phi_{\theta^{(e)}})\right|$
  Backward propagation
  $h^{(e+1)} \leftarrow h^{(e)} - \eta_h \left(\sum_{k=1}^{K} \alpha_k^{(e)} \Delta_h \epsilon_k(h^{(e)})\right)$ ▷ (*)
  $h'^{(e+1)} \leftarrow h'^{(e)} + \eta_h \left(\sum_{k=1}^{K} \alpha_k^{(e)} \Delta_{h'} HDisc(h'^{(e)})\right)$
  $\theta^{(e+1)} \leftarrow \theta^{(e)} - \eta_\theta \left(\sum_{k=1}^{K} \alpha_k^{(e)} \Delta_\theta \epsilon_k(\theta^{(e)}) + \Delta_\theta HDisc(\theta^{(e)})\right)$
  $\alpha_k^{(e+1)} \leftarrow \alpha_k^{(e)} - \eta_\alpha \left(\Delta_{\alpha_k} HDisc(\alpha_k^{(e)}) + 2\lambda\alpha_k^{(e)}\right)$
  Clip weights of $\phi_\theta^{(e+1)}$, $h^{(e+1)}$ and $h'^{(e+1)}$
  $\alpha^{(e+1)} = \alpha^{(e+1)}/||\alpha^{(e+1)}||_1$
end for
(*) For a parameter p and a loss L, we note $\Delta_p L(p_0)$ the gradient of L with respect to p computed at p_0

1. $\mathcal{L}_h = \alpha_k \epsilon_k$ updates h to minimize the source loss
2. $\mathcal{L}_{h'} = -HDisc$ updates h' to maximize discrepancy
3. $\mathcal{L}_\theta = HDisc + \sum_{k=1}^{K} \alpha_k \epsilon_k$ updates ϕ_θ to minimize discrepancy and source loss
4. $\mathcal{L}_\alpha = HDisc + \lambda||\alpha||_2$ updates α to minimize the discrepancy between α-weighted domain and target domain

The predictor h' can be seen as a discriminator in traditional adversarial domain adaptation methods. It is trained to give predictions close to h on one domain and far on another. The representations ϕ_θ are gradually updated against the discriminator. We include a source loss in its update as otherwise extracted features would be meaningless for the final task. The loss $\mathcal{L}_h$ ensures that h is performing well on the source domains.

In the loss $\mathcal{L}_\alpha$, we only included the discrepancy term. Indeed, our goal is to select the domains closer to the source in terms of discrepancy. Including the source loss in $\mathcal{L}_\alpha$ may give too high weights to sources that are "easy" to predict. It is possible to keep the term with a μ parameter to control its influence but in our experiments, it did not bring any improvement. It would also be possible to completely update α at each epoch but we found it sub-performing. Our method allows the weights to smoothly adapt to the representations learnt by ϕ_θ.

We present a pseudo-algorithm in Algorithm 1. The order of the steps did not matter in our experiments. It is possible to include a short pre-training phase where hypothesis-discrepancy is not minimized as in the beginning of the training, representations and h may be meaningless and weights may be updated for unrelated sources. Our algorithm can also be applied in the single-source scenario by setting $K = 1$ and $\alpha = 1$.

4 Related Works

Relation with Other Measures. The hypothesis-discrepancy has several advantages. Firstly, it can be estimated with finite samples (see Appendix C). Moreover, it is dependent on the hypothesis class $\mathcal{H}$ and loss L so it is possible to use in both classification and regression settings. The hypothesis-discrepancy is based on the discrepancy introduced in [14]: the original discrepancy is more conservative than our hypothesis-discrepancy as it is defined by $Disc(P, Q) = \sup_{h \in \mathcal{H}} HDisc(P, Q; h)$. Our bound is tighter than the one of [14] or [8] as it involves only one supremum over $\mathcal{H}$. The idea of discrepancy with only one supremum is also used in [12] with the source-discrepancy which is a specific case of our hypothesis-discrepancy with $h = h_s^*$. While the bound is tighter than the original [14] it does not lead to efficient practical solutions.

The popular $d_{\mathcal{H}}$ introduced in [3] for classification also involves only one supremum but often fails in practice for regression problems. Indeed, minimizing it aligns domains in a sense of classification and one can see on Fig. 2 how in a simple linear regression problem $d_{\mathcal{H}}$ would fail. Our algorithm presented in the next section is general to classification and regression.

Discrepancy Minimization. While our method minimizes the new hypothesis-discrepancy, several methods worked on discrepancy minimization: in the original work [14], authors derive a quadratic formulation for ℓ^2-loss in regression where the goal is to re-weight each sample in the source domain and was later extended with kernels in [7]. Recently, discrepancy with linear regressors was used as a measure of distance between probabilities to train GANs [1]. Most works focusing on discrepancy have used specific values of the loss L and hypothesis class $\mathcal{H}$ to compute it: for instance, the ℓ^2-norm with linear regression was presented in the original work of [14], later extended to kernels [8] or even used to train GANs [1]. For classification, previous works used $d_{\mathcal{H}}$ as it is a special case of discrepancy for binary classification.

Adversarial Domain Adaptation. The first work using adversarial learning for Domain Adaptation was introduced in [10] with the gradient reversal layer. Using

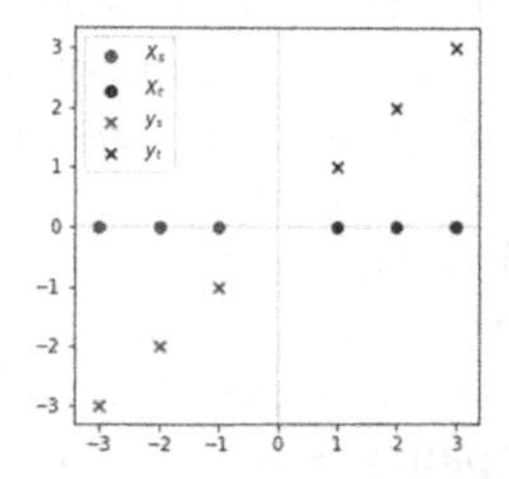

Fig. 2. Basic linear regression problem: source domain is in red and target in blue. Here using linear regressors and ℓ^2-loss, for any h, $HDisc(X_s, X_t; h) = 0$ but using linear classifiers $d_{\mathcal{H}}(X_s, X_t) = 1$ as domains are perfectly separable

the bound of Ben-David [2], authors find a new representation that is discriminative for the task but where domains are confounded. [19] follows a similar idea. The main benefit of our method is that the adversarial structure is directly dependent on the task at hand. Maximum Classifier Discrepancy [18] is the closest idea to our practical solution even though they use a specific instance of discrepancy. Our *HDisc*-based algorithm is more general as it works for both regression and classification. To the best of authors' knowledge, this is the first work proposing an adversarial domain adaptation tailored for regression.

Multi-source Domain Adaptation. Other authors also proposed methods to compute weights for different sources. For instance, in [15], authors assume that the target distribution is a convex combination of the source distributions and derive optimal weights. [16] and [11], authors use the Rényi divergence to derive ideal weights for each of the sources.

Other authors tried to attribute weights to different sources with an adversarial objective. Recently, [22] and [13] extended the adversarial framework of [10] to multi-source domain adaptation. Namely, they both prove extended previous bounds from single source DA based on $d_{\mathcal{H}}$ and Wasserstein distance. [22] gives a bound for regression based on $d_{\mathcal{H}}$, but is based on the generalized VC-dimension for regression which is hard to use in practice. Moreover the adopted weighted scheme in their algorithm sequentially adapts the worst source to the target which we assume to be suboptimal. [13] uses relationships between sources to get weights to adapt to the target using Wasserstein distance, which is not tailored for regression applications. A very recent pre-print [20] proposes a weighting scheme for multi-source DA using the original discrepancy of [14] in the linear regression case, similar to [1].

5 Experiments

In this section, we evaluate our method both in the multi-source (**AHD-MSDA**) and single-source scenario (**AHDA**) on several datasets[1]. It should be noted that unsupervised Domain Adaptation is hard to evaluate as having no labeled samples limits usage of classical comparison tools (cross-validation, hyperparameter tuning, ...). Hence we firstly use a synthetic dataset to test how our algorithm behaves. Secondly, we improve on previous state of the art results on an extended version of the Multi-Domain Amazon Review dataset originally used in [5]. We also applied our algorithm to a digit classification task for which we get comparable results with methods tailored for classification.

5.1 Synthetic Dataset (Friedman)

To generate synthetic data, we use a modified version of the Friedman regression problem [9]. The Friedman dataset uses inputs x of dimension 5 and the

[1] Code is available at https://github.com/GRichard513/ADisc-MSDA.

prediction function is

$$y(x) = 10sin(\pi x_1 x_2) + 20(x_3 - 0.5)^2 + 10x_4 + 5x_5 + \epsilon$$

where $\epsilon \sim \mathcal{N}(0, \sigma_y)$.

To highlight the two contributions of this work, our goal is two-fold: we firstly want to demonstrate the effectiveness of *HDisc* in the single-source DA scenario (K=1 and $\alpha = 1$). Then we run different experiments to show when multi-source DA is expected to bring an improvement.

We generate source data as follows: we define three clusters of sources $\{\mathcal{S}_1, \mathcal{S}_2, \mathcal{S}_3\}$, $\{\mathcal{S}_4, \mathcal{S}_5, \mathcal{S}_6\}$ and $\{\mathcal{S}_7, \mathcal{S}_8, \mathcal{S}_9\}$. For each cluster, and for each feature (1...5), a mean of -1 or 1 is selected at random. Then each mean of each source of a given cluster is shifted by a random shift i.e. for a source k associated to a cluster c $\mu_k^{(f)} \sim \mathcal{N}(\mu_c, \sigma_c)$. Finally, each source sample is randomly chosen with a normal distribution $p_k = \mathcal{N}(\mu_k^{(f)}, \sigma_k)$. In Fig. 3, we display each feature of the generated data.

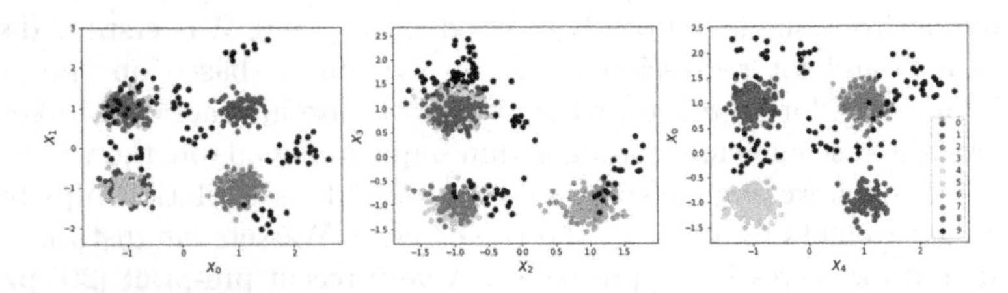

Fig. 3. Data for the single-source Friedman experiment over the 5 features with $\sigma_k = 0.2$, $\sigma_c = 0.2$, $\mu_{shift} = 0.5$, $\sigma_{shift} = 0.5$. From right to left: (x_0, x_1); (x_2, x_3); $(x_4; x_0)$. Each color corresponds to a source, the target is in black (Color figure online).

In order to separate the effects of our two contributions, we split the experiments in two parts: in the first experiment, demonstrate the effectiveness of our hypothesis-discrepancy minimization in the single source scenario (i.e. when $\alpha_k = \frac{1}{K}$) and why the classical $d_{\mathcal{H}}$ fails in this regression scenario. In a second scenario, we show how our weighting scheme helps adaptation in multi-source, especially to select sources that are related to the target.

Hypothesis-Discrepancy. For single-source, we merge all 9 source domains together to create one. The target sample is generated choosing uniformly one of the source domains for each element and adding a noise of the form $\mathcal{N}(\mu_{shift}, \sigma_{shift})$. This experiment helps to understand the purpose of unsupervised domain adaptation. Indeed, as the underlying condition for unsupervised DA to work is that the labelling function is similar in every domain ($\eta_{\mathcal{H}}$ small in our experiments), unsupervised DA is closely related to the issue of generalization. As a consequence, if the algorithm learnt on the source data is able to generalize, domain adaptation will not bring any improvement. As such, one can see unsupervised DA as a data-driven regularization to improve the target risk.

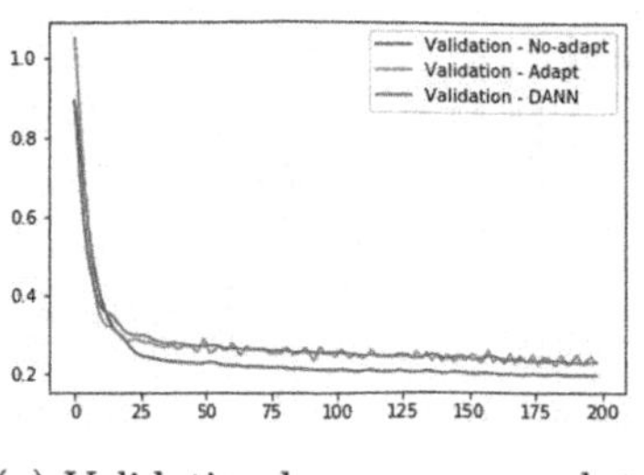

(a) Validation loss on source data

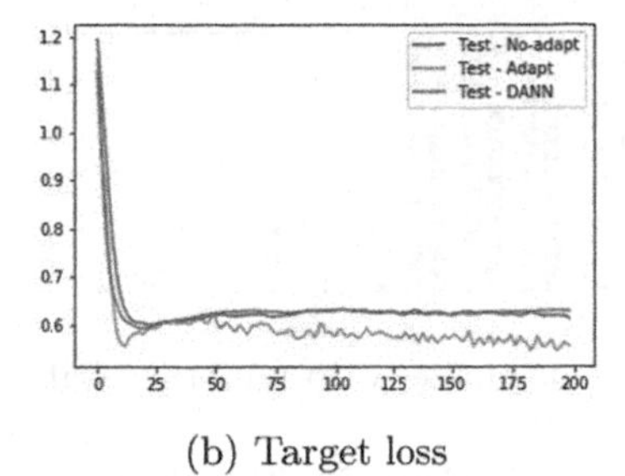

(b) Target loss

Fig. 4. Training curves for Single Source DA: without adaptation, the target loss increases as the validation loss keeps decreasing. **DANN** exposes the same behaviour as the target loss of **AHDA** decreases.

For this experiment, we use a shallow network with 2 layers with 5 neurons and *LeakyRelu* activation for the feature extractor and 1 layer for the final predictor as detailed in Appendix D We keep this architecture for three methods: Multi-Layer Perceptron (**MLP**) without adaptation, Domain-Adversarial Neural Network (**DANN**) minimizing the $d_{\mathcal{H}}$ between domains and our Adversarial Hypothesis-Discrepancy Adaptation (**AHDA**) our method in the case of single source (no weights α). To get the results for **DANN**, we tuned then hyperparameter μ balancing the regression and domain losses: without this tuning, **DANN** always fails to converge because its adversarial scheme is related to classification.

We conducted the experiments with various amount of shift. In Fig. 4, we report the validation loss (computed validation set different from the training set) and target loss for each method for $\sigma_x = 0.2$, $\mu_{shift} = 0.5$ and $\sigma_{shift} = 0.5$, which is the case of target data related but not too close to the source domain plottend on Fig. 5. While **MLP** and **DANN** overfit on the source data, our method is able to decrease the target loss. Hence, our adversarial scheme helps the algorithm to better learn for the target data. We report in Table 1 the average MSE scores over ten runs for the three methods and various shifts. One can notice that the further μ_{shift} is, the more useful the adaptation is. We also report in Appendix D a visualization of the extracted features from **AHDA** and **DANN** which shows that **DANN** tries to align domains only to be able not to separate them while **AHDA** is constrained by the final regression task.

Table 1. Single source domain adaptation: MSE for different amounts of shift.

μ_{shift}	0.0	0.1	0.2	0.3	0.4	0.5	0.6	0.7	0.8	0.9	1.0
MLP	**0.221**	**0.214**	0.260	0.360	0.510	0.695	0.909	1.132	1.322	1.484	1.632
DANN	0.223	0.235	0.296	0.412	0.581	0.784	1.017	1.258	1.462	1.629	1.772
AHDA	0.222	**0.214**	**0.255**	**0.350**	**0.490**	**0.670**	**0.881**	**1.044**	**1.197**	**1.392**	**1.446**

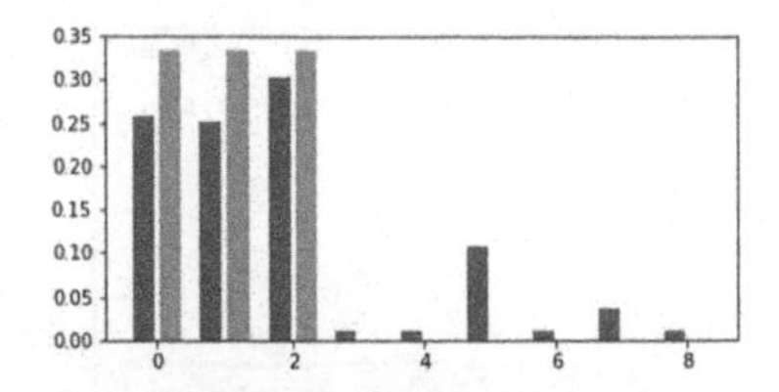

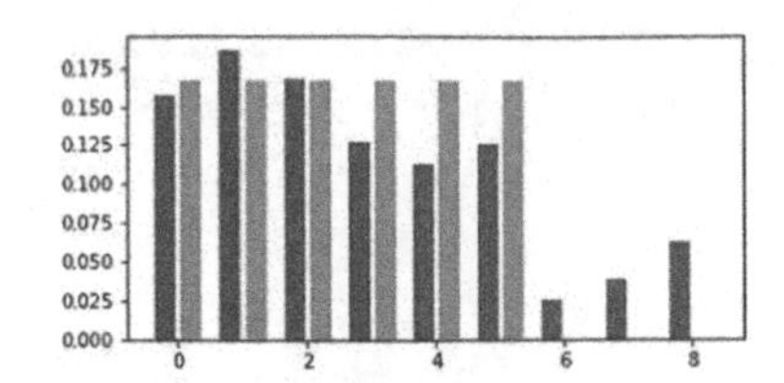

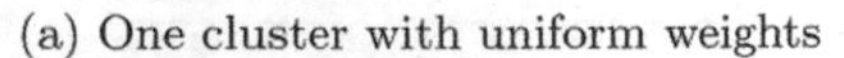
(a) One cluster with uniform weights

(b) Two clusters with uniform weights

Fig. 5. Friedman Multiple Source experiment: α found by **AHD-MSDA** (blue) vs True α (orange) (Color figure online)

Multi-Source DA. We also tried our multi-source algorithm on the previous target domain. As we expected, multi-source domain adaptation did not bring any improvement in the previous scenario as the target was sharing the same relations with every source. But MSDA is particularly interesting in the case where some of the sources are not useful for adaptation. We demonstrate it in an experiment where we control the weights α given to each source in the creation of the target domain.

In a first experiment, we give equal weights $\alpha_{1,2,3} = 1/3$ to every source in one cluster and $\alpha_{4:9} = 0$ for every other cluster. In a second experiment, we gave equal weights to two clusters $\alpha_{1:6} = 1/6$ and $\alpha_{7:9} = 0$ In both experiments, and on different runs, the algorithm was able to retrieve weights close to the real ones. It was translated by a decrease of the target loss. When putting weights to only one source in a cluster, our algorithm struggled to identify a specific source but still gave high weight to sources in the same cluster. In some experiments where sources were very different from each other (no cluster), we noticed a tendency for α to give 1 for one source and 0 for all the others. It can be meaningful as the target may be close to only one source in that case or be balanced using the λ parameter of the ℓ^2 regularization.

Discussion. Based on this toy experiment, we can conclude that our **AHD-MSDA** and **AHDA** perform well for regression under the condition that the labelling functions are the same. The proposed weighting scheme is performing well when sources have really different relations to the output. The main limitation we found to *HDisc* is that since it is dependent on h, its estimation is hard, especially in regression where values are not constrained. As we have seen the weighting scheme mainly helps when the target data is close to only few of the sources. In the next experiment, we show how our algorithm performs on a real world dataset.

5.2 Multi-domain Amazon Dataset

We use the extended version of the Multi-Domain Sentiment Dataset[2] with 25 categories (books, dvd, ...). The dataset is made of textual reviews from Amazon

[2] https://www.cs.jhu.edu/~mdredze/datasets/sentiment/.

and associated ratings. We treat it as a regression problem where the goal is to predict the rating based on the review. As in previous papers [10,22], we transform the data using tf-idf transform and filtering only the 1,000 words with highest coefficients.

Similarly to the reduced dataset for classification [5], we kept 2,000 samples for each category (or less when there was not 2,000 available). We alternate on each category as a target and use every other category as a source.

Table 2. Average Mean Average Error (MAE) over 5 runs for each method and domain of the Amazon Multi-Domain Dataset

Dataset	Apparel	Auto	Baby	Beauty	Books	Camera	Cellphones	Computer	Dvd
MLP	0.897	0.902	0.841	0.880	1.024	0.923	0.871	0.842	0.912
DANN	0.929	1.017	0.893	0.878	1.141	0.944	1.051	1.135	1.052
AHDA	**0.837**	0.887	0.840	0.860	**0.945**	0.893	0.859	0.849	0.882
MDAN	0.980	0.797	0.908	0.973	1.234	0.921	0.954	1.749	1.322
AHD-MSDA	0.859	**0.767**	**0.794**	**0.790**	0.969	**0.876**	**0.825**	**0.770**	**0.868**
Dataset	Electronics	Food	Grocery	Health	Jewelry	Kitchen	Magazines	Music	Musical
MLP	0.833	0.882	0.774	0.869	0.759	0.851	0.960	0.976	0.778
DANN	1.064	1.036	0.793	0.955	0.783	0.856	0.985	1.293	0.727
AHDA	0.844	0.866	0.796	0.851	0.815	0.849	**0.909**	**0.955**	0.895
MDAN	1.041	**0.820**	0.789	0.987	0.755	0.850	1.295	1.330	1.117
AHD-MSDA	**0.815**	0.849	**0.716**	**0.840**	**0.715**	**0.848**	0.911	0.991	**0.723**
Dataset	Office	Outdoor	Software	Sports	Tools	Toys	Video	Average	Avg rank
MLP	0.969	0.831	0.924	0.844	0.823	0.873	0.866	0.892	3.12
DANN	**0.854**	0.803	1.090	0.850	0.888	0.861	1.00	0.955	3.92
AHDA	0.956	0.851	0.880	0.839	0.813	0.864	**0.878**	0.868	2.52
MDAN	1.041	0.789	0.964	0.910	1.656	**0.843**	1.467	1.060	3.96
AHD-MSDA	0.882	**0.755**	**0.858**	**0.814**	**0.741**	0.873	0.883	**0.838**	**1.48**

In order to study the performance of our method **AHD-MSDA**, we compare it to several other baselines on different datasets:

- **MLP** corresponds to merging all sources and applying to the target dataset without adaptation.
- **DANN** corresponds to merging all sources into one and applying the **DANN** from [10].
- **AHDA** corresponds to merging all sources into one and applying our adversarial scheme using discrepancy specific to the task.
- **MDAN** is a multi-source domain adaptation based on the $d_{\mathcal{H}}$ distance proposed in [22]. It gives high weights to sources far from the target (we used the soft version).
- **AHD-MSDA** is our method described in Algorithm 1.

We also experimented the Multi-Domain Matching Network (**MDMN**) that uses Wasserstein distance between sources and the target to give weights but it did not perform well at all for regression as the adversarial structure is not well conditioned for regression. For **DANN** and **MDAN** the predictors have been specified to regression by changing the last layer by a regression layer and loss with mean squared error. The implementation of **MDAN** is inspired from the original implementation from the authors.[3] We tried some methods to pre-compute weights for our multi-source algorithm based on different distances between domains but at best it led to similar results to **AHDA** so we do not report it here.

For fair comparison, the basic architecture is kept the same for every method (see Appendix E). We tried different architectures and came to the same conclusion. For every method, we use *Adam* optimizer. Hyperparameter selection in unsupervised domain adaptation is a hard task as no labeled target data is available. For **DANN** and **MDAN**, we tried different sets of values and only report the ones giving the best results on test data which is very advantageous and normally not possible for unsupervised DA ($\lambda = 0.01$ for **DANN**, $\gamma = 10$, $\mu = 0.1$ for **MDAN** which are the hyperparameters they used in the classification setting). For our algorithm, the only hyperparameter is λ for ℓ^2-regularization of α for which we tried several values and we report results for $\lambda = 1$ here as we did not notice significant differences for values between 10^{-3} and 10.

We ran the experiment 5 times and report the average MAE for each method in Table 2. The standard deviations are around 0.01 for **AHD-MSDA** and **AHDA** (see Appendix E) and for most domains, improvements are statistically significants using a Wilcoxon rank test. While the gains over the **MLP** with no adaptation may seem small, we emphasize that obtaining an improvement is already challenging as the **MLP** can already learn general features with the variety of data it is feeded. For most domains, **AHD-MSDA** obtains the best result with multi-source bringing an improvement over **AHDA** for many domains. One can notice the instability of **DANN** and **MDAN** for a few number of domains where the error becomes very large.

Overall, the **AHDA** achieves the best score and shows its efficiency for adversarial regression domain adaptation. The weights obtained by our method were meaningful for some domains (giving high weight for software to transfer to computer for instance) but not for all. We assume that the efficiency of the method is very dependent on the nature of the dataset as even in this case, for some domains, every adaptation method we tried fails.

5.3 Digit Classification

Finally, we experiment our method on a digit classification task using 5 different domains: *MNIST*, *MNIST-M*, *Synth*, *SVHN* and *USPS*. *SVHN* and *USPS* are known to be the hardest datasets to classify as they are much more diverse. In each experiment, we use one domain as the target and the 4 others as the sources. We resize all domains to images made of 28×28 pixels. For fair comparison we use the same architecture for every network, using a simple convolutional

[3] https://github.com/KeiraZhao/MDAN.

neural network (**CNN** see Appendix F for details). For this experiment, we use a reduced version of the datasets with 10, 000 samples in each domain.

Since we are in a classification setting, the loss used to train **AHDA** and **AHD-MSDA** cannot be the Mean Squared Error. We found that the cross-entropy loss was performing better than the ℓ^1-loss, so we use it to compute *HDisc* (even though it does not follow triangle inequality). Our adversarial structure becomes very close to the MCD introduced in [18] and our weighting scheme extends this structure to a multi-source setting.

We report the accuracy for each dataset and each method in Table 3. One can see that for a classification task, our *HDisc* still gives state of the art results compared to other methods tailored for classification. For MNIST, even without adaptation, the CNN can learn general features from other datasets. **MDAN** fails as it gives high weights to datasets that are not similar (such as **USPS** for **SVHN**) while our **AHD-MSDA** does not suffer from this drawback.

Table 3. Accuracy for the visual adaptations on digit datasets

Dataset	MNIST	MNIST-M	SVHN	Synth	USPS
CNN	0.916	0.450	0.489	0.562	0.624
DANN	0.918	**0.530**	**0.546**	0.710	0.659
AHDA	0.912	0.512	0.510	0.769	**0.667**
MDAN	**0.923**	0.460	0.488	0.671	0.574
AHD-MSDA	**0.923**	0.518	0.501	**0.793**	0.657

6 Conclusion

In this work we proposed a new general domain adaptation bound based on a new hypothesis-discrepancy compatible with classification and regression. We proposed an adversarial domain adaptation algorithm tailored for the task at hand, which is novel for regression. We extended these results to multiple sources to find ideal convex combinations of the sources. We demonstrated the efficiency of our method for a regression task for which we improved on previous state of the art results. For a classification task, **AHD-MSDA** obtains comparable results to other state of the art results. We emphasize that in the multi-source setting, our weighting scheme limits the risk of negative transfer. As we saw in our experiments, we mainly expect improvements of the multi-source method when some domains are really closer to other forming clusters.

The main limitation of our work comes from the assumption of unsupervised domain adaptation that the labelling functions are the same in every domain. In our future work, we intend to investigate the semi-supervised and few-shot learning settings, where a few labeled target data is available.

References

1. Adlam, B., Cortes, C., Mohri, M., Zhang, N.: Learning GANs and ensembles using discrepancy. In: Advances in Neural Information Processing Systems, pp. 5788–5799 (2019)
2. Ben-David, S., Blitzer, J., Crammer, K., Kulesza, A., Pereira, F., Vaughan, J.W.: A theory of learning from different domains. Machine Learn. **79**(1–2), 151–175 (2010)
3. Ben-David, S., Blitzer, J., Crammer, K., Pereira, F.: Analysis of representations for domain adaptation. In: Advances in Neural Information Processing Systems, pp. 137–144 (2007)
4. Ben-David, S., Lu, T., Luu, T., Pál, D.: Impossibility theorems for domain adaptation. In: International Conference on Artificial Intelligence and Statistics, pp. 129–136 (2010)
5. Blitzer, J., Dredze, M., Pereira, F.: Biographies, bollywood, boom-boxes and blenders: domain adaptation for sentiment classification. In: Proceedings of the 45th Annual Meeting of the Association of Computational Linguistics, pp. 440–447 (2007)
6. Cortes, C., Gonzalvo, X., Kuznetsov, V., Mohri, M., Yang, S.: AdaNet: adaptive structural learning of artificial neural networks. In: Proceedings of the 34th International Conference on Machine Learning-Volume 70, pp. 874–883. JMLR.org (2017)
7. Cortes, C., Mohri, M.: Domain adaptation and sample bias correction theory and algorithm for regression. Theoret. Comput. Sci. **519**, 103–126 (2014)
8. Cortes, C., Mohri, M., Medina, A.M.: Adaptation based on generalized discrepancy. J. Machine Learn. Res. **20**(1), 1–30 (2019)
9. Friedman, J.H.: Multivariate adaptive regression splines. Ann. Stat. **19**, 1–67 (1991)
10. Ganin, Y., et al.: Domain-adversarial training of neural networks. J. Mach. Learn. Res. **17**(1), 2030–2096 (2016)
11. Hoffman, J., Mohri, M., Zhang, N.: Algorithms and theory for multiple-source adaptation. In: Advances in Neural Information Processing Systems, pp. 8246–8256 (2018)
12. Kuroki, S., Charoenphakdee, N., Bao, H., Honda, J., Sato, I., Sugiyama, M.: Unsupervised domain adaptation based on source-guided discrepancy. In: Proceedings of the AAAI Conference on Artificial Intelligence, vol. 33, pp. 4122–4129 (2019)
13. Li, Y., Carlson, D.E., et al.: Extracting relationships by multi-domain matching. In: Advances in Neural Information Processing Systems, pp. 6798–6809 (2018)
14. Mansour, Y., Mohri, M., Rostamizadeh, A.: Domain adaptation: learning bounds and algorithms. In: COLT 2009 - The 22nd Conference on Learning Theory, February 2009
15. Mansour, Y., Mohri, M., Rostamizadeh, A.: Domain adaptation with multiple sources. In: Advances in Neural Information Processing Systems, pp. 1041–1048 (2009)
16. Mansour, Y., Mohri, M., Rostamizadeh, A.: Multiple source adaptation and the Rényi divergence. arXiv preprint arXiv:1205.2628 (2012)
17. McDiarmid, C.: Concentration. In: Habib, M., McDiarmid, C., Ramirez-Alfonsin, J., Reed, B. (eds.) Probabilistic Methods for Algorithmic Discrete Mathematics. Algorithms and Combinatorics, vol. 16, pp. 195–248. Springer, Heidelberg (1998). https://doi.org/10.1007/978-3-662-12788-9_6

18. Saito, K., Watanabe, K., Ushiku, Y., Harada, T.: Maximum classifier discrepancy for unsupervised domain adaptation. In: Proceedings of the IEEE Conference on Computer Vision and Pattern Recognition, pp. 3723–3732 (2018)
19. Tzeng, E., Hoffman, J., Saenko, K., Darrell, T.: Adversarial discriminative domain adaptation. In: Proceedings of the IEEE Conference on Computer Vision and Pattern Recognition, pp. 7167–7176 (2017)
20. Wen, J., Greiner, R., Schuurmans, D.: Domain aggregation networks for multi-source domain adaptation. arXiv preprint arXiv:1909.05352 (2019)
21. Xu, Z., Sun, S.: Multi-source transfer learning with multi-view Adaboost. In: Huang, T., Zeng, Z., Li, C., Leung, C.S. (eds.) ICONIP 2012. LNCS, vol. 7665, pp. 332–339. Springer, Heidelberg (2012). https://doi.org/10.1007/978-3-642-34487-9_41
22. Zhao, H., Zhang, S., Wu, G., Moura, J.M., Costeira, J.P., Gordon, G.J.: Adversarial multiple source domain adaptation. In: Advances in Neural Information Processing Systems, pp. 8559–8570 (2018)

Open Set Domain Adaptation Using Optimal Transport

Marwa Kechaou(✉), Romain Herault, Mokhtar Z. Alaya, and Gilles Gasso

Normandie Univ., UNIROUEN, UNIHAVRE, INSA Rouen, LITIS, 76 000 Rouen, France
marwa.kechaou@insa-rouen.fr

Abstract. We present a 2-step optimal transport approach that performs a mapping from a source distribution to a target distribution. Here, the target has the particularity to present new classes not present in the source domain. The first step of the approach aims at rejecting the samples issued from these new classes using an optimal transport plan. The second step solves the target (class ratio) shift still as an optimal transport problem. We develop a dual approach to solve the optimization problem involved at each step and we prove that our results outperform recent state-of-the-art performances. We further apply the approach to the setting where the source and target distributions present both a label-shift and an increasing covariate (features) shift to show its robustness.

Keywords: Optimal transport · Open set domain adaptation · Rejection · Label-Shift

1 Introduction

Optimal Transport (OT) approaches tackle the problem of finding an optimal mapping between two distributions P^s and P^t respectively from a source domain and a target domain by minimizing the cost of moving probability mass between them. Efficient algorithms are readily available to solve the OT problem [18].

A wide variety of OT applications has emerged ranging from computer vision tasks [4] to machine learning applications [1,8]. Among the latter, a body of research work was carried out to apply OT to domain adaptation task [7,8, 20,27]. Domain Adaptation (DA) assumes labelled samples (x, y) in the source domain while only unlabelled (or a few labelled) data are available in the target domain. It intends to learn a mapping so that the prediction model tuned for the source domain applies to the target one in the presence of shift between source and target distributions. The distribution shift may be either a *Covariate-Shift* where the marginal probability distributions $P^s(X)$ and $P^t(X)$ vary across domains while conditional probability distributions is invariant (i.e. $P^s(Y|X) = P^t(Y|X)$) or a *Label-Shift* where label distributions $P(Y)$ for both domains do not match but their conditional probability distributions $P(X|Y)$ are the same. Theoretical works [2,29] have investigated the generalization guarantees on target domain when transferring knowledge from the labeled source data to the target domain.

F. Hutter et al. (Eds.): ECML PKDD 2020, LNAI 12457, pp. 412–435, 2021.
https://doi.org/10.1007/978-3-030-67658-2_24

Courty et al. [8] settled OT to deal with covariate-shift by enforcing samples from a class in the source domain to match with the same subset of samples in the target domain. Follow up works extend OT to asymmetrically-relaxed matching between the distributions $P^s(X)$ and $P^t(X)$ or to joint distribution $P(X, Y)$ matching between source and target domains [3,7]. Recently, Redko et al. [19] focus on multi-source domain adaptation under target shift and aim to estimate the proper label proportions $P^t(Y)$ of the unlabelled target data. Traditional DA methods for classification commonly assume that the source and target domains share the same label set. However in some applications, some source labels may not appear in the target domain. This turns to be an extreme case of label-shift when the related target class proportions drop to zero. The converse case, termed as open set domain adaptation [17], considers a target domain with additional labels which are deemed abnormal as they are *unknown classes* from the source domain standpoint. This results in a substantial alteration in the label distributions as $P^s(y_k) = 0$ and $P^t(y_k) \neq 0$ for some labels y_k not occurring in the source domain. Therefore, aligning the label distributions $P^s(Y)$ and $P^t(Y)$ may lead to a negative transfer. To tackle this issue, open set domain adaptation aims at rejecting the target domain "abnormal samples" while matching the samples from the shared categories [10,17,21].

In this paper, we address the open set DA using optimal transport. The approach we propose consists of the following two steps: 1) rejection of the outlier samples from the unknown classes followed by 2) a label shift adaptation. Specifically, we frame the rejection problem as learning an optimal transport map together with the target marginal distribution $P^t(X)$ in order to prevent source samples from sending probability mass to unknown target samples. After having rejected the outliers from target domain, we are left with a label shift OT-based DA formulation. Contrary to the first step, we fix the resulting target marginal $P^t(X)$ (either to a uniform distribution or to the $P^t(X)$ learned at the first stage) and optimize for a new transport map and the source marginal distribution $P^s(X)$ in order to re-weight source samples according to the shift in the proportions of the shared labels. We also propose a decomposition of $P^s(X)$ and show its advantage to reduce the number of involved parameters. To the best of our knowledge, this is the first work considering open set DA problem using OT approach. The key contributions of the paper are: i) We devise an OT formulation to reject samples of unknown class labels by simultaneously optimizing the transport map and the target marginal distribution. ii) We propose an approach to address the label-shift which estimates the target class proportions and enables the prediction of the target sample labels. iii) We develop the dual problem of each step (rejection and label-shift) and give practical algorithms to solve the related optimization problems. iv) We conduct several experiments on synthetic and real-datasets to assess the effectiveness of the overall method.

The paper is organized as follows: in Sect. 2 we detail the related work. Section 3 presents an overview of discrete OT, our approach, and the dual problem of each step. It further details the optimization algorithms and some implementation remarks. Sect. 4 describes the experimental evaluations.

2 Related Work

Arguably the most studied scenario in domain adaptation copes with the change in the marginal probability distributions $P^s(X)$ and $P^t(X)$.

Only a few dedicated works have considered the shift in the class distributions $P^s(Y)$ and $P^t(Y)$.

To account for the label-shift, Zhang et al. [28] proposed a re-weighting scheme of the source samples. The weights are determined by solving a maximum mean matching problem involving the kernel mean embedding of the marginal distribution $P^s(X)$ and the conditional one $P^s(X/Y)$. In the same vein, Lipton et al. [16] estimated the weights $P^t(y_k)/P^s(y_k)$ for any label y_k using a black box classifier elaborated on the source samples. The estimation relies on the confusion matrix and on the approximated target class proportions via the pseudo-labels given by the classifier.

The re-weighting strategy is also investigated in the JCPOT procedure [19] using OT and under multiple source DA setting. The target class proportions are computed by solving a constrained Wassesrtein barycenter problem defined over the sources. Wu et al. [27] designed DA with asymmetrically relaxed distribution alignment to lift the adversarial DA approach [11] to label-shift setup. Of a particular note is the label distribution computation [22] which hinges on mixture proportion estimation. The obtained class proportions can be leveraged on to adapt source domain classifier to target samples. Finally, JDOT approach [7] addresses both the covariate and label shifts by aligning the joint distributions $P^s(X,Y)$ and $P^t(X,\hat{Y})$ using OT. As in [16] the target predictions $\hat{Y}$ are given by a classifier learned jointly with the related OT map.

Regarding open set DA, the underlying principle of the main approaches resembles the one of multi-class open set recognition or multi-class anomaly rejection (see [13,22–24] and references therein) where one looks for a classifier with a reject option. [17] proposed an iterative procedure combining assignment and linear transformation to solve the open set DA problem. The assignment step consists of a constrained binary linear programming which ensures that any target sample is either assigned to a known source class (with some cost based on the distance of the target sample to the class center) or labelled as outlier. Once the unknown class samples are rejected, the remaining target data are matched with the source ones using a linear mapping. Saito et al. [21] devised an adversarial strategy where a generator is trained to indicate whether a target sample should be discarded or matched with the source domain. Recently, Fang et al. [10] proposed a generalization bound for open set DA and thereon derived a so-called distribution alignment with open difference in order to sort out the unknown and known target samples. The method turns to be a regularized empirical risk minimization problem.

3 The Proposed Approach

We assume the existence of a labeled source dataset $Z^s = \{(x_1^s, y_1^s), .., (x_{n_s}^s, y_{n_s}^s)\}$ where $\{y_i\}_{i=1}^{n_s} \in \{1, .., C\}^{n_s}$ with C the number of classes and n_s the number of source samples. We also assume available a set of n_t unlabeled target samples $Z^t = \{x^t_1, \ldots, x^t_{n_t}\}$. The target samples are assumed to be of labels y^t in $\{1, .., C, C+1\}$ where the class $C+1$ encompasses all target samples from other classes not occurring in the source domain. Moreover we assume that the proportions of the shared classes may differ between source and target domains i.e. $P^s(y) \neq P^t(y)$ for $y \in \{1, .., C\}$.

Let $P^s(x|y)$ and $P^t(x|y)$ be the conditional distributions of source and target respectively with possibly $P^s(x|y) \neq P^t(x|y)$ for $y \in \{1, .., C\}$. Similarly we denote the marginal source and target distributions as $P^s(x)$ and $P^t(x)$. Our goal is to learn a distribution alignment scheme able to reject from Z^t the samples of the unknown class $C+1$ while matching correctly the remaining source and target samples by accounting for the label-shift and possibly a shift in the conditional distributions. For this, we propose a two-step approach (see in Fig. 2 in the Appendix A.6 for an illustration): a rejection step followed by the label shift correction. To proceed we rely on discrete OT framework which is introduced hereafter.

3.1 The General Optimal Transport Framework

This section reviews the basic notions of discrete OT and fixes additional notation. Let Σ_n be the probability simplex with n bins, namely the set of probability vectors in $\mathbb{R}^n_+$, i.e., $\Sigma_n = \{\omega \in \mathbb{R}^n_+ : \|\omega\|_1 := \sum_{j=1}^n \omega_j = 1\}$. Let μ^s and μ^t be two discrete distributions derived respectively from Σ_{n_s} and Σ_{n_t} such that

$$\mu^s = \sum_{i=1}^{n_s} \mu^s_i \delta_{x^s_i} \text{ and } \mu^t = \sum_{j=1}^{n_t} \mu^t_j \delta_{x^t_j},$$

where μ^s_i stands for the probability mass associated to the i-th sample (the same for μ^t_j). Computing OT distance between μ^s and μ^t, referred to as the Monge-Kantorovich or Wasserstein distance [14,26]. amounts to solving the linear problem given by

$$W(\mu^s, \mu^t) = \min_{\gamma \in \Pi(\mu^s, \mu^t)} \langle \zeta, \gamma \rangle_F, \tag{1}$$

where $\langle \cdot, \cdot \rangle_F$ denotes Frobenius product between two matrices, that is $\langle T, W \rangle_F = \sum_{i,j} T_{ij} W_{ij}$. Here the matrix $\zeta = (\zeta_{ij})_{1 \leq i \leq n_s; 1 \leq j \leq n_t} \in \mathbb{R}^{n_s \times n_t}_+$, where each ζ_{ij} represents the energy needed to move a probability mass from x^s_i to x^t_i. In our setting ζ is given by the pairwise Euclidean distances between the instances in the source and target distributions, i.e., $\zeta_{ij} = \|x^s_i - x^t_j\|_2$. The matrix $\gamma = (\gamma_{ij}) \in \mathbb{R}^{n_s \times n_t}_+$ is called a transportation plan, namely each entry γ_{ij} represents the fraction of mass moving from x^s_i to x^t_j. The minimum γ's in problem (1) is taken over the convex set of probability couplings between μ^s and μ^t defined by

$$\Pi(\mu^s, \mu^t) = \{\gamma \in \mathbb{R}^{n_s \times n_t}_+ : \gamma \mathbf{1}_{n_t} = \mu^s, \gamma^\top \mathbf{1}_{n_s} = \mu^t\},$$

where we identify the distributions with their probability mass vectors, i.e. $\mu^s \equiv (\mu^s_1, \ldots, \mu^s_{n_s})^\top$ (similarly for μ^t), and $\mathbf{1}_n \in \mathbb{R}^n$ stands for all-ones vector. The set

$\Pi(\mu^s, \mu^t)$ contains all possible joint probabilities with marginals corresponding to μ^s and μ^t. In the sequel when applied to matrices and vectors, product, division and exponential notations refer to element-wise operators.

Computing classical Wassertein distance is computationally expensive, since its Kantorovich formulation (1) is a standard linear program with a complexity $O(\max(n_s, n_t)^3)$ [15]. To overcome this issue, a prevalent approach, referred to as regularized OT [9], operates by adding an entropic regularization penalty to the original problem and it writes as

$$W_\eta(\mu^s, \mu^t) = \min_{\gamma \in \Pi(\mu^s, \mu^t)} \{\langle \zeta, \gamma \rangle_F - \eta H(\gamma)\} \tag{2}$$

where $H(\gamma) = -\sum_{i=1}^{n_s} \sum_{j=1}^{n_t} \gamma_{ij} \log \gamma_{ij}$ defines the entropy of the matrix γ and $\eta > 0$ is a regularization parameter to be chosen. Adding the entropic term makes the problem significantly more amenable to computations. In particular, it allows to solve efficiently the optimization problem (2) using a balancing algorithm known as Sinkhorn's algorithm [25]. Note that the Sinkhorn iterations are based on the dual solution of (2) (see [18] for more details).

3.2 First Step: Rejection of Unknown Class Samples

In the open set DA setting, a naive application of the preceding OT framework to source set $Z^s = \{(x_1^s, y_1^s), .., (x_{n_s}^s, y_{n_s}^s)\}$ and target dataset $Z^t = \{x_1^t, \ldots, x_{n_t}^t\}$ will lead to undesirable mappings as some source samples will be transported onto the abnormal target samples. To avoid this, we intend to learn a transportation map such that the probability mass sent to the unknown abnormal samples of the target domain will be negligible, hence discarding those samples. A way to achieve this goal is to adapt the target marginal distribution $P^t(X)$ while learning the map.

Therefore, to discard the new classes appearing in the target domain, in a first stage, we solve the following optimization problem:

$$\gamma_{\text{rej}}^\star, \mu^{t^\star} = \underset{\substack{\gamma \in \Pi(\mu^s, \mu^t) \\ \mu^t \in \Sigma_{n_t}}}{\text{argmin}} \{\langle \zeta, \gamma \rangle_F - \eta H(\gamma)\}, \tag{3}$$

where μ^t stands for the target marginal $P^t(X)$ and μ^s for the source one $P^s(X)$.

The first stage of the rejection step as formulated in (3) aims at calculating a transportation plan while optimizing the target marginal μ^t. The rationale for updating μ^t is linked to the new classes appearing in target domain. Therefore, the formulation allows some freedom on μ^t and leads to more accurate matching between known marginal source and unknown marginal target. To solve this optimization problem, we use Sinkhorn iterations [9]. Towards this end, we explicit its dual form in Lemma 1. Hereafter, we set $B(f, g) = \text{diag}(e^f) K \text{diag}(e^g)$ where $K = e^{-\zeta/\eta}$ stands for the Gibbs kernel associated to the cost matrix ζ and where diag denotes the diagonal operator.

Lemma 1. *The dual problem of* (3) *reads as*

$$(f^\star_{\text{rej}}, g^\star_{\text{rej}}) = \underset{f\in\mathbb{R}^{n_s}, g\in\mathbb{R}^{n_t}}{\operatorname{argmin}} \{\mathbf{1}_{n_s}^\top B(f,g)\mathbf{1}_{n_t} - \langle f, \mu^s\rangle + \chi_{-\mathbf{1}_{n_t}}(g)\}, \tag{4}$$

where for all $g \in \mathbb{R}^{n_t}$ *we denote by*

$$\chi_{-\mathbf{1}_{n_t}}(g) = \begin{cases} 0, & \text{if } g = -\mathbf{1}_{n_t}, \text{i.e.}\ g_j = -1, \forall j = 1, \dots, n_t, \\ \infty, & \text{otherwise.} \end{cases}$$

Note that the optimal solutions $\gamma^\star_{\text{rej}}$ and $\mu^{t^\star}$ of the primal problem take the form

$$\gamma^\star_{\text{rej}} = B(f^\star_{\text{rej}}, g^\star_{\text{rej}}), \quad \mu^{t^\star} = \gamma^{\star\top}_{\text{rej}}\mathbf{1}_{n_s}.$$

Once μ^t is learned, the second stage consists in discarding the new classes by relying on the values of $\mu^{t^\star}$. Specifically, we reject the j-th sample in the target set whenever $\mu^{t^\star}_j$ is a neglectable value with respect to some chosen threshold. Indeed, since $\gamma^\star_{\text{rej}}$ satisfies the target marginal constraint $\mu^{t^\star}_j = \sum_{i=1}^{n_s}(\gamma^\star_{\text{rej}})_{ij}$ for all $j = 1, \dots, n_t$, we expect that the row entries $\{(\gamma^\star_{\text{rej}})_{ij} : i = 1, \dots, n_s\}$ take small values for each j-th sample associated to a new class, that is we avoid transferring probability mass from source samples to the unknown target j-th instance. The tuning of the rejection threshold is exposed in Sect. 4.1.

The overall rejection procedure is depicted in Algorithm 1. To grasp the elements of Algorithm 1 and its stopping condition, we derive the Karush-Kuhn-Tucker (KKT) optimality conditions [5] for the rejection dual problem (Eq. 4) in Lemma 2.

Lemma 2. *The couple* $(f^\star_{\text{rej}}, g^\star_{\text{rej}})$ *optimum of problem* (4) *satisfies*

$$(f^\star_{\text{rej}})_i = \log(\mu^s_i) - \log\Big(\sum_{j=1}^{n_t} K_{ij}e^{(g^\star_{\text{rej}})_j}\Big) \tag{5}$$

and

$$\sum_{j=1}^{n_t} e^{(f^\star_{\text{rej}})_i}K_{ij}e^{(g^\star_{\text{rej}})_j}(1 + (g^\star_{\text{rej}})_j) \le 0, \tag{6}$$

for all $i = 1, \dots, n_s$.

The proofs of Lemma 1 and 2 are postponed to Appendix A.1.

We remark that we have a closed form of $f^\star_{\text{rej}}$, see Eq. 5, while it is not the case for $g^\star_{\text{rej}}$ as shown in Eq. 6. This is due to non-differentiability of the objective function defining the couple $(f^\star_{\text{rej}}, g^\star_{\text{rej}})$. Therefore, we tailor Algorithm 1 with a sufficient optimality condition to guarantee Eq. 6, in particular we set $(g^\star_{\text{rej}})_j \le -1$ for all $j = 1, \dots, n_t$. These latter conditions can be tested on the update of the target marginal μ^t for the rejection problem (see Steps 6–9 in Algorithm 1). We use the condition $\|B(f,g)\mathbf{1}_{n_t} - \mu^s\|_1 + \|B(f,g)^\top\mathbf{1}_{n_s} - \mu^t\|_1 \le \varepsilon$ (ε-tolerance) as a stopping criterion for Algorithm 1, which is very natural since it requires that $B(f,g)\mathbf{1}_{n_t}$ and $B(f,g)^\top\mathbf{1}_{n_s}$ are close to the source and target marginals μ^s and μ^t.

Algorithm 1. Rejection (see Equation 3)

require: η: regularization parameter, ζ: cost matrix, Z^t: target samples, n_s: number of source samples, n_t: number of target samples, *tol*: tolerance, *thresh*: threshold;
output: transport matrix: $\gamma_{\text{rej}} = B(f_{\text{rej}}, g_{\text{rej}})$; target marginal: μ^t; rejected samples: X^t_{rej}

1: **initialize:**
2: $err \leftarrow 0$; $f \leftarrow \mathbf{0}_{n_s}$; $g \leftarrow -\mathbf{1}_{n_t}$; $\mu^t \leftarrow \frac{1}{n_t}\mathbf{1}_{n_t}$;
3: **while** $err > tol$ **do**
4: $\quad f \leftarrow \log(\mu^s) - \log(Ke^g)$;
5: $\quad \mu^t \leftarrow B(f,g)^\top \mathbf{1}_{n_s}$;
6: $\quad$ **for all** $j = 1, \dots, n_t$ **do**
7: $\quad\quad$ **if** $\mu^t_j > e^{-1}\sum_{i=1}^{n_s} K_{ij}e^{g_j}$ **then**
8: $\quad\quad\quad \mu^t_j \leftarrow e^{-1}\sum_{i=1}^{n_s} K_{ij}e^{g_j}$;
9: $\quad\quad$ **end if**
10: $\quad$ **end for**
11: $\quad g \leftarrow \log(\mu^t)$ - $\log(K^\top e^f)$;
12: $\quad err \leftarrow \|B(f,g)\mathbf{1}_{n_t} - \mu^s\|_1 + \|B(f,g)^\top \mathbf{1}_{n_s} - \mu^t\|_1$;
13: **end while**
14: $Z^t_{\text{rej}} \leftarrow Z^t[\mu^t \leq thresh]$
15: **return:** $B(f,g)$, μ^t and Z^t_{rej}

3.3 Second Step: Label-Shift Correction

We re-weight source samples to correct the difference in class proportions between source and target domains. Correcting the label shift is formulated as

$$\gamma^\star_{\text{ls}}, \nu^\star = \underset{\substack{\gamma \in \Pi(D\nu, \mu^t) \\ \nu \in \Delta_C}}{\text{argmin}} \{\langle \zeta, \gamma \rangle_F - \eta H(\gamma)\}, \tag{7}$$

where the target marginal μ^t is either a uniform distribution or the one learned at the rejection step and where $D = (d_{ic}) \in \mathbb{R}^{n_s \times C}_+$ is a linear operator, such that for $i = 1, \dots, n_s$ and $c = 1, \dots, C$

$$d_{ic} = \begin{cases} \frac{1}{n^c_s}, & \text{if } y^s_i = c, \\ 0, & \text{otherwise.} \end{cases}$$

Here n^c_s denotes the cardinality of source samples with class c, namely $n^c_s = \#\{i = 1, \dots, n_s : y^s_i = c\}$. The parameter vector $\nu = (\nu_c)^C_{c=1}$ belongs to the convex set

$$\Delta_C = \Big\{\alpha \in \mathbb{R}^C_+ : \sum_{c=1}^{C}\sum_{i=1}^{n_s} d_{ic}\alpha_c = 1\Big\}.$$

In order to estimate the unknown class proportions in the target domain, we set-up the source marginal as $\mu^s = D\nu$ where the entry ν_c expresses the c-class proportion for all $c = 1, \dots, C$. Once we estimate theses proportions, we can get the class proportions in target domain thanks to OT matching. We shall stress that Problem (7) involves the simultaneous calculation of the transportation plan γ_{ls} and the source class re-weighting. Our procedure resembles the re-weighting method of JCPOT [19] except that we do not rely on a Wasserstein barycentric problem required by the multiple source setting addressed in [19]. The estimation $\nu^\star$ can be explicitly calculated using the source marginal constraint satisfied by the transportation plan $\gamma_{\text{ls}}^\star$, i.e.,

$$\nu^\star = (D^\top D)^{-1} D^\top \gamma_{\text{ls}}^\star \mathbf{1}_{n_t}.$$

As for the rejection step, we use Sinkhorn algorithm with an update on the source marginal $\mu^s = D\nu$ to solve the label shift Problem (7) via its dual as stated in Lemma 3.

Lemma 3. *The dual of Problem (7) writes as*

$$(f_{\text{ls}}^\star, g_{\text{ls}}^\star) = \underset{f \in \mathbb{R}^{n_s}, g \in \mathbb{R}^{n_t}}{\operatorname{argmin}} \{\mathbf{1}_{n_s}^\top B(f,g)\mathbf{1}_{n_t} - \langle g, \mu^t \rangle + \chi_{\mathcal{F}}(f)\}, \tag{8}$$

where $\mathcal{F} = \left\{f \in \mathbb{R}^{n_s} : \sum_{i=1}^{n_s} (f_i + 1) d_{ic} = 0, \forall c = 1, \ldots, C\right\}$ *and*

$$\chi_{\mathcal{F}}(f) = \begin{cases} 0, & \text{if } f \in \mathcal{F}, \\ \infty, & \text{otherwise.} \end{cases}$$

Moreover, the closed form of the transportation plan in the Label-Shift step is given by

$$\gamma_{\text{ls}}^\star = B(f_{\text{ls}}^\star, g_{\text{ls}}^\star).$$

The analysis details giving the dual formulation in Eq. (8) in Lemma 3 are presented in the appendices. As for the rejection problem, the optimality conditions of the Label-Shift problem are described in the dedicated Lemma 4 which proof is differed to Appendix A.3.

Lemma 4. *The couple* $(f_{\text{ls}}^\star, g_{\text{ls}}^\star)$ *optimum of problem* (8) *satisfies*

$$(g_{\text{ls}}^\star)_j = \log(\mu_j^t) - \log\Big(\sum_{i=1}^{n_s} K_{ij} e^{(f_{\text{ls}}^\star)_i}\Big) \tag{9}$$

and

$$\sum_{i=1}^{n_s} e^{(f_{\text{ls}}^\star)_i} K_{ij} e^{(g_{\text{ls}}^\star)_j} (1 + (f_{\text{ls}}^\star)_i) \leq 0, \tag{10}$$

for all $j = 1, \ldots, n_t$.

Algorithm 2 shows the related optimization procedure. Similarly to the rejection problem, we see that $g_{\text{ls}}^\star$ admits a close form (Eq. 9), while $f_{\text{ls}}^\star$ does not. As previously, we endow the Algorithm 2 with the sufficient optimality conditions (10) by ensuring $(f_{\text{ls}}^\star)_i \leq -1$ for all $i = 1, \ldots, n_s$. The conditions are evaluated on the source marginal $\mu^s = D\nu^\star$ (see Steps 5–9 in Algorithm 2). Finally we use the same ε-tolerance stopping condition $\|B(f,g)\mathbf{1}_{n_t} - \mu^s\|_1 + \|B(f,g)^\top \mathbf{1}_{n_s} - \mu^t\|_1 \leq \varepsilon$.

Algorithm 2. Label-Shift (see Equation 7)

require: η: regularization term; ζ: cost matrix, Y^s: source labels, n_s: number of source samples; n_t: number of target samples; C: number of classes; tol: tolerance; D: linear operator;

output: transport matrix: $\gamma_{\text{ls}} = B(f_{\text{ls}}, g_{\text{ls}})$; class proportions: ν; Prediction of target labels: $\hat{Y}^t$

1: **initialize:** $err \leftarrow 1$; $\nu \leftarrow \frac{1}{C}\mathbf{1}_C$; $f \leftarrow -\mathbf{1}_{n_s}$; $g \leftarrow \mathbf{0}_{n_t}$; $A \leftarrow (D^\top D)^{-1}D^\top$;
2: **while** $err > tol$ **do**
3: $\quad g \leftarrow \log(\mu^t) - \log(K^\top e^f)$;
4: $\quad \mu^s \leftarrow D\nu$;
5: $\quad$ **for all** $i = 1, \dots, n_s$ **do**
6: $\quad\quad$ **if** $\mu_i^s < e^{-1}\sum_{j=1}^{n_t} K_{ij}e^{g_j}$ **then**
7: $\quad\quad\quad \mu_i^s \leftarrow e^{-1}\sum_{j=1}^{n_t} K_{ij}e^{g_j}$;
8: $\quad\quad$ **end if**
9: $\quad$ **end for**
10: $\quad f \leftarrow \log(\mu^s) - \log(Ke^g)$;
11: $\quad \nu \leftarrow AB(f, g)\mathbf{1}_{n_t}$;
12: $\quad err \leftarrow \|B(f, g)1_{n_t} - \mu^s\|_1 + \|B(f, g)^\top \mathbf{1}_{n_s} - \mu^t\|_1$;
13: **end while**
14: $\hat{Y}^t \leftarrow \text{argmax}(D^\top B(f, g))$;//indices of the max. values of $D^\top B(f, g)$'s columns [19]
15: **return:** $B(f, g)$, ν and $\hat{Y}^t$

3.4 Implementation Details and Integration

Our proposed approach to open set DA performs samples rejection followed by sample matching in order to predict the target labels (either outlier or known source domain label). Hence, at the end of each step, we identify either rejected samples or predict target labels (see Step 14 of Algorithms 1 and 2).

For rejection, we compare the learned target marginal $\mu^{t^\star}$ to some threshold to recognize the rejected samples (See Sect. 4.1 for its tuning). To fix the threshold, we assume that the target samples that receive insufficient amount of probability mass coming from source classes likely cannot be matched to any source sample and hence are deemed outliers.

To predict the labels of the remaining target samples, we rely on the transportation map $\gamma_{\text{ls}}^\star$ given by Algorithm 2. Indeed for Label-Shift, JCPOT [19] suggested a label propagation approach to estimate labels from N transportation maps (corresponding to N source domains). Following JCPOT, the obtained transport matrix $\gamma_{\text{ls}}^\star$ is proportional to the target class proportions. Therefore, we estimate the labels of the target samples based on the probability mass they received from each source class using $\hat{Y}^t = \text{argmax}(D^\top \gamma_{\text{ls}}^\star)$. The term $D^\top \gamma_{\text{ls}}^\star$ provides a matrix of mass distribution over classes.

Finally, we stress that the rejection and Label-Shift steps are separately done allowing us to compare theses approaches with the sate-of-art. Nevertheless, we can make a joint 2-step, that means after rejecting the instances with new classes in the target domain we plug the obtained target marginal $\mu^{t^\star}$ in the Label-Shift step. Experimental evaluations show that similar performances are attained for separate and joint steps.

4 Numerical Experiments

To assess the performance of each step, we first present the evaluations of Rejection and Label-Shift algorithms so that we can compare them to state-of-the-art approaches. Then we present overall accuracy of the joint 2-step algorithm.

4.1 Abnormal Sample Rejection

We frame the problem as a binary classification where common and rejected classes refer respectively to positive and negative classes. Therefore, source domain has only one class (the positive) while target domain includes a mixture of positives and negatives. We estimate their proportions and compare our results to open set recognition algorithms for unknown classes detection.

To reject the target samples, we lay on the assumption that they correspond to entries with a small value in $\mu^{t^\star}$. The applied threshold to these entries is strongly linked to the regularization parameter η of the OT problem (3). We remark that when η increases, the threshold is high and vice versa, making the threshold proportional to η. Also experimentally, we notice that the threshold has the same order of magnitude of $1/(n_s + n_t)$. Therefore, we define a new hyper-parameter α such that the desired threshold is given by $\lambda = \alpha\frac{\eta}{n_s+n_t}$.

In order to fix the hyper-parameters (η, α) of the Rejection algorithm, we resort to Reverse Validation procedure [6,30]. For a standard classification problem where labels are assumed to be only available for source samples, a classifier is trained on $\{X^s, Y^s\}$ in the forward pass and evaluated on X^t to predict $\hat{Y}^t$. In the backward pass, the target samples with the pseudo-labels $\{X^t, \hat{Y}^t\}$ are used to retrain the classifier with the same hyper-parameters used during the first training, to predict $\hat{Y}^s$. The retained hyper-parameters are the ones that provide the best accuracy computed from $\{Y^s, \hat{Y}^s\}$ without requiring Y^t.

We adapt the reverse validation principle to our case. For fixed (η, α), Algorithm 1 is run to get μ^t and to identify abnormal target samples. These samples are removed from X^t leading to X^t_{rej}. Then the roles of X^s and X^t_{rej} are reversed. By running the Rejection algorithm to map X^t_{rej} onto X^s we expect that the yielded marginal μ^s will have entries greater than the threshold λ. This suggests that we did not reject erroneously the target samples during the forward pass. As we may encounter mus-rejection, we select the convenient hyper-parameters (η, α) that correspond to the highest $\frac{\#(\mu^t \leq \lambda)}{n_s}$. Algorithm 3 in Appendix A.5 gives the implementation details of the adapted Reverse Validation approach.

We use a grid search to find optimal hyperparameters (η, α). η was searched in the following set $\{0.001, 0.01, 0.05, 0.1, 0.5, 1, 5, 10\}$ and α in $\{0.1,1,10\}$. We apply Algorithm 3 and get $\eta = 0.1$ and $\alpha = 1$ for synthetic data and $\eta = 0.01$ and $\alpha = 10$ for real datasets.

Experiments on Synthetic Datasets. We use a mixture of 2D Gaussian dataset with 3 classes. We choose 1 or 2 classes to be rejected in target domain as shown in Table 1. We generate 1000 samples for each class in both domains with varying noise levels.

The change of rejected classes at each run induces a distribution shift between shared (Sh) and rejected (Rj) class proportions. Tables 1 and 2 present the recorded F1-score. For a fair comparison, we tune the hyper-parameters of the competitor algorithms and choose the best F1-score for each experiment.

Table 1. F1-score on target domain for the Rejection algorithm applied to synthetic dataset, Noise level = 0.5 and $\eta = 0.1$

Sh classes	{0,1}	{0,2}	{1,2}	{0}	{1}	{2}
Rj classes	{2}	{1}	{0}	{1,2}	{0,2}	{0,1}
% of Rj classes	33%	33%	33%	66%	66%	66%
1Vs (Linear)	0.46	0.5	0	0	0	0
WSVM (RBF)	0.99	**0.99**	0.99	–	–	–
PISVM (RBF)	0.99	**0.99**	0.69	0.5	0.5	0.5
Ours	**1**	**0.99**	**0.99**	**1**	**1**	**0.98**

Table 2. F1-score on target domain for the Rejection algorithm applied to synthetic dataset, Noise level = 0.75 and $\eta = 0.5$

Sh classes	{0,1}	{0,2}	{1,2}	{0}	{1}	{2}
Rj classes	{2}	{1}	{0}	{1,2}	{0,2}	{0,1}
% of Rj classes	33%	33%	33%	66%	66%	66%
1Vs (Linear)	0.5	0.37	0.49	0	0	0
WSVM (RBF)	0.81	0.8	0.79	–	–	–
PISVM (RBF)	0.94	0.83	0.8	0.5	0.5	0.5
Ours	**0.95**	**0.96**	**0.97**	**0.98**	**0.98**	**0.96**

Experiments on Real Datasets. For this step, we first evaluate our rejection algorithm on datasets under Label-Shift and open set classes. We modify the set of classes for each experiment in order to test different proportions of *rejected class*. We use USPS (U), MNIST (M) and SVHN (S) benchmarks. All the benchmarks contain 10 classes. USPS images have single channel and a size of 16×16 pixels, MNIST images have single channel and a size of 28×28 pixels while SVHN images have 3-color channels and a size of 32×32 pixels.

As a first experiment, we sample our source and target datasets from the same benchmark i.e. USPS → USPS, MNIST → MNIST and SVHN → MNIST.

We choose different samples for each domain and modify the set of shared and rejected classes. Then, we present challenging cases with increasing Covariate-Shift as source and target samples are from different benchmarks as shown in Table 3. For each benchmark, we resize the images to 32×32 pixels and split source samples into training and test sets. We extract feature embeddings using the following process: 1) We train a Neural Network (as suggested in [12]) on the training set of source domain, 2) We randomly sample 200 images (except for USPS 72 images instead) for each class from test set of source and target domains, and 3) We extract image embeddings of chosen samples from the last Fc layer (128 units) of the trained model.

We compare our Rejection algorithm to the 1-Vs Machine [24], PISVM [13] and WSVM [23][1] which are based on SVM and require a threshold to provide a decision. For tasks with a single rejected class, we get results similar to PISVM and WSVM when noise is small (Table 1) and outperfom all methods when noise increases (Table 2). These results prove that we are more robust to ambiguous dataset. For tasks with multiple rejected classes, WSVM is not suitable to this case and PISVM and 1Vs performs poorly compared to our approach. In fact, these approaches strongly depend on openness measure [13,23].

As for the case with small noise, we obtain similar results for DA tasks with Label-Shift only as shown in Table 3 while we outperform state-of-art methods for DA tasks combining target and covariate shifts (Table 4) except for last task where WSVM slightly exceeds our method. This confirms the ability of our approach to address challenging shifts. In addition, our proposed approach for the rejection step is based on OT which provides a framework consistent with the Label-Shift step.

Table 3. F1-score of Rejection algorithms applied to target samples of MNIST benchmark

Sh classes	{0,2,4}	{6,8}	{1,3,5}	{7,9}	{0,1,2,3,4}
Rj classes	{6,8}	{0,2,4}	{7,9}	{1,3,5}	{5,6,7,8,9}
% of Rj classes	40%	60%	40%	60%	50%
1Vs (Linear)	0.65 ± 0.01	0	0.74 ± 0.01	0.29 ± 0.04	0.61
WSVM (RBF)	0.97 ± 0.02	0.95 ± 0.0	**0.98** ± 0.01	0.76 ± 0.2	0.96 ± 0.01
PISVM (RBF)	**0.98** ± 0.01	0.96 ± 0.02	**0.98** ± 0.01	0.80 ± 0.16	**0.97** ± 0.01
Ours	**0.98** ± 0.01	**0.99** ± 0.01	**0.98** ± 0.014	**0.97** ± 0.01	0.93 ± 0.02

4.2 Label-Shift

We sample unbalanced source datasets and reversely unbalanced target datasets for both MNIST and SVHN benchmarks in order to create significant Label-Shift as shown in Fig. 1 in Appendix A.4. USPS benchmark is too small (2007 samples for test) and is already unbalanced. Therefore we use all USPS samples for the experiments M→U and U→M.

[1] https://github.com/ljain2/libsvm-openset.

Table 4. F1-score of Rejection algorithms applied to target samples where source domain: MNIST and target domain: USPS

Sh classes	{0,2,4}	{6,8}	{1,3,5}	{7,9}	{0,1,2,3,4}
Rj classes	{6,8}	{0,2,4}	{7,9}	{1,3,5}	{5,6,7,8,9}
% of Rj classes	40%	60%	40%	60%	50%
1Vs (Linear)	0.57 ± 0.04	0	0.62 ± 0.05	0.27 ± 0.06	0.53 ± 0.04
WSVM (RBF)	0.82 ± 0.09	0.69 ± 0.07	0.86 ± 0.05	0.64 ± 0.06	**0.79** ± 0.04
PISVM (RBF)	0.82 ± 0.09	0.68 ± 0.06	0.86 ± 0.05	0.66 ± 0.06	0.77 ± 0.04
Ours	**0.9** ± 0.02	**0.83** ± 0.03	**0.87** ± 0.05	**0.92** ± 0.02	0.74 ± 0.06

We create 5 tasks by increasing Covariate-Shift to evaluate the robustness of our algorithm. We compare our approach to JDOT [7] and JCPOT [19] which predicts target label in two different ways (label propagation JCPOT-LP and JCPOT-PT). We used the public code given by the authors for JDOT[2] and JCPOT[3]. Note that JCPOT is applied to multi-source samples. Consequently,we split $\{X^s, Y^s\}$ into N sources with random class proportions and chose N which gives the best results (N = 5). We present the results on 5 trials. We set $\eta = 0.001$ for all experiments with the Label-Shift algorithm. JCPOT uses a grid search to get its optimal η.

For synthetic dataset, JCPOT and our Label-Shift method give similar results (Table 5). However, for real datasets as shown in Table 6, we widely outperform other state-of-the-art DA methods especially for DA tasks that present covariate shift in addition to the Label-Shift. These results prove that our approach is more robust to high-dimensional dataset as well as to distributions with combined label and covariate shifts.

Table 5. Recorded F1-score for Label-Shift algorithms applied to synthetic datasets.

Setting	JDOT	JCPOT-LP(5)	JCPOT-PT(5)	Ours
Noise = 0.5	0.5	0.997	0.99	0.997
Noise = 0.75	0.45	0.98	0.94	0.98

Table 6. F1-score of Label-Shift algorithms on digits classification tasks.

Methods	M→M	S→S	M→U	U→M	S→M
JDOT	0.52 ± 0.04	0.53 ± 0.01	0.64 ± 0.01	0.87 ± 0.02	0.43 ± 0.01
JCPOT-LP(5)	**0.98** ± 0.002	0.37 ± 0.43	0.56 ± 0.0.026	0.89 ± 0.01	0.21 ± 0.237
JCPOT-PT(5)	0.96 ± 0.004	0.81 ± 0.045	0.40 ± 0.327	0.86 ± 0.013	0.46 ± 0.222
Ours	**0.98** ± 0.001	**0.92** ± 0.006	**0.76** ± 0.019	**0.92** ± 0.006	**0.65** ± 0.017

[2] Code available at https://github.com/rflamary/JDOT.
[3] Code available at https://github.com/ievred/JCPOT.

4.3 Full 2-Step Approach: Rejection and Label-Shift

The same shared and rejected classes from the rejection experiments tasks have been chosen. We also create significant Label-Shift as done for Label-Shift experiments (Unbalanced and Reversely-unbalanced class proportions) for synthetic datasets as well as for MNIST and SVHN real benchmarks. Nevertheless, we keep the initial class proportions of USPS due to the size constraint of the database. This time, we implement a jointly 2-step. Namely, we plug the obtained target marginal in the Label-Shift step after discarding rejected samples. We apply Algorithm 3 to rejection step to get optimal hyperparameters (η, α) and keep the same η for Label-shift step. We obtained $\eta = 0.001$ and $\alpha = 1$.

In Table 7, we show results for synthetic data generated with different noises. When noise increases, i.e., boundary decision between classes is ambiguous, the performance is affected. Table 8 presents F1-score over 10 runs of our 2-step approach applied to real datasets. For DA tasks with only Label-Shift (M→M and S→S), F1-score is high. However it drops when we address both Covariate and Label-Shift (M→U, U→M and S→M). In fact, previous results for each step (Tables 4 and 6) have shown that performance was affected by Covariate-Shift. The final result of our 2-step approach is linked to the performance of each separate step. We present an illustration of the full algorithm in Fig. 2 in Appendix A.6.

Table 7. F1-score across target samples of combined our 2-step approach applied to synthetic data, $\eta = 0.001$, $\alpha = 1$

Sh classes	{0,1}	{0,2}	{1,2}
Rj classes	{2}	{1}	{0}
Noise = 0.5	1	0.99	0.99
Noise = 0.75	0.93	0.87	0.85

Table 8. F1-score across target samples of our combined 2-step approach applied to real datasets features, $\eta = 0.001$, $\alpha = 1$

Benchmarks	M→M	S→S	M→U	U→M	S→M
Sh {0,2,4} Rj {6,8}	0.93 ± 0.005	0.91 ± 0.008	0.65 ± 0.011	0.59 ± 0.014	0.66 ± 0.011
Sh {6,8} Rj {0,2,4}	0.95 ± 0.006	0.89 ± 0.012	0.82 ± 0.013	0.61 ± 0.01	0.53 ± 0.014
Sh {1,3,5} Rj {7,9}	0.93 ± 0.009	0.86 ± 0.01	0.76 ± 0.02	0.58 ± 0.011	0.74 ± 0.018
Sh {7,9} Rj {1,3,5}	0.97 ± 0.009	0.90 ± 0.011	0.75 ± 0.008	0.52 ± 0.007	0.65 ± 0.021
Sh {0,1,2,3,4} Rj {5,6,7,8,9}	0.91 ± 0.01	0.82 ± 0.007	0.73 ± 0.013	0.74 ± 0.01	0.68 ± 0.01

5 Conclusion

In this paper, we proposed an optimal transport framework to solve open set DA. It is composed of two steps solving Rejection and Label-shift adaptation problems. The main idea was to learn the transportation plans together with the marginal distributions. Notably, experimental evaluations showed that applying our algorithms to various datasets lead to consistent outperforming results over the state-of-the-art. We plan to extend the framework to learn deep networks for open set domain adaptation.

Acknowledgements. This work was supported by the National Research Fund, Luxembourg (FNR) and the OATMIL ANR-17-CE23-0012 Project of the French National Research Agency (ANR).

A Appendix

A.1 Proof of Lemma 1

Define the dual Lagrangian function

$$\begin{aligned}&\mathscr{L}_{\text{rej}}(\gamma, \mu^t, \lambda, \beta, \vartheta, \theta)\\&= \langle \zeta, \gamma\rangle_F - \eta H(\gamma) + \langle \lambda, \gamma \mathbf{1}_{n_t} - \mu^s\rangle + \langle \beta, \gamma^\top \mathbf{1}_{n_s} - \mu^t\rangle - \langle \vartheta, \mu^t\rangle + \theta(\|\mu^t\|_1 - 1)\\&= \langle \zeta, \gamma\rangle_F - \eta H(\gamma) + \langle \lambda, \gamma \mathbf{1}_{n_t}\rangle + \langle \beta, \gamma^\top \mathbf{1}_{n_s}\rangle - \langle \beta, \mu^t\rangle + \theta\|\mu^t\|_1 - \langle \vartheta, \mu^t\rangle - \langle \lambda, \mu^s\rangle - \theta\end{aligned}$$

equivalently

$$\mathscr{L}_{\text{rej}}(\gamma, \mu^t, \lambda, \beta, \theta) = E_{\text{rej}}(\gamma) + F_{\text{rej}}(\mu^t) + G_{\text{rej}}(\lambda, \theta),$$

where

$$E_{\text{rej}}(\gamma) = \langle \zeta, \gamma\rangle_F - \eta H(\gamma) + \langle \lambda, \gamma \mathbf{1}_{n_t}\rangle + \langle \beta, \gamma^\top \mathbf{1}_{n_s}\rangle,$$

$$F_{\text{rej}}(\mu^t) = -\langle \beta, \mu^t\rangle - \langle \vartheta, \mu^t\rangle + \theta\|\mu^t\|_1, \text{ and } G_{\text{rej}}(\lambda, \theta) = -\langle \lambda, \mu^s\rangle - \theta.$$

We have

$$\frac{\partial \mathscr{L}_{\text{rej}}(\gamma, \mu^t, \lambda, \beta, \vartheta, \theta)}{\partial \gamma_{ij}} = \frac{\partial E_{\text{rej}}(\gamma)}{\partial \gamma_{ij}} = C_{ij} + \eta(\log \gamma_{ij} + 1) + \lambda_j + \beta_j,$$

and

$$\frac{\partial \mathscr{L}_{\text{rej}}(\gamma, \mu^t, \lambda, \beta, \vartheta, \theta)}{\partial \mu^t_j} = \frac{\partial F_{\text{rej}}(\mu^t)}{\partial \mu^t_j} = -\beta_j - \vartheta_j + \theta.$$

Then the couple $(\gamma^\star_{\text{rej}}, \mu^{t\star})$ optimum of the dual Lagrangian function $\mathscr{L}_{\text{rej}}(\gamma, \mu^t, \lambda, \beta, \theta)$ satisfies the following

$$\begin{cases}\frac{\partial \mathscr{L}_{\text{rej}}(\gamma^\star_{\text{rej}}, (\mu^t)^\star, \lambda, \beta, \vartheta, \theta)}{\partial \gamma^\star_{ij}} = 0\\ \frac{\partial \mathscr{L}_{\text{rej}}(\gamma, (\mu^t)^\star, \lambda, \beta, \vartheta, \theta)}{\partial {\mu^t}^\star_{ij}} = 0\end{cases} \equiv \begin{cases}(\gamma^\star_{\text{rej}})_{ij} = \exp\big(-\frac{C_{ij}+\lambda_i+\beta_j}{\eta} - 1\big),\\ \theta - \beta_j = 0,\end{cases}$$

for all $i = 1, \dots, n_s$ and $j = 1, \dots, n_t$. Now, plugging this solution in the Lagrangian function we get

$$\begin{aligned}\mathscr{L}_{\text{rej}}(\gamma^\star_{\text{rej}}, \mu^{t\star}, \lambda, \beta, \vartheta, \theta) &= \sum_{i=1}^{n_s}\sum_{j=1}^{n_t} C_{ij} \exp\big(-\frac{C_{ij}+\lambda_i+\beta_j}{\eta} - 1\big) \\ &\quad + \eta \sum_{i=1}^{n_s}\sum_{j=1}^{n_t} (-\frac{C_{ij}+\lambda_i+\beta_j}{\eta} - 1) \exp\big(-\frac{C_{ij}+\lambda_i+\beta_j}{\eta} - 1\big) \\ &\quad + \sum_{i=1}^{n_s} \lambda_i \sum_{j=1}^{n_t} \exp\big(-\frac{C_{ij}+\lambda_i+\beta_j}{\eta} - 1\big) \\ &\quad + \sum_{j=1}^{n_t} \beta_j \sum_{i=1}^{n_s} \exp\big(-\frac{C_{ij}+\lambda_i+\beta_j}{\eta} - 1\big) \\ &\quad - \langle \beta, \mu^{t\star}\rangle - \langle \vartheta, \mu^{t\star}\rangle + \theta\|\mu^{t\star}\|_1, -\langle \lambda, \mu^s\rangle - \theta.\end{aligned}$$

Note that $-\langle \beta, \mu^{t\star}\rangle - \langle \vartheta, \mu^{t\star}\rangle + \theta\|\mu^{t\star}\|_1 = \langle -\beta - \vartheta + \theta \mathbf{1}_{n_t}, \mu^{t\star}\rangle$, hence taking into account the constraint $\theta - \beta_j - \vartheta_j = 0$, for all $j = 1, \dots, n_t$, it entails that $-\langle \beta + \vartheta, \mu^{t\star}\rangle + \theta\|\mu^{t\star}\|_1 = 0$. Hence

$$\mathscr{L}_{\text{rej}}(\gamma^\star_{\text{rej}}, \mu^{t\star}, \lambda, \beta, \vartheta, \theta) = -\eta \sum_{i=1}^{n_s}\sum_{j=1}^{n_t} \exp\big(-\frac{C_{ij}+\lambda_i+\beta_j}{\eta} - 1\big) - \langle \lambda, \mu^s\rangle - \theta,$$

subject to $\theta - \beta_j - \vartheta_j = 0$ for all $j = 1, \dots, n_t$. Setting the following variable change $f = -\frac{\lambda}{\eta} - \frac{1}{2}\mathbf{1}_{n_s}$ and $g = -\frac{\beta}{\eta} - \frac{1}{2}\mathbf{1}_{n_t}$ we get

$$\begin{aligned}\mathscr{L}_{\text{rej}}(\gamma^\star_{\text{rej}}, \mu^{t\star}, \lambda, \beta, \vartheta, \theta) &= -\eta \sum_{i=1}^{n_s}\sum_{j=1}^{n_t} \exp\big(-\frac{C_{ij}}{\eta} + f_i + g_j\big) + \eta\langle (f + \frac{1}{2}\mathbf{1}_{n_s}), \mu^s\rangle - \theta \\ &= -\eta \sum_{i=1}^{n_s}\sum_{j=1}^{n_t} \exp\big(-\frac{C_{ij}}{\eta} + f_i + g_j\big) + \eta\langle f, \mu^s\rangle + \eta\frac{1}{2} - \theta \\ &= -\eta \mathbf{1}_{n_s}^\top B(f,g)\mathbf{1}_{n_t} + \eta\langle f, \mu^s\rangle + \eta\frac{1}{2} - \theta.\end{aligned}$$

Then

$$\mathscr{L}_{\text{rej}}(\gamma^\star_{\text{rej}}, \mu^{t\star}, \lambda, \beta, \vartheta, \theta) = -\eta\big\{\mathbf{1}_{n_s}^\top B(f,g)\mathbf{1}_{n_t} - \langle f, \mu^s\rangle - \frac{1}{2} + \frac{\theta}{\eta}\big\},$$

subject to $\theta + \eta(g_j + \frac{1}{2}) = 0$. Putting $\kappa = \frac{\theta}{\eta} - \frac{1}{2}$, then $\theta = \eta(\kappa + \frac{1}{2})$. This gives

$$\mathscr{L}_{\text{rej}}(\gamma^\star_{\text{rej}}, \mu^{t\star}, \lambda, \beta, \vartheta, \theta) \equiv \mathscr{L}_{\text{rej}}(\gamma^\star_{\text{rej}}, \mu^{t\star}, \lambda, \beta, \kappa) = -\eta\big\{\mathbf{1}_{n_s}^\top B(f,g)\mathbf{1}_{n_t} - \langle f, \mu^s\rangle + \kappa\big\},$$

subject to $g_j + \kappa + 1 = 0$, for all $j = 1, \dots, n_t$. We remark that

$$\mathbf{1}_{n_s}^\top B(f,g)\mathbf{1}_{n_t} = \sum_{i=1}^{n_s}\sum_{j=1}^{n_t} e^{f_i - \kappa} K_{ij} e^{g_j + \kappa} = \mathbf{1}_{n_s}^\top B(f - \kappa\mathbf{1}_{n_s}, g + \kappa\mathbf{1}_{n_t})\mathbf{1}_{n_t},$$

then using a variable change $\tilde{f} = f - \kappa\mathbf{1}_{n_s}$ and $\tilde{g} = g + \kappa\mathbf{1}_{n_t}$ we get

$$(f^\star_{\text{rej}}, g^\star_{\text{rej}}) = \underset{\substack{\tilde{f}\in\mathbb{R}^{n_s}, \tilde{g}\in\mathbb{R}^{n_t}, \\ \tilde{g}_j + 1 = 0, \forall j = 1,\dots,n_t}}{\text{argmin}} \{\mathbf{1}_{n_s}^\top B(\tilde{f},\tilde{g})\mathbf{1}_{n_t} - \langle \tilde{f}, \mu^s\rangle\}.$$

Therefore

$$(f^\star_{\text{rej}}, g^\star_{\text{rej}}) = \underset{f\in\mathbb{R}^{n_s}, g\in\mathbb{R}^{n_t}}{\operatorname{argmin}} \{\mathbf{1}_{n_s}^\top B(f,g)\mathbf{1}_{n_t} - \langle f, \mu^s\rangle + \chi_{-\mathbf{1}_{n_t}}(g)\}.$$

A.2 Proof of Lemma 2

Setting

$$\Psi(f,g) = \mathbf{1}_{n_s}^\top B(f,g)\mathbf{1}_{n_t} - \langle f, \mu^s\rangle + \chi_{-\mathbf{1}_{n_t}}(g),$$

Writting the KKT optimality condition for the above problem leads to the following: we have $f \mapsto \Psi(f,g)$ is differentiable, hence we can calculate a gradient with respect to f. However $g \mapsto \Psi(f,g$ is not differentiable, then we just calculate a subdifferentiale as follows:

$$\nabla\Psi(f,g) = \Big\{e^{f_i}\sum_{j=1}^{n_t} K_{ij}e^{g_j} - \mu_i^s\Big\}_{1\le i\le n_s} \in \mathbb{R}^{ns},$$

and

$$\partial_g(\Psi(f,g)) = \Big\{e^{g_j}\sum_{i=1}^{n_s} K_{ij}e^{f_i} + \partial(\chi_{-\mathbf{1}_{n_t}}(g))\Big\}_{1\le j\le n_t},$$

where $\partial_g(\chi_{-\mathbf{1}_{n_t}}(g))$ is the subdifferential of the indicator function $\chi_{-\mathbf{1}_{n_t}}$ at g is known as the normal cone, namely

$$\begin{aligned}\partial(\chi_{-\mathbf{1}_{n_t}}(g)) &= \{\boldsymbol{u}\in\mathbb{R}^{n_t} | \boldsymbol{u}^\top g \ge -\boldsymbol{u}^\top \mathbf{1}_{n_t}\}\\ &= \Big\{\boldsymbol{u}\in\mathbb{R}^{n_t} | \sum_{j=1}^{n_t} u_j g_j \ge -\sum_{j=1}^{n_t} u_j\Big\}.\end{aligned}$$

Therefore, KKT optimality conditions give

$$e^{f^\star_{\text{rej}}} = \frac{\mu}{Ke^{g^\star_{\text{rej}}}} \text{ and } -e^{g^\star_{\text{rej}}}\cdot K^\top e^{f^\star_{\text{rej}}} \in \partial(\chi_{-\mathbf{1}_{n_t}}(g^\star_{\text{rej}})),$$

(the division / and the multiplication $\cdot$ between vectors have to be understood elementwise). So

$$e^{(f^\star_{\text{rej}})_i} = \frac{\mu_i^s}{\sum_{j=1}^{n_t} K_{ij}e^{(g^\star_{\text{rej}})_j}} \text{ and } -\sum_{j=1}^{n_t} e^{(g^\star_{\text{rej}})_j}K_{ij}e^{(f^\star_{\text{rej}})_i}(g^\star_{\text{rej}})_j \ge -(-\sum_{j=1}^{n_t} e^{(g^\star_{\text{rej}})_j}K_{ij}e^{(f^\star_{\text{rej}})_i}),$$

equivalently

$$e^{(f^\star_{\text{rej}})_i} = \frac{\mu_i^s}{\sum_{j=1}^{n_t} K_{ij}e^{(g^\star_{\text{rej}})_j}} \text{ and } \sum_{j=1}^{n_t} e^{(f^\star_{\text{rej}})_i}K_{ij}e^{(g^\star_{\text{rej}})_j}(1+(g^\star_{\text{rej}})_j) \le 0.$$

for all $i = 1,\ldots,n_s$.

A.3 Proof of Lemma 3

First, observe that $\Delta_C = \{\alpha \in \mathbb{R}_+^C : \mathbf{1}_C^\top D\alpha = 1\}$. Then the dual Lagrangian function is given by

$$\begin{aligned}
&\mathscr{L}_{\text{ls}}(\gamma, \nu, \lambda, \beta, \vartheta, \theta) \\
&= \langle \zeta, \gamma \rangle_F - \eta H(\gamma) + \langle \lambda, \gamma \mathbf{1}_{n_t} - D\nu \rangle + \langle \beta, \gamma^\top \mathbf{1}_{n_s} - \mu^t \rangle - \langle \vartheta, \nu \rangle + \theta(\mathbf{1}_C^\top D\nu - 1) \\
&= \langle \zeta, \gamma \rangle_F - \eta H(\gamma) + \langle \lambda, \gamma \mathbf{1}_{n_t} \rangle + \langle \beta, \gamma^\top \mathbf{1}_{n_s} \rangle - \langle \lambda, D\nu \rangle - \langle \beta, \mu^t \rangle - \langle \vartheta, \nu \rangle + \theta \mathbf{1}_C^\top D\nu - \theta,
\end{aligned}$$

equivalently

$$\mathscr{L}_{\text{ls}}(\gamma, \mu^t, \lambda, \beta, \vartheta, \theta) = E_{\text{ls}}(\gamma) + F_{\text{ls}}(\nu) + G_{\text{ls}}(\lambda, \theta),$$

where

$$E_{\text{ls}}(\gamma) = \langle \zeta, \gamma \rangle_F - \eta H(\gamma) + \langle \lambda, \gamma \mathbf{1}_{n_t} \rangle + \langle \beta, \gamma^\top \mathbf{1}_{n_s} \rangle,$$

$$F_{\text{ls}}(\nu) = -\langle \lambda, D\nu \rangle - \langle \vartheta, \nu \rangle + \theta \mathbf{1}_C^\top D\nu, \text{ and } G_{\text{ls}}(\beta, \theta) = -\langle \beta, \mu^s \rangle - \theta.$$

We have

$$\frac{\partial \mathscr{L}_{\text{ls}}(\gamma, \mu^t, \lambda, \beta, \vartheta, \theta)}{\partial \gamma_{ij}} = \frac{\partial E_{\text{ls}}(\gamma)}{\partial \gamma_{ij}} = C_{ij} + \eta(\log \gamma_{ij} + 1) + \lambda_j + \beta_j,$$

and

$$\frac{\partial \mathscr{L}_{\text{rej}}(\gamma, \mu^t, \lambda, \beta, \vartheta, \theta)}{\partial \nu_c} = \frac{\partial F_{\text{ls}}(\nu)}{\partial \nu_c} = -\sum_{i=1}^{n_s} \lambda_i d_{ic} + \theta \sum_{i=1}^{n_s} d_{ic} = \sum_{i=1}^{n_s} (\theta - \lambda_i) d_{ic} - \vartheta_c$$

Then the couple $(\gamma_{\text{ls}}^\star, \nu^\star)$ optimum of the dual Lagrangian function $\mathscr{L}_{\text{ls}}(\gamma, \nu, \lambda, \beta, \theta)$ satisfies the following

$$\begin{cases} \frac{\partial \mathscr{L}_{\text{ls}}(\gamma_{\text{ls}}^\star, \nu^\star, \lambda, \vartheta, \beta, \theta)}{\partial (\gamma_{\text{ls}}^\star)_{ij}} = 0 \\ \frac{\partial \mathscr{L}_{\text{ls}}(\gamma_{\text{ls}}^\star, \nu^\star, \lambda, \beta, \vartheta, \theta)}{\partial \nu_c^\star} = 0 \end{cases} \equiv \begin{cases} (\gamma_{\text{ls}}^\star)_{ij} = \exp\big(-\frac{C_{ij} + \lambda_i + \beta_j}{\eta} - 1\big), \\ \sum_{i=1}^{n_s} (\theta - \lambda_i) d_{ic} - \vartheta_c = 0, \end{cases}$$

for all $i = 1, \ldots, n_s$, $j = 1, \ldots, n_t$, and $c = 1, \ldots, C$. Now, plugging this solution in the Lagrangian function we get

$$\begin{aligned}
\mathscr{L}_{\text{ls}}(\gamma_{\text{ls}}^\star, \nu^\star, \lambda, \vartheta, \beta, \theta) &= \sum_{i=1}^{n_s} \sum_{j=1}^{n_t} C_{ij} \exp\big(-\frac{C_{ij} + \lambda_i + \beta_j}{\eta} - 1\big) \\
&+ \eta \sum_{i=1}^{n_s} \sum_{j=1}^{n_t} \big(-\frac{C_{ij} + \lambda_i + \beta_j}{\eta} - 1\big) \exp\big(-\frac{C_{ij} + \lambda_i + \beta_j}{\eta} - 1\big) \\
&+ \sum_{i=1}^{n_s} \lambda_i \sum_{j=1}^{n_t} \exp\big(-\frac{C_{ij} + \lambda_i + \beta_j}{\eta} - 1\big) \\
&+ \sum_{j=1}^{n_t} \beta_j \sum_{i=1}^{n_s} \exp\big(-\frac{C_{ij} + \lambda_i + \beta_j}{\eta} - 1\big) \\
&- \langle \lambda, D\nu^\star \rangle - \langle \vartheta, \nu^\star \rangle + \theta \mathbf{1}_C^\top D\nu^\star - \langle \beta, \mu^t \rangle - \theta
\end{aligned}$$

Observe that

$$\theta \mathbf{1}_C^\top D\nu^\star - \langle \lambda, D\nu^\star \rangle - \langle \vartheta, \nu^\star \rangle = \sum_{c=1}^{C}\sum_{i=1}^{n_s} d_{ic}\nu_c^\star - \sum_{c=1}^{C}\sum_{i=1}^{n_s} d_{ic}\lambda_i d_{ic}\nu_c^\star - \sum_{c=1}^{C}\vartheta_c \nu_c^\star$$

$$= \sum_{c=1}^{C}\Big(\sum_{i=1}^{n_s}(\theta - \lambda)d_{ic} - \vartheta_c\Big)\nu_c^\star.$$

Taking into account the constraint $\sum_{i=1}^{n_s}(\theta-\lambda_i)d_{ic}-\vartheta_c = 0$, for all $c = 1, \dots, C$, it entails that $\theta \mathbf{1}_C^\top D\nu^\star - \langle \lambda, D\nu^\star \rangle - \langle \vartheta, \nu^\star \rangle = 0$. Hence

$$\mathscr{L}_{\text{ls}}(\gamma_{\text{ls}}^\star, \nu^\star, \lambda, \beta, \theta) = -\eta \sum_{i=1}^{n_s}\sum_{j=1}^{n_t} \exp\big(-\frac{C_{ij}+\lambda_i+\beta_j}{\eta} - 1\big) - \langle \beta, \mu^t \rangle - \theta,$$

subject to $\sum_{i=1}^{n_s}(\theta - \lambda_i)d_{ic} - \vartheta_c = 0$, for all $c = 1, \dots, C$. Setting the following variable change $f = -\frac{\lambda}{\eta} - \frac{1}{2}\mathbf{1}_{n_s}$ and $g = -\frac{\beta}{\eta} - \frac{1}{2}\mathbf{1}_{n_t}$ we get

$$\mathscr{L}_{\text{ls}}(\gamma_{\text{ls}}^\star, \nu^\star, \lambda, \beta, \vartheta, \theta) = -\eta \sum_{i=1}^{n_s}\sum_{j=1}^{n_t} \exp\big(-\frac{C_{ij}}{\eta} + f_i + g_j\big) + \eta\langle (g + \frac{1}{2}\mathbf{1}_{n_t}), \mu^t \rangle - \theta$$

$$= -\eta \sum_{i=1}^{n_s}\sum_{j=1}^{n_t} \exp\big(-\frac{C_{ij}}{\eta} + f_i + g_j\big) + \eta\langle g, \mu^t \rangle + \eta\frac{1}{2} - \theta$$

$$= -\eta \mathbf{1}_{n_s}^\top B(f,g)\mathbf{1}_{n_t} + \eta\langle g, \mu^t \rangle + \eta\frac{1}{2} - \theta.$$

Then

$$\mathscr{L}_{\text{ls}}(\gamma_{\text{ls}}^\star, \nu^\star, \lambda, \beta, \vartheta, \theta) \equiv \mathscr{L}_{\text{ls}}(\gamma_{\text{ls}}^\star, \nu^\star, \lambda, \beta, \theta) = -\eta\{\mathbf{1}_{n_s}^\top B(f,g)\mathbf{1}_{n_t} - \langle g, \mu^t \rangle - \frac{1}{2} + \frac{\theta}{\eta}\},$$

subject to $\sum_{i=1}^{n_s}(\theta + \eta(f_i + \frac{1}{2}))d_{ic} = 0$, for all $c = 1, \dots, C$. Putting $\kappa = \frac{\theta}{\eta} - \frac{1}{2}$, then $\theta = \eta(\kappa + \frac{1}{2})$. This gives

$$\mathscr{L}_{\text{ls}}(\gamma_{\text{ls}}^\star, \nu^\star, \lambda, \beta, \kappa) = -\eta\{\mathbf{1}_{n_s}^\top B(f,g)\mathbf{1}_{n_t} - \langle g, \mu^t \rangle + \kappa\},$$

subject to $\sum_{i=1}^{n_s}(f_i + \kappa + 1)d_{ic} = 0$, for all $c = 1, \dots, C$. We remark that

$$\mathbf{1}_{n_s}^\top B(f,g)\mathbf{1}_{n_t} = \sum_{i=1}^{n_s}\sum_{j=1}^{n_t} e^{f_i - \kappa}K_{ij}e^{g_j + \kappa} = \mathbf{1}_{n_s}^\top B(f - \kappa\mathbf{1}_{n_s}, g + \kappa\mathbf{1}_{n_t})\mathbf{1}_{n_t}^\top,$$

then using a variable change $\tilde{f} = f + \kappa\mathbf{1}_{n_s}$ and $\tilde{g} = g - \kappa\mathbf{1}_{n_t}$ we get

$$(f_{\text{ls}}^\star, g_{\text{ls}}^\star) = \underset{\substack{\tilde{f}\in\mathbb{R}^{n_s}, \tilde{g}\in\mathbb{R}^{n_t}, \\ \sum_{i=1}^{n_s}(\tilde{f}_i+1)d_{ic}=0, \forall c=1,\dots,C}}{\operatorname{argmin}} \{\mathbf{1}_{n_s}^\top B(\tilde{f}, \tilde{g})\mathbf{1}_{n_t} - \langle \tilde{g}, \mu^t \rangle\}.$$

Finally

$$(f_{\text{ls}}^\star, g_{\text{ls}}^\star) = \underset{\substack{f\in\mathbb{R}^{n_s}, g\in\mathbb{R}^{n_t}, \\ \sum_{i=1}^{n_s}(f_i+1)d_{ic}=0, \forall c=1,\dots,C}}{\operatorname{argmin}} \{\mathbf{1}_{n_s}^\top B(f, g)\mathbf{1}_{n_t} - \langle g, \mu^t \rangle\}$$

that is

$$(f_{\text{ls}}^{\star}, g_{\text{ls}}^{\star}) = \operatorname*{argmin}_{f \in \mathbb{R}^{n_s}, g \in \mathbb{R}^{n_t}} \{\mathbf{1}_{n_s}^{\top} B(f, g) \mathbf{1}_{n_t} - \langle g, \mu^t \rangle + \chi_{\mathcal{F}}(f)\}.$$

Remark 1. We omit the proof of Lemma 4 since it follows exactly the same lines as proof of Lemma 2.

A.4 Unbalancement Trade-Off

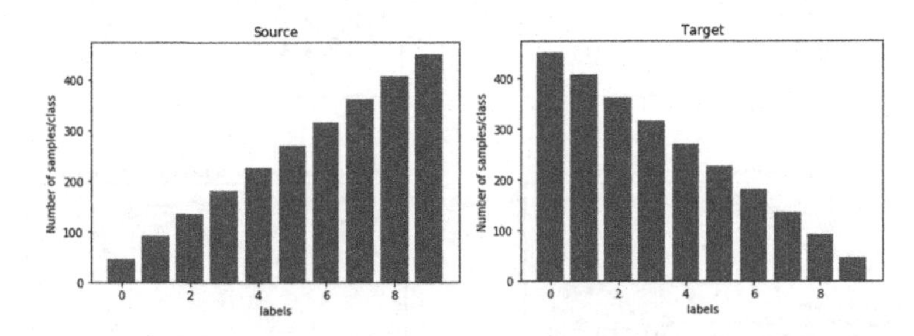

Fig. 1. Unbalanced source and reversely unbalanced target class proportions applied in label shift experiments as shown in Table 6

A.5 Reverse validation details

Algorithm 3. Reverse Validation

---

require: η: list of suggested regularization terms; α: list of values to fix threshold, X^s: source samples; $\{X^t\}$: target samples; n_s: number of source samples; n_t: number of target samples;
output: (η_i, α_j): tuple of hyperparameters;

1: **initialize:**
2: $k \leftarrow 0, errors \leftarrow [\,], hyperparameters \leftarrow [\,]$;
3: **for all** η_i in η **do**
4:     **for all** α_j in α **do**
5:         $thresh \leftarrow \alpha \frac{\eta}{n_s + n_t}$;
6:         $\mu^t \leftarrow$ Rejection$(X^s, X^t, thresh)$;
7:         $X_{sc}^t \leftarrow X^t[\mu^t > thresh]$;
8:         $X_{new}^s \leftarrow X_{sc}^t, X_{new}^t \leftarrow X^s$;
9:         $\mu_{sc}^t \leftarrow$ Rejection$(X_{new}^s, X_{new}^t, thresh)$;
10:         $errors[k] \leftarrow \frac{\#(\mu^t \leq thresh)}{n_s}$;
11:         $hyperparameters[k] \leftarrow (\eta_i, \alpha_j)$;
12:         $k \leftarrow k + 1$;
13:     **end for**
14: **end for**
15: **return:** $hyperparameters[\text{argmax}(errors)]$;

A.6 Algorithm Illustration

See Fig. 3.

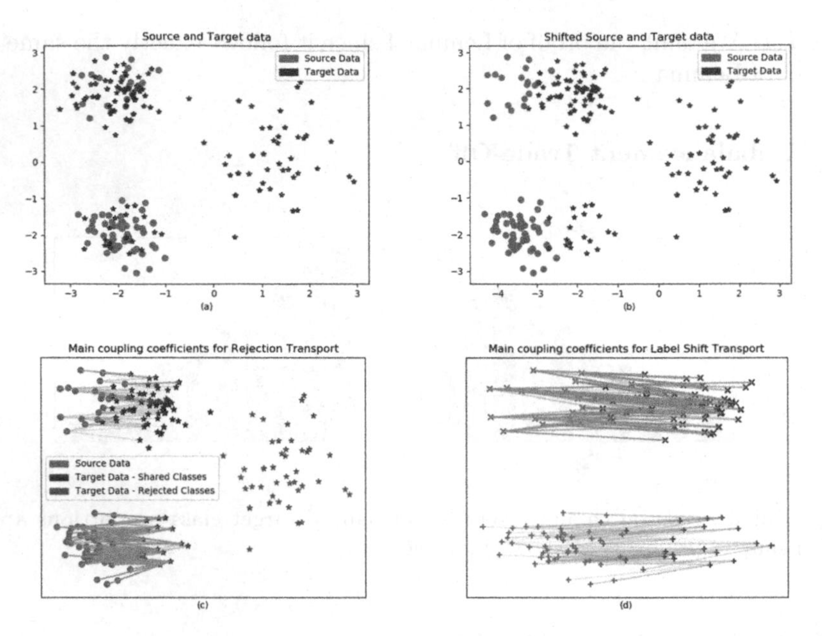

Fig. 2. Illustration of our 2-step approach for open set DA: (a) Mixture of 2D Gaussian data where the classes {0,1} are the common classes between source and target domains and the class {2} is the rejected class in the target domain. The common classes' proportions for the source domain are $[0.25, 0.75]$ while for the target domain they are chosen as $[0.75, 0.25]$; (b) Source data are slightly shifted from target data to ease visualization of mass transportation in figures (c) and (d); (c) Rejection step: Rejected points (in green) correspond to the points that receive a negligible amount of probability mass from source samples (d) Label-shift: mass transportation map obtained by label-shift algorithm. (Color figure online)

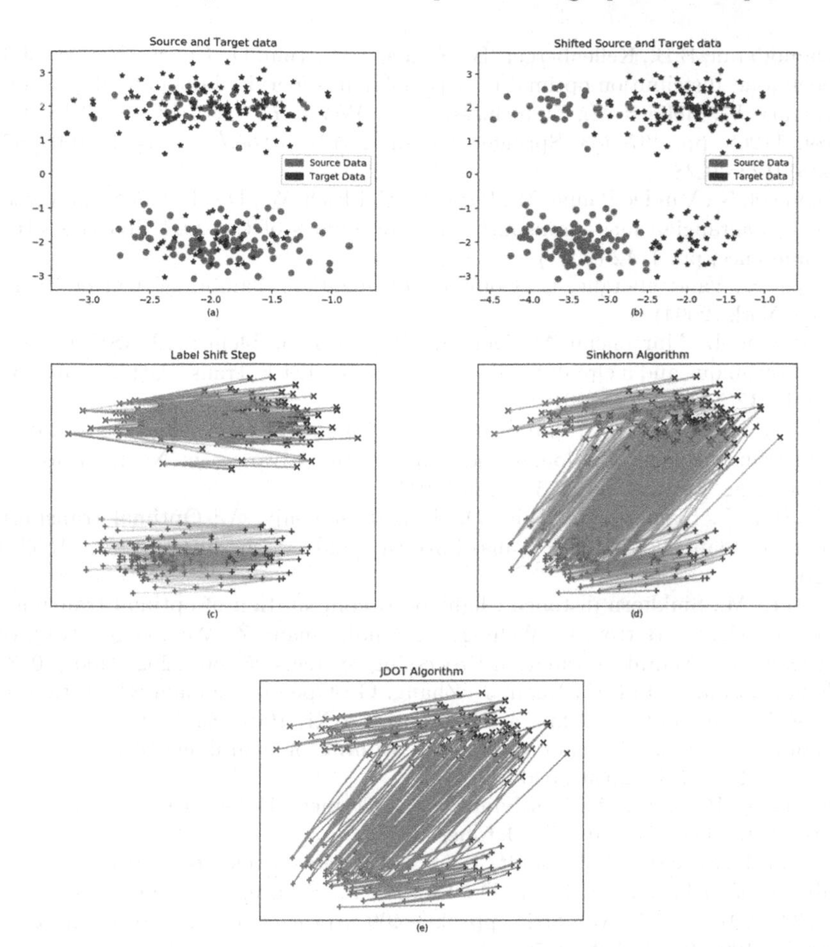

Fig. 3. Illustration of the Label-Shift on mixture of 2D Gaussian data with 2 classes such that the class proportions of the source domain are $[0.25, 0.75]$ while for the target domain are $[0.75, 0.25]$: (a) data; (b) Source data are slightly shifted from target data to ease visualization of mass transportation in figures (c), (d) and (e); (c) Our Label-Shift step; (d) Unsupervised DA solution using uniform source and target marginals. Samples from one source class split their probability mass between target samples of both classes due to the existing label shift; (e) A similar behavior is observed for JDOT algorithm.

References

1. Arjovsky, M., Chintala, S., Bottou, L.: Wasserstein generative adversarial networks. In: Proceedings of the 34th International Conference on Machine Learning, vol. 70, pp. 214–223 (2017)
2. Ben-David, S., Blitzer, J., Crammer, K., Kulesza, A., Pereira, F., Vaughan, J.: A theory of learning from different domains. Machine Learn. **79**, 151–175 (2010)

3. Damodaran, B.B., Kellenberger, B., Flamary, R., Tuia, D., Courty, N.: DeepJDOT: deep joint distribution optimal transport for unsupervised domain adaptation. In: Ferrari, V., Hebert, M., Sminchisescu, C., Weiss, Y. (eds.) ECCV 2018. LNCS, vol. 11208, pp. 467–483. Springer, Cham (2018). https://doi.org/10.1007/978-3-030-01225-0_28
4. Bonneel, N., Van De Panne, M., Paris, S., Heidrich, W.: Displacement interpolation using Lagrangian mass transport. In: Proceedings of the 2011 SIGGRAPH Asia Conference, pp. 1–12 (2011)
5. Boyd, S., Vandenberghe, L.: Convex Optimization. Cambridge University Press, New York (2004)
6. Bruzzone, L., Marconcini, M.: Domain adaptation problems: A DASVM classification technique and a circular validation strategy. IEEE Trans. Pattern Anal. Mach. Intell. **32**(5), 770–787 (2010)
7. Courty, N., Flamary, R., Habrard, A., Rakotomamonjy, A.: Joint distribution optimal transportation for domain adaptation. In: Advances in Neural Information Processing Systems, pp. 3730–3739 (2017)
8. Courty, N., Flamary, R., Tuia, D., Rakotomamonjy, A.: Optimal transport for domain adaptation. IEEE Trans. Pattern Anal. Mach. Intell. **39**(9), 1853–1865 (2016)
9. Cuturi, M.: Sinkhorn distances: lightspeed computation of optimal transport. In: Burges, C.J.C., Bottou, L., Welling, M., Ghahramani, Z., Weinberger, K.Q. (eds.) Advances in Neural Information Processing Systems 26, pp. 2292–2300 (2013)
10. Fang, Z., Lu, J., Liu, F., Xuan, J., Zhang, G.: Open set domain adaptation: theoretical bound and algorithm. arXiv preprint arXiv:1907.08375 (2019)
11. Ganin, Y., et al.: Domain-adversarial training of neural networks. J. Machine Learn. Res. **17**(1), 2030–2096 (2016)
12. Haeusser, P., Frerix, T., Mordvintsev, A., Cremers, D.: Associative domain adaptation. In: The IEEE ICCV, October 2017
13. Jain, L.P., Scheirer, W.J., Boult, T.E.: Multi-class open set recognition using probability of inclusion. In: Fleet, D., Pajdla, T., Schiele, B., Tuytelaars, T. (eds.) ECCV 2014. LNCS, vol. 8691, pp. 393–409. Springer, Cham (2014). https://doi.org/10.1007/978-3-319-10578-9_26
14. Kantorovich, L.: On the transfer of masses. Dokl. Akad. Nauk SSSR **2**, 227–229 (1942). (in Russian)
15. Lee, Y.T., Sidford, A.: Path finding methods for linear programming: solving linear programs in ö(vrank) iterations and faster algorithms for maximum flow. In: 2014 IEEE 55th Annual Symposium on Foundations of Computer Science, pp. 424–433 (2014)
16. Lipton, Z.C., Wang, Y., Smola, A.J.: Detecting and correcting for label shift with black box predictors. In: Proceedings of the 35th International Conference on Machine Learning (2018)
17. Panareda Busto, P., Gall, J.: Open set domain adaptation. In: Proceedings of the IEEE International Conference on Computer Vision, pp. 754–763 (2017)
18. Peyré, G., Cuturi, M.: Computational optimal transport. Found. Trends® Mach. Learn. **11**(5–6), 355–607 (2019)
19. Redko, I., Courty, N., Flamary, R., Tuia, D.: Optimal transport for multi-source domain adaptation under target shift. Proc. Mach. Learn. Res. **89**, 849–858 (2019)
20. Redko, I., Habrard, A., Sebban, M.: Theoretical analysis of domain adaptation with optimal transport. In: Ceci, M., Hollmén, J., Todorovski, L., Vens, C., Džeroski, S. (eds.) ECML PKDD 2017. LNCS (LNAI), vol. 10535, pp. 737–753. Springer, Cham (2017). https://doi.org/10.1007/978-3-319-71246-8_45

21. Saito, K., Yamamoto, S., Ushiku, Y., Harada, T.: Open set domain adaptation by backpropagation. In: Ferrari, V., Hebert, M., Sminchisescu, C., Weiss, Y. (eds.) ECCV 2018. LNCS, vol. 11209, pp. 156–171. Springer, Cham (2018). https://doi.org/10.1007/978-3-030-01228-1_10
22. Sanderson, T., Scott, C.: Class proportion estimation with application to multiclass anomaly rejection. In: Artificial Intelligence and Statistics, pp. 850–858 (2014)
23. Scheirer, W.J., Jain, L.P., Boult, T.E.: Probability models for open set recognition. IEEE Trans. Pattern Anal. Mach. Intell. **36**(11), 2317–2324 (2014)
24. Scheirer, W., Rocha, A., Sapkota, A., Boult, T.: Toward open set recognition. IEEE Trans. Pattern Anal. Mach. Intell. **35**, 1757–72 (2013)
25. Sinkhorn, R.: Diagonal equivalence to matrices with prescribed row and column sums. Am. Math. Month. **74**(4), 402–405 (1967)
26. Villani, C.: Topics in Optimal Transportation. Graduate studies in mathematics, American Mathematical Society (2003)
27. Wu, Y., Winston, E., Kaushik, D., Lipton, Z.: Domain adaptation with asymmetrically-relaxed distribution alignment. In: International Conference on Machine Learning, pp. 6872–6881 (2019)
28. Zhang, K., Schölkopf, B., Muandet, K., Wang, Z.: Domain adaptation under target and conditional shift. In: International Conference on Machine Learning, pp. 819–827 (2013)
29. Zhao, H., Combes, R.T.D., Zhang, K., Gordon, G.: On learning invariant representations for domain adaptation. In: Proceedings of the 36th International Conference on Machine Learning, vol. 97, pp. 7523–7532 (2019)
30. Zhong, E., Fan, W., Yang, Q., Verscheure, O., Ren, J.: Cross validation framework to choose amongst models and datasets for transfer learning. In: Balcázar, J.L., Bonchi, F., Gionis, A., Sebag, M. (eds.) ECML PKDD 2010. LNCS (LNAI), vol. 6323, pp. 547–562. Springer, Heidelberg (2010). https://doi.org/10.1007/978-3-642-15939-8_35

Sketching, Sampling, and Binary Projections

Revisiting Wedge Sampling for Budgeted Maximum Inner Product Search

Stephan S. Lorenzen[1] and Ninh Pham[2(✉)]

[1] University of Copenhagen, Copenhagen, Denmark
lorenzen@di.ku.dk
[2] University of Auckland, Auckland, New Zealand
ninh.pham@auckland.ac.nz

Abstract. Top-k maximum inner product search (MIPS) is a central task in many machine learning applications. This work extends top-k MIPS with a budgeted setting, that asks for the best approximate top-k MIPS given a limited budget of computational operations. We investigate recent advanced sampling algorithms, including wedge and diamond sampling, to solve budgeted top-k MIPS. First, we show that diamond sampling is essentially a combination of wedge sampling and basic sampling for top-k MIPS. Our theoretical analysis and empirical evaluation show that wedge is competitive (often superior) to diamond on approximating top-k MIPS regarding both efficiency and accuracy. Second, we propose *dWedge*, a very simple *deterministic* variant of wedge sampling for budgeted top-k MIPS. Empirically, dWedge provides significantly higher accuracy than other budgeted top-k MIPS solvers while maintaining a similar speedup.

Keywords: Budgeted maximum inner product search · Sampling

1 Introduction

Maximum inner product search (MIPS) is the task of, given a point set $\mathbb{X} \subset \mathbb{R}^d$ of size n and a query point $\mathbf{q} \in \mathbb{R}^d$, finding the point $\mathbf{p} \in \mathbb{X}$ such that,

$$\mathbf{p} = \arg\max_{\mathbf{x} \in \mathbb{X}} \mathbf{x} \cdot \mathbf{q} \ .$$

MIPS and its variant top-k MIPS, which finds the top-k largest inner product points with a query, are central tasks in the retrieval phase of standard collaborative filtering based recommender systems [7,17]. They are also algorithmic ingredients in a variety of machine learning tasks, for instance, prediction tasks on multi-class learning [8,23] and neural network [6,26].

Modern real-world online recommender systems often deal with very large-scale data sets and a limited amount of response time. Such collaborative filtering

Partially supported by the Innovation Fund Denmark through the DABAI project.

F. Hutter et al. (Eds.): ECML PKDD 2020, LNAI 12457, pp. 439–455, 2021.
https://doi.org/10.1007/978-3-030-67658-2_25

based systems often present users and items as low-dimensional vectors. A large inner product between these vectors indicates that the items are relevant to the user preferences. The recommendation is often performed in the *online* manner since the user vector is updated online with ad-hoc contextual information only available during the interaction [3,15,16]. A personalized recommender needs to infer user preferences based on online user behavior, e.g. recent search queries and browsing history, as implicit feedback to return relevant results [13,22]. Since the retrieval of recommended items is only performed online, the result of this task might not be "perfect" given a small amount of waiting time but its accuracy/relevance should be improved given more waiting time. Hence, it is challenging to not only speed up the MIPS process, but to trade the search efficiency for the search quality for performance improvement.

Motivated by the computational bottleneck in the retrieval phase of modern recommendation systems, this work investigates the *budgeted* MIPS problem, a natural extension of MIPS with a computational limit for the search efficiency and quality trade-off. Our budgeted MIPS addresses the following question:

Given a data structure built in $\tilde{\mathcal{O}}(dn)$ time [1] *and budgeted computational operations, can we have an algorithm to return the best approximate top-k MIPS?*

To measure the accuracy of approximate top-k MIPS, we use the search *recall*, i.e. the empirical probability of retrieving the true top-k MIPS. In our budgeted setting, we limit the time complexity of building a data structure to $\tilde{\mathcal{O}}(dn)$ since when a context is used in a recommender system, the learning phase cannot be done entirely offline [3,15]. In other words, the item vectors are also computed online and hence a high cost of constructing the data structure will degrade the performance. Furthermore, since user preferences often change over time, a recommender system needs to frequently update its factorization model to address such *drifting* user preferences. This means that the item vectors and our data structure will be updated frequently.

It is worth noting that the budgeted MIPS has been recently studied in [30] given a budget of $B = \eta n$ inner product computation where η is a small constant, e.g. 5%. Furthermore, such budget constraints on the number of computational operations or on accessing a limit number of data points are widely studied not only on search problems [21] but also on clustering [19,24] and other problems [11, 31] when dealing with large-scale complex data sets.

1.1 Prior Art on Solving MIPS and Its Limit on Budgeted MIPS

It is well-known that due to the "curse of dimensionality", any exact solution for MIPS based on data or space partitioning indexing data structures generally degrades when dimensionality increases. It is no better than a simple sequential scanning when the dimensionality is larger than 10 [18,28]. Hence recent work on solving MIPS focuses on speeding up sequential scanning by pruning the search space [2,18,27]. Though such methods can solve MIPS exactly, they require $\Theta(n)$

[1] Polylogarithmic factors, e.g. $\log d \log n$ is absorbed in the $\tilde{\mathcal{O}}$-notation.

operations and therefore do not fit well to the budgeted MIPS setting where we might need $o(n)$ operations for some query. Furthermore, these methods do not provide any trade-off between the search quality and efficiency for online queries.

Another research direction is investigating approximation solutions that trade accuracy for efficiency. Since *locality-sensitive hashing* (LSH) [12] has emerged as a basic algorithmic tool for similarity search in high dimensions due to the sublinear query time guarantee, several approaches have followed this direction to obtain sublinear solutions for approximate MIPS [14,20,25,29]. Due to the inner product not being a metric, these LSH-based solutions have to convert MIPS to near neighbor search problem by applying order-preserving transformations to exploit the LSH framework.

Although LSH-based approaches can guarantee sublinear query time, the top-k inner product values are often very small compared to the vector norms in high dimensions. This means that the distance gap between "close" and "far apart" points after transformation in the LSH framework is arbitrarily small. That leads to not only the space usage (i.e. the number of hash tables) blow up, but also degrading LSH performance [1]. Furthermore, the LSH trade-off between search quality and efficiency is somewhat "fixed" for any query since it is governed by specific parameters of the LSH data structure, e.g. number of hash functions. We will show these limitations in our experiment section. Note that the learning phase of a recommender system has to be executed in an online manner when a context is used [3,15]. In this case, the significant cost of building LSH tables will be a computational bottleneck for handling online recommendations.

An alternative efficient solution is applying sampling methods to estimate the matrix-vector multiplication derived by top-k MIPS [4,5]. The basic idea is to sample a point $\mathbf{x}$ with probability proportional to the inner product $\mathbf{x} \cdot \mathbf{q}$. The larger inner product values the point $\mathbf{x}$ has, the more occurrences of $\mathbf{x}$ in the sample set. By the end of the sampling process, we extract top-B points ($B > k$) with the largest occurrences in the sample set via a counting histogram. The top-k points with the largest inner product among these B points will be returned as an approximate top-k MIPS. Such sampling schemes naturally fit the budgeted setting since the more samples provide the higher accuracy.

1.2 Our Contribution

This work studies sampling methods for solving the budgeted MIPS since they naturally fit to the class of budgeted problems. Sampling schemes provide not only the trade-off between search quality and search efficiency but also a flexible mechanism to control this trade-off via the number of samples S and the number of inner product computation B. Our contributions are as follows:

1. We revise popular sampling methods for solving MIPS, including basic sampling, wedge sampling [5], and the state-of-the-art diamond sampling methods [4]. We show that diamond sampling is essentially a combination of basic sampling and wedge sampling.

2. Our novel theoretical analysis and empirical evaluation illustrate that wedge is competitive (often superior) to diamond on approximating top-k MIPS regarding both efficiency and accuracy.
3. We propose *dWedge*, a very simple but efficient *deterministic* variant of wedge sampling, with a flexible mechanism to govern the trade-off between search quality and efficiency for the budgeted top-k MIPS.
4. Empirically, dWedge outperforms other competitive budgeted MIPS solvers [20,29,30] on standard recommender system data sets. Especially, dWedge returns the top-10 MIPS with at least 90% accuracy with the speedup between 20x and 180x compared to the brute-force search on our large-scale data sets.

2 Notation and Preliminaries

We present the point set $\mathbb{X}$ as a matrix $\mathbf{X} \subset \mathbb{R}^{n \times d}$ where each point $\mathbf{x}_i$ corresponds to the ith row, and the query point $\mathbf{q}$ as a column vector $\mathbf{q} = (q_1, \ldots, q_d)^T$. We use $i \in [n]$ to index row vectors of $\mathbf{X}$, i.e. $\mathbf{x}_i = (x_{i1}, \ldots, x_{id}) \in \mathbb{R}^d$. Since we will describe our investigated methods using the column-wise matrix-vector multiplication $\mathbf{Xq}$, we use $j \in [d]$ to index column vectors of $\mathbf{X}$, i.e. $\mathbf{y}_j = (x_{1j}, \ldots, x_{nj})^T \in \mathbb{R}^n$. For each column j, we pre-compute its 1-norm $c_j = \|y_j\|_1$.

We briefly review sampling approaches for estimating inner products $\mathbf{x}_i \cdot \mathbf{q}$. For simplicity, we first assume that $\mathbf{X}$ and $\mathbf{q}$ are non-negative. Then we show how to extend these approaches to handle negative inputs with their limits. We consider the column-wise matrix-vector multiplication $\mathbf{Xq}$ as follows.

$$
\begin{aligned}
\mathbf{Xq} &= \begin{bmatrix} x_{11} \\ \vdots \\ x_{n1} \end{bmatrix} q_1 + \begin{bmatrix} x_{12} \\ \vdots \\ x_{n2} \end{bmatrix} q_2 + \ldots + \begin{bmatrix} x_{1d} \\ \vdots \\ x_{nd} \end{bmatrix} q_d \\
&= \mathbf{y}_1 q_1 + \mathbf{y}_2 q_2 + \ldots + \mathbf{y}_d q_d
\end{aligned} \tag{1}
$$

2.1 Basic Sampling

Basic sampling is a very straightforward method to estimate the inner product $\mathbf{x}_i \cdot \mathbf{q}$ for the point $\mathbf{x}_i$. For any row i, we sample a column j with probability $q_j/\|\mathbf{q}\|_1$ and return x_{ij}. Define a random variable $Z_i = x_{ij}$, we have

$$
\mathbf{E}\left[Z_i\right] = \sum_{j=1}^{d} x_{ij} \frac{q_j}{\|\mathbf{q}\|_1} = \frac{\mathbf{x}_i \cdot \mathbf{q}}{\|\mathbf{q}\|_1} .
$$

The basic sampling suffers large variance when most of the contribution of $\mathbf{x}_i \cdot \mathbf{q}$ are from a few coordinates. In particular, the variance will be significantly large when the main contributions of $\mathbf{x}_i \cdot \mathbf{q}$ are from a few coordinates $x_{ij}q_j$ and q_j are very small. Note that this basic sampling approach has been used in [10] as an efficient sampling technique for approximating matrix-matrix multiplication.

Algorithm 1: Wedge sampling

Data: Matrices $\mathbf{X}$, query $\mathbf{q}$, pre-computed values z and c_j for each $j \in [d]$, number of samples S and number of inner products B.
Result: Approximate top-k MIPS for $\mathbf{q}$.
1 **Screening:** Wedge sample S points and increase its counter value.
2 Extract top-B points with the largest values from the counter histogram.
3 **Ranking:** Compute these B inner products and return top-k points with the largest inner product values.

Negative Inputs: To handle the negative cases, one can change the sampling probability to $|q_j|/\|\mathbf{q}\|_1$ and return $Z_i = \mathtt{sgn}(q_j)x_{ij}$ where $\mathtt{sgn}$ is the sign function, i.e. $\mathtt{sgn}(u) = -1$ if $u < 0$ and $\mathtt{sgn}(u) = 1$ if $u \geq 0$. It is clear that $\mathbf{E}[Z_i] = \mathbf{x}_i \cdot \mathbf{q}/\|\mathbf{q}\|_1$. Despite providing the unbiased estimate, this scheme needs $S = \Omega(n)$ samples for estimating n inner product values to answer top-k MIPS.

2.2 Wedge Sampling

Cohen and Lewis [5] proposed an efficient sampling approach, called wedge sampling, to approximate matrix multiplication and to isolate the largest inner products as a byproduct. Wedge sampling needs to pre-compute some statistics, including the sum of all inner products $z = \sum_i z_i$ where $z_i = \mathbf{x}_i \cdot \mathbf{q}$ and 1-norm of column vectors $c_j = \|\mathbf{y}_j\|_1$. Since we can pre-compute c_j *before* querying, computing $z = \sum_j q_j c_j$ clearly takes $\mathcal{O}(d)$ query time. We can think of $q_j c_j / z$ as the contribution ratio of the column j to the sum of inner product values z.

The basic idea of wedge sampling is to randomly sample a row index i corresponding to $\mathbf{x}_i$ with probability z_i/z. Hence, the larger the inner product $z_i = \mathbf{x}_i \cdot \mathbf{q}$, the larger the number of occurrences of i in the sample set. Consider Eq. (1), wedge sampling first samples a column j corresponding to $\mathbf{y}_j$ with probability $q_j c_j / z$, and then samples a row i corresponding to $\mathbf{x}_i$ from $\mathbf{y}_j$ with probability x_{ij}/c_j. By Bayes's theorem, we have

$$\begin{aligned}\mathbf{Pr}[\mathtt{Sampling}\ i] &= \sum_{j=1}^{d} \mathbf{Pr}[\mathtt{Sampling}\ i|\mathtt{Sampling}\ j] \cdot \mathbf{Pr}[\mathtt{Sampling}\ j] \\ &= \sum_{j=1}^{d} \frac{x_{ij}}{c_j} \cdot \frac{q_j c_j}{z} = \frac{\sum_{j=1}^{d} x_{ij} q_j}{z} = \frac{z_i}{z} .\end{aligned}$$

Applying wedge sampling method on $\mathbf{Xq}$, we obtain a sample set where each index i corresponding to $\mathbf{x}_i$ is sampled according to an independent Bernoulli distribution with parameter $p_i = z_i/z$. As a screening phase, a simple counting algorithm will be used to find the points with the largest counters. Given S samples and a constant cost for each sample, such a counting algorithm runs in $\mathcal{O}(S + min(S, n) \log k)$ time to answer approximate top-k MIPS. If we have an additional budget of $B > k$ inner product computation, we can compute the exact inner product values of the top-B points with the largest counter values for ranking. Such ranking (or post-processing) phase with an additional $\mathcal{O}(dB)$ (since $d > \log k$) computational cost will provide higher accuracy for top-k MIPS in practice. Algorithm 1 shows how the wedge sampling works.

We note that since wedge sampling uses the contribution ratio $q_j c_j / z$ to sample the column j, it can alleviate the effect of skewness of $\mathbf{x}_i \cdot \mathbf{q}$ where large contributions are from a few coordinates. Hence wedge sampling achieves lower variance than the basic sampling in practice.

Negative Inputs: Again, we can use the sign trick to deal with negative cases. We note that this trick has been first exploited in the diamond sampling approach [4]. In particular, we execute wedge sampling on absolute values of $\mathbf{X}$ and $\mathbf{q}$, and return $Z_i = \texttt{sgn}(x_{ij})\texttt{sgn}(q_j)$ for the point $\mathbf{x}_i$. It is clear that $\mathbf{E}[Z_i]$ is proportional to $\mathbf{x}_i \cdot \mathbf{q}$. The analysis of this trick under some assumptions of the data distribution can be found in [9].

2.3 Diamond Sampling

Ballard et al. [4] proposed diamond sampling to find the largest *magnitude* elements from a matrix-matrix multiplication $\mathbf{XQ}$ without computing the final matrix directly. The method first presents $\mathbf{XQ}$ as a weighted tripartite graph. Then it samples a diamond, i.e. four cycles from such graph with probability proportional to the value $(\mathbf{XQ})^2_{ij}$, which claims to amplify the focus on the largest magnitude elements.

Consider a vector $\mathbf{q}$ as a one-column matrix $\mathbf{Q}$, it is clear that diamond sampling can be applied to solve MIPS. Indeed, we will show that diamond sampling is essentially a combination of wedge sampling and basic sampling when approximating $\mathbf{Xq}$. In particular, diamond sampling first makes use of wedge sampling to return a random row i corresponding to $\mathbf{x}_i$ with probability z_i/z. Given such row i, it then applies basic sampling to sample a random column j' with probability $q_{j'}/\|\mathbf{q}\|_1$ and return $x_{ij'}$ as a scaled estimate of $(\mathbf{x}_i \cdot \mathbf{q})^2$. Define a random variable $Z_i = x_{ij'}$ corresponding to $\mathbf{x}_i$, using the properties of wedge sampling and basic sampling we have

$$\mathbf{E}[Z_i] = \sum_{j'=1}^{d} x_{ij'} \frac{q_{j'}}{\|\mathbf{q}\|_1} \cdot \frac{z_i}{z} = \frac{(\mathbf{x}_i \cdot \mathbf{q})^2}{z\|\mathbf{q}\|_1} .$$

Since diamond sampling builds on basic sampling, it suffers from the same drawback as basic sampling. To answer top-k MIPS, diamond follows the same procedure as wedge hence shares the same asymptotic running time.

Negative Inputs: Handling negative cases using diamond is similar to wedge. We apply diamond sampling on absolute values of $\mathbf{X}$ and $\mathbf{q}$ then return $Z_i = \texttt{sgn}(q_j)\texttt{sgn}(x_{ij})\texttt{sgn}(q_{j'})x_{ij'}$ where j is the column sampled by wedge sampling and j' is the column sampled by basic sampling. We can verify that $\mathbf{E}[Z_i]$ is proportional to $(\mathbf{x}_i \cdot \mathbf{q})^2$.

Although diamond sampling can deal with negative inputs, its concentration bound only works on non-negative cases. Furthermore, diamond sampling indeed solves a different problem, i.e. $\arg\max_i (\mathbf{x}_i \cdot \mathbf{q})^2$, which will give a completely different result on negative inputs. In practice, the implementation of diamond sampling requires significant query time overhead due to the basic sampling. This sampling process generates random variables from a discrete distribution derived from the query $\mathbf{q}$ and requires expensive *random* access operations to access $x_{ij'}$. We will show that wedge significantly outperforms diamond regarding both accuracy and efficiency on our benchmark data sets.

3 Wedge Sampling for Budgeted Top-k MIPS

This section first presents a new analysis of wedge sampling. Our theoretical concentration bound shows that wedge requires fewer samples than diamond on approximating top-k MIPS. Then we present a drawback of wedge sampling for the budgeted MIPS and propose *dWedge*, a simple deterministic variant to handle such drawback. dWedge can govern the trade-off between search quality and efficiency with two parameters: the number of samples S and the number of B inner product computation.

3.1 Analysis of Wedge Sampling

This subsection shows the analysis of wedge sampling on non-negative inputs. Consider a counting histogram of n counters corresponding to n point indexes, the following theorem states the number of samples required to distinguish between two inner product values τ_1 and τ_2.

Theorem 1. *Fix two thresholds $\tau_1 > \tau_2 > 0$ and suppose $S \geq \frac{3z\ln n}{(\sqrt{\tau_1}-\sqrt{\tau_2})^2}$ where $z = \sum_i \mathbf{x}_i \cdot \mathbf{q}$. With probability at least $1 - \frac{1}{n}$, the following holds for all pairs $i_1, i_2 \in [n]$: if $\mathbf{x}_{i_1} \cdot \mathbf{q} \geq \tau_1$ and $\mathbf{x}_{i_2} \cdot \mathbf{q} \leq \tau_2$, then counter$[i_1] >$ counter$[i_2]$.*

Proof. Define $p_1 = \frac{\mathbf{x}_{i_1}\cdot\mathbf{q}}{z} \geq \frac{\tau_1}{z}$ and $p_2 = \frac{\mathbf{x}_{i_2}\cdot\mathbf{q}}{z} \leq \frac{\tau_2}{z}$. For $l = 1, \ldots, S$, we consider independent pair of random variables (X_l, Y_l) where

$$X_l = \begin{cases} 1 \text{ if } \mathbf{x}_{i_1} \text{is chosen at } l\text{th sample;} \\ 0 \text{ otherwise} \end{cases} \qquad Y_l = \begin{cases} 1 \text{ if } \mathbf{x}_{i_2} \text{is chosen at } l\text{th sample;} \\ 0 \text{ otherwise.} \end{cases}$$

Define $X = \sum_{l=1}^{S} X_l$ and $Y = \sum_{l=1}^{S} Y_l$. We only consider the failure case where $Y - X \geq 0$. Applying Markov inequality for any $\lambda > 0$, we have

$$\begin{aligned}\mathbf{Pr}\left[Y - X \geq 0\right] &= \mathbf{Pr}\left[e^{\lambda(Y-X)} \geq 1\right] \leq \mathbf{E}\left[e^{\lambda(Y-X)}\right] \\ &= \mathbf{E}\left[e^{\lambda(\sum_l Y_l - \sum_l X_l)}\right] = \prod_{l=1}^{S} \mathbf{E}\left[e^{\lambda(Y_l - X_l)}\right].\end{aligned}$$

We also have

$$\mathbf{E}\left[e^{\lambda(Y_l - X_l)}\right] = e^{\lambda}p_2 + (1 - p_1 - p_2) + e^{-\lambda}p_1$$
$$\geq 2\sqrt{p_1 p_2} + 1 - p_1 - p_2 = 1 - \left(\sqrt{p_1} - \sqrt{p_2}\right)^2 .$$

The equality holds when $\lambda = \ln\sqrt{p_1/p_2} > 0$. In other words, by choosing $\lambda = \ln\sqrt{p_1/p_2}$, we have

$$\mathbf{Pr}\left[Y - X \geq 0\right] \leq \left(1 - (\sqrt{p_1} - \sqrt{p_2})^2\right)^S \leq e^{-S\left(\sqrt{p_1}-\sqrt{p_2}\right)^2} .$$

By choosing $S \geq \frac{3z \ln n}{(\sqrt{\tau_1}-\sqrt{\tau_2})^2} \geq \frac{3 \ln n}{\left(\sqrt{p_1}-\sqrt{p_2}\right)^2}$ and the union bound, the theorem holds with probability at least $1 - 1/n$. □

Trade-Off Between Search Quality and Efficiency: By choosing S as Theorem 1, we have $\sqrt{\tau_1} - \sqrt{\tau_2} \geq \sqrt{3z(\ln n)/S}$. Assume that the top-$k$ value is τ_1, it is clear the more samples we use, the smaller gap between the top-k MIPS values and the other values we can distinguish. We note that we can compute and rank B inner products of the top-B points with the largest counter values in the ranking phase. Increasing B corresponds to increasing the gap $Y - X$ in our analysis, and hence decreasing the failure probability. This means that both B and S can be used to control such trade-off.

Comparison to Diamond Sampling: For a fair theoretical comparison, we consider the same setting as in [4, Theorem 4] where we want to distinguish $\mathbf{x}_{i_1} \cdot \mathbf{q} \geq \tau$ and $\mathbf{x}_{i_2} \cdot \mathbf{q} \leq \tau/4$, and all entries in $\mathbf{X}$ and $\mathbf{q}$ are positive[2]. Applying Theorem 1, wedge sampling needs $S_w \geq 12z \ln n/\tau$. Diamond sampling needs $S_d \geq 12K\|\mathbf{q}\|_1 z \ln n/\tau^2$ where all entries in $\mathbf{X}$ are at most K. Since $K\|\mathbf{q}\|_1 \geq \tau$ for any τ, wedge requires strictly less samples than diamond.

General Inputs: For the general cases, Theorem 1 does not hold anymore. We observe that there are several ways to convert both data and query into non-negative forms without changing their inner product order. For instance, by shifting each column $\mathbf{y}_j$ a constant factor (e.g. its minimum or maximum value) dependent on $\mathbf{sgn}(q_j)$, the order of inner products $\mathbf{Xq}$ is preserved and therefore Theorem 1 still holds for MIPS in the non-negative transformation space.We leave this research direction to future work.

3.2 dWedge: A Simple Deterministic Variant for Budgeted MIPS

This subsection presents a significant drawback of wedge sampling to solve top-k MIPS with a $o(n)$ budgeted computation. We then introduce dWedge, a simple deterministic variant to handle such a drawback. For simplicity, we present our approach on non-negative $\mathbf{X}$ and $\mathbf{q}$.

[2] Diamond sampling's analysis only works on non-negative inputs.

Algorithm 2: dWedge Sampling

Data: For each dimension j, sort data in descending order on x_{ij} and store them in the list L_j. Other pre-computed statistics z, c_j, and the query $\mathbf{q}$.

Result: A counting histogram of sampled data points.

1 Compute the number of samples $s_j = Sc_jq_j/z$ for each sorted list L_j.
2 **foreach** *sorted list* L_j **do**
3 Select $\mathbf{x}_i$ in the descending order of x_{ij}.
4 Increase $\mathbf{x}_i$'s counter and the current number of samples used of L_j by $\lceil s_j x_{ij}/c_j \rceil$.
5 If the current number of samples is larger than s_j, stop iterating L_j.
6 **end**

Drawback: We observe that wedge sampling first samples the column j and then samples the point $\mathbf{x}_i$ on this column j. In other words, given a fixed number of samples S, wedge sampling allocates S samples to d columns. Each column j receives s_j samples and $\sum_j s_j = S$. To ensure that wedge provides an accurate estimate, these s_j samples must approximate the discrete distribution $\mathbf{y}_j/\|\mathbf{y}_j\|_1$ well. Given the budget of $S = o(n)$ samples, the number of samples s_j of the column j is $S/d \ll n$ in expectation. Since $s_j \ll n$ and the data set is often dense, it is impossible to approximate the discrete distribution $\mathbf{y}_j/\|\mathbf{y}_j\|_1$ by s_j samples. In a realistic case of the Netflix data set with $n = 17,770$ and $d = 300$, if we use $S = n$ samples, in expectation we only have nearly 60 samples to approximate a discrete distribution with 17,770 values. Hence the performance of wedge (and hence diamond) sampling dramatically degrades in the budgeted setting, as can be seen in the experiment section.

dWedge: Observing that wedge sampling carefully distributes S samples to each dimension. Dimension j receives $s_j = Sc_jq_j/z$ samples and hence the point $\mathbf{x}_i$ on dimension j will receive s_jx_{ij}/c_j samples in expectation. Given $S = o(n)$, we have $s_j \ll n$ and therefore can only sample a few points on the dimension j. Due to this limit, we propose to *greedily* sample $\mathbf{x}_i$ with the largest x_{ij} values in the column j. For each selected $\mathbf{x}_i$, we sample $\lceil s_jx_{ij}/c_j \rceil$ times.

Algorithm 2 presents our simple heuristic solution on non-negative $\mathbf{X}$ and $\mathbf{q}$. Before executing this algorithm, we need a pre-processing step that sorts all data points in descending order for each dimension. We iterate these sorted list and greedily sample $\mathbf{x}_i$ on the descending order. For each list L_j, we stop the iteration when exceeding the number of samples s_j. For the general case, dWedge exploits the sign trick as standard wedge, executing on the absolute values of $\mathbf{X}$ and $\mathbf{q}$. If selected, $\mathbf{x}_i$'s counter is increased by $\texttt{sgn}(x_{ij})\texttt{sgn}(q_j)\lceil s_jx_{ij}/c_j \rceil$.

Time Complexity: It is clear that the pre-processing step takes $\mathcal{O}(dn \log n)$ times and $\mathcal{O}(dn)$ additional space. dWedge sampling takes $\mathcal{O}(S)$ time. If $S \ll n$, we can use a hash table to maintain the counting histogram. Otherwise, we use a vector of size n. The running time of dWedge for answering top-k MIPS with a post-processing B inner products is $\mathcal{O}\left(S + min(S, n) \log B + dB\right)$.

Cost Model: While the cost of extracting top-B points in the ranking phase is larger than the sampling cost in theory, this cost in practice is much smaller due to the optimization of C++ `std::priority_queue`. We observe that dWedge executes two main operations at Step 4–5 for each sample. Given S samples, dWedge's cost is upper bounded by the cost of computing $2S/d$ inner products. Hence we can model the operation cost of dWedge for answering top-k MIPS as $2S/d + B$ inner product computation. Empirically, this cost model is very accurate due to the simplicity of dWedge. We will use this simple cost model to estimate the speedup of dWedge and to tune the parameters S and B for comparing to other budgeted MIPS solvers, as shown in the experiment.

Relation to Greedy-MIPS [30]: This approach exploits the upper bound $\mathbf{x}_i \cdot \mathbf{q} \leq d \max_j \{q_j x_{ij}\}$ to construct B candidates. In principle, it greedily selects B points $\mathbf{x}_i$ with the largest $q_j x_{ij}$ values for each dimension j, then merges them to find top-B candidates with $\max_j \{q_j x_{ij}\}$. While dWedge shares the same preprocessing step and greedy spirit, there is a significant difference between dWedge and Greedy. dWedge greedily estimates and differentiates the largest elements on each dimension j using s_j samples. The more samples used, the more of the largest elements have been considered. This leads to higher quality of top-B candidates and hence top-k MIPS. On the other hand, Greedy-MIPS does not have the sampling step to improve the quality of candidates and hence its performance degrades, as shown in the experiment on the Gist data set.

4 Experiment

We implement the sampling schemes and other competitors in C++[3] using -O3 optimization and conduct experiments on a 2.80 GHz core i5-8400 32 GB of RAM. We present empirical evaluations to verify our claims, including: (1) Wedge is competitive (often superior) to Diamond; (2) dWedge provides better accuracy than both Wedge and Greedy-MIPS; and (3) dWedge is more flexible and often superior to competitive LSH-based solvers [20,29] for the budgeted MIPS.

For measuring the accuracy and efficiency, we used the standard Precision@10 and the speedup over the brute-force algorithm, defined as follows.

$$\text{Precision@10} = |\text{Retrieved top-10} \cap \text{True top-10}|/10\,,$$
$$\text{Speedup} = \text{Running time of brute-force / Running time of algorithm.}$$

4.1 Experiment Setup and Data Sets

Since diamond exploits wedge, applying dWedge to diamond derives a new variant, called dDiamond[4]. The list of all implemented algorithms includes: (1)

[3] https://github.com/NinhPham/MIPS.

[4] This variant is not deterministic due to the randomness from the basic sampling.

The traditional wedge (*Wedge*) and diamond (*Diamond*) sampling and the proposed solutions *dWedge* and *dDiamond*; (2) The Greedy-MIPS approach (*Greedy*) [30]; (3) Representative LSH-based solutions, including *SimpleLSH* [20] and the recently improved algorithm *RangeLSH* [29]; and (4) Brute-force algorithm with the Eigen-3.3.4 library[5] for the extremely fast C++ matrix-vector multiplication.

We conduct experiments on standard real-world data sets[6], including Netflix, Yahoo, and Gist. For the sake of comparison, we use the Netflix-200 ($n = 17,770; d = 200$) from [30], Netflix-300 ($n = 17,770; d = 300$) and Yahoo ($n = 624,961; d = 300$) from [7], and Gist ($n = 1,000,000; d = 960$). For Netflix and Yahoo, the item matrices are used as the data points. We randomly pick 1000 users from the user matrices to form the query sets. All randomized results are the average of 5 runs of the algorithms.

4.2 Comparison Between Wedge and Diamond Schemes

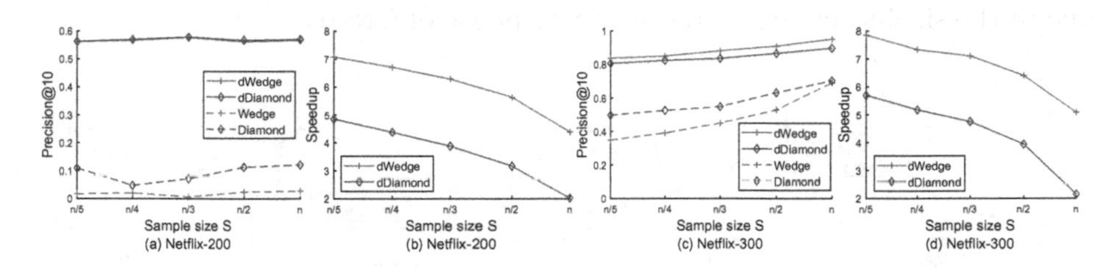

Fig. 1. Comparison of accuracy and speedup between dWedge, dDiamond, Wedge and Diamond on Netflix-200 and Netflix-300 when fixing $B = 100$ and varying S.

In this subsection, we compare the top-k MIPS performance on Netflix between wedge-based and diamond-based schemes using deterministic and randomized generators. For each sampling scheme, we consider two corresponding variants, including Wedge, Diamond, dWedge and dDiamond. We measure their performance on Precision@10 value and speedup over the brute-force search where we varied the sample size S and fix $B = 100$ for post-processing.

Figure 1 reveals that the proposed deterministic generator gives higher accuracy than the randomized one. dWedge and dDiamond outperform Wedge and Diamond, respectively, over a wide range of S. Especially, on Netflix-200, Wedge and Diamond suffer very low accuracy whereas dWedge and dDiamond return more than 55%. Netflix-300 shows the superiority of both dWedge and dDiamond with the accuracy at least 80%. We note that the two Netflix versions are generated by different matrix factorization tools. On Netflix-300, the true top-k points tend to dominate the others on each dimension, hence dWedge and dDiamond achieve higher accuracy than on Netflix-200.

[5] http://eigen.tuxfamily.org/index.php?title=Main_Page.
[6] https://drive.google.com/drive/folders/1BHpiaii6Ur0rKSy5c9AFVwAhfLUQMXFE.

In term of speedup, dWedge runs significantly faster than dDiamond, especially at least twice faster when $S = n$. Wedge variants run faster than diamond variants since diamond requires the basic sampling step which requires expensive cost for random accesses. This gap will be more substantial on large data sets but not reported here due to the lack of space.

4.3 Comparison Between dWedge and Greedy-MIPS

In this subsection, we compare the top-k MIPS performance between dWedge and Greedy on all used data sets. Since the cost of the screening phase of Greedy is implicitly governed by the number of inner product computation B_g in the ranking phase, the larger B_g provides the higher accuracy for Greedy. Note that dWedge's cost is about $2S/d + B$ inner product computation. Since we want to show that dWedge always runs as fast as Greedy but achieves higher accuracy, we often set $B_g > 2S/d + B$ inner product computation for Greedy. In particular, for small Netflix-200 and Netflix-300, we use $B_g = 2S/d + B + 50$ and $B_g = 2S/d + B + 20$, respectively. For the large Yahoo data set, we use $B_g = B$ due to the significant cost of the screening phase of Greedy.

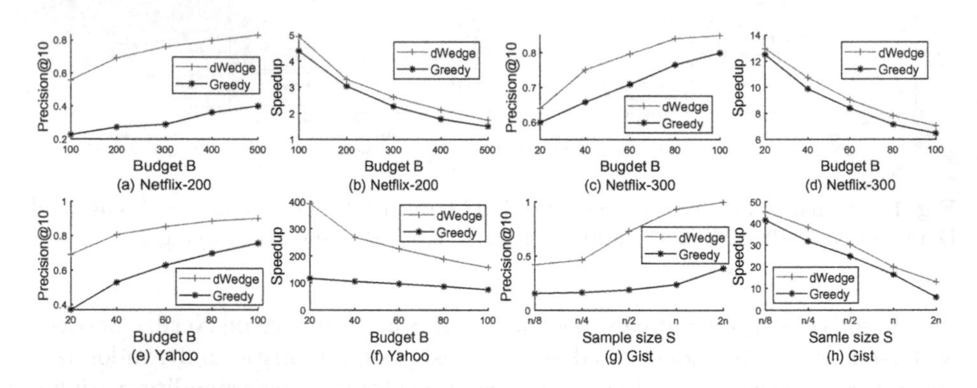

Fig. 2. Comparison of accuracy and speedup between dWedge and Greedy when fixing $S = 10{,}000$ and varying B on Netflix-200 (a, b); fixing $S = 4{,}500$ and varying B on Netflix-300 (c, d); fixing $S = 4{,}500$ and varying B on Yahoo (e, f); and fixing $B = 200$ and varying S on Gist (g, h).

Figure 2 (a–d) shows that, given our setting above, dWedge runs slightly faster and provides dramatically higher accuracy than Greedy on the two Netflix versions. dWedge achieves up to 80% Precision@10 on both versions while Greedy bears at most 40% on Netflix-200. On Yahoo in Fig. 2 (e, f), since the screening phase of Greedy incurs remarkable cost, dWedge yields substantially higher accuracy and speedup. We note that the Eigen's batch inner product computation runs nearly 20 times faster than a standard implementation on our machine. Hence, using our cost model, dWedge's speedup can be estimated as $n/(40S/d + 20B)$.

In order to show the benefit of the sampling phase of dWedge, we measure its performance on Gist while fixing $B = 200$ and varying S. Again for Greedy, we add additional inner product computation into the post-processing phase so that dWedge and Greedy achieve the similar speedup, as can be seen in Fig. 2 (h). Figure 2 (g) shows that Greedy suffers from very small accuracy with at most 40% while dWedge achieves nearly perfect recall, i.e. 99% when $S = 2n$. In general, dWedge with two parameters, i.e. number of samples S and number of post-processing inner products B, governs the trade-off between search efficiency and quality more efficiently than Greedy. Particularly, dWedge offers a least 80% Precision@10 on all 4 data sets with remarkable speedups while Greedy does not.

4.4 Comparison Between dWedge and LSH-based Schemes

This subsection shows experiments comparing the performance between dWedge and LSH-based schemes, including SimpleLSH [20] and RangeLSH [29] on Yahoo and Gist. There are two strategies of using LSH for top-k MIPS: constructing binary codes for efficient estimation and constructing hash tables for efficient lookups. We consider both strategies in our experiments.

SimpleLSH [20] transformed data $\mathbf{x} \mapsto \{\mathbf{x}/m, \sqrt{1 - \|\mathbf{x}\|_2/m}\}$ and query $\mathbf{q} \mapsto \{\mathbf{q}/\|\mathbf{q}\|_2, 0\}$ where m is the maximum 2-norm value of all data points. RangeLSH [29] partitions the data set into several partitions and applies SimpleLSH on each partition. We show empirically that the MIPS values in the transformation space are much smaller and therefore degrades the efficiency of both LSH schemes.

LSH Estimation: For the first strategy, the LSH time complexity is dominated by the hash evaluation, i.e. $\mathcal{O}(dh)$, and the linear cost of Hamming distance computation, i.e. $\mathcal{O}(nh)$, where h is the code length. In practice, we can use the Eigen library with fast matrix-vector multiplication to speed up the hash computation. Furthermore, the Hamming distance computation is also fast due to the `_builtin_popcount` function of GCC compilers. Hence, LSH-based estimation is still much faster than brute-force search. To compare dWedge with LSH schemes, we set $S = 6,000 \approx n/100$ for Yahoo and $S = 2n$ for Gist. Both dWedge and LSH-based methods use $B = 100$ inner products for post-processing. We evaluate SimpleLSH and RangeLSH over a wide range of code length h.

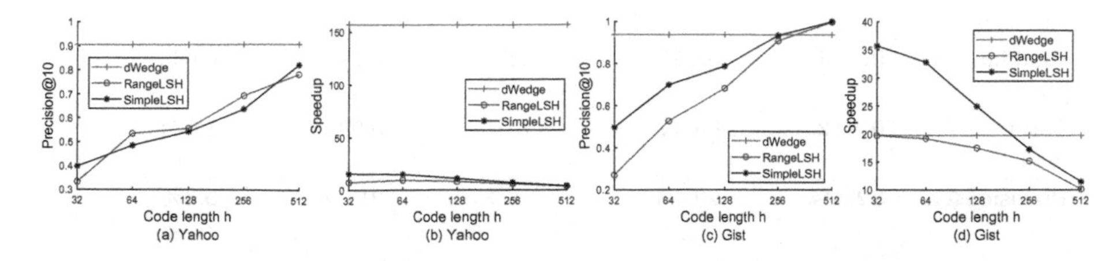

Fig. 3. Comparison of accuracy and speedup between dWedge ($S = n/100$ for Yahoo, $S = 2n$ for Gist), SimpleLSH and RangeLSH when fixing $B = 100$ and varying h.

Figure 3 shows that dWedge is superior to LSH schemes on these two data sets. On Yahoo, dWedge achieves 90% Precision@10 with approximate 180x speedup whereas the LSH schemes give at most 80% with negligible speedup when $h = 512$. On Gist, LSH schemes only achieve slightly higher accuracy but run twice slower than dWedge when $h = 512$. After transformation, both the MIPS values and the gap between "close" and "far apart" points are often very small. That requires significantly large code length, i.e. $h = 512$, to achieve acceptable accuracy. For the other values of h, dWedge always returns higher accuracy and runs dramatically faster than RangeLSH. For the same speedup setting, dWedge still provides higher accuracy than SimpleLSH but not reported here due to the lack of space. We note that dWedge provides significant speedup on Yahoo compared to Gist because of the much smaller number of samples S.

In our experiment, SimpleLSH is competitive with RangeLSH. RangeLSH roughly estimates $\mathbf{x} \cdot \mathbf{q}$ as $m_i \cos \pi(1-p)$ where p is the collision probability and m_i is the maximum 2-norm of the partition of $\mathbf{x}$. While m_i is always non-negative, $\cos \pi(1-p)$ might be negative. A small error induced from LSH functions will change the sign of $\cos \pi(1-p)$ and therefore change the results completely. On the other hand, SimpleLSH does not suffer this issue. Since the comparison of LSH schemes is out of the scope, we do not discuss it in more detail.

LSH Lookups: For the second strategy, since we do not know the collision probability between the query and all data points, it is impossible to tune LSH parameters for any budget cost. Given a fixed number of hash tables L, the number of concatenating hash functions h will control the number of collisions and therefore the candidate set size. In particular, large values of h will decrease the collision probability and hence candidate set size and vice versa. The query complexity will be dominated by $\mathcal{O}(dhL + dB)$ where the first term is from the hash computation and the second term comes from the post-processing phase of computing B inner products. Again, we used the Eigen library for speeding up the hash computation.

Table 1. Comparison of accuracy and running time of the screening and ranking phases between dWedge and SimpleLSH on Yahoo when fixing $B = 40$ and varying h.

Method	dWedge $S = 6{,}000$	SimpleLSH $h = 8$	SimpleLSH $h = 10$	SimpleLSH $h = 12$	SimpleLSH $h = 16$
Screening time (ms)	**0.158**	0.342	0.444	0.649	0.998
Ranking time (ms)	**0.144**	0.255	0.224	0.242	0.043
Total time (ms)	**0.312**	0.646	0.731	0.984	1.171
Precision@10	**82.4%**	20.4%	21.6%	21.8%	5.3%

Since we need to set up these parameters before querying, LSH provides a fixed trade-off between search efficiency and quality. To compare with dWedge, we first fix $L = 512$ as suggested in previous LSH-based MIPS solvers [20,25] and vary the parameter h. Due to the similar results[7], we report the representative results of SimpleLSH on Yahoo. As explained in Sect. 4.3, the screening cost of dWedge with S samples is similar to the cost of computing $2S/d$ inner products. Hence, dWedge with $S = 6,000 \approx n/100$ and $B = 40$ balances the screening time and ranking time on Yahoo with $d = 300$. We also set $B = 40$ as the candidate size of SimpleLSH for the sake of comparison.

Table 1 shows the comparison between dWedge and SimpleLSH in several components, including the screening time, ranking time, total running time, and accuracy. The screening time of dWedge and SimpleLSH are the sampling and hash evaluation time, respectively. It is clear that the screening time and ranking time dominate the total running time. Furthermore, dWedge has not only similar screening and ranking time but also negligible overheads, which expresses the simplicity of dWedge on tuning parameters. Given such a balanced setting, dWedge not only runs at least twice faster but also achieves 82.4% accuracy, which is significantly higher than the maximum 21.8% of SimpleLSH with $h = 12$.

Regarding the screening time, dWedge runs at least twice faster than SimpleLSH with $h = 8$. Regarding the ranking time, dWedge is also faster than SimpleLSH, except for the case $h = 16$, though we use the same $B = 40$. dWedge chooses the top-B with the largest counter values hence has a higher chance to retrieve top-k MIPS earlier. This significantly reduces the number of insertions into the priority queue. On the other hand, such a chance of SimpleLSH is uniform, hence SimpleLSH has no such benefit.

For SimpleLSH, the screening time increases due to the increase of the hash evaluation time. However, the ranking time of SimpleLSH is similar when $h < 12$ and then drops to a negligible amount at $h = 16$. Similarly, Precision@10 slightly increases when h increases and then significantly drops at $h = 16$. This is due to the fact that SimpleLSH has enough candidates (i.e. $B = 40$) when $h \leq 12$. However, it does not have enough collisions on all L hash tables when $h = 16$. We observe that the average value of B is just 6 for all queries. This fact is not surprising. A simple computation indicates that the average maximum MIPS is 0.24 after transformation. Hence, the number of collisions is at most $n(1 - (1 - 0.24^h)^L) \approx 0$ when $h = 16$ and $L = 512$. We note that this observation is also true with RangeLSH. Due to the lack of space, we will not discuss RangeLSH's performance here. In general, LSH with both strategies is inferior to dWedge on our benchmark data sets. LSH schemes cannot govern effectively the trade-off between search quality and efficiency as dWedge.

5 Conclusions

This paper studies top-k MIPS given a limited budget of computational operations and investigates recent advanced sampling algorithms, including wedge and diamond sampling to solve it. We theoretically and empirically show that wedge

[7] RangeLSH provides slightly better than SimpleLSH with small h values. When $h \geq 20$, they will share the same results of very low accuracy on both data sets.

sampling is competitive (often superior) to diamond sampling on approximating top-k MIPS regarding both efficiency and accuracy.

We propose dWedge, a very simple deterministic wedge variant for budgeted top-k MIPS. dWedge runs significantly faster or provides higher accuracy than other competitive budgeted MIPS solvers while maintaining an accuracy of at least 80% on our real-world benchmark data sets.

References

1. Ahle, T.D., Pagh, R., Razenshteyn, I.P., Silvestri, F.: On the Complexity of Inner Product Similarity Join. In: PODS, pp. 151–164 (2016)
2. Abuzaid, F., Sethi, G., Bailis, P., Zaharia, M.: To index or not to index: optimizing exact maximum inner product search. In: ICDE, pp. 1250–1261 (2019)
3. Bachrach, Y., et al.: Speeding up the Xbox recommender system using a Euclidean transformation for inner-product spaces. In: RecSys, pp. 257–264 (2014)
4. Ballard, G., Kolda, T.G., Pinar, A., Seshadhri, C.: Diamond sampling for approximate maximum all-pairs dot-product (MAD) search. In: ICDM, pp. 11–20 (2015)
5. Cohen, E., Lewis, D.D.: Approximating matrix multiplication for pattern recognition tasks. J. Algorithms **30**(2), 211–252 (1999)
6. Covington, P., Adams, J., Sargin, E.: Deep neural networks for Youtube recommendations. In: RecSys, pp. 191–198 (2016)
7. Cremonesi, P., Koren, Y., Turrin, R.: Performance of recommender algorithms on top-n recommendation tasks. In: RecSys, pp. 39–46 (2010)
8. Dean, T.L., Ruzon, M.A., Segal, M., Shlens, J., Vijayanarasimhan, S., Yagnik, J.: Fast, accurate detection of 100,000 object classes on a single machine. In: CVPR, pp. 1814–1821 (2013)
9. Ding, Q., Yu, H., Hsieh, C.: A fast sampling algorithm for maximum inner product search. AISTATS **2019**, 3004–3012 (2019)
10. Drineas, P., Kannan, R., Mahoney, M.W.: Fast monte carlo algorithms for matrices I: approximating matrix multiplication. SIAM J. Comput. **36**(1), 132–157 (2006)
11. Fetaya, E., Shamir, O., Ullman, S.: Graph approximation and clustering on a budget. In: AISTATS (2015)
12. Har-Peled, S., Indyk, P., Motwani, R.: Approximate nearest neighbor: towards removing the curse of dimensionality. Theory Comput. **8**(1), 321–350 (2012)
13. Hu, Y., Koren, Y., Volinsky, C.: Collaborative filtering for implicit feedback datasets. In: ICDM, pp. 263–272 (2008)
14. Huang, Q., Ma, G., Feng, J., Fang, Q., Tung, A.K.H.: Accurate and fast asymmetric locality-sensitive hashing scheme for maximum inner product search. In: KDD, pp. 1561–1570 (2018)
15. Koenigstein, N., Dror, G., Koren, Y.: Yahoo! music recommendations: modeling music ratings with temporal dynamics and item taxonomy. In: RecSys, pp. 165–172 (2011)
16. Koren, Y.: Collaborative filtering with temporal dynamics. In: KDD, pp. 447–456 (2009)
17. Koren, Y., Bell, R.M., Volinsky, C.: Matrix factorization techniques for recommender systems. IEEE Comput. **42**(8), 30–37 (2009)
18. Li, H., Chan, T.N., Yiu, M.L., Mamoulis, N.: FEXIPRO: fast and exact inner product retrieval in recommender systems. In: SIGMOD, pp. 835–850 (2017)

19. Mai, S.T., He, X., Hubig, N., Plant, C., Böhm, C.: Active density-based clustering. In: ICDM, pp. 508–517 (2013)
20. Neyshabur, B., Srebro, N.: On symmetric and asymmetric LSHs for inner product search. In: ICML, pp. 1926–1934 (2015)
21. Ram, P., Lee, P., Gray, A.G.: Nearest-neighbor search on a time budget via max-margin trees. In: SDM, pp. 1011–1022 (2012)
22. Rendle, S., Freudenthaler, C., Gantner, Z., Schmidt-Thieme, L.: BPR: Bayesian personalized ranking from implicit feedback. In: UAI, pp. 452–461 (2009)
23. Russakovsky, O., Deng, J., Su, H., Krause, J., Satheesh, S., Ma, S., Huang, Z., Karpathy, A., Khosla, A., Bernstein, M.S., Berg, A.C., Li, F.: ImageNet large scale visual recognition challenge. IJCV **115**(3), 211–252 (2015)
24. Shamir, O., Tishby, N.: Spectral clustering on a budget. In: AISTATS, pp. 661–669 (2011)
25. Shrivastava, A., Li, P.: Asymmetric LSH (ALSH) for sublinear time maximum inner product search (MIPS). In: NIPS, pp. 2321–2329 (2014)
26. Spring, R., Shrivastava, A.: Scalable and sustainable deep learning via randomized hashing. In: KDD, pp. 445–454 (2017)
27. Teflioudi, C., Gemulla, R.: Exact and approximate maximum inner product search with LEMP. TODS **42**(1), 5:1–5:49 (2017)
28. Weber, R., Schek, H., Blott, S.: A quantitative analysis and performance study for similarity-search methods in high-dimensional spaces. In: VLDB, pp. 194–205 (1998)
29. Yan, X., Li, J., Dai, X., Chen, H., Cheng, J.: Norm-ranging LSH for maximum inner product search. In: NeurIPS, pp. 2956–2965 (2018)
30. Yu, H., Hsieh, C., Lei, Q., Dhillon, I.S.: A greedy approach for budgeted maximum inner product search. In: NIPS, pp. 5459–5468 (2017)
31. Zilberstein, S.: Using anytime algorithms in intelligent systems. AI Mag. **17**(3), 73–83 (1996)

Modeling Winner-Take-All Competition in Sparse Binary Projections

Wenye Li[1,2](✉)

[1] The Chinese University of Hong Kong, Shenzhen, China
wyli@cuhk.edu.cn
[2] Shenzhen Research Institute of Big Data, Shenzhen, China
http://mypage.cuhk.edu.cn/academics/wyli/

Abstract. With both theoretical and practical significance, the study of sparse binary projection models has attracted considerable recent research attention. The models project dense input samples into a higher-dimensional space and output sparse binary data representations after the Winner-Take-All competition, subject to the constraint that the projection matrix is also sparse and binary. Following the work along this line, we developed a supervised-WTA model which works when training samples with both input and output representations are available, from which the optimal projection matrix can be obtained with a simple, efficient, yet effective algorithm after proper relaxation. We further extended the model and the algorithm to an unsupervised setting where only the input representation of the samples is available. In a series of empirical evaluation on similarity search tasks, the proposed models reported significantly improved results over state-of-the-art methods in both search accuracies and running speed. The successful results give us strong confidence that the work provides a useful tool for industrial applications.

Keywords: Sparse binary projection · Winner-take-all · Supervised learning

1 Introduction

Random projection has emerged as a powerful tool in data analysis applications [4]. It is often used to reduce the dimension of data samples in the Euclidean space, and has become well-known for its simplicity and effectiveness. It provides a computationally efficient way to reduce the complexity of the data by trading a controlled amount of representation error [15] for faster processing speed, with quite successful results in practical applications including natural language processing and image processing [4].

Very recently, a novel FLY algorithm was successfully designed, which provided a fundamentally different viewpoint to the study of random projection methods [7]. Instead of performing dimension reduction, the algorithm increases

F. Hutter et al. (Eds.): ECML PKDD 2020, LNAI 12457, pp. 456–472, 2021.
https://doi.org/10.1007/978-3-030-67658-2_26

the input dimension with a random sparse binary projection matrix, which has strong evidence from neural circuits when forming the tag of an input signal for behavioral responses [33]. Then after a process called winner-take-all (WTA) competition, the samples are output in the form of sparse binary vectors. In similarity search tasks, it was reported that such sparse binary vectors outperformed the hashed vectors produced by the classical locality sensitive hashing (LSH) method that is based on the random dense projection.

Following the work along this line, we proposed two sparse binary projection models with the explicit modeling of the WTA competition. Different from the FLY algorithm which resides on the random generation of the projection matrix, our two models seek the *optimal* projection matrix under a supervised setting and an unsupervised setting respectively. Through proper relaxation, we derived a computational approach that is simple, effective, and efficient for each model. In empirical evaluations, the proposed models reported significantly improved results in both similarity search accuracies and running speed over the state-of-the-art approaches, which hence provided a practical tool for data analysis applications with high potential.

A note on notation. Unless specified otherwise, a capital letter, such as W, denotes a matrix. A lower-case letter, with or without a subscript, denotes a vector or a scalar. For example $w_{i.}$ denotes the i-th row, $w_{.j}$ denotes the j-th column, and w_{ij} denotes the (i, j)-th entry of the matrix W.

The paper is organized as follows. Section 2 introduces the necessary background. Section 3 presents our models and algorithms. Section 4 reports the experiments and the results, followed by the conclusion in Sect. 5.

2 Background

2.1 LSH and Sparse Binary Projection Algorithms

LSH is a data-independent hashing method that maps higher-dimensional data points into lower-dimensional spaces [6,9]. The key idea behind the LSH method is provided by the Johnson-Lindenstrauss theorem [15]. The theorem states that data points in a vector space of sufficiently high-dimension can be projected into a suitable lower-dimensional space in a way that approximately preserves the pairwise distances between the data points. To realize the method, a randomly generated projection matrix, such as a random Gaussian matrix, is routinely applied. The dense vectors in the output space are obtained by multiplying the projection matrix with the input vectors.

Different from the classical random projection method that projects data samples from a higher-dimensional space to a lower-dimensional space, the FLY algorithm increases the dimension of the input samples and converts them to be sparse binary vectors in the output space [7].

The motivation of the FLY algorithm was from simulating the fruit fly's olfactory circuit, whose function is to associate similar odors with similar tags. Each odor is initially represented as an input vector of 50-dimensional features of firing rates and will be transformed into a sparse binary vector in an output space of much higher dimensions. To associate each odor with a tag involves three steps. Firstly, a divisive normalization step [24] is necessary to center the mean of the feature vector. Secondly, the dimension of the feature vector is expanded from 50 to 2,000 with a sparse binary connection matrix [5,33], which has the same number of ones in each row. Thirdly, the WTA competition is involved as a result of strong inhibitory feedback coming from an inhibitory neuron. After the competition, all but the highest-firing 5% out of the 2,000 features are silenced [31]. This remaining 5% features just correspond to the tag assigned to the input odor.

In certain sense, the FLY algorithm can be viewed as a special form of the LSH method as it aims to produce similar hashes for similar input signals. Empirically, the FLY algorithm reported improved results over the LSH method in similarity search applications [7].

The success of the FLY algorithm inspired subsequent research attention on sparse binary projection methods, among which one of particular interest to us is the LIFTING algorithm [19] that removes the randomness assumption of the projection matrix, which is partially supported by most recent biological discoveries [33]. In the work, the projection matrix is obtained through supervised learning. Suppose training samples with both dense input representation $X \in \mathcal{R}^{d \times n}$ and sparse output representation $Y \in \{0,1\}^{d' \times n}$ are available. The LIFTING algorithm seeks the projection matrix W by minimizing an objective of $\frac{1}{2}\|WX - Y\|_F^2 + \beta \|W\|_{\frac{1}{2}}$ in the feasible region of sparse binary matrices. To solve the optimization problem, the Frank-Wolfe algorithm, which involves successive solutions of linear programs, was found to have quite good empirical performance [8,13].

A challenge to the LIFTING algorithm is the lack of output codes to the supervised training. When the output representation Y is not available, the algorithm suggests minimizing $\frac{1}{2}\left\|\gamma X^T X - Y^T Y\right\|_F^2 + \alpha \|Y\|_{\frac{1}{2}}$ to obtain the desired sparse binary Y, again with the Frank-Wolfe algorithm. Unfortunately, the scalability of the Frank-Wolfe algorithm may become a nontrivial concern when the number of training samples gets large, due to the heavy computation involved.

2.2 Winner-Take-All Competition

Evidence in neuroscience showed that excitation and inhibition are common activities in neurons [31,32]. Based on the lateral information, some neurons raise to the excitatory state, while the others get inhibited and remain silent. Such excitation and inhibition result in competitions among neurons. Modeling neuron competitions is of key importance, with which useful applications are found in a variety of tasks [2,21]. Specifically in machine learning, the competition mechanism has motivated the design of computer algorithms for a long time, from the early self-organizing map [17] to more recent work in developing novel neural network architectures [20,26].

To model the competition stage, the WTA model is popularly adopted. We are interested in a variant of the WTA model with the following form. For a

d-dimensional input vector x and a given hash length of $k\,(k \ll d)$, a function $WTA_k^d : \mathcal{R}^d \to \{0,1\}^d$ outputs a vector $y = WTA_k^d\,(x)$ that satisfies, for each $1 \le i \le d$,

$$y_i = \begin{cases} 1, \text{ if } x_i \text{ is among top-}k \text{ entries of } (x_1, \cdots, x_d). \\ 0, \text{ otherwise.} \end{cases} \tag{1}$$

Thus the output entries with value 1 just mark the positions of top-k values of x. For simplicity and without causing ambiguity, we do not differentiate whether the input/output vector of the WTA function is a row vector or a column vector throughout this paper.

3 Model

3.1 Supervised Training

A natural extension of the data-independent projection models is to consider a data-dependent setting with supervised information. In the case of sparse binary projections, assume a set of training samples are available and given in the form of $X \in \mathcal{R}^{d \times n}$ and $Y \in \{0,1\}^{d' \times n}$ with $d \ll d'$. Each $x_{.m}\,(1 \le m \le n)$ of X denotes an input sample and $y_{.m}$ of Y denotes the corresponding output representation of $x_{.m}$, satisfying $\|y_{.m}\|_1 = k$ for a given integer k, which is far less than d' and is called the hash length of a sparse binary projection in the sequel.

Let us first assume that, for a fixed positive integer c[1], which is far less than d, there exists a sparse binary projection matrix $W \in \{0,1\}^{d' \times d}$ with $\|w_{i.}\|_1 = c$ for all $1 \le i \le d'$ and $y_{.m} = WTA_k^{d'}\,(Wx_{.m})$ for all m. Then from the WTA function defined in Eq. (1), we have:

$$w_{i.}x_{.m} \ge w_{j.}x_{.m}, \text{ if } y_{im} = 1 \text{ and } y_{jm} = 0 \tag{2}$$

for all $1 \le m \le n$ and $1 \le i, j \le d'$.

Now we are interested in inferring such a projection matrix W from the given data. But unfortunately, seeking such a matrix directly from Eq. (2) is generally hard. A matrix that meets all the constraints may not exist due to the noise in the observed samples. Even if it exists, the computational requirement can be non-trivial. For example, a straightforward formulation of the problem needs to check the feasible region of a linear integer program. Unfortunately, the formulation would involve $d' \times d$ variables and $O\,(nk\,(d'-k)+d')$ constraints, which is infeasible to solve even for moderately small n and d'.

To ensure the tractability, we resort to a relaxation-based approach. For any feasible m, i and j, we define a measure $y_{im}\,(1-y_{jm})\,(w_{i.}x_{.m} - w_{j.}x_{.m})$ to quantify the compliance, of two row vectors $w_{i.}$ and $w_{j.}$, with the condition specified in Eq. (2). When $y_{im} = 1$ and $y_{jm} = 0$, the measure is non-negative if

[1] As in [7], c is set to $\lfloor 0.1 \times d \rfloor$ in this paper.

the condition is met; otherwise, the measure is negative. Naturally, we sum up the values of the measure over all possible m, i and j, and define

$$L_s(W) = \sum_{m=1}^{n}\sum_{i=1}^{d'}\sum_{j=1}^{d'} y_{im}(1 - y_{jm})(w_{i.}x_{.m} - w_{j.}x_{.m}). \tag{3}$$

The value of $L_s(W)$ measures how well a matrix W meets the conditions in Eq. (2). Maximizing L_s with respect to W in the feasible region of sparse binary matrices provides a principled solution to seeking the projection matrix. And we call it the supervised-WTA model.

Considering that

$$\begin{aligned}
\arg_W \max L_s &= \arg_W \max \sum_{m=1}^{n}\sum_{i=1}^{d'}\sum_{j=1}^{d'} y_{im}(1 - y_{jm})(w_{i.}x_{.m} - w_{j.}x_{.m}) \\
&= \arg_W \max \sum_{m=1}^{n}\left[d'\sum_{i=1}^{d'} y_{im}w_{i.}x_{.m} - k\sum_{j=1}^{d'} w_{j.}x_{.m}\right] \\
&= \arg_W \max \sum_{m=1}^{n}\left[\sum_{i=1}^{d'} y_{im}w_{i.}x_{.m} - \frac{k}{d'}\sum_{i=1}^{d'} w_{i.}x_{.m}\right] \\
&= \arg_W \max \sum_{m=1}^{n}\left[\sum_{i=1}^{d'}\left(y_{im} - \frac{k}{d'}\right)w_{i.}x_{.m}\right] \\
&= \arg_W \sum_{i=1}^{d'} \max\left\{w_{i.}\left[\sum_{m=1}^{n} x_{.m}\left(y_{im} - \frac{k}{d'}\right)\right]\right\}
\end{aligned}$$

Therefore, maximizing $L_s(W)$ is equivalent to d' maximization sub-problems. Each sub-problem seeks a row vector $w_{i.}$ $(1 \le i \le d')$ by

$$\max w_{i.}\left[\sum_{m=1}^{n} x_{.m}\left(y_{im} - \frac{k}{d'}\right)\right] \tag{4}$$

subject to:

$$w_{i.} \in \{0,1\}^{1\times d}, \text{ and } \|w_{i.}\|_1 = c. \tag{5}$$

Denote

$$\ell_{.i} = \sum_{m=1}^{n} x_{.m}\left(y_{im} - \frac{k}{d'}\right), \tag{6}$$

and the optimal solution of $w_{i.}$ to Eq. (4) is given by

$$w_{i.}^{*} = WTA_c^d(\ell_{.i}). \tag{7}$$

3.2 Unsupervised Training

The supervised-WTA model utilizes both input and output representations to learn a projection matrix. In practice, it is more often the case that there are only input vectors but no or not enough output vectors for supervised training. To handle the difficulty, we can extend the supervised model to an unsupervised-WTA model, by maximizing the objective:

$$L_u(W,Y) = \sum_{m=1}^{n}\sum_{i=1}^{d'}\sum_{j=1}^{d'} y_{im}(1-y_{jm})(w_{i.}x_{.m} - w_{j.}x_{.m}) \tag{8}$$

subject to the constraints: $w_{i.} \in \{0,1\}^{1\times d}$, $\|w_{i.}\|_1 = c$, $y_{.m} \in \{0,1\}^{d'\times 1}$, and $\|y_{.m}\|_1 = k$ for all $1 \le i \le d'$ and $1 \le m \le n$.

Different from the supervised model which seeks the projection matrix W only, the unsupervised model treats the output representation Y as unknown variables too, and jointly optimizes on both W and Y to maximize the objective L_u.

Here, an alternating algorithm can be applied. We start with random initialization of W as W^1, and solve the model iteration by iteration. In t-th ($t = 1, 2, \cdots$) iteration, we maximize $L_u(W^t, Y)$ with respect to Y and get the optimal Y^t. Then we maximize $L_u(W, Y^t)$ with respect to W and get the optimal W^{t+1}.

In t-th iteration, the optimal Y^t is given by:

$$y_{.m}^t = WTA_k^{d'}(W^t x_{.m}) \tag{9}$$

for all $1 \le m \le n$.

By applying the result from the supervised model in Eq. (7), the optimal W^{t+1} in $(t+1)$-th iteration can be obtained by:

$$w_{i.}^{t+1} = WTA_c^d(\ell_{.i}^t) \tag{10}$$

for all $1 \le i \le d'$, where $\ell_{.i}^t = \sum_{m=1}^{n} x_{.m}\left(y_{im}^t - \frac{k}{d'}\right)$.

The convergence of the proposed training algorithm can be understood via follows. Denote by $L_u^t = L_u(W^t, Y^t)$. Note that $Y = Y^t$ is the maximizer to $L_u(W^t, Y)$ and $W = W^{t+1}$ is the maximizer to $L_u(W, Y^t)$. The value of the sequence $\{L_u^t\}$ monotonically increases for $t = 1, 2, \cdots$. The alternating optimization process stops when the objective value of L_u^t can't be increased anymore [29]. Note, as there might be equal entries in each Wx, the maximizer of $Y = WTA_k^{d'}(WX)$ and hence the projection matrix W are possibly not unique as well.

The unsupervised-WTA model has some interesting characteristics that deserve our attention. It can be studied as a generic clustering method [14]. The model puts n data samples into d' clusters and each sample belongs to k clusters. A special case of $k = 1$ leads to a hard clustering method. Two samples with the element of one in the same output dimension indicate that they have the same cluster membership. If we further switch the constraints on W from $\|w_{i.}\|_1 = c$ to $\|w_{i.}\|_2 = c$ and release the binary constraints on W, then the algorithm is much similar to the classical k-means clustering method [22]. The only difference is that the k-means method seeks a center with the smallest distance to all intra-cluster points, while our algorithm seeks a center with the maximum inner product to its intra-cluster points.

The unsupervised-WTA model can also be treated as a feature selection method [11]. This can be seen from the fact that each output dimension is associated with a subset of c features, instead of all d features in the input space. The model can choose these c features automatically and encode the selection results in the projection matrix W. Therefore, combining the two characteristics, it is straightforward to apply the unsupervised-WTA model as a new clustering algorithm with the ability of automatic feature selection.

3.3 Complexity Analysis

Computing the optimal solution to the supervised-WTA model is straightforward. To obtain each projection vector $w_{i.}$, a naïve implementation needs $O(dn + d\log c)$ operations, among which $O(dn)$ are for the summation operation in Eq. (6) and $O(d\log c)$ are for the sorting operations in Eq. (7) by the Heapsort algorithm [16]. Therefore, computing the whole projection matrix needs $O(d'dn + d'd\log c)$ operations. In fact, by utilizing the sparse structure of the output matrix Y, the computational complexity for W can be further reduced to $O(kdn + d'd\log c)$. As seen in our empirical evaluation (ref. Sect. 4.3), this is a highly efficient result.

To solve the unsupervised-WTA model, we proposed an iterative algorithm. In each iteration the algorithm needs to compute both Y and W. Computing one Y needs $O(cdn + d'd\log k)$ operations by utilizing the sparse structure of W, where $O(cdn)$ are for multiplying W with X and $O(d'd\log k)$ are for the sorting operations required by Eq. (9). Computing one W has the same complexity as in the supervised-WTA model, which is $O(kdn + d'd\log c)$. Therefore, the total complexity per iteration is $O((k+c)dn + d'd\log(kc))$, which is also efficient as seen in our evaluation.

The same as the FLY algorithm, the memory requirement of both WTA models is mainly from the storage of the matrices X, Y and W, and the memory complexity is $O(dn + d'n + d'd)$, which can be further reduced to $O(dn + kn + d'c)$ if sparse matrix representation is adopted.

In addition to the training complexity, it is important to analyze the search complexity, which is a key concern in large-scale similarity search or nearest neighbor search applications [3,23]. The search time includes both the projection time and retrieval time. The projection time is used to compute the hash vector, i.e., to map the input vectors to their output representations. The retrieval time is used to compute the similarities or distances between a query with all search candidates in their output representations.

The LSH method, which projects a d-dimensional input vector to a k-dimensional output vector, needs $O(kd)$ multiplication and $O(kd)$ addition operations. Comparatively, a sparse binary projection method, which projects a d-dimensional input vector to a d'-dimensional sparse output vector with k ones, needs $O(d'c)$ addition operations for increasing the dimension to d', and $O(d' \log k)$ comparison operations for finding the k-largest entries from a d'-dimensional vector. Here c, which is far less than d, is the number of ones in each row of the projection matrix. In our evaluation, the projection time of the LSH method was faster than the sparse binary projection method, but the difference can be safely ignored as the projection time only occupies a very small portion in the search time.

The benefit brought by the sparse binary projection method is most significant in the retrieval time. To compute the distance between two k-dimensional dense vectors typically needs $O(k)$ subtraction operations and $O(k)$ multiplication operations, followed by $O(k)$ addition operations. All these operations are on floating numbers. Comparatively, computing the distance between two d'-dimensional sparse binary vectors with exactly k ones each only needs $O(d')$ logical bit operations and $O(k)$ addition operations, which, as evidenced in our evaluation, can be tens of times faster than computing the distance with the k-dimensional dense representation.

4 Evaluation

4.1 General Settings

To evaluate the performance of the proposed models, we carried out a series of experiments and reported the results below[2].

Application: Similarly to the work of [7], we applied the proposed models in similarity search tasks. Similarity search is a fundamental problem in computer science, with wide industrial applications. It aims to find similar samples to a given query object among potential candidates, according to a certain distance or similarity measure [3]. The complexity of accurately determining similar samples relies heavily on both the number of candidates and the dimension of the data. Computing the pairwise distances between the query and the candidates seems straightforward, but unfortunately, it could often become expensive or even prohibitive if the number of candidates to evaluate is too large or the dimension of the data is too high.

To handle the computing difficulty brought by the high dimension of the input data, the classical hashing method is to reduce the data dimension while approximately preserving their pairwise distances. Yet with the sparse binary projection, we can also consider, as in this paper, to increase the dimension but confining the data in the output space to be sparse and binary, in the hope of achieving significantly improved search speed with the new representation [7].

[2] More details are available at: http://mypage.cuhk.edu.cn/academics/wyli/research/ecml20.html.

Objective and Methodology: Our major objective is to evaluate and compare the similarity search accuracies between our proposed models and existing algorithms. Each sample in a given dataset was used, in turn, as the query object, and the other samples in the same dataset were used as the search candidates. For each query object, we compared its 100 nearest neighbors in the output space with its 100 nearest neighbors in the input space, and recorded the ratio of common neighbors in both spaces. The ratio is averaged over all query objects as the search accuracy of each algorithm. Obviously, a higher similarity search accuracy indicates better preserving locality structures from the input space to the output space by the algorithm.

In addition to the search accuracies, another objective is to evaluate the running time. The running time comes from two parts. The first part is the training time of the proposed models. The second part is the search time needed by different output representations, which is also an important concern in practice.

Datasets: Real datasets and artificially generated datasets were used in the evaluation. The real datasets have the input representation X only; while the artificial datasets have both the input representation X and the output representation Y. The datasets include:

- GLOVE [28]: 100- to 1000-dimensional *GloVe* word vectors trained on a subset of 330 million tokens from wikimedia database dumps[3] with the 50,000 most frequent words.
- ImageNet [30]: a collection of 1.2 million images represented as 1,000-dimensional visual words quantized from SIFT features.
- ARTFC: five sets of 1,000-dimensional dense vectors (X) and 2,000-dimensional sparse binary vectors (Y). For each hash length of $k = 2/4/8/16/32$, firstly a set of 2,000-dimensional sparse binary vectors were randomly generated with the hash length. Then the vectors were projected to 1,000-dimensional dense vectors through principal component analysis. In this way, the samples' pairwise distances are roughly preserved between the input space and the output space, that is, $\|x_{.m} - x_{.m'}\|_2^2 \approx \|y_{.m} - y_{.m'}\|_2^2$ for all pairs of samples in the same set.

Algorithms to Compare: We compared the proposed supervised-WTA (denoted by SUP) model and the unsupervised-WTA (UNSUP) model with the data-independent hashing methods including the LSH algorithm [6,9], the fast Jonson-Lindenstrauss projection (FJL) algorithm [1], and the FLY algorithm [7]. The LSH algorithm maps d-dimensional inputs to k-dimensional dense vectors with a random dense projection matrix. The FJL algorithm is a fast implementation of the LSH algorithm with a sparse projection matrix. The FLY algorithm uses a random sparse binary matrix to map d-dimensional inputs to d'-dimensional vectors.

For data-dependent methods, we compared the proposed WTA models with the LIFTING algorithm [19], which trains a sparse binary projection matrix in

[3] https://dumps.wikimedia.org/.

a supervised manner for the d-dimensional to d'-dimensional projection. Both the FLY and the LIFTING algorithms involve a WTA competition stage in the output space to generate sparse binary vectors for each hash length.

In addition to the LSH method, we further conducted the comparison with several other data-dependent hashing methods, including the iterative quantization (ITQ) algorithm [10], the spherical hashing (SPH) algorithm [12] and the isotrophic hashing (ISOH) algorithms [18]. These algorithms were popularly used in literature to produce sparse binary data embeddings.

Computing Environment: All the algorithms were tested on an 8-way computing server, with which a maximum of 128 threads was enabled for each algorithm. The codes were implemented in MATLAB platform, with intel MKL as the underlying maths library. For the LIFTING algorithm, IBM CPLEX was used as the linear program solver that was needed by the Frank-Wolfe algorithm.

4.2 Similarity Search Accuracy

Table 1. Search accuracies on various datasets with a fixed output dimension ($d' = 2,000$). The proposed WTA models reported consistently improved results in all the settings. Note: on *ImageNet* with one million samples, the results of SUP/LIFTING algorithms are not available due to the prohibitive computation to obtain the output representation.

Datasets	k	*SUP*	*UNSUP*	*LSH*	*FJL*	*FLY*	*LIFTING*	*ITQ*	*SPH*	*ISOH*
	2	**0.1758**	0.1143	0.0174	0.0169	0.0474	0.1748	0.0103	0.0097	0.0101
ARTFC	4	**0.6665**	0.3531	0.0243	0.0237	0.0673	0.6134	0.0175	0.0138	0.0227
$d = 1,000$	8	0.3647	**0.3944**	0.0259	0.0255	0.0376	0.2612	0.0360	0.0173	0.0331
	16	**0.5884**	0.3267	0.0278	0.0282	0.0402	0.1694	0.0367	0.0202	0.0349
	32	**0.3141**	0.1319	0.0324	0.0336	0.0443	0.0832	0.0382	0.0235	0.0375
	2	0.1317	**0.1596**	0.0217	0.0198	0.0511	0.0831	0.0221	0.0195	0.0198
GLOVE	4	0.2310	**0.3251**	0.0356	0.0328	0.0964	0.1458	0.0617	0.0311	0.0594
$d = 300$	8	0.3061	**0.3959**	0.0655	0.0618	0.1073	0.1914	0.1209	0.0591	0.1112
	16	0.4030	**0.4495**	0.1138	0.1081	0.1809	0.2851	0.1939	0.1004	0.1882
	32	**0.4374**	0.4323	0.2039	0.2139	0.2808	0.3917	0.3208	0.1717	0.2727
ImageNet	4	*N.A.*	**0.1863**	0.0502	0.0578	0.1058	*N.A.*	0.0406	0.0389	0.0392
$d = 1,000$	8	*N.A.*	**0.2177**	0.0824	0.0854	0.1519	*N.A.*	0.0925	0.0806	0.0826
	16	*N.A.*	**0.2391**	0.1522	0.1527	0.2122	*N.A.*	0.1679	0.1338	0.1378
	32	*N.A.*	**0.2480**	0.2282	0.2337	0.2430	*N.A.*	0.2311	0.1801	0.2002

We experimented on the artificial datasets and the real datasets. From each ARTFC dataset, we randomly chose 10,000 training samples with both the input (X) and the output (Y) representations, and chose another 10,000 testing samples with the input representation only. For the two proposed WTA models, we trained a sparse binary projection matrix W each based on the training data. Then we generated 2,000-dimensional sparse binary output vectors via the WTA competition after projecting the testing samples with the matrix. For the LIFTING algorithm, the same training and testing procedures were applied. For all

Table 2. Search accuracies on GLOVE dataset with various input dimensions and a fixed output dimension ($d' = 2,000$). The proposed WTA models reported consistently improved results in all the settings.

Dimension	k	*SUP*	*UNSUP*	*LSH*	*FJL*	*FLY*	*LIFTING*	*ITQ*	*SPH*	*ISOH*
$d = 100$	2	0.1007	**0.1210**	0.0208	0.0210	0.0503	0.0982	0.0245	0.0229	0.0230
	4	0.1720	**0.2274**	0.0335	0.0317	0.0787	0.1449	0.0683	0.0452	0.0633
	8	0.2365	**0.2816**	0.0591	0.0602	0.1059	0.1898	0.1509	0.0705	0.1297
	16	0.3113	**0.3572**	0.1096	0.1125	0.1698	0.2279	0.2201	0.1232	0.1995
	32	0.3779	**0.3831**	0.1962	0.2007	0.2581	0.2954	0.3311	0.2398	0.2952
$d = 200$	2	0.0808	**0.1073**	0.0183	0.0177	0.0387	0.0733	0.0231	0.0223	0.1234
	4	0.1432	**0.2030**	0.0275	0.0257	0.0624	0.1037	0.0692	0.0395	0.0212
	8	0.2008	**0.2551**	0.0459	0.0329	0.0786	0.1363	0.1197	0.0624	0.1256
	16	0.2759	**0.3189**	0.0816	0.0798	0.1284	0.1712	0.1804	0.1105	0.1905
	32	0.3254	**0.3331**	0.1442	0.1502	0.1991	0.2391	0.3025	0.2051	0.2782
$d = 500$	2	0.0689	**0.0866**	0.0148	0.0152	0.0226	0.0490	0.0197	0.0173	0.0182
	4	0.1328	**0.1702**	0.0195	0.0183	0.0394	0.0711	0.0522	0.0301	0.0397
	8	0.1878	**0.2252**	0.0278	0.0276	0.0421	0.0892	0.0973	0.0521	0.0885
	16	**0.2768**	0.2696	0.0437	0.0469	0.0710	0.1188	0.1497	0.0995	0.1305
	32	**0.3172**	0.2727	0.0728	0.0804	0.1115	0.1806	0.2119	0.1502	0.2117
$d = 1,000$	2	0.0464	**0.0508**	0.0132	0.0129	0.0172	0.0293	0.0177	0.0166	0.0179
	4	0.0985	**0.1131**	0.0162	0.0175	0.0288	0.0396	0.0356	0.0289	0.0322
	8	0.1447	**0.1615**	0.0214	0.0261	0.0303	0.0490	0.0434	0.0312	0.0365
	16	**0.2115**	0.2071	0.0305	0.0372	0.0474	0.0662	0.0912	0.0787	0.0883
	32	**0.2194**	0.2049	0.0464	0.0511	0.0699	0.0874	0.1507	0.1339	0.1303

other algorithms, we applied each of them on the testing samples to get either dense or sparse binary output vectors. Then the output vectors are used in similarity search and compared against the input vectors, as illustrated in Sect. 4.1.

We repeated the process for fifty runs and recorded the average accuracies. The results are given in Table 1. Each row shows the similarity search accuracies with a specific hash length[4]. Consistent with the results reported in [7], the sparse binary projection algorithms reported improved results over the classical LSH method. Among the algorithms, it is clearly shown that, with the support of the supervised information, the LIFTING and the supervised-WTA algorithms reported further improved results over the FLY algorithm. Most prominently, with the hash length of $k = 4$, the FLY algorithm has an accuracy of 6.73%, while the supervised-WTA model's accuracy reaches 66.7%, almost ten times higher. When comparing the two supervised algorithms, the supervised-WTA model outperformed the LIFTING algorithm with all hash lengths.

Among the unsupervised learning algorithms, the proposed unsupervised-WTA model reported the best performances, significantly better than the results given by the LSH, FJL, FLY, ITQ, SPH, and ISOH algorithms. Its accuracies are even better than the supervised-WTA model with the hash length of $k = 8$.

[4] As in [7,19], the hash length is defined as the number of ones in each output vector for the FLY, LIFTING and WTA algorithms. For other algorithms, it is defined as the output dimension.

On the GLOVE dataset, only the input representation X is available. We randomly chose 10,000 samples for training, and 10,000 samples for testing. We computed $Y^* = \arg_Y \min \frac{1}{2} \left\| \gamma X^T X - Y^T Y \right\|_F^2 + \alpha \left\| Y \right\|_{\frac{1}{2}}$ for the training data via the Frank-Wolfe algorithm[5], and used Y^* as the output representation for training. Then we experimented with the same setting as on ARTFC datasets. Again the two WTA models reported improved results.

When comparing the performances of two WTA models on the GLOVE dataset, the unsupervised-WTA model performed even better than the supervised-WTA model on most tests. On these datasets only the input representation X was known. An approximation of Y was obtained through matrix factorization. The quality of this approximated Y becomes critical to the supervised-WTA model. We believe this is the major reason why the supervised model no longer excels.

Besides, we tested the algorithms' performances on a much larger ImageNet dataset with one million images for training and 10,000 images for testing. Computing the output representation Y^* becomes infeasible on such a large training set, and therefore the results of the supervised-WTA and the LIFTING models were not available. Comparing with the other algorithms with available results, once again the unsupervised-WTA algorithm reported improved search accuracies.

In addition to the experiment on similarity search accuracies, we further investigated the influence of different input/output dimensions on the performance of the proposed models. We fixed the output dimension to $d' = 2,000$ while varying the input dimension from 100 to 1,000 on GloVe word vectors. We recorded the similarity search accuracies from all algorithms. From the results in Table 2, it was also shown that, when the output dimension and the hash length were fixed, the search accuracies of all algorithms decreased when increasing the input dimension. Comparatively, we can see that the two WTA models had improved results.

4.3 Running Speed

As a practical concern, we compared the training time of the proposed WTA models with the LIFTING algorithm. In the experiment, we used the ARTFC datasets with 1,000-dimensional inputs and 2,000-dimensional outputs, and the number of training samples varied from 1,000 to 50,000.

We recorded the training time of each algorithm to compute the sparse binary projection matrix W. On all training sets, the proposed models reported significantly faster speed than the LIFTING algorithm. With 1,000 samples (ref. Fig. 1(a)), the supervised-WTA model took less than 0.2 s to get the optimal solution, hundreds of times faster than the LIFTING algorithm which took around 50 s.

The unsupervised-WTA model needs to solve multiple W^t and Y^t iteratively. It took 10 to 20 s with 1,000 samples, which was slower than the supervised-WTA model but several times faster than the LIFTING algorithm. With 50,000

[5] We set $\gamma = 1$ without fine-tuning the parameter.

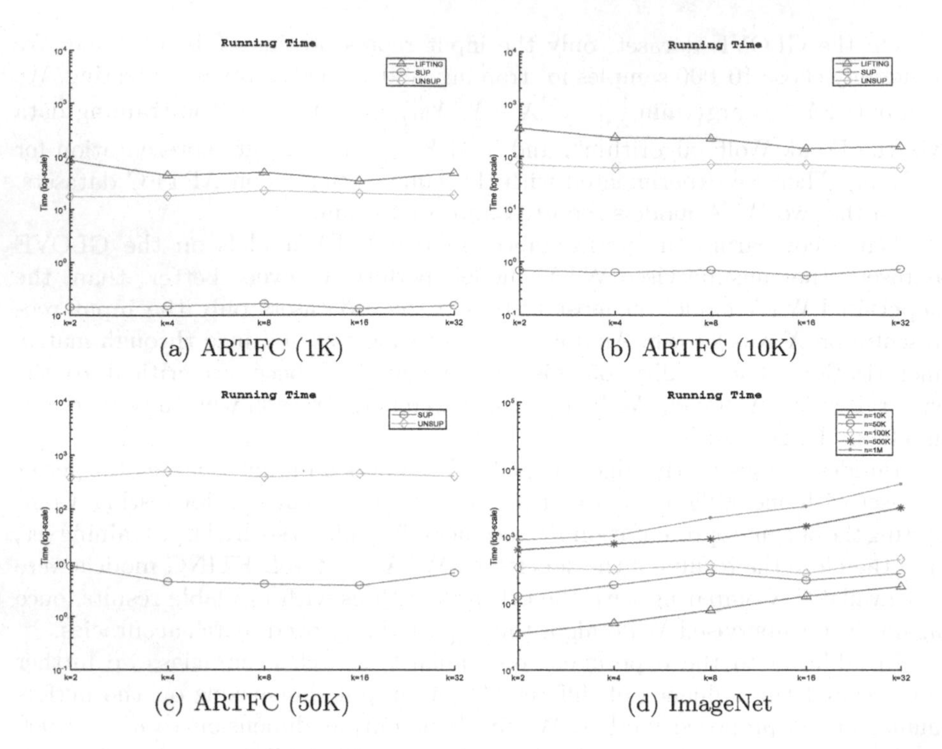

(a) ARTFC (1K) (b) ARTFC (10K)

(c) ARTFC (50K) (d) ImageNet

Fig. 1. Comparison of training time. Horizontal: the hash length ($k = 2/4/8/16/32$). Vertical: training time (seconds) in log-scale. (a)–(c): Comparison of LIFTING/SUP/UNSUP algorithms with $1K$ to $50K$ training samples of the ARTFC dataset (with both input and output representations). In (c), the results of LIFTING are not available due to the prohibitive computation. (d) Training time of the UNSUP algorithm with $10K$ to $1M$ training samples of the ImageNet dataset. The results of SUP/LIFTING algorithms are not available due to the prohibitive computation to obtain the output representations.

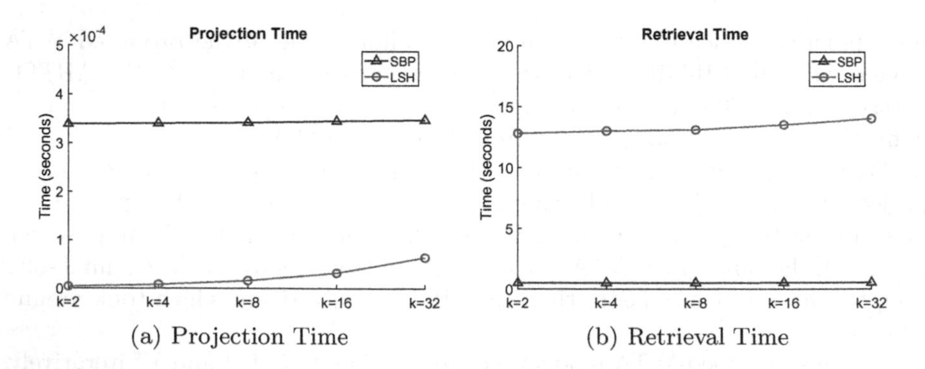

(a) Projection Time (b) Retrieval Time

Fig. 2. Comparison of search speed on 10 million simulated images. Horizontal: the hash length ($k = 2/4/8/16/32$). Vertical: running time in seconds. In the experiment, $d = 1,000$ and $d' = 2,000$, and $c = 100$.

samples (ref. Fig. 1(b)), the supervised-WTA model took less than 10 s, and the unsupervised-WTA model took about 400 s to get the solutions. For the LIFTING algorithm, we didn't finish the execution in our platform within 12 h. All these real results were consistent with the complexity analysis given in Sect. 3.3, and justified the running efficiency of the proposed WTA models.

We further experimented on the much larger ImageNet dataset and reported the results in Fig. 1(d). Due to the prohibitive computation to obtain the output representation, only the results of the UNSUP algorithm are available. From the results we can see, with a hash length of $k = 32$, the algorithm took less than 200 s to train a projection matrix with $10K$ samples, and took around 6,000 s to train with one million samples.

In addition to the training time, we further evaluated the search time under a much larger setting. We simulated an environment with a database of 10 million candidate images with 1,000-dimensional features. Assuming the projection matrix has been given, to obtain the similarity between a query image and the candidate images, the computations are mainly from two phases:

- Projection: to compute the hash vector for each query image with the projection matrix;
- Retrieval: to compute and compare the distances between each query image with all candidate images in the database with the hash vector.

We compared the projection and retrieval time over this database by the dense representation (denoted by LSH) and the sparse binary representation (denoted by SBP). The results are shown in Fig. 2(a) and 2(b). It can be seen that, although the sparse binary projection method took a longer time than the LSH method in the projection phase, it was much faster in the retrieval phase. Summing up the projection time and retrieval time, the sparse binary projection method took about only 0.5 s to find the most similar images for a query over such a large database, which was over 20 times faster than LSH with various hash lengths. Our results verified the benefit of the sparse binary representation which makes it possible to apply much faster logical operations instead of arithmetic operations in distance or similarity calculations.

4.4 Evaluation Summary

In summary, our experimental results revealed a number of key benefits of the proposed models over state-of-the-art methods, which are usually appealing in industrial applications.

1. The proposed WTA models reported improved accuracies over several popular hashing methods.
2. The proposed WTA models reported significantly improved training speed over the LIFTING method.
3. The sparse binary projection representation reported improved search speed over the dense representation with a range of hash lengths.

Considering the improvements in search accuracies, we believe that the computational overhead of training a projection matrix instead of randomizing one should be acceptable in many application scenarios. On the other hand, the search speed is vital in numerous tasks, where the sparse binary representation may find its wide applications. Although the reported results are still preliminary, our work exhibited the high potential of the proposed methods in industrial applications.

5 Conclusion

With strong evidence from biological science, the study of sparse binary projection models has attracted much research attention recently. By mapping lower-dimensional dense data to higher-dimensional sparse binary vectors, the models have reported excellent empirical results and proved to be useful in practical applications.

Sparse binary projections are tightly coupled with WTA competitions. The competition is an important stage for pattern recognition activities that happen in the brain. Accordingly, our work started from the explicit treatment of the competition, and proposed two models to seek the desired projection matrix. Specifically, the supervised-WTA model utilizes both input and output representations of the samples, and trains the projection matrix as a supervised learning problem. The unsupervised-WTA model extends the supervised model and utilizes the input representation only to train the projection matrix in an unsupervised manner, which equips the model with wider application scenarios.

Despite the benefits from the WTA competition, it often brings non-trivial computation burdens to the training algorithms, which limits its application in various tasks. Our work designed a relaxation scheme that greatly simplified the computational burden, with which we were able to develop simple, efficient, and effective computing procedures to the proposed WTA models. The relaxation technique potentially supports the inclusion of the WTA procedure into more learning models.

Our results trigger some topics for further study. Firstly, the computing routines for both models only involve simple vector addition and scalar comparison operations. The computing procedures are highly parallelizable and can be possibly implemented with customized hardware for high-throughput applications [25].

Secondly, the unsupervised-WTA model provides a unified framework that combines the clustering and feature selection techniques. This viewpoint may provide a potential bridge that helps to better understand the WTA mechanism and make it clear why the competition could lead to algorithms that preserve the locality structures of the data well.

Thirdly, the proposed models potentially help to design new artificial neural network architectures. The WTA competition and the relaxation techniques adopted in this paper can be used as an activation function of the neurons in an artificial neural network. We warmly anticipate future work along this direction [20,27].

In addition, another useful topic is on the relationship between the input dimension d and the output dimension d'. From our empirical results, a higher output dimension seems to be needed for a given higher input dimension to keep the search accuracy. Yet a rigorous study on this problem will be of both theoretical and practical values.

Acknowledgment. We thank the anonymous reviewers for their insightful comments and suggestions. The work is supported by Shenzhen Fundamental Research Fund (JCYJ201703061410-38939, KQJSCX20170728162302784, JCYJ20170410172341657).

References

1. Ailon, N., Chazelle, B.: The fast Johnson-Lindenstrauss transform and approximate nearest neighbors. SIAM J. Comput. **39**(1), 302–322 (2009)
2. Arbib, M.: The Handbook of Brain Theory and Neural Networks. MIT Press, Cambridge (2003)
3. Baeza-Yates, R., Ribeiro-Neto, B.: Modern Information Retrieval, vol. 463. ACM Press, Cambridge (1999)
4. Bingham, E., Mannila, H.: Random projection in dimensionality reduction: applications to image and text data. In: Proceedings of the 7th ACM SIGKDD International Conference on Knowledge Discovery and Data Mining, pp. 245–250. ACM (2001)
5. Caron, S., Ruta, V., Abbott, L., Axel, R.: Random convergence of olfactory inputs in the drosophila mushroom body. Nature **497**(7447), 113 (2013)
6. Charikar, M.: Similarity estimation techniques from rounding algorithms. In: Proceedings of the 34th Annual ACM Symposium on Theory of Computing, pp. 380–388. ACM (2002)
7. Dasgupta, S., Stevens, C., Navlakha, S.: A neural algorithm for a fundamental computing problem. Science **358**(6364), 793–796 (2017)
8. Frank, M., Wolfe, P.: An algorithm for quadratic programming. Naval Res. Logist. **3**(1–2), 95–110 (1956)
9. Gionis, A., Indyk, P., Motwani, R.: Similarity search in high dimensions via hashing. In: Proceedings of the 25th International Conference on Very Large Data Bases, vol. 99, pp. 518–529 (1999)
10. Gong, Y., Lazebnik, S., Gordo, A., Perronnin, F.: Iterative quantization: a procrustean approach to learning binary codes for large-scale image retrieval. IEEE Trans. Pattern Anal. Mach. Intell. **35**(12), 2916–2929 (2012)
11. Guyon, I., Elisseeff, A.: An introduction to variable and feature selection. J. Mach. Learn. Res. **3**, 1157–1182 (2003)
12. Heo, J., Lee, Y., He, J., Chang, S., Yoon, S.: Spherical hashing: Binary code embedding with hyperspheres. IEEE Trans. Pattern Anal. Mach. Intell. **37**(11), 2304–2316 (2015)
13. Jaggi, M.: Revisiting Frank-Wolfe: projection-free sparse convex optimization. In: Proceedings of the 30th International Conference on Machine Learning, pp. 427–435 (2013)
14. Jain, A., Murty, N., Flynn, P.: Data clustering: a review. ACM Comput. Surv. **31**(3), 264–323 (1999)
15. Johnson, W., Lindenstrauss, J.: Extensions of Lipschitz mappings into a Hilbert space. Contemporary Mathematics **26**(189–206), 1 (1984)

16. Knuth, D.: The Art of Computer Programming, Sorting and Searching, vol. 3. Addison-Wesley, Reading (1998)
17. Kohonen, T.: The self-organizing map. Proc. IEEE **78**(9), 1464–1480 (1990)
18. Kong, W., Li, W.: Isotropic hashing. In: Advances in Neural Information Processing Systems, pp. 1646–1654 (2012)
19. Li, W., Mao, J., Zhang, Y., Cui, S.: Fast similarity search via optimal sparse lifting. In: Advances in Neural Information Processing Systems, pp. 176–184 (2018)
20. Lynch, N., Musco, C., Parter, M.: Winner-take-all computation in spiking neural networks. arXiv preprint arXiv:1904.12591 (2019)
21. Maass, W.: On the computational power of winner-take-all. Neural Comput. **12**(11), 2519–2535 (2000)
22. MacQueen, J.: Some methods for classification and analysis of multivariate observations. In: Proceedings of the 5th Berkeley Symposium on Mathematical Statistics and Probability, Oakland, CA, USA, vol. 1, pp. 281–297 (1967)
23. Manning, C., Raghavan, P., Schütze, H.: Introduction to Information Retrieval. Cambridge University Press, Cambridge (2008)
24. Olsen, S., Bhandawat, V., Wilson, R.: Divisive normalization in olfactory population codes. Neuron **66**(2), 287–299 (2010)
25. Omondi, A., Rajapakse, J.: FPGA Implementations of Neural Networks, vol. 365. Springer, Heidelberg (2006). https://doi.org/10.1007/0-387-28487-7
26. Panousis, K., Chatzis, S., Theodoridis, S.: Nonparametric Bayesian deep networks with local competition. In: Proceedings of the 36th International Conference on Machine Learning, pp. 4980–4988 (2019)
27. Pehlevan, C., Sengupta, A., Chklovskii, D.: Why do similarity matching objectives lead to Hebbian/anti-Hebbian networks? Neural Comput. **30**(1), 84–124 (2018)
28. Pennington, J., Socher, R., Manning, C.: Glove: global vectors for word representation. In: Proceedings of the 2014 Conference on Empirical Methods in Natural Language Processing, pp. 1532–1543 (2014)
29. Powell, M.: On search directions for minimization algorithms. Math. Program. **4**(1), 193–201 (1973)
30. Russakovsky, O., Deng, J., Su, H., et al.: Imagenet large scale visual recognition challenge. Int. J. Comput. Vision **115**(3), 211–252 (2015)
31. Stevens, C.: What the fly's nose tells the fly's brain. Proc. Natl. Acad. Sci. **112**(30), 9460–9465 (2015)
32. Turner, G., Bazhenov, M., Laurent, G.: Olfactory representations by drosophila mushroom body neurons. J. Neurophysiol. **99**(2), 734–746 (2008)
33. Zheng, Z., Lauritzen, S., Perlman, E., Robinson, C., et al.: A complete electron microscopy volume of the brain of adult drosophila melanogaster. Cell **174**(3), 730–743 (2018)

LOAD: LSH-Based ℓ_0-Sampling over Stream Data with Near-Duplicates

Dingzhu Lurong[1], Yanlong Wen[1](✉), Jiangwei Zhang[2], and Xiaojie Yuan[1]

[1] TKLNDST, College of Computer Science, Nankai University, Tianjin, China
{lulongdingzhu,wenyanlong,yuanxiaojie}@dbis.nankai.edu.cn
[2] National University of Singapore, Singapore, Singapore
A0054808@u.nus.edu

Abstract. Massive amounts of stream data nowadays almost make any real-time analysis impossible. To overcome the challenge of processing this huge amount of data, previous works typically use sampling to extract representatives and conduct analysis on this sampled dataset. In this paper, we propose LOAD, a Locality-Sensitive Hashing (LSH) based ℓ_0-sampling over stream data. Instead of having the same diameter for all dimensions, LOAD utilizes the dimension-specific diameters which could fit the distribution of groups better. Therefore, LOAD always generates a better representative identification result. To facilitate the real-time analysis, we further optimize LOAD by applying LSH. Since nearest items are hashed into the same bucket with high probability, hence distinguishing the representatives becomes lightning fast. Extensive experiments show that LOAD is not only more accurate than other state-of-the-art algorithms, but also faster by an order of magnitude.

Keywords: Stream data · Data sampling · ℓ_0-sampling

1 Introduction

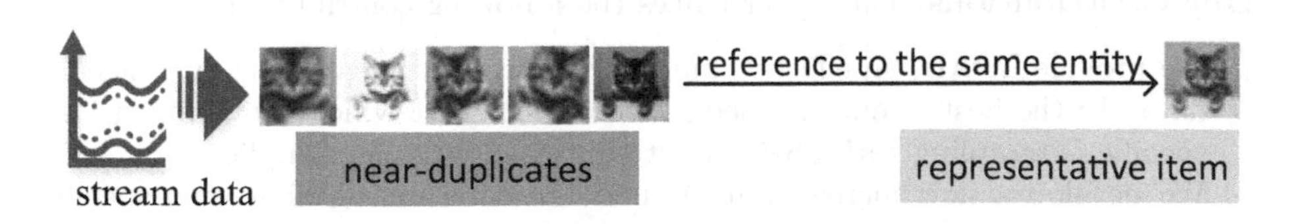

Fig. 1. An illustration of near-duplicates in image stream.

Nowadays, enormous amounts of stream data are being generated, and it's nearly impossible to conduct any real-time analysis over the whole stream data. To allow real-time analysis, one must do sampling. One of the commonly used sampling methods is l_0-sampling, which samples the entities uniformly, in other words, the

F. Hutter et al. (Eds.): ECML PKDD 2020, LNAI 12457, pp. 473–489, 2021.
https://doi.org/10.1007/978-3-030-67658-2_27

entities have the same probability of being sampled. The standard l_0-sampling treats each item from stream data as a distinct entity. But in reality, stream data are always noisy, with much of the data items are near-duplicates, which means those items look differently on the surface, but they may reference to the same entity essentially. For example, Fig. 1 illustrates that images of the same content may regard differently due to diverse color adjustments (brightness, contrast), rotations, random crops, formats, etc. Hence, instead of treating these near-duplicates as distinct entities, we should treat those near-duplicates as one entity, which could dramatically reduce the data size without losing too much useful information. Therefore, the standard ℓ_0-sampling cannot be directly applied here, since the items with more near-duplicates will be selected as the sampled item with a higher probability. Consequently, this imbalanced sampled dataset will give biased result for subsequent analysis. Thus, the question then is how to compute those near-duplicates in a real-time manner, and conduct robust ℓ_0-sampling over stream data with near-duplicates.

Recently, Chen et al. [1] firstly studied robust ℓ_0-sampling over stream data and proposed RSIW, which posts random grid cells with equal sides in the Euclidean spaces. For each arriving item p, RSIW validates whether p is the representative and calculates whether the cell of p is sampled. If so, p is added to sampled dataset, then RSIW continually maintains the sampled dataset with the latest arrived items. At each query, it returns the result from sampled dataset via random sampling. However, RSIW has limitations in effectiveness and efficiency. First of all, RSIW uses the same diameter for all dimensions when clustering the near-duplicates, which is not applicable when the data items distribute differently in individual dimension. One may say that normalizing data along each dimension would estimate this problem. We notice data normalization scales the items to a specific interval, but the distributions of items in each dimensions remain unchanged, thus the group diameters in individual dimensions are still different. Secondly, when computing the near-duplicates, RSIW searches the representatives in a brute force manner, which has a time complexity $O(nm)$, where n is the number of items in stream data, and m is total number of arrived representatives. To overcome the limitations, we propose LOAD.

Our Contributions. This paper makes the following contributions.

- We provide LOAD, an effective and efficient robust l_0-sampling over stream data. To the best of our knowledge, we are the first which allow dimension-specific l_0-sampling with LSH over stream data with near-duplicates.
- We develop a new incremental method for determining dimension-specific diameters, which guarantees a better representative identification result and improves the accuracy of robust l_0-sampling effectively.
- We firstly utilize LSH into robust ℓ_0-sampling on stream data, while it accelerates the identification of representatives among near-duplicates, and improves the execution time by an order of magnitude.

Outline. The rest of the paper is organized as follows. In Sect. 2, we lay out the preliminaries and formally define the problem. In Sect. 3, we present our solution

Table 1. Commonly used notations.

Symbol	Description	Symbol	Description
S	Raw stream data	M	The maximum size of B^{acc}
d	The dimensionality of S	w	The quantization width of LSH
rep	Representative	k	The number of projections of LSH
S^{rep}	All *reps* from S	γ	The array of separation ratios of S
B^{acc}	All accepted *reps* in buckets	α	The array of group diameters
B^{rej}	All rejected *reps* in buckets	β	The array of inter-group distances
$\mathrm{F}_0(S,\alpha)$	The number of groups in S	$Cell(u)$	The cell that item u locates
$\mathrm{F}_0(S,\alpha_i)$	The number of groups in S_i	$Adj(u)$	Adjacent cells with center item u

LOAD in detail, followed by the comprehensive experiment analysis in Sect. 4. Section 5 surveys related work of ℓ_0-sampling and data deduplication. Section 6 concludes the paper.

2 Preliminaries

In this section, we first introduce some concepts, then we formally define our problems. For the ease of reference, we summarize the commonly used notations in this paper in Table 1.

Definition 1. *(α,β)-sparsity stream data. Let $\alpha[\alpha_1,...,\alpha_d]$ be an array of group diameters, where each $\alpha_i \subset [0,\infty), i \in [1,d]$. $\beta[\beta_1,...,\beta_d]$ be an array of inter-group distances which holds $\beta_i > \alpha_i$ $(i \in [1,d])$. $D(.,.) : U \times U \to [0,\infty)$ be the distance function of the Euclidean space. We say S is (α,β)-sparsity if for any two items $u[u_1,...,u_d]$ and $v[v_1,...,v_d]$ from S in dimension i, there are either $D(u_i,v_i) \leq \alpha_i$, or $D(u_i,v_i) \geq \beta_i$, $i \in [1,d]$.*

Take Fig. 2 (a) for example, S is ([2,4]-[6,16])-sparsity. Because for any items $u[u_1,u_2]$ and $v[v_1,v_2]$, there are $D(u_1,v_1) \leq 2$ or $D(u_1,v_1) \geq 6$ in x-axis.

Definition 2. *Separation ratios γ. Let γ $[\gamma_1,...,\gamma_d]$ be the separation ratios of S, and each γ_i holds $\gamma_i = \beta_i/\alpha_i$, $i \in [1,d]$, where β_i, α_i is the maximum, minimum in all the values that $D(u_i,v_i) \geq \beta_i$, $D(u_i,v_i) \leq \alpha_i$ holds in dimension i.*
Take Fig. 2 (a) for example, S has the γ [3,4]. Simply because S is ([2,4]-[6,16])-sparsity, and $\gamma_1 = 6/2 = 3$ in x-axis, $\gamma_2 = 16/4 = 4$ in y-axis.

Definition 3. *Well-shaped stream data. Let S be the (α,β)-sparsity stream data, γ be the separation ratios of S. We say S is well-shaped stream data, if and only if $\gamma_i > 2$ $(i \in [1,d])$. When the $\gamma_i > 2$ $(i \in [1,d])$, the items in S can be partitioned into several disjoint groups in each dimension with the intra-group distance at most α_i and inter-group distance at least β_i.*
Take Fig. 2 (a) for example, S is a well-shaped stream data, simply because the separation ratios γ of S are [3,4], and obviously $\gamma_1 = 3 > 2$, $\gamma_2 = 4 > 2$.

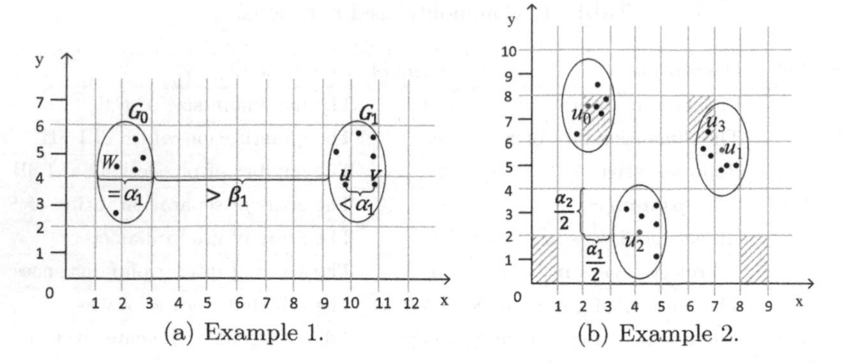

(a) Example 1. (b) Example 2.

Fig. 2. **(a)** An illustration of well-shaped dataset distribution in 2-dimensional space. Each black point is an item from S, where S are ([2,4]–[6,16*])-sparsity and well-shaped (* it can be any number larger than 8, and we take 16 for example). Each black circle represents a group, then items in the same group are near-duplicates of each other. In the x-axis, the intra-group distance of S is smaller than 2 (α_1), and the inter-group distance is larger than 6 (β_1). **(b)** An example of robust ℓ_0-sampling. Each black circle represents an identified group and each point is an item from stream data, while the red points are the *reps* of corresponding groups. Each square with gray edges represents a cell. Cells with blue stripe are sampled cells. The side lengths of each cells are half of $\alpha[\alpha_1 = 2, \alpha_2 = 4]$. Then, u_0 is accepted (add to B^{acc}), because u_0 is a *rep* and the *Cell*(u_0) is sampled. u_1 is rejected (add to B^{rej}), because the *Cell*(u_1) is not sampled, but there is one of adjacent cells *Adj*(u_1) is sampled. u_2 is ignored, because both *Cell*(u_2) and *Adj*(u_2) are not sampled. Note that u_3 locates in a sampled cell, but it is ignored since it's not a *rep* firstly.

Definition 4. *Cell. [2] Let S be the stream data, α be the array of group diameters of S, then the random grids of side length α_i in dimension $i (i \in [1, d])$ are posted on $\mathbb{R}^d$, and we call a grid cell simply a cell.*

The function Cell(u) calculates the cell that item u locates, and function Adj(u) selects the cells which locate in the space of diameters α with a center at u. Take Fig. 2 for example. Each square with gray edges represents a cell.

Definition 5. *Near-duplicates. Let Space(p, α) be the space centered at item p with diameter α_i in each dimension i ($i \in [1, d]$). For items u, v from sequential stream S, if u, v are near-duplicates of each other (denoted as $u \cong v$), we have $v \in$ Space(u, α), which indicates u and v belong to the same group. On the contrary, if u, v are not near-duplicates of each other, then $u \ncong v$.*
Take Fig. 2 (a) for example, the items in G_1, like u and v are near-duplicates. But items u and w are not near-duplicates.

Definition 6. *Representative (abbr rep). Let S^{rep} be the set of reps from stream data S and it is initially empty. We say an item u from S is a rep, if and only if u is the first item arrived in the group, which means $\forall\ p \in S^{rep}$, $p \ncong u$.*

Take Fig. 2 (b) for example, items u_0, u_1, u_2 from S are reps of S, as we can see, there is only one rep for a group.

Definition 7. *ℓ_0-sampling on well-shaped stream data. [1] Let S be well-shaped stream data, α be the array of group diameters of S, v be any rep from S. Then ℓ_0-sampling outputs an item u with uniform distribution, that $\forall u \in S$, $Pr[u \cong v] = 1/F_0(S, \alpha)$, where $F_0(S, \alpha)$ is the number of distinct groups in S.*

Definition 8. *robust ℓ_0-sampling on stream data with near-duplicates. [1] Let α be the group diameters of stream data S, v be any rep from S. Then robust ℓ_0-sampling on S outputs an item u, that $\forall u \in S$, $Pr[u \cong v] = \Theta(1/F_0(S, \alpha))$.*

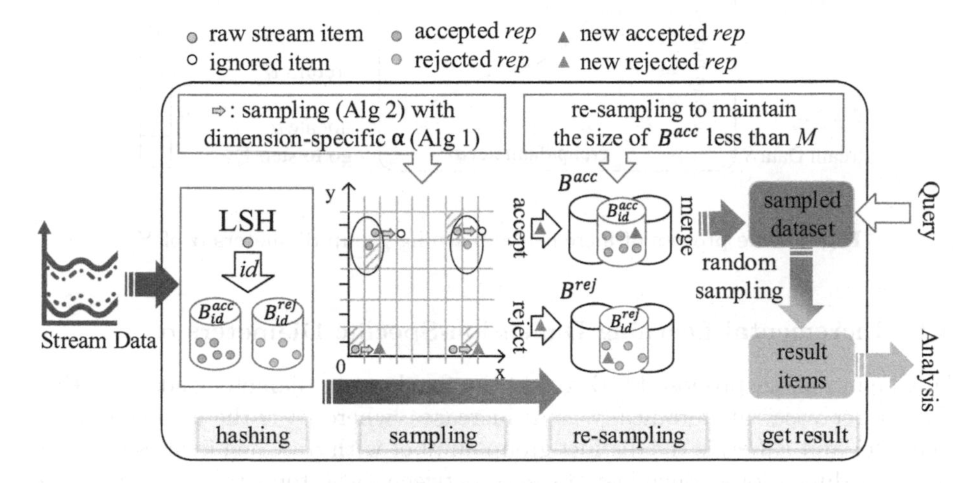

Fig. 3. Overview of the LOAD.

3 LOAD

The overview of LOAD (LSH-based ℓ_0-sampling over stream data) can be found in Fig. 3. LOAD first allocates a bucket id id for each data item u from the raw stream data through LSH. Then it compares u with all the *reps* in the buckets B^{acc}_{id}, B^{rej}_{id} to determine whether the item u is a *rep*. If not, we just ignore it, since there is already a *rep* which is near-duplicate with u. On the other hand, if u is a *rep*, then we use cells with dimension-specific diameters to calculate whether u should be accepted or rejected or ignored. As the LOAD is continuously processing raw stream data, lots of accepted *reps* are adding to B^{acc}_{id}, which are merged as the sampled dataset. Whenever a query comes, the result can be computed from this sampled dataset for later analysis. During the process, the buckets B^{acc} and B^{rej} will be re-sampled whenever the number of items in B^{acc} is greater than the threshold M. In the following section, we will explain in detail of LOAD.

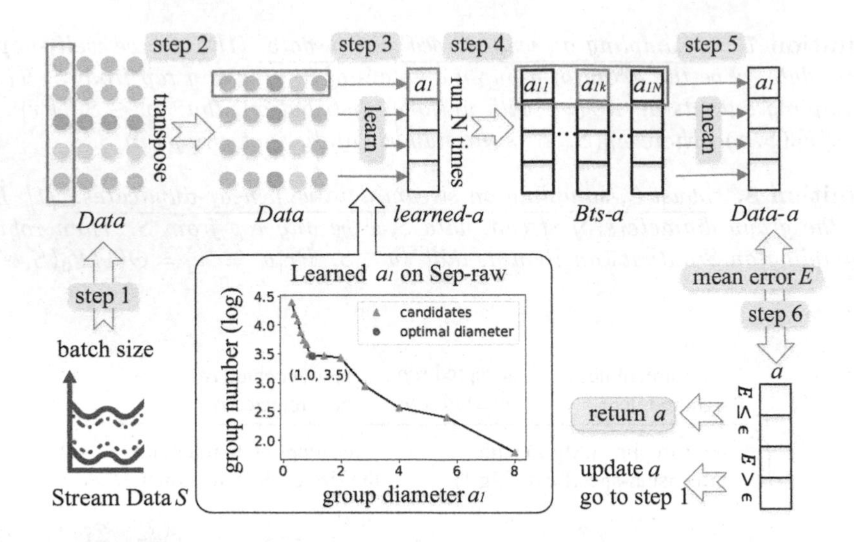

Fig. 4. The process of incremental learning group diameters α of S.

3.1 Incremental Learning Dimension-Specific Diameters α

Di Chen et al. [2] propose FGD to find the fixed group diameter and prove that $F_0(\alpha')$ decreases monotonically as α' increases, where α' is the assigned group diameter and $F_0(\alpha')$ is the distinct group number with α'. Chen focuses on using the high-dimensional Euclidean distance between data items to measure a same diameter on each dimensions. But in this paper, we pay more attention to the information in individual dimensions of dataset. We calculate the array of intra-group distances α (group diameters) and the array of inter-group distances β, where each element of the arrays presents a dimension-specific distance. Therefore the arrays can fit the characteristics of dataset better. especially when the separation ratios γ of dataset are different from each dimensions greatly.

The array of α are made up of α_1,..., α_d, where each α_i means the learned group diameter in dimension i ($i \in [1, d]$) and the group number $\mathrm{F_0}(S, \alpha_i)$ decreases monotonously with growth of α_i in each dimension. As shown in Fig. 4, the group number decreases gradually with increase of α_1 on the first dimension of Sep-raw. Let α^{act} be the array of actual group diameters of S, if $\alpha_i < \alpha_i^{act}$, then $\mathrm{F_0}(S, \alpha_i) \geq \mathrm{F_0}(S, \alpha_i^{act})$ ($i \in [1, d]$), as actual groups will be broken down into smaller groups in each dimension. On the contrary, if $\alpha_i > \alpha_i^{act}$, then $\mathrm{F_0}(S, \alpha_i) \leq \mathrm{F_0}(S, \alpha_i^{act})$ ($i \in [1, d]$), as small and different groups will be merged into larger groups. Subsequently, if dataset S is (α^{act}, β)-sparsity, then the learned group diameters α in dimension i holds $\alpha_i^{act} < \alpha_i < \beta_i$, $i \in [1, d]$, and there are $\mathrm{F_0}(S, \alpha_i^{act}) = \mathrm{F_0}(S, \alpha_i) = \mathrm{F_0}(S, \beta_i)$, which means the group number remain unchanged in this interval. Therefore the basic idea to calculate the group diameter α_i from candidate sequence α_{i0},...,α_{ic} (c is the number of candidates)

Algorithm 1: Compute the Group Diameters α

Input: batch dataset *Data* from S, iteration times N, error threshold ϵ
Output: the learned group diameters α of S

1 Transpose(*Data*);
2 $\{min, max\}$ = Get-Min-Max(*Data*);
3 **for** *repeat N times* **do**
4   **for** *each dimension i of Data* **do**
5   tmp-α = min_i;
6   **while** *tmp-α < max_i* **do**
7   tmp-F_0 = Count-F_0($Data_i$, tmp-α);
8   push {tmp-α, log(tmp-F_0)} into *Dim-F_0*;
9   update tmp-α;
10   **end**
11   tmp-α = Get-Min-Gradient-α(*Dim-F_0*);
12   *learned-α_i* = tmp-α;
13   **end**
14   push *learned-α* into *Bts-α*;
15   Transpose-Shuffle-Transpose(*Data*);
16 **end**
17 *Data-α* = Get-Mean-α (*Bts-α*);
18 **if** *the average error between Data-α and α is larger than ϵ* **then**
19   update α with Data-α;
20   load next batch *Data* from S and go to line 1;
21 **else**
22   **return** α;
23 **end**

is searching for a region with a low gradient ∇ in dimension i.

$$\nabla = \frac{|log(F_0(\alpha_{ij})) - log(F_0(\alpha_{ij-1}))|}{\alpha_{ij} - \alpha_{ij-1}}, \; j \in (0, c). \tag{1}$$

When we get the region with the lower ∇, then we select the smaller value α_{ij-1} of the two alternative values $\alpha_{ij-1}, \alpha_{ij}$ as the approximate value of group diameter α_i in dimension i. Take Fig. 4 for example, we calculate all the candidate diameters and the corresponding group numbers (the green triangles) of α_1 on Sep-raw, then we calculate the interval with minimum gradient and get the smaller value as the learned diameter α_1(red point).

Algorithm 1 Computes the Group Diameters α. After reading batch dataset *Data* from stream data in memory as the step 1 in Fig. 4, they are transposed for easy calculation (Line 1) in step 2. Then, the array of minimum and maximum distances between coordinates in each dimension of *Data* are calculated and denoted as *min, max* (Line 2), which specify the range of candidate group diameters in each dimensions. When calculating the group diameter of

Data (Line 4–13) as the step 3 in Fig. 4, we firstly initialize α_i (denoted as tmp-α) to min_i (Line 5). Then, we do the while loop to look through all the candidates of α_i by factors of $\sqrt{2}$, and calculate the corresponding number of groups, then we save them (tmp-α, log(tmp-F_0)) to *Dim-*F_0 as pairs (Line 6–10). Subsequently, we use Eq. (1) to select the interval with the minimum gradient in *Dim-*F_0 and choose the least diameter in that interval as the learned diameter of α_i (Line 11). After computing α_i in each dimensions, we successfully get the current learned diameters *learned-*α of *Data*. Next, the step 3 will be executed N times (Line 3–16) as the step 4 in Fig. 4. In each execution, we firstly recalculate and save the present *learned-*α to *Bts-*α and secondly, we reorder the items from *Data* for next calculation (Line 15), since using α_i to estimate the number of groups on $Data_i$ is related to the sequential order of items. After that, the step 5 (Line 17) calculates the mean of *Bts-*α in each dimensions as the learned group diameter *Data-*α of *Data*. Finally, the step 6 (Line 18–23) calculates the average error between *Data-*α and α, where α is the learned group diameters of S. Next, if the error is larger than ϵ, then we load next batch dataset from S as new *Data* and recalculate the *Data-*α from step 1. In the end, we return α of S until the error is small than ϵ.

3.2 Accepted Representative Identification

One can simply find that if u is an accepted *rep*, then u must satisfy followings: Firstly, u is a *rep* of S and secondly, the cell of u is sampled. Let's first identify *reps*. As defined before, if item u is a *rep*, then all the items v belong to S^{rep}, that hold $u \not\cong v$. After the group diameters $\alpha[\alpha_1, ..., \alpha_d]$ of S were learned, we can calculate $u \not\cong v$ explicitly as $D(u, v, \alpha) > 1$, where

$$D(u, v, \alpha) = \sum_{i=1}^{d} \frac{(u_i - v_i)^2}{\alpha_i^2}. \tag{2}$$

When the item u arrives, it needs to be compared with all the so-far arrived *reps*. Once there is a *rep* v, that $D(u, v, \alpha) \leq 1$, then u and v are near-duplicates of each other. Hence, u is not a *rep* of S. If there is no such *rep* v, then u is a *rep* of S. Next, we employ hash function $h_R(C) = h(C) \ mod \ R$ to calculate whether the cell C is sampled, where $h : [\Delta^2] \rightarrow \{0, ..., 2^{\lceil 2log\Delta \rceil} - 1\}$ is a fully random hash function and $R = 2^k (k \in \mathbb{N})$. Then, when an item u is identified as a *rep*, we say the *rep* u

- u is accepted (add to B^{acc}). If and only if $Cell(u)$ is sampled, which means $h_R(C) = 0$.
- u is rejected (add to B^{rej}). If $Cell(u)$ is not sampled ($h_R(Cell(u)) \neq 0$), but one or more cells belong to $Adj(u)$ is sampled ($h_R(Adj(u)) = 0$).
- u is ignored. If none of the $Cell(u)$ or $Adj(u)$ is sampled.

All these three cases are shown in Fig. 2 (b) that we post all the items in 2-dimensional Euclidean space as easy to understand. Note that maintaining B^{rej}

is in order to sample items uniformly from stream data with near-duplicates, which is critical importance in robust ℓ_0-sampling over stream data. Now, we formally define accepted representative in this paper.

Definition 9. *Accepted representative. Let S be the stream data, S^{rep} be the set of reps of S, α be the group diameter of S, u be a data item from S, then we say u is an accepted rep of S, if u satisfying the following rules: Firstly, $\forall v \in S^{rep}$, $D(u, v, \alpha) > 1$ and secondly, $h_R(Cell(u)) = 0$.*

Take Fig. 2 (b) for example. There is only item u_0 is an accepted rep. Since u_0 is the first arrived item in the corresponding group, thus it is the rep of that group. Then, the cell of u_0 $Cell(u_0)$ is sampled.

3.3 LSH-Based Robust ℓ_0-Sampling on Stream Data

LSH Optimal Parameters Learning. LSH uses a probabilistic approach to find the nearest neighbors, and it can greatly reduce the retrieval time on large-scale data, since similar data items are hashed into the same buckets with high probability. Then it only needs to traverse a few buckets, rather than the entire dataset. LSH was formally defined as follows:

Definition 10. *Locality-sensitive hashing (LSH). [3] let S be the dataset, U be the item universe, $D(.,.)$ be the distance metric, $\mathcal{H}$ be the hash family $\{h : S \rightarrow U\}$, we say $\mathcal{H}$ be (r_1, r_2, p_1, p_2)-sensitive, if any data items u, v from s, that*

- *$D(u, v) \leq r_1$, then $Pr_{\mathcal{H}}[h(u) = h(v)] \geq p_1$.*
- *$D(u, v) \geq r_2$, then $Pr_{\mathcal{H}}[h(u) = h(v)] \leq p_2$.*

There are many different distance metrics $D(.,.)$, we follow previous works [4], and use the normal Euclidean distance as distance metric, let u, v be the data items from S, d be the dimensionality of S, then

$$D(u, v) = \sqrt{(u_1 - v_1)^2 + ... + (u_d - v_d)^2}. \tag{3}$$

Let w be the quantization width, b be the bias term that randomly selected from a uniform distribution between 0 and w, v be a random vector with the same dimensionality as u, where each coordinate is a Gaussian random variable $N(0, 1)$, the hash function $h(u)$ in hash family $\mathcal{H}$, is

$$h(u) = \lfloor \frac{u \cdot v + b}{w} \rfloor. \tag{4}$$

In this paper, the required parameters for an LSH implementation are the quantization width w, the number of projections k, and the number of repetitions L. Slaney et al. [4] propose the optimization algorithm for parameters of LSH $OA(n, \delta, d_{nn}, d_{any})$, where n is the number of items in dataset, δ is an acceptable error probability that we miss the exact nearest neighbor, d_{nn} is the probability distribution functions (pdf) for the distance between the query item and its nearest neighbor, d_{any} is the pdf for distance between the query item and any random item in the dataset. Given this, the algorithm OA returns the optimal parameters for a LSH implementation: the quantization width w, and the number of projections k. Note that, we do not repeat the whole hash processing, which means L=1.

Performance Improvement with LSH. The efficiency of the ℓ_0-sampling algorithm becomes particularly important, and the basic idea to improve the efficiency is to reduce the number of comparisons between item u and *reps* in S^{rep} when identifying whether u is a *rep*. In this paper, we use LSH to hash similar *reps* into the same bucket with a higher probability, and the buckets of accepted/rejected *reps* are denoted as B^{acc}/B^{rej} separately. Then, whenever data item u arrives, we only need to find the corresponding bucket B^{acc}_{id} and B^{rej}_{id} with the bucket id id of u, and traverse items in those two buckets to determine whether u is a *rep*. Thus we can identify item u is a *rep*, if

$$\forall v \in B^{acc}_{id} \cup B^{rej}_{id}, D(u, v, \alpha) > 1. \tag{5}$$

The steps to obtain the bucket id id corresponding to a data item u are as follows: Firstly, initialization. Generate k random vectors v, p separately with the dimensionality d and each coordinate is a Gaussian random variable $N(0,1)$, denoted as V, P. Then select b from a uniform distribution from 0 to w as the bias item. Secondly, one-line projection. Using the hash function $h(u)$ in hash family $\mathcal{H}$ to project a data item u to an integer i, where $h(u)$ is presented in Eq. (4). Thirdly, multiline projection. Doing k one-line projection to obtain a vector I, which contains k integer i, and each i is the result from one one-line projection for u. Lastly, random projection. For I, doing $\sum_{j=0}^{k-1} I \cdot P_j$, then getting the final result as bucket id id.

Algorithm 2 About LSH-Based Robust ℓ_0-Sampling. When an item u arrives, the bucket id id of u will be calculated with LSH (Line 3), then the buckets B^{acc}_{id} and B^{rej}_{id} corresponding to u can be found with id. Next, u needs to compare with all the items in $B^{acc}_{id} \cup B^{rej}_{id}$ to find out whether u is a *rep* (Line 4). If u is not a *rep*, then the algorithm just ignores u and continues to process the next item. On the contrary, if u is a *rep*, then we need to calculate whether the cell of u $Cell(u)$ is sampled (Line 5). If $Cell(u)$ is sampled, then we add u into B^{acc}_{id} (Line 6) which means u is accepted. On the other hand, if $Cell(u)$ is not sampled, then we calculate whether there is a cell belongs to $Adj(u)$ is sampled (Line 7). If so, we add u into B^{rej}_{id} (Line 8). We continuously process items from stream data, and update B^{acc}_{id}, B^{rej}_{id} synchronously with the latest items. During processing, we ceaselessly maintain the size of B^{acc} under M (Line 11–14). Once the size of B^{acc}_{id} is greater than M, then R is doubled and B^{acc}_{id}, B^{rej}_{id} will be updated with the new h_R. Finally, for each query, result dataset can be randomly sampled from B^{acc} for application-specific analysis (Line 16).

Algorithm 2: LSH-Based Robust ℓ_0-Sampling

```
   Input: raw stream data S, the maximum size of B^acc M
   Output: result dataset
 1 Initialization: R ← 1, B^acc ← ∅, B^rej ← ∅
 2 for each data item u from S do
 3     id = Get-Bucket-ID (u) ;
 4     if ∀v ∈ B^acc_id ∪ B^rej_id, D(u, v, α) > 1 then
 5         if h_R(Cell(u)) = 0 then
 6             push u into B^acc_id;
 7         else if ∃C ∈ Adj(u), h_R(Cell(C)) = 0 then
 8             push u into B^rej_id;
 9         end
10     end
11     if Num-of-items(B^acc) > M then
12         R ← 2R ;
13         re-sample B^acc and B^rej ;
14     end
15 end
16 return Random-Sample(B^acc) ;
```

4 Experiment Results

Datasets. Table 2 presents the datasets used in this paper: We use image dataset cifar10 [5], synthetic dataset Sep-raw, and we follow [1], using datasets Rand20, Seeds[1] and Yacht[2]. Let α be the actual diameter of Sep-raw, and $\alpha_i = \sqrt{2} \times \alpha_{i-1}$ ($i \in (0, 10]$), where $\alpha_0 = 1$. β follows $\beta_i = 5 \times \alpha_i + 0.1$. Thus each coordinate of separation ratios γ on Sep-raw is larger than 2, which means Sep-raw is a well-shaped dataset. Then for each item v already generated in Sep-raw, the generation of new item u applies the following rules: $D(u_i, v_i) < \alpha_i \, or D(u_i, v_i) > \beta_i$. For each image in cifar10, we use image augment (rotation, scaling, shearing, random crops, horizontal flips) to generate 99 near-duplicates, and we encode the images into latent space $\mathbb{R}^{20}$, $\mathbb{R}^{200}$, $\mathbb{R}^{1000}$ using ClusterGAN [6], producing datasets C20, C200, C1000. For the rest four datasets, we generate near-duplicates using two different methods mentioned in [1], and we add '-uniform', '-power' after the name of dataset.

Performance Metrics. We follow [7], the accuracy of ℓ_0-sampling is defined as following: Let S be the dataset, α be the learned diameters, $n = F_0(S, \alpha)$ be the number of groups in S, $f^* = 1/n$ be the target probability, f_i be the empirical sampling probability of the i-th group, then:

[1] https://archive.ics.uci.edu/ml/datasets/seeds.
[2] https://archive.ics.uci.edu/ml/datasets/Yacht+Hydrodynamics.

Table 2. Characteristics of datasets.

Dataset	Num of items	Num of *reps*	Dimensionality
C20	6000000	60000	20
C200			200
C1000			1000
Sep-raw	500	500	10
Sep-uniform	25127		
Sep-power	3678		
Rand20-uniform	26085	500	20
Rand20-power	3678		
Seeds-uniform	11055	210	7
Seeds-power	1360		
Yacht-uniform	15153	308	7
Yacht-power	2115		

- stdDevNm:

$$\sum_{i=0}^{n-1}\left|\frac{f_i - f^*}{f^*}\right|, \ i \in [0, n).$$

- maxDevNm:

$$max_i\left\{\left|\frac{f_i - f^*}{f^*}\right|\right\}, \ i \in [0, n).$$

We follow [1], and defined the execution time as:

- pTime: Processing time per item (millisecond with single thread).

Experiment Environment. We implement LOAD in C++ (gcc version 5.4.0). The experiment is conducted on Ubuntu 16.04.4 LTS with Intel(R) Core(TM) i7-6800k CPU @3.40 GHZ and 62 GB memory.

4.1 Time Consumption

EXP-1: Average Runtime. Figure 5 depicts the execution time about LOAD and RSIW [1] on different datasets. It's very clear that LOAD takes less time than RSIW by an order of magnitude. On Rand20-uniform and Rand20-power, it's clear that LOAD is approximately 10 times faster than RSIW. we run LOAD and RSIW 100 times and record the average time to reduce the measuring error.

EXP-2: Effect of Number of *Reps*. In Fig. 5, for datasets Seeds-uniform, Seeds-power and Yacht-uniform, Yacht-power, LOAD is only 3–4 times faster than RSIW, since the number of *reps* in them are only 210 and 308, thus the improvement of LSH is not huge. But with the number of *reps* increases, for dataset C20, C200, C1000 with 60,000 *reps* in Fig. 6, LOAD is nearly 12–16 times

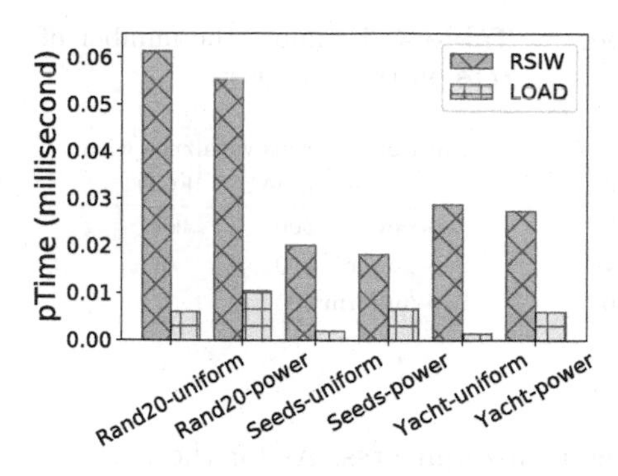

Fig. 5. Time Consumption *w.r.t.* average runtime on six different datasets.

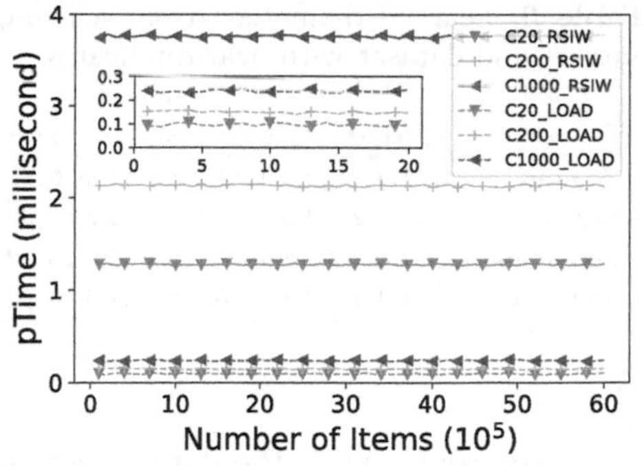

Fig. 6. Time Consumption *w.r.t.* the effect of data dimension.

faster than RSIW, which demonstrates LOAD is more applicable for stream data processing than RSIW, while stream data always contains huge amounts of *reps*.

EXP-3: Effect of Data Dimension. In Fig. 6, we report the runtime of RSIW and LOAD on different dimensional data. It can be noted that LOAD performs pretty well on all datasets, especially on high dimensional dataset, it is nearly 16 times faster than RSIW on C1000. On the other hand, there is an obvious increase in runtime of RSIW with the rise of data dimension. Overall, LOAD shows better performance on high dimensional datasets than RSIW.

4.2 Accuracy of Sampling

EXP-4: Learning Group Diameters α. Table 3 presents the result about algorithm 1 running on three datasets. For well-shaped dataset Sep-raw (Row 1), we can find that learned α are the same as actual α (Row 4), which shows that algorithm 1 works pretty well on the well-shaped dataset. Simply because, when the dataset is well-shaped, all the distances between two coordinates of items u and v are either smaller than α_i or larger than β_i in the i-axis. Thus, the α_i in each dimension on well-shaped dataset can be found properly. For dataset contains near-duplicates Sep-power (Row 2) and Sep-uniform (Row 3), the learned α are approximately same as actual α with the max errors no larger than 0.2 in each dimensions, which demonstrates Algorithm 1 can work on dataset with near-duplicates.

EXP-5: Representative Identification. In Table 4, we report the number of *reps* estimate by LOAD and FGD [2] on several datasets. It's clear that no matter the dataset with or without near-duplicates, LOAD can precisely find all *reps* (Column 2), where the actual number of *reps* in three datasets is 500. On the contrary, FGD find 499 *reps* when the dataset is well-shaped (Row 1, Column 3), but as the number of near-duplicates increases, the total *reps* measured by FGD decreases dramatically from 499 to 3, 2 (Row 2,3, Column 3) that many small groups are merged by other groups, which means FGD can not identify the

Table 3. Learned diameters α on well-shaped dataset and dataset with near-duplicates.

Dataset	α_0	α_1	α_2	α_3	α_4	α_5	α_6	α_7	α_8	α_9
Sep-raw	1	1.4	2	2.8	4	5.7	8	11.3	16	22.6
Sep-power	1	1.3	2	2.9	4	5.6	8	11.3	16	22.7
Sep-uniform	1	1.5	2.2	2.8	4.2	5.7	8.2	11.5	16	22.7
actual α	1	1.4	2	2.8	4	5.7	8	11.3	16	22.6

Table 4. Estimate the number of *reps* on three dataset.

Dataset	*reps* with LOAD	*reps* with FGD
Sep-raw	500	499
Sep-power	500	3
Sep-uniform	500	2

accurate groups when the dataset contains near-duplicates. As for the reasons, on the one hand, FGD just uses a fixed global α to verify groups, which could not detect correct groups adequately due to lack of enough dimension-specific characteristics of dataset. On the other hand, the near-duplicates significantly impact the FGD to find the suitable α, where the fixed global α computed by FGD are 243.9, 1168.1, 1425.1 for Sep-raw, Sep-power, Sep-uniform separately. Besides, the α that learned by LOAD is recorded in Table 3.

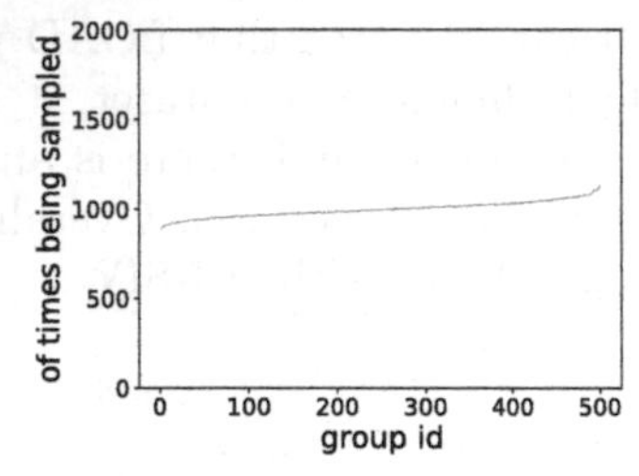

(a) Sep-uniform #run=500,000

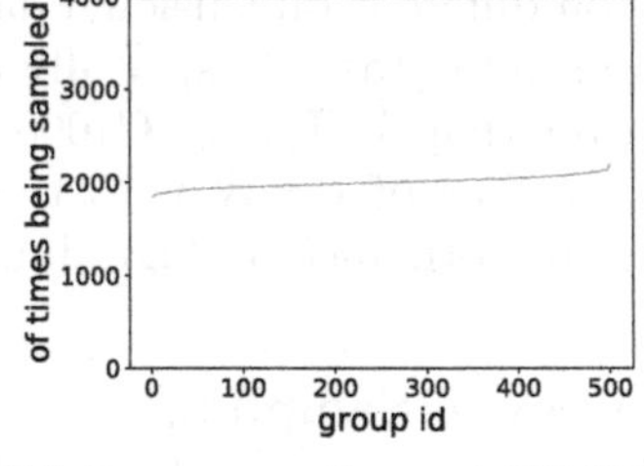

(b) Sep-power #run=1,000,000

Fig. 7. Accuracy *w.r.t.* the times of groups being sampled separately.

EXP-6: Empirical Sampling Distribution of LOAD. In Fig. 7, we report the result about times of groups being sampled separately after sampled 500,000 times on Sep-uniform and 1,000,000 times on Sep-power (denote as #run). It's clear that LOAD can almost uniformly sample *reps* from raw stream data, which presents that LOAD is a qualified robust ℓ_0-sampling with good performance.

EXP-7: Accuracy of LOAD on ℓ_0-Sampling. Interestingly, results in Fig. 8 show that increasing M actually boosts the accuracy of ℓ_0-sampling. The reason is that B^{acc} needs to be updated, when the number of items in B^{acc} is up to M, and this operation will delete nearly half of *reps* in B^{acc}, which may cause the errors that some *reps* can not be sampled uniformly. Overall, even M is small enough, the maxDevNm is still no larger than 0.22, and the stdDevNm is smaller than 0.06, which demonstrates that LOAD is high accuracy in ℓ_0-sampling. In addition, RSIW can not estimate properly *reps* firstly when dataset is not well-shaped, thus it is meaningless to evaluate stdDevNm and maxDevNm on RSIW.

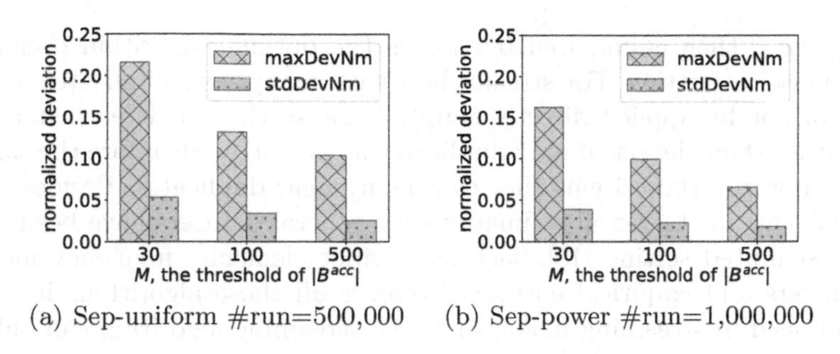

(a) Sep-uniform #run=500,000 (b) Sep-power #run=1,000,000

Fig. 8. Accuracy *w.r.t.* the maxDevNm and stdDevNm with different M.

5 Related Work

The most related works to LOAD are ℓ_0-sampling and data deduplication, where data deduplication [8] (or duplicate detection, record linkage, etc.) is the problem of estimating distinct elements on dataset. Previous works mostly work on datasets without near-duplicates (noiseless setting), and recently, Zhang et al. [2] firstly study about robust data deduplication with near-duplicates, where it proposed a sampling method to overcome the problem of data deduplication with near-duplicates, and it proposes FGD (Finding Group Diameter) to find the fixed global group diameter. Then, Chen et al. [1] firstly investigate robust ℓ_0-sampling on stream data with near-duplicates and propose RSIW (Robust ℓ_0-Sampling-IW), which focuses on sampling the item uniformly from the stream data. This paper extends the line of [1] and proposes LOAD, an effective and efficient robust ℓ_0-sampling over stream data with near-duplicates.

Standard ℓ_0-Sampling. Unlike robust ℓ_0-sampling which selects the representatives as the entities, standard ℓ_0-sampling simply treats each item as an entity, therefore the sampling result may be erroneous if we feed the dataset with near-duplicates to it. Standard ℓ_0-sampling is first formally studied in [9,10], where it returns items (almost) uniformly at random from sampled set as the sampling result. Graham Cormode et al. [7] provided a unifying framework for standard ℓ_0-sampling algorithms, which abstracts three major steps for building ℓ_0-sampler: sampling, recovery and selection. Brian Babcock et al. [11] applied the sequence-based and timestamp-based sliding window technique to standard ℓ_0-sampling. For distributed streaming settings, [12] presented the random ℓ_0-sampling algorithms that continuously maintain a random sample of distinct entities from data stream, whose input items come from multiple distributed sites that communicate via central coordinator. To improve the efficiency of sampling, the tight bound both upper bound and lower bound for approximate ℓ_p-samplers are investigated in [13,14].

Data Deduplication. There was a long research line about data deduplication over stream data without near-duplicates. The first streaming algorithm to estimate distinct elements was introduced in [15], followed by many studies

([16,17], etc.), then culminated by [18] as the optimal algorithm over stream data in noiseless setting. For stream data with many near-duplicates, all these works can not be applied directly simply because they treat each item as an entity rather than the set of near-duplicates as an entity, therefore the sampling is biased towards those items that own many near-duplicates. Besides, several basic problems in statistical estimations for general datasets have been studied in the distributed setting [19], such as distinct elements, frequency moments, heavy hitters and empirical entropy. However all those algorithms in [19] can not be applied in streaming mode, since the streaming algorithm can only scan the dataset once without looking back.

6 Conclusion

In this paper, we present LOAD for robust ℓ_0-sampling over stream data with near-duplicates. LOAD adopts dimension-specific group diameters to better identify *reps*, while the group diameters are learned incrementally with so-far arrived items from stream data. To enhance the performance, LOAD utilizes LSH to reduce computation load when distinguishing *reps*, which accelerates the sampling process significantly. The extensive experiments demonstrate LOAD gains much better results than state-of-the-art methods in terms of efficiency and accuracy.

Acknowledgement. This research is supported by Chinese Scientific and Technical Innovation Project 2030 (No. 2018AAA0102100), National Natural Science Foundation of China (No. 61772289, U1936206). We thank the reviewers for their constructive comments. We also thank Jiecao Chen and Qin Zhang for their generous help.

References

1. Chen, J., Zhang, Q.: Distinct sampling on streaming data with near-duplicates. In: Proceedings of the 37th ACM SIGMOD-SIGACT-SIGAI Symposium on Principles of Database Systems, pp. 369–382. ACM (2018)
2. Chen, D., Zhang, Q.: Streaming algorithms for robust distinct elements. In: Proceedings of the 2016 International Conference on Management of Data, pp. 1433–1447. ACM (2016)
3. Indyk, P., Motwani, R.: Approximate nearest neighbors: towards removing the curse of dimensionality. In: Proceedings of the Thirtieth Annual ACM Symposium on Theory of Computing, pp. 604–613. ACM (1998)
4. Slaney, M., He, J., Lifshits, Y.: Optimal parameters for locality-sensitive hashing. Proc. IEEE **100**(9), 2604–2623 (2012)
5. Krizhevsky, A.: Learning multiple layers of features from tiny images. Technical report (2009)
6. Mukherjee, S., Asnani, H., Lin, E., Kannan, S.: Clustergan: latent space clustering in generative adversarial networks. In: Proceedings of the AAAI Conference on Artificial Intelligence **33**, 4610–4617 (2019)
7. Cormode, G., Firmani, D.: A unifying framework for l0-sampling algorithms. Distrib. Parallel Databases **32**(3), 315–335 (2014)

8. Elmagarmid, A.K., Ipeirotis, P.G., Verykios, V.S.: Duplicate record detection: a survey. IEEE Trans. Knowl. Data Engineering **19**(1), 1–16 (2006)
9. Frahling, G., Indyk, P., Sohler, C.: Sampling in dynamic data streams and applications. In: Symposium on Computational Geometry (2005)
10. Gibbons, P.B., Tirthapura., S.: Estimating simple functions on the union of data streams. In: Proceedings of the Thirteenth Annual ACM Symposium on Parallel Algorithms and Architectures, pp. 281–291. ACM (2001)
11. Babcock, B., Datar, M., Motwani, R.: Sampling from a moving window over streaming data. In: Proceedings of the Thirteenth Annual ACM-SIAM Symposium on Discrete Algorithms, pp. 633–634. Society for Industrial and Applied Mathematics (2002)
12. Chung, Y.-Y., Tirthapura, S.: Distinct random sampling from a distributed stream. In: 2015 IEEE International Parallel and Distributed Processing Symposium, pp. 532–541. IEEE (2015)
13. Ba, K.D., Indyk, P., Price, E., Woodruff, D.P.: Lower bounds for sparse recovery. In: Proceedings of the Twenty-First Annual ACM-SIAM Symposium on Discrete Algorithms, pp. 1190–1197. SIAM (2010)
14. Jowhari, H., Sağlam, M., Tardos, G.: Tight bounds for LP samplers, finding duplicates in streams, and related problems. In: Proceedings of the thirtieth ACM SIGMOD-SIGACT-SIGART Symposium on Principles of Database Systems, pp. 49–58. ACM (2011)
15. Flajolet, P., Martin, G.N.: Probabilistic counting algorithms for data base applications. J. Comput. Syst. Sci. **31**(2), 182–209 (1985)
16. Beyer, K., Haas, P.J., Reinwald, B., Sismanis, Y., Gemulla, R.: On synopses for distinct-value estimation under multiset operations. In: Proceedings of the 2007 ACM SIGMOD International Conference on Management of Data, pp. 199–210. ACM (2007)
17. Ganguly, S.: Counting distinct items over update streams. Theoret. Comput. Sci. **378**(3), 211–222 (2007)
18. Kane, D.M., Nelson, J., Woodruff, D.P.: An optimal algorithm for the distinct elements problem. In: Proceedings of the Twenty-Ninth ACM SIGMOD-SIGACT-SIGART Symposium on Principles of Database Systems, pp. 41–52. ACM (2010)
19. Zhang, Q.: Communication-efficient computation on distributed noisy datasets. In: Proceedings of the 27th ACM Symposium on Parallelism in Algorithms and Architectures, pp. 313–322. ACM (2015)

Spatio-Temporal Tensor Sketching via Adaptive Sampling

Jing Ma(✉), Qiuchen Zhang, Joyce C. Ho, and Li Xiong

Department of Computer Science, Emory University, Atlanta, Georgia
{jing.ma,qzhan84,joyce.c.ho,lxiong}@emory.edu

Abstract. Mining massive spatio-temporal data can help a variety of real-world applications such as city capacity planning, event management, and social network analysis. The tensor representation can be used to capture the correlation between space and time and simultaneously exploit the latent structure of the spatial and temporal patterns in an unsupervised fashion. However, the increasing volume of spatio-temporal data has made it prohibitively expensive to store and analyze using tensor factorization.

In this paper, we propose SkeTenSmooth, a novel tensor factorization framework that uses adaptive sampling to compress the tensor in a temporally streaming fashion and preserves the underlying global structure. SkeTenSmooth adaptively samples incoming tensor slices according to the detected data dynamics. Thus, the sketches are more representative and informative of the tensor dynamic patterns. In addition, we propose a robust tensor factorization method that can deal with the sketched tensor and recover the original patterns. Experiments on the New York City Yellow Taxi data show that SkeTenSmooth greatly reduces the memory cost and outperforms random sampling and fixed rate sampling method in terms of retaining the underlying patterns.

Keywords: Spatio-temporal data · Tensor sketching · Tensor completion

1 Introduction

The increasing availability of spatio-temporal data has brought new opportunities in application domains including urban planning, informed driving, and infectious disease spread modeling [2,8,24]. Unfortunately, the rapid growth in these data streams can be prohibitively expensive to store, communicate and analyze. In addition, the high-dimensional, multi-aspect spatio-temporal data poses analytic challenges due to the correlations in the measurements from both time and space. Moreover, human-intensive and domain-specific supervised models are not tractable due to the constant and evolving deluge of measurements.

Given the high-dimensional, multi-aspect nature of spatio-temporal data, a tensor serves as an efficient way to represent and model such data [5,8,24]. As

F. Hutter et al. (Eds.): ECML PKDD 2020, LNAI 12457, pp. 490–506, 2021.
https://doi.org/10.1007/978-3-030-67658-2_28

an example, each element of the tensor can represent the occurrences of an event at a specific location (encoded as latitude and longitude) within a specific time interval. Compared with the low-dimensional matrix-based methods such as [15], tensors can capture the correlation between each mode. Furthermore, tensor factorization offers an unsupervised, data-driven approach to identify the global structure of the data via a high-order decomposition. It is also more interpretable compared to deep learning methods [30]. Unfortunately, existing models require the full tensor information (i.e., all the data samples must be stored) and do not readily scale to extremely large tensors.

The computation and storage limitations of large tensors motivate the need to approximate such tensors with relatively small "sketches" of the original tensors. Not only are these manageable-sized tensor sketches more readily stored on a single machine, the computationally expensive tasks such as tensor decomposition can be performed on the smaller tensors while still preserving the underlying structure. Although sketching has been proposed as a linear algebra tool for reducing the memory cost, traditional methods involve the linear transformation of the original tensor with a "fat" random projection matrix to reduce the dimensions [21,27]. However, the transformation has inherent shortcomings: 1) the design of the random projection matrix may not capture the evolving patterns without a priori information; and 2) continuous sketching is computationally expensive since it involves the inversion of a fat random projection matrix.

An alternative sketching approach is based on tensor sparsification – subsampling the original tensor while preserving the original tensor structure. Compared with the random projection method, sampling incurs negligible online complexity. Existing tensor sparsification methods include random sampling based on the tensor spectral norm [19], sampling according to the entry values [28], and sampling according to the pre-computed tensor distribution [4]. However, these tensor sparsification algorithms suffer from the following limitations: 1) they cannot deal with streaming data where measurements are not available a priori and the tensor is incrementally updated; 2) there is no formal mechanism to reconstruct the original streams [7]. While tensor reconstruction is often a byproduct of dealing with missing data, reconstruction of the original streams can help track dynamic and abrupt changes at particular locations.

In this paper, we propose SkeTenSmooth, a factorization framework that uses adaptive sampling to generate tensor sketches on-the-fly and preserves the underlying global structure. We explore the problem of serially acquired time slices (measurements appear once they are available). We introduce SkeTen, a tensor sketching method that adaptively adjusts the sampling intervals and samples time slices according to the feedback error between the prior estimate and the true value. Thus it can capture the time slices that are not well modeled by the prior forecasting model, while avoiding the storage of the time slices that contain redundant information (i.e., patterns that are already captured). Furthermore, we propose a novel method SkeSmooth to decompose the tensor sketches and reconstruct the underlying temporal trends. Unlike previous tensor

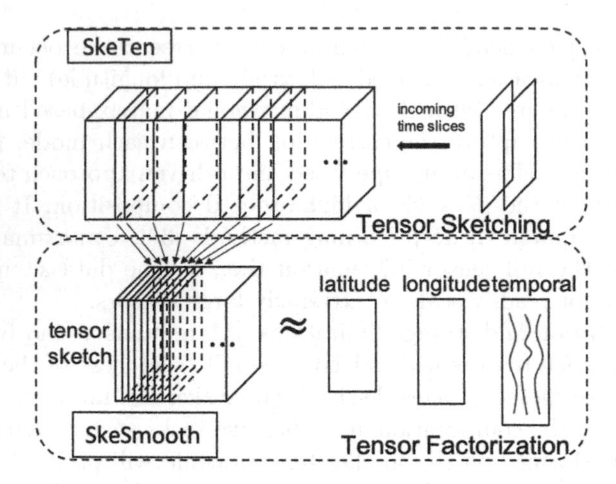

Fig. 1. The overall flow of SkeTenSmooth, including SkeTen and SkeSmooth.

factorization algorithms that deal with randomly missing entries, the sketched tensor has missing time slices (i.e., no data from a specific time point) which poses additional challenges. Thus, we introduce a robust tensor factorization algorithm that incorporates temporal smoothness constraints using auxiliary information about the data gained from SkeTen. Figure 1 shows the overall flow of the proposed framework. We briefly summarize our contributions as:

1) **Tensor sketching with adaptive sampling.** We propose SkeTen, a tensor sketching technique that helps process large volumes of data with low memory requirements and preserves the underlying temporal dynamics.
2) **Tensor factorization with smoothness constraint.** We propose a tensor factorization framework, SkeSmooth, that decomposes the smaller tensor "sketches" with smoothness constraints to achieve robust recovery of the underlying latent structure and the missing entries.
3) **Case study on New York taxi data.** We illustrate the ability to produce small tensor "sketches" and generate smooth temporal factors on real data.

2 Preliminaries and Notations

This section summarizes the notations used in this paper. Note that we use $\mathcal{X}$ to denote a tensor, $\mathbf{X}$ to denote a matrix, and $\mathbf{x}$ to denote a vector. The mode-d matricization of a tensor is denoted by $X_{(n)}$. The row and column vectors are represented by $\mathbf{x}_{i:}, \mathbf{x}_{:r}$ respectively.

2.1 Tensor Factorization

Definition 1. *(Khatri-Rao product). Khatri-Rao product is the "columnwise" Kronecker product of two matrices* $\mathbf{A} \in \mathbb{R}^{I\times R}$ *and* $\mathbf{B} \in \mathbb{R}^{J\times R}$*. The result is a* *matrix of size* $(IJ \times R)$ *and defined by* $\mathbf{A} \odot \mathbf{B} = [\mathbf{a}_1 \otimes \mathbf{b}_1 \cdots \mathbf{a}_R \otimes \mathbf{b}_R]$. $\otimes$ *denotes the Kronecker product. The Kronecker product of two vectors* $\mathbf{a} \in \mathbb{R}^I$, $\mathbf{b} \in \mathbb{R}^J$ *is*

$$\mathbf{a} \otimes \mathbf{b} = \begin{bmatrix} a_1\mathbf{b} \\ \vdots \\ a_I\mathbf{b} \end{bmatrix}$$

Definition 2. *(CANDECOMP-PARAFAC Decomposition). The CANDECOMP-PARAFAC (CP) decomposition is to approximate the original tensor* $\mathcal{Y}$ *by the sum of* R *rank-one tensors where* R *is the rank of tensor* $\mathcal{Y}$*. For a three-mode tensor* $\mathcal{Y} \in \mathbb{R}^{I \times J \times K}$*, the CP decomposition can be represented as*

$$\mathcal{O} \approx \mathcal{X} = \sum_{r=1}^{R} \boldsymbol{a}_{:r} \circ \boldsymbol{b}_{:r} \circ \boldsymbol{c}_{:r}, \tag{1}$$

where $\boldsymbol{a}_{:r} \in \mathbb{R}^I$, $\boldsymbol{b}_{:r} \in \mathbb{R}^J$, $\boldsymbol{c}_{:r} \in \mathbb{R}^K$ *are the* r*-th column vectors within the three factor matrices* $\boldsymbol{A} \in \mathbb{R}^{I \times R}$, $\boldsymbol{B} \in \mathbb{R}^{J \times R}$, $\boldsymbol{C} \in \mathbb{R}^{K \times R}$, $\circ$ *denotes the outer product.*

In this paper, the spatio-temporal tensor has three dimensions. The first dimension is the temporal dimension, while the other two are longitude and latitude dimensions representing the spatial modes.

3 SkeTen: Tensor Sketching via Adaptive Sampling

The main idea of SkeTen is to discard slices along the temporal mode that are "predictable". Intuitively, this adaptive sampling strategy will capture sudden temporal changes of the incoming streams, which is difficult for other methods such as random sampling and fixed rate sampling to capture. Moreover, SkeTen does not require a priori knowledge of the data and can adjust the sampling interval in real-time. SkeTen consists of two main components. The adaptive sampling module uses a prediction model to measure the "predictability" of the data. The second component selects random time series projections to reduce the training time of the prediction models.

3.1 Adaptive Sampling

SkeTen introduces an adaptive sampling method that can adjust according to the detected temporal dynamics and thus can better capture the underlying patterns even with the same amount of data stored. In SkeTen, individual time-series prediction models are developed for each spatial location. The adaptive sampling strategy then analyzes the slice feedback error, or the feedback error across all the locations.

Definition 3. *(Slice Feedback Error). Each tensor fiber along the temporal mode,* $\mathcal{X}(:, j, k)$*, is denoted as a time-series stream,* $\boldsymbol{x}_{j,k}$*. If* $\boldsymbol{x}_{j,k}^{(t)}$ *represents the* *true value at a particular time t, where $0 \leq t \leq T$, then the slice feedback error at time t, E_t is defined as:*

$$E_t = \sum_{k=1}^{K} \sum_{j=1}^{J} \left| \hat{\boldsymbol{x}}_{j,k}^{(t)} - \boldsymbol{x}_{j,k}^{(t)} \right| / \max\{\boldsymbol{x}_{j,k}^{(t)}, \delta\}, \tag{2}$$

where $\hat{\boldsymbol{x}}_{j,k}^{(t)}$ is the estimated value based on the prediction model, and δ is a small sanity bound.

The slice feedback error reflects how well the current time series models fit the current trends for each time series fiber. If there is a sudden increase across multiple spatial locations (i.e., an increase in the slice feedback error), then the sampling interval should be shortened to better capture the evolving trends.

Time Series Prediction Model. SkeTen uses an Autoregressive Integrated Moving Average (ARIMA) model to predict the values of the temporal slices. Given a time-series sequence $\{x_i\}$, $i = 1, \cdots, t$, where i is the time index, ARIMA(p, d, q) is defined as

$$\nabla^d x_t = \sum_{j=1}^{p} \alpha_j x_{t-j} + \varepsilon_t + \sum_{j=1}^{q} \beta_j \varepsilon_{t-j} + c \tag{3}$$

where p is the order of lags of the autoregressive model, d is the degree of differencing, and q is the order of the moving average model. $\alpha_j, j = 1, \cdots, p$ and $\beta_j, j = 1, \cdots, q$ represent the coefficients of AR and MA, respectively, ε_t is the white noise at time t, and c is the bias term. An ARIMA model is trained on each tensor fiber for the initial sampled time slices.

ARIMA has several convenient properties that make it more suitable than other time series prediction models in our setting. (1) The auto-regressive part of ARIMA ($\{\alpha_1, \cdots, \alpha_p\}$) generates the coefficients that are later exploited in SkeSmooth for the smoothness constraint in Sect. 4.1. This is essential for recovering the original temporal patterns with missing time slices. (2) It doesn't require a large number of training samples so the compression can occur much earlier than more complex models. (3) The coefficients can be used to understand the adaptive sampling for the PID controller which we will explain next. In preliminary experiments, we tried several state-of-the-art deep learning models such as ConvLSTM [29], but the compression rates and sampled slices only improved marginally. Thus, ARIMA is an optimal choice for our setting.

Feedback Control. SkeTen uses a controller system to detect rapid changes in the slice feedback error and adaptively adjusts the sampling rate. We adopt a PID controller similar to [6,13] to change the sampling interval over time. The PID controls the *Proportional, Integral* and *Derivative* errors. *Proportional error* is defined as $\Gamma_p = \gamma_p E_t$ to control the sampling intervals by keeping the controller proportional to the current slice feedback error. *Integral error* is

Algorithm 1: SkeTen

Input: ARIMA models for each tensor fiber, incoming tensor slices $\mathcal{X}(:, j, k), (j \leq J, k \leq K)$, set of PID controller parameters $\{\gamma_p, \gamma_i, \gamma_d\}$, next sampling point ns

1 **while** *not reach the end of data streams* **do**
2   **if** *next time slice* $t == ns$ **then**
3   sample the next time slice to the sketched tensor;
4   compute slice feedback error according to eq. (2);
5   compute the PID controller error Γ according to eq. (4);
6   obtain the new interval according to eq. (5);
7   update the next sampling point $ns = ns + newInterval$;
8   **else**
9   move on to the next time slice.

defined as $\Gamma_i = \frac{\gamma_i}{M_t} \sum_{m=0}^{M_t} E_t$, where M_t represents how many errors have been taken until time t. Thus the integral error considers the past errors to eliminate offset. *Derivative error* is defined as $\Gamma_d = \gamma_d \frac{E_t - E_{t-1}}{t_m - t_{m-1}}$ to prevent large errors in the future. Therefore, the full PID controller is defined as

$$\Gamma = \Gamma_p + \Gamma_i + \Gamma_d,$$
$$\text{s.t. } \gamma_p, \gamma_i, \gamma_d \geq 0, \ \gamma_p + \gamma_i + \gamma_d = 1. \tag{4}$$

The PID errors can thus be interpreted as control actions based on the past, the present and the future [3]. We set the PID parameters according to the common practice that *proportional* > *integral* > *derivative* [6]. With the PID error, the sampling interval is adjusted according to

$$newInterval = \max\{1, \theta(1 - e^{\frac{\Gamma - \xi}{\xi}})\}, \tag{5}$$

where θ and ξ are predefined user-specific parameters. The adaptive sampling process is presented in Algorithm 1.

3.2 Random Projection of ARIMA Coefficients

For large scale spatio-temporal data with many spatial locations (countless values of longitude and latitude), training the prediction model (i.e., ARIMA) may be computationally infeasible. To reduce the computational cost and retain the temporal fidelity of the data, we propose a random projection algorithm to train a limited number of ARIMA models, while preserving the competitive performance of the trained models. Inspired by the Locality Sensitive Hashing (LSH) algorithm, which has demonstrated its success in Approximate Nearest Neighbor search, our proposed projection method has the following steps: (1) train L ARIMA models $g_i, i = 1, ..., L$ from L time series $ts_i, i = 1, ..., L$, where each of

Algorithm 2: Random Projection of ARIMA Coefficients

Input: Tensor fiber set $\mathcal{T} = \{ts_1, ..., ts_M\}$, number of models to train L

1 **for** *i=1,...,L* **do**
2 train ARIMA model g_i for the randomly selected time series ts_i
3 **for** *j=1,...,M* **do**
4 **for** *i=1,...,L* **do**
5 map each ts_j into bucket i according to $arg \min_{g_i} ||g_i(ts_j) - ts_j||$

the L time series are chosen at random from the time series set $\mathcal{T}$ containing all tensor fibers $ts_j, j = 1, ..., M$; (2) construct L buckets from the time series set $\mathcal{T}$, each bucket i corresponds to an ARIMA model g_i; (3) map each time series $ts_j, j = 1, ..., M$ from $\mathcal{T}$ into bucket i according to the RMSE between the true and the predicted values using the ARIMA model g_i. In this way, we treat the fitting of each time series to the randomly selected time series as a special "random projection", and then map each time series according to the least RMSE to each bucket i. This process (detailed in Algorithm 2) allows locations that may not have been spatially close to be clustered to the same model and can decrease the training time, which is the bottleneck of the adaptive sampling process.

We compare our proposed random projection algorithm for the ARIMA coefficients with the k-means algorithm where similar time series can also be clustered and thereby reducing training time. Figure 2 (c), (d) demonstrate that despite the high computation cost of the k-means algorithm, it provided limited improvement over the random projection algorithm.

4 SkeSmooth: Smooth Tensor Factorization

We then consider analyzing the sketched tensor using the CP decomposition to capture the underlying multi-linear latent structure of the data and reconstruct the unsampled time slices. Built on the success of the Quadratic Variation (QV) matrix constraint [30,33], we consider the delay effect of the time series and utilize the coefficients from the pre-trained ARIMA model from SkeTen to make the temporal pattern smooth and robust to missing entire time slices.

4.1 Formulation

CP Decomposition with Missing Data. We consider the sketched tensor as an incomplete data problem where the goal is to both learn the underlying patterns of the data and reconstruct the unsampled values. Previous CP-based tensor completion algorithms [1,23] have formulated the completed tensor $\overline{\mathcal{X}}$ as:

$$\overline{\mathcal{X}} = \mathcal{W} * \mathcal{X} + (1 - \mathcal{W}) * [\![\mathbf{A}, \mathbf{B}, \mathbf{C}]\!], \tag{6}$$

where $*$ denotes the element-wise product, the observed tensor is factorized as $\mathcal{X} = [\![\mathbf{A}, \mathbf{B}, \mathbf{C}]\!]$ and the binary weight tensor $\mathcal{W}$ is of the same size as $\mathcal{X}$ such that

$$w_{i,j,k} = \begin{cases} 1, & \text{if } w_{i,j,k} \text{ is known} \\ 0, & \text{if } w_{i,j,k} \text{ is missing.} \end{cases}$$

ARIMA-Based Regularization Matrix. Standard CP decomposition with missing data assumes the entries are missing at random, whereas SkeTen yields a sketched tensor with entire time slices missing. Moreover, they do not account for temporal information which can improve interpretability and robustness. In spatio-temporal datasets, successive observations at the same spatial location (i.e., $x_{j,k}^{(t)}$ and $x_{j,k}^{(t+1)}$) are unlikely to change significantly. Thus, adjacent time slots are generally smooth. SkeSmooth incorporates the coefficients generated by the various ARIMA models as auxiliary information into the regularization matrix $\mathcal{L}$. Based on Eq. (3), we observe that $\mathbf{a}_i$ can be approximated as:

$$\mathbf{a}_i = \alpha_1 \mathbf{a}_{i-1} + \alpha_2 \mathbf{a}_{i-2} + \cdots + \alpha_p \mathbf{a}_{i-p},\ p < i \tag{7}$$

Thus by utilizing the learned predictive models in SkeTen (the autoregressive coefficients and the order of the number of time lags, p), the algorithm can better reconstruct the missing time slices and learn the underlying patterns.

Although each ARIMA model may have different order of time lags, $p_{j,k}$, we use the largest order $p = \max p_{j,k}$ and set the coefficients $\alpha_{p_{j,k}+1}, \cdots, \alpha_p$, to zero if $p_{j,k} < p$. Since each tensor fiber is clustered to the ARIMA model that fits best, we can thus get the weight for each ARIMA model as the number of tensor fibers clustered to it. We then compute the weighted average of the autoregressive coefficients to produce the regularization matrix, $\mathcal{L}$, defined as:

$$\mathcal{L} = \begin{bmatrix} -\alpha_p & 0 & 0 & 0 & \cdots & 0 \\ \vdots & \ddots & \vdots & \vdots & \ddots & \vdots \\ -\alpha_1 & \cdots & -\alpha_p & 0 & \cdots & 0 \\ 1 & -\alpha_1 & \cdots & -\alpha_p & \cdots & 0 \\ \vdots & \ddots & \ddots & \vdots & \ddots & \vdots \\ 0 & \cdots & 1 & -\alpha_1 & \cdots & -\alpha_p \end{bmatrix} \tag{8}$$

SkeSmooth Optimization Problem. The objective function for SkeSmooth after adding the smoothness constraint is:

$$\min_{\mathcal{X}} \frac{1}{2} \|\mathcal{Y} - \mathcal{W} * [\![\mathbf{A}, \mathbf{B}, \mathbf{C}]\!]\|_F^2 + \frac{\rho}{2} \|\mathcal{L}\mathbf{A}\|_F^2, \quad \text{s.t. } \mathcal{Y} = \mathcal{W} * \mathcal{X}, \tag{9}$$

where $\mathcal{Y}$ denotes the sketched tensor with missing entries. Thus higher values of the regularization parameter, ρ, will yield smoother temporal factors. However, this may potentially reduce the overall fit of the tensor.

Algorithm 3: SkeSmooth

Input: Sparse tensor sketch $\mathcal{Y}$, weight tensor $\mathcal{W} \in \mathbb{R}^{I\times J\times K}$, smooth parameter ρ.

1 Randomly initialize the factor matrices $\mathbf{A} \in \mathbb{R}^{I\times R}, \mathbf{B} \in \mathbb{R}^{J\times R}, \mathbf{C} \in \mathbb{R}^{K\times R}$

2 **while** *A, B, C not converge* **do**

3 compute tensor $\mathcal{Z} = \mathcal{W} * [\![\mathbf{A}, \mathbf{B}, \mathbf{C}]\!]$;

4 compute function value $f = \frac{1}{2}\|\mathcal{Y} - \mathcal{Z}\|_F^2$;

5 update gradient G^n according to Eq. (10);

6 update $\mathbf{A}$, $\mathbf{B}$ and $\mathbf{C}$ with l-bfgs.

4.2 Algorithm

We use a first-order method to solve the optimization problem, similar to [1]. The gradient-based method has been shown to be robust to overspecification of the rank. As the regularization matrix is only enforced on the temporal mode (first mode), only the gradient for the temporal mode involves the gradient of the regularization matrix. Thus the updates for all three modes are:

$$\begin{aligned} \mathbf{G}^{(1)} &= -(\mathcal{Y}_{(1)} - \mathcal{Z}_{(1)})(\mathbf{C} \odot \mathbf{B}) + \rho\mathcal{L}^T\mathcal{L}\mathbf{A};\\ \mathbf{G}^{(2)} &= -(\mathcal{Y}_{(2)} - \mathcal{Z}_{(2)})(\mathbf{C} \odot \mathbf{A});\\ \mathbf{G}^{(3)} &= -(\mathcal{Y}_{(3)} - \mathcal{Z}_{(3)})(\mathbf{B} \odot \mathbf{A});\\ \text{s.t. } \mathcal{Z} &= \mathcal{W} * [\![\mathbf{A}, \mathbf{B}, \mathbf{C}]\!], \end{aligned} \tag{10}$$

where $\mathcal{Y}_{(n)}$ is the matricization on the n-th mode of the tensor sketch. Our gradient-based algorithm uses the limited-memory BFGS for the first order optimization. The optimization process is shown in Algorithm 3.

4.3 Complexity Analysis

The time complexity of SkeSmooth is dominated by the matricized tensor times Khatri-Rao product (MTTKRP) operations, which is computed in Eq. (10) as the matricized residual tensor $\mathcal{Y}_{(n)} - \mathcal{Z}_{(n)}$ times the Khatri-Rao product between two factor matrices. For simplicity purpose, we denote an N-th order tensor $\mathcal{X} \in \mathbb{R}^{I\times\cdots\times I}$ as the representation of the residual tensor $\mathcal{Y}_{(n)} - \mathcal{Z}_{(n)}$. For each mode, computing MTTKRP takes $O(nnz(\mathcal{X})R)$, where R is the approximate rank of the tensor, $nnz(\mathcal{X})$ represents the number of non-zero elements of $\mathcal{X}$. While computing the term $\rho\mathcal{L}^T\mathcal{L}\mathbf{A}$ for the first mode takes $O(I^2R^2)$. Thus the overall time complexity for SkeSmooth is $O(nnz(\mathcal{X})NR + I^2R^2)$.

5 Experiments

We evaluate SkeTenSmooth on a large real-world dataset, the New York City (NYC) Yellow Taxi data (see [18] for experiment settings). The goal of our evaluation is to assess both SkeTen and SkeSmooth from three aspects:

1. **Effectiveness of SkeTen:** Analyze memory compression without hurting the performance. Use SkeTen to evaluate the latent structure preservation.
2. **Effectiveness of SkeSmooth:** Evaluate with the sketched tensor on the effectiveness of the smoothness constraint.
3. **Sketching result on prediction tasks:** Analyze how the decomposed factor matrices perform on a downstream prediction task, to indirectly evaluate whether the sketching results are meaningful.

5.1 Dataset

We evaluate SkeTenSmooth on the NYC Yellow Taxi Data[1], which is collected and published by the New York City Taxi and Limousine Commission. To demonstrate the scalability of our methods, we collect data from January, 2009 to June, 2015, which has 1,090,939,222 trips in total, on a daily basis[2]. We investigate the NYC area using the latitude range of $[40.66, 40.86]$ and longitude range of $[-74.03, -73.91]$. We partition the NYC area with a 0.001×0.001 degree grid, where 0.001 degree is roughly 111.32 meters. Then we form a tensor $\mathcal{X}$ of size $2341 \times 120 \times 200$, for time, latitude and longitude modes respectively. The first 60% of the tensor is used for training and is thus considered as offline data, while the latter 40% is used to test SkeTen and SkeSmooth.

5.2 Evaluation Metrics

The tensor decomposition performance is evaluated by two criteria: Factor Match Scores (FMS) and Tensor Completion Scores (TCS) [1]. FMS is defined as:

$$score(\bar{\mathcal{X}}) = \frac{1}{R}\sum_r \left(1 - \frac{|\xi_r - \bar{\xi_r}|}{max\left\{\xi_r, \bar{\xi_r}\right\}}\right) \prod_{\mathbf{x=a,b,c}} \frac{\mathbf{x}_r^T \bar{\mathbf{x}}_r}{\|\mathbf{x}_r\|\|\bar{\mathbf{x}}_r\|},$$

$$\xi_r = \prod_{\mathbf{x=a,b,c}} \|\mathbf{x}_r\|, \bar{\xi_r} = \prod_{\mathbf{x=a,b,c}} \|\bar{\mathbf{x}}_r\|$$

where $\bar{\mathcal{X}} = [\![\bar{\mathbf{A}}, \bar{\mathbf{B}}, \bar{\mathbf{C}}]\!]$ is the estimated factor and $\mathcal{X} = [\![\mathbf{A}, \mathbf{B}, \mathbf{C}]\!]$ is the true factor. $\mathbf{x}_r$ is the rth column of factor matrices. FMS measures the similarity between two tensor decomposition solutions. It ranges from 0 to 1, and the best possible FMS is 1. We choose the factorization result of the complete tensor using CP-OPT [1] as a true factor to compare with.

[1] https://www1.nyc.gov/site/tlc/about/tlc-trip-record-data.page.

[2] We also conducted experiments on hourly basis (a finer granularity), which shows consistent results with the daily basis. Results can be found in Supplement in [18].

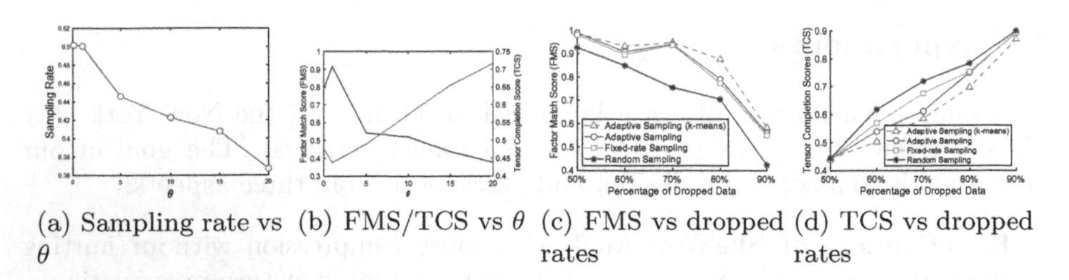

(a) Sampling rate vs θ (b) FMS/TCS vs θ (c) FMS vs dropped rates (d) TCS vs dropped rates

Fig. 2. Overall sampling rates, FMS and TCS over the change of θ ((a) and (b)). FMS and TCS over dropped rates ((c) and (d)).

TCS indicates the relative error of the missing entries, and is defined as

$$TCS = \frac{\left\|(1 - \mathcal{W}) * (\mathcal{X} - \bar{\mathcal{X}})\right\|}{\|(1 - \mathcal{W}) * \mathcal{X}\|},$$

The best possible TCS is 0.

5.3 Parameter Selection

In SkeTen, parameters involved in achieving the adaptive sampling algorithm include: the PID controller parameters $\{\gamma_p, \gamma_i, \gamma_d\}$, θ and ξ for the interval adjustment. The PID parameters are set as $\{\gamma_p, \gamma_i, \gamma_d\} = \{0.7, 0.2, 0.1\}$[3]. Parameter θ determines the magnitude of the sampling interval adaptation, its impact is shown in Fig. 2(a) and (b). From Fig. 2(b), we observe that as θ increases, the reconstruction performance measured by FMS and the TCS decreases due to the enlarged average sampling intervals (shown in Fig. 2(a)). The optimal choice shown in the figure is $\theta = 1$. When $\theta = 0.1$, the reconstruction performance (TCS and FMS) is worse than when $\theta = 1$ due to an insufficient interval adjustment. ξ is determined as 3.5 for the tolerance of PID error.

For SkeSmooth, the smoothness constraint ρ is chosen as 600 after grid search through {500, 600, 700, 800, 900, 1000}, and the rank of the tensor is determined as 5 after grid search in {5, 10, 15, 20, 25} (see Supplement in [18] for more details). The number of lags is determined as $p = 3$ based on the majority of all the trained ARIMA models. We use the number of tensor fibers mapped into the same bucket as weights, and compute the final coefficients for the regularization by weighted average of the coefficients of the randomly trained ARIMA models and is determined as $\boldsymbol{\alpha} = [0.55, -0.19, 0.04]$[4].

---

[3] We experimented with several PID settings, and it hardly affected the sampling rate as long as we set the controller ID according to *proportional* > *integral* > *derivative*. We thus consider our system robust to the control parameters and choose the optimal setting in empirical studies.

[4] The k-means based coefficients are computed the same way using the Dynamic Time Warping (DTW) distance and are determined as $\boldsymbol{\alpha}_{k-means} = [0.51, -0.07, 0.007]$.

5.4 Tensor Sketching

Here we consider a streaming setup where the tensor sketching will be applied to a succession of incoming time slices.

Baselines. We compare SkeTen with two baseline sampling techniques:

- **Fixed rate sampling** samples the data with a predefined, fixed interval.
- **Random sampling** has been proposed as a way to sparsify the tensor [19].

By evaluating the performance of fixed rate sampling and random sampling, we can gain a better understanding of the benefits of adaptive sampling under the setting of time-evolving streams and how it may provide more informative representations of the streams at the same storage cost.

Evaluation. We evaluate the three sampling methods for different levels of dropped data by using the proposed SkeSmooth algorithm. The performance is measured by both FMS and TCS. We explore the sketching level from 50% to 90%. Figure 2 (c) and (d) show the FMS and TCS of different sampling methods when applied to SkeSmooth algorithm. As more data are sampled, FMS gradually approaches 1, which means the extracted temporal patterns are equivalent to the true patterns. Notice that when the level of sampling is 50%, adaptive sampling and fixed-rate sampling method show FMS as high as 0.9886 and 0.9803, which means that with only 50% of the original data, SkeSmooth can successfully recover the original temporal pattern. Adaptive sampling achieves slightly better reconstruction performance than fixed-rate sampling, because the daily taxi data are fairly regular and periodical so the interval adjustments for adaptive sampling helps marginally. Both adaptive sampling and fixed-rate sampling are better than the random sampling, illustrating that carefully designing the sampling strategy is critical for pattern and data reconstruction.

5.5 Tensor Decomposition with Sketches

To evaluate the performance of tensor sketching, we first need to look into how it can preserve the temporal trend and the latent structures on the temporal dimension. We decompose the sketched results to generate the factor matrices as latent patterns, and use these factor matrices to reconstruct the original tensor. We compare SkeSmooth with the following state-of-the-art methods: CP-WOPT [1], HaLRTC [16], SPC [30], BCPF [32], t-SVD [31], t-TNN [12].

Evaluation. Since not all baseline algorithms are CP-based tensor completions, it is infeasible to compare the FMS. Therefore, we use TCS as the evaluation metric. Figure 3 illustrates the TCS as a function of the level of data dropped for different sketching methods. We note that SkeSmooth outperforms the baseline tensor completion algorithms in achieving a lower TCS (relative error), verifying

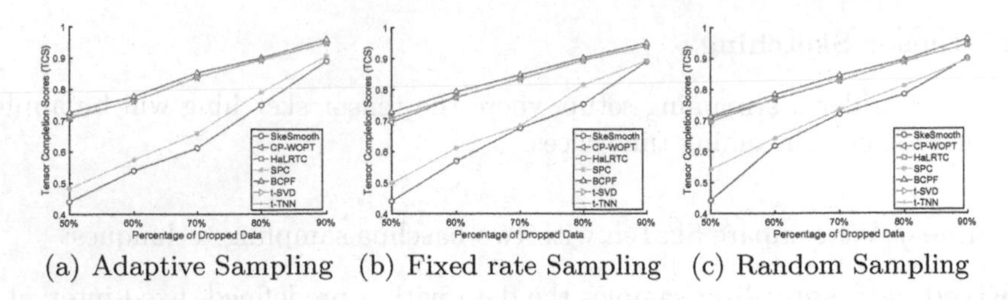

(a) Adaptive Sampling (b) Fixed rate Sampling (c) Random Sampling

Fig. 3. Tensor Completion Scores Comparison between SkeSmooth and the Tensor Completion baselines for different sampling mechanisms: (a) adaptive sampling; (b) fixed rate sampling; (c) random sampling.

its robustness in dealing with the missed time slices in the sketched tensor. In particular, we observe the noticeable improvement using the ARIMA-based coefficients in comparison with the SPC algorithm which adopts the QV smoothness constraint. This observation is consistent for all sampling mechanisms.

5.6 Prediction Task

It is also critical to analyze how meaningful the generated temporal patterns are by performing a downstream prediction task using the decomposed factor matrices. We fit a multivariate Long short-term memory (LSTM) model, which is well-suited for the traffic demand prediction task [26]. For the prediction task, the taxi demand is predicted on a daily basis where each day's number of pickups are treated as the daily demand with a train-test partition of 70-30. The LSTM model uses the factorized temporal mode factor matrix $\mathbf{A}$ of size $I \times R$, where each feature within $\mathbf{A}$ ($\mathbf{A}_{i:}$ of size $1 \times R$) represents a latent transportation pattern. The LSTM model is configured as one hidden layer of 30 LSTM units with ReLu activation function, followed by an output dense layer[5].

Table 1 presents the root-mean-square error (RMSE) differences for adaptive sampling (SkeTen), fixed rate sampling, and random sampling with varying percentage of data dropped from 50% to 90%. Results demonstrate that adaptive sampling outperforms other two sampling methods in maintaining the predictive power with 60%-80% data dropped. As more data are dropped, the test error (RMSE) increases, indicating that it becomes harder to fit the LSTM model.

6 Related Works

We review the previous work regarding the tensor sketching and tensor completion approaches in this section.

[5] The model is trained by the RMSprop optimizer with a fixed learning rate (0.01) and a batch size of 8 for 50 epochs.

Table 1. Predictive performance (test RMSEs) for adaptive sampling, fixed rate sampling, and random sampling with 50% to 90% of data dropped.

Sampling techniques	50%	60%	70%	80%	90%
Adaptive sampling	0.0155	**0.0270**	**0.0303**	**0.02827**	0.0450
Fixed-rate sampling	0.0173	0.0278	0.0309	0.0427	0.0452
Random sampling	**0.0154**	0.0282	0.0317	0.0342	**0.0322**

6.1 Tensor Sketching

Sketching has been proposed and investigated as an indispensable numerical linear algebra tool to help process large volume of data [27]. Thus far, there have been two research directions. Random projection is the predominant form of tensor sketching. [20] developed an approach to randomly compress a big tensor into a much smaller tensor by multiplying the tensor with a random matrix on each mode. [21] extended this work to improve the permutation matching performances of the resulting factor matrices. [10] expanded the work further to the online setting. However, these compression methods cannot avoid the inherent bottleneck of the matrix-tensor multiplication operators.

An alternative for tensor sketching is tensor sparsification. The goal of tensor sparsification is to generate a sparse "sketch" of the large tensor. [19] proposed random sampling based on the tensor spectral norm. Entries are sampled with probability proportional to the magnitude of the entries. [28] extended this work with: 1) the extension from cubic tensors to general tensors; 2) the criteria and treatment of "small" entry values. They proposed to keep all large entries, sample proportionally to moderate entries, and uniformly sample small entries. [4] proposed a new sampling method that samples entries based on a pre-computed tensor distribution. These tensor sparsification algorithms do not consider the data streaming setting and require prior knowledge about tensors.

6.2 Tensor Factorization and Completion

Tensor completion is more considered as a byproduct when dealing with missing data in the tensor factorization process. As a result, CP decomposition and Tucker decomposition are two major types of tensor completion methods [23]. [1] proposed CP-WOPT, a tensor CP decomposition algorithm that can handle missing entries using a binary weight tensor. Tucker-based tensor completion including geomCG [14], HaLRTC/FaLRTC [16], and TNCP [17], etc. There are other approaches besides CP and Tucker decomposition, such as t-SVD [31]. [23] and [22] provide comprehensive surveys on the topic of tensor completion.

Tensor completion has also been widely applied to spatio-temporal data analysis for traffic prediction [11,25], urban mobility pattern mining [24], and urban event detection [5]. [33] applied a similar tensor completion algorithm with smoothness constraints as in [30] and [33] to the imputation of missing internet traffic data, that tries to minimize the difference between adjacent time slots.

[9] explored the similarity for each element within one dimension and used this as auxiliary information to better recover the original tensor. Nevertheless, the above algorithms did not consider the tensor streaming problem. [25] proposed a dynamic tensor completion (DTC) algorithm, where traffic data are represented as a dynamic tensor pattern, but will still require the entire tensor.

7 Conclusions

SkeTenSmooth is a tensor factorization framework that generates tensor sketches in a streaming fashion using adaptive sampling mechanism and decomposes the sampled smaller tensor sketch with an ARIMA-based smoothness constraint. It is well suited for the incrementally increased spatio-temporal data analysis that can greatly reduce the storage cost and sample complexity, while preserving the latent global structure. SkeTenSmooth includes two main components: 1) SkeTen as a tensor sketching algorithm that tackles the memory efficiency issue by adaptively adjusting the sampling interval based on the model prediction error; and 2) SkeSmooth incorporates an ARIMA-based smoothness constraint, which better demonstrates the auto correlation between each time point of a time series. Experiments on the large-scale NYC taxi dataset illustrate that with adaptive sampling strategy, SkeTen can greatly reduce the memory cost by 80% and still preserve the latent temporal patterns and the predictive power. Future work will focus on the distributed setting to improve scalability and efficiency.

Acknowledgment. This work was supported by the National Science Foundation, award IIS-#1838200 and CNS-#2027783.

References

1. Acar, E., Dunlavy, D.M., Kolda, T.G., Mørup, M.: Scalable tensor factorizations for incomplete data. Chemometr. Intell. Lab. Syst. **106**(1), 41–56 (2011)
2. Angulo, J., et al.: Spatiotemporal infectious disease modeling: a BME-SIR approach. PLoS ONE **8**(9), e72168 (2013)
3. Åström, K.J., Wittenmark, B.: Computer-Controlled Systems: Theory and Design. Courier Corporation, North Chelmsford (2013)
4. Bhojanapalli, S., Sanghavi, S.: A new sampling technique for tensors. arXiv preprint arXiv:1502.05023 (2015)
5. Chen, L., Jakubowicz, J., Yang, D., Zhang, D., Pan, G.: Fine-grained urban event detection and characterization based on tensor cofactorization. IEEE Trans. Hum.-Mach. Syst. **47**(3), 380–391 (2016)
6. Fan, L., Xiong, L.: An adaptive approach to real-time aggregate monitoring with differential privacy. IEEE TKDE **26**(9), 2094–2106 (2013)
7. Gama, F., Marques, A.G., Mateos, G., Ribeiro, A.: Rethinking sketching as sampling: linear transforms of graph signals. In: 2016 50th Asilomar Conference on Signals, Systems and Computers, pp. 522–526. IEEE (2016)
8. Gauvin, L., Panisson, A., Cattuto, C.: Detecting the community structure and activity patterns of temporal networks: a non-negative tensor factorization approach. PLoS ONE **9**(1), e86028 (2014)

9. Ge, H., Caverlee, J., Zhang, N., Squicciarini, A.: Uncovering the spatio-temporal dynamics of memes in the presence of incomplete information. In: Proceedings of the 25th CIKM, pp. 1493–1502. ACM (2016)
10. Gujral, E., Pasricha, R., Yang, T., Papalexakis, E.E.: Octen: online compression-based tensor decomposition. arXiv preprint arXiv:1807.01350 (2018)
11. He, L., Qin, Z.T., Bewli, J.: Low-rank tensor recovery for geo-demand estimation in online retailing. Procedia Comput. Sci. **53**, 239–247 (2015)
12. Hu, W., Tao, D., Zhang, W., Xie, Y., Yang, Y.: A new low-rank tensor model for video completion. arXiv preprint arXiv:1509.02027 (2015)
13. King, M., et al.: Process Control: A Practical Approach. Wiley Online Library (2011)
14. Kressner, D., Steinlechner, M., Vandereycken, B.: Low-rank tensor completion by riemannian optimization. BIT Numer. Math. **54**(2), 447–468 (2014)
15. Lakhina, A., Papagiannaki, K., Crovella, M., Diot, C., Kolaczyk, E.D., Taft, N.: Structural analysis of network traffic flows. In: ACM SIGMETRICS Performance Evaluation Review, vol. 32, pp. 61–72. ACM (2004)
16. Liu, J., Musialski, P., Wonka, P., Ye, J.: Tensor completion for estimating missing values in visual data. IEEE Trans. Pattern Anal. Mach. Intell. **35**(1), 208–220 (2012)
17. Liu, Y., Shang, F., Jiao, L., Cheng, J., Cheng, H.: Trace norm regularized CANDECOMP/PARAFAC decomposition with missing data. IEEE Trans. Cybern. **45**(11), 2437–2448 (2014)
18. Ma, J., Zhang, Q., Ho, J.C., Xiong, L.: Spatio-temporal tensor sketching via adaptive sampling. arXiv preprint arXiv:2006.11943 (2020)
19. Nguyen, N.H., Drineas, P., Tran, T.D.: Tensor sparsification via a bound on the spectral norm of random tensors. arXiv preprint arXiv:1005.4732 (2010)
20. Sidiropoulos, N.D., Kyrillidis, A.: Multi-way compressed sensing for sparse low-rank tensors. IEEE Signal Process. Lett. **19**(11), 757–760 (2012)
21. Sidiropoulos, N.D., Papalexakis, E.E., Faloutsos, C.: Parallel randomly compressed cubes: a scalable distributed architecture for big tensor decomposition. IEEE Signal Process. Mag. **31**(5), 57–70 (2014)
22. Sobral, A., Zahzah, E.: Matrix and tensor completion algorithms for background model initialization: a comparative evaluation. Pattern Recogn. Lett. (2016). https://doi.org/10.1016/j.patrec.2016.12.019
23. Song, Q., Ge, H., Caverlee, J., Hu, X.: Tensor completion algorithms in big data analytics. ACM TKDD **13**(1), 6 (2019)
24. Sun, L., Axhausen, K.W.: Understanding urban mobility patterns with a probabilistic tensor factorization framework. Transp. Res. Part B Methodol. **91**, 511–524 (2016)
25. Tan, H., Wu, Y., Shen, B., Jin, P.J., Ran, B.: Short-term traffic prediction based on dynamic tensor completion. IEEE Trans. Intell. Transp. Syst. **17**(8), 2123–2133 (2016)
26. Tsai, T.H., Lee, C.K., Wei, C.H.: Neural network based temporal feature models for short-term railway passenger demand forecasting. Expert Syst. Appl. **36**(2), 3728–3736 (2009)
27. Woodruff, D.P., et al.: Sketching as a tool for numerical linear algebra. Found. Trends® Theor. Comput. Sci. **10**(1–2), 1–157 (2014)
28. Xia, D., Yuan, M.: Effective tensor sketching via sparsification. arXiv preprint arXiv:1710.11298 (2017)

29. Xingjian, S., Chen, Z., Wang, H., Yeung, D.Y., Wong, W.K., Woo, W.C.: Convolutional LSTM network: a machine learning approach for precipitation nowcasting. In: NIPS, pp. 802–810 (2015)
30. Yokota, T., Zhao, Q., Cichocki, A.: Smooth PARAFAC decomposition for tensor completion. IEEE Trans. Signal Process. **64**(20), 5423–5436 (2016)
31. Zhang, Z., Ely, G., Aeron, S., Hao, N., Kilmer, M.: Novel methods for multilinear data completion and de-noising based on tensor-SVD. In: CVPR, pp. 3842–3849 (2014)
32. Zhao, Q., Zhang, L., Cichocki, A.: Bayesian CP factorization of incomplete tensors with automatic rank determination. IEEE Trans. Pattern Anal. Mach. Intell. **37**(9), 1751–1763 (2015)
33. Zhou, H., Zhang, D., Xie, K., Chen, Y.: Spatio-temporal tensor completion for imputing missing internet traffic data. In: 2015 34th IPCCC, pp. 1–7. IEEE (2015)

Graphical Models and Causality

Orthogonal Mixture of Hidden Markov Models

Negar Safinianaini[1](✉), Camila P. E. de Souza[2], Henrik Boström[1], and Jens Lagergren[1,3]

[1] School of Electrical Engineering and Computer Science, KTH Royal Institute of Technology, Stockholm, Sweden
{negars,bostromh}@kth.se

[2] Department of Statistical and Actuarial Sciences, The University of Western Ontario, London, Canada
camila.souza@uwo.ca

[3] Science for Life Laboratory, Solna, Sweden
jens.lagergren@scilifelab.se

Abstract. Mixtures of Hidden Markov Models (MHMM) are widely used for clustering of sequential data, by letting each cluster correspond to a Hidden Markov Model (HMM). Expectation Maximization (EM) is the standard approach for learning the parameters of an MHMM. However, due to the non-convexity of the objective function, EM can converge to poor local optima. To tackle this problem, we propose a novel method, the Orthogonal Mixture of Hidden Markov Models (oMHMM), which aims to direct the search away from candidate solutions that include very similar HMMs, since those do not fully exploit the power of the mixture model. The directed search is achieved by including a penalty in the objective function that favors higher orthogonality between the transition matrices of the HMMs. Experimental results on both simulated and real-world datasets show that the oMHMM consistently finds equally good or better local optima than the standard EM for an MHMM; for some datasets, the clustering performance is significantly improved by our novel oMHMM (up to 55 percentage points w.r.t. the v-measure). Moreover, the oMHMM may also decrease the computational cost substantially, reducing the number of iterations down to a fifth of those required by MHMM using standard EM.

Keywords: Hidden markov models · Mixture models · Mixture of hidden markov models · Expectation maximization · Orthogonality · Regularization · Penalty

1 Introduction

Clustering of sequential data is an important machine learning task with a wide range of applications from biology to finance. Various methods have been applied for this task, including feature-based methods, deep-learning approaches and model-based methods [1,6,19]. Model-based methods have the advantages of

F. Hutter et al. (Eds.): ECML PKDD 2020, LNAI 12457, pp. 509–525, 2021.
https://doi.org/10.1007/978-3-030-67658-2_29

being probabilistic, interpretable, providing measures of uncertainty, as well as allowing for rapid prototyping [7,33]. Motivated by these advantages, we focus on model-based approaches; more specifically on the Hidden Markov Model (HMM), or a Mixture of Hidden Markov Models (MHMM), introduced in [29]. They have been applied to tasks such as time series clustering, music analysis, motion recognition, handwritten digit recognition, as well as to problems within finance and biology [9,12,13,16,23,30,34]. In recent years, various methods have been proposed to enhance an MHMM with respect to, e.g., time complexity and sparsity of the model [24,30], when using Expectation Maximization (EM) [10], which is the standard inference approach when clustering based on an MHMM [6]. However, since EM maximizes a non-convex objective function, it can lead to poor local optima; a proper initialization plays a crucial role here [6,8]. Moreover, singularity in mixture model estimation is also a potential problem; i.e., a cluster can be collapsed into a single data point [6]. To tackle the first problem, it is common to consider several random initializations [6,8]. For the second problem, the inclusion of a penalty or prior on the optimization variable has been suggested as a solution [6]. In general, to enhance the predictive power of a model, an HMM in this context, penalization can be applied on EM [20]; controlling the behavior of the model parameters. Some recently proposed penalties over HMM parameters (either transition or emission distribution parameters) include the pairwise mean difference penalty [21], the smoothing penalty [32], the sparse penalty [31], the self-transition penalty [22] and the determinantal point process [17] penalty [20].

Each of the penalties introduced in previous work is however designed for a single HMM and is consequently not particularly suited for training a mixture of HMMs. In particular, these penalties do not introduce any dependencies between the HMMs. In contrast, we introduce a novel penalty, the *orthogonality penalty*, by using the inner product as a distance measure between pairs of HMMs, or more exactly, their transition matrices. Distance measures between HMMs have been used also previously for the purpose of clustering sequences, however, without using MHMMs [15]. Specifically, the method in [15] do not use HMMs to represent clusters of sequences; instead, each sequence is represented by a unique HMM and the actual clustering is performed on the HMMs, that is, subsequent to the construction of the HMMs. When it comes to incorporating the concept of orthogonality, a similar idea was exploited in [3], where orthogonality among the weights of a neural network was used as a constraint to improve maximum likelihood estimation w.r.t. initialization sensitivity.

In this work, we modify the standard EM to infer an MHMM, where the novel orthogonality penalty is used to push the EM algorithm to avoid a poor local optimum, by increasing the distance between the transition matrices of each pair of HMMs. We term the modified EM procedure as the Orthogonal Mixture of Hidden Markov Models (oMHMM). A key feature of the oMHMM method, which is missing in an MHMM due to the independence of the transition matrices, is that the estimation of each transition matrix may be affected by all other transition matrices (representing different clusters); hence realizing a *global context*. As we aim to improve upon the standard EM when inferring an MHMM, we compare the proposed oMHMM to the standard EM implementation of MHMM, as, to the best of our knowledge, no other approaches to infer MHMMs have been put forward in the literature[1].

In the next section, we provide notation and background on HMMs, MHMMs, and a linear algebraic definition of orthogonality. In Sect. 3, we introduce the oMHMM; our novel approach. In Sect. 4, we evaluate and compare it to standard EM for MHMM. Finally, in Sect. 5, we summarize the main findings and point out directions for future research.

2 Preliminaries

In this section, we first provide definitions concerning clustering of sequences using mixtures of HMMs and, then, a short description of orthogonality.

2.1 Hidden Markov Models

A Hidden Markov Model (HMM) is a probabilistic graphical model [5] in which a sequence of emitted symbols (observation sequence) is observed, but the sequence of states (state sequence) emitting the observation sequence, is hidden and follows a Markov structure. The Markov property implies that the next state in the state sequence only depends on the current state. We assume that the input sequences have length M. The observation sequence is denoted as $Y = \{y_1, \ldots, y_M\}$ and the state sequence is denoted as $C = \{c_1, \ldots, c_M\}$. Each state in C takes a value j for $j = 1, \ldots, J$; we denote each state at step m with value j as $c_{m,j}$ for $m = 1, \ldots, M$. An HMM is parameterized by initial probabilities, $p(c_{1,j})$, transition probabilities, $p(c_{m,j}|c_{m-1,i})$, and emission probabilities, $p(y_m|c_{m,j})$. These sets of parameters are referred to as ρ, A, and O respectively, jointly referred to as θ. Inference of HMM parameters can be conducted by maximizing the likelihood using the Expectation Maximization (EM) algorithm [6], which consists of two steps: the *E-step*, calculating $Q(\theta, \theta^{old}) = E_{C|Y,\theta^{old}}\big[\log P(Y, C|\theta)\big]$, the expected value of $\log P(Y, C|\theta)$, the *complete-data log likelihood* [6], w.r.t. the conditional distribution of the hidden states given the observed data and old parameter estimates (θ^{old}); the *M-step*, maximizing the Q function calculated in the E-step w.r.t. the parameters of interest. The E-step involves calculating the marginal posterior distribution of a latent variable $c_{m,j}$, denoted as $\gamma(c_{m,j})$, and the joint posterior distribution of two successive latent variables, $\varepsilon(c_{m-1,i}, c_{m,j})$. For details of the calculations of these terms, we refer to [6]. In the M-step, ρ, O, and A are updated using $\gamma(c_{m,j})$ and $\varepsilon(c_{m-1,i}, c_{m,j})$. The M-step concerns a maximization problem per parameter set; here we focus on the maximization problem regarding the transition probabilities, A, as the contribution of this work involves this problem.

[1] Spectral learning of MHMM [30] improves the time complexity and does not improve the clustering. Sparse MHMM [24], requiring data coming from a set of entities connected in a graph with a known topology, can be used together with oMHMM.

Therefore, the maximization problem shown in Eq. 1 results from maximizing the Q function w.r.t. A_{ij} which is the transition probability of moving from state i to state j; the optimization is subject to the constraint $\sum_{j=1}^{J} A_{ij} = 1$.

$$\max_{A_{ij}} \sum_{m=2}^{M} \sum_{i=1}^{J} \sum_{j=1}^{J} \varepsilon(c_{m-1,i}, c_{m,j}) \log A_{ij} \tag{1}$$

2.2 Mixtures of Hidden Markov Models

A Mixture of Hidden Markov Models (MHMM) is a probabilistic graphical model where each observation sequence is generated by a *mixture model* [6] with K components, each representing a cluster which corresponds to a unique HMM parameter setting. We denote each observation sequence as $Y_n = \{y_{n1}, \cdots, y_{nM}\}$ and Z_n is the latent variable concerning the cluster assignment of Y_n, for $n = 1, \ldots, N$. An observation sequence Y_n, belonging to the k-th (k takes a value between 1 and K) cluster, arises from an HMM with state sequence $C_n = \{c_1^n, \ldots, c_M^n\}$ parameterized by ρ_k, O_k, A_k. The parameter defining the probability of the observation sequences belonging to component k, the mixture probability, is π_k with $\sum_{k=1}^{K} \pi_k = 1$. Performing EM on the MHMM, all parameters, $\rho_{1:K}$, $O_{1:K}$, $A_{1:K}$, and $\pi_{1:K}$, are updated at each iteration. Note that using these parameters, the posterior probability of each observation sequence belonging to component k can be calculated; i.e. $p(Z_n = k|Y_n, \theta^{old})$. For details on the update equations, we refer to [24]. Similar to the previous section, the E-step concerns calculating $Q(\theta, \theta^{old}) := E_{C,Z|Y,\theta^{old}}\big[\log p(Y, C, Z|\theta)\big]$ and M-step maximizing the Q function. Here we focus again on the M-step, where the maximization problem concerns the transition matrices, i.e., maximizing the Q function w.r.t. A_{kij}; the probability of transition from state i to j concerning component k. The maximization is formulated in Eq. 2 subject to $\sum_{j=1}^{J} A_{kij} = 1$.

$$\max_{A_{kij}} \sum_{k=1}^{K} \sum_{n=1}^{N} \sum_{m=2}^{M} \sum_{i=1}^{J} \sum_{j=1}^{J} \varepsilon_k(c_{m-1,i}^n, c_{m,j}^n) \log A_{kij} \tag{2}$$

Algorithm 1. MHMM

1: **procedure** LEARN($Y_{1:N}$):
2: Initialise $\theta := \{\rho, O, A, \pi\}$
3: **repeat**
4: E-step: calculate $Q(\theta, \theta^{old}) := E_{C,Z|Y_{1:N},\theta^{old}}\big[\log p(Y_{1:N}, C, Z|\theta)\big]$
5: M-step: update A by Eq. 2
6: update ρ, O, π
7: **until** convergence
8: **return** θ

The EM algorithm concerning an MHMM is illustrated in Algorithm 1. For writing simplicity, we use MHMM when referring to this algorithm.

2.3 Orthogonality

In linear algebra, two vectors, $\mathbf{a}$ and $\mathbf{b}$, in a vector space are orthogonal when, geometrically, the angle between the vectors is 90 degrees. Equivalently, their inner product is zero, i.e. $\langle \mathbf{a}, \mathbf{b} \rangle = 0$. Similarly, the inner product of two orthogonal matrices is also zero. For matrices A and B, $\langle A, B \rangle := trace(A^T B)$ [14] where *trace* refers to the sum of the elements of the diagonal of a matrix. The inner product of square matrices, $trace(A^T B)$, is calculated as the following [14]:

$$trace(A^T B) = \sum_{i=1}^{J} \sum_{j=1}^{J} A_{ij} B_{ij} \quad where \quad A, B \in R^{J \times J} \tag{3}$$

3 Orthogonal Mixture of Hidden Markov Models

In this section we describe our solution, Orthogonal Mixture of Hidden Markov Models (oMHMM), to avoid a poor local optimum in the MHMM. In a mixture of HMMs, each HMM, and consequently each transition matrix, corresponds to a different cluster. Therefore, we consider a transition matrix as a cluster representation. The underlying idea of the oMHMM method is to direct the search for a transition matrix solution away from candidate solutions that are very similar, i.e., not fully exploiting the power of the mixture model. This is achieved by increasing the distance between transition matrices—equivalently, the dissimilarity of the clusters—at each iteration of EM. A geometric intuition of the idea is illustrated in Fig. 1, where we use orthogonality as a distance measure between cluster representations (transition matrices). Intuitively, as the angle between the representations of the two clusters increases, the dissimilarity of those clusters increases.

In order to enforce the concept of distant transition matrices, we propose a penalty, *orthogonality penalty*, for the objective function described in Eq. 2 in the M-step of the EM algorithm. Concretely, the penalty is the sum of all the pairwise inner products of the transition matrices. By adding the orthogonality penalty, having a form as in Eq. 3, to the original maximization problem in Eq. 2, we achieve the penalized objective function presented in Eq. 4 subject to $\sum_{j=1}^{J} A_{kij} = 1$. By introducing a subtraction of the penalty, maximizing the objective function implies minimizing the inner product; i.e. maximizing the orthogonality. Note that λ is the hyperparameter for the penalty.

$$\max_{A_{kij}} \sum_{k=1}^{K} \sum_{n=1}^{N} \sum_{m=2}^{M} \sum_{i=1}^{J} \sum_{j=1}^{J} \left[\varepsilon_k(c^n_{m-1,i}, c^n_{m,j}) \log A_{kij} - \lambda \sum_{k' \neq k}^{K} trace(A^T_{kij} A_{k'ij}) \right] \tag{4}$$

Plugging the right hand side of Eq. 3 into Eq. 4, we get the following:

$$\max_{A_{kij}} \sum_{k=1}^{K} \sum_{n=1}^{N} \sum_{m=2}^{M} \sum_{i=1}^{J} \sum_{j=1}^{J} \varepsilon_k(c^n_{m-1,i}, c^n_{m,j}) \log A_{kij} - \lambda \sum_{k' \neq k}^{K} \sum_{i=1}^{J} \sum_{j=1}^{J} A_{kij} A_{k'ij} \tag{5}$$

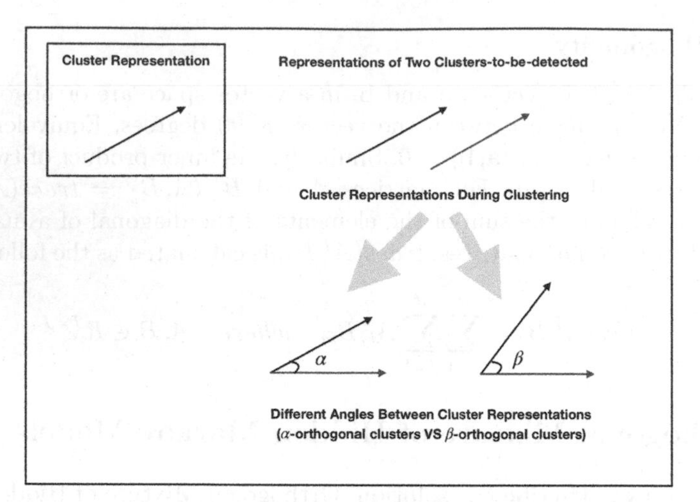

Fig. 1. A geometric view of the orthogonality of representations in clustering: each vector corresponds to a cluster representation, which can be expressed as any linear algebraic subspace. As the angle between representations of the two clusters increases, so does the dissimilarity between those clusters.

We solve the maximization problem stated in Eq. 5 by using the "cvxpy" Python library [2,11], as numerical optimization methods are needed due to the non-closed form solution to this optimization problem. Namely, setting the derivative of Eq. 5 w.r.t. A_{kij}, $\sum_{n=1}^{N}\sum_{m=2}^{M}\frac{\varepsilon(c_{m-1,i}^{k,n},c_{m,j}^{k,n})}{A_{kij}}+\delta-\sum_{k'\neq k}^{K}A_{k'ij}$, to zero cannot be solved by isolating the variable A_{kij}, where δ is the Lagrange multiplier of the constraint $\sum_{j=1}^{J}A_{kij}=1$.

A key feature of the oMHMM, lacking in the MHMM due to the treatment of transition matrices as independent, where the occurrence of one does not affect the probability of the occurrence of another, is that the estimation of a transition matrix, A_k, will be affected by all other transition matrices, $A_{k'}\quad\forall k'\neq k$; hence realizing a *global context*—parameters of *all* clusters are pulled into context.

Algorithm 2 summarizes oMHMM, which is a penalized EM. Note that all of the calculations are identical to the MHMM (Algorithm 1), except for line 5, where the update of transition matrices, A, is affected by Eq. 5.

4 Experiments

Our experimental objective is twofold: firstly, we aim to investigate the relative performance of the MHMM and oMHMM methods (Algorithms 1, 2) on various real-world datasets; we then use simulated datasets, inspired by the real-world datasets, to study the behavior of oMHMM in a more controlled manner, where the ground truth transition matrices are known.

Algorithm 2. oMHMM

1: **procedure** LEARN($Y_{1:N}$):
2: Initialise $\theta := \{\rho, O, A, \pi\}$
3: **repeat**
4: E-step: calculate $Q(\theta, \theta^{old}) := E_{C,Z|Y_{1:N},\theta^{old}}[\log p(Y_{1:N}, C, Z|\theta)]$
5: M-step: update A by Eq. 5
6: update ρ, O, π
7: **until** convergence
8: **return** θ

For the implementation of oMHMM and the complete test results, we refer to https://github.com/negar7918/oMHMM.

4.1 Performance Metrics

When measuring method performance, we use v-measure [26] concerning clustering results (in all of the experiments performed, we have access to the cluster labels). The v-measure, which gives a score between 0% (imperfect) and 100% (perfect), captures the homogeneity and completeness properties, that is, it measures how well a cluster contains only its own members and how many of those members. Note that the rand index [25] provides very similar results as the v-measure, and is for that reason not included here.

For the real-world datasets, to extend the investigation of clustering performance, we use accuracy in addition to v-measure; accuracy is commonly used [15,30] despite the focus on unsupervised learning. Using accuracy, we can show the proportion of correctly estimated clusters. Moreover, we extend the investigation by reporting the number of EM iterations, measuring the computational cost (the elapsed time for each iteration of oMHMM and MHMM are nearly equal). As the goal of experimenting on the simulated datasets is to examine the clustering performance hypothesis, i.e., an increase of the orthogonality between the true transition matrices leads to increased clustering performance, it is sufficient to use v-measure.

The experiments are conducted using different random initializations, where each initialization is used by both of the methods, MHMM and oMHMM. In the tables below, the result of the better performing model is highlighted with a bold font.

4.2 Experiments with Biological Data

We perform experiments on a previously published biological dataset[2] stemming from single-cell whole-genome sequencing from 18 primary colon cancer tumor cells and 18 metastatic cells from matched liver samples for one patient

[2] Available from the NCBI Sequence Read Archive (SRA) under accession number SRP074289; for pre-processing of the data, see [18].

referred to as CRC2 in [18]. In this work, we cluster CRC2 patient data using the genomic sequence of chromosome 4; the sequence comprises 808 genomic regions (sequence length of 808) where each region bears the characteristic of that region by a count number. In other words, the nature of the data is a count number. As mentioned in [18], it is reasonable that the primary tumor cells from the colon and the metastatic cells from the liver should cluster separately; therefore, we consider to cluster the sequence data into primary and metastatic clusters. Note that, for the purpose of evaluation, we know which cell sequence belongs to which cluster according to [18]. In order to perform clustering, we use a mixture of HMMs, where each cell sequence with a length of 808 represents an observation sequence and each HMM comprises hidden variables with three states, where the states correspond to the *copy numbers* for chromosome 4 used in [18]. In Fig. 2, illustrating metastatic and primary cancer cells, we can observe how, on average, the depth of the count data concerning metastatic cells reach lower values than the case of primary cells. Moreover, we can see that the variation of the counts across the sequence position is higher in metastatic than primary cells. The variations occurring throughout the sequence data are commonly used to perform clustering in cancer research [18]. These variations, per cluster, can be modeled as state transitions in an HMM in the mixture model. In Fig. 3, two cells are shown, each belonging to a different cluster: M-67 from the metastatic cluster and P-8 from the primary cluster. Note that, it is harder to see their difference compared to the difference between the average of metastatic and primary cells illustrated in Fig. 2. However, we can still see that they follow the patterns in Fig. 2; i.e., the lower counts belong to the metastatic cell (M-67) and the variation of the counts is higher in M-67 compared to P-8.

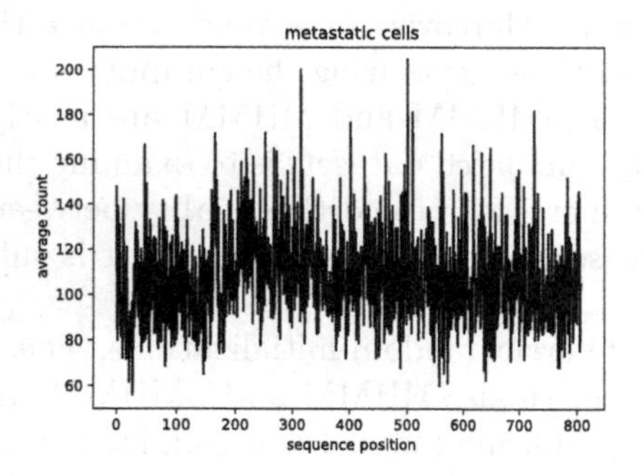

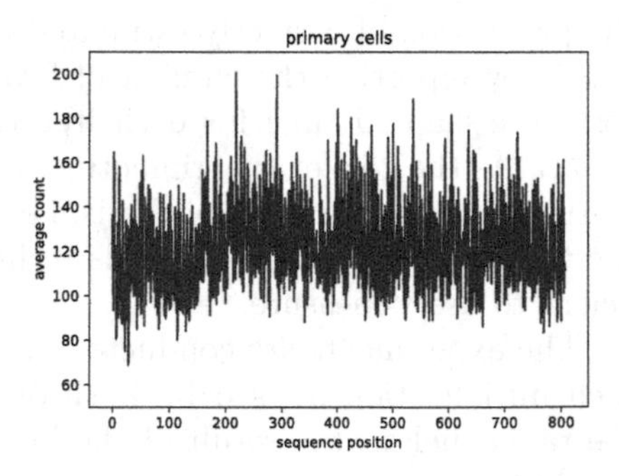

Fig. 2. The average counts of metastatic and primary cancer cells, per sequence position.

After performing the clustering of metastatic- and primary cells, using the MHMM and oMHMM methods, we calculate the resulting v-measure and accuracy measures. Moreover, we give an account of the number of EM iterations for each method. We run the methods on the random initializations A and B concerning transition matrices; initialization A has a Dirichlet distribution prior

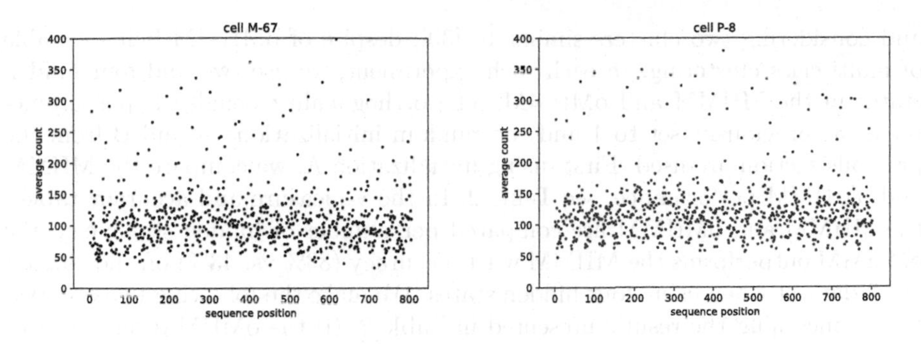

Fig. 3. The count data concerning two cells: M-67 metastatic and P-8 primary.

with parameters set to 0.1 for each row of the transition matrix, and B follows a discrete uniform distribution (the choices of the priors are inspired by [3,30]). We set the orthogonality penalty's hyperparameter to 1, allowing for a full effect of the orthogonality penalty. Table 1 shows that the oMHMM outperforms the MHMM for both initializations w.r.t. v-measure, accuracy, and the number of iterations. The highest v-measure and accuracy are 84% and 97%, respectively, and the maximum improvement is 55 percentage points w.r.t. v-measure and 44 percentage points w.r.t. accuracy, for initialization A. Finally, looking at the table concerning the number of iterations, we can observe that the oMHMM outperforms the MHMM with one iteration fewer required by the oMHMM than that for the MHMM. The fewer number of iterations implies a faster convergence.

Table 1. Accuracy, v-measure, and number of iterations are compared between the MHMM and oMHMM.

% V-measure

initialization	MHMM	oMHMM
A	4%	**59%**
B	69%	**84%**

% Accuracy

initialization	MHMM	oMHMM
A	47%	**91%**
B	94%	**97%**

Iterations

initialization	MHMM	oMHMM
A	5	**4**
B	3	**2**

4.3 Experiments with Handwritten Digit Data

We cluster handwritten digits from the so-called "pen-based recognition of handwritten digits" dataset in the UCI machine learning repository [4]. We repeatedly cluster datasets obtained by restricting the entire dataset to two digits at a time and considering two clusters similar to [30]; despite of oMHMM being capable of multi-class clustering. In each such experiment, we use two and four hidden states in the MHMM and oMHMM. The orthogonality penalty hyperparameter is, as previously, set to 1 and the random initializations A and B from the previous section are used. First, using initialization A, we compare the MHMM and oMHMM performance, see Table 2. In the v-measure and accuracy tables, the MHMM and oMHMM are compared per number of hidden states e.g., the oMHMM outperforms the MHMM w.r.t. accuracy (83% vs. 43%) on the "digit 6 vs 7" dataset when using four hidden states. We make the following two observations concerning the results presented in Table 2: (i) the oMHMM outperforms the MHMM and achieves the v-measures 36%, 29%, and 7%, corresponding to accuracies 83%, 76%, and 59%; (ii) the total number of iterations is reduced approximately to one third for all of the datasets. As shown in Table 3 for initialization B, the oMHMM outperforms the MHMM with v-measures 91%, 38%, 28%, 18%, and 10% corresponding to accuracies 98%, 78%, 73%, 68%, and 60%. For all of the datasets, the total number of iterations for the oMHMM is only 20% of that for the MHMM.

For each of the four datasets ("digit 6 vs 7", "digit 2 vs 9", "digit 4 vs 2", and "digit 5 vs 8"), the oMHMM results in the best v-measure and accuracy: (i) the highest v-measure are 91%, 29%, 18%, and 10%, respectively; (ii) the accuracies are 98%, 76%, 68%, and 60%, respectively. Note that in each of the experiments, the oMHMM results in a v-measure and accuracy that are equal to or greater than those of the MHMM. The greatest v-measure improvement obtained by the oMHMM is 38 percentage points increment for the dataset "digit 6 vs 7" and two states, Table 3. As shown in Table 2, the oMHMM results in the greatest accuracy increase, 40 percentage points increment, for "digit 6 vs 7" and four states. Finally, the oMHMM outperforms the MHMM with respect to the number of iterations, reducing the computational cost to approximately 20–50% of that achieved by the MHMM.

4.4 Experiments with Hand Movement Data

We perform experiments on the Libras movement dataset from the UCI machine learning repository [4], in which there are 15 classes of hand movements. Experiments are performed similarly to the previous section. We repeatedly cluster datasets obtained by restricting the entire dataset to two hand movements at a time.

First, using initialization A, we compare the MHMM and oMHMM performance, see Table 4. We can see that the oMHMM outperforms the MHMM and achieves v-measures 34%, 3%, 34%, 13%, 1%, and 17% corresponding to accuracies 75%, 60%, 75%, 71%, 54%, and 62%. Note that the accuracy achieved by the oMHMM is greater than the one from the MHMM for the "2 vs 1" dataset when using two hidden states; however, the v-measure is the same for both methods. The total number of iterations is reduced for the "13 vs 15" dataset and is almost unchanged for the other datasets.

Table 2. Using initialization A, accuracy, v-measure, and number of iterations are compared between the MHMM and oMHMM. S shows the number of hidden states.

% V-measure

Dataset	MHMM		oMHMM	
	S=2	S=4	S=2	S=4
digit 6 vs 7	5%	2%	5%	**36%**
digit 2 vs 9	21%	0%	**29%**	0%
digit 4 vs 2	0%	0%	**7%**	0%
digit 5 vs 8	1%	0%	1%	0%

% Accuracy

Dataset	MHMM		oMHMM	
	S=2	S=4	S=2	S=4
digit 6 vs 7	37%	43%	37%	**83%**
digit 2 vs 9	73%	47%	**76%**	47%
digit 4 vs 2	50%	50%	**59%**	50%
digit 5 vs 8	44%	50%	44%	50%

Total Iterations

Dataset	MHMM	oMHMM
digit 6 vs 7	16	**5**
digit 2 vs 9	17	**5**
digit 4 vs 2	17	**5**
digit 5 vs 8	16	**6**

Table 3. Using initialization B, accuracy, v-measure, and number of iterations are compared between the MHMM and oMHMM. S shows the number of hidden states.

% V-measure

Dataset	MHMM		oMHMM	
	S=2	S=4	S=2	S=4
digit 6 vs 7	0%	60%	**38%**	**91%**
digit 2 vs 9	0%	0%	**28%**	0%
digit 4 vs 2	0%	0%	**18%**	0%
digit 5 vs 8	0%	0%	**10%**	0%

% Accuracy

Dataset	MHMM		oMHMM	
	S=2	S=4	S=2	S=4
digit 6 vs 7	51%	91%	**78%**	**98%**
digit 2 vs 9	48%	48%	**73%**	48%
digit 4 vs 2	50%	50%	**68%**	50%
digit 5 vs 8	50%	50%	**60%**	50%

Total Iterations

Dataset	MHMM	oMHMM
digit 6 vs 7	28	**15**
digit 2 vs 9	31	**6**
digit 4 vs 2	30	**11**
digit 5 vs 8	35	**7**

As shown in Table 5 for initialization B, the oMHMM outperforms the MHMM with v-measures of 13%, 2%, 13% and 1%; however, w.r.t. accuracy, the oMHMM outperforms the MHMM on more accounts with values 73%, 71%, 58%, 71%, 54%, and 52%. The total number of iterations is reduced for the datasets "15 vs 3" and "3 vs 1", while unchanged for the other two datasets.

For each of the four datasets ("13 vs 15", "15 vs 3", "3 vs 1", and "2 vs 1"), the oMHMM results in the best v-measure and accuracy: (i) the highest

Table 4. Using initialization A, accuracy, v-measure, and number of iterations are compared between the MHMM and oMHMM. S shows the number of hidden states.

% V-measure

Dataset	MHMM		oMHMM	
	S=2	S=4	S=2	S=4
13 vs 15	1%	1%	**3%**	**34%**
15 vs 3	0%	0%	0%	**34%**
3 vs 1	4%	0%	**13%**	**1%**
2 vs 1	5%	0%	5%	**17%**

% Accuracy

Dataset	MHMM		oMHMM	
	S=2	S=4	S=2	S=4
13 vs 15	54%	54%	**60%**	**75%**
15 vs 3	52%	48%	52%	**75%**
3 vs 1	62%	50%	**71%**	**54%**
2 vs 1	60%	50%	**62%**	**62%**

Total Iterations

Dataset	MHMM	oMHMM
13 vs 15	12	**7**
15 vs 3	**8**	9
3 vs 1	**8**	9
2 vs 1	**8**	9

Table 5. Using initialization B, accuracy, v-measure, and number of iterations are compared between the MHMM and oMHMM. S shows the number of hidden states.

% V-measure

Dataset	MHMM		oMHMM	
	S=2	S=4	S=2	S=4
13 vs 15	0%	20%	**13%**	20%
15 vs 3	2%	0%	2%	**2%**
3 vs 1	0%	6%	**1%**	**13%**
2 vs 1	0%	0%	0%	0%

% Accuracy

Dataset	MHMM		oMHMM	
	S=2	S=4	S=2	S=4
13 vs 15	50%	64%	**71%**	**73%**
15 vs 3	58%	54%	58%	**58%**
3 vs 1	52%	65%	**54%**	**71%**
2 vs 1	50%	50%	50%	**52%**

Total Iterations

Dataset	MHMM	oMHMM
13 vs 15	14	14
15 vs 3	24	**8**
3 vs 1	18	**14**
2 vs 1	4	4

v-measures are 34%, 34%, 13%, and 17%, respectively; (ii) the accuracies are 75%, 75%, 71%, and 62%, respectively. Note that in each of the experiments, the oMHMM results in v-measure and accuracy of equal to or greater than those of the MHMM. The greatest improvement obtained by the oMHMM w.r.t. v-measure and accuracy concerns the "15 vs 3" dataset using four hidden states in Table 4, with an increase of 34 and 27 percentage points, respectively.

4.5 Experiments with Simulated Data

The goal of this section is to test the following hypothesis: greater orthogonality between the true transition matrices results in greater improvement achieved by the oMHMM compared to MHMM. This may only be investigated when having access to the ground truth (the true transition matrices), which is why we consider synthetic datasets here. We consider data generated from a given MHMM, comprising ground truth model parameters (we refer to this model by MHMM-gen), and evaluate the clustering obtained by each of the MHMM and the oMHMM from this data. The contribution of the orthogonality penalty is investigated by comparing the v-measures obtained by the oMHMM and the MHMM, based on the ground truth cluster labels from MHMM-gen.

Inspired by the biological data comprising two clusters in Sect. 4.2, we construct the simulated datasets, assuming the two transition patterns where each represents an inclination towards a specific state (we assume 3 vs. 4). Moreover, we consider 50 observation sequences with a length of 800. These sequences form the first dataset, *Scenario 1.* To study the behavior of the oMHMM when having lower orthogonality among the transition matrices in the ground truth model, we create a new dataset, *Scenario 2*, where we design the transition matrices so that they result in lower orthogonality than in Scenario 1. We achieve this by adding a third HMM (cluster) following a transition pattern similar to one of the two transition patterns in Scenario 1; this similarity results in lower orthogonality.

Having the ground truth, we evaluate the clustering results produced by the MHMM and oMHMM. V-measure is used as the main clustering performance metric. We perform the tests considering the transition matrix initializations used in the real-world datasets, A and B, to evaluate the orthogonality hypothesis in the already performed setting. Moreover, we add a third initialization, C, to extend the evaluation. C holds a Dirichlet distribution prior with parameters set to 0.5 for each row of the transition matrix.

Regarding the hyperparameters of the orthogonality penalty, λ in Eq. 5, we define the set of possible values to be $\{0, 0.1, 0.5, 1\}$. For each scenario and each initialization, we tune the value of λ using a separate dataset (with the same size as the test datasets used subsequently in the experiments) which we never use again in the following experiments. For each of those separate datasets, we choose the λ which gives the highest v-measure when performing the oMHMM. In case of ties among the best-performing values of λ, the highest value is selected.

Scenario 1. Using MHMM-gen, we generate 50 observation sequences of length 800, divided into two clusters. The parameters of the MHMM-gen model are set as follows. The number of hidden states is set to 4, the emission distribution is a Poisson distribution with one state-dependent rate per hidden state, randomly chosen between 80 and 100. We use mixture probabilities of 0.5 for the two components with transition matrices, A_1 and A_2. Conceptually, one cluster is inclined to stay at state 3 and the other at state 4 (these are highlighted as column 3 and 4 in the matrices below). In order to compare Scenario 1 to Scenario 2 for testing our hypothesis, we use orthogonality which is defined as one minus

Table 6. V-measure is compared between MHMMs and oMHMMs for Scenario 1 and 2.

% V-measure Scenario 1

initialization	MHMM	oMHMM
A	29%	29%
B	40%	**87%**
C	0%	**11%**

% V-measure Scenario 2

initialization	MHMM	oMHMM
A	72%	72%
B	62%	**72%**
C	66%	**72%**

the normalized inner product, i.e., $1 - \frac{<A_1,A_2>}{\sqrt{<A_1,A_1>}\sqrt{<A_2,A_2>}}$. The orthogonality of the matrices, A_1 and A_2, is 89%.

$$A_1 = \begin{pmatrix} 0 & 0 & .07 & \mathbf{.93} \\ 0 & .003 & .007 & \mathbf{.99} \\ 0 & 0 & .06 & \mathbf{.94} \\ 0 & 0 & .02 & \mathbf{.98} \end{pmatrix} \quad A_2 = \begin{pmatrix} 0 & .002 & \mathbf{.99} & .008 \\ 0 & 0 & \mathbf{.95} & .05 \\ 0 & 0 & \mathbf{.92} & .08 \\ 0 & 0 & \mathbf{.87} & .13 \end{pmatrix}$$

The hyperparameter values of the orthogonality penalty, λ, for initializations A, B, and C are 0, 1, and 1, respectively, after performing hyperparameter tuning as aforementioned. We give account for the v-measure performance of the MHMM and oMHMM concerning different initializations in the left-hand side of Table 6. We can observe that the oMHMM outperforms the MHMM with a 47 and 11 percentage points increase in v-measure for initialization B and C, respectively. Finally, the oMHMM results in the highest v-measure.

Scenario 2. Similar to Scenario 1, we generate observation sequences (a total of 100), forming three clusters. Each HMM is generated similarly to Scenario 1, except that the sequence length is 200 and the four Poisson rates are set to 1, 3, 9, and 27, respectively, for states 1, 2, 3, and 4. The following transition matrices are considered for the three HMMs: A_1, A_2, and A_3. A_1 and A_2 follow Scenario 1 and A_3 is similar to A_1, however with less inclination towards state 4. The orthogonality of 65% is achieved, which is 24 percentage points less than that for Scenario 1 (the orthogonality score is calculated using the same formula as in Scenario 1).

$$A_1 = \begin{pmatrix} 0 & 0 & .07 & \mathbf{.93} \\ 0 & .003 & .007 & \mathbf{.99} \\ 0 & 0 & .06 & \mathbf{.94} \\ 0 & 0 & .02 & \mathbf{.98} \end{pmatrix} \quad A_2 = \begin{pmatrix} 0 & .002 & \mathbf{.99} & .008 \\ 0 & 0 & \mathbf{.95} & .05 \\ 0 & 0 & \mathbf{.92} & .08 \\ 0 & 0 & \mathbf{.87} & .13 \end{pmatrix} \quad A_3 = \begin{pmatrix} 0 & .3 & 0 & .7 \\ 0 & .4 & 0 & .6 \\ 0 & .3 & 0 & .7 \\ 0 & .3 & .02 & .68 \end{pmatrix}$$

The hyperparameter λ is again tuned using the procedure outlined for Scenario 1. For initialization A, $\lambda = 1$ is chosen as it results in the highest v-measure. For initializations B and C, due to the identical v-measure results, $\lambda = 1$ is chosen (following our assumption explained in Sect. 4.5).

Looking at the right-hand side in Table 6, we can observe that the oMHMM outperforms the MHMM with a 10 and 6 percentage points increase in v-measure regarding initialization B and C, respectively. Note that the oMHMM results in the highest v-measure; however, the oMHMM has equal performance to MHMM, using initialization A. Comparing these results to the ones from Scenario 1, we can observe that the contribution of the oMHMM decreases as the orthogonality among transition matrices decreases. This confirms our hypothesis that the more orthogonal the ground truth transition matrices are, the greater improvement can be expected from the oMHMM.

4.6 Discussion of Results

The experiments showed that oMHMM can significantly improve MHMM. Concretely, oMHMM was observed to achieve up to a 55 percentage points increase w.r.t. v-measure, a 44 percentage points increase w.r.t. accuracy, and was observed to reduce the number of iterations down to a fifth. The experiments conducted on simulated data confirmed our hypothesis that the more orthogonal the ground truth transition matrices are, the greater improvement may be obtained by using the oMHMM instead of the standard MHMM. When the orthogonality is less pronounced we expect oMHMM to perform equal to MHMM given that the penalty hyperparameter is properly tuned.

5 Concluding Remarks

The use of EM for clustering of sequential data based on an MHMM may lead to poor local optima. This type of problem is often handled by augmenting the objective function, in the M-step of EM, with a penalty term. Several different penalties have been proposed for the EM algorithm when handling a single HMM. To tackle the problem for an MHMM, we propose a new penalty, the orthogonality penalty, which takes multiple HMMs into account. We call the so obtained EM algorithm oMHMM. The underlying idea is that the clustering can be expected to be improved when increasing the dissimilarity of clusters based on the orthogonality of the corresponding transition matrices of the constituent HMMs. We have presented results from experiments in which the novel algorithm is compared to the standard EM for an MHMM. The results show that the oMHMM performs on par or better than the standard EM for an MHMM with respect to v-measure, accuracy, and the number of iterations. These promising results show that the proposed penalty has a positive effect on sequence clustering using an MHMM.

One direction for future research is to combine the sparse mixture of HMMs [24] with the oMHMM. Another direction concerns investigating the theoretical and statistical properties of the oMHMM, e.g., consistency, efficiency, and convergence rate. MHMMs, in contrast to HMMs, have not received sufficient attention, e.g., various penalties can be investigated concerning EM for an MHMM. Finally, observing the high performance of oMHMM on the biological data in this work, motivates future work on critical biomedical applications. To name some, one can extend the studies on cancer cell clustering [28] and early prediction of a therapy outcome [27] by applying the orthogonality constraint.

Acknowledgments. We thank Johan Fylling, Mohammadreza Mohaghegh Neyshabouri, and Diogo Pernes for their great help during the preparation of this paper.

References

1. Aghabozorgi, S., Seyed Shirkhorshidi, A., Ying Wah, T.: Time-series clustering - a decade review. Inf. Syst. **53**, 16–38 (2015)
2. Agrawal, A., Verschueren, R., Diamond, S., Boyd, S.: A rewriting system for convex optimization problems. J. Control Decision **5**(1), 42–60 (2018)
3. Altosaar, J., Ranganath, R., Blei, D.: Proximity variational inference. AISTATS (2017)
4. Bache, K., Lichman, M.: Uci machine learning repository. UCI machine learning repository (2013)
5. Baum, L., Petrie, T.: Statistical inference for probabilistic functions of finite state markov chains. Ann. Math. Stat. **37**(6), 1554–1563 (1966)
6. Bishop, C.: Pattern recognition and machine learning. Springer, Information science and statistics, New York (2006)
7. Bishop, C.: Model-based machine learning. Philosophical transactions. Series A, Mathematical, physical, and engineering sciences 371 (2012)
8. Blei, D., Kucukelbir, A., Mcauliffe, J.: Variational inference: a review for statisticians. J. Am. Statist. Assoc. **112**(518), 859–877 (2017)
9. Chamroukhi, F., Nguyen, H.: Model based clustering and classification of functional data. Wiley Interdiscip. Rev. Data Mining Knowl. Disc. **9**(4), e1298 (2019)
10. Dempster, A.P., Laird, N.M., Rubin, D.B.: Maximum likelihood from incomplete data via the EM algorithm. J. Roy. Stat. Soc.: Ser. B (Methodol.) **39**(1), 1–22 (1977)
11. Diamond, S., Boyd, S.: CVXPY: a Python-embedded modeling language for convex optimization. J. Mach. Learn. Res. **17**(83), 1–5 (2016)
12. Dias, J., Vermunt, J., Ramos, S.: Mixture hidden markov models in finance research. In: Advances in Data Analysis, Data Handling and Business Intelligence, pp. 451–459 (2009)
13. Esmaili, N., Piccardi, M., Kruger, B., Girosi, F.: Correction: Analysis of healthcare service utilization after transport-related injuries by a mixture of hidden markov models. PLoS One **14**(4), e0206274 (2019)
14. Horn, R.A., Johnson, C.R.: Matrix Analysis. Cambridge University Press, Cambridge (2013)
15. Jebara, T., Song, Y., Thadani, K.: Spectral clustering and embedding with hidden markov models. In: Machine Learning: ECML 2007: 18th European Conference on Machine Learning 4701, pp. 164–175 (2007)
16. Jonathan, A., Sclaroff, S., Kollios, G., Pavlovic, V.: Discovering clusters in motion time-series data. In: CVPR (2003)
17. Kulesza, A., Taskar, B.: Determinantal point processes for machine learning. Found. Trends Mach. Learn. **5**(2–3), 123–286 (2012)
18. Leung, M., et al.: Single-cell DNA sequencing reveals a late-dissemination model in metastatic colorectal cancer. Genome Res. **27**(8), 1287–1299 (2017)

19. Ma, Q., Zheng, J., Li, S., Cottrell, G.: Learning representations for time series clustering. Adv. Neural Inf. Process. Syst. **32**, 3781–3791 (2019)
20. Maoying Qiao, R., Bian, W., Xu, D., Tao, D.: Diversified hidden markov models for sequential labeling. IEEE Trans. Knowl. Data Eng. **27**(11), 2947–2960 (2015)
21. McGibbon, R., Ramsundar, B., Sultan, M., Kiss, G., Pande, V.: Understanding protein dynamics with l1-regularized reversible hidden markov models. In: Proceedings of the 31st International Conference on Machine Learning, vol. 32, no. 2, pp. 1197–1205 (2014)
22. Montanez, G., Amizadeh, S., Laptev, N.: Inertial hidden markov models: modeling change in multivariate time series. In: AAAI Conference on Artificial Intelligence (2015)
23. Oates, T., Firoiu, L., Cohen, P.: Clustering time series with hidden markov models and dynamic time warping. In: IJCAI-99 Workshop on Neural, Symbolic and Reinforcement Learning Methods for Sequence Learning, pp. 17–21 (1999)
24. Pernes, D., Cardoso, J.S.: Spamhmm: sparse mixture of hidden markov models for graph connected entities. In: 2019 International Joint Conference on Neural Networks (IJCNN), pp. 1–10 (2019)
25. Rand, W.: Objective criteria for the evaluation of clustering methods. J. Am. Statist. Assoc. **66**(336), 846–850 (1971)
26. Rosenberg, A., Hirschberg, J.: V-measure: a conditional entropy-based external cluster evaluation measure. In: EMNLP-CoNLL (2007)
27. Safinianaini, N., Boström, H., Kaldo, V.: Gated hidden markov models for early prediction of outcome of internet-based cognitive behavioral therapy. In: Riaño, D., Wilk, S., ten Teije, A. (eds.) AIME 2019. LNCS (LNAI), vol. 11526, pp. 160–169. Springer, Cham (2019). https://doi.org/10.1007/978-3-030-21642-9_22
28. Safinianaini, N., De Souza, C., Lagergren, J.: Copymix: mixture model based single-cell clustering and copy number profiling using variational inference. bioRxiv (2020). https://doi.org/10.1101/2020.01.29.926022
29. Smyth, P.: Clustering sequences with hidden markov models. In: Advances in Neural Information Processing Systems (1997)
30. Subakan, C., Traa, J., Smaragdis, P.: Spectral learning of mixture of hidden markov models. Adv. Neural Inf. Process. Syst. **27**, 2249–2257 (2014)
31. Tao, L., Elhamifar, E., Khudanpur, S., Hager, G., Vidal, R.: Sparse hidden markov models for surgical gesture classification and skill evaluation. In: Proceedings of International Conference on Natural Language Processing and Knowledge Engineering, pp. 167–177 (2012)
32. Wang, Q., Schuurmans, D.: Improved estimation for unsupervised part-of-speech tagging. In: Proceedings of International Conference on Natural Language Processing and Knowledge Engineering, pp. 219–224 (2005)
33. Xing, Z., Pei, J., Keogh, E.: A brief survey on sequence classification. ACM SIGKDD Explor. Newslett. **12**(1), 40–48 (2010)
34. Yuting, Q., Paisley, J., Carin, L.: Music analysis using hidden markov mixture models. IEEE Trans. Signal Process. **55**(11), 5209–5224 (2007)

Poisson Graphical Granger Causality by Minimum Message Length

Kateřina Hlaváčková-Schindler[1,2(✉)] and Claudia Plant[1,3]

[1] Faculty of Computer Science, University of Vienna, Vienna, Austria
{katerina.schindlerova,claudia.plant}@univie.ac.at
[2] Institute of Computer Science, Czech Academy of Sciences, Prague, Czech Republic
[3] ds:UniVie, University of Vienna, Vienna, Austria

Abstract. Graphical Granger models are popular models for causal inference among time series. In this paper we focus on the Poisson graphical Granger model where the time series follow Poisson distribution. We use minimum message length principle for determination of causal connections in the model. Based on the dispersion coefficient of each time series and on the initial maximum likelihood estimates of the regression coefficients, we propose a minimum message length criterion to select the subset of causally connected time series with each target time series. We propose a genetic-type algorithm to find this set. To our best knowledge, this is the first work on applying the minimum message length principle to the Poisson graphical Granger model. Common graphical Granger models are usually applied in scenarios when the number of time observations is much greater than the number of time series, normally by several orders of magnitude. In the opposite case of "short" time series, these methods often suffer from overestimation. We demonstrate in the experiments with synthetic Poisson and point process time series that our method is for short time series superior in precision to the compared causal inference methods, i.e. the heterogeneous Granger causality method, the Bayesian causal inference method using structural equation models LINGAM and the point process Granger causality.

Keywords: Granger causality · Poisson graphical Granger model · Minimum message length · Ridge regression for GLM

1 Introduction

Granger causality is a popular method for causality analysis in time series due to its computational simplicity. Its application to time series with non-Gaussian distribution can be however misleading. Recently, Behzadi et al. in [2] proposed the heterogeneous graphical Granger Model (HGGM) for detecting causal relations among time series having a distribution from the exponential family, which includes a wider class of common distributions. HGGM employs regression in generalized linear models (GLM) with adaptive Lasso as a variable selection method and applies it to time series with a given lag. The approach allows to

F. Hutter et al. (Eds.): ECML PKDD 2020, LNAI 12457, pp. 526–541, 2021.
https://doi.org/10.1007/978-3-030-67658-2_30

apply causal inference among time series with discrete values. Poisson graphical Granger model (PGGM) is a special case of HGGM for detecting Granger-causal relationships among $p \geq 3$ Poisson processes. Each process in the model, represented by time series, is a count. A count process can be e.g. a process of events such as the arrival of a telephone call at a call centre in a time interval. Poisson processes can serve as models of point process data, including neural spike trains, [4]. Poisson graphical Granger model may be appropriate when investigating temporal interactions among processes as e.g. the number of transit passengers of an airport within a time period or in criminology, when temporal relationships among various crimes in some time interval are investigated.

In this paper we approach the inference in the Poisson graphical Granger model by the principle of minimum message length (MML).

- We use minimum message length principle for determination of causal connections in the Poisson graphical Granger model.
- For the highly collinear design matrix of the model we define a corrected form of the Fisher information matrix using the ridge penalty.
- Based on the dispersion coefficient of each time series and on the initial maximum likelihood estimates of the regression coefficients, we propose a minimum message length criterion to select the subset of causally connected time series with each target time series.
- We propose a genetic-type algorithm to find this set.
- To our best knowledge, this is the first work on applying the minimum message length principle to the Poisson graphical Granger model.
- We demonstrate experimentally that our method is superior in precision to the compared causal inference methods, i.e. the heterogeneous Granger causality method, the Bayesian causal inference method using structural equation models LINGAM [20] and the point process Granger causality in the case of short data, i.e. when the number of time observations is approximately of the same order as the number of time series.

The paper is organized as follows. Section 2 presents preliminaries, concretely Granger causality and the Poisson graphical Granger model. Section 3 presents the PGGM as an instance of multiple Poisson regression. The MML code for PGGM is computed and the main theorem is stated in Sect. 4. The algorithm to compute the MML code for PGGM and the genetic algorithm for variable selection in PGGM are explained in Sect. 5. Related work is discussed in Sect. 6. Our experiments are in Sect. 7. Section 8 is devoted to conclusions and the derivation of the criterion can be found in Appendix.

2 Preliminaries

Relevance of Granger Causality. Since its introduction, there has been lead a criticism of Granger causality, since it e.g. does not take into account counterfactuals, [10,15]. As its name implies, Granger causality is not necessarily true causality. In defense of his method, Granger in [5] wrote: "Possible causation is not considered for any arbitrarily selected group of variables, but only for variables for which the researcher has some prior belief that causation is, in some sense, likely." In other words, drawing conclusions about the existence of a causal relation between time series and about its direction is possible only if theoretical knowledge of mechanisms connecting the time series is accessible. Nevertheless as confirmed by a recent Nature publication [14], if the theoretical background of investigated processes is insufficient, methods to infer causal relations from data rather than knowledge of mechanisms (Granger causality including) are helpful. These methods can also make possible to perform credible analyses with large amount of observational time data, e.g. in social networks [9], since they are less costly than common epidemiological or marketing research approaches.

Graphical Granger Model. The (Gaussian) graphical Granger model extend the autoregressive concept of Granger causality to $p \geq 2$ time series and time lag $d \geq 1$ [1]. Let $x_1^t, \ldots, x_p^t$ be p time series, $t = 1, \ldots, n$. Consider the vector autoregressive (VAR) models with lag d for $i = 1, \ldots, p$

$$x_i^t = X_{t,d}^{Lag} \beta_i' + \varepsilon_i^t \tag{1}$$

where $X_{t,d}^{Lag} = (x_1^{t-d}, \ldots, x_1^{t-1}, \ldots, x_p^{t-d}, \ldots, x_p^{t-1})$ and β_i be a matrix of the regression coefficients and ε_i^t be white noise. One can easily show that $X_{t,d}^{Lag} \beta_i' = \sum_{j=1}^{p} \sum_{l=1}^{d} x_j^{t-l} \beta_j^l$. One says the time series x_j Granger–causes the time series x_i for the given lag d, denote $x_j \rightarrow x_i$ for $i, j = 1, \ldots, p$ if and only if at least one of the d coefficients in $j-th$ row of β_i in (1) is non-zero.

Poisson Graphical Granger Model. The Poisson graphical Granger model has the form

$$x_i^t \approx \lambda_i^t = \exp(X_{t,d}^{Lag} \beta_i') = \exp(\sum_{j=1}^{p} \sum_{l=1}^{d} x_j^{t-l} \beta_j^l) \tag{2}$$

for x_i^t, $i = 1, \ldots, p, t = d+1, \ldots, n$ having a Poisson distribution. Applying the HGGM approach to the case, when the link function for each process x_i is function exp, problem (2) can be solved as

$$\hat{\beta}_i = \arg\min_{\beta_i} \sum_{t=d+1}^{n} (x_i^t - \exp(X_{t,d}^{Lag} \beta_i'))^2 + \rho_i R(\beta_i) \tag{3}$$

for a given lag $d > 0$ and all $t = d+1, \ldots, n$ with $R(\beta_i)$ adaptive Lasso penalty function. (The sign $'$ denotes a transpose of a matrix). One says, the time series x_j Granger–causes the time series x_i for the given lag d, denote $x_j \rightarrow x_i$ for $i, j = 1, \ldots, p$ if and only if at least one of the d coefficients in $j-th$ row of $\hat{\beta}_i$ of the solution of (3) is non-zero [2].

3 Poisson Graphical Granger Model as Multiple Poisson Regression

In this section we will derive the Poisson Granger model (2) with a fixed lag d as an instance of a multiple Poisson regression with a fixed design matrix. Consider the full model for p Poisson variables x_i^t and (integer) lag $d \geq 1$ corresponding to the optimization problem (2). To be able to use the maximum likelihood (ML) estimation over the regression parameters, we reformulate the matrix of lagged time series $X_{t,d}^{Lag}$ from (1) into a fixed design matrix form. Assume $n - d > pd$ and denote $x_i = (x_i^{d+1}, x_i^{d+2}, \ldots, x_i^n)$. We construct the $(n-d) \times (d \times p)$ design matrix

$$X = \begin{bmatrix} x_1^d & \ldots & x_1^1 & \ldots & x_p^d & \ldots & x_p^1 \\ x_1^{d+1} & \ldots & x_1^2 & \ldots & x_p^{d+1} & \ldots & x_p^2 \\ \vdots & \vdots & \vdots & \vdots & \vdots & \vdots & \vdots \\ x_1^{n-1} & \ldots & x_1^{n-d+1} & \ldots & x_p^{n-1} & \ldots & x_p^{n-d+1} \end{bmatrix} \tag{4}$$

and a $1 \times (d \times p)$ vector $\beta_i = (\beta_1^1, \ldots, \beta_1^d, \ldots, \beta_p^1, \ldots, \beta_p^d)$. We can see that problem

$$x_i' \approx \lambda_i = \exp(X\beta_i') \tag{5}$$

is equivalent to problem (2) in the matrix form where we mean by exp a function operating on each coordinate $i = d+1, \ldots, n$ and $\lambda_i = (\lambda_i^{d+1}, \ldots, \lambda_i^{d+1})$.

Denote now by $\gamma_i \subset \Gamma = \{1, \ldots, p\}$ the subset of indices of regressor's variables and $k_i := |\gamma_i|$ its cardinality. Let $\beta_i := \beta_i(\gamma_i) \in \mathbb{R}^{1 \times (d \times k_i)}$ be the vector of unknown regression coefficients with a fixed ordering within the γ_i subset. For illustration purposes and without lack of generality we can assume that the first k_i indices out of p vectors belong into γ_i. Considering only the columns from matrix X in (4) corresponding to γ_i, we define the $(n-d) \times (d \times k_i)$ matrix of lagged vectors with indices from γ_i as

$$X_i := X(\gamma_i) = \begin{bmatrix} x_1^d & \ldots & x_1^1 & \ldots & x_{k_i}^d & x_{k_i}^{d-1} & \ldots & x_{k_i}^1 \\ x_1^{d+1} & \ldots & x_1^2 & \ldots & x_{k_i}^{d+1} & x_{k_i}^d & \ldots & x_{k_i}^2 \\ x_1^{d+2} & \ldots & x_1^3 & \ldots & x_{k_i}^{d+2} & x_{k_i}^{d+1} & \ldots & x_{k_i}^3 \\ \vdots & \vdots & \vdots & \vdots & \vdots & \vdots & \vdots & \vdots \\ x_1^{n-1} & \ldots & x_1^{n-d+1} & \ldots & x_{k_i}^{n-1} & x_{k_i}^{n-2} & \ldots & x_{k_i}^{n-d+1} \end{bmatrix} \tag{6}$$

The problem (5) for explanatory variables with indices from γ_i is expressed as

$$x_i' \approx \lambda_i = E(x_i'|X_i) = \exp(X_i\beta_i') \tag{7}$$

or alternatively

$$\log(x_i') \approx \log(\lambda_i) = \log(E(x_i'|X_i)) = X_i\beta_i' \tag{8}$$

with $\beta_i := \beta_i(\gamma_i)$ to be a $1 \times (dk_i)$ matrix of unknown coefficients and log operates on each coordinate. Wherever it it clear from context, we will simplify the notation β_i instead of $\beta_i(\gamma_i)$ and X_i instead of $X(\gamma_i)$.

4 Minimum Message Length for Poisson Granger Model

Denote Γ the set of all subsets of covariates $x_i, i = 1, \ldots, p$. Assume now a fixed set $\gamma_i \in \Gamma$ of covariates with size $k_i \leq p$ and the corresponding design matrix X_i from (6). It is well known that the Poisson regression model can be still used in over- or underdispersed settings. (However the standard error for Poisson would not be correct for the overdispersed situation.) In the Poisson graphical Granger model, it is the case when for the dispersion of at least one time series holds $\phi_i \neq 1$. So using the Poisson regression model, we assume that the likelihood function for x_i in PGGM does not depend on ϕ_i. It is usual to assume that the targets x_i are independent random variables, conditioned on the features given by X_i, so that the likelihood function can be factorized into the product $p(x_i|\beta_i, X_i, \gamma_i) = \prod_{t=1}^{n-d} p(x_i^t|\beta_i, X_i, \gamma_i)$. The log-likelihood function has then the form

$$L_i := \log p(x_i|\beta_i, X_i, \gamma_i) = \sum_{t=1}^{n-d} \log p(x_i^t|\beta_i, X_i, \gamma_i). \quad (9)$$

Since X_i is highly collinear, to make the ill-posed problem for coefficients β_i a well-posed one, one can use regularization by the ridge regression for GLM (see e.g. [19]). Ridge regression requires an initial estimate of β_i which can be as the maximum likelihood estimator of (7) obtained by the iteratively reweighted least square algorithm (IRLS). For a fixed $\rho_i > 0$, for the ridge estimates of coefficients $\hat{\beta}_{i,\rho_i}$ holds

$$\hat{\beta}_{i,\rho_i} = \arg\min_{\beta_i \in \mathbb{R}^+} \{-L_i + \rho_i \beta_i' \Sigma_i \beta_i\}. \quad (10)$$

In our paper however, we will not use the GLM ridge regression in form (10). Instead, we will apply the principle of minimum description length. Ridge regression in the minimum description length framework is equivalent to allowing the prior distribution to depend on a hyperparameter (= ridge regularization parameter). To compute the message length using the MML87 approximation proposed in [17], we need the negative log-likelihood function, prior distribution over the parameters and an appropriate Fisher information matrix. [17] proposed the corrected form of Fisher information matrix for a GLM regression with ridge penalty. In our work, we will use this form of ridge regression and apply it to the Poisson graphical Granger model. In the following, we will construct the MML code for every subset of covariates in PGGM. The derivation of the criterion can be found in Appendix.

The MML Criterion for PGGM. *For each* $d \geq 1$ *assume* $x_i, i = 1, \ldots, p, t = 1, \ldots, n$ *and the estimate of the dispersion parameter* $\hat{\phi}_i$ *be given. Assume* $\hat{\beta}_i$ *be an initial solution of* (7) *achieved as the maximum likelihood estimate.*

(i) *The causal graph of the Poisson Granger problem* (7) *can be inferred from the solutions of* p *variable selection problems, where for each* $i = 1, \ldots, p$, *the set* $\hat{\gamma}_i$ *of Granger-causal variables to* x_i *is found.*

(ii) *For the estimated set* $\hat{\gamma}_i$ *holds*

$$\hat{\gamma}_i = \arg\min_{\gamma_i \in \Gamma}\{I(x_i, \hat{\beta}_i, \hat{\phi}_i, \hat{\rho}_i, X_i, \gamma_i) + I(\gamma_i)\} \text{ where} \tag{11}$$

$I(x_i, \hat{\beta}_i, \hat{\phi}_i, \hat{\rho}_i, X_i, \gamma_i) = \min_{\rho_i \in \mathbb{R}^+}\{MML(x_i, \hat{\beta}_i, \hat{\phi}_i, \rho_i, X_i, \gamma_i)\}$ *where* $MML(x_i, \hat{\beta}_i, \hat{\phi}_i, \rho_i, X_i, \gamma_i)$ *is the minimum message length code of the set* γ_i *and can be expressed as*

$$MML(x_i, \hat{\beta}_i, \hat{\phi}_i, \rho_i, X_i, \gamma_i) = -L_i + \frac{1}{2}\log|X_i'W_iX_i + \rho_i\Sigma_i| - \frac{1}{2}\log|\Sigma_i| \tag{12}$$

$$+\tfrac{k_i}{2}\log(\tfrac{2\pi}{\rho_i}) + (\tfrac{\rho_i}{2\hat{\phi}_i})\hat{\beta}_i'\Sigma_i\hat{\beta}_i + \tfrac{1}{2}\log(n-d) - \tfrac{k_i+1}{2}\log(2\pi) + \tfrac{1}{2}\log((k_i+1)\pi)$$

where $|\hat{\gamma}_i| = k_i$, Σ_i *is the unity matrix of size* $dk_i \times dk_i$, W_i *is a diagonal matrix with entries* $W_i(t) = \lambda_i^t = \exp(X_i\hat{\beta}_i')^t, t = 1, \dots, n-d$ *for Poisson* x_i *and* $W_i(t) = \lambda_i^t = [x_i^{d+t} - \exp(X_i\hat{\beta}_i')]^2$ *for over- or underdispersed Poisson* x_i, $L_i = \log(p(x_i|\hat{\beta}_i, X_i, \gamma_i)) = \sum_{t=d+1}^{n} x_i^t[X_i\hat{\beta}_i']^t - \exp([X_i\hat{\beta}_i']^t) - \log(x_i^t!)$ *and* $I(\gamma_i) = \log\binom{p}{k_i} + \log(p+1)$.

Remark: Schmidt and Makalic in [18] compared AIC_c criterion with MML code for generalized linear models. We constructed the AIC_c criterion also for PGGM. However this criterion requires pseudoinverse of a matrix multiplication which includes matrices X_i. Since X_is are highly collinear, these matrix multiplications had in our experiments very high condition numbers. This consequently lead the AIC_c criterion for PPGM to spurious results and therefore we do not report them in our paper.

5 Variable Selection in Poisson Graphical Granger Model

For both Poisson and overdispersed Poisson cases we consider the family of models $M(\gamma_i) := \{p(x_i|\beta_i, X_i, \gamma_i), \gamma_i \in \Gamma\}$ defined by Poisson densities $p(x_i|\beta_i, X_i, \gamma_i)$. First, we present the procedure in Algorithm 1 which for each x_i computes the MML code for a set $\gamma_i \subset \Gamma$. Then we present Algorithm 2 for computation of $\hat{\gamma}_i$.

In general, the selection of the best structure γ_i amounts to evaluate values of $MML(\gamma_i)$ for all $\gamma_i \subset \Gamma$, i.e. for all 2^p possible subsets and then to pick the subset with which the minimum of the function was achieved. To avoid the exhaustive search approach, we find γ_i with minimum MML by the proposed genetic algorithm type procedure called MMLGA. The idea of MMLGA is as follows. Consider an arbitrary $\gamma_i \subset \Gamma$ with size k_i for a fixed i and $d \geq 1$. Define a Boolean vector Q_i of length p corresponding to a given γ_i in so that it has ones in the positions of the indices of covariates from γ_i, otherwise zeros. Define $I(Q_i) := I(\gamma_i)$ where $I(\gamma_i)$ is from (11). Genetic algorithm MMLGA executes genetic operations on populations of Q_i. In the first step a population of size m (m be an even integer), is generated randomly in the set of all 2^p binary strings (individuals) of length p. Then we select $m/2$ individuals in the current

Algorithm 1. MML Code for γ_i

Input: $\gamma_i \in \Gamma, d \geq 1$, series is the matrix of x_i^t, $\hat{\phi}_i$ dispersion parameter, $i = 1, \ldots, p, t = 1, \ldots, n-d$, Σ_i, H a set of positive numbers;
Output: For each i minimum $I(x_i, \hat{\beta}_i, \hat{\rho}_i, X_i, \gamma_i)$ over H is found;
for all x_i **do**
  // Construct the d-lagged matrix X_i with time series with indices from γ_i.
  //Compute matrix W_i.
  for all $\rho_i \in H$ **do**
    // Compute L_i from (9).
    // Find the initial estimates of $\hat{\beta}_i$.
    //Compute $MML(x_i, \hat{\beta}_i, \rho_i, X_i, \gamma_i)$ from (12).
  end for// to ρ_i
  // Compute $I(x_i, \hat{\beta}_i, \hat{\rho}_i, X_i, \gamma_i) = \min_{\rho_i \in R^+} MML(x_i, \hat{\beta}_i, \rho_i, X_i, \gamma_i)$.
end for// to x_i
return $I(x_i, \hat{\beta}_i, \hat{\rho}_i, X_i, \gamma_i)$ for each i.

population with the lowest value of (11) as the elite subpopulation of parents of the next population. For a predefined number of generated populations n_g, the crossover operation of parents and the mutation operation of a single parent are executed on the elite to create the rest of the new population. A mutation corresponds to a random change in Q_i and a crossover combines the vector entries of a pair of parents. After each run of these two operations on a current population, the current population is replaced with the children with the lowest value of (11) to form the next generation. The algorithm stops after the number of population generations n_g is achieved. The algorithm MMLGA is summarized in Algorithm 2. Our code in Matlab is publicly available at: https://t1p.de/b3gf.

5.1 Computational Complexity of MMLGA

For computation of $I(x_i, \hat{\beta}_i, \hat{\rho}_i, X_i, \gamma_i)$ we used Matlab function *fminsearch*. It is well-known that the upper bound of the computational complexity of a genetic algorithm is of order of the product of the size of an individual, of the size of each population, of the number of generated populations and of the complexity of the function to be minimized. Therefore an upper bound of the computational complexity of MMLGA for p time series, size p of an individual, m the population size and n_g the number of population generations is $\mathcal{O}(pmn_g) \times O(fminsearch) \times p$ where $O(fminsearch)$ can be also estimated. The highest complexity in *fminsearch* has the computation of the Hessian matrix, which is the same as for the Fisher information matrix (our matrix W_i) or the computation of the determinant. The computational complexity of Hessian for i fixed for $(n-d) \times (n-d)$ matrix is $\mathcal{O}(\frac{(n-d)(n-d+1)}{2})$. An upper bound on complexity of determinant in (12) is $\mathcal{O}((pd)^3)$. As before we assume $n - d \geq pd$. Denote $M = \max\{pd, (n-d+1)\}$. Then holds also $M^3 \geq \frac{(M-1)M}{2}$. Since we have p optimization functions, our upper bound on the computational complexity of MMLGA is then $\mathcal{O}(p^2 mn_g M^3)$.

Algorithm 2. MMLGA

Input: Γ, $d \geq 1, p, n_g, m$ an even integer, $z \leq p$ position for off-spring;
series is the matrix of $x_i^t, i = 1, \ldots, p, t = 1, \ldots, n-d$;
Output: Adj := adjacency matrix of the output causal graph;
// For every x_i Q_i with minimum of (11) is found;
for all x_i **do**
  Create initial population $\{Q_i^j, j = 1, \ldots, m\}$ at random;
  Compute $I(Q_i^j) := I(x_i, \hat{\beta}_i, \hat{\rho}_i, X_i, Q_i^j) + \binom{p}{k_i^j} + \log(p+1)$ for each $j = 1, \ldots, m$
  where k_i^j is the number of ones in Q_i^j; v:=1;
  while $v \leq n_g$ **do**
    u:=1;
    while $u \leq m$ **do**
      Sort $I(Q_i^j)$ ascendingly and create the elite population; By crossover of Q_i^j and Q_i^r, $r \neq j$ create children and add them to elite; Compute $I(Q_i^j)$ for each j; Mutate a single parent Q_i^j at a random position; Compute $I(Q_i^j)$ for each j; Add the children with minimum $I(Q_i^j)$ until the new population not filled;
      u:=u+1;
    end while// to u
    v:=v+1;
  end while// to v
end for// to x_i
The $i-th$ row of Adj: $Adj_i := Q_i$ with min of (11)
return (Adj)

---

6 Related Work

The minimum message length (MML) is an information theoretic principle based on the statistical inference and data compression. The key idea is, if a statistical model compresses data, then the model has (with a high probability) captured regularities in the data. The MML principle selects the model which most compresses the data (i.e. the one with the "shortest message length") as the most descriptive for the data. To be able to decompress this representation of the data, the details of the statistical model used to encode the data must also be part of the compressed data string. The calculation of the exact message is an NP hard problem, however the most widely used less computationally intensive is the Wallace-Freeman approximation called MML87 [22].

Compression schemes for Poisson regression have been already studied in the framework of generalized linear models (GLM). Hansen and Yu 2003 in [6] derived objective functions for one-dimensional GLM regression by the minimum description principle. Schmidt and Makalic in [18] used MML87 to derive the MML code of a multivariate GLM ridge regression. The mentioned codes cannot be however directly used for a Granger model due to the lag in the time series and the highly collinear matrix of covariates. To our best knowledge, compression criteria for Poisson graphical Granger model has not been published yet. Other papers inferring Granger causality by MDL are [3,11,12]. The inference in this papers is however done for the bivariate Granger causality and the extension to graphical Granger methods is not straightforward.

Kim et al. in [8] proposed the statistical framework Granger causality (SFGC) that can operate on point processes, including neural-spike trains. The proposed framework uses multiple statistical hypothesis testing for each pair of involved neurons. A pair-wise hypothesis test was used for each pair of possible connections among all time series and the false discovery rate (FDR) applied.

For a fair comparison with our method we selected causal inference methods which are designed for $p \geq 3$ non-Gaussian processes. In our experiments, we used SFGC as a comparison method and the publicly available point process time series provided by the authors. As another comparison method we selected the method LINGAM from Shimizu et al. [20] which estimates a causal structure in Bayesian networks among non-Gaussian time series using structural equation models and independent component analysis. The experiments reported in the papers with comparison methods were done only in scenarios when the number of time observations is by several orders of magnitude greater than the number of time series.

7 Experiments

We performed experiments with MMLGA on synthetically generated Poisson processes and on neural spike train data from [8]. We used the method HGGM [2], the method LINGAM [20] and the point process Granger causality SFGC [8] for comparison. To assess similarity between the target and output causal graphs by all methods, we used the commonly applied F-measure, which takes both precision and recall into account.

7.1 Implementation and Parameter Setting

The comparison method HGGM uses Matlab package *penalized* from [16] with adaptive Lasso penalty. The algorithm in this package employs the Fisher scoring algorithm to estimate the coefficients of regressions. As recommended by the author of *penalized* in [16] and employed in [2] we used adaptive Lasso with $\lambda_{max} = 5$, applying cross validation and taking the best result with respect to F measure from the interval $(0, \lambda_{max}]$. We also followed the recommendation of the authors of LINGAM in [20] and used threshold = 0.05 and number of boots n/2, where n is the length of the time series. In method SFGC we used the setting recommended by the authors, the significance level 0.05 of FDR. The method SFGC is designed for binomial time series so as expected, using the generated Poisson time series as input of this method gave very low or zero F-measure, so we do not report these values in our results. For a fair comparison to MML, HGGM and LINGAM, we will examine the performance of SFGC with input binomial time series in Sect. 7.3.

In MMLGA, the initial estimates of β_i were achieved by the iteratively re-weighted least square procedure implemented in Matlab function *glmfit*, in the same function we obtained also the estimates of the dispersion parameters of time series. (Considering initial estimates of β_i by the IRLS procedure using function *penalized* with ridge gave poor results in the experiments.) The minimization over ρ_i was done by function *fminsearch* which defined set H from Algorithm 1 as positive numbers greater or equal to 0.1.

7.2 Synthetically Generated Poisson Processes

To be able to evaluate the performance of MMLGA and to compare it to other methods, the ground truth, i.e. the target causal graph in the experiments should be known. In this series of experiments we examined randomly generated Poisson processes together with the correspondingly generated target causal graphs. The performance of MML, HGGM and LINGAM depends on various parameters including the number of time series (features), the number of causal relations in Granger causal graph (dependencies), the length of time series and finally the lag parameter. We examined causal graphs with $p = 5$ and with $p = 9$ time series. The length of generated time series was 'short', varying from 100 to 1000. We generated Poisson time series randomly. Concerning the calculation of an appropriate lag for each time series, theoretically it can be done by AIC or BIC. However, the calculation of AIC and BIC assumes that the degrees of freedom are equal to the number of nonzero parameters, which is only known to be true for the Lasso penalty [23] but not known for adaptive Lasso. In our experiments we followed the recommendation of [2] how to select the lag of time series. They observed that varying the lag parameter from 3 to 50 did not influence either the performance of HGGM nor SFGC significantly. Based on that we considered lags 3 and 4 in our experiments. For the causal graphs with $p = 5$ and with $p = 9$ we tested the performance of algorithms for number of dependencies from 6 to 9. The results of our experiments on causal graphs with 5 features ($p = 5$) are presented in Table 1. Each value in Table 1 represents the mean value of all F-measures over 20 random generations of causal graphs for length n and lag d. One can see from Table 1 that MMLGA gave significantly higher precision in terms of F-measure than both comparison methods for n up to 500. On the other hand, HGGM gave the highest F-measure for $n = 1000$ which can be for $p = 5$ considered as a scenario of a large data set. The results of our experiments with causal graphs with $p = 9$ are presented in Table 2. Each value in Table 2 represents the mean value of all F-measures over 20 random generations of causal graphs for length n and lag d.

Similarly as in the experiments with $p = 5$, one can see in Table 2 for $p = 9$ that MMLGA gave significantly higher F-measure than for both comparison methods for n up to 500. HGGM gave higher F-measure than MMLGA for $n = 1000$ which can be for $p = 5$ considered as a scenario of a large data set. In both networks with $p = 5$ and $p = 9$ time series, method LINGAM had the lowest F-measures for all investigated n.

Table 1. $p = 5$, average F-measure for each method, MMLGA, with $n_g = 10$, $m = 50$, HGGM with $\lambda_{max} = 5$, LINGAM with $n/2$ boots.

$d = 3, n =$	50	100	200	300	500	1000
MMLGA	**0.8**	**0.82**	**0.83**	**0.77**	**0.77**	0.73
HGGM	0.67	0.73	0.73	0.73	0.71	**0.8**
LINGAM	0.71	0.71	0.7	0.69	0.65	0.65
$d = 4, n =$	50	100	200	300	500	1000
MMLGA	**0.75**	**0.77**	**0.77**	**0.8**	**0.8**	0.67
HGGM	0.66	0.73	0.71	0.73	0.73	**0.8**
LINGAM	0.64	0.65	0.64	0.63	0.65	0.64

Table 2. $p = 9$, average F-measure for each method, MMLGA, with $n_g = 10$, $m = 50$, HGGM with $\lambda_{max} = 5$, LINGAM with $n/2$ boots.

$d = 3, n =$	50	100	200	300	500	1000
MMLGA	**0.67**	**0.62**	**0.69**	**0.69**	**0.69**	0.67
HGGM	0.6	0.48	0.5	0.62	0.65	**0.82**
LINGAM	0.4	0.28	0.29	0.33	0.35	0.36
$d = 4, n =$	50	100	200	300	500	1000
MMLGA	**0.62**	**0.61**	**0.67**	**0.64**	**0.64**	0.63
HGGM	0.5	0.51	0.54	0.54	0.55	**0.8**
LINGAM	0.4	0.39	0. 39	0.4	0.37	0.38

7.3 Neural Spike Train Data

In this section we examine the performance of MMLGA, SFGC, HGGM and LINGAM on time series representing the spike train data. We used the nine-neuron network with the spike train data from [8] and the corresponding target network in Fig. 1-B of the paper. Based on the experimental settings described in [8], the authors generated 100,000 samples for each neuron, and the total number of spikes for each neuron ranged from 2176 through 2911, i.e. three orders of magnitude more than the number of time series. The target network corresponds thus to the long time series. For fair comparison of all methods, we compared their precision on short time series in terms of F-measure to the target network from Fig. 1-B of the paper.

Spike train data is a special case of a temporal point process. A temporal point process is a stochastic time series of binary events that occur in continuous time. It can only take on two values at each point in time, indicating whether or not an event has actually occurred. When considering the data set of size n, the point process model of a spike train for neuron i can be defined as a counting process, which can be denoted as $\{N(t), 1 \leq t \leq n\}$. A counting process represents the total number of occurrences or events that have happened up to and including time t. It is a Poisson process. We used the point process time series from [8] as input of SFGC and their Poisson representation as described above as input of methods MMLGA, HGGM and LINGAM. We experimented with short time series with n from 100 up to 1000. Table 3 gives the F-measures of the methods. Rephrasing the F-measures into percent, one can see that for $n = 500$ method MMLGA is able to reconstruct 59 % of the target causal network, while the best reconstruction results are for SFGC 12 % (achieved for $n = 900$ and $n = 1000$), for HGGM 45 % (achieved for $n = 300$) and for LINGAM 19 % (achieved for $n = 900$). One can see that MMLGA outperformed significantly the other three methods in precision measured by F-measure for almost all investigated n.

Table 3. $p = 9$, F-measure for each method, MMLGA with $d = 3$, $n_g = 30$, $m = 50$, HGGM with $\lambda_{max} = 5$, LINGAM with $n/2$ boots.

$d = 3, n =$	100	200	300	400	500	600	700	800	900	1000
MMLGA	0.27	0.43	0.45	0.51	**0.59**	0.55	0.47	0.46	0.47	0.46
SFGC	0	0.06	0	0	0	0	0.06	0	**0.12**	**0.12**
HGGM	0.34	0.44	**0.45**	0.41	0.44	0.38	0.39	0.38	0.41	0.39
LINGAM	0.39	0.39	0.39	0.39	0.39	0.39	0.39	0.33	**0.42**	0.39

7.4 Analysis of Chicago Violence Crime Dataset

Chicago's violent crime rate is substantially higher than the US average. Although national crime rates in the US have stayed near historic lows, Chicago had nearly half of 2016's increase in crimes in the US [13]. Thus any research on possible causes of the increased number of crimes is valuable for the law enforce agencies. We used the data set of 5 most frequent crimes in Chicago from [13], i.e. battery, narcotics consumption, criminal damage (= violation of property rights), theft and other offense (e.g. harassment by telephone, a weapon violation). These are yearly measurements from 2001 to 2017. Due to the small data size, this is rather a toy example. We investigated temporal interactions of these time series. No target graph was given. Our goal was to find out whether the resulting causal graphs for each test method support empirical evidence. Statistical distribution fitting test confirmed Poisson distribution of all time series. We investigated causal graphs for lags 1 to 3 (to keep the condition $n - d \geq pd$ as discussed above). Method HGGM gave for each lag a different causal graph, for $d = 1$ it gave the complete graph. So we excluded it from further analysis. Methods LINGAM and MMLGA gave only one causal graph as output for all considered lags and the causal graphs can be found (for $n_g = 10$ and $m = 30$) in Fig. 1. Focusing on crime narcotic consumptions, MMLGA outputs other offenses as causal to narcotics and narcotics causal to crime battery. Both claims support the empirical evidence, in the second case it is known that drug consumption increases the effects generating violence, as stated in the reports of the Bureau of Justice Statistics, US Department of Justice, e.g. [21]. On the other hand, LINGAM outputs battery as a cause of narcotics, which seems unrealistic. So the output of MMLGA gave a more realistic causal graph than LINGAM.

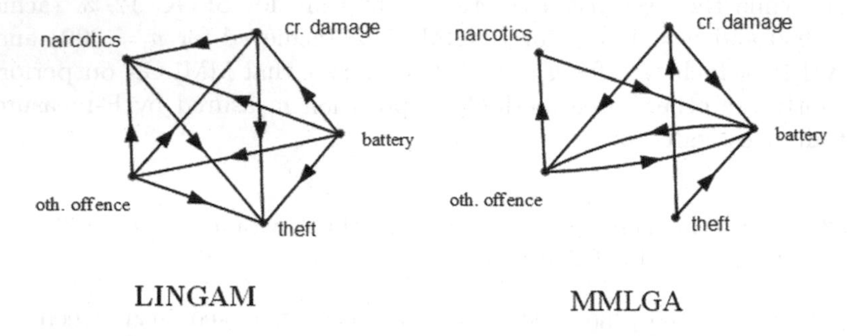

Fig. 1. Results of causal interactions among 5 most frequent crimes in Chicago for LINGAM and MMLGA.

8 Conclusions

Common graphical Granger models are usually applied in scenarios when the number of time observations is by several orders of magnitude greater than the number of time series. In the opposite case of short time series, these methods often suffer from overestimation. In this paper we used minimum message length principle for determination of causal connections in the Poisson graphical Granger model. Based on the dispersion coefficient of each time series and on the initial maximum likelihood estimates of the regression coefficients, we proposed a minimum message length criterion to select the subset of time series causal to each target time series. We used a genetic-type algorithm MMLGA to find this set. We demonstrated in the experiments on synthetic Poisson time series and on point process time series that our method is on short time series superior in precision to the compared causal inference methods, i. e. the heterogeneous Granger causality method, the Bayesian causal inference method using structural equation models LINGAM and the point process Granger causality. Both MMLGA and HGGM use penalization, MMLGA uses ridge, HGGM adaptive Lasso. The superiority of MMLGA with respect to HGGM for short time series can be explained by using the dispersion of the time series in the criterion as additional information with respect to HGGM. To our best knowledge, this is the first work on applying the minimum message length principle to the Poisson graphical Granger model. In our future work we would like to investigate other utilization of the minimum message principle for PGGM, for example for the dispersion parameter of the involved time series.

Acknowledgement. This work was supported by the Czech Science Foundation, project $GA19-16066S$.

9 Appendix

Derivation of the MML Criterion for PGGM

Assume p independent Poisson random variables expressed by time series with lag $d > 0$ $x_i^t, t = d+1, \dots, n$, and the problem (7). Assume the estimate $\hat{\phi}_i$ is given. We consider now γ_i fixed, so for simplicity of writing we omit it from the list of variables of the functions. First we need to express the log-likelihood function in terms of parameters β_i. Since we use Poisson model for x_i having the Poisson distribution or overdispersed Poisson, we omit ϕ_i from the list of parameters which condition function p. For a given set of parameters β_i, the probability of attaining $x_i^{d+1}, \dots, x^n$ is given by $p(x_i^{d+1}, \dots, x_i^n | X_i, \beta_i) = \prod_{t=d+1}^n \frac{(\lambda_i^t)^{x_i^t} \exp(-\lambda_i^t)}{(x_i^t)!} = \prod_{t=d+1}^n \frac{\exp([X_i\beta_i']^t)^{x_i^t} \exp(-\exp([X_i\beta_i']^t))}{x_i^t!}$ where $[X_i\beta_i']^t$ denotes the t-th coordinate of the vector $X_i\beta_i'$. The log-likelihood in terms of β_i is $L_i = ll(\beta_i|x_i, X_i) = \log p(\beta_i|x_i, X_i) = \sum_{t=d+1}^n x_i^t [X_i\beta_i']^t - \exp([X_i\beta_i']^t) - \log(x_i^t!)$. Having function L_i, we can now compute an initial estimate of $\hat{\beta}_i$ from (7) which is the solution to the system of score equations. Since $-ll(\beta_i|x_i, X_i)$ is a convex function, one can use standard convex optimization techniques (e.g. Newton-Raphson method) to solve these equations numerically. (In our code, we use the Matlab implementation of an iteratively reweighted least squares (IRLS) algorithm of the Newton-Raphson method). Assume now we have an initial solution $\hat{\beta}_i$ from (7).

1. Now we derive matrix W_i for x_i with Poisson distribution:
The Fisher information matrix $J_i = J(\beta_i) = -\mathbb{E}_{\beta_i}(\nabla^2 ll(\beta_i|x_i, X_i))$ may be obtained by computing the second order partial derivatives of ll for $r, s = 1, \dots, k_i$. This gives
$\frac{\delta^2 ll(\beta_i|x_i, X_i)}{\delta^2 \beta_i^r \beta_i^s} = \frac{\delta ll}{\delta \beta_i^s} \sum_{t=d+1}^n [x_i^t \sum_{l=1}^d x_r^{t-l} - \exp(\sum_{j=1}^{k_i} \sum_{l=1}^d x_j^{t-l} \beta_j^l) \sum_{l=1}^d x_r^{t-l}]$ $= -\sum_{t=d+1}^n \exp(\sum_{j=1}^{k_i} \sum_{l=1}^d x_j^{t-l} \beta_j^l)(\sum_{l=1}^d x_s^{t-l})(\sum_{l=1}^d x_r^{t-l})$. If we denote $W_i := diag(\exp(\sum_{j=1}^{k_i} \sum_{l=1}^d x_j^{d+1-l} \beta_j^l), \dots, \exp(\sum_{j=1}^{k_i} \sum_{l=1}^d x_j^{n-l} \beta_j^l))$ then we have Fisher information matrix $J(\beta_i) = (X_i)' W_i X_i$.

2. Derivation of matrix W_i for x_i with overdispersed Poisson distribution:
Assume now the dispersion parameter $\phi_i > 0, \neq 1$. The variance of the overdispersed Poisson distribution is $\phi_i \lambda_i$. We know that the Poisson regression model can be still used in overdispersed settings and the function ll is the same as $ll(\beta_i)$ derived above. We use the robust sandwich estimate of covariance of $\hat{\beta}_i$, proposed in [7] for a general Poisson regression. The Fisher information matrix of overdispersed problem is $J_i = J(\beta_i) = (X_i)' W_i X_i$ where W_i is constructed for PGGM based on [7] and has the form $W_i = diag([x_i^{d+1} - \exp(\sum_{j=1}^{k_i} \sum_{l=1}^d x_j^{d+1-l} \beta_j^l)]^2, \dots, [x_i^n - \exp(\sum_{j=1}^{k_i} \sum_{l=1}^d x_j^{n-l} \beta_j^l)]^2)$. Having parameters $\hat{\beta}_i, \hat{\phi}_i, \Sigma_i$ W_i and ρ_i, we still need to construct the function $MML(\gamma_i)$.

Construction of function $MML(\gamma_i)$:
Having these parameters, we use for each $i = 1, \ldots, p$ and regression (7) formula (18) from [18] i.e. for the case when in $\alpha := 0$ and $\beta := \beta_i$ and $X := X_i$, $y := x_i$, $n := n - d$, $k := k_i$, $\theta := \hat{\beta}_i$, $\lambda := \hat{\rho}_i$, $\phi := \phi_i$, $S = \Sigma_i$ is the unity matrix of dimension dk_i, the corrected Fisher information matrix for the parameters β_i is then $J(\beta_i|\phi_i, \rho_i) = (\frac{1}{\phi_i})X_i'W_iX_i + \rho_i\Sigma_i$ where $\lambda_i = \exp(X_i\beta_i)$. Function $c(m)$ for $m := k_i + 1$ is then $c(k_i + 1) = -\frac{k_i+1}{2}\log(2\pi) + \frac{1}{2}\log((k_i + 1)\pi) - 0.5772$ and the constants independent of k_i we omitted from MML code, since the optimization over γ_i is of them independent. Among all subsets $\gamma_i \in \Gamma$, there are $\binom{p}{k_i}$ subsets of size k_i. If nothing is known a priori about the likelihood of any covariate x_i being included in the final model, a prior that treats all subset sizes equally likely $\pi(|\gamma_i|) = 1/(p+1)$ is appropriate [18]. This gives the code length $I(\gamma_i) = \log\binom{p}{k_i} + \log(p+1)$ as in (11).

References

1. Arnold, A., Liu, Y., Abe, N.: Temporal causal modeling with graphical Granger methods. In: ACM SIGKDD, pp. 66–75 (2007)
2. Behzadi, S., Hlaváčková-Schindler, K., Plant, C.: Granger causality for heterogeneous processes. In: Yang, Q., Zhou, Z.-H., Gong, Z., Zhang, M.-L., Huang, S.-J. (eds.) PAKDD 2019. LNCS (LNAI), vol. 11441, pp. 463–475. Springer, Cham (2019). https://doi.org/10.1007/978-3-030-16142-2_36
3. Budhathoki, K., Vreeken, J.: Origo: causal inference by compression. Knowl. Inf. Syst. **56**(2), 285–307 (2018)
4. Brown E.N.: Theory of point processes for neural systems. In: Chow, C., et al. (ed.) Methods and Models in Neurophysics, pp. 691–726. Elsevier, Paris (2005)
5. Granger, C.W.: Some recent development in a concept of causality. J. Econ. **39**(1–2), 199–211 (1988)
6. Hansen, M.H., Yu, B.: Minimum description length model selection criteria for generalized linear models. Lecture Notes-Monograph Series, pp. 145–163 (2003)
7. Huber, P.J.: The behavior of maximum likelihood estimates under nonstandard conditions. In: Proceedings of the Fifth Berkeley Symposium on Mathematical Statistics and Probability, vol. 1, pp. 221–233. University of California Press (1967)
8. Kim, S., Putrino, D., Ghosh, S., Brown, E.N.: A Granger causality measure for point process models of ensemble neural spiking activity. PLOS Comput. Biol. 1–13 (2011)
9. Kwak, H., Lee, C., Park, H., Moon, S.: What is twitter, a social network or a news media? In: Proceedings of the 19th International Conference on World Wide Web, pp. 591–600. ACM (2010)
10. Mannino, M., Bressler, S.L.: Foundational perspectives on causality in large-scale brain networks. Phys. Life Rev. **15**, 107–123 (2015)
11. Marx, A., Vreeken, J.: Telling cause from effect using MDL-based local and global regression. In: IEEE ICDM, pp. 307–316 (2017)
12. Marx, A., Vreeken, J.: Causal inference on multivariate and mixed-type data. In: ECML PKDD 2018, pp. 655–671 (2018)
13. Mangipudi, V.: Analysis of crimes in Chicago 2001–2017. https://rstudio-pubs-static.s3.amazonaws.com/294927b602318d06b74e4cb2e6be336522e94e.html. Accessed 21 Feb 2020

14. Marinescu, I.E., Lawlor, P.N., Kording, K.P.: Quasi-experimental causality in neuroscience and behavioural research. Nat. Hum. Behav. **2**(1), 891–898 (2018)
15. Maziarz, M.: A review of the granger-causality fallacy. J. Philos. Econ. Reflect. Econ. Soc. Issues **8**(2), 86–105 (2015)
16. McIlhagga, W.H.: Penalized: a MATLAB toolbox for fitting generalized linear models with penalties. J. Stat. Softw. **72**(6), 1–21 (2016)
17. Schmidt, D.F., Makalic, E.: MML invariant linear regression. In: Nicholson, A., Li, X. (eds.) AI 2009. LNCS (LNAI), vol. 5866, pp. 312–321. Springer, Heidelberg (2009). https://doi.org/10.1007/978-3-642-10439-8_32
18. Schmidt, D.F., Makalic, E.: Minimum message length ridge regression for generalized linear models. In: Cranefield, S., Nayak, A. (eds.) AI 2013. LNCS (LNAI), vol. 8272, pp. 408–420. Springer, Cham (2013). https://doi.org/10.1007/978-3-319-03680-9_41
19. Segerstedt, B.: On ordinary ridge regression in generalized linear models. Commun. Stat. Theory Methods **21**(8), 2227–2246 (1992)
20. Shimizu, S., et al.: DirectLiNGAM: a direct method for learning a linear non-Gaussian structural equation model. J. Mach. Learn. Res. **12**, 1225–1248 (2011)
21. U.S. Department of Justice, Office of Justice Programs, Bureau of Justice Statistics. https://www.bjs.gov/content/pub/pdf/DRRC.PDF
22. Wallace, C.S., Freeman, P.R: Estimation and inference by compact coding. J. R. Stat. Soc. Ser. B **49**(3), 240–252 (1987)
23. Zou, H., Hastie, T., Tibshirani, R.: On the “degrees of freedom” of the lasso. Annal. Stat. **35**(5), 2173–2192 (2007)

Counterfactual Propagation for Semi-supervised Individual Treatment Effect Estimation

Shonosuke Harada[1(✉)] and Hisashi Kashima[1,2]

[1] Kyoto University, Kyoto, Japan
sh1108@ml.ist.i.kyoto-u.ac.jp, kashima@i.kyoto-u.ac.jp
[2] RIKEN AIP, Tokyo, Japan

Abstract. Individual treatment effect (ITE) represents the expected improvement in the outcome of taking a particular action to a particular target, and plays important roles in decision making in various domains. However, its estimation problem is difficult because intervention studies to collect information regarding the applied treatments (i.e., actions) and their outcomes are often quite expensive in terms of time and monetary costs. In this study, we consider a semi-supervised ITE estimation problem that exploits more easily-available unlabeled instances to improve the performance of ITE estimation using small labeled data. We combine two ideas from causal inference and semi-supervised learning, namely, matching and label propagation, respectively, to propose *counterfactual propagation,* which is the first semi-supervised ITE estimation method. Experiments using semi-real datasets demonstrate that the proposed method can successfully mitigate the data scarcity problem in ITE estimation.

Keywords: Causal inference · Treatment effect estimation · Semi-supervised learning

1 Introduction

One of the important roles of predictive modeling is to support decision making related to taking particular actions in responses to situations. The recent advances of in the machine learning technologies have significantly improved their predictive performance. However, most predictive models are based on passive observations and do not aim to predict the causal effects of actions that actively intervene in environments. For example, advertisement companies are interested not only in their customers' behavior when an advertisement is presented, but also in the causal effect of the advertisement, in other words, the change it causes on their behavior. There has been a growing interest in moving from this passive predictive modeling to more active causal modeling in various domains, such as education [14], advertisement [8,22], economic policy [20], and health care [12].

Taking an action toward a situation generally depends on the expected improvement in the outcome due to the action.

This is often called the *individual treatment effect (ITE)* [32] and is defined as the difference between the outcome of taking the action and that of *not* taking the action. An intrinsic difficulty in ITE estimation is that ITE is defined

F. Hutter et al. (Eds.): ECML PKDD 2020, LNAI 12457, pp. 542–558, 2021.
https://doi.org/10.1007/978-3-030-67658-2_31

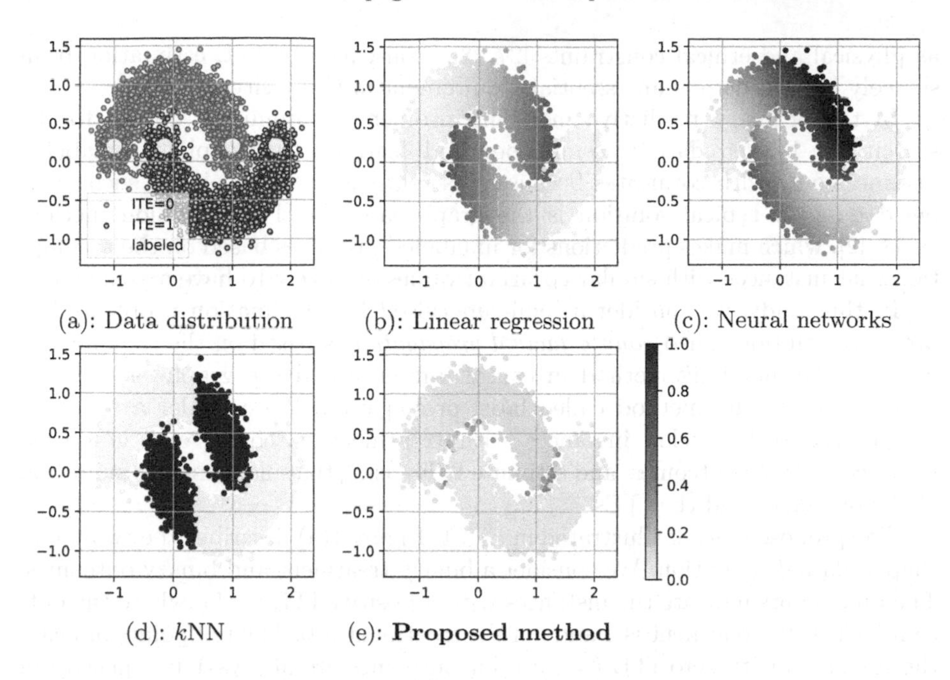

(a): Data distribution (b): Linear regression (c): Neural networks

(d): kNN (e): **Proposed method**

Fig. 1. Illustrative example using (a) two-moon dataset. Each moons has a constant ITE either of 0 and 1. Only two labeled instances are available for each moon, denoted by yellow points, whose observed (treatment, outcome) pairs are $(0, 1), (1, 1), (1, 1), (0, 0)$ from left to right. Figures (b), (c), and (d) show the ITE estimation error (PEHE) by the standard two-model approach using different base models suffered from the lack of labeled data. The deeper-depth color indicates larger errors. The proposed semi-supervised method (e) successfully exploits the unlabeled data to estimate the correct ITEs. (Color figure online)

as the difference between the factual and counterfactual outcomes [21,26,32]; in other words, the outcome that we can actually observe is either of the one when we take an action or the one when we do not, and it is physically impossible to observe both. To address the counterfactual predictive modeling from observational data, various techniques including matching [31], inverse-propensity weighting [29], instrumental variable methods [3], and more modern deep learning-based approaches have been developed [17,33]. For example, in the matching method, matching pairs of instances with similar covariate values and different treatment assignments are determined. The key idea is to consider the two instances in a matching pair as the counterfactual instance of each other so that we can estimate the ITE by comparing the pair.

Another difficulty in ITE estimation is data scarcity. For ITE estimation, we need some labeled instances whose treatments (i.e., whether or not an action was taken on the instance) and their outcomes (depending on the treatments) as well as their covariates are given. However, collecting such labeled instances can be quite costly in terms of time and money, or owing to other reasons, such as physical and ethical constraints [27,28]. Consequently, ITE estimation from scarcely labeled data is an essential requirement in many situations.

In the ordinary predictive modeling problem, a promising option to the scarcity of labeled data is semi-supervised learning that exploits unlabeled instances only with covariates because it is relatively easy to obtain such unlabeled data. A typical solution is the graph-based label propagation method [4,38,42], which makes predictions for unlabeled instances based on the assumption that instances with similar covariate values are likely to have a same label.

In this study, we consider a semi-supervised ITE estimation problem. The proposed solution called *counterfactual propagation* is based on the resemblance between the matching method in causal inference and the graph-based semi-supervised learning method called label propagation. We consider a weighted graph over both labeled instances with treatment outcomes and unlabeled instances with no outcomes, and estimate ITEs using the smoothness assumption of the outcomes and the ITEs.

The proposed idea is illustrated in Fig. 1. Figure 1(a) describes the two-moon shaped data distribution. We consider a binary treatment and binary outcomes. The blue points indicate the instances with a positive ITE ($=1$), where the outcome is 1 if the treatment is 1 and 0 if the treatment is 0. The red points indicate the instances with zero ITE ($=0$); their outcomes are always 1 irrespective of the treatments. We have only four labeled data instances shown as yellow points, whose observed (treatment, outcome) pairs are $(0,1),(1,1),(1,1),(0,0)$ from left to right. Since the amount of labeled data is considerably limited, supervised methods relying only on labeled data fail to estimate the ITEs. Figures 1(b), (c), (d) show the ITE estimation errors by the standard two-model approach using different base learners, which show poor performance. In contrast, the proposed approach exploits unlabeled data to find connections between the red points and those between the blue points to estimate the correct ITEs (Fig. 1(e)).

We propose an efficient learning algorithm assuming the use of a neural network as the base model, and conduct experiments using semi-synthetic real-world datasets to demonstrate that the proposed method estimates the ITEs more accurately than baselines when the labeled instances are limited.

2 Semi-supervised ITE Estimation Problem

We start with the problem setting of the semi-supervised treatment effect estimation problem. Suppose we have N labeled instances and M unlabeled instances. (We usually assume $N \ll M$.) The set of labeled instances is denoted by $\{(\mathbf{x}_i, t_i, y_i^{t_i})\}_{i=1}^{N}$, where $x_i \in \mathbb{R}^D$ is the covariates of the i-th instance, $t_i \in \{0,1\}$ is the treatment applied to instance i, and $y_i^{t_i}$ is its outcome. Note that for each instance i, either $t_i = 0$ or $t_i = 1$ is realized; accordingly, either y_i^0 or y_i^1 is available. The unobserved outcome is called a counterfactual outcome. The set of unlabeled instances is denoted by $\{(\mathbf{x}_i)\}_{i=N+1}^{N+M}$, where only the covariates are available.

Our goal is to estimate the ITE for each instance. Following the Rubin-Neyman potential outcomes framework [32,34], the ITE for instance i is defined as $\tau_i = y_i^1 - y_i^0$ exploiting both the labeled and unlabeled sets. Note that τ_i is not known even for the labeled instances, and we want to estimate the ITEs for both the labeled and unlabeled instances.

We make typical assumptions in ITE estimation in this study. i.e., (i) stable unit treatment value: the outcome of each instance is not affected by the treatment assigned to other instances; (ii) unconfoundedness: the treatment assignment to an instance is independent of the outcome given covariates (confounder variables); (iii) overlap: each instance has a positive probability of treatment assignment.

3 Proposed Method

We propose a novel ITE estimation method that utilizes both the labeled and unlabeled instances. The proposed solution called *counterfactual propagation* is based on the resemblance between the matching method in causal inference and the graph-based semi-supervised learning method.

3.1 Matching

Matching is a popular solution to address the counterfactual outcome problem. Its key idea is to consider two similar instances as the counterfactual instance of each other so that we can estimate the causal effect by comparing the pair. More concretely, we define the similarity w_{ij} between two instances i and j, as that defined between their covariates; for example, we can use the Gaussian kernel.

$$w_{ij} = \exp\left(-\frac{\|\mathbf{x}_i - \mathbf{x}_j\|^2}{\sigma^2}\right). \tag{1}$$

The set of (i, j) pairs with w_{ij} being larger than a threshold and satisfying $t_i \neq t_j$ are found and compared as counterfactual pairs. Note that owing to definition of the matching pair, the matching method only uses labeled data.

3.2 Graph-Based Semi-supervised Learning

Graph-based semi-supervised learning methods assume that the nearby instances in a graph are likely to have similar outputs. For a labeled dataset $\{(\mathbf{x}_i, y_i)\}_{i=1}^N$ and an unlabeled dataset $\{\mathbf{x}_i\}_{i=N+1}^{N+M}$, their loss functions for standard predictive modeling typically look like

$$L(f) = \sum_{i=1}^{N} l(y_i, f(\mathbf{x}_i)) + \lambda \sum_{i,j=1}^{N+M} w_{ij} \left(f(\mathbf{x}_i) - f(\mathbf{x}_j)\right)^2, \tag{2}$$

where f is a prediction model, l is a loss function for the labeled instances, and λ is a hyper-parameter. The second term imposes "smoothness" of the model output over the input space characterized by w_{ij} that can be considered as the weighted adjacency matrix of a weighted graph; it can be seen the same as that used for matching (1).

The early examples of graph-based methods include label propagation [42] and manifold regularization [4]. More recently, deep neural networks have been used as the base model f [38].

3.3 Treatment Effect Estimation Using Neural Networks

We build our ITE estimation model based on the recent advances of deep-learning approaches for ITE estimation, specifically, the treatment-agnostic representation network (TARNet) [33] that is a simple but quite effective model. TARNet shares common parameters for both treatment instances and control instances to construct representations but employs different parameters in its prediction layer, which is given as:

$$f(\mathbf{x}_i, t_i) = \begin{cases} \Theta_1^\top g\left(\Theta^\top x_i\right) & (t_i = 1) \\ \Theta_0^\top g\left(\Theta^\top x_i\right) & (t_i = 0) \end{cases}, \tag{3}$$

where Θ is the parameters in the representation learning layer and Θ_1, Θ_0 are those in the prediction layers for treatment and controlled instances, respectively. The g is a non-linear function such as ReLU. One of the advantages of TARNet is that joint representations learning and separate prediction functions for both treatments enable more flexible modeling.

3.4 Counterfactual Propagation

It is evident that the matching method relies only on labeled data, while the graph-based semi-supervised learning method does not address ITE estimation; however, they are quite similar because they both use instance similarity to interpolate the factual/counterfactual outcomes or model predictions as mentioned in Sect. 3.2. Our idea is to combine the two methods to propagate the outcomes and ITEs over the matching graph assuming that similar instances would have similar outcomes.

Our objective function consists of three terms, $L_\mathrm{s}, L_\mathrm{o}, L_\mathrm{e}$, given as

$$L(f) = L_\mathrm{s}(f) + \lambda_\mathrm{o} L_\mathrm{o}(f) + \lambda_\mathrm{e} L_\mathrm{e}(f), \tag{4}$$

where λ_o and λ_e are the regularization hyper-parameters. We employ TARNet [33] as the outcome prediction model $f(\mathbf{x}, t)$. The first term in the objective function (4) is a standard loss function for supervised outcome estimation; we specifically employ the squared loss function as

$$L_s(f) = \sum_{i=1}^{N} (y_i^{t_i} - f(\mathbf{x}_i, t_i))^2. \tag{5}$$

Note that it relies only on the observed outcomes of the treatments that are observed in the data denoted by t_i.

The second term L_o is the outcome propagation term:

$$L_o(f) = \sum_t \sum_{i,j=1}^{N+M} w_{ij}((f(\mathbf{x}_i, t) - f(\mathbf{x}_j, t))^2. \tag{6}$$

Similar to the regularization term (2) in the graph-based semi-supervised learning, this term encourages the model to output similar outcomes for similar instances by penalizing the difference between their outcomes. This regularization term allows the model to propagate outcomes over a matching graph. If two nearby instances have different treatments, they interpolate the counterfactual outcome of each other, which compares the factual and (interpolated) counterfactual outcomes to estimate the ITE. The key assumption behind this term is the smoothness of outcomes for each treatment over the covariate space. While w_{ij} indicates the adjacency between nodes i and j in the graph-based regularization, it can be considered as a matching between the instances i and j in the treatment effect estimation problem. Even though traditional matching methods have only rely on labeled instances, we combine matching with graph-based regularization which also utilizes unlabeled instances. This regularization enables us to propagate the outcomes for each treatment over the matching graph and mitigate the counterfactual problem.

The third term L_e is the ITE propagation term defined as

$$L_e(f) = \sum_{i,j=1}^{N+M} w_{ij}(\hat{\tau}_i - \hat{\tau}_j)^2, \tag{7}$$

where $\hat{\tau}_i$ is the ITE estimate for instance i:

$$\hat{\tau}_i = f(\mathbf{x}_i, 1) - f(\mathbf{x}_i, 0).$$

$L_e(f)$ imposes the smoothness of the ITE values in addition to that of the outcomes imposed by outcome propagation (6). In comparison to the standard supervised learning problems, where the goal is to predict the outcomes, as stated in Sect. 2, our objective is to predict the ITEs. This term encourages the model to output similar ITEs for similar instances. We expect that the outcome propagation and ITE propagation terms are beneficial especially when the available labeled instances are limited while there is an abundance of unlabeled instances, similar to semi-supervised learning.

3.5 Estimation Algorithm

As mentioned earlier, we assume the use of neural networks as the specific choice of the outcome prediction model f based on the recent successes of deep neural networks in causal inference. For computational efficiency, we apply a sampling approach to optimizing Eq. (4). Following the existing method [38], we employ

Algorithm 1: Counterfactual propagation

Input: labeled instances $\{(\mathbf{x}_i, t_i, y_i^{t_i})\}_{i=1}^{N}$, unlabeled instances $\{(\mathbf{x}_i)\}_{i=N+1}^{N+M}$, a similarity matrix $w = (w_{ij})$, and mini-batch sizes b_1, b_2.

Output: estimated outcome(s) for each treatment $\hat{y}_i^1$ and/or $\hat{y}_i^0$ using Eq. (3).

while *not converged* **do**

 # Approximating the supervised loss
 Sample b_1 instances $\{(\mathbf{x}_i, t_i, y_i)\}$ from the labeled instances
 Compute the supervised loss (5) for the b_1 instances
 # Approximating propagation terms
 Sample b_2 pairs of instances $\{(\mathbf{x}_i, \mathbf{x}_j)\}$
 Compute the outcome propagation terms $\lambda_{\mathrm{o}} L_{\mathrm{o}}$ for b_2 pairs of instances
 Sample b_2 pairs of instances $\{(\mathbf{x}_i, \mathbf{x}_j)\}$
 Compute the ITE propagation terms $\lambda_{\mathrm{e}} L_{\mathrm{e}}$ of b_2 pairs of instances
 Update the parameters to minimize $L_s + \lambda_o L_o + \lambda_e L_e$ for the sampled instances

end

the Adam optimizer [18], which is based on stochastic gradient descent to train the model in a mini-batch manner.

Algorithm 1 describes the procedure of model training, which iterates two steps until convergence. In the first step, we sample a mini-batch consisting of b_1 labeled instances to approximate the supervised loss (5). In the second step, we compute the outcome propagation term and the ITE propagation terms using a mini-batch consisting of b_2 instance pairs. Note that in order to make the model more flexible, we can employ different regularization parameters for the treatment outcomes and the control outcomes. The b_1 and b_2 are considered as hyper-parameters; the details are described in Sect. 4. In practice, we optimize only the supervised loss for the first several epochs, and decrease the strength of regularization as training proceeds, in order to guide efficient training [38].

4 Experiments

We test the effectiveness of the proposed semi-supervised ITE estimation method in comparison with various supervised methods, especially when the available labeled data are strictly limited. We first conduct experiments using two semi-synthetic datasets based on public real datasets. We also design some experiments varying the magnitude of noise on outcomes to explore how the noisy outcomes affect the proposed method. Our implementation is available on Github[1].

[1] https://github.com/SH1108/CounterfactualPropagation.

4.1 Datasets

Owing to the counterfactual nature of ITE estimation, we rarely access real-world datasets including ground truth ITEs, and therefore cannot directly evaluate ITE estimation methods like the standard supervised learning methods using cross-validation. Therefore, following the existing work [17], we employ two semi-synthetics datasets whose counterfactual outcomes are generated through simulations. Refer to the original papers for the details on outcome generations [14,17].

News Dataset is a dataset including opinions of media consumers for news articles [17]. It contains 5,000 news articles and outcomes generated from the NY Times corpus[2]. Each article is consumed on desktop ($t = 0$) or mobile ($t = 1$) and it is assumed that media consumers prefer to read some articles on mobile than desktop. Each article is generated by a topic model and represented in the bag-of-words representation. The size of the vocabulary is 3,477.

IHDP Dataset is a dataset created by randomized experiments called the Infant Health and Development Program (IHDP) [14] to examine the effect of special child care on future test scores. It contains the results of 747 subjects (139 treated subjects and 608 control subjects) with 25 covariates related to infants and their mothers. Following the existing studies [17,33], the ground-truth counterfactual outcomes are simulated using the NPCI package [10].

4.2 Experimental Settings

Since we are particularly interested in the situation when the available labeled data are strictly limited, we split the data into a training dataset, validation dataset, and a test dataset by limiting the size of the training data. We change the ratio of the training to investigate the performance; we use 10%, 5%, and 1% of the whole data from the News dataset, and use 40%, 20%, and 10% of those from the IHDP dataset for the training datasets. The rest 80% and 10% of the whole News data are used for test and validation, respectively. Similarly, 50% and 10% of the whole IHDP dataset are used for test and validation, respectively. We report the average results of 10 trials on the News dataset and 50 trials on the IHDP dataset.

In addition to the evaluation under labeled data scarcity, we also test the robustness against label noises. As pointed out in previous studies, noisy labels in training data can severely deteriorate predictive performance, especially in semi-supervised learning. Following the previous work [14,17], we add the noise $\epsilon \sim \mathcal{N}(0, c^2)$ to the observed outcomes in the training data, where $c \in \{1, 3, 5, 7, 9\}$. In this evaluation, we use 1% of the whole data as the training data for the News dataset and 10% for the IHDP dataset, respectively, since we are mainly interested in label-scarce situations.

[2] https://archive.ics.uci.edu/ml/datasets/Bag+of+Words.

The hyper-parameters are tuned based on the prediction loss using the observed outcomes on the validation data. We calculate the similarities between the instances by using the Gaussian kernel; we select σ^2 from $\{5 \times 10^{-3}, 1 \times 10^{-3}, \ldots, 1 \times 10^2, 5 \times 10^2\}$, and select λ_o and λ_e from $\{1 \times 10^{-3}, 1 \times 10^{-2}, \ldots, 1 \times 10^2\}$. Because the scales of treatment outcomes and control outcomes are not always the same, we found scaling the regularization terms according to them is beneficial; specifically, we scale the regularization terms with respect to the treatment outcomes, the control outcomes, and the treatment effects by $\alpha = 1/\sigma_{y^1}^2, \beta = 1/\sigma_{y^0}^2$, and $\gamma = 1/(\sigma_{y^1}^2 + \sigma_{y^0}^2)$, respectively. We apply principal component analysis to reduce the input dimensions before applying the Gaussian kernel; we select the number of dimensions from $\{2, 4, 6, 8, 16, 32, 64\}$. The learning rate is set to 1×10^{-3} and the mini-batch sizes b_1, b_2 are chosen from $\{4, 8, 16, 32\}$.

As the evaluation metrics, we report the *Precision in Estimation of Heterogeneous Effect (PEHE)* used in the previous research [14]. PEHE is the estimation error of individual treatment effects, and is defined as

$$\epsilon_{\text{PEHE}} = \frac{1}{N+M} \sum_{i=1}^{N+M} (\tau_i - \hat{\tau}_i)^2.$$

Following the previous studies [33,40], we evaluate the predictive performance for labeled instances and unlabeled instances separately. Note that, although we observe the factual outcomes of the labeled data, their true ITEs are still unknown because we cannot observe their counterfactual outcomes.

4.3 Baselines

We compare the proposed method with several existing supervised ITE estimation approaches. (i) Linear regression (Ridge, Lasso) is the ordinary linear regression models with ridge regularization or lasso regularization. We consider two variants: one that includes the treatment as a feature (denoted by 'Ridge-1' and 'Lasso-1'), and the other with two separated models for treatment and control (denoted by 'Ridge-2' and 'Lasso-2'). (ii) k-nearest neighbors (kNN) is a matching-based method that predicts the outcomes using nearby instances. (iii) Propensity score matching with logistic regression (PSM) [29] is a matching-based method using the propensity score estimated by a logistic regression model. We also compared the proposed method with tree models such as (iv) random forest (RF) [6] and its causal extension called (v) causal forest (CF) [37]. In CF, trees are trained to predict propensity score and leaves are used to predict treatment effects. (vi) TARNet [33] is a deep neural network model that has shared layers for representation learning and different layers for outcome prediction for treatment and control instances. (vii) Counterfactual regression (CFR) [33] is a state-of-the-art deep neural network model based on balanced representations between treatment and control instances. We use the Wasserstein distance.

Table 1. The performance comparison of different methods on News dataset. The † indicates that our proposed method (CP) performs statistically significantly better than the baselines by the paired t-test ($p < 0.05$). The bold results indicate the best results in terms of the average.

$\sqrt{\epsilon_{\text{PEHE}}}$	News 1%		News 5%		News 10%	
Method	Labeled	Unlabeled	Labeled	Unlabeled	Labeled	Unlabeled
Ridge-1	$^{\dagger}4.494_{\pm 1.116}$	$^{\dagger}4.304_{\pm 0.988}$	$^{\dagger}4.666_{\pm 1.0578}$	$^{\dagger}3.951_{\pm 0.954}$	$^{\dagger}4.464_{\pm 1.082}$	$^{\dagger}3.607_{\pm 0.943}$
Ridge-2	$2.914_{\pm 0.797}$	$^{\dagger}2.969_{\pm 0.814}$	$^{\dagger}2.519_{\pm 0.586}$	$^{\dagger}2.664_{\pm 0.614}$	$^{\dagger}2.560_{\pm 0.558}$	$^{\dagger}2.862_{\pm 0.621}$
Lasso-1	$^{\dagger}4.464_{\pm 1.082}$	$^{\dagger}3.607_{\pm 0.943}$	$^{\dagger}4.466_{\pm 1.058}$	$^{\dagger}3.367_{\pm 0.985}$	$^{\dagger}4.464_{\pm 1.0822}$	$^{\dagger}3.330_{\pm 0.984}$
Lasso-2	$^{\dagger}3.344_{\pm 1.022}$	$^{\dagger}3.476_{\pm 1.038}$	$^{\dagger}2.568_{\pm 0.714}$	$^{\dagger}2.848_{\pm 0.751}$	$^{\dagger}2.269_{\pm 0.628}$	$^{\dagger}2.616_{\pm 0.663}$
kNN	$^{\dagger}3.678_{\pm 1.250}$	$^{\dagger}3.677_{\pm 1.246}$	$^{\dagger}3.351_{\pm 1.004}$	$^{\dagger}3.434_{\pm 1.018}$	$^{\dagger}3.130_{\pm 0.752}$	$^{\dagger}3.294_{\pm 0.766}$
PSM	$^{\dagger}3.713_{\pm 1.149}$	$^{\dagger}3.662_{\pm 1.127}$	$^{\dagger}3.363_{\pm 0.901}$	$^{\dagger}3.500_{\pm 0.961}$	$^{\dagger}3.260_{\pm 0.734}$	$^{\dagger}3.526_{\pm 0.832}$
RF	$^{\dagger}4.494_{\pm 1.116}$	$^{\dagger}3.691_{\pm 0.878}$	$^{\dagger}4.466_{\pm 1.058}$	$^{\dagger}2.975_{\pm 0.874}$	$^{\dagger}4.464_{\pm 1.082}$	$^{\dagger}2.657_{\pm 0.682}$
CF	$^{\dagger}3.691_{\pm 1.082}$	$^{\dagger}3.607_{\pm 0.943}$	$^{\dagger}3.196_{\pm 0.901}$	$^{\dagger}3.215_{\pm 0.910}$	$^{\dagger}3.101_{\pm 0.806}$	$^{\dagger}3.129_{\pm 0.818}$
TARNET	$^{\dagger}3.166_{\pm 0.742}$	$^{\ddagger}3.160_{\pm 0.722}$	$^{\dagger}2.670_{\pm 0.796}$	$^{\dagger}2.666_{\pm 0.773}$	$^{\dagger}2.589_{\pm 0.894}$	$^{\dagger}2.598_{\pm 0.869}$
CFR	$2.908_{\pm 0.752}$	$2.925_{\pm 0.746}$	$^{\dagger}2.590_{\pm 0.772}$	$^{\dagger}2.546_{\pm 0.796}$	$^{\dagger}2.570_{\pm 0.519}$	$^{\dagger}2.451_{\pm 0.547}$
CP (proposed)	$\mathbf{2.844}_{\pm 0.683}$	$\mathbf{2.823}_{\pm 0.656}$	$\mathbf{2.310}_{\pm 0.430}$	$\mathbf{2.446}_{\pm 0.471}$	$\mathbf{2.003}_{\pm 0.393}$	$\mathbf{2.153}_{\pm 0.436}$

4.4 Results and Discussions

We discuss the performance of the proposed method compared with the baselines by changing the size of labeled datasets, and then investigate the robustness against the label noises.

We first see the experimental results for different sizes of labeled datasets and sensitivity to the choice of the hyper-parameters that control the strength of label propagation. Tables 1 and 2 show the PEHE values by different methods for the News dataset and the IHDP dataset, respectively. Overall, our proposed method exhibits the best ITE estimation performance for both labeled and unlabeled data in both of the datasets; the advantage is more significant in the News dataset. The News dataset is a relatively high-dimensional dataset represented using a bag of words. The two-model methods such as Ridge-2 and Lasso-2 perform well in spite of their simplicity, and in terms of regularization types, the Lasso-based methods perform relatively better due to the high-dimensional nature of the dataset.

The proposed method also performs the best in the IHDP dataset; however, the performance gain is rather moderate, as shown by the no statistical significance against CFR [33] with the largest 40%-labeled data, which is the most powerful baseline method. The reason for the moderate improvements is probably because of the difficulty in defining appropriate similarities among instances, because the IHDP dataset has various types of features including continuous variables and discrete variables. The traditional baselines such as Ridge-1, Lasso-1, k-NN matching, and the tree-based models show limited performance; in contrast, the deep learning based methods such as TARNet and CFR demonstrate remarkable performance. Generally, the performance gain by the proposed method is larger on labeled data than on unlabeled data.

Our proposed method has two different propagation terms, the outcome propagation term and the ITE propagation term, as regularizers for semi-supervised

Table 2. The performance comparison of different methods on IHDP dataset. The † indicates that our proposed method (CP) performs statistically significantly better than the baselines by the paired t-test ($p < 0.05$). The bold results indicate the best results in terms of average.

$\sqrt{\epsilon_{\text{PEHE}}}$	IHDP 10%		IHDP 20%		IHDP 40%	
Method	Labeled	Unlabeled	Labeled	Unlabeled	Labeled	Unlabeled
Ridge-1	$^\dagger 5.484_{\pm 8.825}$	$^\dagger 5.696_{\pm 7.328}$	$^\dagger 5.067_{\pm 8.337}$	$^\dagger 4.692_{\pm 6.943}$	$^\dagger 4.80_{\pm 8.022}$	$^\dagger 4.448_{\pm 6.874}$
Ridge-2	$^\dagger 3.426_{\pm 5.692}$	$^\dagger 3.357_{\pm 5.177}$	$^\dagger 2.918_{\pm 4.874}$	$^\dagger 2.918_{\pm 4.730}$	$^\dagger 2.605_{\pm 4.314}$	$^\dagger 2.639_{\pm 4.496}$
Lasso-1	$^\dagger 6.685_{\pm 10.655}$	$^\dagger 6.408_{\pm 9.900}$	$^\dagger 6.435_{\pm 10.147}$	$^\dagger 6.2446_{\pm 9.639}$	$^\dagger 6.338_{\pm 9.704}$	$^\dagger 6.223_{\pm 9.596}$
Lasso-2	$^\dagger 3.118_{\pm 5.204}$	$^\ddagger 3.292_{\pm 5.725}$	$^\dagger 2.684_{\pm 4.428}$	$^\dagger 2.789_{\pm 4.731}$	$^\dagger 2.512_{\pm 4.075}$	$^\dagger 2.571_{\pm 4.379}$
kNN	$^\dagger 4.457_{\pm 6.957}$	$^\dagger 4.603_{\pm 6.629}$	$^\dagger 4.023_{\pm 6.193}$	$^\dagger 4.370_{\pm 6.244}$	$^\dagger 3.623_{\pm 5.316}$	$^\dagger 4.109_{\pm 5.936}$
PSM	$^\dagger 6.506_{\pm 10.077}$	$^\dagger 6.982_{\pm 10.672}$	$^\dagger 6.277_{\pm 9.708}$	$^\dagger 7.209_{\pm 11.077}$	$^\dagger 6.065_{\pm 9.362}$	$^\dagger 7.181_{\pm 9.362}$
RF	$^\dagger 6.924_{\pm 10.620}$	$^\dagger 5.356_{\pm 8.790}$	$^\dagger 6.854_{\pm 10.471}$	$^\dagger 4.845_{\pm 8.241}$	$^\dagger 6.928_{\pm 10.396}$	$^\dagger 4.549_{\pm 7.822}$
CF	$^\dagger 5.389_{\pm 8.736}$	$^\dagger 5.255_{\pm 8.070}$	$^\dagger 4.939_{\pm 7.762}$	$^\dagger 4.955_{\pm 7.503}$	$^\dagger 4.611_{\pm 7.149}$	$^\dagger 4.764_{\pm 7.448}$
TARNET	$^\dagger 3.827_{\pm 5.315}$	$^\dagger 3.664_{\pm 4.888}$	$^\dagger 2.770_{\pm 3.617}$	$^\dagger 2.770_{\pm 3.542}$	$^\dagger 2.005_{\pm 2.447}$	$^\dagger 2.267_{\pm 2.825}$
CFR	$^\dagger 3.461_{\pm 5.1444}$	$^\dagger 3.292_{\pm 4.619}$	$^\dagger 2.381_{\pm 3.126}$	$^\dagger 2.403_{\pm 3.080}$	$1.572_{\pm 1.937}$	$1.815_{\pm 2.204}$
CP (proposed)	$\mathbf{2.427}_{\pm 3.189}$	$\mathbf{2.652}_{\pm 3.469}$	$\mathbf{1.686}_{\pm 1.838}$	$\mathbf{1.961}_{\pm 2.343}$	$\mathbf{1.299}_{\pm 1.001}$	$\mathbf{1.485}_{\pm 1.433}$

learning. Table 3 investigates the contributions by the different propagation terms. The proposed method using the both propagation terms (denoted by CP) shows better results than the one only with the ITE propagation denoted by CP ($\lambda_o = 0$); on the other hand, the improvement over the one only with the outcome regularization is marginal. This observation implies the outcome propagation contributes more to the predictive performance than the ITE propagation.

We also examine the sensitivity of the performance to the regularization hyper-parameters. Figure 2 reports the results using 40% and 10% of the whole data as the training data of the News and IHDP datasets, respectively. The proposed method seems rather sensitive to the strength of the regularization terms, particularly on the IHDP dataset, which suggests that the regularization parameters should be carefully tuned using validation datasets in the proposed method. In our experimental observations, slight changes in the hyper-parameters sometimes caused significant changes of predictive performance. We admit the hyper-parameter sensitivity is one of the current limitations in the proposed method and efficient tuning of the hyper-parameters should be addressed in future.

Finally, we compare the proposed method with the state-of-the-art methods by varying the magnitude of noises added to the outcomes. Fig 3 shows the performance comparison in terms of $\sqrt{\epsilon_{\text{PEHE}}}$. Note that the results when $c = 1$ correspond to the previous results in Tables 1 and 2 . The proposed method stays tolerant of relatively small magnitude of noises; however, with larger label noises, it suffers more from wrongly propagated outcome information than the baselines. This is consistent with the previous studies reporting the vulnerability of semi-supervised learning methods against label noises [7,11,23,35].

Table 3. Investigation of the contributions by the outcome propagation and the ITE propagation in the proposed method. The upper table shows the results for the News dataset, and the lower for the IHDP dataset. The $\lambda_o = 0$ and $\lambda_e = 0$ indicate the proposed method (CP) with only the ITE propagation and the outcome propagation, respectively, The † indicates that our proposed method (CP) performs statistically significantly better than the baselines by the paired t-test ($p < 0.05$). The bold numbers indicate the best results in terms of the average.

$\sqrt{\epsilon_{PEHE}}$	News 1%		News 5%		News 10%	
Method	Labeled	Unlabeled	Labeled	Unlabeled	Labeled	Unlabeled
CP ($\lambda_o = 0$)	**2.812**$\pm$**.651**	**2.806**$\pm$**0.598**	†2.527$\pm$0.474	†2.531$\pm$0.523	†2.400$\pm$0.347	†2.410$\pm$0.450
CP ($\lambda_e = 0$)	2.879$\pm$0667	2.885$\pm$0.609	2.351$\pm$0.450	2.483$\pm$0.481	**1.996**$\pm$**0.338**	2.221$\pm$0.455
CP	2.844$\pm$0.683	2.823$\pm$0.656	**2.310**$\pm$**0.430**	**2.446**$\pm$**0.471**	2.003$\pm$0.393	**2.153**$\pm$**0.436**
$\sqrt{\epsilon_{PEHE}}$	IHDP 10%		IHDP 20%		IHDP 40%	
Method	Labeled	Unlabeled	Labeled	Unlabeled	Labeled	Unlabeled
CP ($\lambda_o = 0$)	†2.883$\pm$3.708	†3.004$\pm$4.071	†1.972$\pm$1.930	†2.144$\pm$2.465	1.574$\pm$1.392	1.674$\pm$1.874
CP ($\lambda_e = 0$)	2.494$\pm$3.201	2.698$\pm$3.461	1.728$\pm$2.194	1.977$\pm$2.450	1.344$\pm$1.383	1.585$\pm$1.923
CP	**2.427**$\pm$**3.189**	**2.652**$\pm$**3.469**	**1.686**$\pm$**1.838**	**1.961**$\pm$**2.343**	**1.299**$\pm$**1.001**	**1.485**$\pm$**1.433**

5 Related Work

5.1 Treatment Effect Estimation

Treatment effect estimation has been one of the major interests in causal inference and widely studied in various domains. Matching [1,31] is one of the most basic and commonly used treatment effect estimation techniques. It estimates the counterfactual outcomes using its nearby instances, whose idea is similar to that of graph-based semi-supervised learning. Both methods assume that similar instances in terms of covariates have similar outcomes. To mitigate the curse of dimensionality and selection bias in matching, the propensity matching method relying on the one-dimensional propensity score was proposed [29,30]. The propensity score is the probability of an instance to get a treatment, which is modeled using probabilistic models like logistic regression, and has been successfully applied in various domains to estimate treatment effects unbiasedly [24]. Tree-based methods such as regression trees and random forests have also been well studied for this problem [9,37]. One of the advantages of such models is that they can build quite expressive and flexible models to estimate treatment effects. Recently, deep learning-based methods have been successfully applied to treatment effect estimation [17,33]. Balancing neural networks (BNNs) [17] aim to obtain balanced representations of a treatment groups and a control group by minimizing the discrepancy between them, such as the Wasserstein distance [33]. Most recently, some studies have addressed causal inference problems on network-structured data [2,13,36]. Alvari et al. applied the idea of manifold regularization using users activities as causality-based features to detect harmful users in social media [2]. Guo et al. considered treatment effect estimation on social networks using graph convolutional balancing neural networks [13]. In contrast with their work assuming the network structures are readily available, we do not assume them and considers matching network defined using covariates.

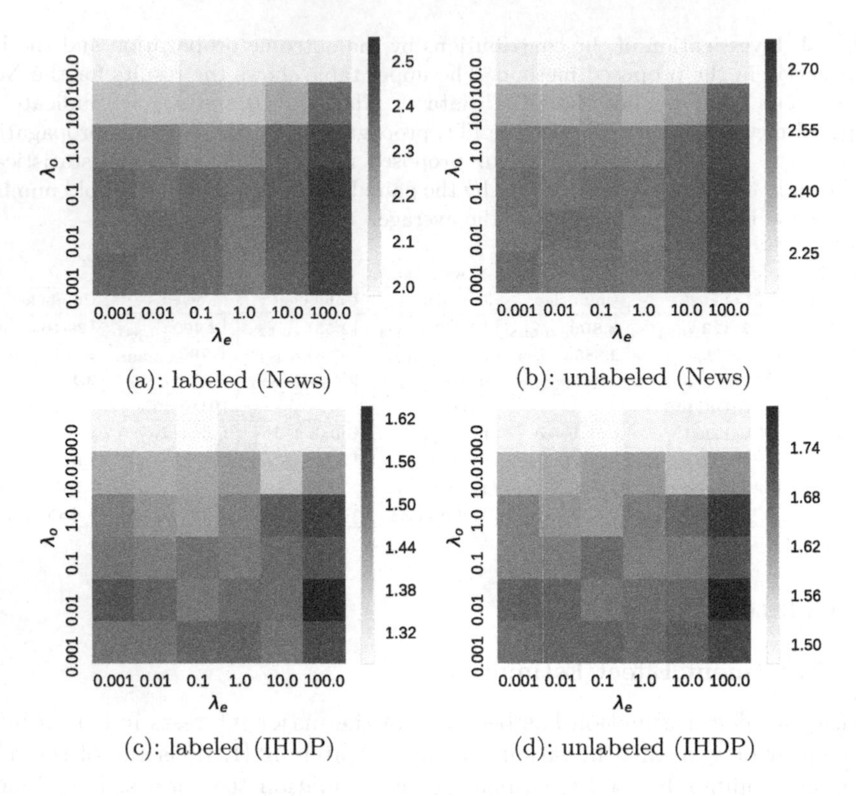

Fig. 2. Sensitivity of the results to the hyper-parameters. The colored bars indicate $\sqrt{\epsilon_{\text{PEHE}}}$ for (a)(b) the News dataset and (c)(d) the IHDP dataset when using the largest size of labeled data. The deeper-depth color indicates larger errors. It is observed that the proposed method is somewhat sensitive to the choice of hyper-parameters, especially, the strength of outcome regularizations (λ_o). (Color figure online)

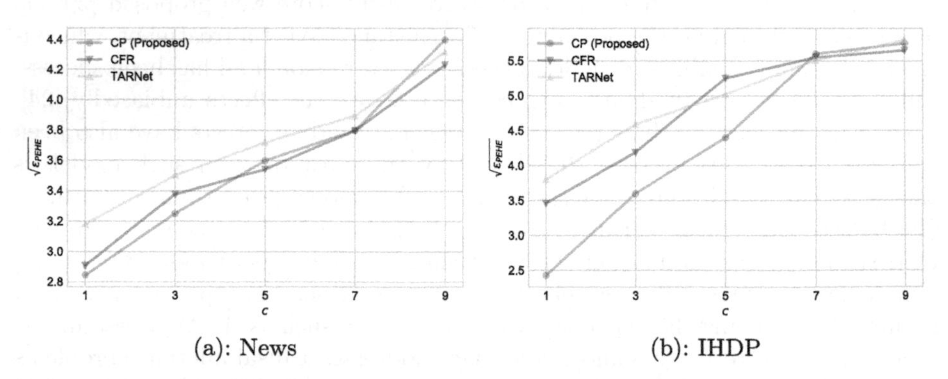

Fig. 3. Performance comparisons for different levels of noise c added to the labels on (a) News dataset and (b) IHDP dataset. Note that the results when $c = 1$ correspond to the previous resuls (Tables 1 and 2).

5.2 Semi-supervised Learning

Semi-supervised learning, which exploits both labeled and unlabeled data, is one of the most popular approaches, especially in scenarios when only limited labeled data can be accessed [5,15]. Semi-supervised learning has many variants, and because it is almost impossible to refer to all of them, we mainly review the graph-based regularization methods, known as label propagation or manifold regularization [4,38,42]. Utilizing a given graph or a graph constructed based on instance proximity, graph-based regularization encourages the neighbor instances to have similar labels or outcomes [4,42]. Such idea is also applied to representation learning in deep neural networks [7,16,19,38,39]; they encourage nearby instances not only to have similar outcomes, but also have similar intermediate representations, which results in remarkable improvements from ordinary methods. One of the major drawbacks of semi-supervised approaches is that label noises in training data can be quite harmful; therefore, a number of studies managed to mitigate the performance degradation [11,23,25,35].

One of the most related work to our present study is graph-based semi-supervised prediction under sampling biases of labeled data [41]. The important difference between this work and ours is that they do not consider intervention and we do not consider the sampling biases of labeled data.

6 Conclusion

We addressed the semi-supervised ITE estimation problem. In comparison to the existing ITE estimation methods that only rely on labeled instances including treatment and outcome information, we proposed a novel semi-supervised ITE estimation method that also utilizes unlabeled instances. The proposed method called counterfactual propagation is built on two ideas from causal inference and semi-supervised learning, namely, matching and label propagation, respectively; accordingly, we devised an efficient learning algorithm. Experimental results using the semi-simulated real-world datasets revealed that our methods performed better in comparison to several strong baselines when the available labeled instances are limited. However, this method had issues related to reasonable similarity design and hyper-parameter tuning.

One of the possible future directions is to make use of balancing techniques such as the one used in CFR [33], which can be also naturally integrated into our model. Our future work also includes addressing the biased distribution of labeled instances. As mentioned in Related work, we did not consider such sampling biases for labeled data. Some debiasing techniques [41] might also be successfully integrated into our framework. In addition, robustness against noisy outcomes under semi-supervised learning framework is still the open problem and will be addressed in the future.

Acknowledgments. This work was partially supported by JSPS KAKENHI Grant Number 20H04244.

References

1. Abadie, A., Imbens, G.W.: Large sample properties of matching estimators for average treatment effects. Econometrica **74**(1), 235–267 (2006)
2. Alvari, H., Shaabani, E., Sarkar, S., Beigi, G., Shakarian, P.: Less is more: semi-supervised causal inference for detecting pathogenic users in social media. In: Proceedings of the 2019 World Wide Web Conference (WWW), pp. 154–161 (2019)
3. Baiocchi, M., Cheng, J., Small, D.S.: Instrumental variable methods for causal inference. Stat. Med. **33**(13), 2297–2340 (2014)
4. Belkin, M., Niyogi, P., Sindhwani, V.: Manifold regularization: a geometric framework for learning from labeled and unlabeled examples. J. Machine Learn. Res. **7**, 2399–2434 (2006)
5. Bengio, Y., Lamblin, P., Popovici, D., Larochelle, H.: Greedy layer-wise training of deep networks. In: Advances in Neural Information Processing Systems (NeurIPS), pp. 153–160 (2007)
6. Breiman, L.: Random forests. Machine Learn. **45**(1), 5–32 (2001)
7. Bui, T.D., Ravi, S., Ramavajjala, V.: Neural graph learning: training neural networks using graphs. In: Proceedings of the 11th ACM International Conference on Web Search and Data Mining (WSDM), pp. 64–71 (2018)
8. Chan, D., Ge, R., Gershony, O., Hesterberg, T., Lambert, D.: Evaluating online ad campaigns in a pipeline: causal models at scale. In: Proceedings of the 16th ACM SIGKDD International Conference on Knowledge Discovery and Data Mining (KDD), pp. 7–16 (2010)
9. Chipman, H.A., George, E.I., McCulloch, R.E., et al.: Bart: Bayesian additive regression trees. Ann. Appl. Stat. **4**(1), 266–298 (2010)
10. Dorie, V.: NPCI: Non-parametrics for causal inference. https://www.github.com/vdorie/npci (2016)
11. Du, B., Xinyao, T., Wang, Z., Zhang, L., Tao, D.: Robust graph-based semisupervised learning for noisy labeled data via maximum correntropy criterion. IEEE Trans. Cyber. **49**(4), 1440–1453 (2018)
12. Glass, T.A., Goodman, S.N., Hernán, M.A., Samet, J.M.: Causal inference in public health. Annual Rev. Public Health **34**, 61–75 (2013)
13. Guo, R., Li, J., Liu, H.: Learning individual causal effects from networked observational data. In: Proceedings of the 13th International Conference on Web Search and Data Mining (WSDM), pp. 232–240 (2020)
14. Hill, J.L.: Bayesian nonparametric modeling for causal inference. J. Comput. Graph. Stat. **20**(1), 217–240 (2011)
15. Hinton, G.E., Osindero, S., Teh, Y.W.: A fast learning algorithm for deep belief nets. Neural Comput. **18**(7), 1527–1554 (2006)
16. Iscen, A., Tolias, G., Avrithis, Y., Chum, O.: Label propagation for deep semi-supervised learning. In: Proceedings of the IEEE Conference on Computer Vision and Pattern Recognition (CVPR), pp. 5070–5079 (2019)
17. Johansson, F., Shalit, U., Sontag, D.: Learning representations for counterfactual inference. In: Proceedings of the 33rd International Conference on Machine Learning (ICML), pp. 3020–3029 (2016)
18. Kingma, D.P., Ba, J.: Adam: A method for stochastic optimization. arXiv preprint arXiv:1412.6980 (2014)
19. Kipf, T.N., Welling, M.: Semi-supervised classification with graph convolutional networks. arXiv preprint arXiv:1609.02907 (2016)

20. LaLonde, R.J.: Evaluating the econometric evaluations of training programs with experimental data. The American Economic Review, pp. 604–620 (1986)
21. Lewis, D.: Causation. J. Philosophy **70**(17), 556–567 (1974)
22. Li, S., Vlassis, N., Kawale, J., Fu, Y.: Matching via dimensionality reduction for estimation of treatment effects in digital marketing campaigns. In: Proceedings of the 25th International Joint Conference on Artificial Intelligence (IJCAI), pp. 3768–3774 (2016)
23. Liu, W., Wang, J., Chang, S.F.: Robust and scalable graph-based semisupervised learning. Proc. IEEE **100**(9), 2624–2638 (2012)
24. Lunceford, J.K., Davidian, M.: Stratification and weighting via the propensity score in estimation of causal treatment effects: a comparative study. Stat. Med. **23**(19), 2937–2960 (2004)
25. Pal, A., Chakrabarti, D.: Label propagation with neural networks. In: Proceedings of the 27th ACM International Conference on Information and Knowledge Management (CIKM), pp. 1671–1674 (2018)
26. Pearl, J.: Causality. Cambridge University Press (2009)
27. Pombo, N., Garcia, N., Bousson, K., Felizardo, V.: Machine learning approaches to automated medical decision support systems. In: Handbook of Research on Artificial Intelligence Techniques and Algorithms, pp. 183–203. IGI Global (2015)
28. Radlinski, F., Joachims, T.: Query chains: learning to rank from implicit feedback. In: Proceedings of the 11th ACM SIGKDD International Conference on Knowledge Discovery in Data Mining (KDD), pp. 239–248. ACM (2005)
29. Rosenbaum, P.R., Rubin, D.B.: The central role of the propensity score in observational studies for causal effects. Biometrika **70**(1), 41–55 (1983)
30. Rosenbaum, P.R., Rubin, D.B.: Constructing a control group using multivariate matched sampling methods that incorporate the propensity score. Am. Stat. **39**(1), 33–38 (1985)
31. Rubin, D.B.: Matching to remove bias in observational studies. Biometrics, pp. 159–183 (1973)
32. Rubin, D.B.: Estimating causal effects of treatments in randomized and nonrandomized studies. J. Educ. Psychol. **66**(5), 688 (1974)
33. Shalit, U., Johansson, F.D., Sontag, D.: Estimating individual treatment effect: generalization bounds and algorithms. In: Proceedings of the 34th International Conference on Machine Learning (ICML), pp. 3076–3085. JMLR.org (2017)
34. Splawa-Neyman, J., Dabrowska, D.M., Speed, T.: On the application of probability theory to agricultural experiments. essay on principles. Section 9. Statistical Science, pp. 465–472 (1990)
35. Vahdat, A.: Toward robustness against label noise in training deep discriminative neural networks. In: Advances in Neural Information Processing Systems (NeurIPS), pp. 5596–5605 (2017)
36. Veitch, V., Wang, Y., Blei, D.: Using embeddings to correct for unobserved confounding in networks. In: Advances in Neural Information Processing Systems (NeurIPS), pp. 13769–13779 (2019)
37. Wager, S., Athey, S.: Estimation and inference of heterogeneous treatment effects using random forests. J. Am. Stat. Assoc. **113**(523), 1228–1242 (2018)
38. Weston, J., Ratle, F., Mobahi, H., Collobert, R.: Deep learning via semi-supervised embedding. In: Montavon, G., Orr, G.B., Müller, K.-R. (eds.) Neural Networks: Tricks of the Trade. LNCS, vol. 7700, pp. 639–655. Springer, Heidelberg (2012). https://doi.org/10.1007/978-3-642-35289-8_34
39. Yang, Z., Cohen, W.W., Salakhutdinov, R.: Revisiting semi-supervised learning with graph embeddings. arXiv preprint arXiv:1603.08861 (2016)

40. Yao, L., Li, S., Li, Y., Huai, M., Gao, J., Zhang, A.: Representation learning for treatment effect estimation from observational data. In: Advances in Neural Information Processing Systems (NeurIPS), pp. 2633–2643 (2018)
41. Zhou, F., Li, T., Zhou, H., Zhu, H., Jieping, Y.: Graph-based semi-supervised learning with non-ignorable non-response. In: Advances in Neural Information Processing Systems (NeurIPS), pp. 7013–7023 (2019)
42. Zhu, X., Ghahramani, Z., Lafferty, J.D.: Semi-supervised learning using gaussian fields and harmonic functions. In: Proceedings of the 20th International conference on Machine learning (ICML), pp. 912–919 (2003)

(Spatio-)Temporal Data and Recurrent Neural Networks

Real-Time Fine-Grained Freeway Traffic State Estimation Under Sparse Observation

Yangxin Lin[1], Yang Zhou[2], Shengyue Yao[3], Fan Ding[4], and Ping Wang[1,5,6](✉)

[1] School of Software and Microelectronic, Peking University, Beijing, China
[2] Department of Civil and Environmental Engineering, University of Wisconsin, Madison, USA
[3] Department of Transportation and Urban Engineering, Technical University of Braunschweig, Braunschweig, Germany
[4] School of Transportation, Southeast University, Nanjing, China
[5] National Engineering Research Center for Software Engineering, Peking University, Beijing, China
[6] Ministry of Education, Key Laboratory of High Confidence Software Technologies (PKU), Beijing, China
pwang@pku.edu.cn

Abstract. Obtaining sufficient traffic state (e.g. traffic flow, density, and speed) data is critical for effective traffic operation and control. Especially for emerging advanced traffic applications, fine-grained traffic state estimation is non-trivial. With the development of advanced sensing and communication technology, connected vehicles provide unprecedented opportunities to sense traffic state and change current estimation methods. However, due to the low penetration rate of connected vehicles, traditional traffic state estimation methods do not work well under fine-grained requirements. To overcome such a problem, a probabilistic approach to estimate fine-grained traffic state of freeway under sparse observation is proposed in this paper. Specifically, we propose Residual Attention Conditional Neural Process (RA-CNP), which is an approximation of Gaussian Processes Regression (GPR) using neural network, to model spatiotemporally varying traffic states. The method can comprehensively extract both constant spatial-temporal and dynamic traffic state dependency from sparse data and have better estimation accuracy. Besides, the proposed method has less computational cost compared with traditional GPR, which makes it applicable to real-time traffic estimation applications. Extensive experiments using real-world traffic data show that the proposed method provides lower estimation error and more reliable results than other traditional traffic estimation methods under sparse observation.

Keywords: Traffic state estimation · Conditional neural process · Sparse observation

F. Hutter et al. (Eds.): ECML PKDD 2020, LNAI 12457, pp. 561–577, 2021.
https://doi.org/10.1007/978-3-030-67658-2_32

1 Introduction

Traffic state estimation, which refers to inference traffic state variables (e.g. traffic density, flow, and speed) using partial observation [21], plays an important role in traffic operations and management, especially under congested traffic conditions. During the past decades, various traffic state estimation methods have been proposed and widely used in transportation systems.

Most traditional traffic state estimation methods highly rely on traffic state data collected from local fixed traffic sensors (e.g. loop detector, video camera) [31]. However, due to financial reasons, fixed sensors cannot be widely deployed [1,2], which results in a blind area (typically greater than 1000 m for freeway) between two fixed sensors. The estimation accuracy and granularity cannot satisfy the demand of advanced transportation applications, such as dynamic traffic control and vehicle automation [23,36]. In order to achieve better performance, those advanced applications usually exhibit fine-grained traffic estimation requirements (e.g. traffic states of a small range of a specific traffic lane (e.g. less than 100 m) during a few seconds [11]). For instance, variable speed limit control in the freeway, which is used to harmonize vehicle speed to mitigate traffic breakdown, highly relies on the estimation of downstream traffic states to make a reliable control [8]. Additionally, automated vehicles need a reliable downstream traffic states estimation to avoid unnecessary maneuvers [37].

Recent advances in connected vehicle technology make vehicles become the mobile traffic sensor using on-board sensors, and provide unprecedented opportunities to collect traffic data in a cooperative manner via Vehicle to Infrastructure (V2I) communication [30]. The beauty of such a connected environment lies in its ability to sense and integrate spatial-temporally correlated data (e.g. trajectory, spacing, and speed) to facilitate advanced transportation applications. However, it is not practical to have a fully connected environment in the very near future, due to the low penetration of connected vehicles, which brings enormous challenges to the fine-grained traffic state estimation. The difficulties lie in the following aspects.

First, fine-grained traffic state estimation suffers from sparse observation. Although connected vehicles can provide detailed trajectory data, the wide adoption of connected vehicles is still decades away. With sparsely observed traffic state data, some estimation approaches (e.g. LSTM [34], DCRNN [14], GaAN [33], and STGRAT [17]) are not directly applicable with incomplete historical data. Other approaches (e.g. linear regression [35], tensor-based methods [29]) always lead to huge estimation errors due to limited information. Therefore, most existing methods do not work well in fine-grained estimation.

Second, the stochasticity in fine-grained traffic state estimation is non-trivial. With the decrement of estimation granularity, drivers' behavioral stochasticities [37] will dominate the fine-grained traffic state. The stochasticity of the traffic state needs to be better described, otherwise, it may result in the estimation error. However, most of the existing methods do not consider the stochasticity in modeling. To address the stochasticity issue, apart from an expectation value, the variance of the estimated traffic state is vital to be provided. In addition, the variance can represent the estimation reliability, which is essential for traffic and vehicle automation robust control [36].

In recent works, Gaussian Processes Regression (GPR) [18] liked methods are proposed to address the aforementioned sparseness and stochasticity problems in many similar fields [4,24]. Among those methods, the Conditional Neural Process (CNP), which combines neural networks with features reminiscent of Gaussian processes, has great potential for regression and data completion [7]. However, those GPR-liked methods have the problem of constant spatial-temporal dependency, which is inconsistent with the actual traffic situation. In an actual traffic scenario, the dependency among traffic states is time-dependent and related to the traffic condition (e.g. congested or not). Therefore, the constant dependency issue will also lead to an estimation error. How to achieve a fine-grained traffic state estimation especially under sparse observation is still not fully studied.

In this work, we propose a novel method, Residual Attention CNP (RA-CNP), for fine-grained freeway traffic state estimation under sparse observation. Unlike aforementioned GPR-liked methods that only consider the spatial-temporal relationship among traffic states, RA-CNP involves the residual attention module to extend the traditional CNP model and uses both spatial-temporal and traffic state attention for efficiently capturing time-dependent dependencies within freeway from sparsely observed data. Besides, the proposed method has less computational consumption compared with the traditional GPR, which makes it applicable to real-time traffic estimation applications. Lastly, we demonstrate the applicability of the proposed method on a real freeway dataset. The experiment shows that the proposed model has a lower estimation error than existing models, especially under sparse observation.

The rest of the paper is organized as follows: Section 2 reviews related works. Section 3 formulizes a general problem of fine-grained traffic state estimation. Section 4 introduces the RA-CNP model for fine-grained traffic state estimation. Section 5 adopts the actual dataset to evaluate the performance of the proposed method in the freeway. Section 6 concludes the research and points out the development direction.

2 Related Works

2.1 Traffic State Estimation Methods

Traffic state estimation methods are grouped into two categories, namely, model-driven and data-driven methods [1].

Model-driven methods utilize the prior knowledge to make a specific assumption about the traffic flow model and estimate missing traffic state using partial observation. Limited by modeling complexity, most existing traffic flow models are simplified to some degree. For instance, Cell Transmission Model (CTM) utilizes a linear method to model the dynamic of traffic flow in both free and congested traffic [3]; other studies combined CTM and Kalman Filter (KF) to estimate traffic states [1,15]. However, these simplified models are incapable of capturing the complex traffic characteristics and always lead to great estimation errors under sparse observation.

Data-driven methods extensively rely on historical-data, instead of physically interpreted traffic flow models. In general, data-driven methods perform better due to the flexible model structure and fit advances. Especially, tensor-based and NN-based methods are commonly used in research. The tensor-based method converts the traffic state estimation problem to the tensor completion problem [25,29]. However, with the percentage of missing data increases, tensor-based methods perform worse without the prior knowledge of traffic characteristics. The NN-based method can automatically extract the dependency among various traffic features without any artificial parameters [27], but it usually cannot handle the issue of missing data under sparse observation [14,17,33,34]. More recently, with the developments in the Gaussian process regression, researchers find that the Bayesian formulation and explicit probabilistic output of Gaussian process regression make it well suited for handling sparseness and stochasticity issues meet in traffic state estimation scenario. Gaussian processes regression has been shown to outperform traditional approaches in various applications, such as traffic speed estimation [20], traffic volume estimation [28], and travel time estimation [12].

2.2 Gaussian Process and Conditional Neural Process

Gaussian process regression model is a fully probabilistic model, which works well on sparse datasets and could output the reliability measurement (i.e. estimation variance) [18]. The probabilistic regression maps input $x_i \in R^{d_x}$ to an output $y_i \in R^{d_y}$ according to $y = f(x)+\epsilon$, where $\epsilon \in N(0, \sigma)$ is Gaussian noise, and $f(x)$ satisfies Gaussian processes prior. The Gaussian processes prior is formulated as $f(x) \sim GP(m(x), k(x, x'))$ using a mean function $m(x)$ and kernel function $k(x, x')$ [13].

For the estimation problem with the observation set O, $(x_O, y_O) = \{(x_i, y_i)\}_{i \in O}$, and target set T, $(x_T, y_T) = \{(x_i, y_i)\}_{i \in T}$, which needs to be estimated, the posterior distribution of points at x_T could be expressed as:

$$p(y_T|x_T, x_O, y_O) = \int p(y_T|x_T, f)p(f|x_O, y_O)df. \tag{1}$$

Although the Gaussian assumption mentioned above makes sure that the posterior distribution is tractable, Gaussian process regression is computationally expensive (the complexity is $O((m + n)^3)$ for making m estimation with n observed data) due to the limitation of kernel functions. Besides, its performance highly relies on the selection of mean and kernel functions [19], which also limits its applications.

Conditional Neural Process (CNP) is a conditional distribution over functions trained to model the empirical conditional distributions of Gaussian processes. CNP gives up the mathematical guarantees associated with Gaussian processes, trading this off for functional flexibility and scalability [7]. The posterior distribution in (1) is approximated using the following expression in CNP:

$$p(y_T|x_T, x_O, y_O) \approx p(y_T|x_T, r_T) \tag{2}$$

where the latent representation vector r_T is used to parameterize the Gaussian processes prior f; both r_T and $p(y_T|x_T, r_T)$ could be calculated using neural networks. Therefore, in CNP, learning a function is fast and uncertainty-aware, while the estimation at test time is still efficient [22]. This also makes related methods cover a broad range of problems from learning RL agents to increasingly more challenging tasks to modeling dynamic 3D scenes [5].

2.3 Attention Mechanisms

An attention function can be described as mapping a query and a set of key-value pairs to an output, where the query, keys, values, and output are all vectors [26]. The output aggregates the weighted sum of the input values. The weight assigned to each value vector is calculated according to a similarity function of the query with the corresponding key [26]. In recent years, attention mechanisms have been widely applied in various areas, such as natural language processing and image recognition.

The typical implementation of attention mechanisms including dot-product attention, multi-head attention [32]. In dot-product attention, queries, keys, and values are packed together into matrix $Q \in R^{m \times d_k}$, $K \in R^{n \times d_k}$, and $V \in R^{n \times d_v}$ respectively [26]. The attention could be calculated as:

$$DotProduct(Q, K, V) = softmax((QK^T)/\sqrt{d_k})V, \tag{3}$$

where $\sqrt{d_k}$ is a scaling factor. Moreover, multi-head attention is an extension of dot-product attention. It uses multiple parallel dot-product attention modules to extract more latent dependencies.

3 Problem Statement

In our work, the fine-grained traffic estimation problem focuses on estimating the traffic state of a specific traffic lane in the freeway. Given a traffic lane (as shown in Fig. 1), we evenly partition it into M sub-segments. Within an observation window with N equal time segments, the time-variant traffic state of sub-segments could be denoted by a $M \times N$ matrix of *spatial-temporal cells*. Each cell could be represented by a tuple (x, y), where $x = (m, n)$ is the index of the cell in the matrix, y is the traffic state of the cell. In this paper, we focus on the estimation of traffic flow q, density k, and speed v of cells (i.e. $y = (q, k, v)$), which is widely applied in traffic engineering.

Given the observation set O, $(x_O, y_O) = \{(x_i, y_i)\}_{i \in O}$, which is cooperatively collected from connected vehicles, and target set T, $(x_T, y_T) = \{(x_i, y_i)\}_{i \in T}$, the fine-grained traffic state estimation problem is equivalent to solving the following optimization problem:

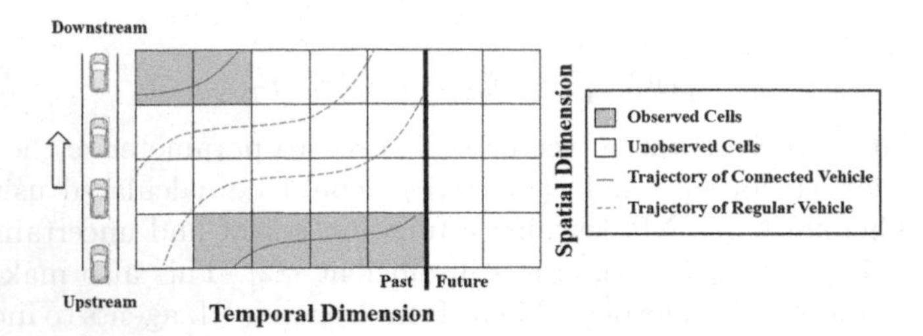

Fig. 1. Typical scenario of fine-grained traffic state estimation. Given a traffic lane, the temporal-spatial space of its traffic state is divided into $M \times N$ cells. There are 4 vehicles on the lane. The first and last vehicles are connected vehicles, who can collect surrounding traffic state (blue cells). White cells' traffic state need to be estimated. (Color figure online)

$$y_T = argmax_{y_T} p(y_T | x_T, x_O, y_O). \tag{4}$$

where $p(\cdot)$ denotes the probability density function.

Figure 1 provides us an illustrative example of fine-grained traffic state estimation. Figure 1 demonstrates the space-time profile of vehicles under a typical traffic scenario, which consists of a consecutive sequence of vehicles. The first and last vehicles are connected vehicles, whose trajectories are denoted by solid curves. Others are regular vehicles, whose trajectories are represented by dashed curves. With the help of the on-board sensors, connected vehicles could detect the traffic state of spatial-temporal cells they passed by (in blue). Our goal is to estimate the traffic state of unobserved cells (in white) under limited observation.

4 Proposed Method

In this section, we will first introduce the basic CNP method for fine-grained traffic state estimation and analyze the reason why CNP cannot fit the traffic field. Then, RA-CNP is proposed to further extract comprehensive dependency on the basic of CNP, in order to improve the estimation accuracy.

4.1 Basic CNP for Estimation

The fine-grained traffic state estimation problem could be solved by CNP directly. The basic CNP adopts an encoder-decoder architecture to implement the above posterior distribution function in (2). The encoder is designed to extracts latent representation vector r_T from the observed data,

$$r_T = Encoder(x_T, x_O, y_O). \tag{5}$$

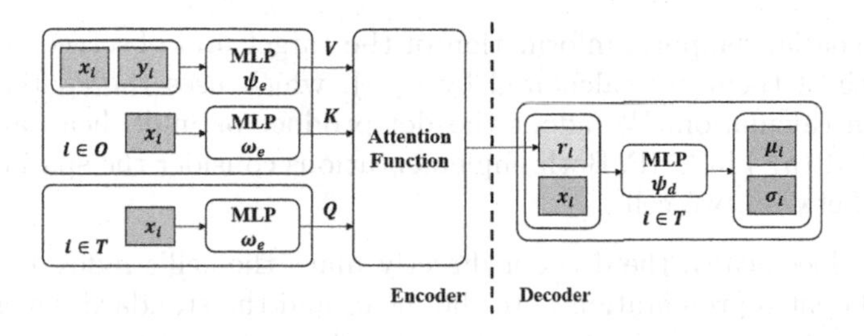

Fig. 2. Structure of CNP model. O and T denotes the observed and target dataset respectively; V, K, and Q denotes value, key and query input of attention function; latent representation r_j is calculated using similarity between K and Q, in order to estimate the mean μ_j and standard deviation σ_j of traffic state.

The decoder outputs the conditional distribution of target data y_T, which is assumed to be a Gaussian random variable $N(y_T; \mu_T, \sigma_T^2)$, given the latent representation r_T,

$$p(y_T|x_T, x_O, y_O) = N(y_T; \mu_T, \sigma_T^2), with\,(\mu_T, \sigma_T) = Decoder(x_T, r_T). \quad (6)$$

The detailed structure of CNP-based estimation is shown in Fig. 2.

Encoder. The encoder outputs the latent representation r_T, which parameterizes the Gaussian processes prior of the traffic state. As we have known, the kernel function in the Gaussian process calculates the covariance between data, which is critical for the representation of Gaussian processes prior. In CNP, the encoder implements such a process using attention mechanism, which adopts key-query similarity (i.e. dependency) to approximate the covariance between data.

To be specific, the encoder can be formulized as the following expressions:

$$r_T = \{r_i\}_{i \in T} = Attn\begin{pmatrix} Q = \{\omega_e(x_i)\}_{i \in T}, \\ K = \{\omega_e(x_i)\}_{i \in O}, \\ V = \{\psi_e(x_i, y_i)\}_{i \in O} \end{pmatrix} \quad (7)$$

where $Attn(\cdot)$ is the attention function. Q, K, and V denotes the queries, keys, and values input for attention function respectively; both ω_e and ψ_e are Multi-layer Perceptron (MLP) [9] functions.

For the attention function in the encoder, the input V is calculated by $\psi_e(\cdot)$, which is designed to extract the latent representation of Gaussian processes prior from every observation input pair $(x_i, y_i)_{i \in O}$ independently. According to the attention mechanism, the weighted sum of V is r_T. The weights of representations are calculated according to the similarity between Q and K. From the perspective of fine-grained traffic state estimation, Q and K represents the extracted spatial-temporal information of the target and observed cells respectively. Both of them are calculated by $\omega_e(\cdot)$, which uses the spatial-temporal index x_i in calculation. We adopt the dot-product or multi-head attention to implement $Attn(\cdot)$ in CNP. Both implementations consider the spatial-temporal similarity between two cells.

Decoder. Specifically, the decoder directly maps the cell's index x_i and corresponding latent representation r_i to the mean and the standard deviation of y_i. Considering the input pairs (x_i, r_i) has already contained the critical information for estimation, we adopt a simple implementation, MLP, in the decoder. The decoder could be expressed as follows:

$$(\mu_i, \sigma_i) = \psi_d(y_i | x_i, r_i),\ \forall i \in T. \tag{8}$$

where $\psi_d(\cdot)$ is the MLP function.

In CNP, the estimated traffic state is calculated according to cells' spatial-temporal dependency (i.e. the similarity between the cell's index x_T and x_O), which will keep constant after training because the index of spatial-temporal cells never changes. However, in the actual traffic environment, the dependency among cells' traffic state is time-dependent and related to the traffic state (e.g. congested or not). For instance, in congested traffic, the state of downstream cells dominates that of upstream cells because traffic jams would propagate backward; in free traffic, such a dependency disappears because free traffic flow would like to move forwards. Therefore, the constant spatial-temporal dependency in CNP could not fully describe the characteristic of traffic flow. Directly using CNP to solve the problem in the traffic field might lead to a huge estimation error.

4.2 Residual Attention CNP (RA-CNP) for Estimation

The aforementioned analysis shows that the dynamic traffic state dependency should also be considered in CNP to comprehensively model the traffic flow. However, due to the limited observation, we cannot directly extract the dependency between the traffic state of observed and target cells. Fortunately, the latent representation r_T extracted from the encoder contains critical information of estimated traffic state, which could be used for further dependency extraction. Therefore, in the proposed RA-CNP, we insert a residual attention module between the encoder and decoder to promote CNP's capability in estimation.

Residual Attention Module. In RA-CNP, the residual attention module serves as the extension of the encoder. Similar to CNP, the residual attention module will adopt another attention function to extract traffic state dependency based on the output of the encoder, r_T and $r_O = \{r_i\}_{i \in O}$ (calculated using the same way as r_T).

$$r_T^* = \{r_i^*\}_{i \in T} = Attn \begin{pmatrix} Q = \{\omega_r(r_i)\}_{i \in T}, \\ K = \{\omega_r(r_i)\}_{i \in O}, \\ V = \{\omega_r(r_i)\}_{i \in O} \end{pmatrix} \tag{9}$$

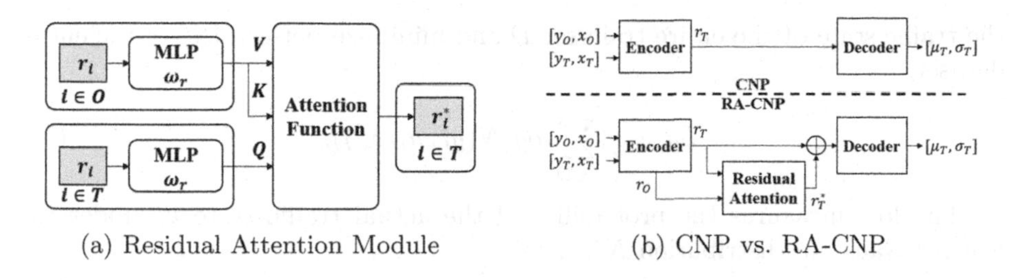

(a) Residual Attention Module (b) CNP vs. RA-CNP

Fig. 3. Structure of RA-CNP model. Compared with CNP, RA-CNP has an extra residual attention module between encoder and decoder.

where $Attn(\cdot)$ is the attention function. Q and K represents the extracted traffic state information of the target and observed cells respectively. Both of them are calculated by $\omega_r(\cdot)$, which independently maps encoder's representation to a new hidden space of traffic state for further information extraction. The values input V is set as the same as K. The output r_T^* is the weighted sum of V, whose weights are determined by traffic state dependency between observed and target cells. Same to CNP, $\omega_r(\cdot)$ is implemented by MLP. The implementation detail of the residual attention module is shown in Fig. 3(a).

Model Structure. Figure 3(b) illustrates the structure of RA-CNP. It is noted that the encoder and decoder in CNP and RA-CNP are the same, except that: (1) The encoder in RA-CNP also outputs the observed cell's representation r_O, which will be used in the residual attention module; (2) We adopt a residual connection [10] to merge the result of the encoder and residual attention module, and use $r_T + r_T^*$ to replace the input of the decoder. The reason for using the residual connection is that we would like RA-CNP to extract comprehensive dependency without increasing network depth, which is easier for model training. In summary, the RA-CNP model incorporates CNP with residential attention.

From the perspective of attention module, both attention functions of the encoder and the residual attention module in RA-CNP can be implemented using the same way. However, the encoder focuses on spatial-temporal dependency; the residual attention module focuses on the traffic state dependency. For the sake of simplicity, the attention module in the encoder is referred to as the *spatial-temporal attention*, and that of the residual attention module is denoted as *state attention*. According to Fig. 3(b), it is clear that the residual connection merges latent representations from both spatial-temporal and state attention. In other words, RA-CNP considers both the spatial-temporal dependency and traffic state dependency when estimating the missing traffic state.

Model Training. Due to the implementation using neural networks, both CNP and RA-CNP model are easy to train via gradient descent. To provide proper input for model training, the actual index x_i and traffic state y_i will be mapped into $[0,1]^2$ and $[0,1]^d$ respectively, where d is the dimension of traffic states.

Given the dataset D, we randomly generate observation set O and target set T, where $O \cup T = D$. During the training, we use the data in O to estimate the traffic state of the entire train set D and minimize per-cell loss of the entire dataset,

$$L_{prob} = \sum_{i \in D} log(N(y_i^*; \mu_i, \sigma_i^2)). \tag{10}$$

The loss measures the probability of the actual traffic state y_i^* under the learned Gaussian distribution $N(\mu_i, \sigma_i^2)$.

Computational Cost. In the basic CNP model, CNP with dot-product attention implementation has less time complexity $O(mn)$ for making m estimation with n observed cells, compared with traditional GPR, $O((m+n)^3)$. The attention function contributes the most in the time complexity because it needs to calculate the similarity between observed and unobserved cells. Similarly, the time complexity of RA-CNP is $O((m+n)n)$, which is a little bit higher than CNP ($O(mn)$), because the encoder in RA-CNP needs to calculate extra representation r_O. In general, RA-CNP has lower complexity than GPR, which makes RA-CNP possible for real-time application, especially under sparse observation condition, where $n \ll m$.

5 Numerical Experiments

5.1 Experimental Setup

Datasets. The data adopted for this paper comes from the Next Generation Simulation (NGSIM) dataset for I-80 freeway in California, which records the trajectories of around 10,000 vehicles from 4:00 pm to 5:30 pm on April 13, 2005 [16]. The study area was approximately 500 m in length. The trajectories of the study area are provided in Fig. 4(a). In our experiment, the studied traffic lane is equally divided into 8 sub-segments (each segment is around 60 m). The observation window has 8 time segments (each segment is 5 s). Spatial-temporal cells are indicated by the red rectangles in Fig. 4(a). The traffic state of the cells could be calculated using trajectories according to the traffic flow theory [21], Fig. 4(b) shows the result of the calculated traffic speed.

Experiments. In the experiment, we focus on the estimation of traffic flow, density, and speed of cells. The first experiment will compare the estimation error of various methods. This experiment will be conducted under two different mode: random observation and trajectory observation. The random observation mode assumes that the observed cells are randomly distributed in the studied area. To quantify the sparseness of the observation in a dataset D (consists of observation and target set, O and T), we define the sparse ratio $s = |T|/|D|$, where $|\cdot|$ indicates the size of the set. We will compare the estimation error of methods in different sparse ratio (from 0.5 to 0.9). The trajectory observation mode simulates the actual traffic state estimation scenario, where we suppose that there is only one connected vehicle is deployed on the target traffic lane. The connected vehicle passes through several cells, whose traffic state is available

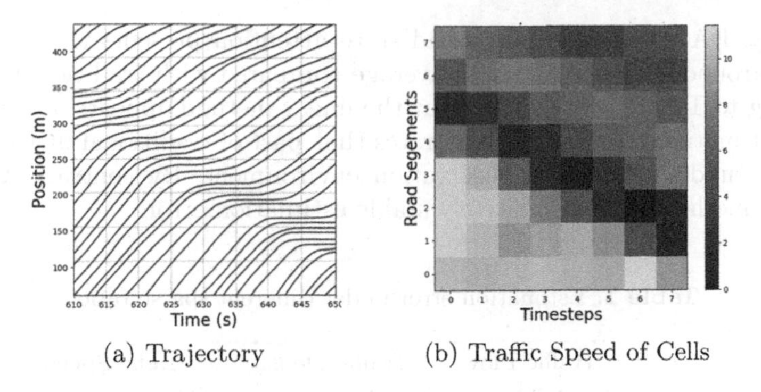

(a) Trajectory (b) Traffic Speed of Cells

Fig. 4. Data in NGSIM dataset. (a) Trajectory data in NGSIM; (b) Traffic speed of cells (m/s). Blue curves in (a) are trajectories; red rectangles in (a) denote the spatial-temporal cells; Darker color in (b) represents lower traffic speed. (Color figure online)

for estimation. In this mode, we will visualize the estimation result and give an illustrative example of methods' performance under congested traffic scenario.

To make a comprehensive comparison between various methods, we will first compare the estimation error between CNP and RA-CNP. We suppose that both the CNP and RA-CNP adopt dot-product attention in the spatial-temporal attention. Besides, the effect of state attention's implementation is also considered in this paper. We will compare the difference between dot-product and multi-head state attention. For demonstration convenience, we name the RA-CNP with dot-product state attention as RA-CNP-DP and that with 8 head state attention as RA-CNP-MH. Moreover, we will also develop several traditional traffic state estimation methods as the baselines: Kalman Filter (KF) [15], LSTM [6], and Gaussian process regression (GPR) with squared exponential kernel [20].

In the second experiment, we will visualize the attention weights extracted by both the spatial-temporal attention and state attention in the trained RA-CNP, and explain the reason why the proposed method has a better performance in traffic state estimation. We will compare the attention weights between the free and congested traffic scenarios.

Metrics. The metric adopted in this paper is the Mean Absolute Error (MAE) between the estimated and actual traffic state of each cell, $MAE = \sum_{i \in D} |y_i - y_i^*|$, where y_i, and y_i^* represents the estimated and actual traffic state respectively.

5.2 Estimation Error

Random Observation. Table 1 demonstrates the changes in estimation error under different sparse ratio (from 0.5 to 0.9). As the sparse ratio rises, the estimation error of all methods increases gradually. CNP and RA-CNP methods outperform other traditional estimation methods under all sparse scenarios.

Especially, RA-CNP with multi-head state attention presents the lowest error and has around 40% less MAE on average compared with traditional methods. According to Table 1, we can see that the linear method appears a huge growth rate in estimation error, which illustrates that part of traditional methods is not applicable under the sparse observation environment. By contrary, CNP and RA-CNP methods have a relatively stable estimation error.

Table 1. Estimation error under different sparse ratio

	Traffic Flow (veh/h)			Traffic Density (veh/km)			Traffic Speed (km/h)		
Sparse Ratio	0.5	0.7	0.9	0.5	0.7	0.9	0.5	0.7	0.9
KF	351	396	546	22.9	28.6	38.4	3.32	4.84	9.06
LSTM	349	381	399	25.8	27.4	29.3	5.15	5.28	6.13
GPR	211	232	273	11.1	13.3	16.2	2.44	3.21	4.50
CNP	202	218	256	10.2	11.8	14.2	2.47	3.00	4.18
RA-CNP-DP	175	198	249	9.31	10.9	13.9	2.31	2.85	4.14
RA-CNP-MH	**169**	**187**	**241**	**8.80**	**10.4**	**13.9**	**2.03**	**2.55**	**3.93**

Trajectory Observation. In this mode, for the sake of simplicity, we only show the traffic speed in the studied area. The typical traffic scenario is demonstrated in Fig. 5(a), where the horizontal and vertical axis indicates the time and position respectively, and the traffic flow moves from bottom to up in the vertical direction (the downstream of the traffic lane is on the top). The congested traffic forms few low-speed spatial-temporal cells (darker area), which propagates along the direction of arrow A. The connected vehicle passes through several cells denoted by the black cells in Fig. 5(e). To better illustrate the result, we also mark the observed cells with red lines in other estimation results in Fig. 5.

The results in Fig. 5 shows that RA-CNP-MH outperforms other methods in MAE, and reproduces the most reasonable result. Besides, the traditional method KF has the worst result. We believe that, due to the stochasticity and sparse observation, it is difficult to design explicit prior model for traditional methods. As a result, those model-based methods will lose the estimation accuracy rapidly when the estimated cell is far away from the observation.

From the perspective of GPR, it can only produce blurred estimation result compared with CNP and RA-CNP models. We believe that the predefined kernel function limits its capability of extracting traffic dependency. As a result, the traffic state of cell B (in Fig. 5(d)), which captures a low-speed area, does not help in the estimation. Although CNP promotes the performance by involving neural network and spatial-temporal attention, it is still limited by the constant dependency. As pointed out by arrow C in Fig. 5(f), CNP presents a different propagation direction. Due to the constant dependency, CNP will always use

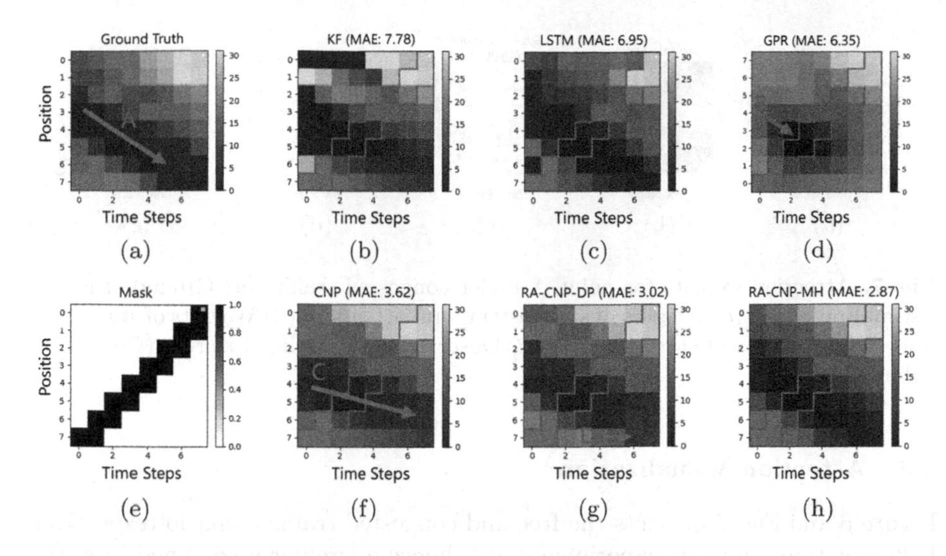

Fig. 5. Estimation with trajectory observation. (a) Trajectory data in NGSIM; (b) Traffic speed of cells (km/h). Blue curves in (a) are trajectories; red rectangles in (a) denote the spatial-temporal cells; Darker color in (b) represents lower traffic speed. (Color figure online)

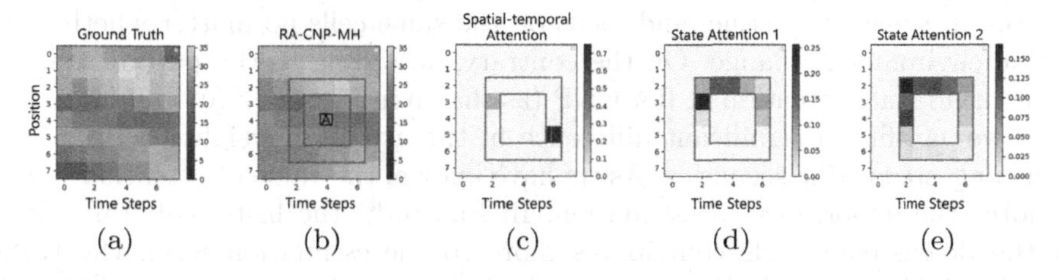

Fig. 6. Attention weights for point A under free traffic. (a) Ground truth; (b) Estimation result; (c) Weights of spatial-temporal attention; (d) Weights of state attention 1; (e) Weights of state attention 2. Observed cells are lined in red. (Color figure online)

the same propagation direction in estimation without considering the traffic conditions (e.g. congested or not). On the contrary, both RA-CNP-DP and RA-CNP-MH get a more reasonable result due to the application of state attention.

In RA-CNP based methods, compared with dot-product attention, multi-head state attention can comprehensively extract latent dependency among cells, due to the multiple projections. As shown in Fig. 5(g), RA-CNP-DP performs worse in estimating the traffic state of cell D, which means that the dependency between cells B and D is not well extracted by RA-CNP-DP. By contrary, RA-CNP-MH has better performance.

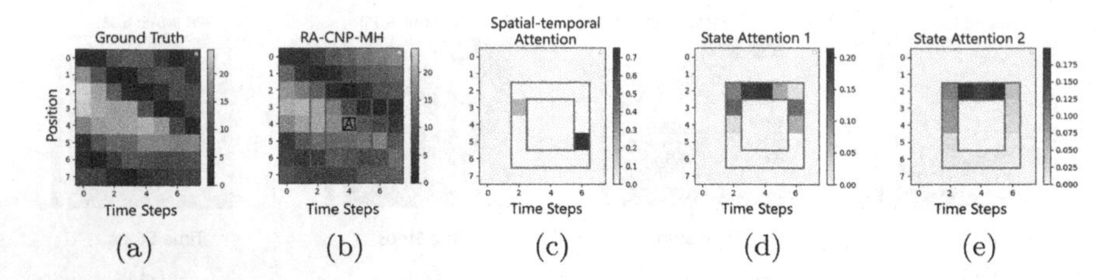

Fig. 7. Attention weights for point A under congested traffic. (a) Ground truth; (b) Estimation result; (c) Weights of spatial-temporal attention; (d) Weights of state attention 1; (e) Weights of state attention 2. Observed cells are lined in red. (Color figure online)

5.3 Attention Visualization

Figure 6 and Fig. 7 presents the free and congested traffic scenario respectively. Different from previous experiments, we choose a circular area (lined in red) as the source of our observation, and analysis which cell contributes more for the estimation of the central cell A (marked in Fig. 6(b)).

Figure 6(c) and Fig. 7(c) visualize the weight of spatial-temporal attention in the encoder. It is obvious that encoder (especially for CNP) has constant spatial-temporal dependency issue, and focus on the same cells no matter whether the traffic environment change. On the contrary, according to the partial result of multi-head state attention in RA-CNP (as shown in Fig. 6(d)–(e) and Fig. 7(d)–(e)), we can find a significant difference in the attention weights between free and congested traffic scenarios. As we have known, the value of attention weight denotes the importance in estimation. In Fig. 6(d), the historical traffic state of the downstream cells contributes more to the estimation when the traffic speed is high; in Fig. 7(d), congested downstream cells receive more attention under jam traffic because the model needs to consider the influence of that. A similar phenomenon could also be found in state attention 2 in Fig. 6(e) and Fig. 7(e). The only difference between them is the value of attention weights. In summary, RA-CNP could solve the constant dependency issue in CNP using residual attention, and improve estimation result.

6 Conclusion

In this paper, we present a real-time fine-grained freeway traffic state estimation method under sparse observation. Inspired by the Conditional Neural Process (CNP), we formulate the distribution of the traffic state estimation function by Residual Attention Conditional Neural Process (RA-CNP). The RA-CNP is an approximation of Gaussian Processes Regression (GPR) using neural networks. It combines the advantage of both GPR and neural networks, which give us a flexible way to solve the sparseness and stochasticity problems meet in fine-grained estimation. Compared with CNP, RA-CNP involves the residual attention module to further extract traffic state dependency, and get a more reliable estimation. At the same time, RA-CNP still keeps a low complexity, which makes it applicable in real-time traffic applications. Numerical test on actual traffic data shows that the proposed method greatly lowers the traffic state estimation error compared with traditional methods and CNP, even under sparse observation.

For the foreseeable future, the proposed method can be used for advanced traffic applications and contribute to the research of traffic flow theory. In addition, due to the flexibility of the RA-CNP model, RA-CNP-liked methods can contribute to the research of traffic flow theory in extracting latent dependency among traffic states.

Acknowledgment. This work is supported by National Key R&D Program of China No. 2017YFB1200700. Authors thank FHWA for making the NGSIM trajectory data available.

References

1. Bekiarisliberis, N., Roncoli, C., Papageorgiou, M.: Highway traffic state estimation with mixed connected and conventional vehicles. IEEE Trans. Intell. Transport. Syst. **17**(12), 3484–3497 (2016)
2. Chen, C., Varaiya, P.: Freeway performance measurement system (pems). PATH Research Report (2002)
3. Daganzo, C.F.: The cell transmission model: a dynamic representation of highway traffic consistent with the hydrodynamic theory. Transport. Res. Part B-Methodol. **28**(4), 269–287 (1994)
4. Datta, A., Banerjee, S., Finley, A.O., Gelfand, A.E.: Hierarchical nearest-neighbor gaussian process models for large geostatistical datasets. J. Am. Stat. Assoc. **111**(514), 800–812 (2016)
5. Eslami, S.M.A., et al.: Neural scene representation and rendering. Science **360**(6394), 1204–1210 (2018)
6. Fu, R., Zhang, Z., Li, L.: Using LSTM and GRU neural network methods for traffic flow prediction **2016**, 324–328 (2016)
7. Garnelo, M., et al.: Conditional neural processes. arXiv: Learning (2018)
8. Han, Y., Chen, D., Ahn, S.: Variable speed limit control at fixed freeway bottlenecks using connected vehicles. Transport. Res. Part B-Methodol. **98**, 113–134 (2017)
9. Hastie, T., Tibshirani, R., Friedman, J.H.: The elements of statistical learning: data mining, inference, and prediction. Math. Intell. **27**(2), 83–85 (2005)
10. He, K., Zhang, X., Ren, S., Sun, J.: Deep residual learning for image recognition, pp. 770–778 (2016)
11. Houenou, A., Bonnifait, P., Cherfaoui, V., Yao, W.: Vehicle trajectory prediction based on motion model and maneuver recognition, pp. 4363–4369 (2013)
12. Ide, T., Kato, S.: Travel-time prediction using gaussian process regression: a trajectory-based approach, pp. 1185–1196 (2009)
13. Kim, H., et al.: Attentive neural processes (2019)

14. Li, Y., Yu, R., Shahabi, C., Liu, Y.: Diffusion convolutional recurrent neural network: data-driven traffic forecasting. arXiv: Learning (2017)
15. Nantes, A., Ngoduy, D., Bhaskar, A., Miska, M., Chung, E.: Real-time traffic state estimation in urban corridors from heterogeneous data. Transport. Res. Part C-Emerg. Technol. **66**, 99–118 (2016)
16. NGSIM: Next generation simulation. https://data.transportation.gov/Automobiles/Next-Generation-Simulation-NGSIM-Vehicle-Trajector/8ect-6jqj
17. Park, C., et al.: Stgrat: a spatio-temporal graph attention network for traffic forecasting. arXiv: Learning (2019)
18. Quinonerocandela, J., Rasmussen, C.E.: A unifying view of sparse approximate gaussian process regression. J. Machine Learn. Res. **6**, 1939–1959 (2005)
19. Rasmussen, C.E.: Gaussian processes in machine learning (2003)
20. Rodrigues, F., Pereira, F.C.: Heteroscedastic gaussian processes for uncertainty modeling in large-scale crowdsourced traffic data. Transport. Res. Part C-Emerging Technol. **95**, 636–651 (2018)
21. Seo, T., Bayen, A.M., Kusakabe, T., Asakura, Y.: Traffic state estimation on highway: a comprehensive survey. Ann. Rev. Control **43**, 128–151 (2017)
22. Singh, G., Yoon, J., Son, Y., Ahn, S.: Sequential neural processes. arXiv: Learning (2019)
23. Smaragdis, E., Papageorgiou, M., Kosmatopoulos, E.B.: A flow-maximizing adaptive local ramp metering strategy. Transport. Res. Part B-Methodol. **38**(3), 251–270 (2004)
24. Tagade, P., Hariharan, K.S., Ramachandran, S., Khandelwal, A., Naha, A., Kolake, S.M., Han, S.H.: Deep gaussian process regression for lithium-ion battery health prognosis and degradation mode diagnosis. J. Power Sources **445**, 227281 (2020)
25. Tan, H., Feng, G., Feng, J., Wang, W., Zhang, Y., Li, F.: A tensor based method for missing traffic data completion. Transport. Res. Part C-Emerging Technol. **28**, 15–27 (2013)
26. Vaswani, A., et al.: Attention is all you need, pp. 5998–6008 (2017)
27. Wu, Y., Tan, H., Qin, L., Ran, B., Jiang, Z.: A hybrid deep learning based traffic flow prediction method and its understanding. Transport. Res. Part C-Emerg. Technol. **90**, 166–180 (2018)
28. Xie, Y., Zhao, K., Sun, Y., Chen, D.: Gaussian processes for short-term traffic volume forecasting. Transport. Res. Record **2165**(2165), 69–78 (2010)
29. Xu, D.W., Dong, H.H., Li, H.J., Jia, L.M., Feng, Y.J.: The estimation of road traffic states based on compressive sensing. Transportmetrica B-Transport Dyn. **3**(2), 131–152 (2015)
30. Yang, F., Wang, S., Li, J., Liu, Z., Sun, Q.: An overview of internet of vehicles. China Commun. **11**(10), 1–15 (2014)
31. Yuan, Y., Van Lint, J.W.C., Wilson, R.E., Van Wageningenkessels, F.L.M., Hoogendoorn, S.P.: Real-time lagrangian traffic state estimator for freeways. IEEE Trans. Intell. Transport. Syst. **13**(1), 59–70 (2012)
32. Zhang, H., Goodfellow, I., Metaxas, D.N., Odena, A.: Self-attention generative adversarial networks. arXiv: Machine Learning (2018)
33. Zhang, J., Shi, X., Xie, J., Ma, H., King, I., Yeung, D.: Gaan: gated attention networks for learning on large and spatiotemporal graphs. arXiv: Learning (2018)
34. Zhao, Z., Chen, W., Wu, X., Chen, P.C.Y., Liu, J.: LSTM network: a deep learning approach for short-term traffic forecast. IET Intell. Transport Syst. **11**(2), 68–75 (2017)

35. Zhong, M., Lingras, P., Sharma, S.: Estimation of missing traffic counts using factor, genetic, neural and regression techniques. Transport. Res. Part C-Emerg. Technol. **12**(2), 139–166 (2004)
36. Zhou, Y., Ahn, S., Chitturi, M., Noyce, D.A.: Rolling horizon stochastic optimal control strategy for ACC and CACC under uncertainty. Transport. Res. Part C-Emerg. Technol. **83**, 61–76 (2017)
37. Zhou, Y., Ahn, S., Wang, M., Hoogendoorn, S.P.: Stabilizing mixed vehicular platoons with connected automated vehicles: An h-infinity approach. Transport. Res. Procedia **38**, 441–461 (2019)

Revisiting Convolutional Neural Networks for Citywide Crowd Flow Analytics

Yuxuan Liang[1(✉)], Kun Ouyang[1], Yiwei Wang[1], Ye Liu[1], Junbo Zhang[2,3,4], Yu Zheng[2,3,4], and David S. Rosenblum[1]

[1] School of Computing, National University of Singapore, Singapore, Singapore
yuxliang@outlook.com, {ouyangk,y-wang,liuye,david}@comp.nus.edu.sg
[2] JD Intelligent Cities Research and JD Intelligent Cities Business Unit, Beijing, China
{msjunbozhang,msyuzheng}@outlook.com
[3] Institute of Artificial Intelligence, Southwest Jiaotong University, Chengdu, China
[4] Xidian University, Xi'an, China

Abstract. Citywide crowd flow analytics is of great importance to smart city efforts. It aims to model the crowd flow (e.g., inflow and outflow) of each region in a city based on historical observations. Nowadays, Convolutional Neural Networks (CNNs) have been widely adopted in raster-based crowd flow analytics by virtue of their capability in capturing spatial dependencies. After revisiting CNN-based methods for different analytics tasks, we expose two common critical drawbacks in the existing uses: 1) inefficiency in learning global spatial dependencies, and 2) overlooking latent region functions. To tackle these challenges, in this paper we present a novel framework entitled DeepLGR that can be easily generalized to address various citywide crowd flow analytics problems. This framework consists of three parts: 1) a local feature extraction module to learn representations for each region; 2) a global context module to extract global contextual priors and upsample them to generate the global features; and 3) a region-specific predictor based on tensor decomposition to provide customized predictions for each region, which is very parameter-efficient compared to previous methods. Extensive experiments on two typical crowd flow analytics tasks demonstrate the effectiveness, stability, and generality of our framework.

1 Introduction

Citywide crowd flow analytics is very critical to smart city efforts around the world. A typical task is citywide crowd flow prediction [12,20,21], which aims to predict the traffic (e.g., inflows and outflows of every region) for the next time slot, given the historical traffic observations. It can help the governors conduct traffic control and avoid potential catastrophic stampede before a special event. Another important task is to infer the fine-grained crowd flows from available coarse-grained data sources, which can reduce the expense of urban systems [11,13]. Other tasks [19,24] are also actively studied by the community due to the vital impact of citywide crowd flow analytics.

F. Hutter et al. (Eds.): ECML PKDD 2020, LNAI 12457, pp. 578–594, 2021.
https://doi.org/10.1007/978-3-030-67658-2_33

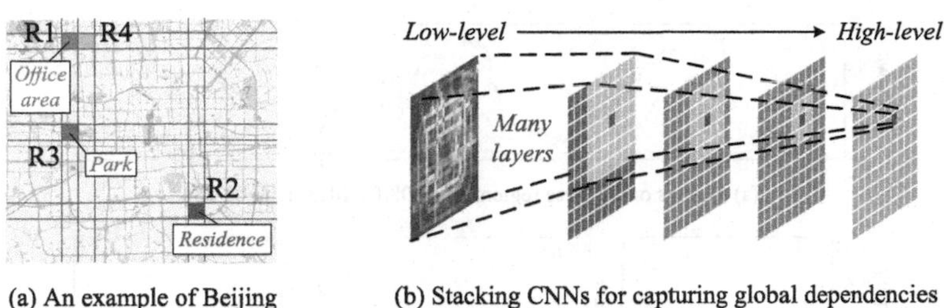

(a) An example of Beijing (b) Stacking CNNs for capturing global dependencies

Fig. 1. Application of CNNs for citywide crowd flow analytics (Better view in color). (Color figure online)

Crowd flow analytics is not trivial as the traffic can be affected by multiple complex factors in spatio-temporal domains. As shown in Fig. 1(a), the inflow of Region R1 is affected by outflows of nearby regions like R4 as well as distant regions, which indicates the spatial dependencies. For the temporal dependencies, crowd flow in a region is affected by recent, daily, and weekly historical traffic. To model the spatio-temporal dependencies, Convolutional Neural Networks (CNNs) have been widely used and achieved promising performance. A pioneering work [21] provided the first CNN-based method (DeepST) for modeling crowd flow, where convolution operators are used to extract spatially near and distant dependencies and the temporal dependencies are considered in different branches of networks. ST-ResNet [20] further enhanced the performance of DeepST using residual structures. Very recently, a novel ConvPlus structure in DeepSTN+ [12] was proposed to learn the long-term spatial dependencies between two arbitrary regions. These CNN-based methods are characterized by two components: a complicated ST feature learner to capture features of the measurements, and a simple task-specific predictor to generate predictions on all regions. However, they have two main drawbacks:

1) *Inefficiency in learning global spatial dependencies.* Take traveling in Beijing (Fig. 1) as an example. When predicting the inflow of R1 during morning hours, the outflow of distant regions like R2 needs to be considered, since it is common that people commute from a distant residence location. As people can travel around a modern city quickly, it becomes crucial to capture global spatial dependencies in this task. To this end, existing arts employ two approaches:

- *Stacking CNNs to increase receptive fields.* Most previous studies like DeepST and ST-ResNet employ CNNs to capture information locally. But to capture global spatial dependencies, they have to stack many layers to increase the receptive field of the network (see Fig. 1(b)). This is very inefficient since relationships between distant regions can only be captured by a near-top layer with a sufficiently large receptive field to cover all the regions of interest.
- *Learning long-range spatial dependencies directly.* Instead of gradually increasing receptive fields, DeepSTN+ attempts to capture global spatial dependencies in *every layer* using ConvPlus structure, which explicitly models all pairwise relationship between regions. However, a single layer of ConvPlus

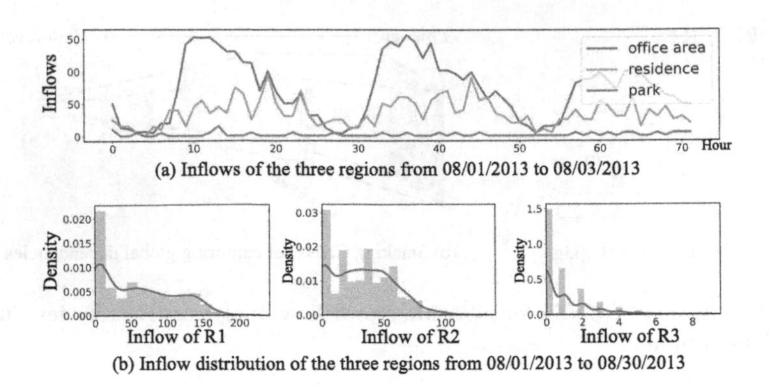

Fig. 2. Illustration of daily patterns and inflow distribution in three regions.

without pooling requires $O(n^2)$ parameters, where n is the number of regions. Constrained by this bloated structure, DeepSTN+ cannot easily go deeper to learn higher-level representations for each region. Thus, how to learn global spatial dependencies more efficiently still remains a major challenge.

2) *Ignoring latent region functions.* Different from pixels in image processing, urban regions have different land functions according to their locations and surrounding POIs [14,23]. Recall that R1, R2 and R3 in Fig. 1 correspond to an office area, a residential area and a park zone respectively. From Fig. 2(a), it can be seen easily that their daily patterns are entirely different. For instance, the office area (R1) usually reaches a traffic peak in the morning, while the residential area (R2) usually exhibits growth after dinner time. The difference between their daily flow distributions can also be seen from Fig. 2(b). However, the aforementioned methods have overlooked such varying latent functions among regions and used a simple predictor with shared parameters to predict flow for all regions, which inevitably resulted in degraded performance.

To address the above problems, we make the following contributions to the community. Primarily, we introduce DeepLGR, the first-ever general framework for raster-based crowd flow analytics. It is named according to how it stratifies a given task into three major procedures: 1) **L**ocal feature extraction to learn representations for each region within small receptive fields; 2) **G**lobal context aggregation to efficiently capture the global spatial dependencies; and 3) **R**egion-specific prediction. Respectively,

- we present the first attempt to extract local region representations using Squeeze-and-Excitation networks (SENet) [5], which excels by including the channel-wise information as additional knowledge;
- we design a global context module that firstly aggregates the region representations using a specific pooling method, and then upsample the global priors back to the original scale to generate global-aware features;
- we introduce a region-specific predictor based on tensor decomposition that factorizes the region-specific parameters of the predictor into a smaller core tensor and adjoint matrices.

In addition, we evaluate our framework on two typical crowd flow analytics tasks: crowd flow forecasting [20,21] and fine-grained crowd flow inference [11, 13]. Extensive experiments demonstrate the state-of-the-art performance and stability achieved by our framework. We have released our code at https://github.com/yoshall/DeepLGR for public use.

2 Formulation

In this section, we introduce several notations and formulate the problem of crowd flow analytics. As shown in Fig. 1(a), we first follow the previous study [21] to partition an area of interest (e.g., a city) evenly into a $H \times W$ grid map based on longitude and latitude where a grid denotes a region. Thus, the crowd flow at a certain time t can be denoted as a 3D tensor $\mathcal{P}_t \in \mathbb{R}^{H \times W \times K}$, where K is the number of different flow measurements (e.g., inflow and outflow). Each entry (i, j, k) denotes the value of the k-th measurement in the region (i, j).

Without loss of generality, we use $\mathcal{X} \in \mathbb{R}^{H \times W \times C}$ and $\mathcal{Y} \in \mathbb{R}^{H' \times W' \times D}$ as the input and output for a crowd flow analytics task, where C and D are the number of channels. For example, in the task of crowd flow prediction [12,20,21], the input is the historical observations $\mathcal{X} = \{\mathcal{P}_i | i = 1, 2, \cdots, \tau\} \in \mathbb{R}^{H \times W \times K\tau}$ and the target is to predict $\mathcal{Y} = \mathcal{P}_{\tau+1} \in \mathbb{R}^{H \times W \times K}$.

3 Methodology

Figure 3 presents the framework of DeepLGR, which can be easily generalized to all kinds of citywide crowd flow. Compared to the previous methods composed of an ST feature learner and a shared predictor for all regions, our framework contains three major components: local feature extraction, global context module and region-specific predictor. In the first component, we employ the SENet to learn representations for each region within small (i.e., local) receptive fields from the input tensor $\mathcal{X}$. To capture global spatial dependencies, we further design the global context module that considers the full region of interest. It first extracts global contextual priors from the learned region representations using a specific pooling method, and then upsamples the priors to the original scale to generate the global features. Once we obtain features from both local view and global view, we concatenate them into a tensor and then feed it to the region-specific predictor to make customized predictions for each region respectively.

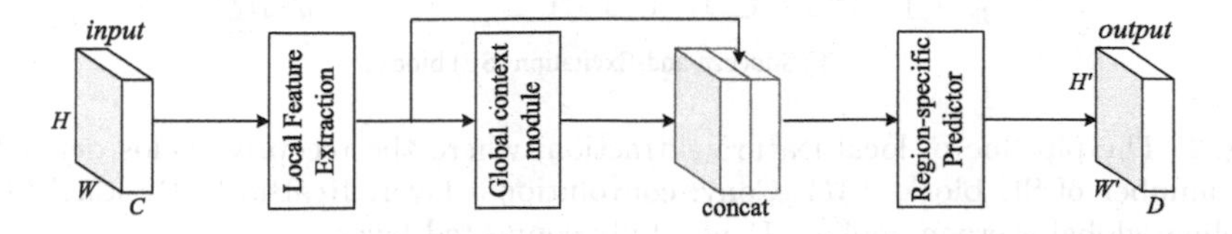

Fig. 3. The pipeline of DeepLGR, which contains three major components.

For spatial dependencies, our framework employs the first two components which strategically capture both local-level (neighborhood) and global-level dependencies between regions. Following the mainstream CNN architectures for citywide crowd flow analytics [12,20,21], the temporal dependencies like closeness (recent), period (daily) and trend (weekly), if any, are considered in the channels of input. These temporal dependencies can interact with each other in the backbone network. Next, we will detail the three components respectively.

3.1 Local Feature Extraction

Recall that both the previous and current state-of-the-arts [12,20] use residual blocks to model the spatial dependencies from nearby regions. However, these methods mainly focus on the spatial dimension and have overlooked the channel-wise information in the feature maps. Thus, we employ SENet to fuse both spatial and channel-wise information within small (i.e., local) receptive fields at each layer, which has proven to be effective in producing compacted and discriminative features of each grid. Figure 4(a) illustrates the pipeline of the module for local feature extraction. The input is fed to a convolutional layer for initialization. Then, we stack M squeeze-and-excitation (SE) blocks in Fig. 4(b) for feature extraction, which is composed of three stages: 1) a residual block [3] for feature learning; 2) a squeeze operation to squeeze global spatial information into a channel descriptor by global average pooling; 3) an excitation operation to fully capture the channel-wise dependencies: it first computes the attention coefficients over each channel via two fully connected layers followed by a sigmoid function, and then rescales the channels of original inputs by these weights. Finally, we use an output convolutional layer to transform the obtained high-level feature maps to the input of the next module. In summary, the SE structure enables this module to learn better representations for each region locally within receptive fields.

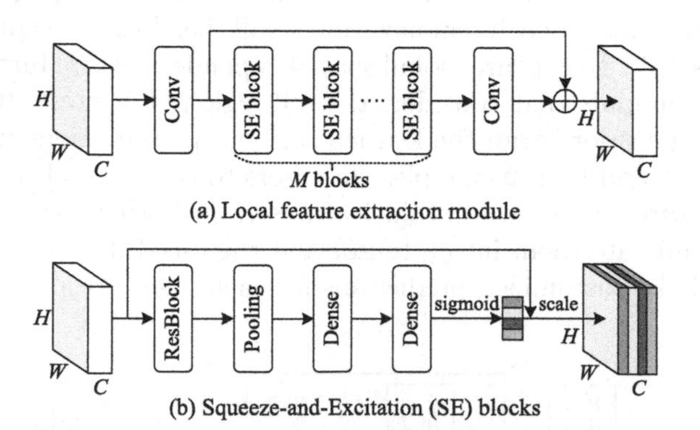

Fig. 4. The pipeline of local feature extraction, where the receptive fields depend on the number of SE blocks (M). Conv: convolutional layer. ResBlock: Residual block. Pooling: global average pooling. Dense: fully connected layer.

3.2 Global Context Module

After local feature extraction, we have designed a specific module that takes the output of the former component as input to generate global contexts for each region, so as to capture global spatial dependencies. As depicted in Fig. 5, we first employ spatial pyramid pooling [2] to generate a set of the global priors, where each prior is a spatially abstract of the original input under different pyramid scales. This operation allows the module to separate the feature map into different sub-regions and build pooled representation for different locations. For example, the 1×1 prior (the red cube) denotes the coarsest level with only one single value at each channel, which is equivalent to global pooling operation that covers the whole image. In our experiments, we use a 4-level pyramid (1×1, 2×2, 4×4 and 8×8) to squeeze the input by average pooling.

Once the global priors are obtained, an 1×1 convolution layer followed by a Batchnorm layer [6] is used for dimension reduction of channels from N to $N/8$. Inspired by the study [11] aiming at inferring fine-grained crowd flow from coarse-grained counterparts, we employ the Subpixel block [15] to upsample the priors to generate new representations with the same size as the original inputs. For example, after the Subpixel block in 4×4 branch, the output feature maps grow $H/4$ and $W/4$ times larger in height and width respectively with the number of channels unchanged. Different from PSPNet [22] using bilinear interpolation for upsampling the priors, the Subpixel block considers the relationship between a super-region and its corresponding sub-regions by introducing a parametric design. Finally, we concatenate the input (i.e., region representations) with all levels of global features (i.e., context) as the output of this module.

In summary, this module first converts the input feature map into priors (e.g., 1×1 prior that encodes the information of all regions) and then upsamples the

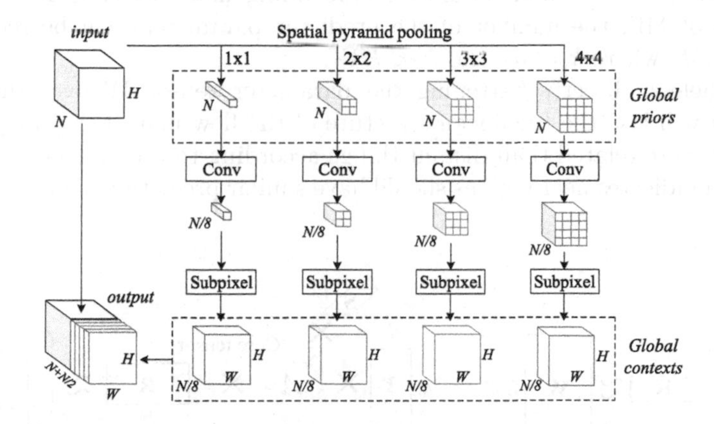

Fig. 5. The pipeline of global context module, where Conv denotes a 1×1 convolutional layer for dimension reduction, and Subpixel contains a convolutional layer and a pixelshuffle operation sequentially to upsample the contextual priors. For simplicity, we use a 4-level pyramid (1×1, 2×2, 3×3 and 4×4) for an illustration.

priors to learn the global-context-to-region influence (i.e., global spatial dependencies). Compared to the previous attempt (ConvPlus layer in DeepSTN+), our solution is more efficient and lightweight. Each ConvPlus layer directly models the pairwise relationships among all regions, thus demanding $O(n^2)$ parameters. With the increase of spatial granularity, it will induce extremely high computational costs due to the massive parameters. Thus, DeepSTN+ can hardly learn higher-level representations by simply increasing network depth. In contrast, as we have separated the procedures of local feature extraction and global context modeling, we can easily increase the network depth to gain better capacity.

3.3 Region-Specific Predictor

As mentioned before, each urban region has its unique land function. Previous studies [12,20,21] mainly employ a single fully connected layer (equivalent to a 1×1 convolution) with shared weights as the predictor for all regions, which fails to capture this critical property. Thus, it is necessary to assign region-specific predictor to each region.

Recall that the high-level feature obtained from last module is $\mathcal{Z} \in \mathbb{R}^{H \times W \times N'}$ and prediction result is $\mathcal{Y} \in \mathbb{R}^{H \times W \times D}$, where $N' = N + N/2$. Conventionally, the number of parameters in a shared fully connected layer is $n_f = N'D$. To achieve region-specific predictor, an intuitive solution is to use a customized fully connected layer for each region. However, it will induce $HW \times n_f$ parameters (denoted as a tensor $\mathcal{W} \in \mathbb{R}^{H \times W \times n_f}$), which can easily bloat up as the granularity increases. Recently, matrix factorization (MF) was used to avoid these drawbacks [14], in which the parameter tensor $\mathcal{W}$ is reshaped to a matrix $\mathbf{W} \in \mathbb{R}^{HW \times n_f}$. As shown in Fig. 6(a), the authors from [14] decompose the weight matrix $\mathbf{W}$ into two *learnable* low-rank matrices, i.e., region embedding matrices $\mathbf{L} \in \mathbb{R}^{HW \times k}$ and parameter embedding matrices $\mathbf{R} \in \mathbb{R}^{k \times n_f}$. With the usage of MF, the number of the predictor parameters can be reduced to $(HW + n_f)k$, where $k \ll n_f$ and $k \ll HW$.

Nonetheless, directly flattening the parameter tensor $\mathcal{W}$ over the region dimension will lose the Euclidean structure of the flow map. For example, near things are more related than distant things according to the first law of geography, which indicates near regions should have similar prediction weights. Instead,

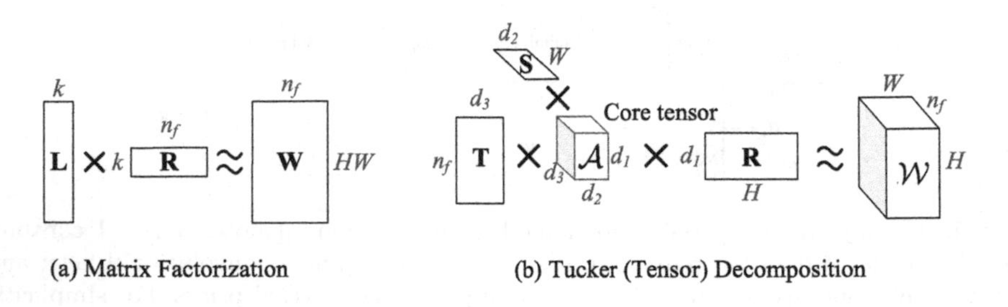

Fig. 6. Illustration of matrix factorization and tensor decomposition.

we present a new idea for decomposing $\mathcal{W}$ using Tensor Decomposition (TD) [17]. It not only preserves the spatial similarity (dependencies) between regions, but also reduces the amount of parameters. As illustrated in Fig. 6(b), tensor $\mathcal{W}$ is decomposed into the multiplication of a core tensor $\mathcal{A} \in \mathbb{R}^{d_1 \times d_2 \times d_3}$ and three adjoint matrices, where d_1, d_2, and d_3 denote the number of latent factors for each matrix. The computation is as follows:

$$\mathcal{W} = \mathcal{A} \times_R \mathbf{R} \times_S \mathbf{S} \times_T \mathbf{T}, \quad (1)$$

where $\times_R$ stands for the tensor-matrix multiplication; the subscript R is the corresponding mode of the multiplication. For instance, $\mathbf{H} = \mathcal{A} \times_R \mathbf{R}$ is $\mathbf{H}_{ijk} = \sum_{i=1}^{d_1} A_{ijk} \times R_{ij}$. By this, we have changed the optimization target from $\mathcal{W}$ to the core tensor $\mathcal{A}$ as well as the three learnable matrices $\mathbf{R}$, $\mathbf{S}$ and $\mathbf{T}$. The core tensor is a low-rank representation summarising both the parametric and spatial information of the origin tensor $\mathcal{W}$. Compared to MF-based solution [14], our tensor decomposition can handle the higher-order relationships within the parameters. In addition, the number of parameters required is $d_1 d_2 d_3 + d_1 H + d_2 W + d_3 n_f$. Since d_1, d_2 and d_3 are usually very small, TD can achieve even much fewer parameters than MF, which is validated in our experiments.

3.4 Optimization

Since our framework is smooth and differentiable everywhere, it can be trained via the back-propagation algorithm. During the training phase, we use Adam optimizer to train our model by minimizing the entry-wise mean absolute error (MAE) between our prediction $\widehat{\mathcal{Y}}$ and the corresponding ground truth $\mathcal{Y}$:

$$\mathcal{L}(\mathbf{\Theta}) = \left\|\mathcal{Y} - \widehat{\mathcal{Y}}\right\|_1 \quad (2)$$

where $\mathbf{\Theta}$ denotes all learnable parameters in our framework.

4 Experiments

To validate the generality of DeepLGR, we conduct experiments on two typical tasks of citywide crowd flow analytics:

- *Crowd flow forecasting*: This task is to forecast the inflow and outflow of each region in a city from historical readings. Following the settings of [20], we consider the temporal dependencies (i.e., closeness, period and trend) in different channels of input, and the output is the prediction of inflow and outflow for the next timestamp. Similar to [20], we set the length of closeness (recent), period (daily) and trend (weekly) to 5, 3, 3.
- *Fine-grained flow inference*: In this task, we aim to infer fine-grained crowd flows throughout a city based on coarse-grained observations. We extend the state-of-the-art method named UrbanFM [11] using our framework. Specifically, we replace the ResNet-based feature extraction of UrbanFM by our first component (SENet). Then, we add the global context module and region-specific predictor after the subpixel blocks in UrbanFM.

4.1 Experimental Settings

Datasets. Two datasets were used in our experiments, including TaxiBJ and HappyValley. The former is the fine-grained version of the ones used by [20] and the latter is provided from [11]. Specifically, TaxiBJ consists of four different time spans (denoted as P1 to P4 with different number of taxicabs and distribution), while HappyValley is the hourly observations of human flow in a theme park in Beijing from ten months. The statistics are detailed in Table 1. We select the flow data between 6 am and 11 pm to conduct our experiments. Using both datasets, we evaluate DeepLGR over the two aforementioned tasks: In the first task, we employ the first 80% data as training set, the next 10% as validation set and the rest for test set; In the second task, we follow all the experiment settings of [11], including training, validation and test set partition. The upscaling factors in TaxiBJ and HappyValley are 4 and 2 respectively.

Evaluation Metrics. We employ two widely-used criteria to evaluate our model from different aspects, including mean absolute error (MAE) and symmetric mean absolute percentage error (SMAPE). They are defined as:

$$\text{MAE} = \frac{1}{z}\sum_{i=1}^{z}|y_i - \tilde{y}_i|, \quad \text{SMAPE} = \frac{1}{z}\sum_{i=1}^{z}\frac{|y_i - \tilde{y}_i|}{|y_i| + |\tilde{y}_i|},$$

where y and $\tilde{y}$ are ground truth and predicted value respectively; z is the total number of all entries. Smaller metric scores indicate better model performance.

Baselines. In the first task, we compare our framework with heuristics, time series methods and CNN-based baselines. Specifically, a naive method (**Last**) simply uses the last observation as the prediction result, and another heuristic (**CA**) leverages the closeness property to predict the future crowds by averaging the values from the previous 5 time steps. **ARIMA** is a well-known model for forecasting future values in a time series. Besides, the CNN-based baselines (including **DeepST** [21], **ST-ResNet** [20], **ConvLSTM** [15] and **DeepSTN+** [12]) have been introduced in Sect. 1.

Table 1. Dataset description.

Dataset	TaxiBJ	HappyValley
Data type	Inflow and outflow	Staying flow
Resolution	(128, 128)	(50,100)
Sampling rate	30 min	1 h
Time Span	P1: 07/01/2013-10/31/2013	01/01/2018-
(mm/dd/yyyy)	P2: 02/01/2014-06/30/2014	10/31/2018
	P3: 03/01/2015-06/30/2015	
	P4: 11/01/2015-03/31/2016	

The second task was introduced only very recently by [11], where the authors presented the state-of-the-art method named **UrbanFM**. It considers the unique characteristics of this task, including the spatial hierarchy and external factors. Other strong baselines included in this work are related to image super-resolution, such as **VDSR** [7] and **SRResNet** [8]. We mainly use these three baselines for model comparison in this task. It is worth noting that all baselines are implemented with their default settings in both tasks.

Training Details and Hyperparameters. Our framework, as well as the above baselines, are fully implemented by Pytorch 1.1.0 with one GTX 2080TI. During the training phase, the learning rate is 0.005 and the batch size is 16. For the number of stacked SE blocks (denoted as M) in the first component, we conduct a grid search over $\{3, 6, 9, 12\}$. For simplicity, we use the same hidden dimension (i.e., number of channels) at each 3×3 convolutional layer in SE blocks, and conduct a grid search over $F = \{32, 64, 128\}$.

Table 2. Prediction results on TaxiBJ over different time spans (P1-P4), where the bold number indicates the best performance of the column. We train and test each method five times, and present results using the format: "mean ± standard deviation".

Method	P1		P2	
	MAE	SMAPE	MAE	SMAPE
CA	3.43	0.290	4.23	0.288
Last	3.39	0.242	4.09	0.241
ARIMA	3.08	0.403	3.53	0.385
DeepST	2.59 ± 0.05	0.41 ± 0.01	2.94 ± 0.05	0.39 ± 0.01
ST-ResNet	2.53 ± 0.05	0.38 ± 0.05	2.93 ± 0.06	0.34 ± 0.07
ConvLSTM	2.42 ± 0.02	0.41 ± 0.01	2.77 ± 0.01	0.39 ± 0.01
DeepSTN+	2.33 ± 0.04	0.35 ± 0.08	2.67 ± 0.02	0.32 ± 0.05
DeepLGR	**2.15 ± 0.00**	**0.19 ± 0.00**	**2.46 ± 0.00**	**0.18 ± 0.00**
Method	P3		P4	
	MAE	SMAPE	MAE	SMAPE
CA	4.17	0.286	2.81	0.286
Last	4.07	0.240	2.82	0.239
ARIMA	3.68	0.363	2.61	0.420
DeepST	2.97 ± 0.04	0.39 ± 0.01	2.16 ± 0.04	0.43 ± 0.02
ST-ResNet	2.91 ± 0.06	0.33 ± 0.05	2.15 ± 0.04	0.32 ± 0.06
ConvLSTM	2.87 ± 0.01	0.39 ± 0.01	2.09 ± 0.02	0.43 ± 0.02
DeepSTN+	2.82 ± 0.04	0.38 ± 0.05	2.05 ± 0.01	0.34 ± 0.05
DeepLGR	**2.56 ± 0.02**	**0.19 ± 0.04**	**1.84 ± 0.01**	**0.19 ± 0.00**

4.2 Results on Crowd Flow Forecasting

Model Comparison. Here, we compare our framework with the baselines over the two datasets. We report the result of DeepLGR with $M = 9$ and $F = 64$ as our default setting. Further results regarding different M will be discussed later.

Table 2 shows the experimental results over P1 to P4 in TaxiBJ. We can observe that our framework clearly outperforms all baselines over both metrics. For instance, DeepLGR shows 10.2% and 44.1% improvements on MAE and SMAPE beyond the state-of-the-art method (DeepSTN+) in P4. The conventional model ARIMA performs much worse than deep learning models in these datasets, since it only considers the temporal dependencies among time series. Apart from the CNN-based methods, ConvLSTM advances DeepST and ST-ResNet because of the positive effect of its LSTM structure. However, it overlooks the global spatial dependencies between regions, which leads to inferiority compared to DeepSTN+ and DeepLGR. Another interesting observation is that the heuristics including CA and Last achieves much less SMAPE than previous CNN-based methods. Recall that SMAPE prefers to penalize the errors in regions with lower flow volumes. This observation reveals the importance of the temporal dependencies in such regions since CA and Last only consider the temporal closeness for forecasting. Only our method performs better than the heuristics on SMAPE with the usage of tensor decomposition, which will be detailed in the ablation study. Last but not least, DeepLGR is also more stable than the baselines according to the standard deviation observations.

Compared to TaxiBJ with a citywide scale, HappyValley focuses on a local area with a highly skewed flow distribution, where only a few regions contain dense populations. Table 3 presents a comprehensive comparison of each model over this dataset. First, it can be seen easily that our framework shows great superiority against the CNN-based methods and slightly outperforms ConvLSTM in terms of both metrics, while using as little as 6.2% of the amount of parameters required in the state-of-the-art method (DeepSTN+). This fact demonstrates that our model is more practical than other CNN-based solutions

Table 3. Prediction results of various methods on the HappyValley dataset, where #Params is the number of parameters and M denotes million.

Method	#Params	MAE	SMAPE
CA	x	2.23	0.46
Last	x	2.20	**0.38**
ARIMA	0.00M	2.14	0.47
DeepST	0.59M	2.02 ± 0.05	0.56 ± 0.05
ST-ResNet	2.73M	1.98 ± 0.05	0.53 ± 0.04
ConvLSTM	5.98M	1.86 ± 0.01	0.48 ± 0.10
DeepSTN+	15.70M	1.92 ± 0.01	0.54 ± 0.06
DeepLGR	0.97M	$\mathbf{1.84 \pm 0.01}$	0.40 ± 0.02

in real-world systems. Second, similar to the results in TaxiBJ, DeepLGR performs more stable than the baselines according to the standard deviation in multiple experiments. Third, the heuristic method (Last) achieves the lowest SMAPE but the second-highest MAE, which can prove the skew distribution of this dataset. Last, the fact that DeepLGR and DeepSTN+ outperform ST-ResNet verifies the necessity of modeling global context in such a small area.

Ablation Study. To further investigate the effectiveness of each component, we compare DeepLGR with its variants over TaxiBJ-P1. For simplicity, we use the terms as local, global and TD to denote the three components in our framework respectively. Based on them, DeepLGR and its variants can be denoted as:

- **local+global+TD**: The original implementation of DeepLGR.
- **local+global+MF**: To show the effectiveness and lightweight property of TD against MF, we replace TD in the region-specific predictor by MF.
- **local+global**: Similar to the CNN-based baselines [12,20,21], this variant uses shared parameters (i.e., not region-specific) as the predictor.
- **local+TD**: The variant of DeepLGR without global context module.
- **local+MF**: We first remove global context module from DeepLGR and then replace TD in region-specific predictor by MF.
- **local+bilinear**: We employ bilinear interpolation rather than Subpixel block to upsample the global priors, so as to obtain new global representations.
- **local**: The last two components are removed from DeepLGR.

Table 4 illustrates the variant comparison over TaxiBJ-P1. We discuss the effects of each model component as follows:

- *Local feature extraction*: A powerful ST feature extractor enables the capability of extracting useful representations for each region. Compared to previous attempts like ST-ResNet based on residual blocks, our feature extraction module largely improves the performance (e.g., local vs. ST-ResNet in Tables 2 and 4). We further investigate the effects of the number of SE blocks in this module. As shown in Fig. 7, it achieves the best performance when $M = 6$ in the test set. Noted that we choose $M = 9$ as the default setting of DeepLGR because of its best performance on the validation set rather than the test set. Besides, we replace the SE blocks in this module by residual blocks to show the advantages of SE blocks, where the results are also in Fig. 7.

Table 4. Results of different variants over TaxiBJ-P1 (trained/tested five times).

Variants	#Params	MAE	SMAPE
Local	0.72M	2.21 ± 0.01	0.37 ± 0.03
Local+MF	0.89M	2.19 ± 0.02	0.36 ± 0.03
Local+TD	0.74M	2.19 ± 0.01	0.32 ± 0.03
Local+bilinear	0.73M	2.20 ± 0.02	0.35 ± 0.03
Local+global	2.30M	2.17 ± 0.02	0.29 ± 0.03
Local+global+MF	2.46M	2.15 ± 0.00	0.27 ± 0.01
Local+global+TD	2.31M	$\mathbf{2.15 \pm 0.00}$	$\mathbf{0.19 \pm 0.00}$

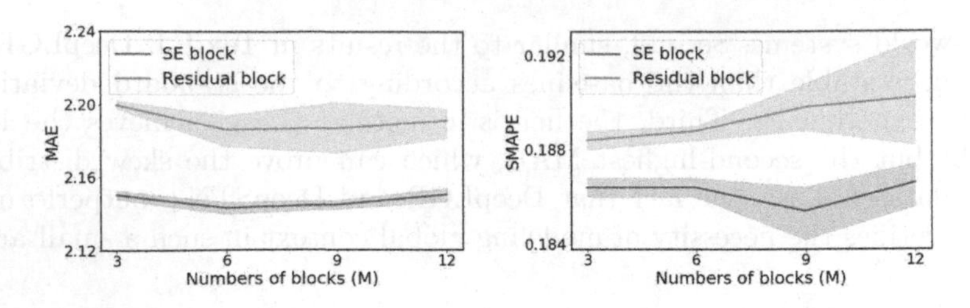

Fig. 7. SE vs. residual block over P1, where the shade area is the standard deviation.

- *Global context module*: As a vital component in our framework, this module provides the global information to boost the performance. As illustrated in Table 4, the comparison between local and local+global (also local+TD and local+global+TD) can verify the effectiveness of this module. With the usage of Subpixel block with a parametric design, local+global brings an improvement beyond local+bilinear.
- *Region-specific predictor*: This module is used to determine the region-specific parameters for predictions. Thus, we compare it with a shared fully connected layer with n_f parameters (local+global), and the matrix decomposition method. From the last three rows of Table 4, we observe that TD demonstrates very competitive accuracy while using as little as 6.3 % of the number of parameters required in MF (i.e., 0.01 M vs. 0.16 M). Moreover, TD significantly outperforms MF over SMAPE since it allows the model to capture spatial dependencies between regions.

4.3 Results on Fine-Grained Flow Inference

Experimental results on the second task have demonstrated the superiority of our framework again. From Table 5, we have the following observations: 1) UrbanFM

Table 5. Results of various models for fine-grained flow inference. We train/test each method five times, and present results using the format: "mean ± standard deviation".

Method	TaxiBJ-P1		HappyValley	
	MAE	SMAPE	MAE	SMAPE
VDSR	2.23 ± 0.05	0.54 ± 0.03	2.13 ± 0.04	0.61 ± 0.02
SRResNet	2.20 ± 0.05	0.52 ± 0.03	1.89 ± 0.05	0.61 ± 0.03
UrbanFM	2.07 ± 0.03	0.25 ± 0.02	1.80 ± 0.02	0.41 ± 0.02
Local	1.98 ± 0.01	0.20 ± 0.01	1.83 ± 0.01	0.43 ± 0.01
Local+global	1.96 ± 0.00	0.20 ± 0.01	1.78 ± 0.01	0.38 ± 0.01
Local+global+TD	**1.95 ± 0.00**	**0.18 ± 0.01**	**1.76 ± 0.01**	**0.35 ± 0.00**

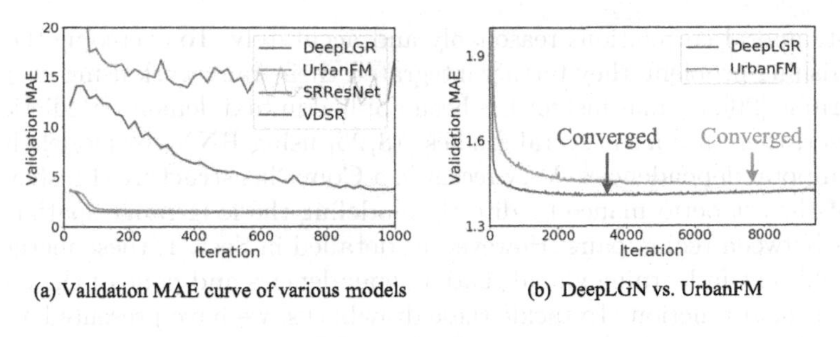

(a) Validation MAE curve of various models (b) DeepLGN vs. UrbanFM

Fig. 8. Convergence speed of various methods over P1.

equipped with our framework (denoted as local+global+TD) shows considerable improvements against its original version on both datasets, validating its great generality in different applications. For example, DeepLGR achieves 5.8% lower MAE and 28.0% lower SMAPE than UrbanFM in the TaxiBJ-P1 dataset. 2) The three components of DeepLGR are effective according to the advancement of performance (only except local vs. UrbanFM in HappyValley). 3) Compared to VDSR and SRResNet for image-resolution , UrbanFM outperforms them by considering the domain knowledge, i.e., spatial hierarchy and external influence [11]. From above discussions, we can see that existing approaches like UrbanFM can be easily integrated with our framework.

We further investigate the efficiency of DeepLGR. Figure 8 plots the MAE on the validation set during the training phase using TaxiBJ-P1. Remarkably, UrbanFM and DeepLGR converge much smoother and faster than the others as shown in Fig. 8(a). A more detailed comparison between UrbanFM and DeepLGR lies in Fig. 8(b). From this figure, we can see that DeepLGR converges at iteration 3540 (epoch 37) while UrbanFM early-stops at iteration 7720 (epoch 81). This fact demonstrates that our framework can also accelerate the training phase of existing method.

5 Related Work

Citywide crowd flow analytics has attracted considerable attention of researchers in recent years. A series of studies have explored forecasting millions or even billions of individual mobility traces [1,16]. Different from analyzing crowd behaviors on an individual level, several works started to forecast citywide crowd flow by aggregating the crowds into corresponding regions [4,10]. Among them, statistical learning was employed to capture inter-region relationship. With interest in obtaining fine-grained regional data, several studies [11,13,24] presented techniques to recover fine-grained crowd flow from coarse-grained data.

Recently, there have been many attempts focusing on end-to-end deep learning solutions such as CNNs for citywide crowd flow analytics. A pioneering study by [21] presented a general framework based on CNNs for citywide crowd flow prediction. By using a CNN architecture, their method can capture the spatio-temporal correlations reasonably and accurately. To overcome the gradient vanishing problem, they further integrated their framework using deep residual learning [20]. Similar insight has been applied in taxi demand prediction [19]. Moreover, there are also several studies [18,25] using RNNs to model the periodic temporal dependencies. Very recently, a ConvPlus structure [12] showed the state-of-the-art performance by directly modeling the long-range spatial dependencies between region pairs. However, as detailed in Sect. 1, these methods are very inefficient in learning global spatial dependencies and none of them considers latent land function. To tackle these drawbacks, we have presented a general framework that can be easily generalized to all kinds of crowd flow data.

6 Conclusion and Future Work

In this paper, we have carefully investigated existing CNN-based methods for citywide crowd flow analytics, and exposed their inefficiency in capturing global spatial dependencies and incapability in generating region-specific predictions. Based on our discovery, we have presented the DeepLGR framework which decouples the local feature extraction and global context modeling, and provides a parameter-efficient solution for customizing regional outputs. We have evaluated DeepLGR over two real-world citywide crowd flow analytics tasks. In the prediction task, DeepLGR outperforms the state-of-the-art (DeepSTN+) by average 8.8% and 45.9% on TaxiBJ dataset, and 4.2% and 25.9% on HappyValley dataset in terms of MAE and SMAPE metrics respectively. Moreover, our framework is more lightweight than the state-of-the-art methods, which is very important in real practice. In the second task, we have verified that the existing approach can be easily integrated with our framework to boost its performance. In the future, we will explore two directions. First, we notice that manually designing neural networks requires amount of expert efforts and domain knowledge. To overcome this problem, we can follow a very recent study [9] to study Neural Architecture Search (NAS), which can automatically construct a general neural network for diverse spatio-temporal tasks in cities. Second, we will extend our framework to a much broader set of spatio-temporal tasks by using graph convolutions.

Acknowledgement. We thank all reviewers for their constructive and kind suggestions. This work was supported by the National Key R&D Program of China (2019YFB2101805) and Beijing Academy of Artificial Intelligence (BAAI).

References

1. Fan, Z., Song, X., Shibasaki, R., Adachi, R.: Citymomentum: an online approach for crowd behavior prediction at a citywide level. In: Proceedings of the ACM International Joint Conference on Pervasive and Ubiquitous Computing (2015)
2. He, K., Zhang, X., Ren, S., Sun, J.: Spatial pyramid pooling in deep convolutional networks for visual recognition. IEEE Trans. Pattern Anal. Machine Intell. **37**(9), 1904–1916 (2015)

3. He, K., Zhang, X., Ren, S., Sun, J.: Deep residual learning for image recognition. In: CVPR, pp. 770–778 (2016)
4. Hoang, M.X., Zheng, Y., Singh, A.K.: FCCF: forecasting citywide crowd flows based on big data. In: SIGSPATIAL, p. 6 (2016)
5. Hu, J., Shen, L., Sun, G.: Squeeze-and-excitation networks. In: Proceedings of the IEEE Conference on Computer Vision and Pattern Recognition, pp. 7132–7141 (2018)
6. Ioffe, S., Szegedy, C.: Batch normalization: Accelerating deep network training by reducing internal covariate shift. arXiv preprint arXiv:1502.03167 (2015)
7. Kim, J., Kwon Lee, J., Mu Lee, K.: Accurate image super-resolution using very deep convolutional networks. In: Proceedings of the IEEE Conference on Computer Vision and Pattern Recognition, pp. 1646–1654 (2016)
8. Ledig, C., et al.: Photo-realistic single image super-resolution using a generative adversarial network. In: Proceedings of the IEEE Conference on Computer Vision and Pattern Recognition, pp. 4681–4690 (2017)
9. Li, T., Zhang, J., Bao, K., Liang, Y., Li, Y., Zheng, Y.: Autost: efficient neural architecture search for spatio-temporal prediction. In: Proceedings of the 26th ACM SIGKDD International Conference on Knowledge Discovery & Data Mining (2020)
10. Li, Y., Zheng, Y., Zhang, H., Chen, L.: Traffic prediction in a bike-sharing system. In: SIGSPATIAL, pp. 1–10 (2015)
11. Liang, Y., et al.: Urbanfm: Inferring fine-grained urban flows. In: Proceedings of the 25th ACM SIGKDD International Conference on Knowledge Discovery & Data Mining, p. 3132–3142 (2019)
12. Lin, Z., Feng, J., Lu, Z., Li, Y., Jin, D.: Deepstn+: context-aware spatial-temporal neural network for crowd flow prediction in metropolis. Proc. AAAI Conf. Artif. Intell. **33**, 1020–1027 (2019)
13. Ouyang, K., Liang, Y., Liu, Y., Tong, Z., Ruan, S., Zheng, Y., Rosenblum, D.S.: Fine-grained urban flow inference. arXiv preprint arXiv:2002.02318 (2020)
14. Pan, Z., Wang, Z., Wang, W., Yu, Y., Zhang, J., Zheng, Y.: Matrix factorization for spatio-temporal neural networks with applications to urban flow prediction. In: CIKM, pp. 2683–2691 (2019)
15. Shi, W., et al.: Real-time single image and video super-resolution using an efficient sub-pixel convolutional neural network. In: Proceedings of the IEEE Conference on Computer Vision and Pattern Recognition, pp. 1874–1883 (2016)
16. Song, X., Zhang, Q., Sekimoto, Y., Shibasaki, R.: Prediction of human emergency behavior and their mobility following large-scale disaster. In: Proceedings of the 20th ACM SIGKDD International Conference on Knowledge Discovery and Data Mining, pp. 5–14 (2014)
17. Tucker, L.R.: Some mathematical notes on three-mode factor analysis. Psychometrika **31**(3), 279–311 (1966)
18. Yao, H., Tang, X., Wei, H., Zheng, G., Li, Z.: Revisiting spatial-temporal similarity: a deep learning framework for traffic prediction. In: AAAI (2019)
19. Yao, H., et al.: Deep multi-view spatial-temporal network for taxi demand prediction. In: AAAI (2018)
20. Zhang, J., Zheng, Y., Qi, D.: Deep spatio-temporal residual networks for citywide crowd flows prediction. In: Thirty-First AAAI Conference (2017)
21. Zhang, J., Zheng, Y., Qi, D., Li, R., Yi, X.: DNN-based prediction model for spatio-temporal data. In: SIGSPATIAL, p. 92 (2016)

22. Zhao, H., Shi, J., Qi, X., Wang, X., Jia, J.: Pyramid scene parsing network. In: Proceedings of the IEEE Conference on Computer Vision and Pattern Recognition, pp. 2881–2890 (2017)
23. Zheng, Y., Capra, L., Wolfson, O., Yang, H.: Urban computing: concepts, methodologies, and applications. ACM Trans. Intell. Syst. Technol. (TIST) **5**(3), 1–55 (2014)
24. Zong, Z., Feng, J., Liu, K., Shi, H., Li, Y.: DeepDPM: Dynamic population mapping via deep neural network. Proc. AAAI Conf. Artif. Intell. **33**, 1294–1301 (2019)
25. Zonoozi, A., Kim, J.j., Li, X.L., Cong, G.: Periodic-CRN: a convolutional recurrent model for crowd density prediction with recurring periodic patterns. In: IJCAI, pp. 3732–3738 (2018)

RLTS: Robust Learning Time-Series Shapelets

Akihiro Yamaguchi(✉), Shigeru Maya, and Ken Ueno

System AI Lab., Corporate R&D Center, Toshiba Corporation, Kawasaki, Japan
akihiro5.yamaguchi@toshiba.co.jp

Abstract. Shapelets are time-series segments effective for classifying time-series instances. Joint learning of both classifiers and shapelets has been studied in recent years because such a method provides both superior classification performance and interpretable results. For robust learning, we introduce Self-Paced Learning (SPL) and adaptive robust losses into this method. The SPL method can assign latent instance weights by considering not only classification losses but also understandable shapelet discovery. Furthermore, the adaptive robustness introduced into feature vectors is jointly learned with shapelets, a classifier, and latent instance weights. We demonstrate the superiority of AUC and the validity of our approach on UCR time-series datasets.

Keywords: Time-series shapelets · Self-paced learning · Robust losses

1 Introduction

The Internet of Things (IoT) has spurred development of time-series classification technologies using machine learning. Time-series classifications differ from general classifications in that attribute ordering and shapes in the time series are important, and the time series may include phase shifts. Among time-series classifications, learning classifiers by discovering shapelets (time-series segments effective for classifying time-series instances) has attracted considerable interest [7,18]. The idea is that discriminative features appear only in small segments, not throughout entire time series. Shapelet methods quickly predict class labels once learning models are complete. These methods typically achieve high accuracy while providing interpretability for domain experts.

The study of shapelets originally started from search-based methods [18], and several learning-based shapelet methods have been proposed [7,15,17]. These learning-based methods use Stochastic Gradient Descent (SGD) algorithms in nonconvex settings to learn both shapelets and classifiers. Learned shapelets are thus not restricted to being subseries in training time-series instances, and may not be understandable. However, they reduce algorithmic complexity as compared with search-based methods [7] and improve classification performance in terms of accuracy, F-measure, and partial AUC (pAUC) [7,15,17].

F. Hutter et al. (Eds.): ECML PKDD 2020, LNAI 12457, pp. 595–611, 2021.
https://doi.org/10.1007/978-3-030-67658-2_34

Robust learning aims at reducing the influence of noise and outliers, and has been attracting increasing attention in machine learning. One promising research direction is Self-Paced Learning (SPL) [12]. SPL was inspired by the learning processes of humans and animals—first, more reliable and easier concepts are learned and then gradually noisier and more confusing instances are incorporated into training. Unlike in conventional curriculum learning [4], learners dynamically generate their own curriculum according to what they have already learned. SPL is beneficial for avoiding bad local minima and achieves better generalization results [10,12,13,20]. Another important research direction is to learn with robust or truncated loss functions. Robust losses are less influenced by outliers than by inliers without hard thresholds between inliers and outliers, although robustness parameters need to be tuned according to data in most robust loss functions [2,9].

Real-world time-series data are often noisy and confusing due to environmental variability, measurement noise, etc. In addition, some manually assigned class labels may be incorrect. Robustness in the presence of confusing data is thus inherently important in time-series classification. In particular, it is important to discover shapelets that are robust to abnormal fluctuation in shapelet methods. SPL can be naturally applied to SGD-based shapelet methods [7,15] to avoid bad local minima and confusion of unreliable instances. Moreover, we believe that it is better to measure distances between shapelets and time-series instances by using robust losses if noises and outliers are meaningless for discriminating different classes. Despite the potential for shapelet methods to improve performance, to our knowledge, there are no previous studies in the field of time-series classification.

We propose "Robust Learning Time-series Shapelets" (RLTS) as a shapelet method to address two challenges. The first is to introduce SPL into a shapelet method. It is important that shapelets not only discriminate classes but also match reliable time-series instances of appropriate classes while avoiding noisy and confusing instances. This can be useful if, for example, we want to discover understandable shapelets. In the proposed SPL, shapelets can optionally match reliable time-series instances of the classes that shapelets latently belong to, just like imitating their time-series segments.

The second challenge is to introduce adaptive and efficient robust losses to the feature vectors of shapelet methods and to determine robustness. Automatic adjustment of the robustness parameter reduces tuning costs. Feature vectors of shapelet methods are derived from distances between time-series segments and shapelets. Because these distances are calculated many times during training, it is necessary to reduce the computational cost. We thus define a feature vector as lightweight robust distances, and the robustness parameter can be jointly and efficiently learned with shapelets and a classifier by using SGD.

Our main contributions are summarized as follows:

- We propose RLTS, which jointly learns not only shapelets and a classifier but also a robustness parameter, latent instance weights (instance reliability) in SPL, and classes that shapelets latently belong to (Fig. 1).

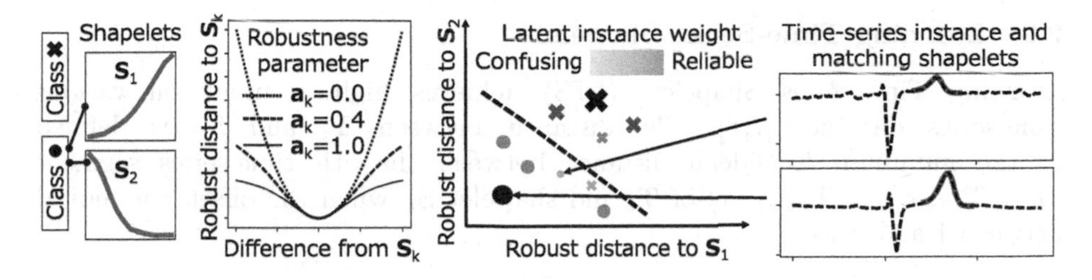

Fig. 1. RLTS jointly learns shapelets **S** and their class assignments, a model parameter of a classifier, a robustness parameter **a**, and latent instance weights in SPL.

- To discover understandable shapelets as an option, we introduce a SPL method that considers not only classification losses but also shapelet matching to reliable time-series instances of appropriate classes.
- We introduce adaptive robust losses to shapelet-based feature vectors. Even so, the algorithmic complexity is linear in time-series length by using subdifferentials, as in Ref. [17], in contrast with the quadratic or higher complexities in most existing shapelet methods [7,15,18,19].
- Using UCR time-series datasets, we demonstrate the superiority of our approach in terms of AUC as compared with state-of-the-art learning-based shapelet methods [7,15,17] and its validity through case studies.

2 Background

2.1 Preliminaries

The following describes binary classification problems where class labels $\mathcal{Y} = \{1, 0\}$. Let $\mathbf{y}_i \in \mathcal{Y}$ be the true class label of the ith instance. Given I instances, logistic classification loss with an ℓ_2 regularization term is

$$\sum_{i=i}^{I} \mathcal{L}_i, \quad \mathcal{L}_i = -\mathbf{y}_i \ln\sigma(\hat{\mathbf{y}}_i) - (1-\mathbf{y}_i)\ln\left(1-\sigma(\hat{\mathbf{y}}_i)\right) + \frac{\alpha}{I}\sum_{k=1}^{K}\mathbf{w}_k^2, \tag{1}$$

where σ is a sigmoid function defined as $\sigma(\hat{\mathbf{y}}_i) = \left(1+e^{-\hat{\mathbf{y}}_i}\right)^{-1}$, and $\alpha \geq 0$ is a regularization parameter. Here, $\hat{\mathbf{y}}_i$ is expressed via model parameter (plus bias) $\mathbf{w} \in \mathbb{R}^{K+1}$ and feature vector $\mathbf{X}_i \in \mathbb{R}^K$ for the ith instance in

$$\hat{\mathbf{y}}_i = \sum_{k=1}^{K} \mathbf{w}_k \mathbf{X}_{i,k} + \mathbf{w}_0. \tag{2}$$

As shown in Fig. 1, a time-series instance and a shapelet are respectively long and short sequences of ordered values. Let $\mathbf{T} \in \mathbb{R}^{I \times Q}$ be I time-series instances of length Q, and $\mathbf{S} \in \mathbb{R}^{K \times L}$ be K shapelets of length L. We denote the jth value of the ith time-series instance $\mathbf{T}_i$ as $\mathbf{T}_{i,j}$, and the lth value of the kth shapelet $\mathbf{S}_k$ as $\mathbf{S}_{k,l}$. There is a total of $J := Q - L + 1$ segments of length L for each time-series instance.

2.2 Learning Time-Series Shapelets

Learning Time-series Shapelets (LTS) achieves high accuracy on various time-series datasets [7,8]. The distance between $\mathbf{T}_i$ and $\mathbf{S}_k$ is defined as the minimum Euclidean distance between the jth time-series segment $(\mathbf{T}_{i,j}, \mathbf{T}_{i,j+1}, \cdots, \mathbf{T}_{i,j+L-1})$ of $\mathbf{T}_i$ and shapelet $\mathbf{S}_k$ when the offset j is moved between 1 and J as

$$\mathbf{X}_{i,k} = \min_{j=1,2,\cdots,J} \frac{1}{L} \sum_{l=1}^{L} (\mathbf{T}_{i,j+l-1} - \mathbf{S}_{k,l})^2. \tag{3}$$

For the ith instance, feature vector $\mathbf{X}_i$ is defined as $(\mathbf{X}_{i,1}, \mathbf{X}_{i,2}, \cdots, \mathbf{X}_{i,K})$. The formulation jointly optimizes shapelets $\mathbf{S}$ and model parameter $\mathbf{w}$ in

$$\underset{\mathbf{S}\in\mathbb{R}^{K\times L},\ \mathbf{w}\in\mathbb{R}^{K+1}}{\text{minimize}} \quad \sum_{i=1}^{I} \mathcal{L}_i, \tag{4}$$

where $\mathcal{L}_i$ is the per-instance regularized classification loss in Eq. (1) via Eq. (2) and feature vectors based on shapelets are defined in Eq. (3).

An extended method [8] learns where shapelets appear in a time series by an alternating direction method, focusing on reduced training time at the expense of classification accuracy as compared with LTS. Another method [19] can be extended to unsupervised clustering. Furthermore, extended methods that outperform binary classification performance in specific conditions have been proposed. Cost-Sensitive Learning Time-series Shapelets (CSLTS) [15] extends LTS by introducing an autotuned asymmetric cost-sensitive logistic loss function and outperforms the F-measures in imbalanced binary classification. Learning Time-series Shapelets for optimizing Partial AUC (LTSpAUC) [17] jointly learns shapelets and a scoring function to optimize pAUC or AUC, and outperforms the pAUC in a low False Positive Rate (FPR) range. However, prior studies have investigated introducing neither SPL nor robust losses to shapelet methods.

2.3 Self-paced Learning

SPL can automatically and dynamically choose the order in which training instances are processed for solving nonconvex optimization problems. In SPL, latent instance weights $\mathbf{v} \in [0,1]^I$ are introduced into the learning objective to indicate whether the ith instance is easy. The formulation jointly optimizes model parameter $\mathbf{w}$ and latent instance weights $\mathbf{v}$ in

$$\underset{\mathbf{w}\in\mathbb{R}^{K+1},\ \mathbf{v}\in[0,1]^I}{\text{minimize}} \quad \sum_{i=1}^{I} \left(\mathbf{v}_i \mathcal{L}_i + \tilde{\lambda} \tilde{f}(\mathbf{v}_i) \right), \quad \tilde{f}(\mathbf{v}_i) = -\mathbf{v}_i, \tag{5}$$

where $\tilde{\lambda} > 0$ is the age parameter, and $\tilde{f}(\mathbf{v}_i)$ is the SPL regularizer. Age parameter $\tilde{\lambda}$ gradually increases the learning pace. When $\tilde{\lambda}$ is small, only easy instances with small classification losses will be considered in training. As $\tilde{\lambda}$ grows, more instances with larger classification losses will be gradually appended to train a more mature model [10,12,13].

Extended SPL regularizers have been proposed with their efficient solutions. Although the SPL regularizer in Eq. (5) can either select ($\mathbf{v}_i = 1$) or unselect ($\mathbf{v}_i = 0$) in a hard-weighting scheme [12], soft-weighting methods for assigning real-valued weights have also been proposed [10,20]. More recently, a dynamic SPL regularizer that changes function $\tilde{f}$ according to the progress of the learning process has been proposed [13]. SPL has been successfully applied to various tasks, including matrix factorization [20], multimedia retrieval [10], and convolutional neural networks [13]. However, no existing methods introduce SPL to time-series classification. In particular, we use SPL that considers not only classification losses but also interpretability to discover understandable shapelets.

2.4 Robust Loss for Time-Series Distance

Some studies have considered various invariances in time-series distance. Most shapelet methods can intrinsically deal with invariances in phase shift, offset, and scaling (linear stretching). Dynamic Time Warping (DTW) is a well-known method for dealing with invariance to warping, and invariance based on time-series complexity has been proposed [3]. In Ref. [16], DTW is introduced into LTS [7] although the algorithmic complexity is cubic in time-series length. However, such approaches are different from the introduction of robust losses.

There are various nonconvex robust or truncated losses such as Tukey's biweight and Cauchy losses. We can apply these to shapelet-based feature vectors by considering the distance between a shapelet and a time-series instance as error. In most robust losses, it is tedious and expensive to tune robustness parameters according to data, although there is no need to define a hard threshold between inliers and outliers [9]. To overcome this problem, a general and adaptive robust loss function to learn a robustness parameter has been proposed [2]. However, the gradient-based learning process in Ref. [2] needs normalization so that the loss function becomes the negative log-likelihood of the probability density function, and the calculation is time-consuming. In contrast, we do not need such normalization when introducing adaptive robust losses not to objective functions but to feature vectors, and we adopt simple and efficient robust losses as described in Sect. 3.1.

3 RLTS: Robust Learning Time-Series Shapelets

3.1 Shapelet-Based Feature Vector via Robust Losses

We introduce robust losses to a shapelet-based feature vector by considering distances between a time-series instance and shapelets as losses. Let $\mathbf{a} \in \mathbb{R}^K_{\geq 0}$ be a robustness parameter. For the ith instance, we define the kth element of the feature vector as a robust distance in

$$\mathbf{X}_{i,k} = \min_{j=1,2,\cdots,J} \frac{1}{L} \sum_{l=1}^{L} \frac{\mathbf{D}^2_{i,j,k,l}}{1 + \mathbf{a}_k \mathbf{D}^2_{i,j,k,l}/2}, \quad \mathbf{D}_{i,j,k,l} = \mathbf{S}_{k,l} - \mathbf{T}_{i,j+l-1}, \tag{6}$$

where $\mathbf{a}_k \geq 0$. As shown in Fig. 1, increasing $\mathbf{a}_k$ increases robustness by reducing the influence of outliers. Clearly, our feature vector is reduced to the existing shapelet-based feature vector in Eq. (3) when $\mathbf{a}_k = 0$. Unlike existing shapelet methods [7,15,17] that use the feature vectors in Eq. (3), RLTS learns robustness parameter $\mathbf{a}$. Nevertheless, the simple formulation in Eq. (6) allows for efficient calculation of not only $\mathbf{X}_{i,k}$ but also gradients ($\partial \mathbf{X}_{i,k}/\partial \mathbf{a}_k$ and $\partial \mathbf{X}_{i,k}/\partial \mathbf{S}_{k,l}$) without increasing algorithmic complexity, as described in Sect. 3.4.

3.2 Shapelet Assignments

To support shapelet discovery, we introduce a shapelet-matching regularizer (shapelet regularizer for short). The regularizer aims for shapelets to approximate the best-matching segments of reliable time-series instances in the classes that the shapelets latently belong to. Let $\mathbf{b}_k \in \mathcal{Y}$ be the class that the kth shapelet $\mathbf{S}_k$ latently belongs to, and we call $\mathbf{b} := (\mathbf{b}_1, \mathbf{b}_2, \cdots, \mathbf{b}_K)$ the shapelet assignments. Given shapelets $\mathbf{S} \in \mathbb{R}^{K \times L}$, model parameter $\mathbf{w} \in \mathbb{R}^{K+1}$, robustness parameter $\mathbf{a} \in \mathbb{R}^K_{\geq 0}$, latent instance weights $\mathbf{v} \in [0,1]^I$, and imitation parameter $\gamma_y \geq 0$ for class $y \in \mathcal{Y}$, we assign each shapelet of $\mathbf{S}$ to a class of $\mathcal{Y}$ so as to minimize the shapelet regularizer in

$$\underset{\mathbf{b} \in \mathcal{Y}^K}{\text{minimize}} \quad \sum_{i=1}^{I} \gamma_{\mathbf{y}_i} \mathbf{v}_i \sum_{k=1}^{K} \mathbb{I}(\mathbf{b}_k = \mathbf{y}_i) \mathbf{X}_{i,k}, \tag{7}$$

where $\mathbf{X}_{i,k}$ is a feature vector element in Eq. (6), and $\mathbb{I}$ is an indicator function with value 1 if its argument is true and 0 otherwise. Shapelets should not be assigned to classes where their instances have low latent instance weights. We thus set a large value of γ_y for the unreliable class y, and vice versa. The subproblem in Eq. (7) is optionally incorporated into our optimization formulation in order to make shapelets imitate reliable segments, as described in Sect. 3.3.

3.3 Optimization Formulation

We can formulate our optimization problem to jointly learn shapelets $\mathbf{S}$, model parameter $\mathbf{w}$, robustness parameter $\mathbf{a}$, latent instance weights $\mathbf{v}$, and shapelet assignments $\mathbf{b}$ as

$$\underset{\substack{\mathbf{S} \in \mathbb{R}^{K \times L},\ \mathbf{w} \in \mathbb{R}^{K+1} \\ \mathbf{a} \in \mathbb{R}^K_{\geq 0},\ \mathbf{v} \in [0,1]^I \\ \mathbf{b} \in \mathcal{Y}^K}}{\text{minimize}} \sum_{i=1}^{I} G_i, \ G_i = \mathbf{v}_i \mathcal{L}_i + \lambda_{\mathbf{y}_i} f(\mathbf{v}_i) + \gamma_{\mathbf{y}_i} \mathbf{v}_i \sum_{k=1}^{K} \mathbb{I}(\mathbf{b}_k = \mathbf{y}_i) \mathbf{X}_{i,k}, \tag{8}$$

where $\mathbf{X}_{i,k}$ is the kth element of the robust feature vector (i.e., the kth robust distance) in Eq. (6), $\mathcal{L}_i$ is the per-instance classification loss in Eq. (1) via Eq. (2), and $f(\mathbf{v}_i)$ is the SPL regularizer, defined as

$$f(\mathbf{v}_i) = \frac{q}{1+q} \mathbf{v}_i^{\frac{1}{q}+1} - \mathbf{v}_i, \tag{9}$$

where $q > 0$ changes during learning as in Ref. [13]. To solve this problem, we apply an alternate optimization algorithm. In each iteration, one block of variables is optimized while keeping the other blocks fixed. At each iteration, we increase age parameter $\lambda_y > 0$ to gradually involve more difficult instances while decreasing $q > 0$, and decrease imitation parameter $\gamma_y \geq 0$ to gradually reduce the effect of the shapelet regularizer just as humans imitate their environment before learning to create new things. We omit use of the shapelet regularizer by setting $\gamma_y = 0$. Section 3.7 describes how to decide these paces.

Our SPL regularizer f has the following important properties.

Proposition 1. *$f(v)$ in Eq. (9) satisfies the following when $0 \leq v \leq 1$:*

(a) $f(v)$ is a strictly monotonically decreasing function (i.e., f always satisfies $f(v) > f(\tilde{v})$ for any v and $\tilde{v}$, where $0 \leq v < \tilde{v} \leq 1$).
(b) $f(v)$ is a convex function.

Proof. Noting that $q > 0$, because $\partial f(v)/\partial v = v^{\frac{1}{q}} - 1$, $\partial f(v)/\partial v < 0$ where $0 \leq v < 1$ and $\partial f(v)/\partial v = 0$ where $v = 1$, so item (a) holds. Because $\partial^2 f(v)/\partial v^2 = \frac{1}{q} v^{\frac{1}{q}-1} \geq 0$ where $0 \leq v \leq 1$ and $q > 0$, item (b) holds.

Proposition 1(a) indicates that the optimal value of $\mathbf{v}_i$ in Eq. (8) goes to 1 if $\lambda_{\mathbf{y}_i} \to \infty$ for any ith instance. In addition, because shapelet assignments $\mathbf{b}$ only affect the third term of G_i in Eq. (8), we can neglect optimization of $\mathbf{b}$ when $\gamma_{\mathbf{y}_i} = 0$. Our formulation is therefore reduced to that of LTS in Eq. (4) with $\lambda_y \to \infty$, $\gamma_y = 0$, and $\mathbf{a}_k = 0$ for $k = 1, 2, \cdots, K$ and $y \in \mathcal{Y}$.

To alternately optimize Eq. (8), we divide the variables into three disjoint blocks. In the first block, RLTS simultaneously learns shapelets $\mathbf{S}$, model parameter $\mathbf{w}$, and robustness parameter $\mathbf{a}$. In the second and third blocks, RLTS learns latent instance weights $\mathbf{v}$ and shapelet assignments $\mathbf{b}$, respectively. The following three sections describe how to learn these blocks.

3.4 Learning Shapelets, and Model and Robustness Parameters

We optimize shapelets $\mathbf{S}$, model parameter $\mathbf{w}$, and robustness parameter $\mathbf{a}$ while fixing the other variables $\mathbf{v}$ and $\mathbf{b}$. Although the subproblem remains nonconvex with respect to $\mathbf{S}$ and $\mathbf{a}$, our objective function is decomposable into the sum of each instance contribution G_i. Therefore, we apply a SGD algorithm to jointly optimize $\mathbf{S}$, $\mathbf{w}$, and $\mathbf{a}$.

The gradients of per-instance objective function G_i with respect to shapelets $\mathbf{S}$ are expressed via the chain rule for derivatives by noting that SPL regularizer f does not depend on $\mathbf{S}_{k,l}$ in

$$\frac{\partial G_i}{\partial \mathbf{S}_{k,l}} = \mathbf{v}_i \frac{\partial \mathcal{L}_i}{\partial \hat{\mathbf{y}}_i} \frac{\partial \hat{\mathbf{y}}_i}{\partial \mathbf{X}_{i,k}} \frac{\partial \mathbf{X}_{i,k}}{\partial \mathbf{S}_{k,l}} + \gamma_{\mathbf{y}_i} \mathbf{v}_i \mathbb{I}(\mathbf{b}_k = \mathbf{y}_i) \frac{\partial \mathbf{X}_{i,k}}{\partial \mathbf{S}_{k,l}}. \tag{10}$$

The calculation of $\partial \mathcal{L}_i / \partial \hat{\mathbf{y}}_i$ and $\partial \hat{\mathbf{y}}_i / \partial \mathbf{X}_{i,k}$ is the same as in Ref. [7], namely,

$$\frac{\partial \mathcal{L}_i}{\partial \hat{\mathbf{y}}_i} = \sigma(\hat{\mathbf{y}}_i) - \mathbf{y}_i, \quad \frac{\hat{\mathbf{y}}_i}{\partial \mathbf{X}_{i,k}} = \mathbf{w}_k. \tag{11}$$

We derive $\partial \mathbf{X}_{i,k}/\partial \mathbf{S}_{k,l}$ as

$$\frac{\partial \mathbf{X}_{i,k}}{\partial \mathbf{S}_{k,l}} = \frac{\mathbf{D}_{i,j^*,k,l}}{L\left(1+\mathbf{a}_k \mathbf{D}_{i,j^*,k,l}^2/2\right)}\left(2-\frac{\mathbf{a}_k \mathbf{D}_{i,j^*,k,l}^2}{1+\mathbf{a}_k \mathbf{D}_{i,j^*,k,l}^2/2}\right),$$
$$j^* = \underset{j=1,2,\cdots,J}{\operatorname{argmin}} \frac{1}{L}\sum_{l=1}^{L} \frac{\mathbf{D}_{i,j,k,l}^2}{1+\mathbf{a}_k \mathbf{D}_{i,j,k,l}^2/2}, \quad \mathbf{D}_{i,j,k,l} = \mathbf{S}_{k,l} - \mathbf{T}_{i,j+l-1}. \tag{12}$$

Here, we use the subdifferential for the differential of the minimum function without approximating the minimum function, as in Ref. [17].

The gradients of per-instance objective function G_i with respect to robustness parameter $\mathbf{a}$ are expressed by noting that f does not depend on $\mathbf{a}_k$ in

$$\frac{\partial G_i}{\partial \mathbf{a}_k} = \mathbf{v}_i \frac{\partial \mathcal{L}_i}{\partial \hat{\mathbf{y}}_i}\frac{\partial \hat{\mathbf{y}}_i}{\partial \mathbf{X}_{i,k}}\frac{\partial \mathbf{X}_{i,k}}{\partial \mathbf{a}_k} + \gamma_{\mathbf{y}_i}\mathbf{v}_i \mathbb{I}(\mathbf{b}_k = \mathbf{y}_i)\frac{\partial \mathbf{X}_{i,k}}{\partial \mathbf{a}_k}. \tag{13}$$

We derive $\partial \mathbf{X}_{i,k}/\partial \mathbf{a}_k$ by using subdifferentials as in Eq.(12):

$$\frac{\partial \mathbf{X}_{i,k}}{\partial \mathbf{a}_k} = -\frac{1}{2L}\sum_{l=1}^{L} \frac{\mathbf{D}_{i,j^*,k,l}^4}{\left(1+\mathbf{a}_k \mathbf{D}_{i,j^*,k,l}^2/2\right)^2}. \tag{14}$$

We thus reduce the algorithmic complexity to linear for time-series length despite introducing robust losses, by using the subdifferential-based formulation.

Since model parameter $\mathbf{w}$ only depends on $\mathcal{L}_i$ in Eq. (8), the gradients of per-instance objective function G_i with respect to $\mathbf{w}$ are the same as in Ref. [7] except for latent instance weight $\mathbf{v}_i$:

$$\frac{\partial G_i}{\partial \mathbf{w}_k} = \begin{cases} \mathbf{v}_i\left((\sigma(\hat{\mathbf{y}}_i) - \mathbf{y}_i)\,\mathbf{X}_{i,k} + \frac{2\alpha}{I}\mathbf{w}_k\right) & \text{if } k = 1, 2, \cdots, K, \\ \mathbf{v}_i\left(\sigma(\hat{\mathbf{y}}_i) - \mathbf{y}_i\right) & \text{if } k = 0. \end{cases} \tag{15}$$

3.5 Learning Latent Instance Weights

We optimize latent instance weights $\mathbf{v}$ while fixing the other variables. Because our objective function is decomposable as $\sum_{i=1}^{I} G_i$, it is sufficient to optimize G_i individually. We summarize the properties of G_i as follows.

Proposition 2. *Per-instance objective function G_i satisfies the following:*

(a) G_i is convex with respect to $\mathbf{v}_i$.
(b) $\mathbf{v}_i^* := \operatorname{argmin}_{\mathbf{v}_i \in [0,1]} G_i$ is monotonically increasing with respect to $\lambda_{\mathbf{y}_i}$.

Proof. *Term $\lambda_{\mathbf{y}_i} f(\mathbf{v}_i)$ is convex with respect to $\mathbf{v}_i$* where *$\lambda_{\mathbf{y}_i} > 0$ from Proposition 1 (b). Other terms $\mathbf{v}_i \mathcal{L}_i$* and *$\gamma_{\mathbf{y}_i}\mathbf{v}_i \sum_{k=1}^{K}\mathbb{I}(\mathbf{b}_k = \mathbf{y}_i)\mathbf{X}_{i,k}$ are also convex with respect to $\mathbf{v}_i$. Because a sum of convex functions is convex, item (a) holds. Although $\lambda_{\mathbf{y}_i} f(\mathbf{v}_i)$ is monotonically decreasing with respect to $\mathbf{v}_i$ from Proposition 1 (a), other terms $\mathbf{v}_i \mathcal{L}_i$* and *$\gamma_{\mathbf{y}_i}\mathbf{v}_i \sum_{k=1}^{K}\mathbb{I}(\mathbf{b}_k = \mathbf{y}_i)\mathbf{X}_{i,k}$ are monotonically increasing with respect to $\mathbf{v}_i$. This tradeoff relation indicates that increasing $\lambda_{\mathbf{y}_i}$ never decreases $\mathbf{v}_i^*$,so item (b) holds.*

From Proposition 2, we can solve the optimal value of latent instance weight $\mathbf{v}_i$ in a closed form as follows.

Proposition 3. *When variables* $\mathbf{S}$, $\mathbf{w}$, $\mathbf{a}$, *and* $\mathbf{b}$ *are fixed, the optimal value of latent instance weight* $\mathbf{v}_i$ *in Eq. (8) for the ith instance is*

$$\mathbf{v}_i = \begin{cases} \left(1 - \frac{\mathcal{L}_i + \gamma_{\mathbf{y}_i}\sum_{k=1}^{K}\mathbb{I}(\mathbf{b}_k = \mathbf{y}_i)\mathbf{X}_{i,k}}{\lambda_{\mathbf{y}_i}}\right)^q & \text{if } \mathcal{L}_i + \gamma_{\mathbf{y}_i}\sum_{k=1}^{K}\mathbb{I}(\mathbf{b}_k = \mathbf{y}_i)\mathbf{X}_{i,k} \leq \lambda_{\mathbf{y}_i}, \\ 0 & \text{if } \mathcal{L}_i + \gamma_{\mathbf{y}_i}\sum_{k=1}^{K}\mathbb{I}(\mathbf{b}_k = \mathbf{y}_i)\mathbf{X}_{i,k} > \lambda_{\mathbf{y}_i}. \end{cases} \tag{16}$$

Proof. Under the first condition, we can derive the closed form by solving the following equation:

$$\frac{\partial G_i}{\partial \mathbf{v}_i} = \mathcal{L}_i + \lambda_{\mathbf{y}_i}\left(\mathbf{v}_i^{\frac{1}{q}} - 1\right) + \gamma_{\mathbf{y}_i}\sum_{k=1}^{K}\mathbb{I}(\mathbf{b}_k = \mathbf{y}_i)\mathbf{X}_{i,k} = 0.$$

From Proposition 2 (a), the solution is globally optimal, so the first case holds. When $\lambda_{\mathbf{y}_i}$ *is equal to* $\mathcal{L}_i + \gamma_{\mathbf{y}_i}\sum_{k=1}^{K}\mathbb{I}(\mathbf{b}_k = \mathbf{y}_i)\mathbf{X}_{i,k}$, $\mathbf{v}_i$ *is zero. From Proposition 2 (b), when* $\lambda_{\mathbf{y}_i}$ *decreases from the value,* $\mathbf{v}_i$ *does not increase from zero. The second case therefore also holds.*

We call term $\gamma_{\mathbf{y}_i}\sum_{k=1}^{K}\mathbb{I}(\mathbf{b}_k = \mathbf{y}_i)\mathbf{X}_{i,k}$ in Eq. (16) the per-instance shapelet regularizer. Figure 2 shows the values of latent instance weight $\mathbf{v}_i$ when we change the per-instance shapelet regularizer and per-instance classification loss.

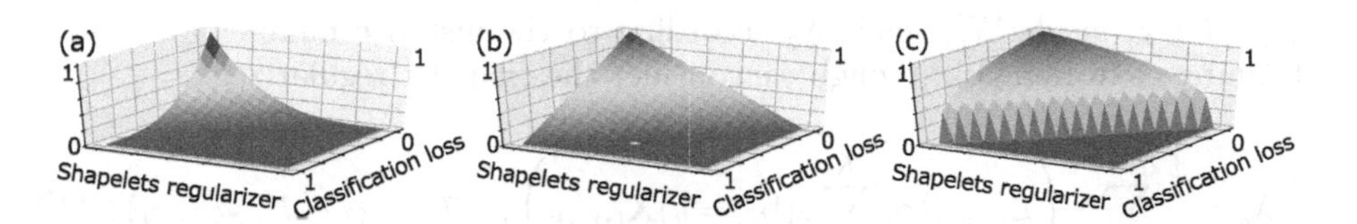

Fig. 2. Latent instance weight $\mathbf{v}_i$ in Eq. (16) with changing per-instance shapelet regularizer and classification loss where $\lambda_{\mathbf{y}_i} = 1$, while decreasing q in from (a) to (c).

3.6 Learning Shapelet Assignments

We optimize shapelet assignments $\mathbf{b}$ while fixing the other variables. Because $\mathbf{b}$ only affects the third term of G_i in Eq. (8), we can reduce the optimization problem in Eq. (8) to the subproblem in Eq. (7). We can efficiently find the optimal class that each shapelet belongs to as follows.

Proposition 4. *Optimal assignment in Eq. (7) for the kth shapelet is:*

$$\mathbf{b}_k^* = \operatorname*{argmin}_{y\in\mathcal{Y}}\sum_{i=1}^{I}\gamma_{\mathbf{y}_i}\mathbf{v}_i\mathbb{I}(y = \mathbf{y}_i)\mathbf{X}_{i,k}. \tag{17}$$

Proof. Because the objective function in Eq. (7) *is decomposable into the sum of each shapelet contribution* $\sum_{i=1}^{I} \gamma_{\mathbf{y}_i} \mathbf{v}_i \mathbb{I}(\mathbf{b}_k = \mathbf{y}_i) \mathbf{X}_{i,k}$ *for* $k = 1, 2, \cdots, K$, *it is sufficient to minimize each contribution individually. Therefore,* $\mathbf{b}^* := (\mathbf{b}_1^*, \mathbf{b}_2^*, \cdots, \mathbf{b}_K^*)$ *in Eq.* (17) it is a global minimizer in Eq. (7).

3.7 Implementation

RLTS repeatedly optimizes each block in an outer loop while performing SGD in an inner loop. In the outer loop, age parameter λ_y increases while imitation parameter γ_y decreases for class $y \in \mathcal{Y}$. In practice, it is difficult to decide those paces from absolute values. We first decide imitation parameter γ_y according to the ratio between weighted classification losses and the shapelet regularizer as

$$\gamma_y = \frac{\beta}{\sum_{i=1}^{I} \mathbb{I}(\mathbf{y}_i = y)\mathbf{v}_i} \frac{\tilde{M} - \tilde{m}}{\tilde{M} - 1} \frac{\sum_{i=1}^{I} \mathbb{I}(\mathbf{y}_i = y)\mathbf{v}_i \mathcal{L}_i}{\sum_{i=1}^{I} \mathbb{I}(\mathbf{y}_i = y)\mathbf{v}_i \sum_{k=1}^{K} \mathbb{I}(\mathbf{b}_k = y)\mathbf{X}_{i,k}}, \tag{18}$$

where $\tilde{M}$ and $\tilde{m}$ are respectively the numbers of outer and current iterations, and $\tilde{m}$ increases from 1 to $\tilde{M}$ in the outer loop. Here, imitation weight $\beta \geq 0$ is a single meta-parameter for tuning the pace. We omit use of the shapelet regularizer by setting $\beta = 0$. Since Eq. (18) normalizes by the total latent instance weights of class y, imitation parameter γ_y increases if class y is unreliable, and vice versa.

Next, we choose age parameter λ_y for each class $y \in \mathcal{Y}$ to balance selected instances among classes. Let $(i)_y$ be the ith index among the I instances when we sort the class y instances in $\{\mathcal{L}_{\tilde{i}} + \gamma_y \sum_{k=1}^{K} \mathbb{I}(\mathbf{b}_k = y)\mathbf{X}_{\tilde{i},k}\}_{\tilde{i} \in \mathcal{I}_y}$ where $\mathcal{I}_y = \{\tilde{i} \in I \mid \mathbf{y}_{\tilde{i}} = y\}$. We decide λ_y according to the instance number involved in training as in Ref. [13], though we consider the shapelet regularizer as

$$\lambda_y = \sum_{i=1}^{\lceil \frac{\tilde{m}}{\tilde{M}} |\mathcal{I}_y| \rceil} \left(\mathcal{L}_{(i)_y} + \gamma_y \sum_{k=1}^{K} \mathbb{I}(\mathbf{b}_k = y)\mathbf{X}_{(i)_y,k} \right), \quad \mathcal{I}_y = \{\tilde{i} \in I \mid \mathbf{y}_{\tilde{i}} = y\}, \tag{19}$$

where $|\cdot|$ denotes cardinality, and $\lceil \cdot \rceil$ is the ceiling function.

Algorithm 1 shows pseudocode for RLTS. Variables are initialized at line 1. We use a k-means++ clustering algorithm [1] to acquire initial shapelets as the cluster centroids, as in Ref. [7]. The outer loop (lines 2–11) is iterated $\tilde{M}$ times in the alternate optimization. After updating age and imitation parameters and q in Eq. (9) at line 3, RLTS learns shapelets $\mathbf{S}$, model parameter $\mathbf{w}$, and robustness parameter $\mathbf{a}$ in the inner SGD loop (lines 4–9). At lines 7–8, we apply a projected sub-gradient method to satisfy $\mathbf{a}_k \geq 0$. We update latent instance weights $\mathbf{v}$ at line 10, and shapelet assignments $\mathbf{b}$ at line 11.

Algorithmic Complexity: As in Ref. [7], we omit K and L from our calculations because they take small values. The complexity in the inner loop is $O\left(\frac{MIQ}{\tilde{M}}\right)$, and we need to sort instances to calculate Eq. (19) in the outer loop.

Algorithm 1: RLTS

Input: Training time-series instances: $\mathbf{T} \in \mathbb{R}^{I \times Q}$; Class labels: $\mathcal{Y}^I$; Shapelet length: L; Number of shapelets: K; ℓ_2 regularization parameter: α; Imitation weight: β; Learning rate: η; Numbers of total and outer iterations: $(M, \tilde{M})$

Output: Shapelets: $\mathbf{S} \in \mathbb{R}^{K \times L}$; Model parameter: $\mathbf{w} \in \mathbb{R}^{K+1}$; Robustness parameter: $\mathbf{a} \in \mathbb{R}^K_{\geq 0}$; Latent instance weights: $\mathbf{v} \in [0,1]^I$; Shapelet assignments: $\mathbf{b} \in \mathcal{Y}^K$

1 Initialize $\mathbf{S}$ and $\mathbf{b}$, and set $\mathbf{w} \leftarrow \mathbf{0}$, $\mathbf{a} \leftarrow \mathbf{0}$, and $\mathbf{v} \leftarrow \mathbf{1}$.
2 **for** $\tilde{m} = 1, 2, \cdots, \tilde{M}$ **do**
3   Set γ_y and λ_y for $y \in \mathcal{Y}$ in Eqs. (18)–(19), and $q \leftarrow 100^{\frac{1}{2} - \frac{\tilde{m}-1}{\tilde{M}-1}}$.
4   **for** $m = 1, 2, \cdots, M/\tilde{M}$ **do**
5   **for** $i = 1, 2, \cdots, I$ **do**
6             $\mathbf{S}_{k,l} \leftarrow \mathbf{S}_{k,l} - \eta \frac{\partial G_i}{\partial \mathbf{S}_{k,l}}$ for $k = 1, 2, \cdots, K$ and $l = 1, 2, \cdots, L$ in Eqs. (10)–(12).
7             $\mathbf{a}_k \leftarrow \mathbf{a}_k - \eta \frac{\partial G_i}{\partial \mathbf{a}_k}$ for $k = 1, 2, \cdots, K$ in Eqs. (11) and (13)–(14).
8             $\mathbf{a}_k \leftarrow \max\{0, \mathbf{a}_k\}$ for $k = 1, 2, \cdots, K$.
9             $\mathbf{w}_k \leftarrow \mathbf{w}_k - \eta \frac{\partial G_i}{\partial \mathbf{w}_k}$ for $k = 0, 1, \cdots, K$ in Eq. (15).
10     $\mathbf{v} \leftarrow \operatorname{argmin}_{\mathbf{v} \in [0,1]^I} G$ in Eq. (16).
11     $\mathbf{b} \leftarrow \operatorname{argmin}_{\mathbf{b} \in \mathcal{Y}^K} G$ in Eq. (17).

The total complexity is thus $O\left(MIQ + \tilde{M} \sum_{y \in \mathcal{Y}} (|\mathcal{I}_y| \log |\mathcal{I}_y|)\right)$. As described in Sect. 3.4, RLTS can linearly scale to time-series length Q despite using adaptive robust distances.

4 Experiments

4.1 Experimental Setup

We compared the proposed method (**RLTS**) with the following state-of-the-art learning-based shapelet methods. **RLTSNoAd** does not use robust distances (i.e., $\mathbf{a} = \mathbf{0}$) by modifying RLTS. **RLTSNoMa** does not use the shapelet regularizer (i.e., $\beta = 0$) by modifying RLTS. **LTS** [7] is a well-known learning-based shapelet method that achieves high accuracy in time-series classification [7,8,14]. **CSLTS** [15] was proposed as a shapelet method that extends LTS by introducing cost-sensitive learning and achieves high F-measures in imbalanced binary classification[1]. **LTSpAUC** [17] predicts not class labels but ranking scores optimized in AUC, and achieves high AUC[2].

[1] CSLTS more severely penalizes error in class 1 than in class 0. We therefore swap class labels when class 1 is larger than class 0.

[2] LTSpAUC was originally proposed to optimize pAUC in any FPR range. Here, we use it to optimize AUC.

We use 30 UCR time-series datasets [5] for binary classification after removing datasets with missing values. We use the default training and testing splits, and repeat the experiments ten times. We use training data to tune meta-parameters. Learning rate η for each method is chosen from $\{0.01, 0.1, 1, 10, 100\}$. The regularization parameter based on ℓ_2 constraint is chosen from $\{1, 10, 100\}$ for LTSpAUC, while ℓ_2 regularization parameter α is chosen from $\{0.01, 1, 100\}$ for other methods. RLTS and RLTSNoAd have an additional imitation weight, which is chosen from $\beta \in \{0.1, 1, 10\}$. CSLTS has two additional meta-parameters, which are chosen from $\theta \in \{1, 10, 100\}$ and $D \in \{0.1, 10\}$. To reduce computation times, we set the following common meta-parameters: shapelet length L and number of shapelets K are $0.1 \times Q$, and the numbers of total iterations M and outer iterations $\tilde{M}$ are respectively 600 and 20.

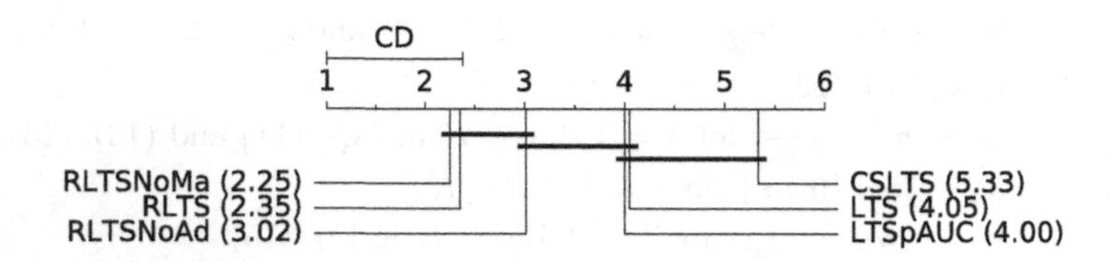

Fig. 3. Critical difference diagram for AUC during testing.

4.2 Experimental Results

We compare the AUC of the six methods for the test datasets, because AUC is a threshold-independent measure and is appropriate for both balanced and imbalanced binary classification. In their place, we present a Critical Difference (CD) diagram [6] to summarize the comparison. Figure 3 shows the CD diagram for the 30 datasets. Values in the figure are the respective rank means (lower is better). Methods connected by a bold bar have no significant differences to each other at the 95% confidence level.

The figure shows that the proposed methods RLTSNoMa, RLTS, and RLTSNoAd are better than the existing methods, and that RLTSNoMa and RLTS significantly outperform the existing methods. RLTSNoAd performed worst among the proposed methods. This indicates that introducing both SPL and adaptive robust losses effectively improves AUC. In addition, by comparing RLTS and RLTSNoMa we find that the shapelet regularizer decreases AUC, although the difference is not significant. However, the shapelet regularizer can discover understandable shapelets that match reliable time series, as described in Sect. 5.

4.3 Scalability in Time-Series Length

Figure 4(a) shows the RLTS (Algorithm 1) runtime while we change the length of the synthetic time series, as in Ref. [17]. As that figure shows, our algorithm

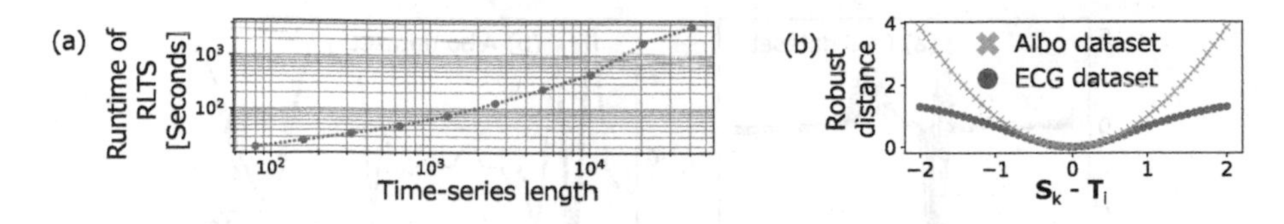

Fig. 4. (a) Scalability in time-series length. (b) Learned robustness parameter.

scales linearly with time-series length Q when Q is large. These results are consistent with the analysis of algorithmic complexity in Sect. 3.7.

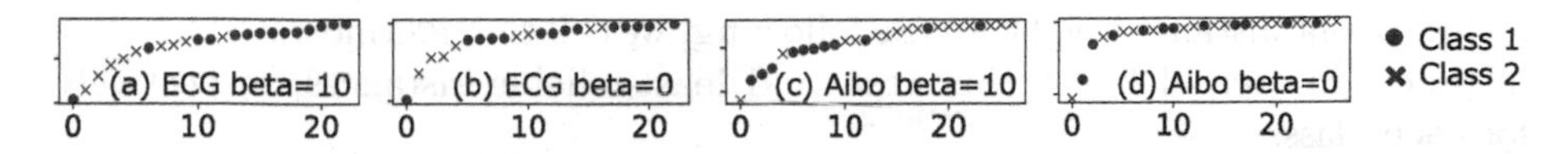

Fig. 5. Sorted latent instance weights $\mathbf{v}$ on each dataset when imitation weight β is 10 or 0. In each figure, the rightmost and leftmost instances are the most- and least-reliable, respectively, for each class in terms of SPL.

5 Case Studies

We evaluate learning of the robustness parameter and latent instance weights as well as the shapelet regularizer through case studies. The experimental setting is the same as in Sect. 4.1. We measure the extent to which shapelet $\mathbf{S}_k$ contributes to the classification in $|\mathbf{w}_k|$. In other words, shapelet $\mathbf{S}_{k^*}$ where $k^* = \text{argmax}_{k=1,\cdots,K}|\mathbf{w}_k|$ most contributes to the classification.

5.1 Medical Diagnosis Using ECG

ECG time series allows noninvasive and inexpensive access for many patients, so their analyses play an important role in medical diagnosis. In the ECG Five Days dataset, the only medically significant difference is cardiologist-identified T waves [11]. Because the discriminative difference is not a sharp change in waveform but a smoothed change of T waves, the learned robustness parameter $\mathbf{a}$ should have larger-than-zero values to enhance robustness. In this dataset, the mean value of robustness parameter $\mathbf{a}$ is 0.52, and Fig. 4(b) shows the robust distances in Eq. (6) for the mean parameter value when we change the difference between shapelet $\mathbf{S}_k$ and time-series instance $\mathbf{T}_i$. We find that RLTS learned to enhance robustness, as expected.

Figure 5(a) shows latent instance weights $\mathbf{v}$ during training when we strengthen the shapelet regularizer (i.e., $\beta = 10$). We plot class 1 and class 2 as "○" and "×", respectively. In Fig. 6(a), black curves are training time-series instances of class 1, and the red bold curve is the instance of the lowest latent instance weight in class 1 (i.e., the leftmost instance in Fig. 5(a)). We find that

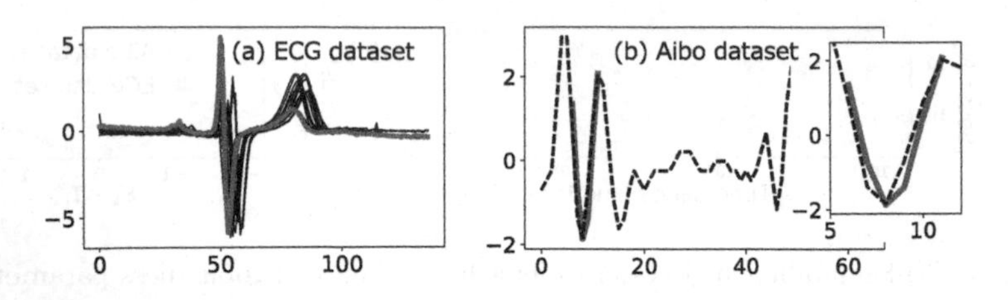

Fig. 6. Time-series instances of the lowest latent instance weights.

the red curve is different from other class 1 training instances and that the T wave of the red curve is smaller than theirs. This abnormality is consistent with its low latent instance weight. In the following, we call the rightmost and leftmost instances in each of Fig. 5 the most- and least-reliable instances, respectively, for each class.

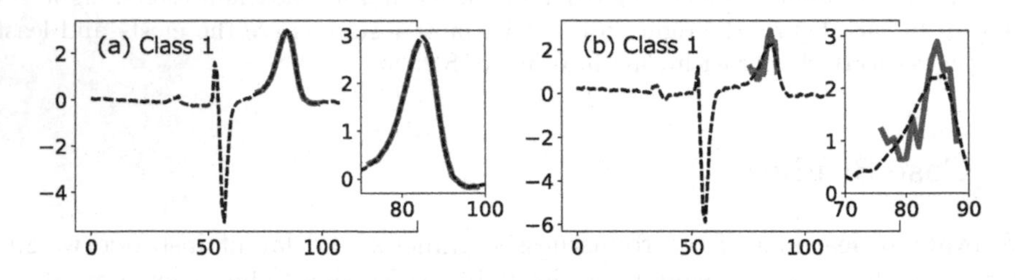

Fig. 7. Shapelets by RLTS for the ECG dataset, with (a) $\beta = 10$ and (b) $\beta = 0$.

We study the effect of the shapelet regularizer by comparing imitation weight β settings between 0 and 10. When we enhance the shapelet regularizer (i.e., $\beta = 10$), RLTS learned that the most-contributing shapelet $\mathbf{S}_{k^*}$ and the next-most contributing shapelet latently belong to class 1. Figure 7(a) shows the two most-contributing shapelets at the best matching position for the most reliable time-series instance of class 1. In contrast, Figs. 5(b) and 7(b) respectively show latent instance weights $\mathbf{v}$ and shapelet $\mathbf{S}_{k^*}$ when we neglect the shapelet regularizer (i.e., $\beta = 0$). In this case, RLTS learned that shapelet $\mathbf{S}_{k^*}$ latently belongs to class 1. We plot $\mathbf{S}_{k^*}$ at the best-matching position for the most reliable time-series instance of class 1. From Figs. 7(a) and (b), we find that those shapelets capture T waves and that the results are consistent with medical knowledge. Furthermore, comparing these figures shows that enhancing the shapelet regularizer causes the most-contributing shapelets to well match reliable time-series instances and to improve interpretability.

5.2 Surface Detection Using Sony Aibo Robot

Robot calibration for a given environment is a tedious task that usually requires human intervention, so detecting specific environmental states using robotic sensors is an interesting topic. The Sony Aibo Robot Surface II dataset contains time series collected from the accelerometer sensor of a small robot. The robot walks on carpeted and cement surfaces, and we want to classify these surfaces. Cemented floors are harder than carpets, and there are clear, sharp changes in acceleration on cement floors. In contrast to the ECG case, the learned robustness parameter **a** should have small (nearly zero) values to capture this sharpness. As Fig. 4(b) shows, the mean value of the robustness parameter **a** is zero, so RLTS learned to reduce the robustness, as expected.

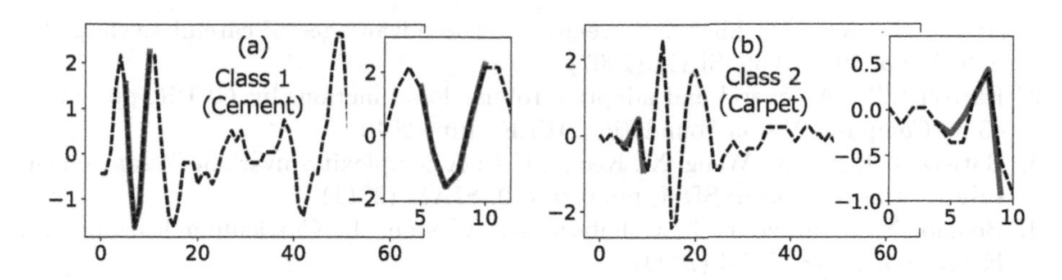

Fig. 8. Shapelets by RLTS for the Aibo dataset, with (a) $\beta = 10$ and (b) $\beta = 0$.

Figure 5(c) shows latent instance weights $\mathbf{v}$ during training when we strengthen the shapelet regularizer (i.e., $\beta = 10$). RLTS learned that the most-contributing shapelet $\mathbf{S}_{k^*}$ latently belongs to class 1. Figure 6(b) shows that shapelet $\mathbf{S}_{k^*}$ at the best-matching position for the least reliable time-series instance of class 1. As shown in the enlarged view of the figure, the instance has a left protrusion. In contrast, shapelet $\mathbf{S}_{k^*}$ and most class 1 training instances have right protrusions. This abnormality is consistent with the low latent instance weight.

We study the effects of the shapelet regularizer by comparing imitation weight β settings between 0 and 10. Figure 8(a) shows shapelet $\mathbf{S}_{k^*}$ at the best-matching position for the most reliable time-series instance of class 1 when we enhance the shapelet regularizer (i.e., $\beta = 10$). In contrast, Figs. 5(d) and 8(b) show latent instance weights $\mathbf{v}$ and shapelet $\mathbf{S}_{k^*}$, respectively, when we neglect the shapelet regularizer (i.e., $\beta = 0$). In this case, RLTS learned that shapelet $\mathbf{S}_{k^*}$ latently belongs to class 2. We plot $\mathbf{S}_{k^*}$ at the best-matching position for the most-reliable time-series instance of class 2. Shapelet $\mathbf{S}_{k^*}$ of Fig. 8(b) does not match the reliable time series, while that of Fig. 8(a) matches. These results confirm the effect of the shapelet regularizer.

6 Conclusion

For robust learning, we proposed RLTS by introducing SPL and adaptive robust losses into a learning-based shapelet method. RLTS jointly learns shapelets, a classifier, a robustness parameter, latent instance weights, and classes that shapelets latently belong to. The SPL method can optionally consider shapelet matching to reliable time-series instances. The adaptive losses are applied to shapelet-based feature vectors to efficiently learn robustness without increasing algorithmic complexity. Using UCR datasets, we confirmed that for AUC, RLTS significantly outperforms LTS, CSLTS, and LTSpAUC, which are state-of-the-art learning-based shapelet methods. Case studies demonstrated the validity of our approach in terms of the learned robustness parameter and latent instance weights as well as the shapelet regularizer.

References

1. Arthur, D., Vassilvitskii, S.: K-means++: the advantages of careful seeding. In: SODA, pp. 1027–1035. SIAM (2007)
2. Barron, J.T.: A general and adaptive robust loss function. In: CVPR, pp. 4331–4339. Computer Vision Foundation. IEEE, June 2019
3. Batista, G.E.A.P.A., Wang, X., Keogh, E.J.: A complexity-invariant distance measure for time series. In: SDM, pp. 699–710. SIAM (2011)
4. Bengio, Y., Louradour, J., Collobert, R., Weston, J.: Curriculum learning. In: ICML, pp. 41–48. ACM (2009)
5. Dau, H.A., et al.: Hexagon-ML: the UCR Time Series Classification Archive, October 2018. https://www.cs.ucr.edu/~eamonn/time_series_data_2018/
6. Demšar, J.: Statistical comparisons of classifiers over multiple data sets. J. Mach. Learn. Res. **7**, 1–30 (2006)
7. Grabocka, J., Schilling, N., Wistuba, M., Schmidt-Thieme, L.: Learning time-series shapelets. In: KDD, pp. 392–401. ACM (2014)
8. Hou, L., Kwok, J.T., Zurada, J.M.: Efficient learning of timeseries shapelets. In: AAAI, pp. 1209–1215. AAAI Press (2016)
9. Huber, P.J.: Robust Statistics, pp. 1248–1251. Springer, Heidelberg (2011). https://doi.org/10.1007/978-3-642-04898-2_594
10. Jiang, L., Meng, D., Mitamura, T., Hauptmann, A.G.: Easy samples first: self-paced reranking for zero-example multimedia Search. In: MM, pp. 547–556. ACM (2014)
11. Keogh, E., Rakthanmanon, T.: Fast shapelets: a scalable algorithm for discovering time series shapelets. In: SDM, pp. 668–676. SIAM (2013)
12. Kumar, M.P., Packer, B., Koller, D.: Self-paced learning for latent variable models. In: International Conference on Neural Information Processing Systems, pp. 1189–1197. Curran Associates Inc. (2010)
13. Li, H., Gong, M.: Self-paced convolutional neural networks. In: IJCAI, pp. 2110–2116. AAAI Press (2017)
14. Li, X., Lin, J.: Evolving separating references for time series classification. In: SDM, pp. 243–251. SIAM (2018)
15. Roychoudhury, S., Ghalwash, M., Obradovic, Z.: Cost sensitive time-series classification. In: Ceci, M., Hollmén, J., Todorovski, L., Vens, C., Džeroski, S. (eds.) ECML PKDD 2017. LNCS (LNAI), vol. 10535, pp. 495–511. Springer, Cham (2017). https://doi.org/10.1007/978-3-319-71246-8_30
16. Shah, M., Grabocka, J., Schilling, N., Wistuba, M., Schmidt-Thieme, L.: Learning DTW-shapelets for time-series classification. In: CODS, pp. 1–8. ACM (2016)

17. Yamaguchi, A., Maya, S., Maruchi, K., Ueno, K.: LTSpAUC: learning time-series shapelets for optimizing partial AUC. In: SDM, pp. 1–9. SIAM (2020)
18. Ye, L., Keogh, E.: Time series shapelets: a new primitive for data mining. In: KDD, pp. 947–956. ACM (2009)
19. Zhang, Q., Wu, J., Yang, H., Tian, Y., Zhang, C.: Unsupervised feature learning from time series. In: IJCAI, pp. 2322–2328. AAAI Press (2016)
20. Zhao, Q., Meng, D., Jiang, L., Xie, Q., Xu, Z., Hauptmann, A.G.: Self-paced learning for matrix factorization. In: AAAI, pp. 3196–3202. AAAI Press (2015)

Disentangled Sticky Hierarchical Dirichlet Process Hidden Markov Model

Ding Zhou[1(✉)], Yuanjun Gao[2], and Liam Paninski[1]

[1] Department of Statistics, Columbia University, New York, USA
dz2336@columbia.edu, liam@stat.columbia.edu
[2] Jane Street, New York, USA

Abstract. The Hierarchical Dirichlet Process Hidden Markov Model (HDP-HMM) has been used widely as a natural Bayesian nonparametric extension of the classical Hidden Markov Model for learning from sequential and time-series data. A sticky extension of the HDP-HMM has been proposed to strengthen the self-persistence probability in the HDP-HMM. However, the sticky HDP-HMM entangles the strength of the self-persistence prior and transition prior together, limiting its expressiveness. Here, we propose a more general model: the disentangled sticky HDP-HMM (DS-HDP-HMM). We develop novel Gibbs sampling algorithms for efficient inference in this model. We show that the disentangled sticky HDP-HMM outperforms the sticky HDP-HMM and HDP-HMM on both synthetic and real data, and apply the new approach to analyze neural data and segment behavioral video data.

Keywords: Bayesian nonparametrics · Time series · Hierarchical Dirichlet Process · Hidden Markov Model

1 Introduction

Hidden Markov models (HMMs) provide a powerful set of tools for modeling time series data. In the HMM we assume that the time series observations are modulated by underlying latent time-varying variables which take a discrete set of states. This model class is useful in its own right and can also be incorporated as a building block for more complicated models. It has been widely used in speech recognition [9], musical audio analysis [12,23], acoustic-phonetic modeling [16], behavior segmentation [1,8,28], sequential text modeling [11,31], financial time series data analysis [8,30], computational biology [15], and many other fields.

Selecting the number of HMM states is an important question for practitioners. Classical model selection techniques can be used, but these methods can be computationally intensive and are sometimes unreliable in practice [3,26]. Also,

Electronic supplementary material The online version of this chapter (https://doi.org/10.1007/978-3-030-67658-2_35) contains supplementary material, which is available to authorized users.

F. Hutter et al. (Eds.): ECML PKDD 2020, LNAI 12457, pp. 612–627, 2021.
https://doi.org/10.1007/978-3-030-67658-2_35

for real datasets, it is often reasonable to assume that the number of latent states may be unbounded, violating classical assumptions needed to establish consistency results for model selection. Based on previous work in [2,31] proposed the Hierarchical Dirichlet process HMM (HDP-HMM), a Bayesian nonparametric framework. In the HDP-HMM the transition matrix follows a hierarchical Dirichlet process (HDP) prior. [9] noted that the HDP-HMM tends to rapidly switch among redundant states, and proposed the sticky HDP-HMM (S-HDP-HMM), which strengthens the self-persistence probability. This modification often leads to significant improvements in modeling real data.

In the HMM, it is important to distinguish three features: 1, the similarity of the rows of the transition matrix; 2, the average self-persistence probability of the latent states (controlled by the mean of the diagonal of the transition matrix); and 3, the strength of the self-persistence prior across states (i.e., the inverse prior variance of the diagonal elements of the transition matrix). In the HDP-HMM, there is only one parameter controlling feature 1. The sticky HDP-HMM adds one more parameter to control feature 2, but still entangles features 1 and 3 with only one parameter, thus limiting the expressiveness of the prior.

We show that we can add one additional hyperparameter to generalize the sticky HDP-HMM formulation, obtaining three degrees of freedom to model the three features discussed above. We call this new model the disentangled sticky HDP-HMM (DS-HDP-HMM).

The rest of the paper is organized as follows. In Sect. 2, we provide a brief introduction to Bayesian HMM, HDP-HMM, and sticky HDP-HMM. In Sect. 3, we discuss the limitations of these models. In Sect. 4, we introduce disentangled sticky HDP-HMM, and in Sect. 5 we develop efficient Gibbs sampling inference methods for this new model. Section 6 demonstrates the effectiveness of the disentangled sticky HDP-HMM on both synthetic and real data, including applications to analyzing neural data and segmenting behavior video. The notation table can be found in the supplementary material [33] section A.

2 Background on Bayesian HMM and HDP-HMM

Our goal here is to fit an HMM to time series data. On its face, this would seem to be a solved problem; after all, we can compute the HMM likelihood easily, and the basic expectation-maximization algorithm for HMM fitting is textbook material [27]. Nonetheless, a fully Bayesian solution to this problem has remained elusive. Specifically, we would like to be able to compute a posterior over all of the unknown HMM parameters (including the number of latent states). Quantification of posterior uncertainty is critical in many applications: for example, given short time series data, often we do not have enough data to sufficiently identify the HMM parameters. Even for longer time series data, we might want to fit richer models as we collect more data. Here "richer models" correspond to more latent states, and since the number of parameters in the HMM grows quadratically with the number of states, we may be left again with some irreducible uncertainty about the model parameters.

The HDP-HMM [31] provides a useful starting point for fully Bayesian HMM inference. The basic idea here is to sample a global transition distribution prior from a Dirichlet process (described below), and then for each latent state we sample a transition distribution from this shared (random) global prior distribution. To develop the details of this HDP-HMM idea we first need to define some notation for the Dirichlet process (DP). Given a base distribution H on a parameter space Θ and a positive concentration parameter γ, a Dirichlet process $G \sim \mathrm{DP}(\gamma, H)$ (sometimes also denoted by $\mathrm{DP}(\gamma H)$) can be constructed by the following stick-breaking procedure [29]: let

$$\beta \sim \mathrm{GEM}(\gamma),\ \theta_i \overset{iid}{\sim} H,\ i = 1, 2, \cdots, \tag{1}$$

where $\beta \sim \mathrm{GEM}(\gamma)$ is a random probability mass function (p.m.f.) defined on a countably infinite set as follows:

$$v_i \sim \mathrm{Beta}(1, \gamma),\ \beta_i = v_i \prod_{l=1}^{i-1}(1 - v_l),\ i = 1, 2, \cdots. \tag{2}$$

Then the discrete random measure $G = \sum_i \beta_i \delta_{\theta_i}$ is a sample from $\mathrm{DP}(\gamma H)$, where δ_{θ_i} denotes the Dirac measure centered on θ_i.

The HDP-HMM [31] uses the DP to define a prior on the rows of the HMM transition matrix in a setting where the number of latent states is unbounded. The HDP-HMM is defined as

$$\begin{aligned}
\text{DP shared global prior}: &\quad \beta \sim \mathrm{GEM}(\gamma)\\
&\quad \theta_j \overset{iid}{\sim} H,\ j = 1, 2, \cdots\\
\text{Transition matrix prior}: &\quad \pi_j \overset{iid}{\sim} \mathrm{DP}(\alpha\beta),\ j = 1, 2, \cdots\\
\text{Latent states}: &\quad z_t \sim \pi_{z_{t-1}},\ t = 1, \cdots, T\\
\text{Observations}: &\quad y_t \sim f(y|\theta_{z_t}),\ t = 1, \cdots, T
\end{aligned} \tag{3}$$

Here, β and $\{\theta_j\}_{j=1}^{\infty}$ are defined as in the DP described above, and then each transition distribution π_j for state j is defined as a random sample of a second DP with base measure β and concentration parameter α. Here α controls how similar π_j is to the global transition distribution β. Finally, as usual, z_t denotes the state of a Markov chain at time t, and the observation y_t is independently distributed given the latent state z_t and parameters $\{\theta_j\}_{j=1}^{\infty}$, with emission distribution $f(\cdot)$.

The sticky HDP-HMM from [9] modifies the transition matrix prior by adding a point mass distribution with stickiness parameter κ to encourage self-persistence:

$$\text{Transition matrix prior}: \quad \pi_j \sim \mathrm{DP}(\alpha\beta + \kappa\delta_j),\ j = 1, 2, \cdots, \tag{4}$$

where δ_j denotes the Dirac measure centered on j. Figure 1(a) provides the graphical model for the sticky HDP-HMM.

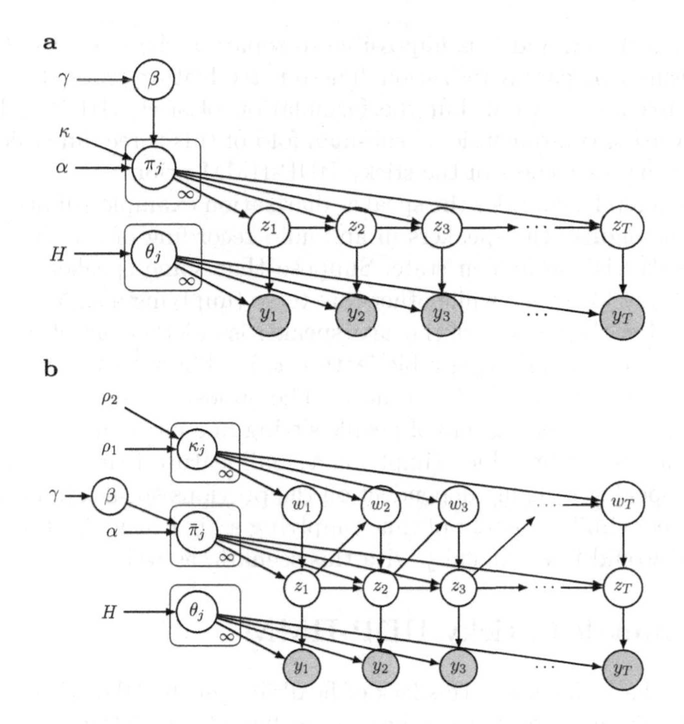

Fig. 1. Graphical models for sticky HDP-HMM (a) and disentangled sticky HDP-HMM (b).

3 Limitations of the HDP-HMM and Sticky HDP-HMM

The HDP-HMM uses the concentration parameter α to control the following feature of the HMM:

- feature 1: the strength of the transition matrix prior, or the similarity of the rows of the transition matrix.

In other words, a large value of α here means that the transition probability for each state is close to the global transition distribution β.

A flexible model should have additional parameters to control two additional features:

- feature 2: the average self-persistence probability, or the mean of the diagonal of the transition matrix.
- feature 3: the strength of the self-persistence prior, or the similarity of the diagonal elements of the transition matrix.

The sticky HDP-HMM adds just one parameter κ compared to the HDP-HMM. feature 2 is controlled by $\kappa/(\alpha + \kappa)$, while both feature 1 and feature 3 are controlled by $\alpha + \kappa$. Note that a strong prior (large $\alpha + \kappa$) means that both the self-persistence probability and the transition probability are quite similar across states, and it is impossible to separate the strength of these two elements using this parameterization. These three features should occupy three degrees of freedom in total, but the formulation of sticky HDP-HMM is only able to traverse a two-dimensional sub-manifold of this three-dimensional space, limiting the expressiveness of the sticky HDP-HMM prior.

More concretely, consider the speaker diarization example studied in [9]. The task is to distinguish the speakers in an audio recording of a conversation; the current speaker is the hidden state. Suppose that some speakers are likely to speak for a very long time while others are terse (implying a small feature 3, i.e. small $\alpha+\kappa$), but the identity of the next speaker is independent of the identity of the previous speaker (implying a big feature 1, i.e. big $\alpha+\kappa$). The sticky HDP-HMM would have trouble in this scenario. The opposite case is plausible as well: as an example, consider a group of people sitting in a circle and expressing their opinions in a clock-wise fashion (implying a small feature 1 since the distribution of the next speaker is highly dependent on the previous speaker); if each speaker talks for a very similar amount of time (implying a big feature 3), then the sticky HDP-HMM would have difficulty with this scenario as well.

4 Disentangled Sticky HDP-HMM

Now that we have diagnosed this lack of flexibility in the HDP-HMM and sticky HDP-HMM, we can construct a new more flexible model that separates the strength of the self-persistence from the similarity of the transition probabilities. Specifically, we modify the transition matrix prior as

$$\begin{aligned}\text{Transition matrix prior}: \quad & \kappa_j \overset{iid}{\sim} \text{beta}(\rho_1, \rho_2) \\ & \bar{\pi}_j \overset{iid}{\sim} \text{DP}(\alpha\beta) \\ & \pi_j = \kappa_j \delta_j + (1-\kappa_j)\bar{\pi}_j, \; j = 1, 2, \cdots ,\end{aligned} \tag{5}$$

where the transition distribution π_j is a mixture distribution. A sample from π_j has the self-persistence probability of κ_j to come from a point mass distribution at j, and has probability $1-\kappa_j$ to come from $\bar{\pi}_j$, a sample from the DP with base measure β. We call this new model the disentangled sticky HDP-HMM.

Here the beta(ρ_1, ρ_2) prior has the flexibility to control both the expectation of self-persistence (feature 2), and the variability of self-persistence (feature 3). Meanwhile α is free to control the variability of the transition probability around the mean transition β (feature 1). In short, we use 3 parameters (ρ_1, ρ_2, α) rather than the 2 parameters (κ, α) to separate the strength of the self-persistence and the transition priors.

When $\rho_1 = 0$ and $\rho_2 > 0$, all the $\kappa_j = 0$, and the disentangled sticky HDP-HMM reduces to HDP-HMM. Importantly, the sticky HDP-HMM is also a special case of the disentangled sticky HDP-HMM, as shown in Theorem 1. See the supplementary material [33] section B for a proof.

Theorem 1. *The sticky HDP-HMM formulation in Eq. 4 is a special case of the disentangled sticky HDP-HMM by setting* $(\rho_1, \rho_2) = (\kappa, \alpha)$.

An equivalent formulation of $z_t \sim \pi_{z_{t-1}}$ in the disentangled sticky HDP-HMM is as follows:

$$\begin{aligned} \text{Latent states}: \quad & w_t \sim \text{Ber}(\kappa_{z_{t-1}}) \\ & z_t \sim w_t \delta_{z_{t-1}} + (1 - w_t)\bar{\pi}_{z_{t-1}}, \; t = 1, \cdots, T, \end{aligned} \tag{6}$$

where we add binary auxiliary variables w_t, which decide whether the next step is self-persistent or switching. See Fig. 1(b) for the corresponding graphical model. We use this formulation to facilitate inference in Sect. 5.

5 Gibbs Sampling Inference

In this section, we introduce a new direct assignment Gibbs sampler (Algorithm 1) (similar to [31]) and a new weak-limit Gibbs sampler (Algorithm 2) (similar to [9]) for inference in the disentangled sticky HDP-HMM. The direct assignment sampler generates samples from the true posterior of the disentangled sticky HDP-HMM when the Gibbs chains converge. The weak-limit sampler uses finite approximation of the HDP-HMM to accelerate the mixing rate of the Gibbs chains and can be easily adapted to parallel computing. [8] noted that the weak-limit sampler was useful for observation models with dynamics such as auto-regressive HMM (ARHMM) or switching linear dynamic system (SLDS). For detailed derivations of these two algorithms, see the supplementary material [33] section C.

5.1 Direct Assignment Sampler

The direct assignment sampler for the HDP-HMM marginalizes transition distributions π_j and parameters θ_j and sequentially samples z_t given all the other states $z_{\backslash t}$, observations $\{y_t\}_{t=1}^T$, and the global transition distribution β. The main difference between our direct assignment sampler and the corresponding HDP-HMM sampler is that instead of only sampling z_t, we sample $\{z_t, w_t, w_{t+1}\}$ in blocks. We sample α, β, γ only using z_t that switch to other states by $\bar{\pi}_{z_{t-1}}$ ($w_t = 0$), and sample $\{\kappa_j\}_{j=1}^{K+1}, \rho_1, \rho_2$ only using z_t that stick to state z_{t-1} ($w_t = 1$).

Algorithm 1. Direct assignment sampler for disentangled sticky HDP-HMM

1: Sequentially sample $\{z_t, w_t, w_{t+1}\}$ for $t = 1, \cdots, T$.
2: Sample $\{\kappa_j\}_{j=1}^{K+1}$. K is defined as number of unique states in $\{z_t\}_{t=1}^T$.
3: Sample β. Same as HDP-HMM.
4: Optionally, sample hyperparameter $\alpha, \gamma, \rho_1, \rho_2$.

For step 1, we sequentially compute the probability for each possible case of the posterior $p(z_t, w_t, w_{t+1} | z_{\backslash t}, w_{\backslash \{t,t+1\}}, \{y_t\}_{t=1}^T, \alpha, \beta, \{\kappa_j\}_{j=1}^{K+1})$, and sample

$\{z_t, w_t, w_{t+1}\}$ from the corresponding multinomial distribution. If $z_t = K + 1$, i.e. a new state appears, we will increment K, sample self-persistence probability κ_{K+1} for a new state from the prior, and update β using stick-breaking. For step 2, given w_{t+1} whose corresponding z_t is j, we can sample κ_j using beta-binomial conjugacy. For step 3, by introducing auxiliary variables $\{m_{jk}\}_{j,k=1}^{K}$, we sample β using Dirichlet-multinomial conjugacy. For step 4, we compute the empirical transition matrix $\{n_{jk}\}_{j,k=1}^{K}$, where n_{jk} is the number of transitions from state j to k with $w_t = 0$ in $\{z_t\}_{t=1}^{T}$, and introduce additional auxiliary variables. Then the posterior of α and γ are gamma-conjugate, given the auxiliary variables. We approximate the posterior of ρ_1, ρ_2 by finite grids. The complexity for each step in Algorithm 1 is $\mathcal{O}(TK)$, $\mathcal{O}(K)$, $\mathcal{O}(K)$, and $\mathcal{O}(K)$ respectively, so the total complexity per iteration is $\mathcal{O}(TK)$.

It is worth noting that instead of modeling $\{\kappa_j\}_{j=1}^{K+1}$ as samples from a beta distribution, it is natural to consider any distribution on the $[0, 1]$ interval. The Gibbs algorithm here is easily adaptable to cases where we have extra prior information on the self-persistence probability.

5.2 Weak-Limit Sampler

The weak-limit sampler for the sticky HDP-HMM constructs a finite approximation to the HDP prior based on the fact that

$$\begin{aligned} \beta|\gamma &\sim \text{Dir}\left(\gamma/L, \cdots, \gamma/L\right) \\ \pi_j|\alpha, \beta &\sim \text{Dir}\left(\alpha\beta_1, \cdots, \alpha\beta_L\right), \ j = 1, \cdots, L \end{aligned} \tag{7}$$

converges to the HDP prior when L goes to infinity. Using this approximation, one can jointly sample latent variables $\{z_t\}_{t=1}^{T}$ with the HMM forward-backward procedure [27], which accelerates the mixing rate of the Gibbs sampler.

The main difference between our weak-limit Gibbs sampler and the corresponding sticky HDP-HMM sampler is that we now have two dimensional latent variables $\{z_t, w_t\}_{t=1}^{T}$ to sample.

Algorithm 2. Weak-limit sampler for disentangled sticky HDP-HMM

1: Jointly sample $\{z_t, w_t\}_{t=1}^{T}$.
2: Sample $\{\kappa_j\}_{j=1}^{L}$.
3: Sample $\{\beta_j\}_{j=1}^{L}, \{\bar{\pi}_j\}_{j=1}^{L}$. Same as HDP-HMM.
4: Sample $\{\theta_j\}_{j=1}^{L}$.
5: Optionally, sample hyperparameter $\alpha, \gamma, \rho_1, \rho_2$.

For step 1, we apply the forward-backward procedure to jointly sample the two dimensional latent variables $\{z_t, w_t\}_{t=1}^{T}$. Step 2 is the same as in Algorithm 1. For step 3, we sample β and $\bar{\pi}$ based on Dirichlet-multinomial conjugacy, given auxiliary variables $\{m_{jk}\}_{j,k=1}^{L}$, the empirical transition matrix $\{n_{jk}\}_{j,k=1}^{L}$, and the approximate prior in Eq. 7. For step 4, we place a conjugate prior on θ_j and use conjugacy to sample from the posterior. Step 5 is the same as in Algorithm 1. The complexity for each step in Algorithm 2 is $\mathcal{O}(TL^2)$, $\mathcal{O}(L)$, $\mathcal{O}(L)$, $\mathcal{O}(L)$, and $\mathcal{O}(L)$ respectively, with total complexity $\mathcal{O}(TL^2)$.

By jointly sampling the full latent sequence $\{z_t, w_t\}_{t=1}^T$, the weak-limit sampler greatly improves the mixing rate. The correlated observations in ARHMM and SLDS further slows the mixing rate of the direct assignment sampler, so jointly sampling is especially important for those models with dynamics [8].

6 Empirical Results

In this section, we apply the disentangled sticky HDP-HMM to both simulated and real data. We compared the performance of our disentangled sticky HDP-HMM with the two baseline models: sticky HDP-HMM and HDP-HMM. We evaluated the model performance according to two metrics: normalized Hamming distance (defined as the element-wise difference between the inferred states and the true underlying states) on training data, and the predictive negative log-likelihood on held-out test data. To compute the Hamming distance, we used the Munkres algorithm [19] to map the indices of the estimated state sequence to the set of indices that maximize the overlap with the true sequence. To compute the predictive negative log-likelihood, we used the set of parameters inferred every 10th Gibbs iteration after it converges, and ran the forward algorithm of [27].

For all the experiments, we adopted a full Bayesian approach, and used the following hyperpriors. For all three models, we placed a Gamma(1, 0.01) prior on the concentration parameters α (and for the sticky HDP-HMM on $\alpha + \kappa$) to cover a wide range of α values. We placed a Gamma(2, 1) prior on γ to avoid extremely small and large γ samples, because small γ will cause numerical instability when sampling β, while large γ will generate too many states. For our model and sticky HDP-HMM, we placed non-informative priors on the self-persistence parameters. We placed a Unif([0, 1]) prior on the self-persistence proportion parameter $\phi = \frac{\rho_1}{\rho_1+\rho_2}$. For our model, we placed a Unif([0, 2]) on the self-persistence scale parameter $\eta = (\rho_1 + \rho_2)^{-1/3}$, and cut $[0, 1] \times [0, 2]$ (the support of ϕ, η) into 100×100 grids (for simulated data) or 30×30 (for real data) to numerically compute the posterior for ϕ, η.

We used the direct assignment sampler to fit the simulated data, because the simulated data here has a relatively small sample size, and the direct assignment sampler generates samples from the true posterior of the model. We used the weak-limit sampler to fit real data (with a larger sample size) in parallel. We have also applied the direct assignment sampler to a short version of the hippocampal data; the results obtained are qualitatively similar.

6.1 Simulated Data

In the simulation studies, we focus on two settings that serve to clearly illustrate the differences between the sticky versus disentangled sticky models. In both

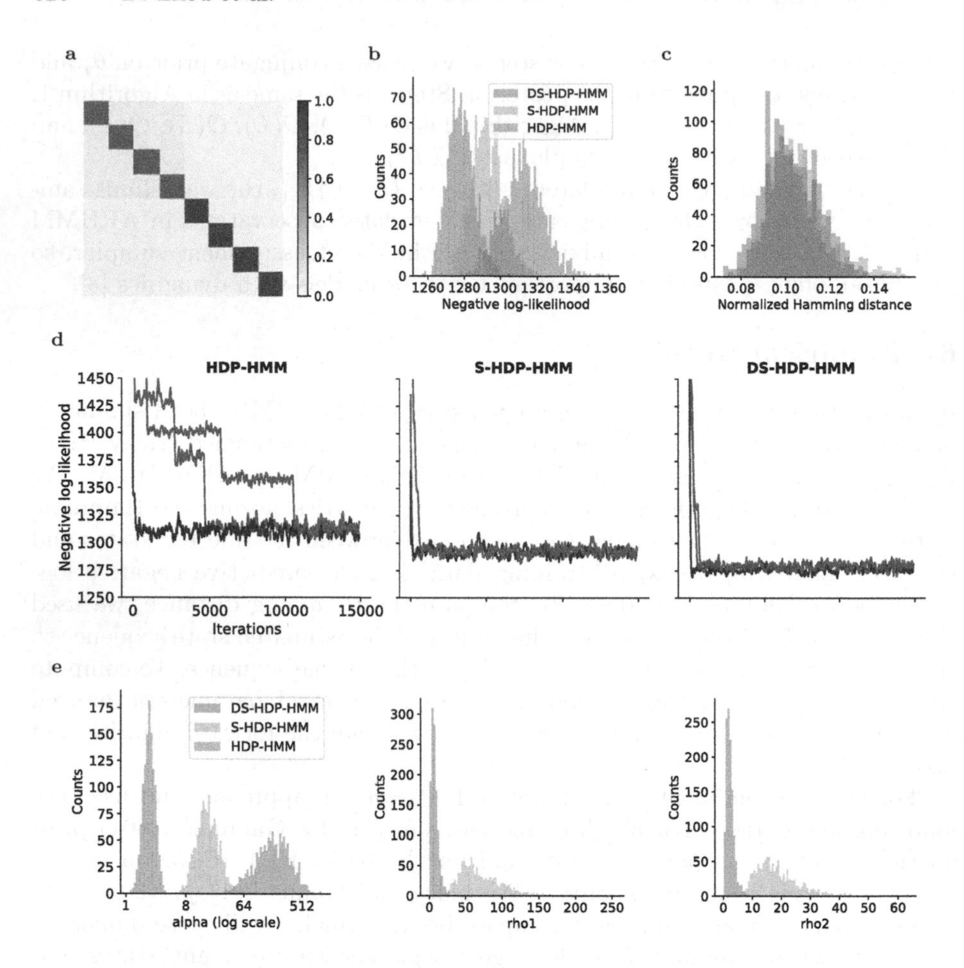

Fig. 2. The DS-HDP-HMM provides good fits to the simulated data with different self-persistence (ρ_1, ρ_2 small), same transition (α large), and multinomial emission. (a) True transition matrix. (b) Histogram of negative log-likelihood on test data. (c) Histogram of normalized Hamming distance between the estimated states and true states on training data. (d) Negative log-likelihood on test data over 15000 direct assignment Gibbs samples from 3 chains for each model. (e) Histogram of hyperparameters α, ρ_1, ρ_2. Note that we plot κ, α for sticky HDP-HMM in the histogram of ρ_1, ρ_2, since κ, α would be equal to ρ_1, ρ_2 respectively if we treat sticky HDP-HMM as a special case of our model. Our model learns a big α and small ρ_1, ρ_2, which is consistent with the data.

settings, we experimented with both multinomial and Gaussian emissions. See the Gaussian emission results in the supplementary material [33] section E.

Different Self-persistence, Same Transition. We simulated data using a transition matrix (Fig. 2(a)) with different κ_j and the same $\bar{\pi}_j$ across states, which corresponds to a large transition concentration parameter α (big feature 1), and small self-persistence parameters ρ_1, ρ_2 (small feature 3). We assigned the multinomial observation to be equal to the latent state with probability 0.9 and to other states with equal chance with probability 0.1.

We placed a symmetric Dir$(1, \cdots, 1)$ prior on the multinomial parameters. We ran 3 MCMC chains, each with 15000 iterations, with 11000 iterations as burn-in (Fig. 2(d)).

Results of three models fit on multinomial emission are shown in Fig. 2. As shown in Fig. 2(e), our model learns a big α and small ρ_1, ρ_2, which is consistent with the data, while sticky HDP-HMM model entangles these two parameters and learns something in the middle. Better recovery of the hyperparameters leads to better fits on the data: our model outperforms the two baseline models in terms of negative log-likelihood and Hamming distance (Fig. 2(b)(c)). The advantages of our model are even more clear under the Gaussian emission (see figure in the supplementary material [33] section E). The three models generally learn similar γ, which means that they infer similar numbers of states.

For each emission model, we compared these three models on 10 datasets generated from the same model using different random seeds. The conclusions are consistent across the 10 different datasets.

Same Self-persistence, Different Transition. We simulated data using a transition matrix (Fig. 3(a)) with the same κ_j and different $\bar{\pi}_j$ across states, which corresponds to small α (small feature 1), and large ρ_1, ρ_2 (large feature 3). We assigned the multinomial observation to be equal to the latent state with probability 0.8 and to other states with equal chance with probability 0.2.

Again, consistently across 10 replications, our model learns hyperparameters consistent with the data (Fig. 3(d)) and outperforms the two baseline models (Fig. 3(b)(c)), though the advantages are not as big as in the previous scenario. This is likely because in the previous example, our model can learn the hyperparameters based on the similarity among rows of the transition matrix, while in this scenario, it can learn hyperparameters mostly from the similarity among diagonal elements of the transition matrix, which contain much less information.

6.2 Inferring Rat Hippocampal Population Codes

Next we applied our model to a public electrophysiological hippocampus dataset [24,25][1]. In the experiment, a rat freely explored in an open square environment (~50 cm × 40 cm), while neural activity in the hippocampal CA1 area was recorded using silicon probes. See Fig. 4(a) for an example trace illustrating the position of the rat over the course of the experiment.

We selected the 100 most active putative pyramidal neurons and binned the ensemble spike activities with a frame rate 10 Hz. The dataset consists of ~ 36k frames. We used the first 8k frames, cut it into blocks of 500 frames, and

[1] https://buzsakilab.nyumc.org/datasets/PastalkovaE/i01/i01_maze15_MS.001/.

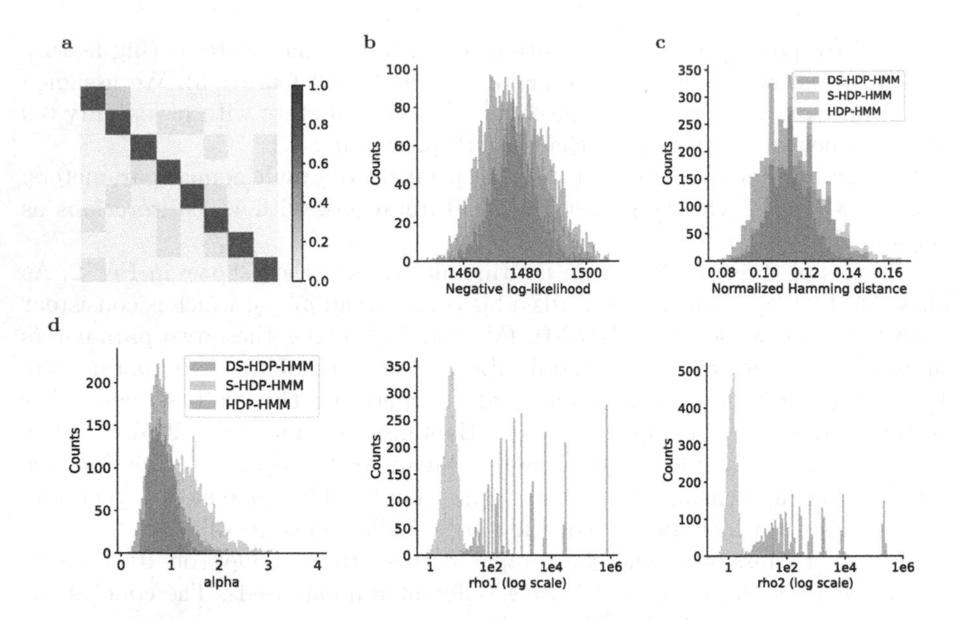

Fig. 3. Results for the simulated data with same self-persistence (ρ_1, ρ_2 large), different transition (α small) and multinomial emission. We ran 3 chains, each with 30000 iterations, with 20000 iterations as burn-in for this data. Conventions as in Fig. 2. Our model learns a small α and big ρ_1, ρ_2, which is consistent with the data. Note that the "spikiness" in the histogram of ρ_1, ρ_2 comes from the approximation of the posterior of ρ_1, ρ_2 using finite grids. An alternative way to avoid the discretization would be to use Metropolis-Hasting sampling [10].

randomly took 8 blocks as training data (4k frames), and the remaining 8 blocks as test data. The spike count at time t for cell c is modeled as Poisson with rate $\lambda_{z_t,c}$, i.e. $y_{t,c} \sim \text{Poisson}(\lambda_{z_t,c})$. As in [17], we used a conjugate gamma(a_c, b_c) prior for the firing rate $\lambda_{j,c}$, $j = 1, 2, \cdots, c = 1, \cdots, 100$. We fixed the shape parameter $a_c = 1$ across cells, and placed a gamma$(1, 1)$ prior for the scale parameter b_c.

We set $L = 200$ and ran 7 MCMC chains, each with 15000 iterations, with 11000 iterations as burn-in (Fig. 4(d)). Our model achieves the smallest negative log-likelihood on test data (Fig. 4(c)). A two-sample t-test for the mean difference between negative log-likelihood of DS-HDP-HMM and S-HDP-HMM for 7 chains is significant (p-value < 0.05). The inferred states are correlated with the spatial locations of the rat (Fig. 4(b)). Compared to sticky HDP-HMM, our model infers a bigger α and smaller ρ_1, ρ_2 (Fig. 4(e)). Bigger α implies smaller variability of the switching transition, consistent with the rat quickly going to other locations since it has fast running speed. Smaller ρ_1, ρ_2 imply bigger variability of the self-persistence probability, consistent with the rat spending different durations at different locations. Note that the disentangled sticky HDP-HMM and sticky HDP-HMM perform similarly on the rat hippocampal data; bigger differences are seen in the mouse behavior video data in the next section.

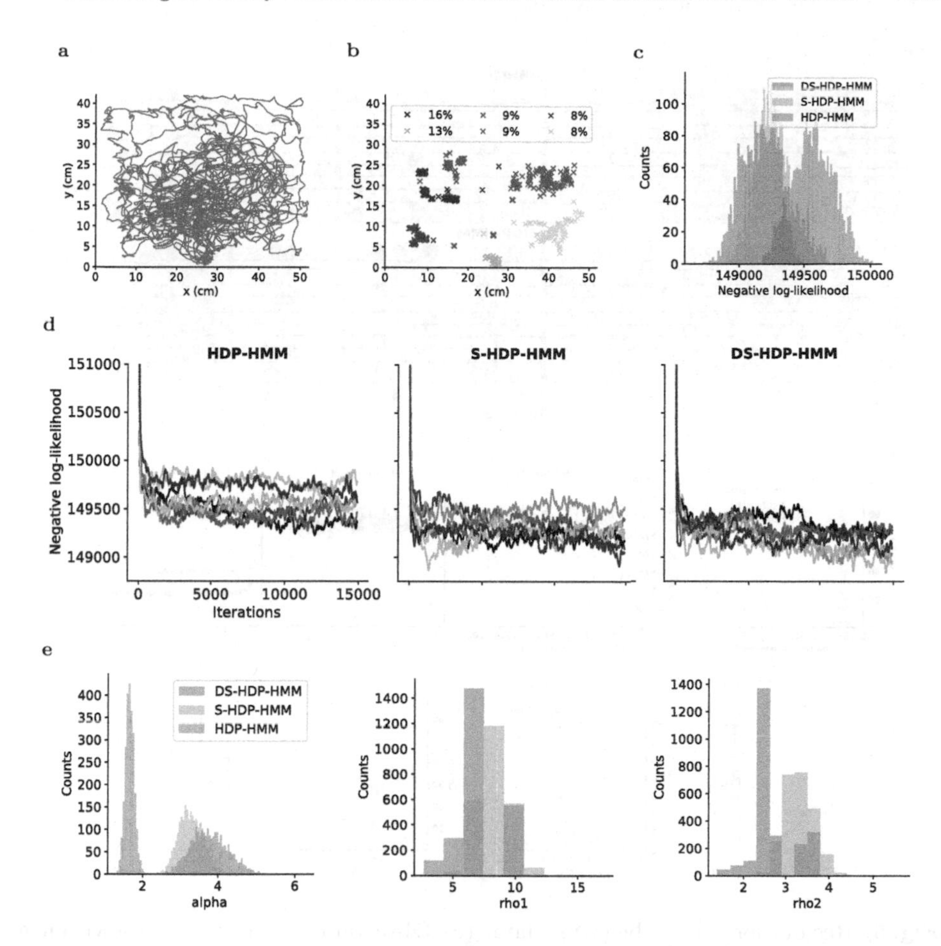

Fig. 4. Results for rat hippocampal data. (a) The rat's moving trajectory in training and test data. (b) The spatial locations of the rat corresponding to 6 most frequent states inferred by DS-HDP-HMM. The figure legend shows the percentage of each of the 6 states. (c)(d)(e) Conventions as Fig. 2, though with more traces and using the weak-limit sampler in (d). Our model infers a bigger α than ρ_1, ρ_2 (big feature 1, small feature 3).

6.3 Segmenting Mouse Behavior Video

We also applied our model to a public mouse behavior dataset [4,21]. In this experiment, a head-fixed mouse performed a visual decision task while neural activity across dorsal cortex was optically recorded using widefield calcium imaging. We only used the behavior video data (128×128 pixels each grayscale video frame), which was recorded using two cameras (one side view and one bottom view). The behavior video is high dimensional, so we directly adopted the dimension reduction result in [1], which output 9-dimensional continuous variables estimated using a convolutional autoencoder.

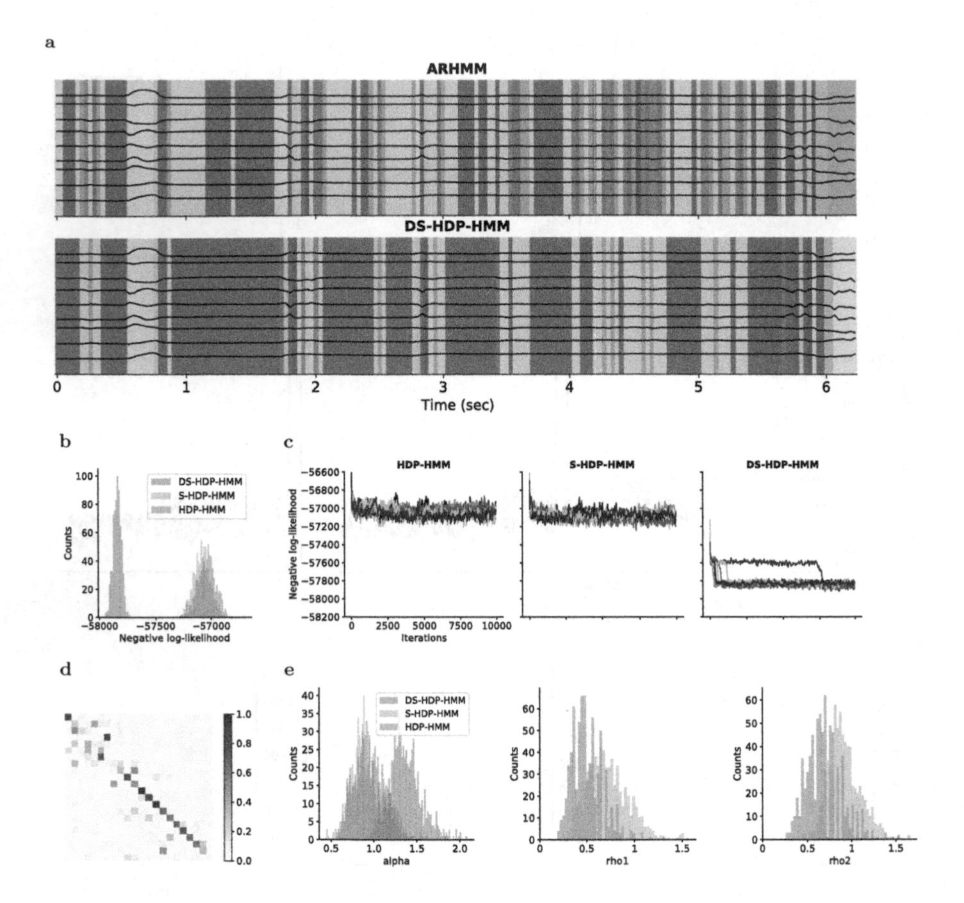

Fig. 5. Results for mouse behavior data. (a) Observations (black) are shown on a sample trial over time, with background colors indicating the discrete state inferred for that time step using the ARHMM (number of states = 16) and our model (inferred number of states = 22) (colors are chosen to maximize the overlap between these two results). (b)(c)(e) Conventions as in the corresponding plots in Fig. 4. Note the significant loglikelihood gap between the DS-HDP-HMM and the baseline models here. Our model infers a bigger α than ρ_1, ρ_2 (big feature 1, small feature 3). (d) Transition matrix estimated by DS-HDP-HMM. Note that the diagonal elements are strongly variable across states (small feature 3). (Color figure online)

The dataset consists of 1126 trials across two sessions, with 189 frames each trial (30 Hz frame rate). We randomly chose 100 trials as training data ($\sim$ 20k frames), and 30 trials as test data ($\sim$ 6k frames). We assumed the observation follows an ARHMM, i.e. $y_t \sim \mathcal{N}(A_{z_t} y_{t-1}, \Sigma_{z_t})$. As in [8], we standardized the observations, and assumed $\{A_j, \Sigma_j\}$ follows a conjugate matrix-normal inverse-Wishart (MNIW) prior. The details of this prior are in the supplementary material [33] section D. We initialized the states $\{z_t\}_{t=1}^{T}$ of the three Bayesian nonparametric models using the state assignments result of a 32-states parametric ARHMM in [1].

We set $L = 40$ and ran 7 MCMC chains, each with 10000 iterations, with 9000 iterations as burn-in (Fig. 5(c)). From Fig. 5(a), we can see that our model has less rapid switches among states than the parametric ARHMM even if we have more states. The diagonal elements are strongly variable across states (small feature 3) (Fig. 5(d)). For this dataset, α, ρ_1, and ρ_2 are all small, but they still have different values. Our model infers a bigger α than ρ_1, ρ_2 (big feature 1, small feature 3), as shown in Fig. 5(e). Again, our model achieves much smaller negative predictive log-likelihood on test data than the other two Bayesian nonparametric models (Fig. 5(b)), due to disentanglement of hyperparameters, improving the expressiveness of the HDP prior.

7 Discussion and Conclusion

In this paper, we propose an extension of the sticky HDP-HMM, to decouple the strength of the self-persistence prior (or the similarity of the diagonal elements of the transition matrix) and transition prior (or the similarity of the rows of the transition matrix). We develop two novel Gibbs samplers for performing efficient inference. We also show in simulated and real data that our extension outperforms existing approaches.

The work [14] proposed a Bayesian nonparametric prior to Hidden Semi-Markov Model [20] which also extends the sticky HDP-HMM. The idea is to break the Markovian assumption and instead model the distribution of duration of self-persistence for each state explicitly. The flexibility of the choice of duration distribution provides a rich set of possible priors, though the choice of duration distribution can be challenging for practitioners. Our formulation can be thought as a special case of their model, but with a few important advantages. First, we clarify that the limitations of the sticky HDP-HMM might not necessarily come from the Markovian assumption but could instead be due to the entanglement of the prior. Second, our extension adds a single hyperparameter, making it easy for practitioners to use. Third, our formulation still maintains the Markovian structure, which is useful in many applications. For example, there are applications of HDP-HMM prior to Markov decision processes [5–7], in which the Markovian structure is critical. Also, by harnessing the Markovian structure, our model enjoys scalable $O(TK^2)$ message passing algorithms, while the message passing algorithm of HSMM has complexity of $O(T^2K + TK^2)$ in general.

In the future we hope to explore alternative computational approaches (specifically, stochastic variational [13,32] and amortized inference [22] methods), which may further improve the scalability of the model introduced here. Another important direction is to adapt the mixture of finite mixtures (MFM) model [18] to the HMM setting, as a replacement for the HDP prior; as in the mixture modeling setting discussed in [18], we expect that the MFM prior may lead to better estimates of the number of latent states.

Open source code is available at https://github.com/zhd96/ds-hdp-hmm.

Acknowledgements. We thank John Paisley, Xue-Xin Wei, Kenneth Kay, Matthew R Whiteway, Pengcheng Zhou and Ari Pakman for helpful discussions. We thank the authors of [1] for generously sharing their data. We acknowledge computing resources from Columbia University's Shared Research Computing Facility project, which is supported by NIH Research Facility Improvement Grant 1G20RR030893-01, and associated funds from the New York State Empire State Development, Division of Science Technology and Innovation (NYSTAR) Contract C090171, both awarded April 15, 2010.

References

1. Batty, E., et al.: Behavenet: nonlinear embedding and Bayesian neural decoding of behavioral videos. In: Advances in Neural Information Processing Systems, pp. 15680–15691 (2019)
2. Beal, M.J., Ghahramani, Z., Rasmussen, C.E.: The infinite hidden Markov model. In: Advances in Neural Information Processing Systems, pp. 577–584 (2002)
3. Celeux, G., Durand, J.B.: Selecting hidden Markov model state number with cross-validated likelihood. Comput. Statist. **23**(4), 541–564 (2008). https://doi.org/10.1007/s00180-007-0097-1
4. Churchland, A., Musall, S., Kaufman, M.T., Juavinett, A., Gluf, S.: Single-trial neural dynamics are dominated by richly varied movements: dataset, vol. 10, no. 14224/1, p. 38599 (2019)
5. Doshi-Velez, F.: The infinite partially observable Markov decision process. In: Advances in Neural Information Processing Systems, pp. 477–485 (2009)
6. Doshi-Velez, F., Pfau, D., Wood, F., Roy, N.: Bayesian nonparametric methods for partially-observable reinforcement learning. IEEE Trans. Pattern Anal. Mach. Intell. **37**(2), 394–407 (2013)
7. Doshi-Velez, F., Wingate, D., Roy, N., Tenenbaum, J.B.: Nonparametric Bayesian policy priors for reinforcement learning. In: Advances in Neural Information Processing Systems, pp. 532–540 (2010)
8. Fox, E., Sudderth, E.B., Jordan, M.I., Willsky, A.S.: Nonparametric Bayesian learning of switching linear dynamical systems. In: Advances in Neural Information Processing Systems, pp. 457–464 (2009)
9. Fox, E.B., Sudderth, E.B., Jordan, M.I., Willsky, A.S.: A sticky HDP-HMM with application to speaker diarization. Ann. Appl. Statist. **5**, 1020–1056 (2011)
10. Hastings, W.K.: Monte Carlo sampling methods using Markov chains and their applications. Biometrika **57**(1), 97–109 (1970)
11. Heller, K., Teh, Y.W., Gorur, D.: Infinite hierarchical hidden Markov models. In: Artificial Intelligence and Statistics, pp. 224–231 (2009)
12. Hoffman, M.D., Cook, P.R., Blei, D.M.: Data-driven recomposition using the hierarchical Dirichlet process hidden Markov model. In: ICMC. Citeseer (2008)
13. Hughes, M., Kim, D.I., Sudderth, E.: Reliable and scalable variational inference for the hierarchical Dirichlet process. In: Artificial Intelligence and Statistics, pp. 370–378 (2015)
14. Johnson, M.J., Willsky, A.S.: Bayesian nonparametric hidden semi-Markov models. J. Mach. Learn. Res. **14**, 673–701 (2013)
15. Krogh, A., Brown, M., Mian, I.S., Sjolander, K., Haussler, D.: Hidden Markov models in computational biology: applications to protein modeling. J. Mol. Biol. **235**(5), 1501–1531 (1994)

16. Lee, C.y., Glass, J.: A nonparametric Bayesian approach to acoustic model discovery. In: Proceedings of the 50th Annual Meeting of the Association for Computational Linguistics: Long Papers-Volume 1, pp. 40–49. Association for Computational Linguistics (2012)
17. Linderman, S.W., Johnson, M.J., Wilson, M.A., Chen, Z.: A Bayesian nonparametric approach for uncovering rat hippocampal population codes during spatial navigation. J. Neurosci. Methods **263**, 36–47 (2016)
18. Miller, J.W., Harrison, M.T.: Mixture models with a prior on the number of components. J. Am. Statist. Assoc. **113**(521), 340–356 (2018)
19. Munkres, J.: Algorithms for the assignment and transportation problems. J. Soc. Ind. Appl. Math. **5**(1), 32–38 (1957)
20. Murphy, K.P.: Hidden semi-markov models (hsmms). unpublished notes 2 (2002)
21. Musall, S., Kaufman, M.T., Juavinett, A.L., Gluf, S., Churchland, A.K.: Single-trial neural dynamics are dominated by richly varied movements. Nat. Neurosci. **22**(10), 1677–1686 (2019)
22. Pakman, A., Wang, Y., Mitelut, C., Lee, J., Paninski, L.: Discrete neural processes. arXiv preprint arXiv:1901.00409 (2018)
23. Pardo, B., Birmingham, W.: Modeling form for on-line following of musical performances. In: Proceedings of the National Conference on Artificial Intelligence, Menlo Park, CA, Cambridge, MA, London, vol. 20, p. 1018. AAAI Press; MIT Press 1999 (2005)
24. Pastalkova, E., Wang, Y., Mizuseki, K., Buzsáki, G.: Simultaneous extracellular recordings from left and right hippocampal areas ca1 and right entorhinal cortex from a rat performing a left/right alternation task and other behaviors. CRCNS.org (2015). https://doi.org/10.6080/K0KS6PHF
25. Pastalkova, E., Itskov, V., Amarasingham, A., Buzsáki, G.: Internally generated cell assembly sequences in the rat hippocampus. Science **321**(5894), 1322–1327 (2008)
26. Pohle, J., Langrock, R., van Beest, F., Schmidt, N.M.: Selecting the number of states in hidden Markov models-pitfalls, practical challenges and pragmatic solutions. arXiv preprint arXiv:1701.08673 (2017)
27. Rabiner, L.R.: A tutorial on hidden Markov models and selected applications in speech recognition. Proc. IEEE **77**(2), 257–286 (1989)
28. Saeedi, A., Hoffman, M., Johnson, M., Adams, R.: The segmented ihmm: a simple, efficient hierarchical infinite hmm. In: International Conference on Machine Learning, pp. 2682–2691 (2016)
29. Sethuraman, J.: A constructive definition of Dirichlet priors. Statistica sinica **4**, 639–650 (1994)
30. Song, Y.: Modelling regime switching and structural breaks with an infinite hidden Markov model. J. Appl. Econ. **29**(5), 825–842 (2014)
31. Teh, Y.W., Jordan, M.I., Beal, M.J., Blei, D.M.: Hierarchical Dirichlet processes. J. Am. Statist. Assoc. **101**(476), 1566–1581 (2006). https://doi.org/10.1198/016214506000000302
32. Zhang, A., Gultekin, S., Paisley, J.: Stochastic variational inference for the HDP-HMM. In: Artificial Intelligence and Statistics, pp. 800–808 (2016)
33. Zhou, D., Gao, Y., Paninski, L.: Disentangled sticky hierarchical dirichlet process hidden markov model. arXiv preprint arXiv:2004.03019 (2020)

Predicting Future Classifiers for Evolving Non-linear Decision Boundaries

Kanishka Khandelwal(✉), Devendra Dhaka, and Vivek Barsopia

NEC Corporation, Tokyo, Japan
{k_khandelwal,barsopiav}@nec.com, deven.dhaka@gmail.com

Abstract. In streaming data applications, the underlying concept often changes with time which necessitates the update of employed classifiers. Most approaches in the literature utilize the arriving labeled data to continually update the classifier system. However, it is often difficult/expensive to continuously receive the labels for the arriving data. Moreover, in domains such as embedded sensing, resource-aware classifiers that do not update frequently are needed. To tackle these issues, recent works have proposed to predict classifiers at a future time instance by additionally learning the dynamics of changing classifier weights during the initial training phase. This strategy bypasses the need to retrain/relearn the classifiers, and thus the additional labeled data is no longer required. However, the current progress is limited to the prediction of linear classifiers. As a step forward, in this work, we propose a probabilistic model for predicting future non-linear classifiers given time-stamped labeled data. We develop a variational inference based learning algorithm and demonstrate the effectiveness of our approach through experiments using synthetic and real-world datasets.

Keywords: Concept drift · Data streams · Classification

1 Introduction

In dynamically changing environments, the underlying concept that determines the response variable y, for the covariate $\mathbf{x}$, may vary over time. This variation is reflected as a change in the conditional distribution $p(y|\mathbf{x})$ which makes the environment non-stationary. For example, a user's preference for online news articles varies over time owing to the geopolitical/social factors. Consequently, the user's click pattern for news articles will change, even if the topics of published articles remain the same.

In applications where data arrives sequentially, this phenomenon is common and is referred to as *real* concept drift [10]. In a classification context, this would mean the underlying decision boundary is evolving over time. A classifier employed in such an environment should be updated to avoid a deterioration in the performance. In the above example, the news provider must update its

D.Dhaka—Contributed while working at NEC Corporation.

F. Hutter et al. (Eds.): ECML PKDD 2020, LNAI 12457, pp. 628–643, 2021.
https://doi.org/10.1007/978-3-030-67658-2_36

classifier to continue recommending articles in line with the user's preference; else the churn rate increases. To achieve this, most approaches in the literature propose to update the decision rule of classifier (or classifier system) in an online manner using the arriving labeled data. Of these, some are *passive* approaches that continuously update the decision rule while others are *active* approaches that use change detectors to trigger an update [2].

However, there are two issues associated with this online update strategy when the decision boundary is evolving. *Firstly*, it is often expensive/difficult to continuously acquire labels for the sequentially arriving data since labeling is done by domain experts in most cases. And if labels are unavailable, it is not possible to detect changes in $p(y|\mathbf{x})$ without taking any assumption [22, 24]. *Secondly*, it could be impractical to allow such online updates in computationally constrained applications. For example, adaptive classifiers are needed in wearable embedded systems that collect physiological sensor data to perform tasks such as physical activity recognition [1, 34]. Updating the decision rule in an online fashion becomes prohibitive in such systems with limited memory and processing capabilities [2, 23].

The above issues motivate the research on classifier models that can avoid the online update step using labeled data yet cope with the situation of evolving (underlying) decision boundary. In this work, we propose a model to predict future classifier (i.e. weights of the classifier at a future time instance) to maintain the classification performance. We achieve this by jointly learning the dynamics of an evolving boundary along with the classifier weights using time-stamped labeled data during the training phase. The learned dynamics and past values of classifier weights can be used to predict the weights at a future time instance thus avoiding the need for an online update of the classifier weights using labeled data. The advantages of predicting future classifier are twofold: *first*, effective use of available training data makes it possible to (reasonably) approximate the evolving boundary for some future time instances which reduces the cost of labeling the arriving data; *second*, the online update step is no longer necessary and thus the classifier system can be deployed in resource-constrained applications.

The current progress in this direction is limited to the prediction of future linear classifier [20–22] which is expected to perform poorly if the (evolving) underlying decision boundary is non-linear (i.e. the classes are not linearly separable). As a step forward, we propose a probabilistic model to predict a non-linear classifier at a future time instance. We utilize the non-parametric framework of Dirichlet process mixture model [8] to model the joint distribution of label y and covariates x, in the form $p(x, y) = p(y|x)p(x)$ and allow $p(y|x)$ to change over time. Further, we develop a variational inference based learning algorithm and demonstrate the effectiveness of our approach through experiments using both synthetic and real datasets.

In the next section, we discuss the related work in this domain and highlight the challenges in designing methods for future non-linear classifier prediction. In Sect. 3, the task is formally defined which is followed by a description of our model. Next, we develop a variational-inference based learning algorithm in Sect. 4. We present the evaluation results of our approach on synthetic and real-world datasets in Sect. 5 and summarize this paper in Sect. 6.

2 Related Work

Changing data distribution in streaming data applications is a well-studied problem and many methods have been proposed to maintain the classifier performance in such non-stationary conditions. Online algorithms [6,33], forgetting algorithms [17,19,27], ensemble methods [5,18], evolving fuzzy systems [3,30] etc. usually capitalize on arriving labeled data for updating the classifier. However, these methods are not well suited for applications where acquiring labeled data is difficult/expensive. Active learning algorithms [36,37] look to optimize the labeling cost by querying labels for specific instances. Some methods have been proposed [13–15] that use unlabeled data with/without the labeled data. However, all of the above methods look to update the decision rule in an online manner using the arriving data. As explained in the previous section, this strategy could be prohibitive in applications where resources are limited.

To alleviate these issues, a possible *proactive* strategy is to predict future classifiers using the labeled training data at hand. In this direction, Kumagai and Iwata [20–22] have proposed probabilistic models for prediction of future classifiers by learning the dynamics of evolving decision boundary during the initial training phase. The intuition behind their approach is as follows. The weights of a classifier employed in an environment where the decision boundary is evolving ought to change with time to maintain the performance. Using the time-stamped labeled data available in the initial training phase, the classifier can be trained at different time instances in the past. A time series model fitted over these weights can capture the dynamics of the decision boundary. Thus, the past values of classifier weights and the time series model can be used to predict the classifier at future time instances.

In their initial work [20], logistic regression is used as a classifier model and autoregressive (AR) process as a time series model over classifier weights. Since the AR process could only capture the linear dynamics of the decision boundary, in [21] the authors proposed to use Gaussian Process (GP) as the time series model over the weights in logistic regression model for capturing the non-linear dynamics of the decision boundary. The GP-based model was further extended in [22] to utilize the unlabeled data available at test (future) time instances. Low-density separation criterion, i.e. decision boundary should pass through low density regions, is used to regularize the posterior of the classifier parameters which may improve the prediction of future classifiers in some scenarios. On the other hand, waiting for a few unlabeled batches of data (belonging to test time instances) makes this approach inappropriate for the real-time prediction task.

In all of these works, logistic regression is used for the classification task which limits their applicability to linearly separable settings and is bound to perform poorly in scenarios where the decision boundary is non-linear. Extending these models using a non-linear classifier in place of logistic regression classifier is difficult. A non-parametric non-linear classification algorithm (like GP classification, nearest neighbors, SVM, etc.) learns a data-dependent classifier. Since the data samples can be different at different time instances, establishing a correspondence between the learned weights across time instances is not obvious; thus the time series model cannot be applied straightforwardly. On the other hand, with parametric classifiers (such as Multilayer perceptron) there is an issue of the huge parameter space. The model formulation and training would be difficult since the model should also identify the subset of parameters to be modeled using time series.

3 Proposed Method

3.1 Task

We consider a binary classification task with an available set $\mathcal{D}$ of datasets $\mathcal{D}_t := \{(\mathbf{x}_{t,i}, y_{t,i})\}_{i=1}^{N_t}$ collected at regular intervals with time-stamp denoted using $t \in \{1, 2 \ldots T\}$. Here, $\mathbf{x}_{t,i} \in \mathbb{R}^D$ is the D-dimensional covariate vector of the i-th sample at time t, with $y_{t,i} \in \{0, 1\}$ as the class label and N_t is the number of samples available at time t. Note that sequential time-stamped data can be represented in this format by discretizing at regular intervals and considering the data falling within same interval to have same time-stamp. Given the set of training data $\mathcal{D} = \{D_t\}_{t=1}^{T}$, our aim is to predict binary classifier $h_t : \mathbb{R}^D \rightarrow \{0, 1\}$ at future time instances $t \in \{T + 1, T + 2, ...\}$ to precisely classify the future data.

3.2 Model

The non-parametric mixture model based on Dirichlet process (DP) [8] and its extensions form a well-known family of probabilistic models used for tasks such as topic modelling [11], regression [35], etc. The DP framework assumes that a countably infinite set of mixture components is needed to generate the available data, but, the DP prior (with a concentration parameter α) exhibits a clustering property [31]. Our model is among the DP-GLM family [12,29] and is explicitly developed for a non-stationary environment wherein $p(y|\boldsymbol{x})$ is changing. Each component within a mixture locally models the distribution of covariates as well as a linear (classifying) relationship between the response variable and covariates, separately at all time instances. Additionally, within a component, a time series model is used to capture the dynamics of changing classifier weights. Globally, a mixture of such components is expected to model a time-varying non-linear relationship between the response variable and covariates.

Figure 1 shows the graphical representation of our model and below we explain the generative process of a sample pair $(\mathbf{x}_{t,i}, y_{t,i})$. Our model first generates π_k independently for each component $k \in \{1, 2, \cdots \infty\}$ with the distribution $Beta(1, \alpha)$. The mixture weight given to each component, π'_k, is determined from $\pi = \{\pi_k\}_{k=1}^{\infty}$ using the stick breaking process representation of DP [28] as,

$$\pi'_k = \pi_k \Pi_{k'=1}^{k-1}(1 - \pi_{k'}) \tag{1}$$

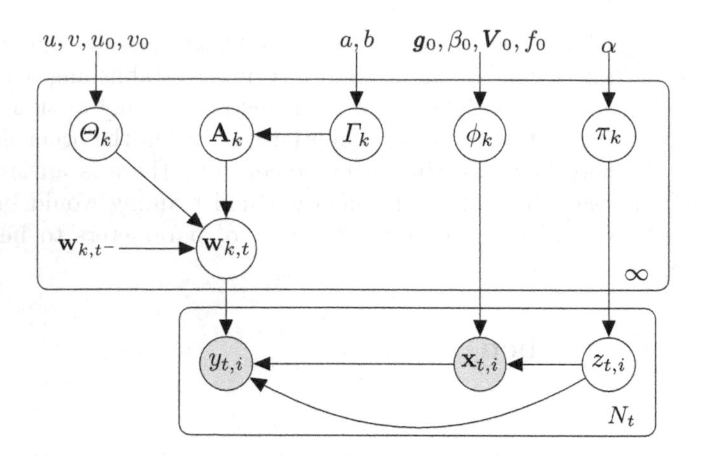

Fig. 1. Graphical model representation of the proposed model at time t. Here, $\mathbf{w}_{k,t^-} := \{\mathbf{w}_{k,t-m}\}_{m=1}^{M}$, $\Theta_k := \{(\theta_k, \theta_{0,k})\}$, $\Gamma_k := \{\gamma_{k,m}\}_{m=0}^{M}$. Nodes with gray and empty circles are observed and latent variables, respectively. Nodes with deterministic value are without circles. Note that $\mathbf{w}_{k,t^-}$ is already determined by the time t

The component assignment variable, $z_{t,i}$, for our sample is obtained from a multinomial distribution as,

$$z_{t,i}|(\pi'_1, \pi'_2, \ldots) \sim Mult(\pi'_1, \pi'_2, \ldots) \tag{2}$$

Here, $z_{t,i}$ is represented as an infinite length indicator vector with the entry corresponding to the selected component being 1, and 0 for all others. Given $z_{t,i}$, the covariate $\mathbf{x}_{t,i}$ is obtained from a multivariate normal distribution as,

$$\mathbf{x}_{t,i}|z_{t,i}, \{(\mathbf{m}_k, \mathbf{R}_k^{-1})\}_{k=1}^{\infty} \sim \Pi_{k=1}^{\infty} \left(\mathcal{N}(\mathbf{m}_k, \mathbf{R}_k^{-1})\right)^{z_{t,i,k}} \tag{3}$$

The mean $\mathbf{m}_k \in \mathbb{R}^D$ and precision matrix $\mathbf{R}_k \in \mathbb{R}^{D\times D}$ for each component are drawn from a Normal-Wishart distribution with mean vector $g_0 \in \mathbb{R}^D$, scale matrix $V_0 \in \mathbb{R}^{D\times D}$, scale parameter $\beta_0 \in \mathbb{R}$ and degree of freedom $f_0 \in \mathbb{R}$ as,

$$\phi_k = \{\mathbf{m}_k, \mathbf{R}_k\} \sim \mathcal{NW}(\boldsymbol{g}_0, \beta_0, \boldsymbol{V}_0, f_0) \tag{4}$$

Note that ϕ_k doesn't vary with time since we assume the covariate distribution $p(x)$ to be stationary. The binary label $y_{t,i}$ is drawn from a Bernoulli (Ber) distribution with the "success" parameter given by standard logistic regression of $\mathbf{x}_{t,i}$ as,

$$y_{t,i}|z_{t,i}, \mathbf{x}_{t,i}, \{\mathbf{w}_{k,t}\}_{k=1}^{\infty} \sim \Pi_{k=1}^{\infty} \left(Ber(\sigma(\mathbf{w}_{k,t}^T \mathbf{x}_{t,i}))\right)^{z_{t,i,k}} \tag{5}$$

Here, $\sigma(\cdot)$ is the sigmoid function and $\mathbf{w}_{k,t}$ are the classifier weights for the component k at time t. Same as [20], we use simple AR process [26] with lag order M to model the dynamics of classifier weights $\mathbf{w}_{t,k}$, and their values are obtained as,

$$\mathbf{w}_{k,t}|\mathbf{A}_k, \theta_k \sim \mathcal{N}(\sum_{m=1}^{M} \mathbf{A}_{k,m}\mathbf{w}_{k,t-m} + \mathbf{A}_{k,0},\ \theta_k^{-1}\mathbf{I}_D) \quad t > M$$
$$\mathbf{w}_{k,t}|\theta_{0,k} \sim \mathcal{N}(\mathbf{0},\ \theta_{0,k}^{-1}\mathbf{I}_D) \quad t \leq M \tag{6}$$

where $\mathbf{A}_{k,1}, \mathbf{A}_{k,2}, \ldots \mathbf{A}_{k,M} \in \mathbb{R}^{D\times D}$ define the classifier dynamics, $\mathbf{A}_{k,0} \in \mathbb{R}^D$ is the bias term, and $\theta_{0,k}, \theta_k \in \mathbb{R}^+$ are precision parameters. In the above equation, the classifier weights at time t depends on the weights of the same component at past M time instances, through linear dynamics. Since the classifier weights $\mathbf{W_k} := \{\mathbf{w}_{k,t}\}_{t=1}^T$ and time series parameters $\mathbf{A_k} := \{\mathbf{A}_{k,m}\}_{m=0}^M$ are not shared across the components, our model has the flexibility to learn separate dynamics for each component. For the sake of simplicity, we restrict $\mathbf{A}_{k,m}$ to be a diagonal matrix and for convenience use the same notation for the vector of diagonal elements of $\mathbf{A}_{k,m}$. This implies the dth dimension of classifier weight $w_{k,t,d}$, where $d \in \{1, \ldots, D\}$, depends only on its own past values $w_{k,t-1,d}, \ldots w_{k,t-M,d}$. For $t \leq M$ we assume each $\mathbf{w}_{k,t}$ is generated from $\mathcal{N}(\mathbf{0}, \theta_{0,k}^{-1}\mathbf{I}_D)$. The time series parameters $\mathbf{A}_{k,m}$ for $m = 0, \ldots, M$ are generated from multivariate normal distribution as,

$$\mathbf{A}_{k,m}|\gamma_{k,m} \sim \mathcal{N}(\mathbf{0}, \gamma_{k,m}^{-1}\mathbf{I}_D) \tag{7}$$

The precision parameters $\theta_{0,k}, \theta_k, \gamma_{k,m}$ are sampled from the Gamma priors as,

$$\theta_{0,k} \sim \Gamma(u_0, v_0), \quad \theta_k \sim \Gamma(u, v), \quad \gamma_{k,m} \sim \Gamma(a, b) \tag{8}$$

where u_0, v_0, u, v, a, b are fixed hyperparameters.

Correspondingly, the joint distribution of the labeled data $\mathcal{D}$, location parameters $\mathbf{\Phi} := \{\phi_k\}_{k=1}^\infty$, classifier weights $\mathbf{W} := \{\mathbf{W}_k\}_{k=1}^\infty$ and their precision parameters $\mathbf{\Theta} := \{(\theta_k, \theta_{0,k})\}_{k=1}^\infty$, dynamics' parameters $\mathbf{A} := \{\mathbf{A_k}\}_{k=1}^\infty$ and their precision parameters $\Gamma := \{\{\gamma_{k,m}\}_{m=0}^M\}_{k=0}^\infty$, assignment variables $\mathbf{Z} := \{\{z_{t,i}\}_{i=1}^{N_t}\}_{t=1}^T$ and stick components π can be written as,

$$p(\mathcal{D}, \mathbf{W}, \mathbf{A}, \mathbf{Z}, \mathbf{\Phi}, \mathbf{\Theta}, \Gamma, \pi) = p(\mathcal{D}|\mathbf{\Phi}, \mathbf{W}, \mathbf{Z})p(\mathbf{W}|\mathbf{A}, \mathbf{\Theta})$$
$$\times\ p(\mathbf{A}|\Gamma)p(\mathbf{Z}|\pi)p(\Gamma)p(\mathbf{\Theta})p(\mathbf{\Phi})p(\pi) \tag{9}$$

We have omitted the hyperparameters from the notation.

4 Inference

In this section we develop a learning algorithm to compute the posterior distribution of hidden variables i.e. $p(\mathbf{W}, \mathbf{A}, \mathbf{Z}, \mathbf{\Phi}, \mathbf{\Theta}, \Gamma, \pi|\mathcal{D})$. Since exact posterior inference is computationally intractable because of evidence calculation, we take a variational inference-based approach and approximate the model posterior using the distribution $q(\mathbf{W}, \mathbf{A}, \mathbf{Z}, \Phi, \mathbf{\Theta}, \Gamma, \pi)$ (shortly referred as $q(\cdot)$).

4.1 Evidence Lower Bound (ELBO)

In a standard fashion, we aim to minimize the reverse KL divergence of the model posterior with respect to the variational distribution $q(\cdot)$ [4]. Since the model evidence $p(\mathcal{D})$ is constant with respect to $q(\cdot)$, the minimization task is equivalent to the maximization of the lower bound of log evidence $L(q)$ given as,

$$L(q) := \mathbb{E}_q\big[\log p(\mathcal{D}, \mathbf{W}, \mathbf{A}, \mathbf{Z}, \Phi, \boldsymbol{\Theta}, \Gamma, \pi)\big] + H(q) \tag{10}$$

with respect to free variational parameters of the distribution $q(\cdot)$ where $H(q)$ denotes the entropy of distribution $q(\cdot)$. The logistic regression introduces non-conjugacy into our model. To tackle this issue, we use the following Gaussian lower bound [16]:

$$p(y_{t,i}|\mathbf{x}_{t,i}, \mathbf{w}_{k,t}) \;\geq\; \sigma(\xi_{k,t,i}) \exp\left(\frac{2y_{t,i}b - b - \xi_{k,t,i}}{2} + h(\xi_{k,t,i})(b^2 - \xi^2_{k,t,i})\right)$$

where $b := \mathbf{w}^T_{k,t}\mathbf{x}_{t,i}$ and $h(\xi_{k,t,i}) = \frac{2\sigma(\xi_{k,t,i})-1}{4\xi_{k,t,i}}$. This inequality introduces an extra set of parameters ξ for all samples in $\mathcal{D}$ to approximate the logistic expression in each component mixture. This gives us a new variational objective function $L(q, \xi)$, a lower bound to $L(q)$, with additional parameters ξ to estimate. Maximizing the newly obtained variational objective function $L(q, \xi)$ is equivalent to minimizing the KL divergence between $q(\cdot)$ and model posterior $p(\cdot)$, with accuracy of the approximation given by parameters ξ.

4.2 Variational Distributions and Update Equations

For analytical purposes, we take a mean-field approximation and restrict the distribution $q(\cdot)$ to a fully factorized form. We utilize truncated representation of DP's stick-breaking process [32] in our variational distribution and fix a truncation level K by enforcing $q(\pi_K = 1) = 1$. That is, the component $k = K$ takes all of the stick left after $K - 1$ splits and the remaining (infinite) components are assigned zero probability of being chosen by the assignment variable. The value of K can be chosen before the start of the learning process. Putting it all together, $q(\cdot)$ can be represented as,

$$q(\cdot) := \prod_{k=1}^{K}\prod_{t=1}^{T} q(\mathbf{w}_{k,t}) \prod_{k=1}^{K}\prod_{m=0}^{M} q(A_{k,m})q(\gamma_{k,m}) \\ \times \prod_{t=1}^{T}\prod_{i=1}^{N_t} q(z_{t,i}) \prod_{k=1}^{K} q(\phi_k)q(\theta_k)q(\theta_{0,k})q(\pi_k) \tag{11}$$

The log of the optimal solution for each factor in $q(\cdot)$ in the above equation is obtained by taking the expectation of log of the joint distribution given in (9) with respect to all other factors [4]. Using this method, the optimal forms are derived as,

$$q(\pi'_k) = Beta(a^{\pi}_k, b^{\pi}_k), \qquad q(z_{t,i}) = Mult(\nu_{t,i,1} \dots \nu_{t,i,K})$$
$$q(\phi_k) = \mathcal{NW}(\boldsymbol{g}_k, \beta_k, \boldsymbol{V}_k, f_k), \qquad q(\mathbf{w}_{k,t}) = \prod_{d=1}^{D} \mathcal{N}(\eta_{k,t,d}, \lambda^{-1}_{k,t,d})$$
$$q(\theta_{0,k}) = \Gamma(u^{\theta_0}_k, v^{\theta_0}_k), \qquad q(\mathbf{A}_{k,m}) = \mathcal{N}(\boldsymbol{\mu}_{k,m}, \boldsymbol{S}^{-1}_{k,m})$$
$$q(\theta_k) = \Gamma(u^{\theta}_k, v^{\theta}_k), \qquad q(\gamma_{k,m}) = \Gamma(a^{\gamma}_{k,m}, b^{\gamma}_{k,m}) \tag{12}$$

where $\eta_{k,t,d} \in \mathbb{R}$, $\boldsymbol{g}_k, \boldsymbol{\mu}_{k,m} \in \mathbb{R}^D$ and $\boldsymbol{V}_k, \boldsymbol{S}_{k,m} \in \mathbb{R}^{D \times D}$ and the remaining parameters belong to $\mathbb{R}$. Corresponding update equations for the parameters in (12) are provided below. The variational posterior distribution $q(\cdot)$ is obtained by iterating through these equations in a cyclic fashion until a convergence criterion is satisfied.

$$a^{\pi}_k = 1 + \sum_{t=1}^{T}\sum_{i=1}^{N_t} \nu_{k,t,i}, \quad b^{\pi}_k = \alpha + \sum_{k'=k+1}^{K}\sum_{t=1}^{T}\sum_{i=1}^{N_t} \nu_{k,t,i}, \quad \boldsymbol{g}_k = \frac{1}{\beta_k}(\beta_0 \boldsymbol{g}_0 + \bar{\nu}_k \bar{\mathbf{x}}_k)$$

$$\beta_k = \beta_0 + \bar{\nu}_k, \quad f_k = f_0 + \bar{\nu}_k, \quad \mathbf{V}^{-1}_k = \boldsymbol{V}^{-1}_0 + \bar{\nu}_k \bar{\sigma}_k + \frac{\beta_0 \bar{\nu}_k}{\beta_0 + \bar{\nu}_k}(\bar{\mathbf{x}}_k - \boldsymbol{g}_0)(\bar{\mathbf{x}}_k - \boldsymbol{g}_0)^T$$

$$u^{\theta}_k = u + \frac{1}{2}(T - M)D, \quad u^{\theta_0}_k = u_0 + \frac{1}{2}MD, \quad v^{\theta_0}_k = v_0 + \frac{1}{2}\sum_{t=1}^{T} ||\boldsymbol{\eta}_{k,t}||^2 + Tr(\boldsymbol{\Lambda}^{-1}_{k,t})$$

$$v^{\theta}_k = v + \frac{1}{2}\sum_{t=M+1}^{T}\Bigg(Tr(\boldsymbol{S}^{-1}_{0,k} + \boldsymbol{\Lambda}^{-1}_{k,t}) + ||\boldsymbol{\eta}_{k,t} - \sum_{l=1}^{M} dg(\boldsymbol{\mu}_{k,l})\boldsymbol{\eta}_{k,t-l} - \boldsymbol{\mu}_{k,0}||^2$$
$$+ \sum_{l=1}^{M} \boldsymbol{\eta}^T_{k,t-l}\boldsymbol{S}^{-1}_{k,l}\boldsymbol{\eta}_{k,t-l} + \sum_{l=1}^{M}\Big\{Tr(dg(\boldsymbol{\mu}_{k,l})^2)\boldsymbol{\Lambda}^{-1}_{k,t-l} + Tr(\boldsymbol{S}^{-1}_{k,l}\boldsymbol{\Lambda}^{-1}_{k,t-l})\Big\}\Bigg)$$
$$\forall k = 1 \dots K$$

$$\log \nu_{t,i,k} \propto \sum_{k'=1}^{k-1} \psi(b^{\pi}_{k'}) - \psi(a^{\pi}_{k'} + b^{\pi}_{k'}) + \psi(a^{\pi}_k) - \psi(a^{\pi}_k + b^{\pi}_k) + \frac{1}{2}\Psi(f_k/2)$$
$$+ \frac{1}{2}(D\log 2 + \log|\boldsymbol{V}_k|) + (y_{t,i} - 0.5)\boldsymbol{\eta}^T_{k,t}\mathbf{x}_{t,i} - \mathbf{x}^T_{t,i}[\boldsymbol{\eta}_{k,t}\boldsymbol{\eta}^T_{k,t} + \boldsymbol{\Lambda}^{-1}_{k,t}]\mathbf{x}_{t,i}h(\xi_{t,i,k})$$
$$- \frac{1}{2}\left(\frac{D}{\beta^{-1}_k} + f_k(\mathbf{x}_{t,i} - \boldsymbol{g}_k)^T \boldsymbol{V}_k(\mathbf{x}_{t,i} - \boldsymbol{g}_k)\right) \forall k = 1 \dots K, i = 1 \dots N_t, t = 1 \dots T$$

$$\eta_{k,t,d} = \lambda^{-1}_{k,t,d}\Bigg[\sum_{i=1}^{N_t} \nu_{k,t,i}\Big((y_{t,i} - 0.5)x_{t,i,d} - 2h(\xi_{t,i,k})\sum_{l \neq d} \eta_{k,t,l}\mathbf{x}_{t,i,l}\mathbf{x}_{t,i,d}\Big)$$
$$+ \frac{u^{\theta}_k}{v^{\theta}_k}\sum_{l=1}^{K^t_1}\Big((\eta_{k,m+l,d} - \mu_{k,0,d})\mu_{k,m+l-t,i} - \sum_{l' \neq m+l-t}^{m} \eta_{k,m+l-l',d} \cdot \mu_{k,l',d} \cdot \mu_{k,m+l-t,d}\Big)\Bigg]$$

$$\boldsymbol{\Lambda}_{k,t} = \frac{u_k^\theta}{v_k^\theta} \sum_{l=1}^{K_1^t} \left(dg(\boldsymbol{\mu}_{k,m-t+l})^2 + \boldsymbol{S}_{k,m-t+l}^{-1} \right) + \frac{u_k^{\theta_0}}{v_k^{\theta_0}} \mathbf{I}_d + 2 \sum_{i=1}^{N_t} h(\xi_{t,i,k} dg(\mathbf{x}_{t,i})^2)$$

$$\forall k = 1, \ldots K, \quad t = 1, \ldots M, \quad d = 1, \ldots D$$

$$\eta_{k,t,d} = \lambda_{k,t,d}^{-1} \left(\frac{u_k^\theta}{v_k^\theta} \Big[\sum_{l=1}^{K_2^t} \Big((\eta_{k,t+l,d} - \mu_{k,0,d}) \mu_{k,l,d} - \sum_{l' \neq l}^{m} \eta_{k,t+l-l',d} \cdot \mu_{k,l',d} \cdot \mu_{k,l,d} \Big) \right.$$

$$\left. + \sum_{l=1}^{m} \eta_{k,t-l,d} \mu_{k,l,d} + \mu_{k,0,d} \Big] + \sum_{i=1}^{N_t} \nu_{k,t,d} \mathbf{x}_{t,i,d} \Big\{ y_{t,i} - \frac{1}{2} - 2h(\xi_{t,i,k}) \sum_{l \neq d} \eta_{k,t,l} \mathbf{x}_{t,i,l} \Big\} \right)$$

$$\boldsymbol{\Lambda}_{k,t} = \frac{u_k^\theta}{v_k^\theta} \sum_{l=1}^{K_2^t} (dg(\boldsymbol{\mu}_{k,l})^2 + \boldsymbol{S}_{k,l}^{-1}) + \frac{u_k^{\theta_0}}{v_k^{\theta_0}} \mathbf{I}_d + 2 \sum_{i=1}^{N_t} h(\xi_{t,i,k}) dg(\mathbf{x}_{t,i})^2$$

$$k = 1, \ldots K, \quad t = M+1, \ldots T, \quad d = 1, \ldots D$$

$$(\xi_{t,i,k})^2 = \mathbf{x}_{t,i}^T (\boldsymbol{\Lambda}_{t,k}^{-1} + \boldsymbol{\eta}_{t,k} \boldsymbol{\eta}_{t,k}^T) \mathbf{x}_{t,i} \qquad k = 1, \ldots K, t = 1, \ldots T, i = 1, \ldots N_t$$

$$\boldsymbol{S}_{k,m} = \frac{a_{k,m}^\gamma}{b_{k,m}^\gamma} \mathbf{I}_d + \frac{u_k^\theta}{v_k^\theta} \sum_{t=m+1}^{T} (dg(\boldsymbol{\eta}_{k,t-m})^2 + \boldsymbol{\Lambda}_{k,t-m}^{-1}), \boldsymbol{S}_{k,0} = \Big(\frac{a_{k,0}^\gamma}{b_{k,0}^\gamma} + (T-M) \frac{u_k^\theta}{v_k^\theta} \Big) \mathbf{I}_D$$

$$a_{k,m}^\gamma = a + \frac{1}{2} D, \quad \boldsymbol{\mu}_{k,m} = \boldsymbol{S}_{k,m}^{-1} \frac{u_k^\theta}{v_k^\theta} \sum_{t=m+1}^{T} dg\Big(\boldsymbol{\eta}_{k,t} - \sum_{l \neq m} dg(\boldsymbol{\mu}_{k,l}) \boldsymbol{\eta}_{k,t-l} - \boldsymbol{\mu}_0 \Big) \boldsymbol{\eta}_{k,t-m}$$

$$b_{k,m}^\gamma = b + \frac{1}{2} (||\boldsymbol{\mu}_{k,m}||^2 + Tr(\boldsymbol{S}_{k,m}^{-1})), \ \boldsymbol{\mu}_{k,0} = \boldsymbol{S}_{k,0}^{-1} \frac{u_k^\theta}{v_k^\theta} \sum_{t=1}^{T} (\boldsymbol{\eta}_{k,t} - \sum_{m=1}^{M} dg(\boldsymbol{\mu}_{k,m}) \boldsymbol{\eta}_{k,t-m})$$

$$k = 1....K, \quad m = 1...M$$

where $\bar{\sigma}_k = \frac{1}{\bar{\nu}_k} \sum_t^T \sum_i^{N_t} \nu_{t,i,k} (\mathbf{x}_{t,i} - \bar{\mathbf{x}}_k)(\mathbf{x}_{t,i} - \bar{\mathbf{x}}_k)^T$, $\bar{\nu}_k = \sum_t^T \sum_i^{N_t} \nu_{t,i,k}$, $\bar{\mathbf{x}}_k = \frac{1}{\bar{\nu}_k} \sum_t^T \sum_i^{N_t} \nu_{t,i,k} \mathbf{x}_{t,i}$, $\boldsymbol{\Lambda}_{k,t} \in \mathcal{R}^{D \times D}$ is a diagonal matrix with diagonal elements as $(\lambda_{k,t,1}, \ldots \lambda_{k,t,D})$, $\psi(\cdot)$ and $\Psi(\cdot)$ are digamma and polygamma functions, $K_1^t := min(t, T-m)$, $K_2^t := min(m, T-t)$, $dg(\mathbf{x})$ is a diagonal matrix whose diagonal elements are $\mathbf{x} = (x_1, \ldots x_d)$, $h(x) = \frac{1}{2x}(\sigma(x) - \frac{1}{2})$, $||\cdot||$ is l_2-norm, and Tr represents the trace.

4.3 Prediction

For unlabeled time-stamped data sample $\mathbf{x}_{T',i}$ arrived at time $T' > T$, the probability of its label being 1 can be obtained as,

$$Pr(y_{T',i} = 1 | \mathbf{x}_{T',i}) = \sum_{k=1}^{K} \omega_{k,T',i} c_{k,T',i} \tag{13}$$

where $c_{k,T',i}$ is the prediction by classifier at time T' of component k obtained using AR process and $\omega_{k,T',i} = P(z_{T',i,k} = 1 | \mathbf{x}_{T',i})$ is the probabilistic weight of the component in the mixture. Here,

$$\begin{aligned}\omega_{k,T',i} &\propto Pr(z_{T',i,k}=1)p(\mathbf{x}_{T',i}|\mathbf{\Phi}, z_{T',i,k}=1)\\ &= Pr(z_{T',i,k}=1)\mathcal{N}(\mathbf{x}_{T',i}|\mathbf{m}_k, \mathbf{R}_k^{-1})\end{aligned} \quad (14)$$

Note that, $\sum_{k=1}^{K}\omega_{k,T',i}=1$. The prior weight of a component k in the mixture $Pr(z_{T',i,k}=1)$ can be approximated as $\frac{\sum_{t=1}^{T}\sum_{i=1}^{N_t}\nu_{k,t,i}}{N}$, where N denotes total number of samples in the training data $\mathcal{D}$.

5 Experiments

In this section, we first describe the methods used for comparison and the experimental settings that were used to evaluate the performance of the proposed method. Next, we describe the synthetic and real-world data used for experiments and present the obtained results.

Comparison Methodology. We compare our method's performance against AAAI16 [20], AAAI17 [21], KDD18 [22], RF-Batch and RF-Present. AAAI17 and AAAI16 are the baseline approaches for predicting future linear classifiers. KDD18 is an extension to AAAI17 which utilizes the unlabeled data at test time instances for the training purpose. RF is a random forest classifier trained using the labeled data available at training time instances. RF-Batch uses all of the available training data $\mathcal{D}_{1:T}$ to learn the classifier while RF-Present uses the data available at the latest time instance i.e. $\mathcal{D}_T$. We use the area under the ROC curve (AUC) as an evaluation metric, a well-used measure for classification tasks, which was also used in the previous studies [21,22]. The performance results presented are obtained by aggregating the AUC values at test time instances and are averaged over five independent runs for each experiment.

Settings. For the proposed method, we set the value of hyperparameters as follows. We choose same values as [20] for Gamma prior hyperparameters: $u_0 = u = a = 1, v_0 = v = b = 0.1$. Since we normalize the data, the mean vector $\mathbf{g}_0$ and scale matrix $\mathbf{V}_0$ of Normal-Wishart distribution are set as $\mathbf{0}$ and $\mathbf{I}_D$, respectively. The scale parameter β_0 and degree of freedom f_0 should be greater than 0 and $D-1$, respectively. α, the concentration parameter of the Dirichlet Process, should be a positive real. As $\alpha \to 0$, the concentration property of the Dirichlet Process increases, and the number of inferred components decreases. K is the truncation parameter which determines the maximum number of inferred components and is user-specified. M is the lag order of the AR process and should be less than T in our model. We set their values as $\beta_0 = 2, f_0 = D+2, \alpha = 2, K = 6, M = 5$ from preliminary experiments. In AAAI16, AAAI17, and KDD18, we set the hyperparameters as suggested by authors in [21] and vary the AR lag order for AAAI16 from $1, \ldots 9$. All of these methods are variational inference-based and we run each of them till convergence.

5.1 Datasets

Synthetic Data. We consider two non-stationary settings (datasets) to test our method's competence in the prediction of future non-linear classifiers. Each setting is described briefly followed by the generative process for a sample pair $(\mathbf{x}_{t,i}, y_{t,i}) \in \mathcal{D}_t$.

Multimodal: In this experiment, covariate vector is sampled from a two-dimensional bimodal Gaussian mixture distribution. Decision boundary in one mode is rotating while in the second mode it is horizontal and moving vertically. We expect our method to infer two components one for each mode and learn the two different dynamics separately.

1. $z_{t,i} \sim Bernoulli(0.5)$
2. $\mathbf{x}_{t,i}|z_{t,i} \sim z_{t,i}\mathcal{N}(\mathbf{m}_1, 0.2\mathbf{I}_2) + (1 - z_{t,i})\mathcal{N}(\mathbf{m}_2, 0.2\mathbf{I}_2)$ where $\mathbf{x}_{t,i} \in \mathbb{R}^2, \mathbf{m}_1 = (0,0), \mathbf{m}_2 = (2,0)$
3. If $z_{t,i} = 1$, $y_{t,i} = 1$ if $\mathbf{x}_{t,i}^T(\cos\psi, \sin\psi) \geq 0$ else 0; $\psi = \pi(t-1)/4$ If $z_{t,i} = 0$, $y_{t,i} = 1$ if $x_{t,i,2} + (t - 15)/150 \geq 0$ else 0

XOR: We conduct this experiment to emulate the classic example of XOR classification where covariates in the opposite quadrants have same class labels. At any given time instance, we can approximate the underlying decision boundary using just two linear classifiers. We take uniform stationary distribution for the covariates, and gradually rotate the decision boundary.

1. $\mathbf{x}_{t,i,d} \sim Uniform(-2, 2)$ where $d = \{1, 2\}$
2. $z'_{t,i} = sgn(x_{t,i,1}\cos\psi + x_{t,i,2}\sin\psi)$; $\psi = \pi(t-1)/8$
   $z''_{t,i} = sgn(-x_{t,i,1}\sin\psi + x_{t,i,2}\cos\psi)$; $\psi = \pi(t-1)/8$
3. $y_{t,i} = 1$ if $z'_{t,i}z''_{t,i} > 0$ else 0

Real Data. We conduct experiments on two real-world datasets that have been used as benchmarks in previous studies on concept drift.

NOAA[1]: This dataset contains approximately 50 years of meteorological data containing measurements of temperature, pressure, etc. on a daily basis and the data is shown to be cyclical in nature [7]. It has 18, 159 samples and 8 features. The task is to predict whether it rained or not.

ONP[2]: This is a dataset about the news articles published on a digital media website for a period of 2 years. It consists of 39, 797 samples and 61 features. The regression task is converted into a classification problem by setting a threshold on the number of shares of an article [9].

Preprocessing. For the study of temporal variation of the classification performance in our experiments, we utilize the data until a certain time unit for

[1] https://www.ncdc.noaa.gov/cdo-web/datasets.
[2] https://archive.ics.uci.edu/ml/datasets/Online+News+Popularity.

Table 1. Average and standard deviation of the mean of AUC at all test time instances for five independent runs

Method	Multimodal	XOR	ONP2	ONP4	NOAA
RF-Batch	0.648 ± 0.003	0.464 ± 0.008	$\mathbf{0.615 \pm 0.008}$	$\mathbf{0.633 \pm 0.002}$	0.712 ± 0.003
RF-Present	0.722 ± 0.001	0.541 ± 0.004	0.579 ± 0.010	0.607 ± 0.006	0.690 ± 0.006
AAAI16	0.717 ± 0.004	0.516 ± 0.002	0.503 ± 0.011	0.524 ± 0.061	0.634 ± 0.049
AAAI17	0.710 ± 0.009	0.508 ± 0.009	0.569 ± 0.025	0.584 ± 0.006	0.772 ± 0.086
KDD18	0.732 ± 0.010	0.514 ± 0.010	0.568 ± 0.025	0.584 ± 0.009	–
Proposed	$\mathbf{0.990 \pm 0.003}$	$\mathbf{0.976 \pm 0.005}$	$\mathbf{0.616 \pm 0.025}$	$\mathbf{0.630 \pm 0.014}$	$\mathbf{0.828 \pm 0.034}$

training, and the remaining for testing. For Multimodal and XOR, we obtain 300 samples randomly per time instance, i.e. $N_t = 300$ for $t \in \{1, \ldots 30\}$. We split the data into train and test data by taking $T = 20$. Since the data is generated in a discretized manner, it can be directly used in the proposed and the other three competing methods (i.e. KDD18, AAAI17, and AAAI16). The real datasets that we use have sequential data. As a result, data within an interval of length ΔT is binned together and considered to have arrived at the same timestamp. The value of ΔT may have a significant effect on the performance of the four methods and a grid search approach should be used to determine its value. However, its optimum value may be different for each of the three methods. Thus, we do not tune this parameter to compare the performance over the same sequential data. Moreover, we randomly sample 80% of the data at each time instance and train the models using it. For NOAA, we discard the last 159 samples and set $\Delta T = 600$ days which splits the data into 30 units and keep $T = 20$. The remaining 10 units are used as testing data. For ONP, we use the same setting as [22] i.e. one month as ΔT and $T = 15$ while the remaining 10 units are used as testing data. ONP2 and ONP4 datasets are generated by further subsampling the data at training time instances by 20% and 40%, respectively.

Complexity of Datasets. We perform a short study on the classification complexity of the prepared datasets by training a linear Support Vector Machine (SVM) (as described in [25]). The mean m of the distances of incorrectly classified samples from the decision boundary is used to find a measure of linearity of the dataset as follows:

$$L = 1 - \frac{1}{1 + m}$$

As a result, we found the values of L for ONP2, ONP4, NOAA are respectively 0.37, 0.37, and 0.44, which shows the non-linearity of decision boundary in these datasets. Note that low values for L (bounded in $[0, 1)$) indicate that the problem is close to being linearly separable.

5.2 Results

Table 1 shows the average and standard deviation of the mean of AUCs over test time instances and Fig. 2 shows the average and standard deviation of AUCs for

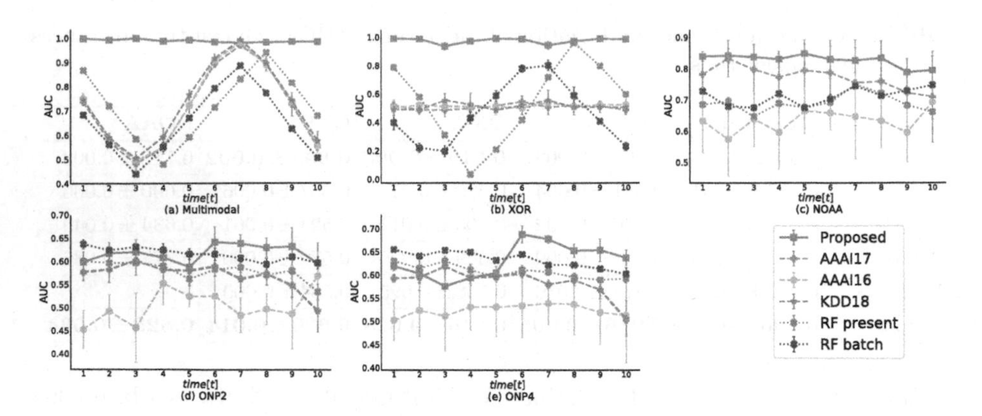

Fig. 2. Average and standard deviations of AUCs for all test time units over five runs

all test time instances. In synthetic data experiments, the proposed method consistently outperformed all other approaches by a significant margin. Competing approaches (i.e. KDD18, AAAI17, and AAAI16) that predict linear classifiers in the future performed poorly since the aggregated decision boundary was non-linear in the dataset. RF-Batch and RF-Present showed poor performances since the classifiers are not updated at test time instances in these approaches.

In Multimodal, the proposed method was able to identify the two modes and learn the dynamics separately for each mode to predict future classifiers, as validated by the AUC scores. KDD18, AAAI17, and AAAI16 did not predict classifiers based on the dynamics of the rotating boundary (corresponding to the left mode) and thus showed an oscillating AUC performance (see Fig. 2). RF-Batch and RF-Present showed a similar trend with RF-Present achieving the best AUC value at $t_{Test} = 8$, which corresponds to the period of rotation.

The nature of the XOR dataset exposes the limitations of linear approaches since any orientation of a linear classifier will separate the input space into two regions with (almost) an equal number of positive and negative samples. As a result, all three competing approaches had average AUC values near 0.5 i.e. as poor as a random guess (see Table 1). The proposed method showed robust performance even if the underlying covariate distribution is not a Gaussian mixture, and the decision boundary is non-linear. Figure 3 shows the predicted future classifiers by the proposed method for the XOR dataset in a single run. It can be observed from the figure that the proposed method identified the location of the mixture components in the input space such that the evolving non-linear decision boundary can be approximated well with the mixture of linear classifiers at all time-instances.

On real datasets, the proposed method achieved better average AUC values (see Table 1) than the competing linear approaches. In ONP2, the proposed method consistently showcased better AUC values at all the test time instances. In ONP4, the proposed method showcased similar or better performance than competing approaches at all time instances except at $t_{Test} = 3$. In ONP experiments, the results of AAAI16 were poor compared to AAAI17. One of the reasons for this drop is the difficulty in modeling the dynamics of the decision

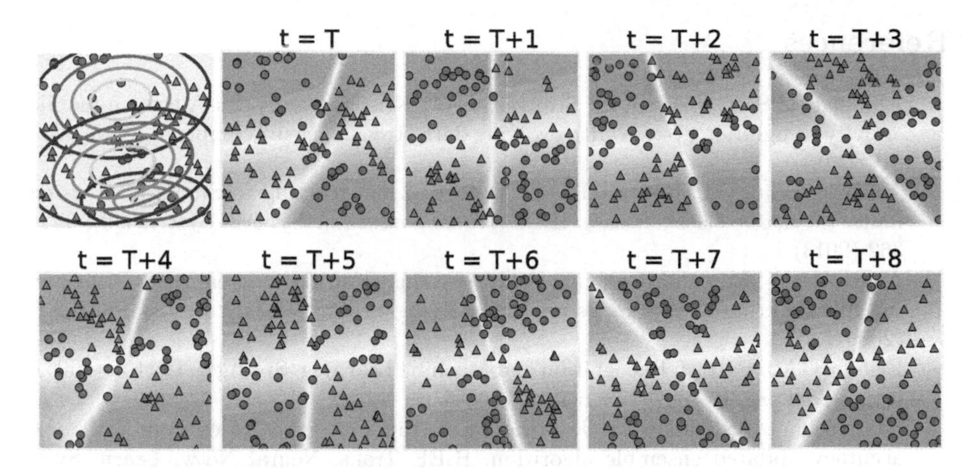

Fig. 3. For XOR dataset, contour plot of covariate distribution of the 3 inferred components followed by filled contour plots of (global) class probabilities at last training instance $t = T$ and future predictions for $t = T + 1, \ldots T + 8$. Red to blue scale represents class probabilities varying from 0 to 1. Ground truth positive (blue dots) and negative samples (red triangles) are shown too (Color figure online)

boundary using the AR process compared to GP. The proposed method uses the AR process as well for modeling the dynamics within each component. However, we believe it is not a limitation in our method since the mixture approach could help capture the complex dynamics. However, a more rigorous study on the hypothesis space of such evolutionary models is required, which is out of the scope of current work. For ONP experiments, the proposed method showed a similar result to the RF-Batch method and RF is considered as one of the best non-linear classification algorithms. At last, the proposed method was able to outperform all other approaches on the NOAA dataset. The result of KDD on the NOAA dataset is not presented since we faced convergence issues in the maximization step for mean and variances of classifier weights for the test time instances (refer [22]). The results validate that the proposed method can predict future classifiers when the decision boundary, possibly non-linear, is evolving and maintain it for some time.

6 Conclusion

We proposed a probabilistic model for predicting future non-linear classifiers given time-stamped labeled data collected until the current time. At each time instance, our model approximates the non-linear decision boundary using a mixture of linear classifiers. We used the Dirichlet process as prior in our model to automatically determine the number of required linear classifiers and developed a variational inference-based learning algorithm. Through synthetic and real data experiments we confirmed the effectiveness of the proposed method.

References

1. Abdallah, Z.S., Gaber, M.M., Srinivasan, B., Krishnaswamy, S.: Adaptive mobile activity recognition system with evolving data streams. Neurocomputing **150**, 304–317 (2015)
2. Alippi, C., Liu, D., Zhao, D., Bu, L.: Detecting and reacting to changes in sensing units: the active classifier case. IEEE Trans. Syst. Man Cybern. Syst. **44**(3), 353–362 (2013)
3. Angelov, P., Zhou, X.: Evolving fuzzy systems from data streams in real-time. In: 2006 International Symposium on Evolving Fuzzy Systems, pp. 29–35. IEEE (2006)
4. Bishop, C.M.: Pattern Recognition and Machine Learning. Springer, Boston (2006). https://doi.org/10.1007/978-1-4615-7566-5
5. Brzezinski, D., Stefanowski, J.: Reacting to different types of concept drift: the accuracy updated ensemble algorithm. IEEE Trans. Neural Netw. Learn. Syst. **25**(1), 81–94 (2014)
6. Crammer, K., Even-Dar, E., Mansour, Y., Vaughan, J.W.: Regret minimization with concept drift. In: Proceedings of the 23rd Annual Conference on Learning Theory (COLT) (2010)
7. Ditzler, G.: Incremental learning of concept drift from imbalanced data (2011)
8. Ferguson, T.S.: A bayesian analysis of some nonparametric problems. The annals of statistics pp. 209–230 (1973)
9. Fernandes, K., Vinagre, P., Cortez, P.: A proactive intelligent decision support system for predicting the popularity of online news. In: Pereira, F., Machado, P., Costa, E., Cardoso, A. (eds.) EPIA 2015. LNCS (LNAI), vol. 9273, pp. 535–546. Springer, Cham (2015). https://doi.org/10.1007/978-3-319-23485-4_53
10. Gama, J., Žliobaitė, I., Bifet, A., Pechenizkiy, M., Bouchachia, A.: A survey on concept drift adaptation. ACM Comput. Surv. (CSUR) **46**(4), 44 (2014)
11. Griffiths, T.L., Jordan, M.I., Tenenbaum, J.B., Blei, D.M.: Hierarchical topic models and the nested Chinese restaurant process. In: Advances in Neural Information Processing Systems, pp. 17–24 (2004)
12. Hannah, L.A., Blei, D.M., Powell, W.B.: Dirichlet process mixtures of generalized linear models. J. Mach. Learn. Res. **12**, 1923–1953 (2011)
13. Haque, A., Khan, L., Baron, M.: Sand: semi-supervised adaptive novel class detection and classification over data stream. In: THIRTIETH AAAI Conference on Artificial Intelligence (2016)
14. Haque, A., Khan, L., Baron, M., Thuraisingham, B., Aggarwal, C.: Efficient handling of concept drift and concept evolution over stream data. In: 2016 IEEE 32nd International Conference on Data Engineering (ICDE), pp. 481–492. IEEE (2016)
15. Hofer, V.: Adapting a classification rule to local and global shift when only unlabelled data are available. Eur. J. Oper. Res. **243**(1), 177–189 (2015)
16. Jaakkola, T.S., Jordan, M.I.: Bayesian parameter estimation via variational methods. Stat. Comput. **10**(1), 25–37 (2000)
17. Klinkenberg, R.: Learning drifting concepts: example selection vs. example weighting. Intell. Data Anal. **8**(3), 281–300 (2004)
18. Kolter, J.Z., Maloof, M.A.: Using additive expert ensembles to cope with concept drift. In: Proceedings of the 22nd International Conference on Machine Learning, pp. 449–456. ACM (2005)
19. Koychev, I.: Gradual forgetting for adaptation to concept drift. In: Proceedings of ECAI 2000 Workshop on Current Issues in Spatio-Temporal ... (2000)

20. Kumagai, A., Iwata, T.: Learning future classifiers without additional data. In: Thirtieth AAAI Conference on Artificial Intelligence (2016)
21. Kumagai, A., Iwata, T.: Learning non-linear dynamics of decision boundaries for maintaining classification performance. In: Thirty-First AAAI Conference on Artificial Intelligence (2017)
22. Kumagai, A., Iwata, T.: Learning dynamics of decision boundaries without additional labeled data. In: Proceedings of the 24th ACM SIGKDD International Conference on Knowledge Discovery & Data Mining, pp. 1627–1636. ACM (2018)
23. Lee, K.H., Verma, N.: A low-power processor with configurable embedded machine-learning accelerators for high-order and adaptive analysis of medical-sensor signals. IEEE J. Solid-State Circuits **48**(7), 1625–1637 (2013)
24. Lindstrom, P., Delany, S.J., Mac Namee, B.: Handling concept drift in a text data stream constrained by high labelling cost. In: Twenty-Third International FLAIRS Conference (2010)
25. Lorena, A.C., Garcia, L.P., Lehmann, J., Souto, M.C., Ho, T.K.: How complex is your classification problem? A survey on measuring classification complexity. ACM Comput. Surv. (CSUR) **52**(5), 1–34 (2019)
26. Lütkepohl, H.: Vector Autoregressive Models. Springer, Heidelberg (2011). https://doi.org/10.1007/978-3-642-04898-2_609
27. Martínez-Rego, D., Pérez-Sánchez, B., Fontenla-Romero, O., Alonso-Betanzos, A.: A robust incremental learning method for non-stationary environments. Neurocomputing **74**(11), 1800–1808 (2011)
28. Sethuraman, J.: A constructive definition of Dirichlet priors. Statistica Sinica, 639–650 (1994)
29. Shahbaba, B., Neal, R.: Nonlinear models using Dirichlet process mixtures. J. Mach. Learn. Res. **10**, 1829–1850 (2009)
30. Škrjanc, I., Iglesias, J.A., Sanchis, A., Leite, D., Lughofer, E., Gomide, F.: Evolving fuzzy and neuro-fuzzy approaches in clustering, regression, identification, and classification: a survey. Inf. Sci. **490**, 344–368 (2019)
31. Teh, Y.W., Jordan, M.I., Beal, M.J., Blei, D.M.: Sharing clusters among related groups: Hierarchical Dirichlet processes. In: Advances in Neural Information Processing Systems, pp. 1385–1392 (2005)
32. Wang, C., Paisley, J., Blei, D.: Online variational inference for the hierarchical Dirichlet process. In: Proceedings of the Fourteenth International Conference on Artificial Intelligence and Statistics, pp. 752–760 (2011)
33. Wang, J., Zhao, P., Hoi, S.C.H.: Exact soft confidence-weighted learning. In: Proceedings of the 29th International Conference on Machine Learning, pp. 121–128 (2012)
34. Yang, X., Dinh, A., Chen, L.: Implementation of a wearerable real-time system for physical activity recognition based on Naive Bayes classifier. In: 2010 International Conference on Bioinformatics and Biomedical Technology, pp. 101–105. IEEE (2010)
35. Yuan, C., Neubauer, C.: Variational mixture of gaussian process experts. In: Advances in Neural Information Processing Systems, pp. 1897–1904 (2009)
36. Zhu, X., Zhang, P., Lin, X., Shi, Y.: Active learning from stream data using optimal weight classifier ensemble. IEEE Trans. Syst. Man Cybern. Part B (Cybern.) **40**(6), 1607–1621 (2010)
37. Žliobaitė, I., Bifet, A., Pfahringer, B., Holmes, G.: Active learning with drifting streaming data. IEEE Trans. Neural Netw. Learn. Syst. **25**(1), 27–39 (2013)

Parameterless Semi-supervised Anomaly Detection in Univariate Time Series

Oleg Iegorov(✉) and Sebastian Fischmeister

University of Waterloo, Waterloo, Canada
{oiegorov,sfischme}@uwaterloo.ca

Abstract. Anomaly detection algorithms that operate without human intervention are needed when dealing with large time series data coming from poorly understood processes. At the same time, common techniques expect the user to provide precise information about the data generating process or to manually tune various parameters. We present SIM-AD: a semi-supervised approach to detecting anomalies in univariate time series data that operates without any user-defined parameters. The approach involves converting time series using our proposed Sojourn Time Representation and then applying modal clustering-based anomaly detection on the converted data. We evaluate SIM-AD on three publicly available time series datasets from different domains and compare its accuracy to the PAV and RRA anomaly detection algorithms. We conclude that SIM-AD outperforms the evaluated approaches with respect to accuracy on trendless time series data.

Keywords: Anomaly detection · Time series · Motifs · Modal clustering

1 Introduction

Time series data arise when any data generating process is observed over time. These data are used in diverse fields of research, such as intrusion detection for cyber-security, medical surveillance, economic forecasting, fault detection in safety-critical systems, and many others. One of the main tasks performed on time series data is anomaly detection (AD). Anomalies appear when the underlying process deviates from its normal behavior. For this reason, anomalies can help analysts better understand the data generating process or safely steer it towards nominal operation. As Aggarwal points out in [5], virtually all AD algorithms train a model of the normal patterns in the data and then compute anomaly scores for new observations based on the deviation from these patterns.

With the ever-growing rate of data generation, data mining (DM) and machine learning (ML) tasks need to be performed with minimal human intervention, and anomaly detection is no exception [37]. This task usually requires a combination of expertise in the target application domain as well as in ML and DM algorithms. Experts are rare, and their work is costly. Companies thus need

F. Hutter et al. (Eds.): ECML PKDD 2020, LNAI 12457, pp. 644–659, 2021.
https://doi.org/10.1007/978-3-030-67658-2_37

AD tools that do not require manual parameter tweaking and that can work on raw, unprocessed data without jeopardizing accuracy of the results. Consequently, the demand for parameterless solutions has recently given rise to the field of automated ML (AutoML) [24].

There exist numerous algorithms for anomaly detection in time series data; for example, the surveys in [12] and [20] point to dozens of AD approaches aiming at time series data. Indeed, different application domains typically have their own AD problem formulations that require dedicated techniques. An anomaly detection problem can be described with a set of largely independent *facets*, some of which we list below. The underlined facets define the anomaly detection problem that we address in this paper:

- time series: univariate/multivariate
- anomaly detection type: unsupervised/semi-supervised/supervised
- parameters: none/training/anomaly detection
- anomaly type: subsequence/whole time series
- anomaly score: binary/numeric
- anomaly detection mode: online/offline.

This way, we are interested in *online* anomaly detection in *univariate* time series data. Moreover, we expect at least one time series capturing normal (i.e., non-anomalous) system behavior to be available for training: this scenario is often referred to as *semi-supervised* anomaly detection in the literature [12,13, 32]. Also, we want to detect anomalous *subsequences* and have *binary* yes/no anomaly scores for subsequences in the target time series. Finally, we want both the training and anomaly detection stages to be completely *parameterless*.

One of the application domains that can benefit from our problem formulation is safety-critical embedded systems used in automotive, avionics, medical device, and other mission-critical industries. Anomaly detection in such contexts is a crucial task since a system malfunction can cause catastrophic damages or worse put human lives in danger. As an example, consider an embedded system running a water treatment plant [18]. The hardware setup consists of various sensors and actuators, each generating a stream of univariate time series data. A malfunctioning system component can incur damage to the facility (e.g., by causing a tank overflow) or contaminate treated water[1]. It is possible to collect data during normal system operation and train a system model on these data. However, the large amount of generated data makes any manual pre-processing or visual analysis of the data complicated or not feasible. Therefore, a parameterless training on unprocessed data is much preferable. Another requirement is to detect anomalies in an online mode allowing preventative measures to be taken immediately. Finally, it is equally important to detect momentary and slowly-evolving malfunctions, both of which can be captured by anomalous subsequences in a time series data stream. Chapter 7 in [12] provides other examples of application domains where our problem formulation is relevant.

[1] https://www.theregister.co.uk/2016/03/24/water_utility_hacked.

In this paper, we propose a solution to the anomaly detection problem defined above: the SIM-AD approach. Since our goal is to make anomaly detection parameterless, we have to assume that all the knowledge about the normal operation of the target system is captured in a single or multiple time series available for training. In data mining, one obtains knowledge about a system by mining data patterns [21]. Therefore, we focus on mining patterns from time series data generated by the target system during normal operation and then look for the mined patterns in the new data that may capture anomalous system behavior. With SIM-AD, we represent a time series pattern as a typical number of timesteps that the time series spends (or *sojourns*) in a value range. To facilitate using SIM-AD, we have developed a web application available at https://sim-ad.herokuapp.com/.

The rest of the paper is organized as follows. In Sects. 2 and 3, we provide an overview of SIM-AD and review the related work. In Sects. 4 and 5, we present the main components of SIM-AD, including our new time series representation and the SIM algorithm. We then compare the accuracy of SIM-AD with the RRA and PAV anomaly detectors in Sect. 6.

2 Terminology and Overview

A univariate time series T of length N is a sequence of N measurements, or *observations*, of some variable collected chronologically: $T = \langle t_1, t_2, \ldots, t_N \rangle$. Observations are usually made at regular time intervals. A subsequence s of a time series T is a contiguous sampling of observations $t_i, \ldots, t_{i+m}$ of length $m < N$, where i is an arbitrary position, such that $1 \leq i \leq N - m + 1$.

With SIM-AD, we view a time series T as a sequence of temporal *states*. A state can be one of two types. A *motif-state* is a state that the system visits multiple times; it corresponds to a time series pattern, or a *motif* [31], i.e., a subsequence of T that looks similar to one or more other subsequences of T. A *nonmotif-state*, on the other hand, is visited only once; it corresponds to a subsequence of T that does not look similar to any other subsequence of T. Detection of motif-states corresponds to the problem of motif discovery, which is an active area of research in the data mining community [40]. Unfortunately, mining motifs of all possible lengths in raw time series poses a scalability problem [17]. A common approach to tackle this problem involves transforming the original time series T into some representation that has fewer data points than T [7]. SIM-AD uses a new time series representation which we call Sojourn Time Representation (STR). An STR of a time series T is a sequence of *sojourn times*. A sojourn time (ST) represents the number of timesteps that T sojourns in one of two contiguous and non-overlapping value ranges called Bin 1 and Bin 2 (see Fig. 1). *With SIM-AD, we assume that the first and last observations of a motif of a time series T correspond to consecutive intersection points between T and the line separating Bin 1 and Bin 2.* By making this assumption, we can map a motif or a nonmotif of a time series T to an ST in the STR of T. However, the same motif in T may correspond to slightly different STs in the STR of T due

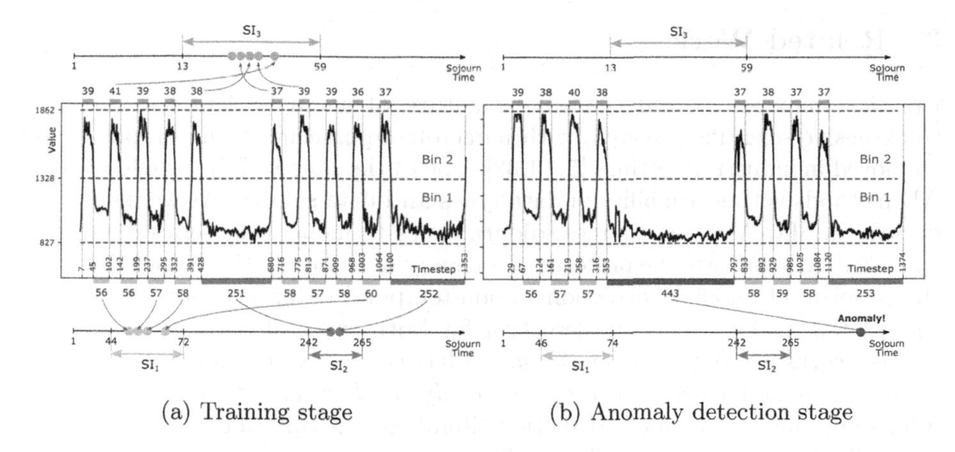

(a) Training stage (b) Anomaly detection stage

Fig. 1. Overview of SIM-AD (the shown time series are excerpts from the Power dataset introduced in Sect. 6.1) (Color figure online)

to random fluctuations in real-world time series data. Moreover, having a group of STs corresponding to the same motif in T and given a new ST, we would like to be able to decide whether the new ST belongs to that group; this way, we could conclude whether a subsequence of some time series T' corresponds to one of the states of T. We propose the Sojourn Interval Miner (SIM) algorithm that addresses the two problems mentioned above by clustering STs and finding outlier limits around the clusters. SIM mines *sojourn intervals* from the STR of a time series T. A sojourn interval (SI) is a pair of integers defining the outlier limits around a cluster of STs. Conceptually, an SI plays the role of an STR-equivalent of a time series state. Finally, given a set of SIs and a test time series T', SIM-AD detects anomalies in T' by finding subsequences of T' whose STs fall outside of SIs mined from the training time series T.

Figure 1 presents an overview of the training and anomaly detection stages of SIM-AD on a pair of excerpts from the Power time series introduced in Sect. 6.1. During the training stage (Fig. 1a), SIM-AD first computes the STR of the training time series T using Bin 1 and Bin 2. It then clusters STs from each bin independently, obtaining three clusters: two (blue and yellow) for STs from Bin 1 and one (orange) for the STs from Bin 2. This way, SIM-AD represents T as a sequence of three motif-states. For example, a subsequence between timesteps 237 and 295 belongs to one motif-state while a subsequence between timesteps 428 and 680 belongs to another motif-state. Next, SIM-AD mines an SI for each cluster of STs in T. For example, $\text{SI}_3 = [13, 59]$ denotes a range of lengths of subsequences in a test time series falling into Bin 2 that will map to the motif-state represented by the orange STs extracted from the training time series T. Finally, SIM-AD extracts STs from the test time series T' and finds a single ST whose value of 443 falls outside of SIs mined from T (Fig. 1b). Therefore, SIM-AD reports the corresponding subsequence of T' (highlighted in red) as an anomaly.

3 Related Work

Practical importance and challenging nature of anomaly detection have led to extensive scientific research, with numerous approaches being proposed and employed in industrial settings [5,13,32]. One of the causes of the proliferation of AD methods is the variability in data types and data sources. As we mentioned in Sect. 2, SIM-AD uses discrete sojourn times to model time series data. This way, SIM-AD converts the problem of anomaly detection in time series data into the problem of anomaly detection in non-temporal data. Therefore, we review the related work in anomaly detection for both types of data. Also, we only discuss works whose problem formulation has the largest overlap with ours.

It is common to see the terms *anomaly*, *outlier*, *novelty*, or *discord* used interchangeably in anomaly detection literature. In this work, we distinguish anomalies from outliers as in [32]: outlier subsequences contaminate normal time series data, and the goal is to cope with their presence during the training stage. We abstain from using the term discord since it was introduced in the context of unsupervised anomaly detection [26]. Finally, we do not make any distinction between anomalies and novelties.

3.1 Anomaly Detection in Time Series Data

Regression-based anomaly detection techniques, such as Long Short-Term Memory (LSTM) networks [19,23], use a window of w consecutive observations t_i, $t_{i+1}, \dots, t_{i+w-1}$ to predict the values of the subsequent n observations $t_{i+w}, \dots, t_{i+w+n-1}$. They then declare observations whose predicted value significantly deviates from the true value as anomalies. These methods require substantial user involvement in preprocessing raw data and tuning various parameters.

Discord mining algorithms, e.g., RRA [36], address the problem of unsupervised anomaly detection in univariate time series data. The term *discord* denotes the most unusual subsequence within a time series. Early discord mining algorithms [26] require the user to specify the length of discords. Later works allow for mining variable-length discords [36,41] but still require the user to choose the time series discretization parameters. This way, discord mining algorithms have good accuracy when the approximate length of motifs is chosen correctly [37]. Moreover, these algorithms output a ranked list of discords, and the user must choose a threshold on the length of this list to distinguish discords from normal subsequences.

Similarly to SIM-AD, segmentation-based anomaly detectors, e.g., PAV [14] and Gecko [35], segment a time series and consider each segment as a state. They then either learn a Finite State Automaton (FSA) of the mined states [35] or simply memorize the frequency of each state [14] and then detect anomalies as segments that do not match any state in the learned FSA or that appear infrequently. Unlike SIM-AD, the Gecko algorithm requires a database of similar time series for training [12] and takes the minimum length of a segment as a parameter.

Another relevant anomaly detection technique uses Numenta's Hierarchical Temporal Memory (HTM) [6]. This method addresses the problem of online and unsupervised anomaly detection in univariate time series data streams. Based on our understanding of the HTM theory, and confirmed by Numenta's engineers[2], this technique assumes that the first timestamps of motif occurrences are known.

3.2 Anomaly Detection in Non-temporal Data

Statistical anomaly detection techniques fit a mathematical model to the given data instances and then apply an inference test to determine whether a new instance belongs to this model. Instances that have a low probability to be generated from the learned model are declared as anomalies [13]. Parametric approaches assume that normal data are generated from a parametric distribution fully defined by parameters θ [32]. These parameters are then estimated from the training data. Nonparametric approaches, e.g., the ones based on Kernel Density Estimation (KDE) [38] (also called parzen window estimation), do not make assumptions about the distribution from which data are sampled. However, they require a threshold on probability density that separates normal instances from the anomalous ones [15].

Nearest neighbor (NN) techniques detect anomalies by considering distances between data instances. The distances are used either directly, by assigning anomaly scores to instances based on the distance to their kth nearest neighbor [33], or indirectly, by computing local densities of data instances and then declaring instances with smaller densities as anomalies [10]. In both cases, NN approaches require the user to specify the size of the local neighborhood.

Semi-supervised clustering techniques group normal data instances into clusters and then detect anomalies based on the distances between new data instances and the clusters. Some approaches adopt distance-based clustering that requires the number of clusters [16] or a similarity threshold [22]. Moreover, distance-based clustering assumes that clusters have a particular shape [28]. In contrast, density-based clustering does not require the number of clusters nor does it make assumptions about the shape of clusters [28]. However, this type of clustering requires the user to choose a smoothing factor. In DBSCAN-like clustering [11], the smoothing factor is expressed as the minimum cluster size, while in modal clustering [30] it is the kernel bandwidth that is used to compute the KDE of the data. An advantage of modal clustering is that the kernel bandwidth can be estimated automatically using, for example, the Improved Sheather-Jones plug-in rule [8]. We look closer at modal clustering in Sect. 5.1. Regardless of the underlying algorithm, clustering-based anomaly detection techniques need an anomaly threshold to determine whether a new data instance belongs to some cluster or constitutes an anomaly.

[2] https://discourse.numenta.org/t/3141.

4 Sojourn Time Representation (STR)

Transforming a time series T into a representation having fewer data points than T is often a prerequisite for scalable motif discovery. At the same time, we want to reduce the size of T without requiring any parameters. Common time series representations, e.g., PLR [25] and SAX [29], require the user to set at least one parameter, such as the length of a sliding window. This motivated us to propose Sojourn Time Representation (STR): a new type of time series representation, similar to the clipped representation [34], that we construct from a time series in a parameterless way.

An STR of a time series T is the run-length encoding of a sequence T_d, where T_d is the result of discretization of T into two bins, Bin 1 and Bin 2. Although the number of bins can be considered as a hard-coded parameter, we justify using two discretization bins in Sect. 6.5. We define Bin 1 and Bin 2 using non-overlapping intervals $[v_1, v_2)$ and $[v_2, v_3]$ correspondingly, where v_1 is the minimum value in T, v_3 is the maximum value in T, and v_2 is the median value among the unique values in T. This way, $T_d = \langle d_1, d_2, \ldots, d_N \rangle$, where $d_i = 1$ if $t_i \in [v_1, v_2)$ and $d_i = 2$ if $t_i \in [v_2, v_3]$, for $\forall i = 1, 2, \ldots, N$. Next, we apply run-length encoding to T_d. As a result, we obtain a sequence of tuples (r, b), where r refers to the run length, that is, the number of consecutive elements in T_d having the same value, and b refers to the bin number. We call this run-length encoding of T_d the Sojourn Time Representation of T, STR(T). Also, we refer to the run lengths in STR(T) as the *sojourn times* of T with respect to Bin 1 and Bin 2. As an example, the first few elements of the STR of the time series shown in Fig. 1a are $\langle (39, 2), (56, 1), (41, 2), (56, 1), (39, 2), \ldots \rangle$.

5 Sojourn Interval Miner (SIM)

The SIM algorithm lies at the core of SIM-AD and consists of two steps briefly mentioned in Sect. 2: clustering sojourn times and mining sojourn intervals given an STR of a time series T. The general problem addressed by SIM is the following: given a set of integers J (i.e., sojourn times extracted from training data) and a new integer j' (i.e., a sojourn time extracted from test data), find whether j' is anomalous with respect to all $j \in J$.

5.1 Parameterless Modal Clustering of Sojourn Times

In Sect. 3.2, we explained why modal clustering is the most relevant approach to clustering outlier-contaminated univariate data: it is possible to make it parameterless without assuming that data are sampled from some exponential family. We next consider modal clustering in detail and then show how we use it for anomaly detection.

In modal clustering, clusters correspond to densely-populated regions of the sample space [30]. A probability density function (PDF) is estimated from a given data sample using a nonparametric technique and then clusters are formed around the modes in the estimated PDF (hence the name *modal* clustering). The most popular nonparametric approach to density estimation is Kernel Density Estimation (KDE) [8,38]. In KDE, an estimate $\hat{f}$ of the true PDF f at a point x is computed by placing a kernel function K, usually a Gaussian of particular variance h, on each observation X_i ($i \in 1, \ldots, n$) and then summing up the values of all kernels at point x:

$$\hat{f}(x) = \frac{1}{nh} \sum_{i=1}^{n} K\left(\frac{x - X_i}{h}\right). \tag{1}$$

Each mode in the estimated PDF $\hat{f}$ then defines a cluster, and observations are assigned to clusters based on the distance to their nearest mode.

Note that the variance of Gaussian kernels, h, commonly called the *bandwidth*, is a parameter that must be estimated separately or chosen manually. Its value greatly influences the number and location of modes in $\hat{f}$ [38]. There exist various methods to estimate h from a data sample, the most popular being the rule-of-thumb and plug-in estimators. These methods find a value of h that maximizes the estimation accuracy of $\hat{f}$, usually based on the asymptotic mean integrated squared error (AMISE). Computing the AMISE requires approximating the functionals of the unknown density f. Most of the bandwidth estimation methods approximate the functionals of f assuming that f is normal, which is usually not the case. One exception is the Improved Sheather-Jones (ISJ) plug-in rule [8]. This method does not use the normal reference rule and thus is completely data-driven. ISJ was shown to accurately estimate densities of unimodal and multimodal distributions from samples of various sizes. Moreover, the numerical procedure used in ISJ is fast when implemented using the Discrete Cosine Transform [8]. We, therefore, use the ISJ method to automatically estimate the bandwidth h from Eq. 1. Interestingly, the literature on modal clustering does not focus on the problem of estimating h [30], leaving it to be set manually or using a rule-of-thumb approach which is known to significantly oversmooth multimodal densities. Instead, these works consider multivariate data and focus on finding the modes of $\hat{f}$ in multidimensional spaces, which is a non-trivial task.

Gaussian KDE is known to be sensitive to outliers, resulting in the appearance of "spurious" modes in $\hat{f}$ centered at outlier data instances [8,9]. This behavior is undesirable when the goal is to minimize the AMISE of $\hat{f}$ with respect to the true density f. Therefore, various methods have been proposed to make KDE more robust to outliers by smoothing out the tails of $\hat{f}$ [8]. We, however, use this peculiarity of the Gaussian KDE as a feature. It allows to get more accurate clustering results by assigning outliers to their own, distinct clusters instead of "stretching out" the existing clusters to accommodate the outliers.

5.2 Mining Sojourn Intervals

Once we grouped the STs from a set J into a set of clusters C, how can we decide whether a new ST $j' \notin J$ is anomalous with respect to J? We provide an answer to this question using the property of the Gaussian kernel density estimator mentioned above: if an ST j' is an anomaly, then the KDE of $\{J \cup j'\}$ will have a mode centered at j'. Indeed, the convex shape of the Gaussian kernel guarantees that the KDE of $\{J \cup j'\}$ will have a "bump" around j' when the distance from j' to the closest cluster $c \in C$ is sufficiently large given the kernel bandwidth h estimated from J.

When we perform online anomaly detection with SIM-AD, we must analyze STs of the incoming time series T' in real-time. We avoid computing the KDE of $\{J \cup j'\}$ for each ST j' from T' by finding sojourn intervals (SIs) around clusters C during the training stage. Given a cluster $c \in C$, we apply a binary search to find the closest ST g, such that the KDE of $\{c \cup g\}$ has a bump on g, using the bandwidth h estimated from normal STs J. We perform this search on both sides of the cluster c and define an SI using the found pair of outlier STs g_1 and g_2. This allows us to detect anomalies by simply comparing the values of the incoming STs to the SIs and reporting the subsequences of T' that correspond to STs that fall outside of the SIs.

6 Experiments

We compare the accuracy of SIM-AD, PAV, and RRA anomaly detectors on three publicly-available time series datasets previously used in the anomaly detection literature. We evaluate the accuracy of the detectors with a pair of F-scores.

6.1 Datasets

Power. The dataset contains 35,040 measurements of power consumption of a Dutch research facility during the entire year of 1997 [4]. The measurements capture the aggregate power consumption during fifteen-minute time intervals. The dataset was used for anomaly detection in [26,36,39]. There are 44 normal and 8 anomalous weeks in the dataset. Figure 2 shows power consumption during a normal week and an anomalous week. During a normal week, we can observe five consecutive peaks and valleys corresponding to the consumption on workdays, while the longer valleys are observed on weekends when relatively little power was consumed.

ECG. The electrocardiogram (ECG) dataset from the MIT-BIH Arrhythmia Database [1] was previously used for anomaly detection in [26,36,39]. The dataset contains 21,600 measurements of electrical potential between two points on the body surface of a patient during 55 heartbeats. Individual heartbeats have a similar shape, but their duration varies slightly. There are three anomalous heartbeats in the dataset reported by a cardiologist [26]. Figure 3 shows examples of three normal consecutive heartbeats and an anomalous heartbeat.

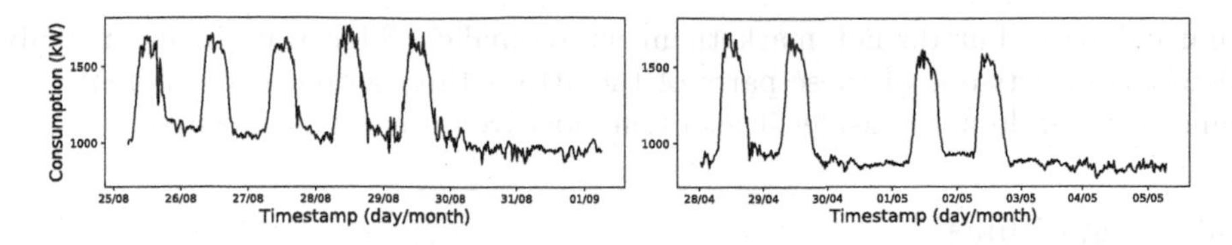

Fig. 2. A normal week (left) and an anomalous week (right) in the Power dataset

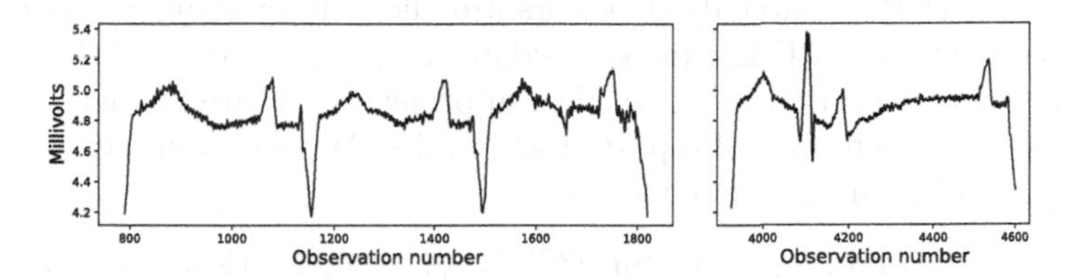

Fig. 3. Three normal heartbeats (left) and an anomalous heartbeat (right) in the ECG dataset

LIT101. Finally, we consider time series data generated by the LIT101 sensor of the Secure Water Treatment (SWaT) testbed [3] (version 0). Datasets from this testbed were used for anomaly detection in [19,27]. SWaT is a scaled down version of a real-world industrial water treatment plant [18] producing five gallons/minute of double-filtered water. It consists of 51 sensors and actuators that control its six-stage filtration process. The LIT101 sensor measures the raw water tank level (in millimeters). The data collected from the testbed comprise eleven days of continuous operation. The researchers who had built the SWaT testbed ran it normally during the first seven days and then launched attacks on different components during the remaining four days. This way, the LIT101 dataset consists of a normal time series with 496,800 measurements and an attack time series with 449,919 measurements. A total of 36 attacks were launched during the last four days of SWaT operation. Only five attacks targeted the LIT101 sensor. Figure 4 shows a part of the normal LIT101 time series, where the raw water tank was filled and emptied seven times, as well as two anomalous parts during the attacks on the LIT101 sensor. Some of the other 31 attacks also affected the LIT101 measurements. Moreover, the time series sometimes contains unusual subsequences before or after the launched attack. We manually annotate these

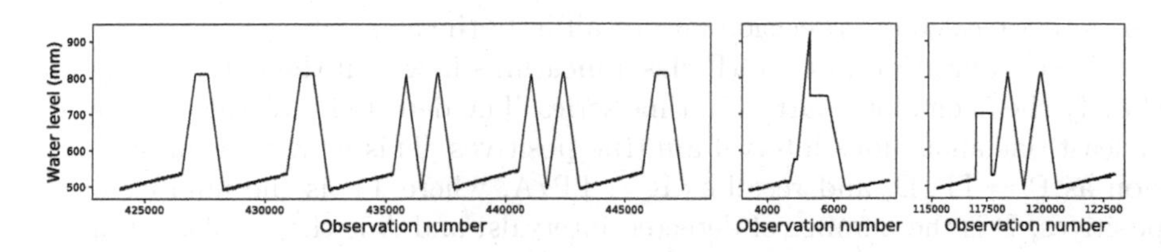

Fig. 4. Normal (left) and two anomalous (center and right) subsequences from the LIT101 dataset

unusual parts but do not mark them as anomalies. This way, if an anomaly detector reports one of these parts of the attack time series as an anomaly, we ignore it and do not consider it as a true positive or as a false positive.

6.2 Algorithms

To the best of our knowledge, the Pattern Anomaly Value (PAV) and Rare Rule Anomaly (RRA) anomaly detectors are the only existing algorithms that support parameterless AD in time series data. However, unlike SIM-AD, both of them aim at unsupervised AD and both approaches require the user to choose a threshold on the number of reported anomalies. We next explain how we use these detectors in our experiments.

RRA is a discord-mining algorithm [36]. Its implementation is freely available as part of the GrammarViz 3.0 tool [2,37]. RRA mines variable-length discords and requires setting three SAX-discretization parameters: sliding window size w, PAA word size p, and alphabet size a. We set these parameters using the semi-automated method from [37]. This method requires choosing a *parameter learning interval*, that is, a training subsequence that captures the normal behavior of the generative process. The user must also set the ranges of acceptable parameter values. The authors of RRA propose setting the range for p from 2 to 50, for a from 2 to 15, and for w from 10 to the doubled length of time series motifs. We followed these guidelines and set the upper value of the range for w to 500. We also configured RRA to use the Re-Pair grammar inference algorithm, EXACT numerocity reduction strategy, and set the normalization threshold to 0.01.

PAV mines *linear patterns* in univariate time series data. If observations are equally spaced in time (which is true for our experimental data), then a linear pattern is simply the value difference between consecutive observations. PAV reports the n linear patterns with the smallest number of occurrences as anomalies, where n is a user-specified threshold.

6.3 Accuracy Metrics

We evaluate the accuracy of anomaly detectors using a pair of F-scores, where $\text{F-score} = 2 \cdot \text{P} \cdot \text{R}/(\text{P} + \text{R})$, P is the precision, and R is the recall. The first F-score measures the classification accuracy while the second one measures the coverage accuracy of a detector. We next explain why two F-scores are necessary and how we calculate precision and recall in both cases.

The classification F-score (F-class) measures how well the detected intervals classify the anomalous parts of a time series. The detected intervals that overlap at least one anomalous interval are true positives. This way, we compute precision as $\text{P} = \text{TP}/\text{D}$, and recall as $\text{R} = \text{TP}/\text{A}$, where TP is the number of true positives, D is the number of detected intervals, and A is the number of anomalous intervals. F-class, however, does not take into account the lengths of the detected intervals. Indeed, a detector which reports an interval that covers the entire time series will have a perfect F-class = 1, since this interval necessarily overlaps all anomalous intervals.

The coverage F-score (F-cover) measures how well the detected intervals cover the anomalous intervals. In this case, a true positive is a timestamp of an anomalous interval covered by one or more detected intervals. We thus have precision P = TP/C and recall R = TP/A, where TP is the number of true positives, C is the number of timestamps in all detected intervals, and A is the number of timestamps in all anomalous intervals. F-cover alone, however, does not say how many anomalies were detected. Indeed, a detector that reports an interval that fully covers a long anomalous interval but leaves shorter anomalies uncovered will still have a large F-cover. Therefore, we need both F-scores to evaluate the accuracy of an anomaly detector.

6.4 Results

We report the accuracies of the SIM-AD, RRA, and PAV anomaly detectors applied on the three datasets presented in Sect. 6.1. For each dataset, we show classification and coverage F-scores of the detectors on two plots. The X-axis on these plots indicates the number of top discords for RRA and the number of smallest unique frequencies of linear patterns for PAV that the user chooses as an anomaly threshold. SIM-AD has constant F-class and F-cover since its number of detected anomalies does not depend on any threshold.

As a preliminary step, we split the Power and ECG datasets into a training set and a test set, such that the training set does not include any labeled anomalies. The LIT101 dataset comes already partitioned into such two sets. We use the entire training set as a parameter learning interval for the RRA detector and run the PAV detector only on the test set.

Power. SIM-AD dominates other detectors both in classification and in coverage accuracies (Fig. 5a). RRA performs much worse on the Power dataset than it has been reported in [36]. In fact, the parameter learning method returns clearly not optimal values of $w = 10$, $p = 2$, $a = 7$ when the entire training set is used as the parameter learning interval. Finally, the anomalies reported by PAV appear to be random. Indeed, PAV aims at detecting *point anomalies* [12] (i.e., unusual spikes/dips in observed values), while the anomalies in the Power time series are of *collective* type [12] (i.e., unusual sequences of observed values).

ECG. Both RRA and PAV detectors show better classification F-score than SIM-AD for some values of the threshold. For the RRA detector, the parameter learning method returns a very low value for w, and the top discords do not match the ones reported in [36], where the learning interval was chosen manually. SIM-AD detects only one out of three anomalies and does not report any false positives. The inferior classification F-score of SIM-AD on this dataset is due to a "noisy" trailing subsequence in the ECG time series where the value range of the heartbeats changes drastically. Incidentally, removing the noisy subsequence

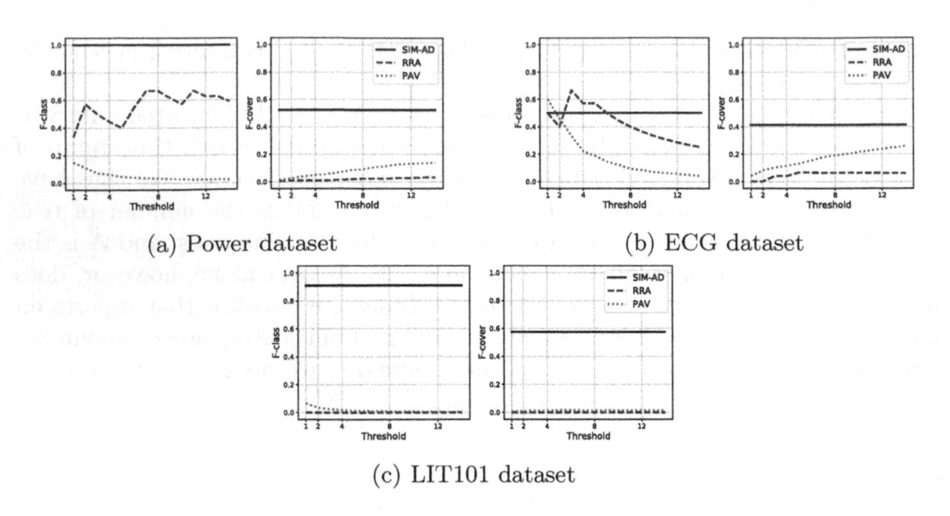

(a) Power dataset (b) ECG dataset

(c) LIT101 dataset

Fig. 5. Accuracy results for the SIM-AD, RRA, and PAV anomaly detectors

from the training set allows SIM-AD to achieve a perfect F-class of 1. Although having an inferior F-class, SIM-AD covers the anomalous intervals in this dataset much better than the other two detectors, as shown by the F-cover plot in Fig. 5b.

LIT101. SIM-AD outperforms both the RRA and PAV detectors on this dataset. Our approach detects all five anomalies and a single short false positive (Fig. 5c). In the case of the RRA detector, none of the top 24 discords overlap anomalous subsequences in the attack LIT101 dataset. Similarly to the Power dataset, the parameter selection method learned a very small window length $w = 10$, and the detector failed to combine observations from short windows into longer patterns. Although PAV correctly detects all five anomalies starting from the threshold of 1, it also returns numerous false positives that bring its F-scores down.

6.5 Discussion

Experimental results show that SIM-AD outperforms both the RRA and PAV anomaly detectors on two out of three considered time series datasets. We can explain the inferior accuracy of SIM-AD on the ECG dataset by the presence of a trend towards the end of the training time series. Indeed, SIM-AD assumes that a motif's occurrences are not shifted in the value range. Notice that without additional knowledge about a system, our approach has no way of knowing whether two subsequences mapped to different sojourn intervals correspond to the same, although shifted in value, motif. We can summarize this limitation in the following way: SIM-AD performs poorly on time series data having a trend. In future work, we will address this limitation by automatically fitting the line that defines discretization bins to the time series trend.

Another limitation of SIM-AD concerns time series motifs that map to a *sequence* of sojourn intervals. As an example, consider the weekly motif in the Power dataset (see Fig. 1). The motif consists of a sequence of peaks and valleys corresponding to a sequence of sojourn intervals mined by SIM-AD: $\langle SI_3, SI_1, SI_3, SI_1, SI_3, SI_1, SI_3, SI_1, SI_3, SI_2 \rangle$. The current version of our approach does not mine such composite motifs, and we leave this task as future work.

Why did we choose to use two discretization bins to compute the STR of a time series? Notice that identical subsequences will have identical STRs, hence, will be represented with the same set of SIs. If similar subsequences have small variations in them, their STRs will be different, but their set(s) of SIs are likely to be the same. Indeed, the more occurrences a motif has in a time series, the more STs it produces, the more robust to variations the mined SIs become. It is possible, however, that SIM-AD maps structurally-different subsequences that spend similar amounts of time in Bin 1 and Bin 2 to the same set of SIs. If one of these subsequences corresponds to an instance of a motif while the other one does not, the anomaly detection stage of SIM-AD will incorrectly report both subsequences as instances of the same motif, i.e., it will produce a false negative. We plan to address this issue in the future work by recursively mining fine-grained structure of subsequences mapped to the same set of SIs.

The computational complexity of the training stage of SIM-AD includes the time needed to (a) construct an STR of the training time series; (b) find the optimal bandwidth of the KDE kernel; (c) compute the KDE of the sojourn times. The first step has linear complexity with respect to the length of the training time series, while the other two operations depend on the length of the STR and have efficient implementations that use Discrete Cosine Transform [8] and Fast Fourier Transform [38]. Indeed, training each of the considered time series datasets took SIM-AD less than a second on a single core of an Intel Core i5-6200U CPU. The anomaly detection stage of SIM-AD needs to run only two relational operators on each incoming observation. Therefore, it is possible to perform anomaly detection on data streams in real-time.

7 Conclusion

In this paper, we addressed the problem of parameterless anomaly detection in univariate time series data. To this end, we proposed and evaluated SIM-AD: a semi-supervised anomaly detection approach that does not require any parameters. At the core of SIM-AD lies a modal clustering-based anomaly detection approach that uses kernel density estimation and the Improved Sheather-Jones plug-in bandwidth estimator. We showed that SIM-AD outperforms the relevant anomaly detection algorithms on trendless time series data. In future work, we plan to address the three limitations of SIM-AD mentioned in Sect. 6.5.

References

1. ECG Dataset. http://www.cs.ucr.edu/~eamonn/discords/mitdbx_mitdbx_108.txt. (2nd column)
2. Grammarviz 3.0. https://grammarviz2.github.io/grammarviz2_site/
3. LIT101 Dataset. https://itrust.sutd.edu.sg/itrust-labs_datasets/
4. Power Dataset. http://www.cs.ucr.edu/~eamonn/discords/power_data.txt
5. Aggarwal, C.C.: Outlier analysis. Data Mining, pp. 237–263. Springer, Cham (2015). https://doi.org/10.1007/978-3-319-14142-8_8
6. Ahmad, S., Lavin, A., Purdy, S., Agha, Z.: Unsupervised real-time anomaly detection for streaming data. Neurocomputing **262**, 134–147 (2017)
7. Bettaiah, V., Ranganath, H.S.: An analysis of time series representation methods: data mining applications perspective. In: Proceedings of the 2014 ACM Southeast Regional Conference, pp. 16:1–16:6 (2014)
8. Botev, Z., Grotowski, J., Kroese, D.: Kernel density estimation via diffusion. Ann. Stat. **38**(5), 2916–2957 (2010)
9. Breiman, L., Meisel, W., Purcell, E.: Variable kernel estimates of multivariate densities. Technometrics **19**(2), 135–144 (1977)
10. Breunig, M.M., Kriegel, H.P., Ng, R.T., Sander, J.: LOF: identifying density-based local outliers. In: Proceedings of the International Conference on Management of Data, vol. 29, pp. 93–104. ACM (2000)
11. Campello, R.J., Moulavi, D., Zimek, A., Sander, J.: Hierarchical density estimates for data clustering, visualization, and outlier detection. ACM Trans. Knowl. Discov. Data (TKDD) **10**(1), 5:1–5:51 (2015)
12. Chandola, V.: Anomaly detection for symbolic sequences and time series data. Ph.D. thesis, University of Minnesota (2009)
13. Chandola, V., Banerjee, A., Kumar, V.: Anomaly detection: a survey. ACM Comput. Surv. (CSUR) **41**(3), 15 (2009)
14. Chen, X., Zhan, Y.: Multi-scale anomaly detection algorithm based on infrequent pattern of time series. J. Comput. Appl. Math. **214**(1), 227–237 (2008)
15. Chow, C.: Parzen-window network intrusion detectors. In: Proceedings of the 16th International Conference on Pattern Recognition (ICPR), pp. 385–388 (2002)
16. Clifton, D., Bannister, P., Tarassenko, L.: A framework for novelty detection in jet engine vibration data. Key Eng. Mater. **347**, 305–310 (2007)
17. Gao, Y., Lin, J.: HIME: discovering variable-length motifs in large-scale time series. Knowl. Inf. Syst. **61**(1), 513–542 (2019)
18. Goh, J., Adepu, S., Junejo, K.N., Mathur, A.: A dataset to support research in the design of secure water treatment systems. In: Havarneanu, G., Setola, R., Nassopoulos, H., Wolthusen, S. (eds.) CRITIS 2016. LNCS, vol. 10242, pp. 88–99. Springer, Cham (2017). https://doi.org/10.1007/978-3-319-71368-7_8
19. Goh, J., Adepu, S., Tan, M., Lee, Z.S.: Anomaly detection in cyber physical systems using recurrent neural networks. In: 18th IEEE International Symposium on High Assurance Systems Engineering (HASE), pp. 140–145 (2017)
20. Gupta, M., Gao, J., Aggarwal, C.C., Han, J.: Outlier detection for temporal data: a survey. IEEE Trans. Knowl. Data Eng. **26**(9), 2250–2267 (2014)
21. Han, J., Pei, J., Kamber, M.: Data Mining: Concepts and Techniques. Elsevier, Amsterdam (2011)
22. He, Z., Xu, X., Deng, S.: Discovering cluster-based local outliers. Pattern Recognit. Lett. **24**(9–10), 1641–1650 (2003)

23. Hundman, K., Constantinou, V., Laporte, C., Colwell, I., Soderstrom, T.: Detecting spacecraft anomalies using LSTMs and nonparametric dynamic thresholding. In: Proceedings of the 24th ACM International Conference on Knowledge Discovery & Data Mining (KDD), pp. 387–395. ACM (2018)
24. Hutter, F., Kotthoff, L., Vanschoren, J.: Automated Machine Learning. Springer, Heidelberg (2019). https://doi.org/10.1007/978-3-030-05318-5
25. Keogh, E., Chu, S., Hart, D., Pazzani, M.: Segmenting time series: a survey and novel approach. In: Data Mining in Time Series Databases, pp. 1–21 (2004)
26. Keogh, E., Lin, J., Fu, A.: HOT SAX: efficiently finding the most unusual time series subsequence. In: Fifth IEEE International Conference on Data Mining (ICDM), pp. 226–233 (2005)
27. Kravchik, M., Shabtai, A.: Detecting cyber attacks in industrial control systems using convolutional neural networks. In: Proceedings of the 2018 Workshop on Cyber-Physical Systems Security and Privacy, pp. 72–83. ACM (2018)
28. Kriegel, H.P., Kroger, P., Sander, J., Zimek, A.: Density-based clustering. Wiley Interdiscip. Rev. Data Min. Knowl. Discov. **1**(3), 231–240 (2011)
29. Lin, J., Keogh, E., Wei, L., Lonardi, S.: Experiencing SAX: a novel symbolic representation of time series. Data Min. Knowl. Discov. **15**(2), 107–144 (2007)
30. Menardi, G.: A review on modal clustering. Int. Stat. Rev. **84**(3), 413–433 (2016)
31. Patel, P., Keogh, E.J., Lin, J., Lonardi, S.: Mining motifs in massive time series databases. In: Proceedings of the 2002 IEEE International Conference on Data Mining (ICDM), pp. 370–377 (2002)
32. Pimentel, M.A., Clifton, D.A., Clifton, L., Tarassenko, L.: A review of novelty detection. Signal Process. **99**, 215–249 (2014)
33. Ramaswamy, S., Rastogi, R., Shim, K.: Efficient algorithms for mining outliers from large data sets. In: Proceedings of the International Conference on Management of Data, vol. 29, pp. 427–438. ACM (2000)
34. Ratanamahatana, C., Keogh, E., Bagnall, A.J., Lonardi, S.: A novel bit level time series representation with implication of similarity search and clustering. In: Ho, T.B., Cheung, D., Liu, H. (eds.) PAKDD 2005. LNCS (LNAI), vol. 3518, pp. 771–777. Springer, Heidelberg (2005). https://doi.org/10.1007/11430919_90
35. Salvador, S., Chan, P.: Learning states and rules for detecting anomalies in time series. Appl. Intell. **23**(3), 241–255 (2005)
36. Senin, P., et al.: Time series anomaly discovery with grammar-based compression. In: Proceedings of the 18th International Conference on Extending Database Technology (EDBT), pp. 481–492 (2015)
37. Senin, P., et al.: GrammarViz 3.0: interactive discovery of variable-length time series patterns. ACM Trans. Knowl. Discov. Data (TKDD) **12**(1), 10 (2018)
38. Silverman, B.W.: Density Estimation for Statistics and Data Analysis. Chapman & Hall, London (1986)
39. Singh, A.: Anomaly detection for temporal data using long short-term memory (LSTM). Master's thesis, KTH Information and Communication Technology, Sweden (2017)
40. Torkamani, S., Lohweg, V.: Survey on time series motif discovery. Wiley Interdiscip. Rev. Data Min. Knowl. Discov. **7**(2), e1199 (2017)
41. Wang, X., Lin, J., Patel, N., Braun, M.: Exact variable-length anomaly detection algorithm for univariate and multivariate time series. Data Min. Knowl. Discov. **32**(6), 1806–1844 (2018)

The Temporal Dictionary Ensemble (TDE) Classifier for Time Series Classification

Matthew Middlehurst(✉), James Large, Gavin Cawley, and Anthony Bagnall

School of Computing Sciences, University of East Anglia, Norwich, UK
M.Middlehurst@uea.ac.uk

Abstract. Using bag of words representations of time series is a popular approach to time series classification (TSC). These algorithms involve approximating and discretising windows over a series to form words, then forming a count of words over a given dictionary. Classifiers are constructed on the resulting histograms of word counts. A 2017 evaluation of a range of time series classifiers found the bag of symbolic-Fourier approximation symbols (BOSS) ensemble the best of the dictionary based classifiers. It forms one of the components of hierarchical vote collective of transformation-based ensembles (HIVE-COTE), which represents the current state of the art. Since then, several new dictionary based algorithms have been proposed that are more accurate or more scalable (or both) than BOSS. We propose a further extension of these dictionary based classifiers that combines the best elements of the others combined with a novel approach to constructing ensemble members based on an adaptive Gaussian process model of the parameter space. We demonstrate that the Temporal Dictionary Ensemble (TDE) is more accurate than other dictionary based approaches. Furthermore, unlike the other classifiers, if we replace BOSS in HIVE-COTE with TDE, HIVE-COTE becomes significantly more accurate. We also show this new version of HIVE-COTE is significantly more accurate than the current top performing classifiers on the UCR time series archive. This advance represents a new state of the art for time series classification.

Keywords: Time series · Classification · Bag of words · HIVE-COTE

1 Introduction

Dictionary based approaches adapt the bag of words model commonly used in signal processing, computer vision and audio processing for time series classification (TSC). A comparison of TSC algorithms, commonly known as the bake off [2], formed a taxonomy of approaches based on representations of discriminatory features, with dictionary approaches being one of these. From the bake off the bag of Symbolic-Fourier-Approximation symbols (BOSS) [16] ensemble was found to be the most accurate dictionary classifier by a significant amount.

F. Hutter et al. (Eds.): ECML PKDD 2020, LNAI 12457, pp. 660–676, 2021.
https://doi.org/10.1007/978-3-030-67658-2_38

BOSS was found to be the third most accurate algorithm out of the 20 compared. This highlights the utility of dictionary methods for TSC.

This performance lead to BOSS being incorporated into the hierarchical vote collective of transformation-based ensembles (HIVE-COTE) [14], a heterogeneous ensemble encompassing multiple time series representations. The inclusion of BOSS and the subsequent significant improvement in accuracy places HIVE-COTE in the state of the art for TSC among three other algorithms proposed more recently. These are the time series combination of heterogeneous and integrated embeddings forest (TS-CHIEF) [19], which also a hybrid of multiple representations including BOSS, the random convolutional kernel transform (ROCKET) [6], and the deep learning approach InceptionTime [9].

Since the bake off a number of dictionary algorithms have been published, focusing on improving accuracy [12,18], prediction time efficiency [18], train time and memory efficiency [15]. These algorithms are mostly extensions of BOSS, making alterations to different parts of the original algorithm. Word extraction for time series classification (WEASEL) [18] abandons the ensemble structure in favour of feature selection and changes the method of word discretisation. Spatial BOSS (S-BOSS) [12] introduces temporal information and additional features using spatial pyramids. Contractable BOSS (cBOSS) [15] changes the method used by BOSS to form its ensemble to improve efficiency and allow for a number of usability improvements.

Each of these methods constitutes an improvement to the dictionary representation from BOSS. Our contribution is to combine design features of these four classifiers (BOSS, WEASEL, S-BOSS and cBOSS) to make a new algorithm, the Temporal Dictionary Ensemble (TDE). Like BOSS, TDE is a homogeneous ensemble of nearest neighbour classifiers that use distance between histograms of word counts and injects diversity through parameter variation. TDE takes the ensemble structure from cBOSS, which is more robust and scalable. The use of spatial pyramids is adapted from S-BOSS. From WEASEL, TDE uses bi-grams and an alternative method of finding word breakpoints.

We found the simplest way of combining these components did not result in significant improvement. We speculate that the massive increase in the parameter space made the randomised diversity mechanism result in too many poor learners in the ensemble. We propose a novel mechanism of base classifier model selection based on an adaptive form of Gaussian process (GP) modelling of the parameter space. Through extensive evaluation with the UCR time series classification repository [5], we show that TDE is significantly more accurate than WEASEL and S-BOSS while retaining the usability and scalability of cBOSS. We further show that if TDE replaces BOSS in HIVE-COTE, the resulting classifier is significantly more accurate than HIVE-COTE with BOSS and all three competing state of the art classifiers.

The rest of this paper is structured as follows. Section 2 provides background information for the four dictionary based algorithms relevant to TDE. Section 3 describes the TDE algorithm, including the GP based parameter search. Section 4 presents the performance evaluation of TDE. Conclusions are drawn in Sect. 5 and future work is discussed.

2 Dictionary Based Classifiers

Dictionary based classifiers have the same broad structure. A sliding window of length w is run across a series. For each window, the real valued series of length w is converted through approximation and discretisation processes into a symbolic string of length l, which consists of α possible letters. The occurrence in a series of each 'word' from the dictionary defined by l and α is counted, and once the sliding window has completed the series is transformed into a histogram. Classification is based on the histograms of the words extracted from the series, rather than the raw data.

The bag of Symbolic-Fourier-Approximation symbols (BOSS) [16] was found to be the most accurate dictionary based classifier in a 2017 study [2]. Hence, it forms our benchmark for new dictionary based approaches. BOSS is described in detail in Sect. 2.1. A number of extensions and alternatives to BOSS have been proposed.

- One of the problems with BOSS is that it can be memory and time inefficient, especially on data where many transforms are accepted into the final ensemble. cBOSS (Sect. 2.2) addresses the scalability issues of BOSS [15] by altering the ensemble structure.
- BOSS ignores the temporal location of patterns. Rectifying this led to an extension of BOSS based on spatial pyramids, called S-BOSS [12], described in Sect. 2.3.
- WEASEL [18] is a dictionary based classifier by the same team that produced BOSS. It is based on feature selection from histograms for a linear model (see Sect. 2.4).

We propose a dictionary classifier that merges these extensions and improvements to the core concept of BOSS, called the Temporal Dictionary Ensemble (TDE). It lends from the sped-up ensemble structure of cBOSS, the spatial pyramid structure of S-BOSS, and the word and histogram forming improvements of WEASEL. TDE is fully described in Sect. 3.

2.1 Bag of Symbolic-Fourier-Approximation Symbols (BOSS) [16]

Algorithm 1 gives a formal description of the bag forming process of an individual BOSS classifier. Words are created using symbolic Fourer approximation (SFA) [17]. SFA first finds the Fourier transform of the window (line 8), then discretises the first l Fourier terms into α symbols to form a word, using a bespoke supervised discretisation algorithm called multiple coefficient binning (MCB) (line 13). It has an option to normalise each window or not by dropping the first Fourier term (lines 6–7). Lines 14–16 encapsulates the process of not counting trivially self similar words: if two consecutive windows produce the same

Algorithm 1. baseBOSS(A list of n time series of length m, $\mathbf{T} = (\mathbf{X}, \mathbf{y})$)

Parameters: the word length l, the alphabet size α, the window length w, normalisation parameter p

1: Let $\mathbf{H}$ be a list of n histograms $(\mathbf{h}_1, \ldots, \mathbf{h}_n)$
2: Let $\mathbf{B}$ be a matrix of l by α breakpoints found by MCB
3: **for** $i \leftarrow 1$ to n **do**
4:   **for** $j \leftarrow 1$ to $m - w + 1$ **do**
5:     $\mathbf{s} \leftarrow x_{i,j} \ldots x_{i,j+w-1}$
6:     **if** p **then**
7:       $s \leftarrow$normalise(s)
8:     $\mathbf{q} \leftarrow$ DFT($\mathbf{s}, l, \alpha$,p) { $\mathbf{q}$ *is a vector of the complex DFT coefficients*}
9:     **if** p **then**
10:       $\mathbf{q}' \leftarrow (q_2 \ldots q_{l/2+1})$
11:     **else**
12:       $\mathbf{q}' \leftarrow (q_1 \ldots q_{l/2})$
13:     $\mathbf{r} \leftarrow$ SFAlookup($\mathbf{q}', \mathbf{B}$)
14:     **if** $\mathbf{r} \neq \mathbf{p}$ **then**
15:       $pos \leftarrow$index($\mathbf{r}$)
16:       $h_{i,pos} \leftarrow h_{i,pos} + 1$
17:     $\mathbf{p} \leftarrow \mathbf{r}$

word, the second occurrence is ignored. This is to avoid a slow-changing pattern relative to the window size being over-represented in the resulting histogram.

BOSS uses a non-symmetric distance function in conjunction with a nearest neighbour classifier. Only the words contained in the test instance's histogram (i.e. the word's count is above zero) are used in the distance calculation, but it is otherwise the Euclidean distance.

The final classifier is an ensemble of individual BOSS classifiers (parameterised transform plus nearest neighbour classifier) found through first fitting and evaluating a large number of individual classifiers, then retaining only those within 92% accuracy of the best classifier. The BOSS ensemble (also referred to as just BOSS), evaluates and retains the best of all transforms parameterised in the range $w \in \{10 \ldots m\}$ with $m/4$ values where m is the length of the series, $l \in \{16, 14, 12, 10, 8\}$ and $p \in \{true, false\}$. α stays at the default value of 4.

2.2 Contractable BOSS (cBOSS) [15]

Due to its grid-search and method of retaining ensemble members BOSS is unpredictable in its time and memory resource usage, and is impractical for larger problems. cBOSS significantly speeds up BOSS while retaining accuracy by improving how the transform parameter space is evaluated and the ensemble is formed. The main change from BOSS to cBOSS is that it utilises a filtered random selection of parameters to find its ensemble members. cBOSS allows the user to control the build through a time contract, defined as the maximum amount

Algorithm 2. cBOSS(A list of n cases length m, $\mathbf{T} = (\mathbf{X}, \mathbf{y})$)

Parameters: the number of parameter samples k, the max ensemble size s,

1: Let w be window length, l be word length, p be normalise/not normalise and α be alphabet size.
2: Let $\mathbf{C}$ be a list of s BOSS classifiers $(\mathbf{c}_1, \ldots, \mathbf{c}_s)$
3: Let $\mathbf{E}$ be a list of s classifier weights $(\mathbf{e}_1, \ldots, \mathbf{e}_s)$
4: Let $\mathbf{R}$ be a set of possible BOSS parameter combinations
5: $i \leftarrow 0$
6: $lowest\_acc \leftarrow \infty, lowest\_acc\_idx \leftarrow \infty$
7: **while** $i < k$ AND $|\mathbf{R}| > 0$ **do**
8: $[l, a, w, p] \leftarrow random\_sample(\mathbf{R})$
9: $\mathbf{R} = \mathbf{R} \setminus \{[l, \alpha, w, p]\}$
10: $\mathbf{T}' \leftarrow$ subsample_data($\mathbf{T}$, 0.7)
11: $cls \leftarrow$ baseBOSS($\mathbf{T}', l, \alpha, w, p$)
12: $acc \leftarrow$ LOOCV(cls) { *train data accuracy*}
13: **if** $i < s$ **then**
14: **if** $acc < lowest\_acc$ **then**
15: $lowest\_acc \leftarrow acc$, $lowest\_acc\_idx \leftarrow i$
16: $c_i \leftarrow cls$, $e_i \leftarrow acc^4$
17: **else if** $acc > lowest\_acc$ **then**
18: $c_{lowest\_acc\_idx} \leftarrow cls$, $e_{lowest\_acc\_idx} \leftarrow acc^4$
19: $[lowest\_acc, lowest\_acc\_idx] \leftarrow$ find_new_lowest_acc($\mathbf{C}$)
20: $i \leftarrow i + 1$

of time spent constructing the classification model. Algorithm 2 describes the decision procedure for search and maintaining individual BOSS classifiers for cBOSS.

A new parameter k (default 250) for the number of parameter combinations samples is introduced (line 7), of which the top s with the highest accuracy are kept for the final ensemble (lines 13–19). The k parameter is replaceable with a time limit t through contracting. Each ensemble member is built on a subsample of the train data, (line 10) using random sampling without replacement of 70% of the whole training data. An exponential weighting scheme for the predictions of the base classifiers is introduced, to produce a tilted distribution (line 18).

cBOSS was shown to be an order of magnitude faster than BOSS on both small and large datasets from the UCR archive while showing no significant difference in accuracy [15].

2.3 BOSS with Spatial Pyramids (S-BOSS) [12]

BOSS intentionally ignores the locations of words in series, classifying based on the frequency of patterns rather than their location. For some datasets we know that the locations of certain discriminatory subsequences are important, however. Some patterns may gain importance only when in a particular location, or a mutually occurring word may be indicative of different classes depending on when it occurs. Spatial pyramids [13] bring some temporal information back into the bag-of-words paradigm.

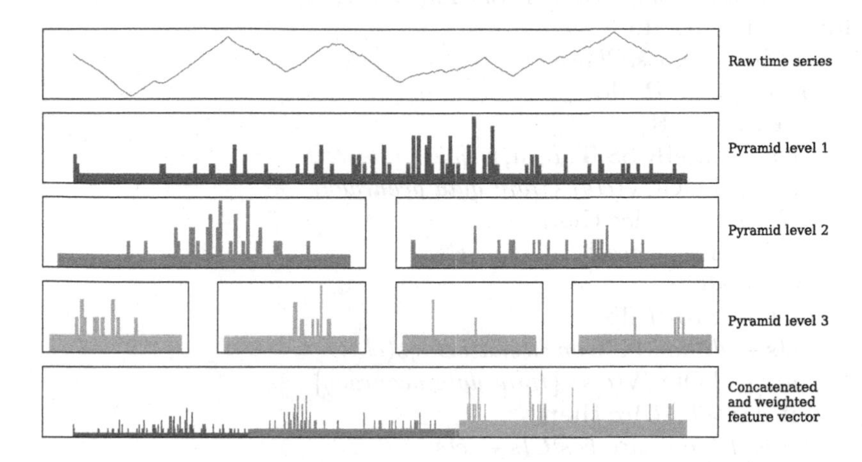

Fig. 1. An example transformation of an OSULeaf instance to demonstrate the additional steps to form S-BOSS from BOSS. Note that each histogram is represented in a sparse manner; the set of words along the x-axis of each histogram at higher pyramid levels may not be equal.

S-BOSS, described in Algorithm 3 and illustrated in Fig. 1, incorporates the spatial pyramids technique into the BOSS algorithm. S-BOSS creates a standard BOSS transform at the global level (line 6), which constitutes the first level of the pyramid. An additional degree of optimisation is then performed to find the best pyramid height $h \in \{1, 2, 3\}$ (lines 11–16). Height defines the importance of localisation for this transform. Creating the next pyramid level involves creating additional histograms each sub-region of the series at the next scale. Histograms are weighted to give more importance to similarities in the same locations than global similarity, and are concatenated to form an elongated feature vector per instance. The histogram intersection distance measure, more commonly used for approaches using histograms, replaces the BOSS distance for the nearest neighbour classifiers. S-BOSS retains the BOSS ensemble strategy (line 17), such that each S-BOSS ensemble member is a BOSS transform with its own spatial pyramid optimisation plus nearest neighbour classifier.

2.4 Word Extraction for Time Series Classification (WEASEL) [18]

Like BOSS, WEASEL performs a Fourier transform on each window, creates words by discretisation, and forms histograms of words counts. It also does this for a range of window sizes and word lengths. However, there are important differences. WEASEL is not an ensemble nearest neighbour classifiers. Instead, WEASEL constructs a single feature space from concatenated histograms for

Algorithm 3. S-BOSS(A list of n cases length m, $\mathbf{T} = (\mathbf{X}, \mathbf{y})$)

Parameters: the set of possible $[\alpha, w, p]$ parameter combinations $\mathbf{R}$, the set of possible $[l]$ parameter values $\mathbf{L}$, the maximum pyramid height H

1: Let $\mathbf{C}$ be a list of s BOSS classifiers $(\mathbf{c}_1, \ldots, \mathbf{c}_s)$
2: **for** $i \leftarrow 1$ to $|\mathbf{L}|$ **do**
3: $bestAcc \leftarrow 0, bestCls \leftarrow \varnothing$
4: **for** $j \leftarrow 1$ to $|\mathbf{R}|$ **do**
5: $[\alpha, w, p] \leftarrow \mathbf{R_j}$
6: $cls \leftarrow$ baseBOSS$(\mathbf{T}, L_i, a, w, p)$
7: $acc \leftarrow$ LOOCV(cls) {*train data accuracy*}
8: **if** $acc > bestAcc$ **then**
9: $bestAcc \leftarrow acc, bestCls \leftarrow cls$
10: $cls \leftarrow bestCls$
11: **for** $h \leftarrow 1$ to H **do**
12: $cls \leftarrow divideAndConcatenateBags(cls)$
13: $acc \leftarrow$ LOOCV(cls) {*train data accuracy*}
14: **if** $acc > bestAcc$ **then**
15: $bestAcc \leftarrow acc, bestCls \leftarrow cls$
16: $\mathbf{C}_i \leftarrow bestCls$
17: $keepWithinBest(\mathbf{C}, 0.92)$ {*keep those cls with train accuracy within 0.92 of the best*}

different parameter values, then uses logistic regression and feature selection. Histograms of individual words and bigrams of the previous non-overlapping window for each word are used. Fourier terms are selected for retention by the application of an F-test. The retained values are then discretised into words using information gain binning (IGB), similar to the MCB step in BOSS. The number of features is further reduced using a chi-squared test after the histograms for each instance are created, removing any words which score below a threshold. It performs a parameter search for p (whether to normalise or not) and over a reduced range of l, using a 10-fold cross-validation to determine the performance of each set. The alphabet size α is fixed to 4 and the *chi* parameter is fixed to 2. Algorithm 4 gives an overview of WEASEL, although the formation and addition of bigrams is omitted for clarity.

3 Temporal Dictionary Ensemble (TDE)

The easiest algorithms to combine are cBOSS and S-BOSS. cBOSS speeds up BOSS through subsampling training cases and random parameter selection. The number of levels parameter introduced by S-BOSS can be included in the random parameter selection used by cBOSS. For comparisons we call this naive hybrid of algorithms cS-BOSS. We use this as a baseline to justify the extra complexity we introduce in TDE. Algorithm 5 provides an overview of the ensemble build process for TDE, which follows the general structure and weighting scheme of

Algorithm 4. WEASEL(A list of n cases of length m, $\mathbf{T} = (\mathbf{X}, \mathbf{y})$)

Parameters: the word length l, the alphabet size α, the maximal window length w_{max}, mean normalisation parameter p

1: Let $\mathbf{H}$ be the histogram $\mathbf{h}$
2: Let $\mathbf{B}$ be a matrix of l by α breakpoints found by MCB using information gain binning
3: **for** $i \leftarrow 1$ to n **do**
4:     **for** $w \leftarrow 2$ to w_{max} **do**
5:         **for** $j \leftarrow 1$ to $m - w + 1$ **do**
6:             $\mathbf{o} \leftarrow x_{i,j} \ldots x_{i,j+w-1}$
7:             $\mathbf{q} \leftarrow \text{DFT}(o, w, p)$ { $\mathbf{q}$ *is a vector of the complex DFT coefficients*}
8:             $\mathbf{q}' \leftarrow \text{ANOVA-F}(q, l, y)$ { *use only the* $\mathbf{l}$ *most discriminative ones*}
9:             $\mathbf{r} \leftarrow \text{SFAlookup}(\mathbf{q}', \mathbf{B})$
10:            $pos \leftarrow \text{index}(\mathbf{w}, \mathbf{r})$
11:            $h_{i,pos} \leftarrow h_{i,pos} + 1$
12: $h \leftarrow \chi^2(h, y)$ { *feature selection using the chi-squared test* }
13: fitLogistic(h, y)

cBOSS. The classifier returned by improvedBaseBOSS includes spatial pyramids and also includes the following enhancements taken from both S-BOSS and WEASEL. Like S-BOSS, it uses the histogram intersection distance measure which has been shown to be more accurate than BOSS distance [12]. It uses bigram frequencies in the same way as WEASEL. Base classifiers can use either IGB from WEASEL or MCB from BOSS in the discretisation. TDE samples parameters from the range given in Table 1 using a method sampleParameters. While expensive in general, leave-one-out cross-validation can be used for estimating train accuracy due to the use of nearest neighbour classifiers. At first thought using the leftover data from subsampling would seem a better choice. However, this data would first have to be transformed before it can be used to make an estimation, while the subsampled data is already transformed.

Table 1. Parameter ranges for TDE base classifier selection.

Parameter	Range
Word lengths	$l = \{16, 14, 12, 10, 8\}$
Window lengths	$w = \{10...m\}$
Normalise	$p = \{true, false\}$
Alphabet size	$\alpha = \{4\}$
No. pyramid levels	$h = \{1, 2, 3\}$
Discretisation	$b = \{MCB, IGB\}$

Algorithm 5. TDE(A list of n cases length m, $\mathbf{T} = (\mathbf{X}, \mathbf{y})$)

Parameters: the number of parameter samples k, the max ensemble size s

1: Let w be window length, l be word length, p be normalise/not normalise, α be alphabet size, h be number of pyramid levels and b be MCB or IGB discretisation.
2: Let $\mathbf{C}$ be a list of s BOSS classifiers $(\mathbf{c}_1, \ldots, \mathbf{c}_s)$
3: Let $\mathbf{E}$ be a list of s classifier weights $(\mathbf{e}_1, \ldots, \mathbf{e}_s)$
4: Let $\mathbf{G}$ be a list of k BOSS parameter and accuracy pairs $(\mathbf{g}_1, \ldots, \mathbf{g}_k)$
5: Let $\mathbf{R}$ be a set of possible BOSS parameter combinations
6: $i \leftarrow 0$
7: $lowest\_acc \leftarrow \infty, lowest\_acc\_idx \leftarrow \infty$
8: **while** $i < k$ AND $|\mathbf{R}| > 0$ **do**
9: $\quad [l, \alpha, w, p, h, b] \leftarrow$ chooseParameters($\mathbf{R}$,$\mathbf{G}$,i)
10: $\quad \mathbf{R} = \mathbf{R} \setminus \{[l, \alpha, w, p, h, b]\}$
11: $\quad \mathbf{T}' \leftarrow$ subsampleData($\mathbf{T}$, 0.7)
12: $\quad cls \leftarrow$ improvedBaseBOSS($\mathbf{T}', l, \alpha, w, p, h, b$)
13: $\quad acc \leftarrow$ LOOCV(cls) { *train data accuracy*}
14: $\quad$ **if** $i < s$ **then**
15: $\quad\quad$ **if** $acc < lowest\_acc$ **then**
16: $\quad\quad\quad lowest\_acc \leftarrow acc, lowest\_acc\_idx \leftarrow i$
17: $\quad\quad c_i \leftarrow cls,\ e_i \leftarrow acc^4$
18: $\quad$ **else if** $acc > lowest\_acc$ **then**
19: $\quad\quad c_{lowest\_acc\_idx} \leftarrow cls,\ e_{lowest\_acc\_idx} \leftarrow acc^4$
20: $\quad\quad [lowest\_acc, lowest\_acc\_idx] \leftarrow$ findNewLowestAcc($\mathbf{C}$)
21: $\quad g_i \leftarrow \{[l, a, w, p, h, b], acc\}$
22: $\quad i \leftarrow i + 1$

The increase in the parameter search space caused by the inclusion of pyramid and IGB parameters makes the random parameter selection used by cBOSS less effective. Instead, TDE uses a guided parameter selection for ensemble members inspired by Bayesian optimisation [20]. A Gaussian process model is built over the regressor parameter space $\mathbf{R}$ for parameters $[l, a, w, p, h, b]$ ($\boldsymbol{x}$) to predict the accuracy (y), using n previously observed ($\boldsymbol{x}$,y) pairs $\mathbf{G}$ from previous classifiers.

A Gaussian Process [21] describes a distribution over functions, $f(\boldsymbol{x}) \sim \mathcal{GP}(m(\boldsymbol{x}, k(\boldsymbol{x}, \boldsymbol{x}')))$, characterised by a mean function, $m(\boldsymbol{x})$, and a covariance function, $k(\boldsymbol{x}, \boldsymbol{x}')$, such that

$$m(\boldsymbol{x}) = \mathbb{E}\left[f(\boldsymbol{x})\right],$$
$$k(\boldsymbol{x}, \boldsymbol{x}') = \mathbb{E}\left[(f(\boldsymbol{x}) - m(\boldsymbol{x}))(f(\boldsymbol{x}') - m(\boldsymbol{x}'))\right],$$

where any finite collection of values has a joint Gaussian distribution. Commonly the mean function is constant, $m(\boldsymbol{x}) = \gamma$, or even zero, $m(\boldsymbol{x}) = 0$. The covariance function $k(\boldsymbol{x}, \boldsymbol{x}')$ encodes the expected similarity of the function evaluated at pairs of input-space vectors, $\boldsymbol{x}$ and $\boldsymbol{x}'$. For example, the squared exponential covariance function,

Algorithm 6. chooseParameters(**R**,**G**)

Parameters: the number of classifiers built n

1: **if** $n < 50$ **then**
2: $[l, \alpha, w, p, h, b] \leftarrow$ randomSample(**R**)
3: **else**
4: $gp \leftarrow$ buildGaussianProcesses(**G**)
5: $[l, \alpha, w, p, h, b] \leftarrow$ bestPredictedParameters(**R**, gp)
6: **return** $[l, \alpha, w, p, h, b]$

$$k(\boldsymbol{x}, \boldsymbol{x}') = \sigma_f^2 \exp\left\{-\frac{(\boldsymbol{x} - \boldsymbol{x}')^2}{2\ell^2}\right\},$$

encodes a preference for smooth functions, where ℓ is a hyper-parameter that specifies the characteristic length-scale of the covariance functions (large values yield smoother functions) and σ_f governs the magnitude of the variance.

Typically in a regression setting the response variables of the training samples, $\mathcal{D} = \{(\boldsymbol{x}_i, y_i) \mid i = 1, 2, \ldots, n\}$, are assumed to be realisations of a deterministic function that have been corrupted by additive Gaussian noise, i.e.

$$y_i = f(\boldsymbol{x}_i) + \varepsilon_i, \qquad \text{where} \qquad \varepsilon_i \sim \mathcal{N}\left(0, \sigma_n^2\right).$$

In that case, the joint distribution of the training sample, and a single test point, $\boldsymbol{x}_*$, is given by,

$$\begin{bmatrix} \boldsymbol{y} \\ f_* \end{bmatrix} \sim \mathcal{N}\left(\boldsymbol{0}, \begin{bmatrix} \boldsymbol{K} + \sigma_n^2 \boldsymbol{I} & \boldsymbol{k}_* \\ \boldsymbol{k}_*^T & k(\boldsymbol{x}_*, \boldsymbol{x}_*) \end{bmatrix}\right),$$

where $\boldsymbol{K}$ is the matrix of pairwise evaluation of the covariance function for all points belonging to the training sample and $\boldsymbol{k}_*$ is a column vector of the evaluation of the covariance function for the test point and each of the training points. The Gaussian predictive distribution for the test point is then specified by

$$\bar{f}_* = \boldsymbol{k}_*^T \left(\boldsymbol{K} + \sigma_n^2 \boldsymbol{I}\right)^{-1} \boldsymbol{y},$$
$$\mathbb{V}[f_*] = k(\boldsymbol{x}_*, \boldsymbol{x}_*) - \boldsymbol{k}_*^T \left(\boldsymbol{K} + \sigma_n^2 \boldsymbol{I}\right)^{-1} \boldsymbol{k}_*.$$

The hyper-parameters of the Gaussian process can be handled by tuning them, often via maximisation of the marginal likelihood, or by full Bayesian marginalisation, using an appropriate hyper-prior distribution. For further details, see Williams and Rasmussen [21]. We use a basic form of GP and treat all the regressors (TDE parameters) as continuous. The bestPredictedParameters operation in line 5 of Algorithm 6 is limited to the same parameter ranges used for random search given in Table 1.

4 Results

Our experiments are run on 112 datasets from the recently expanded UCR/UEA archive [5], removing any datasets that are unequal length or contain missing values. We also remove the dataset Fungi as it only provides a single train case for each class. For each classifier-dataset combination we run 30 stratified resamples, with the first sample being the original train test split. For reproducibility each dataset resample and classifier is seeded to its resample number. All experiments were run single threaded on a high performance computing cluster with a run time limit of 7 days. We used an open source Weka compatible code base called tsml for experimentation[1]. Implementations of BOSS, cBOSS, S-BOSS, WEASEL, HIVE-COTE and TS-CHIEF provided by the algorithm inventors are all available in tsml. InceptionTime and ROCKET experiments were run using the Python based package sktime and a deep learning extension thereof[2].

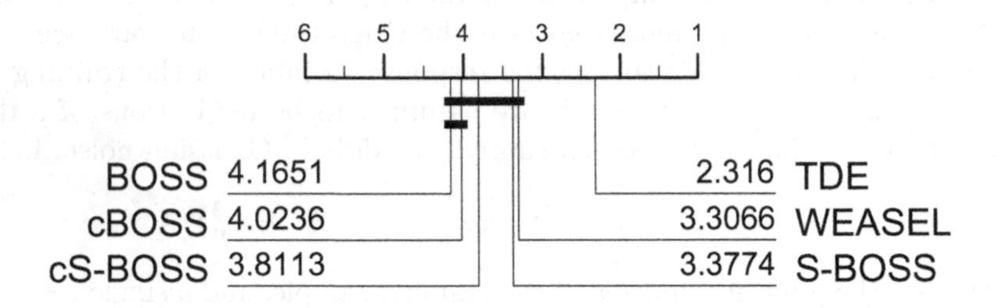

Fig. 2. Critical difference diagram for six dictionary based classifiers on 106 UCR time series classification problems. Full results are available on the accompanying website.

Guidance on how to recreate the resamples and code to reproduce the results is available on the accompanying website[3]. We also provide results and parameter settings for all classifiers used in experimentation.

Our experiments are designed to test whether TDE is better in terms of predictive performance and run time than other dictionary based classifiers, and whether it improves HIVE-COTE when it replaces BOSS in the meta ensemble HIVE-COTE.

4.1 TDE vs Other Dictionary Classifiers

For the dictionary classifiers, we were only able to obtain complete results for 106 of the 112 datasets. This was due to the long run time of S-BOSS and WEASEL. The missing problems are: ElectricDevices; FordA; FordB; HandOutlines; NonInvasiveFetalECGThorax1; and NonInvasiveFetalECGThorax2.

---

[1] https://github.com/uea-machine-learning/tsml.

[2] https://github.com/sktime.

[3] http://timeseriesclassification.com/TDE.php.

Figure 2 shows a critical difference diagram [7] for the six dictionary based classifiers considered. The number on each line is the average rank of an algorithm over 106 UCR datasets (lower is better). The solid bars are cliques. There is no detectable significant difference between classifiers in the same clique. Comparison of the performance of classifiers is done using pairwise Wilcoxon signed rank tests and cliques are formed using the Holm correction, following recommendations from [3] and [11].

TDE is significantly more accurate than all other classifiers. There are then two cliques: cBOSS is significantly worse than S-BOSS, cS-BOSS and WEASEL. BOSS and cBOSS show no significant difference. This also confirms that the simple hybrid cS-BOSS is no better than S-BOSS in terms of accuracy.

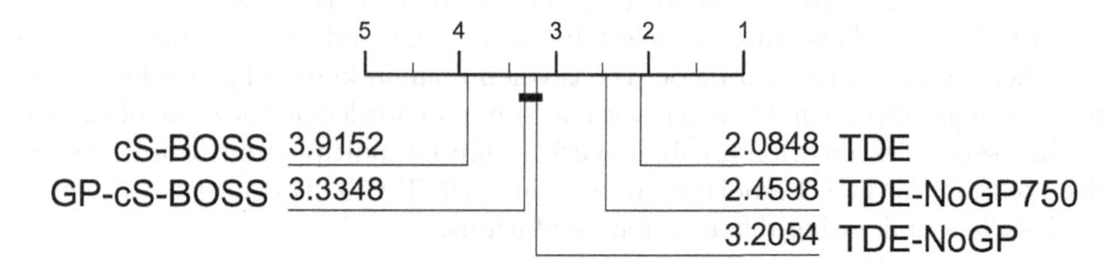

Fig. 3. Critical difference diagram for TDE variations and cS-BOSS on 112 UCR time series classification problems. Full results are available on the accompanying website.

Table 2. Summary of run time for six dictionary based classifiers over 106 UCR problems. The median time over 30 resamples is used for each dataset.

Classifier	Max run time (hrs)	Total run time (hrs)
BOSS	11.33	52.10
cBOSS	0.63	3.74
S-BOSS	34.11	148.82
cS-BOSS	2.91	12.62
WEASEL	4.86	28.50
TDE	5.67	21.73

While TDE provides a significant boost in accuracy, it is not readily visible how each change contributes to this. Figure 3 compares cS-BOSS and TDE to: cS-BOSS using TDE's GP parameter search; TDE without the GP parameter search but all else equal; and TDE without the GP parameter search once more but with triple the k parameter value, 750. Removing either the GP or additional algorithm features from WEASEL results in a significantly worse classifier, though still significantly better than cS-BOSS. Randomly selecting 750 parameter sets is also significantly worse than selecting 250 using GP, as well as being nearly twice as slow to train on average over the 112 datasets.

Table 2 summarises the run time of each dictionary based classifier. All experiments are conducted sequentially on a single processor. TDE completes all the problems in under a day. It is faster than WEASEL and considerably faster than S-BOSS. The max run timings for BOSS and S-BOSS demonstrate the problem with the traditional BOSS algorithm addressed by cBOSS and cS-BOSS: the ensemble design means they may require a long runtime, and it is not very predictable when this will happen. Figure 4 shows the scatter plot of runtime for BOSS vs TDE and demonstrates that TDE scales much better than BOSS.

4.2 TDE with HIVE-COTE

TDE is significantly more accurate than all other dictionary based time series classification algorithms, and faster than the best of the rest, S-BOSS and WEASEL. We believe there is merit in finding the best single representation classifier because there will be occasions when domain knowledge would recommend a single approach. However, with no domain knowledge, the state of the art in time series classification involves hybrids built on multiple representations, or deep learning to fit a bespoke representation. HIVE-COTE is a meta ensemble of classifiers built using different representations.

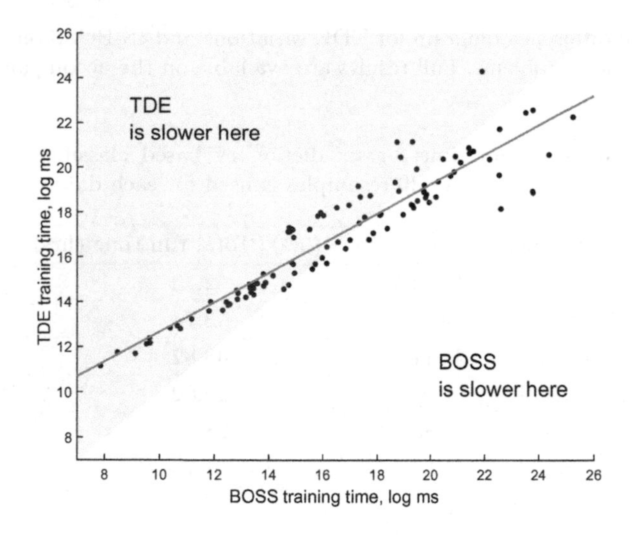

Fig. 4. Pairwise scatter diagram, on log scale, of TDE and BOSS training times. TDE has larger overheads which make it slower on smaller problems, but it scales much better towards larger problems.

For our experiments we use a recent and more efficient version of HIVE-COTE, HC 1.0 [1]. All HIVE-COTE variants used in our experiments are built with four components: the random interval spectral ensemble (RISE) [10], shapelet transform classifier (STC) [4] with a 4 hour contract and time series forest (TSF) [8] plus one other dictionary based classifier.

With these other settings fixed, we have reconstructed HIVE-COTE using TDE instead of BOSS. We call this HC-TDE for differentiation purposes.

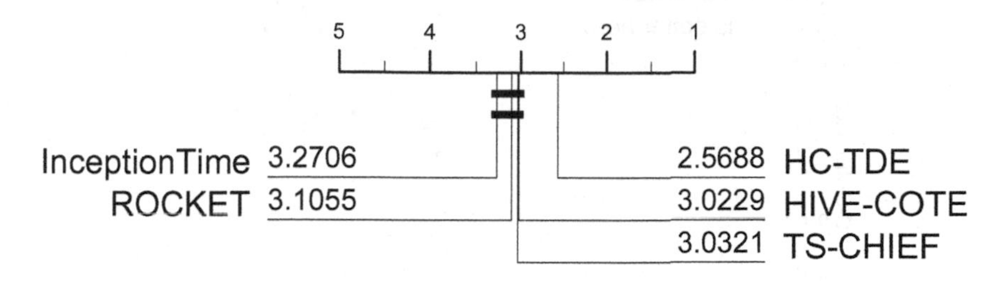

Fig. 5. Critical difference diagram for six dictionary based classifiers on 109 UCR time series classification problems. Full results are available on the accompanying website.

TS-CHIEF [19] is a tree ensemble that embeds dictionary, spectral and distance based representations, and is set to build 500 trees. InceptionTime [9] is a deep learning ensemble, combining 5 homogeneous networks each with random weight initialisations for stability. ROCKET [6] uses a large number, 10,000, of randomly parameterised convolution kernels in conjunction with a linear ridge regression classifier. We use the configurations of each classifier described in their respective publications.

Figure 5 shows the ranked performance of HC-TDE against HIVE-COTE, TS-CHIEF, InceptionTime and ROCKET on 109 problems. We are missing three datasets, HandOutlines, NonInvasiveFetalECGThorax1 and NonInvasiveFetalECGThorax2 because TS-CHIEF could not complete them within the seven day limit.

HC-TDE is significantly better than all four algorithms currently considered state of the art. The actual differences between HIVE-COTE and HC-TDE are understandably small, given their similarities. However, they are consistent: replacing BOSS with TDE improves HIVE-COTE on 69 problems, and makes it worse on just 32 (with 8 ties). HC-TDE does show significant variation to TS-CHIEF (see Fig. 6) and is on average over 1% more accurate. HC-TDE is the top performing algorithm using a range of performance measures such as AUROC, F1 and balanced accuracy (see accompanying website). The improvement over InceptionTime is even greater: it is on average 1.5% more accurate.

It is worth considering whether replacing BOSS with either S-BOSS or WEASEL would give as much improvement to HIVE-COTE as TDE does. We replaced BOSS with WEASEL (HC-WEASEL) and S-BOSS (HC-S-BOSS). Figure 7 shows the performance of these relative to HC-TDE, InceptionTime and TS-CHIEF. Whilst it is true that HC-S-BOSS is not significantly worse than HC-TDE, it is also not significantly better than the current state of the art. HC-WEASEL does not perform well. We speculate that this is because the major differences in WEASEL mean that its improvement is at problems better suited to other representations, and this improvement comes at the cost of worse performance at problems suited to dictionary classifiers.

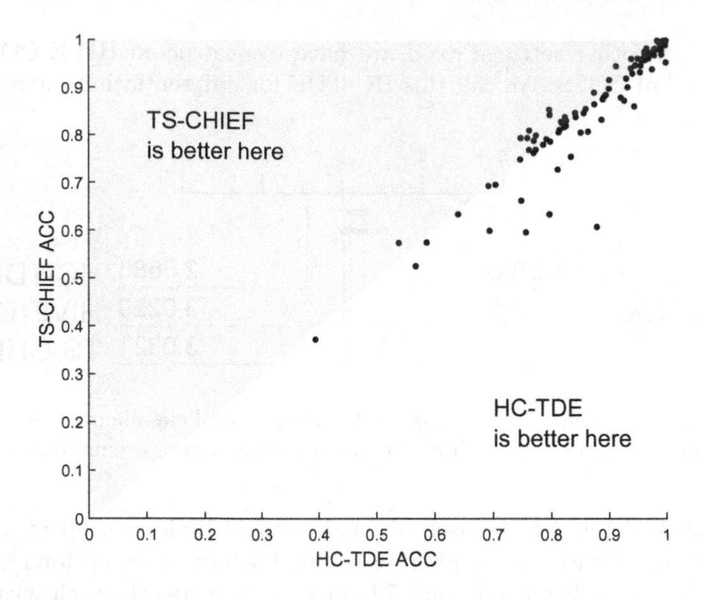

Fig. 6. Scatter plot of TS-CHIEF against HC-TDE. HC-TDE wins on 62, draws on 6 and loses on 41 data sets.

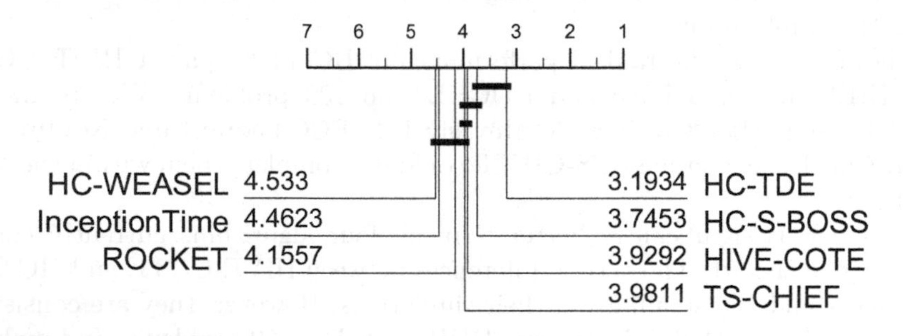

Fig. 7. Critical difference diagram for seven classifiers on 106 UCR time series classification problems. Full results are available on the accompanying website.

5 Conclusion

TDE combines the best elements of existing dictionary based classifiers with a novel method of improving the ensemble through a Gaussian process model for parameter selection. TDE is more accurate and scalable than current top performing dictionary algorithms. When we replace BOSS with TDE in HIVE-COTE, the resulting classifier is significantly more accurate than the current state of the art. TDE has some drawbacks. It is memory intensive. It requires about three times more memory than BOSS, and the maximum memory required was 10 GB for ElectricDevices. Like all nearest neighbour classifiers, TDE is relatively slow to classify new cases. If fast predictions are required, WEASEL may be preferable. Future work will focus on making TDE more scalable.

Acknowledgements. This work is supported by the UK Engineering and Physical Sciences Research Council (EPSRC) iCASE award T206188 sponsored by British Telecom. The experiments were carried out on the High Performance Computing Cluster supported by the Research and Specialist Computing Support service at the University of East Anglia.

References

1. Bagnall, A., Flynn, M., Large, J., Lines, J., Middlehurst, M.: On the usage and performance of the hierarchical vote collective of transformation-based ensembles version 1.0 (hive-cote 1.0). arXiv preprint arXiv:2004.06069 (2020)
2. Bagnall, A., Lines, J., Bostrom, A., Large, J., Keogh, E.: The great time series classification bake off: a review and experimental evaluation of recent algorithmic advances. Data Min. Knowl. Discov. **31**(3), 606–660 (2017)
3. Benavoli, A., Corani, G., Mangili, F.: Should we really use post-hoc tests based on mean-ranks? J. Mach. Learn. Res. **17**(1), 152–161 (2016)
4. Bostrom, A., Bagnall, A.: Binary shapelet transform for multiclass time series classification. In: Hameurlain, A., Küng, J., Wagner, R., Madria, S., Hara, T. (eds.) Transactions on Large-Scale Data- and Knowledge-Centered Systems XXXII. LNCS, vol. 10420, pp. 24–46. Springer, Heidelberg (2017). https://doi.org/10.1007/978-3-662-55608-5_2
5. Dau, H.A., et al.: The UCR time series archive. IEEE/CAA J. Automatica Sinica **6**(6), 1293–1305 (2019)
6. Dempster, A., Petitjean, F., Webb, G.I.: Rocket: exceptionally fast and accurate time series classification using random convolutional kernels. arXiv preprint arXiv:1910.13051 (2019)
7. Demšar, J.: Statistical comparisons of classifiers over multiple data sets. J. Mach. Learn. Res. **7**, 1–30 (2006)
8. Deng, H., Runger, G., Tuv, E., Vladimir, M.: A time series forest for classification and feature extraction. Inf. Sci. **239**, 142–153 (2013)
9. Fawaz, H.I., et al.: InceptionTime: finding AlexNet for time series classification. arXiv preprint arXiv:1909.04939 (2019)
10. Flynn, M., Large, J., Bagnall, T.: The contract random interval spectral ensemble (c-RISE): the effect of contracting a classifier on accuracy. In: Pérez García, H., Sánchez González, L., Castejón Limas, M., Quintián Pardo, H., Corchado Rodríguez, E. (eds.) HAIS 2019. LNCS (LNAI), vol. 11734, pp. 381–392. Springer, Cham (2019). https://doi.org/10.1007/978-3-030-29859-3_33
11. Garcia, S., Herrera, F.: An extension on "statistical comparisons of classifiers over multiple data sets" for all pairwise comparisons. J. Mach. Learn. Res. **9**, 2677–2694 (2008)
12. Large, J., Bagnall, A., Malinowski, S., Tavenard, R.: On time series classification with dictionary-based classifiers. Intell. Data Anal. **23**(5), 1073–1089 (2019)
13. Lazebnik, S., Schmid, C., Ponce, J.: Beyond bags of features: spatial pyramid matching for recognizing natural scene categories. In: 2006 IEEE Computer Society Conference on Computer Vision and Pattern Recognition (CVPR 2006), vol. 2, pp. 2169–2178. IEEE (2006)
14. Lines, J., Taylor, S., Bagnall, A.: Time series classification with HIVE-COTE: the hierarchical vote collective of transformation-based ensembles. ACM Trans. Knowl. Discov. Data (TKDD) **12**(5), 52 (2018)

15. Middlehurst, M., Vickers, W., Bagnall, A.: Scalable dictionary classifiers for time series classification. In: Yin, H., Camacho, D., Tino, P., Tallón-Ballesteros, A.J., Menezes, R., Allmendinger, R. (eds.) IDEAL 2019. LNCS, vol. 11871, pp. 11–19. Springer, Cham (2019). https://doi.org/10.1007/978-3-030-33607-3_2
16. Schäfer, P.: The boss is concerned with time series classification in the presence of noise. Data Min. Knowl. Discov. **29**(6), 1505–1530 (2015)
17. Schäfer, P., Högqvist, M.: SFA: a symbolic Fourier approximation and index for similarity search in high dimensional datasets. In: Proceedings of the 15th International Conference on Extending Database Technology, pp. 516–527 (2012)
18. Schäfer, P., Leser, U.: Fast and accurate time series classification with weasel. In: Proceedings of the 2017 ACM on Conference on Information and Knowledge Management, pp. 637–646 (2017)
19. Shifaz, A., Pelletier, C., Petitjean, F., Webb, G.I.: TS-CHIEF: a scalable and accurate forest algorithm for time series classification. Data Min. Knowl. Discov. **34**, 1–34 (2020)
20. Snoek, J., Larochelle, H., Adams, R.P.: Practical Bayesian optimization of machine learning algorithms. In: Advances in Neural Information Processing Systems, pp. 2951–2959 (2012)
21. Williams, C.K., Rasmussen, C.E.: Gaussian Processes for Machine Learning, vol. 2. MIT Press, Cambridge (2006)

Incremental Training of a Recurrent Neural Network Exploiting a Multi-scale Dynamic Memory

Antonio Carta[1](✉), Alessandro Sperduti[2], and Davide Bacciu[1]

[1] Department of Computer Science, University of Pisa, Pisa, Italy
{antonio.carta,bacciu}@di.unipi.it
[2] Department of Mathematics, University of Padova, Padova, Italy
sperduti@math.unipd.it

Abstract. The effectiveness of recurrent neural networks can be largely influenced by their ability to store into their dynamical memory information extracted from input sequences at different frequencies and timescales. Such a feature can be introduced into a neural architecture by an appropriate modularization of the dynamic memory. In this paper we propose a novel incrementally trained recurrent architecture targeting explicitly multi-scale learning. First, we show how to extend the architecture of a simple RNN by separating its hidden state into different modules, each subsampling the network hidden activations at different frequencies. Then, we discuss a training algorithm where new modules are iteratively added to the model to learn progressively longer dependencies. Each new module works at a slower frequency than the previous ones and it is initialized to encode the subsampled sequence of hidden activations. Experimental results on synthetic and real-world datasets on speech recognition and handwritten characters show that the modular architecture and the incremental training algorithm improve the ability of recurrent neural networks to capture long-term dependencies.

Keywords: Recurrent Neural Networks · Linear Dynamical Systems · Incremental Learning

1 Introduction

Time series, such as speech and music sound waves [9,17], and raw sensor data from several domains are sampled at high frequencies, generating large datasets of long and fast-flowing sequences. Recurrent neural networks must dynamically extract information from these samples at different frequencies. Unfortunately, these data present two challenges that limit the application of recurrent neural networks. First, vanilla RNN [5] and LSTM [11] do not take into account the

Electronic supplementary material The online version of this chapter (https://doi.org/10.1007/978-3-030-67658-2_39) contains supplementary material, which is available to authorized users.

F. Hutter et al. (Eds.): ECML PKDD 2020, LNAI 12457, pp. 677–693, 2021.
https://doi.org/10.1007/978-3-030-67658-2_39

importance of different frequencies when processing a sequence, making it difficult to capture high and low frequencies together. Furthermore, low-frequency information requires capturing long-term temporal dependencies in the sequence, which are difficult to learn by stochastic gradient descent (SGD) due to the vanishing gradient problem [11,20]. The most popular solutions to address this problem is to subsample the sequence, discarding useful data, or to process the sequence to extract hardcoded features that reduce the sampling rate. Both solutions circumvent the problem by avoiding to learn features for short-term dependencies while shortening the long-term dependencies. Both approaches simplify the learning problem but have the drawback that useful information is discarded when it could be used to improve the quality of network predictions.

In this paper, we propose a new model, called the MultiScale LMN (MS-LMN), a recurrent neural network designed to easily represent and learn short-term and long-term dependencies. The architecture of the MS-LMN partitions the memory state of the model into separate modules, and each module is updated with a different frequency. Therefore, high frequency and low frequency information is separated such that the short-term dependencies do not interfere with the long-term ones.

To improve the learning of long-term dependencies we propose an incremental training algorithm. Since short-term dependencies are easier to learn, the model can be trained to learn them before the long-term ones, by learning only the parameters relevant to the high-frequency memory state. Incrementally, new modules can be added to the MS-LMN to learn longer dependencies. Each module is initialized to memorize the subsampled sequence and finetuned with SGD. The complexity of the model grows according to the complexity of the learning task, which is given by the maximum dependency length.

Synthetic and real world time series datasets are used to assess the proposed model. In particular, we present empirical evidence that the model is able to generate a long sequence, learning to model both short-term and long-term dependencies in the data. Moreover, experiments on speech recognition and handwritten character recognition show that the proposed architecture and associated incremental training algorithm can effectively learn on real-world datasets when using high-frequency features.

In summary, the main contributions of the paper are as follows:

- the proposal of a novel hierarchical RNN architecture designed to model low frequency and high frequency features separately;
- the proposal of a novel incremental training algorithm for RNNs designed to incrementally learn long-term dependencies;
- an experimental assessment of the proposed approach that shows the benefits of incremental training on synthetic and real-world datasets.

2 Background

In this section we describe two models, the Linear Memory network (LMN) [1], and the linear autoencoder for sequences (LAES) [21]. The proposed architecture

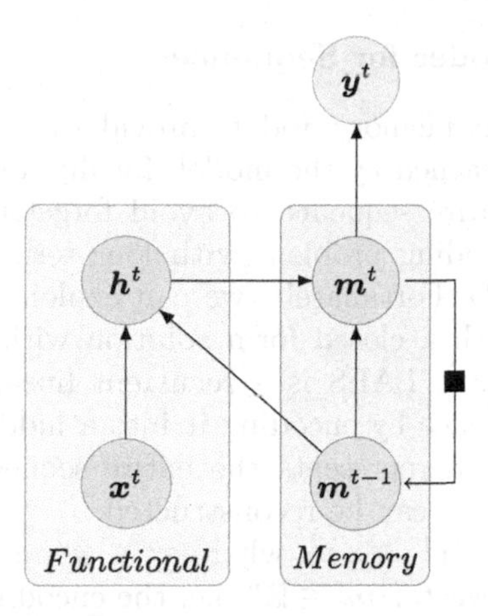

Fig. 1. A Linear Memory Network, where $\boldsymbol{x}^t, \boldsymbol{h}^t, \boldsymbol{m}^t, \boldsymbol{y}^t$ are the input, hidden, memory and output vector, respectively. The edge with the small black square represents a time delay.

extends the LMN with a modular design. The LAES training algorithm is a fundamental step of the incremental training.

2.1 Linear Memory Network

The dynamic multiscale architecture put forward in this paper is built on the Linear Memory Network (LMN) [1] concepts. The LMN is a recurrent neural network that updates its memory state using a nonlinear mapping of the input and a linear mapping of its memory state. The LMN computes a hidden state $\boldsymbol{h}^t$ and a memory state $\boldsymbol{m}^t$ as follows:

$$\boldsymbol{h}^t = \sigma(\mathbf{W}^{xh}\boldsymbol{x}^t + \mathbf{W}^{mh}\boldsymbol{m}^{t-1}),$$
$$\boldsymbol{m}^t = \mathbf{W}^{hm}\boldsymbol{h}^t + \mathbf{W}^{mm}\boldsymbol{m}^{t-1},$$

where $\mathbf{W}^{xh} \in \mathbb{R}^{N_h \times N_x}$, $\mathbf{W}^{mh} \in \mathbb{R}^{N_h \times N_m}$, $\mathbf{W}^{hm} \in \mathbb{R}^{N_m \times N_h}$, $\mathbf{W}^{mm} \in \mathbb{R}^{N_m \times N_m}$ are the model parameters matrices, with N_x, N_h, N_m the input size, hidden size and memory size respectively, and σ is a non-linear activation function ($tanh$ for the purpose of this paper). The memory state $\boldsymbol{m}^t$ is the output of the entire layer, which is passed to the next layer. The output $\boldsymbol{y}^t$, is computed by a linear layer:

$$\boldsymbol{y}^t = \mathbf{W}^{my}\boldsymbol{m}^t. \tag{1}$$

A schematic view of the LMN is shown in Fig. 1. Notice that the linearity of the recurrence does not limit the expressive power of the entire model since given an RNN such that $\boldsymbol{h}^t = \sigma(\tilde{\mathbf{W}}^{xh}\boldsymbol{x}^t + \tilde{\mathbf{W}}^{hh}\boldsymbol{h}^{t-1})$, it is possible to initialize an equivalent LMN that computes the same hidden activations by setting $\mathbf{W}^{xh} = \tilde{\mathbf{W}}^{xh}$, $\mathbf{W}^{mh} = \tilde{\mathbf{W}}^{hh}$, $\mathbf{W}^{hm} = \mathbb{I}$, $\mathbf{W}^{mm} = 0$. The linearity of the memory update plays a key role during the initialization of a new module in the incremental training algorithm.

2.2 Linear Autoencoder for Sequences

The linearity of the LMN memory update provides an opportunity to explicitly optimize the encoding learned by the model. Ideally, we would like to train the model to encode the entire sequence to avoid forgetting past elements. However, this is an autoencoding problem with long-term dependencies, which is difficult to solve by SGD. Fortunately, we can exploit the linearity to find the optimal autoencoder with a closed form solution with the linear autoencoder for sequences (LAES) [21]. LAES is a recurrent linear model that is able to memorize an input sequence by encoding it into a hidden memory state vector recursively updated which represents the entire sequence. Given the memory state, the original sequence can be reconstructed.

Given a sequence $s = \boldsymbol{x}^1, \ldots, \boldsymbol{x}^l$, where $\boldsymbol{x}^i \in \mathbb{R}^a$, a linear autoencoder computes the memory state vector $\boldsymbol{m}^t \in \mathbb{R}^p$, i.e. the encoding of the input sequence up to time t, using the following equations:

$$\boldsymbol{m}^t = \mathbf{A}\boldsymbol{x}^t + \mathbf{B}\boldsymbol{m}^{t-1}, \tag{2}$$

$$\begin{bmatrix} \boldsymbol{x}^t \\ \boldsymbol{m}^{t-1} \end{bmatrix} = \mathbf{C}\boldsymbol{m}^t, \tag{3}$$

where p is the memory state size, $\mathbf{A} \in \mathbb{R}^{p \times a}$, $\mathbf{B} \in \mathbb{R}^{p \times p}$ and $\mathbf{C} \in \mathbb{R}^{(a+p) \times p}$ are the model parameters. Equation (2) describes the encoding operation, while Eq. (3) describes the decoding operation.

Training Algorithm. The linearity of the LAES allows us to derive the optimal solution with a closed-form equation, as shown in [22]. For simplicity, let us assume that the training set consists of a single sequence $\{\boldsymbol{x}_1, \ldots, \boldsymbol{x}_l\}$ and define $\mathbf{M} \in \mathbb{R}^{l \times p}$ as the matrix obtained by stacking by rows the memory state vectors of the LAES at each timestep. From Eq. (2) it follows that:

$$\underbrace{\begin{bmatrix} \boldsymbol{m}^{1\top} \\ \boldsymbol{m}^{2\top} \\ \boldsymbol{m}^{3\top} \\ \vdots \\ \boldsymbol{m}^{l\top} \end{bmatrix}}_{\mathbf{M}} = \underbrace{\begin{bmatrix} \boldsymbol{x}^{1\top} & 0 & \ldots & 0 \\ \boldsymbol{x}^{2\top} & \boldsymbol{x}^{1\top} & \ldots & 0 \\ \vdots & \vdots & \ddots & \vdots \\ \boldsymbol{x}^{l\top} & \boldsymbol{x}^{l-1\top} & \ldots & \boldsymbol{x}^{1\top} \end{bmatrix}}_{\boldsymbol{\Xi}} \underbrace{\begin{bmatrix} \mathbf{A}^\top \\ \mathbf{A}^\top\mathbf{B}^\top \\ \vdots \\ \mathbf{A}^\top\mathbf{B}^{l-1\top} \end{bmatrix}}_{\boldsymbol{\Omega}}. \tag{4}$$

The matrix $\boldsymbol{\Xi} \in \mathbb{R}^{l \times la}$ contains the reversed subsequences of s, while $\boldsymbol{\Omega} \in \mathbb{R}^{la \times p}$ contains the matrices to encode the input elements for up to l timesteps. The encoder parameters $\mathbf{A}$ and $\mathbf{B}$ can be identified by exploiting the singular value decomposition (SVD) $\boldsymbol{\Xi} = \mathbf{V}\boldsymbol{\Sigma}\mathbf{U}^\top$, where imposing $\mathbf{U}^\top\boldsymbol{\Omega} = \mathbf{I}$ yields $\boldsymbol{\Omega} = \mathbf{U}$. Given this additional constraint, we can then exploit the structure of $\boldsymbol{\Xi}$ to

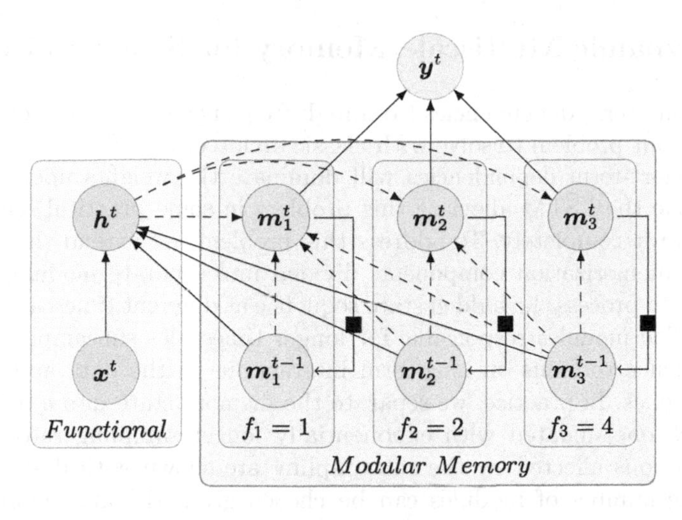

Fig. 2. Architecture of the MultiScale Linear Memory Network with 3 memory modules. Dashed connections are affected by the subsampling, while connections with small black squares represent time delays.

recover $\mathbf{A}$, $\mathbf{B}$, and the decoder parameters $\mathbf{C} = \begin{bmatrix}\mathbf{A}^\top \\ \mathbf{B}^\top\end{bmatrix}$. Specifically, $\mathbf{\Omega} = \mathbf{U}$ is satisfied by using the matrices

$$\mathbf{P} \equiv \begin{bmatrix} \mathbf{I}_a \\ \mathbf{0}_{a(l-1)\times a} \end{bmatrix}, \quad \text{and} \quad \mathbf{R} \equiv \begin{bmatrix} \mathbf{0}_{a\times a(l-1)} & \mathbf{0}_{a\times a} \\ \mathbf{I}_{a(l-1)} & \mathbf{0}_{a(l-1)\times a} \end{bmatrix},$$

to define $\mathbf{A} \equiv \mathbf{U}^\top \mathbf{P}$ and $\mathbf{B} \equiv \mathbf{U}^\top \mathbf{R} \mathbf{U}$, where $\mathbf{I}_u$ is the identity matrix of size u, and $\mathbf{0}_{u\times v}$ is the zero matrix of size $u \times v$. The algorithm can be easily generalized to multiple sequences by stacking the data matrix Ξ_q for each sequence s^q and padding with zeros to match sequences length, as shown in [21].

The optimal solution reported above allows encoding the entire sequence without errors with a minimal number of memory units $p = rank(\Xi)$. Since we fix the number of memory units before computing the LAES, we approximate the optimal solution using the truncated SVD, which introduces some errors during the decoding process. The computational cost of the training algorithm is dominated by the cost of the truncated SVD decomposition, which for a matrix of size $n \times m$ is $\mathcal{O}(n^2 m)$. Given a dataset of sequences with lengths $l^1, \ldots, l^S$, with $l = \sum_{i=1}^{S} l^i$, $l_{max} = \max_i\{l^i\}$, we have $\Xi \in \mathbb{R}^{l \times l_{max} a}$, which requires $\mathcal{O}(l_{max} l a)$ memory for a dense representation. The memory usage can be reduced for sparse inputs, such as music in a piano roll representation, by using a sparse representation. To mitigate this problem in a more general setting, we approximate the SVD decomposition using the approach proposed in [19], which computes the SVD decomposition with an iterative algorithm that decomposes Ξ in slices of size $l \times a$ and requires $\mathcal{O}(la)$ memory.

3 A Dynamic Multiscale Memory for Sequential Data

Learning long-term dependencies from high-frequency data with recurrent models is a difficult problem to solve with classic architectures. Due to the vanishing gradient, short-term dependencies will dominate the weights update. Popular solutions like the LSTM alleviate this problem in some practical scenarios but do not solve it completely. To address this problem, we extend the LMN with a modular memorization component, divided into separate modules, each one responsible to process the hidden state sequence at different timescales, as shown in Fig. 2. The modules responsible for longer timescales subsample the hidden states sequence to focus on long-term interactions in the data and ignore the short-term ones. In practice, we separate the memory state into g different substates, each one updated with exponentially longer sampling rates $1, 2, ..., 2^g$. The connections affected by the subsampling are shown with dashed edges in Fig. 2. The number of modules can be chosen given the maximum length of the sequences in the training set l_{max}. Given l_{max}, the maximum number of different frequencies is $g = \lfloor \log l_{max} \rfloor$, which means the model only needs a logarithmic number of modules. Each memory module is connected only to slower modules, and not vice-versa, to avoid interference of the faster modules with the slower ones. The organization of the memory into separate modules with a different sampling rate is inspired by the Clockwork RNN [14], which is an RNN with groups of hidden units that work with different sampling frequencies. Differently from the Clockwork RNN, we apply this decomposition only to the memory state. Furthermore, by adopting a linear recurrence we can achieve better memorization of long sequences, as we will see in the experimental results in Sect. 5.

The model update computes at each timestep t an hidden state $\boldsymbol{h}^t$ and g memory states $\boldsymbol{m}_1^t, \ldots, \boldsymbol{m}_g^t$ as follows:

$$\boldsymbol{h}^t = \sigma(\mathbf{W}^{xh}\boldsymbol{x}^t + \sum_{i=1}^{g} \mathbf{W}^{m_i h}\boldsymbol{m}_i^{t-1}), \tag{5}$$

$$\boldsymbol{m}_k^t = \begin{cases} \boldsymbol{m}_{new}^t & if\ t \mod 2^{k-1} = 0 \\ \boldsymbol{m}_k^{t-1} & otherwise \end{cases} \qquad \forall k \in 1, \ldots, g, \tag{6}$$

$$\boldsymbol{m}_{new}^t = \mathbf{W}^{hm_k}\boldsymbol{h}^t + \sum_{i=k}^{g} \mathbf{W}^{m_i m_k}\boldsymbol{m}_i^{t-1}, \tag{7}$$

where $\boldsymbol{x}^t \in \mathbb{R}^{N_x}$, $\boldsymbol{h}^t \in \mathbb{R}^{N_h}$, $\boldsymbol{m}_k^t \in \mathbb{R}^{N_m}$. The subsampling of the hidden state sequence is performed by choosing when to update the memory state using the modulo operation. The network output can be computed from the memory modules' output as follows:

$$\boldsymbol{y}^t = \sum_{i=1}^{g} \mathbf{W}^{m_i y}\boldsymbol{m}_i^t. \tag{8}$$

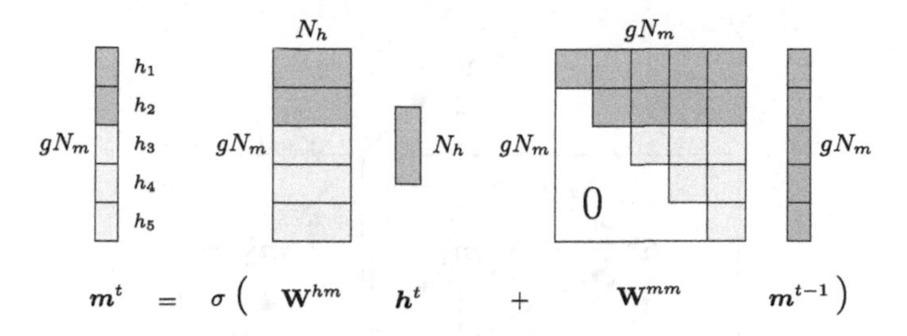

Fig. 3. Representation of the memory update with block matrices showing the size for $g = 5$, assuming that only the first two modules are active at time t. Darker blocks represent the active weights.

Figure 2 shows a schematic view of the architecture. For a more efficient implementation, more amenable to parallel architectures like GPUs, we can combine all the operations performed by Eqs. (5)–(8) for each module into a single matrix multiplication. In the following, we show the procedure for Eq. (6) since the same approach can be applied to Eqs. (5) and (8). First, we notice that the memory modules are ordered by frequency, from fastest to slowest, and their sampling frequencies are powers of 2. As a consequence, if module i is active, then all the modules j with $j < i$ are also active since $t \mod 2^i = 0$ implies $t \mod 2^j = 0$ whenever $j \geq 0$ and $j < i$. Therefore, we only need to find the maximum index of the active modules $i^t_{max} = max\{i \mid t \mod 2^{i-1} = 0 \wedge i \leq g\}$ to know which memory modules must be updated. We can combine the activations and parameters of the modules together as follows:

$$\boldsymbol{m}^t = \begin{bmatrix} \boldsymbol{m}_1^{t\top} \\ \vdots \\ \boldsymbol{m}_k^{t\top} \end{bmatrix}, \mathbf{W}^{hm} = \begin{bmatrix} \mathbf{W}^{h_1 m} \\ \vdots \\ \mathbf{W}^{h_g m} \end{bmatrix},$$

$$\mathbf{W}^{mm} = \begin{bmatrix} \mathbf{W}^{m_1 m_1} & & \dots & \mathbf{W}^{m_g m_1} \\ 0 & \mathbf{W}^{m_2 m_2} & \dots & \mathbf{W}^{m_g m_2} \\ \vdots & \ddots & \ddots & \vdots \\ 0 & \dots & 0 & \mathbf{W}^{m_g m_g} \end{bmatrix}.$$

Equation (6) becomes:

$$\boldsymbol{m}^t[: i_{max}] = \mathbf{W}^{hm}[: i_{max}]\boldsymbol{h}^t[: i_{max}] + \mathbf{W}^{mm}[: i_{max}]\boldsymbol{m}^t[: i_{max}], \tag{9}$$

$$\boldsymbol{m}^t[i_{max} :] = \boldsymbol{m}^{t-1}[i_{max} :], \tag{10}$$

where the slicing operator $\boldsymbol{m}[i_{start} : i_{end}]$ returns a vector $\boldsymbol{m}[i_{start} : i_{end}] \in \mathbb{R}^{i_{end}-i_{start}}$ with $\boldsymbol{m}[i_{start} : i_{end}]_k = \boldsymbol{m}[i_{start} + k]$, which is used to select the vector of the currently active memory modules. Using Eqs. (9)–(10), the subsampling is performed by finding i_{max} which determines which slice of $\boldsymbol{m}^t$ must

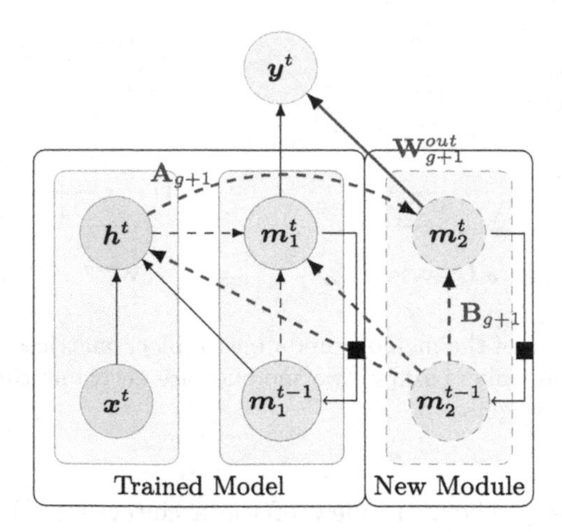

Fig. 4. MS-LMN pre-training: *(right side of picture)* the $g + 1$th new memory model is added. New connections are shown in red. The LAES training algorithm is used to pre-train the weight matrices $\mathbf{A}_{g+1}$, $\mathbf{B}_{g+1}$, and $\mathbf{W}_{g+1}^{out}$, while the remaining new connections are initialized with zeros.

be updated or remain constant. Notice that $\mathbf{W}_{mm}$ is a block diagonal matrix, therefore it has fewer parameters than a corresponding LMN with gN_m hidden units. Figure 3 shows the block structure of the MS-LMN parameters for the memory state update, with the darker blocks being the currently active modules.

4 Incremental Training of the Multiscale Memory

The MS-LMN is designed to model long-term dependencies at different frequencies. To better exploit the modular architecture, we propose a constructive algorithm that incrementally trains the network by adding a new module at each step. During an incremental step the algorithm performs the following operations:

1. Add a new module with a lower sampling rate.
2. Initialize the new module to encode the hidden state into the memory state.
3. Finetune the model by SGD.

Incremental training helps to gradually learn longer dependencies, while the initialization ensures that the new module is able to encode a long sequence of activations, which helps to extract long-term dependencies from the data. In this section we show how to train the MS-LMN by describing the incremental construction of the model and the initialization procedure of the new module.

4.1 MS-LMN Incremental Training

The training algorithm for the MS-LMN works incrementally. During the first iteration, a MS-LMN with a single module is trained by SGD. After a fixed number of epochs[1], an additional memory module with a slower frequency is added, and it is initialized to encode the hidden activations of the current model using the LAES training algorithm (see Fig. 4). The resulting model is trained with SGD and the process is repeated until all the modules have been added.

The addition of a new module works as follows. Let us assume to have already trained an MS-LMN with g modules as defined in Eqs. (5)–(6). We collect the sequences of hidden activations $\boldsymbol{h}^1, \ldots, \boldsymbol{h}^t$ computed by the model for each sample in the training set, subsample them with frequency 2^g, which is the subsampling rate of the new module, and train a LAES with parameters $\mathbf{A}_{g+1}$ and $\mathbf{B}_{g+1}$. A new module with sampling frequency 2^g is added to the model and the new connections are initialized as follows:

$$\mathbf{W}^{hm_{g+1}} = \mathbf{A}_{g+1}, \quad \mathbf{W}^{m_{g+1}m_{g+1}} = \mathbf{B}_{g+1}, \quad \mathbf{W}^{m_{g+1}o} = \mathbf{W}^{out}_{g+1},$$

where $\mathbf{W}^{out}_{g+1}$ is obtained by training a linear model (via pseudoinverse) to predict the desired output $\boldsymbol{y}^t$ from the memory state of the entire memory $\boldsymbol{m}^t_1, \ldots, \boldsymbol{m}^t_{g+1}$. The remaining new parameters are initialized with zeros, while the old parameters remain unchanged. After the initialization of the new connections, the entire model is trained end-to-end by gradient descent. Notice that during this phase the entire model is trained, which means that the connections of the old modules can still be updated. This process is repeated until all the memory modules have been added and the entire model has been trained for the last time. Algorithm 1 shows the pseudocode for the entire training procedure.

Algorithm 1. MS-LMN training

1: **procedure** MS-LMNTRAIN($Data$, N_h, N_m, g)
2:     **for** $i \in \{0, \ldots, g-1\}$ **do**
3:         *ms-lmn.add-module*($Data, i$)
4:         *ms-lmn.fit*($Data$)
5:     **return** *ms-lmn*
6: **procedure** ADD-MODULE($self$, $Data$, g)
7:     $\mathbf{H}_g = [\,]$
8:     **for** $seq \in Data$ **do**
9:         $\boldsymbol{h}_{seq}, \boldsymbol{m}_{seq} \leftarrow self(seq)$
10:         $\boldsymbol{h}_{seq}.subsample(2^g)$
11:         $\mathbf{H}_g.append(\boldsymbol{h}_{seq})$
12:     $laes \leftarrow train\text{-}laes(N_m)$
13:     $laes.fit(\mathbf{H}_g)$
14:     $self.\mathbf{W}^{hm_{g+1}} \leftarrow laes.\mathbf{A}$
15:     $self.\mathbf{W}^{m_{g+1}m_{g+1}} \leftarrow laes.\mathbf{B}$
16:     $self.\mathbf{W}^{m_{g+1}o} \leftarrow fit\_readout(self, laes, Data)$ ▷ computed with the pseudoinverse

[1] This is a simple but suboptimal strategy that has been adopted to show the effectiveness of the proposed model. This strategy can easily be improved so to optimize the performance on the validation set.

The computational cost of the incremental training process is the sum of the cost of training the autoencoders and the cost of the SGD. In practice, despite the worse computational cost, training the LAES requires a negligible amount of time compared to the SGD, due to the large number of iterations needed by the latter to reach convergence. It is important to notice that incremental training is less expensive than training the full model with all memory modules from scratch by SGD since during the first phases the model has a lower number of modules, and therefore it has a lower computational cost.

5 Experimental Analysis

In the following experiments, we study the ability of the modular memorization component of the MS-LMN to capture long-term dependencies in three different sequential problems. The model is compared against LSTM, LMN, and CW-RNN, including all datasets used in the original CW-RNN publication [14]. Each model is a single layer recurrent neural network with *tanh* activation function (only in the functional component for LMN and MS-LMN). We tested the MS-LMN on audio signals since they are a natural choice to assess the modular memory, which works by subsampling its input sequence. The first task, *Sequence Generation*, is a synthetic problem that requires the model to output a given signal without any eternal input. The second task, *Spoken Word Classification*, is a sequence classification task that uses a subset of spoken words extracted from TIMIT [6]. The task is designed to have long-term dependencies by only considering a restricted subset of words that have a common suffix. The third experiment performs handwritten character recognition on the IAM-OnDB online dataset [8]. Throughout all the experiments we use the Adam optimizer [13] with L2 weight decay. The source code is available online[2].

5.1 Sequence Generation

The Sequence Generation task is a synthetic problem that requires the model to output a given signal without any eternal input. We extracted a sequence of 300 data points generated from a portion of a music file, sampled at 44.1 kHz starting from a random position. Sequence elements are scaled in the range $[-1, 1]$. The task stresses the ability to learn long-term dependencies by requiring the model to encode the entire sequence without any external input. Notice that this task is used only to assess the effectiveness of the modular architecture and therefore we do not use the incremental pretraining. The CW-RNN [14] is the state-of-the-art on this task, where it reaches better results than comparable RNNs and LSTMs. We tested each model with 4 different numbers of parameters in $\{100, 250, 500, 1,000\}$, by varying the number of hidden neurons. The models are trained to optimize the Normalized MSE. Table 1 shows the most relevant hyperparameters of the best configuration for each model which were found with a random search.

[2] https://github.com/AntonioCarta/mslmn.

Table 1. Number of parameters and corresponding number of hidden units for different configurations trained on the Sequence Generation task. For CW-RNN and MS-LMN we show the total number of units and the number of units for each module (the latter between parenthesis).

#parameters	RNN	LSTM	CW-RNN	LMN	MS-LMN
	N_h	N_h	$N_h, (\frac{N_h}{g})$	N_h, N_m	$N_h, N_m, (\frac{N_m}{g})$
100	9	4	9, (1)	4, 6	1, 9, (1)
250	15	7	18, (2)	7, 10	1, 18, (2)
500	22	10	27, (3)	11, 13	1, 27, (3)
1,000	31	15	36, (4)	2, 29	1, 36, (4)

Table 2. Results for the Sequence Generation task. Performance computed using NMSE (lower is better).

	RNN	LSTM	CW-RNN	LMN	MS-LMN
Hidden units	31	15	36	2	1
Memory units	/	/	/	29	36
Learning rate	1×10^{-3}	1×10^{-2}	5×10^{-5}	5×10^{-4}	5×10^{-3}
# parameters	1,000	1,000	1,000	1,000	1,000
Epochs	6,000	12,000	2,000	5,000	8,000
NMSE (10^{-3})	79.5	20.7	12.5	38.4	**0.116**

The CW-RNN and MS-LMN use 9 modules. Notice that adding more modules would be useless since the sequence length is 300. The number of hidden units for the CW-RNN and of memory units for the MS-LMN is the number of modules times the number of units per module. For the LSTM, we obtained the best results by initializing the forget gate to 5, as suggested in [4].

The results of the experiments are reported in Table 2. Figure 5 shows the reconstructed sequence for each model. The results confirm those found in [14] for the CW-RNN and show a clear advantage of the CW-RNN over the RNN and LSTM. The LMN provides an approximation of the sequence, slightly worse than that of the LSTM, which closely follows the global trend of the sequence, but it is not able to model small local variations. The MS-LMN obtains the best results and closely approximates the original sequence. The model combines the advantages of a linear memory, as shown by the LMN performance, with the hierarchical structure of the CW-RNN. The result is that the MS-LMN is able to learn to reproduce both the short-term and long-term variations of the signal.

5.2 Common Suffix TIMIT

TIMIT [6] is a speech corpus for training acoustic-phonetic models and automatic speech recognition systems. The dataset contains recordings from different speak-

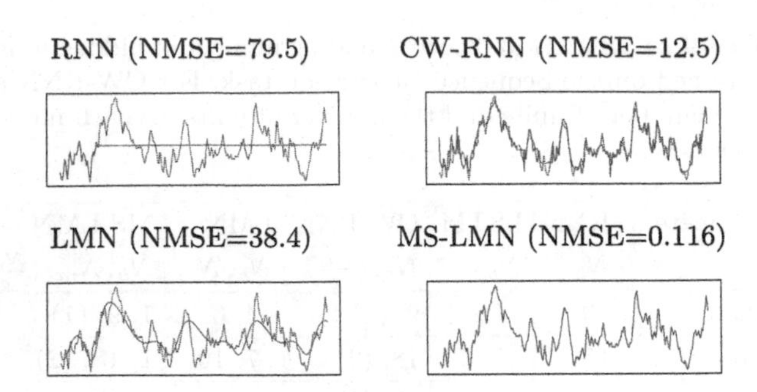

Fig. 5. Generated output for the Sequence Generation task. The original sequence is shown with a dashed blue line while the generated sequence is a solid green line. (Color figure online)

ers in major American dialects, and it provides a train-test split and information about words and phonemes for each recorded audio file.

Since we are interested in the ability of the assessed models to capture long-term dependencies, we extract from TIMIT a subset of words that have a common suffix, following the preprocessing of the original CW-RNN paper [14]. We took 25 words pronounced by 7 different speakers, for a total of 175 examples. The chosen words can be categorized into 5 clusters based on their suffix:

- *Class 1:* making, walking, cooking, looking, working;
- *Class 2:* biblical, cyclical, technical, classical, critical;
- *Class 3:* tradition, addition, audition, recognition, competition;
- *Class 4:* musicians, discussions, regulations, accusations, conditions;
- *Class 5:* subway, leeway, freeway, highway, hallway.

The common suffix makes the task more difficult since the model is forced to remember the initial part of the sequence to correctly classify each word.

Each file in the TIMIT dataset is a WAV containing the recording of a sentence, therefore we trimmed it to select a single word using the segmentation metadata provided with the dataset. The words extracted from this procedure are preprocessed to extract a sequence of MFCC coefficients using a window length of 25 ms, a step of 1 ms, preemphasis filter of 0.97, 13 cepstrum coefficients, where we replaced the zeroth cepstral coefficient with the *energy* (the log of total frame energy). As a result, we obtain 13 features for each timestep. We normalized each feature to have mean 0 and variance 1.

To allow a direct comparison with the work in [14], we split the dataset by taking 5 words for training and 2 for test from each class. This split ensures a balanced train and test set. During training we added gaussian noise to the sequence with a standard deviation of 0.6. Due to the small size of the dataset, we do not use an additional train-validation split and we use the clean version of the training set as a validation set, as done in [14]. Unfortunately, [14] does not provide the exact split used in their experiments. Given the small size of the dataset, in our experiments we have found a great variance between different splits. Therefore, we cannot directly compare against the results in [14] and we decided to train the CW-RNN with our train-test split. To ensure the reproducibility of our work and a fair comparison for future work, we provide the splits in the supplementary material.

We found the best hyperparameters with a random search on the batch size in $\{1, 25\}$, l2 decay in $\{0, 10^{-3}, 10^{-4}, 10^{-5}\}$, learning rate in $\{10^{-3}, 10^{-4}\}$ and hidden units per module in $[5, 40]$. Each model is trained to minimize the cross-entropy loss. When using a batch size equal to 25 we keep the classes balanced by taking one sample per class. The CW-RNN and MS-LMN use 7 modules, since the longer sequence has 97 data points. We initialize the LSTM forget gate to 5, as suggested in [4].

Table 3 shows the results of the experiment. The CW-RNN and MS-LMN obtain much better results than the LMN due to their ability to learn long-term dependencies. The MS-LMN shows superior results compared to the CW-RNN, and the incremental training improves the performance of the LMN substantially.

5.3 Handwritten Character Recognition

IAM-OnDB [8] is a dataset of handwritten digits. The dataset is composed of 12,179 handwritten sentences. Each sentence is represented as a sequence of pen strokes, where each stroke is a sequence of temporal positions of the pen on the whiteboard. Each measurement is represented as a quadruple <*time*, x, y, *end_stroke*>, where *time* is the timestamp of the measurement, x and y the coordinates of the pen on the whiteboard, and *end_stroke* is 1 for the last measurement of each stroke and 0 otherwise. The labels are the corresponding character, lower case or upper case, digits, and some special characters. We extracted a different sample from each line in the dataset and we split the data into a train, validation and test set, of 5,364, 3,859, and 2,956 sentences respectively. We normalized each stroke separately such that the timestamp and positions lie in the range $[0, 1]$. Given a sequence of pen measurements, the model must predict the corresponding sequence of characters.

A single layer bidirectional network is trained with connectionist temporal classification (CTC) loss [7] for each model. In order to perform a fair compar-

Table 3. Test set accuracy on Common Suffix TIMIT. Variance computed by training with different random seeds for 5 times using the best hyperparameters found during the model selection.

Model	n_h	n_m	Accuracy
LSTM	41	–	55.9 ± 5.3
LMN	52	–	55.0 ± 1.0
CW-RNN	13	–	74.4 ± 2.9
MS-LMN	25	25	78.0 ± 3.4
pret-MS-LMN	25	25	**79.6 ± 3.8**

Table 4. Best path accuracy computed on IAM-OnDB online test set.

	n_h	n_m	Accuracy
LSTM	110		**77.3**
CW-RNN	33		32.9
MS-LMN	144	16	26.5
pret-MS-LMN	144	16	66.8

ison, all models use a number of hidden units corresponding to approximately 100,000 parameters, with 9 memory modules for the CW-RNN and MS-LMN. For the incremental training, we add a new module every 50 epochs. We stop the training once the validation loss does not improve for 50 epochs. In our preliminary experiments all the models were highly sensitive to the random seed and initialization. Therefore, we decided to train each model 5 times and report the test set results of the best model for each configuration (selected on the validation set).

Table 4 shows the results of the experiment. On this dataset, the incremental training is fundamental to achieve a good performance on the Multiscale LMN, which allows to double the accuracy of the model. However, the best accuracy is obtained by the LSTM. We hypothesize that the use of gates in a multiscale architecture like the CW-RNN and the MS-LMN could further improve the performance of these models. The result also suggests that the LSTM's gates are effectively adding expressive power to the model in addition to their role in reducing the vanishing gradient since the multiscale architecture is already able to reduce the vanishing gradient. While the maximum number of modules for the hierarchical models is set to 9, the incremental training stops at 8 modules due to the early stopping criterion. This result shows another advantage of the incremental training, which is the ability to adaptively select the model complexity.

6 Related Work

Recurrent neural networks such as the vanilla RNN [5] and the LSTM [11] are state-of-the-art models on many sequence classification problems, for example speech recognition [9] or handwritten character recognition [8]. However, due to the vanishing gradient problem [10,20], it is difficult to learn long-term dependencies with these models. Hierarchical RNNs, such as the Clockwork RNN [14], Phased LSTM [16], and Hierarchical Multiscale RNN [3], solve this limitation by modifying the architecture to easily encode long-term dependencies in the hidden state. Most of the effort of the literature focus on architectural modifications [3,15,16]. Another line of research explores the use of online algorithms to train RNNs [12,18,23]. Instead, in this paper, we additionally propose an incremental training algorithm that exploits the modularity of the hierarchical architecture during training.

The MS-LMN is based on the Linear Memory Network [1], and the initialization procedure exploits the training algorithm for the LAES [22]. While [1] proposes a pretraining algorithm for the LMN that also exploits the LAES, the procedure proposed in this paper for the MS-LMN represents an improvement under several aspects. First, it does not require the expensive pretraining of the unrolled model. Furthermore, the incremental approach is less computationally expensive for long sequences.

The incremental training can be seen as a form of curriculum training [2] since the model is trained on gradually more difficult problems by increasing the complexity of the network. The main difference is that traditionally in the curriculum learning scenario the model is fixed while the data is changing, while in the incremental training it is the model which is changed to model the more difficult long-term dependencies.

7 Conclusion

Time series datasets with high-frequency samples provide challenging environments for learning with RNNs. The MS-LMN provides a natural solution for the problem by separating the memory state of the model into several modules, each one updated at different frequencies. Furthermore, the incremental training algorithm helps to learn long-term dependencies by incrementally adding new modules to the network. The experimental results show that the modular network can effectively learn features at different frequencies, such as modelling low-frequency and high-frequency changes in a generated sequence. Furthermore, the incremental training provides a consistent improvement to the final performance compared to traditional stochastic gradient descent.

In the future, we plan to apply the incremental training algorithm to novel architectures and learning settings and to optimize it to improve the convergence speed, for example by considering more sophisticated criteria for the dynamic addition of a module. The modular architecture and incremental learning can be also exploited in different learning settings, such as multi-task or continual learning scenarios, where the modular separation can be evinced from the data.

Acknowledgments. This work has been supported by MIUR under project SIR 2014 LIST-IT (RBSI14STDE) and by the DEEPer project, University of Padova.

References

1. Bacciu, D., Carta, A., Sperduti, A.: Linear Memory Networks. In: ICANN (2019)
2. Bengio, Y., Louradour, J., Collobert, R., Weston, J.: Curriculum learning. In: Proceedings of the 26th Annual International Conference on Machine Learning, pp. 41–48. ACM (2009)
3. Chung, J., Ahn, S., Bengio, Y.: Hierarchical multiscale recurrent neural networks. In: ICLR (2017)

4. Cummins, F., Gers, F.A., Schmidhuber, J.: Learning to forget: continual prediction with LSTM. Neural Comput. **2**, 850–855 (2000). https://doi.org/10.1197/jamia.M2577
5. Elman, J.L.: Finding structure in time. Cognit. Sci. **14**(2), 179–211 (1990)
6. Garofolo, J.S., Fisher, W.M., Fiscus, J.G., Pallett, D.S., Dahlgren, N.L.: DARPA TIMIT: Acoustic-phonetic continuous speech corpus CD-ROM, NIST speech disc 1-1.1 (1993)
7. Graves, A., Fernández, S., Gomez, F., Schmidhuber, J.: Connectionist temporal classification: labelling unsegmented sequence data with recurrent neural networks. In: Proceedings of the 23rd International Conference on Machine Learning (2006)
8. Graves, A., Liwicki, M., Bunke, H., Schmidhuber, J., Fernández, S.: Unconstrained on-line handwriting recognition with recurrent neural networks. In: Advances in Neural Information Processing Systems, pp. 577–584 (2008)
9. Graves, A., Mohamed, A.r., Hinton, G.: Speech recognition with deep recurrent neural networks. In: Department of Computer Science, University of Toronto. IEEE International Conference (3), pp. 6645–6649 (2013). https://doi.org/10.1093/ndt/gfr624
10. Hochreiter, S.: The vanishing gradient problem during learning recurrent neural nets and problem solutions. Int. J. Uncertain. Fuzz. Knowl.-Based Syst. **6**(02), 107–116 (1998)
11. Hochreiter, S., Schmidhuber, J.: Long short-term memory. Neural Comput. **9**(8), 1–32 (1997). https://doi.org/10.1144/GSL.MEM.1999.018.01.02
12. Ke, N.R., Alias Parth Goyal, A.G., Bilaniuk, O., Binas, J., Mozer, M.C., Pal, C., Bengio, Y.: Sparse attentive backtracking: temporal credit assignment through reminding. In: Bengio, S., Wallach, H., Larochelle, H., Grauman, K., Cesa-Bianchi, N., Garnett, R. (eds.) Advances in Neural Information Processing Systems, vol. 31, pp. 7640–7651. Curran Associates, Inc. (2018). http://papers.nips.cc/paper/7991-sparse-attentive-backtracking-temporal-credit-assignment-through-reminding.pdf
13. Kingma, D.P., Ba, J.: Adam: A Method for Stochastic Optimization. arXiv preprint arXiv:1412.6980 pp. 1–15 (2014). https://doi.org/10.1145/1830483.1830503
14. Koutnik, J., Greff, K., Gomez, F., Schmidhuber, J.: A clockwork RNN. In: Proceedings of the 31 St International Conference on Machine Learning, Beijing, China, vol. 32, pp. 1–9 (2014)
15. Mali, A., Ororbia, A., Giles, C.L.: The neural state pushdown automata. ArXiv abs/1909.05233 (2019)
16. Neil, D., Pfeiffer, M., Liu, S.C.: Phased LSTM: accelerating recurrent network training for long or event-based sequences. In: Advances in Neural Information Processing Systems, pp. 3882–3890 (2016)
17. van den Oord, A., et al.: WaveNet: a generative model for raw audio. arXiv:1609.03499 [cs], September 2016
18. Ororbia, A., Mali, A., Giles, C.L., Kifer, D.: Continual learning of recurrent neural networks by locally aligning distributed representations. IEEE Trans. Neural Netw. Learn. Syst. **31**, 1–12 (2020)
19. Pasa, L., Sperduti, A.: Pre-training of recurrent neural networks via linear autoencoders. Adv. Neural Inf. Process. Syst. **27**, 3572–3580 (2014)
20. Pascanu, R., Mikolov, T., Bengio, Y.: On the difficulty of training Recurrent Neural Networks. In: International Conference on Machine Learning, November 2013
21. Sperduti, A.: Exact solutions for recursive principal components analysis of sequences and trees. In: Kollias, S.D., Stafylopatis, A., Duch, W., Oja, E. (eds.) ICANN 2006, Part I. LNCS, vol. 4131, pp. 349–356. Springer, Heidelberg (2006). https://doi.org/10.1007/11840817_37

22. Sperduti, A.: Efficient computation of recursive principal component analysis for structured input. In: Kok, J.N., Koronacki, J., Mantaras, R.L., Matwin, S., Mladenič, D., Skowron, A. (eds.) ECML 2007. LNCS (LNAI), vol. 4701, pp. 335–346. Springer, Heidelberg (2007). https://doi.org/10.1007/978-3-540-74958-5_32
23. Tallec, C., Ollivier, Y.: Unbiased online recurrent optimization. In: International Conference on Learning Representations, February 2018

Flexible Recurrent Neural Networks

Anne Lambert, Françoise Le Bolzer, and François Schnitzler(✉)

InterDigital Inc., Cesson-Sévigné, France
{anne.lambert,francoise.lebolzer,francois.schnitzler}@interdigital.com

Abstract. We introduce two methods enabling recurrent neural networks (RNNs) to trade off accuracy for computational cost during the analysis of a sequence. This opens up the possibility to adapt RNNs in real time to changing computational constraints, such as when running on shared hardware with other processes or in mobile edge computing nodes. The first approach makes minimal changes to the model. Therefore, it avoids loading new parameters from slow memory. In the second approach, different models can replace one another within a sequence analysis. The latter works on more data sets. We evaluate these two approaches on permuted MNIST, adding task and a human activity recognition task. We demonstrate that changing the computational cost of a RNN with our approaches leads to sensible results. Indeed, the resulting accuracy and computational cost is typically a weighted average of the corresponding metrics of the models used. The weight of each model also increases with the number of time steps a model is used.

Keywords: Recurrent neural networks · Flexibility · Edge computing

1 Introduction

This paper introduces flexible recurrent neural networks (FRNNs): recurrent neural networks (RNNs) whose computational cost can be dynamically modified during the analysis of a sequence. RNNs are a class of deep learning architecture designed to process sequential data. They have been extensively used to analyze audio [33], video [29] or other sensor data [7]. As deep learning models are increasingly run on user devices rather than in the cloud, approaches to reduce their computational cost have recently received a lot of attention. Examples applied to RNNs include reducing computation for some inputs [2], reducing model size [23] or parallelization [4]. These diverse approaches lead to substantial savings in memory, energy and/or computational costs. However, they do not allow to adjust these costs to respond to changes in the computational environment the process runs in, which is the focus of our work.

Our work is motivated in particular by two computational environments: edge computing and shared hardware. Edge computing refers to the use of computing resources physically located close to end users, at or close to the end of the network, as opposed to a more remote cloud [19]. Several versions of edge computing have been proposed, under various names. A recent overview of the

F. Hutter et al. (Eds.): ECML PKDD 2020, LNAI 12457, pp. 694–709, 2021.
https://doi.org/10.1007/978-3-030-67658-2_40

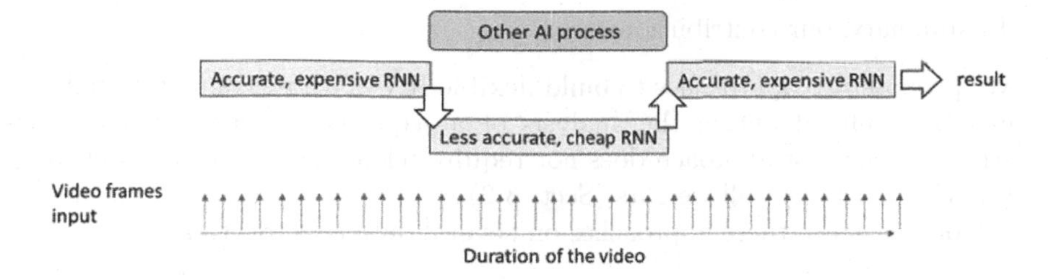

Fig. 1. Illustration of our methods: an accurate but expensive RNN analyzing an incoming video is modified during inference to be temporarily less accurate but cheaper when another process running on the same hardware limits the resources available.

interaction of deep learning and edge computing is [26]. Running applications on edge nodes rather than on cloud servers improves latency and reduce network congestion while still leveraging hardware potentially more powerful than user devices and without energy constraints. Edge nodes might not be available everywhere and, as a mobile device moves around, some computations may need to be scaled down or moved back to the mobile device. In such a scenario, computational and energy constraints are likely to change, and a RNN would need to adapt to the new constraints.

The second environment is shared hardware. With the advent of computer chips dedicated to neural networks, user devices with limited computational resources increasingly run deep learning models [6,18,27,30,31,33]. Such devices may have several processes, including deep learning ones, running at the same time. Some processes might need to free computational resources for other, more important processes. Without flexibility, it may mean stopping a recurrent model. With our approach, it may instead mean a temporary reduction in accuracy. In addition, reducing computation will also typically save energy, offering more flexibility for battery management.

To address these challenges, this paper introduces two methods to construct RNNs whose computational cost can be adjusted during the analysis of a sequence and traded off for accuracy. To the best of our knowledge, this work is the first to enable such an adaptation at runtime for RNNs. The first method uses a single model and controls the cost through a single parameter. It avoids loading a new model into memory. The second method constructs a family of models. Models can replace one another within a sequence analysis to modify the computational cost. The resulting accuracy and computational cost are typically a weighted average of the corresponding metrics of the models used. These two methods leverage and rely on existing architectures that speed-up RNN by conditional computation, such as skip-RNN [2]. We illustrate our methods with skip-RNN, but it has the potential to be used with other architectures as well. These methods are evaluated on permuted MNIST, adding task and a video-based human activity recognition task.

In summary, our contributions are:

- We propose two approaches to build flexible RNNs whose computational cost can be modified during the analysis of a sequence to meet dynamic constraints. The first approach does not require reloading the model (Sect. 3.1), the second changes all weights (Sect. 3.2).
- We demonstrate these approaches on several data sets (Sect. 4).

2 Related Work

A Recurrent Neural Network (RNN) can be seen as a function taking a sequence of inputs $\mathbf{x} = (\mathbf{x}_1, \ldots, \mathbf{x}_T)$ and recursively computing a set of states $\mathbf{s} = (\mathbf{s}_1, \ldots, \mathbf{s}_T)$. Each state vector $\mathbf{s}_t \in \mathbb{R}^m$ is computed from $\mathbf{s}_{t-1}$ and $\mathbf{x}_t \in \mathbb{R}^n$ by a cell S of the RNN. Frequently used neural architectures include long short-term memory (LSTM) networks [10] and gated recurrent units (GRU) [3]. Enabling these models to run on constrained hardware, in terms of computational power, memory or energy, has received a lot of attention lately.

Accuracy-Efficiency Trade-Offs. A few architectures have been proposed to enable trade-offs for convolutional neural networks (CNNs). Such trade-offs are achieved by varying the quantization level [6], the width of the network [30, 31] or pruning a mixture of CNNs [18]. These approaches allow CNNs running on shared hardware to be flexible. To the best of our knowledge, we are the first to propose trade-offs during inference for RNNs. In addition, we also cover transferring computation to another hardware, which is less relevant for CNNs.

Conditional Computation. A first class of methods speed up RNNs by constructing several versions of a cell S with different computational costs and selecting one version for each input, conditionally on previous inputs. The skip-RNN architecture [2] constructs a cell S that may skip processing of an input and keep the hidden state unchanged. This architecture is further described in Sect. 2.1. A recent work has also shown that an existing model can be modified to skip inputs without retraining S [22]. It is also possible to skip updates for individual variables of the hidden state [15] or to consider the current input in addition to the hidden state to decide to skip that input [32]. Approaches similar to skip-RNN were developed in [21, 28] but rather than iteratively deciding whether to skip the current input or not, the network directly selects the next input to process. Structural-Jump-LSTM [8] specializes this idea for text. When a new input word is received, the model can decide to a) update the state b) skip the word or c) jump to the next punctuation mark.

Skim-RNN [20] constructs two or more cells S that may update only part of the state. For example, one "big" S updates the whole state and a "small" S updates only half of it. VCRNN [11] is another method that updates a subset of the state. The model decides on the number d of variables to update in the hidden state and uses a subset of size $d \times d$ of a single parameter matrix to compute the updated values. In both cases, the decision to update at time t is also based on the current input $\mathbf{x}_t$.

G-LSTM [25] learns different clocks for each value in the hidden state vector. These clocks then induce soft cyclic updates on these neurons.

We also perform conditional computation. However, these works focus on improving computational cost while keeping almost the same accuracy. Furthermore, they create single models with a given accuracy/computational cost trade-off. This trade-off is not modified during the analysis of a sequence. We, on the other hand a) sacrifice accuracy to achieve higher computational savings and b) can modify the trade-off dynamically.

Small Architecture. The size of the model can have a large influence on the run-time of the algorithm. It directly impacts the number of operations per time-step. It can also create additional delay, for example if the model does not fit in the memory cache [24]. Therefore, some works have looked at developing small recurrent architectures. FastGRNN [12] is an architecture that relies on weighted residual connections and sparse, low-rank representation of parameter matrices. KPRNNs [23] are RNNs whose parameter matrices are approximated by Kronecker Products.

Parallelization. ShaRNN increases the parallelization of the computation of RNNs by using two layers [4]. The first layer operates on windows of inputs, without using previous inputs. This layer is thus parallelizable. A second layer takes as input the output of each window.

These two lines of research are complementary to our works. The common objective is to facilitate the execution of RNNs on hardware with constraints. Like conditional computation methods however, they optimize a single model with fixed performance characteristics. We, on the other hand, want to modify the behavior of the model dynamically to adapt to changing constraints.

2.1 Skip-RNN

The FRNNs we develop are based on the skip-RNN architecture [2]. In this section we take a closer look at it. Skip-RNN augments S to allow the model to skip some inputs. When an input is processed, an update gate computes a quantity $\Delta\tilde{u}_t$ that controls how many inputs will be skipped. In practice, subsequent inputs $\mathbf{x}_t$ are skipped ($u_t = 0$) and $\Delta\tilde{u}_t$ is accumulated to $\tilde{u}_t$ every time step until $\tilde{u}_t \geq 0.5$. The resulting architecture can be described as follows:

$$u_t = f_{binarize}(\tilde{u}_t) \tag{1}$$

$$\mathbf{s}_t = u_t S_t(\mathbf{s}_{t-1}, \mathbf{x}_t) + (1 - u_t)\mathbf{s}_{t-1} \tag{2}$$

$$\Delta\tilde{u}_t = \sigma(W\mathbf{s}_t + b) \tag{3}$$

$$\tilde{u}_{t+1} = u_t \Delta\tilde{u}_t + (1 - u_t)(\tilde{u}_t + \min(\Delta\tilde{u}_t, 1 - \tilde{u}_t)) \ . \tag{4}$$

In these equations, $f_{binarize}(z)$ denotes a binarization function: $f_{binarize}(z)$ is 0 if $z < 0.5$ and 1 otherwise, σ a non-linear function and W and b parameters. As $f_{binarize}$ is not differentiable, it is approximated by the straight-through estimator [1,9] during training.

The loss used to train this model contains two terms:

$$L_{skip-RNN} = L_{acc}(x, y) + \lambda L_{budget} \ . \tag{5}$$

The first term measures the accuracy of the task (for example cross-entropy for classification or Euclidian loss for regression), and the second one penalizes computational operations: $L_{budget} = \sum_t u_t$. The hyperparameter λ controls the strength of the penalty.

3 Flexible RNNs

For ease of exposition, we describe the two FRNN architectures we propose based on skip-RNN. Like skip-RNN, the methods are generic and can be applied to several RNN architectures including LSTM and GRU. The first method, ThrRNN, is described in Sect. 3.1 and is relatively simple. It modifies the architecture slightly and no retraining of a skip-RNN is necessary. Adapting the computational cost does not require reloading the model. The method is well suited for modifying the cost of the model in place. However, as shown in the experimental section, it only works on some data sets. The second method, SwitchRNN is described in Sect. 3.2. SwitchRNN does require training the models specifically to be flexible. As it creates several models, it also increases the memory footprint. This is not an issue when moving the RNN computation to other hardware, such as in edge computing.

3.1 ThrRNN

The ThrRNN architecture makes minimal change to the architecture of the original conditional computation architecture. It changes the threshold of the gate that selects the computation to be executed for a new input. Therefore, the model is almost the same as the skip-RNN model. Only $f_{binarize}$ is different. The function accepts an additional parameter, *thr*. The resulting architecture is the same as in Sect. 2.1, except for Eq. 1:

$$u_t = f_{binarize}(\tilde{u}_t, thr) = \begin{cases} 0 \text{ if } \tilde{u}_t < thr, \\ 1 \text{ otherwise} \ . \end{cases} \tag{6}$$

Varying *thr* changes the behavior of the model. When it increases, the model will skip more inputs. At the same time, the model still adapts its computation to the data. While this is a simple change, our experiments suggest that training a model with a constant value for *thr* and then modifying *thr* during inference gracefully trades off accuracy for computational cost on several data sets.

3.2 SwitchRNN

The second approach, SwitchRNN, relies on the construction of a set (or family) of RNNs $\mathcal{M} = \{M_i\}_{i \in 1,\ldots,K}$ of at least $K \geq 2$ recurrent neural networks M_i with different computational cost and accuracy but trained for the same task. This type of trade-offs can be achieved using any existing method, for example skip-RNN. Figure 6 in the experimental section illustrates such a trade-off. Typically, such a family will only contain models that belong to a Pareto frontier over computational cost and accuracy. In other words, for any $M_i \in \mathcal{M}$, there is no model $M_k \in \mathcal{M}$ that has both a better accuracy and lower computational cost than M_i.

Modifying the trade-off is then as simple as changing the model. This is trivial when the model can be changed between sequences. However, we propose to build a family of models $\mathcal{M}$ that can be switched during the analysis of a sequence. This opens up the possibility to update the computational cost of a RNN for long or infinite sequences. Such a family $\mathcal{M}$ can be used as follows.

- An initial model $M_1 \in \mathcal{M}$ matching the available computational resources is selected to begin the analysis.
- Inputs are processed as they arrive.
- When the computational resources available to the process change, another model $M_2 \in \mathcal{M}$ of the family may be selected. The hidden state of M_1 is transferred to M_2. Input processing can then continue.

This idea is illustrated in Fig. 1.

Training a Family of Switchable Models. As using unrelated models does not work, this method is designed to train K models $\mathcal{M} = \{M_i\}_{i \in 1,\ldots,K}$ whose internal states are similar. These models are trained one by one, iteratively from the previous model in the family. We first discuss how to build a family of $K = 2$ models and then how to expand it. The intuition behind this method is to learn two different RNN models that a) have a different accuracy/computational cost trade-off while b) forcing these models to compute similar internal states for the same input sequence. This is achieved as follows:

1. Train one model M_1 with a given accuracy/computational cost trade-off by parameterizing its loss function $L_{acc}(x, y) + \lambda_1 L_{budget}$, where the value of the hyperparameter λ_1 is selected to achieve a desirable trade-off.
2. Train a second model M_2 with the following loss:

$$L_{SwitchRNN} = L_{acc}(x, y) + \lambda_2 L_{budget} + \mu L_{align}(\mathbf{s}^1, \mathbf{s}^2) \ . \quad (7)$$

In this loss, L_{align} is chosen to enforce that the internal states $\mathbf{s}^2$ computed by the model M_2 have values close to the internal states $\mathbf{s}^1$ computed by M_1 on any input sequence $\mathbf{x}$. Examples include the l1 or l2 losses:

$$L_{align}(\mathbf{s}^1, \mathbf{s}^2) = \sum_{t=1}^{T} |\mathbf{s}_t^1 - \mathbf{s}_t^2| \quad (8)$$

or

$$L_{align}(\mathbf{s}^1, \mathbf{s}^2) = \sum_{t=1}^{T} (\mathbf{s}_t^1 - \mathbf{s}_t^2)^2 \ . \quad (9)$$

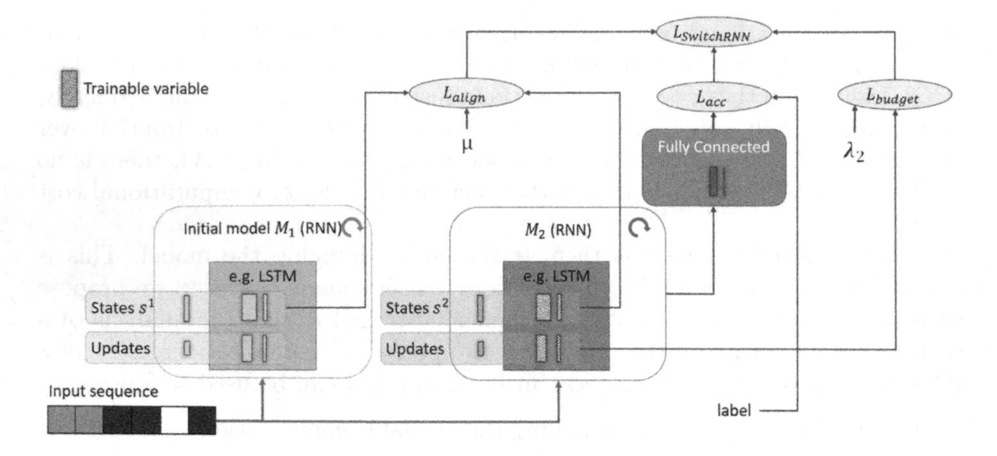

Fig. 2. Training procedure for SwitchRNN. Only the second model is trained.

In these losses, μ is an additional hyperparameter that controls the importance of the alignment term. The value of λ_2 must be chosen to enforce a different accuracy/computational cost trade-off than the model M_1.

The family can be further extended by iteratively training an additional model M_i from model M_{i-1} or any combination of models previously trained.

Implementation Details. An illustration of the computational graph for training M_2 is displayed in Fig. 2. Here only M_2 is trained: the weights of the fully connected layer are fixed to the ones found during training of M_1.

Training often seems to work better when model M_2 is allowed a larger computational budget than M_1. We conjecture that this is because optimizing for an additional objective (state alignment) is harder. Hence, relaxing the constraint on the computational cost rather than tightening it makes learning easier. We used the l2 loss as L_{align} in our experiments.

4 Experiments

We demonstrate that our approaches can adapt the computational cost of a RNN during inference and that the resulting accuracy and computational cost is typically a weighted average of the corresponding metrics of the adapted models used. This means that using a faster RNN to free up computational resources for a short time relative to the length of a sequence will not have a large impact on the accuracy of the result. We report here our results on three data sets:

- **Permuted MNIST** (pMNIST): a fixed random permutation is applied to the pixels of each MNIST image [14] and the result is presented sequentially (one pixel at a time) to the recurrent network. The goal is to identify the digit in the image. This task is thought to be useful for testing long-range

dependencies in RNN models [13]. For this task, we trained a 110 cell single-layer RNN with batches of 256 images of length 784 pixels.
- **Human Activity Recognition** (HAR): The HAR dataset consists of 2D-pose time series extracted from videos. The videos belong to a subset of the Berkeley MHAD (multimodal Human Activity database) [17] recorded on 12 subjects with stationary cameras. Six actions are considered. There is a total of 1438 videos. The 2D-poses are vectors of 36. A pose is extracted from each frame and estimated by 18 body joints expressed in a 2D coordinate system. Each action must then be identified based on a sequence of 32 frames. Each video gives several sequences by applying an overlapping ratio. Following the parameter setting defined in [5], the dataset is split into 2 independent partitions: 22625 sequences for training and 5751 for validation. The training was performed with batches of 4096. The architecture consists in a fully connected layer and 2 RNN layers. Each of the 2 layers contains 40 cells. The output layer is a softmax classifier with 6 cells.
- **Adding Task**: This dataset consists of synthetic data obtained from the adding problem, commonly used to evaluate the performance of RNN models [2,10,16]. The input data is a sequence of (value, marker) tuples where the value elements are uniformly sampled in the range $(-0.5, 0.5)$ and where the marker elements are equal to 1 or 0. The task output is the sum of all values with a marker element equal to 1. To generate the dataset, we reproduced the experimental setup defined in [16]. Only two values were marked for addition. The first marker was randomly placed among the first 10% of the sequence (drawn with uniform probability). Similarly, the second marker was placed in the last half of the sequence.

We detail the results on the first two tasks but only summarize those on Adding Task, as they are similar to HAR.

For each data set, we evaluate our methods when adapting the computational cost of a RNN cell during the analysis of the sequence. This is the main focus of this work as it offers maximum flexibility. We also analyze model adaptation between sequences for ThrRNN. This setting is still interesting as ThrRNN allows adapting the model in memory. For SwitchRNN, this is not the case. Models that are switched between sequences could be any RNN as state coherence is not necessary. So, between sequences, SwitchRNN has no benefit over skip-RNN. We use skip-RNN in that context as a baseline.

We evaluate our approach using skip-RNN [2] as the conditional computation RNN approach. We experimented with LSTM and GRU cells on pMNIST. Their behavior was very similar, so we report results for LSTM only. Computational complexity is quantified by the average number of inputs that are not skipped.

4.1 Permuted MNIST

Flexibility Between Sequence Analysis. We first show that *ThrRNN*, while very easy to use, does not work well on all data sets. Figure 3 illustrates the typical behavior for ThrRNN on pMNIST. Minimal changes in *thr* in both

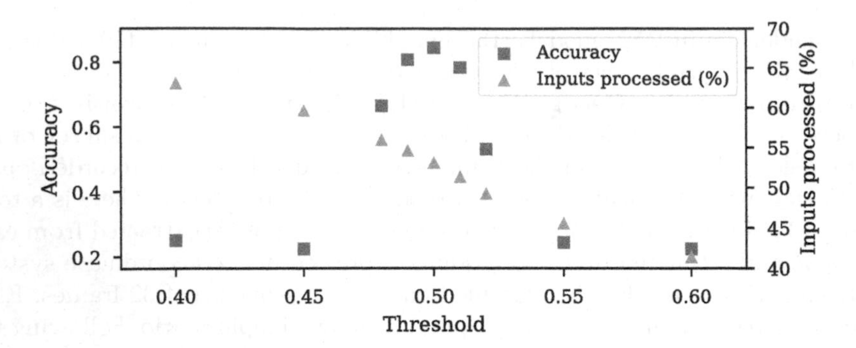

Fig. 3. Examples of accuracy/number of updates trade-offs for one ThrRNN on pMNIST. The model is a ThrLSTM model learned with $\lambda = 5e{-}4$ and $thr = 0.5$.

directions cause a sharp decrease in accuracy. Computational complexity on the other hand changes slowly. Therefore, decreasing thr does not improve the model as it becomes both worse in accuracy and computational cost. Increasing thr still leads to models that could be useful as they have worse accuracy but lower computational cost. However, the decrease in accuracy is so rapid that such a model is probably not worth the small gain in computation. We conjecture this is due to the fact that ThrRNN does not handle long-range relationships well.

It is nevertheless possible to obtain different models that exhibit an interesting set of different trade-offs on this challenging data set. Figure 6 illustrates such a set of trade-offs for a skip-RNN model using Long Short-Term Memory (LSTM) architecture for each cell. As the value of λ is changed, the models achieve different behaviors: the number of updates is reduced, but so is accuracy. The latter is at first reduced only slightly but gradually the gap widens and, when the model is constrained to update its internal states only 5% of the time, the accuracy goes down to 77%. The comparison to SwitchRNN will be discussed later.

Flexibility During Sequence Analysis. SwitchRNN can generate models similar to these shown in the previous paragraph and that, in addition, can be switched during the analysis of a sequence. Furthermore, the resulting accuracy and computational cost tend to be a weighted average of the corresponding metrics for the models used. Figure 4 and 5 provide examples of this behavior.

Figure 4 displays the impact of one model switch during the analysis of a sequence. The average accuracy and computational cost of the analysis of a sequence in the test set are reported as a function of the position of the switch from M_1 to M_2 during the analysis of the sequence. Hence, $x = 0$ and $x = 784$ correspond to the performance of individual models M_2 and M_1 respectively. Values $0 < x < 784$ correspond to a change from M_1 to M_2 at position x. As $\lambda_2 > \lambda_1$, M_2 has a worse accuracy but a lower computational cost than M_1, when these models analyze the sequence individually. When the model is switched during the analysis ($0 < x < 784$), the resulting accuracy and computational

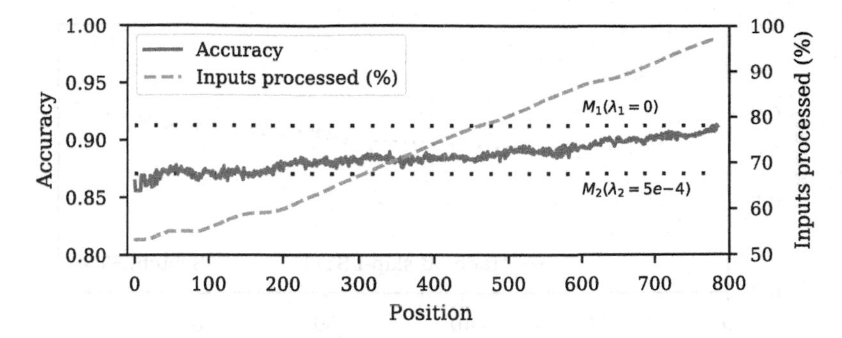

Fig. 4. Average accuracy and number of updated states for SwitchLSTM ($\mu = 1$) on pMNIST (y axis) obtained when replacing model M_1 ($\lambda_1 = 0$) by another model M_2 ($\lambda_2 = 5e-4$) at various points (x axis) during the analysis of a sequence of length 784. Dotted lines signal the accuracy of individual models.

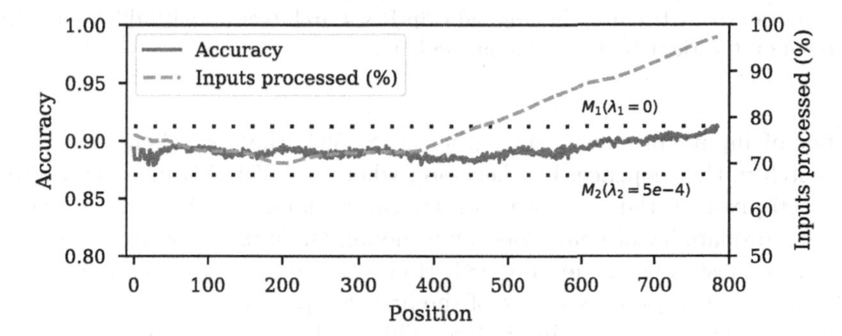

Fig. 5. Average accuracy and number of updated states for SwitchLSTM ($\mu = 1$) on pMNIST when running M_1 ($\lambda_1 = 0$) up to x, using M_2 ($\lambda_2 = 5e-4$) up to $x + 400$ then switching back to M_1. Dotted lines signal the accuracy of individual models.

cost increase almost linearly from the values of M_2 to the values of M_1. This suggests that *SwitchRNN* can be used in practice to adapt the computational cost of a RNN.

More than one switch can also be made during the analysis of a single sequence, as illustrated in Fig. 5 using the same models as in Fig. 4. Figure 5 shows the average accuracy and computational cost of the analysis of the sequence, reported as a function of the position in the sequence of the first switch. The second model switch was performed $\Delta = 400$ inputs after the first one. No second switch took place if the end of the sequence was reached before. Results with $\Delta = 100$, 200 and 300 were similar. Hence, $x = 784$ corresponds to the performance of model M_1. Values $0 \leq x < 784$ correspond to a change from M_1 to M_2 at position x and back to M_1 at input $t = x + \Delta$, provided $t < 784$.

The first point to take from this figure is that the accuracy is typically between the accuracy of both models. Accuracy also rises towards the end when a larger part of the sequence is analyzed by M_1. The second point is that the

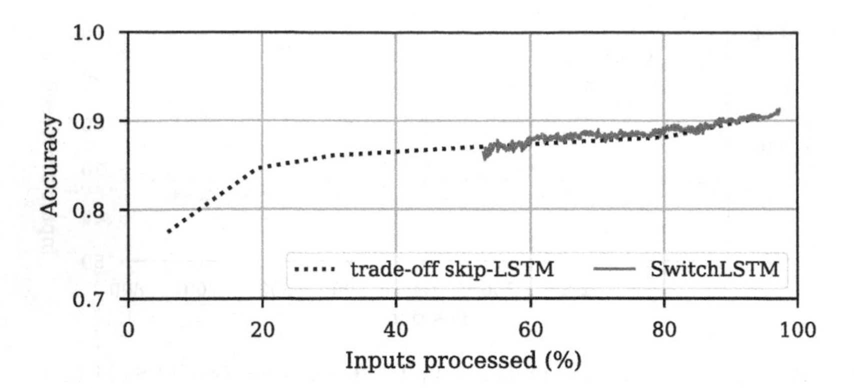

Fig. 6. Average accuracy as a function of the number of updated states for Switch-LSTM ($\mu = 1$) on pMNIST when switching from M_1 ($\lambda_1 = 0$) to M_2 ($\lambda_2 = 5e{-}4$) (same results as Fig. 4). The dotted line corresponds to the performance of individual models obtained by training the original skip-RNN architecture with different λ (from $\lambda = 1e{-}4$ on the right to $\lambda = 3e{-}3$ on the left).

number of inputs processed rises when $x > 384$, that is, when there is only one switch in the sequence. Furthermore, when $x \leq 384$, M_2 may choose to see up to 400 pixels of the sequence and M_1 the remaining 384. If every pixel had the same probability of being used by a model, the number of inputs processed would be almost constant for $x < 384$. However, this is not the case for M_2: the probability that M_2 processes one of the first 200 pixels is close to 65%, whereas for the next 400 this probability is 45%. This explains why we find the minimum number of inputs processed around the position 200.

Finally, Fig. 6 allows an easier comparison between skip-LSTM and Switch-LSTM with one switch in the sequence. It can be seen that the latter can match the performance of the former even when dynamically updating the trade-off during inference.

4.2 Activity Recognition

Flexibility Between Sequence Analysis. This set of experiments evaluates our methods on a more realistic problem. The first observation is that ThrRNN works well on this data set. Figure 7 displays the accuracy/computational cost trade-off of a ThrLSTM model on HAR. Using a model trained with $thr = 0.5$, increasing thr leads to a slow decrease in accuracy and large gains in computational complexity. As opposed to the results on pMNIST, this diminution is progressive on HAR, although it accelerates when the number of frames gets really low ($\lambda = 0.1$). Therefore, this architecture is interesting on HAR.

Trade-offs achieved using ThrLSTM with $\lambda = 0.01$ in Fig. 7 are similar to these achieved by skip-LSTM. In particular, ThrLSTM trained with $\lambda = 0.01$ and run with $thr = 0.8$ is very close in performance to skip-LSTM trained with

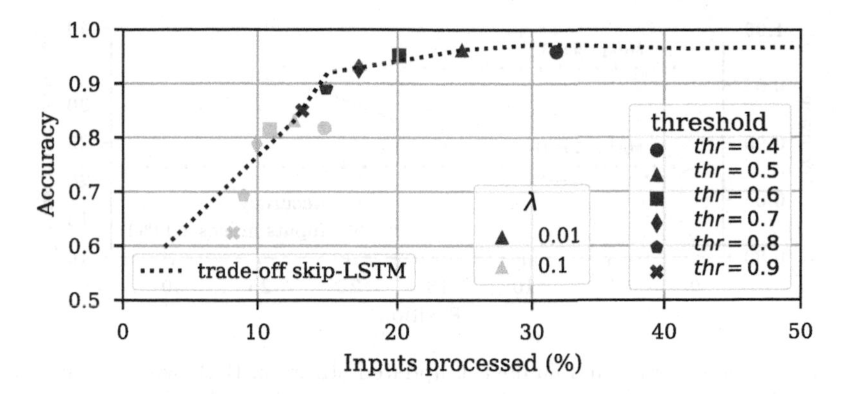

Fig. 7. Example of accuracy/number of updates trade-offs on HAR for two ThrLSTMs trained with $thr = 0.5$. The dotted line corresponds to the performance of individual models obtained by training the original skip-RNN architecture with different λ (from $\lambda = 3e{-}3$ on the right to $\lambda = 5e{-}1$ on the left). Increasing thr at inference leads to a decrease in accuracy and computational cost.

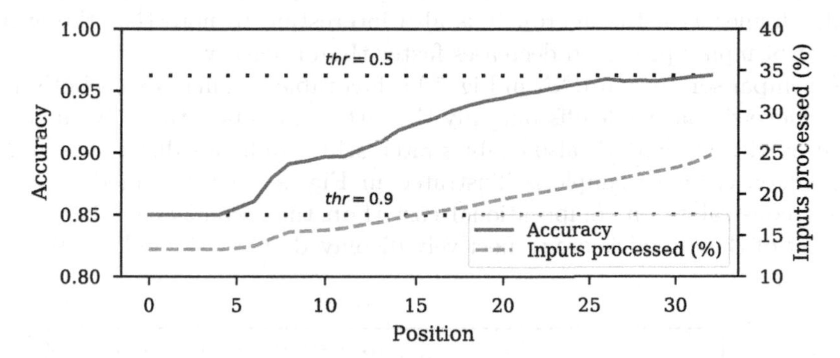

Fig. 8. Average accuracy and number of updated states on HAR with ThrLSTM trained with $thr = 0.5$ when using $thr = 0.5$ up to x and then $thr = 0.9$. The model was trained with $\lambda = 0.01$. Dotted lines indicate the accuracy when keeping thr constant.

$\lambda = 0.1$. This suggests that ThrRNN yields on this data set trade-offs of a similar quality to skip-RNN.

Flexibility During Sequence Analysis. On HAR, the behavior displayed by ThrRNN during inference is similar to what was observed on pMNIST with SwitchRNN. Figure 8 provides an example of this behavior in terms of accuracy/computational cost trade-offs when modifying thr once during the analysis of a sequence. As in previous figures, these metrics are provided as a function of the index of the input where thr was modified. In this example, thr is low (0.5) in the first model and higher (0.9) in the second model. Therefore, accuracy and computational costs are low for low values of x in the figure, as the model with $thr = 0.9$ is processing most of the sequence for these values. The accuracy

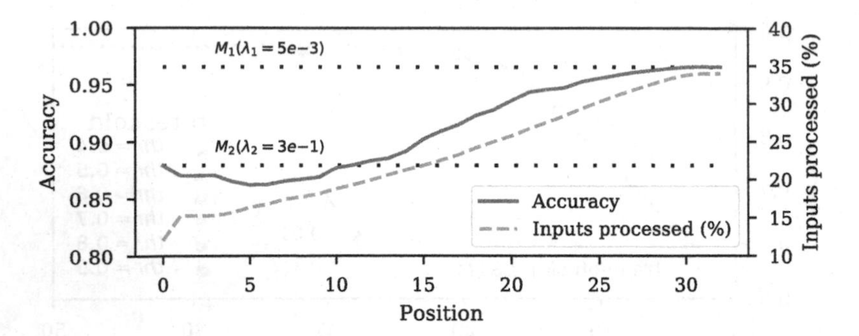

Fig. 9. Average accuracy and number of updated states on HAR with a SwitchLSTM ($\mu = 1e-4$) when using M_1 ($\lambda_1 = 5e-3$) up to x and then M_2 ($\lambda_2 = 3e-1$). Dotted lines correspond to the accuracy of individual models.

resulting from a change in *thr* is between the accuracy of individual models and gradually increases as a model with a lower *thr* processes a larger number of inputs (higher x in the figure). It is also interesting to note that the average number of inputs processed decreases faster than accuracy.

A comparison to skip-RNN in Fig. 10 is favorable to ThrRNN, as both methods achieve similar trade-offs but only the latter can adapt these dynamically.

On HAR, SwitchRNN also enables interesting trade-offs during the analysis of a sequence. One example is illustrated in Fig. 9. The first model used has weaker constraints on computational cost than the second one. The accuracy and cost of these models are respectively displayed at $x = 32$ and $x = 0$. In this

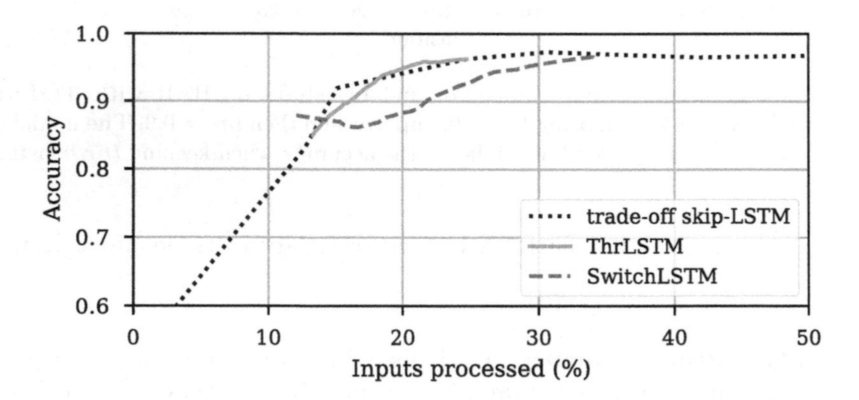

Fig. 10. Average accuracy as a function of the number of updated states on HAR for SwitchLSTM ($\mu = 1e-4$) when switching from M_1 ($\lambda_1 = 5e-3$) to M_2 ($\lambda_2 = 3e-1$) (same results as Fig. 9) and for ThrLSTM trained with $thr = 0.5$ and $\lambda = 0.01$ when switching from $thr = 0.5$ to $thr = 0.9$ (same results as Fig. 8). The dotted line corresponds to the performance of individual models obtained by training the original skip-RNN architecture with different λ (from $\lambda = 3e-3$ on the right to $\lambda = 5e-1$ on the left).

example, the accuracy resulting from the switch can be lower than the accuracy of the most computationally efficient model (M2). That being said, the largest drop in accuracy is only 1.7% below M2. For the majority of the sequence, switching the models results in an accuracy that is between the accuracy of the two models. Risking a lower accuracy if the switch takes place at a bad time might be acceptable in exchange for the flexibility afforded by our method.

In Fig. 10, SwitchLSTM achieves trade-offs that are a bit worse than ThrRNN and skip-LSTM. This is to be expected considering that switching models led to a slightly worse accuracy for some of the sequence on this data set.

4.3 Adding Task

Results obtained on Adding Task were mostly similar to the previous section. Both ThrRNN and SwitchRNN achieve interesting trade-offs. When modifying accuracy and computational cost during the analysis of a sequence, the average of these metrics on the whole sequence lies between the values of these metrics for the configurations used. However, these metrics change faster as a function of the position of the model modification than on pMNIST or HAR. We believe this is due to the nature of the task: the only important elements of the sequence are the two pairs where the marker is one.

5 Conclusion

We have proposed the first two methods to trade off accuracy for computational cost during the analysis of a sequence for recurrent neural networks. The first one only modifies one parameter of a model, hence minimizing time spent loading a new model. The parameter could also be provided as input to the model. This method is therefore also interesting to adapt computational cost between sequences. It is well suited to running in a shared environment with other processes. The second method replaces all parameters of a model. It works on a broader set of tasks than the first method but requires either reloading the model or keeping two models in the working memory. This method is in particular interesting when a process moves between computational environments, such as in edge computing where it follows a user.

Both methods can gracefully trade off accuracy and computational cost during the analysis of a sequence, although computational cost is reduced only in expectation. Furthermore, the resulting accuracy is typically better than the accuracy obtained by running the most computationally efficient model on its own. Therefore, our methods can also be used to leverage temporarily available computational resources.

Our approaches rely on and are complementary to conditional computation methods to speed-up RNNs, such as skip-RNN, which we have used in this work. In our opinion skip-RNN is particularly interesting as skipping an input also skips any preprocessing, such as applying a CNN on an image. Whether another conditional computation method would work better is an open question.

References

1. Bengio, Y., Léonard, N., Courville, A.: Estimating or propagating gradients through stochastic neurons for conditional computation. arXiv preprint arXiv:1308.3432 (2013)
2. Campos, V., Jou, B., i Nieto, X.G., Torres, J., Chang, S.F.: Skip RNN: learning to skip state updates in recurrent neural networks. In: International Conference on Learning Representations (2018)
3. Cho, K., et al.: Learning phrase representations using RNN encoder-decoder for statistical machine translation. In: Proceedings of the 2014 Conference on Empirical Methods in Natural Language Processing (EMNLP), pp. 1724–1734 (2014)
4. Dennis, D., et al.: Shallow RNN: accurate time-series classification on resource constrained devices. In: Advances in Neural Information Processing Systems, pp. 12896–12906 (2019)
5. Eiffert, S.: RNN for human activity recognition - 2D pose input. https://github.com/stuarteiffert/RNN-for-Human-Activity-Recognition-using-2D-Pose-Input
6. Guerra, L., Zhuang, B., Reid, I., Drummond, T.: Switchable precision neural networks. arXiv preprint arXiv:2002.02815 (2020)
7. Hammerla, N.Y., Halloran, S., Plötz, T.: Deep, convolutional, and recurrent models for human activity recognition using wearables. In: Proceedings of the Twenty-Fifth International Joint Conference on Artificial Intelligence, pp. 1533–1540 (2016)
8. Hansen, C., Hansen, C., Alstrup, S., Simonsen, J.G., Lioma, C.: Neural speed reading with structural-jump-LSTM. In: International Conference on Learning Representations (2019)
9. Hinton, G.: Neural networks for machine learning. Coursera Video Lect. **264**(1) (2012)
10. Hochreiter, S., Schmidhuber, J.: Long short-term memory. Neural Comput. **9**(8), 1735–1780 (1997)
11. Jernite, Y., Grave, E., Joulin, A., Mikolov, T.: Variable computation in recurrent neural networks. In: International Conference on Learning Representations (2017)
12. Kusupati, A., Singh, M., Bhatia, K., Kumar, A., Jain, P., Varma, M.: FastGRNN: a fast, accurate, stable and tiny kilobyte sized gated recurrent neural network. In: Advances in Neural Information Processing Systems, pp. 9017–9028 (2018)
13. Le, Q.V., Jaitly, N., Hinton, G.E.: A simple way to initialize recurrent networks of rectified linear units. arXiv preprint arXiv:1504.00941 (2015)
14. LeCun, Y., Bottou, L., Bengio, Y., Haffner, P.: Gradient-based learning applied to document recognition. Proc. IEEE **86**(11), 2278–2324 (1998)
15. Li, Y., Yu, T., Li, B.: Recognizing video events with varying rhythms. arXiv preprint arXiv:2001.05060 (2020)
16. Neil, D., Pfeiffer, M., Liu, S.C.: Phased LSTM: accelerating recurrent network training for long or event-based sequences. In: Advances in Neural Information Processing Systems, pp. 3882–3890 (2016)
17. Ofli, F., Chaudhry, R., Kurillo, G., Vidal, R., Bajcsy, R.: Berkeley MHAD: a comprehensive multimodal human action database. In: 2013 IEEE Workshop on Applications of Computer Vision (WACV), pp. 53–60. IEEE (2013)
18. Ruiz, A., Verbeek, J.J.: Adaptative inference cost with convolutional neural mixture models. In: 2019 IEEE/CVF International Conference on Computer Vision (ICCV), pp. 1872–1881 (2019)
19. Satyanarayanan, M., Bahl, P., Caceres, R., Davies, N.: The case for VM-based cloudlets in mobile computing. IEEE Pervasive Comput. **8**(4), 14–23 (2009)

20. Seo, M., Min, S., Farhadi, A., Hajishirzi, H.: Neural speed reading via skim-RNN. In: International Conference on Learning Representations (2018)
21. Song, I., Chung, J., Kim, T., Bengio, Y.: Dynamic frame skipping for fast speech recognition in recurrent neural network based acoustic models. In: 2018 IEEE International Conference on Acoustics, Speech and Signal Processing (ICASSP), pp. 4984–4988. IEEE (2018)
22. Tao, J., Thakker, U., Dasika, G., Beu, J.: Skipping RNN state updates without retraining the original model. In: Proceedings of the 1st Workshop on Machine Learning on Edge in Sensor Systems, pp. 31–36 (2019)
23. Thakker, U., et al.: Compressing RNNs for IoT devices by 15–38x using kronecker products. arXiv preprint arXiv:1906.02876 (2019)
24. Thakker, U., Dasika, G., Beu, J., Mattina, M.: Measuring scheduling efficiency of RNNs for NLP applications. arXiv preprint arXiv:1904.03302 (2019)
25. Thornton, M., Anumula, J., Liu, S.C.: Reducing state updates via gaussian-gated LSTMs. arXiv preprint arXiv:1901.07334 (2019)
26. Wang, X., Han, Y., Leung, V.C., Niyato, D., Yan, X., Chen, X.: Convergence of edge computing and deep learning: a comprehensive survey. IEEE Commun. Surv. Tutor. **22**(2), 869–904 (2020)
27. Wu, C.J., et al.: Machine learning at Facebook: understanding inference at the edge. In: 2019 IEEE International Symposium on High Performance Computer Architecture (HPCA), pp. 331–344. IEEE (2019)
28. Yu, A.W., Lee, H., Le, Q.: Learning to skim text. In: Proceedings of the 55th Annual Meeting of the Association for Computational Linguistics (vol. 1: Long Papers), Vancouver, Canada, pp. 1880–1890, July 2017
29. Yu, H., Wang, J., Huang, Z., Yang, Y., Xu, W.: Video paragraph captioning using hierarchical recurrent neural networks. In: Proceedings of the IEEE Conference on Computer Vision and Pattern Recognition, pp. 4584–4593 (2016)
30. Yu, J., Huang, T.S.: Universally slimmable networks and improved training techniques. In: Proceedings of the IEEE International Conference on Computer Vision, pp. 1803–1811 (2019)
31. Yu, J., Yang, L., Xu, N., Yang, J., Huang, T.: Slimmable neural networks. In: International Conference on Learning Representations (2019)
32. Zhang, S., Loweimi, E., Xu, Y., Bell, P., Renals, S.: Trainable dynamic subsampling for end-to-end speech recognition. In: Proceedings of INTERSPEECH 2019, pp. 1413–1417 (2019)
33. Zhang, Y., Suda, N., Lai, L., Chandra, V.: Hello edge: keyword spotting on microcontrollers. CoRR abs/1711.07128 (2017)

Z-Embedding: A Spectral Representation of Event Intervals for Efficient Clustering and Classification

Zed Lee[1,2](✉), Šarūnas Girdzijauskas[2], and Panagiotis Papapetrou[1]

[1] DSV, Stockholm University, Stockholm, Sweden
{zed.lee,panagiotis}@dsv.su.se
[2] EECS, KTH Royal Institute of Technology, Stockholm, Sweden
{zed,sarunasg}@kth.se

Abstract. Sequences of event intervals occur in several application domains, while their inherent complexity hinders scalable solutions to tasks such as clustering and classification. In this paper, we propose a novel spectral embedding representation of event interval sequences that relies on bipartite graphs. More concretely, each event interval sequence is represented by a bipartite graph by following three main steps: (1) creating a hash table that can quickly convert a collection of event interval sequences into a bipartite graph representation, (2) creating and regularizing a bi-adjacency matrix corresponding to the bipartite graph, (3) defining a spectral embedding mapping on the bi-adjacency matrix. In addition, we show that substantial improvements can be achieved with regard to classification performance through pruning parameters that capture the nature of the relations formed by the event intervals. We demonstrate through extensive experimental evaluation on five real-world datasets that our approach can obtain runtime speedups of up to two orders of magnitude compared to other state-of-the-art methods and similar or better clustering and classification performance.

Keywords: Event intervals · Bipartite graph · Spectral embedding · Clustering · Classification

1 Introduction

The problem of sequential pattern mining has been studied in many application areas, and the main goal is to extract frequent patterns from event sequences

This work was partly supported by the VR-2016-03372 Swedish Research Council Starting Grant, as well as the EXTREME project funded by the Digital Futures framework.

Electronic supplementary material The online version of this chapter (https://doi.org/10.1007/978-3-030-67658-2_41) contains supplementary material, which is available to authorized users.

F. Hutter et al. (Eds.): ECML PKDD 2020, LNAI 12457, pp. 710–726, 2021.
https://doi.org/10.1007/978-3-030-67658-2_41

[10,22] or cluster instances that are characterized by similar patterns [13]. However, a major limitation of traditional sequential pattern mining algorithms is their assumption that events occur instantaneously. As a result, they fail to capture temporal relations that may occur between events, in application areas where events have a time duration. To overcome this limitation, a wide body of research has been developed where the notion of event is extended to that of event interval. The advantage of this representation is that it can model event durations and hence the relationships between different event intervals based on their relative time positions, providing further insights into the nature of the underlying events. A wide range of application areas employ event interval representations, including including medicine [18], geoinformatics [14], or sign language [11].

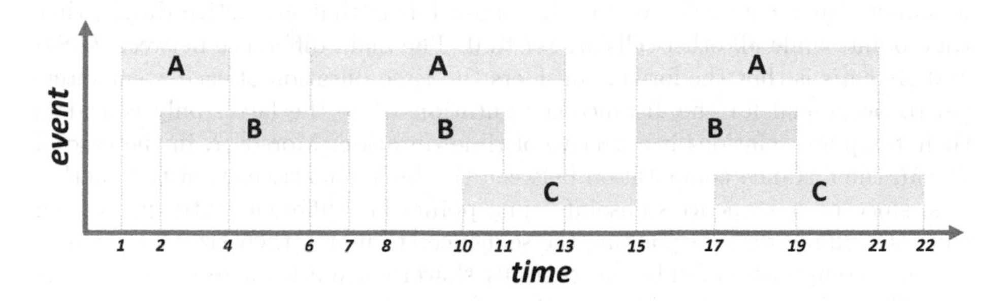

Fig. 1. Example of an sequence of eight event intervals of three event labels (A, B, and C). The time-span of the sequence is 22 time points.

Example. Figure 1 represents a sequence of nine event intervals defined over an alphabet of three labels, A, B, and C. Each event interval is characterized by a specific label, as well as its start and end time points. Over time, the same event label can appear multiple times, and a different type of temporal relation can occur between each pair of event labels.

A multi-set of event intervals arranged in chronological order constitutes an *event interval sequence*. Relative relations between many events that occur within a sequence can lead to various forms of temporal compositions, e.g., by considering the temporal relations between the events using Allen's temporal logic [1]. In this setting, one challenging problem is finding robust and computationally cheap distance or similarity functions that can effectively capture the commonalities between pairs of such complex sequences. Such functions can then be used to provide scalable solutions to the problems of clustering and classification.

1.1 Related Work

Early research on classification and clustering of sequences of temporal intervals has been focusing mostly on defining proper distance functions between the sequences. One such distance metric is *Artemis* [7], which calculates the distance between two event interval sequences by computing the ratio of temporal relations of specific pairs that both sequences have in common. In this case, the absolute time duration of each sequence is ignored, which makes *Artemis* oblivious to the length of the sequences. However, this method requires a substantial amount of computation time for checking all temporal relations occurring before and after each and every event interval in a sequence. The application of this metric to k-NN classification demonstrated promising predictive performance.

An alternative measure, called *IBSM* [8], calculates the distance between two event interval sequences by constructing a binary matrix per sequence that is used to monitor the active event labels at each time point without explicitly considering any temporal relations between the events. Each time point is represented by a binary vector of size equal to the number of event labels in the alphabet, having its cells set to 1 for those labels that are active during that time point, while all other cells are set to 0. The main difference between *IBSM* and *Artemis* is that the former considers the time duration of each event interval to be crucial for the distance computation while the latter only considers their temporal relations irrespective of time duration. Moreover, in the case of *IBSM*, the distance computation between the binary matrices is fast. Nonetheless, since each sequence's absolute time points are different, extra processing time is required for interpolating the sequences to match their size. As a result, the time computation can be substantially slower when it involves sequence pairs with the highly disproportional number of event labels and time durations.

Recently, *STIFE* [2] has been proposed for classifying sequences of event intervals by using a combination of static and temporal features extracted from the sequences. The temporal features include pairs of event intervals that can achieve high class-separation (e.g., with respect to information gain). Nonetheless, the feature extraction time can make the algorithm even slower than *IBSM*. Furthermore, the extracted features are calibrated for feature-based classification and cannot be directly applied to other tasks, such as clustering.

Finally, the connection between graphs and temporal intervals is demonstrated by converting dynamic graphs to event intervals [6]. In this paper, we demonstrate that our proposed bipartite graph representation can be used to define a feature space at a substantially lower computational cost by using *graph spectral embeddings*, hence addressing the aforementioned scalability deficiencies of the three main competitors, i.e., *ARTEMIS*, *IBSM*, and *STIFE*. Graph spectral embeddings constitute a common clustering technique in graph mining for capturing community structure in graphs [12,20,21]. For bipartite graphs, bi-spectral clustering is introduced to speed up the embedding process using the bi-adjacency matrix, which removes the space for edges between instances in the same set from adjacency matrix [9], and its variants have been introduced on stochastic block model [24]. In recent years, the technique of regularizing this affinity matrix has been actively studied [3,15] and shown to work well in terms of eigenvector perturbation on stochastic block model [5], conductance [23], and sensitivity to outliers [4]. This embedding space of an affinity matrix can also be used as a feature space for classification, showing better performance than previous distance metrics [17].

1.2 Contributions

In this paper, we propose a time-efficient approach for mapping event interval sequences to the spectral feature space of a bipartite graph representing the original sequences. We additionally introduce space pruning techniques for achieving further speedups for both classification and clustering. The main contributions of this paper are summarized as follows:

- **Novelty.** We propose a novel three-step framework for representing sequences of event intervals by (1) constructing a search-efficient data structure (`G-HashTable`), (2) mapping the e-sequences to bipartite graphs, and (3) exploiting the bipartite graph representation to eventually map them to their corresponding spectral graph embedding space (`Z-Embedding`);
- **Efficiency.** The proposed three-step mapping results in a new feature space constructed with substantially less computation time (up to a factor of 200) compared to existing state-of-the-art representations;
- **Flexibility.** the proposed representation additionally exploits vertical and horizontal supports to represent the nature of the interval data space better, as well as three pruning parameters {`maxSup`, `minSup`, `gap`} for adding flexibility to the targeted features;
- **Clustering performance.** We demonstrate that the proposed representation can achieve higher clustering purity values than earlier methods for clustering event interval sequences on five real-world datasets;
- **Classification performance.** Using the same datasets, we demonstrate that the proposed feature space can achieve comparable classification performance against state-of-the-art classification methods for event interval sequences while maintaining substantially low computation time.

2 Background

Let $\Sigma = \{e_1, \ldots, e_m\}$ be a set of m event labels. An event that occurs during a specific time duration is called *event interval.* A set of event intervals for the same entity (e.g., a patient in a medical dataset) forms an *event-interval sequence* or *e-sequence*. Next, we provide more formal definitions for these two concepts.

Definition 1 (event interval). *An event interval $s = (e, t_s, t_e)$ is defined as a tuple of three elements, where $s.e \in \Sigma$ and $s.t_s$, $s.t_e$ define the start and end time of the interval, with $s.t_s \leq s.t_e$.*

In the special case where $s.t_s = s.t_e$, the event interval is *instantaneous.*

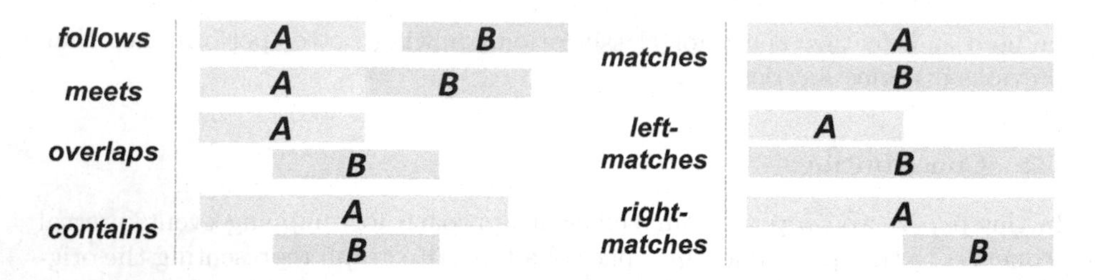

Fig. 2. Seven temporal relations that two event intervals can have, as defined in Allen's temporal logic [1].

Definition 2 (e-sequence). *An e-sequence* $\mathcal{S}=\{s_1,\ldots,s_n\}$ *is a collection of event intervals. Event intervals in an e-sequence are sorted chronologically and can contain the same event label multiple times. More concretely, they first follow an ascending order based on their start time. If the start times are the same, the end times are arranged in ascending order. If the end times are also the same, they follow the order in which event labels are lexicographically sorted.*

We consider seven temporal relations (see Fig. 2), defined in the following set, $\mathcal{I}$ = {*follows, meets, overlaps, matches, contains, left-matches, right-matches*}. Formally, a *temporal relation* $\mathcal{R}$ between two event intervals s_a and s_b, with $s_a.e = e_i, s_b.e = e_j$, is defined as a triplet $<e_i, e_j, r>$, with $r \in \mathcal{I}$. In [1], the number of relations is 13, including six inverse relations, but it can be reduced to seven by forcing the order by start time and removing six inverse relations except for *matches*, which does not have an inverse form. In several applications, we may not be interested in the absolute time values of event intervals but rather in the temporal relations between them. Hence, a simplified representation may be used.

Definition 3 (vertical support). *Given an e-sequence* $\mathcal{S} = \{s_1,\ldots,s_n\}$ *and a temporal relation* $\mathcal{R} =< e_i, e_j, r >$ *defined by event labels* $e_i, e_j \in \Sigma$, *with* $r \in \mathcal{I}$, *the* vertical support *of* $\mathcal{R}$ *is defined as the number of e-sequences where event labels* e_i, e_j *occur with relation* r.

While there can be multiple occurrences of $\mathcal{R}$ in the same e-sequence, the relation is counted only once.

Let function $occ(\cdot)$ indicate a single occurrence of a temporal relation in an e-sequence, such that $occ(\mathcal{R},\mathcal{S}) = 1$ if $\mathcal{R}$ occurs in $\mathcal{S}$, and 0 otherwise. We define a frequency function $\mathcal{F} : [0, |\mathcal{D}|] \rightarrow [0, 1]$ that computes the *relative* vertical support of a temporal relation $\mathcal{R}$ in an e-sequence database $\mathcal{D}$ as follows:

$$\mathcal{F}(\mathcal{R}) = \frac{1}{|\mathcal{D}|} \sum_{\mathcal{S}_i \in \mathcal{D}}^{|\mathcal{D}|} occ(\mathcal{R}, \mathcal{S}_i) .$$

Definition 4 (horizontal support). *Given an e-sequence* $\mathcal{S}$ *and a temporal relation* $\mathcal{R}$, *the* horizontal support *of* $\mathcal{R}$ *is defined as the number of occurrences of* $\mathcal{R}$ *in* $\mathcal{S}$.

For horizontal support, multiple occurrences of $\mathcal{R}$ in the same e-sequence are counted.

Problem 5 **(relation-preserving embedding).** Given an e-sequence $\mathcal{S}$ with u being the total number of temporal relations in $\mathcal{S}$, define a mapping function to an embedding vector space $\mathbb{R}^d$, with $d \leq u$ such that $f : \mathcal{S} \longrightarrow \mathbb{R}^d$, where the underlying temporal relations in $\mathcal{S}$ are preserved.

By construction, the proposed embedding space achieves substantially scalable solutions to the problems of **clustering** and **classification of e-sequences.**

3 Z-Embedding: A Spectral Embedding Representation of Event Interval Sequences

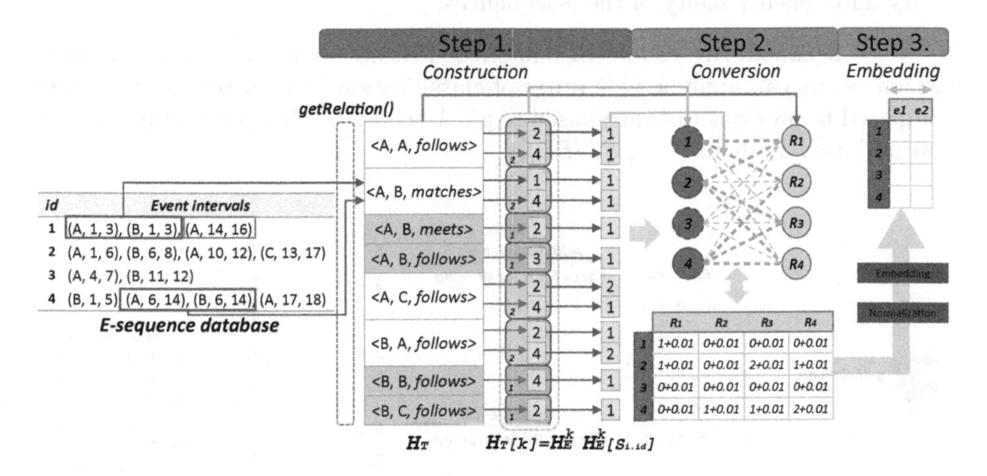

Fig. 3. An example of the process of `Z-Embedding` with the parameters $\{\texttt{minSup} : 0.5, \texttt{maxSup} : 1, \texttt{gap} : 0.5.\}$

`Z-Embedding` is an efficient three-step framework for converting an e-sequence database into a spectral embedding vector space representation, where important structural information regarding the temporal relations in the e-sequences is preserved by pruning techniques, hence facilitating scalable clustering and classification. The first two steps of the framework convert the original e-sequence database into a bipartite graph, while at the third step, the bipartite graph is converted into a spectral embedding space. The final space representation can then be readily used by off-the-shelf clustering or classification algorithms. These steps, also outlined in Fig. 3 and Algorithm 1, are described below:

1. **Construction of `G-HashTable`**: This is a data structure that efficiently stores the information needed to create a bipartite graph after scanning an e-sequence database. We can apply various pruning processes based on temporal relations to the table for better graph representation.
2. **Conversion to a bipartite graph**: The pruned table is converted to a weighted bipartite graph with two vertex sets of e-sequences and temporal relations. The bipartite graph is represented as a form of a bi-adjacency matrix. We represent the bipartite graph with the two interestingness factors defined in Sect. 2, i.e., *vertical support* and *horizontal support* [18]. We use vertical support as a pruning factor because it is a measure of how prevalent the temporal relation is across the entire database, while horizontal support is used as a weight of the edge of the graph since it represents the strength of a specific temporal relation in different e-sequences.
3. **Spectral embedding of the bipartite graph**: After generating the bi-adjacency matrix, the feature vector of each e-sequence is generated through regularization and singular value decomposition, hence reducing the complexity and dimensionality of the e-sequences.

Since `Z-Embedding` results in numerical feature vector representation of e-sequences, we can apply a wide range of classification and clustering algorithms compared to previous distance-based (e.g., *Artemis*, *IBSM*) and non-numerical-feature-based methods (e.g., *STIFE*).

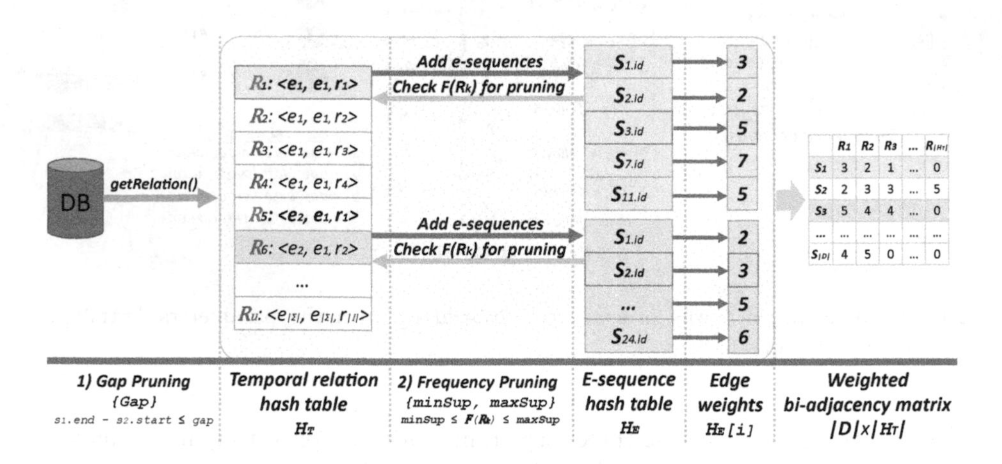

Fig. 4. An instantiation of `G-HashTable`.

3.1 Construction of `G-HashTable`

`G-HashTable` is a hash table composed of three layers constructed from the e-sequence database for facilitating its conversion to a bipartite graph. It efficiently maintains all information for the conversion and occurrence-based pruning by

Algorithm 1: Z-Embedding

```
   Data: D: E-sequence database, d: dimension factor
         constraints: predefined constraints {minSup, maxSup, gap}
   Result: U: Row embedding of regularized bi-adjacency matrix
 1 // Step 1: Construction of G-HashTable
 2 H_T = {};
 3 for S_i ∈ D do
 4     for s_a, s_b[s_a < s_b] ∈ S_i do
 5         r ← getRelation(s_a, s_b, constraints.gap);
 6         if r ≠ None then
 7             R ← (s_a.e, s_b.e, r);
 8             if R ∉ H_T then
 9                 H_T.index(R);
10             if S_i.id ∉ H_T[R] then
11                 H_T[R].index(S_i.id);
12             H_T[R][S_i.id].addHorizontalSupport();
13 for R_k ∈ H_T do
14     if F(R) < constraints.minSup ∨ F(R) > constraints.maxSup then
15         remove H_T[R_k]
16 // Step 2: Conversion to a bipartite graph
17 B = 0^{|D|×|H_T|};
18 for R_j ∈ H_T do
19     for S_i.id ∈ H_T[R_k] do
20         B[S_i.id][hash(R_k, |H_T|)] = H_T[R_k][S_i.id];
21 // Step 3: Spectral embedding of the bipartite graph
   B_S = spectralEmbedding(B, d);
22 return B_S
```

scanning temporal relations only once in the database. There are two main steps to creating the hash table: (1) construction step (blue arrows in Fig. 4), (2) pruning step (orange arrows in Fig. 4).

Construction Step. First, we traverse all event intervals in the e-sequence database in chronological order. For target event interval s_a, we make a pair with all event intervals s_b that occur after s_a. Thereafter, we check the temporal relation between event intervals s_a and s_b (lines 1–5, Algorithm 1). Then, a temporal relation between them, $\mathcal{R}_k = <s_a.e, s_b.e, r>$, with $r \in \mathcal{I}$ is formed and stored as a key in the first hash table $\mathcal{H}_T$, which we call *temporal relation hash table* (lines 6–9). Whenever a relation $\mathcal{R}_k$ is found, we identify the e-sequence id containing it, and use it as a key in the second hash table $\mathcal{H}_E$, called *e-sequence* *hash table* (lines 10–11). We note that each record $\mathcal{H}_T[\mathcal{R}_k] \in \mathcal{H}_T$ is mapped to its respective e-sequence hash table, denoted as $\mathcal{H}_E^k$. The keys of this hash table are the e-sequence ids where $\mathcal{R}_k$ occurs, while the values are the edge weights of the bipartite graph quantifying the occurrence $\mathcal{R}_k$ in the e-sequence. When we firstly create a specific key, we set the value in $\mathcal{H}_E^k$ equal to one, which corresponds to the horizontal support of the temporal relation in the e-sequence. Thus, if the same temporal relation $\mathcal{R}_k$ occurs more than once in the same e-sequence $\mathcal{S}_i$, we add the count to $\mathcal{H}_E^k[i]$ to update its horizontal support (line 12).

Pruning Step. The pruning step helps limit the unnecessary formation of relations and helping graph only to represent necessary information; This step consists of two sub-steps, which occur at different times:

1. **Gap pruning:** A gap constraint limits the maximum distance of *follows* relations between intervals. This pruning process eliminates unnecessary relations that occur just because they are far apart rather than having a meaningful relation. The gap is checked when checking the temporal relation while scanning the database (line 5, Algorithm 1). We receive the gap constraint with a value in the range [0, 1], meaning the ratio of the average time duration of e-sequences and prune the *follows* relations having a distance above that ratio.
2. **Frequency pruning:** Frequency pruning is a step of removing temporal relations $\mathcal{R}_k$, whose relative vertical supports $\mathcal{F}(\mathcal{R}_k)$ are below or above the predefined criteria after the table is completely formed (lines 13–15). To do this, we impose the following two constraints:
   - Minimum support constraint: corresponding to the minimum occurrence frequency of each temporal relation. This helps increase the cluster's purity by limiting the small size temporal relations that can be different within a cluster.
   - Maximum support constraint: corresponding to the maximum occurrence frequency of each temporal relation. This limits the temporal relations spanning almost all e-sequences, allowing the embedding space to represent the e-sequence space holistically.

Example. Consider an e-sequence database of size 4 (Fig. 3). For this example, we will use the following parameter settings: $\texttt{minSup} : 0.5, \texttt{maxSup} : 1, \texttt{gap} : 0.5$. Hence, we need to find relation pairs with absolute vertical supports from 2 to 4. Moreover, the gap constraint of 0.5 implies that the longest span of a *follows* relation can be at most half the average time length of all e-sequences, which is $\frac{16+17+12+18}{4} \times 0.5 = 7.875$. First, we scan the database to get all temporal relations, which is accomplished by checking the temporal relations between the event intervals in the database. In this example, we see that (A, 1, 3), (B, 1, 3) in the first e-sequence and (A, 6, 14), (B, 6, 14) in the fourth e-sequence form temporal relation: $<A, B, matches>$. Then, we place the relation as the key for the first layer of the hash table $\mathcal{H}_T$. After that, since the same temporal relation occurs in both the first and fourth e-sequences, we can store their ids into the second layer along with their corresponding vertical supports (left square boxes in the *e-sequence hash table* $\mathcal{H}_E$). Finally, we compute the horizontal support by counting how often the temporal relation has occurred in each stored e-sequence ({1, 4} in the example). We only have a single horizontal occurrence in both e-sequences. Hence we add ones to the values of the *e-sequence hash table* (right square boxes). The gap constraint pruning is applied together with the `G-HashTable` construction. Actually, (A, 1, 3) and (A, 14, 16) in the first e-sequence must have formed a *follows* relation without the gap constraint. However, since the distance between the two event intervals is 11 (>7.875), we skip creating a record in the `G-HashTable`. The same pruning holds for the fourth e-sequence and event intervals (B, 1, 5), (A, 17, 18) having a distance equal to 12. After constructing the `G-HashTable`, frequency pruning is performed by applying {`minSup, maxSup`}. Since $minSup = 0.5$ (or support count of 2), temporal relations with vertical support equal to 1 are subsequently excluded from the first layer of the table (gray triplets in $\mathcal{H}_T$ in the example).

The advantage of `G-HashTable` is that we can easily consider two types of frequencies and apply pruning techniques by scanning the e-sequence database only once. Moreover, we can directly convert the table to its corresponding bi-adjacency matrix weighted by the frequencies in the e-sequences. All that is required is to scan the database once and scan the first layer of the table to apply pruning and scan the first and second layers of the table to convert it into the bi-adjacency matrix.

Hence, given an e-sequence database $\mathcal{D} = \{\mathcal{S}_1, \ldots, \mathcal{S}_{|\mathcal{D}|}\}$, the set of possible relations $\mathcal{I}$, and the alphabet of event labels Σ, the time complexity for creating the bi-adjacency matrix is quadratic in the worst case as follows:

$$(\sum_{\mathcal{S}_i \in \mathcal{D}}^{|\mathcal{D}|} |\mathcal{S}_i|^2 \times |\mathcal{I}|) + (|\Sigma|^2 \times |\mathcal{I}|) + (|\Sigma|^2 \times |\mathcal{I}| \times |\mathcal{D}|) \,.$$

3.2 Conversion to a Weighted Bipartite Graph

In this step, we use the notion of a weighted bipartite graph.

Definition 6 (bipartite graph). *A bipartite graph $G = \{U, V, E\}$ is a special form of a graph, having vertices divided into two disjoint sets U and V, meaning that $U \cap V = \emptyset$, and a set of edges $E = \{e_{u,v} | u \in U, v \in V\}$.*

A bipartite graph consists of edges that can only lead from the vertex set U to the other vertex set V, while vertices belonging to the same set cannot be connected. A *weighted bipartite graph* is trivially an extension of G, where each $e_{u,v} \in E$ equals to the corresponding edge weight between u and v, or to 0 if no edge exists between u and v.

After the construction and pruning steps resulting into `G-HashTable`, we create the corresponding weighted bipartite graph by directly using each layer of the `G-HashTable`. The temporal relations in the first layer of `G-HashTable` are used as the right-hand side nodes, while the e-sequence ids of the second layer are used as the left-side nodes. Furthermore, the edges are created to link each e-sequence id (left-hand side nodes) to the corresponding temporal relations (right-hand side nodes) it contains. Horizontal supports are used as weights (having applied {`minSup`, `maxSup`} thresholds). The resulting graph is a weighted bipartite graph $G = \{U, V, E\}$, with

$$U : \{i \mid \mathcal{S}_i \in \mathcal{D}, i \in [1, |\mathcal{D}|]\},$$
$$V : \{\mathcal{H}_T.keys \mid \forall \mathcal{R} \in \mathcal{H}_T : minSup < \mathcal{F}(\mathcal{R}) < maxSup\}\} \text{ and}$$
$$E : \{e_{i,j} = \mathcal{H}_E^i[j] \mid i \in [1, |\mathcal{D}|], j \in \mathcal{H}_E^i\}.$$

Using G, we construct its bi-adjacency matrix B (lines 17–20, Algorithm 1). The bi-adjacency matrix $B \in \mathbb{R}^{|U| \times |V|}$ is a two-dimensional matrix representing G, with the dimensions corresponding to vertex sets U and V, and edges between sets U and V are defined as the elements of the matrix, with

$$B_{u,v} = \begin{cases} e_{u,v} > 0, & \text{if and only if } e_{u,v} \in E \\ 0, & \text{otherwise} \end{cases}$$

A bi-adjacency matrix is computationally efficient as it reduces the matrix size from $|U + V| \times |U + V|$ to $|U| \times |V|$, while storing the same information. In Fig. 4, we see an example of a conversion from the set of temporal relations to a bipartite graph G and its corresponding weighted bi-adjacency matrix B.

Example. After the construction and pruning step, we have four temporal relations {<A, A, *follows*>, <A, B, *matches*>, <A, C, *follows*>, <B, A, *follows*>} and four e-sequence ids {1, 2, 3, 4} that meet all the constraints. Then we can create a 4 × 4 bi-adjacency matrix and fill the values of the third layers of the `G-HashTable` as key of the second layer, key of the first layer in the matrix shown in Step 2 in Fig. 3. For example, since the horizontal support of pair <B, A, *follows*> and e-sequence 4 is 2, we can insert the value 2 into the matrix with key {<B, A, *follows*>, 4}, which is the right bottom value in the matrix. If no edge occurs between an e-sequence and a relation pair, we can set that value to zero by following the definition of the adjacency matrix. The third e-sequence will have all zeros in the matrix since all of its relations are pruned.

3.3 Spectral Embedding of `Z-Embedding`

After constructing the bipartite graph and its bi-adjacency matrix, we proceed with defining a reduced-rank spectral embedding [24]. First, we apply regularization with a regularization factor α to ensure noise and outlier robustness of the spectral embedding. The factor α is determined by prior knowledge based on the properties of the datasets. We used the most recent technique introduced in [4], which is adding a constant α equally to all elements of the bi-adjacency matrix (line 1, Algorithm 2). From a graph perspective, this means adding small-weight edges to every pair of nodes between the sets (green edges in Fig. 3).

Next, using the bi-adjacency matrix we create the *normalized Laplacian matrix* N_B. We only calculate N_B^{UR}, the top-right part of N_B (line 2).

Algorithm 2: spectralEmbedding

Data: B: a bi-adjacency matrix of intervals where $B \in \mathbb{R}^{|U| \times |V|}$
d: dimension factor
Result: U: Embedding of the rows
1 $B_R = B + \alpha * 1^{|U| \times |V|}$
2 $N_B^{UR} = D_1^{-\frac{1}{2}} B_R D_2^{-\frac{1}{2}}$
3 calculate SVD $N_B^{UR} = M \Sigma W^T$
4 pick leading d singular values and corresponding d columns from M
5 **return** $M[: U, : d]$

Definition 7 (Normalized Laplacian matrix). *We define a normalized Laplacian matrix* N *of a graph as* $N = D^{-\frac{1}{2}} L D^{-\frac{1}{2}} = I_{|U|} - D^{-\frac{1}{2}} A D^{-\frac{1}{2}}$, *where* $D_U \in \mathbb{R}^{|U| \times |U|}$ *is a diagonal degree matrix where* $D_{ii} = deg(u_i)$, *with* $i \in [1, |U|]$.

Note that we can only use the bi-adjacency matrix part (top-right), since N can be expressed as a bi-adjacency matrix as follows:

$$N = \begin{bmatrix} I_{|U|} & -D_U^{-\frac{1}{2}} B D_V^{-\frac{1}{2}} \\ -D_V^{-\frac{1}{2}} B^T D_U^{-\frac{1}{2}} & I_{|V|} \end{bmatrix} \quad (1)$$

The normalized Laplacian matrix provides an approximate solution for finding the sparsest cut of the graph, providing a good graph partition [19].

The next step is to define the spectral embedding space of N_B^{UR}, hence reducing the horizontal dimension of the matrix, which can have a maximum size of $|\Sigma|^2 * |\mathcal{I}|$, to create reduced-size feature vectors that can be processed at a faster speed. The spectral embedding space is achieved by constructing a new embedding space and obtaining the leading eigenvectors of the adjacency matrix. Since the bi-adjacency matrix is not square, we apply SVD as an equivalent process of eigendecomposition [16]. Then we sort the singular values and choose the d leading values, where $d \leq min(|U|, |V|)$, and the corresponding columns from M. The target dimension parameter k is set based on prior knowledge and the dataset properties. Finally, we return the selected d columns of M (size $U \times d$), which defines the spectral embedding space of each e-sequence.

The intuition behind these three steps is that pruning in the spectral space will result in e-sequences of the same class label having similar but unique distributions of pairwise relations.

4 Experiments

We demonstrated the applicability of the `Z-Embedding` representation on five real-world datasets for clustering and classification, and compared it against two state-of-the-art competitors for the task of clustering and three for classification. For **repeatability** purposes, our datasets and code can be found on github[1].

[1] https://github.com/zedshape/zembedding.

Table 1. A summary of the properties of the real-world datasets.

Dataset	# of e-seq.	# of event labels	# of event intervals	Avg. interval length	Avg. e-seq. length	# of unique temp. rel.	# of total temp. rel.
BLOCKS	210	8	1,207	5.75	54.13	174	3,245
PIONEER	160	92	8,949	55.93	57.19	26,429	252,986
CONTEXT	240	54	19,355	80.65	191.85	6,723	804,504
SKATING	530	41	23,202	43.78	1,916.08	4,844	516,272
HEPATITIS	498	63	53,921	108.28	3,193.55	20,865	3,785,167

4.1 Setup

Datasets. We used five public datasets collected from different application domains. Table 1 summarizes the properties of datasets. Detailed information for each dataset can be found in earlier works (e.g., [7]).

Competitor Methods. We demonstrated the runtime efficiency of `Z-Embedding` and its applicability to the clustering tasks and classification tasks. More concretely, for clustering, we benchmarked k-means and k-medoids under the Euclidean distance in the `Z-Embedding` space, and compared them against using two alternative state-of-the-art distance functions, *Artemis* and *IBSM*. Moreover, for classification, we benchmarked four different classifiers using the `Z-Embedding` feature space, i.e., 1-Nearest Neighbor (1-NN), Random Forests (RF), Support Vector Machine (SVM) with the Radial Basis Function (RBF) kernel (SVM_RBF), and SVM with the 3-degree polynomial kernel (SVM_Poly). These were compared against 1-NN using *Artemis* and *IBSM*. For completeness, we additionally compared against *STIFE*, a RF feature-based classifier for e-sequences. All clustering and classification algorithms were implemented using the scikit-learn library.

4.2 Results

All algorithms were implemented in Python 3.7 and run on an Ubuntu 18.10 system with Intel i7-8700 CPU 3.20 GHz and 32 GB main memory. All results contain the average values of 10-fold cross-validation (for classification) and 100 trials (for clustering). If the algorithm required hyperparameters, we followed the parameter setup defined by the authors of each paper for a fair comparison. For the dimension factor d for spectral embedding, we chose $d = 4$ for BLOCKS dataset as it has comparably smaller in terms of the number of temporal relations compared to other datasets (Table 1), while for the rest of the datasets we set $d = 8$. Throughout this process, the resulting feature vectors provided a compressed version of the original space by almost 99% for all datasets, which has contributed to the high computation speedups obtained. Using Z-Embedding , we could achieve speedups of up to a factor of 292 compared to the competitors[2].

Clustering Results. We set the expected number of clusters to the actual number of class labels in the dataset, and computed the total runtime and purity values required for all the algorithms. Since *Artemis* is only calculating distances, k-means was inapplicable. K-medoids was generally faster than k-means because it could be run after pre-calculating the pairwise distances between the data e-sequences. To construct the embedding space for Z-Embedding, we used the same regularization factor, $\alpha = 0.001$, for every dataset.

Table 2. Clustering results for Z-Embedding and all competitors in terms of clustering purity (%) and runtime (seconds).

Dataset	*Artemis*		*IBSM*				Z-Embedding			
	K-medoids		K-medoids		K-means		K-medoids		K-means	
	Purity	Time	Purity	Time	Purity	Time	Purity	Time	Purity	Time
BLOCKS	85.62	1.20	95.30	0.71	99.09	10.57	93.81	**0.02**	99.82	0.04
PIONEER	66.13	15.64	63.94	4.41	64.09	74.13	74.75	**0.89**	83.12	0.91
CONTEXT	65.13	122.23	75.22	5.19	82.66	204.82	77.54	**1.99**	82.36	2.02
SKATING	36.52	180.48	70.21	286.10	-	>1 h	62.45	**1.48**	74.40	1.52
HEPATITIS	-	>1 h	67.91	444.77	-	>1 h	71.70	**9.60**	70.08	9.63

Table 2 shows the results in terms of clustering purity and runtime. For each method, we set a one-hour time limit to its runtime. We firstly applied each algorithm to create the feature vectors, and then k-means and k-medoids were applied, respectively. In terms of runtime, Z-Embedding was faster in both cases compared to two competitors. In particular, *Artemis* did not complete the calculation within an hour on the HEPATITIS dataset. *IBSM* showed deficient runtime performance for the datasets with long e-sequences, such as SKATING or HEPATITIS. Specifically, when k-medoids was used on SKATING, the speed was even slower than that of *Artemis*. Moreover, k-means could not complete within an hour on SKATING and HEPATITIS, while Z-Embedding with k-means completed in 1.52 s on SKATING and 9.63 s on HEPATITIS.

In terms of purity, Z-Embedding also showed remarkable results. In the k-medoids trials, *Artemis* had the lowest purity values on all data sets except for PIONEER. *IBSM* showed the highest purity only on SKATING with k-medoids, but it was about 193 times slower than Z-Embedding. On the other datasets, Z-Embedding showed the fastest runtime performance and achieved the highest purity. In the k-means experiment, Z-Embedding showed the highest purity values, except for CONTEXT. *IBSM* led by a slight difference of 0.3 percent on CONTEXT but was also about 101 times slower than Z-Embedding.

[2] This is even an underestimate as for the cases where competitors that did not finish within the one-hour execution time limit, our approach is at least 300 times faster.

Classification Results. For each competitor method, we used the classifiers suggested by the authors in the corresponding papers. Since *Artemis* and *IBSM* are distance-based algorithms, the number of applicable algorithms is highly limited. Therefore, for these two competitors, the 1-NN classifier was applied. On the other hand, since *STIFE* generates non-numeric feature vectors, distance-based algorithms cannot be applied, and in this case, RF was applied. For *STIFE*, we applied the recommended optimal parameters [2]. In order to adjust the parameters of `Z-Embedding` for each dataset, we performed a grid search on 1-NN classification accuracy within the range of [0, 1] for each of the three parameters {`maxSup, minSup, gap`}, in increments of 0.1. The top 10 parameter settings and the experimental results are available in the supplementary materials.

Table 3. Classification results for all competitors in terms of classification accuracy (%) and runtime (seconds).

Dataset	*Artemis*		*IBSM*		*STIFE*	
	1-NN		1-NN		RF	
	Acc	Time	Acc	Time	Acc	Time
BLOCKS	99.00	1.43	100	0.77	100	2.96
PIONEER	97.50	19.27	93.75	4.43	98.75	8.51
CONTEXT	90.00	130.22	97.08	5.32	98.33	12.1
SKATING	84.00	208.79	97.74	286.24	96.42	21.4
HEPATITIS	–	>1 h	77.91	445.83	82.13	83.7

Table 4. Classification results for `Z-Embedding` in terms of classification accuracy (%) and runtime (seconds).

Dataset	`Z-Embedding`										
	Constraints			1-NN		RF		SVM_RBF		SVM_Poly	
	`minSup`	`maxSup`	`gap`	Acc	Time	Acc	Time	Acc	Time	Acc	Time
BLOCKS	0.0	0.4	0.0	100	**0.02**	100	0.12	100	**0.02**	100	**0.02**
PIONEER	0.0	0.7	0.1	100	**1.49**	100	1.62	100	1.50	100	**1.49**
CONTEXT	0.4	0.5	0.2	95.00	**1.35**	96.25	1.46	97.50	1.36	97.08	1.36
SKATING	0.5	0.6	0.1	91.32	**0.98**	92.07	1.10	93.58	0.99	92.45	0.99
HEPATITIS	0.0	1.0	0.1	76.30	10.83	82.13	11.27	83.73	**10.82**	83.34	11.04

Unlike existing algorithms, `Z-Embedding` can apply a wide range of algorithms as it forms numeric feature vectors. In this experiment, we applied the ones that previous methods used, such as 1-NN and RF, and we also ran two SVM with RBF kernel and polynomial kernel. Table 3 shows the classification accuracy and runtime for each competitor method, while Table 4 shows the results for Z-Embedding. 1-NN under *Artemis* had the longest runtime and lowest accuracy for all datasets, while on HEPATITIS it failed to complete within the 1-h runtime limit. On the other hand, *IBSM* achieved the best performance on SKATING, but it is 13 times slower than *STIFE* and up to 292 times slower than Z-Embedding . Finally, *STIFE* was the algorithm with the highest speed and accuracy performance (except for SKATING) among the other competitors. It even achieved better performance than Z-Embedding on CONTEXT and SKATING, but it was about up to 9 times slower than Z-Embedding on CONTEXT, and 21 times on SKATING.

5 Conclusion

We proposed a novel representation of event interval sequences using a bipartite graph for efficient clustering and classification. We benchmarked our representation on five real-world datasets against several competitor algorithms. Our experimental benchmarks showed that the proposed spectral embedding representation can achieve substantially lower runtimes compared to earlier competitors and even higher values of purity (for clustering) and classification accuracy (for classification) than some of its competitors. Future work includes extending the bipartite graph representation to tripartite or higher multipartite by calculating higher orders of temporal relations, investigating the usage of our framework for providing scalable solutions to other machine learning problems, and also exploring alternative link-analysis ranking methods such as rooted PageRank.

References

1. Allen, J.F.: Maintaining knowledge about temporal intervals. CACM **26**(11), 832–843 (1983)
2. Bornemann, L., Lecerf, J., Papapetrou, P.: STIFE: a framework for feature-based classification of sequences of temporal intervals. In: Calders, T., Ceci, M., Malerba, D. (eds.) DS 2016. LNCS (LNAI), vol. 9956, pp. 85–100. Springer, Cham (2016). https://doi.org/10.1007/978-3-319-46307-0_6
3. Chaudhuri, K., Chung, F., Tsiatas, A.: Spectral clustering of graphs with general degrees in the extended planted partition model. In: COLT, p. 35-1 (2012)
4. De Lara, N., Bonald, T.: Spectral embedding of regularized block models. In: ICLR (2020)
5. Joseph, A., Yu, B., et al.: Impact of regularization on spectral clustering. Ann. Stat. **44**(4), 1765–1791 (2016)
6. Kostakis, O., Gionis, A.: On mining temporal patterns in dynamic graphs, and other unrelated problems. In: Cherifi, C., Cherifi, H., Karsai, M., Musolesi, M. (eds.) COMPLEX NETWORKS 2017 2017. SCI, vol. 689, pp. 516–527. Springer, Cham (2018). https://doi.org/10.1007/978-3-319-72150-7_42
7. Kostakis, O., Papapetrou, P.: On searching and indexing sequences of temporal intervals. Data Min. Knowl. Disc. **31**(3), 809–850 (2017). https://doi.org/10.1007/s10618-016-0489-3
8. Kotsifakos, A., Papapetrou, P., Athitsos, V.: IBSM: interval-based sequence matching. In: SDM, pp. 596–604. SIAM (2013)

9. Kunegis, J.: Exploiting the structure of bipartite graphs for algebraic and spectral graph theory applications. Internet Math. **11**(3), 201–321 (2015)
10. Lam, H.T., Mörchen, F., Fradkin, D., Calders, T.: Mining compressing sequential patterns. SADM **7**(1), 34–52 (2014)
11. Liu, L., Wang, S., Hu, B., Qiong, Q., Wen, J., Rosenblum, D.S.: Learning structures of interval-based Bayesian networks in probabilistic generative model for human complex activity recognition. PR **81**, 545–561 (2018)
12. Ng, A.Y., Jordan, M.I., Weiss, Y.: On spectral clustering: analysis and an algorithm. In: NIPS, pp. 849–856 (2002)
13. Perera, D., Kay, J., Koprinska, I., Yacef, K., Zaïane, O.R.: Clustering and sequential pattern mining of online collaborative learning data. TKDE **21**(6), 759–772 (2008)
14. Pissinou, N., Radev, I., Makki, K.: Spatio-temporal modeling in video and multimedia geographic information systems. GeoInformatica **5**(4), 375–409 (2001)
15. Qin, T., Rohe, K.: Regularized spectral clustering under the degree-corrected stochastic blockmodel. In: NIPS, pp. 3120–3128 (2013)
16. Ramasamy, D., Madhow, U.: Compressive spectral embedding: sidestepping the SVD. In: NIPS, pp. 550–558 (2015)
17. Schmidt, M., Palm, G., Schwenker, F.: Spectral graph features for the classification of graphs and graph sequences. CompStat **29**(1–2), 65–80 (2014)
18. Sheetrit, E., Nissim, N., Klimov, D., Shahar, Y.: Temporal probabilistic profiles for sepsis prediction in the ICU. In: KDD, pp. 2961–2969 (2019)
19. Shi, J., Malik, J.: Normalized cuts and image segmentation. TPAMI **22**(8), 888–905 (2000)
20. Von Luxburg, U.: A tutorial on spectral clustering. Stat. Comput. **17**(4), 395–416 (2007)
21. Von Luxburg, U., Belkin, M., Bousquet, O.: Consistency of spectral clustering. ANN STAT **36**, 555–586 (2008)
22. Wang, J., Han, J.: Bide: efficient mining of frequent closed sequences. In: ICDE, p. 79. IEEE (2004)
23. Zhang, Y., Rohe, K.: Understanding regularized spectral clustering via graph conductance. In: NIPS, pp. 10631–10640 (2018)
24. Zhou, Z., Amini, A.A.: Analysis of spectral clustering algorithms for community detection: the general bipartite setting. JMLR **20**(47), 1–47 (2019)

Collaborative Filtering and Matrix Completion

Neural Cross-Domain Collaborative Filtering with Shared Entities

M. Vijaikumar(✉), Shirish Shevade, and M. N. Murty

Department of Computer Science and Automation,
Indian Institute of Science, Bangalore, India
{vijaikumar,shirish,mnm}@iisc.ac.in

Abstract. Cross-Domain Collaborative Filtering (CDCF) provides a way to alleviate data sparsity and cold-start problems present in recommendation systems by exploiting the knowledge from related domains. Existing CDCF models are either based on matrix factorization or deep neural networks. Independent use of either of the techniques in isolation may result in suboptimal performance for the prediction task. Also, most of the existing models face challenges particularly in handling diversity between domains and learning complex non-linear relationships that exist amongst entities (users/items) within and across domains. In this work, we propose an end-to-end neural network model – NeuCDCF, to address these challenges in a cross-domain setting. More importantly, NeuCDCF is based on a wide and deep framework and learns the representations jointly using both matrix factorization and deep neural networks. We perform experiments on four real-world datasets and demonstrate that our model performs better than state-of-the-art CDCF models.

Keywords: Cross-domain collaborative filtering · Deep learning · Neural networks · Wide and deep framework · Recommendation system

1 Introduction

Personalized recommendation systems play an important role in extracting relevant information for user requirements, for example, product recommendation from Amazon[1], event recommendation from Meetup[2], scholarly references recommendation from CiteULike[3]. Collaborative Filtering (CF) is one of the widely used techniques [25] in recommendation systems that exploits the interactions between users and items for predicting unknown ratings.

The performance of the CF techniques mainly depends on the number of interactions the users and items have in the system. However, in practice, most of the users interact with a very few items. For instance, even the popular and

[1] www.amazon.com.
[2] www.meetup.com.
[3] www.citeulike.org.

F. Hutter et al. (Eds.): ECML PKDD 2020, LNAI 12457, pp. 729–745, 2021.
https://doi.org/10.1007/978-3-030-67658-2_42

well-preprocessed datasets such as Netflix[4] and MovieLens-20M[5] have 1.18% and 0.53% available ratings respectively. Furthermore, new users and items are added to the system continuously. Due to this, the new entities (users/items) may have a very few or even no ratings associated with them. These issues are referred to as data sparsity problem and cold-start problem [25]; and the entities with a very few or no ratings are called cold-start entities. When the above-mentioned issues exist, learning parameters in CF techniques becomes a daunting task.

Cross-Domain Collaborative Filtering (CDCF) [2] is a promising technique in such scenarios that alleviates data sparsity and cold-start problems by leveraging the knowledge extracted from user-item interactions in related domains[6], where information is rich, and transfers appropriately to the target domain where a relatively smaller number of ratings are known. In this work, we study the problem of Cross-Domain Recommendation (CDR) with the following assumption: users (or items) are shared across domains and no side information is available.

Most of the existing models proposed for CDR are based on Matrix Factorization (MF) techniques [5,8,14,15,23,24]. They may not handle complex non-linear relationships that exist among entities within and across domains very well. More recently, attempts have been made to exploit the transfer learning ability of the deep networks in cross-domain settings [4,11,17,19,33,34]. Although they have achieved performance improvement to some extent, they have the following shortcomings. First, they do not consider the diversity among the domains. That is, they learn shared representations from the source and the target domain together instead of learning domain-specific and domain-independent representations separately. Second, they do not consider both the wide (MF based) and deep (deep neural networks based) representations together. We explain why addressing these shortcomings are essential in the following:

Motivation for the Use of Domain-Specific Embeddings to Handle Diversity. When users (or items) are shared between different domains one can ignore the diversity among domains and treat all the items (or users) as members of the same domain, and apply the single domain CF models. However, it loses important domain-specific characteristics that are explicit to different domains. For example, let us assume that we have two domains: movie and book. Suppose the movie domain is defined by the features: *genre*, *language* and *visual effects*; and the book domain is defined by the features: *genre*, *language* and *number of pages*. Though the features *genre* and *language* are shared by both the domains, *visual effects* and *number of pages* are specific to the movie and book domains, respectively.

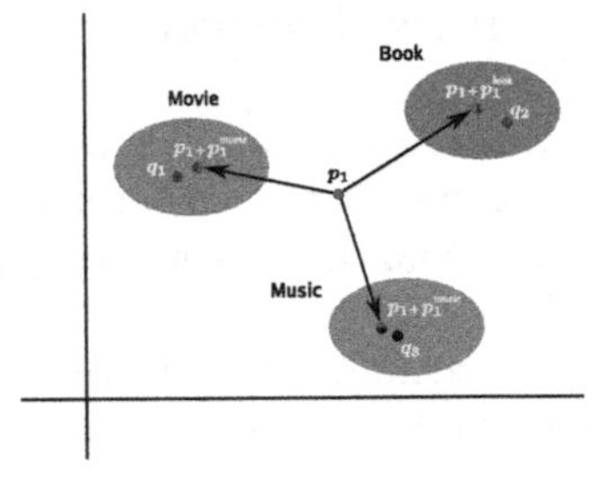

Fig. 1. Illustration of domain-specific preferences to handle diversity.

[4] https://www.kaggle.com/netflix-inc/netflix-prize-data.
[5] http://files.grouplens.org/datasets/movielens.
[6] We follow the domain definition as in [15,17,19].

When users and items are embedded in some latent space, learning only common preferences of users may not be sufficient to handle diversity among domains. This is illustrated in Fig. 1. Suppose, user $u1$ rates high on items $j1, j2$ and $j3$ belonging to movie, book and music domains, respectively. Let p_1, q_1, q_2 and q_3 be embeddings for user $u1$ and items $j1, j2$ and $j3$, respectively. In order to maintain higher similarity values with all the items $j1, j2$ and $j3$, the existing models [9,13,20,24] embed p_1 for user $u1$ as shown in Fig. 1. This may deteriorate the performance of the prediction task for CDR. This is because we need to account for the domain-specific embeddings p_1^{movie}, p_1^{book} and p_1^{music} as illustrated in Fig. 1 to get the appropriate preferences for movie, book and music domains, respectively.

Importance of Learning Non-linear Relationships Across Domains. Learning non-linear relationships present across domains is important in transferring knowledge across domains. For example[7], let us assume that a movie production company releases a music album in advance to promote a movie. Let one feature for 'music' be *popularity*, and the two features for 'movie' be *visibility* (reachability to the people) and *review-sentiment*. As the *popularity* of the music album increases, the *visibility* of the movie also increases, and they are positively correlated. However, the movie may receive positive reviews due to the music *popularity* or it may receive negative reviews if it does not justify the affiliated music album's *popularity*. So, the movie's *review-sentiment* and music album's *popularity* may have a non-linear relationship, and it has been argued [9,18,21] that such relationships may not very well be handled by matrix factorization models. Thus, it is essential to learn non-linear relationships for the entities in cross-domain systems using deep networks.

Necessity of Fusing Wide and Deep Networks. The existing models for CDR are either based on MF techniques or neural network-based techniques. However, following either approach leads to suboptimal performance. That is, deep networks are crucial to capture the complex non-linear relationships that exist among entities in the system. At the same time, we cannot ignore the complementary representations provided by wide networks such as matrix factorization [3].

Contributions. In this work, we propose a novel end-to-end neural network model for CDR, Neural Cross-Domain Collaborative Filtering (NeuCDCF), to address the above-mentioned challenges. In particular, our novelty lies in the following places:

- We extend matrix factorization ideas to the cross-domain recommendation, which helps learn wide representations. This extension is referred to as Generalized Collective Matrix Factorization (GCMF). We propose an extra component in GCMF that learns the domain-specific embeddings. This helps significantly in addressing diversity issues.

[7] This example is inspired from [18].

- Inspired from the success of encoder-decoder architecture in machine translation [1], we propose Stacked Encoder-Decoder (SED) to obtain deep representations of shared entities. These deep representations learn the complex non-linear relationships that exist among entities. This is obtained by constructing target domain ratings from the source domain.
- More importantly, we fuse GCMF and SED to get the best of both wide and deep representations. All the parameters of both networks are learned in an end-to-end fashion. To our best knowledge, NeuCDCF is the first model that is based on a wide and deep framework for the CDR setting.

We conduct extensive experiments on four real-world datasets – two from Amazon and two from Douban, in various sparse and cold-start settings, and demonstrate the effectiveness of our model compared to the state-of-the-art models. Our implementation is available at https://github.com/mvijaikumar/NeuCDCF.

2 Problem Formulation

We denote the two domains, source and target, by the superscripts S and T, respectively. Ratings from the source and the target domains are denoted by the matrices $R^S = [r^S_{uj}]_{m \times n_S}$ and $R^T = [r^T_{uj}]_{m \times n_T}$ where $r^S_{uj}, r^T_{uj} \in [\gamma_{min}, \gamma_{max}] \cup \{0\}$, u and j respectively denote user and item indices, 0 represents unavailability of a rating for the pair (u, j), and γ_{min} and γ_{max} denote minimum and maximum rating values in rating matrices R^S and R^T, respectively. Further, m, n_S and n_T represent the number of users, the number of items in source domain and the number of items in target domain, respectively. Let, $n = n_S + n_T$. We indicate the available ratings in the source domain by $\Omega^S = \{(u, j)|r^S_{uj} \neq 0\}$ and the target domain by $\Omega^T = \{(u, j)|r^T_{uj} \neq 0\}$. We drop the superscripts S and T from r^S_{uj} and r^T_{uj} wherever the context is clear through Ω^S and Ω^T.

Further, $p \odot q$ and $p'q$ denote element-wise multiplication and dot product of two column vectors p and q of the same size, $a(\cdot)$ denotes an activation function, and $\|x\|$ denotes the l_2 norm of a vector x. Let $R_{u,:}$ ($R_{:,j}$) be u^{th} row (j^{th} column) of the matrix R corresponding to the user u (item j).

A set of users, shared across domains, is denoted by $\mathcal{U}$ and the sets of items for the source and the target domains are denoted by $\mathcal{V}^S$ and $\mathcal{V}^T$ respectively. Let $\mathcal{V} = \mathcal{V}^S \cup \mathcal{V}^T$ and $\mathcal{V}^S \cap \mathcal{V}^T = \varnothing$.

Problem Statement: Given the partially available ratings for the source and the target domains $r_{uj}, \forall (u, j) \in \Omega^S \cup \Omega^T$, our aim is to predict target domain ratings r_{uj}, $\forall (u, j) \notin \Omega^T$. We assume that no side information is available.

3 The Proposed Model

In this section, we explain our proposed model – NeuCDCF in detail. First, we explain the individual networks, GCMF and SED, and then discuss about how these two are fused to get NeuCDCF.

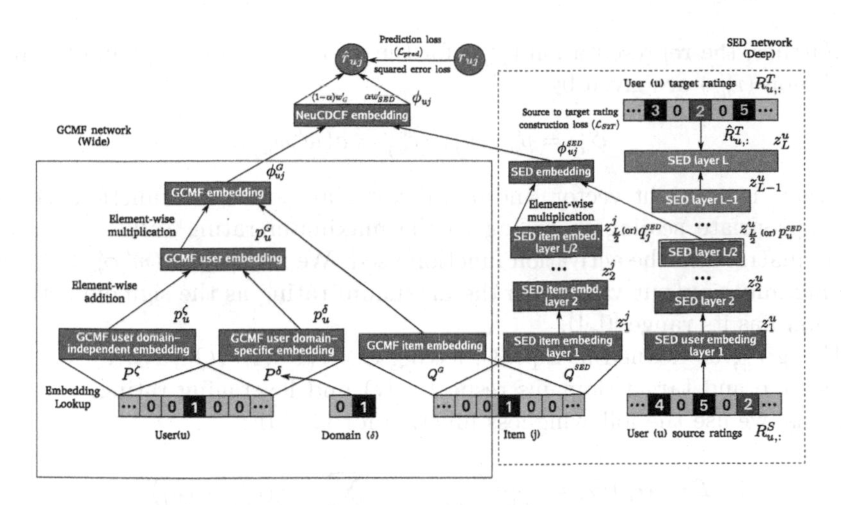

Fig. 2. The architecture of our proposed NeuCDCF Model.

3.1 Generalized Collective Matrix Factorization (GCMF)

The proposed GCMF network is shown inside the solid-lined box in Fig. 2. Let $P^{\delta} \in \mathbb{R}^{m \times k}$ and $P^{\zeta} \in \mathbb{R}^{m \times k}$ respectively be the domain-specific and domain-independent user embedding matrices, $Q^{G} \in \mathbb{R}^{n \times k}$ be the item embedding matrix and k be the embedding dimension. Note that the matrix Q^{G} contains the embedding of all the source and the target domain items. Note also that, we have a single P^{ζ} shared across the domains and separate P^{δ} for the source ($\delta = 0$) and target ($\delta = 1$) domains. Having a separate user embedding matrix P^{δ} for each domain δ helps in handling diversity across domains. We define the domain-specific embedding (p_u^{δ}) and domain-independent embedding (p_u^{ζ}) for user u as follows:

$$p_u^{\delta} = P^{\delta} x_u, \quad p_u^{\zeta} = P^{\zeta} x_u, \tag{1}$$

where x_u and y_j are one-hot encodings of user u and item j respectively. The user and item embeddings, p_u^{G} and q_j^{G} for the user-item pair (u, j) is obtained as

$$p_u^{G} = p_u^{\zeta} + p_u^{\delta}, \quad q_j^{G} = Q^{G} y_j. \tag{2}$$

We combine the domain-specific and domain independent embeddings to get the user embeddings. The embedding of item j is denoted by q_j^{G}. During training, p_u^{ζ} gets updated irrespective of its interactions with the items from different domains, whereas p_u^{δ} gets updated only with respect to its specific domain.

It is possible to define p_u^{G} in different ways, instead of using Eq. (2). For example, one can concatenate them or use the element-wise product ($p_u^{\delta} \odot p_u^{\zeta}$) to obtain p_u^{G}. However, intuitively we can think of p_u^{δ} as an offset to p_u^{ζ} to account for the domain-specific preferences (e.g., embeddings p_1^{movie}, p_1^{music} and p_1^{book} for movie, music and book in Fig. 1) that provide the resultant user representation p_u^{G} according to the domain. This also encourages the model to flexibly learn the item embeddings suitable for the domains.

Further, the representation (ϕ^G_{uj}) for the user-item pair (u, j) and the rating prediction ($\hat{r}^G_{uj}$) are given by

$$\phi^G_{uj} = p^G_u \odot q^G_j, \quad \hat{r}^G_{uj} = a(w'_G \phi^G_{uj}), \tag{3}$$

where w_G is a weight vector, and $a(\cdot)$ denotes an activation function. One can use appropriate scaling according to the maximum rating value γ_{max} in the rating matrix and the activation function used. We use $\gamma_{max}.\sigma(w'_G \phi^G_{uj})$ to match the maximum output value with the maximum rating as the sigmoid activation function has its range $[0, 1]$.

Using $\hat{r}^G_{uj}$ as defined in Eq. (3), letting $\Theta = \{P^\delta, P^\varsigma, Q^G, w_G\}$ for $\delta \in \{0, 1\}$ (for source and target domains respectively) and leveraging ratings from both domains, we use the following loss function for GCMF:

$$\mathcal{L}_{GCMF}(\Theta) = \frac{1}{|\Omega^S \cup \Omega^T|} \sum_{(u,j) \in \Omega^S \cup \Omega^T} (\hat{r}^G_{uj} - r_{uj})^2. \tag{4}$$

GCMF is a wide network and it is a generalization of many well-known matrix factorization models such as PMF [20], CMF [24] and GMF [9]. That is, we obtain CMF by assigning w_G to the vector of all ones and setting $p^\delta = 0$; we get GMF by setting $p^\delta = 0$ and using only single domain ratings; and we obtain PMF by assigning w_G to the vector of all ones, setting $p^\delta = 0$ and using only single domain ratings.

3.2 Stacked Encoder-Decoder (SED)

Inspired by the ability to construct target language sentences from the source language in neural machine translation [1], we adopt an encoder-decoder network for cross-domain recommendation tasks. The proposed SED network is illustrated in Fig. 2 inside the dashed-lined box. For the sake of brevity, we include the feed-forward network that provides the item embedding (q^{SED}_j) from the one-hot encoding vector (y_j), and the SED embedding (ϕ^{SED}_{uj}) layer as a part of the SED network.

As we discussed earlier, it is essential to capture the non-linear relationships that exist among entities within and across domains. Hence, while constructing the target domain ratings from the source domain ratings, non-linear embeddings of the shared users and items are obtained as illustrated in Fig. 2.

The encoder part of the SED network is a feed-forward network that takes partially observed ratings for users from the source domain, $R^S_{u,:} \in \mathbb{R}^{n_S} \forall u \in \mathcal{U}$, projects them into a low dimensional ($z^u_{\frac{L}{2}} \in \mathbb{R}^k$) embedding space as follows:

$$z^u_{\frac{L}{2}} = a(W^U_{\frac{L}{2}}(a(W^U_{\frac{L}{2}-1}(\ldots a(W^U_1 R^S_{u,:} + b^U_1) \ldots) + b^U_{\frac{L}{2}-1})) + b^U_{\frac{L}{2}}), \tag{5}$$

where z^u_l, W^U_l and b^U_l denote the user representation, weight matrix and bias at layer l, respectively. At layer $\frac{L}{2}$, the embedding vector $z^u_{\frac{L}{2}}$ acts as a low-dimensional compact representation (p^{SED}_u) for the user u. Here, p^{SED}_u contains the transferable user behavior component for the user u from the source to the target domain.

Further, the decoder is also a feed-forward network that takes this low dimensional embedding ($z^u_{\frac{L}{2}}$) and constructs the user ratings for the target domain ($\hat{R}^T_{u,:} \in \mathbb{R}^{n_T}$) as follows:

$$z^u_L = a(W^U_L(a(W^U_{L-1}(\dots a(W^U_1 z^u_{\frac{L}{2}} + b^U_{\frac{L}{2}})\dots) + b^U_{L-1})) + b^U_L), \tag{6}$$

where, z^u_L is treated as the constructed ratings ($\hat{R}^T_{u,:}$) of the user u for the target domain. Hence, we have the following loss function for the source to the target domain rating construction:

$$\mathcal{L}_{S2T}(\Theta) = \frac{1}{|\Omega^T|} \sum_{u \in \mathcal{U}} \|(z^u_L - R^T_{u,:}) \odot \mathbb{1}(R^T_{u,:} > 0)\|^2_2, \tag{7}$$

where Θ is the set of parameters that contains all the weight matrices and bias vectors, and $\mathbb{1}(\cdot)$ denotes indicator function.

In addition to this, we get the item representation from the SED network as

$$q^{SED}_j = z^j_{\frac{L}{2}}, \quad \text{where } z^j_{\frac{L}{2}} = a(W^V_{\frac{L}{2}}(a(W^V_{\frac{L}{2}-1}(\dots a(W^V_1 Q^{SED} y_j + b^V_1)\dots) + b^V_{\frac{L}{2}-1})) + b^V_{\frac{L}{2}}). \tag{8}$$

Here $Q^{SED} \in \mathbb{R}^{n \times k}$ denotes the item embedding matrix and q^{SED}_j represents the embedding vector for the item j in the SED network. q^{SED}_j is further used along with p^{SED}_u to predict the target domain ratings as shown in Fig. 2. Note here that, in order to obtain low dimensional deep representations from the SED network and wide representations from the GCMF network for the items, we use respectively two separate embedding matrices Q^{SED} and Q^G.

Here, p^{SED}_u ($z^u_{\frac{L}{2}}$) acts as a component that transfers the user behavior to the target domain. Besides, p^{SED}_u is particularly helpful to obtain the representations for the users when they have a very few or no ratings in the target domain. Further, stacked layers help in extracting the complicated relationships present among the entities.

Similar to GCMF, we learn the representations ϕ^{SED}_{uj} and predict the rating $\hat{r}^{SED}_{uj}$ for the user-item pair (u, j) as follows:

$$\hat{r}^{SED}_{uj} = a(w'_{SED}\phi^{SED}_{uj}), \quad \text{where } \phi^{SED}_{uj} = p^{SED}_u \odot q^{SED}_j. \tag{9}$$

Here w_{SED} denotes the weight vector. The loss function of the SED network ($\mathcal{L}_{SED}$) contains two terms: the rating prediction loss ($\mathcal{L}_{pred}$) and the source to the target rating construction loss ($\mathcal{L}_{S2T}$). Let $\Omega = \Omega^S \cup \Omega^T$. The loss function is given as follows:

$$\mathcal{L}_{SED}(\Theta) = \frac{1}{|\Omega|} \sum_{(u,j)\in\Omega} (\hat{r}^{SED}_{uj} - r_{uj})^2 + \frac{1}{|\Omega^T|} \sum_{u\in\mathcal{U}} \|(\hat{R}^T_{u,:} - R^T_{u,:}) \odot \mathbb{1}(R^T_{u,:} > 0)\|^2_2. \tag{10}$$

3.3 Fusion of GCMF and SED

We fuse the GCMF and SED networks to get the best of both wide and deep representations. The embeddings for NeuCDCF, and the rating prediction ($\hat{r}_{uj}$) are obtained as follows:

$$\hat{r}_{uj} = a(w'\phi_{uj}), \quad \text{where } \phi_{uj} = \begin{bmatrix} \phi_{uj}^{G} \\ \phi_{uj}^{SED} \end{bmatrix}, \quad w = \begin{bmatrix} (1-\alpha)w_G \\ \alpha w_{SED} \end{bmatrix}. \tag{11}$$

Here $\alpha \in [0, 1]$ is a trade-off parameter between the GCMF and SED networks and it is tuned using cross-validation. When α is explicitly set to 0 or 1, NeuCDCF considers only the GCMF network or the SED network respectively.

The objective function for NeuCDCF is the sum of three individual loss functions and it is optimized in an end-to-end fashion.

$$\begin{aligned} \mathcal{L}(\Theta) &= \mathcal{L}_{pred} + \mathcal{L}_{S2T} + \mathcal{L}_{reg}, \\ \text{where } \mathcal{L}_{pred} &= \frac{1}{|\Omega^S \cup \Omega^T|} \sum_{(u,j)\in\Omega^S\cup\Omega^T} (\hat{r}_{uj} - r_{uj})^2, \\ \mathcal{L}_{S2T} = \frac{1}{|\Omega^T|} \sum_{u\in\mathcal{U}} &\|(\hat{R}^T_{u,:} - R^T_{u,:}) \odot \mathbb{1}(R^T_{u,:} > 0)\|_2^2, \quad \text{and } \mathcal{L}_{reg} = \mathcal{R}(\Theta). \end{aligned} \tag{12}$$

Here, $\mathcal{L}_{pred}$ is the prediction loss from both the source and the target domains together, $\mathcal{L}_{S2T}$ is the source to the target rating construction loss between the predicted ratings $\hat{R}^T_{u,:}$ and the available true ratings $R^T_{u,:}$ for all the users $u \in \mathcal{U}$, and $\mathcal{L}_{reg}$ is the regularization loss derived using all the model parameters. Here, $\mathcal{L}_{S2T}$ also acts as a regularizer that helps in properly learning p_u^{SED} and ϕ_{uj}^{SED}.

Note that the proposed architecture can easily be adapted to the implicit rating settings where only the binary ratings are available (e.g., clicks or likes) by changing the loss function from the squared error loss to cross-entropy [9] or pairwise loss [22] with appropriate negative sampling strategy. Since the datasets we experiment with contain real-valued ratings, the choice of squared error loss is natural.

4 Experiments

Here, we describe the conducted experiments to address the following questions:

RQ1 How do GCMF, SED and NeuCDCF perform under the different sparsity and cold-start settings?

RQ2 Does fusing wide and deep representations boost the performance?

RQ3 Is adding domain-specific component helpful in handling diversity and improving the performance of GCMF?

RQ4 Does capturing non-linear relationship by SED provide any benefits?

We first present the experimental settings and then answer the above questions.

Table 1. Statistics of the Amazon and Douban datasets

Dataset		# users	# items	# ratings	Density	Role
Amazon	Movie (M)	3677	7549	222095	0.80 %	Source
	Book (B)	3677	9919	183763	0.50 %	Target
	Movie (M)	2118	5099	223383	2.07 %	Source
	Music (Mu)	2118	4538	91766	0.95 %	Target
Douban	Movie (M)	5674	11676	1863139	2.81 %	Source
	Book (B)	5674	6626	432642	1.15 %	Target
	Music (Mu)	3628	8118	554075	1.88 %	Source
	Book (B)	3628	4893	293488	1.65 %	Target

4.1 Experimental Setup

Datasets. We adopt four widely used datasets: two from Amazon – an e-commerce product recommendation system, and two from Douban [32] – a Chinese online community platform for movies, books and music recommendations. All the datasets are publicly available[8,9] and have the ratings from multiple domains. We do the following preprocessing similar to the one done in [5,19]: we retain (1) only those ratings associated with the users who are common between both the domains and (2) the users and items that have at least 10 and 20 ratings (different thresholds are used here to have diversity in density across datasets) for Amazon and Douban datasets, respectively. Statistics of the datasets are given in Table 1.

Evaluation Procedure. We randomly split the target domain ratings of the datasets into training (65%), validation (15%), and test (20%) sets. In this process, we make sure that at least one rating for every user and every item is present in the training set. For the CDCF models, we include the source domain ratings in the above-mentioned training set. Each experiment is repeated five times for different random splits of the dataset and we report the average test set performance and standard deviation with respect to the best validation error as the final results. We evaluate our proposed models (GCMF, SED and NeuCDCF) against baseline models under the following 3 experimental settings:

1. **Sparsity:** In this setting, we remove $K\%$ of the target domain ratings from the training set. We use different values of K from $\{0, 20, 40, 60, 80\}$ to get different sparsity levels.
2. **Cold-start:** With the above setting, we report the results obtained for only those users who have less than five ratings in the target domain of the training set. We call them cold-start users. The similar setting was defined and followed in [7]. We do not include 0% and 20% sparsity levels here as the respective splits have a very few cold-start users.

[8] http://jmcauley.ucsd.edu/data/amazon (we rename CDs-and-Vinyl as Music).
[9] https://sites.google.com/site/erhengzhong/datasets.

Table 2. Sparsity – Performance of different models on four real-world datasets at different sparsity levels. Here, $K\%$ indicates the percentage of the target domain ratings removed from the training set. The bold-faced values indicate the best performance for every dataset.

Sparsity setting						
Dataset	Model	$K = 0$	$K = 20$	$K = 40$	$K = 60$	$K = 80$
Amazon (M-B)	PMF [20]	0.7481 ± 0.0042	0.7828 ± 0.0061	0.8449 ± 0.0178	0.9640 ± 0.0035	1.2223 ± 0.0145
	GMF [9]	0.7029 ± 0.0110	0.7108 ± 0.0056	0.7254 ± 0.0078	0.7805 ± 0.0151	0.8878 ± 0.0207
	GMF-CD [9]	0.6961 ± 0.0012	0.7085 ± 0.0056	0.7186 ± 0.0018	0.7552 ± 0.0236	0.8009 ± 0.0325
	CMF [24]	0.7242 ± 0.0022	0.7434 ± 0.0044	0.7597 ± 0.0019	0.7919 ± 0.0039	0.8652 ± 0.0032
	MV-DNN [4]	0.7048 ± 0.0024	0.7160 ± 0.0047	0.7192 ± 0.0035	0.7255 ± 0.0019	0.7624 ± 0.0019
	GCMF (ours)	0.6705 ± 0.0019	0.6862 ± 0.0052	0.7023 ± 0.0025	0.7254 ± 0.0023	0.7873 ± 0.0030
	SED (ours)	0.6880 ± 0.0033	0.7024 ± 0.0060	0.7094 ± 0.0030	0.7194 ± 0.0023	0.7428 ± 0.0024
	NeuCDCF (ours)	**0.6640 ± 0.0018**	**0.6786 ± 0.0052**	**0.6852 ± 0.0025**	**0.6916 ± 0.0010**	**0.7291 ± 0.0060**
Amazon (M-Mu)	PMF [20]	0.7135 ± 0.0105	0.7366 ± 0.0050	0.7926 ± 0.0118	0.9324 ± 0.0174	1.2080 ± 0.0186
	GMF [9]	0.6583 ± 0.0056	0.6731 ± 0.0090	0.7008 ± 0.0083	0.7268 ± 0.0194	0.8331 ± 0.0046
	GMF-CD [9]	0.6628 ± 0.0025	0.6666 ± 0.0047	0.6883 ± 0.0175	0.6997 ± 0.0061	0.7566 ± 0.0277
	CMF [24]	0.6923 ± 0.0041	0.7071 ± 0.0013	0.7298 ± 0.0103	0.7594 ± 0.0071	0.8279 ± 0.0046
	MV-DNN [4]	0.6843 ± 0.0017	0.6827 ± 0.0046	0.6898 ± 0.0077	0.6905 ± 0.0037	0.7018 ± 0.0031
	GCMF (ours)	0.6173 ± 0.0053	0.6261 ± 0.0042	0.6527 ± 0.0111	0.6809 ± 0.0040	0.7436 ± 0.0038
	SED (ours)	0.6406 ± 0.0032	0.6430 ± 0.0040	0.6597 ± 0.0114	0.6770 ± 0.0039	0.7006 ± 0.0044
	NeuCDCF (ours)	**0.6095 ± 0.0043**	**0.6170 ± 0.0044**	**0.6318 ± 0.0101**	**0.6535 ± 0.0050**	**0.6937 ± 0.0061**
Douban (M-B)	PMF [20]	0.5733 ± 0.0019	0.5811 ± 0.0018	0.5891 ± 0.0020	0.6136 ± 0.0016	0.7141 ± 0.0066
	GMF [9]	0.5730 ± 0.0016	0.5747 ± 0.0017	0.5781 ± 0.0033	0.5971 ± 0.0071	0.6133 ± 0.0072
	GMF-CD [9]	0.5803 ± 0.0026	0.5836 ± 0.0029	0.5879 ± 0.0031	0.6002 ± 0.0089	0.6117 ± 0.0078
	CMF [24]	0.5771 ± 0.0006	0.5821 ± 0.0014	0.5862 ± 0.0012	0.5979 ± 0.0025	0.6188 ± 0.0016
	MV-DNN [4]	0.5956 ± 0.0005	0.6009 ± 0.0019	0.6039 ± 0.0011	0.6127 ± 0.0023	0.6224 ± 0.0015
	GCMF (ours)	0.5608 ± 0.0009	0.5664 ± 0.0023	0.5709 ± 0.0016	0.5849 ± 0.0023	0.6118 ± 0.0018
	SED (ours)	0.5822 ± 0.0003	0.5865 ± 0.0021	0.5905 ± 0.0018	0.6011 ± 0.0029	0.6124 ± 0.0022
	NeuCDCF (ours)	**0.5603 ± 0.0009**	**0.5647 ± 0.0021**	**0.5704 ± 0.0014**	**0.5800 ± 0.0023**	**0.5957 ± 0.0019**
Douban (Mu-B)	PMF [20]	0.5750 ± 0.0022	0.5800 ± 0.0016	0.5894 ± 0.0033	0.6146 ± 0.0037	0.7319 ± 0.0099
	GMF [9]	0.5745 ± 0.0033	0.5768 ± 0.0036	0.5765 ± 0.0018	0.5900 ± 0.0051	0.6241 ± 0.0153
	GMF-CD [9]	0.5825 ± 0.0023	0.5847 ± 0.0040	0.5883 ± 0.0040	0.5962 ± 0.0067	0.6137 ± 0.0037
	CMF [24]	0.5827 ± 0.0017	0.5881 ± 0.0015	0.5933 ± 0.0024	0.6035 ± 0.0017	0.6231 ± 0.0019
	MV-DNN [4]	0.5892 ± 0.0015	0.5918 ± 0.0013	0.5946 ± 0.0016	0.6039 ± 0.0022	0.6180 ± 0.0022
	GCMF (ours)	0.5675 ± 0.0012	0.5707 ± 0.0013	0.5768 ± 0.0018	0.5905 ± 0.0020	0.6187 ± 0.0021
	SED (ours)	0.5769 ± 0.0024	0.5782 ± 0.0019	0.5839 ± 0.0020	0.5934 ± 0.0023	0.6090 ± 0.0025
	NeuCDCF (ours)	**0.5625 ± 0.0015**	**0.5646 ± 0.0013**	**0.5688 ± 0.0017**	**0.5794 ± 0.0021**	**0.5954 ± 0.0024**

3. **Full-cold-start:** We remove all the ratings of $K\%$ of users from the target domain and call them as full-cold-start users since these users have no ratings at all in the target domain of the training set. We use different values of K from {10, 20, 30, 40, 50} to get different full-cold-start levels. This particular setting was followed in [19].

Following [2,7,14,15], we employ Mean Absolute Error (MAE) for performance analysis for real-valued rating setting. Here, smaller values of MAE indicate better prediction.

Comparison of Different Models. To evaluate the performance of NeuCDCF and its individual networks (GCMF and SED) in the CDR setting, we compare our models with the representative models from the following three categories:

(1) Deep neural network-based CDCF models (Deep):

- **MV-DNN** [4]: It is one of the state-of-the-art neural network models for the cross-domain recommendation. It learns the embeddings of the shared entities

Table 3. Cold-start – Performance of different models on four real-world datasets at different sparsity levels. Here, $K\%$ indicates the percentage of the target domain ratings removed from the training set. The bold-faced values indicate the best performance for every dataset.

Cold-start setting				
Dataset	Model	$K = 40$	$K = 60$	$K = 80$
Amazon (M-B)	PMF [20]	1.1063 ± 0.0501	1.1476 ± 0.0250	1.5210 ± 0.0316
	GMF [9]	0.9040 ± 0.0704	1.0721 ± 0.0791	1.0735 ± 0.0250
	CMF [24]	0.8617 ± 0.0327	0.8734 ± 0.0126	0.9362 ± 0.0129
	MV-DNN [4]	0.8052 ± 0.0293	0.8135 ± 0.0167	0.8347 ± 0.0243
	GCMF (ours)	0.8086 ± 0.0296	0.8220 ± 0.0158	0.8792 ± 0.0144
	SED (ours)	0.7925 ± 0.0301	0.8002 ± 0.0157	0.8169 ± 0.0118
	NeuCDCF (ours)	**0.7830 ± 0.0304**	**0.7791 ± 0.0165**	**0.8015 ± 0.0157**
Amazon (M-Mu)	PMF [20]	1.0552 ± 0.0315	1.1734 ± 0.0577	1.5235 ± 0.0330
	GMF [9]	0.8425 ± 0.0687	0.8490 ± 0.0345	1.0013 ± 0.0155
	CMF [24]	0.8359 ± 0.0320	0.8245 ± 0.0139	0.8746 ± 0.0194
	MV-DNN [4]	0.7320 ± 0.0253	**0.7075 ± 0.0313**	**0.7241 ± 0.0157**
	GCMF (ours)	0.7537 ± 0.0262	0.7600 ± 0.0252	0.8295 ± 0.0147
	SED (ours)	0.7281 ± 0.0314	0.7176 ± 0.0265	0.7437 ± 0.0156
	NeuCDCF (ours)	**0.7144 ± 0.0232**	0.7134 ± 0.0267	0.7390 ± 0.0161
Douban (M-B)	PMF [20]	0.8654 ± 0.0593	0.7527 ± 0.0199	0.8275 ± 0.0135
	GMF [9]	0.6205 ± 0.0530	0.7290 ± 0.0740	0.6625 ± 0.0371
	CMF [24]	0.5490 ± 0.0711	0.5998 ± 0.0156	0.6171 ± 0.0069
	MV-DNN [4]	0.5624 ± 0.0807	0.6110 ± 0.0165	0.6206 ± 0.0113
	GCMF (ours)	0.5495 ± 0.0625	0.5983 ± 0.0172	0.6220 ± 0.0037
	SED (ours)	0.5462 ± 0.0671	0.5983 ± 0.0159	0.6105 ± 0.0050
	NeuCDCF (ours)	**0.5372 ± 0.0699**	**0.5911 ± 0.0157**	**0.6031 ± 0.0033**
Douban (Mu-B)	PMF [20]	0.9201 ± 0.0951	0.7629 ± 0.0237	0.8451 ± 0.0161
	GMF [9]	0.5489 ± 0.0594	0.6873 ± 0.0731	0.6964 ± 0.0634
	CMF [24]	0.5875 ± 0.0543	0.6081 ± 0.0208	0.6194 ± 0.0060
	MV-DNN [4]	0.5485 ± 0.0367	0.6131 ± 0.0198	0.6089 ± 0.0049
	GCMF (ours)	0.5220 ± 0.0717	0.6129 ± 0.0212	0.6205 ± 0.0047
	SED (ours)	0.5423 ± 0.0380	0.6011 ± 0.0239	0.5999 ± 0.0046
	NeuCDCF (ours)	**0.5200 ± 0.0482**	**0.5963 ± 0.0246**	**0.5954 ± 0.0046**

from the ratings of the source and the target domains combined using deep neural networks.

- **EMCDR** [19]: It is one of the state-of-the-art models for the full-cold-start setting. EMCDR is a two-stage neural network model for the cross-domain setting. In the first stage, it finds the embeddings of entities by leveraging matrix factorization techniques with respect to its individual domains. In the second stage, it learns a transfer function that provides target domain embeddings for the shared entities from its source domain embeddings.

(2) Matrix factorization based single domain models (Wide):

- **PMF** [20]: Probabilistic Matrix Factorization (PMF) is a standard and well-known baseline for single domain CF.

Table 4. Full-cold-start – Performance of different models on four real-world datasets for complete cold-start users at different cold-start levels. Here, $K\%$ indicates the percentage of users whose all ratings in the target domain are removed from the training set. The bold-faced values indicate the best performance.

Full-cold-start setting						
Dataset	Model	$K = 10$	$K = 20$	$K = 30$	$K = 40$	$K = 50$
Amazon (M-B)	CMF [24]	0.7870 ± 0.0079	0.7856 ± 0.0115	0.8017 ± 0.0286	0.8214 ± 0.0170	0.8590 ± 0.0327
	EMCDR [19]	0.7340 ± 0.0042	0.7324 ± 0.0089	0.7436 ± 0.0195	0.7831 ± 0.0167	0.8067 ± 0.0211
	MV-DNN [4]	0.7401 ± 0.0038	0.7366 ± 0.0086	0.7328 ± 0.0161	0.7415 ± 0.0047	0.7483 ± 0.0075
	GCMF (ours)	0.7233 ± 0.0076	0.7208 ± 0.0080	0.7226 ± 0.0177	0.7371 ± 0.0081	0.7590 ± 0.0034
	SED (ours)	0.7181 ± 0.0040	0.7108 ± 0.0062	0.7115 ± 0.0151	0.7178 ± 0.0060	0.7277 ± 0.0044
	NeuCDCF (ours)	**0.7096 ± 0.0073**	**0.7012 ± 0.0077**	**0.6983 ± 0.0171**	**0.7091 ± 0.0077**	**0.7215 ± 0.0043**
Amazon (M-Mu)	CMF [24]	0.7467 ± 0.0138	0.7413 ± 0.0123	0.7600 ± 0.0220	0.7865 ± 0.0146	0.8348 ± 0.0309
	EMCDR [19]	0.6781 ± 0.0147	0.6910 ± 0.0104	0.7349 ± 0.0212	0.7682 ± 0.0172	0.7917 ± 0.0272
	MV-DNN [4]	0.6848 ± 0.0270	0.6923 ± 0.0170	0.6833 ± 0.0274	0.7037 ± 0.0116	0.7178 ± 0.0135
	GCMF (ours)	0.6658 ± 0.0237	0.6766 ± 0.0154	0.6635 ± 0.0237	0.6943 ± 0.0118	0.7280 ± 0.0104
	SED (ours)	0.6523 ± 0.0281	0.6544 ± 0.0181	0.6452 ± 0.0214	0.6693 ± 0.0115	0.6864 ± 0.0098
	NeuCDCF (ours)	**0.6418 ± 0.0257**	**0.6471 ± 0.0183**	**0.6384 ± 0.0222**	**0.6650 ± 0.0107**	**0.6856 ± 0.0095**
Douban (M-B)	CMF [24]	0.5919 ± 0.0052	0.5987 ± 0.0062	0.5970 ± 0.0055	0.5983 ± 0.0037	0.6004 ± 0.0012
	EMCDR [19]	0.5942 ± 0.0069	0.5926 ± 0.0071	0.5973 ± 0.0064	0.5996 ± 0.0032	0.6112 ± 0.0031
	MV-DNN [4]	0.6004 ± 0.0071	0.6106 ± 0.0069	0.6062 ± 0.0062	0.6068 ± 0.0025	0.6075 ± 0.0032
	GCMF (ours)	0.5852 ± 0.0070	0.5923 ± 0.0074	0.5919 ± 0.0044	0.5930 ± 0.0021	0.5932 ± 0.0022
	SED (ours)	0.5897 ± 0.0072	0.5998 ± 0.0094	0.5955 ± 0.0066	0.5965 ± 0.0026	0.5978 ± 0.0032
	NeuCDCF (ours)	**0.5807 ± 0.0065**	**0.5877 ± 0.0072**	**0.5858 ± 0.0048**	**0.5864 ± 0.0019**	**0.5869 ± 0.0020**
Douban (Mu-B)	CMF [24]	0.6063 ± 0.0112	0.5971 ± 0.0048	0.5995 ± 0.0025	0.6000 ± 0.0019	0.5997 ± 0.0021
	EMCDR [19]	0.6058 ± 0.0072	0.6021 ± 0.0046	0.6083 ± 0.0034	0.6102 ± 0.0033	0.6092 ± 0.0042
	MV-DNN [4]	0.6145 ± 0.0112	0.6053 ± 0.0033	0.6072 ± 0.0031	0.6059 ± 0.0025	0.6035 ± 0.0035
	GCMF (ours)	0.6142 ± 0.0054	0.6044 ± 0.0030	0.6115 ± 0.0037	0.6082 ± 0.0025	0.6078 ± 0.0039
	SED (ours)	0.6037 ± 0.0119	0.5936 ± 0.0045	0.5962 ± 0.0033	0.5940 ± 0.0023	0.5927 ± 0.0025
	NeuCDCF (ours)	**0.5978 ± 0.0098**	**0.5880 ± 0.0042**	**0.5927 ± 0.0029**	**0.5898 ± 0.0017**	**0.5885 ± 0.0030**

- **GMF**[9]: Generalized Matrix Factorization (GMF) is a state-of-the-art model for single domain recommendation settings proposed as a part of NeuMF [9].

(3) Matrix factorization based CDCF models (Wide):

- **GMF-CD**[9]: It is the same model as the GMF model for CDR where both the source and the target domain ratings are used for training.
- **CMF** [24]: Collective Matrix Factorization (CMF) is a standard baseline for CDR when entities are shared across the domains.

Parameter Settings and Reproducibility. We implemented our models using Tensorflow 1.12. We use the squared error loss as an optimization loss across all models to have a fair comparison [9,17]. We use $L2$ regularizer for matrix factorization models (PMF and CMF) and dropout regularizer for neural network models. Hyperparameters were tuned using cross-validation and test error corresponding to the best performing model on the validation set is reported. In particular, different values used for different parameters are: λ for $L2$ regularization from {0.0001, 0.0005, 0.005, 0.001, 0.05, 0.01, 0.5}, dropout from {0.1, 0.2, 0.3, 0.4, 0.5, 0.6}, embedding dimension (i.e., number of factors k) from {8, 16, 32, 48, 64, 80} and α value from {0.05, 0.1, 0.2, 0.4, 0.6, 0.8, 0.9, 0.95}. We train the models for a maximum of 120 epochs with early stopping criterion. We randomly initialize the model parameters with normal distribution with 0 mean and 0.002 standard deviation. We adopt RMSProp [6] with mini-batch optimization procedure. We tested the learning rate (η) of {0.0001, 0.0005, 0.001, 0.005, 0.01}, batch size of {128, 256, 512, 1024}. From the validation set performance, we set η to 0.002 for PMF and CMF, 0.001 for MV-DNN, 0.005 for other models; dropout to 0.5; batch size to 512. In the SED network, for the encoder, we use 3 hidden layers with the number of neurons: $4k \rightarrow 2k \rightarrow k$ and the sigmoid activation function, and for the decoder, we use the same as the encoder with the order of the number of hidden layer neurons reversed, and k is tuned using the validation set.

4.2 Results and Discussion

Tables 2, 3 and 4 detail the comparison results of our proposed models GCMF, SED and NeuCDCF, and baseline models at different sparsity and cold-start levels on four datasets. In addition, we conduct a paired t-test and the improvements of NeuCDCF over each of the compared models passed with significance value $p < .01$. The findings are summarized below.

Sparsity and Cold-Start Settings (RQ1). As we see from Table 2, NeuCDCF outperforms the other models at different sparsity levels. In particular, the performance of GCMF is better when the target domain is less sparse, but, the performance of the SED model improves eventually when sparsity increases. This phenomenon is more obvious on the Amazon datasets. Furthermore, adding the source domain helps in recommendation performance. This is evident from the performance of GCMF as compared to its counterparts – PMF and GMF (single-domain models).

We have two different settings for the cold-start analysis: cold-start and full-cold-start, and the performance of the recommendation models is given in Tables 3 and 4. The overall performance of NeuCDCF, in particular the SED model, is better than the other comparison models. Since the SED model adapts and transfers the required knowledge from the source domain to the target domain rating construction, it helps the cold-start entities to learn better representations. This demonstrates that a network such as SED is important in learning the non-linear representation for cold-start entities.

GCMF and SED Integration (RQ2). NeuCDCF consistently performs better than all the models that are either based on wide or deep framework including the proposed GCMF and SED networks. This result can be inferred from Tables 2, 3 and 4. We thus observe that the individual networks – GCMF (wide) and SED (deep) provide different complementary knowledge about the relationships that exist among the entities within and across domains.

Domain-Specific Embeddings (RQ3). Adding source domain blindly, without taking into account of domain-specific embedding, might deteriorate the performance. This is illustrated in Figs. 3(a) and 3(b) using Douban datasets. When sparsity is less, surprisingly the performance of the single-domain models (PMF and GMF) is better than that of the cross-domain models (CMF and

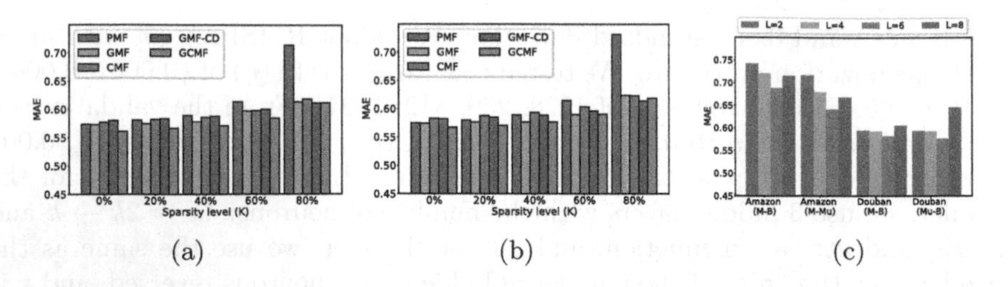

(a) (b) (c)

Fig. 3. Performance of GCMF model and its counterparts – PMF, GMF, GMF-CD and CMF in MAE with respect to the various sparsity levels on (a) Douban (Movie-Book) dataset, (b) Douban (Music-Book) dataset. (c) SED performance in MAE with respect to the number of layers (L) on all the datasets.

GMF-CD). However, GCMF performs better than the counterparts – PMF, GMF, GMF-CD and CMF in almost all the cases. These results demonstrate that GCMF provides a more principled way to understand the diversity among the domains using domain-specific embeddings.

Deep Non-linear Representations (RQ4). Having deep representations helps in improving the performance when the sparsity increases in the target domain. This is demonstrated in Table 2 by the SED model and supported by MV-DNN particularly on the Amazon datasets. Despite both being deep networks, the SED model performs better than MV-DNN. This is because MV-DNN learns the embeddings of shared users from the source and the target domain together, whereas, in the SED model, the embeddings are learned appropriately to obtain the target domain ratings using the source domain ratings. Further, we show the performance of the SED model for the different number of layers (L) on all the datasets in Fig. 3(c). This shows that the performance improves when we increase the number of layers (L) from 2 to 6. In other words, using deep representations in the SED model helps in boosting the performance. When $L = 8$ the performance decreases because of the overfitting due to the high complexity of the model.

5 Related Work

In the literature of CDR, early works [2,5,14,15,24] mainly adopt matrix factorization models. In particular, [14] constructs a cluster-level rating matrix (codebook) from user-item rating patterns and through which it establishes links to transfer the knowledge across domains. A similar approach with an extension to soft-membership was proposed in [15]. Collective matrix factorization (CMF) [24] was proposed for the case where the entities participate in more than one relation. However, as many studies pointed out, MF models may not handle non-linearity and complex relationships present in the system [9,18,30].

On the other hand, recently, there has been a surge in methods proposed to explore deep learning networks for recommendation systems [30]. Most of the models in this category focus on utilizing neural network models for extracting embeddings from side information such as reviews [31], descriptions [12], content information [26] and knowledge graphs [29]. Nevertheless, many of these models are traces to matrix factorization models, that is, in the absence of side information, these models distill to either MF [13] or PMF [20].

Further, to combine the advantages of both matrix factorization models and deep networks such as multi-layer perceptron (MLP), some models have been proposed [3,9] for learning representations from the available ratings with no side information. These models combine both the wide and deep networks together to provide better representations. Stacked denoising autoencoder [28], variational autoencoder [16], and recurrent neural networks [27] have also been exploited for recommendation systems. However, the above neural network models use only the interaction between users and items from a single domain. Hence, they suffer from the aforementioned issues such as sparsity and cold-start.

In recent years, attempts have been made to utilize neural network models for cross-domain recommendation [4,11,17,19,33,34]. In particular, MV-DNN [4] uses an MLP to learn shared representations of the entities common across multiple domains. A factorization based multi-view neural network was proposed in CCCFNet [17], where the representations learned from multiple domains are coupled with the representations learned from content information. A two-stage approach was followed in [11,19,34], wherein the first stage, user embeddings are learned, and in the second stage, a function is learned to map the user from the source domain to the target domain.

While the models [4,11,17,19,34] consider learning embeddings together, they completely ignore the domain-specific representations of the shared users or items. The performance of these models [4,17] is highly dependent on the relatedness of the domains. In contrast, our proposed model learns domain-specific representations that significantly improves the prediction performance. Further, [17] rely on content information to bridge the source and the target domains. Besides, all of these models [4,11,17,19,33,34] are either based on wide or deep networks but not both. We are also aware of the models proposed for cross-domain settings [10,17,33]. However, they differ from the research scope of ours because they bridge the source and the target domains using available side information.

6 Conclusion

In this work, we proposed a novel end-to-end neural network model, NeuCDCF which is based on a wide and deep framework. NeuCDCF addresses the main challenges in CDR – diversity and learning complex non-linear relationships among entities in a more systematic way. Through extensive experiments, we showed the suitability of our model for various sparsity and cold-start settings.

The proposed framework is general and can be easily extended to a multi-domain setting as well as the setting where a subset of entities is shared. Further, it is applicable to ordinal and top-N recommendation settings with the only modification in the final loss function. NeuCDCF is proposed for rating only settings when no side information is available. If side information is available, it can easily be incorporated as a basic building block in place of matrix factorization to extract effective representations from user-item interactions.

References

1. Bahdanau, D., Cho, K., Bengio, Y.: Neural machine translation by jointly learning to align and translate. In: ICLR (2015)
2. Cantador, I., Fernández-Tobías, I., Berkovsky, S., Cremonesi, P.: Cross-domain recommender systems. In: Recommender Systems Handbook, pp. 919–959 (2015)
3. Cheng, H.T., et al.: Wide & deep learning for recommender systems. In: RecSys, pp. 7–10. ACM (2016)
4. Elkahky, A.M., Song, Y., He, X.: A multi-view deep learning approach for cross domain user modeling in recommendation systems. In: WWW, pp. 278–288 (2015)
5. Gao, S., Luo, H., Chen, D., Li, S., Gallinari, P., Guo, J.: Cross-domain recommendation via cluster-level latent factor model. In: ECML PKDD (2013)
6. Goodfellow, I., Bengio, Y., Courville, A., Bengio, Y.: Deep learning, vol. 1. MIT Press, Cambridge (2016)
7. Guo, G., Zhang, J., Yorke-Smith, N.: TrustSVD: collaborative filtering with both the explicit and implicit influence of user trust and of item ratings. In: AAAI (2015)
8. He, M., Zhang, J., Yang, P., Yao, K.: Robust transfer learning for cross-domain collaborative filtering using multiple rating patterns approximation. In: WSDM, pp. 225–233. ACM (2018)
9. He, X., Liao, L., Zhang, H., Nie, L., Hu, X., Chua, T.S.: Neural collaborative filtering. In: WWW, pp. 173–182 (2017)
10. Kanagawa, H., Kobayashi, H., Shimizu, N., Tagami, Y., Suzuki, T.: Cross-domain recommendation via deep domain adaptation. In: ECIR, pp. 20–29 (2019)
11. Kang, S., Hwang, J., Lee, D., Yu, H.: Semi-supervised learning for cross-domain recommendation to cold-start users. In: CIKM, pp. 1563–1572 (2019)
12. Kim, D., Park, C., Oh, J., Lee, S., Yu, H.: Convolutional matrix factorization for document context-aware recommendation. In: RecSys, pp. 233–240. ACM (2016)
13. Koren, Y., Bell, R., Volinsky, C.: Matrix factorization techniques for recommender systems. Computer **8**, 30–37 (2009)
14. Li, B., Yang, Q., Xue, X.: Can movies and books collaborate? Cross-domain collaborative filtering for sparsity reduction. In: IJCAI, vol. 9, pp. 2052–2057 (2009)
15. Li, B., Yang, Q., Xue, X.: Transfer learning for collaborative filtering via a rating-matrix generative model. In: ICML, pp. 617–624 (2009)
16. Li, X., She, J.: Collaborative variational autoencoder for recommender systems. In: SIGKDD, pp. 305–314 (2017)
17. Lian, J., Zhang, F., Xie, X., Sun, G.: CCCFNet: a content-boosted collaborative filtering neural network for cross domain recommender systems. In: WWW (2017)
18. Liu, Y.F., Hsu, C.Y., Wu, S.H.: Non-linear cross-domain collaborative filtering via hyper-structure transfer. In: ICML, pp. 1190–1198 (2015)
19. Man, T., Shen, H., Jin, X., Cheng, X.: Cross-domain recommendation: an embedding and mapping approach. In: IJCAI, pp. 2464–2470. AAAI Press (2017)
20. Mnih, A., Salakhutdinov, R.R.: Probabilistic matrix factorization. In: NeurIPS, pp. 1257–1264 (2008)
21. Rendle, S.: Factorization machines. In: ICDM, pp. 995–1000. IEEE (2010)

22. Rendle, S., Freudenthaler, C., Gantner, Z., Schmidt-Thieme, L.: BPR: Bayesian personalized ranking from implicit feedback. In: UAI (2009)
23. Sahebi, S., Brusilovsky, P.: It takes two to tango: an exploration of domain pairs for cross-domain collaborative filtering. In: RecSys, pp. 131–138. ACM (2015)
24. Singh, A.P., Gordon, G.J.: Relational learning via collective matrix factorization. In: SIGKD, pp. 650–658. ACM (2008)
25. Su, X., Khoshgoftaar, T.M.: A survey of collaborative filtering techniques. In: Advances in Artificial Intelligence (2009)
26. Wang, H., Wang, N., Yeung, D.Y.: Collaborative deep learning for recommender systems. In: SIGKDD, pp. 1235–1244. ACM (2015)
27. Wu, C.Y., Ahmed, A., Beutel, A., Smola, A.J., Jing, H.: Recurrent recommender networks. In: WSDM, pp. 495–503. ACM (2017)
28. Wu, Y., DuBois, C., Zheng, A.X., Ester, M.: Collaborative denoising auto-encoders for top-n recommender systems. In: WSDM, pp. 153–162. ACM (2016)
29. Zhang, F., Yuan, N.J., Lian, D., Xie, X., Ma, W.Y.: Collaborative knowledge base embedding for recommender systems. In: SIGKDD, pp. 353–362. ACM (2016)
30. Zhang, S., Yao, L., Sun, A., Tay, Y.: Deep learning based recommender system: a survey and new perspectives. CSUR **52**(1), 5 (2019)
31. Zheng, L., Noroozi, V., Yu, P.S.: Joint deep modeling of users and items using reviews for recommendation. In: WSDM, pp. 425–434. ACM (2017)
32. Zhong, E., Fan, W., Yang, Q.: User behavior learning and transfer in composite social networks. TKDD **8**(1), 6 (2014)
33. Zhu, F., Chen, C., Wang, Y., Liu, G., Zheng, X.: DTCDR: a framework for dual-target cross-domain recommendation. In: CIKM, pp. 1533–1542 (2019)
34. Zhu, F., Wang, Y., Chen, C., Liu, G., Orgun, M.A., Wu, J.: A deep framework for cross-domain and cross-system recommendations. In: IJCAI, pp. 3711–3717 (2018)

NoisyCUR: An Algorithm for Two-Cost Budgeted Matrix Completion

Dong Hu(✉), Alex Gittens(✉), and Malik Magdon-Ismail(✉)

Rensselaer Polytechnic Institute, Troy, NY 12180, USA
{hud3,gittea}@rpi.edu, magdon@cs.rpi.edu

Abstract. Matrix completion is a ubiquitous tool in machine learning and data analysis. Most work in this area has focused on the number of observations necessary to obtain an accurate low-rank approximation. In practice, however, the cost of observations is an important limiting factor, and experimentalists may have on hand multiple modes of observation with differing noise-vs-cost trade-offs. This paper considers matrix completion subject to such constraints: a budget is imposed and the experimentalist's goal is to allocate this budget between two sampling modalities in order to recover an accurate low-rank approximation. Specifically, we consider that it is possible to obtain low noise, high cost observations of individual entries or high noise, low cost observations of entire columns. We introduce a regression-based completion algorithm for this setting and experimentally verify the performance of our approach on both synthetic and real data sets. When the budget is low, our algorithm outperforms standard completion algorithms. When the budget is high, our algorithm has comparable error to standard nuclear norm completion algorithms and requires much less computational effort.

Keywords: Matrix completion · Low-rank approximation · Nuclear norm minimization

1 Introduction

Matrix completion (MC) is a powerful and widely used tool in machine learning, finding applications in information retrieval, collaborative filtering, recommendation systems, and computer vision. The goal is to recover a matrix $\boldsymbol{A} \in \mathbb{R}^{m \times n}$ from only a few, potentially noisy, observations $\mathbf{y} \in \mathbb{R}^d$, where $d \ll mn$.

In general, the MC problem is ill-posed, as many matrices may give rise to the same set of observations. Typically the inversion problem is made feasible by assuming that the matrix from which the observations were generated is in fact low-rank, $\text{rank}(\boldsymbol{A}) = r \ll \min\{m, n\}$. In this case, the number of degrees of

---

Electronic supplementary material The online version of this chapter (https://doi.org/10.1007/978-3-030-67658-2_43) contains supplementary material, which is available to authorized users.

F. Hutter et al. (Eds.): ECML PKDD 2020, LNAI 12457, pp. 746–761, 2021.
https://doi.org/10.1007/978-3-030-67658-2_43

freedom in the matrix is $(n+m)r$, so if the observations are sufficiently diverse, then the inversion process is well-posed.

In the majority of the MC literature, the mapping from the matrix to its observations, although random, is given to the user, and the aim is to design algorithms that minimize the sample complexity under these observation models. Some works have considered modifications of this paradigm, where the user designs the observation mapping themselves in order to minimize the number of measurements needed [6,14,21].

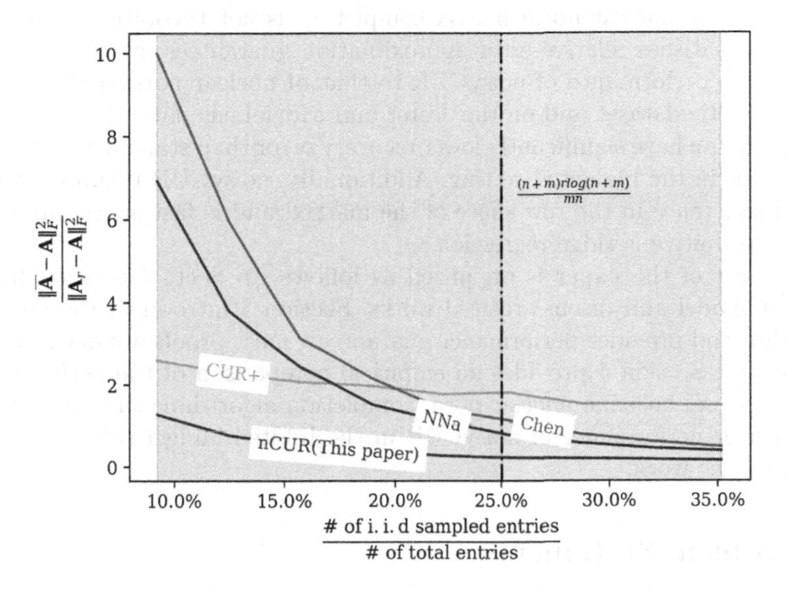

Fig. 1. noisyCUR, a regression-based MC algorithm (CUR+), and two nuclear norm MC algorithms applied to the budgeted MC problem on a synthetic incoherent matrix $\boldsymbol{A} \in \mathbb{R}^{80\times 60}$. See Sect. 4 for more details on the experimental setup. The performance of each method at each budget level is averaged over 10 runs, with hyper-parameters selected at each value of d by cross-validation. Within the red region the budget is small enough that not enough entries can be sampled for nuclear norm methods to have theoretical support. In this regime, noisyCUR significantly outperforms the baseline algorithms.

This paper considers a budgeted variation of noisy matrix completion with competing observation models: the goal is to allocate a finite budget between the observation models to maximize the accuracy of the recovery. This setup is natural in the context of experimental design: where, given competing options, the natural goal of an experimenter is to spend their budget in a way that maximizes the accuracy of their imputations. Specifically, this work considers a two-cost model where either entire columns can be observed with high per-entry error but low amortized per-entry cost, or entries can be observed with low per-entry error but high per-entry cost. This is a natural model in, for instance, recommender systems applications, where one has a finite budget to allocate between incentivizing users to rate either entire categories of products or individual items. The former corresponds to inexpensive, high-noise measurements of entire columns and the latter corresponds to expensive, low-noise measurements of individual entries.

The noisyCUR algorithm is introduced for this two-cost budgeted MC problem, and guarantees are given on its recovery accuracy. Figure 1 is illustrative of the usefulness of noisyCUR in this setting, compared to standard nuclear norm MC approaches and another regression-based MC algorithm. Even in low-budget settings where nuclear norm matrix completion is not theoretically justifiable, noisyCUR satisfies relative-error approximation guarantees. Empirical comparisons of the performance of noisyCUR to that of nuclear norm MC approaches on a synthetic dataset and on the Jester and MovieLens data sets confirm that noisyCUR can have significantly lower recovery error than standard nuclear norm approaches in the budgeted setting. Additionally, noisyCUR tolerates the presence of coherence in the row-space of the matrix, and is fast as its core computational primitive is ridge regression.

The rest of the paper is organized as follows. In Sect. 2 we introduce the two-cost model and discuss related works. Section 3 introduces the noisyCUR algorithm and provides performance guarantees; most proofs are deferred to the supplement. Section 4 provides an empirical comparison of the performance of noisyCUR and baseline nuclear norm completion algorithms that demonstrates the superior performance of noisyCUR in the limited budget setting. Section 5 concludes the work.

2 Problem Statement

2.1 Notation

Throughout this paper, scalars are denoted by lowercase letters (n), vectors by bolded lowercase letters ($\mathbf{x}$), and matrices by bolded uppercase letters $\boldsymbol{A}$. The spectral, Frobenius, and nuclear norms of $\boldsymbol{A} \in \mathbb{R}^{m \times n}$ are written $\|\boldsymbol{A}\|_2$, $\|\boldsymbol{A}\|_F$, and $\|\boldsymbol{A}\|_\star$ respectively, and its singular values are ordered in a decreasing fashion: $\sigma_1(\boldsymbol{A}) \geq \cdots \geq \sigma_{\min\{m,n\}}(\boldsymbol{A})$. The smallest nonzero singular value of $\boldsymbol{A}$ is denoted by $\sigma_{\min}(\boldsymbol{A})$. The condition number of $\boldsymbol{A}$ is taken to be the ratio of the largest singular value and the smallest *nonzero* singular value, $\kappa_2(\boldsymbol{A}) = \sigma_1(\boldsymbol{A})/\sigma_{\min}(\boldsymbol{A})$. The orthogonal projection onto the column span of $\boldsymbol{A}$ is denoted by $\boldsymbol{P_A}$. Semi-definite inequalities between the positive-semidefinite matrices $\boldsymbol{A}$ and $\boldsymbol{B}$ are written, e.g., as $\boldsymbol{A} \preceq \boldsymbol{B}$. The standard Euclidean basis vectors are written as $\mathbf{e}_1$, $\mathbf{e}_2$, and so on.

2.2 Problem Formulation

Given a limited budget B with which we can pay to noisily observe a matrix $\boldsymbol{A} \in \mathbb{R}^{m \times n}$, our goal is to construct a low-rank matrix $\overline{\boldsymbol{A}}$ that approximates $\boldsymbol{A}$ well.

Algorithm 1. noisyCUR algorithm for completion of a low-rank matrix $\boldsymbol{A} \in \mathbb{R}^{m \times n}$

Require: d, the number of column samples; s, the number of row samples; σ_e, the noise level of the column samples; σ_c, the noise level of the row samples; and λ, the regularization parameter
Ensure: $\overline{\boldsymbol{A}}$, approximation to $\boldsymbol{A}$
1: $\tilde{\boldsymbol{C}} \leftarrow \boldsymbol{C} + \boldsymbol{E}_c$, where the d columns of $\boldsymbol{C}$ are sampled uniformly at random with replacement from $\boldsymbol{A}$, and the entries of $\boldsymbol{E}_c$ are i.i.d. $\mathcal{N}(0, \sigma_c^2)$.
2: $\boldsymbol{U} \leftarrow$ an orthonormal basis for the column span of $\tilde{\boldsymbol{C}}$
3: $\ell_i \leftarrow \frac{1}{2}\|\mathbf{e}_i^T \boldsymbol{U}\|_2^2 / \|\boldsymbol{U}\|_F^2 + \frac{1}{2m}$ for $i = 1, \ldots, m$
4: $\boldsymbol{S} \leftarrow$ SamplingMatrix($\boldsymbol{\ell}, m, s$), the sketching matrix[1] used to sample s rows of $\boldsymbol{A}$
5: $\boldsymbol{Y} = \boldsymbol{S}^T \boldsymbol{A} + \boldsymbol{E}_e$, where the entries of $\boldsymbol{E}_e$ are i.i.d. $\mathcal{N}(0, \sigma_e^2)$.
6: $\boldsymbol{X} \leftarrow \arg\min_{\boldsymbol{Z}} \|\boldsymbol{Y} - \boldsymbol{S}^T \tilde{\boldsymbol{C}} \boldsymbol{Z}\|_F^2 + \lambda \|\boldsymbol{Z}\|_F^2$
7: $\overline{\boldsymbol{A}} \leftarrow \tilde{\boldsymbol{C}} \boldsymbol{X}$
8: **return** $\overline{\boldsymbol{A}}$

There are two modes of observation: very noisy and cheap samples of entire columns of $\boldsymbol{A}$, or much less noisy but expensive samples of individual entries of $\boldsymbol{A}$.

The following parameters quantify this two-cost observation model:

- Each low-noise sample of an individual entry costs p_e.
- Each high-noise sample of a column costs $p_c > 0$. Because columns are cheap to sample, $p_c < mp_e$.
- The low-noise samples are each observed with additive $\mathcal{N}(0, \sigma_e^2)$ noise.
- Each entry of the high-noise column samples is observed with additive $\mathcal{N}(0, \sigma_c^2)$ noise. Because sampling columns is noisier than sampling entries, $\sigma_c^2 > \sigma_e^2$.

2.3 Related Work

To the best of the authors' knowledge, there is no prior work on budgeted low-rank MC. Standard approaches to matrix completion, e.g. [11,12,16] assume that the entries of the matrix are sampled uniformly at random, with or without noise. The most related works in the MC literature concern adaptive sampling models that attempt to identify more informative observations to reduce the sample complexity [3,6,13,14,21]. One of these works is [6], which estimates non-uniform sampling probabilities for the entries in $\boldsymbol{A}$ and then samples according to these to reduce the sample complexity. The work [14] similarly estimates the norms of the columns of $\boldsymbol{A}$ and then samples entries according to these to reduce the sample complexity. The work [21] proposes a regression-based algorithm for noiseless matrix completion that uses randomly sampled columns, rows, and entries of the matrix to form a low-rank approximation.

[1] SamplingMatrix($\boldsymbol{\ell}, m, s$) returns a matrix $\boldsymbol{S} \in \mathbb{R}^{m \times s}$ such that $\boldsymbol{S}^T \boldsymbol{A}$ samples and rescales, i.i.d. with replacement, s rows of $\boldsymbol{A}$ with probability proportional to their shrinked leverage scores. See the supplement for details.

3 The noisyCUR Algorithm

The noisyCUR (nCUR) algorithm, stated in Algorithm 1, is a regression-based MC algorithm. The intuition behind noisyCUR is that, if the columns of $\boldsymbol{A}$ were in general position and $\boldsymbol{A}$ were exactly rank-r, then one could recover $\boldsymbol{A}$ by sampling exactly r columns of A and sampling r entries from each of the remaining columns, then using regression to infer the unobserved entries of the partially observed columns.

noisyCUR accomodates the presence of noise in the column and entry observations, and the fact that $\boldsymbol{A}$ is not low-rank. It samples d noisy columns from the low-rank matrix $\boldsymbol{A} \in \mathbb{R}^{m \times n}$, with entry-wise error that has variance σ_c^2 and collects these columns in the matrix $\tilde{\boldsymbol{C}} \in \mathbb{R}^{m \times d}$. It then uses $\tilde{\boldsymbol{C}}$ to approximate the span of $\boldsymbol{A}$ and returns an approximation of the form $\overline{\boldsymbol{A}} = \tilde{\boldsymbol{C}}\boldsymbol{X}$.

The optimal coefficient matrix $\boldsymbol{X}$ would require $\boldsymbol{A}$ to be fully observed. The sample complexity is reduced by instead sampling s rows noisily from $\boldsymbol{A}$, with entry-wise error that has variance σ_e^2, to form the matrix $\boldsymbol{Y}$. These rows are then used to estimate the coefficient matrix $\boldsymbol{X}$ by solving a ridge-regression problem; ridge regression is used instead of least-squares regression because in practice it better mitigates the noisiness of the observations. The rows are sampled according to the shrinked leverage scores [15,18] of the rows of $\tilde{\boldsymbol{C}}$.

The cost of observing the entries of $\boldsymbol{A}$ needed to form the approximation $\overline{\boldsymbol{A}}$ is $dp_c + snp_e$, so the number of column observations d and the number of row observations s are selected to use the entire budget, $B = dp_c + snp_e$.

Our main theoretical result is that when $\boldsymbol{A}$ is column-incoherent, has low condition number and is sufficiently dense, noisyCUR can exploit the two sampling modalities to achieve additive error guarantees in budget regimes where the relative error guarantees of nuclear norm completion do not apply.

First we recall the definition of column-incoherence.

Definition 1. *If $\boldsymbol{A} \in \mathbb{R}^{m \times n}$ and $\boldsymbol{V} \in \mathbb{R}^{m \times r}$ is an orthonormal basis for the row space of $\boldsymbol{A}$, then the column leverage scores of $\boldsymbol{A}$ are given by*

$$\ell_i = \|\mathbf{e}_i^T \boldsymbol{V}\|_2^2 \quad \text{for } i = 1, \ldots, n.$$

The column coherence of $\boldsymbol{A}$ is the maximum of the column leverage scores of $\boldsymbol{A}$, and $\boldsymbol{A}$ is said to have a β-incoherent column space if its column coherence is smaller than $\beta\frac{r}{n}$.

Theorem 1. *Let $\boldsymbol{A} \in \mathbb{R}^{m \times n}$ be a rank-r matrix with β-incoherent column space. Assume that $\boldsymbol{A}$ is dense: there is a $c > 0$ such that at least half[1] of the entries of $\boldsymbol{A}$ satisfy $|a_{ij}| \geq c$.*

[1] This fraction can be changed, with corresponding modification to the sample complexities d and s.

Fix a precision parameter $\varepsilon \in (0,1)$ *and invoke Algorithm 1 with*

$$d \geq \max\left\{\frac{6+2\varepsilon}{3\varepsilon^2}\beta r \log\frac{r}{\delta}, \frac{8(1+\delta)^2}{c^2(1-\varepsilon)\varepsilon} r\kappa_2(\boldsymbol{A})^2\sigma_c^2\right\} \text{ and } s \geq \frac{6+2\varepsilon}{3\varepsilon^2} 2d\log\frac{d}{\delta}.$$

The returned approximation satisfies

$$\|\boldsymbol{A}-\overline{\boldsymbol{A}}\|_F^2 \leq \left(\gamma+\varepsilon+40\frac{\varepsilon}{1-\varepsilon}\right)\|\boldsymbol{A}\|_F^2 + 12\varepsilon\left(\frac{\sigma_e^2}{\sigma_c^2}\right) d\sigma_r^2(\boldsymbol{A})$$

with probability at least $0.9 - 2\delta - 2\exp(\frac{-(m-r)\delta^2}{2}) - \exp(\frac{-sn}{32})$. *Here,*

$$\gamma \leq 2\left(\frac{1+\varepsilon}{1-\varepsilon}\right)\left[\frac{\lambda}{(1+\varepsilon)\left(\frac{1}{2}\sqrt{m-r}-\sqrt{d}\right)^2\sigma_c^2+\lambda}\right]^2.$$

The proof of this result is deferred.

Comparison of Nuclear Norm and noisyCUR Approximation Guarantees. Theorem 1 implies that, if $\boldsymbol{A}$ has low condition number, is dense and column-incoherent, and the regularization parameter λ is selected to be $o((\sqrt{m-r}-\sqrt{d})^2\sigma_c)$, then a mixed relative-additive bound of the form

$$\begin{aligned}\|\boldsymbol{A}-\overline{\boldsymbol{A}}\|_F^2 &\leq \varepsilon'\|\boldsymbol{A}\|_F^2 + \varepsilon''\tilde{O}(r)\sigma_r^2(\boldsymbol{A}) \\ &= \varepsilon'\|\boldsymbol{A}\|_F^2 + \varepsilon''\tilde{O}(\|\boldsymbol{A}\|_F^2)\end{aligned} \quad (1)$$

holds with high probability for the approximation returned by Algorithm 1, where ε' and ε'' are $o(1)$, when d and s are $\tilde{\Omega}(r)$.

By way of comparison, the current best guarantees for noisy matrix completion using nuclear norm formulations state that if $d = \Omega((n+m)r\log(n+m))$ entries are noisily observed with noise level σ_e^2, then a nuclear norm formulation of the matrix completion problem yields an approximation $\overline{\boldsymbol{A}}$ that satisfies

$$\|\boldsymbol{A}-\overline{\boldsymbol{A}}\|_F^2 = O\left(\frac{\sigma_e^2}{\sigma_r^2(\boldsymbol{A})}\frac{nm}{r}\right)\|\boldsymbol{A}\|_F^2$$

with high probability [2,7]. The conditions imposed to obtain this guarantee [7] are that $\boldsymbol{A}$ has low condition number and that both its row and column spaces are incoherent. If we additionally require that $\boldsymbol{A}$ is dense, so that the assumptions applied to both algorithms are comparable, then $\|\boldsymbol{A}\|_F^2 = \Omega(mn)$ and the guarantee for nuclear norm completion becomes

$$\|\boldsymbol{A}-\overline{\boldsymbol{A}}\|_F^2 = O(\sigma_e^2\kappa_2^2(\boldsymbol{A}))\|\boldsymbol{A}\|_F^2. \quad (2)$$

Comparison of Budget Requirements for Nuclear Norm Completion and Noisy CUR. For nuclear norm completion approaches to assure guarantees of the form (2) it is necessary to obtain $\Omega((n+m)r\log(n+m))$ high precision noisy samples [4], so the budget required is

$$B_{NN} = \Omega((n+m)r\log(n+m)p_e).$$

When B_{NN} exceeds the budget B, there is no theory supporting the use of a mix of more expensive high and cheaper low precision noisy measurements.

The noisyCUR algorithm allows exactly such a mix: the cost of obtaining the necessary samples is

$$B_{nCUR} = dp_c + snp_e = \tilde{\Omega}(r)p_c + \tilde{\Omega}_r(nr)p_e,$$

where the notation $\tilde{\Omega}_r(\cdot)$ is used to indicate that the omitted logarithmic factors depend only on r. It is evident that

$$B_{nCUR} < B_{NN},$$

so the noisy CUR algorithm is applicable in budget regimes where nuclear norm completion is not.

3.1 Proof of Theorem 1

Theorem 1 is a consequence of two structural results that are established in the supplement.

The first result states that if $\operatorname{rank}(\boldsymbol{C}) = \operatorname{rank}(\boldsymbol{A})$ and the bottom singular value of $\boldsymbol{C}$ is large compared to σ_c, then the span of $\tilde{\boldsymbol{C}}$ will contain a good approximation to $\boldsymbol{A}$.

Lemma 1. *Fix an orthonormal basis $\boldsymbol{U} \in \mathbb{R}^{m\times r}$ and consider $\boldsymbol{A} \in \mathbb{R}^{m\times n}$ and $\boldsymbol{C} \in \mathbb{R}^{m\times d}$ with factorizations $\boldsymbol{A} = \boldsymbol{U}\boldsymbol{M}$ and $\boldsymbol{C} = \boldsymbol{U}\boldsymbol{W}$, where both $\boldsymbol{M}$ and $\boldsymbol{W}$ have full row rank. Further, let $\tilde{\boldsymbol{C}}$ be a noisy observation of $\boldsymbol{C}$, that is, let $\tilde{\boldsymbol{C}} = \boldsymbol{C} + \boldsymbol{G}$ where the entries of $\boldsymbol{G}$ are i.i.d. $\mathcal{N}(0,\sigma_c^2)$. If $\sigma_{\min}(\boldsymbol{C}) \geq 2(1+\delta)\sigma_c\sqrt{m/\varepsilon}$, then*

$$\|(\boldsymbol{I} - \boldsymbol{P}_{\tilde{\boldsymbol{C}}})\boldsymbol{A}\|_F^2 \leq \varepsilon\|\boldsymbol{A}\|_F^2$$

with probability at least $1 - \exp\left(\frac{-m\delta^2}{2}\right)$.

Recall the definition of a $(1 \pm \varepsilon)$-subspace embedding.

Definition 2 (Subspace embedding [20]). *Let $\boldsymbol{A} \in \mathbb{R}^{m\times n}$ and fix $\varepsilon \in (0,1)$. A matrix $\boldsymbol{S} \in \mathbb{R}^{m\times s}$ is a $(1\pm\varepsilon)$-subspace embedding for $\boldsymbol{A}$ if*

$$(1-\varepsilon)\|\mathbf{x}\|_2^2 \leq \|\boldsymbol{S}^T\mathbf{x}\|_2^2 \leq (1+\varepsilon)\|\mathbf{x}\|_2^2$$

for all vectors $\mathbf{x}$ in the span of $\boldsymbol{A}$, or equivalently, if

$$(1-\varepsilon)\boldsymbol{A}^T\boldsymbol{A} \preceq \boldsymbol{A}^T\boldsymbol{S}\boldsymbol{S}^T\boldsymbol{A} \preceq (1+\varepsilon)\boldsymbol{A}^T\boldsymbol{A}.$$

Often we will use the shorthand "subspace embedding" for $(1\pm\varepsilon)$-subspace embedding.

The second structural result is a novel bound on the error of sketching using a subspace embedding to reduce the cost of ridge regression, when the target is noisy.

Corollary 1. *Let* $\tilde{\boldsymbol{C}} \in \mathbb{R}^{m \times d}$, *where* $d \leq m$, *and* $\tilde{\boldsymbol{A}} = \boldsymbol{A} + \boldsymbol{E}$ *be matrices, and let* $\boldsymbol{S}$ *be an* $(1 \pm \varepsilon)$*-subspace embedding for* $\tilde{\boldsymbol{C}}$. *If*

$$\boldsymbol{X} = \arg\min_{\boldsymbol{Z}} \|\boldsymbol{S}^T(\tilde{\boldsymbol{A}} - \tilde{\boldsymbol{C}}\boldsymbol{Z})\|_F^2 + \lambda \|\boldsymbol{Z}\|_F^2,$$

then

$$\|\boldsymbol{A} - \tilde{\boldsymbol{C}}\boldsymbol{X}\|_F^2 \leq \|(\boldsymbol{I} - \boldsymbol{P}_{\tilde{\boldsymbol{C}}})\boldsymbol{A}\|_F^2 + \gamma \|\boldsymbol{P}_{\tilde{\boldsymbol{C}}}\boldsymbol{A}\|_F^2 + \frac{4}{1-\varepsilon}\|\boldsymbol{S}^T\boldsymbol{E}\|_F^2 + \frac{4}{1-\varepsilon}\|\boldsymbol{S}^T(\boldsymbol{I} - \boldsymbol{P}_{\tilde{\boldsymbol{C}}})\boldsymbol{A}\|_F^2,$$

where $\gamma = 2\left(\frac{1+\varepsilon}{1-\varepsilon}\right)\left(\frac{\lambda}{(1+\varepsilon)\sigma_d(\tilde{\boldsymbol{C}})^2 + \lambda}\right)^2$.

Corollary 1 differs significantly from prior results on the error in sketched ridge regression, e.g. [1,18], in that: (1) it bounds the *reconstruction error* rather than the *ridge regression objective*, and (2) it considers the impact of noise in the target. This result follows from a more general result on sketched noisy proximally regularized least squares problems, stated as Theorem ?? in the supplement.

Together with standard properties of Gaussian noise and subspace embeddings, these two results deliver Theorem 1.

Proof (Proof of Theorem 1). The noisyCUR algorithm first forms the noisy column samples $\tilde{\boldsymbol{C}} = \boldsymbol{C} + \boldsymbol{E}_c$, where $\boldsymbol{C} = \boldsymbol{A}\boldsymbol{M}$. The random matrix $\boldsymbol{M} \in \mathbb{R}^{n \times d}$ selects d columns uniformly at random with replacement from the columns of $\boldsymbol{A}$, and the entries of $\boldsymbol{E}_c \in \mathbb{R}^{m \times d}$ are i.i.d. $\mathcal{N}(0, \sigma_c^2)$. It then solves the sketched regression problem

$$\boldsymbol{X} = \arg\min_{\boldsymbol{Z}} \|\boldsymbol{S}^T(\tilde{\boldsymbol{A}} - \tilde{\boldsymbol{C}}\boldsymbol{Z})\|_F^2 + \lambda \|\boldsymbol{Z}\|_F^2,$$

and returns the approximation $\overline{\boldsymbol{A}} = \tilde{\boldsymbol{C}}\boldsymbol{X}$. Here $\tilde{\boldsymbol{A}} = \boldsymbol{A} + \boldsymbol{E}_e$, where $\boldsymbol{E}_e \in \mathbb{R}^{m \times n}$ comprises i.i.d $\mathcal{N}(0, \sigma_e^2)$ entries, and the sketching matrix $\boldsymbol{S} \in \mathbb{R}^{m \times s}$ samples s rows using the shrinked leverage scores of $\tilde{\boldsymbol{C}}$.

By [18, Appendix A.1.1], $\boldsymbol{S}$ is a subspace embedding for $\tilde{\boldsymbol{C}}$ with failure probability at most δ when s is as specified. Thus Corollary 1 applies and gives that

$$\begin{aligned}\|\boldsymbol{A} - \tilde{\boldsymbol{C}}\boldsymbol{X}\|_F^2 &\leq \|(\boldsymbol{I} - \boldsymbol{P}_{\tilde{\boldsymbol{C}}})\boldsymbol{A}\|_F^2 + \gamma' \|\boldsymbol{P}_{\tilde{\boldsymbol{C}}}\boldsymbol{A}\|_F^2 \\ &\quad + \frac{4}{1-\varepsilon}\|\boldsymbol{S}^T\boldsymbol{E}\|_F^2 + \frac{4}{1-\varepsilon}\|\boldsymbol{S}^T(\boldsymbol{I} - \boldsymbol{P}_{\tilde{\boldsymbol{C}}})\boldsymbol{A}\|_F^2 \\ &= T_1 + T_2 + T_3 + T_4,\end{aligned}$$

where $\gamma' = 2\left(\frac{1+\varepsilon}{1-\varepsilon}\right)\left(\frac{\lambda}{(1+\varepsilon)\sigma_d(\tilde{\boldsymbol{C}})^2 + \lambda}\right)^2$. We now bound the four terms T_1, T_2, T_3, and T_4.

To bound T_1, note that by [19, Lemma 13], the matrix $\sqrt{\frac{n}{d}}\boldsymbol{M}$ is a subspace embedding for $\boldsymbol{A}^T$ with failure probability at most δ when d is as specified. This gives the semidefinite inequality $\frac{n}{d}\boldsymbol{C}\boldsymbol{C}^T = \frac{n}{d}\boldsymbol{A}\boldsymbol{M}\boldsymbol{M}^T\boldsymbol{A}^T \succeq (1-\varepsilon)\boldsymbol{A}\boldsymbol{A}^T$, which in turn gives that

$$\begin{aligned}\sigma_r^2(\boldsymbol{C}) &\geq (1-\varepsilon)\frac{d}{n}\sigma_r^2(\boldsymbol{A}) \geq \frac{8(1+\delta)^2}{c^2\varepsilon}\frac{r}{n}\|\boldsymbol{A}\|_2^2\sigma_c^2 \\ &\geq \frac{8(1+\delta)^2}{c^2\varepsilon n}\|\boldsymbol{A}\|_F^2\sigma_c^2 \geq 4(1+\delta)^2\frac{m}{\varepsilon}\sigma_c^2.\end{aligned}$$

The second inequality holds because

$$d \geq \frac{8(1+\delta)^2}{c^2(1-\varepsilon)\varepsilon} r\kappa_2(\boldsymbol{A})^2\sigma_c^2 \quad \text{implies} \quad \sigma_r^2(\boldsymbol{A}) \geq \frac{8(1+\delta)^2}{c^2(1-\varepsilon)\varepsilon}\frac{r}{d}\|\boldsymbol{A}_2\|^2\sigma_c^2 \tag{3}$$

The third inequality holds because $r\|\boldsymbol{A}\|_2^2$ is an overestimate of $\|\boldsymbol{A}_F\|_2^2$. The final inequality holds because the denseness of $\boldsymbol{A}$ implies that $\|\boldsymbol{A}\|_F^2 \geq \frac{1}{2}c^2mn$.

Note also that the span of $\boldsymbol{C} = \boldsymbol{A}\boldsymbol{M}$ is contained in that of $\boldsymbol{A}$, and since $\frac{n}{d}\boldsymbol{C}\boldsymbol{C}^T \succeq (1-\varepsilon)\boldsymbol{A}\boldsymbol{A}^T$, in fact $\boldsymbol{C}$ and $\boldsymbol{A}$ have the same rank and therefore span the same space. Thus the necessary conditions to apply Lemma 1 are satisfied, and as a result, we find that

$$T_1 \leq \varepsilon\|\boldsymbol{A}\|_F^2$$

with failure probability at most $\exp(-\frac{m\delta^2}{2})$.

Next we bound T_2. Observe that $\|\boldsymbol{P}_{\tilde{\boldsymbol{C}}}\boldsymbol{A}\|_F^2 \leq \|\boldsymbol{A}\|_F^2$. Further, by Lemma ?? in the supplement,

$$\sigma_d(\tilde{\boldsymbol{C}}) \geq \left(\frac{1}{2}\sqrt{m-r} - \sqrt{d}\right)^2 \sigma_c^2$$

with failure probability at most $\exp(\frac{-(m-r)\delta^2}{2})$. This allows us to conclude that

$$T_2 \leq \gamma\|\boldsymbol{A}\|_F^2,$$

where γ is as specified in the statement of this theorem.

To bound T_3, we write

$$\begin{aligned}T_3 &= \frac{4}{1-\varepsilon}\|\boldsymbol{S}^T\boldsymbol{P}_{\boldsymbol{S}}\boldsymbol{E}\|_F^2 \leq \frac{4}{1-\varepsilon}\|\boldsymbol{S}\|_2^2\|\boldsymbol{P}_{\boldsymbol{S}}\boldsymbol{E}\|_F^2 \\ &\leq \frac{8}{1-\varepsilon}\frac{m}{s}\|\boldsymbol{Q}^T\boldsymbol{E}\|_F^2,\end{aligned}$$

where $\boldsymbol{Q}$ is an orthonormal basis for the span of $\boldsymbol{S}$. The last inequality holds because [18, Appendix A.1.2] shows that $\|\boldsymbol{S}\|_2^2 \leq 2\frac{m}{s}$ always. Finally, note that $\boldsymbol{Q}$ has at most s columns, so in the worst case $\boldsymbol{Q}^T\boldsymbol{E}$ comprises sn i.i.d. $\mathcal{N}(0, \sigma_e^2)$ entries. A standard concentration bound for χ^2 random variables with sn degrees of freedom [17, Example 2.11] guarantees that

$$\|\boldsymbol{Q}^T\boldsymbol{E}\|_F^2 \leq \frac{3}{2}sn\sigma_e^2$$

with failure probability at most $\exp(\frac{-sn}{32})$. We conclude that, with the same failure probability,

$$T_3 \leq \frac{12}{1-\varepsilon} mn\sigma_e^2.$$

Now recall (3), which implies that

$$\begin{aligned}\varepsilon(1-\varepsilon)d\sigma_r^2(\boldsymbol{A}) &\geq \frac{8(1+\delta)^2}{c^2} r\|\boldsymbol{A}_2\|^2\sigma_c^2 \geq \frac{8(1+\delta)^2}{c^2}\|\boldsymbol{A}\|_F^2\sigma_c^2 \\ &\geq 4(1+\delta)^2 mn\sigma_c^2 \geq mn\sigma_c^2.\end{aligned}$$

It follows from the last two displays that

$$T_3 \leq 12\varepsilon\left(\frac{\sigma_e^2}{\sigma_c^2}\right) d\sigma_r^2(\boldsymbol{A}).$$

The bound for T_4 is an application of Markov's inequality. In particular, it is readily verifiable that $\mathbb{E}[\boldsymbol{S}\boldsymbol{S}^T] = \boldsymbol{I}$, which implies that

$$\mathbb{E}T_4 = \frac{4}{1-\varepsilon}\|(\boldsymbol{I}-\boldsymbol{P}_{\tilde{C}})\boldsymbol{A}\|_F^2 = \frac{4}{1-\varepsilon}T_1 \leq \frac{4\varepsilon}{1-\varepsilon}\|\boldsymbol{A}\|_F^2.$$

The final inequality comes from the bound $T_1 \leq \varepsilon\|\boldsymbol{A}\|_F^2$ that was shown earlier. Thus, by Markov's inequality,

$$T_4 \leq \frac{40\varepsilon}{1-\varepsilon}\|\boldsymbol{A}\|_F^2$$

with failure probability at most 0.1.

Collating the bounds for T_1 through T_4 and their corresponding failure probabilities gives the claimed result.

4 Empirical Evaluation

In this section we investigate the performance of the noisyCUR method on a small-scale synthetic data set and on the Jester and MovieLens data sets. We compare with the performance of three nuclear norm-based algorithms in a low and a high-budget regime.

4.1 Experimental Setup

Four parameters are manipulated to control the experiment setup:

1. The budget, taken to be of the size $B = c_0 mrp_e$ for some constant positive integer c_0. This choice ensures that the $O((n+m)r\log(n+m))$ high precision samples needed for nuclear norm completion methods cannot be obtained.
2. The ratio of the cost of sampling a column to that of individually sampling each entry in that column, $\alpha = \frac{p_c}{mp_e}$. For all three experiments, we set $\alpha = 0.2$.

3. The entry sampling noise level σ_e^2.
4. The column sampling noise level σ_c^2.

Based on the signal-to-noise ratio between the matrix and the noise level of the noisiest observation model, σ_c^2, we classify an experiment as being high noise or low noise. The entry-wise signal-to-noise ratio is given by

$$SNR = \frac{\|\boldsymbol{A}\|_F^2}{mn\sigma_c^2}.$$

High SNR experiments are said to be low noise, while those with low SNR are said to be high noise.

4.2 Methodology: noisyCUR and the Baselines

We compare to three nuclear norm-based MC algorithms, as nuclear norm-based approaches are the most widely used and theoretically investigated algorithms for low-rank MC. We additionally compare to the CUR+ algorithm of [21] as it is, similarly to noisyCUR, a regression-based MC algorithm.

To explain the baselines, we introduce some notation. Given a set of indices Ω, the operator $\mathcal{P}_\Omega : \mathbb{R}^{m\times n} \to \mathbb{R}^{m\times n}$ returns a matrix whose values are the same as those of the input matrix on the indices in Ω, and zero on any indices not in Ω. The set Ω_s below comprises the indices of entries of $\boldsymbol{A}$ sampled with high accuracy, while Ω_c comprises the indices of entries of $\boldsymbol{A}$ sampled using the low accuracy column observation model.

(nCUR). Given the settings of the two-cost model, the noisyCUR algorithm is employed by selecting a value for d, the number of noisy column samples; the remaining budget then determines s, the number of rows that are sampled with high precision. Cross-validation is used to select the regularization parameter λ.

(CUR+). The CUR+ algorithm is designed for noiseless matrix completion [21]; it is adapted in a straightforward manner to our setting. Now d is the number of noisy row and column samples, and $d/2$ columns and $d/2$ rows are sampled uniformly with replacement from $\boldsymbol{A}$ and noisily observed to form column and row matrices $\boldsymbol{C}$ and $\boldsymbol{R}$. The remaining budget is used to sample entries to form Ω_e and $\boldsymbol{A}_{\text{obs}}$, the partially observed matrix which contains the observed noisy entry samples and is zero elsewhere. The CUR+ algorithm then returns the low-rank approximation $\overline{\boldsymbol{A}} = \boldsymbol{CUR}$, where $\boldsymbol{U}$ is obtained by solving

$$\boldsymbol{U} = \arg\min \|\mathcal{P}_{\Omega_e}(\boldsymbol{A}_{\text{obs}} - \boldsymbol{CUR})\|_F^2.$$

(NNa). The first of the nuclear norm baselines is the formulation introduced in [5], which forms the approximation

$$\begin{aligned} \overline{\boldsymbol{A}} = \arg\min\nolimits_{\boldsymbol{Z}} & \|\boldsymbol{Z}\|_\star \\ \text{s.t. } & \|\mathcal{P}_{\Omega_e}(\boldsymbol{Z} - \boldsymbol{A}_{\text{obs}})\|_F \leq \delta, \quad (i,j) \in \Omega_e. \end{aligned} \tag{4}$$

All of the budget is spend on sampling entries to form Ω_e and $\boldsymbol{A}_{\text{obs}}$, the partially observed matrix which contains the observed noisy entry samples and is zero elsewhere. Thus the performance of this model is a constant independent of d in the figures. The hyperparameter δ is selected through cross-validation. This baseline is referred to as NNa (nuclear norm minimization for all entries) in the figures.

(NNs). The second nuclear norm baseline is a nuclear norm formulation that penalizes the two forms of observations separately, forming the approximation

$$\begin{aligned} \overline{\boldsymbol{A}} &= \arg\min_{\boldsymbol{Z}} \|\boldsymbol{Z}\|_\star \\ &\text{s.t. } \|\mathcal{P}_{\Omega_c}(\boldsymbol{Z} - \boldsymbol{A}_{\text{obs}})\|_F^2 \leq C_1 d m \sigma_c^2 \\ &\quad\;\; \|\mathcal{P}_{\Omega_e}(\boldsymbol{Z} - \boldsymbol{A}_{\text{obs}})\|_F^2 \leq C_2 f \sigma_e^2 \end{aligned} \tag{5}$$

where C_1 and C_2 are parameters, and again $\boldsymbol{A}_{\text{obs}}$ is the partially observed matrix which contains the observed noisy column samples and entry samples and is zero elsewhere. As with the noisyCUR method, given a value of d, the remaining budget is spent on sampling s rows with high precision. The hyperparameters C_1 and C_2 are selected through cross-validation. This baseline is referred to as NNs (nuclear norm split minimization) in the figures.

(Chen). The final nuclear norm baseline is an adaptation of the two-phase sampling method of [6]. This method spends a portion of the budget to uniformly at random sample entries to estimate leverage scores, then uses the rest of the budget to sample entries according to the leverage score and reconstructs from the union of these samples using the same optimization as NNa. The performance is therefore independent of d. This baseline is referred to as Chen in the figures.

The details of cross-validation of the parameters for the nuclear norm methods are omitted because there are many relevant hyperparameters: in addition to the constraint parameters in the optimization formulations, there are important hyperparameters associated with the ADMM solvers used (e.g., the Lagrangian penalty parameters).

4.3 Synthetic Dataset

Figure 2 compares the performance of the baseline methods and noisyCUR on an incoherent synthetic data set $\boldsymbol{A} \in \mathbb{R}^{80\times 60}$ generated by sampling a matrix with i.i.d. $\mathcal{N}(5,1)$ entries and taking its best rank four approximation. For each value of d, the regularization parameter λ of noisyCUR is selected via cross-validation from 500 logarithmically spaced points in the interval $(10^{-4}, 10)$.

4.4 Jester

Figure 3 compares the performance of the baseline methods and noisyCUR on a subset of the Jester dataset of [9]. Specifically the data set was constructed by extracting the submatrix comprising the 7200 users who rated all 100 jokes. For each value of d, the regularization parameter λ of noisyCUR is selected via cross-validation from 200 points logarithmically spaced in the interval $(10, 10^5)$.

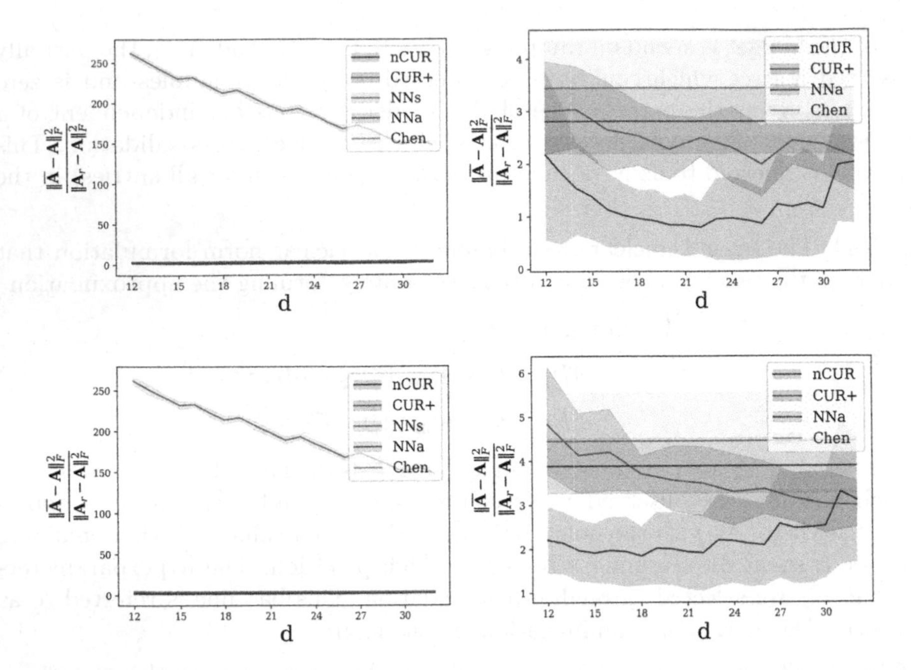

Fig. 2. Performance on the synthetic dataset. $\beta = 15\%$ for all plots. The noise level is low in the top plots: $\sigma_e^2 = 0.01, \sigma_c^2 = 0.05$. The upper left plot shows all methods, while the upper right plot removes the NNs method to facilitate comparison of the better performing methods. In the bottom two plots, the noise level is higher: $\sigma_e^2 = 0.04, \sigma_c^2 = 0.2$. Similarly, the bottom left plot shows all methods, while the bottom right removes the NNs method. Each point in the plots is the average of 100 runs.

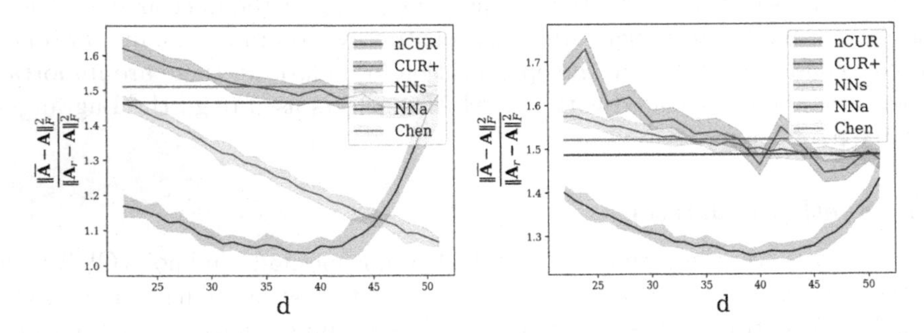

Fig. 3. Performance on the Jester data set. $\beta = 11\%$ for all plots. The noise level is low in the top plots: $\sigma_e^2 = 0.04, \sigma_c^2 = 2$. In the bottom two plots, the noise level is higher: $\sigma_e^2 = 0.25$, $\sigma_c^2 = 12.5$. As in Fig. 2, the plots to the left contain the NNs baseline while those to the right do not. Each point in the plots is the average over 10 runs.

4.5 Movielens-100K

Figure 4 compares the performance of the baseline methods and noisyCUR on the Movielens-100K dataset of [10]. The original 1682×943 data matrix is quite

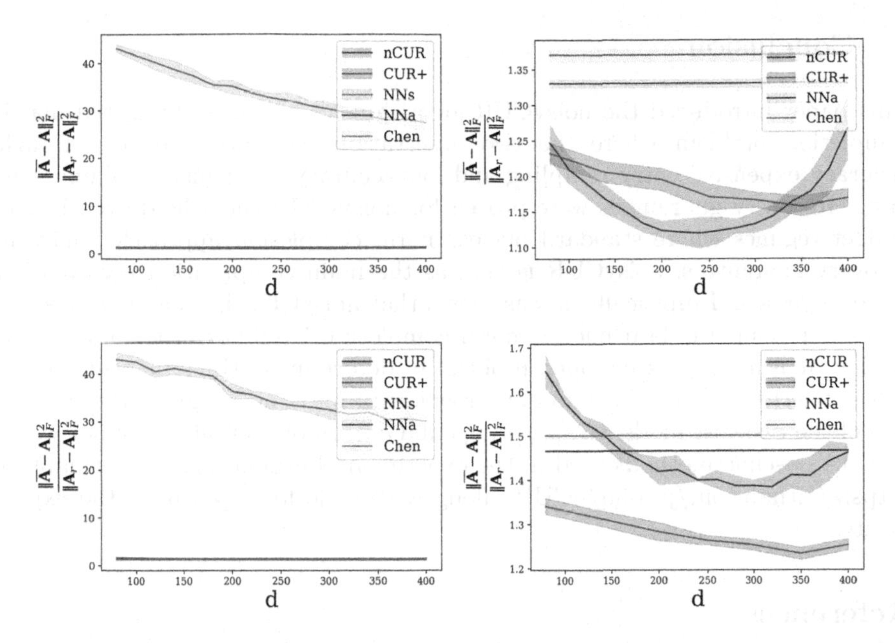

Fig. 4. Performance on the MovieLens-100K data set. $\beta = 10.6\%$ for all plots. The noise level is low in the upper two plots: $\sigma_e^2 = 0.003, \sigma_c^2 = 0.06$. In the bottom two plots, the noise level is higher: $\sigma_e^2 = 0.0225, \sigma_c^2 = 0.65$. As in Fig. 2, the plots to the left show the NNs baseline while the plots to the right do not. Each point in the plots is the average of 10 runs.

sparse, so the iterative SVD algorithm of [8] is used to complete it to a low-rank matrix before applying noisyCUR and the three baseline algorithms. For each value of d, the parameter λ of noisyCUR is cross-validated among 200 points linearly spaced in the interval $[1, 200]$.

4.6 Observations

In both the lower and higher noise regimes on all the datasets, noisyCUR exhibits superior performance when compared to the nuclear norm baselines, and in all but one of the experiments, noisyCUR outperforms CUR+, the regression-based baseline. Also, noisyCUR produces a v-shaped error curve, where the optimal approximation error is achieved at the bottom of the v. This convex shape of the performance of noisyCUR with respect to d suggests that there may be a single optimal d given a dataset and the parameters of the two-cost model.

We note a practical consideration that arose in the empirical evaluations: the noisyCUR method is much faster than the nuclear norm algorithms, as they invoke an iterative solver (ADMM was used in these experiments) that computes SVDs of large matrices during each iteration, while noisyCUR solves a single ridge regression problem.

5 Conclusion

This paper introduced the noisyCUR algorithm for solving a budgeted matrix completion problem where the two observation modalities consist of high-accuracy expensive entry sampling and low-accuracy inexpensive column sampling. Recovery guarantees were proven for noisyCUR; these hold even in low-budget regimes where standard nuclear norm completion approaches have no recovery guarantees. noisyCUR is fast, as the main computation involved is a ridge regression. Empirically, it was shown that noisyCUR has lower reconstruction error than standard nuclear norm completion baselines in the low-budget setting. It is an interesting open problem to determine optimal or near-optimal values for the number of column and row samples (d and s), given the parameters of the two-cost model (B, p_e, p_c, σ_e, and σ_c). Implementations of noisyCUR and the baseline methods used in the experimental evaluations are available at https://github.com/jurohd/nCUR, along with code for replicating the experiments.

References

1. Avron, H., Clarkson, K.L., Woodruff, D.P.: Sharper bounds for regularized data fitting. In: Approximation, Randomization, and Combinatorial Optimization. Algorithms and Techniques (APPROX/RANDOM 2017), vol. 81, pp. 27:1–27:22. Schloss Dagstuhl-Leibniz-Zentrum fuer Informatik (2017)
2. Balcan, M.F., Liang, Z., Song, Y., Woodruff, D.P., Zhang, H.: Non-convex matrix completion and related problems via strong duality. J. Mach. Learn. Res. **20**(102), 1–56 (2019)
3. Balcan, M.F., Zhang, H.: Noise-tolerant life-long matrix completion via adaptive sampling. In: Advances in Neural Information Processing Systems 29, pp. 2955–2963. Curran Associates, Inc. (2016)
4. Candès, E.J., Tao, T.: The power of convex relaxation: near-optimal matrix completion. IEEE Trans. Inf. Theory **56**(5), 2053–2080 (2010)
5. Candes, E.J., Plan, Y.: Matrix completion with noise. Proc. IEEE **98**(6), 925–936 (2010)
6. Chen, Y., Bhojanapalli, S., Sanghavi, S., Ward, R.: Coherent matrix completion. In: Proceedings of the 31st International Conference on Machine Learning (ICML), pp. 674–682 (2014)
7. Chen, Y., Chi, Y., Fan, J., Ma, C., Yan, Y.: Noisy matrix completion: understanding statistical guarantees for convex relaxation via nonconvex optimization. arXiv preprint, arXiv:1902.07698 (2019)
8. Cho, K., Reyhani, N.: An iterative algorithm for singular value decomposition on noisy incomplete matrices. In: The 2012 International Joint Conference on Neural Networks (IJCNN), pp. 1–6 (2012)
9. Goldberg, K., Roeder, T., Gupta, D., Perkins, C.: An iterative algorithm for singular value decomposition on noisy incomplete matrices. Inf. Retrieval **6**(2), 133–151 (2001)
10. Harper, F.M., Konstan, J.A.: The movielens datasets: history and context. ACM Trans. Interact. Intell. Syst. (TiiS) **5**, 1–19 (2015)

11. Hastie, T., Mazumder, R., Lee, J.D., Zadeh, R.: Matrix completion and low-rank SVD via fast alternating least squares. J. Mach. Learn. Res. **16**(1), 3367–3402 (2015)
12. Keshavan, R., Montanari, A., Oh, S.: Matrix completion from noisy entries. In: Advances in Neural Information Processing Systems, pp. 952–960 (2009)
13. Krishnamurthy, A., Singh, A.: Low-rank matrix and tensor completion via adaptive sampling. In: Advances in Neural Information Processing Systems 26, pp. 836–844. Curran Associates, Inc. (2013)
14. Krishnamurthy, A., Singh, A.R.: On the power of adaptivity in matrix completion and approximation. arXiv preprint arXiv:1407.3619 (2014)
15. Ma, P., Mahoney, M.W., Yu, B.: A statistical perspective on algorithmic leveraging. J. Mach. Learn. Res. **16**(1), 861–911 (2015)
16. Recht, B.: A simpler approach to matrix completion. J. Mach. Learn. Res. **12**(12), 3413–3430 (2011)
17. Wainwright, M.J.: High-Dimensional Statistics: A Non-asymptotic Viewpoint. Cambridge University Press, Cambridge (2019)
18. Wang, S., Gittens, A., Mahoney, M.W.: Sketched ridge regression: optimization perspective, statistical perspective, and model averaging. J. Mach. Learn. Res. **18**(218), 1–50 (2018)
19. Wang, S., Zhang, Z., Zhang, T.: Towards more efficient SPSD matrix approximation and CUR matrix decomposition. J. Mach. Learn. Res. **17**(1), 7329–7377 (2016)
20. Woodruff, D.P.: Sketching as a tool for numerical linear algebra. Found. Trends® Theor. Comput. Sci. **10**(1–2), 1–157 (2014)
21. Xu, M., Jin, R., Zhou, Z.H.: CUR algorithm for partially observed matrices. In: Proceedings of the 32nd International Conference on Machine Learning. Proceedings of Machine Learning Research, vol. 37, pp. 1412–1421. PMLR (2015)

Author Index

Printed in the United States
By Bookmasters